Advanced Statistics with Applications in R

Advanced Statistics with Applications in R

Eugene Demidenko

Dartmouth College

Registered Office
John Wiley & Sons, Inc., 111 River Street, Hoboken, NJ 07030, USA

Editorial Office
111 River Street, Hoboken, NJ 07030, USA

For details of our global editorial offices, customer services, and more information about Wiley products visit us at www.wiley.com.

Wiley also publishes its books in a variety of electronic formats and by print-on-demand. Some content that appears in standard print versions of this book may not be available in other formats.

Library of Congress Cataloging-in-Publication Data
Names: Demidenko, Eugene, 1948- author.
Title: Advanced statistics with applications in R / Eugene Demidenko
    (Dartmouth College).
Description: Hoboken, NJ : Wiley, 2020. | Series: Wiley series in probability
    and statistics | Includes bibliographical references and index. |
    Identifiers: LCCN 2019015124 (print) | LCCN 2019019543 (ebook) | ISBN
    9781118594131 (Adobe PDF) | ISBN 9781118594612 (ePub) | ISBN 9781118387986
    (hardback)
Subjects: LCSH: Mathematical statistics–Data processing–Problems,
    exercises, etc. | Statistics–Data processing–Problems, exercises, etc. |
    R (Computer program language)
Classification: LCC QA276.45.R3 (ebook) | LCC QA276.45.R3 D4575 2019 (print)
    | DDC 519.5–dc23
LC record available at https://lccn.loc.gov/2019015124

Cover design by Wiley
Cover image: Courtesy of Eugene Demidenko

Set in 11/14pt Computer Modern by SPi Global, Chennai, India

Printed in the United States of America

10 9 8 7 6 5 4 3 2 1

Contents

To my family

Why I Wrote This Book

My favorite part of the recent American Statistical Association (ASA) statement on the p-value [103] is how it starts: "Why do so many people still use $p = 0.05$ as a threshold?" with the answer "Because that's what they were taught in college or grad school." Many problems in understanding and the interpretation of statistical inference, including the central statistical concept of the p-value arise from the shortage of textbooks in statistics where theoretical and practical aspects of statistics fundamentals are put together. On one hand, we have several excellent theoretical textbooks including Casella and Berger [17], Schervish [87], and Shao [94] without single real-life data example. On the other hand, there are numerous recipe-style statistics textbooks where theoretical considerations, assumptions, and explanations are minimized. This book fills that gap.

Statistical software has become so convenient and versatile these days that many use it without understanding the underlying principles. Unfortunately, R packages do not explain the algorithms and mathematics behind computations, greatly contributing to a superficial understanding making statistics too easy. Many times, to my question "How did you compute this, what is the algorithm," I hear the answer, "I found a program on the Internet." Hopefully, this book will break the unwanted trend of such statistics consumption.

I have often been confronted with the question comparing statistics with driving a car: "Why do we need to know how the car works?" Well, because statistics is not a car: the chance of the car breaking is slim, but starting with the wrong statistical analysis is almost guaranteed without solid understanding of statistics background and implied limitations. In this book, we look at what is under the hood.

Each term I start my first class in statistics at Dartmouth with the following statement:

"Mathematics is the queen and statistics is the king of all sciences"

Indeed, mathematics is the idealistic model of the world: one line goes through a pair of points, the perimeter of a polygon converges to $2\pi r$ when the number of edges goes to infinity, etc. Statistics fills mathematics with life. Due to an unavoidable measurement error, one point turns into a cloud of points. How does one draw a line through two clouds of points? How does one measure π in real life? This book starts with a motivating example "Who said π?" in which I suggest to measuring π by taking the ratio of the perimeter of the tire to its diameter. To the surprise of many, the average ratio does not converge to π even if the measurement error is very small. The reader will learn how this seemingly easy problem of estimating π turns into a formidable statistical problem. Statistics is where the rubber meets the road. It is difficult to name a science where statistics is not used.

Examples are a big deal in this book (there are 442 examples in the book). I follow the saying: "Examples are the expressway to knowledge." Only examples show how to use theory and how to solve a real-life problem. Too many theories remain unusable.

Today statistics is impossible without programming: that is why R is the language statisticians speak. The era of statistics textbooks with tables of distributions in an appendix is gone. Simulations are a big part of probability and statistics: they are used to set up a probabilistic model, test the analytical answer, and help us to study small-sample properties. Although the speed of computations with the for loop have improved due to 64-bit computing, vectorized simulations are preferable and many examples use this approach.

Regarding the title of the book, "Advanced Statistics" is not about doing more mathematics, but an advanced understanding of statistical concepts from the perspective of applications. Statistics is an applied science, and this book is about statistics in action. Most theoretical considerations and concepts are either introduced or applied to examples everybody understands, such as mortgage failure, an oil spill in the ocean, gender salary discrimination, the effect of a drug treatment, cancer distribution in New Hampshire, etc.

I again turn the reader's attention to the p-value. This concept falls through the crack of statistical science. I have seen many mathematical statisticians who work in the area of asymptotic expansions and are incapable of explaining the p-value in layman's terms. I have seen many applied statisticians who mostly use existing statistical packages and describe the p-value incorrectly. The goal of this book is to rigorously explain statistical concepts, including the p-value, and illustrate them with concrete examples dependent on the purpose of statistics applications (I suggest an impatient reader jump to Section 7.10 and then Section 8.5). I emphasize the difference between parameter- and individual-based statistical inference. While classical statistics is concerned with parameters, in real-life applications, we are mostly concerned with individual prediction. For example, given a random sample of individual incomes in town, the classical statistics is concerned with estimation of the town mean income (phantom parameter) and the respective confidence interval, but often we are interested in a more practical question. In what range does the income of a randomly asked resident belong with given probability? This distinction is a common theme of the book.

This book is intended for graduate students in statistics, although some sections are accessible for senior undergraduate statistics students with a solid mathematical background in multivariate calculus and linear algebra along with some courses in elementary statistics and probability. I hope that researchers will find this book useful as well to clarify important statistical concepts.

I am indebted to Steve Quigley, former associate publisher at Wiley, for pursuing me to sign the contract on writing a textbook in statistics. Several people read parts of the book and made helpful comments: Senthil Girimurugan; my Dartmouth students James Brofos, Michael Downs, Daniel Kang; and my colleagues, Dan Rockmore, Zhigang Li, James O'Malley and Todd MacKenzie, among others. I am thankful to anonymous reviewers for their thoughts and corrections that improved the book. Finally, I am grateful to John Morris of *Editide* (http://www.editide.us/) for his professional editorial service.

Data sets and R codes can be downloaded at my website:

www.dartmouth.edu/~eugened

I suggest that they be saved on the hard drive in the directory`C:\StatBook\`. The codes may be freely distributed and modified .

I would like to hear comments, suggestions and opinions from readers. Please e-mail me at `eugened@dartmouth.edu`.

Dartmouth College   *Eugene Demidenko*
Hanover, New Hampshire
August 2019

Chapter 1

Discrete random variables

Two types of random variables are distinguished: discrete and continuous. Theoretically, there may be a combination of these two types, but it is rare in practice. This chapter covers discrete distributions and the next chapter will cover continuous distributions.

1.1 Motivating example

In univariate calculus, a variable x takes values on the real line and we write $x \in (-\infty, \infty)$. In probability and statistics, we also deal with variables that take values in $(-\infty, \infty)$. Unlike calculus, we do not know *exactly* what value it takes. Some values are more likely and some values are less likely. These variables are called *random*. The idea that there is uncertainty in what value the variable takes was uncomfortable for mathematicians at the dawn of the theory of probability, and many refused to recognize this theory as a mathematical discipline. To convey information about a random variable, we must specify its distribution and attach a probability or density for each value it takes. This is why the concept of the distribution and the density functions plays a central role in probability theory and statistics. Once the density is specified, calculus turns into the principal tool for treatment.

Throughout the book we use letters in uppercase and lowercase with different meaning: X denotes the random variable and x denotes a value it may take. Thus $X = x$ indicates the event that random variable X takes value x. For example, we may ask what is the chance (probability) that X takes value x. In mathematical terms, $\Pr(X = x)$. For a continuous random variable, we may be interested in the probability that a random variable takes values less or equal to x or takes values from the interval $[x, x + \Delta]$.

A complete coverage of probability theory is beyond the scope of this book –

Advanced Statistics with Applications in R, First Edition. Eugene Demidenko.
© 2020 John Wiley & Sons, Inc. Published 2020 by John Wiley & Sons, Inc.

rather, we aim to discuss only those features of probability theory that are useful in statistics. Readers interested in a more rigorous and comprehensive account of the theory of probability are referred to classic books by Feller [45] or Ross [83], among many others.

In the following example, we emphasize the difference between calculus, which assumes that the world is deterministic, and probability and statistics, which assume that the world is random. This difference may be striking.

Example 1.1 *Who said* π? *The ratio of a circle's circumference to its diameter is* π. *To test this fact, you may measure the circumference of tires and their diameters from different cars and compute the ratios. Does the average ratio approach* π *as the number of measured tires goes to infinity?*

Perhaps to the reader's surprise, even if there is a slight measurement error of the diameter of a tire, the average of empirically calculated π's does not converge to the theoretical value of π; see Examples 3.36 and 6.126. In order to obtain a consistent estimator of π, we have to divide the sum of all circumferences by the sum of all diameters. This method is difficult to justify by standard mathematical reasoning because tires may come from different cars.

This example amplifies the difference between calculus and probability and statistics. The former works in an ideal environment: no measurement error, a unique line goes through two points, etc. However, the world we live in is not perfect: measurements do not produce exactly a theoretically expected result, points do not fall on a straight line, people answer differently to the same question, patients given the same drug recover and some not, etc. All laws of physics including the Newton's free fall formula $S(t) = 0.5gt^2$ (see Example 9.4) do not exactly match the empirical data. To what extent can the mismatch can be ignored? Do measurements confirm the law? Does the Newton's theory hold? These questions cannot be answered without assuming that the measurements made (basically all data) are intrinsically random. That is why statistics is needed every time data are analyzed.

1.2 Bernoulli random variable

The Bernoulli random variable is the simplest random variable with two outcomes, such as *yes* and *no*, but sometimes referred to as *success* and *failure*. Nevertheless, this variable is a building block of all probability theory (this will be explained later when the central limit theorem is introduced).

Generally, we divide discrete random variables into two groups with respect to how we treat the values they take:

- *Cardinal (or numerical).* The variables take numeric values and therefore can be compared (inequality $<$ is meaningful), and the arithmetic is allowed. Examples of cardinal discrete random variables include the number

of children in the family and the number of successes in a series of independent Bernoulli experiments (the binomial random variable). If a random
variable takes values 0, 1, and 2, then $1 - 0 = 2 - 1$; the arithmetic mean
is meaningful for the cardinal random variables.

- *Nominal (or categorical)*. These variables take values that are not numeric
  but merely indicate the name/label or the state (category). For example, if
  we are talking about three categories, we may use quotes "1," "2," or "3"
  if names are not provided. An example of a nominal discrete random variable is the preference of a car shopper among car models "Volvo," "Jeep,"
  "VW," etc. Although the probabilities for each category can be specified,
  the milestone probability concepts such as the mean and the cumulative
  distribution function make no sense. Typically, we will be dealing with cardinal random variables. Formally, the Bernoulli random variable is nominal,
  but with only two outcomes, we may safely code *yes* as 1 and *no* as 0. Then
  the average of Bernoulli outcomes is interpreted as the proportion of having
  a *yes* outcome. Variables may take finite or infinite number of values. An
  example of a discrete random variable that may take an infinite number of
  values is a Poisson random variable, discussed in Section 1.7. Sometimes,
  it is convenient to assume that a variable takes an infinite number of values
  even in cases when the number of cases is bounded, such as in the case of
  the number of children per family.

An example of a binary (or dichotomous) random variable is the answer to
a question such as "Do you play tennis?" (it is assumed that there are only two
answers, *yes* and *no*). As was noted earlier, without loss of generality, we can
encode *yes* as 1 and *no* as 0. If X codes the answer, we cannot predict the answer –
that is why X is a random variable. The key property of X is the probability that
a randomly asked person plays tennis (clearly, the probability that a randomly
asked person does not play tennis is complementary). Mathematically we write
$\Pr(X = 1) = p$. The distribution of a binary random variable X is completely
specified by p. An immediate application of the probability is that, assuming
that a given community consists of N people, we can estimate the number of
tennis players as Np.

We refer to this kind of binary variable as a Bernoulli random variable named
after the Swiss mathematician Jacob Bernoulli (1654–1705). We often denote
$q = 1 - p$ (complementary probability), so that $\Pr(X = 0) = q$. A compact way
to write down the Bernoulli probability of possible outcomes is

$$\Pr(X = y) = p^y(1 - p)^{1-y}, \tag{1.1}$$

where y takes fixed values, 1 or 0. This expression is useful for deriving the
likelihood function for statistical purposes that will be used later in the statistics
part of the book.

The next example applies the Bernoulli random variable to a real-world problem.

Example 1.2 *Safe driving.* *Fred is a safe driver: he has a 1/10 chance each year of getting a traffic ticket. Is it true that he will get at least one traffic ticket over 20 years of driving?*

Solution. Many people say yes. Indeed, since the probability for one year is $1/10$, the probability that he will get a traffic ticket over 20 years is more than 1 and some people would conclude that he will definitely get a ticket. First, this naive computation is suspicious: How can a probability be greater than 1? Second, if he is lucky, he may never get a ticket over 20 years because getting a ticket during one year is just a probability, and the event may never occur this year, next year, etc. To find the probability that Fred will get at least one ticket, we use the method of complementary probability and find the probability that Fred gets no ticket over 20 years. Since the probability to get no ticket each year is $1-1/10$ the probability to get no tickets over 20 years is $(1-1/10)^{20}$. Finally, the probability that Fred gets at least one ticket over 20 years is $1 - (1 - 1/10)^{20} = 1 - (9/10)^{20} = 0.88$. In other words, the probability to be ticket-free over 20 years is greater than 10%. This is a fun problem and yet reflects an important phenomenon in our life: things may happen at random and scientific experiments may not be reproducible with positive probability.

Problems

1. Check formula (1.1) by examination. [Hint: Evaluate the formula at $y = 0$ and $y = 1$.]

2. Demonstrate that the naive answer in Example 1.2 can be supported by the approximation formula $1 - (1 - x)^n \simeq nx$ for small x and $n > 1$. (a) Derive this approximation using the L'Hôpital's rule, and (b) apply it to the probability of getting at least one ticket.

3. Provide an argumentation for the infinite monkey theorem: a monkey hitting keys at random on a computer keyboard for an infinite amount of time will almost surely type a given text, such as "Hamlet" by William Shakespeare (make the necessary assumptions). [Hint: The probability of typing the text starting from any hit is the same and positive; then follow Example 1.2.]

1.3   General discrete random variable

Classical probability theory uses cardinal (numeric) variables: these variables take numeric values that can be ordered and manipulated using arithmetic operations such as summation. For a discrete numeric random variable, we must specify the probability for each unique outcome it takes. It is convenient to use a table to specify its distribution as follows.

Value of X	x_1	x_2	x_3	$\cdots$	x_{n-1}	x_n
Probability	p_1	p_2	p_3	$\cdots$	p_{n-1}	p_n

It is assumed that x_i are all different and the n events $\{X = x_i, i = 1, ..., n\}$ are mutually exclusive; sometimes the set $\{x_i\}$ is called the sample space and particular x_i the outcome or elementary event. Without loss of generality, we will always assume that the values are in ascending order, $x_1 < x_2 < \cdots < x_n$. Indeed, if some x are the same, we sum the probabilities. As follows from this table, X may take n values and

$$\Pr(X = x_i) = p_i, \quad i = 1, 2, ..., n,$$

sometimes referred to as the *probability mass function* (pmf). Since p_i are probabilities and $\{x_i\}$ is an exhaustive set of values, we have

$$\sum_{i=1}^{n} p_i = 1, \quad p_i \geq 0.$$

For $n = 2$, a categorical random variable can be interpreted as a Bernoulli random variable. An example of a categorical random variable with a number of outcomes more than two is a voter's choice in an election, assuming that there are three or more candidates. This is not a cardinal random variable: the categories cannot be arranged in a meaningful order and arithmetic operations do not apply.

An example of a discrete random variable that may take any nonnegative integer value, at least hypothetically, is the number of children in a family. Although practically this variable is bounded (for instance, one may say that the number of children is less than 100), it is convenient to assume that the number of children is unlimited. It is customary to prefer convenience over rigor in statistical applications.

Sometimes we want to know the probability that a random variable X takes a value less or equal to x. This leads to the concept of the *cumulative distribution function* (cdf).

Definition 1.3 *The cumulative distribution function is defined as*

$$F_X(x) = \Pr(X \leq x) = \sum_{x_i \leq x} p_i.$$

The cdf is a step-wise increasing function; the steps are at $\{x_i, i = 1, 2, ..., n\}$. The cdf is convenient for finding the probability of an interval event. For example,

$$\Pr(u < X \le U) = F_X(U) - F_X(u),$$

where u and U are fixed numbers ($u \le U$). We will discuss computation of the cdf in R for some specific discrete random variables later in this chapter.

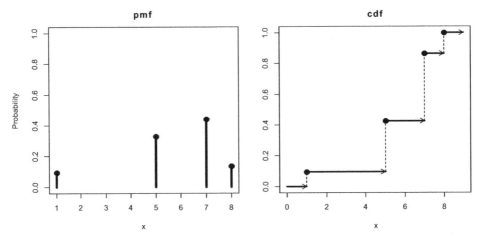

Figure 1.1: *The probability mass function (pmf) and the cumulative distribution function (cdf) of a typical discrete distribution. Note that the cdf is discrete from the left and continuous from the right.*

The pmf and the respective cdf of a typical discrete distribution are shown in Figure 1.1. At each jump the cdf is continuous from the right (indicated by a filled circle) and discrete from the left (indicated by an arrow). This means that $\lim F_X(x) = F(x_i)$ when x approaches x_i from above ($x > x_i$), but $\lim F_X(x) < F(x_i)$ when x approaches x_i from below ($x < x_i$).

Problems

1. (a) Prove that the cdf is a non-decreasing function on $(-\infty, \infty)$. (b) Prove that the cdf approaches 1 when $x \to \infty$ and approaches 0 when $x \to -\infty$.

2. Express p_i in terms of cdf.

3. Express the continuity of a cdf at x_i using notation $\lim_{x\downarrow}$.

1.4   Mean and variance

Expectation or the mean value (mean) is one of the central notions in probability and statistics. Typically, it is difficult to specify the entire distribution of a random variable, but it is informative to know where the center of the distribution lies, the *mean*. The arithmetic average of observations as an estimator of the mean is one of summary statistics characterizing the center of the distribution of a random variable. Another summary statistic, variance, will be discussed in the next section.

We use E to denote the expectation of a random variable. For a discrete random variable X that takes value x_i with probability $p_i = \Pr(X = x_i)$, the mean is defined as

$$E(X) = \sum_{i=1}^{n} x_i \Pr(X = x_i) = \sum_{i=1}^{n} x_i p_i. \tag{1.2}$$

The mean, $E(X)$, can be interpreted as the weighted average of $\{x_i, i = 1, 2, ..., n\}$, where the weights are the probabilities. It is easy to see that for a Bernoulli random variable, the mean equals the probability of occurrence (success), $E(X) = p$. Indeed, as follows from the previous definition of the mean, $E(X) = 1 \times p + 0 \times (1 - p) = p$. This explains why it is convenient to assume that a dichotomous random variable takes the values 0 and 1.

Following standard notation, the Greek letter μ (*mu*) is used for the mean; when several random variables are involved, we may use notation μ_X to indicate that the expectation is of the random variable X.

The mean acts as a linear function: the mean of a linear combination of random variables is the linear combination of the means; in mathematical terms $E(aX) = aE(X)$, $E(X+Y) = E(X)+E(Y)$. The first property is easy to prove; the proof of the second property requires the concept of the bivariate distribution and is deferred to Chapter 3.

1.4.1 Mechanical interpretation of the mean

The mean can be interpreted as the center of mass using the notion of *torque* in physics. Imagine a stick with n masses of weights W_i attached at n locations, $\{x_i, i = 1, 2, ..., n\}$. See Figure 1.2 for a geometric illustration. We want to find the support point, μ, where the stick is in balance; thus μ is the center of mass (center of gravity). From physics, we know that a weight W_i located at x_i with respect to μ creates the torque $(x_i - \mu)W_i$. The stick is in balance if the sum of these torques is zero:

$$\sum_{i=1}^{n} (x_i - \mu)W_i = 0. \tag{1.3}$$

This balance condition leads to the solution

$$\mu = \frac{\sum_{i=1}^{n} x_i W_i}{\sum_{i=1}^{n} W_i}. \tag{1.4}$$

For the example depicted in Figure 1.2, the balance point (center of masses) is $\mu = 4.38$,

$$\mu = \frac{1 \times 5 + 2 \times 5 + 3 \times 5 + 4 \times 2 + 5 \times 1 + 6 \times 2 + 7 \times 4 + 8 \times 1 + 9 \times 4}{5 + 5 + 5 + 2 + 1 + 2 + 4 + 1 + 4}.$$

One can interpret solution (1.4) as the weighted average because we may rewrite $\mu = \sum_{i=1}^{n} x_i w_i$, where w_i is the relative weight. As with probability, p_i, $w_i = W_i / \sum_{j=1}^{n} W_j$. In a special case when weights are the same, $W_i = $ const we arrive at the arithmetic mean, $\overline{x} = \sum_{i=1}^{n} x_i / n$. In summary, the mean is the balance point, or the center of gravity, where probabilities $\{p_i, i = 1, 2, ..., n\}$ act as the relative weights at locations $\{x_i, i = 1, 2, ..., n\}$. A similar interpretation of the mean as a center of the gravity is valid for two and three dimensions.

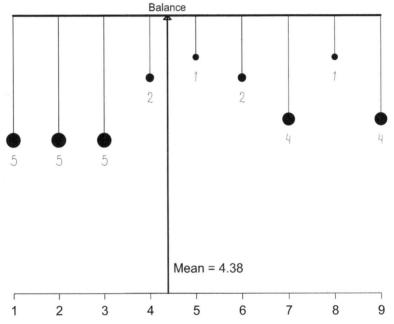

Figure 1.2: *Mechanical interpretation of the mean using hanging weights: the balance is where the resultant torque is zero, $\sum_{i=1}^{n} (x_i - \mu) W_i = 0$. For this example, the support point is $\mu = 4.38$. The mean is the center of gravity.*

The mean is meaningful only for cardinal/numeric random variables, but the mean is not the only way to define the center of the distribution. There are other characteristics of the center, such as mode or median. Sometimes, depending on the subject and the purpose of the study, they may offer a better interpretation than the mean. The mode is defined as the most frequent observation/value: the mode is x_i for which $\Pr(X = x_i) = \max$. Hence one can refer to the mode as the most probable value of the random variable. The mode is applied for categorical random variables where the mean does not make sense, such as when reporting results of the poll among presidential candidates. In another example, we know that the most frequent number of children in American families (mode) is 2 and the average (mean) is 2.2. In other words, 2 is the most probable number of children in the family. If you invite a family with children, you expect to see two kids, not 2.2.

The median is defined as the value x for which the probability of observing $X < x$ is equal to the probability of observing $X > x$. The median does not apply to categorical random variables because it requires ordering.

Sometimes, the weighted mean emerges naturally as a conditional probability, the following is an example.

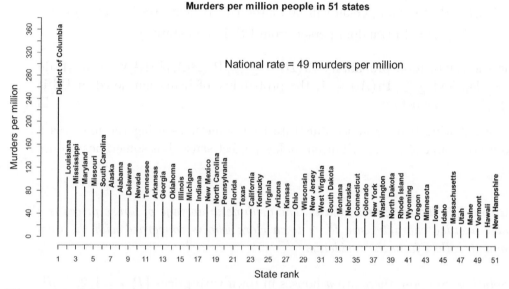

Figure 1.3: *Murder rates in the United States by state. The national murder rate is the weighted mean with the ith weight equal to the number of people living in state i.*

Example 1.4 *National murder rate. Figure 1.3 shows the number of murders per million people in the District of Columbia and all 50 states of the United States. (a) Express the national murder rate as the weighted mean. (b) Interpret the national murder rate as a conditional probability.*

Solution. (a) By definition, the murder rate, r_i is the number of murders divided by the number of people living in the state (state population), $i = 1, 2, ..., 51$. We associate r_i with the probability that a random person from state i will be murdered. If n_i denotes the state population (in millions), the number of murders is $r_i n_i$ and the total number of murders in the country is $\sum_{i=1}^{51} r_i n_i$. To compute the national murder rate, we need to divide the total number of murders by the total number of people in the United States,

$$r = \frac{\sum_{i=1}^{51} r_i n_i}{\sum_{i=1}^{51} n_i} = \sum_{i=1}^{51} r_i w_i$$

where $w_i = n_i / \sum_{i=1}^{51} n_i$ is the proportion of people living in state i. Using this formula and data presented in Figure 1.3, the national murder rate is $r = 49$

murders per million. Looking back into formula (1.2), one can interpret w_i as probabilities. Therefore one can interpret r as the probability of murdering of a random person in the United States. Note that it would be incorrect to compute the national murder rate as the simple average, $\sum_{i=1}^{51} r_i/51$. If less-populated states have high murder rates their contribution would be the same as large states. (b) Now we formulate the problem as a conditional probability. Define

$$\Pr(B|A_i) = \Pr\,(\text{random person will be murdered} \mid \text{person lives in state } i) = r_i$$
$$\Pr(A_i) = \Pr\,(\text{random person from US lives in state } i) = w_i.$$

Using the law of total probability $\Pr(B) = \sum_{i=1}^{n} \Pr(B|A_i)\Pr(A_i)$, where A_i do not overlap and $\sum_{i=1}^{n} \Pr(A_i) = 1$, the probability of being murdered in US is $r = \sum_{i=1}^{51} r_i w_i$, as before.   □

The following two examples illustrate that sometimes using the mean as a measure of center of the distribution makes perfect sense, but sometimes it does not.

Example 1.5 *Town clerk mean.* *The arithmetic average of house prices is a suitable average characteristic for a town clerk who is concerned with the total amount to collect from the residents.*

Solution. Suppose there are n houses in town with prices $\{P_i, i = 1, 2, ..., n\}$. When reporting the average house price, town officials prefer to use the arithmetic average:

$$\overline{P} = \frac{1}{n}\sum_{i=1}^{n} P_i.$$

Indeed, if r is the property tax rate, the town collects $r\sum_{i=1}^{n} P_i$ dollars and $r\overline{P}$ is the average property tax in the town.

Example 1.6 *Buyer's house median.* *A real estate agent shows houses to a potential buyer. What is a suitable average house price for the buyer, the mean or the median?*

Solution. While the mean price makes sense for a town or state official (the property tax is proportional to the mean), it is not useful for the buyer who is thinking of the chance of affording a house he/she likes. Instead, the median means that 50% of the houses he/she saw will have price lower than the median and 50% of the houses will have higher price. In this case, the median has a much more sensible interpretation from the buyer's perspective.   □

Another situation where the mean and median (or mode) depends on the subject of application is the salary distribution in a company. For the company's

CEO, the mean, which is the ratio of the personnel cost to the number of employees, is the most meaningful quantity because it directly affects the profit = revenue minus cost (including cost of labor). For an employee, the median (or maybe the mode) is the most informative parameter of the company salary distribution because he or she can assess if he/she is underpaid. In general, mean has a meaningful interpretation if and only if the sum of observations or measurements has an interpretation. This is the case for both examples: the total wealth of properties and the total labor cost are what the town and the CEO are concerned with, respectively.   □

In the previous discussion, we compared mean with median. The following example underscores the difference between median and mode.

Example 1.7 *Mode for the manager of a shoe store and median for a shoe buyer. Explain why mode is a more appropriate characteristic of the center of the shoe-size distribution for the manager of a shoe store but median is more appropriate for a shoe buyer.*

Solution. The manager is concerned with the most popular shoe size because it tells him/her about the order to make from a shoe factory. The buyer wants to know if the store has the sufficient stock of the popular shoe size.   □

In conclusion, we should not stick with the mean as the most popular and the easiest parameter to characterize the center of the distribution. We must also consider the median or the mode, depending on the application.

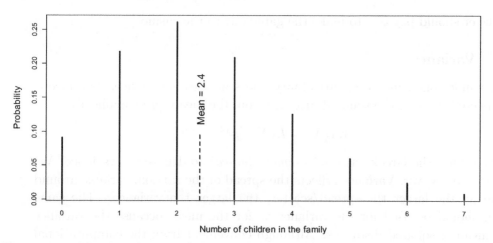

Figure 1.4: *The distribution of the number of children in 100 families. According to the mean you are expected to see 2.4 children.*

Example 1.8 *Number of children in the family. You are invited for dinner to a family and you want to bring presents to each child (you do not know the* *number of children). To make an educated guess on how many presents to buy you find on Internet data on the number of children in 100 families, see Figure 1.4. How many presents do you buy?*

Solution. The typical number of children in the family is 2, the mode. According to the mean you expect to see 2.4 children, a somewhat uncomfortable number. See Example 1.17 where the number of toys is solved assuming that the number of children in the family follows a Poisson distribution.   □

The following example illustrates that some random variables do not have finite mean.

Example 1.9 *St. Petersburg paradox. The game involves a single casino player and consists of a series of coin tosses. The pot starts at $1 and casino doubles the pot every time a head appears. When a tail appears, the game ends and the player wins whatever is in the pot. What would be a fair price to pay the casino for entering the game?*

Solution. We define X as the dollar amount paid to the player. With probability 1/2, the player wins one dollar; with probability 1/4 the player wins $2, with probability 1/8 the player wins $4, and so on. Thus, $\Pr(X = 2^{k-1}) = 1/2^k$ for the number of tosses, $k = 1, 2, ...$ We calculate the expected value to determine a fair price for a player to pay as

$$E(X) = \frac{1}{2} \times 1 + \frac{1}{4} \times 2 + \frac{1}{8} \times 4 + \cdots = \sum_{k=1}^{\infty} \frac{1}{2^k} 2^{k-1} = \sum_{k=1}^{\infty} \frac{1}{2} = \infty.$$

The player should pay ∞ to make the game fair to the casino.

1.4.2 Variance

Another milestone concept of probability and statistics is variance. Variance is the expected value of the squared distance from the mean, or symbolically,

$$\mathrm{var}(X) = E(X - \mu)^2.$$

Usually we use the Greek letter σ^2 (*sigma-squared*) to denote the variance. We write $\sigma^2 = \sigma_X^2 = \mathrm{var}$. Variance reflects the spread of the random variable around the mean: the larger the spread/scatter, the larger the variance. The same caution should be used for the variance as for the mean because the variance is the mean of squared distances. Although convenient from the computational standpoint, it may not appropriate for a particular application.

If X is a discrete random variable, $(X - \mu)^2$ can be viewed as another discrete random variable, so its expectation can be computed as

$$\sigma^2 = \sum_{i=1}^{n} (x_i - \mu)^2 p_i. \tag{1.5}$$

In the following theorem we provide an alternative computation of the variance.

Theorem 1.10 *The following formula holds:*

$$\sigma^2 = E(X^2) - \mu^2. \tag{1.6}$$

Proof. Expanding $(x_i - \mu)^2$ in (1.5), we obtain

$$\sum_{i=1}^{n}(x_i - \mu)^2 p_i = \sum_{i=1}^{n}(x_i^2 - 2\mu x_i + \mu^2)p_i = \sum_{i=1}^{n} x_i^2 p_i - 2\mu \sum_{i=1}^{n} x_i p_i + \mu^2 \sum_{i=1}^{n} p_i.$$

But $\sum_{i=1}^{n} x_i^2 p_i = E(X^2)$ and $\sum_{i=1}^{n} x_i p_i = \mu$. Because $\sum_{i=1}^{n} p_i = 1$, we finally obtain $\sigma^2 = E(X^2) - 2\mu \times \mu + \mu^2 = E(X^2) - \mu^2$. $\square$

Sometimes $E(X^2)$ is called the *noncentral* second moment and σ^2 is called the *central* second moment. These moments are connected as shown in formula (1.6). We shall see later that formula (1.6) holds for continuous random variables as well.

To illustrate formula (1.6), we derive the variance of the Bernoulli random variable, we first find $E(X^2) = 0 \times \Pr(X = 0) + 1 \times \Pr(X = 1) = p$. Then, using the fact that $E(X) = p$, the variance of the Bernoulli variable is $\text{var}(X) = p(1-p)$.

Using formula (1.6) we obtain an explicit expression for the variance, as an alternative formula to (1.5),

$$\sigma^2 = \sum_{i=1}^{n} x_i^2 p_i - \left(\sum_{i=1}^{n} x_i p_i\right)^2. \tag{1.7}$$

Variance is always nonnegative and equals to zero if and only if X takes one value, the mean value.

It is easy to prove that $\text{var}(a + bX) = b^2 \text{var}(X)$ where a and b are numbers. In Chapter 3 we prove that if X and Y are independent random variables, then $\text{var}(X + Y) = \text{var}(X) + \text{var}(Y)$.

Standard deviation (SD) is the square root of variance, $\sigma = \sqrt{\text{var}}$. SD also reflects how random variable deviates from the mean. In contrast with the variance, SD does so on the original scale, compared with the variance, which is more convenient for interpretation. For example, if X is measured in feet, variance is measured in square feet but SD is measured in feet as well. For this reason, SD is often reported in applications to specify the scatter of the variable.

The mean and the variance of a discrete random variable are computed by the same formulas (1.2) and (1.5) when the number of outcomes is infinite, $n = \infty$.

The probability distribution completely specifies X. Mean and SD are integral features – where the random variable values concentrate and how wide the spread is.

The expected value is often used in finance to quantify the expected return, and variance is used to quantify the risk (volatility); the following is a typical example. Here we use the fact that the variance of the sum of two independent random variables equals the sum of two variances.

Example 1.11 *Grant problem. A person applies for a large grant in amount $100K with the probability of getting funded 1/5. There is an alternative: to apply for two small grants in amount of $50K each with the same probability of funding (it is assumed that the probabilities of funding are independent). What strategy is better?*

Solution. We define the expected return (funding) as the sum of possible amounts weighted with respect to their odds/probabilities. Let X define the binary random variable that takes value 100 with probability 1/5 and 0 with probability 4/5. Then the expected return in the first strategy is $100/5 + 0 \times (4/5) = 20$. Similarly, the second strategy leads to the expected return $50/5 + 50/5 = 20$. Thus in terms of the expected funding, the two strategies are equivalent. Now we look at these options from the risk perspective (this is typical for finance calculations). Clearly, between two strategies with the same expected return, we choose the strategy with a smaller risk/variance. Using formula (1.5) the variance of the large grant is $(0 - 20)^2 \times (4/5) + (100 - 20)^2 \times (1/5) = 1600$. Since getting funded from two small grants is independent, the total variance is twice the variance from each grant, $2 \times \left[(0 - 10)^2 \times (4/5) + (50 - 10)^2 \times (1/5) \right] = 800$. We conclude that, although the two options are equivalent in terms of the expected return, the second option is less risky. This fact, known as *diversification*, is the pillar of financial decision making and investment risk management analysis. We will return to the problem of diversification when considering the optimal portfolio selection in Section 3.8.

Problems

1. Introduce a function $M(a) = E(X - a)^2$. (a) Express this function through var and μ. (b) Prove that this function takes the minimal value at $a = \mu$. (c) Find minimum of M.

2. Prove that the mean is a linear function of scale, $E(aX) = aE(X)$, where a is a constant. Prove that variance is a quadratic function of scale, $\mathrm{var}(aX) = a^2 \mathrm{var}(X)$.

3. Plot the cdf of a Bernoulli random variable.

4. Prove that $F_{aX}(x) = F_X(x/a)$ for a positive a.

5. How would you report the average grade in class: mean, mode, or median? Justify using the concept of the students' standing in the class.

6. (a) What would state officials report in the summary statement: mean, median, or mode of income? (b) If state officials want to attract new business to the area, what would they appeal to: mean, mode, or median of company revenue? Explain.

7. Suppose that in Example 1.9 the pot increases by $q\%$. What would be the fair price to enter the game?

8. Generalize the grant problem to an arbitrary number of grants with equal probability.

1.5   R basics

The R programming language will be used throughout the text, so we introduce it at the very beginning.

R is a public domain statistical package. It is freely available for many computational platforms, such as Windows, Macintosh and Unix at

$$http://www.r-project.org/$$

The goal of this section is to give the reader the very basics of R programming. More detail will be given throughout the book. A comprehensive yet succinct description of R is found at

$$https://cran.r-project.org/doc/manuals/R-intro.pdf$$

There are two ways to use R: command line and script (function). We use the command line when a single-formula computation is needed, like a scientific calculator. When computations involve several steps, we combine them in a script. A distinctive feature of R is the assignment operator `<-`. We however prefer the usual = symbol.

For example, to compute 2×2, we write after > (the command prompt) `2*2` and press <Enter>. We can store the result using an identification (name of a variable), say `four<-2*2`. This means that computer computes 2×2 and stores the result in variable named `four`. Other operations are `+`, `-`, `/`, `^`. Many standard mathematical functions are available: `log`, `log10`, `exp`, `sin`, `cos`, `tan`, `atan`. Vectors and matrices are easy to handle in R. For example, to create a ten-dimensional vector of ones, we issue `one10<-rep(1,10)`. Then `one10` is a vector with each component 1; to see this we type `one10` and press <Enter>. Function `rep` is a special function of R, which is short of *repetition*. It has two arguments and in general form is written as `rep(what,size)`. If you do not want to assign special values to components of a vector, use `NA`, which means that component values are *not available*.

Similarly, one can create a matrix. For example, `mat.pi.20.4<-matrix(pi, nrow=20,ncol=4)` will create a 20×4 matrix with the name `mat.pi.20.4` with all entries π (`.` may be a part of a name). Again, to see the results, you type `mat.pi.20.4` and press <Enter>. It is easy to see the history of the commands you issued using arrow keys ↑ or ↓. Thus, instead of retyping `mat.pi.20.4`, you pick the previous command and remove the unwanted parts.

R distinguishes lowercase and uppercase, thus `mat.pi.20.4` and `mat.Pi.20.4` mean different things.

If `a` and `b` are two one-dimensional arrays (vectors) of the same size, `a*b` computes the vector whose values are the component-wise products. The same rule holds for matrices of the same dimensions. In contrast to the component-wise multiplication, the symbol `%*%` is used for vector/matrix multiplication. For example, if $\mathbf{A}$ is a $n \times m$ matrix and $\mathbf{a}$ is an m-dimensional vector (the boldface is used throughout the book to indicate vectors and matrices), then `A%*%a` computes a vector $\mathbf{Aa}$ of dimension n. If $\mathbf{B}$ is a $m \times k$ matrix, then `A%*%B` computes an $n \times k$ matrix. Function `t` is used for vector/matrix transposition. For example, if $\mathbf{A}$ is a $n \times m$ matrix, `t(A)` is $\mathbf{A}'$. If $\mathbf{a}$ is a vector, then `t(a)%*%a`, `sum(a^2)`, and `sum(a*a)` all give the same result. There are a few rules when adding or multiplying matrices and vectors that do not comply with mathematics but are convenient from a computational standpoint. For example, if $\mathbf{A}$ is a $n \times m$ matrix and $\mathbf{a}$ is an m-dimensional vector, then `A+a` is acceptable in R and computes a matrix with columns that are columns of matrix `A` plus `a`. If `a` is an m-dimensional vector, then `A+a` computes a matrix with rows that are rows of `A` plus `a`. When `A` is a square matrix operation is on columns. The same rule works for multiplication (or division). For example, if $\mathbf{A}$ is a $n \times m$ matrix and $\mathbf{a}$ is an n-dimensional vector, then `A*a` produces a $n \times m$ matrix with the ith row as the product of the ith row of matrix $\mathbf{A}$ times a_i. It is important that `a` is a vector, not the result of `A%*%b` that produces a $n \times 1$ matrix. To make `A%*%b` a vector, use `as.vector(A%*%b)`.

1.5.1   Scripts/functions

Typically statistical computations involve many lines of code you want to keep and edit. In this case, you want to write user-defined scripts (functions). Functions have arguments and a body. For example, if you want to create a function by the name `my1`, you would type `my1<-function(){}`. This means that so far this function has no arguments in ( ) and no body in { }. To add this, we type `my1<-edit(my1)` and a text editor window appears where we can add operators. It is good style to use comments in your program. In R everything after # is a comment up to the next carriage return.

For example, let us say we want to write a program that multiplies two user-defined numbers. We start with program creation `twoprod<-function(){}` and then add the text using command `twoprod<-edit(twoprod)`.

Here is a version of the R program:

```
twoprod<-function(t1,t2)
{
    prod<-t1*t2
    return(prod)
}
```

Here t1 and t2 are the arguments. The user can use any numbers. For example, if one enters on the command line twoprod(0.3,24), the answer should be 7.2. Another, more explicit way to run the function is to use twoprod(t1=0.3,t2 =24). A convenient feature of R is that one can define default values for the arguments. For example, one may use t1=10 as default. Then the function should be

```
twoprod<-function(t1,t2=10)
{
    prod<-t1*t2
    return(prod)
}
```

To compute 2.4×10 it suffices to run twoprod(t1=2.4), but this does not mean that t2=10 always. You still can use any t2 even though the default is specified. Each function should return something, in this case the product of numbers. To run a function, it has to have ( ) even if no arguments are specified; otherwise, it prints out the text of the function.

1.5.2 Text editing in R

You have to use a text editor to edit functions (R codes). The default text editor in R is primitive – it does not even number the lines of the code. For example, when R detects errors in your code you may see a message like this:

```
Error in .External2(C_edit, name, file, title, editor) :
unexpected ',' occurred on line 521
use a command like
x <- edit()
to recover
```

This means that you have to count 521 lines yourself! There are plenty of public domain Windows text editors used by programmers worldwide, such as notepad++ or sublime. For example, to make notepad++ your R editor, you have to add the line (for Windows PC)

```
options(editor="c:\\Program Files (x86)\\Notepad++\\notepad++.exe")
```

to the file Rprofile.site, which, for example, is located in the folder C:\Program Files\R\R-3.2.2\etc\.

Alternatively, you may issue this command in the R console every time you start a new session. The notepad++ software can be downloaded from the site https://notepad-plus-plus.org/. Here we assume that the program notepad ++.exe is saved in the folder c:\\Program Files (x86)\\Notepad++ and you use R-3.2.2 version of R; you have to make slight modifications otherwise. The sublime editor can be downloaded from https://www.sublimetext.com/3.

1.5.3   Saving your R code

We strongly recommend saving your code/function/script as a text file every time you run it. For example, assuming that you have a Windows computer and you want to save your function `myfun` in the existing folder/directory, `C:\StatBook`, your first statement in this function may look like this `dump("myfun","C:\\Stat Book\\myfun.r")`. Note that double backlashes are used. If you want to read this function, issue `source("C:\\StatBook\\myfun.r")` in the R console. Saving as a text file has a double purpose: (i) You can always restore the function in the case you forgot to save the R session. (ii) You can keep the R code in the same folder where other documents for your project are kept. Throughout the book, all R codes are saved in the folder `C:\\StatBook\\`.

If you work on a Mac computer, the syntax is slightly different: single forward slashes are used, and there is no reference to the hard drive letter. For example, to save `myfun`, you use `dump("myfun","/Users/myname/StatBook/myfun.r")` with the `source` command being modified accordingly.

1.5.4   for loop

Repeated computation expressed in a loop is the most important method in computer programming. The simplest example of the loop is

```
for(i in 1:10)
{
    #body of the loop
}
```

which repeats operators between { and } ten times by letting i=1, i=2, ..., i=10. Here `1:10` creates a sequence of numbers from 1 to 10. There is a more general way to create sequences `seq(from=,to=,by=)` or `seq(from=,to=,length=)`. For example, `s1<-seq(from=1,to=10,by=1)` and `s2<-seq(from=1,to=10,length=10)` both produce the same sequence from 1 to 10. Let us write a program that computes the sum of all even numbers from 2 to n, where n is user-defined. First, we specify the function and edit it: `twonsum<-function(){};` `twonsum<-edit(twonsum)`. The code may look like this:

```
twonsum<-function(n)
{
    sn<-0 # initialization
    nseq<-seq(from=2,to=n,by=2)
    for(n in nseq)
        {
            sn<-sn+n # summation
        }
    return(sn)
}
```

If there is only one operator in the loop body, the braces are not required, so a shorter version is

```
twonsum<-function(n)
{
    sn<-0 # initialization
    nseq<-seq(from=2,to=n,by=2)
    for(n in nseq)
        sn<-sn+n # summation
    return(sn)
}
```

This function can be significantly shortened because there is a built-in summation function `sum(x)`, where x is a an array. Similarly, `prod(x)` computes the product. Thus, `twonsum` function can be shortened as `sum(seq(from=2,to=n,by=2))`.

1.5.5 Vectorized computations

Many, but not all, loop operations can be vectorized. Then, instead of loops we use operations with vectors. The vectorized versions usually are more compact but sometimes require matrix algebra skills. More importantly, the vectorized approach is more efficient in R – loops are slower, but they may need less RAM. Vectorized operations are faster because they are written in C or FORTRAN. Using the C language, vectorized computations pass the array pointer to C, but the loop communicates with C at every single iteration. Some vectorized operations, like `romSums`, `colMeans`, etc., are already built in.

Compute the mean, variance, and SD of a discrete random variable specified by n-dimensional vectors of values and probabilities.

```
mvsd<-function(x,p)
{
    # x is the vector of values
    # p is the vector of probabilities
    n<-length(x) # recover size
    mux<-sum(x*p) # mean
    ex2<-sum(x^2*p) #E(X^2)
    s2x<-ex2-mux^2 # alternative variance
    sdx<-sqrt(s2x) # SD
    return(c(mux,s2x,sdx)) # return the triple
}
```

For example, if we run `mvsd(x=c(0,1),p=c(.25,.75))` it will give us 0.75000 0.1875000 0.4330127. Indeed, since we specified a Bernoulli distribution, we should have $E(X) = 0.75$, $\mathrm{var}(X) = 0.25 \times 0.75 = 0.1875$, SD $= \sqrt{\mathrm{var}(X)} = 0.433$. R has `mean`, `var` and `sd` as built-in functions with a vector as the argument. For example, if `a<-c(0.1,0.4,0.5)` then `mean(a)` and `sd(a)` return 0.33333 and 0.2081666, respectively. Summation, subtraction, and multiplication of vectors with the same length are component-wise and produce a vector of the same length.

To illustrate the built-in vectorized function consider the problem of computing the SD for each row of a $n \times m$ matrix X, where n is big, say, $n = 500000$ as in genetic applications. Of course, we could do a loop over $500,000$ rows and even use the `mvsd` function. The following is a much faster version.

```
SDbig=function(X)
{
    mrow=rowMeans(X) #averaging for each row across columns
    mrow2=rowMeans((X-mrow)^2)
    SDs=sqrt(mrow2)
    return(SDs)
}
```

We make two comments: (i) `rowMeans` returns a vector with length equal to the number of rows in matrix `X`, and each component of this vector is the mean in the row across columns. (ii) Although `X` is a matrix and `mrow` is a vector, `R` does not complain and subtracts `mrow` from each row when we compute `X-mrow`.

Five vectorized computations for double integral approximation are compared in the following example.

Example 1.12 *Comparison of five vectorized computations. Use vectorized computations for numerical approximation of the double integral*

$$A = \int \int_{x^2+y^2<1} e^{-(3x^2-4xy+2y^2)}dxdy.$$

Solution. The exact integral value can be obtained by rewriting the integral in a form suitable for symbolic algebra software, such as *Maple* or *Mathematica*:

$$A = \int_{-1}^{1}\left(\int_{-\sqrt{1-x^2}}^{\sqrt{1-x^2}} e^{-(3x^2-4xy+2y^2)}dy\right) dx = 1.374.$$

To approximate the integral, we replace the integral with the sum of integrand values over the grid for x and y. Let the grid for x and y be an array from -1 to 1 of `length=N` (the rectangular grid must contain the integration domain). First, compute the $N \times N$ matrix $\mathbf{M}$ of values $3x^2 - 4xy + 2y^2$, and second, compute the sum of values $e^{-(3x^2-4xy+2y^2)}$ multiplied by the step in each grid for which $x^2+y^2 < 1$. The `R` code is found in the file `vecomp.r` and the time of computation by each algorithm is shown in the following table.

Method	Double loop	Single loop	Matrix algebra	rep	expand.grid	outer
job	job=1	job=2	job=3	job=4	job=5	job=6
Time (s)	19	4	3	6	7	5

The first method (job=1) uses the brute-force double loop matrix computation and fills in the matrix M in an element-wise fashion, instead of a vectorized algorithm. The second algorithm (job=2) is a vectorized algorithm using a single loop over the x grid and filling in the matrix M row by row. The vectorized algorithm in job=3 uses matrix algebra to compute matrix M as $\mathbf{M} = 3\mathbf{x}^2\mathbf{1}' - 4\mathbf{xy}' + 21\mathbf{y}^{2'}$, where $\mathbf{x}$, $\mathbf{y}$, and $\mathbf{1}$ are $N \times 1$ vectors ($\mathbf{1}$ is a vector of ones). The fourth algorithm (job=4) computes M on the grid of values (x_i, y_j) for $i, j = 1, ..., N$ using the rep function with two options: (i) to repeat vector x with the option times=N and (ii) to repeat each element of vector x with the option each=N. The fifth method (job=5) is similar to the previous one, but instead of the rep function, another R function, expand.grid, is used (they produce the same grids for two dimensions). This function is especially convenient with multiple dimension grids, say, for three-dimensional integration. Finally, the sixth method (job=6) uses a built-in function outer. This function returns a $N \times N$ matrix of values $f(x_i, y_j)$ and is especially convenient for our purposes. Function f should be specified in R before computing. The time of computation in seconds is assessed using the date() command. A more precise way to find the time of computation is by calling the Sys.time() function before and after the operation and taking the difference. Not surprisingly, double loop takes a long time. Although the third method is the fastest, it may be difficult to generalize to nonlinear functions and integrals of higher dimension. The R function that implements the five methods can be accessed by issuing source("c:\\StatBook\\vecomp.r"). A few remarks on the script: (i) It is easy to lose the R code; it is a good idea to save the code as a text file in a safe place on the hard disk. Command dump("vecomp","c:\\StatBook\\vecomp.r") saves the R code/object in the folder c:\\StatBook under the name vecomp.r with the full name c:\\StatBook\\vecomp.r. (ii) In job=6 the command sum(M[x2y2<1]) computes the sum of elements of matrix M for which condition x2y2<1 holds. An advantage of these integral approximations is that any domain of integration may be used if expressed as an inequality condition. The condition should be a matrix with the same dimension as matrix M. The double integral A is approximated in Example 3.60 via simulations.

apply

Besides rowMeans and the like, such as colMeans, colSums, rowSums, or pmax, R has a built-in capability to do vectorized computations in general way. Here we illustrate this feature by the apply function. This function has three arguments: X, MARGIN, and FUN. The first argument, X, specifies a matrix. The second argument, MARGIN, specifies row or column, and the third argument, FUN, specifies the function to be performed in the vectorized fashion. For example, instead of SDbig, one can use apply(X,1,sd). If the second argument is 2, it returns SD of the columns. Of course, one could use other functions for FUN, such as sum,

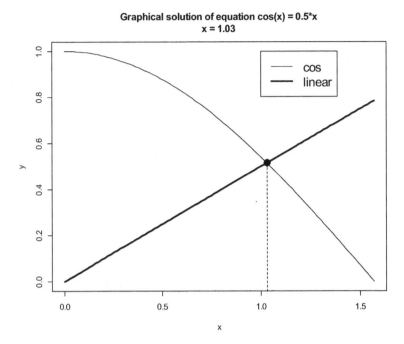

Figure 1.5: *This plot is the result of function* **my1sr()**. *This function returns* **1.031**, *an approximate solution to the equation* $\cos(x) = x/2$.

max, etc. Note that this is the easiest example of **apply**; there are others, more sophisticated examples of this function as well as other vectorized functions, such as **lapply** and **tapply**. The function specified in **FUN** may be any function of the row data and may return a vector. See Example 6.88 where **apply** is used for vectorized simulations.

Example 1.13 *Apply command. Fill a 100 by 12 matrix using numbers from 1 to 1200 and compute (a) the normalized matrix (subtract the mean and divide by SD in each row) using* **apply**, *and (b) two numbers in each row such that 25% and 75% of the data is less than those numbers (the first and third quartiles).*

Solution. The matrix is filled with integers from 1 to 1200 using command **X=matrix(1:1200, ncol=12)**. (a) To normalize the matrix, we write a function with a vector argument that may be thought of as a typical row of the matrix: **normal=function(x) (x-mean(x))/sd(x)**. Now the normalized matrix is computed as **apply(X,1,FUN=normal)**. (b) First, the function orders the array of numbers in each row and then picks the $(3n/4)$th element: **tdqurt=function(x) {xo=x[order(x)]; n=length(xo);c(xo[n/4],xo[3*n/4])}**. Note that the call **apply(X,1, FUN=tdqurt)** returns a matrix 2 by 100.

1.5.6   Graphics

Versatile graphics is one of the most attractive features of R. A picture is worth a thousand words and a good graph may likely become the endpoint of your statistical analysis. Let us start with an example of a graphical solution of a transcendent equation, $\cos x = \alpha x$, where α is a positive user-defined parameter. Specifically, we want to (i) plot two functions, $y_1 = \cos x$ and $y_2 = \alpha x$, on an interval where they intersect and (ii) find the point of intersection. Since y_1 is a decreasing function and y_2 is an increasing function on $(0, \pi/2)$, we plot these functions on this interval. In this function, `alpha` is the used-defined parameter with default value 3. The graphical output of the R function `my1sr` is shown in Figure 1.5. This function can be accessed by issuing `source("c:\\StatBook\\my1sr.r")`. The command `indmin<-which.min(abs(y1-y2))` returns the index for which the absolute difference between function values is minimum.

We make several comments:

1. In the `plot` function, the first argument is an array of x-values, and the second argument is an array of the corresponding y-values; the arrays/vectors should have the same length. The second argument of the function `type` specifies the type of plot. For example, `type="l"` will produce lines and `type="p"` will produce points; `xlab` and `ylab` are the axis labels; `main` specifies the title of the plot.

2. We add the line to the plot using function `lines`. As with the `plot` function, the first and second arguments are the x- and y- values; `lwd` specifies the line width (regular width is 1). The line style can be specified as well (see below), with default `lty=1`.

3. A legend is a must in almost all plots to explain what is plotted. The first argument is the x-coordinate and the second argument is the y-coordinate of the upper-left corner of the legend rectangle; the third argument is the message in the legend, which, in our function, consists of two words; `lty` specifies the line style: 1 is solid, 2 is dotted, and 3 is dashed.

4. In this code, `which.min` returns the index with the minimal value. In our case, this built-in function returns the index where two lines are close to each other.

5. Parameters `cex` and `pch` control the type of the point and its size. The default values are `cex=1`(small) and `pch=0` (empty circle); `cex=2` produces a circle of the size twice large as the default.

6. Title combines text and numbers using the `paste` command. Text and characters need quotes; `\n` forces text to the next line.

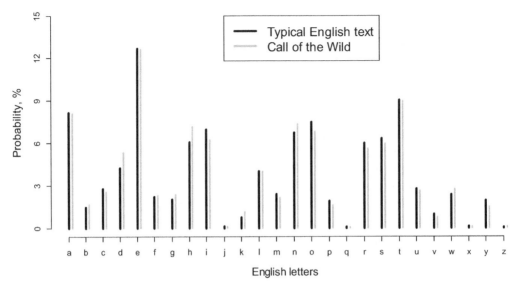

Figure 1.6: *The probability plot of English letters in a typical English text and "Call of the Wild" by Jack London. The distribution of letters in this novel will be further studied in Example 7.66.*

In the following example, we illustrate how two discrete distributions can be visually compared in R.

Example 1.14 *Frequency of English letters in a Jack London novel. Write an R function that plots the frequency of English letters in a typical text and the frequency of letters in "The Call of the Wild" by Jack London side by side. Do the frequencies look similar?*

Solution. The text file `Jack_London_Call_of_the_Wild_The_f1.char` contains *Call of the Wild* character by character. The frequency of English letters in a typical text, as the average over a large number of English texts, can be found on the Web. The R code is found in the file `frlJLET.r`, which creates Figure 1.6. We make a few comments: (i) It is useful to save the R code in a text file using the `dump` command every time we run the function. It is easy to restore the function later by issuing `source("c:\\StatBook\\frlJLET.r")`. (ii) The char file is uploaded using the built-in function `scan`; option `what=""` means that letters are characters. (iii) The operator `ch[ch==freqlet[i,1]]` returns a part of array `ch` for which the condition `ch==freqlet[i,1]` holds, or in other words, it returns characters specified by `freqlet[i,1]`.

We conclude that, on average, the frequencies of the letters in a typical English text match those of a famous Jack London novel; however, for some letters such as "d" and "h," the discrepancy is noticeable. Later we shall test that the frequencies of letters in the novel are the same as in a typical English text using the Pearson chi-square test. The entropy of English texts is computed in Section 2.15.

1.5.7   Coding and help in R

Coding in R, as any other programming language, may be a frustrating experience even for a mature programmer. There are two types of error messages: syntax and run-time errors. Syntax errors are displayed after the window with the code is closed. Unfortunately, while R reports the line where the error may be, (i) the actual error may be not in this line but below or above, and (ii) the built-in editor in R does not number lines. More sophisticated text editors can be used to get line numbers. Examples include Rstudio (official site https://www.rstudio.com) and Notepad++ (official site https://notepad-plus-plus.org). In fact, Rstudio provides a different Graphical User Interface (GUI) for R with multiple windows for the command line, graphs, etc.

There is no unique advice to understand why your code does not work when you get run-time errors. Printing out values of variables or arrays may help to figure out what is going on and where exactly the error occurs.

It is easy to lose the code in R. I recommend saving the R code in a text file via the dump command as in the code fr1JLET in Example 1.14. To restore this code, issue source("c:\\Stat Book\\fr1JLET.r") in the R console.

To get help on a known function, say, legend issue ?legend on the command line. Documentation is the Achilles' heel of R. You are likely to find that the help too concise and sometimes without examples. Typically, options are not explained well. Google it!

Problems

1. Write an R function with argument x that computes e^x using Taylor series expansion $1 + \sum_{n=1}^{\infty} x^n/n!$ and compare with exp(x). Report the output of your function with x=-1 and compare it with exp(-1). [Hint: Use for loop until the next term is negligible, say, eps< 10^-7.]

2. Write an R function with n and m as arguments, matrix dimensions. Generate a matrix, say, matrix(1:(n*m),ncol=m). Compute a vector of length n as the sum of values in each row using rowSums. Do the same using the function apply. Then use sum with a loop over rows and compare the results.

3. Modify function vecomp to approximate integral $\int\int_{x^2+x+y^2<2}(1 + 2x^2 + y^2)^{-1}dxdy$. [Hint: The true value is 2.854; use grid for x from -2 to 1 and for y from $-3/2$ to $3/2$.]

4. Modify Example 1.13 to compare apply with the traditional for loop in terms of computation time using a large matrix X (use date() before and after computation. A more accurate time can be computed using the Sys.time() call.

5. Demonstrate that `apply` is equivalent to `colMeans`, `rowMeans`, `colSums`, and `rowSums`, but slower. [Hint: Use the `for` loop to obtain the time in the order of minutes.]

6. Write a formula for a straight line that goes through two points and verify this formula graphically using R code. The code should have four arguments as coordinate of the first and second point. Use `points` to display the points. [Hint: The equation of the straight line that goes through points (x_1, y_1) and (x_2, y_2) is $(y - y_1)/(y_2 - y_1) = (x - x_1)/(x_2 - x_1)$.]

7. Find the minimum of the function $ax + e^{-bx}$ graphically where a and b are positive user-defined parameters. Plot the function and numerically find where it takes a minimum using the `which` command. Display the minimum point using `points` and display the minimal value. Find the minimum analytically by taking the derivative, and compare the results.

8. Rearrange the plot in function `my1sr` starting with the most frequent letter 'e'. [Hint: Use `[order(-numfr)]` to rearrange the rows of the `freqlet` matrix.]

9. In the analysis of letter frequency, capital letters were reduced to lowercase using the `tolower` command. Does the conclusion remain the same if only lowercase letters are compared (without using `tolower`)?

10. Find a text on the Internet and save it as a txt file. Use the `for` loop over the words and the loop over the number of characters in the word (`nchar`). Then use `substring` to parse words into characters, compute the frequencies, and plot them as in Figure 1.6 using `green` bars.

11. (a) Confirm by simulation the results of Example 1.11 using the `for` loop. (b) Use vectorized simulations. [Hint: Use `X=runif(n=1)<0.2` to generate a Bernoulli random variable.]

12. (a) Is the probability of having a boy and a girl in the family the same as having two boys or two girls? (b) Use vectorized simulations to confirm the analytical answer. [Hint: Generate Bernoulli random variables as in the previous problem.]

1.6   Binomial distribution

The binomial distribution is the distribution of successes in a series of independent Bernoulli experiments (trials). Hereafter we use the word *success* just for the occurrence of the binary event and *failure* otherwise. More precisely, let $\{X_i, i = 1, 2, ..., n\}$ be a series of independent identically distributed (iid) Bernoulli random variables with the probability of success in a single experiment p, meaning that

$\Pr(X_i = 1) = p$. Consequently, the probability of failure is $\Pr(X_i = 0) = 1 - p$. Note that sometimes the notation $q = 1 - p$ is used.

Table 1.1. Four R functions for binomial distribution (`size=n`, `prob=p`)

R function	Formula	Returns	Explan.
`dbinom(x,size,prob)`	$\binom{n}{m} p^m (1-p)^{n-m}$	(1.8)	`x = m`
`pbinom(q,size,prob)`	$\sum_{m=0}^{K} \binom{n}{m} p^m (1-p)^{n-m}$	(1.10)	`q = K`
`rbinom(n,size,prob)`	$X_1, X_2, ..., X_N \overset{\text{iid}}{\sim} B(n,p)$	rand numb	`n = N`
`qbinom(p,size,prob)`	$\sum_{k=0}^{K} \binom{n}{m} p^m (1-p)^{n-m} = P$	quantile	`p = P`

We want to find the probability that in n independent Bernoulli experiments, m successes occur. Since $X_i = 1$ encodes success and $X_i = 0$ encodes failure, we can express the number of successes as the sum, $X = \sum_{i=1}^{n} X_i$. We use the notation $X \sim \mathcal{B}(n,p)$ to indicate that X is a binomial random variable with n trials and the probability of success p in a single trial. Clearly, X can take values $0, 1, ..., m, ..., n$. The celebrated *binomial probability* formula gives the distribution of X, namely,

$$\Pr(X = m) = \binom{n}{m} p^m (1-p)^{n-m}, \quad m = 0, 1, ..., n, \qquad (1.8)$$

where the coefficient $\binom{n}{m}$ is called the *binomial coefficient* (we say "n choose m") and can be expressed through the factorial

$$\binom{n}{m} = \frac{n!}{m!(n-m)!}.$$

We let $\binom{n}{0} = 1$; obviously, $\binom{n}{n} = 1$. An algebraic application of the binomial coefficient is the expansion of the nth power of the sum of two numbers:

$$(a+b)^n = \sum_{m=0}^{n} \binom{n}{m} a^m b^{n-m}. \qquad (1.9)$$

In some simple cases, we do not have to use the binomial coefficient. For example, the probability that in n experiments there is no one single success is $\Pr(X = 0) = (1-p)^n$. Similarly, $\Pr(X = n) = p^n$. However, all these probabilities can be derived from the general formula (1.8).

The binomial distribution is built into R. There are four different functions for each distribution in R including the binomial distribution. The R function `pbinom` computes the cdf, the probability that $X \leq K$,

$$F(K) = \Pr(X \leq K) = \sum_{m=0}^{K} \binom{n}{m} p^m (1-p)^{n-m}, \quad K = 0, 1, ..., n. \qquad (1.10)$$

The R function qbinom is the inverse of pbinom and computes K such that the cdf equals the specified probability. All four R functions for the binomial distribution are presented in Table 1.1. Arguments may be vectors. For example, dbinom(0:size,size,prob) computes probabilities for each outcome $m = 0, ..., n$, where size $= n$ and prob $= p$.

Note that the quantile of a discrete distribution (K) may be not exactly defined given p due to discreteness of the cdf. The call qbinom returns the smallest integer for which the cdf is greater or equal to p. For example, qbinom(p=0.2,size =5, prob=0.5) and qbinom(p=0.4,size=5,prob=0.5) return the same 2. In contrast, pbinom(q=2,size=5,prob=0.5)=0.5.

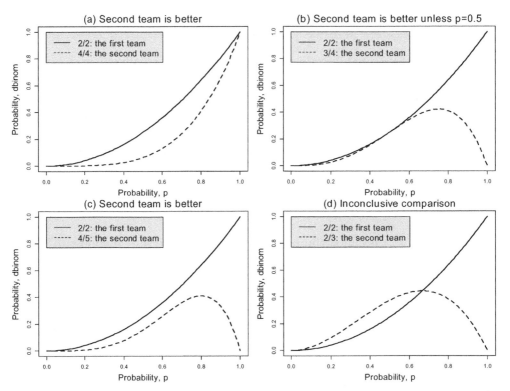

Figure 1.7: *Illustration for Example 1.15. Four answers to the question what team is better based on the number of wins. The second team is better if the probability of the outcome is smaller for all $0 < p < 1$.*

Example 1.15 *What team is better?* *Two teams, say, soccer teams, are compared. The first team won two out of two games, and the second team won three out of four games. What team is better? In other words, what team has a better chance/probability of winning in a single game?*

Solution. Before advancing to the solution, we make a couple of assumptions: (i) The teams did not play against each other, that is, 2/2 and 3/4 are scores of the games with other teams. (ii) The probability of winning against other teams is the same for each team under consideration. These assumptions imply that the number of wins follows the binomial distribution with Bernoulli probabilities to win in a single game for team 1 and 2, p_1 and p_2, respectively. Thus, we ask, is $p_1 < p_2$, $p_1 > p_2$, or is the answer inconclusive?

First, consider the case when two teams won all games played (all wins). Let the number of games played by the first team be n_1 and the number of games played by the second team be n_2. Since all games have been won, one may deduce that $p_1 = 1$ and $p_2 = 1$, and the naive answer is that the two teams are the same. However, consider a specific case when $n_1 = 1$ and $n_2 = 100$. Clearly, the second team should be claimed better. In general, under the "all wins" scenario, the second team is better if $n_1 < n_2$. Now we derive this intuitive rule by comparing the probabilities of winning all games. For the first team this probability is $p_1^{n_1}$, and for the second team this probability is $p_2^{n_2}$. How would probabilities p_1 and p_2 be related if the teams won all games they played, n_1 and n_2, respectively? To find p_2 as a function of p_1, we need to solve the equation $p_1^{n_1} = p_2^{n_2}$ for p_2; that gives $p_1 = p_2^{n_2/n_1} < p_2$. This means that the second team is better, coinciding with our intuition. We do not have to solve the equation for p_1, just plot p^{n_1} and p^{n_2} for $0 < p < 1$ on the same graph; see the upper-left panel in Figure 1.7 with $n_1 = 2$ and $n_2 = 4$, denoted as 2/2 and 4/4, respectively. If the second curve is below, the second team is better.

Second, consider the case where the first team played n_1 games and won all of them, as before, and the second team played n_2 games and won m_2 ($m_2 < n_2$). A naive answer is that the first team is better because $p_1 = n_1/n_1 = 1$ and $p_2 < m_2/n_2 < 1$. But consider the case when $n_1 = 2$ and $m_2 = 99$ and $n_2 = 100$. In this case, the second team is better. As in the previous case, find p_1, which leads to the same result of winning for the second team, $p_1^{n_1} = \binom{n_2}{m_2} p_2^{m_2}(1 - p_2)^{n_2 - m_2}$. If

$$p_1 = \left[\binom{n_2}{m_2} p_2^{m_2}(1 - p_2)^{n_2 - m_2} \right]^{1/n_1} < p_2$$

for all $0 < p_2 < 1$, then the second team is better. Again, it is convenient to plot the curves p^{n_1} and $\binom{n_2}{m_2} p^{m_2}(1 - p)^{n_2 - m_2}$ on the same graph. If the second curve is below the first one for all p the second team is better. If $n_1 = 2$ and $m_2 = 3, n_2 = 4$, the second team is not worse if $\binom{4}{3} p^3(1 - p)^{4-3} \leq 1$ for all $0 < p < 1$. But $\binom{4}{3} = \frac{4!}{3!(4-3)!} = \frac{4 \times 3!}{3!(4-3)!} = 4$ and indeed $4p^3(1 - p) < 1$. Note that the inequality becomes an equality when $p = 0.5$. Hence the second team is better unless the chance of winning in each game is 50/50 for both teams; see panel (b) of Figure 1.7.

Third, in general, we say that the second team is better if its outcome probability is smaller than the outcome of the first team:

$$\binom{n_2}{m_2} p^{m_2}(1 - p)^{n_2 - m_2} < \binom{n_1}{m_1} p^{m_1}(1 - p)^{n_1 - m_1}, \quad 0 < p < 1.$$

The two bottom graphs depict $m_2 = 4$ and $n_2 = 5$ and $m_2 = 2$ and $n_2 = 3$. The last outcome leads to an inconclusive comparison. We show in Example 3.39 how to solve the problem of an inconclusive comparison using a noninformative prior for p.

Example 1.16 *Birthday problem. What is the probability that at least two students in a class of 23 students have the same birthday (assume that there are 365 days/year and birthdays are independent)?*

Solution. It is reasonable to assume that the birth rate is constant over the year. Let the students names be John, Catherine, Bill, etc. Instead of finding the probability that at least two students have the same birthday, we find the probability that all 23 students have birthdays on different days. The probability that John and Catherine do not have the same birthday is $(365 - 1)/365$. The probability that John, Catherine, and Bill do not have birthday on the same day is $(365 - 1)(365 - 2)/365^2$, etc. Finally, the probability that 23 students have at least one shared birthday is

$$1 - \frac{(365 - 1)(365 - 2) \cdots (365 - 22)}{365^{22}}. \tag{1.11}$$

An R code that computes this probability is `1-prod(seq(from=364,to=365-22, by=-1))/365^22`, which gives the answer `0.5072972`. Some may find this probability surprisingly high.

Simulations for the birthday problem

It is instructive to do simulations in R to verify formula (1.11). Imagine that you can go from class to class with the same number of students and ask if at least two students have the same birthday. Then the empirical probability is the proportion of classes where at least two students have the same birthday.

The following built-in R functions are used.

`ceiling(x)` returns the minimum integer, which is greater or equal to `x`. There are two similar functions, `round` and `floor`.

`unique(a)` where `a` is a vector, returns another vector with the vector's distinct (unique) values. For example, `unique(c(1,2,1,2,1))` returns a vector with components 1 and 2.

`length(a)` returns the length (dimension) of the vector.

Function `birthdaysim` simulates this survey. The key to this code is the fact that if a vector has at least two of the same components, the number of unique elements is smaller than the length of the vector. Using the default value Nexp $= 100,000$, this program estimates the proportion to be 0.507. It is important to understand that each run comes with a different number, but all are around 0.507, a close match with the theoretical value.

Unlike theoretical solution (1.11), this simulation program, after a small modification, allows estimating this probability under the assumption that the distribution of birthdays is not uniform throughout the year. In fact, data from different countries confirm that the distribution is not uniform as discussed by Borja [12]: For many countries in the northern hemisphere, the probability of birth on a specific day looks like a sinusoid with the maximum around September. For example, we can model the probability as $\Pr(\text{birthday} = d) = 0.9 - 0.2\sin(2\pi d/365)$ for $d = 1, 2, ..., 365$. The R function is `birthdaysim.sin`. Simulations may take several minutes (use, say, `Nexp=10000` to reduce the time). The probability estimate is around 0.516, slightly higher than under the uniform-birthday distribution. We make a few comments on the code: (i) `cumsum` computes the cumulative sum of an array; this is the shortest way to compute a cdf. (ii) `runif(1)` generates a random number on the interval (0,1); the uniform distribution will be covered in detail in the next chapter.

Problems

1. Derive the formula $\sum_{m=0}^{n} \binom{n}{m} = 2^n$ from the binomial probability (1.8).

2. Use formula (1.8) to prove that $(a+b)^3 = a^3 + 3ab^2 + 3ba^2 + b^3$.

3. Find m for which the binomial probability is maximum.

4. Erica tosses a fair coin n times and Fred tosses $n+1$ times. What is the probability that Fred gets more heads than Erica. Solve the problem theoretically and then write an R function with simulations (n is the argument of the function). Compare the results. [Hint: Use the fact that the probability of getting m_F heads in Fred's tosses and m_E heads in Erica's tosses is the product of the probabilities. Apply command `mean(X>Y)` to compute the proportion of elements of array X greater than Y.]

5. In the birthday example, estimate the probability using simulations that *exactly* two students have the same birthday. (Before computing, what probability is less, *exactly two* or *at least two*?) Plot the theoretical and empirical (from simulations) probabilities that at least two students have the same birthday in the class of n students versus n. Explain the comparison.

6. Sixteen players of two genders sign up for a tournament. What is the probability that there will be eight men and women? Provide a theoretical answer and confirm via simulations.

7. What is the probability that in n fair coin tosses, there will be m head streaks? Give a theoretical formula and verify by simulations. Use two methods to compute the theoretical probability: with or without `dbinom`. Write R code to compute the simulated probability with the number of

simulations equal to 1000K. [Hint: Your solution should be one line of code.]

8. The chance that a person will develop lung cancer in his/her lifetime is about 1 in 15. What is the probability that in a small village with 100 people, (a) there are no individuals with lung cancer, and (b) there is only one case? In both cases check your answer with simulations. [Hint: Use `rbinom`.]

9. Early mathematicians believed that in a fair coin tossing after a long streak of heads, a tail is more likely. Here is one of the proofs: make 100 tosses and consider the longest streak of heads. Tails always occurs after the streak. Give pros and cons for this statement. Do you agree with this statement?

10. Conjecture: Any nonuniform distribution may only increase the probability that at least two people have the same birthday. In other words, (1.11) is the lower bound over all possible distributions of birthdays on $1, 2, ..., 365$. [Hint: Modify function `birthdaysim.sin` to check using simulations; run the function with user-defined birthday probabilities within each month.]

1.7   Poisson distribution

Poisson distribution is useful for modeling a distribution of counts. This distribution specifies the probability that a discrete random variable takes value k, where k may be one of $0, 1, 2, ...$ Simeon Denis Poisson (1781–1840), a famous and powerful French mathematician and physicist, introduced this distribution. We say that the random variable X has a Poisson distribution with a positive parameter λ if

$$\Pr(X = k; \lambda) = \frac{1}{k!} e^{-\lambda} \lambda^k, \quad k = 0, 1, 2, ... \quad (1.12)$$

The presence of the factorial in the denominator (the normalizing coefficient) can be seen from the following calculus formula: $e^x = 1 + x + \frac{1}{2!} x^2 + \cdots + \frac{1}{k!} x^k + \cdots$. Using this formula it is clear that the probabilities add to one, $\sum_{k=0}^{\infty} \Pr(X = k) = 1$, as in the case with all probability distributions. We will use the symbolic notation $X \sim \mathcal{P}(\lambda)$ to indicate that X has a Poisson distribution with a positive parameter λ. Sometimes λ is called the Poisson rate.

In Table 1.2, we show four R functions associated with the Poisson distribution. As is the case with other built-in distributions in R, these functions take vector arguments. For example, if `lambda` is a scalar and `x` is a vector, then `dpois(x,lambda)` returns a vector of the same length as `x` keeping `lambda` the same. If `lambda` is a vector of the same length as `x`, this function returns a vector of probabilities computed using the corresponding pairs of `x` and `lambda`. Even though the support of the Poisson distribution is infinite, there are many examples when this distribution may serve as a good probabilistic model despite the fact that a random variable takes a finite set of values when the upper limit is difficult to specify: the number of children in the family, the number of people visiting a specific website, minutes between consecutive telephone calls, the number of earthquakes in 10 years, the number of accidents in town, etc.

Table 1.2. Four R functions for Poisson distribution (`lambda=`λ)

R function	Formula	Returns	Explan.
`dpois(x,lambda)`	$\frac{1}{k!}e^{-\lambda}\lambda^k$	(1.12)	`x = `k
`ppois(q,lambda)`	$\sum_{k=0}^{K}\frac{1}{k!}e^{-\lambda}\lambda^k$	cdf	`q = `K
`rpois(n,lambda)`	$X_1, X_2, ..., X_n \overset{iid}{\sim} P(\lambda)$	rand. numbers	`n = `n
`qpois(p,lambda)`	$\sum_{k=0}^{K}\frac{1}{k!}e^{-\lambda}\lambda^k = p$	quantile, K	`p = `p

A continuous analog of the Poisson distribution is the gamma distribution; the two distributions take a similar shape; see Section 2.6. The following is a continuation of Example 1.8.

Example 1.17 *How many toys to buy?* *Assuming that the number of children in the family follows a Poisson distribution with $\lambda = 2.4$, how many toys should one buy so that every child gets a present.*

Solution. Strictly speaking, whatever number of toys you buy, there is a chance that at least one child will not get a toy because theoretically the number of children is unbounded. As is customary in probability and statistics, we define a probability, close to 1, that each child gets a present. Let this probability be $p = 0.9$. Then the minimum number of toys to buy is the quantile of the Poisson distribution with $\lambda = 2.4$ and is computed in R as `qpois(p=0.9,lambda=2.4)=4`.

The Poisson distribution possesses a very unique property: the mean and the variance are the same,

$$E(X) = \text{var}(X) = \lambda.$$

To prove this fact, we refer to the definition of the mean of a discrete random variable:

$$E(X) = \sum_{k=0}^{\infty} k \times \Pr(X = k; \lambda).$$

Using the expression for the probability (1.12) and the fact $\sum_{k=0}^{\infty} \frac{e^{-\lambda}}{k!}\lambda^k = 1$, we obtain

$$E(X) = \lambda \sum_{k=1}^{\infty} \frac{e^{-\lambda}}{(k-1)!}\lambda^{k-1} = \lambda \sum_{k=0}^{\infty} \frac{e^{-\lambda}}{k!}\lambda^k = \lambda.$$

Analogously, one can prove that $E(X^2) = \lambda + \lambda^2$, which, in conjunction with formula (1.6), gives $\text{var}(X) = \lambda$.

The binomial and Poisson distributions are relatives. Loosely speaking, the Poisson distribution is a limiting case of the binomial distribution. This fact is formulated rigorously as follows.

Theorem 1.18 *A binomial distribution with an increasing number of trials, n, and decreasing probability of successes, $p = p(n)$, converges to a Poisson distribution with parameter, $\lambda = \lim_{n \to \infty} np(n)$.*

Proof. Expressing p through λ and n and substituting it into the formula for binomial probability (1.8), we obtain

$$\Pr(X = k) = \frac{n!}{k!(n-k)!} \left(\frac{\lambda}{n}\right)^k \left(1 - \frac{\lambda}{n}\right)^{n-k}$$

$$= \frac{\lambda^k n!}{k!(n-k)!\, n^k} \left(1 - \frac{\lambda}{n}\right)^n \left(1 - \frac{\lambda}{n}\right)^{-k}.$$

From the famous limit $\lim_{n \to \infty} \left(1 + \frac{a}{n}\right)^n = e^a$ for any fixed a, we have

$$\lim_{n \to \infty} \left(1 - \frac{\lambda}{n}\right)^n = e^{-\lambda}.$$

But when $n \to \infty$, we have $\lambda/n \to 0$. In addition, using Stirling's formula, it is possible to prove that $\frac{n!}{(n-k)!} \frac{1}{n^k} \to 1$. Therefore,

$$\lim_{n \to \infty} \Pr(X = k) = \frac{1}{k!} \lambda^k e^{-\lambda},$$

the Poisson probability.   □

As an example, consider the distribution of the number of stolen credit cards a credit company experiences over the year. The probability that a credit card will be stolen $m \geq 1$ times over the year for a specific customer (binomial variable) is very small, say, p. But the credit company may have many customers, n. Then the average number of stolen cards is $\lambda = pn$. Thus, from the Theorem 1.18, one may infer that the number of stolen cards follows a Poisson distribution. An advantage of using the Poisson distribution over the binomial distribution is that the upper limit of stolen cards is not specified (technically speaking, it is infinity). It is important to estimate λ with the assumption that the data on the number of stolen cards is available for several years – a common statistical problem. As we shall learn later, λ may be estimated simply as an average of the number stolen cards over the years.

Example 1.19 *Raisin in the cookie.* *n raisins are well mixed in the dough and m cookies are baked. What is the probability that a particular cookie has at least one raisin? Provide the exact formula using the binomial distribution and the Poisson approximation.*

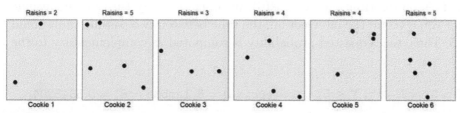

Figure 1.8: *Illustration of the distribution of the number of raisins in a cookie,* $m = 6$, $n = 30$.

Solution. Let X denote the number of raisins in a cookie. See Figure 1.8 for an illustration; since the shape of the dough and cookies does not matter, we depict cookies to have a rectangular shape for simplicity of display. Assuming that raisins are well mixed and the volumes of cookies are equal there will be n/m raisins in the cookie on average with the probability of one raisin $p = 1/m$. The exact distribution of the number of raisins in the cookie X follows a binomial distribution with probability $\Pr(X = k) = \binom{n}{k}p^k(1-p)^{n-k}$. Thus, the exact probability that the cookie has at least one raisin is

$$\Pr(X \geq 1) = 1 - \Pr(X = 0) = 1 - \left(1 - \frac{1}{m}\right)^n. \tag{1.13}$$

If m and n are large, we can approximate the binomial distribution with the Poisson distribution letting $\lambda = n/m$. For large m, the probability, $1/m$, is small, and X can be interpreted as the repetition of n experiments with the mean $\lambda = pn = n/m$. Thus, the probability that a particular cookie has k raisins can be approximated as

$$\Pr(X = k) = \frac{1}{k!}e^{-n/m}\left(\frac{n}{m}\right)^k. \tag{1.14}$$

Then, the probability that at least one raisin will be in a cookie is $\Pr(X > 0) = 1 - \Pr(X = 0) = 1 - e^{-n/m}$. We shall show that (1.14) and (1.13) are close for large m. Indeed, using the approximation $(1 - 1/m)^n \simeq e^{-n/m}$ and letting $\lambda = n/m$, we obtain $\lim_{m \to \infty} \left(1 - \frac{1}{m}\right)^n = e^{-\lambda} \simeq e^{-n/m}$. Thus, for large m, both formulas give close answers. $\square$

In applied probability and statistics, we often need to translate a vaguely formulated real-life question to a rigorously defined problem. Typically, we need to make some assumptions – the next example illustrates this point.

Example 1.20 *Probability of a safe turn.* *Consider turning from a side street onto a busy avenue. There are, on average, 10 cars per minute passing your side street. Assuming that the time between cars follows a Poisson distribution and it takes five seconds to enter the traffic stream safely, what is the probability that you will be able to enter without waiting for a break in traffic?*

Solution. Let X denote the time in seconds between two passing cars. On average, the number of seconds between two passing cars is $60/10 = 6$. Since X

follows a Poisson distribution, $X \sim \mathcal{P}(6)$, in order to safely turn on the avenue right after you come to the intersection, the time between two passing cars should be $X > 5$. Thus, the requested probability is computed as complementary to the cdf:

$$\Pr(X > 5) = 1 - \Pr(X \leq 5) = 1 - \texttt{ppois(q = 5, lambda = 6)} = 0.5543204.$$

Do not make an instant turn unless you are chased by a person who wants to kill you. See Example 2.17, where we compute the wait required for a safe turn. $\square$

An important property of the Poisson distribution is that the sum of independent Poisson variables follows a Poisson distribution with the rate equal the sum of individual rates. Namely, if $X_i \sim \mathcal{P}(\lambda_i)$, $i = 1, 2, ..., n$ and X_i are independent, then

$$\sum_{i=1}^{n} X_i \sim \mathcal{P}\left(\sum_{i=1}^{n} \lambda_i\right). \tag{1.15}$$

We will prove this property using the moment generating function in Section 2.5 of the next chapter. The next example uses this property.

Example 1.21 *No typos. The distribution of the number of typographical errors per 100 pages of a document follows a Poisson distribution with a mean value 4. (a) What is the probability that a 300-page book will have no typos? (b) What is the probability that it will have more than 5 typos? (c) What is the probability that a book of 157 pages has no typos? (d) Run simulations that confirm your analytic answer with L as an argument in the R code.*

Solution. (a) Let X_1 be the number of typographical errors on the first 100 pages of the book. We know that $X_1 \sim \mathcal{P}(4)$. Analogously, let X_2 and X_3 be the number of errors on the next 100 pages and the last 100 pages, respectively. Assuming that the locations of errors are independent, the number of errors in 300 pages, $X = X_1 + X_2 + X_3$, has a Poisson distribution with $\lambda = 3 \times 4 = 12$. Now, to compute the probability that there are no errors on 300 pages, we simply let $k = 0$, so the answer is $\Pr(X = 0) = \frac{1}{0!e^{12}}12^0 = e^{-12} = 6.1 \times 10^{-6}$. Alternatively, we can compute this probability in R as `dpois(0,lambda=12)`. In the previous solution, we assume that the distribution of typos follows a Poisson distribution. Alternatively, one can assume that the distribution of typos follows a binomial distribution with $n = 300$ and the probability of typo per page $p = 4/100 = 1/25$. Then the probability that there will be no typos in 300 pages is $(1 - 1/25)^{300} = 4.8 \times 10^{-6}$. (b) The probability that a 300-page book has 5 typos or less is the cumulative probability and can be computed using `ppois` function. The probability that the book has more than 5 typos is the complementary probability, `1-ppois(5,lambda=12)`. The answer is 0.9872594. (c) The previous solution used in (a) and (b) only works when the number of pages in the book is multiple of 100. What if the number of pages in the book is not multiple of 100, like

157? Then the probability that there are $X = k$ typos on L pages can be modeled using a Poisson distribution with $\lambda = 4L/100 = 0.04L$. Following the previous argument, the probability that there are no typos in a 157-page book is $(0!e^{0.04 \times 157})^{-1}(0.04 \times 157)^0 = e^{-0.04 \times 157} = 1.87 \times 10^{-3}$. (d) Imagine a large number of books with L pages with the number typos distributed according to the Poisson distribution $\mathcal{P}(0.04L)$. If X counts the number of typos in each book, the probability that there are no typos is estimated as the proportion of books with $X = 0$. The probability of this setup can be estimated using the following line of code, `mean(rpois(n=10000000,lambda=0.04*157)==0)`, where `n` is the number of books/simulations and $L = 157$ (of course, `n` and `lambda` may take any values). This command gives the result `0.0018742`, close to our analytic answer.

Problems

1. (a) Prove that, for $\lambda < 1$, probabilities of the Poisson distribution decrease with k. (b) Prove that for $\lambda > 1$ the maximum probability occurs around $\lambda - 1$. [Hint: Consider the ratio $\Pr(X = k+1)$ to $\Pr(X = k)$.]

2. Explain why the distribution of rare diseases, such as cancer, follows a Poisson distribution.

3. Find the minimum number of toys in Example 1.17 that ensures each child gets a toy with probability 99%.

4. (a) Write an R program that computes and plots Poisson distributions of the fertility rates (births per woman) in the United States and India. See https://data.worldbank.org/indicator/sp.dyn.tfrt.in. [Hint: Use `plot` with option `type="h"` and depict the two bars with different color slightly shifted for better visibility, use `legend`.]

5. Prove that for the Poisson distribution, $\text{var}(X) = \lambda$ using formula (1.6).

6. What is the probability that one cookie will have all m raisins?

7. It is well known that the interval mean $\pm 2 \times$ SD covers about 95% of the distribution. Test this statement for the Poisson distribution by generating observations using the `ppois` function in R: plot the Poisson probabilities on the y-axis versus a grid of values for λ on the x-axis, say, `lambda=seq(from=.1,to=3,by=.1)`, and plot the 0.95 horizontal line.

8. Illustrate Theorem 1.18 by plotting Poisson and binomial probabilities side by side for large n and small p (make them arguments of your R function).

9. 10% of families have no children. Assuming that the number of children in the family follows a Poisson distribution, estimate the average number of children in the family.

10. What is the probability that a randomly chosen family with the number of children two or more has children of the same gender? Run simulations to check the answer. [Hint: Assume a Poisson distribution and use conditional probability.]

11. Referring to problem 3 from Section 1.2, how many monkeys is required so that at least one monkey types the novel with probability 0.9? Use an approximation similar to that from Example 1.19.

12. (a) Write R code that simulates Example 1.20. (b) Confirm the probability of a safe turn using simulations.

13. The number of children in the family follows a Poisson distribution with $\lambda = 1.8$. Six families with kids are invited to a birthday party. What is the probability that more than 15 kids come to the party. Give the answer under two scenarios: (a) family brings all kids they have, and (b) the probability that they bring a child is 0.8. Write a simulation program to check your answer.

1.8 Random number generation using sample

1.8.1 Generation of a discrete random variable

A general finite-value discrete random variable, X, takes values $\mathbf{x} = \{x_i, i = 1, 2, ..., n\}$ with probabilities $\mathbf{p} = \{p_i, i = 1, 2, ..., n\}$. In this section, we address the problem of generating a random sample from this distribution. In R, this can be done using a built-in function sample: to generate a random sample of size 10K we issue sample(x=x, size=10000, replace = T, prob = p). Note that arrays x and p must have the same length; replace means that values are drawn with replacement (otherwise, if size is greater than n, the sample cannot be generated).

In the following example, we use the function sample to generate 10,000 Poisson random numbers and test the sample by matching the empirical and theoretical probabilities. Note that since Poisson random variable is unbounded, we must truncate the probabilities. Of course, sample is used here solely for illustrative purposes. A better way to generate Poisson numbers is to use rpois.

Example 1.22 R *sample. Generate 10,000 observations from a Poisson distribution with parameter λ using* sample *and compare with theoretical probabilities by plotting them on the same graph.*

Solution. We need to set an upper bound on the outcomes x_i since the Poisson distribution is unbounded. For example, we may set $\max(x) = \lambda + 5\sqrt{\lambda}$; this guarantees that the right-tail probability is very small. The R code is found in file sampP.r. Option replace=T implies that some observations may repeat. We may use `replace=F` (the default option) if distinct observations are required. This option is used when a subsample with no repeat values is needed (`size` < n). For example, if one needs a subsample from a survey of 1,000 families, option `replace=F` must be used because otherwise one may get repeated observations as if the same family was asked twice. Note that the empirical probabilities do not exactly match the theoretical probabilities (bars) even for a fairly large number of simulated values, `nSim=10000`. However `sampP(nSim=100000)` produces points landing almost on the top of the bars.   □

One can sample from a character vector `x`. For example, `sample(x=c("John",` `"Tom","Rebecca"),size=100,rep=T,prob=c(.25,.5,.25))` produces an array of size 100 with components `"John"`,`"Tom"`,and `"Rebecca"`. The function `sample` is very useful for resampling such as bootstrap – see Example 5.7.

1.8.2 Random Sudoku

Sudoku is a popular mathematical puzzle that originated in France almost two hundred years ago. The objective is to fill blank cells of a 9×9 grid with digits so that each column, each row, and each of the nine 3×3 subsquares contains all of the digits from 1 to 9. We say Sudoku is complete (solved) if there are no blanks; see Figure 1.9 for some *complete Sudoku* puzzles. By contrast, a Sudoku with blank cells are called *puzzle Sudoku*. The goal of this section is to illustrate how `sample` command can be employed to generate and display Sudoku puzzles.

The key to the following theorem is permutation of indices $\{1, 2, 3, 4, 5, 6, 7, 8, 9\}$. For example, a permutation vector $\mathbf{p} = (3, 9, 5, 4, 8, 7, 2, 6, 1)$ means that 1 is replaced with 3, 2 is replaced with 9, etc. In words, this operation may be viewed as index relabeling.

Theorem 1.23 *Let* $\mathbf{A}$ *be a complete Sudoku and* $\mathbf{p}$ *be any permutation of* $\{1, 2, 3, 4, 5, 6, 7, 8, 9\}$. *Then* $\mathbf{B} = \mathbf{A}(\mathbf{p})$ *is also complete Sudoku.*

According to our definition, $\mathbf{A}(\mathbf{p})$ leads to another complete Sudoku where 1 is replaced with p_2, 2 is replaced with p_2, etc. We refer the reader again to Figure 1.9 where the daughter Sudoku ($\mathbf{B}$) is derived from the mother Sudoku ($\mathbf{A}$) by the permutation vector $\mathbf{p} = (3, 9, 5, 4, 8, 7, 2, 6, 1)$. $9! = 362880$ daughter Sudokus can be derived from a mother Sudoku. The connection to `sample` is that a random permutation vector can be obtained as `sample(1:9,size=9,prob=rep(1/9,9))`. This means that from one mother Sudoku, we can create many other puzzles (daughter Sudokus) using the `sample` command.

As was mentioned earlier, in a puzzle Sudoku, some cells are blank, and Sudoku solver needs to fill in the blanks to arrive at a complete Sudoku; see Figure 1.10. The number of blanks determines how difficult the Sudoku puzzle is. For example, with only few blanks the Sudoku is easy. Similarly, a Sudoku with almost all its cells blank is not difficult to solve as well. To create a puzzle

Mother Sudoku

7	9	8	5	3	2	1	6	4
1	4	6	8	9	7	5	3	2
5	3	2	4	6	1	8	7	9
9	5	1	3	4	6	2	8	7
4	8	3	2	7	5	9	1	6
2	6	7	9	1	8	3	4	5
3	1	4	7	5	9	6	2	8
8	7	9	6	2	3	4	5	1
6	2	5	1	8	4	7	9	3

Daughter Sudoku

2	1	6	8	5	9	3	7	4
3	4	7	6	1	2	8	5	9
8	5	9	4	7	3	6	2	1
1	8	3	5	4	7	9	6	2
4	6	5	9	2	8	1	3	7
9	7	2	1	3	6	5	4	8
5	3	4	2	8	1	7	9	6
6	2	1	7	9	5	4	8	3
7	9	8	3	6	4	2	1	5

Figure 1.9: *Mother and daughter Sudokus. The daughter Sudoku is created from the mother Sudoku upon permutation vector* $\mathbf{p} = (3, 9, 5, 4, 8, 7, 2, 6, 1)$.

Sudoku from a complete Sudoku, one needs to blank out some cells, for example, at random locations. Such a puzzle Sudoku is called a random Sudoku. If k is the number of blank cells, then the number of combinations of empty cells out of 81 is $\binom{81}{k}$. Therefore the total number of possible empty cells is

$$\sum_{k=1}^{81} \binom{81}{k} = 2^{81} = 2\,417\,851\,639\,229\,258\,349\,412\,351,$$

enough to keep a Sudoku lover busy!

The R code for displaying, testing, and generating a random puzzle Sudoku is found in file `sudoku.r`. The internal function `test.sudoku` tests whether a 9×9 matrix composed of digits from 1 to 9 is a complete Sudoku. It returns 1 if the Sudoku is complete and 0 otherwise. In the latter case, the number of unique values in each row or column is less than 9, or the number of unique values in each small square is less than 9. The internal function `display.sudoku` plots the 9×9 square and the digit in each cell. When the cell is blank, it is specified as missing (`NA`), and therefore is not displayed.

The `sudoku` function does three jobs: The option `job=1` plots the mother Sudoku and tests whether it is complete (of course other mother Sudoku may be used). The option `job=2` creates Figure 1.9. To create a random daughter Sudoku, a random permutation is computed using the `sample` command (the result is vector `i9`). The random number generation is controlled by `setrand`; different values of `setrand` will produce different daughter Sudokus.

The option `job=3` produces a puzzle Sudoku. First, it generates a random daughter Sudoku from the mother Sudoku using a random permutation, and second, the specified number of cells are blanked. In the Sudoku at left in Figure

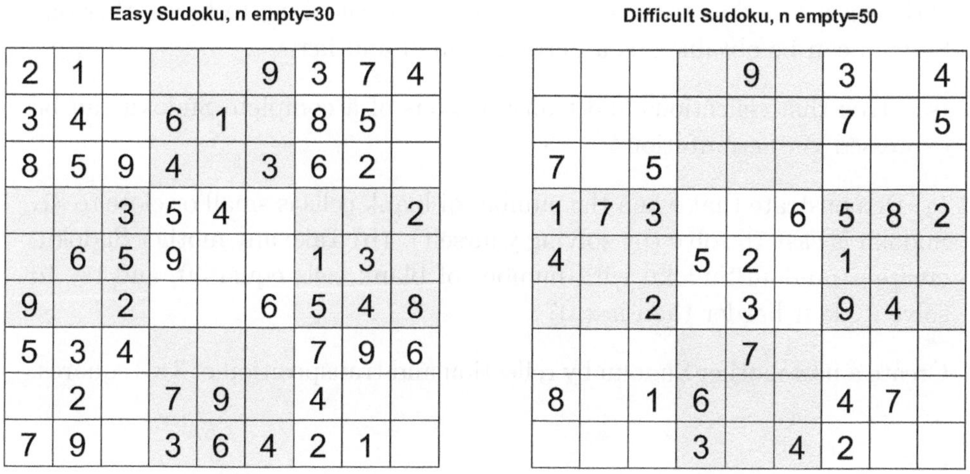

Easy Sudoku, n empty=30

2	1				9	3	7	4
3	4		6	1		8	5	
8	5	9	4		3	6	2	
		3	5	4				2
	6	5	9			1	3	
9		2		6	5	4	8	
5	3	4			7	9	6	
	2		7	9		4		
7	9		3	6	4	2	1	

Difficult Sudoku, n empty=50

				9		3		4
						7		5
7		5						
1	7	3			6	5	8	2
4			5	2		1		
	2			3		9	4	
			7					
8			1	6		4	7	
			3		4	2		

Figure 1.10: *Two puzzle Sudokus with different number of blank cells (n empty). The Sudoku at left is easier than Sudoku at right. This figure is generated by issuing* sudoku(job=3).

1.10, this number is n.blank=30, and in the Sudoku at right, this number is n.blank=50. Once the number of blank cells is given, the random cells get blank/missing again using the sample command applied to 81 pairs of digits (i, j) where $i = 1, 2, ..., 9$ and $j = 1, 2, ..., 9$. These 81 pairs are generated using the rep command with two options times and each. Using this function, random puzzle Sudokus of various levels of difficulty can be generated.

Problems

1. Use sample to generate binomial random numbers by modifying the aforementioned function sampP. Test the sample by plotting the empirical and exact probabilities side by side.

2. Generate $N = 100000$ of $m = 5$-element arrays $(X_1, X_2, ...X_m)$ of Bernoulli variables with probability p using sample and compute the proportion of simulated samples when $\sum_{i=1}^{m}(2X_i^2 - X_i) = 1$. Plot this proportion as a function of $p = 0.1, 0.2, ..., 0.9$. Explain the result. [Hint: Generate Nm binary variables and form an $N \times m$ matrix X; then use rowSums.]

3. Randomly select characters from Example 1.14 and plot the letter frequency in the two samples side by side as in Figure 1.6. Do the frequencies look alike? [Hint: Generate random characters using random index arrays ir1=sample(1:N,size= N/2,prob=rep(1/N,N)) and to exclude those in ir1 as ir2=(1:N)[-ir1] where N=length(ch).]

4. Write an R function that, given two complete Sudokus, tests whether one
Sudoku can be obtained as a permutation of another.

5. Is it true that reflections and transpositions of a complete Sudoku can be
expressed via permutation?

6. (a) Demonstrate that when the number of blank cells is small or close to 81,
Sudoku is easy to solve (by solving yourself). (b) Take any mother Sudoku,
create a random Sudoku with number of blank cells equal 40, and try to
solve it. Is it harder than in (a)?

7. Create a new mother Sudoku by reflection and transposition of 3×3 squares.

Chapter 2

Continuous random variables

Unlike discrete random variables, continuous random variables may take *any* value on a specified interval. Usually, we deal with continuous random variables that take any value on an infinite or semi-infinite interval, denoted as $(-\infty, \infty)$ or $(0, \infty)$, respectively. Some continuous random variables, such as uniform or beta-distributed random variables, belong to an interval. A characteristic property of a continuous random variable is that $\Pr(X = x) = 0$ for any value x. Although it is possible that a random variable is a combination of a discrete and a continuous variable, we ignore this possibility. Thus, only continuous random variables are considered in this chapter, with calculus as the major mathematical tool for investigating of distributions of these variables. Regarding the notation, typically, random variables are denoted as uppercase, like X, and the values it takes or the argument of the density function as lowercase, x. This notation rule will be followed throughout the book.

2.1 Distribution and density functions

While the cumulative distribution function (cdf) is defined for both types of variables, the probability density function (pdf), or density function, is defined only for the continuous type.

2.1.1 Cumulative distribution function

The cumulative distribution function (cdf), or distribution function of a random variable X, is defined as

$$F_X(x) = \Pr(X \leq x), \quad -\infty < x < \infty. \tag{2.1}$$

In words, the cdf is the probability that the random variable X takes a value less than or equal to x, where x may be any number. Since the distribution function

Advanced Statistics with Applications in R, First Edition. Eugene Demidenko.
© 2020 John Wiley & Sons, Inc. Published 2020 by John Wiley & Sons, Inc.

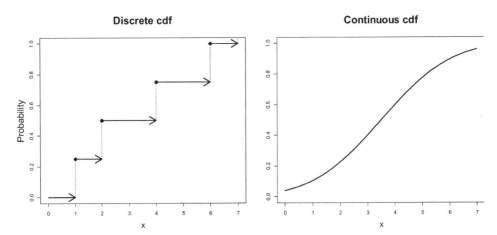

Figure 2.1: *Typical distribution functions for discrete and continuous random variable. Discrete cdf is discontinuous and takes the step p_i and x_i. On the other hand, the cdf of a continuous random variable is a continuous function.*

is a cumulative probability, it is an increasing function of x and takes nonnegative values in the interval $[0,1]$.

A cdf has the following properties:

1. The cdf is an increasing function: $x_1 \leq x_2$ implies $F(x_1) \leq F(x_2)$.

2. The upper limit of the cdf is 1: $F(+\infty) = 1$, or, more precisely, $\lim_{x\to\infty} F(x) = 1$.

3. The lower limit of cdf is 0: $F(-\infty) = 0$, or, more precisely, $\lim_{x\to-\infty} F(x) = 0$.

The first property follows directly from the definition of the distribution function; the other two properties are obvious. A function F is a distribution function if it satisfies properties 1, 2, and 3. For any $a \leq b$ the probability that the random variable belongs to the interval $(a, b]$ can be expressed via a cdf as $\Pr(a < X \leq b) = F(b) - F(a)$. The cdf is defined for discrete or continuous random variable.

If a discrete random variable takes n distinct values $\{x_i, i = 1, 2, ..., n\}$ with respective probabilities $\{p_i, i = 1, 2, ..., n\}$, the distribution function takes a step of height p_i at x_i. This function is continuous from the right but is discontinuous from the left. For a continuous random variable, the distribution function is continuous on the entire real line. See Figure 2.1.

The cdf completely defines the distribution of a continuous random variable. The distribution function is a mathematical concept – in applications, we have to estimate it by the empirical cdf.

2.1.2 Empirical cdf

Let random variable X with unknown cdf, F, take values $x_1, x_2, ..., x_n$. We say that these values are the realization (or observations) of X or a random sample from a general population specified by cdf F.

The procedure to compute the empirical cdf has two steps:

1. List observations in ascending order, $x_{(1)} \leq x_{(2)} \leq \cdots \leq x_{(n)}$, so that $x_{(1)} = \min x_i$ and $x_{(n)} = \max_i x$. $\{x_{(i)}\}$ are called order statistics.

2. Compute the empirical cdf treating the sample as a discrete random variable that takes values $x_{(i)}$ with probability $1/n$. This means that the cdf is computed as

$$\widehat{F}(x) = \frac{1}{n}\#(x_i \leq x), \qquad (2.2)$$

where sign $\#$ means the number of observations equal to or less than x. Sometimes the notation $I(x_i \leq x)$ is used, where I is an indicator function with values 0 or 1; then $\#(x_i \leq x) = \sum_{i=1}^{n} I(x_i \leq x)$. We use "hat" $\widehat{()}$ to indicate that this is an *estimator* of F, a common notation is statistics. It is easy to see that (2.2) can be expressed as a step-wise function with step i/n at $x_{(i)}$. This means that the empirical cdf can plotted as i/n on the y-axis versus $x_{(i)}$ on the x-axis.

For each x, one can treat the numerator in $\widehat{F}(x)$ as the outcome of the binomial random variable with Bernoulli probability $F(x)$ in n trials, where F is the true cdf. Since the expected number of successes is $nF(x)$, we have $E(\widehat{F}(x)) = F(x)$. We say that the empirical cdf is an unbiased estimator of the true cdf. Moreover, since $\text{var}(\widehat{F}(x)) = F(x)(1 - F(x)/n$, the empirical cdf approaches F when $n \to \infty$. In statistics, theses two properties of an estimator are called unbiasedness and consistency, respectively.

It is easy to plot the empirical distribution function in R using the command `plot()` with option `type="s"` after the original sample is ordered using the `order()` or `sort()` command. This method is illustrated in the following example.

Example 2.1 *Web hits cdf. An Internet company analyzes the distribution of the number of visitor hits during the day. The dataset* `compwebhits.dat` *contains the times recorded by a computer server for 100 hits during a typical day. Plot the empirical cdf and interpret its pattern.*

Solution. The R code below produces Figure 2.2. First, we order observations. Second, we create values for the cdf $\{1/n, 2/n, ...1\}$, and finally plot. Sometimes, we use a continuity correction and plot $(i - 0.5)/n$ on the y-axis,

```
webhits=function()
{
```

```
dump("webhits","c:\\statbook\\webhits.r")
x<-scan("c:\\statbook\\comwebhits.dat") # read the data
x<-x[order(x)];n<-length(x) # order observations
Fx<-(1:n)/n # values for cdf
plot(x,Fx,type="s",xlab="Time of the website hit, h",
                            ylab="Probability, cdf")
text(17,.2,paste("Number of hits during 24 hours =",n))
}
```

Looking at the figure, one may notice that $\widehat{F}(x) = 0$ for $x < 5$ a.m. – people are sleeping. The intensity of visitor hits picks up at around 8 a.m. as people wake up. Hits have a steady rate until 3 p.m. and then slow down. Better insights into the intensity of the hits can be drawn from plotting a histogram; see the next section.

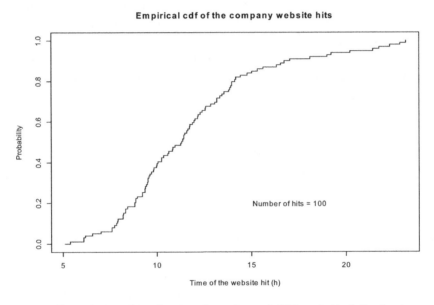

Figure 2.2: *Cumulative distribution function of 100 website hits during a typical day.*

2.1.3 Density function

As we mentioned before, $\Pr(X = x) = 0$ for a continuous random variable. In words, the probability that a continuous random variable takes a specific value is zero. Therefore instead of the point probability, we consider the infinitesimal probability when the length of the interval goes to zero:

$$\frac{\Pr(x < X \le x + h)}{h}. \tag{2.3}$$

Since $\Pr(x < X \leq x+h) = F(x+h) - F(x)$, we define the infinitesimal probability as

$$f(x) = \lim_{h \to 0} \frac{F(x+h) - F(x)}{h}, \qquad (2.4)$$

called the probability density function (pdf), or shortly the density. As follows from the Fundamental Theorem of Calculus, the pdf is the first derivative of the distribution function,

$$f(x) = F'(x),$$

and conversely

$$F(x) = \int_{-\infty}^{x} f(t)dt.$$

Density has the following properties:

1. $f(x) \geq 0$ for all x.

2. $\int_{-\infty}^{\infty} f(x)dx = 1$.

The first property follows from the fact that cdf is an increasing function. Hence pdf cannot be negative. The second property follows from the fact that $F(\infty) - F(-\infty) = 1$. It is easy to see that $f(\pm\infty) = 0$ because otherwise the second property would not hold.

The density reaches its maximum at the *mode*. It is fair to refer to the mode as the most probable value of the random variable because the density at the mode is maximum. We say that the distribution is *unimodal* if there is one mode, i.e. $f(x)$ is (strictly) increasing to the left of the mode and (strictly) decreasing to the right of the mode. We say that the random variable is symmetric if its density is symmetric around the mode. The support of the density is where it is positive. A continuous random variable may be defined on a finite or semi-infinite interval; we call this interval the *support* of the density.

Example 2.2 *Pdf=cdf. Can the same function be the cdf and the pdf at the same time?*

Solution. Yes. Let $g(x)$ be a cdf and pdf at the same time. Then this function satisfies the ordinary differential equation (ODE) $g' = g$ with a solution $g(x) = e^x$. Of course, e^x cannot be a cdf for all x because its values goes beyond 1, but if $x \in (-\infty, 0)$, function $g(x) = e^x$ satisfies the properties of a cdf. Moreover, since the general solution to ODE $g' = g$ is of the form e^{x+c}, where c is any constant, we conclude that the only function that is both a cdf and a density is of the form $g(x) = e^{x+c}$ on the interval $-\infty < x < -c$. This distribution emerges in Section 2.10 as the limiting distribution of the maximum of observations.

Problems

1. Why can one treat the numerator in (2.2) as the outcome of the binomial random variable with Bernoulli probability $F(x)$ in n trials? [Hint: Use the indicator function $1(X \leq x)$ as a Bernoulli random variable.]

2. A linear combination of distributions is called a mixture distribution; see Section 3.3.2. (a) Prove that if F and G are two cdfs and $0 \leq \lambda \leq 1$ is a fixed number, then $\lambda F + (1 - \lambda)G$ is a cdf (mixture). (b) Prove that a similar statement holds for two densities. (c) In the general case, prove that $\sum_{i=1}^{n} \lambda_i F_i(x)$ and $\sum_{i=1}^{n} \lambda_i f_i(x)$ are mixture cdfs and densities, where F_i and f_i are component cdfs and densities, $\lambda_i \geq 0$ and $\sum_{i=1}^{n} \lambda_i = 1$.

3. Let X be a continuous random variable and F be its strictly increasing cdf. Prove that random variables $F(X)$ and $1 - F(X)$ have the same distribution. [Hint: Prove that the cdf of $F(X)$ is x.]

4. (a) Plot the cdf in Example 2.1 with the continuity correction using a different color on the same plot. Does it make any difference? (b) Plot $(F(x_{(i+1)}) - F(x_{(i)}))/(x_{(i+1)} - x_{(i)})$ versus $x_{(i)}$. Make a connection to density.

5. Let random variable X have cdf $F_X(x)$ and density $f_X(x)$. Find the cdf and density of $Y = a + bX$. [Hint: Consider cases when $b > 0$ and $b < 0$ separately.]

6. F is a cdf. Is F^p a cdf?

7. Let F and G be cdfs. Is their product a cdf? Does an analogous statement hold for densities? [Hint: If the answer is positive prove it, otherwise provide a counterexample.]

8. (a) Is it possible that for two cdfs $F(x) < G(x)$ for all x? (b) Is it possible that that for two densities, $f(x) < g(x)$? (c) Is it true that two cdfs always intersect? (d) Is it true that two pdfs always intersect?

9. If $f(x)$ is a density with support on the entire line, is $g(x) = f(x^2)$ a density? The same question but support of f is positive numbers.

10. The density of a random variable is defined as $c \times \sin x$ for $0 < x < \pi$ and 0 elsewhere. Find c and the cdf.

2.2 Mean, variance, and other moments

The mean and variance of a continuous random variable are defined as for a discrete distribution with the sum replaced by an integral. The mean of a continuous distribution with density $f(x)$ is defined as

$$\mu = \int_{-\infty}^{\infty} x f(x) dx, \tag{2.5}$$

and the variance is defined as

$$\sigma^2 = \int_{-\infty}^{\infty} (x - \mu)^2 f(x) dx.$$

We often shorten the notation for mean and variance to $E(X)$ and $\text{var}(X)$, respectively. Sometimes, we use the notation for the standard deviation, $\text{SD}(X) = \sigma$. The following formula describes the relationship between the mean and variance,

$$\sigma^2 = E(X^2) - \mu^2, \tag{2.6}$$

as the counterpart of (1.7). Since variance is nonnegative, we conclude

$$|\mu| \le \sqrt{E(X^2)}. \tag{2.7}$$

The mean, μ, can be interpreted as the center of gravity of a stick with mass density $f(x)$, similar to the discrete case as discussed in Section 1.4.1. By definition, μ is the center of gravity defined as the point at which the resultant torque is zero,

$$\int_{-\infty}^{\infty} (x - \mu) f(x) dx = 0,$$

which leads to definition (2.5).

Example 2.3 *Expectation via cdf*. *Prove that*

$$E(X) = -\int_{-\infty}^{0} F(x) dx + \int_{0}^{\infty} (1 - F(x)) dx \tag{2.8}$$

where F is the cdf of X.

Proof. We have

$$\int_{-\infty}^{\infty} x f(x) dx = \int_{-\infty}^{0} x f(x) dx + \int_{0}^{\infty} x f(x) dx.$$

Using integration by parts in the first integral, define $u = x$, $dv = f(x) dx$, which implies $du = dx$ and $v = F(x)$. Consequently, the first integral can be expressed as

$$\int_{-\infty}^{0} x f(x) dx = x F(x)|_{-\infty}^{0} - \int_{-\infty}^{0} F(x) dx = -\int_{-\infty}^{0} F(x) dx.$$

In the second integral, let $y = -x$. That implies

$$\int_{0}^{\infty} x f(x) dx = -\int_{-\infty}^{0} y f(-y) dy.$$

Use integration by parts again by letting $u = y, dv = f(-y)dy$, which implies $du = dy$ and $v = 1 - F(-y)$. Consequently,

$$\int_{-\infty}^{0} yf(-y)dy = y(1 - F(-y))|_{-\infty}^{0} - \int_{-\infty}^{0} (1 - F(-y))dy = -\int_{0}^{\infty} (1 - F(x))dx.$$

Combining the two integrals yields

$$\int_{-\infty}^{\infty} xf(x)dx = -\int_{-\infty}^{0} F(x)dx + \int_{0}^{\infty} (1 - F(x))dx.$$

The identity (2.8) is proved.   □

    When two random variables are measured on different scales, their variation is easier to compare using the coefficient of variation (CV):

$$\text{CV} = \frac{\sigma}{\mu}.$$

Sometimes CV is expressed as a percent to eliminate units. Indeed, CV does not change when applying the transformation $X \to \alpha X$, where α is a positive coefficient. This coefficient naturally emerges in the lognormal distribution; see Section 2.11.

    The mean and variance may not exist for certain distributions. For example, they do not exist for the Cauchy distribution defined by the density

$$f(x) = \frac{1}{\pi} \frac{1}{1 + x^2}, \quad -\infty < x < \infty. \tag{2.9}$$

Indeed, the integral $\int_{-\infty}^{\infty} x/(1 + x^2)dx$ does not exist.

    The kth noncentral (ν_k) and central (μ_k) moments are defined as

$$\nu_k = \int_{-\infty}^{\infty} x^k f(x)dx, \quad \mu_k = \int_{-\infty}^{\infty} (x - \mu)^k f(x)dx.$$

In another notation,

$$\nu_k = E(X^k), \quad \mu_k = E((X - \mu)^k).$$

The first noncentral moment is the mean and the second central moment is the variance. We say that the distribution is symmetric if the density is an even function around μ, i.e. $f(x - \mu) = f(\mu - x)$. For symmetric distributions all odd central moments are zero and the mean and mode coincide.

    To characterize how skewed a distribution is, we use the *skewness* coefficient:

$$\text{Skewness} = \frac{\mu_3}{\sigma^3}. \tag{2.10}$$

This coefficient is scale independent, so random variables can be compared on the relative scale. If the skewness coefficient is less than zero, we say that the

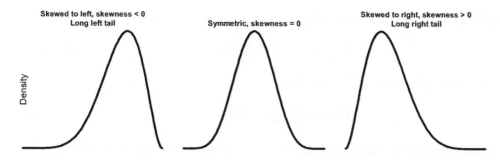

Figure 2.3: *Three types of distributions with respect to the skewness.*

distribution is left skewed (long left tail); if the skewness coefficients greater than zero, we say that the distribution is right skewed (long right tail; see Figure 2.3). The skewness coefficient is zero for a symmetric distribution, but skewness = 0 does not imply that the distribution is symmetric.

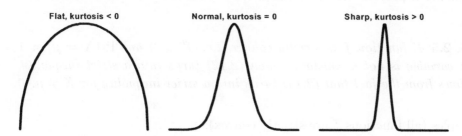

Figure 2.4: *Three types of distributions with respect to kurtosis.*

Kurtosis is used to characterize how flat the distribution is and is computed by the formula

$$\text{kurtosis} = \frac{\mu_4}{\sigma^4} - 3. \tag{2.11}$$

Alternatively, we may say that kurtosis characterizes the sharpness of the density. This coefficient is scale independent as well. The 3 is subtracted to make the normal distribution the reference with kurtosis = 0. For densities with a sharp peak, the kurtosis is positive; for flat densities, the kurtosis is negative. All three cases are illustrated in Figure 2.4.

We calculate skewness and kurtosis for some continuous distributions later in this chapter.

The expectation is defined for any function g of the random variable as

$$E(g(X)) = \int_{-\infty}^{\infty} g(x)f(x)dx.$$

Of course, the expectation may not exist.

Example 2.4 *Jensen's inequality.* *If f is a convex function ($f'' \geq 0$), then*

$$E(f(X)) \geq f(E(X)). \tag{2.12}$$

If f is a concave function ($f'' \leq 0$), then

$$E(f(X)) \leq f(E(X)).$$

Solution. Since f is convex, we have $f(x) \geq f(x_0) + (x - x_0)f'(x_0)$ for all x. Replace x with X and let $x_0 = \mu = E(X)$:

$$f(X) \geq f(\mu) + (X - \mu)f'(\mu). \tag{2.13}$$

Taking the expectation of both sides, we obtain

$$E(f(X)) \geq E(f(\mu)) + E(X - \mu)f'(\mu) = f(E(X)).$$

The inequality is proved. We prove the inequality similarly for the concave function.

Remark 2.5 *If function f is strictly convex, i.e. $f'' > 0$ and $\Pr(X = \mu) < 1$ (random variable is not a constant), then (2.12) turns into a strict inequality. This follows from the fact that (2.13) turns into a strict inequality for $X \neq \mu$.*

Examples (all functions f are strictly convex):

1. $f(x) = x^2 : E(X^2) > E^2(X)$.

2. $f(x) = \ln x : E(\ln X) < \ln E(X)$ if $X > 0$.

3. $f(x) = 1/x : E(1/X) > 1/E(X)$ if $X > 0$.

Example 2.6 *Cauchy inequality.* *Prove that for any random variables X and Y we have*

$$E^2(XY) \leq E(X^2)E(Y^2), \tag{2.14}$$

and the equality holds if and only if $Y = \alpha X$ where α is a constant (all expectations are finite).

Solution. Define $Z = Y - \alpha X$ and express the second noncentral moment through α as

$$E(Z^2) = E(Y^2) - 2\alpha E(XY) + \alpha^2 E(X^2).$$

The right-hand side is a quadratic function of α. Since $E(Z^2) \geq 0$, the discriminant should be nonpositive, $E^2(XY) - E(X^2)E(Y^2) \leq 0$. The equality is true if and only if the discriminant is zero, which happens if $Y = \alpha X$ with probability 1. $\qquad\square$

We make several comments: The Cauchy inequality is sometimes called the Cauchy–Schwarz inequality. The Cauchy inequality holds for discrete random variables as well. Inequality (2.14) can be expressed in terms of any functions f and g:

$$E^2(f(X)g(Y)) \leq E(f^2(X))E(g^2(Y)). \qquad (2.15)$$

One may just think of $f(X)$ and $g(Y)$ as new random variables. Letting $g(Y) = 1$, we get

$$|E(f(X))| \leq \sqrt{E(f^2(X))},$$

an analog of inequality (2.7). Also, it is easy to apply the Cauchy inequality to random variables with the means subtracted:

$$E^2[(X - \mu_X)(Y - \mu_Y)] \leq \sigma_X^2 \sigma_Y^2. \qquad (2.16)$$

This inequality proves the correlation coefficient takes values within the interval $[-1, 1]$. Note that this inequality turns into an equality if and only if random variables are proportionally related, $Y - \mu_Y = \beta(X - \mu_X)$.

The Cauchy inequality can be rewritten in discrete or integral fashion. The discrete version is well known in linear algebra as the inequality between the scalar product and the squared norm,

$$(\mathbf{x'y})^2 \leq \|\mathbf{x}\|^2 \|\mathbf{y}\|^2,$$

where $\mathbf{x'y}$ is the scalar (inner, dot) product, or in the index form

$$\left(\sum_{i=1}^n x_i y_i \right)^2 \leq \left(\sum_{i=1}^n x_i^2 \right) \left(\sum_{i=1}^n y_i^2 \right). \qquad (2.17)$$

The integral version of the inequality is obtained by expressing the expected values in (2.15) via integrals,

$$\left(\int_{-\infty}^{\infty} f(x)g(x)dx \right)^2 \leq \int_{-\infty}^{\infty} f^2(x)dx \times \int_{-\infty}^{\infty} g^2(x)dx$$

provided the integrals on the right-hand side exist. This inequality can be proved using the same method as in Example 2.6.

Under mild conditions, it can be proven that a distribution is uniquely defined by all its moments. We can demonstrate this fact on a discrete random variable: If a random variable takes n distinct values x_i, the noncentral moments ν_k for $k = 1, 2, ..., n - 1$ uniquely define probabilities p_i. We want to prove that the system of linear equations given by $n - 1$ equations,

$$\sum_{i=1}^n p_i x_i^k = \nu_k,$$

has a unique solution for p_i. Note that we have $n-1$ equations because $\sum_{i=1}^{n} p_i = 1$. This system of equations has a unique solution for p_i because its determinant, called the Vandermonde determinant, is not zero.

The following result illustrates how instrumental calculus is in probability and statistics.

Example 2.7 *Stein's identity. Let f be a differentiable density of a continuous random variable X (without loss of generality it is assumed that the support is the entire line) and g be a differentiable function on $(-\infty, \infty)$ such that $E(g(X))$ is finite. Prove that*

$$E_X\left[g(X)(\ln f(X))' + g'(X)\right] = 0,$$

where ' means derivative.

Solution. Represent the expectation as an integral. Since $(\ln f(x))' = f'(x)/f(x)$, we obtain

$$\int_{-\infty}^{\infty} [g(x)(\ln f(x))' + g'(x)]f(x)dx = \int_{-\infty}^{\infty} g(x)f'(x)dx + \int_{-\infty}^{\infty} g'(x)f(x)dx.$$

Now we apply integration by parts, $\int_{-\infty}^{\infty} u\,dv = uv|_{-\infty}^{\infty} - \int_{-\infty}^{\infty} v\,du$, by letting $u = g(x)$ and $dv = f'(x)dx$, so that $du = g'(x)dx$ and $v = f(x)$. This implies

$$\int_{-\infty}^{\infty} g(x)f'(x)dx = g(x)f(x)|_{-\infty}^{\infty} - \int_{-\infty}^{\infty} g'(x)f(x)dx.$$

Since $E(g(X)) < \infty$, we have $g(x)f(x)|_{-\infty}^{\infty} = 0$. Therefore, the Stein's identity follows.

2.2.1   Quantiles, quartiles, and the median

The pth quantile, q_p, is the solution to the equation

$$F(q_p) = p, \quad 0 < p < 1,$$

where F is the cdf. In the language of the inverse cdf, we can define the pth quantile as $q_p = F^{-1}(p)$. The $\frac{1}{4}$ quantile is called the lower (or the first) quartile and is denoted as $q_{1/4}$ and the $\frac{3}{4}$th quantile is called the upper (or the third) quartile and is denoted as $q_{3/4}$. The median is the $\frac{1}{2}$ quantile (or the second quartile):

$$F(\text{median}) = \frac{1}{2}. \tag{2.18}$$

Percentile is a quantile expressed as a percent: 75th percentile, 25th percentile, etc. Quantiles and quartiles can be used to characterize the range of the distribution. We may characterize the range of the distribution using $q_{1/4}$ and $q_{3/4}$, also called the interquartile range, which indicates that 50% of values fall between them. In other words, the interval $[q_{1/4}, q_{3/4}]$ contains 50% of the random variable values.

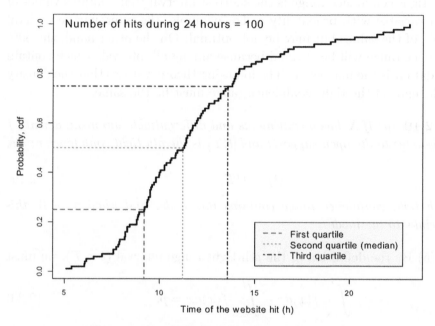

Figure 2.5: *Quartiles for the webhits data.*

Example 2.8 Web hits. *Compute and display $q_{1/4}, q_{1/2},$ and $q_{3/4}$ for the web hits example. Find the interquartile range.*

Solution. The requested quantities are simply computed as x[n/4], x[n/2], and x[3*n/4], the R function is in the file webhitsQ.r. The interquartile range interval is [1] 9.21461 13.56446. About 50% of the web hits happen between 9:30 a.m. and 2 p.m. The R function webhitsQ produces Figure 2.5.

2.2.2 The tight confidence range

Quantification of the range of a random variable that represents the general population is an important task of applied statistics. In many applications, the standard deviation, σ, serves as a characteristic of the range, and sometimes authors use $\mu \pm \sigma$ to report the scatter of the random variable. Using the interval $\mu \pm k\sigma$, where k is a constant, silently assumes that data are symmetric around the mean. When a distribution is not symmetric, a symmetric interval around the mean is not optimal because the same coverage probability can be obtained using an interval with a smaller width. Thus, we arrive at the concept of the *tight confidence range*. We emphasize the difference between confidence range and confidence interval in statistics that covers the true unknown parameter, see Section 7.8.

Definition 2.9 *The pth tight confidence range for a random variable X is the interval $[t, T]$ such that (a) $\Pr(t < X \leq T) = p$, and (b) $T - t = \min$.*

The pth tight confidence range is the shortest interval that contains values of the random variable with probability p. For example, the interquartile interval contains 50% of the values but may be not optimal. On the other hand, the 50% tight confidence range will be optimal because among all intervals which contain 50% of the data it is the narrowest. The following theorem states that the density values at the ends of the tight confidence range must be the same.

Theorem 2.10 *(a) If X has a continuous and differentiable unimodal density f (the mode belongs to the open support) and $[t, T]$ is the pth tight confidence range, then*

$$f(t) = f(T). \tag{2.19}$$

(b) The pth tight confidence range contains the mode, and when $p \to 0$, this interval shrinks to the mode.

Proof. (a) By the definition of the pth tight range interval $(t < T)$, we must have

$$\int_{-\infty}^{T} f(x)dx - \int_{-\infty}^{t} f(x)dx = p. \tag{2.20}$$

Find t and T by solving the optimization problem $T - t = \min$ under constraint (2.20). Introduce the Lagrange function

$$\mathcal{L}(t, T; \lambda) = T - t - \lambda \left(\int_{-\infty}^{T} f(x)dx - \int_{-\infty}^{t} f(x)dx - p \right),$$

where λ is the Lagrange multiplier. The necessary conditions for the minimum are

$$\frac{\partial \mathcal{L}}{\partial t} = -1 + \lambda f(t) = 0, \quad \frac{\partial \mathcal{L}}{\partial T} = 1 - \lambda f(T) = 0.$$

These equations imply (2.19). Since f is a unimodal equation $f(t) = c$ has two solutions, t and T, for each $c \in (0, \max f(x))$. (b) Interval $[t, T]$ contains the mode for each $p > 0$ and it shrinks to the mode because f is a continuous function. $\square$

The above theorem helps compute the pth tight confidence range. To find t and T, we need to solve a system of equations:

$$F(T) - F(t) = p, \quad f(t) = f(T), \tag{2.21}$$

where F is the cdf. We will illustrate computation of the tight confidence range later in the chapter. For symmetric distributions, this interval takes the regular form $\mu \pm k\sigma$. The tight confidence range is especially advantageous for asymmetric distributions, such as the gamma distribution in Section 2.6 or the lognormal distribution in Section 2.11.

The pth tight confidence range can be defined for discrete distributions as well as (t, T) such that $\Pr(t < X \leq T) \geq p$ and $T - t = \min$. Note that we use the inequality sign here because one cannot find t and T such that $\Pr(t < X \leq T) = p$ due to discreteness of the distribution. Since the probability is a discrete function, we cannot rely on calculus. Instead, we use direct computation to find t and T, as in the following example.

Example 2.11 *Confidence range for the binomial distribution.* *Write an R code to compute the $100\lambda\%$ tight confidence range for the binomial random variable.*

Solution. The R code is found in the text file `tr.binom.r`; see Section 1.5.3 to read the file. To find the optimal range, we use double `for` loop over m and M. If, for certain m and M, we have $\Pr(m < X \leq M) \geq \lambda$, we check if the length of the interval is smaller than the current one. Specifically, if `minDIF`. If `M-m<minDIF`, we save m and M as `m.opt` and `M.opt`, respectively. For example, the call

> `tr.binom(n=100,p=.3,lambda=.75)`

`[1] 23.0000000 34.0000000 0.7616109`

gives the 75% range $(23, 34)$, which has the minimal width among all intervals that contain the binomial random variable with probability equal to or greater than λ. In fact, the range $(23, 34)$ gives the probability $0.762 > 0.75$. $\qquad\square$

In applied statistics, we shall use the confidence range to determine the range of individual values as opposed to the traditional confidence interval that is intended to determine the range of an unknown parameter, such as the mean, μ. Example 2.29 illustrates the computation of the tight confidence range for a distribution of house prices on the real estate market.

Problems

1. Prove that $\operatorname{var}(X) \leq E(X^2)$. Is it true that $E(X - \mu)^k \leq E(X^k)$ for any even positive integer k? Specify distributions for which this is true.

2. (a) Prove that $|E(X)| \leq E^{1/p}(|X|^p)$ for $p \geq 1$. (b) Prove that $E(X^3) \geq E^3(X)$ if $X > 0$.

3. (a) Prove formula (2.6). (b) Derive (2.6) from the Jensen's inequality.

4. The central moments can be expressed via noncentral moments using polynomial expansion of $(x - \mu)^k$. Provide a recursive formula for the coefficients of $E(X - \mu)^{k+1}$ as a linear combination of $\nu_1, ..., \nu_{k+1}$. [Hint: Use formula (1.9).]

5. Prove that skewness and kurtosis are scale independent. In other words, skewness and kurtosis are the same for X and aX, where $a > 0$.

6. Prove that the cdf of the Cauchy distribution with density (2.9) is $\frac{1}{\pi} \arctan x +$ $\frac{1}{2}$. Find the median and the $\frac{1}{4}$ and the $\frac{3}{4}$ quantiles. Show that they are symmetric around zero. Define the *shifted* Cauchy distribution and find its cdf.

7. The density of a random variable is defined as $c(1 - x^2)$ for $|x| \leq 1$ and 0 elsewhere. Find c, the cdf, the mean, and the median. Plot the density and cdf in R using `mfrow=c(1,2)`.

8. Express the fourth central moment as a linear combination of noncentral moments. Express the fourth noncentral moment as a linear combination of central moments. [Hint: Use formula (1.9).]

9. The support of X is $(-1, 1)$. Is it true that the kth central moments vanish when $k \to \infty$? Is it true for noncentral moments?

10. Prove that the Poisson distribution is skewed to the right. [Hint: Prove that the third central moment is positive.]

11. Prove that $E\left(g^p(X)\right)$ is a convex function of $p > 0$ where g is a nonnegative function. Derive the respective Jensen's inequality.

12. $p > 0$ and f is a pdf. Is f^p a pdf? [Hint: Use Jensen's inequality.]

13. Plot in R two densities for which median is greater that mode and mode is greater than median. Describe the distributions for which this is true using the language of *tail*.

14. Prove that $E(F(X)) = 1/2$ where F is the cdf of X. [Hint: Assume continuous distribution and apply change of variable.]

15. Prove that the minimum of $\int_{-\infty}^{a} F(x)dx + \int_{a}^{\infty}(1 - F(x))dx$ attains when a is the median.

16. Let a continuous random variable have a positive continuously differentiable density with unique maximum, or specifically, $(\ln f)' = 0$ has unique solution, the mode. Is it true that the following inequality holds: mode $\leq$ median $\leq$ mean? Prove that this inequality turns into equality for symmetric distributions. Is it true that this inequality holds for densities such that $(\ln f)'' < 0$? More about this inequality can be learned from Abadir [1].

17. Prove that, for a symmetric distribution, the pth tight confidence range takes the form $\mu \pm k\sigma$.

18. Write an R program to compute the $100\lambda\%$ tight population confidence range for the Poisson distribution similarly to Example 2.11.

2.3 Uniform distribution

The uniform (or rectangular) distribution is the simplest continuous distribution. Following its name, the uniform distribution is categorized by the fact that the probability of falling within an interval is proportional to its length. We write $X \sim \mathcal{R}(a, b)$ to indicate that random variable X has a uniform distribution on the finite interval (a, b). Sometimes, one can encounter the notation U instead of $\mathcal{R}$. The density for this distribution is constant and the cdf is a linear function (see Figure 2.6), namely,

$$f(x) = \begin{cases} \frac{1}{b-a} \text{ if } a \leq x \leq b \\ 0 \text{ elsewhere} \end{cases}, \quad F(x) = \begin{cases} \frac{x-a}{b-a} \text{ if } a \leq x \leq b \\ 0 \text{ if } x < a \\ 1 \text{ if } x > b \end{cases}.$$

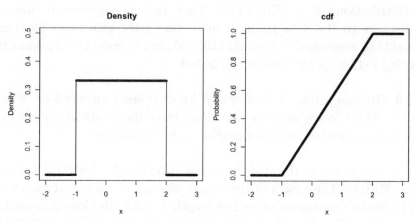

Figure 2.6: *Density and cdf for* $X \sim \mathcal{R}(-1, 2)$. *Sometimes this distribution is called* **rectangular** *because of the shape of the density.*

The mean is $E(X) = (a + b)/2$. Although the result seems obvious, we derive it formally via integration as follows:

$$E(X) = \frac{1}{b-a} \int_a^b x dx = \frac{1}{b-a} \frac{1}{2} x^2 \Big|_a^b = \frac{1}{b-a} \frac{1}{2} (b^2 - a^2) = \frac{a+b}{2}.$$

To be specific, we indicate the distribution of X as a subscript in the mathematical expectation:

$$E_{X \sim \mathcal{R}(a,b)}(X) = \frac{1}{2}(a + b). \tag{2.22}$$

By integration over x^2 and using formula (2.6), we obtain the variance and SD:

$$\text{var}(X) = \frac{(b-a)^2}{12}, \quad \text{SD}(X) = \frac{b-a}{2\sqrt{3}}.$$

The uniform distribution on (a, b) can be derived from the standard uniform distribution on $(0, 1)$ by simple linear transformation. Symbolically we write $\mathcal{R}(a, b) = a + (b - a)\mathcal{R}(0, 1)$. To generate n independent observations from $\mathcal{R}(a, b)$, use `runif(n,min=a,max=b)` where the default values are `a=0` and `b=1`.

Example 2.12 *Waiting time. The shuttle bus comes every 30 minutes. What is the expected waiting time if you come to the bus stop at random time? Use simulations to confirm your answer.*

Solution. Clearly, the wait time is less than 30 minutes. The arrival time can be modeled as a random variable X distributed as $\mathcal{R}(0, 30)$. As follows from (2.22),

$$\text{the expected waiting time} = E(X) = 30/2 = 15 \text{ minutes.}$$

To simulate we imagine that the bus comes at 0:30, 1, 1:30,...,24:00 or on the minute scale $m = \{30, 60, 90, ..., 60 \times 24\}$ and your arrival can be modeled as the uniform distribution, $X \sim \mathcal{R}(0, 1440)$. Then the average arrival time is the average distance to the next point in m to the right, which can be computed using `ceiling` command: `X=runif(100000,min=0,max=1440);mean(30 *ceiling((X/30))-X)` with the output `15.00843`.

Example 2.13 *Broken stick. A stick of unit length is broken at random. What is the probability that a longer piece is more than twice the length of the shorter piece. Provide a theoretical answer and confirm it by simulations.*

Solution. The point where the stick is broken is uniformly distributed along its length, $X \sim \mathcal{R}(0, 1)$. The length of the longer piece, $X_L > 0.5$, and therefore $X_L \sim \mathcal{R}(0.5, 1)$. Since the shorter piece has length $1 - X_L$, the length must be such that $X_L > 2(1 - X_L)$, i.e. $X_L > 2/3$. Using the fact that $X_L \sim \mathcal{R}(0.5, 1)$, the asked probability is $(1 - 2/3) \times 2 = 2/3$.

The R code that estimates the probability via simulations is shown below.

```
longpiece=function(nExp=100000)
{
dump("longpiece","c:\\StatBook\\longpiece.r")
X=runif(nExp)
X.long=X #initialization
X.long[X<0.5]=1-X[X<0.5] # if X is short take the other part
X.short=1-X.long # by definition
pr=mean(X.long>2*X.short) # proportion of experiments
pr
}
```

This function gives the probability `0.66858`, very close to the theoretical answer. We make several comments on the code: (i) The default number of experiments/simulations is 100K. (ii) It is a good idea to save the code as a text file in a safe place. In fact, it is easy to lose code in R: you may forget to save the workspace on exit, your code may be overwritten by another program, etc.. The `dump` command saves the program as a text file under the name `c:\\StatBook\\longpiece.r`. (iii) `X.long[X<0.5]` means that only components of `X.long` for which `X<0.5` are used. The same is true for the right-hand side of the respective line. (iv) Command `X.long>2*X.short` creates a logical vector with `TRUE` if the inequality is true and `FALSE` otherwise. When a numeric operator such as `mean` is applied, `TRUE` is replaced with 1 and `FALSE` is replaced with 0, so that `mean` returns the proportion when the inequality is true.

Alternatively, one can compute the longest and the shortest piece using commands `pmax()` and `pmin()`. These commands compute maximum and minimum in a vectorized fashion, so no loop is required. For example, if `v1`, `v2`, and `v3` are vectors of the same length, `pmax(v1,v2,v3)` returns a vector of the same length with the ith component equal to the maximum of ith components of these vectors. Then the longest and the shortest piece are computed as `X.long=pmax(X,1-X)` and `X.short=pmin(X,1-X)`, respectively.

Example 2.14 *Kurtosis for the uniform distribution. Find the kurtosis of the uniform distribution on* $(0,1)$.

Solution. The formula for the kurtosis is given by (2.11). Find the fourth central moment:

$$\mu_4 = \int_0^1 (x - 0.5)^4 dx = \int_{-1/2}^{1/2} z^4 dz = 2 \int_0^{1/2} z^4 dz = \frac{1}{5 \times 2^4}.$$

But $\sigma^2 = 1/12$ so that kurtosis $= 12^2/5 \times 2^4 - 3 = -1.2$, meaning that the distribution is flatter than normal, as expected. ☐

Below we use the uniform distribution for raison cookies from Example 1.19.

Example 2.15 *Raisins in a cookie simulations. Use simulations to confirm that the probability that one cookie has all n raisins is* $m^{-(n-1)}$.

Solution. Since the probability that there is one raisin in the cookie is $1/m$ the probability that a particular cookie has all the raisins is m^{-n}. Then, the probability that one cookie has all the raisins is $m \times m^{-n} = m^{-(n-1)}$. The simulations presented in the R code `simCookie` confirm this probability. We put m cookies side by side and replace them with unit length segments $[0,1], [1,2], .., [m-1,m]$. Then we throw n raisins on the segment $[0,m]$ at random and observe if all raisins fell into one unit segment; repeat this process `nSim` times (see function `simCookie`). The variable `pr` counts the number of simulations when one cookie has n raisins. The call `simCookie()` with default values `n=3,m=4` gives `0.06304` and the analytical answer is $4^{-(3-1)} = 0.0625$.

Problems

1. Derive a formula for the variance of the uniform distribution. [Hint: Derive the noncentral second moment and then use formula (2.6).]

2. Find the median and the first and third quartiles of the uniform distribution. Use `qunif` function in R to test your answers using $a = -2$ and $b = 3$.

3. Use `pmax` and `pmin` in function `longpiece` to compute the longest and the shortest piece, as suggested above. Test that the new version yields the same answer.

4. Demonstrate that the distribution of the long and short pieces in the previous problem is uniform by plotting the empirical cdfs on one graph side by side using `par(mfrow=c(1,2))`. Prove that the distributions are uniform. [Hint: Use the fact that $\max(X, 1-X) \leq y$ is equivalent to $X \leq y$ and $1 - X \leq y$.]

5. Two ways to get the shortest piece of a stick with two random breaks are suggested: (1) break the first time at random and brake the shortest piece at random again, taking the shortest piece and (2) break the stick at two random places and take the shortest piece. What way produces a shorter piece on average? Modify the `longpiece` function to get the answer via simulations. [Hint: Use `pmin`; find the shortest piece in the second method as $\min(x_1, x_2, 1 - x_1, 1 - x_2, |x_1 - x_2|)$.]

6. (a) Plot the empirical cdf of the shortest piece using the two methods in the previous problem on the same graph. Use different colors and `legend`. Are the distributions uniform? (b) Prove that $X_L \sim \mathcal{R}(0.5, 1)$ by finding the theoretical cdf.

7. Find the kurtosis for $\mathcal{R}(a, b)$.

8. Two friends come to the bus stop at random times between 1 and 2 p.m. The bus arrives every 15 minutes. Use simulations to estimate the probability that the friends end up on the same bus.

9. Modify function `simCookie` to estimate that a particular cookie (your cookie) gets all the raisins, derive an analytical answer and compare the results. [Hint: Use `cookie.id` as an argument in the R function to specify your cookie.]

10. Besides m raisins, l chocolate chips are added to the dough. Use simulations to answer the following questions: (a) What is the probability that at least one cookie gets all raisins and all chocolate chips? (b) What is the probability that a particular cookie gets all the chips but no raisins?

2.4 Exponential distribution

The exponential distribution is commonly used for modeling waiting time and survival analysis. It describes the distribution of a positive continuous random variable with the cdf defined as $F(x; \lambda) = 1 - e^{-\lambda x}$ for $x \geq 0$ and 0 for $x < 0$, where λ is a positive parameter, typically referred to as the *rate*. See Example 2.18 below for why λ is called the rate. The density of the exponential distribution is derived through differentiation:

$$f(x; \lambda) = \lambda e^{-\lambda x}, \quad x \geq 0. \tag{2.23}$$

We use notation $X \sim \mathcal{E}(\lambda)$ to indicate that X has an exponential distribution with parameter λ. We bring the reader's attention to the notation: parameter λ is separated from the argument by a semicolon. The indication that the distribution depends on λ is especially important in statistics where we estimate λ using a sample of observations from this distribution. The density is a decreasing function of x; see Figure 2.7.

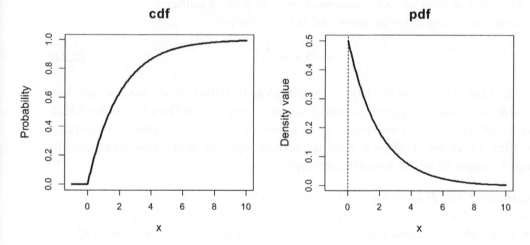

Figure 2.7: *The cdf and pdf of an exponential random variable with the rate parameter $\lambda = 1/2$. The former is computed in R as* pexp *and the latter as* dexp.

The exponential distribution frequently emerges in survival analysis. Then we interpret $F(x)$ as the proportion of dead by time x, and the complementary function $S(x) = 1 - F(x)$ is interpreted as the survival function, so that $F(x + \Delta x) - F(x)$ is the proportion who died between x and $x + \Delta x$. The mortality rate is defined as $(F(x + \Delta x) - F(x))/\Delta x$, the death rate per unit of time. On the scale of survivors, this rate takes the form

$$\frac{(F(x + \Delta x) - F(x))/\Delta x}{1 - F(x)},$$

the death rate per unit of time per proportion of alive at time x. Letting Δx go to zero we arrive at the definition of the *hazard function*,

$$H(x) = \frac{F'(x)}{1 - F(x)} = \frac{-S'(x)}{S(x)},$$

as the instantaneous relative mortality rate: the proportion of dead among survivors. For the exponential distribution, the hazard function is constant $H(x) = \lambda e^{-\lambda x}/e^{-\lambda x} = \lambda$. In words, exponential distribution yields a constant hazard function.

To find the mean of the exponential distribution, we apply integration by parts, $\int u dv = uv - \int v du$, which after letting $u = x$ and $dv = e^{-\lambda x} dx$ gives

$$E(X) = \lambda \int_0^\infty x e^{-\lambda x} dx = -\lambda \times \left. \frac{1}{\lambda} e^{-\lambda x} \right|_0^\infty + \frac{\lambda}{\lambda} \int_0^\infty e^{-\lambda x} dx = \frac{1}{\lambda}.$$

An alternative parametrization uses $\theta = 1/\lambda$, so that the density takes the form $f(x; \theta) = \theta^{-1} e^{-x/\theta}$. Parameter θ is called the *scale* parameter. An advantage of this parametrization is that θ has the same scale unit as X, and, moreover, $E(X) = \theta$. The rate parametrization is used in R by default.

Repeated integration by parts yields the variance

$$\mathrm{var}(X) = \frac{1}{\lambda^2}. \tag{2.24}$$

The exponential distribution is the simplest distribution for positive random variables and may be applied to model the occurrence of random events for which the probability drops monotonically. In particular, it may be applied to describe the time to an event after a meeting arrival, appointment, announcement, or project launch as in the following example.

Example 2.16 *Telephone call.* *Bill said that he will call after 10 a.m. Assuming that the time of his call follows an exponential distribution with parameter $\lambda = 1/10$, which decision maximizes your probability of talking with Bill: (a) wait the first 10 minutes, or (b) wait for the call from 10:10 to 11:00?*

Solution. Let X denote the time elapsed before Bill calls. It is believable that the density of calls is decreasing with time such that the maximum density occurs right after 10 a.m. ($X = 0$). Consequently, the exponential distribution is a good candidate in this case. Specifically, the cdf takes the form

$$F(x) = \Pr(\text{Bill calls within } (10, 10+x) \text{ time interval}) = \Pr(X \le x) = 1 - e^{-x/10}.$$

The first probability (a) is $\Pr(X \le 10) = F(10) = 1 - e^{-10 \times 0.1} = 1 - 1/e = 0.632$. The second probability (b) is $\Pr(10 < X \le 60) = F(60) - F(10) = (1 - e^{-0.1 \times 60}) - (1 - e^{-0.1 \times 10}) = 0.365$. Therefore, it is better to wait the call first 10 minutes after 10 a.m. than to wait the 50 minutes after 10:10 a.m. □

We prove in Section 3.3 that the exponential distribution is memoryless. The following example illustrates how to use simulations with the exponential distribution.

Example 2.17 _Safe turn wait time._ _A truck requires 10 seconds to turn at an intersection onto a busy street. Assuming that the time between two passing cars follows an exponential distribution with the average time three seconds, what is the median time required to make a safe turn? Use simulations to find the answer._

Solution. The R code is realized in function `truck.turn`. The arguments of the function with default values are `tr.tu=10`, `car.time=3`, `Nmax=1000`, `nSim= 10000`. Simulations are carried out using the loop over `isim`. For each `isim` we generate an array of times between two passing cars using command `X=rexp(Nmax, rate=1/ car.time)` because the rate, λ, is the reciprocal of the mean; `Nmax` is the maximum number of cars (it should be big enough but is irrelevant). The command `cX=cumsum(X)` computes the cumulative time, the time elapsed after the truck comes to the intersection. Since the truck turns at the first 10-second gap in traffic, we compute this time as `nsec[isim]=min(cX[X>tr.tu])`. After all simulations are done, we plot the cdf and compute the median wait time using `median` command. With the default parameters, the truck must wait about one minute to make a safe turn. □

The following example shows the connection between the Poisson and exponential distributions.

Example 2.18 _Bathroom break._ _The number of customers arriving at a bank per time t follows the Poisson distribution_ $\mathcal{P}(\lambda t)$, _where_ λ _is the rate of customer flow. If the bank teller just served a customer and needs to take a bathroom break, (a) what is the distribution of time until the next customer walks in? (b) Does he/she have enough time to go to the bathroom if he/she needs five minutes and two customers arrive every 10 minutes on average?_

Solution. First of all, we explain why parameter λ can be interpreted as the rate of customer flow. Because if X denotes the number of customers walking in during time units t, we have $E(X) = \lambda t$. Therefore, $\lambda = E(X)/t$ is the number of expected customers per minute. (a) The probability that no customers arrive within t time units can be modeled as $\Pr(X = 0) = \frac{(\lambda t)^0}{0!} e^{-\lambda t} = e^{-\lambda t}$. Therefore, the cdf of time until the next customer walks in within t time units is the complementary probability, $F(t) = 1 - e^{-\lambda t}$, the cdf of the exponential distribution. (b) In our case, the time unit = 10 minutes, so the probability that no customer walks in within 5 minute bathroom break is computed by formula above using $\lambda = 2$ and $t = 1/2$: $e^{-2/2} = e^{-1} = 0.37$. □

The exponential distribution has an important connection with the chi-square distribution to be discussed later in Section 4.2.

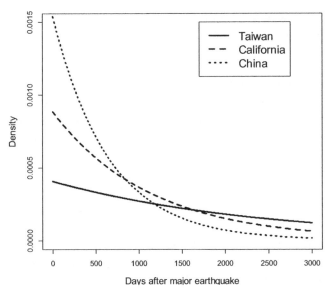

Figure 2.8: *The densities of time between two consecutive earthquakes in three regions.*

Example 2.19 *Earthquake occurrence.* *Three seismological active regions, namely, Taiwan, China, and California, have different patterns of earthquake occurrence. In particular, Wang and Kuo [101] confirmed that the interoccurrence (time between two consecutive earthquakes) nearly follows an exponential distribution with a scale parameter for the three regions of 2465, 649, and 1131, respectively. (a) Depict the three densities, (b) find the median and the third quartile (in years), and (c) compute the probability that no earthquake occurs within one year after a past quake.*

Solution. (a) The three densities are depicted in Figure 2.8. (b) The median is found from the equation $1 - e^{T/\theta} = 0.5$ and the third quartile is found from equation $1 - e^{T/\theta} = 0.75$. This quantile means the time between consecutive earthquake with probability 0.75. (c) The probability that no earthquakes occur within a year is $e^{-365/\theta}$. The results are presented in the following table.

	Taiwan	China	California
Scale, θ	2465 days	649 days	1131 days
Rate, λ	0.148 years	0.562 years	0.322 years
Median	4.68 years	1.23 years	2.15 years
3d quartile	7.42 years	1.95 years	3.40 years
Pr(no earthquake within year)	0.862	0.570	0.724

2.4.1 Laplace or double-exponential distribution

The density of the *Laplace* or *double-exponential* (sometimes called biexponential) distribution is defined as

$$f(x; \lambda) = \frac{\lambda}{2} e^{-\lambda |x|}, \quad -\infty < x < \infty, \tag{2.25}$$

with notation $X \sim \mathcal{L}(\lambda)$.

The cdf is easy to obtain by integration:

$$F(x; \lambda) = \begin{cases} \frac{1}{2} e^{\lambda x} & \text{for } x < 0 \\ 1 - \frac{1}{2} e^{-\lambda x} & \text{for } x \geq 0 \end{cases},$$

see Figure 2.9. This distribution is symmetric, and therefore the mean, mode, and median are the same, 0. The density does not have a derivative at $x = 0$. The variance is $2/\lambda^2$. The density of the shifted Laplace distribution is defined as $(\lambda/2) e^{-\lambda |x - \mu|}$. This distribution is a good model for studying robust statistical inference; as we shall learn later (Section 6.11), this distribution gives rise to the median. See Example 6.161 where statistical properties of the mean and median are compared.

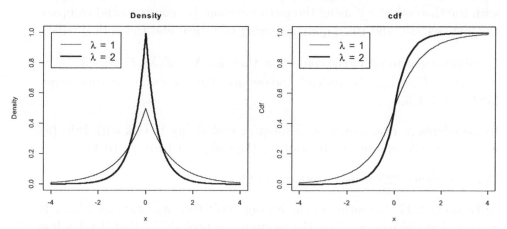

Figure 2.9: *Density and cdf of the Laplace (double-exponential) distribution. The greater λ, the higher the density peak, and the steeper the slope of the cdf at zero.*

2.4.2 R functions

The exponential distribution has a built-in R function. For example, `rexp(n=100, rate=0.5)` will produce 100 random numbers with density $\lambda e^{-\lambda x}$ where $\lambda = 0.5$. There are three other functions associated with the exponential distribution: `qexp`, `pexp`, and `dexp`. For instance, quantiles can be computed as `-log(1-p)/ lambda`.

The Laplace distribution is symmetric; we can use this property to generate random numbers using the following two lines of code: `s=sample(x=c(-1,1), replace=T,size=n,prob=c(.5,.5)); x=s*rexp(n=n,rate=lambda)`.

Problems

1. Prove that if $X \sim \mathcal{R}(0,1)$, then $Y = -\ln X$ has an exponential distribution. Write R code to check your answer: simulate 100K uniformly distributed random numbers and plot the theoretical and empirical cdfs of Y on the same graph.

2. Prove that the only survival function with a constant hazard function is the exponential survival function, $S(x) = e^{-\lambda x}$.

3. Justify why θ is called the scale parameter. [Hint: How does θ change when the scale of x changes from days to weeks, or from months to years?]

4. Show that the 0.5 tight confidence range for the exponential distribution is (0,median). [Hint: The conditions of Theorem 2.10 do not hold.]

5. Write an R program to do the following: (a) generate 10K observations from $\mathcal{E}(2)$ using the function `rexp`; (b) plot the empirical cdf and superimpose with the theoretical cdf using the `pexp` function; (c) compute and compare the first, second, and third quartiles using empirical and theoretical cdfs.

6. Use simulations to reject the conjecture that, for $X \sim \mathcal{E}(\lambda)$, $E\left|X - \lambda^{-1}\right|^k = \lambda^{-k}$. [Hint: Plot the "theoretical" and empirical moments on the same graph for $k = 2, 3, 4, 5$.]

7. In the telephone example, what is a better probability to talk with Bill: (a) wait the first 5 minutes or (b) wait for the call from 10:10 to 10:40?

8. Derive variance (2.24) using formula (2.6).

9. In Example 2.17, (a) estimate the average wait time and explain why it is greater than the median time, (b) estimate the probability that it takes less than 30 seconds for a safe turn, and (c) estimate the median wait time if the truck driver misses the first gap and turns on the second.

10. Compute the probability that two earthquakes happen within a week in each three regions.

11. Prove that if $X \sim \mathcal{L}(\lambda)$, then $E(X) = 0$ and $\mathrm{var}(X) = 2/\lambda^2$.

12. (a) Generate 10K random observations from a shifted Laplace distribution with $\lambda = 2$ and $m = -1$. (b) Plot the empirical and theoretical cdfs on the same graph.

2.5 Moment generating function

Calculating moments of a distribution can be a laborious task. The moment generating function (MGF) is defined as

$$M(t) = E(e^{tX}) = \int_{-\infty}^{\infty} e^{tx} f(x)dx.$$

This function makes the calculation of moments considerably easier especially when expressed in closed form. Once the the MGF is obtained, computation of moments is fairly easy and reduces to evaluation of derivatives of the MGF at $t = 0$. It should be said that the MGF may not exist for some t (the integral may diverge). However, it is sufficient for it to exist in a neighborhood of zero because the derivatives are evaluated at $t = 0$. The MGF is defined for discrete random variables as well; then integral is replaced with a sum. The relationship between the kth moment and the derivative of order k is shown in the following theorem.

Theorem 2.20 *The kth noncentral moment $\nu_k = E(X^k)$ is equal to the kth derivative of the MGF evaluated at zero:*

$$\nu_k = \left. \frac{d^{(k)} M(t)}{dt^k} \right|_{t=0}. \tag{2.26}$$

Proof. Under mild regularity conditions, differentiation may be done under the integral,

$$\frac{d^{(k)} M(t)}{dt^k} = \int_{-\infty}^{\infty} x^k e^{tx} f(x)dx,$$

so that

$$\left. \frac{d^{(k)} M(t)}{dt^k} \right|_{t=0} = \int_{-\infty}^{\infty} x^k f(x)dx = \nu_k.$$

□

This theorem implies that once the MGF is derived, all noncentral moments are fairly straightforward to find. Unfortunately, the MGF does not allow computation of central moments. When noncentral moments up to the kth order are known, the binomial formula may be used to find the kth central moment,

$$E(X - \mu)^k = \sum_{i=0}^{k} (-1)^{k-i} \binom{k}{i} E(X^i) \mu^{k-i}, \tag{2.27}$$

as follows from formula (1.9).

Example 2.21 *Moments of the exponential distribution.* *Find the MGF of the exponential distribution. Find its variance and all noncentral moments using the MGF. Find the third central moment using the binomial formula (2.27) and confirm that the distribution is skewed to right.*

Solution. We have

$$M(t) = \int_0^\infty \lambda e^{-\lambda x} e^{xt} dx = \lambda \int_0^\infty e^{-(\lambda - t)x} dx.$$

We observe that for $\lambda \le t$ the MGF does not exist, but it exists when $t < \lambda$:

$$\int_0^\infty e^{-(\lambda - t)x} dx = \frac{1}{\lambda - t}.$$

Hence, the MGF for the exponential distribution is

$$M(t) = \frac{\lambda}{\lambda - t}, \quad t < \lambda. \tag{2.28}$$

Although the MGF does not exist for all t, the derivatives can be evaluated at zero because the MGF is defined on $(-\infty, \lambda)$. According to the above theorem, the first and the second moments are calculated as

$$E(X) = \left. \frac{dM(t)}{dt} \right|_{t=0} = \lambda (\lambda - t)^{-2} \big|_{t=0} = \frac{1}{\lambda},$$

$$E(X^2) = \left. \frac{d^2 M(t)}{dt^2} \right|_{t=0} = 2\lambda (\lambda - t)^{-3} \big|_{t=0} = \frac{2}{\lambda^2}$$

Using the alternative formula for the variance, we obtain

$$\text{var}(X) = E(X^2) - E^2(X) = \frac{2}{\lambda^2} - \frac{1}{\lambda^2} = \frac{1}{\lambda^2}.$$

Differentiating $M(t)$ k times and evaluating the derivative at zero we obtain the formula for noncentral moments:

$$E(X^k) = k! \lambda^{-k}. \tag{2.29}$$

The third central moment is found from (2.27) as follows:

$$\begin{aligned} E(X - \mu)^3 &= E(X^3) - 3\mu E(X^2) + 3\mu^2 E(X) - \mu^3 \\ &= 6\lambda^{-3} - 6\lambda^{-3} + 3\lambda^{-3} - \lambda^{-3} = 2\lambda^{-3}, \end{aligned}$$

since $\mu = \lambda^{-1}$ and $E(X^3) = 6\lambda^{-3}$. Since the third central moment is positive, we deduce that exponential distribution is skewed to the right.

$\square$

For discrete random variables, the MGF is defined as

$$M(t) = E(e^{tX}) = \sum_{k=1}^\infty e^{tx_k} \Pr(X = x_k),$$

assuming that X takes values $x_1, x_2, \dots$. Again, the kth noncentral moment is equal to the kth derivative of the MGF evaluated at zero.

Example 2.22 *Mean and variance of the Poisson distribution.* *Find the MGF of the Poisson distribution and use it to compute the mean and variance. Use simulations to check the MGF by plotting the empirical and theoretical MGF on the same graph.*

Solution. We have

$$M(t) = e^{-\lambda} \sum_{k=1}^{\infty} \frac{1}{k!} e^{tk} \lambda^k = e^{-\lambda} \sum_{k=1}^{\infty} \frac{1}{k!} (e^t \lambda)^k,$$

where $e^t \lambda$ may be considered "another" parameter λ. Hence $\sum_{k=1}^{\infty} \frac{1}{k!} (e^t \lambda)^k = e^{e^t \lambda}$. Finally, the MGF of the Poisson distribution is

$$M(t) = e^{\lambda(e^t - 1)}.$$

The mean is

$$E(X) = \frac{dM(t)}{dt} \bigg|_{t=0} = \lambda$$

and the second noncentral moment is

$$E(X^2) = \frac{d^2 M(t)}{dt^2} \bigg|_{t=0} = \lambda^2 + \lambda.$$

The variance is

$$\mathrm{var}(X) = E(X^2) - E^2(X) = \lambda.$$

The empirical MGF is computed as the mean of $\exp(tX)$, where $X \sim \mathcal{P}(\lambda)$. See the R code `poisMGF`. □

As we mentioned above, the MGF may not exist at $t = 0$. For example, for a uniformly distributed random variable, $X \sim \mathcal{R}(0, 1)$, we have

$$M(t) = \int_0^1 e^{tx} dx = \frac{1}{t} e^{tx} \bigg|_0^1 = \frac{e^t}{t} - \frac{1}{0},$$

so $M(0)$ formally does not exist. On the other hand, using the limit definition, we obtain

$$\lim_{t \to 0} \frac{e^t - 1}{t} = 1. \tag{2.30}$$

This example emphasizes that some caution should be used when finding the MGF. The noncentral moments of the uniform distribution are easy to find by straightforward integration.

The following theorem expresses the MGF upon linear transformation. The proof is left as an exercise to the reader.

Theorem 2.23 *If a and b are constants and X has MGF $M_X(t)$ then*

$$M_{(X+a)/b}(t) = e^{(a/b)t} M_X(t/b). \tag{2.31}$$

An important property of the MGF is that the MGF of a sum of independent random variables is the product of individual MGFs; this follows from the fact that, for independent random variables, the expected value of the product is equal to the product of expected values (3.16). This property is key in studying the distribution of sums and particularly proving the central limit theorem (CLT). We use this property to prove that the sum of independent Poisson distributions is a Poisson distribution expressed in equation (1.15). Indeed, the ith MGF is $M_i(t) = e^{\lambda_i(e^t-1)}$, and the MGF of the sum is

$$\prod_{i=1}^{n} M_i(t) = \prod_{i=1}^{n} e^{\lambda_i(e^t-1)} = e^{(e^t-1)\sum_{i=1}^{n} \lambda_i},$$

which is the MGF of $\mathcal{P}\left(\sum_{i=1}^{n} \lambda_i\right)$.

2.5.1   Fourier transform and characteristic function

If f is an integrable function on the real line then $\int_{-\infty}^{\infty} f(x)dx$ exists and the Fourier transform is defined as

$$g(t) = \int_{-\infty}^{\infty} e^{-i2\pi xt} f(x)dx,$$

where i is the imaginary unit, $i = \sqrt{-1}$. Compactly, we write $\mathcal{F}(f) = g$, where $\mathcal{F}$ is the Fourier operator that transforms a real function f into a complex-valued function g. Function f can be restored from its Fourier transform as

$$f(x) = \int_{-\infty}^{\infty} e^{i2\pi xt} g(t)dt.$$

Geometrically, g defines a curve on the plane. Using Euler's formula

$$e^{ix} = \cos x + i\sin x,$$

we can separate the Fourier transform into real and imaginary part:

$$g(t) = \int_{-\infty}^{\infty} \cos(2\pi xt)f(x)dx - i\int_{-\infty}^{\infty} \sin(2\pi xt)f(x)dx.$$

This decomposition explains why the Fourier transform exists for every integrable function: $|\cos(2\pi xt)| \leq 1$ and $|\sin(2\pi xt)| \leq 1$, unlike the MGF which, as we know, may not exist for some t. The Fourier transform is popular in applied mathematics and particularly for solving ordinary or partial differential equations where f is not necessarily a density.

In probability theory, the characteristic function is used instead:

$$c(t) = \int_{-\infty}^{\infty} e^{ixt} f(x)dx,$$

where f is the pdf. As follows from the above, the characteristic function always exists as in the case with the MGF. Another important property of the characteristic function is that there is a one-to-one correspondence between the distribution (discrete or continuous) and the characteristic function. if two characteristic functions are the same, then their two distributions (pdf or cdf) are the same, and vice versa. This statement holds for multivariate distributions as well; see Section 3.10.2.

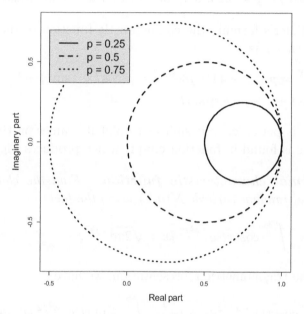

Figure 2.10: *The characteristic functions of the Bernoulli random variable for* $p = 0.25$, 0.5, *and* 0.75.

Alternatively, the characteristic function can be expressed as

$$c(t) = E e^{itX}.$$

Using Euler's formula, we express the characteristic function via trigonometric functions as

$$c(t) = \int_{-\infty}^{\infty} \cos(xt) f(x) dx + i \int_{-\infty}^{\infty} \sin(xt) f(x) dx.$$

Note the obvious relation between the Fourier transform and the characteristic function,

$$c(t) = \mathcal{F} f\left(-\frac{x}{2\pi}\right).$$

Similarly to the MGF, it is easy to obtain the central moments of the distribution by evaluating the derivatives of the characteristic function at zero. The major property of the characteristic function is that the characteristic function of the sum of independent random variables is the product of their characteristic functions. The characteristic function is the major mathematical tool to study the distribution of a sum of independent variables and particularly the rate of convergency to the normal distribution in the central limit theorem (Section 2.10).

Example 2.24 *Characteristic function. Express the characteristic function of the Bernoulli random variable in terms of trigonometric functions. Plot three characteristic functions for $p = 0.25$, 0.5, and 0.75 on the same graph.*

Solution. Using Euler's formula the characteristic function of the Bernoulli distribution can be written as

$$c(t) = Ee^{itX} = pe^{it \times 1} + (1-p)e^{it \times 0} = p(\cos t + i \sin t) + 1 - p$$
$$= (p \cos t + 1 - p) + ip \sin t.$$

The above function c is periodic, so it suffices to plot its values on the segment $t \in [0, 2\pi]$. The R code is found in function `charf`, which produces Figure 2.10.

Example 2.25 *Normal characteristic function. Find the characteristic function of the normal random variable $\mathcal{N}(0,1)$ using the fact*

$$\int_{-\infty}^{\infty} \cos(xt)e^{-x^2/2}dx = \sqrt{2\pi}e^{-\frac{1}{2}t^2}.$$

Solution. From the trigonometric representation, we have

$$c(t) = \frac{1}{\sqrt{2\pi}} \int_{-\infty}^{\infty} \cos(xt)e^{-x^2/2}dx + i\frac{1}{\sqrt{2\pi}} \int_{-\infty}^{\infty} \sin(xt)e^{-x^2/2}dx = e^{-t^2/2}$$

because $\sin(xt)e^{-x^2/2}$ is an odd function of x for every t. In general, the characteristic function is real if the density is an even function (the random variable is symmetric).

Problems

1. Let X be a uniformly distributed random variable on $(-1, 1)$. Find the kth noncentral moment through straightforward integration and using the MGF. Check the equality. [Hint: Use a limit similar to (2.30).]

2. Find all moments of the Bernoulli random variable as a function of $p = \Pr(X = 1)$.

3. (a) Derive the MGF of the binomial distribution using the MGF of the Bernoulli distribution and the fact that the MGF of the sum of independent random variables is the product of individual MGFs. (b) Demonstrate by simulations that your formula is correct by plotting the theoretical and empirical MGFs on the same graph.

4. Prove that the sum of independent binomial random variables with the same probability is again a binomial random variable.

5. Prove that the Poisson distribution is skewed to the right.

6. Use the MGF derived in Example 2.22 to prove that $E(X^3) = \lambda(3\lambda^2+3\lambda+1)$ and $E(X^4) = \lambda(\lambda^3 + 6\lambda^2 + 7\lambda + 1)$.

7. Prove that if $M(t)$ is the MGF of a random variable X, then the MGF of $a + bX$ is $e^{at}M(bt)$.

8. Prove that the MGF of the Laplace distribution is $(1 - (t/\lambda)^2)^{-1}$, where $|t| < \lambda$, and prove that $\mathrm{var}(X) = 2/\lambda^2$.

9. Check formula (2.31) for an exponential distribution with $a = 0$. [Hint: Use the fact that aX has an exponential distribution and formula (2.28).]

10. Find the characteristic function of $\mathcal{N}(\mu, \sigma^2)$.

2.6 Gamma distribution

The gamma distribution is a generalization of the exponential distribution. The gamma distribution is suited for modeling positive continuous random variables such as wait time, time to event, and reliability with applications in science and engineering. The gamma distribution is also important in statistics because its special case, the chi-square distribution, has a wide range of applications in statistical hypothesis testing.

Recall that the density of the exponential distribution is a decreasing function. The density of the gamma distribution, on the other hand, has a positive mode, the point where the density reaches its maximum value. This property is similar to the Poisson distribution, and as we shall see later, these two distributions are closely related.

The density of the gamma distribution is defined as

$$f(x; \alpha, \lambda) = \frac{\lambda^\alpha}{\Gamma(\alpha)} x^{\alpha-1} e^{-\lambda x}, \quad x > 0, \tag{2.32}$$

where α and λ are positive parameters referred to as the shape and rate parameters, respectively, and Γ is the gamma function,

$$\Gamma(\alpha) = \int_0^\infty u^{\alpha-1} e^{-u} du, \quad \alpha > 0.$$

We list some properties of the gamma function, which can be viewed as a continuous counterpart of the factorial: (i) $\Gamma(1) = \Gamma(2) = 1$, (ii) $\Gamma(\alpha+1) = \alpha\Gamma(\alpha)$, (iii) $\Gamma(n) = (n-1)!$, (iv) $\Gamma(0.5) = \sqrt{\pi}$. The gamma function and its logarithm are

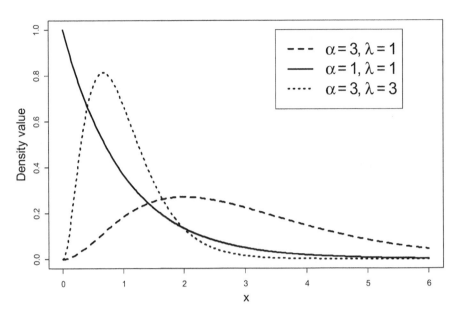

Figure 2.11: *Three gamma densities. The exponential density is a special case with $a = 1$. Parameter λ controls the rate at which the density approaches the x-axis when $x \to \infty$, and parameter α controls how far to the right the density is shifted.*

built-in functions in R, gamma(x) and lgamma(x), respectively. For large values of the argument, the gamma function takes very large values that may cause computation errors; computing the gamma function on the log scale is much more robust, so lgamma is preferable. For example, to compute the log "n choose m" value for large n from Section 1.6, we use lgamma(n+1)-lgamma(m+1)-lgamma(n-m+1).

If X has a gamma distribution with the shape parameter α and the rate parameter λ, we write $X \sim \mathcal{G}(\alpha, \lambda)$. Three densities of the gamma distribution are shown in Figure 2.11.

When $\alpha = 1$, the gamma distribution reduces to the exponential distribution. For $\alpha > 1$, the density of the gamma distribution reaches its maximum at $(\alpha - 1)/\lambda$, the mode (the point of maximum density). For $0 < \alpha \leq 1$, the density is a decreasing function of x.

The MGF for the gamma distribution is given by

$$M(t) = \frac{\lambda^\alpha}{\Gamma(\alpha)} \int x^{\alpha-1} e^{-\lambda x} e^{tx} dx = \frac{\lambda^\alpha}{\Gamma(\alpha)} \int x^{\alpha-1} e^{-(\lambda-t)x} dx \qquad (2.33)$$

$$= \frac{\lambda^\alpha}{\Gamma(\alpha)} \frac{\Gamma(\alpha)}{(\lambda - t)^\alpha} = \frac{1}{(1 - t/\lambda)^\alpha}.$$

The mean and the variance of the gamma distribution are easily derived from the MGF:

$$E(X) = \frac{\alpha}{\lambda}, \quad \text{var}(X) = \frac{\alpha}{\lambda^2}.$$

As follows from Section 2.5, the MGF of the sum of independent random variables is equal to the product of individual MGFs. We apply this result to prove that the sum of independent gamma-distributed random variables with parameters α_i and λ is a gamma-distributed random variable with parameters $\sum_{i=1}^{n} \alpha_i$ and λ (note that the rate parameter is the same). Indeed, as follows from (2.33), the MGF of the sum is

$$\prod_{i=1}^{n} \frac{1}{(1 - t/\lambda)^{\alpha_i}} = \frac{1}{(1 - t/\lambda)^{\sum_{i=1}^{n} \alpha_i}},$$

which proves the statement. In particular, the sum of n exponentially distributed random variables with the same rate is a gamma-distributed random variable with shape parameter $\alpha = n$.

2.6.1   Relationship to Poisson distribution

If events happen according to the Poisson distribution with the mean rate λ per unit time, then the distribution of the waiting time until the αth event can be approximated by the gamma distribution with parameters λ and α. This statement is justified by the following connection between the two cdfs:

$$F(k; \lambda) = 1 - F(\lambda; k + 1, 1), \quad k = 0, 1, 2, ..., \tag{2.34}$$

where F on the left-hand side denotes the cdf of the Poisson distribution with parameter λ and F on the right-hand side denotes the cdf of the gamma distribution evaluated at λ with the shape and rate parameters $k + 1$ and 1, respectively. In words, the cdf of a Poisson distributed random variable with the rate parameter λ evaluated at k is equal to the complementary cdf of the gamma-distributed random variable with the shape parameter $k + 1$ and the rate parameter 1 evaluated at λ. One can get a better approximation if parameter λ in the gamma distribution is augmented by 0.5; see below.

Example 2.26 *Gamma and Poisson distributions. Write R code that confirms the relationship between the Poisson and gamma distributions through simulations. (a) Plot the empirical cdf of the number of event occurrences until α and superimpose the gamma cdf on the same graph (job=1). (b) Plot the Poisson and two gamma cdfs on the same graph (job=2).*

Solution. (a) Imagine that each minute independent events occur at a rate following a Poisson distribution with parameter $\lambda = 5$. We want to find the distribution of times over the day until the first $\alpha = 300$ events occur. The following algorithm is suggested to simulate the times: (i) Compute Nmax=24*60 as the number of minutes a day. (ii) Generate random Poisson-distributed numbers of events for each minute; this is array x of integers of length Nmax (the ith component is the number of events that happened at the ith minute). (iii) Compute the

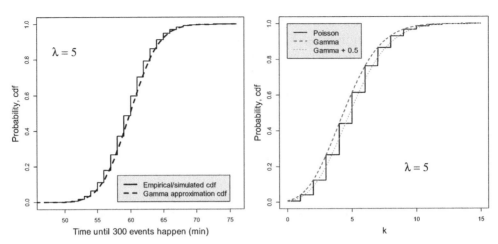

Figure 2.12: *Left: Approximation of the number of Poisson events using the gamma distribution. Right: Approximation of the Poisson cdf with the gamma cdf from (2.34).*

number of minutes elapsed until α events happened (use cumulative sum `cumsum` and find when `cumsum` $\leq \alpha$). The R code is in file `gampois.r`, which produces Figure 2.12 using `job=1`. The empirical cdf is a stepwise function because the time in minutes is discrete integers while the gamma cdf is continuous. (b) The plot at right displays the three cdfs: the step-wise line depicts the Poison cdf. The continuous dashed line depicts the gamma cdf as the right-hand side of equation (2.34) which goes through the corners of the Poisson cdf. The dotted line depicts $1 - F(\lambda + 0.5; k + 1, 1)$, which goes through the center of the Poisson cdf. Note that, instead of computing the complementary cdf, we use the R option `lower.tail=F`. This option produces more accurate results for a large argument.

Example 2.27 *You are fired.* *You work at the college telephone board as an operator and step out for three minutes. The rule is that if you miss two calls in a row, you will be fired. Assuming that calls arrive with the rate $\lambda = 2$ per minute according to the Poisson distribution, what is the probability that you will be fired?*

Solution. Since $\lambda = 2$ and $\alpha = 2$, the probability is

$$\Pr(X \leq 3) = \frac{2^2}{\Gamma(2)} \int_0^3 x e^{-2x} dx.$$

Using integration by parts, $\int u\,dv = uv - \int v\,du$ with $u = x$ and $dv = e^{-2x} dx$, we compute the probability:

$$\Pr(X \leq 3) = \frac{2^2}{1!} \left(-\frac{1}{2} x e^{-2x} \Big|_0^3 + \frac{1}{2} \int_0^3 e^{-2x} dx \right) = 0.983.$$

2.6.2 Computing the gamma distribution in R

Much like other distributions in R, there are four functions for the gamma distribution, as shown in Table 2.1.

Table 2.1. Four R functions for the gamma distribution (shape $= \alpha$, rate $= \lambda$)

R function	Returns/formula	Explanation
dgamma(x,shape,rate)	density (2.32)	x=array x
pgamma(q,shape,rate)	cdf	q=x-value
rgamma(n,shape,rate)	random sample	n=sample size
qgamma(p,shape,rate)	inverse cdf	p=probability

Example 2.28 *Earthquake occurrence. Using the data presented in Example 2.19, compute the probability that three consecutive earthquakes occur in California within five years after the last earthquake assuming that earthquakes are independent? Use simulations based on the exponential distribution and compute the theoretical probability using the gamma distribution.*

Solution. Since the sum of three exponentially distributed random variables with the same scale parameter is a gamma-distributed random variable with $\alpha = 3$, the probability is computed as `pgamma(5,shape=3,scale=1131/365)=` 0.2201662. Simulations give `mean(rexp(100000, rate=365/1131)+rexp(100000 ,rate=365/1131) +rexp(100000,rate=365/1131)<5)=0.22045`. The probabilities are close.

2.6.3 The tight confidence range

The tight confidence range was introduced in Section 2.2.2 as the shortest interval containing values of the random variable with given probability, p. When the distribution is symmetric, the equal-tail probability interval around the mean/mode is the shortest. But the gamma distribution is not symmetric: therefore finding the pth tight confidence range is not trivial. As was proven in Section 2.2.2, if t and T are the lower and the upper limits of the interval, the density values must be the same to provide the shortest width. Let $F(x) = \int_0^x f(u)du$ denote the cdf of the gamma distribution (pgamma). Then the problem of finding t and T reduces to the solution of a system of two equations:

$$(\alpha - 1)(\ln T - \ln t) - \lambda(T - t) = 0, \quad F(T) - F(t) - p = 0, \qquad (2.35)$$

where the first equation is equivalent to $f(t) = f(T)$. We solve this system by
Newton's algorithm as follows:

$$
\begin{bmatrix} t_{s+1} \\ T_{s+1} \end{bmatrix} = \begin{bmatrix} t_s \\ T_s \end{bmatrix} - \begin{bmatrix} -\frac{\alpha-1}{t_s} + \lambda & \frac{\alpha-1}{T_s} - \lambda \\ -f(t_s) & f(T_s) \end{bmatrix}^{-1} \begin{bmatrix} (\alpha-1)\ln\frac{T_s}{t_s} - \lambda(T_s - t_s) \\ F(T_s) - F(t_s) - p \end{bmatrix},
$$

where $s = 0, 1, \dots$ is the iteration index. The elements in the first row of the 2×2
matrix are the derivatives of the first equation with respect to t and T, and the
second row are the derivatives of the second equation (2.35). A good strategy
is to start iterations from the equal-tail probabilities, $t_0 = F^{-1}((1-p)/2)$ and
$T_0 = F^{-1}((1+p)/2)$, where F^{-1} is the inverse cdf (quantile, qgamma). The R code
for computing t and T is found in function pti.gamma. This algorithm is applied
in the following example to find the shortest probability interval for house prices.

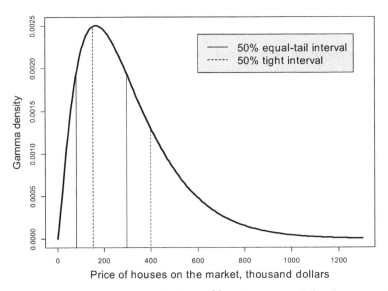

Figure 2.13: *Computation of the 50% tight interval for house price.*

Example 2.29 *House price confidence range.* *House prices on the real estate market follow a gamma distribution with mean = $300K and SD = $200K.
Report the 50% tight confidence range.*

Solution. Find α and λ from the equations $\alpha/\lambda = 300$ and $\alpha/\lambda^2 = 200^2$. We
have $\alpha = (300/200)^2$ and $\lambda = 300/200^2$. The 50% tight confidence range is computed by calling pti.gamma(p=.5, alpha=(300/200)^2, lambda=300/200^2),
which returns $(81, 298)$. This means that the 50% equal-tail confidence range extends from $81 to $298. However, the 50% tight confidence is $(153, 400)$, called
the interquartile range; see Figure 2.13. Since $298 - 81 < 400 - 153$ the tight
confidence range is smaller, as it should be. The difference is striking. It is

interesting that the mean, \$300K, does not even belong to the 50%-equal tail confidence range. Characterizing the center of this distribution using the mean is not as informative as the mode $= (\alpha - 1)/\lambda = \$167K$, the most frequent/popular house price. The mode belongs to every tight confidence range.

Example 2.30 *The fall of the Roman empire.* *Long before cell phones, smoke on the post towers or the top of the mountain was used to warn of an approaching enemy. An army of one million Barbarians invades the Roman empire and moves toward the capital. (a) Assuming that such a warning system consists of three lined up posts, five miles apart, and the average time to notice the enemy and start a big smoke fire is 30 minutes, compute the probability that Rome will be prepared for the siege if it takes six hours after the Barbarians, moving three miles per hour, have been noticed five miles ahead of the first post (assume that, at the time when the Barbarians have been first seen they were 20 miles from Rome). (b) Compute the most probable time (mode) when the Rome guards see smoke at the nearest post and compute the 0.75th tight confidence range (modify function* pti.gamma*).*

Solution. It is plausible to assume that the guards start the fire as soon as possible, and therefore the time when the smoke will be visible from another post follows an exponential distribution with the density $\lambda e^{-\lambda t}$, where t is in hours and $\lambda = 2$ because the expected value is $1/\lambda = 1/2$. The key to the solution is the fact that the sum of exponential distributions follows a gamma distribution. Therefore, the time when the Rome guards see the smoke from the nearest post follows a gamma distribution with the shape parameter $\alpha = 3$ and the rate parameter $\lambda = 2$, and it will take $20/3 = 6.7$ hours to get to Rome. (a) The probability that Rome's people will be prepared for the siege is computed as pgamma(20/3-6,shape=3,rate=2)=0.15. It is unlikely that Rome will be ready for the Barbarian invasion. (b) The modified R function is pti.gamma.Rome. The most probable time (mode) until the Rome guards see the smoke is $(\alpha - 1)/\lambda = 1$ hour and the 0.75 tight confidence range is 0.27–2.1 hours.

Problems

1. (a) Prove that $\Gamma(\alpha) = (\alpha - 1)\Gamma(\alpha - 1)$ using integration by parts. (b) Prove that $\Gamma(1) = \Gamma(2) = 1$. (c) Prove that the density of the gamma distribution reaches its maximum at $(\alpha - 1)/\lambda$ if $\alpha > 1$.

2. The quality-control department investigates the reliability of a product whose time to failure follows a gamma distribution. It was observed that the mean was six years with the standard deviation of three years. What is the probability that after 10 years the product will be still functioning?

3. (a) Following Example 2.27, what is the probability of being fired if you are late by three minutes? Use pgamma to get the answer. (b) How late can

you be to be fired with probability 0.01 (round to seconds)? Use qgamma.

4. The time between earthquakes follows an exponential distribution with an average of one earthquake in two years. (a) Use integration to find the probability that no earthquakes happen in a three-year period and use rexp to check your analytic answer through simulations. (b) Use pgamma to find the probability that three earthquakes happen in a 10-year period and use rgamma to check your answer via simulations.

5. Explain in layman's terms why the answer in Example 2.28 could not be estimated as mean(3*rexp(100000,365/1131)<5).

6. (a) Find $\lim t$ and $\lim T$ in the pth tight confidence range when $p \to 0$. (b) Modify function pti.gamma to illustrate the theoretical results.

7. (a) Plot the density of house prices from Example 2.29. (b) Verify that the tight confidence range $(81, 298)$ contains 50% of houses. (c) Verify that the 50% equal-tail interval is $(153, 400)$. (d) Display the 50% tight and equal-tail intervals using segments with different colors. (e) Use simulated numbers to check (b) and (c). [Hint: Use qgamma to compute the equal-tail interval.]

8. The time to failure of a hard drive (years) follows a gamma distribution with $\alpha = 4$ and $\lambda = 0.2$. Compute the 0.9 tight confidence range for time of failure.

9. There are five friends. The first friend heard a rumor and he/she called the second friend and that friend called the third friend, and so on. Assume that each friend calls once and the average time to call is one hour. What is the most probable time when the last friend hears the rumor? Compute the 90% tight time interval.

2.7   Normal distribution

The normal distribution is the most important distribution in probability and statistics. One of the major arguments for the wide application of the normal distribution is the central limit theorem to be discussed in Section 2.10. The normal-distribution assumption is common in statistics when data are continuous. Sometimes, even when the original distribution is not normal and positive, one can reduce the data to fit a normal distribution by applying a suitable transformation, such as square root or logarithm. In Section 2.13 we show that any continuous distribution can be transformed into a normal distribution by some transformation. In applications to engineering and science, the normal distribution is called *Gaussian* after the great German mathematician Karl Gauss (1777–1855).

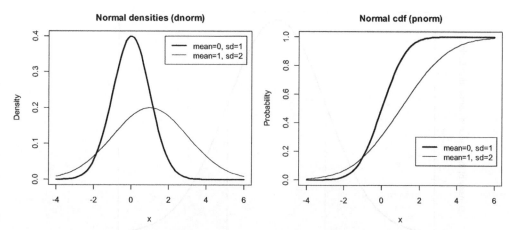

Figure 2.14: *Densities and cdfs of the normal distribution with different* **mean**$=\mu$ *and* **sd**$=\sigma$.

The density of the *standard* normal distribution has the form

$$\phi(x) = \frac{1}{\sqrt{2\pi}}e^{-\frac{1}{2}x^2}, \quad -\infty < x < \infty.$$

We use notation ϕ throughout the book as the standard normal distribution with zero mean and unit variance. Since the area under the density must be 1, we have

$$\int_{-\infty}^{\infty} e^{-\frac{1}{2}x^2}\,dx = \sqrt{2\pi}. \tag{2.36}$$

A more general normal distribution has mean μ and variance σ^2 with the density

$$\phi(x; \mu, \sigma) = \frac{1}{\sqrt{2\pi\sigma^2}}e^{-\frac{1}{2\sigma^2}(x-\mu)^2}, \tag{2.37}$$

where μ and σ are parameters (if μ and σ are not specified, we assume the standard normal density). Notice, we use ; to separate the argument x from the parameters – this is a typical notation in statistics. The normal distribution is symmetric around μ with maximum $1/\sqrt{2\pi\sigma^2}$ at $x = \mu$. Parameter μ determines the center of the distribution; parameter σ determines how wide the distribution is (see Figure 2.14). We say that the normal density is bell shaped. We use notation

$$X \sim \mathcal{N}(\mu, \sigma^2)$$

to indicate that random variable X has $E(X) = \mu$ and $\text{var}(X) = \sigma^2$.

To prove that $E(X) = \mu$, we apply the change of variable, $y = x - \mu$, which yields

$$
\begin{aligned}
E(X) &= \frac{1}{\sigma\sqrt{2\pi}} \int_{-\infty}^{\infty} x e^{-\frac{1}{2\sigma^2}(x-\mu)^2}\,dx = \frac{1}{\sigma\sqrt{2\pi}} \int_{-\infty}^{\infty} (y+\mu)e^{-\frac{1}{2\sigma^2}y^2}\,dy \\
&= \frac{1}{\sigma\sqrt{2\pi}} \int_{-\infty}^{\infty} y e^{-\frac{1}{2\sigma^2}y^2}\,dy + \frac{\mu}{\sigma\sqrt{2\pi}} \int_{-\infty}^{\infty} e^{-\frac{1}{2\sigma^2}y^2}\,dy.
\end{aligned}
$$

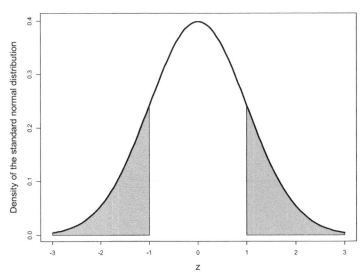

Figure 2.15: *Illustration of identity (2.40) for the standard normal distribution with $c = 1$. The shaded area at left is $\Phi(-c)$ and the shaded area at right is $1 - \Phi(c)$.*

The first integral in the second line of the above expected value is zero because the integrand is an odd function; the second integral is μ, which proves $E(X) = \mu$. To prove $\text{var}(X) = \sigma^2$, integration by parts should be used (the proof is omitted).

One special property of the normal distribution is that the linear function of a normal distribution is again a normal distribution. Precisely, if $X \sim \mathcal{N}(\mu, \sigma^2)$, then

$$a + bX \sim \mathcal{N}(a + b\mu, b^2\sigma^2). \tag{2.38}$$

Reversely, if $X \sim \mathcal{N}(\mu, \sigma^2)$, the *normalized* random variable is

$$Z = \frac{X - \mu}{\sigma} \sim \mathcal{N}(0, 1).$$

The cdf of the normal distribution, Φ, does not admit a closed form and is evaluated via integration:

$$\Phi(x) = \int_{-\infty}^{x} \phi(t)dt = \int_{-\infty}^{x} \frac{1}{\sqrt{2\pi}} e^{-\frac{1}{2}t^2} dt.$$

Function Φ will be referred to as the standard normal cdf. By change of variable it is easy to get that the cdf for a general normal variable $X \sim \mathcal{N}(\mu, \sigma^2)$ can be expressed via the standard normal cdf:

$$\Pr(X \leq x) = \Phi(x; \mu, \sigma^2) = \int_{-\infty}^{x} \phi(t; \mu, \sigma^2)dt$$

$$= \int_{-\infty}^{x} \frac{1}{\sqrt{2\pi\sigma^2}} e^{-\frac{1}{2\sigma^2}(t-\mu)^2} dt = \Phi\left(\frac{x - \mu}{\sigma}\right). \tag{2.39}$$

In statistics, we often use the identity

$$\Pr(|Z| < c) = 1 - 2\Phi(-c), \tag{2.40}$$

where $Z \sim \mathcal{N}(0,1)$ and $c > 0$. This is illustrated in Figure 2.15 with $c = 1$.

When Φ is used with no arguments after the semicolon, it assumes the standard normal cdf ($\mu = 0$ and $\sigma = 1$); the inverse cdf Φ^{-1} defines the quantile. For example, if the left-hand side of (2.40) is $1 - \alpha$, we obtain $c = \Phi^{-1}(1 - \alpha/2)$.

We frequently express the cdf for a general $X \sim \mathcal{N}(\mu, \sigma^2)$ through Φ as

$$\Pr(X \leq a) = \Phi\left(\frac{a - \mu}{\sigma}\right). \tag{2.41}$$

Example 2.31 *MGF and moments of normal distribution.* *Find the MGF of a standard normal distribution and use it to find all moments.*

Solution. If $Z \sim \mathcal{N}(0,1)$, then we have

$$M(t) = \frac{1}{\sqrt{2\pi}} \int_{-\infty}^{\infty} e^{xt - x^2/2} dx.$$

Using elementary algebra, we have $xt - x^2/2 = -(x-t)^2/2 + t^2/2$ and thus

$$M(t) = \frac{e^{t^2/2}}{\sqrt{2\pi}} \int_{-\infty}^{\infty} e^{-(x-t)^2/2} dx = e^{t^2/2} \frac{1}{\sqrt{2\pi}} \int_{-\infty}^{\infty} e^{-z^2/2} dz = e^{t^2/2}.$$

We note that the kth derivative of M can be expressed as $P_k(t)e^{t^2/2}$, where $P_k(t)$ is a polynomial of the kth order. By sequential differentiation, we obtain

$$M'(t) = te^{t^2/2}, \quad M''(t) = (1 + t^2)e^{t^2/2},$$
$$M'''(t) = t(3 + t^2)e^{t^2/2}, \quad M''''(t) = (3 + 6t^2 + t^4)e^{t^2/2}.$$

For example, for $k = 1$, we have $P_1(t) = t$. By induction, one can show that

$$E(Z^k) = \frac{k!}{(k/2)!2^{k/2}}, \quad k = \text{even}, \tag{2.42}$$

and $E(Z^k) = 0$ for $k = $ odd. For the first four k, we have

$$E(Z) = E(Z^3) = 0, \; E(Z^2) = 1, \; E(Z^4) = 3. \tag{2.43}$$

□

From (2.43), we derive a useful formula,

$$\text{var}(Z^2) = 2, \tag{2.44}$$

where $Z \sim \mathcal{N}(0,1)$ is the standard normal variable. If $X \sim \mathcal{N}(0, \sigma^2)$, we have

$$\text{var}(X^2) = \text{var}(\sigma(X/\sigma)) = \sigma^2 \text{var}(X/\sigma) = 2\sigma^2.$$

Combination of the MGF for the standard normal variable above and formula (2.31) yields the MGF for a general normal random variable $\mathcal{N}(\mu, \sigma^2)$:

$$M(t; \mu, \sigma^2) = e^{\mu t + \frac{1}{2}\sigma^2 t^2}. \tag{2.45}$$

Computing the normal distribution in R

There are four built-in functions for the normal distribution in R, as shown in Table 2.2. The default values of the arguments in all four functions are $\mu = 0$ and $\sigma = 1$. Function `rnorm` produces random normal numbers. Use `set.seed(your.number)` where `your.number` is any positive integer (seed number) to get the random numbers you want. During the same R session, you will obtain different random numbers when you repeat `rnorm`, and the numbers will continue be different at next session if the previous session was saved as "workspace image." Using a fixed seed number is convenient for debugging. If the previous session was not saved, random numbers will start the same. Function `qnorm` computes the pth quantile as the solution of equation $\Phi(x; \mu, \sigma) = p$, the inverse cdf.

Table 2.2. Normal distribution in R (`mean` $= \mu$, `sd` $= \sigma$; default values: `mean=0` and `sd=1`)

R function	Returns/formula	Explanation
`dnorm(x,mean,sd)`	density $\phi(x; \mu, \sigma)$	equation (2.37)
`pnorm(q,mean,sd)`	cdf $\Phi(x; \mu, \sigma)$	q = x, equation (2.39)
`rnorm(n,mean,sd)`	$X_1, X_2, ..., X_n \overset{\text{iid}}{\sim} \mathcal{N}(\mu, \sigma^2)$	random sample
`qnorm(p,mean,sd)`	$x = \Phi^{-1}(p; \mu, \sigma)$	quantile

Example 2.32 *The rule of two sigma. Compute the probability* $\Pr(|X - \mu| < 2\sigma)$, *where* $X \sim \mathcal{N}(\mu, \sigma^2)$ *with* $\mu = 1$ *and* $\sigma = \sqrt{2}$ *using the **pnorm** function. Prove that this probability does not depend on* μ *and* σ.

Solution. There are two ways to compute the probability: (a) specify `mean` and `sd` in `pnorm`, or (b) reduce the probability to the standard normal cdf with defaults `mean=0` and `sd=1` using (2.39). Since

$$\Pr(|X - \mu| < 2\sigma) = \Pr(X - \mu < 2\sigma) - \Pr(X - \mu < -2\sigma)$$
$$= \Phi(\mu + 2\sigma; \gamma, \sigma^2) - \Phi(\mu - 2\sigma; \gamma, \sigma^2),$$

we issue `pnorm(1+2*sqrt(2),mean=1,sd=sqrt(2))-pnorm(1-2*sqrt(2),mean =1,sd=sqrt(2))`, which gives 0.9544997. Now, we use (2.39) and reduce computation to the standard normal cdf:

$$\Pr(X < \mu + 2\sigma) - \Pr(X < \mu - 2\sigma) = \Phi\left(\frac{\mu + 2\sigma - \mu}{\sigma}\right) - \Phi\left(\frac{\mu - 2\sigma - \mu}{\sigma}\right)$$
$$= \Phi(2) - \Phi(-2) = 1 - 2\Phi(-2).$$

We get the same answer by issuing `1-2*pnorm(-2)`. Since $\Pr(|X - \mu| < 2\sigma) = 1 - 2\Phi(-2)$, it does not depend on μ and σ. □

Simulations are very useful for checking analytical derivations in probability and statistics, as in the example below.

Example 2.33 *Variance simulation.* *Find* $\mathrm{var}(X^2)$, *where* $X \sim \mathcal{N}(\mu, \sigma^2)$, *and check your derivations with simulations in R using* μ *and* σ^2 *as arguments.*

Solution. Represent $X = \mu + \sigma Z$, where $Z \sim \mathcal{N}(0, 1)$. Then $X^2 = \mu^2 + 2\mu\sigma Z + \sigma^2 Z^2$. Use formula $\mathrm{var}(Y) = E(Y^2) - E^2(Y)$, where $Y = X^2$. Using the fact that $E(Z^3) = 0$ and $E(Z^4) = 3$, we obtain

$$
\begin{aligned}
E(X^4) &= E(\mu^4 + 4\mu^2\sigma^2 Z^2 + \sigma^4 Z^4 + 4\mu^3\sigma Z + 2\mu^2\sigma^2 Z^2 + 4\mu\sigma^3 Z^3) \\
&= \mu^4 + 4\mu^2\sigma^2 + 3\sigma^4 + 2\mu^2\sigma^2 = \mu^4 + 6\mu^2\sigma^2 + 3\sigma^4.
\end{aligned}
$$

Now find $E^2(X^2) = \mu^2 + \sigma^2$. Finally,

$$
\mathrm{var}(X^2) = \mu^4 + 6\mu^2\sigma^2 + 3\sigma^4 - (\mu^2 + \sigma^2)^2 = 4\mu^2\sigma^2 + 2\sigma^4.
$$

The R code is below.

```
varX2=function(mu=1,s2=2,nSim=1000000)
{
dump("varX2","c:\\StatBook\\varX2.r")
X=rnorm(nSim,mean=mu,sd=sqrt(s2))
theor.var=4*mu^2*s2+2*s2^2
emp.var=var(X^2)
cat("empirical variance=",emp.var," theoretical variance=",
        theor.var,"\n")
}
```

The call `varX2()` gives

```
empirical variance = 16.11535 theoretical variance = 16
```

Example 2.34 *Volume of the ball.* *The radius of a ball is* $\rho = 2.5$ *cm, but it is measured with error* $\sigma = 0.001$ *cm. Assuming that the measurement is normally distributed, find the probability that the volume of the ball (i) is greater than the nominal volume* $(4/3)\pi\rho^3 = 65.45$ *cm^3 and (ii) is within* ± 0.01 *of the nominal volume. (iii) Is an estimator of the volume,* $(4/3)\pi\bar{r}^3$, *where* $\bar{r}$ *is an unbiased normally distributed measurement of the radius, unbiased?*

Solution. Of course, with no measurement error, the volume will be exactly 65.45. But measurement errors are random, $r \sim \mathcal{N}(2.5, (0.001)^2)$, so for question (i) we have

$$
\Pr((4/3)\pi r^3 > (4/3)\pi\rho^3) = \Pr(r > \rho) = 1 - \Phi(r \leq \rho) = 1 - \Phi\left(\frac{0}{\sigma}\right) = \frac{1}{2}.
$$

(ii) Express the limits for the volume as the limits for the radius,

$$\Pr\left((4/3)\pi\rho^3 - 0.01 < (4/3)\pi r^3 < (4/3)\pi\rho^3 + 0.01\right)$$

$$= \Pr\left(\sqrt[3]{\frac{(4/3)\pi\rho^3 - 0.01}{(4/3)\pi}} < r < \sqrt[3]{\frac{(4/3)\pi\rho^3 + 0.01}{(4/3)\pi}}\right)$$

$$= \Pr(2.499873 < r < 2.500127) = 0.1.$$

To compute in R we issue

`pnorm(2.500127,mean=2.5,sd=0.001)-pnorm(2.499873,mean=2.5,sd=0.001)`

(iii) Express $\bar{r} = \rho + \sigma Z$, where Z is the standard normal random variable:

$$E(\bar{r}^3) = E(\rho + \sigma Z)^3 = \rho^3 + 2\sigma\rho^2 E(Z) + 2\sigma^2\rho E(Z^2) + \sigma^3 E(Z^3) = \rho^3 + 2\sigma^2\rho > \rho^3.$$

Therefore, the volume estimated based on $\bar{r}$ is positively biased despite $\bar{r}$ being unbiased for the radius. □

The following fact will be used later in connection with the binormal receiver operating characteristic curve (ROC) in Section 5.1.1 and the probit regression in Section 8.8.2.

Theorem 2.35 *The expected value of the normal cdf can be expressed in closed form as*

$$E_{X\sim\mathcal{N}(\mu,\sigma^2)}\Phi(X) = \Phi\left(\frac{\mu}{\sqrt{1+\sigma^2}}\right). \tag{2.46}$$

Proof. Let $Y \sim \mathcal{N}(0,1)$ be independent of X. If X is held fixed $\Phi(X) = \Pr(Y \leq X)$. That is, the left-hand side can be rewritten as

$$E_{X\sim\mathcal{N}(\mu,\sigma^2), Y\sim\mathcal{N}(0,1)}\Pr(Y \leq X) = \Pr(Z \leq 0),$$

where $Z = Y - X \sim \mathcal{N}(-\mu, 1 + \sigma^2)$. Thus, as a result of (2.41), we get

$$\Pr(Z \leq 0) = \Phi\left(\frac{\mu}{\sqrt{1+\sigma^2}}\right).$$

The key to the proof is that $Z = Y - X$ has a normal distribution as the difference of normal distributions is normally distributed.

Example 2.36 *Finite integral via density.* Prove that

$$\int_a^b x\phi(x)dx = \phi(a) - \phi(b). \tag{2.47}$$

Solution. Using the change of variable $y = x^2$ in the integral at the left-hand side, we obtain

$$\frac{1}{\sqrt{2\pi}} \int_a^b x e^{-\frac{1}{2}x^2} dx = \frac{1}{2\sqrt{2\pi}} \int_{a^2}^{b^2} e^{-\frac{1}{2}y} dy = \frac{1}{2\sqrt{2\pi}} \left(-2e^{-\frac{1}{2}y} \Big|_{a^2}^{b^2} \right)$$

$$= \frac{1}{\sqrt{2\pi}} \left(e^{-\frac{1}{2}a^2} - e^{-\frac{1}{2}b^2} \right) = \phi(a) - \phi(b).$$

$\square$

The normal distribution is symmetric around μ. Sometimes we want to model distributions that are asymmetric/skewed distributions. The following definition can be used to specify a distribution on $(-\infty, \infty)$, where parameter λ controls how skewed the distribution is.

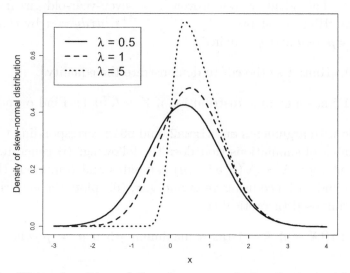

Figure 2.16: *Three densities of the skew-normal distribution with $\mu = 0$ and $\sigma = 1$ defined by $2\phi(x)\Phi(\lambda x)$.*

Definition 2.37 *A continuous distribution on $(-\infty, \infty)$ is called skew-normal if its density is defined as*

$$f(x; \mu, \sigma^2, \lambda) = \frac{2}{\sigma} \phi \left(\frac{x - \mu}{\sigma} \right) \Phi \left(\lambda \frac{x - \mu}{\sigma} \right). \tag{2.48}$$

Note that function Φ has the same argument as ϕ multiplied by coefficient, λ. This parameter drives the skewness of the distribution, and will be further referred to as the skewness coefficient. If $\mu = 0, \sigma^2 = 1, \lambda = 1$, this distribution is called the standard skew-normal distribution. See Figure 2.16 with three skewness coefficients. Note that distributions like gamma or chi-square are also skewed but defined only for positive values. In contrast, the support of the skew-normal distribution is the entire real line. See Example 3.58 for a generalization.

Theorem 2.38 *If random variable X has the skew-normal distribution, then*

$$E(X) = \mu + \sqrt{\frac{2}{\pi}} \frac{\lambda}{\sqrt{1 + \lambda^2}}, \quad \operatorname{var}(X) = \left(1 - \frac{2\lambda}{\pi(1 + \lambda^2)}\right) \sigma^2.$$

In addition, if X has the standard skew-normal distribution, then the following representation holds: $X = \delta |Y| + \sqrt{1 - \delta^2} Z$, where Y and Z are independent normally distributed random variables and $\delta = \lambda / \sqrt{1 + \lambda^2}$.

Problems

1. Prove that $E(X) = \mu$ and $\operatorname{var}(X) = \sigma^2$ by applying Theorem 2.20 to (2.45).

2. Prove (2.36). This identity was proven by a seven-year-old girl from the movie *Gifted*. [Hint: First prove $\int_{-\infty}^{\infty} \int_{-\infty}^{\infty} e^{-(x^2 + y^2)} dx dy = \pi$ by change of variables $x = \rho \cos \theta$ and $y = \rho \sin \theta$.]

3. Prove (2.38). [Hint: Use the cdf to demonstrate the equality.]

4. Prove that Y^2 has a gamma distribution if $Y \sim \mathcal{N}(0, 1)$. Find α and γ.

5. Write R code with arguments `mu`, `sigma2`, and `nSim` corresponding to μ, σ^2, and the number of simulations that does the following: (i) generates `nSim` random observation $X \sim \mathcal{N}(\mu, \sigma^2)$, (ii) computes and prints out the empirical mean and variance using `mean` and `var`, (iii) plots the empirical cdf with a superimposed theoretical cdf.

6. Prove that $\Pr(|X - a| < \tau)$ takes maximum when $a = \mu$ where $X \sim \mathcal{N}(\mu, \sigma^2)$.

7. Show that the skewness and kurtosis equal zero for the normal distribution.

8. Write an R program that verifies the formula for $\operatorname{var}(X^2)$ using simulations with $X \sim \mathcal{N}(\mu, \sigma^2)$. Use σ as an argument and plot the theoretical and simulation-based variance for the range of values of μ from 0 to 1.

9. Repeat derivations and computations of Example 2.32 with three sigma.

10. In Example 2.34, find and compute the probability that the volume of the ball is within ± 0.1 of the nominal volume.

11. Express $E_{X \sim \mathcal{N}(\mu, \sigma^2)} \Phi(a + bX)$ through Φ as in (2.46).

12. The probability of winning a tennis point is $\Phi(\mu)$. Appeal to formula (2.46) to argue that in the presence of poor weather or quality of the court, the chance of winning a point or the entire match becomes closer to 50/50 (make necessary assumptions).

13. Check formula (2.47) for $a = -b$.

14. Generalize formula (2.47) to the normal density with any μ and σ^2.

15. Use representation of the skew-normal random variable from the above theorem to derive formulas for $E(X)$ and $\text{var}(X)$ with $\mu = 0$ and $\sigma = 1$.

16. Generate 10K random numbers from the standard skew-normal distribution using Theorem 2.38 and check the result using the formulas for the mean and variance.

2.8 Chebyshev's inequality

Pafnuty Chebyshev was a famous Russian mathematician. He contributed to numerical mathematics (Chebyshev polynomials) and probability theory. His student Andrey Markov was also famous – he discovered Markov chains.

Chebyshev's inequality provides an upper bound for the probability of being outside the interval around the mean on the scale of the standard deviation of a random variable X. Remarkably, the distribution of X is not involved, so we say that this inequality is distribution free.

Theorem 2.39 *Chebyshev's inequality.* *Let random variable X have mean μ and standard deviation σ. Then for every positive λ, we have*

$$\Pr\left(|X - \mu| \geq \lambda\sigma\right) \leq \frac{1}{\lambda^2}. \tag{2.49}$$

Proof. We will prove Chebyshev's inequality for a continuous random variable by integration. Note that it holds for discrete random variables as well. If $f(x)$ denotes the density of X, by the definition of the variance, we have

$$\sigma^2 = \int_{-\infty}^{\infty} (x - \mu)^2 f(x)dx \geq \int_{|x-\mu|\geq\lambda\sigma} (x - \mu)^2 f(x)dx$$

$$= \int_{|t|\geq\lambda\sigma} t^2 f(t + \mu)dt \geq (\lambda\sigma)^2 \int_{|t|\geq\lambda\sigma} f(t + \mu)dt$$

$$= (\lambda\sigma)^2 \int_{|x-\mu|\geq\lambda\sigma} f(x)dx = (\lambda\sigma)^2 \Pr\left(|X - \mu| \geq \lambda\sigma\right).$$

We arrive at Chebyshev's inequality by dividing both sides by $(\lambda\sigma)^2$. $\qquad\square$

For example, using this inequality, we can say that the probability of being outside the interval mean plus/minus three standard deviations is less or equal $1/9 = 0.11$.

Example 2.40 *Exact probability and Chebyshev's inequality.* *Compute the exact probability $\Pr\left(|X - \mu| \geq \lambda\sigma\right)$ for (a) a uniform random variable, $X \sim$*

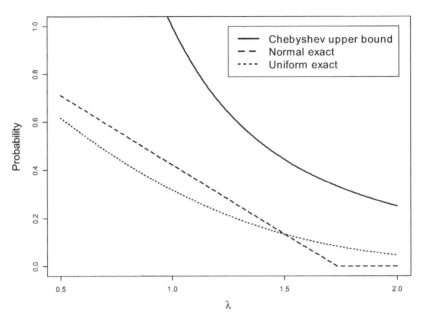

Figure 2.17: *The exact probability* $\Pr(|X - \mu| \geq \lambda\sigma)$ *and Chebyshev's upper bound for uniform* $\mathcal{R}(0,1)$ *and normal* $\mathcal{N}(0,1)$ *distributions.*

$\mathcal{R}(0,1)$ *and (b) a standard normal random variable* $X \sim \mathcal{N}(0,1)$, *and compare with Chebyshev's inequality. Plot the three probabilities against* λ *in the interval* *[0.5,2].*

Solution. For $X \sim R(0,1)$ we have $\mu = 1/2$ and $\sigma = (2\sqrt{3})^{-1}$, so that the exact probability is

$$\Pr(|X - \mu| \geq \lambda\sigma) = \begin{cases} 1 - \lambda/\sqrt{3} \text{ if } \lambda \leq \sqrt{3} \\ 0 \text{ if } \lambda > \sqrt{3} \end{cases}.$$

For normal $X \sim \mathcal{N}(0,1)$, we have $\Pr(|X - \mu| \geq \lambda\sigma) = 2\Phi(-\lambda)$. The exact probabilities and their lower bounds are displayed in Figure 2.17. □

As follows from Chebyshev's inequality, the probability that X is within 1.96σ from the mean is $1 - 1/1.96^2 = 0.74$, while for a normally distributed X, this probability is 0.95.

□

There are several variations on Chebyshev's inequality. In fact, Markov's inequality,

$$\Pr(Z > t) \leq \frac{E(Z)}{t}, \quad t > 0, \tag{2.50}$$

is a generalization of Chebyshev's inequality for a positive random variable Z. Indeed, letting $Z = (X - \mu)^2$ and $t = \lambda^2\sigma^2$, we obtain (2.49) from (2.50). The following is also easy to derive from Markov's inequality:

$$\Pr(|X| > t) \le \frac{E(X^2)}{t^2}. \tag{2.51}$$

If X is a random variable, symmetric and unimodal around μ, then

$$\Pr(|X - \mu| \ge \lambda\sigma) \le \begin{cases} 1 - \lambda/\sqrt{3} \text{ if } 0 < \lambda < 2/\sqrt{3} \\ 4/(9\lambda^2) \text{ if } \lambda \ge 2/\sqrt{3} \end{cases}. \tag{2.52}$$

Problems

1. Let $X \sim \mathcal{N}(\mu, \sigma^2)$. Plot $\Pr\left(|X - \mu| \ge \lambda\sigma\right)$ as a function of λ on the segment $[0, 3]$ using (1) Chebyshev right-hand side inequality (2.49), (2) the exact value using `pnorm`, and (3) the right-hand side of (2.52) on the same graph. Use μ and σ as the program arguments. Display the values using `legend`.

2. Repeat the above for the uniform distribution.

3. (a) Is Chebyshev's inequality sharp, i.e. is there X for which the inequality turns into equality for some λ? (b) Is there X for which the inequality turns into equality for all $\lambda \ge 1$?

4. (a) Prove (2.51) following the line of the proof of Theorem 2.39. (b) Prove (2.51) using (2.49).

5. Derive the left- and the right-hand side of inequality (2.50) for the exponential distribution. Plot both sides as functions of t for $\lambda = 0.1$ and $\lambda = 1$, as in Example 2.40, using `par(mfrow=c(1,2))`.

6. Use the same analysis as in Example 2.40 to investigate how sharp inequality (2.52) is. Prove that inequality (2.52) is sharp for a uniform distribution for $0 < \lambda \le \lambda_0$. [Hint: Find λ_0.]

2.9 The law of large numbers

The law of large numbers (LLN) rules the world. Consider the number of customers in a popular grocery store at a specific time on a specific day of the week. From day to day, this number is remarkably constant. Why? People come shopping independently, and yet the number of shoppers, say, at 2 p.m. Wednesday is very consistent from week to week.

As another example from statistical mechanics, consider a hanging sheet of paper: the sheet is randomly bombarded by millions of air molecules on both sides of the paper and yet it does not move.

These phenomena can be explained by the LLN. The LLN is about convergence of the arithmetic average, $\overline{X}_n$, to its expected value. Before formulating the LLN, it is worthwhile to consider various types of convergence of random variables. These convergences are used to study the asymptotic properties of estimators in statistics, or more specifically, the *consistency* of estimators. When the sample size increases, an asymptotic estimator converges to the true value.

2.9.1   Four types of stochastic convergence

Let X_n be (dependent or independent) sequence of random variables and $n \to \infty$.

1. *Convergence in probability* or *weak convergence*: We say that X_n weakly converge to μ if for every $\varepsilon > 0$,

$$\lim_{n\to\infty} \Pr\left(|X_n - \mu| > \varepsilon\right) = 0.$$

   To indicate convergence in probability, the notation $p\lim_{n\to\infty} X_n = \mu$ is used.

2. *Quadratic convergence*: We say that X_n converges to μ in the quadratic sense if $\lim_{n\to\infty} E(X_n - \mu)^2 = 0$, assuming that $E(X_n - \mu)^2$ exists (is finite) for each n.

3. *Convergence with probability 1* or *strong convergence* (almost surely, a.s.): We say that X_n converges to μ a.s. if $\Pr\left(\lim_{n\to\infty} X_n = \mu\right) = 1$, where $\lim_{n\to\infty} X_n = \mu$ defines the set where the convergence takes place, or equivalently, $\lim_{n\to\infty} \Pr\left(\cup_{m=n}^{\infty} |X_m - \mu| > \varepsilon\right) = 0$.

4. *Convergence in distribution* of X_n to X: we say that the distribution of X_n converges to the distribution of X if

$$\lim_{n\to\infty} \Pr(X_n \leq x) = \Pr(X \leq x)$$

   or $\lim_{n\to\infty} F_n(x) = F(x)$, where F_n is the cdf of X_n, F is the cdf of X, and x is the continuity point of $F(x)$. This type of convergence is used in the formulation of the central limit theorem, see below. Sometimes we use notation $X_n \simeq F$ to indicate that X_n converges to a random variable with cdf F.

There is a parallel between weak and strong convergence one hand and the pointwise and uniform convergence of functions on the other hand. For a comprehensive discussion of these convergences, consult classic books on the asymptotic probability theory, such as Doob [34], Loeve [70], or Feller [45]. We will revisit these definitions in Section 6.4.4 where the asymptotic properties of estimators are discussed. The most popular type of convergence in statistics is convergence in probability; to indicate that $\widehat{\theta}_n$ is a consistent estimator of θ, we write $p\lim_{n\to\infty} \widehat{\theta}_n = \theta$. Under mild conditions, a linear or continuous nonlinear transformation preserves the convergence. Below is a typical result.

Theorem 2.41 *If a and b are constants and $p \lim_{n \to \infty} X_n$ and $p \lim_{n \to \infty} Y_n$ exist, then $p \lim_{n \to \infty} (aX_n + bY_n) = a \times p \lim_{n \to \infty} X_n + a \times p \lim_{n \to \infty} Y_n$. If $f(x)$ is a continuous function, then $p \lim_{n \to \infty} f(X_n) = f(p \lim_{n \to \infty} X_n)$.*

The first statement is often referred to as the Slutsky theorem and the proof follows from the inequality

$$
\begin{aligned}
|(aX_n + bY_n) - (a\mu_X + b\mu_Y)| &= |(aX_n - a\mu_X) + (bY_n - b\mu_Y)| \\
&\leq |a| \times |X_n - \mu_X| + |b| \times |Y_n - \mu_Y|,
\end{aligned}
$$

where $p \lim_{n \to \infty} X_n = \mu_X$ and $p \lim_{n \to \infty} Y_n = \mu_Y$. The proof of the second statement uses the ε-δ definition of the continuous function. □

The four types of convergences are related. Quadratic convergence implies convergence of mathematical expectations. Denoting $\mu_n = E(X_n)$, we prove that $\lim_{n \to \infty} \mu_n = \mu$. Indeed,

$$
E(X_n - \mu)^2 = E[(X_n - \mu_n) - (\mu_n - \mu)]^2 = E(X_n - \mu_n)^2 + (\mu_n - \mu)^2.
$$

If the left-hand side goes to zero, both terms in the right-hand side go to zero as well. In particular, $\mu_n \to \mu$ when $n \to \infty$. In statistics, convergence of mathematical expectations of an estimator is called *asymptotic unbiasedness*.

Theorem 2.42 *If X_n converges to μ in a quadratic sense, then $p \lim_{n \to \infty} X_n = \mu$.*

Proof. First, we prove that if $\lim_{n \to \infty} E(X_n - \mu)^2 = 0$, then $\lim_{n \to \infty} E(X_n - \mu_n)^2 = 0$ where $E(X_n) = \mu_n$. This follows from the fact that $E(X_n - \mu_n)^2 \leq E(X_n - \mu)^2$ for any μ. Second, we use Chebyshev's inequality. Let $\varepsilon > 0$. Applying Chebyshev's inequality (2.49) to $X = X_n$ we obtain

$$
\begin{aligned}
\lim_{n \to \infty} \Pr\left(|X_n - \mu_n| > \varepsilon\right) &= \lim_{n \to \infty} \Pr\left(|X_n - \mu_n| > \sigma(\varepsilon/\sigma)\right) \leq \varepsilon^{-2}\sigma^2 \\
&= \varepsilon^{-2} E(X_n - \mu_n)^2,
\end{aligned}
$$

since $E(X_n - \mu_n)^2 = \sigma^2$, the variance of X_n. We proved that $E(X_n - \mu_n)^2 \to 0$. That is, the left-hand side goes to zero, so $p \lim_{n \to \infty} X_n = \mu$. □

Other relationships between the four types of convergencies exist: (i) strong convergence implies weak convergence, (ii) when X degenerates to a constant, convergence in distribution and weak convergence are equivalent.

There are two formulations of the LLN, the weak and strong formulation, with respect to the implied convergence. The weak formulation requires the absence of correlation, and the strong formulation requires independence.

Theorem 2.43 *The law of large numbers (LLN).* *If $X_1, X_2, X_3, \ldots$ are uncorrelated random variables with the same mean $E(X_i) = \mu$ and variance $\mathrm{var}(X_i) = \sigma^2$ then $p \lim_{n \to \infty} \overline{X}_n = \mu$, the weak LLN. If $X_1, X_2, X_3, \ldots$ are iid and have finite mean μ, then $\lim_{n \to \infty} \overline{X}_n = \mu$ a.s., the strong LLN.*

Proof. First, we make several clarifying comments on the formulation of the theorem. The first part of the theorem says that the arithmetic average, $\overline{X}_n = n^{-1} \sum_{i=1}^{n} X_i$ converges to the mean weakly and the second part says that $\overline{X}_n$ converges to the mean strongly when $n \to \infty$. In fact, the LLN holds under even milder conditions. For example, X_i may have different distributions and be correlated, but the correlation between X_i and X_j should vanish when $|i - j| \to \infty$. See Example 3.86 with correlated X_i. The strong LLN does not work for the Cauchy distribution because it does not have finite mean; see Example 2.77. More technical details on the LLN can be found in the books cited above. Second, the weak LLN follows from the quadratic convergence of the average to the mean because $E(\overline{X}_n - \mu)^2 = \mathrm{var}(\overline{X}_n) = \sigma^2/n \to 0$ when $n \to \infty$. Due to Theorem 2.42, quadratic convergence implies weak convergence.   □

Now consider more carefully two examples from the beginning of the section from the perspective of the LLN theorem. We referred to the grocery store as "popular," implying that the number of potential buyers, n is large. In terms of Theorem 2.43, X_i is an indicator variable ($X_i = 1$ means shopping at 2 p.m. and $X_i = 0$ means otherwise), where i codes the potential buyer, say, the town resident. This random variable follows a Bernoulli distribution with $\Pr(X_i = 1) = p \, (= \mu)$, the probability of shopping at 2 p.m. on Wednesday. Each person decides to shop independently, so X_i are iid. Due to LLN, the average, $\overline{X}_n$, is close to the probability p that a shopper will come to the store around 2 p.m. on Wednesday. Thus the number of shoppers in the store is very stable with the mean np. In fact, since p is small and n is large, but fixed, the number of shoppers in the store may be well described by the Poisson distribution.

In the paper example, the number of air molecules that hit the paper sheet is on the order of millions, $n \simeq \infty$. The hits are independent with the distribution of velocities and angles equal on both sides of the paper (Brownian motion). If X_i and Y_j are the forces of the ith and the jth molecule at left and right, respectively, the force that pushes the paper to the right is proportional to $\sum_{i=1}^{n} X_i/n$, and the force that pushes the paper to the left is proportional to $\sum_{i=1}^{m} Y_j/m$. Since the distribution of forces is the same on both sides of the paper, $\mu = E(X_i) = E(Y_j)$, and due to LLN, we have

$$\lim_{n \to \infty} \sum_{i=1}^{n} X_i/n - \lim_{m \to \infty} \sum_{i=1}^{m} Y_j/m = \mu - \mu = 0.$$

This means that the paper does not move. Every time the sheet moves, the conditions of the LLN are violated, such as the movement of air.

Example 2.44 *Consistency of the ruler.* *The length of an object, L, is measured by the ruler by recording the closest tick mark. Build a probabilistic model and explain under what conditions the average of these imprecise measurements is unbiased and converges to the true length, L.*

Solution. Let L be the unknown length such that $r \leq L \leq r + 1$, where r is the closest tick mark on the ruler to the left of L. The result of the measurement may be encoded through a binary random variable with the outcome either r or $r + 1$. We assume that the probability of the reading is proportional to the distance to the nearest tick mark, r or $r + 1$. More precisely,

$$X = \begin{cases} r \text{ with probability } (r + 1) - L \\ r + 1 \text{ with probability } L - r \end{cases}. \qquad (2.53)$$

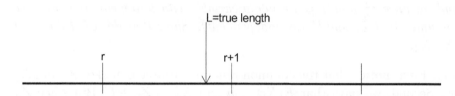

Figure 2.18: *Consistency of the ruler. Each measurement is assigned to the nearest tick mark on the ruler. The model (2.53) implies an unbiased and consistent estimate of L, the true length.*

See Figure 2.18. We calculate the expectation of this Bernoulli random variable as

$$E(X) = r[(r + 1) - L] + (r + 1)(L - r) = r^2 + r - rL + rL - r^2 + L - r = L.$$

We say that, on average, the imprecise ruler unbiasedly estimates the true length L (see more detail on unbiased estimation in Section 6.4.1). According to the LLN, the average of n iid measurements converges to the length of the object, $p \lim_{n \to \infty} \overline{X}_n = L$. □

The following is a similar example.

Example 2.45 *Distance to work.* *I go to the same workplace by car for ten years. The trip is exactly 5 miles and 1387 feet, but the odometer shows the distance in miles. Can the exact distance to work almost exactly be estimated as the average of miles over ten years? Use simulations to demonstrate.*

Solution. The odometer shows the distance covered in miles. For example, if it shows 42673, the car, in fact, might have traveled somewhere between 42673 and 42674 miles. It is safe to assume that the difference between the exact distance to work and the exact distance covered by the car in miles, X, is a uniformly distributed random variable on $[-1/2, 1/2]$. Then the exact distance covered by the car from home to work would be $X + 5 + 1387/5280$ because there are 5280 feet in one mile, but the odometer would show **round**, the closest integer. In the R code **mile**, **n** is the number of times I go to work and come back over 10 years

(52 is the number of weeks in the year and I work five days a week), the result of a run is shown below. The answer is positive. The LLN works its magic: even though the distance is roughly recorded in miles I can reconstruct the distance with amazing precision if averaged over a large number of trips:

```
>mile()
Exact distance to work = 5.262689 miles
Estimated distance to work over ten years = 5.264038 miles      □
```

When variables are dependent, however, the LLN may not hold.

Example 2.46 *When the LLN fails.* Let $X_i = Z_i + U$, where Z_i are iid with mean μ and variance σ_Z^2 and U is a random variable with zero mean and variance τ^2 (it is assumed that Z_i and U are independent). Show that the LLN does not apply to $X_1, X_2, \ldots$.

Solution. First, notice that the common random variable U makes X_i and X_j dependent. Second, we have that $E(X_i) = \mu$ and $\overline{X}_n = \overline{Z}_n + U$. But since $\overline{Z}_n$ and U are independent we have

$$E(\overline{X}_n - \mu)^2 = \mathrm{var}(\overline{X}_n) = \mathrm{var}(\overline{Z}_n) + \mathrm{var}(U) = \sigma^2/n + \tau^2.$$

This implies that $E(\overline{X}_n - \mu)^2 \geq \tau^2$ and therefore the quadratic convergence of $\overline{X}_n$ to μ fails. The fact that $\overline{X}_n$ has variances bounded from below implies that $\overline{X}_n$ does not converge to μ in probability.   □

In statistics, we want to estimate parameters using a random sample. The commonly used estimates of mean μ and variance σ^2 are defined below.

Definition 2.47 *Sample mean and variance.* Let $X_1, X_2, \ldots, X_n$ be iid or a random sample from a distribution with mean μ and variance σ^2. The sample mean and variance are defined as

$$\overline{X}_n = \frac{1}{n}\sum_{i=1}^{n} X_i, \quad \widehat{\sigma}_n^2 = \frac{1}{n-1}\sum_{i=1}^{n}(X_i - \overline{X}_n)^2,$$

respectively. The square root of the sample variance is called the sample standard deviation (SD), $\widehat{\sigma}_n = \sqrt{\widehat{\sigma}_n^2}$, and $\widehat{\sigma}_n/\sqrt{n}$ is called the standard error (SE).

Note that SE can be viewed as the SD of the mean, $\overline{X}_n$. The sum of squares in $\widehat{\sigma}_n^2$ is divided by $n-1$, not n, to make the estimator unbiased; see more detail in Section 6.6.

Example 2.48 *Consistency of the sample mean, variance, and SD.* Prove that the sample mean, variance, and SD are consistent estimators of the mean and variance:

$$p\lim_{n\to\infty} \overline{X}_n = \mu, \quad p\lim_{n\to\infty} \widehat{\sigma}_n^2 = \sigma^2, \quad p\lim_{n\to\infty} \widehat{\sigma}_n = \sigma.$$

Solution. The first limit follows directly from the LLN. The second limit follows from the following decomposition: $n^{-1}\sum_{i=1}^{n}(X_i - \overline{X}_n)^2 = n^{-1}\sum_{i=1}^{n}X_i^2 - \overline{X}_n^2$. Using Theorem 2.41 in combination with the LLN for $\{X_i^2\}$, we obtain

$$p\lim_{n\to\infty}\widehat{\sigma}_n^2 = \lim_{n\to\infty}\frac{n}{n-1} \times \left[p\lim_{n\to\infty}\left(\frac{1}{n}\sum_{i=1}^{n}X_i^2\right) - p\lim_{n\to\infty}\left(\overline{X}_n^2\right)\right]$$

$$= (\sigma^2 + \mu^2) - \left(p\lim_{n\to\infty}\overline{X}_n\right)^2 = (\sigma^2 + \mu^2) - \mu^2 = \sigma^2$$

since $E(X_i^2) = \sigma^2 + \mu^2$ and the quadratic function is continuous. The sample SD is consistent again due to Theorem 2.41 because the square root is a continuous function. Consistency of the sample mean and variance explains why we called them consistent estimators: they converge to the true values when the sample size, n, goes to infinity. □

Sometimes independent random variables X_i have different means and variances and yet the LLN holds.

Theorem 2.49 *Extended LLN.* *If $X_1, X_2, X_3, \ldots$ are pairwise uncorrelated with finite mean $E(X_i) = \mu_i$ and variance $\mathrm{var}(X_i) = \sigma_i^2$ such that*

$$\lim_{n\to\infty}\frac{1}{n}\sum_{i=1}^{n}\mu_i = \mu, \quad \lim_{n\to\infty}\frac{1}{n^2}\sum_{i=1}^{n}\sigma_i^2 = 0, \tag{2.54}$$

then $p\lim_{n\to\infty}\overline{X}_n = \mu$.

Proof. Express

$$\overline{X}_n - \mu = \frac{1}{n}\sum_{i=1}^{n}(X_i - \mu_i) + \left(\frac{1}{n}\sum_{i=1}^{n}\mu_i - \mu\right).$$

The first term converges to zero due to the second condition in (2.54) and the second term converges to zero due to the first condition in (2.54). Here, we use the fact that the variance of the sum of pairwise uncorrelated random variables is equal to the sum of the variances. □

More discussion on stochastic convergence is found in Section 6.4.4.

2.9.2   Integral approximation using simulations

The LLN is the basis for integral approximation by simulations, sometimes called Monte Carlo simulations. Let us start with proper integrals, when the integration limits are finite, $\int_a^b f(x)dx$. We express this integral as the expected value over a uniformly distributed random variable using simple algebra:

$$\int_a^b f(x)dx = (b-a)\int_a^b f(x)\frac{1}{b-a}dx = (b-a)E_{X\sim\mathcal{R}(a,b)}(f(X)).$$

In words, the integral is the expected value of $f(X)$, where X is a uniformly distributed random variable on $[a, b]$. Now let $\{X_1, X_2, ..., X_n\}$ be iid and simulated from $\mathcal{R}(a, b)$. Then, by the LLN, the integral can be approximated as follows:

$$\frac{b-a}{n} \sum_{i=1}^{n} f(X_i) \simeq \int_{a}^{b} f(x)dx. \tag{2.55}$$

The larger the n, the better approximation.

Now consider the approximation of an improper integral (when the lower or upper limit, or both, is infinite). To be specific, suppose that the lower limit is 0 and the upper limit is ∞, that is, the integral to be approximated takes the form $\int_0^\infty f(x)dx$. The idea is the same. Choose a familiar distribution with the same support as the domain of integration and represent the integral as an expectation of a positive random variable. The easiest distribution with positive values is the exponential distribution (see Section 2.4), so we rewrite the integral as

$$\int_0^\infty f(x)dx = \lambda^{-1} \int_0^\infty e^{\lambda x} f(x)(\lambda e^{-\lambda x})dx = \lambda^{-1} E_{X \sim \mathcal{E}(\lambda)}(e^{\lambda X} f(X)).$$

Again, provided the right-hand side exists, due to the LLN for large n, we have

$$\frac{1}{\lambda n} \sum_{i=1}^{n} e^{\lambda X_i} f(X_i) \simeq \int_0^\infty f(x)dx,$$

where each X_i is exponentially distributed with the rate parameter λ. Obviously, the choice of λ matters. The best λ would minimize $\text{var}\left(e^{\lambda X} f(X)\right)$. Since the minimum variance occurs when $e^{\lambda X} f(X) = \text{const}$, we need to choose λ such that $f(x)$ is proportional to $e^{-\lambda x}$ especially when x is large. It is a good idea to plot $f(x)$ and to see how well it can be approximated with an exponential decay function by plotting $-\ln |f(x)|$ and choosing λ as the slope for large x.

Finally, consider evaluation of an improper integral of the type $\int_{-\infty}^{\infty} f(x)dx$. Along the same lines, we may approximate

$$\frac{\sigma\sqrt{2\pi}}{n} \sum_{i=1}^{n} e^{\frac{(X_i-\mu)^2}{2\sigma^2}} f(X_i) \simeq \int_{-\infty}^{\infty} f(x)dx,$$

where $X_i \sim \mathcal{N}(\mu, \sigma^2)$. To make this approximation effective, $e^{\frac{(x-\mu)^2}{2\sigma^2}}$ should be as close to $f(x)$ as possible. The double-exponential distribution can be used as well. The choice of an appropriate distribution is studied in *importance sampling*, a well-developed area especially useful in Bayesian statistics. More detail is found in the book by Rubinstein and Kroese [85].

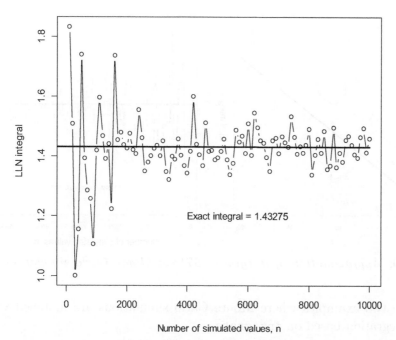

Figure 2.19: *Approximation of an integral (2.56) using simulations with different n.*

The R integrate function

R has a built-in numerical integration function `integrate`. The user must specify the integrand as an R function along with the limits of integration. Indefinite integration is possible by using `Inf`. Below are a few examples.

Problem 1. Find the integral $\int_0^2 e^{-2x} dx$. The exact value is 0.49084 and the R function is listed below.

```
exp2=function(x) exp(-2*x)
integrate(f=exp2,lower=0,upper=2)$value
```

gives 0.4908422

Problem 2. Find the integral $\int_0^2 e^{-\lambda x} dx$, where λ is user-defined.

```
exp2=function(x,lambda) exp(-2*x)
integrate(f=exp2,lower=0,upper=2,lambda=3)$value
```

Note that `lambda` is passed to function `exp2` through the argument in `integrate`.

Problem 3. Find the integral $\int_0^\infty e^{-\lambda x} dx$, where λ is user-defined.

```
exp2=function(x,lambda) exp(-lambda*x)
integrate(f=exp2,lower=0,upper=Inf,lambda=3)$value
```

Problem 4. Compute `pnorm` via integration.

```
pnorm.my=function(x,mu,sigma) dnorm(x,mean=mu,sd=sigma)
integrate(f=pnorm.my,lower=-Inf,upper=1,mu=0,sigma=1)$value
```

produces 0.8413448; `pnorm(1)` produces 0.8413447

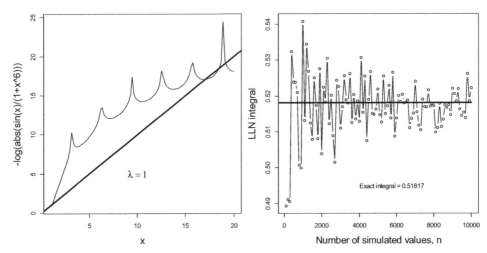

Figure 2.20: *Approximation of integral (2.57) via Monte Carlo simulations.*

Below are two examples where Monte Carlo simulations are matched with numerical integration based on `integrate`.

Example 2.50 *Integral approximation via simulations 1.* *Approximate the integral*

$$\int_{-10}^{5} e^{-0.123x^6} \ln(1 + x^8)dx \qquad (2.56)$$

using simulations, and illustrate the convergence graphically with an increasing number of simulations.

Solution. First, we use analytic software to evaluate this integral for the benchmarking purposes. For example, the symbolic algebra software *Maple* gives

$$\int_{-10}^{5} e^{-0.123x^6} \ln(1 + x^8)dx = 1.4328.$$

Second, to evaluate the integral using simulations, we view this integral as the expected value of a random variable. According to the LLN, the average of the simulated values converges to (2.56). Since the limits of integration are finite, it is convenient to use uniformly distributed random values on $(-10, 5)$. Thus, we view (2.56) as $15 \times E_X[e^{-0.123X^6} \ln(1 + X^8)]$, where X is uniformly distributed on $[-10, 5]$. The R code is found in function `LLNintegral`. We use `integrate` to numerically evaluate the integral and compare with the Monte Carlo estimate. The argument of this function controls the seed number in the case when the integral value needs to be reproduced for debugging. As follows from Figure 2.19, the convergence of the average to its mean (the exact integral) is slow – it's a stochastic convergence, but not the monotonic convergence we used to see in calculus.

Example 2.51 *Integral approximation via simulations 2.* *Approximate integral*

$$\int_0^\infty \frac{\sin(x)}{1+x^6}dx. \tag{2.57}$$

Solution. To choose λ, we plot the negative log integrand shown in Figure 2.20 at left. See the R function `LLNintegral2`. It seems that $\lambda = 1$ is appropriate. In the right plot, we show how the average converges to the true value with increasing n. The exact integral was evaluated numerically using *Maple*. □

The Monte Carlo simulations for approximating a double integral on a finite domain are discussed in Example 3.60.

Problems

1. (a) Generalize definitions 1–3 of the convergence of X_n to random variable X. (b) Prove that convergence in distribution is equivalent to convergence in probability.

2. Does $\lim_{n\to\infty} E(X_n) = a$ imply $p\lim_{n\to\infty} X_n = a$? Does $\lim_{n\to\infty} \text{var}(X_n) = 0$ imply $p\lim_{n\to\infty} X_n = \text{const}$? Prove or provide a counterexample.

3. Prove Theorem 2.49 under assumption that X_i are uncorrelated.

4. Does LLN apply to $X_i = 1/Y_i$ where $Y_i \overset{iid}{\sim} \mathcal{N}(0,1)$? Plot the average for increasing n as in Figure 2.19.

5. Does LLN apply to dependent random variables $\{X_i, i = 1, 2, ...\}$ defined as $X_i = \rho X_{i-1} + Z_i$, where $\{Z_i\}$ are iid with zero mean and constant variance, and $|\rho| < 1$ and $X_0 = 0$?

6. In Example 2.44, derive model (2.53) using the following justification. Let the probabilities of reading the closest tick mark be reciprocal to the distance, $\Pr(X = r+1) = a/((r+1) - L)$ and $\Pr(X = r) = a/(L-r)$, where a is the coefficient such that $\Pr(X = r+1) + \Pr(X = r) = 1$. Show that these assumptions lead to model (2.53).

7. Use a nonuniform distribution in Example 2.45, say, a triangular distribution on $(-1/2, 1/2)$. Can the exact distance be recovered? Prove that the exact distance can be recovered if the distribution on $(-1/2, 1/2)$ is symmetric. Can the exact distance be recovered if a symmetric distribution is defined on $(-a, a)$, where $a > 1/2$?

8. Use Monte Carlo simulations to estimate the integral $\int_0^1 \ln(x + 1)dx$ and compare the estimate with exact value. Refer to the LLN to prove that the simulation value converges to the true integral value when the number of simulations increases.

9. (a) Estimate integral $\int_0^\infty e^{-x/2}\sin x\,dx$ using random numbers with the exponential distribution (use `integrate` to obtain the exact value) and create the plot as in Figure 2.20. Estimate the number of simulations to achieve the probability of the approximation at least 0.99. (b) The same, but for integral $\int_{-\infty}^\infty (1+x^2)e^{-(x-1)^2}\sin x\,dx$.

2.10   The central limit theorem

The central limit theorem (CLT) is one of the most remarkable results of mathematics. This theorem says that the average (upon standardization) of n iid random variables is normally distributed when n goes to infinity.

Theorem 2.52 *Central limit theorem. If $X_1, X_2, X_3, \ldots$ are iid random variables with mean $E(X_i) = \mu$ and variance $\mathrm{var}(X_i) = \sigma^2$ then the distribution of the standardized sum converges to the standard normal distribution:*

$$\frac{\sum_{i=1}^n (X_i - \mu)}{\sigma\sqrt{n}} \to \mathcal{N}(0,1), \quad n \to \infty. \tag{2.58}$$

Proof uses the MGF, see Section 2.5. First, we note that "standardization" means that the random variable on the left has zero mean and unit variance. Next, we express the left-hand side of (2.58) as

$$\frac{\sqrt{n}(\overline{X}_n - \mu)}{\sigma} = \frac{1}{\sqrt{n}}\sum_{i=1}^n Z_i,$$

where $Z_i = (X_i - \mu)/\sigma$ are iid standardized random variables with $E(Z_i) = 0$ and $\mathrm{var}(Z_i) = 1$. Denote $M_Z(t)$ as the MGF of Z_i and $M(t)$ the MGF of $\overline{Z}_n = \sum_{i=1}^n Z_i/\sqrt{n}$. Using the fact that the MGF of the sum of independent random variable is the product of individual MGFs, we obtain

$$M(t) = M_Z^n\left(t/\sqrt{n}\right).$$

We note that $M(0) = 1$, $M'(0) = 0$, and $M''(0) = 1$ because $M_Z(0) = M_Z'(0) = 0$ and $M_Z''(0) = 1$ due to $E(Z_i) = 0$ and $E(Z_i^2) = 1$. Expand $M_Z(t/\sqrt{n})$ up to the second order around $t = 0$:

$$M_Z(t/\sqrt{n}) = 1 + \frac{(t/\sqrt{n})^2}{2!} + o((t/\sqrt{n})^2).$$

Thus, if t is fixed, we have

$$\lim_{n\to\infty} M(t) = \lim_{n\to\infty}\left(1 + \frac{(t/\sqrt{n})^2}{2!} + o((t/\sqrt{n})^2)\right)^n = \lim_{n\to\infty}\left(1 + \frac{a_n}{n}\right)^n,$$

where $a_n = t^2/2 + n \times o((t/\sqrt{n})^2)$. It is not difficult to prove that the second term in a_n vanishes. That is, $\lim_{n\to\infty} a_n = t^2/2$. From calculus, we also know that

$$\lim_{n\to\infty}\left(1+\frac{a_n}{n}\right)^n = \left(\lim_{n\to\infty}\left(1+\frac{a_n}{n}\right)^{n/a_n}\right)^{a_n} = \lim_{n\to\infty} e^{a_n} = e^{t^2/2}.$$

Hence, we deduce that the MGF of the left-hand side of (2.58) converges to $e^{t^2/2}$, but this is the MGF of the standard normal distribution (we rely on the fact that the MGF uniquely identifies the distribution). □

In its explicit form, the CLT states that

$$\lim_{n\to\infty} \Pr(Z_n < z) = \Phi(z)$$

for every z, where

$$Z_n = \sqrt{n}\frac{\overline{X}_n - \mu}{\sigma}$$

is the standardized average. The CLT approximation becomes exact when X_i are normally distributed. Indeed, if $X_i \sim \mathcal{N}(\mu, \sigma^2)$, then $X_i - \mu \sim \mathcal{N}(0, \sigma^2)$, and since the sum of independent normally distributed random variables is normally distributed, we have $\sum_{i=1}^{n}(X_i - \mu) \sim \mathcal{N}(0, n\sigma^2)$, which implies $Z_n \sim \mathcal{N}(0, 1)$ for each n. In other words, for a normal distribution, (2.58) is exact.

The remarkable result of the CLT is that the limiting distribution is a normal distribution regardless of the individual distributions of X_i, even when X_i is a discrete random variable. In layman's terms, a quantity has normal distribution if it is an additive result of many independent factors with uniform contributions.

As follows from the CLT, the sum of iid random variables with mean μ and variance σ^2 can be approximated as

$$\sum_{i=1}^{n} X_i \simeq \mathcal{N}(n\mu, n\sigma^2).$$

This method of approximation is illustrated in an example below.

Example 2.53 CLT for the binomial cdf. *Approximate the binomial cdf with a normal cdf. Write an R script and plot the two cdfs to see how they agree.*

Solution. The binomial distribution is the distribution of $m = \sum_{i=1}^{n} X_i$, where X_i are Bernoulli iid with $\mu = p$ and $\sigma = \sqrt{p(1-p)}$. Therefore, the normalized sum takes the form

$$Z_n = \frac{m - np}{\sqrt{np(1-p)}} \simeq \mathcal{N}(0, 1)$$

due to CLT, sometimes referred to as the de Moivre–Laplace theorem. Thus, the cdf of m can be approximated as

$$\Pr(m \leq k) \simeq \Phi\left(\frac{k - np}{\sqrt{np(1-p)}}\right), \quad k = 0, 1, ..., n \tag{2.59}$$

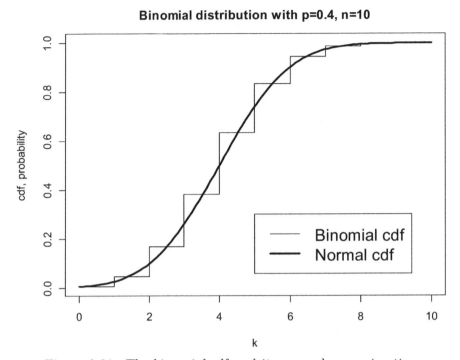

Figure 2.21: *The binomial cdf and its normal approximation.*

Function LLNintegral3 realizes this approximation; the plot with two cdfs is shown in Figure 2.21. The normal cdf is plotted with continuous values of k for better visualization.

Example 2.54 *CLT before your eyes.* *Write R code that illustrates the CLT using (a) binomial, and (b) exponential distribution. Save the plots for each n in one folder and scroll the plots to visualize the convergence to the standard normal density.*

Solution. (a) The probability p and the number of plots N are the arguments of the R function clt.binom with default values 0.1 and 200, respectively. We do not generate binomial random variables because the distribution is a built-in function dbinom. The normalized random variable is (x-p*n)/seb where x=0:n and seb=sqrt(p*(1-p)*n) is the standard error of the binomial distribution. To give the saved filenames the same length, zeros are inserted in front of the file number. The default folder is c:\\StatBook\\clt.binom\\ and must be created before calling the function. The bars with probabilities are displayed in the range $(-3, 3)$. (b) The R function is clt.binom with arguments and their default values given by lambda=1,N=50,nprob=50,nSim=1000000. A nSim by n matrix X with exponentially distributed random variables is created, and the sum is computed using the rowSums command. Two versions for normalization can be used: use theoretical mean and SD of the sum as $n\lambda$ and $\sqrt{n/\lambda^2}$ or empirical by using `mean` and `sd`. Neither version has a significant impact on the distribution of the normalized sum. The probabilities of falling within `nprob` intervals in the range from -3 to 3 are shown as an estimate of the empirical density of the sum. Instead of saving the files, the plots may be shown in R with some time delay using function `Sys.sleep()`.

Example 2.55 ***When the CLT does not hold.*** *Let $X_{\max} = \max X_i$, where X_i are iid uniformly distributed on (0,1). Then the standardized random variable*

$$Z_n = \frac{X_{\max} - E(X_{\max})}{\text{SD}(X_{\max})}$$

does not converge to Φ as $n \to \infty$. Prove and illustrate this result using cdf.

Solution. First, we note that $E(Z_n) = 0$ and $\text{var}(Z_n) = 1$ as in the CLT, but $X_{\max}$ is not the sum, so the CLT does not apply in the first place. Yet, we want to show this explicitly by deriving the limiting distribution of Z_n and demonstrating that it is different from Φ. The cdf of $X_{\max}$ is

$$\Pr(X_1 \le x, X_2 \le x, ..., X_n \le x) = (\Pr(X_1 < x))^n = x^n$$

and the density is

$$f_{\max}(x) = \begin{cases} nx^{n-1} \text{ if } 0 < x < 1 \\ \\ 0 \text{ elsewhere} \end{cases}.$$

The mean and variance can be obtained by direct integration of the density:

$$E(X_{\max}) = \frac{n}{n+1}, \quad \text{var}(X_{\max}) = \frac{n}{(n+1)^2(n+2)}.$$

Rewrite Z_n in a more explicit form as

$$Z_n = \frac{X_{\max} - \frac{n}{n+1}}{\sqrt{\frac{n}{(n+1)^2(n+2)}}}.$$

By construction, Z_n has zero mean and unit variance. Find the cdf as

$$F_n(z) = \left(z\sqrt{\frac{n}{(n+1)^2(n+2)}} + \frac{n}{n+1} \right)^n, \quad \sqrt{n(n+2)} \le z \le \sqrt{\frac{n+2}{n}}. \quad (2.60)$$

Obviously, $F_n(z)$ does not approach Φ because the support of this distribution is $(-\infty, 1)$ when $n \to \infty$. Using the elementary approximation $\ln(1 + a_n) \simeq a_n$ for $a_n \simeq 0$, we obtain the limiting distribution of Z_n as

$$\lim_{n \to \infty} F_n(z) = e^{z-1}, \quad -\infty < z \le 1.$$

The cdfs are shown in Figure 2.22.

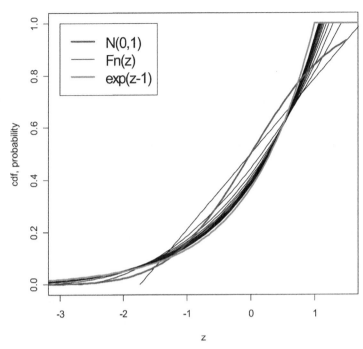

Figure 2.22: *The cdfs (2.60) for $n = 1, 2, ..., 15$. The normal cdf and the limiting cdf e^{z-1} are different.*

Example 2.56 *James Bond chase problem.* *James Bond is chased by an enemy in the sea: the speed of his boat is 1 unit of length per minute, but due to rough sea the actual speed is a uniformly distributed random variable in the interval $(1 - \delta, 1 + \delta)$, where $0 < \delta < 1$ is known. τ minutes after James Bond leaves the dock, the enemy starts chasing him in a faster boat with the speed $1 + v$, but again due to rough sea, the actual speed is a uniformly distributed random variable in the interval $(1 + v - \delta, 1 + v + \delta)$. What is the chance that James Bond will not be caught during a one hour chase?*

Solution. The chase is illustrated in Figure 2.23. The distance covered by James Bond after t minutes is $D_t^{JB} = \sum_{i=1}^{t} V_i$, where $V_i \overset{iid}{\sim} \mathcal{R}(1 - \delta, 1 + \delta)$, and the distance covered by the enemy is $D_t^E = \sum_{i=\tau+1}^{t} U_{i-\tau}$, where $U_i \overset{iid}{\sim} \mathcal{R}(1 + v - \delta, 1 + v + \delta)$ for $t > \tau$. The two simulated distances as functions of t are depicted in the figure. James Bond will be caught if the two curves intersect.

We estimate the distance covered using the normal approximation. Since V_i are iid, we approximate the distribution of D_t^{JB} as $\mathcal{N}(t, t\delta^2/3)$, because the variance of the uniform distribution is $(2\delta)^2/12 = \delta^2/3$. Similarly, the distance D_t^E covered by the enemy can be approximated as $\mathcal{N}((t - \tau)(1 + v), (t - \tau)\delta^2/3)$, so the difference at time t can be approximated as

$$\Delta_t \simeq \mathcal{N}(t - (t - \tau)(1 + v), (2t - \tau)\delta^2/3), \quad t = \tau + 1, \tau + 2, ..., T,$$

where in our case $T = 60$. James Bond will not be caught if $\Delta_t > 0$ for all t. One may be tempted to take the product of these probabilities to estimate the chance that James Bond will not be caught but Δ_t and Δ_{t+1} are not independent. Instead, we assume that James Bond will not be caught if the distance between him and the enemy is positive at the last moment, i.e. $\Delta_T > 0$, which gives

$$\Pr(\text{James Bond is not caught}) \simeq \Pr(\Delta_T > 0) = \Phi\left(\frac{T - (T - \tau)(1 + v)}{\delta\sqrt{(2T - \tau)/3}}\right).$$
(2.61)

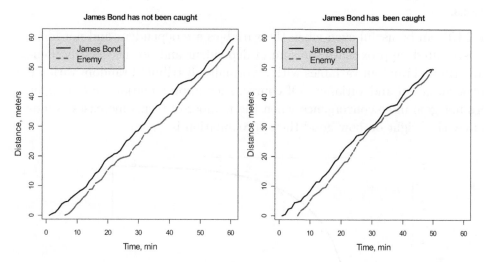

Figure 2.23: *Two outcomes of the chase. The enemy starts the chase 5 minutes after James Bond leaves the dock. He will be caught if the two curves intersect.*

The R code jb(job=1) plots Figure 2.23 and jb(job=2) estimates the probability via vectorized simulations. This code can be accessed by issuing source("c:\\StatBook\\jb.r"). Below, we show only the simulation part.

```
XJB=matrix(runif(nSim*T,min=1-delta,max=1+delta),ncol=T)
XJB=apply(XJB,1,cumsum)
XEN=matrix(runif(nSim*(T-tau),min=1+v-rv,max=1+v+rv),
                ncol=T-tau)
XEN=cbind(matrix(0,ncol=late,nrow=nSim),XEN)
XEN=apply(XEN,1,cumsum)
g=colSums(XJB>XEN)
pr.sim=mean(g==T)
```

We make a few comments: (i) The first line generates matrix XJB with all distances covered by James Bond in one minute. (ii) The second line overwrites this matrix with cumulative distances, so it contains distances covered in t minutes (it should be noted that matrix XJB has dimension T by nSim). (iii) The next three lines computes distances covered by the enemy boat (note that the first τ

minutes are zeros). (iv) The ith component of array g contains the number of times the distance covered by James Bond is greater than the distance covered by the enemy in the ith simulation. (v) The last line computes the proportion of simulations where $\Delta_t > 0$ for all t.

To estimate the probability that James Bond will not be caught, we run
jb(job=2)
Probability that JB will not be caught from simulations = 0.3335
Probability normal approximation = 0.4642508
The normal approximation (2.61) is not very precise. The vectorized simulations are amazingly fast: it takes less than a second to carry out nSim=100000 simulations.   □

The CLT can be useful even when the variables are dependent and not identically distributed: approximate the sum of dependent and not necessarily identically distributed random variables with a normally distributed random variable with the same mean and variance. Of course, such approximation may be poor and does not guarantee convergence when the number of terms increases: simulations may shed light on how good the approximation is.

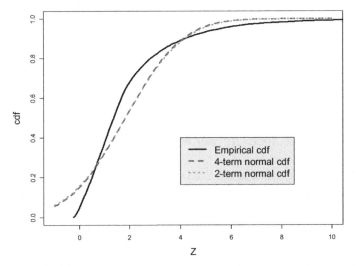

Figure 2.24: *The empirical cdf and two normal cdfs for $Z = X + X^2 + Y + Y^2$.*

Example 2.57 *CLT with violated assumptions*. *Let random variables $X \sim \mathcal{R}(0,1)$ and $Y \sim \mathcal{N}(0,1)$ be independent. Approximate the cdf of $Z = X + X^2 + Y + Y^2$ using the CLT and check your answer using simulations. Rearrange the terms to apply the CLT to the sum of two independent random variables. Plot the three cdfs to compare.*

Solution. Obviously, the terms in the sum are not identically distributed and dependent (some of them are functions of others). Find $E(Z)$ using the formula

$E(U^2) = \text{var}(U) + E^2(U)$ and the fact that $E(X^p) = (1+p)^{-1}$:

$$E(Z) = \frac{1}{2} + \frac{1}{3} + 0 + 1 = \frac{11}{6}.$$

To approximate $\text{var}(Z)$, we use $\text{var}(U^2) = E(U^4) - E^2(U^2)$, so $\text{var}(X^2) = 1/5 - 1/9 = 4/45$. To find $\text{var}(Y^2)$ we use formula (2.43), yielding $\text{var}(Y^2) = 3 - 1 = 2$. Thus applying the CLT to the sum of four terms, we can approximate $Z \simeq \mathcal{N}(11/6, 571/180)$. Now we express $Z = Z_1 + Z_2$, where $Z_1 = X + X^2$ and $Z_2 = Y + Y^2$. Find

$$E(Z_1) = \frac{5}{6}, \quad \text{var}(Z_1) = \left(\frac{1}{3} + \frac{2}{4} + \frac{1}{5}\right) - \left(\frac{5}{6}\right)^2 = \frac{61}{180},$$
$$E(Z_2) = 1, \quad \text{var}(Z_2) = (1+3) - 1^2 = 3,$$

using the fact that $E(Y^4) = 3$. Thus, applying the CLT to $Z_1 + Z_2$, we deduce that the two-term approximation takes the form $Z \simeq \mathcal{N}\left(\frac{5}{6} + 1, \frac{61}{180} + 3\right)$. Note that both normal approximations have the same mean but different variances, $571/180 = 3.172$ and $\frac{61}{180} + 3 = 3.339$, respectively. The plot of the empirical and normal cdfs is depicted in Figure 2.24 produced by function cltP. The four-term and two-term approximations are close to each other (the cdf curves nearly coincide), but are fairly poor approximations as a whole. □

The CLT applies to independent random variables with different distributions under the condition there is no dominant contribution from any term.

Theorem 2.58 *Extended CLT (Lindeberg–Feller Condition).* *Let X_i be independent random variables with mean μ_i and variance σ_i^2. Then*

$$\frac{\sum_{i=1}^n (X_i - \mu_i)}{\sqrt{\sum_{i=1}^n \sigma_i^2}} \simeq \mathcal{N}(0,1)$$

if the Lindeberg–Feller condition on the variances holds:

$$\frac{\max(\sigma_i^2, i = 1, 2, ..., n)}{\sum_{i=1}^n \sigma_i^2} \to 0, \quad n \to \infty. \tag{2.62}$$

The Lindeberg–Feller condition defines what is meant by "no dominant contribution from each term." The CLT can be used to approximate complex numerical approximation, as illustrated in the following example. More details on asymptotic probability theory can be found in Serfling [93] and the books cited in the previous section.

Example 2.59 *CLT for numeric approximation.* *Approximate the volume under the hyperplane $\sum_{i=1}^n p_i x_i = c$ within the n-dimensional cube $[0,1]^n$, where all p_i and c are positive, and, without loss of generality, $\sum p_i^2 = 1$. Use simulations in R to check the answer for $n = 3$ and $n = 10$ with arbitrary (randomly chosen) coefficients p.*

Solution. The exact volume is difficult to compute, especially for large n, say, $n \geq 10$. Instead, we interpret the volume as the cdf of $\sum_{i=1}^{n} p_i x_i$ evaluated at c assuming that $\{x_i, i = 1, ..., n\}$ are iid uniformly distributed random variables on $[0, 1]$. Indeed, approximate the distribution of $\sum_{i=1}^{n} p_i x_i$ by a normal distribution with mean $\mu = \sum_{i=1}^{n} p_i E(x_i)$ and variance $\sigma^2 = \sum_{i=1}^{n} p_i \text{var}(x_i)$. Since $\mu = 0.5 \sum_{i=1}^{n} p_i$ and $\sigma^2 = \frac{1}{12} \sum_{i=1}^{n} p_i^2 = \frac{1}{12}$, we have

$$\text{Vol} = \Pr\left(\sum p_i x_i \leq c\right) \simeq \Phi\left(\left(2c - \sum_{i=1}^{n} p_i\right)\sqrt{3}\right).$$

This approximation is valid due to the extended CLT under the assumption that there are no dominant p_i^2. See function `volcube`. Note that since the number of simulations is large, the empirical cdf/volume is very close to the exact value.

2.10.1 Why the normal distribution is the most natural symmetric distribution

The CLT gives rise to the normal distribution as the limiting distribution of a normalized sum of independent random variables.

We can justify the use of the normal distribution from a different perspective as the simplest approximation of any symmetric unimodal distribution as follows.

Let $f(x)$ be the density of a symmetric unimodal distribution. Suppose that f is differentiable up to the second order on $(-\infty, \infty)$.

1. Since density is positive we can represent $f(x) = e^{h(x)}$.

2. Since $f(x)$ is symmetric, its mean and mode are the same, μ. Expand $h(x)$ as a Taylor series around μ

$$h(x) = h(\mu) + (x - \mu)h'(\mu) + \frac{1}{2}(x - \mu)^2 h''(\mu) + o((x - m)^2),$$

and approximate by dropping the o term:

$$h(x) \simeq h(\mu) + (x - \mu)h'(\mu) + \frac{1}{2}(x - \mu)^2 h''(\mu).$$

3. Prove that $h'(\mu) = 0$ and $h''(\mu) < 0$. Indeed, since μ is the maximum point of f, it is also the maximum point of h, such that we have $h'(\mu) = 0$ and $h''(\mu) < 0$. This means that letting $a = h(\mu)$ and $\sigma^2 = 1/h''(\mu)$, we can approximate

$$h(x) \simeq h(\mu) - \frac{1}{2\sigma^2}(x - \mu)^2.$$

4. The area under $f(x)$ must be 1, so we obtain

$$e^{h(\mu)} = \frac{1}{\int_{-\infty}^{\infty} e^{-\frac{1}{2\sigma^2}(x-\mu)^2} dx} = \frac{1}{\sqrt{2\pi\sigma^2}}.$$

Finally, we arrive at the normal density

$$f(x; \mu, \sigma^2) = \frac{1}{\sqrt{2\pi\sigma^2}} e^{-\frac{1}{2\sigma^2}(x-\mu)^2}.$$

This derivation very much resembles the derivation of the CLT using characteristic functions, Billingsley [9].

2.10.2 CLT on the relative scale

In many situations, contributing factors, X_i, are not additive, but multiplicative. This phenomenon gives rise to the lognormal distribution (discussed further in the next section). In this section, we explain why many real-life quantities follow this distribution due to the multiplicative effect.

The traditional CLT assumes additivity: the difference between two consecutive sums, $S_{i+1} = \sum_{j=1}^{i+1} X_j$ and $S_i = \sum_{j=1}^{i} X_j$, does not depend on S_i because

$$S_{i+1} - S_i = X_{i+1}. \tag{2.63}$$

We say that CLT relies on additivity. One of the consequences of the additivity is the symmetry of the normal distribution. Consequently, the normal distribution may take negative values with positive probability. In fact, violations of additivity are seen in almost every distribution we deal with in real life; below are just a few examples. In these examples, and in many others, contributing factors act on the relative (or equivalently, multiplicative) scale.

1. *The weight of humans, or other biological growth.* Of course genetics play an important role, but a fat-filled diet and unhealthy lifestyle contribute to obesity on the relative scale $Q_{i+1} = Q_i(1 + X_i)$. Obese people gain more weight compared with people with low weight, where X_i is positive or negative but relatively small. The same holds for any other biological growth: factors apply to the current state and the change is proportional to Q_i.

2. *Salary and income.* Salary rises on the relative (or percent scale), $Q_{i+1} = Q_i(1 + X_i/100)$, where X_i is relatively small and represents the percent salary increase. The same rule applies to wealth.

The contributing factors, X_i, are small so that $1 + X_i$ is always positive and can be expressed as $e^{\sigma Z_i}$, where Z_i is a contributing factor on the relative scale expressed as a random variable with zero mean and unit variance, due to the first-order approximation $e^{\sigma Z_i} \simeq 1 + \sigma Z_i$. The relative scale implies that

$$Q_{i+1} = Q_i e^{\sigma Z_i},$$

where $Q_0 = e^\mu$. After standardization, we have

$$Q_n' = Q_0 \left(e^{\sum_{i=1}^n \sigma Z_i} \right)^{1/\sqrt{n}} = e^{\mu + \frac{\sigma}{\sqrt{n}} \sum_{i=1}^n Z_i} \simeq e^{\mathcal{N}(\mu, \sigma^2)},$$

due to the CLT applied to the argument of the exponential function. The right-hand side in the above equation means the distribution of e^U, where $U \sim \mathcal{N}(\mu, \sigma^2)$. This distribution is called *lognormal*; see the next section. This derivation explains why many distributions involving positive data encountered in real life can be reduced to the normal distribution after applying the log transformation.

Problems

1. Approximate the Poisson cdf with the normal cdf similarly to Example 2.53. Plot the two cdfs with $\lambda = 2$ and $\lambda = 10$ using `par(mfrow=c(1,2))`.

2. Prove that if $X \sim \mathcal{P}(\lambda_n)$, then $(X - \lambda_n)/\sqrt{\lambda_n}$ converges in distribution to $\mathcal{N}(0,1)$ when $\lambda_n \to \infty$ using the fact that the sum of independent Poisson random variables is a Poisson random variable. [Hint: Use $\lambda_n = \lambda n$ where $n \to \infty$ and λ is fixed.]

3. It is claimed that the binomial distribution with n trials and probability p can be well approximated by the normal distribution $\mathcal{N}(np, np(1-p))$ when $np > 10$. Compute the difference between the distributions using the maximum distance between the two cdfs and plot it versus np. [Hint: Use `n=1:100` and `np=seq(from=11,to=20,length=100)`.]

4. It is claimed that the Poisson distribution with $\lambda > 10$ can be well approximated with the normal distribution $\mathcal{N}(\lambda, \lambda)$. (a) Compute the difference between the distributions using the maximum distance between the two cdfs and plot it versus λ. (b) What is the difference for $\lambda = 11$? (c) What λ gives the difference 0.05? [Hint: Use `lambda=seq(from=11,to=50, length=100)` and find the difference between cdfs for `x=seq(from=0,to=100,length =10000)`.]

5. Illustrate the CLT using function `LLNintegral3` by showing four binomial distributions with increasing n.

6. Is the CLT true for $\max X_i$ when $n \to \infty$ where X_i are iid and have an exponential distribution?

7. (a) Modify the R function in Example 2.54 to visualize the CLT applied to a Poisson distribution. (b) Use the exponential distribution with random $\lambda_i \sim \mathcal{R}(0,1)$ to see if the normalized sum of variables obeys the CLT. (c) The same as the previous problem, but letting $\lambda_i = 1/i$.

8. Find a sequence of random variables $X_1, X_2, \ldots$ such that $\sum_{i=1}^{n}(X_i - \overline{X}) / \sqrt{\sum_{i=1}^{n}(X_i - \overline{X})^2}$ does not converge to the standard normal distribution.

9. (a) Explain why the normal approximation overestimates the probability that James Bond will not be caught in Example 2.56. Run jb(job=2) under several scenarios to demonstrate this. (b) Plot the probability that James Bond will not be caught as a function of the roughness of the sea (parameter v).

10. Use a normal instead a uniform distribution to model the boat speed at each minute in Example 2.56. Does the normal distribution considerably affect the probability?

11. Random walk on the line: a drunk walker takes step left or right with probability 1/2. Assuming that one step is 3 ft and he takes a step every second, what is the probability that he will be at a distance 100 ft or more from the place he started after one hour? Use vectorized simulations and the CLT to assess the probability. Compare the results.

12. Random walk on the plane: a drunk walker takes a step left, right, up, or down with probability 1/4. Assuming that one step is 3 ft and he takes a step every second, what is the probability that he will stay within the square ± 100 ft centered on the place he started after one hour? Use simulations and the CLT to assess the probability. Compare the results.

13. Approximate the distribution of $\sum_{i=1}^{n} X_i$, where one half of the random variables are iid with mean μ_1 and SD σ_1 and another half of the random variables are iid with mean μ_2 and SD σ_2 (the variables from the two groups are independent). [Hint: Use the fact that the sum of two independent normally distributed random variables is a normally distributed normal variable with the mean and variance equal to the sum of individual means and variances.]

14. Does the CLT apply to a sequence $\{X_i\}$ where $X_i = X_{i-1} + \varepsilon_i$ and ε_i are iid with zero mean and constant variance σ^2 ($X_0 = 0$)? [Hint: Start with normal distribution of ε_i and then apply Theorem 2.58.]

15. Find a normal approximation to the distribution of $\sqrt{n}\sum_{i=1}^{n}(X_i - \overline{X})^2$, where $X_i \overset{\text{iid}}{\sim} \mathcal{R}(0, 1)$.

16. Does the CLT apply to a sequence $X_i = \sum_{j=1}^{i} U_{ij}$ where U_{ij} are iid random variables with mean μ and variance σ^2.

17. Does the CLT apply to identically distributed dependent observations $\{y_i\}$, where y_i and y_{i+k} do not correlate for $k > 1$ but y_i and y_{i+1} correlate with the same correlation coefficient ρ.

18. Investigate under what conditions the normal approximation in Example 2.59 is not valid. [Hint: Express the dominance of p_i in terms of $\max p_i^2$.]

2.11   Lognormal distribution

As discussed in the previous section, a random variable X is said to have a lognormal distribution if it can be expressed as $X = e^U$, where U is a normal random variable with mean μ and variance σ^2. The density of the lognormal distribution can be derived from the cdf

$$F(x; \mu, \sigma^2) = \Pr(X \leq x) = \Phi\left(\frac{\ln x - \mu}{\sigma}\right)$$

by differentiation:

$$f(x; \mu, \sigma^2) = \frac{1}{x\sqrt{2\pi\sigma^2}} e^{-\frac{1}{2\sigma^2}(\ln x - \mu)^2}, \quad x > 0. \tag{2.64}$$

Two densities of the lognormal distribution with $\mu = 1$ and $\mu = 2$, and the same $\sigma = 1$ are shown in Figure 2.25.

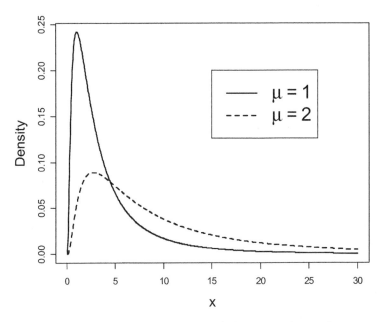

Figure 2.25: *Two lognormal densities* ($\sigma = 1$).

Note that, although x is in the denominator, we can safely assume that $f(0; \mu, \sigma^2) = 0$ because, by letting $h = -\ln x$, we have

$$\lim_{x \to 0} \ln f(x; \mu, \sigma^2) = -\frac{1}{2}\ln(2\pi\sigma^2) + \lim_{h \to \infty}\left(h - \frac{1}{2\sigma^2}(h + \mu)^2\right) = -\infty,$$

which implies that the lognormal density starts at the origin.

The lognormal distribution is skewed to the right. Consequently, it has a heavy right tail meaning that one may encounter large values more often than small values. In rigorous form, the fact that the lognormal distribution has a heavy right tail, compared with the normal distribution, can be expressed as

$$\lim_{x \to \infty} \frac{\phi(x)}{f(x; \mu, \sigma^2)} = \sigma \lim_{x \to \infty} e^{-\frac{1}{2}x^2 + \frac{1}{2\sigma^2}(\ln x - \mu)^2 + \ln x} = 0,$$

because, in vague terms, the quadratic function goes to infinity faster than the log function. That is, the normal density approaches the x-axis faster than the density of the lognormal distribution when $x \to \infty$. This explains why the lognormal distribution is a good candidate to model real-life data, such as weight, income, survival, etc.

The median and the mode of the lognormal distribution are e^μ and $e^{\mu - \sigma^2}$, respectively. It is easy to prove that (a) since the median is invariant with respect to any monotonic transformation, and if m is the median of $\{\ln X_i, i = 1, ..., n\}$, then e^m is the median of $\{X_i, i = 1, ..., n\}$ and (b) the geometric mean calculated on the original data, $(\Pi_{i=1}^n X_i)^{1/n}$ estimates the median of $\{X_i, i = 1, ..., n\}$ because $\overline{\ln X}$ estimates μ.

The expected value and variance are

$$E(X) = e^{\mu + \sigma^2/2}, \quad \text{var}(X) = e^{2\mu + \sigma^2}(e^{\sigma^2} - 1).$$

These formulas can be derived analytically by integration, but we use simulations below.

Example 2.60 *Simulations for the lognormal distribution. Write an R program to check formulas for the expected value and the variance of the lognormal distribution via simulations.*

Solution. The code is found in the text file `lognDS.r`. We make a few comments: (a) The function `cat()` is similar to `paste()` but does not print out quotes. (b) The fact that the empirical and theoretical means are close to each other for large nSim follows from the LLN. The call `lognDS()` gives the following output:

```
Theoretical mean= 12.18249 variance= 255.0156
Empirical mean= 12.18871 variance= 256.5376
```

Note that the result varies slightly from call to call because random numbers do not repeat. To repeat random numbers one needs to have `set.seed(sn)` before simulations start where `sn` is user-defined and fixed. Also, due to the possibility of a large outlier, a fairly large number of simulated values is required to obtain a satisfactory estimator of the theoretical variance (here the default value is 1 million).   □

The lognormal distribution has a close connection with the CV, calculated as the ratio of the SD to the mean, and is an appropriate measure of variation on the *relative scale*, see Section 2.2. For the lognormal distribution, the CV does not depend on μ and is given by

$$\mathrm{CV} = \sqrt{e^{\sigma^2} - 1}.$$

Using the elementary approximation $e^x - 1 \simeq x$ for small x, we obtain that for the lognormal distribution with small σ,

$$\mathrm{CV} \simeq \sigma. \tag{2.65}$$

Roughly speaking, the CV of a lognormally distributed random variable X is equal SD of $\ln X$. For example, if we want to model the distribution of salaries in a firm with median \$90K and variation 20%, we set $\mu = \ln 90 = 4.5$ and $\sigma = 0.2$.

A characteristic feature of the lognormal distribution is that its variance depends on the mean. Many distributions in real life possess this property including those listed in Section 2.10.2.

The lognormal distribution is a part of base R. For example, the density function (2.64) can be evaluated as `dlnorm(x, meanlog=`μ`,sdlog=`σ`)`. Of course, the other three functions `plnorm`, `qlnorm`, `rlnorm` are available as well.

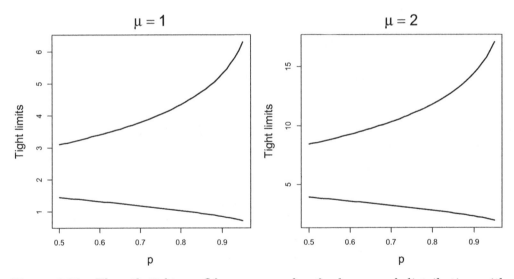

Figure 2.26: *The pth tight confidence range for the lognormal distribution with $\mu = 1$ and $\mu = 2$ and $\sigma = 0.5$ (SD = 50% of the mean) computed using the Newton-Raphson algorithm.*

2.11.1 Computation of the tight confidence range

As mentioned previously, the lognormal distribution is an asymmetric distribution. Typically, we characterize the range of a symmetric distribution as $\mu \pm$

$k \times$SD, where μ is the mean and k is a scaling coefficient. Reporting a symmetric interval for an asymmetric distribution is not an optimal. In Section 2.2.2, we introduced the pth tight confidence range as the interval of the smallest length that contains the random variable with the desired probability p. Applying this notion to the lognormal distribution, as follows from equations (2.21), the lower (t) and the upper (T) limits of the pth tight confidence range are found from

$$\frac{1}{2\sigma^2}(l - \mu)^2 + l - \frac{1}{2\sigma^2}(L - \mu)^2 - L = 0, \tag{2.66}$$

$$\Phi\left(\frac{L - \mu}{\sigma}\right) - \Phi\left(\frac{l - \mu}{\sigma}\right) - p = 0, \tag{2.67}$$

where $L = \ln T$ and $l = \ln t$ are the interval log limits. These equations are solved iteratively by Newton's algorithm:

$$\begin{bmatrix} l \\ L \end{bmatrix}_{\text{new}} = \begin{bmatrix} l \\ L \end{bmatrix} - \begin{bmatrix} \frac{l-\mu}{\sigma^2} + 1 & -\frac{L-\mu}{\sigma^2} - 1 \\ -\frac{1}{\sigma}\phi\left(\frac{l-\mu}{\sigma}\right) & \frac{1}{\sigma}\phi\left(\frac{L-\mu}{\sigma}\right) \end{bmatrix}^{-1} \begin{bmatrix} E_1 \\ E_2 \end{bmatrix},$$

where E_1 and E_2 are the left-hand sides of equations (2.66) and (2.67). The elements of the matrix are the derivatives of E_1 and E_2 with respect to l and L (see Section 10.6.3). The right-hand side of Newton's iterations is computed at the current values of l and L. To start iterations, we use the equal-tail log limits, $l_0 = \mu + \sigma\Phi^{-1}((1 - p)/2)$ and $L_0 = \mu + \sigma\Phi^{-1}((1 + p)/2)$. See function logLnr, which creates Figure 2.26. Newton's iterations are carried out by the local function ptNR. As we can see from Figure 2.26, the upper limit increases rapidly with increasing p because the right tail of the lognormal distribution is heavy compared with the left one.

Example 2.61 *Confidence range for salary. Salary in the firm follows a lognormal distribution. Find the 50% tight confidence range with median $90K and variation 20% using function ptNR.*

Solution. The median of the lognormal distribution is e^μ and σ is the relative variation. The 50% range of salary is found by calling ptNR(), which applies Newton's algorithm described above

```
ptNR(mu=log(90),sigma=.2,p=.5)
[1] 75.36911 99.24609
```

Thus, 50% of employees earn from $75K to $99K. In other words, the chance that a random employee earns from 75 to 99 thousand dollars is 1/2.

Problems

1. (a) Prove that the mode of the lognormal distribution is $e^{\mu - \sigma^2}$ and find the maximum value of the density. (b) Prove that the median of the lognormal distribution is e^μ. [Hint: Express $\ln f$ via $y = \ln x$.]

2. Prove that the mean of the lognormal distribution is $0.5e^{\mu+\sigma^2}$. [Hint: Find Ee^U where $U \sim \mathcal{N}(\mu, \sigma^2)$.]

3. Random variable X has a lognormal distribution. (a) Prove that aX has a lognormal distribution for $a > 0$. (b) Prove that X^p has a lognormal distribution for $p > 0$. (c) Does $\ln X$ has a normal distribution?

4. Prove that μ in the lognormal distribution (2.64) can be interpreted as $\ln(M)$ where M is the median of X.

5. Prove that the right tail of the lognormal distribution is heavier than the left one, $\Pr(X < e^{\mu} - z) < \Pr(X > e^{\mu} + z)$ for all $z < e^{\mu}$.

6. Modify function `lognDS` to investigate how fast the empirical variance converges to the theoretical variance by plotting the empirical variance as a function of `nSim`. Approximate the required `nSim` such that SE $< 0.1\%$.

7. Write R code to illustrate approximation (2.65) for CV in the range from 10% to 100%. Prove that CV $> \sigma$. [Hint: Use CV on the x-axis and σ on the y-axis, display the 45° line.]

8. Write R code to compare how close the equal-tail range limits are close to the tight confidence range limits. Plot the two limits as functions of p as in Figure 2.26.

9. Provide another mathematical formulation of the statement "the lognormal distribution has a heavy right tail, compared with the normal distribution" by finding the limit of $\Pr(Y > x)/\Pr(X > x)$ when $x \to \infty$, where Y is normally distributed, and X is lognormally distributed. [Hint: Use L'Hôpital's rule.]

10. Devise an alternative algorithm for solving equations (2.66) and (2.67). Use a fine grid of c on the interval from 0 to max f (see Problem 1). For each c find l and L such that $f(l) = f(L) = c$. Compute the left-hand side of (2.67) and find c for which its value is closest to 0. Compare the two algorithms on a set of values for μ and σ^2.

11. Find a symmetric 50% interval for salary in Example 2.61 defined as $(e^{\mu} - z, e^{\mu} + z)$ where z is such that $\Pr(|X - e^{\mu}| < z) = 0.5$. Why is the length of this interval smaller than 24?

2.12   Transformations and the delta method

Sometimes, we want to know the distribution of a continuous random variable upon transformation. In this section, we derive the cdf and pdf of a transformed random variable. The concept of the inverse function is the centerpiece in this derivation.

Let X be a continuous random variable with cdf $F_X(x)$ and density $f_X(x)$. Let $g = g(x)$ be a strictly increasing function, that is, $x_1 < x_2$ implies $g(x_1) < g(x_2)$. We will also assume that g is differentiable. From calculus we know that a differentiable function g is a strictly increasing function if $g' > 0$. We want to find the distribution function of the random variable Y upon transformation g : $Y = g(X)$. We say that g^{-1} is the *inverse function* of g if $x = g^{-1}(y)$ is the solution of the equation $g(x) = y$. Since g is a strictly increasing function, there is at most one solution of this equation for every y. This means that the inverse function is well defined. (The notation g^{-1} does not mean $1/g$; treat g^{-1} as a symbol.) For example, the inverse of e^x is $\ln x$ and vice versa.

Two properties of the inverse function will be used: (i) inequalities $g(x) \leq y$ and $x \leq g^{-1}(y)$ are equivalent, and (ii) the derivative of the inverse function is the reciprocal of the of the derivative of the original function, $(g^{-1})' = 1/g'$. [The latter property follows from applying the chain rule to the identity function, $g^{-1}(g) = x$.]

Now we are ready to derive the cdf of Y. As follows from the definition of the cdf,

$$F_Y(y) = \Pr(Y \leq y) = \Pr(g(X) \leq y) = \Pr(X \leq g^{-1}(y)) = F_X(g^{-1}(y)). \quad (2.68)$$

Hence the cdf of the transformed random variable can be expressed through the inverse function as $F_Y(y) = F_X(g^{-1}(y))$. To find the pdf, differentiate $F_Y(y)$ using the chain rule, $f_Y(y) = f_X(g^{-1}(y))/g'(y)$. This formula assumes that g is a strictly increasing function. In the general case, when g is a strictly increasing or decreasing function, the absolute value of the derivative is used,

$$f_Y(y) = \left|(g^{-1}(y))'\right| f_X(g^{-1}(y)) = \frac{1}{|g'(y)|} f_X(g^{-1}(y)). \quad (2.69)$$

Obviously, it suffices to assume that g is strictly increasing (or decreasing) on the support of the random variable X, where $f_X > 0$. In a special case when the transformation is linear, $Y = a + bX$, we have $f_Y(y) = f_X((y-a)/b)/|b|$.

Example 2.62 *Squared uniform distribution. Find the cdf and density of the squared uniform distribution $\mathcal{R}(0,1)$.*

Solution. In this example, $g(x) = x^2$ is a strictly increasing function on the support of $\mathcal{R}(0,1)$. Since $F_X(x) = x$ and $g^{-1}(y) = \sqrt{y}$, the cdf of $Y = X^2$ is $\sqrt{y}$ for $0 < y < 1$. Sometimes it is easier to derive cdf (2.68) and then differentiate to obtain the pdf. Indeed, $F_Y(y) = \Pr(X^2 \leq y) = \Pr(X \leq \sqrt{y}) = \sqrt{y}$. Now to find the density, differentiate cdf to obtain $f_Y(y) = (2\sqrt{y})^{-1}$. Alternatively, we apply formula (2.69): since $g'(x) = 2x$, we obtain $g'(y) = 2\sqrt{y}$, and since $f_X = 1$, we arrive at the same $f_Y(y)$. $\square$

If function g is nonmonotonic, formulas (2.68) and (2.69) do not work. To find the distribution with a nonmonotonic transformation, we need to split the real line into intervals where g is monotonic and then apply formula (2.69) for each interval as illustrated in the following example.

Example 2.63 *Squared normal distribution.* *Find the cdf and density of* X^2 *where* $X \sim \mathcal{N}(\mu, \sigma^2)$.

Solution. First we find the cdf and then differentiate it to find the density. By definition, the cdf of $Y = X^2$ is

$$
\begin{aligned}
F_Y(y) &= \Pr(X^2 < y) = \Pr(|X| < \sqrt{y}) = \Pr(X < \sqrt{y}) - \Pr(X < -\sqrt{y}) \\
&= \Phi\left(\frac{\sqrt{y} - \mu}{\sigma}\right) - \Phi\left(\frac{-\sqrt{y} - \mu}{\sigma}\right), \quad y \geq 0,
\end{aligned}
$$

where Φ is the standard normal cdf. Differentiate with respect to y to find the density using the fact that $\Phi' = \phi$:

$$
f_Y(y) = \frac{1}{2\sigma\sqrt{y}} \phi\left(\frac{\sqrt{y} - \mu}{\sigma}\right) + \frac{1}{2\sigma\sqrt{y}} \phi\left(\frac{\sqrt{y} + \mu}{\sigma}\right), \tag{2.70}
$$

since ϕ is an even function.

Example 2.64 *Two methods to compute the mean.* *Let* $g(x)$ *be an increasing function and* X *have density* $f(x)$. *There are two methods to derive the expected value* $E(g(X))$: *(1) via integration,* $\int_{-\infty}^{\infty} g(x)f(x)dx$, *or (2) find the density of* $Y = g(X)$ *using formula (2.69) and then derive* $E(Y) = \int_{g(-\infty)}^{g(\infty)} y f_Y(y) dy$. *Show that both methods give the same result.*

Solution. In method (2), we explicitly write

$$
E(Y) = \int_{g(-\infty)}^{g(\infty)} y(g^{-1}(y))' f_X(g^{-1}(y)) dy. \tag{2.71}
$$

Make the change of variable $x = g^{-1}(y)$. Then, the limits of integration become $-\infty$ and ∞, and this integral becomes $\int_{-\infty}^{\infty} g(x)f(x)dx$. The two methods are equivalent although the first method is more general because it works even when function g is not monotonic.

Example 2.65 *Normalizing transformation.* *Is it true that upon a nonlinear transformation, any continuous distribution can be transformed into a normal distribution?*

Solution. Yes. The existence of a normalizing transformation for a continuous random variable with density $p(x)$ follows from the theory of ordinary differential equations (ODE). Indeed, let $g(\cdot)$ be an increasing inverse transformation function and ϕ be the density of the standard normal distribution. If g is the normalizing transformation, then it satisfies the functional equation $\phi(x) = g'(x)p(g(x))$. This means that g obeys the ODE

$$g'(x) = \frac{\phi(x)}{p(g(x))}, \qquad (2.72)$$

which is a first-order nonlinear ODE. As follows from Picard's existence theorem (Forsyth [51]), a normalizing monotonic transformation exists for any continuous random variable. Moreover, this transformation is unique. After the ODE is solved, we need to find the inverse function g^{-1}, which transforms the original continuous variable the into normal distribution.

Example 2.66 *Normalizing transformation of the uniform distribution. Find the normalizing transformation of the uniform distribution on $(0,1)$.*

Solution. Symbolically, we express this as $g^{-1} : \mathcal{R} \to \mathcal{N}$. Therefore, the inverse transformation must satisfy the following properties: (a) g is an increasing function on $-\infty < x < \infty$. (b) It takes values from $(0,1)$. (c) It obeys the ODE (2.72). Since the uniform density is 1 we have $p(g(x)) = 1$, that is, $g'(x) = \phi(x)$, and therefore $g(x) = \int_{-\infty}^{x} \phi(t)dt = \Phi(x)$. The transformation that makes a uniform distribution normal is Φ^{-1}, the inverse cdf. This result will be used later in Section 2.13.

Example 2.67 *Normalizing transformation of the exponential distribution. Find the normalizing transformation of the exponential distribution with the rate parameter λ. Write R code to check the analytical derivation via simulations.*

Solution. An increasing inverse transformation function, g^{-1}, is defined on $x > 0$ with values from $(-\infty, \infty)$. Function g is obtained from the ODE $g'(x) = \lambda^{-1}\phi(x)e^{-\lambda g(x)}$ with the solution $g(x) = -\lambda^{-1}\ln(1-\Phi(x))$. It is easy to check that the inverse transformation function takes the form $g^{-1}(x) = \Phi^{-1}(1-e^{-\lambda x})$, $x > 0$. This means that if X has an exponential distribution with parameter λ, then the distribution of $\Phi^{-1}(1-e^{-\lambda X})$ is standard normal. The R function **normex** confirms this statement by (i) generating a large number of exponentially distributed random values, (ii) transforming them using function g^{-1}, and (iii) plotting the empirical cdf of the transformed values and the standard normal cdf on the same graph. The cdfs practically coincide.   □

The fact that any continuous distribution can be transformed into a normal distribution is useful for confidence interval and hypothesis testing to be studied in Chapter 7. Many statistical techniques are developed for normal distribution. Therefore, to comply with the normality assumption, we transform the distribution to normal, build a confidence interval, and then transform back to get the confidence interval on the original scale with the same confidence level. See Chapter 7.

2.12.1   The delta method

The delta method approximates the variance of a nonlinear function of a random variable. This method only requires evaluating the first derivative – this simplicity explains its popularity. Let $Y = g(X)$ and we want to approximate σ_Y^2. If the density of X is $f(x)$, we can compute the exact variance using integration,

$$\sigma_Y^2 = \int_{-\infty}^{\infty} g^2(x)f(x)dx - \left(\int_{-\infty}^{\infty} g(x)f(x)dx \right)^2 .$$

To approximate σ_Y^2, we use the first-order Taylor series expansion around $\mu = E(X)$:

$$g(X) - g(\mu) \simeq (X - \mu)g'(\mu).$$

Taking the square of both sides and the expectation, we obtain the delta method approximation of the variance

$$\sigma_Y^2 \simeq (g'(\mu))^2 \sigma_X^2, \tag{2.73}$$

or, in terms of the SD,

$$\sigma_Y \simeq |g'(\mu)| \, \sigma_X. \tag{2.74}$$

An advantage of the delta method is that this approximation is valid for any distribution of X.

Many real-life problems are formulated in ambiguous language. The statistician should interpret the wording, make reasonable assumptions and suggest a statistical model, as in the example below.

Example 2.68 *Error of the area of the circle. The radius of the circle is measured with 1% error. What percentage measurement error does it imply for the area?*

Solution. The formulation of the problem lacks clarity: we need to explain how to interpret "measured with 1% error." Here is one of the possible interpretations: Let the true radius be ρ. The measured radius, r is a random variable with mean ρ and SD $= \rho/100$. The area of the circle is estimated as πr^2. Thus, the percentage measurement error for the area is $\text{SD}(\pi r^2)/(\pi \rho^2) \times 100$. Since $\sigma_r = \rho/100$, by formula (2.74), we approximate $\text{SD}(\pi r^2) \simeq (2\pi\rho)\sigma_r = 0.02\pi\rho^2$. Thus, a 1% measurement error in the radius implies approximately a 2% measurement for the area. Although imprecise, this result applies to a circle of any radius regardless of the distribution of the measurement error. If the distribution were known, the variance could be found precisely, but the beauty of the delta method is that it produces a simple answer.

Problems

1. Demonstrate identity (2.71) for the random variable $Y = X^k$, where X is an exponentially distributed random variable and $k > 0$.

2. Let a random variable be defined as $Y = g(X)$ where g is a strictly increasing function. (a) Prove that the median of Y is $g(m_X)$ where m_X is the median of X. (b) Does the same statement hold for the mode, namely, $M_Y = g(M_X)$, where M_Y and M_X are modes of Y and X, respectively, assuming that the distribution of X is unimodal? (c) Is the mode of the lognormal distribution e^μ?

3. Find the cdf and density of the quadratic transformation of a uniform distribution on the interval $[-1, 1]$. Verify the answer by plotting the simulation-derived cdf.

4. Find the expected value of the random variable with the density specified by (2.70). Explain the result.

5. Find a function that transforms a normally distributed random variable with mean μ and variance σ^2 into a random variable (a) uniformly distributed on $(0, 1)$ and (b) exponentially distributed with given λ. Verify the answer by plotting the simulation-derived cdf.

6. Define a hyperbolic distribution of a random variable X with the density reciprocal of x on (a, b), where $a < b$ are positive numbers. (a) Prove that $\log X$ has a uniform distribution. (b) Prove that an exponential transformation of a uniform distribution yields a hyperbolic distribution.

7. Redo Example 2.64 when function $g(x)$ is increasing on $(-\infty, x_0)$ and decreasing on (x_0, ∞) and apply this result to Example 2.63.

8. Find an alternative solution in Example 2.67 by transforming the exponential distribution to $\mathcal{R}(0, 1)$.

9. Is it true that the expected value of the random variable $Y = f(X)$ is greater than $f(E(X))$ if f is a convex function? [Hint: Use the Jensen's inequality (2.12).]

10. Is it true that the variance approximated by the delta method is smaller than the original if f is a convex function? [Hint: Start with the quadratic function applied to the uniform and normal distributions.]

11. The height of an object (such as tree) is estimated by measuring the horizontal distance from a point on the ground to the base of the object and the angle at which the top of the object is seen from that point. The distance is measured with 1% error. What percentage error does this imply assuming that the angle is measured precisely?

12. The radius of the sphere is measured with 1% error. What percentage measurement error does this imply for the volume?

13. Approximate the variance of $1/X$ by the delta method. Compute the variance of $1/X$ exactly for $X \sim \mathcal{R}(1, 1+\delta)$, where $\delta > 0$. Plot the two variances as a function of δ and draw conclusions.

2.13   Random number generation

In many statistical packages, including R, a sequence of random numbers is generated numerically by truncation of the last digit starting from the *seed number*. To specify the seed number, we use function `set.seed()`. For example, `set.seed(8)` means that the seed number is 8. It does not mean that the first random number is 8. Sometimes, such as for debugging purposes, it is advantageous to use the same random numbers. In this case it is convenient to have the seed number as an argument of your R function. Also it is important to analyze how the choice of the seed number affects simulations. By default in R, without using the `set.seed` function, different random numbers will be used every time the program is invoked. However, if you quit R and start over, the seed number will be the same.

Every built-in distribution in R has a random number generator (the function name starts with `r`). But sometimes, we deal with a distribution that is not on the list. There is an elegant way to generate continuous random numbers with an arbitrary distribution using the inverse cdf.

Theorem 2.69 *Inverse cdf random number generation. Let F be a strictly increasing continuous distribution function, and let $X \sim \mathcal{R}(0,1)$. Then*

$$Y = F^{-1}(X)$$

has cdf F, where F^{-1} is the inverse cdf.

Proof. Let y be any positive number less than 1. Since F is a strictly increasing continuous function the solution of the equation $y = F(x)$ exists and is unique. We want to prove that $\Pr(Y \leq y) = F(y)$. We use the following facts on the inverse of a strictly increasing continuous function: (a) the inverse function exists, (b) the inverse function is strictly increasing, and $F^{-1}(X) \leq y$ is equivalent to $X \leq F(y)$. Using these facts, we obtain

$$\Pr(Y \leq y) = \Pr(F^{-1}(X) \leq y) = \Pr(X \leq F(y)).$$

But X is a uniform random variable on [0,1]. That is, $\Pr(X \leq F(y)) = F(y)$. This proves the theorem. □

Note that the above method does not work for a discrete random variable because the equation $X = F(Y)$ may have no solution. A modification of this method is discussed in Example 2.76.

Example 2.70 *Generating a logistic distribution.* Write an R program that generates random numbers according to the cdf of a logistic distribution $e^{\alpha+\beta x}(1+e^{\alpha+\beta x})^{-1}$, where α and β are positive parameters. Test the method by plotting the empirical and theoretical cdfs on the same plot.

Solution. The inverse cdf is the solution of the equation $y = e^{\alpha+\beta x}(1+e^{\alpha+\beta x})^{-1}$, i.e. $x = -(\alpha + \ln(y^{-1} - 1))/\beta$. The R code is found in file **expr**. □

The following result is reverse to Theorem 2.69.

Theorem 2.71 *Let random variable X have cdf F. (a) If X is a continuous random variable then $Y = F(X)$ has a uniform distribution on $(0,1)$. (b) If X is a discrete random variable and takes values $\{x_i, i = 1, 2, ..., n\}$, then $Y = F(X)$ takes values $\{F(x_i), i = 1, 2, ..., n\}$ with $\Pr(Y = F(x_i)) = 1/n$, the uniform distribution on $\{x_i, i = 1, 2, ..., n\}$.*

Proof. (a) By definition, the cdf of Y is

$$\Pr(F(X) \leq y) = \Pr(X \leq F^{-1}(y)) = F(F^{-1}(y)) = y,$$

which is the cdf of the uniform distribution where $0 < y < 1$. (b) The cdf of Y evaluated at $F(x)$ is the proportion of $x_i \leq x$, which equals i/n. This means that $\Pr(Y = F(x_i)) = 1/n$. □

To calculate F^{-1}, we solve the equation $F(y) = x$ for each $0 < x < 1$. Sometimes, there is no closed-form solution; then iterations are required as in the example below.

Example 2.72 *Normal/Gaussian mixture random number generation.* *Generate n iid observations from a mixture of two normal distributions specified by $F(x) = p\Phi(x - \mu_1) + (1 - p)\Phi(x - \mu_2)$, where $0 < p < 1$. Plot the empirical and theoretical cdfs to verify that they match.*

Solution. The problem of generation boils down to solving

$$p\Phi(x - \mu_1) + (1 - p)\Phi(x - \mu_2) - Y = 0, \tag{2.75}$$

where $Y \sim \mathcal{R}(0,1)$. Using the fact that $\Phi' = \phi$. Newton's iterations take the form

$$x_{k+1} = x_k - \frac{p\Phi(x - \mu_1) + (1 - p)\Phi(x - \mu_2) - Y}{p\phi(x - \mu_1) + (1 - p)\phi(x - \mu_2)}$$

starting from $x_0 = p\left(\Phi^{-1}(Y) + \mu_1\right) + (1-p)\left(\Phi^{-1}(Y) + \mu_2\right)$. Typically, iterations converge quickly – see function **rugf2**. A distinctive feature of the code is that iterations are carried out in the vectorized fashion by storing all n observations in array **x**. We set the maximum iterations at 10 and the convergence tolerance criterion at 0.001 as default values. An alternative method is described in Example 2.73.

The uniroot function

R has a built-in function `uniroot()` to solve nonlinear equation of the form $f(x) = 0$. Finding a root of a function is a frequent task in statistics. The user must specify the function following the same rule as for `integrate` (see Section 2.9.2) and an interval within which the solution is sought. The function must have different signs at the ends of the interval, so that a continuous function will have at least one root.

Problem 1. Find the solution to the equation $e^{a+bx}/(1+e^{a+bx}) = p$, where a, b, and p are user-defined and $0 < p < 1$.

```
expf=function(x,a,b,p) exp(a+b*x)/(1+exp(a+b*x))-p
uniroot(f=expf,interval=c(-10,10),a=-1,b=2,p=.2)$root
```

Note that specification of the interval may require some additional computation or analytic research to ensure that the function values have different signs at the ends of the interval.

Problem 2. Find the solution to equation (2.75).

```
gnMixN=function(x,p,mu1,mu2,Y) p*pnorm(x-mu1)+(1-p)* pnorm(x-mu2)-Y
X=uniroot(f=gnMixN,interval=c(-3,6),p=.2,mu1=1,mu=2,Y=.3)$root
```

For the mixture distribution specified in Example 2.72, we can generate random numbers much more easily, as follows from the following example.

Example 2.73 *Gaussian mixture random number generation continued.* *To generate n random numbers from the Gaussian mixture distribution, generate np numbers from $\mathcal{N}(\mu_1, 1)$ and $n(1-p)$ numbers from $\mathcal{N}(\mu_2, 1)$, and combine the two samples (the sample can be reshuffled afterword). Write R code with arguments* `mu1,mu2,p,n` *and confirm the empirical and theoretical cdfs match.*

Solution. See function `genmixN`. It has arguments `mu1,mu2,p,N` with default values `1,2,.3,1000`, respectively. There is a close match between the cdfs. □

Many statistical models and tests are built under an assumption of the normal distribution. Below, we show that any continuous distribution can be transformed into a normal distribution by a suitable transformation. The proof relies on one simple fact:

$$\Phi^{-1}_{\mu,\sigma^2} = \mu + \sigma\Phi^{-1}, \qquad (2.76)$$

where Φ^{-1}_{μ,σ^2} and Φ^{-1} are inverse cdfs for $\mathcal{N}(\mu, \sigma^2)$ and $\mathcal{N}(0, 1)$, respectively. In words, the inverse normal cdf can be expressed as a linear function of the inverse standard normal cdf.

Theorem 2.74 *Let X be any random variable with a strictly increasing continuous cdf, F. Then the random variable*

$$Z = \Phi^{-1}_{\mu,\sigma^2}(F(X))$$

has a normal distribution with mean μ and variance σ^2. The above transformation, denoted $\Phi^{-1} \circ F$, is referred to as the normalizing transformation.

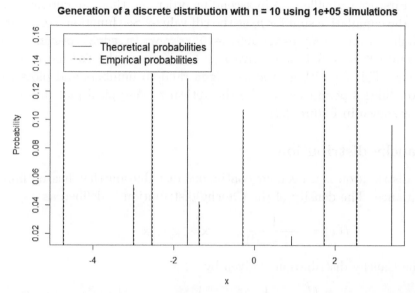

Figure 2.27: *Generation of random numbers from a discrete distribution using inverse cdf. This plot is produced by the R function* `discr.gen` *from Example 2.76.*

Proof. Let F_Z be the cdf of Z given above. We have

$$F_Z(z) = \Pr(Z \leq z) = \Pr\left(\Phi_{\mu,\sigma^2}^{-1}(F(X)) \leq z\right) = \Pr\left(F(X) \leq \Phi_{\mu,\sigma^2}(z)\right).$$

As we did in the proof of Theorem 2.69, we assert that $F(X) \sim \mathcal{R}(0,1)$. Hence $F_Z(z) = \Phi_{\mu,\sigma^2}(z)$. This completes the proof.   □

This result implies that we can comply with the normal assumption with a normalizing transformation that always exists.

Example 2.75 *Exponential via normal distribution. Find the transformation that makes an exponentially distributed random variable a standard normal variable.*

Solution. The standard normal random variable is $\Phi^{-1}(1-e^{-\lambda X})$, where X has an exponential distribution with rate λ.   □

Recall that the `sample` function, introduced in Section 1.8 generates random numbers from a general discrete distribution. The following example illustrates an alternative way using the inverse cdf.

Example 2.76 *Generating a discrete distribution. Apply the inverse cdf to a discrete distribution defined as $\{(x_i, p_i), i = 1, 2, ..., n\}$ and test the algorithm via simulations. Display the theoretical and empirical probabilities side by side.*

Solution. First we reshuffle $\{x_i\}$ in ascending order and then use `cumsum()` in R to compute the cumulative sum for p_i as the cdf values. See function `discr.gen` with the default arguments `Ngen=100000,n=10,x=rnorm(n,sd=3),p=runif(n)`. Then we generate $Y \sim \mathcal{R}(0,1)$ and solve $F(x_i) = Y_i$ using the `for` loop over n by conditioning. This algorithm produces `Ngen` random numbers and plots the empirical probabilities `pemp` close to the theoretical `p`. A typical plot produced by this code is shown in Figure 2.27.

2.13.1   Cauchy distribution

The Cauchy distribution is a peculiar continuous distribution that has no finite mean and variance. The density of the Cauchy distribution is defined as

$$f(x) = \frac{1}{\pi(1+x^2)}, \quad -\infty < x < \infty.$$

The cdf of the Cauchy distribution is given by

$$F(x) = \int_{-\infty}^{x} \frac{1}{\pi(1+t^2)} dt = \frac{1}{\pi}\arctan x + \frac{1}{2}. \tag{2.77}$$

A surprising fact about the Cauchy distribution is that the average of Cauchy distributions is the same Cauchy distribution. This is surprising because the SD of the average is supposed to be square root n smaller. Indeed, the average of n iid random variables with a finite variance σ^2 converges to zero because $\sigma^2/n \to 0$ when $n \to \infty$, but for the Cauchy distribution this rule does not work because the SD is infinite.

Example 2.77 *Average of Cauchy distributions. Prove by simulations that the average of any number of iid Cauchy distributed random variables has the same distribution.*

Solution. To generate random numbers from the Cauchy distribution, we invert the cdf: $U = \tan(\pi(X-0.5))$, where $X \sim \mathcal{R}(0,1)$. The Cauchy distribution is a built-in function R. Alternatively, one could use `rcauchy()` to generate Cauchy random numbers. We use the `rowMeans` function to compute `nSim` means in each row – see the R function `meanC`. When plotting the cdfs, we may use option `xlim` because otherwise the plot will be squashed due to possible outliers. The average of any number of Cauchy-distributed random variables has the same cdf (2.77), and therefore the formula $\mathrm{var}(\overline{X}) = \sigma^2/n$ does not apply here.

Problems

1. Use an inverse cdf to generate observations from the exponential distribution with the rate parameter λ. Use `lambda` as an argument of your function. Check your generation algorithm by comparing the empirical and theoretical cdfs.

2. Confirm Theorem 2.71 by simulations for the following distributions: (i) normal, (ii) binomial, (iii) Poisson. [Hint: Use `job` to parse the tasks and use arguments of the R function to pass the parameters of the distribution. Match the empirical and theoretical cdfs.]

3. (a) Use simulations to check whether the cdf of the random variable defined as $\Phi(X + c)$, where $X \sim \mathcal{N}(0, 1)$ with c as a constant, has a uniform distribution. (b) Prove the result using Theorem 2.71.

4. Generalize the algorithm for generating random numbers (a) from the Gaussian mixture of k normal distributions as in Example 2.72 and (b) as in Example 2.73. Test the generation algorithms through cdf. Which algorithm is faster?

5. Generalize the algorithm for generating of random numbers from the Gaussian mixture when normal components have difference variances σ_1^2 and σ_2^2. Modify function `rugf2` and check the algorithm through cdfs.

6. In Example 2.73, (a) use `qnorm` (not cdf) to confirm that generated random numbers follow the desired mixture distribution, (b) generalize to the case when variances are not equal, (c) generate random numbers that follow the distribution $\sum_{i=1}^{n} \lambda_i Z_i^2$, where $\lambda_i > 0$ and Z_i are iid standard normal random variables; test your algorithm for the case when $\lambda_i = \lambda$ using the chi-square distribution (`pchisq`).

7. Use function `discr.gen` to generate random numbers from a binomial distribution with n and p as arguments. Check the code by plotting empirical and theoretical probabilities side by side as in Figure 2.27.

8. (a) Find the normalizing transformation for the Cauchy distribution. (b) Use simulations to check the answer by plotting the empirical and theoretical cdfs on the same graph.

9. The same as above, but apply the normalizing transformation to the uniform distribution on $(0, 2)$.

10. Is it true that the average of generalized Cauchy distributions defined by the density $(\pi\sigma)^{-1}[1 + ((x - \mu)/\sigma)^2]^{-1}$ has the same distribution, where μ and $\sigma > 0$ are parameters? Modify function `meanC` to check your answer via simulations (use μ and σ as arguments).

11. Box–Muller transform: use simulations to verify that if $U_1, U_2 \overset{\text{iid}}{\sim} \mathcal{R}(0, 1)$, then $\sqrt{-2 \ln U_1} \times \cos(2\pi U_2) \sim \mathcal{N}(0, 1)$ and $\sqrt{-2 \ln U_1} \sin(2\pi U_2) \sim \mathcal{N}(0, 1)$.

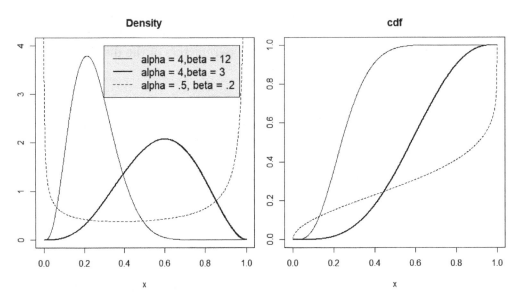

Figure 2.28: *The density and cdf for three beta distributions with ∩ shape density when $\alpha > 1$ and $\beta > 1$, and ∪ shape when $\alpha < 1$ and $\beta < 1$.*

2.14   Beta distribution

The beta distribution specifies the distribution of a random variable on the interval $(0, 1)$ with the pdf given by

$$f(x; \alpha, \beta) = \frac{1}{B(\alpha, \beta)} x^{\alpha-1}(1 - x)^{\beta-1}, \quad 0 < x < 1, \alpha > 0, \beta > 0, \qquad (2.78)$$

where α and β are called the *shape* parameters. We will use notation $X \sim \mathcal{B}(\alpha, \beta)$ to indicate that X has the beta distribution. The beta distribution is a flexible distribution on $(0, 1)$ with special cases as uniform or triangular distributions. In particular, this distribution will be used later to study the distribution of the correlation coefficient in Section 7.7. The beta distribution is a built-in function in R; see Table 2.3. $B(\alpha, \beta)$ is the beta function and is expressed through the Γ function as $B(\alpha, \beta) = \Gamma(\alpha)\Gamma(\beta)/\Gamma(\alpha + \beta)$.

An obvious special case is when $\alpha = \beta = 1$, the uniform distribution on $(0, 1)$. This case is a threshold, which divides the family of beta distributions into two classes: those for which $\alpha < 1$ and $\beta < 1$ with the density having ∪ shape and those for which $\alpha > 1$ and $\beta > 1$ having ∩ shape density; see Figure 2.28.

The moments of the beta distribution are easy to derive due to the relationship between the gamma function evaluated at α and $\alpha + 1$. For example, using

$\Gamma(\alpha + 1) = \alpha\Gamma(\alpha)$ and $\Gamma(\alpha + 2) = \alpha(\alpha + 1)\Gamma(\alpha)$, the mean is found as

$$
\begin{aligned}
E(X) &= \frac{1}{B(\alpha, \beta)} \int_0^1 x x^{\alpha-1}(1-x)^{\beta-1}dx = \frac{1}{B(\alpha, \beta)} \int_0^1 x^{\alpha}(1-x)^{\beta-1}dx \\
&= \frac{B(\alpha+1, \beta)}{B(\alpha, \beta)} = \frac{\Gamma(\alpha+\beta)\Gamma(\alpha+1)}{\Gamma(\alpha)\Gamma(\alpha+\beta+1)} = \frac{\alpha}{\alpha+\beta}.
\end{aligned}
$$

Analogously, we can find the variance of the beta distribution as

$$
\text{var}(X) = \frac{\alpha\beta}{(\alpha + \beta + 1)(\alpha + \beta)^2}.
$$

We can extend the beta distribution to the interval (a, b) with the density proportional to $(x - a)^{\alpha-1}(b - x)^{\beta-1}$.

The binomial and the beta cdfs are closely related:

$$
\int_0^p f(x; k, n - k + 1)dx = 1 - \sum_{m=0}^{k-1} \binom{n}{m} p^m (1 - p)^{n-m}
$$

or, in R terms, `pbeta(p,k,n-k+1)=1-pbinom(k-1,n,p)`. This fact will be used in Section 7.6.3 for to test the binomial proportion.

Table 2.3. Four beta distribution functions in R (`shape1`=α, `shape2`=β)

R function	Returns/formula	Explanation
`dbeta(x,shape1,shape2)`	density (2.78)	x=array x
`pbeta(q,shape1,shape2)`	cdf, $\int_0^q f(x; \alpha, \beta)dx$	cdf at $x = $ q
`rbeta(n,shape1,shape2)`	random numbers	n=sample size
`qbeta(p,shape1,shape2)`	quantile, inverse cdf	p=probability

Problems

1. Show that the mode of the beta distribution is $(\alpha - 1)/(\alpha + \beta - 2)$ if $\alpha > 1$ and $\beta > 1$. Show that the beta distribution is symmetric around 0.5 if $\alpha = \beta$.

2. Express the parameters α and β in terms of mean and variance. Check your answer with simulation: use α and β as arguments, generate a large number of beta-distributed observations, and show that the empirical and theoretical means and variances match.

3. Prove that if $0 < \alpha < 1$, then $\lim_{x \to 0} f(x; \alpha, \beta) = \infty$, and if $0 < \beta < 1$, then $\lim_{x \to 1} f(x; \alpha, \beta) = \infty$.

4. Demonstrate by simulation that if X and Y are gamma-distributed independent random variables, then $X/(X+Y)$ has a beta distribution.

5. Apply Chebyshev's inequality to the beta distribution and plot the probabilities as in Figure 2.17.

6. Find the upper bound for the variance of the beta distribution.

7. Generalize the beta distribution to the interval $0 < x < A$. Derive the cdf, density, and random number generation and test them through simulations using A as the argument of your R function.

2.15   Entropy

The concept of entropy, well known in physics, was developed for engineering purposes by Claude Shannon in 1948 to measure the amount of information in a communication. For example, if a binary message consists of all 0s (or all 1s), the information (entropy) is zero. On the other hand, intuition says that if the proportion of appearance of 0s and 1s in the message is the same, the entropy is maximum (we will prove this statement below). Entropy is used in statistical thermodynamics: the second law of thermodynamics tells that the entropy of a closed system is increasing and eventually leads to chaos.

For a discrete distribution on $\{x_1, ..., x_n\}$ with $p_i = \Pr(X = x_i)$, the Shannon entropy is defined as

$$\mathcal{E} = -\sum_{i=1}^{n} p_i \ln p_i, \tag{2.79}$$

and for a continuous distribution with density $f(x)$, it is defined as

$$\mathcal{E} = -\int_{-\infty}^{\infty} f(x) \ln f(x) dx.$$

These two definitions can be combined through expectation,

$$\mathcal{E} = -E(\ln f(X)).$$

For a discrete random variable, the entropy is always nonnegative because $\ln p_i \leq 0$. But for a continuous density, $\mathcal{E}$ may be negative. For example, it is negative for a normal distribution with a small variance; see Table 2.4. Note that we may omit outcomes for which $p_i = 0$ because $\lim_{p \to 0} p \ln p = 0$. Likewise, the integration above is over the density support, i.e. where $f(x) > 0$.

Let us return to the example about the binary message and prove that, indeed, the entropy reaches its maximum when the proportion of 0s and 1s is the same. Let X be a Bernoulli random variable with $\Pr(X = 1) = p$ and $\Pr(X = 0) = 1-p$. Then the entropy in X is

$$\mathcal{E} = -\left[p \ln p + (1-p) \ln(1-p)\right].$$

The entropy in the binary message/sequence consisting of n iid Bernoulli outcomes is $n\mathcal{E}$. Consider $\mathcal{E}$ as a function of $p \in [0,1]$ and find its maximum by differentiation, $d\mathcal{E}/dp = -\ln p - 1 + \ln(1-p) + 1 = 0$. That is, the maximum of $\mathcal{E}$ occurs when $\ln[p/(1-p)] = 0$ or when $p = \frac{1}{2}$. In words, maximum entropy in the binary message is when the proportion of 0s and 1s is the same. In the next example we generalize this statement.

Table 2.4. Continuous distributions with maximum entropy

Support	Distribution	Entropy
1. Finite, $a < X < b$	Uniform	$\ln(b-a)$
2. Positive, $X > 0$	Exponential	$1 - \ln \lambda$
3. Infinite, $-\infty < X < \infty$	Normal/Gaussian	$\frac{1}{2}\left(1 + \ln(2\pi\sigma^2)\right)$

Example 2.78 *Maximum entropy.* *Prove that the maximum entropy for a discrete distribution is reached when all probabilities are the same, $p_i = 1/n$, i.e. $\mathcal{E}_{\max} = \ln n$.*

Solution. Consider maximization of $\mathcal{E} = \mathcal{E}(p_1, ..., p_n)$ defined by (2.79) under constraint $\sum_{i=1}^{n} p_i = 1$ using the Lagrange multiplier technique:

$$\mathcal{L}(p_1, ..., p_n, \lambda) = -\sum_{i=1}^{n} p_i \ln p_i - \lambda\left(\sum_{i=1}^{n} p_i - 1\right).$$

Differentiating with respect to p_i, we obtain a necessary condition for the maximum, $-\ln p_i - 1 - \lambda = 0$ or $p_i = $ const. This can be true only if $p_i = 1/n$. □
This result says that the maximum entropy occurs when symbols in the message repeat with the same probability.

Example 2.79 *Entropy of an English text.* *Compute the entropy of an average English text and compare with Jack London's "Call of the Wild" novel.*

Solution. We view the text as a sequence of English letters; see Section 2.15. The probability of appearance of English letters is well documented. See the R function found english.r. The maximum entropy in a typical English text with 26 letters is $\ln 26 = 3.26$. We compute the efficiency of the English language with respect to the maximum entropy. The output of this function is English entropy=2.894416 Max entropy=3.258097 %English efficiency=88.83764. The R function frlJLET from Section 2.15 plots the distribution of letters in Jack London's novel (not shown). Using the frequency of letters from that novel, we obtain the entropy 2.892645, very close to the entropy of an average English text. Obviously, rare letters reduce the efficiency of the language. For example, letters

(or any other symbols in messages) with zero probability are redundant and can
be safely removed without loss of information.   □

This may be a surprise to some, but the distribution of digits in π and e looks
uniform. This means that π and e are maximum entropy irrational numbers.
Since digits of rational numbers repeat, they are not maximum entropy numbers.
The problem whether the distribution of digits of any irrational number is actually
uniform is open.

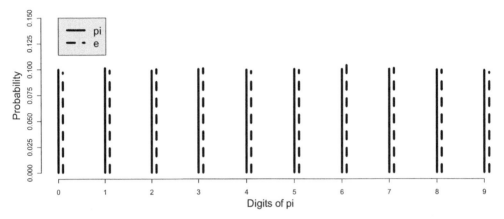

Figure 2.29: *The probability distribution of digits in π and e looks uniform.*

Example 2.80 π *and* e *digit distribution. File* `pi100000digits.csv` *con-*
tains first 100K digits of π *and file e10000digits.csv contains first 10K digits of*
e. *Plot the distribution of digits side by side using the* `plot` *command with option*
`type="h"` *and* `segments` *command.*

Solution. The code in function `pidistr` produces Figure 2.29. We make a
few remarks on the program: (i) `read.csv` command reads the specified file;
option `col.names=F` means that no column names are provided. (ii) Command
`pidig==(i-1)` returns a logical vector of the same length as `pidig` with component `T` if the element equals `i-1` and `F` otherwise; the `mean` command interprets
this logical vector as numeric with `T` replaced with 1 and `F` replaced with 0;
then it computes the mean. The result is the proportion of digits `i-1` in the
array `pidig`. (iii) We use `segments` to plot the probabilities; bars are shifted by
0.1 to avoid an overlap with the previously plotted bars. (iv) The axis labels are
specified by options `xlab` and `ylab`. In this code, we plot the labels using `mtext`
and therefore the labels for the axes are empty; `mtext` has an advantage that
you can easily control the size of labels using `cex` option and the distance from
the axis using the `line` option. (v) We use `gray` background for the legend; `0.9`
gives a nice gray hue (values closer to 1 yield light gray).

The distribution of digits in the two famous irrational numbers is close to
uniform: each digit occurs with the probability close 1/10. As follows from Ex-

```
> pidistr1010()
          0           1           2           3           4           5           6           7           8           9
0 0.009980200 0.010270205 0.009620192 0.009930199 0.009680194 0.010070201 0.010090202 0.010170203 0.010010200 0.010170203
1 0.010420208 0.010340207 0.009920198 0.010100202 0.010300206 0.009710194 0.009820196 0.010080202 0.010440209 0.010240205
2 0.009740195 0.010640213 0.009710194 0.009660193 0.009420188 0.010040201 0.010480210 0.010170203 0.009550191 0.009670193
3 0.009790196 0.009740195 0.009940199 0.010080202 0.010410208 0.010750215 0.009650193 0.010090202 0.009820196 0.009980200
4 0.010200204 0.010020200 0.009870197 0.010140203 0.009710194 0.009620192 0.010310206 0.010080202 0.009750195 0.010000200
5 0.009660193 0.009660193 0.010290206 0.010310206 0.010200204 0.010150203 0.010170203 0.010040201 0.010080202 0.009710194
6 0.010540211 0.010120202 0.010480210 0.009970199 0.010190204 0.009980200 0.009530191 0.009720194 0.009490190 0.010250205
7 0.009480190 0.010460209 0.009810196 0.010310206 0.009450189 0.009610192 0.011000220 0.010120202 0.010400208 0.009610192
8 0.010110202 0.009960199 0.009920198 0.009460189 0.010010200 0.010200204 0.009700194 0.010440209 0.010270205 0.009710194
9 0.010070201 0.010160203 0.009520190 0.010290206 0.010330207 0.010140203 0.009520190 0.009340187 0.009970199 0.009680194
```

Figure 2.30: *The output of the* `pidistr1010` *program. The probability that digit i follows digit j is the same for all i and j.*

ample 2.78, the uniform distribution of probabilities implies maximum entropy – a rather remarkable result. The next result demonstrates that the occurrence of neighboring digits is random.

Example 2.81 π **digits are random.** *Compute the* 10×10 *matrix of probabilities that digit j follows digit i using the file* ***pi100000digits.csv***.

Solution. The same idea is used as in the previous example, but (a) a double for loop is used, and (b) the array of digits is shifted by 1 to compare the current digit with the previous one; see the code `pidistr1010` with the output displayed in Figure 2.30. All 100 probabilities are very close to 0.01. This demonstrates that the digits in π are random and their occurrence is independent. □

More statistical analysis of digits of π can be found at http://blogs.sas.com/content/iml/2015/03/12/digits-of-pi.html.

Now we turn our attention to continuous distributions. There are continuous distributions with maximum entropy for varying support: finite, semi-infinite, and infinite; see Table 2.4. For example, for the uniform distribution on (a, b) we have

$$\mathcal{E} = -\int_a^b \frac{1}{b-a} \ln \frac{1}{b-a} dx = \frac{\ln(b-a)}{b-a}(b-a) = \ln(b-a).$$

Note that for the exponential distribution, the entropy is negative if $\lambda > e$. The possibility of being negative is a drawback of entropy.

These distributions are frequently used in Bayesian statistics to specify non-informative priors.

Problems

1. Explain why when computing the entropy of a continuous distribution integration over $(-\infty, \infty)$ may be replaced with $\{x : f(x) > 0\}$.

2. Prove that the entropy does not change upon a shift. That is, the entropy of $f(x)$ and $f(x + c)$ is the same. Show how the entropy of the original random variable and the linearly transformed $ax + c$ are related.

3. Verify that the entropy of Jack London's novel is 2.892645 by modifying function `frlJL` from Section 2.15.

4. Derive the entropy in Table 2.4 for exponential and normal distributions.

5. Find the file with first digits of $\sqrt{2}$ on the Internet, and add to program `pidistr` and Figure 2.29. Is it uniform as well?

6. Modify function `pidistr1010` from Example 2.81 to compute the 10×10 matrix of probabilities that digit j follows digit i for k locations apart. Print out matrices for $k = 2$ and $k = 7$. [Hint: Use k as an argument of your function.]

2.16 Benford's law: the distribution of the first digit

Almost 150 years ago, the American astronomer Newcomb [76] noticed that first pages of a reference book containing a table of logarithms are more worn out than others. He concluded that 1 occurs more frequently as the first digit than others. Specifically, based on the assumption that "every part of a table of anti-logarithms is entered with equal frequency," he provided the probabilities as we know today. Later Benford [8] rediscovered this phenomenon and suggested the probability distribution for the first significant digit, D, of a positive number written in a simple form as

$$\Pr(D = d) = \log_{10}(1 + 1/d), \quad d = 1, 2, ..., 9. \tag{2.80}$$

This formula for the probability of the first digit will be referred to as Benford's law. Sometimes it is called the *first significant digit law,* meaning that, if the number is positive but less than 1 and all preceding zeros are omitted, D is the first nonzero digit. The fact that the common, not natural, logarithm appears in (2.80) is not surprising because our concern is the digit in the base of 10. It is easy to show that the sum of probabilities adds to 1:

$$\sum_{d=1}^{9} \Pr(D = d) = \sum_{d=1}^{9} \log_{10}(d + 1) - \sum_{d=1}^{9} \log_{10}(d)$$

$$= \sum_{d=2}^{9} \log_{10}(d) + \log_{10}(10) - \log_{10}(1) - \sum_{d=2}^{9} \log_{10}(d)$$

$$= \log_{10}(10) - \log_{10}(1) = 1.$$

As follows from Benford's law, the probability that a number starts with 1 has the highest probability, and the probability that a number starts with 9 is the lowest. The probabilities computed in R by issuing `log10(1+1/1:9)` are shown below

0.30103000 0.17609126 0.12493874 0.09691001 0.07918125 0.06694679
0.05799195 0.05115252 0.04575749

According to Benford's law, 30% of all numbers start with 1, about 18%, start with 2, etc. Many first-digit numbers we deal with follow Benford's law defined by the probability distribution (2.80). For example, Benford in his famous paper, presented the table with twenty data sets of various origins, such as area of rivers, population, atomic weight, cost, addresses, death rate, etc., with distribution of the first digit following the law with little deviation. Surprisingly, these diverse data combined will also follow Benford's law! Many other facts and applications of the Benford's law can be found in a recently published book by Miller [74].

One of the most interesting applications of Benford's law is finance. Consider, the problem of auditing income tax returns on the national level. According to Benford's law, the most suspicious returns would report incomes starting with 8 or 9. It is plausible that the person who makes up his/her income, unfamiliar with Benford's law, wants to have an "average" return and assumes that the first digit has a uniform distribution with equal probability 1/9, which is not true. See an example below where we analyze the data on *ok*, *fraudulent*, and *unknown* transactions. Benford's law was applied to unveil possible fraud in the 2009 Iranian presidential election although not without some controversy (see the Wikipedia article "Benford's law").

Although Benford's law may work for some data sets, it is not universal, as follows from examples below.

Example 2.82 *Distribution of the first digit of birthdays.* *Derive the distribution of the first digits of birthdays in the month. Make necessary assumptions.*

Solution. Assume that birthdays are uniformly distributed within the months and there are 30 days in a month (just to simplify). The first digit of the birthday being 1 has probability 11/30 (birthdays on the day of the month $1, 10, 11, ..., 19$). Being 2 has the same probability, 11/30 (birthdays on the day of the months $2, 20, 21, 22, ..., 29$). Being 3 has probability 2/30 (birthdays on the day 3 and 30), and $4, ..., 9$ is 1/30. Although the probability of the first digit decreases, it does not follow Benford's law. □

Before advancing, we provide the formula for computation of the first significant digit of a positive number. The integer part of number X is denoted as $\lfloor X \rfloor$ and computed via the `floor` function, defined as the largest integer smaller than or equal to X. The fractional part is defined as $X - \lfloor X \rfloor$. For example, $\lfloor 1.8 \rfloor = 1, \lfloor 21 \rfloor = 21, \lfloor 5.2 \rfloor = 5$. Using this notation the first significant digit of a positive number X can be computed as

$$D = \left\lfloor \frac{X}{10^{\lfloor \log_{10} X \rfloor}} \right\rfloor . \tag{2.81}$$

In R, it is computed as `floor(X/10^floor(log10(X)))`. If $X \geq 1$, then D is the first digit of number X. If $0 < X < 1$, then D is the first nonzero digit (the first significant digit): if $X = 0.0326$, we get $D = 3$.

Although several authors have tried to derive Benford's law in rigorous form, it remains an open problem as to why Benford's law works for many diverse data sets. For example, in a recent paper, Goodman [41] writes: "Most proposed explanations for the Benford patterns relate to the properties of numbers themselves and various mathematical sequences. Nevertheless, even the widely acknowledged authority, the mathematician Hill [55], concedes that the law has not been precisely derived, nor is it understood why some Benford-suitable data sets still do not conform to it."

Below, we derive the distribution of the first digit of a random number X with cdf F starting from the distribution of the fractional part defined as $X - \lfloor X \rfloor$. Obviously, the fractional part is nonnegative and less than 1. The following fact expresses the distribution of the fractional part of a random variable X through its cdf.

Lemma 2.83 *Let a continuous random variable X have cdf $F(x)$. Then the fractional part has the following cdf:*

$$F_{X-\lfloor X \rfloor}(z) = \Pr\left(X - \lfloor X \rfloor \leq z\right) = \sum_{k=-\infty}^{\infty} \left(F(k+z) - F(k)\right), \quad 0 \leq z < 1.$$

(2.82)

Proof. The formula simply collects the probabilities $X - \lfloor X \rfloor \leq z$ over the set of all possible values of $\lfloor X \rfloor$, which is the set of all integers. $\qquad \square$

Now we are ready to formulate the main result.

Theorem 2.84 *First-digit distribution.* *Let random variable $X > 0$ have cdf $F(x)$. The probability distribution of the first significant digit is given by*

$$\Pr(D = d) = \sum_{k=-\infty}^{\infty} \left[F\left((d+1)10^k\right) - F(d10^k)\right], \quad d = 1, 2, ..., 9.$$

On the $\log_{10}$ scale, if $Y = \log_{10} X$ has cdf $H(y)$, the above probability can be expressed as

$$\Pr(D = d) = \sum_{k=-\infty}^{\infty} \left[H(k + \log_{10}(d+1)) - H(k + \log_{10}(d))\right].$$

Proof. Let d be any digit from 1 to 9. Using the formula for the first digit

(2.81) combined with the previously derived (2.82), we obtain

$$\Pr(D = d) = \Pr\left(\left\lfloor \frac{X}{10^{\lfloor \log_{10} X \rfloor}} \right\rfloor = d\right) = \Pr\left(d \leq \frac{X}{10^{\lfloor \log_{10} X \rfloor}} < d+1\right)$$

$$= \Pr\left(d10^{\lfloor \log_{10} X \rfloor} \leq X < (d+1)10^{\lfloor \log_{10} X \rfloor}\right)$$

$$= \sum_{k=-\infty}^{\infty} \left[F\left((d+1)10^k\right) - F(d10^k)\right].$$

The proof follows after taking the log:

$$\Pr(D = d)$$
$$= \Pr\left(\lfloor \log_{10} X \rfloor + \log_{10}(d) \leq \log_{10} X < \lfloor \log_{10} X \rfloor + \log_{10}(d+1)\right)$$
$$= \Pr\left(\log_{10}(d) \leq Y - \lfloor Y \rfloor < +\log_{10}(d+1)\right)$$
$$= \sum_{k=-\infty}^{\infty} \left[H(k + \log_{10}(d+1)) - H(k + \log_{10}(d))\right].$$

Example 2.85 *Distribution of the first digit.* *(a) Derive the distribution of the first digit when $Y = \log_{10} X$ follows an exponential distribution with rate λ. (b) Prove that the distribution converges to Benford's law when $\lambda \to 0$. (c) Prove that X follows a Pareto distribution. (d) Use $\lambda = 1$, 0.1, and 0.01 to plot the Pareto-originated and Benford distributions of the first digit.*

Solution. (a) Since for exponential distribution the cdf is $H(y) = 1 - e^{-\lambda y}$, the probability that the first digit, D, of an observation is d is given by

$$\Pr(D = d) = \sum_{k=0}^{\infty} \left(e^{-\lambda \log_{10}(d)} - e^{-\lambda \log_{10}(d+1)}\right) e^{-\lambda k}$$

$$= \left(e^{-\lambda \log_{10}(d)} - e^{-\lambda \log_{10}(d+1)}\right) \sum_{k=0}^{\infty} e^{-\lambda k}$$

$$= \left(\frac{1}{d^{\lambda/\ln 10}} - \frac{1}{(d+1)^{\lambda/\ln 10}}\right) \frac{1}{1 - e^{-\lambda}}.$$

(b) Using L'Hôpital's rule, we obtain

$$\lim_{\lambda \to 0} \frac{e^{-\lambda \log_{10}(d)} - e^{-\lambda \log_{10}(d+1)}}{1 - e^{-\lambda}}$$

$$= \lim_{\lambda \to 0} \frac{-\log_{10}(d)e^{-\lambda \log_{10}(d)} + \log_{10}(d+1)e^{-\lambda \log_{10}(d+1)}}{e^{-\lambda}}$$

$$= \log_{10}(d+1) - \log_{10}(d),$$

Benford's distribution. (c) As follows from Section 2.12, the density of $X = 10^Y$ is the Pareto distribution density:

$$\frac{\lambda}{x^{\lambda/\ln 10 + 1} \ln 10}, \quad x \geq 1.$$

(d) The R code is function `benfordEXP`. It produces the distribution of the first digit according to Benford's law and the distribution with the Pareto distribution for X (exponential distribution for $Y = \log_{10} X$).

Example 2.86 *Fraudulent transactions.* *Data set `sales.Rdata` contains 401,146 sales transactions of undisclosed source studied in Chapter 4 by Torgo [99]. It contains five columns. Compute the first significant digits of the reported monetary value of the sale stratified by* ok*,* fraud *and* unkn*, and compare with Benford's distribution.*

Solution. The R code is found in file `benfordFT.r`. Several conclusions can be made: (i) In general, the shape of the distribution of the first digit is similar to Benford's law, although the probability of 1 is considerably higher, 0.45 versus 0.3. (ii) The proportion of fraudulent transactions starting with 1 is slightly lower than for prudent (ok) transactions, 0.45 versus 0.55. (iii) The proportion of fraudulent transactions starting with 4 and 5 is higher than for ok transactions.

2.16.1   Distributions that almost obey Benford's law

We provide two examples of distributions that almost obey the Benford's law. We want to find a distribution with a positive support such that the distribution of the first significant digit follows the law (2.80). As one infers from Theorem 2.84, a successful cdf, H, on the $\log_{10}$ scale must obey

$$\sum_{k=-\infty}^{\infty} [H(k+y) - H(k)] = y \qquad (2.83)$$

for all $0 \leq y < 1$. Differentiation with respect to y yields an equivalent condition:

$$\sum_{k=-\infty}^{\infty} h(k+y) = 1, \qquad (2.84)$$

where $h = H'$ is the density of Y, and is positive on $(-\infty, \infty)$. Obviously, a uniform distribution on (a, b) satisfies this property, but only when $a \to -\infty$ and $b \to \infty$. Practically, we need to use a uniform distribution of $Y = \log_{10} X$ on a large range. Then on the original scale, the distribution of $X = 10^Y$ turns into a hyperbolic distribution with the density reciprocal of x; see the example below.

Another surprising example for which (2.83), or (2.84), almost holds is the standard normal distribution with $h = \phi$:

$$\frac{1}{\sqrt{2\pi}} \sum_{k=-\infty}^{\infty} e^{-\frac{1}{2}(k+y)^2} = 1 \text{ for } 0 \leq y < 1.$$

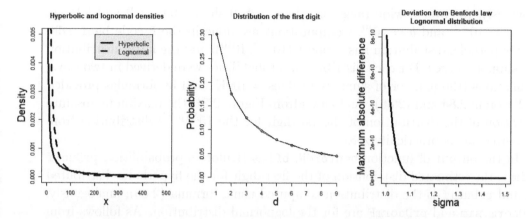

Figure 2.31: *Distributions that almost obey Benford's law. The maximum absolute difference of the first digit probability for the lognormal distribution is about 8×10^{-10} for $\sigma = 1$ and even smaller for $\sigma > 1$. See function* benfordN.

Consequently, the distribution of the first significant digit of the positive random variable X with the lognormal density

$$f_X(x) = \frac{\log_{10} e}{x\sqrt{2\pi}} e^{-\frac{1}{2}\log_{10}^2 x}, \quad x > 0 \tag{2.85}$$

almost exactly obeys Benford's law. It is easy to prove that the distribution of digits does not change if X is replaced with $X + \mu$. The standard normal density is not the only pdf that obeys condition (2.84) almost exactly. One can confirm numerically that a normal density with mean μ and standard deviation greater than one produces Benford's law very closely. More detail is found in the example below.

Example 2.87 *Almost exact Benford's law.* *Show numerically that the distribution of the first significant digit of a random variable X drawn from the lognormal distribution $10^{\mathcal{N}(0,1)}$ and the hyperbolic distribution on (a, b) almost exactly obey Benford's law when a is a small and b is a large positive number. Plot the deviation from Benford's law for the lognormal distribution $10^{\mathcal{N}(\mu,\sigma^2)}$ for given μ and an array of $\sigma^2 \geq 1$.*

Solution. If Y has a uniform distribution on $(\log_{10} a, \log_{10} b)$ then the density of $X = 10^Y$ has the hyperbolic distribution $f_X(x) = [(\ln b - \ln a)x]^{-1}$ for $0 < a < x < b$. The probability that the first significant digit of X is d is given by

$$\Pr(D = d) = \frac{1}{\ln b - \ln a} \left[\sum_{k:a < (d+1)10^k \leq b} \ln \frac{(d+1)10^k}{a} - \sum_{k:a < d10^k < b} \ln \frac{d10^k}{a} \right]. \tag{2.86}$$

Obviously, to cover a wide range of values, $a > 0$ should be small and b large, say, $a = 10^{-10}$ and $b = 10^{10}$. Computations are done in the R code `benfordN`. The hyperbolic distribution has support $(10^{-10}, 10^{10})$, and the latter distribution has support $(0, \infty)$. For each distribution, probabilities were derived in two ways: simulations (the number of simulated values = 100K) and by formulas provided in Theorem 2.84 and (2.86). As follows from Figure 2.31, the maximum absolute deviation of the distribution of the 1st digit for the $10^{\mathcal{N}(\mu, \sigma^2)}$ distribution from Benford's law is practically zero.

In the output of function `benfordN`, `bfl` is Benford's probabilities, `prDunif` `.sim` is the estimated distribution of the first digit for the hyperbolic-distributed X, and `prDunif` is the distribution computed using formula (2.86); analogously, `prDnorm.sim` and `prDnormF` are for the lognormal distribution. As follows from these results, the formulas reproduce probabilities (2.80), and simulation results are very close to exact. Note that the distribution of the first digit of a mixture of these distributions follows Benford's law as well.

Problems

1. (a) Derive the distribution of the first digit of the birthday of the year assuming that they are uniformly distributed on $1, 2, ..., 365$. (b) Check your formula via simulations. (c) Plot Benford's law, the probability formula, and simulated results on the same graph to see how close they match. [Hint: Use `sample` to generate uniform distribution.]

2. Generate t-distributed random numbers in function `fracksim` to see if Benford's law holds. [Hint: Use degrees of freedom as an argument.].

3. Using the results of Section 2.16.1, explain why exponential distribution with smaller λ in Example 2.85 approaches Benford's law.

4. Combine `fraud` and `unkn` transactions and redo the fraudulent-transaction analysis. Does this combination make the probability of the first digit closer to Benford's law?

5. Verify equations (2.83) and (2.84) using $H = \Phi$ and $h = \phi$ by computation.

6. Run function `benfordN` with a narrow range for the hyperbolic distribution, $(10^{-5}, 10^5)$, $(10^{-3}, 10^3)$, $(10^{-1}, 10)$, and $(1, 5)$. Draw conclusions.

7. As a continuation of the previous problem, plot the maximum absolute deviation of the first significant digit distribution from the Benford's law using the hyperbolic distribution on the interval $(10^{-p}, 10^p)$ versus $p = 1, 2, 3, ..., 10$.

8. Given proportion/probabilities p_1, p_2, p_3, and p_4 such that $p_1 + p_2 + p_3 + p_4 = 1$, simulate numbers from the mixture of four distributions used in `benfordN`

and demonstrate that the distribution of the first digit follows Benford's law. [Hint: Use p_1, p_2, and p_3 as arguments of the function and append a column to the output table.]

9. Check numerically that a normal density with mean μ and standard deviation greater than one satisfies (2.83) and (2.84) and produces Benford's law very closely. [Hint: Modify function `benfordN`.]

10. Derive the closed-form expression of function (2.84) when h is the Laplace pdf. Plot this function on $[0, 1]$ for several λs in the range from 0.01 to 2. [Hint: Use the formula for geometric progression $\sum_{k=1}^{\infty} e^{-\lambda k} = (e^{\lambda} - 1)^{-1}$.]

11. Show that if distributions H_k on the $\log_{10}$ scale follow Benford's law, then a linear combination, $\sum \lambda_k H_k$, follows Benford's law where $\lambda_k > 0$ and $\sum \lambda_k = 1$.

2.17 The Pearson family of distributions

It is desirable to generate various continuous distributions from one equation. Almost a hundred years ago, the British statistician Karl Pearson (1857–1936) suggested the following Ordinary Differential Equation (ODE) to describe a rich family of densities:

$$\frac{f'(x)}{f(x)} = \frac{a - x}{b + cx + gx^2}, \tag{2.87}$$

where $f(x)$ is the density and a, b, c, and g are parameters. Note that the density support is not specified beforehand. Neither the fact that $f(x)$ is positive nor the area under $f(x)$ equals 1 are guaranteed, so one has to take care of that after a density candidate is derived (these conditions apply some restrictions on the parameters). One might say that the Pearson equation (2.87) describes the *shape* of the density, not exactly the density.

Example 2.88 *The normal distribution is a member of the Pearson family.* *Derive the Pearson density setting $c = g = 0$, $b > 0$, and show that equation (2.87) gives birth to a normal distribution.*

Solution. When $c = g = 0$, we obtain $f'(x)/f(x) = (a - x)/b$. Letting $h(x) = \ln f(x)$ we note that $h'(x) = f'(x)/f(x)$. That is, in terms of function h, we simplify this ODE as $h'(x) = (a - x)/b$. Express $h'(x) = dh/dx$ and separate h and x by moving the differentials dh and dx to the left- and right-hand sides. Through integration, $\int dh = b^{-1} \int (a - x)dx$, one obtains

$$h(x) = -(x - a)^2/(2b) + C_*,$$

where C_* is a constant. Finally, we exponentiate this solution and derive the normal density,

$$f(x) = Ce^{-\frac{(x-a)^2}{2b}},$$

where C is the normalizing coefficient to ensure that the area under $f(x)$ is 1. $\square$

The denominator in (2.87), as a quadratic function may have no roots, one root, or two roots. It is possible to show that the one-root case produces the gamma distribution and two-root case produces the beta distribution.

Now we consider in detail the case when the denominator has no roots, $b + cx + gx^2 > 0$ for all x and $g > 0$ with the density support $-\infty < x < \infty$. Then in a slightly different parametrization, we redefine the ODE (2.87) as follows:

$$h'(x) = r\frac{a - x}{1 + g(x - m)^2},$$

where $h(x) = \ln f(x)$ and $r > 0$. It is possible to show that

$$\int \frac{a - x}{1 + g(x - m)^2} dx = C_* - \frac{1}{2g} \ln\left(1 + g(x - m)^2\right) + \frac{a - m}{\sqrt{g}} \arctan \sqrt{g}(x - m).$$

Finally we conclude that the four-parameter Pearson density (g and r are positive) takes the form

$$f(x; a, g, m, r) = C\frac{e^{r\frac{a-m}{\sqrt{g}} \arctan \sqrt{g}(x-m)}}{[1 + g(x - m)^2]^{r/(2g)}}, \quad -\infty < x < \infty, \tag{2.88}$$

where normalizing coefficient $C > 0$ is such that the area under the density is 1. Since the numerator is a bounded function of x, we can say that the Pearson density is a skewed Cauchy-type density because it yields the Cauchy density when $r = 2g$ and $a = m$. This density is a good candidate to study the properties of statistical models with heavy asymmetric tails.

Problems

1. Prove that the gamma and beta distributions belong to the Pearson-type families. Take the derivative of the log density and show that it can be expressed as the right-hand side of equation (2.87). What combination of parameters in the ODE yield gamma and beta distributions?

2. Does the exponential distribution belong to the Pearson family?

3. Does the Laplace (double-exponential) or lognormal distribution belong to the Pearson family?

4. Does the Pareto distribution with density $f(x; \alpha) = \alpha/x^{1+\alpha}$ for $x > 1$ belong to the Pearson family ($\alpha > 0$)?

5. Use `integrate` to compute the normalizing coefficient for density (2.88) and plot several densities on the grid of values for g, keeping other parameters fixed. Do the same on the grid of r values. Use `par=mfrow(1,2)` to illustrate how g and r change the shape of the density curve.

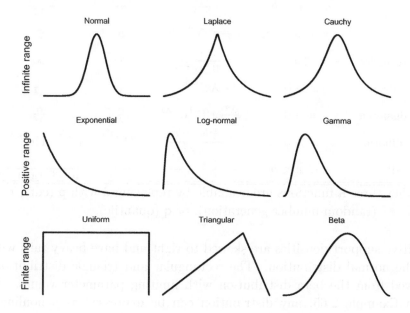

Figure 2.32: *Nine major continuous distributions with infinite, positive, and finite support.*

2.18 Major univariate continuous distributions

Most important continuous distributions are summarized in Table 2.5. We refer the reader to an authoritative book by Johnson et al. [60] for other distributions not covered in this chapter (see also [64]). Nine densities with infinite, semi-infinite, and finite support are depicted in Figure 2.32. All distributions except Laplace are built-in functions in R.

The support of the infinite-range distributions is the entire line, $(-\infty, \infty)$. All three continuous distributions in the top row are unimodal and symmetric around zero. The Laplace density is sharp at zero – the derivative at zero does not exist. This distribution has heavy tails compared to the normal distribution. The Cauchy distribution is unique because it does not have finite mean and variance (the integrals diverge).

All distributions in the second row of the figure are skew-normal distribution skewed to right (the skewness coefficient is positive). The distributions in the third row have support on $[a, b]$. It is easy to generalize the beta distribution with support on $[a, b]$ by letting the density proportional to $(x - a)^{\alpha-1}(b - x)^{\beta-1}$.

Table 2.5. Eight major continuous distributions.

Name	R*	Range	Density	Mean	Variance
Normal	dnorm	$-\infty < x < \infty$	$\dfrac{e^{-\frac{1}{2\sigma^2}(x-\mu)^2}}{\sigma\sqrt{2\pi}}$	μ	σ^2
Uniform		$a < x < b$	$\dfrac{1}{b-1}$	$\dfrac{a+b}{2}$	$\dfrac{(b-a)^2}{12}$
Laplace		$-\infty < x < \infty$	$\dfrac{1}{2\lambda}e^{-\frac{1}{\lambda}|x-\mu|}$	μ	$\dfrac{2}{\lambda^2}$
Cauchy	dcauchy	$-\infty < x < \infty$	$\dfrac{1}{\pi(1+x^2)}$	∞	∞
Exponen.	dexp	$x > 0$	$\lambda e^{-\lambda x}$	$\dfrac{1}{\lambda}$	$\dfrac{1}{\lambda^2}$
Gamma	dgamma	$x > 0$	$\dfrac{\lambda^\alpha}{\Gamma(\alpha)}x^{\alpha-1}e^{-\lambda x}$	$\dfrac{\alpha}{\lambda}$	$\dfrac{\alpha}{\lambda^2}$
Lognorm.	dlnorm	$x > 0$	$\dfrac{e^{-\frac{1}{2}(\ln x-\mu)^2}}{x\sigma\sqrt{2\pi}}$	$e^{\mu+\sigma^2/2}$	$(e^{\sigma^2}-1)e^{2\mu+\sigma^2}$
Beta	dbeta	$a < x < b$	$\dfrac{x^{\alpha-1}(1-x)^{\beta-1}}{B(\alpha,\beta)}$	$\dfrac{\alpha}{\alpha+\beta}$	$\dfrac{\alpha\beta}{(\alpha+\beta+1)(\alpha+\beta)^2}$

*Note: other three R functions are available by replacing d with p (cdf), r (random number generation), or q (quantile).

The positive support densities are skewed to right and have heavy tails with respect to the normal distribution. The rectangular and triangle distributions can be derived from the beta distribution with limiting parameter values. As follows from Example 2.65, any distribution can be expressed as a nonlinear transformation of another.

Problems

1. Explain why the Laplace and Cauchy distributions cannot be approximated with the normal distribution such that the absolute difference between densities is smaller than ε for any $\varepsilon > 0$.

2. As a continuation of the previous problem, find the optimal μ and σ^2 that approximate the density of Laplace ($\lambda = 1$) and Cauchy densities using computations and report maximum absolute difference.

3. The same as above, but for positive range densities.

4. Derive the rectangular and triangular distributions from the beta distribution?

5. Generalize Example 2.65 to transform a random variable with pdf $f(x)$ to a random variable with pdf $p(x)$ and conclude that any distribution from Table 2.5, except the Laplace distribution, can be expressed through another. [Hint: Define the ODE and apply Picard's existence theorem.]

Chapter 3

Multivariate random variables

In the previous chapters, the objects of our attention were univariate random variables. When two random variables have been discussed, the silent assumption was that they were independent. Indeed, the concept of independence is easy to understand although no formal definition has been given yet. The goal of this chapter is to give a formal definition of independence, but more importantly describe how dependence can be rigorously defined and studied using the joint and conditional distributions. The conditional mean and bivariate normal distribution are studied in detail because they are fundamental concepts for describing the relationship between random variables. Multivariate distributions of random vectors are conveniently handled using vector and matrix algebra. Thus, linear algebra techniques and concepts such as matrix algebra manipulations, eigenvectors, and eigenvalues will be used in this chapter. Multiple examples of multivariate continuous distributions can be found in an authoritative book by Kotz et al. [67].

3.1   Joint cdf and density

When two random variables are studied, their *joint* (multivariate) distribution should be defined. We shall use the term *bivariate* when only two random variables are involved, (X, Y). Knowing individual (marginal) distributions is not enough to completely specify the joint distribution because the value of Y may depend on X. For a univariate random variable, the cumulative distribution function (cdf) is the probability of falling within a semi-infinite interval. If two random variables are involved, the interval is replaced with the semi-infinite rectangle on the plane. Following our convention, random variables are denoted using upper case, X or Y, and the arguments of cdf, density, integration, etc. are denoted using lower case letters, x or y.

Advanced Statistics with Applications in R, First Edition. Eugene Demidenko.
© 2020 John Wiley & Sons, Inc. Published 2020 by John Wiley & Sons, Inc.

Definition 3.1 *Let X and Y be two random variables. The joint (bivariate) cdf is defined as the joint probability, $F(x, y) = \Pr(X \leq x, Y \leq y)$.*

In this definition, events $X \leq x$ and $Y \leq y$ occur simultaneously, or in other notation, $(X \leq x, Y \leq y) = (X \leq x) \cap (Y \leq y)$. See the left plot in Figure 3.1 for a geometric illustration: the cdf is the probability that the random point (X, Y) falls within the semi-infinite (shaded) rectangle.

Properties of the bivariate cdf:

1. $0 \leq F(x, y) \leq 1$.

2. $F(-\infty, y) = F(x, -\infty) = 0$, and $F(\infty, \infty) = 1$, or in other notation $\lim_{x \to -\infty} F(x, y) = \lim_{y \to -\infty} F(x, y) = 0$ and $\lim_{x \to \infty} \lim_{y \to \infty} F(x, y) = 1$.

3. If $x_1 \leq x_2$, then $F(x_1, y) \leq F(x_2, y)$ for every y; if $y_1 \leq y_2$ then $F(x, y_1) \leq F(x, y_2)$ for every x. In other words, the cdf is an increasing function in both arguments.

4. If $x_1 \leq x_2$ and $y_1 \leq y_2$, then $F(x_2, y_2) - F(x_1, y_2) - F(x_2, y_1) + F(x_1, y_1) \geq 0$.

One can prove that, if function $F = F(x, y)$ satisfies properties 1–4, then there exists a bivariate random variable (X, Y) with cdf F; see Figure 3.1 for illustration. The left plot depicts the probability of falling within a semi-infinite rectangle. The right plot depicts the probability of falling within the finite rectangle 1 specified in Property 4 with vertices $(x_1, y_1), (x_1, y_2), (x_2, y_2)$, and (x_2, y_1). This probability is expressed through F as follows: $F(x_2, y_2) - F(x_1, y_2) - F(x_2, y_1) + F(x_1, y_1)$ or, symbolically, $1 - 3 - 2 + 4 \geq 0$.

Definition 3.2 *The marginal cdf of X is the limit of the joint cdf when $y \to \infty$: $F_X(x) = F(x, \infty) = \lim_{y \to \infty} F(x, y)$. Similarly, the marginal cdf of Y is defined as $F_Y(y) = F(\infty, y) = \lim_{x \to \infty} F(x, y)$.*

Geometrically, the marginal cdf of X is the probability of (X, Y) of falling to the left of the vertical line at x. Similarly, the marginal cdf of Y is the probability of (X, Y) of falling below the horizontal line at y. The above definition of cdf works for discrete and continuous random variables. If the cdf is twice differentiable, the joint density or the joint probability density function (pdf) can be introduced.

Definition 3.3 *The joint density or pdf of (X, Y) is the mixed partial derivative of the cdf:*

$$f(x, y) = \frac{\partial^2 F(x, y)}{\partial x \partial y}.$$

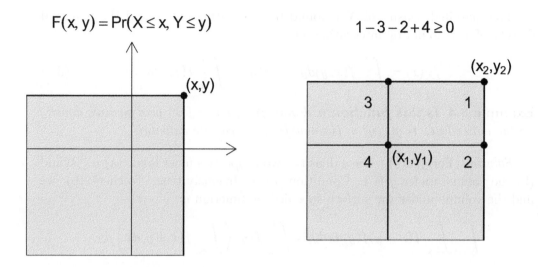

Figure 3.1: *Left: the cdf of the bivariate random variable (X, Y) is the probability of falling within the semi-infinite shaded rectangle specified by the upper-right corner (x, y). Right: property 4 says that the probability of falling within rectangle 1 is nonnegative.*

Using the fundamental theorem of calculus, the cdf is expressed via the density integrated over the semi-infinite rectangle:

$$F(x, y) = \int_{-\infty}^{x} \int_{-\infty}^{y} f(u, v) du dv.$$

Properties of the bivariate pdf (density):

1. The density is a nonnegative function, $f(x, y) \geq 0$ (the support is defined as the set of points where the density is positive).

2. The volume under the density surface in R^3 is 1:

$$\int_{-\infty}^{\infty} \int_{-\infty}^{\infty} f(x, y) dx dy = 1.$$

Property 4 of the bivariate cdf in Definition 3.1 can be derived from the bivariate density by integrating over the rectangular specified by the corners (x_1, y_1) and (x_2, y_2):

$$F(x_2, y_2) - F(x_1, y_2) - F(x_2, y_1) + F(x_1, y_1) = \int_{x_1}^{x_2} \int_{y_1}^{y_2} f(x, y) dx dy \geq 0,$$

where $x_1 \leq x_2$ and $y_1 \leq y_2$.

The marginal density of X is found by integrating out y, and the marginal density of Y is found by integrating out x:

$$f_X(x) = \int_{-\infty}^{\infty} f(x,y)dy, \quad f_Y(y) = \int_{-\infty}^{\infty} f(x,y)dx. \qquad (3.1)$$

Example 3.4 *Is this function a density?* *Let $f = f(x)$ be a positive density on the entire line. Is $g(x,y) = f(x+y)f(y)$ a bivariate density?*

Solution. For $g(x,y)$ to be a density, two properties must hold: (a) $g \geq 0$ and (b) the volume under g is 1. Condition (a) is obviously true. To check (b), we find the volume under the surface specified by function g:

$$\int_{-\infty}^{\infty} \int_{-\infty}^{\infty} f(x+y)f(y)dxdy = \int_{-\infty}^{\infty} f(y) \left(\int_{-\infty}^{\infty} f(x+y)dx \right) dy.$$

By changing variable $z = x + y$ in the inner integral, we obtain $\int_{-\infty}^{\infty} f(x+y)dx = \int_{-\infty}^{\infty} f(z)dz = 1$ for every x. Therefore $\int_{-\infty}^{\infty} \int_{-\infty}^{\infty} f(x+y)f(y)dxdy = \int_{-\infty}^{\infty} f(y)dy = 1$. Thus, the answer is positive. $\qquad \square$

theta=15   theta=60

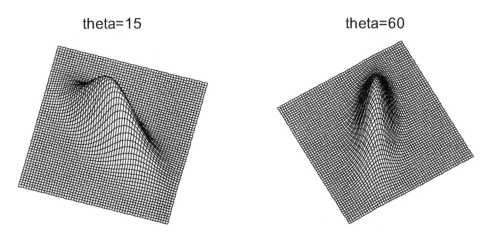

Figure 3.2: *The density surface viewed at different angles using the* `persp` *command in* R.

An example of a bivariate density at two angles is shown in Figure 3.2. We use the joint density from the example above letting $f = \phi$, the standard normal density. This graph was created using the R function `persp`. We make several comments on the code: (i) The script is saved in the folder `c:\\StatBook\\` as an ASCII file under the name `sbmultD.r` every time we run the program. The code may be saved in any folder/directory. (ii) To speed up the computation, the `rep` command is used (other methods of vectorized computation on the grid are discussed in Example 1.12. (iii) The arguments `theta` and `phi` specify the angles of the 3D view with respect to the x-axis and z-axis, respectively. (iv) Option box=F suppresses the (x, y, z) box around the surface for better visualization because it is somewhat irrelevant for our purpose.

Example 3.5 ***A function that is not a density.*** *Suppose that $f(x)$ is a density with support $(-\infty, \infty)$. Prove that $g(x, y) = f(x-y)$ cannot be a bivariate density.*

Solution. The problem with $g(x, y)$ is that it remains constant on the straight line $x - y = c$ for any constant c. This implies that the volume under g is infinite. Algebraically, making the change of variable $z = x - y$ and using the fact that $\int_{-\infty}^{\infty} f(z)dz = 1$, we obtain

$$\int_{-\infty}^{\infty} \int_{-\infty}^{\infty} f(x - y)dxdy = \int_{-\infty}^{\infty} \left(\int_{-\infty}^{\infty} f(z)dz \right) dx = \int_{-\infty}^{\infty} dx = \infty;$$

see Figure 3.3. It is crucial that $f(z) > 0$ for every z because the support is the entire line. The R code that produces the surface density is found in the file statbookgxy.r.

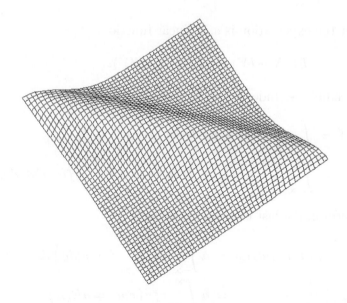

Figure 3.3: *Function $g(x, y) = f(x - y)$ cannot be a density because it is constant along the line $x - y = \text{const}$. Since the volume under this surface $= \infty$, function g cannot be a density.*

In this example, the Gaussian density, $f = \phi$ is used, which is a built-in function, dnorm. We used a straightforward way to compute g through a double loop, which may be inefficient for larger dimensions. There are several ways to carry out faster computations in R. For example, we can used rep

command to speed up computations: `X=rep(x,times=N); Y=rep(x,each=N);`
`gxy=matrix(dnorm(X-Y),ncol=N)`. Other vectorized methods are discussed in
Example 1.12.

3.1.1   Expectation

If $g(X, Y)$ is a function of jointly distributed X and Y, we define the expected
value in a straightforward manner:

$$E(g(X,Y)) = \int_{-\infty}^{\infty} \int_{-\infty}^{\infty} g(x,y)f(x,y)dxdy. \tag{3.2}$$

It is easy to see that this definition complies with the univariate definition of
expectation, namely, $E(g(X,Y)) = E(X)$ if $g(x,y) = x$. Indeed, using (3.1) in
this case:

$$
\begin{aligned}
E(g(X,Y)) &= \int_{-\infty}^{\infty} \int_{-\infty}^{\infty} xf(x,y)dxdy = \int_{-\infty}^{\infty} x\left(\int_{-\infty}^{\infty} f(x,y)dy\right)dx \\
&= \int_{-\infty}^{\infty} xf_X(x)dx = E(X).
\end{aligned}
$$

Now we prove that the expectation is as a linear function:

$$E(aX + bY) = aE(X) + bE(Y),$$

where a and b are numbers. Indeed,

$$
\begin{aligned}
E(aX + bY) &= \int_{-\infty}^{\infty} \int_{-\infty}^{\infty} (ax + by)f(x,y)dxdy \\
&= \int_{-\infty}^{\infty} \int_{-\infty}^{\infty} axf(x,y)dxdy + \int_{-\infty}^{\infty} \int_{-\infty}^{\infty} byf(x,y)dxdy. \tag{3.3}
\end{aligned}
$$

But for the first term in the last expression, we have

$$
\begin{aligned}
\int_{-\infty}^{\infty} \int_{-\infty}^{\infty} axf(x,y)dxdy &= a\int_{-\infty}^{\infty} x\left(\int_{-\infty}^{\infty} f(x,y)dy\right)dx \\
&= a\int_{-\infty}^{\infty} xf_X(x)dx = aE(X)
\end{aligned}
$$

because $\int_{-\infty}^{\infty} f(x,y)dy = f_X(x)$ is the marginal density. Similarly, we prove that
the second term in (3.3) is equal to $bE(Y)$.

3.1.2   Bivariate discrete distribution

Let X take values $\{x_i, i = 1, 2, ..., m\}$ and Y take values $\{y_j, j = 1, 2, ..., n\}$.
To specify the joint distribution, we need to provide probabilities for mn pairs

$\{(x_i, y_j),\ i = 1, 2, ..., m, j = 1, 2, ..., n\}$ by defining the probabilities of the mutual occurrence:

$$p_{ij} = \Pr(X = x_i, Y = y_j).$$

Probabilities p_{ij} sum to 1, and the joint cdf is defined similarly to the continuous case, but the integral is replaced with the sum,

$$F(x, y) = \sum_{i:x_i \le x} \sum_{j:y_j \le y} p_{ij}.$$

The marginal cdf of X is defined by summing out the Y probabilities:

$$F_X(x) = \sum_{i:x_i \le x} \sum_{j=1}^{n} p_{ij}.$$

The expectation of $g(X, Y)$ is defined via the double sum:

$$E(g(X, Y)) = \sum_{i=1}^{m} \sum_{j=1}^{n} g(x_i, y_j) p_{ij}.$$

Problems

1. Show that the marginal cdfs defined in Property 4 of the bivariate cdf satisfy the properties of the univariate cdf as defined in Section 2.1.

2. Prove that (a) Property 4 of the bivariate cdf implies 3, and (b) Property 1 is implied by 2 and 3.

3. Let F_X and F_Y be cdfs. (a) Is $F(x, y) = 2F_X(x)F_Y(y)/(F_X(x) + F_Y(y))$ a joint cdf? (b) Is $F(x, y) = F_X(x)F_Y(y)$ a joint cdf? (c) Let $F(x, y)$ be a joint cdf. Is $F(x, y) = 2F(x, y)/(F_X(x) + F_Y(y))$ a joint cdf?

4. Let $H(x, y) = \max(x, y)/(x + y)$ for $x > 0$ and $y > 0$ and 0 elsewhere. Can H be a cdf? [Hint: First prove that if F is a cdf, then $\lim_{x \to \infty} F(x, x) = 1$.]

5. Describe a bivariate random variable that has cdf $H(x, y) = \max(0, x) \times \max(0, y)$ for $x < 1$ and $y < 1$, and 1 elsewhere.

6. Prove that if F_X, and F_Y are univariate cdfs and $|\alpha| < 1$, then $F_X(x)F_Y(y)[1 + \alpha(1 - F_X(x))\,(1 - F_Y(y))]$ is a bivariate cdf.

7. Let G and H be bivariate cdfs. Is $F = \lambda G + (1 - \lambda)H$ a cdf $(0 \le \lambda \le 1)$? Answer this question by reducing to bivariate density.

8. Display densities in the R function `sbmultD` at angles from $1°$ to $360°$ using a loop over `theta`. Use `par(mfrow=c(1,1))`.

9. (a) If g is a density, can $g(x - y)$ be a density if g has a finite support? (b) If g is a density with the support on positive numbers, can $g(x - y)$ be a density on $0 < y < x$? Give an example or counterexample.

10. Suppose that $f(x)$ is a density. Are (a) $f(x)f(y)$, (b) $0.5(f(x) + f(y))$, (c) $\lambda f(x) + (1 - \lambda)f(y)$ where $0 \le \lambda \le 1$, bivariate densities?

11. Find the marginal cdfs in the "Uniform distribution on the triangle" example.

12. Use `outer` from Section 1.5 to compute the function in the script `sbmultD`.

13. Define a uniform distribution on the unit disk with center at the origin, and find its marginal cdf and densities. [Hint: Use a geometric approach by referring to the area.]

14. (a) Following the definition in Section 2.2, define the center of gravity of an area with the density of the material specified by a pdf. (b) Find the center of gravity of the triangle with vertices $(0,0)$, $(1,1)$, and $(1,0)$ with uniform density. (c) Find the center of gravity of a general triangle with side length A, B, and C, and check the result with simulations. (d) Cut a triangle from a wooden board or plywood, do measurements of the sides, compute the center of the gravity, put the nail at the center, and hang the triangle to test whether your analytic solution is true.

3.2　Independence

From elementary probability, we know that two random events, A and B, are independent if and only if

$$\Pr(A \cap B) = \Pr(A) \times \Pr(B),$$

where $A \cap B$ means that A and B occur simultaneously. Since the joint distribution function $F(x, y)$ is the probability of two simultaneous events, $X \le x$ and $Y \le y$, the random variables X and Y are independent if and only if

$$F(x, y) = F_X(x) \times F_Y(y),$$

where $F_X(x)$ and $F_Y(y)$ are the marginal cdfs. In words, two randomvariables (X, Y) are independent if and only if their joint cdf equals the product of their marginal cdfs.

The same product rule holds for densities. Indeed, using elementary calculus, for independent X and Y, we obtain

$$f(x, y) = \frac{\partial^2 F(x, y)}{\partial x \partial y} = \frac{\partial F_X(x)}{\partial x} \times \frac{\partial F_Y(y)}{\partial y} = f_X(x) \times f_Y(y).$$

Thus, two continuous random variables, X and Y, are independent if and only if the joint density is the product of marginal densities:

$$f(x, y) = f_X(x) \times f_Y(y).$$

Example 3.6 *Variance of sum.* *Prove that if X and Y are independent random variables, then*

$$\operatorname{var}(X + Y) = \operatorname{var}(X) + \operatorname{var}(Y).$$

Solution. Let $E(X) = \mu_X$ and $E(Y) = \mu_Y$. Then

$$\begin{aligned}
\operatorname{var}(X + Y) &= E[(X - \mu_X) + (Y - \mu_Y)]^2 \\
&= E(X - \mu_X)^2 + 2E[(X - \mu_X)(Y - \mu_Y)] + E(Y - \mu_Y)^2 \\
&= \operatorname{var}(X) + \operatorname{var}(Y) + 2E[(X - \mu_X)(Y - \mu_Y)].
\end{aligned}$$

It suffices to prove that the last term is zero. Indeed, we have

$$\begin{aligned}
E[(X - \mu_Y)(X - \mu_Y)] &= \int_{-\infty}^{\infty} \int_{-\infty}^{\infty} (x - \mu_X)(y - \mu_Y) f_X(x) f_Y(y) dx dy \\
&= \int_{-\infty}^{\infty} (x - \mu_X) f_X(x) dx \times \int_{-\infty}^{\infty} (y - \mu_Y) f_Y(y) dy = 0 \times 0.
\end{aligned}$$

Example 3.7 *Mary and John talking.* *Mary and John participate in a teleconference call set up at 10 a.m. The bivariate density of the times they join the group is given by $e^{-(x+y)}$, where x stands for Mary and y stands for John (minutes after 10 a.m.). (a) Prove that Mary and John join the meeting independently. (b) Calculate the probability that John joins the group two minutes after Mary and check the answer via simulations. (c) Estimate the required number of simulations to achieve standard error < 0.001.*

Solution. (a) Find the marginal densities:

$$f_X(x) = \int_0^{\infty} e^{-(x+y)} dy = e^{-x} \int_0^{\infty} e^{-y} dy = e^{-x} \times e^{-y}\big|_0^{\infty} = e^{-x} \times 1 = e^{-x}.$$

Similarly we find that $f_Y(y) = e^{-y}$. Since $f_X(x) \times f_Y(y) = e^{-(x+y)}$, we deduce that X and Y are independent. (b) The probability that John joins the group two minutes after Mary is expressed via a double integral

$$\begin{aligned}
\int_0^{\infty} \left(\int_{x+2}^{\infty} e^{-(x+y)} dy \right) dx &= \int_0^{\infty} \left(e^{-x} \int_{x+2}^{\infty} e^{-y} dy \right) dx \\
&= \int_0^{\infty} \left(e^{-x} \left(-e^{-y} \right)\big|_{x+2}^{\infty} \right) dx \\
&= \int_0^{\infty} \left(e^{-x} e^{-(x+2)} \right) dx = \int_0^{\infty} e^{-2x-2} dx = e^{-2} \left(\frac{1}{2} e^{-2x}\big|_0^{\infty} \right) \\
&= \frac{1}{2} e^{-2} = 0.0677.
\end{aligned}$$

To do simulations, we generate the call times (X_i, Y_i) with cdf $e^{-(x+y)}$ and count the proportion of times $Y_i > X_i + 2$ for $i = 1, 2, ..., N$ (the number of simulated values). Independence means that X_i and Y_i can be generated separately as exponentially distributed random variables. The code is below. We generate exponentially distributed random variables using inverse cdf, alternatively one can use rxep.

```
MJcall=function(nExp=100000)
{ # nExp = number of simulated pairs
    X=-log(1-runif(nExp)) # exponential cdf 1-exp(-x)
    Y=-log(1-runif(nExp)) # exponential cdf 1-exp(-x)
                          # X and Y are independent
    mean(Y>X+2)
}
```

Note that, in the last line, condition Y>X+2 is converted to a numeric array with 0 if the condition does not hold and 1 if the condition holds. Therefore **mean** computes the proportion of simulated values for which Y>X+2. (c) Now we find how many simulated pairs, N, is needed to achieve the desired standard error. First, we run this program with fairly small nExp=10000 that gives the probability $p = 0.066$. Second, we use this estimate to obtain an estimate for N_{exp} by solving the equation $\sqrt{p(1-p)/N_{\text{exp}}} = 0.001$ that yields $N_{\text{exp}} = 0.066 \times (1 - 0.066)/0.001^2 = 61,644$. Thus, nExp=100000 guarantees that the standard error of the estimated probability less than 0.001.

Example 3.8 *Average distance between points.* *Find the average distance between two random points on the interval of unit length. Use simulations to check the answer.*

Solution. Since the density of the uniform distribution is $f(x) = 1$ for $0 < x < 1$ and the joint density is the product of individual densities, $f(x, y) = 1$ if $0 < x < 1$ and $0 < y < 1$, the average distance is

$$\int_0^1 \int_0^1 |x - y|\, dx dy = \int_0^1 \left(\int_0^1 |x - y|\, dx \right) dy = \int_0^1 \left(\frac{1}{2} - y + y^2 \right) dy = \frac{1}{3}.$$

Note that we express $\int_0^1 |x - y|\, dx = -\int_0^y (x - y)dx + \int_y^1 (x - y)dx$. The R code mean(abs(runif(1000000)-runif(100000))) produces [1] 0.3330896.

Example 3.9 *Probability of $X > Y$.* *Two random variables, X and Y, are iid. Find the probability that $X > Y$.*

Solution. Intuition says that this probability is $1/2$. Let us prove it rigorously. The joint density is $g(x, y) = f(x)f(y)$ where f is the density of X and Y. Thus,

$$\Pr(X > Y) = \int_{x>y} g(x, y)dx dy = \int_{-\infty}^{\infty} \left(\int_y^{\infty} g(x, y)dx \right) dy$$

$$= \int_{-\infty}^{\infty} f(y) \left(\int_y^{\infty} f(x)dx \right) dy.$$

But since $\int_y^\infty f(x)dx = 1 - \int_{-\infty}^y f(x)dx = 1 - F(y)$, we have

$$\Pr(X > Y) = \int_{-\infty}^\infty f(y)(1 - F(y))dy = \int_{-\infty}^\infty f(y) - \int_{-\infty}^\infty f(y)F(y)dy$$

$$= 1 - \int_{-\infty}^\infty f(y)F(y)dy.$$

To evaluate the latter integral, make a change of variable, $z = F(y)$. Then $dz = f(y)dy$ and

$$\int_{-\infty}^\infty f(y)F(y)dy = \int_0^1 zdz = \frac{1}{2} z^2\big|_0^1 = \frac{1}{2}.$$

Finally, $\Pr(X > Y) = 1/2$, an expected result due to symmetry.

3.2.1 Convolution

Sometimes, we are interested in the distribution of the sum of two random variables. Specifically, let the joint density of X and Y be given, $f(x, y)$. What is the density of $Z = X + Y$? It is easier to find the cdf first and then take the derivative to obtain the density of Z. Indeed, the cdf of the sum can be expressed as the double integral

$$F_Z(z) = \Pr(X + Y \le z) = \int_{-\infty}^\infty \left(\int_{-\infty}^{z-x} f(x, y)dy \right) dx. \qquad (3.4)$$

This integral is the volume under the joint density over the area defined by the inequality $x + y \le z$. The above integral reads as follows: First, fix x and find the internal integral by integrating $f(x, y)$ over y from $-\infty$ to $z - x$. Second, find the outer integral by integrating the result from the first integration as a function of x over $-\infty$ to ∞. To find the density of Z, we differentiate (3.4) with respect to z using the calculus rule $d\left(\int_{-\infty}^t g(h)dh \right)/dt = g(t)$, yielding the density of Z:

$$f_Z(z) = \int_{-\infty}^\infty f(x, z - x)dx.$$

In a special case when X and Y are independent, the density $f_Z(z)$ is called a *convolution*. Specifically, if $f_X(x)$ and $f_Y(y)$ are densities of X and Y, the convolution is defined as

$$f_Z(z) = \int_{-\infty}^\infty f_X(x)f_Y(z - x)dx. \qquad (3.5)$$

Example 3.10 *Sum of two uniformly distributed random variables. Find the cdf and pdf of the sum of two independent uniformly distributed random variables on (0,1). Check the theoretical cdf via simulations.*

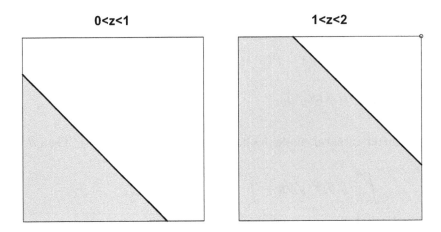

Figure 3.4: *Illustration to calculation of the cdf for $X + Y$ in Example 3.10.*

Solution. Let $X, Y \sim \mathcal{R}(0,1)$ be independent and $Z = X + Y$. First of all, we note that Z takes values on the interval $(0,2)$, and therefore outside of this interval, the density is zero (we say that the support of Z is $(0,2)$). Let z be any fixed value from this interval; then the distribution function of the sum is $F_Z(z) = \Pr(X + Y \le z)$. To compute this probability, we appeal to a geometrical interpretation. The pair (X, Y) may be viewed as a random point on the unit square $[0,1] \times [0,1]$. Consequently, the probability of a point of falling inside of a closed figure is equal to the area of this figure. For $0 \le z \le 1$ the inequality $x + y \le z$ defines a triangle, and therefore $\Pr(X + Y \le z)$ is the area of this triangle; see the left panel of Figure 3.4. Since this area is $z^2/2$, we immediately obtain that $F_Z(z) = z^2/2$, for $0 \le z \le 1$. For $1 < z \le 2$ the area of the figure defined by inequality $x + y \le z$ is easier to compute as complementary to the area of the triangle defined by inequality $x + y > z$. It is easy to see that the latter area is $(2 - z)^2/2$, so finally the cdf and the pdf of $Z = X + Y$ are

$$F_Z(z) = \begin{cases} 0 \text{ if } z \le 0 \\ \frac{1}{2}z^2 \text{ if } 0 < z \le 1 \\ 1 - \frac{1}{2}(2 - z)^2 \text{ if } 1 < z \le 2 \\ 1 \text{ if } z > 2 \end{cases}, \qquad f_Z(z) = \begin{cases} 0 \text{ if } z \le 0 \\ z \text{ if } 0 < z \le 1 \\ 2 - z \text{ if } 1 < z \le 2 \\ 0 \text{ if } z > 2 \end{cases}.$$

The R code that confirms that $F_Z(z)$ is the cdf of $X + Y$ is found in the file `twounif.r`.

Example 3.11 *The shortest piece.* *A stick of unit length is broken at two random points. (a) Find the cdf of the shortest piece (see Section 2.3, Problem 5). (b) Check the theoretical cdf via simulations. (c) Find the length of the shortest piece and confirm your answer by simulations.*

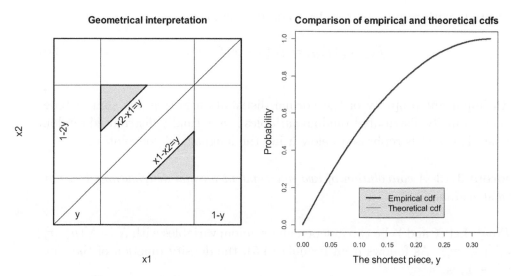

Figure 3.5: *Geometrical illustration and comparison of cdfs for Example 3.11.*

Solution. See Figure 3.5 for a geometric illustration. (a) If x_1 and x_2 are the break points the length of the shortest piece is $X = \min(x_1, x_2, 1 - x_1, 1 - x_2, |x_1 - x_2|)$. Express $\Pr(X \le y)$ via the complementary probability as $1 - \Pr(X > y)$. Since x_1 and x_2 are iid on $[0,1]$, the joint distribution of (x_1, x_2) is uniform on the unit square. Therefore, $\Pr(X > y)$ is the area defined by five inequalities $x_1 > y$, $x_2 > y$, $1 - x_1 > y$, $1 - x_2 > y$, and $|x_1 - x_2| > y$ depicted as the shaded triangles. Since the area of one triangle is $(1 - 3y)^2/2$, the cdf is $F(y) = 1 - (1 - 3y)^2$, where $0 < y < 1/3$ (the length of the shortest piece $< 1/3$). (b) The length of the shortest piece is simulated as follows using command `pmin` to compute the element-wise minimum of the vectors (`nExp=100000`):

```
X1=runif(nExp);X2=runif(nExp)
X.short=pmin(X1,X2,1-X1,1-X2,abs(X1-X2))
```

The empirical and theoretical cdfs are practically the same. (c) To compute the average length, find the density $f(y) = 6(1 - 3y)$, and take the integral $E(X) = \int_0^{1/3} 6y(1 - 3y)dy = 1/9$. Command `mean(X.short)` returns 0.1111338, very close to $1/9 = 0.1111111$.

Example 3.12 *Bivariate density*. *Express the joint density of $(X, X + Y)$ in terms of marginal densities of independent X and Y.*

Solution. The joint cdf, similarly to (3.4), is

$$F_{X, X+Y}(x, y) = \Pr(X \le x, X + Y \le y) = \int_{-\infty}^{x} \left(\int_{-\infty}^{y-u} f_Y(v)dv \right) f_X(u)du$$

$$= \int_{-\infty}^{x} F_Y(y - u) f_X(u)du.$$

Differentiate with respect to x and y to derive the joint density:

$$f_{X,X+Y}(x,y) = f_Y(y-x)f_X(x). \tag{3.6}$$

$\square$

An important property of the normal distribution is that the sum of independent normally distributed random variables is a normally distributed random variable. This fact is rephrased below using the language of convolution.

Theorem 3.13 *A convolution of two independent normal distributions is a normal distribution.*

Proof. Let X and Y be two independent random variables with $X \sim \mathcal{N}(\mu_x, \sigma_x^2)$, $Y \sim \mathcal{N}(\mu_y, \sigma_y^2)$. As follows from formula (3.5), the density function of the sum, $Z = X + Y$, is

$$f_Z(z) = \int_{-\infty}^{\infty} \left(\frac{1}{\sigma_x\sqrt{2\pi}} e^{-\frac{1}{2\sigma_x^2}(x-\mu_x)^2} \right) \left(\frac{1}{\sigma_y\sqrt{2\pi}} e^{-\frac{1}{2\sigma_y^2}(z-x-\mu_y)^2} \right) dx$$

$$= \frac{1}{2\pi\sigma_x\sigma_y} \int_{-\infty}^{\infty} e^{-\frac{1}{2\sigma_y^2}(x-\mu_x)^2 - \frac{1}{2\sigma_y^2}(z-x-\mu_y)^2} dx.$$

Do some algebra: represent the argument of the exponential function as

$$-\frac{1}{2\sigma_x^2}(x-\mu_x)^2 - \frac{1}{2\sigma_y^2}(z-x-\mu_y)^2 = -\frac{1}{2}\left(Ax^2 - 2Bx + C\right),$$

where

$$A = \frac{1}{\sigma_x^2} + \frac{1}{\sigma_y^2}, \quad B = \frac{\mu_x}{\sigma_x^2} + \frac{z-\mu_y}{\sigma_y^2}, \quad C = \frac{\mu_x^2}{\sigma_x^2} + \frac{(z-\mu_y)^2}{\sigma_y^2}.$$

Now, by completing the square, we express $Ax^2 - 2Bx + C = A\left(x - \frac{B}{2A}\right)^2 + \left(C - \frac{B^2}{4A}\right)$. Some algebra confirms that $C - \frac{B^2}{4A} = \frac{(z-\mu_x-\mu_y)^2}{\sigma_x^2+\sigma_y^2}$. Noting that $\int_{-\infty}^{\infty} e^{-\frac{A}{2}(x-\frac{B}{2A})^2} dx = \sqrt{\frac{2\pi}{A}}$, we obtain

$$\frac{1}{2\pi\sigma_x\sigma_y} \int_{-\infty}^{\infty} e^{-\frac{1}{2\sigma_y^2}(x-\mu_x)^2 - \frac{1}{2\sigma_y^2}(z-x-\mu_y)^2} dx = \frac{1}{2\pi\sigma_x\sigma_y} \int_{-\infty}^{\infty} e^{-\frac{A}{2}(x-\frac{B}{2A})^2 - \frac{4AC-B^2}{8A}} dx$$

$$= \frac{1}{2\pi\sigma_x\sigma_y} e^{-\frac{4AC-B^2}{8A}} \int_{-\infty}^{\infty} e^{-\frac{A}{2}(x-\frac{B}{2A})^2} dx = \frac{1}{2\pi\sigma_x\sigma_y} e^{-\frac{4AC-B^2}{2A}} \sqrt{\frac{2\pi}{A}}$$

But

$$\sigma_x\sigma_y\sqrt{A} = \sigma_x\sigma_y\sqrt{\frac{\sigma_x^2+\sigma_y^2}{\sigma_x^2\sigma_y^2}} = \sqrt{\sigma_x^2+\sigma_y^2},$$

so we finally arrive at

$$f_Z(z) = \frac{1}{\sqrt{2\pi(\sigma_x^2 + \sigma_y^2)}} e^{-\frac{1}{2(\sigma_x^2+\sigma_y^2)}(z-\mu_x-\mu_y)^2},$$

the density of the normal distribution $\mathcal{N}(\mu_x + \mu_y, \sigma_x^2 + \sigma_y^2)$. □

In short, $X + Y \sim \mathcal{N}(\mu_x + \mu_y, \sigma_x^2 + \sigma_y^2)$ for independent normally distributed $X \sim \mathcal{N}(\mu_x, \sigma_x^2)$ and $Y \sim \mathcal{N}(\mu_y, \sigma_y^2)$. By induction,

$$\sum_{i=1}^{n} X_i \sim \mathcal{N}\left(\sum_{i=1}^{n} \mu_i, \sum_{i=1}^{n} \sigma_i^2\right),$$

where $X_i \overset{iid}{\sim} \mathcal{N}(\mu_i, \sigma_i^2)$, $i = 1, 2, ..., n$.

Convolution has physics and engineering applications among others. Sometimes, the sign $\circledast$ is used to denote convolution. Symbolically, the above theorem can be written as Normal$\circledast$Normal=Normal. An important fact that explains a wide use of convolution is that the Fourier transform ($\mathcal{F}$) of a convolution is the product of Fourier transforms (see Section 2.5.1). Symbolically, $\mathcal{F}(f \circledast g) = \mathcal{F}(f)\mathcal{F}(g)$.

Theorem 3.14 *If X and Y are independent random variables then any functions of these variables are independent as well.*

Proof. Let $f(X)$ and $g(Y)$ be the transformed random variables. For simplicity we shall assume that f and g are strictly increasing functions. Denote f^{-1} and g^{-1} as the respective inverse functions. Then the joint cdf of the pair $(f(X), g(Y))$ is the product of marginal cdfs as can be seen from the following:

$$\begin{aligned}
F_{f(X),g(Y)}(x,y) &= \Pr\left(f(X) \leq x, g(Y) \leq y\right) = \Pr\left(X \leq f^{-1}(x), Y \leq g^{-1}(y)\right) \\
&= \Pr\left(X \leq f^{-1}(x)\right) \times \Pr(Y \leq g^{-1}(y)) = F_{f(X)}(x) \times F_{g(Y)}(y).
\end{aligned}$$

Note that the final equality follows from the independence of X and Y. □

The following result will be used throughout the book.

Theorem 3.15 *If X and Y are independent random variables, then*

$$E(XY) = E(X)E(Y). \tag{3.7}$$

Proof. Using Definition 3.2 for $g(X, Y) = XY$ and the fact that $f(x,y) = f_X(x)f_Y(y)$ due to independence, we have

$$\begin{aligned}
E(XY) &= \int_{-\infty}^{\infty}\int_{-\infty}^{\infty} xy f_X(x) f_Y(y) dx dy = \int_{-\infty}^{\infty}\int_{-\infty}^{\infty} [x f_X(x)][y f_Y(y)] dx dy \\
&= \left[\int_{-\infty}^{\infty} x f_X(x) dx\right]\left[\int_{-\infty}^{\infty} y f_Y(y) dy\right] = E(X)E(Y).
\end{aligned}$$

In words, if two random variables are independent, then the expectation of the product equals the product of expectations.   □

Using the previous two theorems, it is easy to prove that

$$E(f(X) \times g(Y)) = E(f(X)) \times E(g(Y)) \tag{3.8}$$

for independent X and Y.

Corollary 3.16 *The moment generating function (MGF) of the sum of two independent random variables is the product of individual MGFs, $M_{X+Y}(t) = M_X(t)M_Y(t)$.*

Proof. We have $M_{X+Y}(t) = Ee^{(X+Y)t} = E\left(e^{Xt}e^{Yt}\right)$. But e^{Xt} and e^{Yt} are independent because X and Y are independent. Therefore, $M_{X+Y}(t) = Ee^{Xt} \times Ee^{Yt} = M_X(t)M_Y(t)$.   □

By induction, the MGF of the sum of independent random variables is the product of the individual MGFs. This result yields an alternative proof of Theorem 3.13: if $X_1 \sim \mathcal{N}(\mu_1, \sigma_1^2)$ and $X_2 \sim \mathcal{N}(\mu_2, \sigma_2^2)$ are independent, then the MGF of the sum is

$$M_{X+Y}(t) = e^{\mu_1 t + \frac{1}{2}\sigma_1^2 t^2} e^{\mu_2 t + \frac{1}{2}\sigma_2^2 t^2} = e^{\mu_1 t + \frac{1}{2}\sigma_1^2 t^2 + \mu_2 t + \frac{1}{2}\sigma_2^2 t^2} = e^{(\mu_1 + \mu_2)t + \frac{1}{2}(\sigma_1^2 + \sigma_2^2)t^2},$$

which is the MGF of $\mathcal{N}(\mu_1 + \mu_2, \sigma_1^2 + \sigma_2^2)$.

Proving that two random variables are dependent is not as easy as one might think as follows from an example below.

Example 3.17 Uniform pairs. *U and V are independent random variables uniformly distributed on $(0,1)$. (a) Prove that $X = \min(U,V)$ and $Y = \max(U,V)$ are dependent random variables. (b) Generate and plot 10K random (X, Y).*

Solution. (a) First derive the marginal distributions:

$$\begin{aligned}
F_X(x) &= \Pr(\min(U,V) \le x) = 1 - \Pr(\min(U,V) > x) \\
&= 1 - \Pr(U > x, V > x) = 1 - (1-x)^2, \\
F_Y(y) &= \Pr(\max(U,V) \le y) = \Pr(U \le y, V \le y) = y^2, \quad 0 < x < 1, 0 < y < 1.
\end{aligned}$$

Second, to prove that X and Y are dependent, it is sufficient to show that $F_{X,Y}(x,y) \ne F_X(x)F_Y(y)$ at least for one pair (x,y). Consider the case when $x > y$. Then

$$\begin{aligned}
F_{X,Y}(x,y) &= \Pr(\min(U,V) \le x, \max(U,V) \le y) = \Pr(\max(U,V) \le y) \\
&= y^2 \ne y^2(1 - (1-x)^2),
\end{aligned}$$

where the right-hand side of the inequality is the product of the marginal cdfs. Therefore X and Y are dependent. (b) The R code to generate (X, Y) is found in the file `deprunif.r`. We use the built-in functions `pmin` and `pmax` to compute min and max in parallel (for arrays). The dependence between X and Y follows from an obvious observation that the range of values of Y depends on the value of X – the generated points fill up the triangle. A more formal justification follows from the fact that there no points below the line $x = y$. Indeed, for $x > y$ we have $\Pr(X > x)\Pr(Y > y) = (1-x)^2(1-y^2)$, but $\Pr(X > x, Y > y) = 0$ because if $\min(U, V) > x$ and $x > y$, we cannot have $\max(U, V) > y$. □

The following problem arises from a frequent question about ranking. The problem "Who is better?" was discussed earlier using the binomial distribution as a probabilistic model; see Example 1.15. In the following example, we are given the probability of an individual winning, but the question of who is better remains.

Example 3.18 *Who is better?* *The probability that team (or player) A wins against team B is 0.6 and the probability that team C wins against team A is 0.7. What is the probability that team C wins against team B?*

Solution. Obviously, the chance (a lay-language term) that team C wins against B is greater than 50%. The question about the probability is closely related to ranking the teams. A convenient way to rank teams is by using a continuous performance score system. An advantage of this approach is that once teams get scores, the ranking becomes obvious: the greater the performance score the better the team. Here, we suggest to transforming the probabilities onto the score line using quantiles of the normal distribution. Imagine that the performance score of team i is μ_i. Denote the performance of teams A, B, and C as μ_A, μ_B, and μ_C, respectively. Let σ represent the performance scale unit. Then using the cdf of the standard normal cdf, Φ, we express the probabilities A against B and C against A as

$$\Phi\left(\frac{\mu_A - \mu_B}{\sigma}\right) = 0.6, \quad \Phi\left(\frac{\mu_C - \mu_A}{\sigma}\right) = 0.7.$$

Now the question on the probability of team C winning against team B formulates as follows: what is $\Phi((\mu_C - \mu_B)/\sigma)$? On the quantile scale, we have $(\mu_A - \mu_B)/\sigma = \Phi^{-1}(0.6)$ and $(\mu_C - \mu_A)/\sigma = \Phi^{-1}(0.7)$. Simple algebra yields

$$(\mu_C - \mu_B)/\sigma = (\mu_C - \mu_A)/\sigma + (\mu_A - \mu_B)/\sigma = \Phi^{-1}(0.7) + \Phi^{-1}(0.6)$$
$$= 0.2533 + 0.5244 = 0.7777.$$

This implies that the probability of team C winning against team B is

$$\Phi\left(\frac{\mu_C - \mu_B}{\sigma}\right) = \Phi(0.7777) = 0.7816.$$

Surprisingly, the probability of team C winning against team B is not much greater that winning against team A. Also, notice that the performance scale unit, σ, does not affect the result. Thus, without loss of generality, one can assume that $\sigma = 1$.

Problems

1. See Example 3.7. (a) What is the probability that Mary joins the group after John? (b) What is the probability that both John and Mary will join the group after 10:02 a.m.? (c) What is the average time between John's and Mary's calls? Give (i) an answer using nonalgebraic considerations, (ii) using integral evaluations, and (iii) simulations.

2. Is it generally true that if the support of (X, Y) is a triangle, then X and Y are dependent random variables?

3. Random variables X and Y are independent. Is it true that generally (a) the cdf of $F_{X+Y} = 0.5(F_X + F_Y)$, (b) $f_{X+Y} = 0.5(f_X + f_Y)$? Either prove or provide a counterexample.

4. Prove that the mean of the random variable with density (3.5) is equal to $E(X) + E(Y)$.

5. Prove that the result of Example 3.9 does not hold if X and Y are independent but not identically distributed. [Hint: Use uniform distribution.]

6. Repeat calculations and simulations in Example 3.8 if the length of the intervals is 2.

7. Following the derivation in Example 3.9, express $\Pr(X > Y + 1)$ in terms of f and F.

8. Using formula (3.5) prove that $X + Y$, where X and Y are independent exponentially distributed random variables, is also an exponentially distributed random variable.

9. Derive a similar formula to (3.5) for a convolution of discrete random variables. Apply this formula to prove that the convolution of two independent Poisson distributed random variables is again a Poisson distributed random variable. [Hint: Use the induction method of proof.]

10. See Example 3.10. Approximate the cdf of $X + Y$ using the central limit theorem with $n = 2$.

11. (a) Derive the density of the sum of two independent exponentially distributed random variables. (b) Derive the density of a linear combination of two independent exponentially distributed random variables with positive coefficients.

12. Prove that the sum of two independent gamma distributed random variables with the same rate is again a gamma distributed random variable.

13. Provide an example of dependent X and Y for which (3.8) does not hold. [Hint: Use Y as a function of X.]

14. (a) Prove (3.8) if X and Y are independent. (b) Random variables X and Y are independent with known means and variances. Derive the variance of XY using (3.8).

15. Repeat Example 3.11 but now take the longest piece.

16. (a) Apply the result of Example 3.12 to exponentially distributed X and Y. (b) Prove that the volume under the density is 1.

17. Find the distribution and the density function of three independent random variables uniformly distributed on (0,1). Check your formula by writing R code that generates the sum of three $\mathcal{R}(0,1)$ variables, and plot a empirical distribution function with an superimposed analytical curve.

18. Two points are chosen at random on the square $[0,1]^2$. (a) Find the cdf and density of the distance between the points. (b) Use simulations to check the cdf. [Hint: (1) Find the density of $(X_1 - X_2)^2$ for $X_1, X_2 \sim \mathcal{R}(0,1)$. (2) Use a convolution of the two independent densities to get the final answer.]

19. Random variables X and Y are independent and uniformly distributed on (0,1). Find the cdf of $Z = X/(X+Y)$. [Hint: Use a geometric approach.]

20. Random variables $\{X_i, i = 1, ..., n\}$ are independent and have the same distribution. (a) Prove that $\min X_i$ and $\max X_i$ are dependent. (b) Prove that $\min X_i + \max X_i$ and $\max X_i - \min X_i$ are dependent. Illustrate with a scatterplot for simulated values from $\mathcal{R}(0,1)$. [Hint: Use `apply(X,1,min)` and `apply(X,1,max)` for vectorized simulations.]

21. An oil spill happened in the ocean. Due to wind the oil spill moves toward the shore making 1 ft/s in the x-direction and 2 ft/s in the y-direction independently with probability 2/3 and 4/5, respectively (the oil spill happened at the origin). How many hours it will take before the oil spill reaches the shoreline specified by the equation $x + y = 52800$ with probability 0.9? Make and state necessary assumptions.

22. Using a MGF, prove that the sum of independent (a) Poisson distributed random variables is a Poisson distributed random variable, and (b) exponentially distributed variables is an exponentially distributed random variable.

23. (a) Generalize Example 3.17 to any distribution of U and V. (b) Modify function `deprunif` for normally distributed U and V.

24. Let the probability that team A wins against team B be p and the probability that team C wins against team A be $1-p$. Is it true that the probability that team C wins against team B is 0.5?

25. The choice of Φ in Example 3.18 was arbitrary. Compute the probability that team C wins against team B using the cdf $F(x) = e^x/(1+e^x)$.

26. Random points on a unit disk are defined as $(r\cos\theta, r\sin\theta)$, where $r \sim \mathcal{R}(0,1)$ and $\theta \sim \mathcal{R}(0,2\pi)$ are independent. (a) Generate 100K random points on the square $[-1,1]^2$ and display only points that fall within unit radius to see if the they do not have a uniform distribution on the unit disk (use `pch="."` option). (b) Prove that the distribution is indeed not uniform. (c) Find the cdf of r and use the inverse cdf to generate random points on the disk.

3.3 Conditional density

The conditional distribution is an important concept having roots in the conditional probability defined as $\Pr(A|B) = \Pr(A \cap B)/\Pr(B)$. This formula is employed below.

Example 3.19 ***The exponential distribution is memoryless.*** *(a) Let random variable X have an exponential distribution. Prove that $\Pr(X > s+t|X > s) = \Pr(X > t)$, where s and t are positive numbers. (b) Interpret this formula using the telephone Example 2.16.*

Solution. (a) Following the definition of the conditional probability, we have $(s, t > 0)$

$$\Pr(X > s+t|X > s) = \frac{\Pr(X > s+t \cap X > s)}{\Pr(X > s)} = \frac{\Pr(X > s+t)}{\Pr(X > s)}$$

$$= \frac{e^{-\lambda(s+t)}}{e^{-\lambda s}} = e^{-\lambda t} = \Pr(X > t).$$

(b) Assume that Bill did not call before $10 + s$. Find the probability that he will call after $10 + s + t$. Due to the memorylessness of the exponential distribution, this probability will be the same regardless of the information that Bill did not call earlier than s. In other words, condition $X > s$ does not affect the answer – exponential distribution has no memory.   □

The conditional density of Y given $X = x$ can be viewed as the joint density fixed at x and normalized by the marginal density $f_X(x)$:

$$f_{Y|X=x}(y) = \frac{f(x,y)}{f_X(x)}. \tag{3.9}$$

Here, we treat x as a parameter; thus, the conditional density defines a family of densities (sometimes we write simply $f_{Y|X}$). Whenever $f_X(x) = 0$, we set $f_{Y|X=x}(y) = 0$ to comply with $f(x,y) = 0$. Geometrically, the conditional density $f_{Y|X=x}(y)$ is the profile of surface $f(x,y)$ cut through the plane parallel to the

(y, z)-plane passing through point $(x, 0, 0)$, divided by $f_X(x)$; see Figure 3.6. It easy to see that the area under the conditional density is 1 because

$$\int_{-\infty}^{\infty} \frac{f(x, y)}{f_X(x)} dy = \frac{1}{f_X(x)} \int_{-\infty}^{\infty} f(x, y) dy = \frac{1}{f_X(x)} f_X(x) = 1.$$

The animation of Figure 3.6 can be viewed by double clicking on the file CondDens ity.gif using Internet Explorer (alternatively, PowerPoint *insert picture* may be used). Note that the shape and the variance of conditional density does not change because the bivariate normal distribution is used; see Section 3.5.1. In words, conditional density is the normalized profile of the bivariate density.

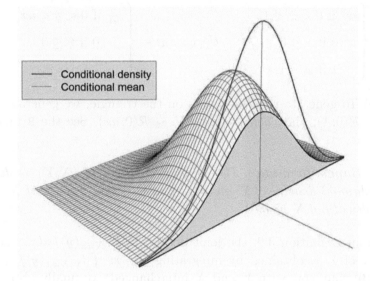

Figure 3.6: *Conditional density and conditional mean. The shaded area under $f(x, y)$ is not 1 but the area under $f(x, y)/f_X(x)$ is 1.*

Theorem 3.20 *Random variables X and Y are independent if and only if the conditional density equals the marginal density.*

Proof. If X and Y are independent, then $f(x, y) = f_X(x)f_Y(y)$. Therefore

$$f_{Y|X=x}(y) = \frac{f_X(x)f_Y(y)}{f_X(x)} = f_Y(y).$$

The same is true for $X|Y = y$, that is, $f_{X|Y=y}(x) = f_X(x)$ if X and Y are independent. Reversely, if $f_{Y|X=x}(y) = f_Y(y)$ (or $f_{X|Y=y}(x) = f_X(x)$), then $f_X(x)f_Y(y) = f(x, y)$. □

The following example illustrates how to generate random pairs (X, Y) from a bivariate distribution using univariate and conditional distributions. First, generate X according to its marginal distribution and, second, generate Y from the conditional distribution $Y|X$.

Example 3.21 *Conditional distribution on a triangle.* *Consider a uniform distribution on the triangle ABC with vertices $A = (0,0), B = (1,a), C = (1,0)$, where a is a positive parameter. (a) Find the marginal cdf for X and the conditional cdf $Y|X = x$. (b) Generate and plot $n = 10,000$ random points on the triangle with a as an argument in the R code.*

Solution. (a) Since the distribution is uniform on the triangle, the marginal cdf is the ratio of the area to the left of x to the entire area, and conditional cdf of Y at $X = x$ is a uniform distribution $Y|(X = x) \sim \mathcal{R}(0, ax)$ with the cdfs given by

$$F_X(x) = \begin{cases} x^2 \text{ if } 0 < x < 1 \\ 0 \text{ if } x \leq 0 \\ 1 \text{ if } x \geq 1 \end{cases}, \qquad F_{Y|X=x}(y) = \begin{cases} \frac{y}{ax} \text{ if } 0 < y < ax \\ 0 \text{ if } y \leq 0 \\ 1 \text{ if } y \geq ax \end{cases},$$

respectively. (b) To generate random points on the triangle, we generate X as $\sqrt{Z}$ where $Z \sim \mathcal{R}(0, 1)$, and then generate $Y \sim \mathcal{R}(0, ax)$. See the R function `rantr`.

Example 3.22 *Bayes formula.* *The joint distribution of (X, Y) is defined through the conditional density of Y given X and marginal density of X. Find the conditional density of X given Y.*

Solution. From Definition 3.9, the joint density is $f_{Y|X=x}(y)f_X(x)$ and the marginal density of Y is obtained by integrating out x: $\int f_{Y|X=x}(y)f_X(x)dx$. Again, using (3.9), but now with Y and X interchanged, we finally obtain the famous Bayes formula

$$f_{X|Y=y}(x) = \frac{f_{Y|X=x}(y)f_X(x)}{\int_{-\infty}^{\infty} f_{Y|X=x}(y)f_X(x)dx}. \qquad (3.10)$$

Example 3.23 *Bayes formula for the normal distribution.* *The conditional density of Y given X is a normal distribution $\mathcal{N}(x, 1)$, and the marginal distribution of X is $\mathcal{N}(0, 1)$. Find the conditional density of X given Y using the Bayes formula (3.10).*

Solution. The numerator is

$$\frac{1}{\sqrt{2\pi}} e^{-\frac{1}{2}(y-x)^2} \times \frac{1}{\sqrt{2\pi}} e^{-\frac{1}{2}x^2} = \frac{1}{2\pi} e^{-\frac{1}{2}[(y-x)^2+x^2]}.$$

The denominator is the integral of the numerator over x:

$$\frac{1}{2\pi} \int_{-\infty}^{\infty} e^{-\frac{1}{2}[(y-x)^2+x^2]} dx.$$

By completing the square, similarly to the proof of Theorem 3.13, we obtain $(y-x)^2 + x^2 = 2x^2 - 2xy + y^2 = 2(x - y/2)^2 + y^2/2$. Using this result, we derive

$$\frac{1}{2\pi} \int_{-\infty}^{\infty} e^{-\frac{1}{2}[(y-x)^2 + x^2]} dx = \frac{e^{-y^2/4}}{\sqrt{2\pi}} \int_{-\infty}^{\infty} \frac{1}{\sqrt{2\pi}} e^{-(x-y/2)^2} dx$$

$$= \frac{e^{-y^2/4}}{\sqrt{2\pi}\sqrt{2}} \int_{-\infty}^{\infty} \frac{\sqrt{2}}{\sqrt{2\pi}} e^{-z^2} dz = \frac{1}{\sqrt{4\pi}} e^{-y^2/4}$$

because $e^{-z^2}/\sqrt{\pi}$ is the density of $\mathcal{N}(0, 1/2)$ and, therefore, the denominator of (3.10) equals 1. Note that $e^{-y^2/4}/\sqrt{4\pi}$ is the density of the normal distribution with zero mean and variance 2. Finally, the conditional density of X given Y is $\pi^{-1/2} e^{-\frac{1}{4}(y-2x)^2} \sim \mathcal{N}(y/2, 1/2)$.

3.3.1 Conditional mean and variance

The conditional mean (expectation), sometimes referred to as a *regression*, is an important concept because it helps to understand how Y is related to X. This task is especially important in statistics where we want to describe the relationship between Y and X. The conditional mean and variance are easy to define using conditional density. The conditional mean is a function of fixed x and defined in a straightforward manner as

$$\mu(x) = E(Y|X = x) = \int_{-\infty}^{\infty} y f_{Y|X}(y) dy. \tag{3.11}$$

The conditional variance is defined as

$$\text{var}(Y|X = x) = E\left((Y - \mu(x))^2 | X = x\right).$$

Alternatively, we can express

$$\text{var}(Y|X = x) = E(Y^2|X = x) - E^2(Y|X = x). \tag{3.12}$$

The conditional mean is illustrated in Figure 3.6, the green line. For this particular bivariate density the conditional mean is a linear function of x, but this is generally not the case.

Theorem 3.24 *If X and Y are independent, then the conditional mean and variance are constant.*

Proof. Indeed,

$$E(Y|X = x) = \int_{-\infty}^{\infty} y f_{Y|X}(y) dy = \int_{-\infty}^{\infty} y f_Y(y) dy = E(Y) = \text{const.}$$

Similarly,

$$E(Y^2|X = x) = \int_{-\infty}^{\infty} y^2 f_Y(y) dy = E(Y^2) = \text{const}$$

and var = const as follows from (3.12). The same is true for the conditional mean and variance for X given $Y = y$. These attributes of independence can be used in graphical assessment based on the cloud of generated pairs. If the cloud is not parallel to the x-axis or the standard deviation for each x is not constant, the variables are dependent.

Example 3.25 *Conditional mean and variance on the uniform triangle.* *Find the conditional mean and variance for a uniform distribution on the triangle ABC with vertices $A = (0,0)$, $B = (1,2)$, $C = (1,0)$, and prove that X and Y are dependent. Display the mean $\pm$ SD lines.*

Solution. From the example above, we know that the conditional distribution of Y given $X = x$ is uniform, $Y|X = x \sim \mathcal{R}(0, 2x)$ for any $x \in (0, 1)$. Therefore, $E(Y|X = x) = x$, the median. Since $Y|X = x$ has a uniform distribution on $(0, x)$, we have $\text{var}(Y|X = x) = x^2/3$. Since $E(Y|X = x) \neq$ const, X and Y are dependent. □

Prediction of Y given $X = x$ is one of the most important applications of probability and statistics. Regression as conditional mean has an important optimal property for prediction as the prediction with minimum variation.

Example 3.26 *Regression is an optimal predictor.* *Prove that conditional mean/regression is the optimal predictor of Y given $X = x$.*

Solution. This property relies on one elementary fact: if Y is a random variable with mean $E(Y) = \mu$, then, for any other number c, we have $E(Y - \mu)^2 \leq E(Y - c)^2$ and the inequality turns into an equality if and only if $\mu = c$. Indeed, using elementary algebra, we obtain

$$
\begin{aligned}
E(Y - c)^2 &= E[(Y - \mu) + (\mu - c)]^2 = E(Y - \mu)^2 + 2E[(Y - \mu)(\mu - c)] \\
+(\mu - c)^2 &= E(Y - \mu)^2 + (\mu - c)^2
\end{aligned}
$$

because $E(Y - \mu) = 0$. Thus, $E(Y - c)^2 \geq E(Y - \mu)^2$ with equality if and only if $\mu = c$. Now, we simply view Y as the random variable conditional on $X = x$ with $\mu = \mu(x) = E(Y|X = x)$, the regression of Y on X. □

Below are important results that connect the (marginal) mean and variance to the conditional mean and variance. They provide a useful way to derive the marginal mean and variance when the joint distribution is derived via the conditional distribution.

Theorem 3.27 *The rule of repeated expectation.* *The expected value of Y can be obtained in two steps. (i) Find the regression as conditional mean, and (ii) find the expected value of the regression:*

$$ E(Y) = E_X(E(Y|X)). \tag{3.13} $$

Proof. We must clarify that the inner expectation is a function of X and the outer expectation is the expectation over X. Replace in (3.11) $f_{Y|X}(y)$ with $f(x,y)/f_X(x)$. Then

$$E_X(E(Y|X)) = \int_{-\infty}^{\infty} \left(\int_{-\infty}^{\infty} y \frac{f(x,y)}{f_X(x)} dy \right) f_X(x) dx = \int_{-\infty}^{\infty} \int_{-\infty}^{\infty} y f(x,y) dy dx$$

$$= \int_{-\infty}^{\infty} y f_Y(y) dy = E(Y).$$

Formula (3.13) is derived.

Example 3.28 *Expectation inequality.* (a) *Prove that*

$$E_{X,Y}(g(X,Y)) \neq E_X(E_Y(g(X,Y))), \qquad (3.14)$$

where g is a function of random variables X and Y, and $E_Y(g(X,Y))$ is derived as if X is a fixed number. (b) Prove that the inequality turns into equality if X and Y are independent. (c) Express $E_{X,Y}(g(X,Y))$ through the conditional mean in the general case using the rule of repeated expectation.

Solution. (a) Let $X = Y$ and $g(X,Y) = XY$. Then the left-hand side of (3.14) is $E(X^2)$. The right-hand side is

$$E_X(E_Y(g(X,Y))) = E_X(XE(X)) = E^2(X).$$

But $E(X^2) > E^2(X)$ unless X is a fixed number. (b) First, note that if X and Y are independent, then the example $X = Y$ in (a) does not work. Second, if X and Y are independent, then the joint density is $f_X(x)f_Y(y)$ and

$$E_{X,Y}(g(X,Y)) = \int \int g(x,y) f_X(x) f_Y(y) dx dy$$

$$= \int \left(\int g(x,y) f_Y(y) dy \right) f_X(x) dx = E_X(E_Y(g(X,Y))).$$

(c) Following the rule of repeated expectation, we express $E_{X,Y}(g(X,Y)) = E_X(E(g(X,Y)|X))$. Note the right-hand side of the above equation differs from the right-hand side of (3.14): the former is the marginal mean holding X fixed, and the latter is the conditional mean. If the above formula is applied to the example in (a), we get the right answer because $E(X^2|X) = X^2$. □

Obviously, the rule of repeated expectation (3.13) can be generalized to any function $f(Y)$

$$E(f(Y)) = E(E(f(Y)|X)). \qquad (3.15)$$

This generalization for the case $f(Y) = (Y - \mu_Y)^2$ is used for the proof of the following theorem.

Theorem 3.29 *Variance decomposition.* *The variance of Y can be decomposed into the expected conditional variance and the variance of the conditional expectation, or*

$$\text{var}(Y) = E(\text{var}(Y|X)) + \text{var}(E(Y|X)). \qquad (3.16)$$

Proof. Apply (3.15) letting $f(Y) = (Y - \mu_Y)^2$. Then denoting $\mu(X) = E(Y|X)$, the inner expectation can be rewritten as

$$E((Y - \mu_Y)^2|X) = E((Y - \mu(X) + \mu(X) - \mu_Y)^2|X)$$
$$= E((Y - \mu(X))^2|X) + 2E((Y - \mu(X))(\mu(X) - \mu_Y)|X) + E((\mu(X) - \mu_Y)^2|X).$$

The second term vanishes because $E(Y|X) = \mu(X)$ and

$$E((Y - \mu(X))(\mu(X) - \mu_Y)|X) = (E(Y|X) - \mu(X))(\mu(X) - \mu_Y) = 0.$$

The first and the second term can be expressed as

$$E_X[E((Y - \mu(X))^2|X)] = E(\text{var}(Y|X)), \quad E((\mu(X) - \mu_Y)^2|X) = \text{var}(E(Y|X))$$

because $E(\mu(X)) = \mu_Y$. Formula (3.16) is proved. $\qquad\square$

The variance decomposition (3.16) gives rise to the coefficient of determination (squared correlation coefficient): $\text{var}(Y)$ is referred to as the total variance, $E(\text{var}(Y|X))$ is referred to as explained variance, and $\text{var}(E(Y|X))$ is referred to as unexplained variance.

Definition 3.30 *Coefficient of determination.* *In the framework of conditional expectation/regression, the ratio of explained variance to total variance is called the coefficient of variation,*

$$\rho^2(x) = \frac{\text{var}(E(Y|X))}{\text{var}(Y)}.$$

As follows from variance decomposition (3.15), $0 \leq \rho^2(x) \leq 1$. When $\rho^2(x) = 1$ for all x as follows from the proof of the above theorem, Y and X are functionally related and $\text{var}(E(Y|X)) = 0$; see Example 3.45. A popular interpretation of the coefficient of determination is the proportion of variance of Y explained by X. Discussion of further properties of the coefficient of determination is deferred to Section 3.5.2, where the bivariate normal distribution is introduced. In particular, we will prove that $\rho^2(x)$ does not depend on x and is equal to the squared correlation coefficient, ρ, as defined in Section 3.4; therefore the notation.

Example 3.31 *Variance decomposition on the uniform triangle.* *Illustrate formulas (3.13) and (3.16) on the triangle distribution from Example 3.25.*

Solution. Formula (3.13): The cdf of X is x^2, the area of the triangle to the left of x, with the marginal density $f_X(x) = 2x$. Therefore, $E(E(Y|X)) = \int_0^1 2x^2 dx = 2/3$. Analogously, the cdf of Y is $1 - (2-y)^2/4$ and $f_Y(y) = (2-y)/2$. The expected value is $E(Y) = 0.5 \int_0^2 y(2-y)dy = 2/3$. Formula (3.16): We have $\text{var}(Y) = 0.5 \int_0^2 y^2(2-y)dy - E^2(Y) = 2/9$ and

$$E(\text{var}(Y|X)) = \int_0^1 \frac{1}{3} x^2 f_X(x) dx = \frac{2}{3} \int_0^1 x^3 dx = \frac{1}{6},$$

$$\text{var}(E(Y|X)) = \text{var}(X) = 2 \int_0^1 x^3 dx - \left(2 \int_0^1 x^2 dx\right)^2 = \frac{1}{18}$$

so that $E(\text{var}(Y|X)) + \text{var}(E(Y|X)) = 1/6 + 1/18 = 2/9 = \text{var}(Y)$.

Example 3.32 *Conditional density and expectation.* (a) *Find the conditional density of $(Y+X)|X = x$ and the conditional expectation, where X and Y are independent.* (b) *Prove that if X and Y are independent and $h(x,y)$ is any function, then*

$$E(h(X,Y)|X = x) = E_Y(h(x,Y)). \tag{3.17}$$

(c) *Derive (a) from (b).*

Solution. (a) Using joint density (3.6), we find

$$f_{Y+X|X=x}(y) = \frac{f_{X,X+Y}(x,y)}{f_X(x)} = f_Y(y-x).$$

Therefore, the conditional expectation is

$$E(X+Y|X = x) = \int_{-\infty}^{\infty} y f_Y(y-x)dy = \int_{-\infty}^{\infty} (x+z)f_Y(z)dz$$

$$= x \int_{-\infty}^{\infty} f_Y(z)dz + \int_{-\infty}^{\infty} z f_Y(z)dz = x + \mu_Y,$$

an intuitively expected result.

(b) If f denotes the joint pdf, by the definition of the conditional density,

$$E(h(X,Y)|X = x) = \int_{-\infty}^{\infty} h(x,y) \frac{f(x,y)}{f_X(x)} dy.$$

Since X and Y are independent, we have $f(x,y) = f_Y(y)f_X(x)$ and

$$E(h(X,Y)|X = x) = \int_{-\infty}^{\infty} h(x,y)f_Y(y)dy = E_Y(h(x,Y)).$$

(c) In case (a) we have $h(x,y) = x+y$ and, therefore, $E(X+Y|X = x) = E_Y(x+Y) = x + \mu_Y$.

Example 3.33 *Conditional mean under inequality.* (a) Express $E(X|X \leq c)$ using the density of X. (b) Find this conditional mean as a function of c for $X \sim \mathcal{N}(0,1)$, and (c) test your answer via simulations.

Solution. (a) First, find the conditional cdf of $X|X \leq c$ as

$$\Pr(X \leq x | X \leq c) = \frac{\Pr(X \leq x \cap X \leq c)}{\Pr(X \leq c)} = \frac{\Pr(X \leq x)}{\Pr(X \leq c)} = \frac{F(x)}{F(c)},$$

where F is the cdf of X since we may assume that $x \leq c$ (otherwise $F(x) = 1$). Second, the density of $X|X \leq c$ is $f(x)/F(c)$ for $-\infty < x \leq c$, where f is the density of X. Finally,

$$E(X|X \leq c) = \frac{1}{\int_{-\infty}^{c} f(x)dx} \int_{-\infty}^{c} xf(x)dx.$$

(b) Now apply this formula to the case when $f = \phi$. Find

$$\int_{-\infty}^{c} x\phi(x)dx = \frac{1}{\sqrt{2\pi}} \int_{-\infty}^{c} xe^{-x^2/2}dx = \frac{1}{\sqrt{2\pi}} \int_{-\infty}^{c} e^{-x^2/2}d(x^2/2) = -\frac{1}{\sqrt{2\pi}}e^{-\frac{1}{2}c^2},$$

so that $E_{X \sim \mathcal{N}(0,1)}(X|X \leq c) = -\phi(c)/\Phi(c)$. (c) The R code `exc` plots theoretical expectations and those from simulations (not shown). Simulations match the analytical result perfectly. □

The conditional distribution can be used to specify the joint distribution using the formula

$$f(x,y) = f_{Y|X=x}(y)f_X(x).$$

According to this formula, we need to specify the marginal and conditional distributions. Then the joint distribution is the product of the two. This representation can be used to generate random numbers with given bivariate density $f(x,y)$ in two steps: (i) generate X according to $f_X(x)$ and (ii) generate Y according to $f_{Y|X=x}(y)$. This idea is used in the following example.

Example 3.34 *Simulations of the bivariate exponential distribution.* The joint density is defined as $f(x,y) = e^{-x}$ for $0 \leq y \leq x < \infty$ and 0 elsewhere. Generate and plot N pairs from this distribution.

Solution. The marginal density of X is

$$f_X(x) = \int_0^x f(x,y)dy = \int_0^x e^{-x}dy = e^{-x}\int_0^x dy = xe^{-x}, \quad x \geq 0.$$

The conditional density of Y given $X = x$ is uniform, $\mathcal{R}(0,x)$. Thus, to generate (X_i, Y_i), we (i) generate X_i with density xe^{-x} and (ii) generate Y_i uniformly distributed on $[0, X_i]$. Note that xe^{-x} is the density of the gamma distribution with the shape parameter $\alpha = 2$ and the rate parameter $\lambda = 1$. Hence, N values X_i can be generated using the R function `rgamma`. See the R code `exexpg`.   □

In the following example, we compute the conditional mean for a discrete distribution.

Example 3.35 *Bivariate Bernoulli distribution.* *Let Y and X be two Bernoulli random variables, distributed as specified by the 2×2 elementary probability table:*

$$\Pr(X = 0, Y = 0) = p_{00} \quad \Pr(X = 0, Y = 1) = p_{01}$$
$$\Pr(X = 1, Y = 0) = p_{10} \quad \Pr(X = 1, Y = 1) = p_{11}$$

Find the conditional mean $E(Y|X = x)$ where $x = 0$ or $x = 1$.

Solution. Obviously, $p_{00} + p_{01} + p_{10} + p_{11} = 1$. Marginal probabilities can be expressed through elementary probabilities $\{p_{ij}, i = 0, 1, j = 0, 1\}$ as follows

$$\Pr(X = 1) = \Pr(X = 1, Y = 0) + \Pr(X = 1, Y = 1) = p_{10} + p_{11}$$

and analogously

$$\Pr(Y = 1) = \Pr(X = 0, Y = 1) + \Pr(X = 1, Y = 1) = p_{01} + p_{11}.$$

Since Y is binary, the expected value reduces to probability, $E(Y|X) = \Pr(Y = 1|X)$. Thus, it suffices to find the conditional probability for $X = 0$ and $X = 1$ as

$$E(Y|X = x) = \begin{cases} p_{01}/(1 - p_{10} - p_{11}) & \text{if } x = 0 \\ p_{11}/(p_{01} + p_{11}) & \text{if } x = 1 \end{cases}.$$

□

The following demonstrates that the naive estimator of π in Example 1.1 is unacceptable in the presence of measurement error. Specifically, the average of empirical πs systematically overestimates π. On the other hand, the ratio of the sum of all circumferences to the sum of all diameters converges to π when the number of measured tires increases to infinity.

Example 3.36 *Who said π?* *Let iid measurement errors of the tire circumference and the diameter be ε and δ, respectively, and let n independent measurements be available. (a) Prove that the average of the ratios of the measured circumferences to the diameter (referred to as the naive estimator of π) has a systematic positive bias when both measurement errors are uniformly distributed on a small interval. (b) Using Jensen's inequality, prove that the naive estimator of π is positively biased for any distribution, not necessarily uniform. (c) Prove that the ratio of the averages of the measured circumferences to diameters is also positively biased but converges to π when the number of measured tires goes to infinity.*

Solution. In short, we aim to prove that the average of ratios is biased and inconsistent, but the ratio of averages is consistent although positively biased.

(a) To be specific, let us assume that measurement error of the diameter and circumference is uniform on the interval $[-h, h]$, where $h > 0$ is a small number. The average of ratios (the naive estimator of π) is

$$\widehat{\pi}_N = \frac{1}{n} \sum_{i=1}^{n} \frac{C_i}{D_i} = \frac{1}{n} \sum_{i=1}^{n} \frac{\pi d_i + \varepsilon_i}{d_i + \delta_i}, \tag{3.18}$$

where d_i is the true (nonrandom) diameter, $D_i = d_i + \delta_i$ is the measured diameter, and $C_i = \pi d_i + \varepsilon_i$ is the measured circumference (h is small enough such that $d_i > h$ for all i). The expectation of the ith term is

$$E\left(\frac{\pi d_i + \varepsilon_i}{d_i + \delta_i}\right) = \frac{1}{(2h)^2} \int_{-h}^{h} \int_{-h}^{h} \frac{\pi d_i + \varepsilon}{d_i + \delta} d\varepsilon d\delta = \pi \frac{d_i}{2h} \ln \frac{d_i + h}{d_i - h}.$$

Let $x = h/d_i$. Introduce function $f(x) = \ln[(1+x)/(1-x)] - 2x$ for $0 < x < 1$. We have $f(0) = 0$ and

$$f'(x) = \frac{1}{1+x} + \frac{1}{1-x} - 2 = \frac{2x^2}{1-x^2} > 0.$$

Therefore, $f(x) > 0$ for $0 < x < 1$, and

$$\frac{d_i}{2h} \ln \frac{d_i + h}{d_i - h} > 1 \tag{3.19}$$

for every i. Finally,

$$E(\widehat{\pi}_N) = \frac{1}{n} \sum_{i=1}^{n} E\left(\frac{\pi d_i + \varepsilon_i}{d_i + \delta_i}\right) > \pi. \tag{3.20}$$

In words, the naive estimator systematically overestimates π.

(b) Now we generalize the above result to any distribution of the measurement error using Jensen's inequality (2.12): $E(1/X) > 1/E(X)$ for a positive random variable X assuming that $E(X)$ exists. Indeed, the expectation of the ith term in (3.18) can be rewritten as

$$E\left(\frac{\pi d_i + \varepsilon_i}{d_i + \delta_i}\right) = E\left(\pi d_i + \varepsilon_i\right) \times E\left(\frac{1}{d_i + \delta_i}\right) = \pi d_i \times E\left(\frac{1}{d_i + \delta_i}\right)$$

because the numerator and the denominator are independent random variables. Letting $X = d_i + \delta_i$, we obtain $E\left(d_i + \delta_i\right)^{-1} > E^{-1}(d_i + \varepsilon_i) = 1/d_i$ and the bias remains.

(c) Now we prove that the ratio of the total circumference to total diameter

$$\widehat{\pi}_R = \frac{\sum_{i=1}^{n} C_i}{\sum_{i=1}^{n} D_i} = \frac{\sum_{i=1}^{n} (\pi d_i + \varepsilon_i)}{\sum_{i=1}^{n} (d_i + \delta_i)}$$

is consistent, i.e. $p \lim_{n\to\infty} \widehat{\pi}_R = \pi$ under assumption that the average of true diameters, d_i, exists when $n \to \infty$, namely, $\bar{d} = \lim_{n\to\infty} \sum_{i=1}^{n} d_i/n > 0$. Indeed, by the law of large numbers (LLN) formulated in Theorem 2.43, we have

$$\bar{\varepsilon} = p \lim_{n\to\infty} \frac{1}{n} \sum_{i=1}^{n} \varepsilon_i = 0, \quad \bar{\delta} = p \lim_{n\to\infty} \frac{1}{n} \sum_{i=1}^{n} \delta_i = 0$$

and, therefore,

$$p \lim_{n\to\infty} \widehat{\pi}_R = p \lim_{n\to\infty} \frac{\sum_{i=1}^{n} (\pi d_i + \varepsilon_i)/n}{\sum_{i=1}^{n} (d_i + \delta_i)/n} = \frac{\pi \bar{d}}{\bar{d}} = \pi.$$

Now, we prove that $\widetilde{\pi}$ is positively based, again using Jensen's inequality:

$$E(\widehat{\pi}_R) = E\left(\sum_{i=1}^{n} C_i\right) E\left(\frac{1}{\sum_{i=1}^{n} D_i}\right) > \pi \frac{n^{-1}\sum_{i=1}^{n} d_i}{n^{-1}\sum_{i=1}^{n} d_i} = \pi.$$

Although biased, the ratio of the average circumferences to the average diameters converges to π when $n \to \infty$. The problem of estimating π is continued in Example 6.126.

3.3.2 Mixture distribution and Bayesian statistics

The mixture distribution emerges in statistics when studying multigroup or multicluster populations. A popular probability model for such a distribution is the *mixture distribution* model. Often mixture distributions have multimodal densities, but not always. In short, a mixture distribution is a linear combination of disruptions. A more rigorous definition is (i) define the distribution of the continuous variable X for each value of the discrete distribution Y treated as the *conditional* distribution, and (ii) define the probability distribution for the discrete variable Y, treated as the *marginal* distribution. To illustrate, we assume that there are two groups, corresponding to the binary random variable outcomes $Y = 1$ and $Y = 0$, respectively, connected to the continuous densities $f_1(x)$ and $f_0(x)$. The conditional density of X given $Y = y$ is

$$f_{X|Y}(x|Y = y) = yf_1(x) + (1 - y)f_0(x),$$

where $y = 0$ or $y = 1$. If the marginal probabilities are defined as $\Pr(Y = 1) = p$ and $\Pr(Y = 0) = 1 - p$, then the marginal density of X is given by

$$f_X(x) = pf_1(x) + (1 - p)f_0(x),$$

the *mixture density*. Sometimes, we want to inverse the probability, find the probability of Y given observation $X = x$. This conditional probability is found from the general Bayes formula,

$$\Pr(Y = 1|X = x) = \frac{pf_1(x)}{pf_1(x) + (1 - p)f_0(x)},$$

$$\Pr(Y = 0|X = x) = \frac{(1 - p)f_0(x)}{pf_1(x) + (1 - p)f_0(x)},$$

and will be referred to as the *Bayesian classifier* because they use the Bayesian formula for conditional probability. Below, we illustrate how the Bayesian classifier can be used to predict gender based on a person's height.

Example 3.37 *Man or woman?* *The height distribution of men and women is defined via normal densities with the means 70 and 64, and standard deviations 4 and 3 in., respectively. (a) Plot the mixture density under assumption that $p = 0.5$ (b) Using surveillance camera it was estimated that the height of a suspect was 68 in. What was the gender of the person (estimate the probability)? (c) Using the 50% probability for gender classification prove that the height threshold is when the densities intersect, and find the threshold for this problem.*

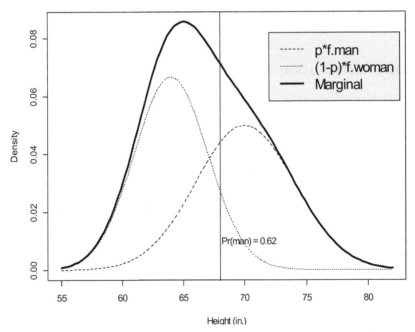

Figure 3.7: *Gaussian mixture distribution of heights. The vertical bar corresponds to the height 68 in. The probability that the person was a man is 0.62.*

Solution. (a) The R function `mixedDens` does the computation and plots Figure 3.7. When the densities f_1 and f_2 are normal, the marginal density f_X is called the *Gaussian mixture* density:

$$f_X(x) = 0.5\phi(x, \text{mean} = 70, \text{sd} = 4) + 0.5\phi(x, \text{mean} = 64, \text{sd} = 3),$$

where ϕ denotes the normal density with respective μ and σ. Note that the marginal density is unimodal but does not have a familiar bell shape. (b) We estimate the probability that the person with the height 68 in. is a man as

$$\Pr(Y = 1|X = 68) = \frac{0.5\phi(68, \text{mean} = 70, \text{sd} = 4)}{0.5\phi(68, \text{mean} = 70, \text{sd} = 4) + 0.5\phi(68, \text{mean} = 64, \text{sd} = 3)}$$
$$= 0.62.$$

(c) The Bayesian classifier computes the probability. Instead, to facilitate classification, we can find the height threshold, x that leads to the probability $p = 0.5$. The implied equation takes the form $f_1(x)/[f_1(x) + f_2(x)] = 0.5$ which implies $f_1(x) = f_2(x)$. For our Gaussian mixture this equation reduces to a quadratic equation $4^{-1}e^{-\frac{1}{2\times 4^2}(x-70)^2} = 3^{-1}e^{-\frac{1}{2\times 3^2}(x-64)^2}$ or

$$-\frac{1}{2 \times 4^2}(x - 70)^2 - \ln 4 = -\frac{1}{2 \times 3^2}(x - 64)^2 - \ln 3$$

with the solution

$$x = \frac{394}{7} + \frac{12}{7}\sqrt{36 + 14\ln 4 - 14\ln 3} = 67.1.$$

Thus, we say that a person is a woman if the height < 67.1 in. and otherwise the person is a man.

Example 3.38 *Mortgage default. The decision on granting or declining a mortgage application is very important. The financial officer has records on 1000 applications. Nine hundred applicants did not default with family income following a lognormal distribution, $\mu_1 = 5$ and $\sigma_1 = 0.4$ (thousand dollars);10 applicants defaulted with family income lognormally distributed, $\mu = 3.5$ and $\sigma = 0.3$ (also see Section 2.11). Determine the income threshold based on the family income to discriminate defaulters from nondefaulters.*

Solution. First, we consider income distribution on the log scale $x = \ln z$, where z is expressed in thousands of dollars, with the marginal density

$$f_X(x) = 0.9\phi(x, \text{mean} = 5, \text{sd} = 1) + 0.1\phi(x, \text{mean} = 3.5, \text{sd} = 0.5).$$

Following the line of argumentation from the previous example, we grant the application if $x > x_*$ where x_* is the solution to the equation

$$0.9\phi(x, \text{mean} = 5, \text{sd} = 1) = 0.1\phi(x, \text{mean} = 3.5, \text{sd} = 0.5),$$

or explicitly

$$\frac{9}{10}\frac{1}{\sqrt{2\pi\sigma_1^2}}e^{-\frac{1}{2\sigma_1^2}(x-\mu_1)^2} = \frac{1}{10}\frac{1}{\sqrt{2\pi\sigma_2^2}}e^{-\frac{1}{2\sigma_2^2}(x-\mu_2)^2},$$

where $\mu_1 = 5$, $\mu_2 = 3.5$, $\sigma_1 = 0.4$, $\sigma_2 = 0.3$. The solution reduces to a quadratic equation after taking the log transformation, $x_* = 3.985$. Therefore, if family income is less than $z_* = e^{3.985} = \$53,785$ the application should be declined. $\square$

The following example illustrates Bayesian statistics where, unlike the frequentist approach studied in this book, parameters are treated as random with distribution given a priori. We do not even try to discuss Bayesian statistics here and refer interested readers to established texts, Gelman and Carlin [36] or Gill [39].

Recall that in Example 1.15 we tried to answer the question, what team is better based on the comparison of probabilities of the number of wins as a function of a win in a single competition. Often, this method leads to an inconclusive answer when probability curves overlap. In contrast, the Bayesian approach, when the binomial probabilities are treated as conditional probabilities, becomes fruitful because it reduces to comparing numbers.

Example 3.39 *What team is better (continued)?* *Answer this question by treating p as a random variable uniformly distributed on $(0, 1)$ with binomial probabilities treated as conditional probabilities. Predict the UFC fight between two fighters, one who previously won all fights out of 13 (Fighter A) and one who won 26 out of 31 (Fighter B).*

Solution. For Fighter A, the conditional probability of winning all 13 fights is p^{13} and, since $p \sim \mathcal{R}(0, 1)$, the marginal probability of winning is $\int_0^1 p^{13} dp = 0.071$. We call this uniform distribution the *noninformative prior*. For Fighter B, this probability is $\int_0^1 \binom{31}{5} p^5 (1-p)^{31-5} dp = 0.031$. We conclude that that Fighter A will win.

3.3.3 Random sum

Sometimes, we want to find the expectation of a sum of random variables with a random number of terms, $\sum_{i=1}^N X_i$, where N and $X_1, ..., X_N$ are random. To simplify, we shall assume that $\{X_i\}$ are iid conditional on N with distribution independent of N. The rule of repeated expectation formulated in Theorem 3.27 can be used to find the expected value of the random sum:

$$E\left(\sum_{i=1}^N X_i\right) = E_N\left(E_X\left(\sum_{i=1}^n X_i \,\middle|\, n = N\right)\right) = E_N\left(E(X_1) \times N\right) = E(N)E(X_1).$$

The following example illustrates an immediate application of this result.

Example 3.40 *Car accident claims.* *An auto-insurance company wants to know the expected total amount of claims per month. It is known that the average number of claims is 1,128 and the average claim is \$3,500. What is the expected total amount?*

Solution. The total amount can be viewed as a random sum with $E(N) = 1,128$ and $E(X_1) = \$3,500$. Under the assumption that claims are iid and do not depend on N, the answer is $1,128 \times \$3,500 = \$3,948$. $\square$

The next example elaborates on the variance and the distribution of the random sum in the case when N follows a nonzero Poisson distribution.

Example 3.41 *Random sums.* *(a) Derive the mean of the random sum with N following the nonzero Poisson distribution and with a normal distribution for X_i. (b) Derive the marginal variance using the variance decomposition formula (3.16). (c) Derive the distribution of the random sum. (Is it normal?). (d) Write R code to check the formulas via simulations, and check the distribution by plotting the theoretical and empirical cdfs on the same graph.*

Solution. (a) We cannot use the regular Poisson distribution for N because it starts with zero; see Section 1.7. The nonzero Poisson distribution takes values $n = 1, 2, \ldots$ with the adjustment coefficient $(1-e^{-\lambda})^{-1}$ to ensure that the probabilities add to one:

$$\Pr(N = n) = \frac{1}{1 - e^{-\lambda}} \frac{\lambda^n}{n!} e^{-\lambda}, \quad n = 1, 2, \ldots$$

The mean of a nonzero Poison distribution is easy to derive, $E(N) = \lambda/(1-e^{-\lambda})$, and therefore the marginal mean of the random sum is given by $E\left(\sum_{i=1}^{N} X_i\right) = \lambda\mu/(1-e^{-\lambda})$. (b) To derive the variance, we note that $E(N^2) = (\lambda+\lambda^2)/(1-e^{-\lambda})$, and, using $\text{var}(N) = E(N^2) - E^2(N)$, we obtain

$$\text{var}(N) = \frac{\lambda + \lambda^2}{1 - e^{-\lambda}} - \left(\frac{\lambda}{1 - e^{-\lambda}}\right)^2.$$

Now using formula (3.16), we arrive at the marginal variance of the random sum

$$\text{var}\left(\sum_{i=1}^{N} X_i\right) = \sigma^2 \frac{\lambda + \lambda^2}{1 - e^{-\lambda}} + \mu^2 \left[\frac{\lambda + \lambda^2}{1 - e^{-\lambda}} - \left(\frac{\lambda}{1 - e^{-\lambda}}\right)^2\right].$$

(c) Conditional on $N = n$, the distribution of the random sum is $\mathcal{N}(\mu n, \sigma^2 n)$. However it is wrong to assume that the marginal distribution of the random sum is normal $\mathcal{N}(\mu E(N), \sigma^2 \times\text{var}(N))$, where the mean and variance of N are given above – we refer to this distribution as naive. To obtain the marginal cdf of the random sum, we have to sum N out over all possible values $n = 1, 2, \ldots$ with weights $\Pr(N = n)$:

$$\Pr\left(\sum_{i=1}^{N} X_i \leq s\right) = \sum_{n=1}^{\infty} \Phi\left(\frac{s - \mu n}{\sigma\sqrt{n}}\right) \Pr(N = n)$$

$$= \frac{1}{1 - e^{-\lambda}} \sum_{n=1}^{\infty} \Phi\left(\frac{s - \mu n}{\sigma\sqrt{n}}\right) \frac{\lambda^n}{n!} e^{-\lambda}. \qquad (3.21)$$

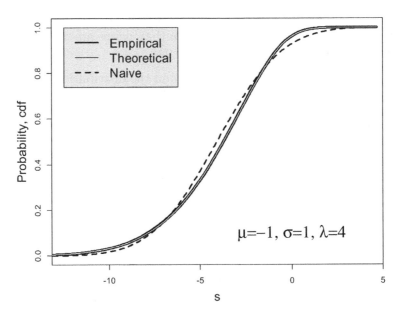

Figure 3.8: *Three cdfs: empirical based on* `nSim`*=100,000 simulations, theoretical, based on equation (3.21), and the naive normal cdf.*

Obviously, the random sum does not follow a normal distribution.

(d) The R code is found in file `randS.r`. We make a few remarks: (i) To generate N from a nonzero Poisson distribution, we first generate N from the regular Poisson distribution and then exclude zeros ($N = 0$). (ii) We use `xlim` following the rule ± three sigma to plot the cdfs. A run of the function is listed below and the plot is depicted in Figure 3.8. Empirical and theoretical values are in good agreement. As follows from the figure, the empirical and theoretical cdfs practically coincide, but the naive cdf does not match the empirical cdf as well as the theoretical one.

```
> randS(nSim=1000000,lambda=2)
Empirical N mean = 2.313256 , theoretical N mean = 2.313035
Empirical var = 1.586594 , theoretical var = 1.588974[1] -2.000000 -2.314956
Empirical RS mean = -2.314956 , theoretical RS mean = -2.313035
Empirical RS var = 3.903632 , theoretical RS mean = 3.902009>
```

Note that a new run will produce a different empirical sum because different values are generated. To produce the same result, one must fix the seed number, e.g. `set.seed(5)`. However, when `nSim` is large, results do not vary much from run to run.

3.3.4　Cancer tumors grow exponentially

Cancer cells are cells with a inflated/uncontrolled rate of proliferation. The uncontrolled cell division leads to rapid tumor growth. Let D_i, at specific time, be the binary event that the ith mother cancer cell divides in two daughter cells over h hours (we assume that h is small enough so that cell cannot divide twice) and let t denote time on the h-scale. That is, $D_i = 1$ means the ith cell divided and $D_i = 0$ did not divide in h hours. To simplify, we assume that cells that do not divide die. Let Y_t be the number of cells at time t and $\Pr(D_{it} = 1) = p$, which implies that the probability of cell proliferation is the same across cells and does not change with time. The number of cells at time $t + 1$ can be expressed as a random sum $Y_{t+1} = 2\sum_{i=1}^{Y_t} D_{it}$. Under assumption that cell division (D_{it}) does not depend on the number of cells (Y_t), we obtain the recursive formula $E(Y_{t+1}) = E(Y_t)(2p)$ with the explicit solution $E(Y_t) = y_0(2p)^{t-1}$, $t = 1, 2, ...$, where y_0 is the initial number of cells at time 1. Cancer tumor grows exponentially if $p > 0.5$.

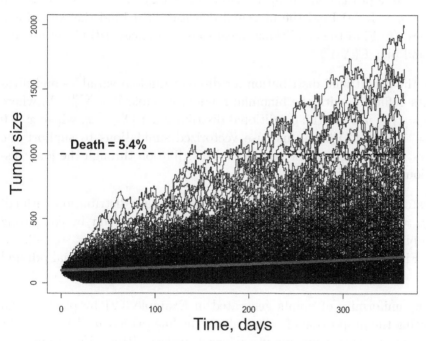

Figure 3.9: *Simulations of 500 tumors generated by code* cancgr. *The bold red line is the expected number of cells (it is assumed that 1 billion cancer cells kills the person bearing the tumor). 5.4% of random curves go beyond 1 billion threshold. That is, 5.4% cancer patients die within one year.*

The R code with simulations is found in the file cancgr.r. This function models cell growth on the scale of millions and creates Figure 3.9 under assumption $h = 24$ h over the course of one year (365 days). The default probability of division is small, $p = 0.501$ and yet 5.4% of people die within a year assuming that death occurs when the tumor reaches 1 billion cancer cells. Note that the variance grows with time as well due to the multiplicative effect.

Problems

1. (a) Prove that the uniform and normal distributions are not memoryless.
   (b) Demonstrate by simulations that the gamma distribution is not memoryless unless $\alpha = 1$.

2. Generate 10K uniformly distributed random points on the half-disk of unit radius defined at $[-1, 1]$ by modifying function `rantr`. Test the uniformity of the generated points by plotting and computing the proportion of points above the line $y = a$ and comparing with your analytical/exact answer.

3. (a) Is $F(x, y)/F_X(x)$ the conditional cdf? (b) Is it the conditional cdf if X and Y are independent?

4. The bivariate pdf is defined by $f(x, y) = (1 - \ln 2)^{-1} e^{-x - x/y}$ where $x > 0$ and $0 < y < 1$. (a) Find the marginal densities. (b) Find the conditional densities. (c) Find the conditional means and variances. (d) Generate 10K pairs and plot $E(X|Y)$.

5. Define the conditional distribution for discrete random variables and illustrate its definition with the binomial random variable $Y = \sum_{i=1}^{n} X_i$ where $\Pr(X_i = 1) = p$. Find the conditional distribution $Y|X_n = x$, where $x = 0$ or $x = 1$, find $E(Y|X_n)$, and use vectorized simulations to support the answers. [Hint: Express the conditional distribution through the joint distribution $p_{ij} = \Pr(X = x_i, Y = y_j)$.]

6. Generate and plot n iid observations from a bivariate distribution with cdf $F(x, y) = 2e^{x+y}(1+e^{x+y})^{-1}$ for $x, y \leq 0$. Check your result by comparing the empirical probabilities $\Pr(X \leq u)$ and $\Pr(Y \leq u)$ for $u = -3, -2, -1$. [Hint: Find the marginal cdf and density and then the conditional pdf and cdf.]

7. Test the uniformity of points generated in Example 3.21 by plotting and computing the proportion of points above the line (a) $y = a$, (b) $y = a + bx$ and comparing with your analytical/exact answer. [Hint: Use a and b as the arguments of the R code.]

8. The bivariate distribution is defined as the marginal pdf $X \sim \mathcal{E}(\lambda)$ and conditional pdf $Y|X = x \sim \mathcal{E}(\eta x)$. (a) Find and plot contours of the joint density and (b) find and plot $E(X|Y = y)$.

9. Find conditional the mean and variance of Y given X, where (X, Y) are uniformly distributed on the parallelogram $A_1 = (0, 0), A_2 = (a, 1), A_3 = (a + 1, 1), A_4 = (1, 0)$, where $a > 0$.

10. Why does a distribution of (X, Y) defined on a finite nonrectangular area, such as polygon, imply that X and Y are not independent?

11. Demonstrate that the reverse to Theorem 3.24 is not true: define a uniformly distributed and dependent pair (X, Y) on a triangle for which the conditional mean is constant. Define a uniformly distributed and dependent pair (X, Y) on a parallelogram for which conditional variance is constant.

12. Is it true that $E(g(X, Y)|X = x) = E_Y(g(x, Y))$? Is it true when function g is (a) linear or (b) quadratic function, or (c) X and Y are independent random variables? [Hint: The quadratic function is defined as $g(x, y) = Ax^2 + By^2 + Cxy + Dx + Ey + F$.]

13. Erica tosses a fair coin n times and Fred tosses $n+1$ times. Prove that Fred gets more heads than Erica with probability 0.5 using the rule of repeated expectation. Use simulations to demonstrate that the probability is 0.5. [Hint: Condition on the number of heads in n tosses.]

14. Does inequality (3.14) turn into equality for $g(x, y) = x + y$?

15. Generalize Example 3.22 to general normal distributions: the conditional density of Y given X is a normal distribution $\mathcal{N}(a+bx, \sigma^2)$ and the marginal distribution of X is $\mathcal{N}(\mu, \tau^2)$.

16. Redo Example 3.25 for $B = (0, 2)$.

17. Prove that inequality (3.19) can be improved by rewriting the right-hand side as $1 + r(h, d_i)$, where $r > 0$ and $r(0, d_i) = 0$ (sign $>$ changes to $\geq$).

18. The optimal prediction of Y given $X = x$ is described as a strictly increasing function $f(x)$. Is it true that the optimal prediction of X given $Y = y$ is described by $f^{-1}(y)$, where f^{-1} is the inverse function? Either prove or provide a counterexample.

19. Random variables X and Y are uniformly distributed on the half disk bounded by $\sqrt{1 - x^2}$, where $|x| < 1$. (a) Predict Y given $X = x$ and derive the conditional SD of the prediction. (b) Generate 100K random observations (X_i, Y_i) from this distribution using the pch="." option and plot the prediction at x=seq(from=-1,to=1,length=100) along with the $\pm$SD lines. (c) Illustrate formula (3.13) by estimating the three terms from simulated values. [Hint: Start by generating uniformly distributed pairs on the rectangle.]

20. Following Example 3.31, illustrate formulas (3.13) and (3.16) analytically using the uniform distribution of the half disk from the previous problem.

21. (a) Develop an algorithm to generate random pairs from the uniform distribution on the half disk using conditional distribution of Y given $X = x$ and marginal distribution of X. (b) Generate and plot 100K pairs. [Hint: Use the inverse cdf to generate X and $\int_{-1}^{z} \sqrt{1 - x^2}dx = 0.5z\sqrt{1 - z^2} + 0.5 \arcsin z + 0.25\pi$. Use uniroot to inverse the cdf; see Section 2.13.]

22. Prove that $\text{var}(E(Y|X)) \leq \text{var}(Y)$. When does this inequality turn into equality?

23. The density of the bivariate exponential distribution of (X_1, X_1) is defined as $[(1 + \theta x_1)(1 + \theta x_2) - \theta)]e^{-(x_1 + x_2 + \theta x_1 x_2)}$. (a) Prove that the marginal distributions are exponential. (b) Prove that the conditional density of X_2 given $X_1 = x_1$ is $[(1 + \theta x_1)(1 + \theta x_2) - \theta)]e^{-(1 + \theta x_1)x_2}$.

24. Prove that (3.17) generally does not hold if X and Y are dependent in Example 3.32 by providing a counterexample.

25. (a) Develop an algorithm to generate random pairs from the uniform distribution on the unit disk. (b) Generate and plot 100K pairs.

26. (a) Predict Y (0 or 1) given $X = 0$ using the result on the conditional mean from Example 3.35. When is the prediction the least precise (probability $= 0.5$)? (b) Do the same, but given $X = 1$.

27. Write R code to simulate tire measurements and illustrate the "Who said π" example using uniform measurements $\mathcal{R}(-h, h)$.

28. Random variables X and Y are independent and uniformly distributed on $(0, 1)$. Define $Z = X/(X + Y)$. Find $E(Z|Y = y)$ and use simulations to check your result. [Hint: Simulate a large number of X and Y, and replace the condition $Y = y$ with $|Y - y| < \varepsilon$ where ε is a small number.]

29. Bid times on an item for sale on e-bay follow an exponential distribution with the expected average time at one hour. What is the expected time between two bids (bids are placed independently)?

30. The rumor about an accident on campus follows an exponential distribution with pdf e^{-x} where $x > 0$ is time in hours after the accident. After Mary heard the rumor, she called her friend according to a conditional pdf $Y|X = x \sim e^{-(y-x)}$, where X stands for the time when Mary learned the news and Y stands for the time she reached John $(y > x)$. Given that John received the call 2 hours after the accident, find the expected time after the accident Mary knew about the accident.

31. Suppose that in the "Who said π" example, $d_i > d_*$ and δ_i are distributed with a positive density on the interval $[-h, h]$ with $h < d_*$. Prove that $\nu > 0$ exists such that $E(\widehat{\pi}_N) > \pi + \nu$.

32. Generalize example *Man or woman?* to the case when the proportion of men in the sample, p, is not known. Plot the height threshold as a function of p on the range $[0.1, 0.99]$ using the Bayesian classifier rule.

33. Derive $E(N)$ and $\text{var}(N)$ in Example 3.41.

34. Redo Example 3.41 with the assumption that $N-1$ follows the binomial distribution with parameters p and n (Section 1.6).

35. In connection with Figure 3.9, how many cancer patients will die within two years? [Hint: run function `cancgr` with a different argument.]

36. Approximate the distribution of the number of cancer cells Y_t using the method of Example 3.41 with the cdf expressed as in equation (3.21).

37. Provide argumentation for modeling the tumor size on the log scale. [Hint: Consult Section 2.11.]

3.4   Correlation and linear regression

An important property of the joint distribution is the *covariance* defined as the expected value of the centered cross-product; the formal definition follows.

Definition 3.42 *The covariance between random variables X and Y is defined as*

$$\mathrm{cov}(X, Y) = E[(X - \mu_x)(Y - \mu_y)].$$

We say that X and Y do not correlate if $\mathrm{cov}(X, Y) = 0$.

The following are elementary properties of the covariance (a and b are numbers):

1. $\mathrm{cov}(X, X) = \mathrm{var}(X)$,

2. $\mathrm{cov}(X, Y) = \mathrm{cov}(Y, X)$,

3. $\mathrm{cov}(X + a, Y) = \mathrm{cov}(X, Y + b) = \mathrm{cov}(X, Y)$,

4. $\mathrm{cov}(X, Y + Z) = \mathrm{cov}(X, Y) + \mathrm{cov}(X, Z)$,

5. $\mathrm{cov}(aX, Y) = a \times \mathrm{cov}(X, Y), \quad \mathrm{cov}(X, bY) = b \times \mathrm{cov}(X, Y)$,

6. $\mathrm{var}(X + Y) = \mathrm{var}(X) + 2\mathrm{cov}(X, Y) + \mathrm{var}(Y)$,

7. if X and Y do not correlate then $\mathrm{var}(X + Y) = \mathrm{var}(X) + \mathrm{var}(Y)$,

8. $\mathrm{cov}(X, Y) = E(XY) - \mu_x \mu_y = E[X(Y - \mu_y)] = E[(X - \mu_x)Y]$,

9. If X and Y are independent, then $\mathrm{cov}(X, Y) = 0$.

Proofs are easy. For example, to prove 6, we use 1 and 4,

$$
\begin{aligned}
\mathrm{var}(X + Y) &= \mathrm{cov}(X + Y, X + Y) = \mathrm{cov}(X, X) + \mathrm{cov}(X, Y) + \mathrm{cov}(Y, X) \\
&+ \mathrm{cov}(Y, Y) = \mathrm{var}(X) + 2\mathrm{cov}(X, Y) + \mathrm{var}(Y).
\end{aligned}
$$

To prove 8, we have

$$\text{cov}(X, Y) = E[XY - X\mu_y - \mu_x Y + \mu_x \mu_y] = E(XY) - \mu_x \mu_y - \mu_x \mu_y + \mu_x \mu_y$$
$$= E(XY) - \mu_x \mu_y.$$

To prove 9, we have

$$E[X(Y - \mu_y)] = E(XY) - \mu_x \mu_y = \text{cov}(X, Y),$$

as follows from 8. Continuing, we have

$$E[(X - \mu_x)(Y - \mu_y)] = \text{cov}(X, Y) = E(X - \mu_x) \times E(Y - \mu_y) = 0 \times 0 = 0,$$

due to (3.7) because of independence.

It is easy to prove that, if random variables are pairwise uncorrelated, then the variance of the sum equals the sum of the variances.

Example 3.43 *Zero correlation does not imply independence.* *Find two random variables that are functionally related and yet have zero covariance.*

Solution. Let X be a random variable such that $E(X) = E(X^3) = 0$ and let $Y = X^2$. Due to Property 6, we have

$$\text{cov}(X, Y) = E((X - \mu_x)X^2) = E(X^3) = 0$$

since $\mu_x = 0$. In words, if two random variables do not correlate, they may be dependent.

Definition 3.44 *The correlation coefficient, sometimes called the Pearson correlation coefficient, is defined as*

$$\rho = \text{cor}(X, Y) = \frac{\text{cov}(X, Y)}{\sqrt{\text{var}(X)\text{var}(Y)}}. \tag{3.22}$$

Unlike covariance, ρ does not depend on the scale of X and Y. We may say that correlation coefficient is the normalized covariance. It is easy to prove that correlation coefficient is invariant with respect to linear transformations

$$\text{cor}(X, Y) = \text{sign}(bd)\text{cor}(a + bX, c + dY),$$

where $a, b, c,$ and d are numbers.

As follows from the Cauchy inequality (2.16),

$$|\text{cov}(X, Y)| \leq \sqrt{\text{var}(X)\text{var}(Y)}.$$

Therefore, the correlation coefficient takes values in the interval $[-1, 1]$, or in other words,

$$|\rho| \leq 1.$$

Moreover, $\rho = \pm 1$ if and only if Y and X are linearly dependent:

$$Y = \mu_y + \beta(X - \mu_x).$$

If $\beta > 0$ then $\rho = 1$; if $\beta < 0$ then $\rho = -1$. Zero correlation is equivalent to zero covariance. Thus, when $\rho = 0$, we say that Y and X are uncorrelated. Zero correlation is a necessary condition for independence; as we shall learn later, it turns into a sufficient condition for the normal distribution. In other words, if two random variables are independent, they are uncorrelated, but the reverse may not be true in general, as shown in example below. Therefore correlation is a measure of *linear dependence*.

Example 3.45 *Coefficient of determination.* *Consider $X \sim \mathcal{N}(0,1)$ and $Y = 1 - X^2$. Prove that the coefficient of determination as defined through the conditional mean in (3.30) is 1, but equals 0 as defined in (3.22).*

Solution. Since Y and X are functionally related, we have $E(Y|X) = 1 - X^2$ and, therefore, $\text{var}(E(Y|X)) = \text{var}(1 - X^2) = \text{var}(Y)$. This implies that $\rho^2 = 1$ in Definition 3.30. Now find $\text{cov}(X,Y) = \text{cov}(X, 1 - X^2) = \text{cov}(X) - \text{cov}(X^3) = 0 - 0 = 0$. This implies that $\rho = 0$ as defined in (3.22). Although Y and X are nonlinearly dependent, they are linearly independent.   □

When $\rho > 0$, we say that correlation is positive; when $\rho < 0$, we say that correlation is negative. Sometimes independence and zero correlation are equivalent as in the following example (item d). Another important case when correlation and independence are equivalent is when two random variables are normally distributed (Section 3.5).

The following is another example when zero correlation and independence are equivalent.

Example 3.46 *Correlation between Bernoulli random variables.* *Using the notation of Example 3.35, (a) express the correlation coefficient, ρ, between two Bernoulli random variables X and Y in terms of elementary probabilities p_{ij}; (b) derive the independence condition between X and Y in terms of p_{ij}; (c) show that independence between two binary random variables implies $\rho = 0$; (d) prove that $\rho = 0$ implies independence.*

Solution. (a) Following the notation of Example 3.35, we have $p_{ij} = \Pr(X = i, Y = j)$, where $i, j = 0, 1$. Consequently, we have $E(XY) = p_{11}$, so that the correlation coefficient between two binary random variables is written as

$$\rho = \frac{p_{11} - (p_{10} + p_{11})(p_{01} + p_{11})}{\sqrt{(p_{10} + p_{11})(1 - p_{10} - p_{11})(p_{01} + p_{11})(1 - p_{01} - p_{11})}}.$$

(b) X and Y are independent if and only if four conditions hold:

$$p_{00} = (1 - p_{10} - p_{11})(1 - p_{01} - p_{11}), \quad p_{01} = (1 - p_{10} - p_{11})(p_{01} + p_{11}),$$
$$p_{10} = (p_{10} + p_{11})(1 - p_{01} - p_{11}), \quad p_{11} = (p_{10} + p_{11})(p_{01} + p_{11}).$$

(c) The fact that independence implies zero correlation follows from Property 9 listed above, but we will prove this directly in the language of probabilities $\{p_{ij}\}$. Indeed, if X and Y are independent, then as shown in (b) $p_{11} = (p_{10} + p_{11})(p_{01} + p_{11})$, which, after some elementary algebra, implies $\rho = 0$. (d) It suffices to show that the fourth condition above implies the three former conditions. Rewrite the first condition as

$$p_{10} = (p_{10} + p_{11}) - (p_{10} + p_{11})(p_{01} + p_{11})$$

or $p_{10} = (p_{10} + p_{11}) - p_{11}$, which is true. Next, rewrite $p_{01} = (1 - p_{10} - p_{11})(p_{01} + p_{11})$ as $p_{01} = p_{01} + p_{11} - (p_{10} + p_{11})(p_{01} + p_{11})$, which is also true. Finally,

$$\begin{aligned} p_{00} &= (1 - (p_{10} + p_{11}))(1 - (p_{01} + p_{11})) = 1 - (p_{10} + p_{11}) - (p_{01} + p_{11}) + p_{11} \\ &= 1 - p_{10} - p_{11} - p_{01} - p_{11} + p_{11} = 1 - p_{10} - p_{11} - p_{01} \end{aligned}$$

which is true as well. $\square$

The concept of correlation is closely related to the concept of linear regression as a linear approximation to the relationship between Y and X.

Definition 3.47 *The least squares linear regression between Y and X is defined as $a + bX$, where a and b are the solution of the following optimization problem:*

$$\min_{a,b} E(Y - a - bX)^2, \tag{3.23}$$

referred to as the least squares (LS) solution.

The function to minimize is referred to as the least squares criterion. The intercept a and slope b are referred to as the least squares estimates. The linear function $y = a + bx$ is referred to as the least squares linear regression.

To find the optimal constants a and b, differentiate $E(Y - a - bX)^2$ with respect to a under expectation and set it to zero,

$$-2E(Y - a - bX) = 0,$$

that implies $a = \mu_y - b\mu_x$. Plugging this back into the squared function we arrive at the minimization of

$$E((Y - \mu_y)) - b(X - \mu_x))^2.$$

Differentiating with respect to b and setting it to zero, we obtain

$$-2E((X - \mu_x)\left[(Y - \mu_y)) - b(X - \mu_x)\right]) = 0.$$

Solving for b, we finally arrive at the least squares formula for the slope:

$$b = \frac{E[(X - \mu_x)(Y - \mu_y)]}{E[(X - \mu_x)^2]} = \frac{\mathrm{cov}(X, Y)}{\mathrm{var}(X)} = \rho \frac{\sigma_y}{\sigma_x}. \tag{3.24}$$

Given b, the LS intercept is obtained as

$$a = \mu_y - b\mu_x. \tag{3.25}$$

As we can see from (3.24), the slope is proportional to the correlation coefficient. In particular, b and ρ are both either positive or negative.

An obvious but important observation is that if X and Y are independent then linear regression = constant = μ_y because $\rho = 0$ implies $b = 0$. Moreover, if X and Y are uncorrelated ($\rho = 0$), the linear regression is constant as well. Reversely, if linear regression is not flat, then X and Y are dependent.

Example 3.48 *Variance decomposition and linear coefficient of determination. Show that the following variance decomposition holds: $\sigma_y^2 = \text{var}(a + bX) + \text{var}(Y - a - bX) = \rho^2 \sigma_y^2 + \sigma^2$, where $\text{var}(a + bX)$ and $\sigma^2 = \text{var}(Y - a - bX)$ are referred to the variances explained and unexplained by the LS regression.*

Solution. Using formula (3.25), we have

$$\begin{aligned}
\text{var}(Y) &= E(Y - \mu_y)^2 = E[(bX - b\mu_x) + (Y - a - bX)]^2 \\
&= E(bX - b\mu_x)^2 + E(Y - a - bX)^2 + E[(bX - b\mu_x)(Y - a - bX)].
\end{aligned}$$

It is elementary to prove that, as follows from (3.24), the first term is $\text{var}(a + bX)$, the second term is $\text{var}(Y - a - bX)$ and the last term vanishes. This variance decomposition gives rise to the squared correlation coefficient ρ^2, called the determination coefficient, as the proportion of variance of Y explained by LS regression on X. This fact underscores once again why the correlation coefficient reflects linear association. More discussion is found in Section 3.5.2. □

In general, the conditional mean $E(Y|X = x)$ defined in (3.11) and the least squares linear regression $a + bx$ with coefficients (3.24) and (3.25) are not the same. But, sometimes, they coincide, as in the following example. Another important example when the two coincide is the bivariate normal distribution; see Section 3.5.

Example 3.49 *Regression on a triangle. Prove that the conditional mean and linear regression are the same for the uniform distribution on a triangle (see Example 3.25).*

Solution. We have

$$E(XY) = \int_0^1 \left(\int_0^{2x} xy \, dy \right) dx = \int_0^1 x \left(\int_0^{2x} y \, dy \right) dx = 2 \int_0^1 x^3 \, dx = \frac{1}{2}.$$

Next we need to compute the marginal means. Since $F_X(x) = x^2$ and $F_Y(y) = 1 - (2 - y)^2/4$ we have

$$E(X) = \int_0^1 2x x \, dx = \frac{2}{3}, \quad E(Y) = \frac{1}{2} \int_0^2 y(2 - y) \, dy = \frac{2}{3}$$

and $\operatorname{var}(X) = \int_0^1 2x^3 dx - E^2(X) = 1/2 - 4/9 = 1/18$. Thus, the slope $b = [E(XY) - E(X)E(Y)]/\operatorname{var}(X) = (1/2 - 4/9)/(1/18) = 1$ and the intercept $a = E(Y) - bE(X) = 0$. These values define the $45°$ line $y = x$, the same as the conditional mean in Example 3.25.   □

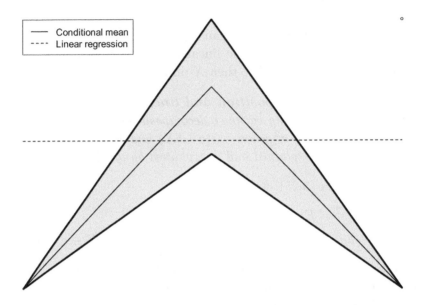

Figure 3.10: *The uniform distribution on the symmetric tetragon. The conditional mean and linear regression are different. Although the linear regression is flat ($\rho = 0$), the random variables are dependent.*

As before, the conditional mean and linear regression are different. To illustrate, in Figure 3.10, we show the two for a uniform distribution on a symmetric tetragon. The conditional mean is a piece-wise linear function (the connected median in the adjacent triangles); the linear regression (the dotted line) is parallel to the x-axis (see Example 3.43), and therefore $\rho = 0$.

The conditional mean and the least squares linear regression can be used for prediction. This feature is illustrated below.

Example 3.50 *Prediction problem.* *The bivariate pdf of a pair of random variables (X, Y) is proportional to xy and defined on the triangle with vertices $A = (0, 0), B = (0, 1),$ and $C = (1, 0)$. (a) Find the normalizing coefficient. (b) Find the marginal density for X. (c) Find the conditional density, cdf, expectation, and SD of Y given $X = x$. (d) Find the LS linear regression. (e) Find the marginal cdf for X. (f) Generate 10K iid pairs of (X, Y) by first generating X and second generating Y given $X = x$ (use **uniroot** to compute the inverse cdf). (g) Predict Y given $X = 0.5$ using the conditional mean and least squares regression (plot the lines and the prediction).*

Solution. (a) To determine the normalizing coefficient, we find the volume under the function xy:

$$\int_{x>0,y>0,x+y<1} xydxdy = \int_0^1 x\left(\int_0^{1-x} ydy\right)dx = \frac{1}{2}\int_0^1 x(1-x)^2 dx$$

$$= \frac{1}{2}\int_0^1 (x - 2x^2 + x^3)dx = \frac{1}{24}.$$

Thus, the normalizing coefficient is the reciprocal of $1/24$, and so the joint density is $f(x,y) = 24xy$. (b) Integrate out y to find the marginal density of X as $f_X(x) = 24\int_0^{1-x} xydy = 12x(1-x)^2$ for $0 < x < 1$. (c) Use general formula (3.9) to find the conditional density, $f_{Y|X=x}(y) = (2y)/(1-x^2)$ for $0 < y < 1-x$. The conditional mean is found using formula (3.11):

$$E(Y|X = x) = \frac{2}{(1-x)^2}\int_0^{1-x} y^2 dy = \frac{2(1-x)^3}{3(1-x)^2} = \frac{2}{3}(1-x), \qquad (3.26)$$

a linear function. To find the conditional variance, use formula (3.12),

$$E(Y^2|X = x) = \frac{2}{(1-x)^2}\int_0^{1-x} y^3 dy = \frac{1}{2}(1-x)^2$$

so that

$$\text{var}(Y|X = x) = E(Y^2|X = x) - E^2(Y|X = x)$$
$$= \frac{1}{2}(1-x)^2 - \frac{4}{9}(1-x)^2 = \frac{1}{18}(1-x)^2,$$
$$\text{SD}(Y|X = x) = \frac{1}{3\sqrt{2}}(1-x).$$

(d) To determine the intercept and slope of the LS linear regression, we need marginal means and variance:

$$E(X) = \int_0^1 12x^2(1-x)^2 dx = \frac{2}{5}, \quad E(X^2) = \int_0^1 12x^3(1-x)^2 dx = \frac{1}{5},$$
$$\text{var}(X) = E(X^2) - E^2(X) = \frac{1}{25}.$$

To find $E(Y)$, we need to find the marginal density of Y (it is expected that the marginal densities for X and Y will be the same due to symmetry),

$$f_Y(y) = 24\int_0^{1-y} xydx = 12y(1-y)^2, \quad 0 < y < 1,$$

so that $E(Y) = 2/5$.

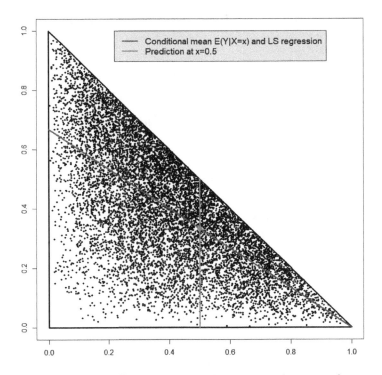

Figure 3.11: *Conditional prediction on the triangle.*

Finally, we need to find $cov(X, Y)$. First, find

$$E(XY) = 24 \int_{x>0,y>0,x+y<1} x^2 y^2 dx dy = \frac{24}{3} \int_0^1 x^2 (1-x)^3 dx = \frac{2}{15}.$$

Then $cov(X, Y) = E(XY) - E(X)E(Y) = 2/15 - (2/5) \times (2/5) = -2/75$, and the least squares slope and intercept are

$$b = \frac{cov(X, Y)}{var(X)} = -\frac{2/75}{1/25} = -\frac{2}{3}, \quad a = E(Y) - bE(X) = \frac{2}{3}.$$

Hence, the least squares linear regression is $a + bx = 2(1-x)/3$, the same as the conditional mean (3.26).

(e) Find the marginal cdf of X by integration of the density obtained above:

$$F_X(x) = 12 \int_0^x (t - 2t^2 + t^3) dt = 6x^2 - 8x^3 + 3x^4.$$

(f) To generate random numbers, find the conditional cdf by integration of the conditional density,

$$F_{Y|X=x}(y) = \frac{2}{(1-x)^2} \int_0^y t \, dt = \frac{y^2}{(1-x)^2}, \quad 0 < y < 1 - x.$$

Random numbers are generated in two steps: (i) generate X by inverse cdf as the solution of the equation $6x^2 - 8x^3 + 3x^4 = r$, where $r \sim \mathcal{R}(0, 1)$, (ii) generate Y as the solution of $y^2/(1 - X)^2 = u$, where $r \sim \mathcal{R}(0, 1)$ and X from the first step, or $Y = (1 - X)\sqrt{u}$.

(g) The prediction lines for conditional mean and linear regression coincide, $\frac{2}{3}(1 - x)$. Therefore, the predicted values of Y at $x = 0.5$ is $1/3$. The R code is found in the file preprob, which produces Figure 3.11. The R function uniroot was discussed in Section 2.13.

Problems

1. Express $\text{cov}(a+bX, c+dY)$ via $\text{cov}(X, Y)$, where a, b, c, and d are constants.

2. Express $\text{cov}(X, aZ+bU)$ using $\text{cov}(X, Z)$ and $\text{cov}(X, U)$, where a and b are constants.

3. The random variables X and Y are uncorrelated and have unit variance. Find the correlation coefficient between X and $X+aY$. Find the correlation coefficient between X and $Y + aX$ as a function of a.

4. The random variables X and Y with unit variance have correlation coefficient ρ. Define a new random variable as a linear combination of X and Y that is uncorrelated with X.

5. A characteristic property of independence of X and Y is that any functions of X and Y are independent as well. Show that it is not true when X and Y are correlated (provide an example). Do X and Y remain uncorrelated upon a linear transformation?

6. The Bland–Altman technique is used in applied statistics to assess the bias between two samples of measurements by plotting $Y - X$ on the y-axis and $(Y + X)/2$ on the x-axis (X and Y may correlate); see a review paper [38]. The bias is judged by the distance of the horizontal line with the height $E(Y) - E(X)$ from the x-axis. Show that this technique is valid if the variances of X and Y are the same (equal-variance assumption).

7. The joint density on the plane is defined as $f(x, y) = e^{-x-y}$ for $x, y > 0$. Find the least squares linear regression.

8. (a) Random variable X is transformed into $X' = c + dX$. Express the regression of Y on X' in terms of the regression Y on X. (b) Random variable Y is transformed into $Y' = c+dY$. Express the regression of Y' on X in terms of the regression Y on X.

9. Prove that $E(Y - a - bX)^2$ as a function of a and b is a convex function. [Hint: Prove that the 2×2 Hessian is positive definite.]

10. Prove that $\text{var}(Y - a - bX) \leq \text{var}(Y)$ where a and b are the LS solution.

11. Using the results from Examples 3.35 and 3.46, predict $E(Y)$ given $X = x$, where $x = 0, 1$ using the least squares regression and conditional mean. Compare the predictions: are they the same?

12. (a) Derive the LS regression for (X, Y) uniformly distributed on a half-disk $|x| < 1$ and $0 < y < \sqrt{1 - x^2}$. (b) Do the conditional mean and linear regression coincide? (c) Is the correlation coefficient between X and Y zero?

13. (a) Derive the LS regression for random variables with zero mean. (b) Derive the formula for the slope.

14. Repeat Example 3.50 with $B = (1, 1)$.

15. A lady arrives at a place at random time between 1 p.m. and 2 p.m. After she arrives, she places a telephone call at a random time before 2 p.m. It is known that she placed the call at 1:45 p.m. Estimate the time she arrived at the place, and provide a geometric illustration. Use simulations to verify the answer. [Hint: Use the conditional mean to estimate the time analytically and then use `mean(x[abs(y-3/4)<eps])` to estimate the time from simulations.]

3.5   Bivariate normal distribution

When random variables X_1 and X_2 are independent and normally distributed with mean μ_1, μ_2 and variances σ_1^2, σ_2^2, respectively, the joint density is the product of their densities. The key parameter of the bivariate normal distribution with *dependent* X_1 and X_2 is the correlation coefficient, ρ. The bivariate normal density depends on five parameters and takes the form

$$f(x_1, x_2) = \frac{1}{(2\pi)\sigma_1\sigma_2\sqrt{1-\rho^2}} e^{-\frac{1}{2(1-\rho^2)}\left[\left(\frac{x_1-\mu_1}{\sigma_1}\right)^2 - 2\rho\left(\frac{x_1-\mu_1}{\sigma_1}\right)\left(\frac{x_2-\mu_2}{\sigma_2}\right) + \left(\frac{x_2-\mu_2}{\sigma_2}\right)^2\right]},$$

(3.27)

where $|\rho| < 1$ is the correlation coefficient defined by (3.22). Indeed, in the theorem to follow, we prove that density (3.27) implies that

$$\rho = \frac{\text{cov}(X_1, X_2)}{\sqrt{\text{var}(X_1)\text{var}(X_2)}} = \frac{\sigma_{12}}{\sigma_1\sigma_2}.$$

When $\rho = 0$, the middle term in the square brackets of (3.27) vanishes the density factors into marginal densities, $X_1 \sim \mathcal{N}(\mu_1, \sigma_1^2)$ and $X_2 \sim \mathcal{N}(\mu_2, \sigma_2^2)$. That is, the components X_1 and X_2 are independent. We know that independence implies zero correlation. For a normal distribution, if two random variables are uncorrelated, they are independent.

Theorem 3.51 *If the pair of random variables (X_1, X_2) has the bivariate normal density specified by (3.27), then (1) the marginal distributions of X_1 and X_2 are $\mathcal{N}(\mu_1, \sigma_1^2)$ and $\mathcal{N}(\mu_2, \sigma_2^2)$, respectively; (2) $E(X_1) = \mu_1$ and $E(X_2) = \mu_2$; (3) $\mathrm{var}(X_1) = \sigma_1^2$ and $\mathrm{var}(X_2) = \sigma_2^2$; (4) $\mathrm{cov}(X_1, X_2) = \sigma_{12} = \rho\sigma_1\sigma_2$ and $\mathrm{cor}(X_1, X_2) = \rho$.*

Proof. (1) To prove that $X_1 \sim \mathcal{N}(\mu_1, \sigma_1^2)$, we need to show that integrating out out x_2 from $f(x_1, x_2)$ yields marginal density of X_1:

$$\int_{-\infty}^{\infty} f(x_1, x_2)dx_2 = \frac{1}{\sqrt{2\pi}\sigma_1}e^{-\frac{1}{2}\left(\frac{x_1-\mu_1}{\sigma_1}\right)^2}.$$

Using change of variable $z_2 = (x_2 - \mu_2)/\sigma_2$, we obtain

$$\int_{-\infty}^{\infty} f(x_1, x_2)dx_2 = \frac{1}{(2\pi)\sigma_1\sqrt{1-\rho^2}} \int_{-\infty}^{\infty} e^{-\frac{1}{2(1-\rho^2)}\left[z_2^2 - 2\rho\frac{x_1-\mu_1}{\sigma_1}z_2 + \left(\frac{x_1-\mu_1}{\sigma_1}\right)^2\right]}dz_2.$$

Express

$$z_2^2 - 2\rho\frac{x_1-\mu_1}{\sigma_1}z_2 = \left(z_2 - \rho\frac{x_1-\mu_1}{\sigma_1}\right)^2 - \rho^2\left(\frac{x_1-\mu_1}{\sigma_1}\right)^2$$

by completing the square. Then the argument of the exponential function of the integrand takes the form

$$-\frac{1}{2(1-\rho^2)}\left[\left(z_2 - \rho\frac{x_1-\mu_1}{\sigma_1}\right)^2 + (1-\rho^2)\left(\frac{x_1-\mu_1}{\sigma_1}\right)^2\right]$$

$$= -\frac{1}{2(1-\rho^2)}\left(z_2 - \rho\frac{x_1-\mu_1}{\sigma_1}\right)^2 - \frac{1}{2}\left(\frac{x_1-\mu_1}{\sigma_1}\right)^2.$$

Finally, using another change of variable,

$$u = \frac{1}{\sqrt{1-\rho^2}}\left(z_2 - \rho\frac{x_1-\mu_1}{\sigma_1}\right),$$

we obtain

$$\int_{-\infty}^{\infty} f(x_1, x_2)dx_2 = \frac{1}{\sqrt{2\pi}\sigma_1}e^{-\frac{1}{2}\left(\frac{x_1-\mu_1}{\sigma_1}\right)^2} \times \frac{1}{\sqrt{2\pi}}\int_{-\infty}^{\infty} e^{-\frac{1}{2}u^2}du = \frac{1}{\sqrt{2\pi}\sigma_1}e^{-\frac{1}{2}\left(\frac{x_1-\mu_1}{\sigma_1}\right)^2}$$

to prove item (1). Items (2) and (3) follow from (1). To prove (4), we show that

$$\int_{-\infty}^{\infty}\int_{-\infty}^{\infty} (x_1 - \mu_1)(x_2 - \mu_2)f(x_1, x_2)dx_1 dx_2 = \rho\sigma_1\sigma_2.$$

Using change of variables $z_1 = (x_1 - \mu_1)/\sigma_1$ and $z_2 = (x_2 - \mu_2)/\sigma_2$, it is easy to see that the above equation is equivalent to

$$\frac{1}{(2\pi)\sqrt{1-\rho^2}} \int_{-\infty}^{\infty} \int_{-\infty}^{\infty} z_1 z_2 e^{-\frac{1}{2(1-\rho^2)}(z_1^2 - 2\rho z_1 z_2 + z_2^2)} dz_1 dz_2 = \rho.$$

To prove this we use change of variable similar to above, $u = (z_2 - \rho z_1)/\sqrt{1-\rho^2}$.

Even if the marginal distributions are normal the joint distribution may not be bivariate normal, as follows from Section 3.5.4. However, Theorem 3.89 says that if every linear combination of X_1 and X_2 has a normal distribution, then the pair (X_1, X_2) has a bivariate normal distribution.

The fact that a couple of random variables has a bivariate normal distribution is symbolically written as

$$\begin{bmatrix} X_1 \\ X_2 \end{bmatrix} \sim \mathcal{N}\left(\begin{bmatrix} \mu_1 \\ \mu_2 \end{bmatrix}, \begin{bmatrix} \sigma_1^2 & \rho\sigma_1\sigma_2 \\ \rho\sigma_1\sigma_2 & \sigma_2^2 \end{bmatrix} \right),$$

or shortly

$$\mathbf{X} \sim \mathcal{N}(\boldsymbol{\mu}, \boldsymbol{\Omega}),$$

where

$$\mathbf{X} = \begin{bmatrix} X_1 \\ X_2 \end{bmatrix}, \quad \boldsymbol{\mu} = \begin{bmatrix} \mu_1 \\ \mu_2 \end{bmatrix}, \quad \boldsymbol{\Omega} = \begin{bmatrix} \sigma_1^2 & \rho\sigma_1\sigma_2 \\ \rho\sigma_1\sigma_2 & \sigma_2^2 \end{bmatrix}.$$

Vector $\boldsymbol{\mu}$ is called the mean vector and matrix $\boldsymbol{\Omega}$ is called the covariance matrix. This notation is easy to generalize to multivariate normal distributions; see the next chapter.

The contour of the bivariate density is an ellipse with the center at (μ_1, μ_2) defined as the set of points $(x_1, x_2) \in R^2$ for which

$$\left(\frac{x_1 - \mu_1}{\sigma_1}\right)^2 - 2\rho\left(\frac{x_1 - \mu_1}{\sigma_1}\right)\left(\frac{x_2 - \mu_2}{\sigma_2}\right) + \left(\frac{x_2 - \mu_2}{\sigma_2}\right)^2 = \text{const.}$$

From analytical geometry, it follows that the area of this ellipse is proportional to the determinant of the covariance matrix $\boldsymbol{\Omega}$. That is why $\det(\boldsymbol{\Omega})$ is sometimes called the *generalized variance*. Upon rescaling $z_j = (x_j - \mu_j)/\sigma_j$ the ellipse simplifies to

$$z_1^2 - 2\rho z_1 z_2 + z_2^2 = \text{const}, \tag{3.29}$$

with the density

$$f(z_1, z_2) = \frac{1}{(2\pi)\sqrt{1-\rho^2}} e^{-\frac{1}{2(1-\rho^2)}(z_1^2 - 2\rho z_1 z_2 + z_2^2)},$$

which is referred to as the *standard bivariate normal density*. To prove that equation (3.29) indeed generates an ellipse, we have to prove that the quadratic form is positive definite (see Appendix), or equivalently, to prove that the 2×2 matrix

$$\mathbf{R} = \begin{bmatrix} 1 & \rho \\ \rho & 1 \end{bmatrix}$$

is positive definite for every $|\rho| < 1$. This follows from the fact that the diagonal elements of matrix $\mathbf{R}$ are positive and $\det(\mathbf{R}) = 1 - \rho^2 > 0$. Alternatively, we can prove that the left-hand side of (3.29) is positive for $z_1^2 + z_2^2 > 0$ and $|\rho| < 1$. This follows from a simple algebraic representation:

$$z_1^2 - 2\rho z_1 z_2 + z_2^2 = (|z_1| - |z_2|)^2 + 2|z_1||z_2|(1 \pm \rho) > 0.$$

To find the principal axes of ellipse (3.29), the characteristic equation is solved (see the Appendix on matrix algebra):

$$\det \begin{bmatrix} 1 - \lambda & \rho \\ \rho & 1 - \lambda \end{bmatrix} = 0,$$

which yields the eigenvalues $\lambda = 1 \pm \rho$. Now we find the principal axes as the eigenvectors of matrix $\mathbf{R}$. Without loss of generality, we can find the eigenvectors in the form $(\cos\theta, \sin\theta)$. The main principal axis, corresponding to the maximum eigenvalue, is the solution to the equation

$$\begin{bmatrix} 1 & \rho \\ \rho & 1 \end{bmatrix} \begin{bmatrix} \cos\theta \\ \sin\theta \end{bmatrix} = (1 + \rho) \begin{bmatrix} \cos\theta \\ \sin\theta \end{bmatrix},$$

yielding $\theta = \pi/4$ (the 45° line). The angle of the axis that corresponds to the minimal eigenvector is $\theta = -\pi/4$ or $\theta = 3\pi/4$ (the $-45°$ line). When variables do not correlate ($\rho = 0$) the contours are circles. Note that, in the general case when $\sigma_1 \neq \sigma_2$, the principal axes do not have angles $\pi/4$ and $-\pi/4$, but they are still orthogonal.

Example 3.52 *Normal ellipse. Plot the ellipse determined by equation (3.29) and show its principal axes.*

Solution. Solve the quadratic equation $z_1^2 - 2\rho z_1 z_2 + z_2^2 = c$ for z_2 in terms of z_1 where $c = \text{const} > 0$ is fixed. The solution is $z_2 = \rho z_1 \pm \sqrt{c - (1 - \rho^2)z_1^2}$, where $|z_1| \leq \sqrt{c/(1 - \rho^2)}$. The maximum distance from the origin to the ellipse is attained at the eigenvector, the 45° line defined by equation $z = z_1 = z_2$ and is found from the equation $z^2 - 2\rho zz + z^2 = c$ with an obvious solution

$$z_{\max} = \pm\sqrt{\frac{c}{2(1 - \rho)}}, \quad d_{\max} = \sqrt{\frac{c}{1 - \rho}}.$$

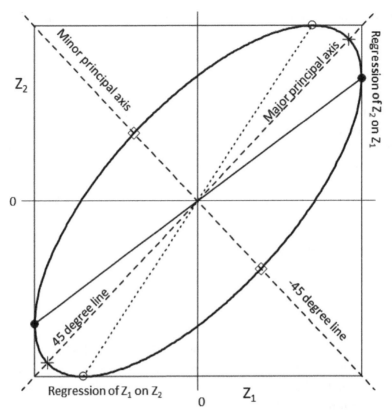

Figure 3.12: *Standard bivariate normal distribution ellipse with major and minor principal axes.*

The minimum distance from the origin attains at the $-45°$ line ($z_1 = -z_2$) and is found from the equation $z^2 + 2\rho zz + z^2 = c$ with solution

$$z_{\min} = \pm\sqrt{\frac{c}{2(1+\rho)}}, \quad d_{\min} = \sqrt{\frac{c}{1+\rho}}.$$

The geometry of the standard bivariate normal distribution is illustrated in Figure 3.12 and the R code for plotting the ellipse along with principal axes is found in the file ell2.r. Argument ro, the correlation coefficient, controls how ellipse is stretched along the $45°$ line; argument const is the scale parameter. Regression of Z_2 on Z_1 has the slope $\tan\theta = \rho$. Regression of Z_1 on Z_2 has the slope $\tan(\pi/2 - \theta) = c\tan\theta = 1/\tan\theta = 1/\rho$. In words, the tangent slope of the regression of X_2 on X_1 with the same variance is ρ, and the tangent slope of the regression of X_1 on X_2 is $1/\rho$. Points on the ellipse with maximum and minimum x-coordinate ($45°$ line) are shown as filled circles. The two most distant points are on the $45°$ line and shown as crosses; points nearest to the origin ($-45°$ line) are shown as squares. $\qquad\square$

The bivariate normal distribution often emerges in statistical applications. For example, Figure 3.13 displays the scatterplot of weight versus height for 25K Korean individuals (the R function `heightweight`). The elliptical shape is an indicator that the joint distribution is binormal. Note that the model for prediction of weight given height is very different from prediction of height given weight. These two regressions correspond to "regression of Z_2 on Z_1" and "regression of Z_1 on Z_2" in Figure 3.12, respectively. See Example 6.137 where the two regressions are estimated by maximum likelihood.

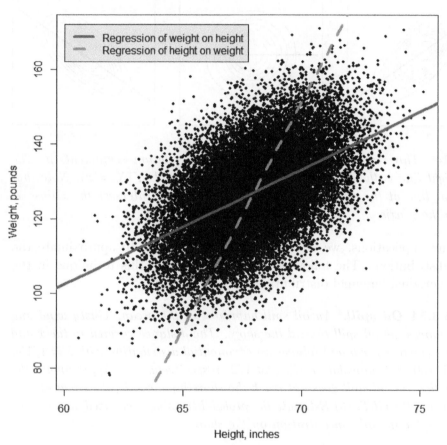

Figure 3.13: *A perfect binormal distribution. The scatterplot weight versus height of 25,000 Korean individuals, 18 years old and younger.*

Contours of the bivariate density, or its estimate, is a convenient geometric representation. This method can be applied to any density; see more detail in Section 5.6. Here, we apply the R function `contour` to the binormal distribution.

Example 3.53 *Contours of the bivariate normal density.* *Plot contours of the general bivariate normal distribution using the R function `contour` for $\rho = -0.7$, 0, and $\rho = 0.7$.*

Solution. The R code **dn2** computes density (3.27) and plots contours for given five parameters (Figure 3.14). As expected, in all three cases the contours are ellipses. When $\rho < 0$ the main principal axis has a negative slope; when $\rho = 0$ the main principal axis is parallel to the x-axis; when $\rho > 0$ main principal axis has a positive slope. If observation points create a cloud with an elliptical shape, the distribution is normal; otherwise, the distribution is not normal.   □

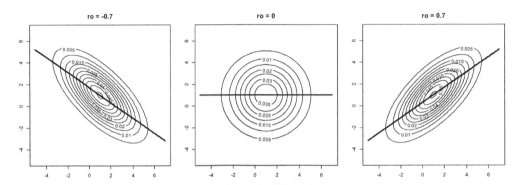

Figure 3.14: *Three contour plots with different correlation coefficient, ρ. The solid straight line is the conditional mean (regression), $E(Y|X = x)$. Note that the tangent line at the point where the regression lines intersect the ellipse is parallel to the y-axis.*

In many applications, we have to appeal to the CLT to approximate the bivariate distribution. The following example illustrates this technique in the case of the original binomial distribution.

Example 3.54 *Oil spill. An oil spill happened in the ocean. Gusty wind and rough sea moves the oil spill toward the shore. The distance covered in the x and y directions (meters per hour) follows an exponential distribution with $\lambda = 1/150$ and $\lambda = 1/100$ with probability 3/5 and 1/2, respectively. (a) Approximate the probability that the oil spill does not reach the shoreline given by the equation $x + y = 3000$ using the CLT. (b) Estimate the probability using vectorized simulations in R. (c) Display the oil concentration on the shore.*

Solution. Without loss of generality, we can assume that the oil spill happened at the origin. Each hour, the spill moves X meters to the right with probability 3/5 where X follows and exponential distribution $\mathcal{E}(1/150)$. Thus, after n hours the x-coordinate can be expressed as $D_{nx} = \sum_{i=1}^{n} U_i X_i$ where U_i is a Bernoulli random variable with $\Pr(U_i = 1) = 3/5$ and $\Pr(U_i = 0) = 2/5$ and $X_i \sim \mathcal{E}(1/150)$. Similarly, the y-coordinate can be expressed as $D_{ny} = \sum_{i=1}^{n} V_i Y_i$ where $\Pr(V_i = 1) = 1/2$ and $\Pr(V_i = 0) = 1/2$ and $Y_i \sim \mathcal{E}(1/100)$. All random variables, including D_{nx} and D_{ny}, are independent.

(a) The asked probability can be written as $\Pr(D_{nx} + D_{ny} < 3000)$. To find this probability, approximate the distribution of D_{nx} with the normal distribution using the CLT, $D_{nx} \simeq \mathcal{N}(E(D_{nx}), \text{var}(D_{nx}))$.

Since U_i and X_i are iid, we have

$$E(D_{nx}) = nE(U_1 X_1) = nE(U_1)E(X_1) = n \times \frac{3}{5} \times 150 = 90n.$$

$$\text{var}(D_{nx}) = n \times \text{var}(U_1 X_1) = n \left[E(U_1^2 X_1^2) - E^2(U_1 X_1) \right]$$

$$= n \left[E(U_1^2)E(X_1^2) - E^2(U_1 X_1) \right] = n \left[\frac{3}{5} \times 2 \times 150^2 - 90^2 \right] = 18900n.$$

Analogously, $E(D_{ny}) = n \times \frac{1}{2} \times 100 = 50n$ and $\text{var}(D_{ny}) = 7500n$. Due to independence of D_{nx} and D_{ny}, we have $D_{nx} + D_{ny} \simeq \mathcal{N}(140n, 26400n)$, and the probability of not reaching the shore within $n = 24$ hours is

$$\Pr(D_{nx} + D_{ny} < 3000) \simeq \Phi\left(\frac{3000 - 140n}{\sqrt{26400n}} \right) = 0.326.$$

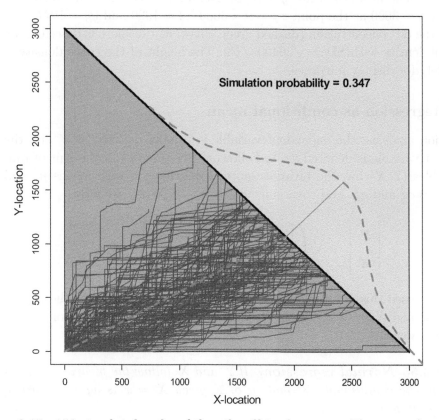

Figure 3.15: *150 simulated paths of the oil spill in the ocean. The normal approximation probability that the oil does not reach the shoreline is 0.326. The green curve depicts the normal density of the oil at the shore.*

(b) The 150 simulated oil spill paths are shown in Figure 3.15. Note that the simulation probability differs from the normal approximation probability because of the low number of simulations; `job=1` creates the graph and `job=2` does vectorized simulations. To create a path for each simulation (the loop over `isim`), we simulate `nSim` by `n` matrices of Bernoulli and exponentially distributed random variables and multiply them. The distances in the x and y directions (D_{nx} and D_{ny}) are found using the `rowSums` command – see the code `oilspill`. The output is below:

```
oilspill(job=2,nSim=100000)
Simulated probability not to reach the shore = 0.34354
Normal approximation probability not to reach the shore = 0.3255
```

The difference between the two probabilities, although not very large, is due to the normal approximation.

(c) The density of the oil on the shoreline is approximated as the conditional density of the bivariate normal distribution with independent $D_{nx} \simeq \mathcal{N}(90n, 18900n)$ and $D_{ny} \simeq \mathcal{N}(50n, 7500n)$ conditional on $D_{nx} + D_{ny} = 3000$. The mean path is the straight line $y = (5/9)x$, shown in green in Figure 3.15, and reaches $x + y = 3000$ at the point $x_S = 9 \times 3000/14 = 1929$ and $y_S = 1071$. The density of the oil concentration (dashed green curve) at the shore is described by the normal density with SD $= \sqrt{26400n}/\sqrt{2}$. The height of the curve density is chosen solely for display purposes.

3.5.1   Regression as conditional mean

In regression analysis, the dependent variable is usually denoted as Y and the independent variable is denoted as X. Following this practice, if the pair of random variables (Y, X) has the bivariate normal distribution with means μ_y and μ_x, variances σ_y^2 and σ_x^2, and covariance $\sigma_{yx} = \sigma_{xy} = \rho\sigma_x\sigma_y$, we write

$$\begin{bmatrix} Y \\ X \end{bmatrix} \sim \mathcal{N}\left(\begin{bmatrix} \mu_y \\ \mu_x \end{bmatrix}, \begin{bmatrix} \sigma_y^2 & \rho\sigma_x\sigma_y \\ \rho\sigma_x\sigma_y & \sigma_x^2 \end{bmatrix} \right).$$

Note that we used this notation earlier, here we just treat $X_1 = Y$ as the dependent variable and $X_2 = X$ as the independent variable following the convention of regression analysis.

Theorem 3.55 *Normal regression.* *If Y and X follow the bivariate normal distribution, the conditional distribution of Y given $X = x$ is again a normal distribution given by*

$$Y|(X = x) \sim \mathcal{N}(\mu_y + \rho\frac{\sigma_y}{\sigma_x}(x - \mu_x), \sigma_{y|x}^2),$$

where

$$E(Y|X = x) = \mu_y + \rho\frac{\sigma_y}{\sigma_x}(x - \mu_x) \tag{3.30}$$

is the conditional mean/regression and $\sigma^2_{y|x}$ is the conditional, or residual, variance:

$$\sigma^2_{y|x} = (1 - \rho^2)\sigma^2_y. \tag{3.31}$$

Proof. The conditional density is found using formula (3.9),

$$f(y|X = x) = \frac{\frac{1}{(2\pi)\sigma_x\sigma_y\sqrt{1-\rho^2}}e^{-\frac{1}{2(1-\rho^2)}\left[\left(\frac{x-\mu_x}{\sigma_x}\right)^2 - 2\rho\left(\frac{x-\mu_x}{\sigma_x}\right)\left(\frac{y-\mu_y}{\sigma_y}\right) + \left(\frac{y-\mu_y}{\sigma_y}\right)^2\right]}}{\frac{1}{\sqrt{2\pi}\sigma_x}e^{-\frac{1}{2}\left(\frac{x-\mu_x}{\sigma_x}\right)^2}}$$

$$= \frac{1}{\sqrt{2\pi}\sigma_y\sqrt{1-\rho^2}}e^{-\frac{1}{2(1-\rho^2)}\left[\rho^2\left(\frac{x-\mu_x}{\sigma_x}\right)^2 - 2\rho\left(\frac{x-\mu_x}{\sigma_x}\right)\left(\frac{y-\mu_y}{\sigma_y}\right) + \left(\frac{y-\mu_y}{\sigma_y}\right)^2\right]}$$

Rewrite the expression in the brackets as

$$\frac{1}{\sigma^2_y}\left[(y-\mu_y)^2 - 2\rho\frac{\sigma_y}{\sigma_x}(x-\mu_x)(y-\mu_y) + \rho^2\frac{\sigma^2_y}{\sigma^2_x}(x-\mu_x)^2\right]$$

$$= \frac{1}{\sigma^2_y}\left(y-\mu_y - \rho\frac{\sigma_y}{\sigma_x}(x-\mu_x)\right)^2.$$

Thus, finally, the conditional density takes the form

$$f(y|X = x) = \frac{1}{\sqrt{2\pi}\sigma_{y|x}}e^{-\frac{1}{2\sigma^2_{y|x}}(y-\mu_y-\rho\frac{\sigma_y}{\sigma_x}(x-\mu_x))^2},$$

where $\mu_y + \rho\frac{\sigma_y}{\sigma_x}(x-\mu_x) = E(Y|X = x)$ and $\sigma_y\sqrt{1-\rho^2} = \sigma_{y|x}$. □

Two characteristic properties of the bivariate normal distribution are that (a) the regression is a linear function, $\alpha + \beta x$, where the slope is related to the regression coefficient as

$$\beta = \rho\frac{\sigma_Y}{\sigma_X} \tag{3.32}$$

and the intercept

$$\alpha = \mu_Y - \beta\mu_X, \tag{3.33}$$

and (b) the conditional variance is constant. Only the normal distribution holds these two properties (Kagan et al. [61]). The bivariate distribution is not normal if at least one property fails.

Another important property of the bivariate normal distribution is that the regression as conditional mean defined by (3.30) and the least squares regression defined by (3.23) are the same, as follows from the equality of slopes and intercepts. In short, for binormal distribution, the LS regression and conditional mean are the same. This is a fundamental fact that raises the bivariate normal distribution above all.

3.5.2   Variance decomposition and coefficient of determination

When the value of X is not specified, the regression as a linear function of X can be expressed as $\mu_y + \rho\frac{\sigma_y}{\sigma_x}(X - \mu_x)$ with the variance of the predicted value (explained variance) given by

$$\mathrm{var}\left(\mu_y + \rho\frac{\sigma_y}{\sigma_x}(X - \mu_x)\right) = \rho^2\frac{\sigma_y^2}{\sigma_x^2}\sigma_x^2 = \rho^2\sigma_y^2. \tag{3.34}$$

Therefore, as follows from (3.31), the following *variance decomposition* holds

$$\sigma_y^2 = \rho^2\sigma_y^2 + \sigma_{y|x}^2, \tag{3.35}$$

where $\sigma_{y|x}^2$ is unexplained or residual variance, $\rho^2\sigma_y^2$ is explained variance and σ_y^2 is total variance. In words,

Total variance = Explained variance + Unexplained variance.

The squared correlation coefficient, ρ^2, is called the *(pointwise) coefficient of determination*. Using identity (3.35), we represent

$$\rho^2 = 1 - \frac{\sigma_{y|x}^2}{\sigma_y^2}. \tag{3.36}$$

Therefore, the coefficient of determination has the following interpretation.

Coefficient of determination = proportion of variance of Y explained by X. (3.37)

As we know from Section 3.3.1, ρ^2 as defined in (3.36) is a function of x, but a bivariate normal distribution is independent of x. Sometimes, the squared correlation coefficient is computed for a nonnormal bivariate distribution. According to Example 3.48, it is also interpreted as the proportion of variance of Y explained by X. The difference is that for the binormal distribution this interpretation is valid for each point x (that is why *pointwise*), but for a linear coefficient of determination, this interpretation applies globally with $\sigma_{y|x}^2$ replaced by $E(Y - a - bX)^2$, where a and b are the LS coefficients from Example 3.48. This fact underscores the special nature of the binormal distribution.

Example 3.56 *Repeated expectation.* *Check the repeated expectation rule (3.13) and decomposition variance (3.16) from Section 3.3.1 for normal regression.*

Solution. To check (3.13), one must prove that

$$E\left(\mu_y + \rho\frac{\sigma_y}{\sigma_x}(X - \mu_x)\right) = \mu_y,$$

which is obvious. To prove (3.16), we use (3.34) in combination with $E(\text{var}(Y|X)) = \sigma^2_{y|x} = (1 - \rho^2)\sigma^2_y$. Therefore, $\text{var}(Y) = \text{var}(E(Y|X)) + E(\text{var}(Y|X))$. $\square$

One may say that the variance decomposition (3.35) is a special case of the general result (3.16), which is valid for any, not necessarily, linear regression. Then, ρ^2 is a function of x, as defined in Definition 3.30.

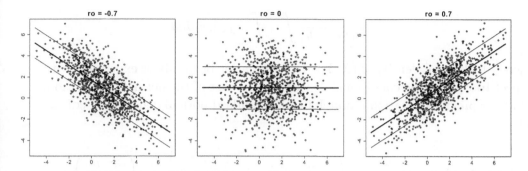

Figure 3.16: *Three bivariate normally distributed pairs generated by the* R *code* dn3 *from Example 3.57.*

3.5.3 Generation of dependent normal observations

It is easy to generate normal pairs using Theorem 3.55 in two steps: (i) generate X using its normal marginal distribution and (ii) generate Y given X using the conditional normal distribution. The following illustrates the procedure.

Example 3.57 *Generation from the bivariate normal distribution. Write an R program for generation of random pairs from the general bivariate normal distribution with $\rho = -0.7$, 0, and $\rho = 0.7$ (use the three-panel plot). Display the normal regression and $\pm$ regression SD line.*

Solution. The R function dn3 is a modification of dn2. The simulated points are displayed in Figure 3.16. A few comments on the code are warranted: (i) The rule of three sigma is used to determine the range for x and y (x.range and y.range). (ii) To increase the size of the font for the axes and the main title we use options cex.axis=1.5 and cex.main=2, respectively. $\square$

The following example gives rise to a generalization of the skew-normal distribution (the traditional skew-normal distribution was introduced in Definition 2.37).

Example 3.58 *Skew-normal distribution via conditional inequality. The bivariate normal distribution of the pair (X,Y) is defined via the conditional distribution $Y|X = x \sim \mathcal{N}(\alpha + \beta x, \sigma^2_{y|x})$ and the marginal distribution $X \sim \mathcal{N}(\mu_x, \sigma^2_x)$. Find the conditional densities $X|(Y < h)$ and $X|(Y > h)$, where h is a constant. Write an R program that checks the result using simulations by superimposing the kernel density on the theoretical density.*

Solution. It is easier to start with the cdf when the condition is expressed in terms of inequality; then the density is simply the derivative of the cdf. In general terms,

$$F_{X|(Y<h)}(x) \;=\; \Pr(X < x | Y < h) = \frac{\Pr(X < x \cap Y < h)}{\Pr(Y < h)}$$

$$= \frac{1}{\Pr(Y < h)} \int_{-\infty}^{x} \int_{-\infty}^{h} f(u,v)dudv,$$

where f is the joint density of (X,Y). Differentiation of the above expression with respect to x yields the conditional density. In the case of the normal distribution, using our previous notation ϕ and Φ for the standard normal density and cdf, we obtain

$$f_{X|(Y<h)}(x) \;=\; \frac{1}{\Pr(Y < h)} \int_{-\infty}^{h} f(x,v)dv = \frac{\Phi\left(\frac{h-\alpha-\beta x}{\sigma_{y|x}}\right)}{\Phi\left(\frac{h-\mu_y}{\sigma_y}\right)} \frac{1}{\sigma_x}\phi\left(\frac{x-\mu_x}{\sigma_x}\right),$$

$$f_{X|(Y>h)}(x) \;=\; \frac{\Phi\left(\frac{\alpha+\beta x-h}{\sigma_{y|x}}\right)}{\Phi\left(\frac{\mu_y-h}{\sigma_y}\right)} \frac{1}{\sigma_x}\phi\left(\frac{x-\mu_x}{\sigma_x}\right).$$

These are generalizations of the skew-normal densities; see Definition 2.37. The R function for simulations and kernel density is found in the file `twocd.r` (the plot is in Figure 3.17). The difference between the theoretical and empirical densities is barely visible.

Example 3.59 *Gaussian mixture. The bivariate distribution of (X_1, X_2) is a mixture of two normal distributions with means $\boldsymbol{\mu}_1$, $\boldsymbol{\mu}_2$ and unit variance, or symbolically $p\mathcal{N}(\boldsymbol{\mu}_1, \mathbf{I}_2) + (1-p)\mathcal{N}(\boldsymbol{\mu}_1, \mathbf{I}_2)$, where p is the proportion of mixture $(0 < p < 1)$. (a) Find the marginal densities. (b) Find the conditional density of $X_1 | X_2 = x$, (c) Find the conditional mean, $E(X_1 | X_2 = x)$. (d) Find the limits of the conditional mean when $x \to \pm\infty$. (e) Plot the contours of the mixture normal distribution density with p and $\boldsymbol{\mu}_1$, $\boldsymbol{\mu}_2$ as parameters, and superimpose with the conditional mean. (f) Check the computation of the conditional mean via simulations by averaging within a running window.*

Solution. (a) Using the notation ϕ, the joint density is expressed as

$$f(x_1, x_2) = p\phi(x_1 - \mu_{11})\phi(x_2 - \mu_{12}) + (1-p)\phi(x_1 - \mu_{21})\phi(x_2 - \mu_{22}),$$

where $\boldsymbol{\mu}_1 = (\mu_{11}, \mu_{12})'$ and $\boldsymbol{\mu}_2 = (\mu_{21}, \mu_{22})'$. Therefore, the marginal densities,

$$f_1(x_1) \;=\; p\phi(x_1 - \mu_{11}) + (1-p)\phi(x_1 - \mu_{21})$$
$$f_2(x_1) \;=\; p\phi(x_2 - \mu_{12}) + (1-p)\phi(x_2 - \mu_{22}),$$

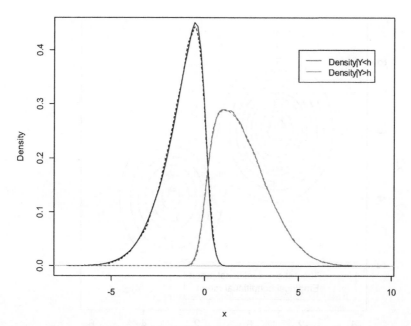

Figure 3.17: *Two skew-normal conditional densities. The dashed lines are empirical kernel densities; they practically overlap with the theoretical densities.*

are mixtures of the univariate normal distributions. (b) The conditional density is found from the general formula

$$f_{1|2}(x_1|x) = \frac{f(x_1, x)}{\int_{-\infty}^{\infty} f(x_1, x) dx_1}.$$

In our particular case, the denominator becomes

$$\int_{-\infty}^{\infty} f(x_1, x) dx_1 = p\phi(x - \mu_{12}) + (1 - p)\phi(x - \mu_{22}),$$

so that

$$f_{1|2}(x_1|x) = \frac{p\phi(x_1 - \mu_{11})\phi(x - \mu_{12}) + (1 - p)\phi(x_1 - \mu_{21})\phi(x - \mu_{22})}{p\phi(x - \mu_{12}) + (1 - p)\phi(x - \mu_{22})}.$$

(c) The conditional (regression) is the expectation of the conditional density,

$$E(X_1|X_2 = x) = \int_{-\infty}^{\infty} x_1 f_{1|2}(x_1|x) dx_1 = \frac{p\mu_{11}\phi(x - \mu_{12}) + (1 - p)\mu_{21}\phi(x - \mu_{22})}{p\phi(x - \mu_{12}) + (1 - p)\phi(x - \mu_{22})}.$$

(d) Assuming that $\mu_{12} < \mu_{22}$, we have $\phi(x - \mu_{12})/\phi(x - \mu_{22}) \to 0$ when $x \to -\infty$ so that

$$\lim_{x \to -\infty} E(X_1|X_2 = x) = \lim_{x \to -\infty} \frac{p\mu_{11}\phi(x - \mu_{12})/\phi(x - \mu_{22}) + (1 - p)\mu_{21}}{p\phi(x - \mu_{12})/\phi(x - \mu_{22}) + (1 - p)} = \mu_{21},$$

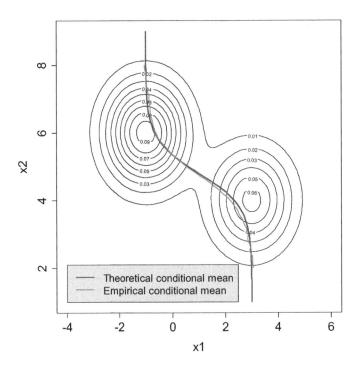

Figure 3.18: *Conditional mean for a mixture of two normal distributions.*

and, analogously, $\lim_{x \to \infty} E(X_1 | X_2 = x) = \mu_{22}$. (e) The R function is `mi2n`. We make a few comments: (i) Densities `f1` and `f2` are computed in the vectorized fashion as vectors and then rearranged as N by N matrices. We use the fast grid computation based on the `rep` command; the first grid is computed as `rep(x1,N)`, and the second grid is computed as `rep(x2,each=N)`. (ii) Since the vector components are independent, the simulated values are generated separately. (iii) The first mixture component has `Nsim*p` values and the second mixture component has `Nsim-Nsim*p` values, so that the proportion of the first component is p. (iv) To estimate the conditional mean from simulations, we cut a window of length `2*w` in the center of the current `x2s` value and average the first component over `X12` values; see Example 2.73. The theoretical and empirical conditional means are very close as can be seen from Figure 3.18.

Example 3.60 *Double integral approximation via simulations.* *Approximate the integral from Example 1.12 via Monte Carlo simulations using uniform and normal distributions. Plot the integral approximation versus SD of the normal distribution.*

Solution. First, we discuss how to approximate integrals with integrand $g(x, y)$ over domain $H(x, y) < c$ in general terms. Introduce the indicator function $I = I(x, y)$ such that $I = 1$ if $H(x, y) < c$ and $I = 0$ otherwise. Then if

(X, Y) have density f, we express

$$
\begin{aligned}
A &= \int\!\!\int_{H(x,y)<c} g(x,y)\,dx\,dy = \int_{-\infty}^{\infty}\int_{-\infty}^{\infty} I(H(x,y)<c)g(x,y)\,dx\,dy \\
&= \int_{-\infty}^{\infty}\int_{-\infty}^{\infty} \frac{I(H(x,y)<c)g(x,y)}{f(x,y)} f(x,y)\,dx\,dy \\
&= E_{(X,Y)}\left(\frac{I(H(X,Y)<c)g(X,Y)}{f(X,Y)} \right).
\end{aligned}
$$

To approximate A via simulations, we generate a large number of pairs (X_i, Y_i) with the bivariate density f and compute the average as indicated above. Clearly, f must be positive on the set $H(x,y) < c$. Usually, several bivariate distributions may be suggested with various performances.

Now we apply this technique for approximation A. First, consider uniformly distributed independent $X \sim \mathcal{R}(-1,1)$ and $Y \sim \mathcal{R}(-1,1)$ with $f = 1/4$ on $[-1,1]^2$; see the R code dint. Note that $[-1,1]^2$ contains $x^2 + y^2 < 1$. The uniform distribution gives the answer close to the exact one, 1.375. Notice that the result varies from run to run even for a large number of simulations, such as 1 million.

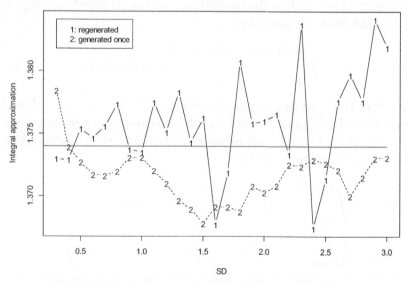

Figure 3.19: *Integral approximation using normal distribution in Example 3.60.*

Second, consider normally distributed X and Y with zero mean and SD varying from 0.3 to 3; see function dint for job=2. The plot of the estimate of A versus SD is shown in Figure 3.19. Two line estimates are shown: regenerated and generated once. In the former case, random normal variables are regenerated for each SD (rnorm in the loop) and in the second case random variables are generated once before the loop. These two methods help to separate the variation in estimation due to the choice of SD from the randomness due to generation of X and Y. Obviously, the second method has smaller variation.

Several lessons can be learned: (i) This may require a large number of simulations to achieve satisfactory approximation. (ii) The approximations have a stochastic nature and, unlike calculus, approximate the integral with certain probability. (iii) This probability is usually difficult to estimate without more simulations.

3.5.4   Copula

Copula is the method to create a multivariate distribution with dependent components using marginal distributions. Here we consider bivariate distributions. Let two marginal cdfs, $F_X(x)$ and $F_Y(y)$, be given. How do we create a bivariate distribution with dependent components such that their marginal distributions are $F_X(x)$ and $F_Y(y)$? Take the standard bivariate normal density with zero means, unit variances, and correlation coefficient ρ,

$$\phi(u, v; \rho) = \frac{1}{(2\pi)\sqrt{1 - \rho^2}} e^{-\frac{1}{2(1-\rho^2)}(u^2 - 2\rho uv + v^2)}, \qquad (3.38)$$

as the basis of our creation. Define the *normal (or Gaussian) copula* through the double integral as the joint cdf given by

$$F(x, y; \rho) = \int_{-\infty}^{\Phi^{-1}(F_X(x))} \int_{-\infty}^{\Phi^{-1}(F_Y(y))} \phi(u, v; \rho) du dv, \qquad (3.39)$$

where Φ is the cdf of the standard normal distribution. It is easy to see that the marginal cdfs of F are $F_X(x)$ and $F_Y(y)$ for each ρ because the marginal density of the bivariate ϕ is $\mathcal{N}(0, 1)$. Also, we prove that if $\rho = 0$, then the copula cdf turns into the product of marginal cdfs:

$$
\begin{aligned}
F(x, y; 0) &= \int_{-\infty}^{\Phi^{-1}(F_X(x))} \int_{-\infty}^{\Phi^{-1}(F_Y(y))} \phi(u)\phi(v) du dv \\
&= \left(\int_{-\infty}^{\Phi^{-1}(F_X(x))} \phi(u) du \right) \left(\int_{-\infty}^{\Phi^{-1}(F_Y(y))} \phi(v) dv \right) \\
&= \Phi(\Phi^{-1}(F_X(x))) \times \Phi(\Phi^{-1}(F_Y(y))) = F_X(x) F_Y(y).
\end{aligned}
$$

To derive the copula density, differentiate $F(x, y; \rho)$ with respect to x and y,

$$f(x, y; \rho) = \frac{\phi\left(\Phi^{-1}(F_X(x)), \Phi^{-1}(F_Y(y)); \rho\right)}{\phi(\Phi^{-1}(F_X(x)))\phi(\Phi^{-1}(F_Y(y)))} f_X(x) f_Y(y),$$

where $f_X(x) = F'_X(x)$ and $f_Y(y) = F'_Y(y)$ are the respective marginal densities. Obviously, when $\rho = 0$, the numerator and denominator cancel out and the copula density turns into the product of marginal densities in agreement with the previous result regarding the cdfs. It is easy to see that, if the marginal densities are Gaussian, then copula f turns into a bivariate normal density (3.38).

The conditional densities can easily be derived from the copula density. For example, the conditional density of Y given $X = x$

$$f(y|x; \rho) = \frac{\phi\left(\Phi^{-1}(F_X(x)), \Phi^{-1}(F_Y(y)); \rho\right)}{\phi(\Phi^{-1}(F_X(x))\phi(\Phi^{-1}(F_Y(y)))} f_Y(y).$$

The conditional mean and variance obtained by integration.

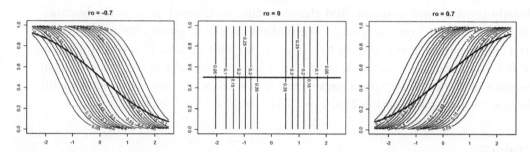

Figure 3.20: *Normal copula contours of the bivariate density with standard normal and uniform marginal distributions.*

Example 3.61 *Copula of normal and uniform. Create a bivariate distribution on $(-\infty, \infty) \times (0, 1)$ as a normal copula with standard normal and uniform marginal distributions. Display the copula density using contours for $\rho = -0.7$, 0, and 0.7 and superimpose with the conditional mean (regression) of Y given $X = x$.*

Solution. The marginal cdf and density of the first component are $\Phi(x)$ and $\phi(x)$, respectively. The marginal cdf and density of the second component are y and 1, respectively. Therefore,

$$\Phi^{-1}(F_X(x)) = \Phi^{-1}(\Phi(x)) = x, \quad \Phi^{-1}(F_Y(y)) = \Phi^{-1}(y)$$

and the copula density becomes

$$\begin{aligned} f(x, y; \rho) &= \phi\left(x, \Phi^{-1}(y); \rho\right) \\ &= \frac{1}{\sqrt{2\pi(1 - \rho^2)}} e^{\frac{1}{2}\left(\Phi^{-1}(y)\right)^2 - \frac{1}{2(1-\rho^2)}\left(x^2 - 2\rho x \Phi^{-1}(y) + \left(\Phi^{-1}(y)\right)^2\right)}. \end{aligned}$$

The conditional mean of Y given $X = x$ is expressed as the integral

$$E(Y|X = x) = \frac{1}{\sqrt{1 - \rho^2}} \int_0^1 y e^{\frac{1}{2}x^2 + \frac{1}{2}\left(\Phi^{-1}(y)\right)^2 - \frac{1}{2(1-\rho^2)}\left(x^2 - 2\rho x \Phi^{-1}(y) + \left(\Phi^{-1}(y)\right)^2\right)} dy.$$

$$(3.40)$$

One can simplify the above integral a bit by change of variable $z = \Phi^{-1}(y)$. Then $\Phi(z) = y$ and $dy = \phi(z)$ and we arrive at

$$E(Y|X = x) = \frac{1}{\sqrt{1 - \rho^2}} \int_{-\infty}^{\infty} \Phi(z) e^{\frac{1}{2}x^2 + \frac{1}{2}z^2 - \frac{1}{2(1-\rho^2)}(x^2 - 2\rho xz + z^2)} dz. \qquad (3.41)$$

This integral cannot be expressed in closed form; numerical integration is required. The function `copRN` plots the contours and conditional mean for three values of the correlation coefficient; see Figure 3.20. Note that $\rho < 0$ implies a decreasing and $\rho > 0$ implies an increasing conditional mean. When $\rho = 0$ the distributions are independent, and therefore the conditional mean is the horizontal line passing 0.5 on the y-axis. We make a couple of comments on the code: (i) The R function `outer` is used to compute the joint density in a timely fashion (it is faster than using a double loop). (ii) Computation of the conditional mean requires numerical integration using `integrate`. Here, we use expression (3.40) defined as the local function `EYx`, but (3.41) could be used as well.

Problems

1. Prove by integration that the volume under the bivariate normal density is 1. [Hint: Apply change of variables to reduce to the standard bivariate normal density.]

2. (a) Write an R function that displays the bivariate density using the `persp` command with the angles specified by the arguments of the function. The five parameters of the density are specified as arguments, define default values. (b) Display the density over a 360° loop of `theta` in a movie fashion. Use optimal `phi` angle and the distance parameter `r` of `persp` to make the view as informative as possible.

3. Prove that matrix Ω is positive definite if and only if $|\rho| < 1$.

4. Write R code that plots contours of the bivariate density using the `contour` function (modify function `ell2` and follow the instructions of Problem 2 above).

5. The concentration of radioactive particles in the air, due to explosion at a nuclear plant, follows the bivariate normal density with unit variances and correlation coefficient $\rho = -0.5$. (a) Derive the radiation density along the main street in town defined by equation $x + y = 1$ (assume that the plant is at the origin). (b) Derive the (x, y) location on the line where the radiation density is maximum. (c) Estimate the amount of radiation, as an integral over the density, for houses on the main street located at the distance 2 or farther to north from the maximum point.

6. Write R code to approximate the double integral $\int_{-\infty}^{a} \int_{-\infty}^{b} f(x_1, x_2) dx_1 dx_2$ based on the `outer` function where f is the bivariate density specified by five parameters. Check your code using $\rho = 0$.

7. Prove that $\Pr(X \leq x, Y \leq y) > \Pr(X \leq x) \Pr(Y \leq y)$ where the pair (X, Y) has a bivariate normal distribution with a positive correlation coefficient. Prove that the reverse inequality holds if the correlation coefficient is negative. [Hint: Reduce the problem to standard normal X and Y.]

8. In the Example 3.52, (a) explain why the point regression of Z_2 on Z_1 touches the line parallel to the y-axis; (b) explain why the point regression of Z_1 on Z_2 touches the line parallel to the x-axis. (c) Is it true that the angle between the regression of Z_2 on Z_1 and the major principal axis equals the angle between the regression of Z_1 on Z_2 and the major principal axis?

9. Modify function `dn2` to (a) display the regression of Z_1 on Z_2 in all three plots as in Figure 3.14 and (b) show three contours with levels $M/4, M/2$, and $2M/3$ where M is the maximum value of the bivariate density using colors `red`, `green`, and `blue`, respectively.

10. Add contours to the cloud of points in Figure 3.13. [Hint: Compute estimates of the mean and covariance matrix using `colMeans` and `cov`, respectively.]

11. Using Theorem 3.55, (a) give a rigorous definition to what we mean when we say that normal regression goes through the center of random variables; (b) prove that Y and X are uncorrelated if and only if regression is a horizontal line; (c) formulate and prove the theorem when X is regressed on Y.

12. In Example 3.50, (a) compute the pointwise coefficient of determination as a function of x and provide an interpretation. (b) What assumption is violated to make interpretation (3.37) valid? (c) Compute the global coefficient of determination by replacing $\sigma_{y|x}^2$ with $E(Y - a - bX)^2$, and provide an interpretation. (d) Is it true that the global coefficient of determination is the expected pointwise coefficient of determination?

13. Modify function `dn3` to generate and plot 100K normally distributed random pairs (use `pch="."`); display the regression line and its $\pm 1.96\sigma$ band. Count the number of data points covered by the band, and explain why the proportion of covered points is very close to 95%.

14. In Example 3.58, (a) check that density $f_{X|(Y<h)}(x)$ turns into a regular normal density when $h \to \infty$; (b) check that density $f_{X|(Y>h)}(x)$ turns into a regular normal density when $h \to -\infty$; and (c) prove that $F_{X|(Y<h)}(x) \leq F_{X|(Y>h)}(x)$ for all x and h.

15. Repeat all steps in the Gaussian mixture example with a general 2×2 covariance matrix, $p\mathcal{N}(\boldsymbol{\mu}_1, \boldsymbol{\Omega}) + (1-p)\mathcal{N}(\boldsymbol{\mu}_1, \boldsymbol{\Omega})$. [Hint: The three parameter values that define the covariance matrix should be passes to the R code as arguments.]

16. Repeat all steps in Example 3.60 with domain of integration $x^2 - xy + 2y^2 < 1$.

17. Box–Muller transform: Prove that if $U_1, U_2 \overset{\text{iid}}{\sim} \mathcal{R}(0,1)$, then $Z_1 = \sqrt{-2\ln U_1} \times \cos(2\pi U_2)$, $Z_1 = \sqrt{-2\ln U_1}\sin(2\pi U_2) \overset{\text{iid}}{\sim} \mathcal{N}(0,1)$. Use simulations to verify this result.

18. Prove that if marginal densities are Gaussian, then copula f turns into a bivariate normal density.

19. Prove that the marginal cdfs in (3.39) are F_X and F_Y, regardless of ρ.

20. (a) Create a bivariate distribution on $(-\infty, \infty) \times (0, \infty)$ using a normal copula with standard normal and exponential distribution with parameter λ as marginal distributions. (b) Create a bivariate distribution on $(0, \infty) \times (0, \infty)$ using two exponential distributions with different rate parameters.

21. Let $F_1(x)$ and $F_2(y)$ be two cdfs with $\theta \geq 1$. Define the joint cdf as $F(x, y) = e^{-(L_1^\theta(x) + L_2^\theta(y))^{1/\theta}}$ where $L_1(x) = -\ln F_1(x)$ and $L_2(y) = -\ln F_2(y)$; this F is called the Gumbel–Hougaard (G-H) copula. (a) Prove that $\theta = 1$ defines the cdf of independent random variables. (b) Derive the G-H copula for standard uniform distributions and plot contours for cdf and pdf for $\theta = 1$ and $\theta = 2$. (c) Do the same, but for standard normal distributions.

3.6 Joint density upon transformation

In Section 2.12, we derived the density (pdf) of a univariate random variable upon linear or nonlinear transformation. In this section, we will extend this result to the bivariate joint density. Although here we assume that $m = 2$, the formula below stays valid for anyrandom vector $\mathbf{X}$ of size m with density f transformed into an vector $\mathbf{Y}$ of size m by the rule

$$\mathbf{Y} = \mathbf{h}(\mathbf{X}). \tag{3.42}$$

What is the density of $\mathbf{Y}$? Sometimes, we refer to $\mathbf{h}$ as a mapping from R^m to R^m and write $\mathbf{h}: R^m \rightarrow R^m$. We make some assumptions on this mapping: (a) $\mathbf{h}$ is a continuously differentiable; (b) $\mathbf{h}$ is one-to-one, so there exists an inverse mapping, $\mathbf{g} = \mathbf{h}^{-1}$; and (c) the Jacobian $\partial \mathbf{g}/\partial \mathbf{y}$ has full rank, so that $\det(\partial \mathbf{g}/\partial \mathbf{y}) \neq 0$. Then the joint density of $\mathbf{Y}$ is

$$f_Y(\mathbf{y}) = \left| \det\left(\frac{\partial \mathbf{g}}{\partial \mathbf{y}}\right) \right| f_X(\mathbf{g}(\mathbf{y})), \tag{3.43}$$

where the bars mean absolute value of the determinant of $\partial\mathbf{g}/\partial\mathbf{y}$ as an $m \times m$ matrix. Note that, in this formula, the inverse $\mathbf{g}$ is used, not the original transformation. Thus, finding the inverse and computing the determinant of its Jacobian are the tasks required to derive the joint density of $\mathbf{Y}$. The above formula is generalizable to the case when the inverse mapping exists on a partition of R^m, see more detail in Casella and Berger [17], Theorem 2.1.3. Sometimes it is easier to derive the Jacobian for $\mathbf{h}$; then use the fact that

$$\left| \det\left(\frac{\partial\mathbf{g}}{\partial\mathbf{y}} \right) \right| = \frac{1}{\left| \det\left(\frac{\partial\mathbf{h}}{\partial\mathbf{y}} \right) \right|}.$$

It is instructive to write down the formula for the density in the coordinate form for a bivariate random variable:

$$f_Y(y_1, y_2) = \left| \det\begin{pmatrix} \frac{\partial g_1}{\partial y_1} & \frac{\partial g_1}{\partial y_2} \\ \frac{\partial g_2}{\partial y_1} & \frac{\partial g_2}{\partial y_2} \end{pmatrix} \right| f_X(g_1(y_1, y_2), g_2(y_1, y_2)).$$

Example 3.62 *Joint density of sum and difference.* *Random variables* X_1 *and* X_2 *are iid with density* $f(x)$. *Find the joint density of* $Y_1 = X_1 + X_2$ *and* $Y_2 = X_1 - X_2$.

Solution. In this case, the transformation is linear:

$$\mathbf{h}(x_1, x_2) = \begin{bmatrix} 1 & 1 \\ 1 & -1 \end{bmatrix} \begin{bmatrix} x_1 \\ x_2 \end{bmatrix}$$

with inverse

$$\mathbf{g}(y_1, y_2) = \begin{bmatrix} 1 & 1 \\ 1 & -1 \end{bmatrix}^{-1} \begin{bmatrix} y_1 \\ y_2 \end{bmatrix} = \frac{1}{2}\begin{bmatrix} 1 & -1 \\ -1 & -1 \end{bmatrix} \begin{bmatrix} y_1 \\ y_2 \end{bmatrix} = \begin{bmatrix} (y_1 - y_2)/2 \\ -(y_1 + y_2)/2 \end{bmatrix}.$$

We have

$$\left| \det\begin{pmatrix} \frac{\partial g_1}{\partial y_1} & \frac{\partial g_1}{\partial y_2} \\ \frac{\partial g_2}{\partial y_1} & \frac{\partial g_2}{\partial y_2} \end{pmatrix} \right| = \left| \det\left(\frac{1}{2}\begin{bmatrix} 1 & -1 \\ -1 & -1 \end{bmatrix} \right) \right| = 1.$$

Since variables X_1 and X_2 are independent, the joint density of the pair $Y = (Y_1, Y_2)$ is

$$f_Y(y_1, y_2) = f\left(\frac{y_1 - y_2}{2} \right) f\left(-\frac{y_1 + y_2}{2} \right).$$

□

Transformation of independent random variables can be used to define dependent bivariate distributions; below is an example.

Example 3.63 *Bivariate exponential distribution. Random variables X_1 and X_2 have independent exponential distributions with rate parameters λ_1 and λ_2. Find the bivariate density of $Y_1 = X_1$ and $Y_2 = X_2/(aX_1)$, where $a > 0$.*

Solution. Since the joint density of (X_1, X_2) is $\lambda_1\lambda_2 e^{-\lambda_1 x_1 - \lambda_2 x_2}$ and $x_2 = ay_1y_2$, the joint density of (Y_1, Y_2) is $\lambda_1\lambda_2 ay_1 e^{-\lambda_1 y_1 - a\lambda_2 y_1 y_2}$. □

The joint density under transformation given by expression (3.43) may work even when the condition on differentiability does not hold, as in the following example.

Example 3.64 *Ratio of two normal RVs has a Cauchy distribution. Prove that the ratio of two independent standard normal variables has a Cauchy distribution using formula (3.43).*

Solution. Transform independent X and Y with standard normal distribution into the new pair of random variables: $U = X/Y$ and $V = Y$. In notation (3.42) $h_1(x, y) = x/y$ and $h_2(x, y) = y$. The inverse transformation is $g_1(u, v) = uv$ and $g_2(u, v) = v$ with the Jacobian

$$\begin{pmatrix} \frac{\partial g_1}{\partial u} & \frac{\partial g_1}{\partial v} \\ \frac{\partial g_2}{\partial u} & \frac{\partial g_2}{\partial v} \end{pmatrix} = \begin{pmatrix} v & u \\ 0 & 1 \end{pmatrix}.$$

The absolute value of the determinant is $|v|$. Since the joint density of (X, Y) is $f(x, y) = (2\pi)^{-1} e^{-\frac{1}{2}(x^2 + y^2)}$, the joint density for (U, V), as follows from (3.43), is

$$f(u, v) = (2\pi)^{-1} |v| \, e^{-\frac{1}{2}v^2(1 + u^2)}.$$

Now we find the marginal distribution of U by integrating out V (let $a = 1 + u^2$ to shorten the notation):

$$\int_{-\infty}^{\infty} f(u, v) dv = \frac{1}{2\pi} \int_{-\infty}^{\infty} |v| \, e^{-\frac{a}{2}v^2} dv = \frac{1}{2\pi} \int_0^{\infty} v e^{-\frac{a}{2}v^2} dv^2$$

$$= \frac{1}{2\pi} \int_0^{\infty} e^{-\frac{a}{2}t} dt = \frac{1}{\pi(1 + u^2)},$$

the Cauchy density. □

A slight modification of the above example leads to quite a different distribution: if all conditions of the example hold but Y has mean μ, the density of the ratio is

$$f_{X/Y}(z) = e^{-\frac{\mu^2 z^2}{2(z^2 + 1)}} \frac{\mu}{\sqrt{2\pi}(z^2 + 1)^{3/2}} \left[2\Phi\left(\frac{\mu}{\sqrt{z^2 + 1}}\right) - 1 \right] + \frac{1}{\pi(z^2 + 1)} e^{-\frac{\mu^2}{2}},$$

$$(3.44)$$

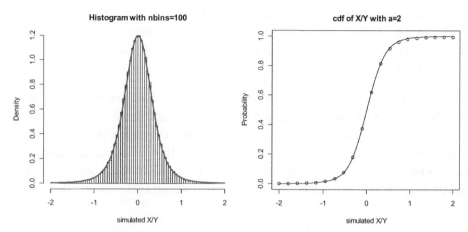

Figure 3.21: *Left: The histogram of simulated X/Y and the theoretical density (3.44) with $\mu = 3$. Right: The empirical (solid line) and theoretical (circle) cdf as the integral of density (3.44).*

where Φ is the standard normal cdf. The key in deriving this density is the fact

$$\int_a^b u\phi(u)du = \phi(a) - \phi(b),\qquad(3.45)$$

where ϕ is the standard normal density. This density was derived by Fieller [46] almost one hundred years ago. Neither expected value nor variance exists for this density.

Example 3.65 *Simulations for Fieller density. Use simulations to confirm the Fieller density (3.44) via histogram and cdf.*

Solution. The R code is `file1XY`, which produces Figure 3.21. We make several comments: (i) Since the variance of X/Y is infinite, the simulated values may be large and the possibility of getting such values increases with the number of simulations (`nExp`). The existence of extreme values creates difficulties in displaying the histogram and cdf. Therefore to truncate the range we use `a` as an argument: values `r=X/Y` with absolute value larger than `a` are removed. (ii) To compute the cdf, we compute the density (3.44) as function `XYdens` and then use `integrate`. (iii) Argument `nbins` controls the number of bins (`nclass`) in the histogram. Several attempts may be required to get the right `nbins`. (iv) The red/gray line depicts the theoretical Cauchy density. There is a slight difference between theoretical and empirical cdfs due to the truncation.

Problems

1. Check the joint density from Example 3.62 with X_1 and X_2 being independent standard normally distributed random variables using simulations: (a)

generate and plot 10K random pairs of (Y_1, Y_2) using the `pch="."` option and plot `contour` with the option `add=T`. Then compute a matrix of the density values for f_Y, and use `contour` with the option `add=T` again with the same `levels` but different color.

2. (a) Find the pdf and cdf of $X - Y$, where X and Y are independent standard normally distributed random variables using the joint density transformation approach as in Example 3.64. (b) Check your derivation by finding the pdf and cdf using the fact that a linear combination of normally distributed random variable is a normally distributed random variable. (c) Use simulations in R to confirm your result by plotting the two cdfs on the same graph.

3. Similarly to Example 3.62, find the joint pdf of $X_1 + X_2$ and X_1 where X_1 and X_2 are independent.

4. Random variables X and Y are independent and distributed as $\mathcal{N}(0, 1)$. Find the density of random $\rho > 0$ and $0 \le \theta < 2\pi$ defined as $X = \rho \cos \theta$ and $Y = \rho \sin \theta$. Use simulations to verify the answer.

5. Find the cdf and pdf of $U = X/Y$, where X and Y are independent random variables uniformly distributed on $[0, 1]$. Use simulations in R to confirm your result by modifying function `file1XY`.

6. Use integration by parts to prove identity (3.45).

7. Derive density (3.44) and plot several densities on the same graph for several μ, say, 0, 1, 2, and 3.

8. Can the uniform distribution on the unit disk be defined as $X = R \cos \Theta$ and $Y = R \sin \Theta$, where random variables $R \sim \mathcal{R}(0, 1)$ and $\Theta \sim \mathcal{R}(0, 2\pi)$ are independent? (a) Derive the marginal cdf of X and compare with the empirical cdf from simulations. (b) Prove that (X, Y) are not uniformly distributed on the disk by finding the cdf of $X^2 + Y^2$. (c) Do the same as in the previous item, but derive the bivariate density upon transformation. [Hint: Use the fact that $\int_{-1}^{x} \sqrt{1 - t^2} dt = 0.5x\sqrt{1 - x^2} + 0.5 \arcsin x + 0.25\pi$.]

9. Find the density of $U = X/Y$ where $X \sim \mathcal{N}(\mu_x, \sigma_x^2)$ and $Y \sim \mathcal{N}(\mu_y, \sigma_y^2)$ are independent. Use simulations in R to confirm your result. [Hint: Split the real line into two pieces, $Y < 0$ and $Y > 0$.]

10. Let $f(x)$ and $g(y)$ be two densities. Define the joint density proportional to $f(a_1 + b_1 x + c_1 y)g(a_1 + b_2 x + c_2 y)$, where all coefficients are fixed numbers. (a) Find the coefficient of the proportionality. (b) Under what condition is the product of the densities a joint density? (c) Under what condition does the joint density specify the density of independent random variables?

11. The distribution of a unit disk is defined as $X = r\cos\theta$ and $Y = r\sin\theta$, where $r \sim \mathcal{R}(0,1)$ and $\theta \sim \mathcal{R}(0, 2\pi)$ are independent. (a) Generate and plot 100K points using `pch="."` option. (b) Prove that the distribution is not uniform. (c) Modify the distribution of r to make the distribution uniform; replot the points to visually check the uniformity. [Hint: Find $\Pr(X^2 + Y^2 \leq z)$.]

12. Prove that the Laplace distribution from Section 2.4.1 can be obtained as $(\sqrt{2}/\lambda)\sqrt{Y}Z$, where Y is the exponential distribution with a unit rate and Z is the standard normal variable. First use simulations and then prove theoretically using the results of this section.

3.7 Geometric probability

Sometimes it is easy to find probabilities using a geometric illustration. When the answer is difficult to obtain analytically, simulations may help; often, we use simulations to verify our mathematical derivation. In fact, it is always desirable to check complex derivations with simulations. In this section, we demonstrate this approach with two types of examples: the meeting problem and various problems involving random objects such as random lines on the plane. However, careful consideration is required to define *throwing an object at random*.

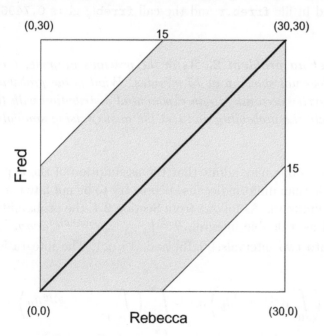

Figure 3.22: *Geometric probability for the meeting problem, where, without loss of generality, we can assume that the y-axis represents the Fred's arrival and the x-axis represents the Rebecca's arrival in minutes.*

3.7.1   Meeting problem

Example 3.66 *Meeting problem 1. Fred and Rebecca decided to meet at 5 p.m. Neither of them is punctual, so they arrive between 5 p.m. and 5:30 p.m., wait for 15 minutes, and if the friend does not show up, leave. Assuming that the arrival time is independent and distributed uniformly, what is the probability that they meet? Use simulations in R to check the answer.*

Solution. Denote X and Y as the random time arrival after 5 p.m. for Rebecca and Fred, respectively, $0 \le X \le 30$ and $0 \le Y \le 30$. Since X and Y are independent and uniformly distributed on $[0, 30]$, the random pair (X, Y) has a uniform distribution on the square $[0, 30] \times [0, 30]$. Fred and Rebecca meet if $|X - Y| < 15$, when a randomly chosen point on the square falls near the $45°$ line ± 15; see the shaded area in Figure 3.22. Algebraically, the condition on the meeting is expressed as a set of points $\{(x, y) \in [0, 30]^2 : |x - y| \le 15\}$. Since the distribution is uniform, the meeting probability is easy to compute as the ratio of the shaded area to the area of the square or, as the complementary probability from the two unshaded triangles,

$$\Pr(\text{friends meet}) = 1 - \frac{2 \times 15 \times 15/2}{30 \times 30} = 1 - \frac{1}{4} = \frac{3}{4}.$$

Simulations are found in file `frreb.r` and the call `frreb()` gives 0.74966, close to 0.75.

Example 3.67 *Meeting problem 2. As in the previous example, the person leaves if the friend does not show up in 15 minutes. What is the probability that friends meet if they arrive according to an exponential distribution with the scale parameter θ? Compute the probability and test the answer using simulations for an array of values θ.*

Solution. First of all one may admit that the assumption of the exponential arrival is more realistic than uniform because people try to be not late (we assume that friends wait indefinitely). As follows from Section 2.4, the probability of the meeting is expressed as a double integral, $\theta^{-2} \int_{|x-y| \le 15} e^{-(x+y)/\theta} dx\, dy$. Split the integration over x into two intervals, $[0, 15]$ and $[15, \infty)$. The integration is as follows:

$$2\theta^{-2} \left[\int_0^{15} \left(\int_0^x e^{-(x+y)/\theta} dy \right) dx + \int_{15}^{\infty} \left(\int_{x-15}^x e^{-(x+y)/\theta} dy \right) dx \right]$$

$$= 2\theta^{-1} \int_0^{15} \left(e^{-\frac{x}{\theta}} - e^{-\frac{2x}{\theta}} \right) dx + 2\theta^{-1} \int_{15}^{\infty} \left(e^{-\frac{2x-15}{\theta}} - e^{-\frac{2x}{\theta}} \right) dx$$

$$= 1 + e^{-30/\theta} - 2e^{-15/\theta} + e^{-15/\theta} - e^{-30/\theta}$$

$$= 1 - e^{-15/\theta}.$$

The R code with simulations is found in function `metpr2`. Exponentially distributed random numbers are generated using the built-in function `rexp` with the rate parameter $\lambda = 1/\theta$. R converts the condition `abs(X-Y)<15` into a binary array (1 if condition is true and 0 otherwise); therefore `mean` computes the proportion when the meeting condition holds. Simulations and the mathematically derived answer match perfectly. As follows from the graph (not shown), if both arrive on average 10 minutes after 5 p.m. ($\theta = 10$), the probability of the meeting is about 80%.

3.7.2 Random objects on the square

The following problem was posed and solved in the eighteenth century by a French mathematician de Buffon.

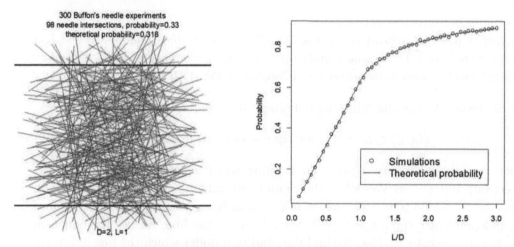

Figure 3.23: *Geometric illustration of the Buffon's needle problem and probability of crossing. The theoretical probability is computed by formula (3.46). See function* **buffon**.

Example 3.68 *Buffon's needle.* *A needle of length L is dropped at random on a plane ruled with parallel lines D units apart. What is the probability that the needle crosses a line? The answer is*

$$
p = \begin{cases} \frac{2L}{\pi D} & \text{if } L < D \\ \frac{2}{\pi} \arccos \frac{D}{L} + \frac{2L}{\pi D}\left[1 - \sqrt{1 - \left(\frac{D}{L}\right)^2}\right] & \text{if } L \geq D \end{cases} . \tag{3.46}
$$

Confirm this result via simulations.

Solution. The R function is `buffon`; `job=1` does the plot (see below) and `job=2` carries out simulations in vectorized fashion. See Figure 3.23. Without loss of generality, the center of the needle belongs to the interval $[0, D]$ with the random orientation angle from $(0, \pi)$. The x-position of the center of the needle does not matter; we use a random position solely for illustration purpose. The needle intersects the lower or the upper line if and only if one of the y-coordinates of the end of the needle is outside the interval $[0, D]$; variable `ninter` counts the number of times the needle intersects a line (bar | in R means the logical operator *or*). In `job=1`, we use the `for` loop over simulations, and we use the vectorized computation of the probability in `job=2`. In this problem, the probability can be expressed as the ratio L/D, that is, D may be set to 1 without loss of generality, as we do in `job=2`. Simulations and formula for p are in good agreement. □

The inverse/statistical problem, namely, estimation of L/D given proportion of intersections is considered later in the statistics part of the book, Example 6.115.

Example 3.69 *Random lines on the disk.* *Define the random straight line on the square* $[-1, 1]^2$. *Using simulations estimate the probability that a random line intersects a disk with radius $r < 1$ located in the center of the square.*

Solution. We use the following definition of a random line on the plane:

$$\{(x, y) \in R^2 : (x - x_0)\cos\theta + (y - y_0)\sin\theta = 0\}, \qquad (3.47)$$

where x_0 and y_0 specify a point on the line and θ defines the angle (the line direction), $0 \leq \theta < \pi$. We define the random straight line on the unit square by choosing (i) x_0 and y_0 independently from the uniform distribution on $[-1, 1]$ and (ii) independently the angle $\theta \sim \mathcal{R}(0, \pi)$. To draw the line, we use the equation $y = \tan(\theta)(x - x_0) + y_0$. Now we find the condition under which the line intersects the center disk of radius r. In specification (3.47), the distance from the line to the center (the distance from the origin to the closest point on the line) is $|x_0 \cos\theta + y_0 \sin\theta|$. Since the disk is defined as $\{(x, y) \in R^2 : x^2 + y^2 \leq r^2\}$, the line intersects the disk if and only if $|x_0 \cos\theta + y_0 \sin\theta| \leq r$. See the R code `ranl`. Lines that intersect the disk are shown in green (the plot is not shown).

Example 3.70 *Random segment.* *A segment of length L is thrown at random on the square* $[0, 2]^2$. *What is the probability that the segment entirely belongs to the square? Use vectorized simulations to get the probability estimate.*

Solution. As in the previous example, the key is to define what the random segment on $[0, 2]^2$ is. As in the previous examples, we define the random segment as a segment with a randomly chosen center from $[0, 2]^2$ and a randomly chosen angle $\theta \in (0, \pi)$. Thus the random center of length L has the ends $(x_0 \pm 0.5L\cos\theta, y_0 \pm 0.5L\sin\theta)$, where (x_0, y_0) is the center of the segment; see function `stickSQ`. As in Example 3.68, `job=1` plots random segments and `job=2` performs vectorized simulation.

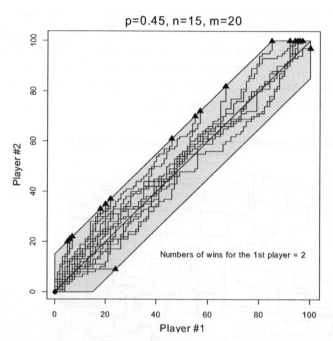

Figure 3.24: *Trajectories of the point-by-point game from Example 3.71 generated by the program* sbE. *Black triangles indicate wins.*

Example 3.71 *Who wins the game? Two players play a point-by-point game until one of the players accumulates n points more than the other. If nobody accumulates more than n points by the 200th trial, the player with more points wins. Assuming that in each trial the first player gets the point with probability p, represent the game graphically (display the point trajectory on the 100 by 100 square). Write an R program to plot and estimate the probability of winning via simulations.*

Solution. The R code, sbE, generates Figure 3.24. The probability that the first player wins in each trial, p, and the number of points, n, are arguments to the function (the default values are $p = 0.45$ and $n = 15$). Argument m is the number of games played, and ss is the seed number to control simulated values (the default values are m=20 and ss=3). To estimate the probability of winning of the first player via simulations, we run the for loop with large m, say, m=10000 over a range of p and n values.

Example 3.72 *Throwing random squares. A square of length d < 1 is thrown K times at random on the unit square. Display random squares and compute coverage probability using simulations.*

Solution. By *throwing at random*, we mean that the center of the square (x, y) is a random point on $[0, 1]^2$, so that the square has coordinates $x \pm d/2$ and

$y \pm d/2$. To compute the probability and display, we represent the unit square as a $N \times N$ matrix $\mathbf{M}$, where N is sufficiently large, say, $N = 1000$. Then the center of the random square is chosen at random on the segment $[0, N]$. Initially, matrix $\mathbf{M}$ has all zero elements. After each throw the elements of matrix $\mathbf{M}$ which are within a random square are set to 1. Then the coverage probability is estimated as $\sum M_{ij}/N^2$. See the code s47 and Figure 3.25.

200 throws of random squares of length 0.05 on the unit square
coverage=82.2%

Figure 3.25: $K = 200$ random throws of a square of length 0.05 on the unit square; see the R code s47.

Problems

1. The conditions of Example 3.66 hold, but Fred waits until 5:30 p.m. What is the probability that the friends meet? Illustrate geometrically and perform simulations to confirm your theoretical answer.

2. In Example 3.67, what is the probability that the friends meet if they arrive according to the triangular distribution with the density $1 - x/30$? Perform simulations to confirm the theoretical answer.

3. Provide a justification that, in Example 3.68, the probability, p, depends on the ratio L/D.

4. What is the probability that a disk of diameter L dropped at random on a plane ruled with parallel lines D units apart will cross a line? Perform simulations to confirm the theoretical answer.

5. What is the probability that a square with diagonal L dropped at random on a plane ruled with parallel lines D units apart will cross a line? Perform simulations to confirm the theoretical answer.

6. Derive the theoretical probability that the random line intersects the disk in Example 3.69.

7. As in Example 3.69, what is the probability of intersection if the disk is replaced with a square of diagonal $r < 1$ (the sides of the small square are parallel to the sides of the unit square)? Use simulations to confirm the answer.

8. As in Example 3.70, a stick of length 1 is thrown at random on the square $[0, 2]^2$. (a) What is the probability that the stick intercepts the main diagonal? (b) What is the probability that the stick does not intersect the diagonal in 1000 throws? Do simulations to verify your answer.

9. A stick of length 1 is thrown at random on the disk of radius 1. Define the "random stick." Use simulations to assess the probability that the stick belongs entirely to the disk. [Hint: Produce a figure similar to Figure 3.23.]

10. Use vectorized simulations to estimate (a) the probability in Example 3.69 and (b) the average distance between random points on the unit square.

11. (a) Can one guarantee that the number of throws until complete coverage is finite in Example 3.72? Run s47 several times until compete coverage and report K. (b) Plot K until complete coverage as a function of d (modify the code to omit graphics for speed).

12. The diameter line is drawn in the bottom of a can of diameter D. A quarter is thrown into the can. What is probability that it intersects the diameter line? (a) Estimate the probability using simulations, and (b) provide an analytical answer by reducing to an integral with arcsin as an integrand (the diameter of the quarter is 0.955 in.).

13. A man lost his way back to a truck he parked in the desert. Assume that (1) he can see his truck from a distance r yards (known) and (2) the point where he got lost is random with independent components X and Y that follow a normal distribution with variance σ^2 around the location of the truck (origin) conditional on $X^2 + Y^2 > r^2$, where $\sigma > r$. To find his truck, he decided to go straight following a path with a random angle, $x(t) = X + t \cos \theta$ and $y(t) = Y + t \sin \theta$, where $t \geq 0$ and $\theta \sim \mathcal{R}(0, 2\pi)$. What is the probability that he finds his truck? Give an analytical answer, provide a geometric illustration, and check your answer through simulations.

14. Do the same as above, but the path is a spiral specified by equations $x(t) = X + \pi^{-1} rt \cos t$ and $y(t) = Y + \pi^{-1} rt \sin t$. Provide a geometric

illustration and prove that he will find his truck with probability 1. Suggest a rectangular path that leads the man to the truck with probability 1. Carry out simulations to estimate the distribution of t_{see} for which $x^2(t_{\text{see}}) + y^2(t_{\text{see}}) = r^2$.

3.8   Optimal portfolio allocation

Diversification is a financial version of the common-sense advice: "Do not put all eggs into one basket." In this section, we prove the benefits of investment diversification by statistical means. The benefit of diversification has been seen earlier in Example 1.11, where it was shown that applying for several small grants leads to a smaller volatility of return than applying for a single big grant although, the expected return is the same. What follows is a continuous version of this phenomenon applied to two stocks. The foundation of portfolio theory was established by Harry Markowitz in his Nobel Prize work of 1952. His concept, the "Markowitz bullet" is a fixture in financial mathematics. In recent studies, *Markowitz optimal portfolio* theory sometimes termed as the mean – variance approach.

To simplify, we start with two stocks or, more generally, two financial products (or two types of investment, assets). We assume here that they have the same expected performance (return) but different risk or, in financial language, *volatility*. The performance of a *stock* (or an investment asset) is judged by its rate of return, or shortly *return*, as the ratio of the difference between stock prices at two time points to the price at the earlier time or, shortly,

$$r_t = \frac{p_t - p_{t-1}}{p_{t-1}}.$$

Since stock prices and the associated returns are unpredictable, it is useful to view the returns as random variables X and Y. By assumption, these stocks have the same expected return or, in probability notation,

$$E(X) = E(Y) = \mu.$$

However, returns have different volatility or, in the probability terms,

$$\text{var}(X) = \sigma_X^2, \quad \text{var}(Y) = \sigma_Y^2.$$

Question: if you want to invest \$1,000 in these stocks would you buy each stock in equal proportion? Let α represent the proportion of the first stock in the portfolio; thus the proportion of the second stock is $1 - \alpha$. To diversify, we want to choose α that minimizes the overall risk – this is the problem of the optimal portfolio allocation. The return of the portfolio is

$$Z = \alpha X + (1 - \alpha)Y, \tag{3.48}$$

where $0 \leq \alpha \leq 1$. Since both stocks yield the same expected return the portfolio yields the same expected return for any α because

$$
\begin{aligned}
E(Z) &= E\left(\alpha X + (1 - \alpha)Y\right) = E\left(\alpha X\right) + E\left((1 - \alpha)Y\right) \\
&= \alpha E\left(X\right) + (1 - a)E(Y) = \alpha\mu + (1 - \alpha)\mu = \mu.
\end{aligned}
$$

Since portfolio (3.48) yields the same expected return, we choose the portfolio with minimal volatility (variance).

3.8.1 Stocks do not correlate

We start with the assumption that stocks do not correlate, $\text{cov}(X, Y) = 0$. The variance/volatility of the portfolio is

$$
\begin{aligned}
\text{var}(Z) &= \text{var}(\alpha X + (1 - \alpha)Y) = \text{var}\left(\alpha X\right) + \text{var}\left((1 - \alpha)Y\right) \\
&= \alpha^2 \text{var}(X) + (1 - \alpha)^2 \text{var}(Y) = \alpha^2 \sigma_X^2 + (1 - \alpha)^2 \sigma_Y^2.
\end{aligned}
$$

We need to find α which minimizes the risk, $\min_\alpha \text{var}(Z)$, or

$$
\min_{0 \leq \alpha \leq 1} \left(\alpha^2 \sigma_X^2 + (1 - \alpha)^2 \sigma_Y^2\right).
$$

The risk function, $R(\alpha) = \alpha^2 \sigma_X^2 + (1 - \alpha)^2 \sigma_Y^2$ is a quadratic function of α. Find the minimum of $R(\alpha)$ by differentiation:

$$
\frac{dR}{d\alpha} = 2\alpha\sigma_X^2 - 2(1 - \alpha)\sigma_Y^2 = 0,
$$

which gives

$$
\alpha_{\text{opt}} = \frac{\sigma_Y^2}{\sigma_X^2 + \sigma_Y^2}. \tag{3.49}
$$

The optimal portfolio yields the minimal risk, $R_{\min} = R(\alpha_{\text{opt}})$. Indeed,

$$
R_{\min} = \frac{\sigma_Y^4}{(\sigma_X^2 + \sigma_Y^2)^2}\sigma_X^2 + \frac{\sigma_X^4}{(\sigma_X^2 + \sigma_Y^2)^2}\sigma_Y^2 = \frac{\sigma_X^2\sigma_Y^2}{(\sigma_X^2 + \sigma_Y^2)^2}\left(\sigma_X^2 + \sigma_Y^2\right) = \frac{\sigma_X^2\sigma_Y^2}{\sigma_X^2 + \sigma_Y^2}. \tag{3.50}
$$

It is easy to prove that $R_{\min} < \sigma_X^2$ and $R_{\min} < \sigma_Y^2$ (assuming that both variances are positive). For example, $\sigma_X^2\sigma_Y^2/(\sigma_X^2 + \sigma_Y^2) < \sigma_X^2$ is equivalent to $\sigma_Y^2/(\sigma_X^2 + \sigma_Y^2) < 1$, which is, of course, true. In words, the optimal portfolio has smaller volatility than either stock.

As seen from (3.49), the ratio of stock X to stock Y in the optimal portfolio equals the ratio of the reciprocal variances, $\alpha_{\text{opt}}/(1 - \alpha_{\text{opt}}) = (1 - \alpha_{\text{opt}}) = (1/\sigma_X^2)/(1/\sigma_Y^2)$. In a special case when the volatilities of the two stocks are the same, one should buy stocks in a 50/50 proportion. The smaller volatility of the stock, the larger proportion it has in the portfolio.

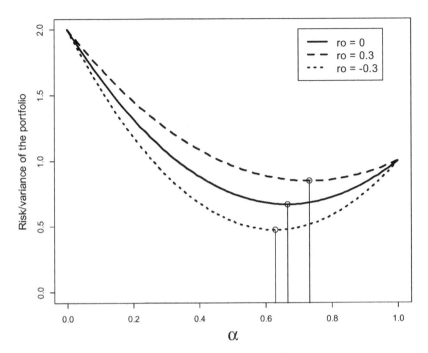

Figure 3.26: *Risk function for uncorrelated ($\rho = 0$) and correlated stocks ($\rho = \pm 0.3$) with $\sigma_X = 1$ and $\sigma_Y = 2$.*

3.8.2   Correlated stocks

Now consider the case when stocks do correlate with the correlation coefficient ρ assuming that $|\rho| < 1$. Then the risk function (the variance of the portfolio) takes the form

$$R(\alpha) = \alpha^2 \sigma_X^2 + 2\alpha(1 - \alpha)\rho\sigma_X\sigma_Y + (1 - \alpha)^2 \sigma_Y^2. \qquad (3.51)$$

Differentiating with respect to α and setting the derivative to zero yields the optimal proportion,

$$\alpha_{\text{opt}} = \frac{\sigma_Y^2 - \rho\sigma_X\sigma_Y}{\sigma_X^2 - 2\rho\sigma_X\sigma_Y + \sigma_Y^2} \qquad (3.52)$$

with the minimum risk

$$R_{\text{min}} = \frac{(1 - \rho^2)\sigma_X^2\sigma_Y^2}{\sigma_X^2 + \sigma_Y^2 - 2\rho\sigma_X\sigma_Y}. \qquad (3.53)$$

It is easy to verify that this optimal proportion turns into the uncorrelated case (3.49) when $\rho = 0$. In the uncorrelated case, $0 < \alpha_{\text{opt}} < 1$. This is not true for correlated stocks. It is easy to see that the denominator of (3.52) is positive since

$$\sigma_X^2 - 2\rho\sigma_X\sigma_Y + \sigma_Y^2 = (\sigma_X - \sigma_Y)^2 + 2\sigma_X\sigma_Y(1 - \rho) > 0.$$

Thus, $\alpha_{\text{opt}} < 0$ when the numerator of (3.52) is negative, when $\rho > \sigma_Y/\sigma_X$. It is easy to prove that $\alpha_{\text{opt}} > 1$ if $\rho > \sigma_X/\sigma_Y$. Thus, to have $0 < \alpha_{\text{opt}} < 1$, one has to have $\rho < \min(\sigma_X/\sigma_Y, \sigma_Y/\sigma_X)$.

To understand how the correlation between stocks affects the minimum risk, we take the difference between (3.50) and (3.53). Simple algebra shows that the difference is positive if and only if

$$\rho\left(\rho - \frac{2\sigma_X\sigma_Y}{\sigma_X^2 + \sigma_Y^2}\right) > 0.$$

It is easy to see that the above inequality holds if $\rho < 0$. In words, if two pairs of stocks have the same return and the same variance and stocks in the first pair are uncorrelated and stocks in the second pair are negatively correlated, then the second optimal portfolio is less volatile. This result is easy to explain: if stocks change in opposite directions they compensate the volatility by combination. Thus, for minimization of the volatility, one has to find a pair of negatively correlated stocks; however, most stocks are positively correlated.

Three cases of correlation between stocks are presented in Figure 3.26. The minimum variance/risk points are computed by formulas (3.52) and (3.53). The risk of a negatively correlated pair of stocks is uniformly lower.

3.8.3 Markowitz bullet

The *Markowitz bullet* depicts the relationship between the return of the portfolio (y-axis) and the associated volatility/variance (x-axis) as a function of α. As before, consider two correlated stocks with expected returns μ_X and μ_Y. The expected return of the portfolio $Z = \alpha X + (1 - \alpha)Y$ is

$$\mu(\alpha) = \alpha\mu_X + (1 - \alpha)\mu_Y$$

with volatility, $\text{var}(Z)$, given by formula (3.51). The Markowitz bullet is a parametrically defined curve with the variance of the portfolio, $R(\alpha)$ on the x-axis and $\mu(\alpha)$ on the y-axis for $\alpha \in [0, 1]$. A typical Markowitz bullet is depicted in Figure 3.27 for $\rho = 0, -0.3$, and 0.3; other parameters are $\mu_X = 0.03$, $\mu_Y = 0.05$, $\sigma_X = 0.005$, and $\sigma_Y = 0.008$. In financial terms, the first stock yields the annual return 3%, and the second stock yields return 5%. In a way, the Markowitz bullet is the $90°$ rotated risk function as depicted in Figure 3.26. The difference is that in Figure 3.26 the x-axis is α and in Figure 3.27 the y-axis is $\alpha\mu_X + (1 - \alpha)\mu_Y$, a linear function of α.

The low branch of the bullets starts at the right-low point (σ_X^2, μ_X), corresponding to $\alpha = 1$, and the upper branch starts at the right-upper point (σ_Y^2, μ_Y), corresponding to $\alpha = 0$. The filled circle on the bullet indicates the portfolio with the lowest volatility. The upper part of the bullet is referred to as the *efficient frontier*, and the lower part of the bullet is referred to as the *inefficient frontier*. The right-upper branch of the bullet characterizes the high risk/high return

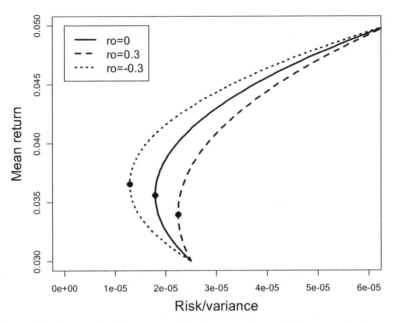

Figure 3.27: *Markowitz bullet curves for three values of the correlation coefficient,* ρ, *between stocks.*

investment strategy, and left-upper branch of the bullet characterizes the low risk/low return strategy. Interestingly, the smaller the correlation between the stocks, the better, i.e. the same return is obtained with smaller volatility (the bullet curve moves to the left when ρ gets smaller). This follows common sense: when two stocks are highly correlated, there is no advantage to a portfolio. On the other hand, if stocks are negatively correlated, the optimal portfolio compensates the risk.

3.8.4   Probability bullet

The variance of the portfolio may not be an optimal characteristic of the risk simply because a portfolio return smaller than the expected is typically more harmful than the gain associated with the greater-than-expected return. In other words, the risk of deviation from the expected return is asymmetric, but the Markowitz bullet is symmetric. A better way to look at the risk of investment is to compare of the portfolio with some guaranteed return, say, if the money were put into a savings account. To be specific, let us operate with an annual return on the percent scale assuming that the savings guarantee the lowest interest rate, r_0 percent. Instead of variance, we measure the risk associated with portfolio $\alpha X + (1 - \alpha)Y$ as the probability of obtaining the return less than the lowest rate r_0. Shortly, the risk of investment portfolio of obtaining a return less than the guaranteed rate is expressed as the probability $\Pr(\alpha X + (1 - \alpha)Y < r_0)$.

Under the simplifying assumption that returns are normally distributed and independent, we write the above probability through the standard normal cdf (pnorm) as

$$\Pr(\alpha X + (1-\alpha)Y < r_0) = \Phi\left(\frac{r_0 - \alpha\mu_X - (1-\alpha)\mu_Y}{\sqrt{\alpha^2\sigma_X^2 + (1-\alpha)^2\sigma_Y^2}}\right).$$

The *probability bullet* is a parametrically defined curve with probability of getting a portfolio return less than the lowest rate on the x-axis and the expected return $\alpha\mu_X + (1-\alpha)\mu_Y$ on the y-axis. It is easy to prove that the probability bullet has a bullet shape if $\mu_X > r_0$ and $\mu_Y > r_0$, which is expected because r_0 is the lowest return:

$$\alpha_{\text{opt}} = \frac{\mu_X - r_0}{(\mu_X\sigma_Y^2 + \mu_Y\sigma_X^2) - (\sigma_X^2 + \sigma_Y^2)r_0}\sigma_Y^2.$$

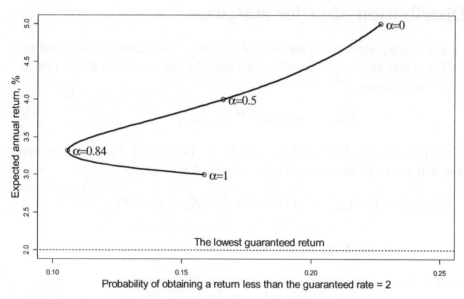

Figure 3.28: *The probability bullet plotted by function **prmb** with the default parameters: $\mu_X = 3\%$, $\sigma_X = 1$, $\mu_Y = 5\%$, $\sigma_Y = 4$, and $r_0 = 2$. The safest portfolio has $\alpha = 0.84$ with the safest mean return 3.32%. The probability that the actual return will be smaller than 3.32% is 0.11.*

Figure 3.28 depicts a typical probability bullet with the annual lowest guaranteed return $r_0 = 2\%$, the returns of the first and second stock $\mu_X = 3$ and $\mu_Y = 5$, and $\sigma_X = 1$ and $\sigma_Y = 4$, respectively (see the R code prmb). The above formula gives the optimal (safest) portfolio $\alpha_{\text{opt}} = 0.84$, with the mean return $\alpha_{\text{opt}}\mu_X + (1 - \alpha_{\text{opt}})\mu_Y = 3.32\%$. The probability that the actual return will be smaller that r_0 is 0.11.

Problems

1. Define the individual risks/volatilities as $R_X = \text{var}(X)$ and $R_Y = \text{var}(Y)$. Show directly that $R_{\min} < R_X$ and $R_{\min} < R_Y$, where α is defined by (3.49).

2. Prove formula (3.53).

3. (a) Reproduce Figure 3.26. (b) Reproduce Figure 3.27.

4. (a) Prove if, in the probability bullet, $\mu_X > r_0$ and $\mu_Y > r_0$, then $0 < \alpha_{\text{opt}} < 1$. (b) Download and run the R code `prmb` to confirm that `aopt` is within $[0, 1]$ under these conditions.

5. Generalize the probability bullet to the case when stocks correlate. Write an R program that computes and plots the probability bullet as does `prmb`.

3.9 Distribution of order statistics

In some applications, we work with the minimum or maximum of iid random variables. The distribution and densities are easy to derive in this case. Let us start with the maximum:

$$X_{\max} = \max(X_1, X_2, ..., X_n),$$

where X_i are iid with cdf $F(x)$ and density $f(x)$. The cdf of $X_{\max}$, denoted as $F_{\max}$, is the nth power of the original cdf. Indeed, due to independence, we have

$$
\begin{aligned}
F_{\max}(x) &= \Pr(X_{\max} \leq x) = \Pr(X_1 \leq x, X_2 \leq x, ..., X_n \leq x) \\
&= \Pr(X_1 \leq x) \times \Pr(X_2 \leq x) \times \cdots \times \Pr(X_n \leq x) \\
&= [\Pr(X_1 \leq x)]^n = F^n(x).
\end{aligned}
$$

Once the cdf is known, the density is obtained by differentiation:

$$f_{\max}(x) = nF^{n-1}(x)f(x). \tag{3.54}$$

Now we find the cdf and density of the minimum. Using the complementary probability, we obtain

$$
\begin{aligned}
F_{\min}(x) &= \Pr(X_{\min} \leq x) = 1 - \Pr(X_{\min} > x) \\
&= 1 - \Pr(X_1 > x, X_2 > x, ..., X_n > x) = 1 - (1 - F(x))^n,
\end{aligned}
$$

and, by differentiation,

$$f_{\min}(x) = n(1 - F(x))^{n-1}f(x). \tag{3.55}$$

Minimum and maximum are the special cases of order statistics. In general, reorder the original observations $\{X_i, i = 1, 2, ..., n\}$ in ascending order $X_{(1)} \leq X_{(2)} \leq \cdots \leq X_{(n)}$. Observations $X_{(i)}$ are referred to as order statistics. In particular, $X_{(1)} = X_{\min}$, $X_{(n)} = X_{\max}$, and $X_{(n+1)/2} = $ median, assuming that n is an odd number. The joint density of order statistics is

$$f(x_1, x_2, ..., x_n) = \begin{cases} n! \displaystyle\prod_{i=1}^{n} f(x_i) \text{ if } x_1 \leq x_2 \leq ... \leq x_n \\ \\ 0 \text{ elsewhere} \end{cases}. \qquad (3.56)$$

It can be shown that the density of the rth order statistic, $X_{(r)}$, is

$$f_r(x) = \frac{n!}{(r-1)!(n-r)!} F^{r-1}(x)[1 - F(x)]^{n-r} f(x). \qquad (3.57)$$

It is easy to see that this formula gives the previously derived density for the maximum (3.54) and minimum (3.55) letting $r = n$ and $r = 1$ in formula (3.57), respectively.

The cdf of the order statistic is the beta distribution (Section 2.14) evaluated at F. Indeed, integrate (3.57) by change of variable $t = F(u)$,

$$\begin{aligned} F_r(x) &= \frac{n!}{(r-1)!(n-r)!} \int_{-\infty}^{x} F^{r-1}(u)[1 - F(u)]^{n-r} f(u) du \\ &= \frac{n!}{(r-1)!(n-r)!} \int_{0}^{F(x)} t^{r-1}(1-t)^{n-r} dt, \end{aligned}$$

which is the cdf of the beta distribution evaluated at $F(x)$ with the shape parameters r and $n - r + 1$.

In the special case of the uniform distribution, $X_i \sim \mathcal{R}(0, 1)$, we obtain the standard beta density

$$f_r(x) = \frac{n!}{(r-1)!(n-r)!} x^{r-1}(1-x)^{n-r} \sim \mathcal{B}(r, n-r+1) \qquad (3.58)$$

because $F(x) = x$.

Example 3.73 *Distribution of the median. Find the cdf of the median of n iid observations from a Laplace distribution with mode μ and parameter λ (n is odd). Use simulations to verify the answer.*

Solution. Since the cdf of an individual observation is

$$F(x; \lambda) = \begin{cases} \frac{1}{2} e^{-\lambda(\mu - x)} \text{ if } x < \mu \\ \\ 1 - \frac{1}{2} e^{-\lambda(x - \mu)} \text{ if } x \geq \mu \end{cases},$$

the cdf of the median is the cdf of the beta distribution with the argument $F(x; \lambda)$ and shape parameters r and $n - r + 1$, where r is the median of $\{1, 2, ..., n\}$. The R code with simulations that match the theoretical and empirical cdf is found in the file cdfMED.r.   □

The following result is a continuation of density (3.58).

Theorem 3.74 *Let* $X_1, X_2, ..., X_n \overset{iid}{\sim} \mathcal{R}(0, 1)$. *Then*

$$X_{(p)} - X_{(q)} \sim \mathcal{B}(p - q, n - p + q + 1), \quad p > q.$$

The following two special cases are of interest:

1. $p = n$ and $q = 1$ (the range):

$$X_{\max} - X_{\min} \sim \mathcal{B}(n - 1, 2). \tag{3.59}$$

2. $q = p - 1$ (difference between neighboring order statistics):

$$X_{(p)} - X_{(p-1)} \sim \mathcal{B}(1, n). \tag{3.60}$$

Example 3.75 *Vectorized simulations.* *Use vectorized simulations to check (3.59) by plotting the empirical and theoretical cdfs on the same graph.*

Solution. See the R function pmaxmin. Functions apply (X,1,max) and apply(X,1,min) compute maximum and minimum in each row. If matrix X has N rows these functions return an array of length N.

Example 3.76 *Throwing random segments.* *Segments of length δ are thrown at random on the interval of unit length ($\delta < 1$). (a) Approximate the probability that the segments will cover the interval after n throws. (b) Carry out simulations to test the precision of the probability.*

Solution. Without loss of generality, we can assume that the interval is $(0, 1)$. By random segment of length δ, we mean that the center of the interval is a random variable X uniformly distributed on $(0, 1)$, so that the random segment of length δ is $(X - \delta/2, X + \delta/2)$. Denote X_i as the center of the segment in the ith throw. Order centers to obtain the order statistics $X_{(i)}$. The n segments cover $(0, 1)$ if and only if $X_{(i+1)} - X_{(i)} < \delta$ for all $i = 1, 2, ..., n - 1$. (a) Hence the probability that n intervals cover $(0, 1)$ can be approximated as $p^{n-1}(\delta)$, where $p(\delta)$ is the value of the cdf of $\mathcal{B}(1, n)$ evaluated at δ. This probability can be found in closed form as

$$p(\delta) = \frac{\Gamma(n + 1)}{\Gamma(1)\Gamma(n)} \int_0^\delta (1 - x)^{n-1} dx.$$

Since $\Gamma(n+1) = n\Gamma(n)$ and $\Gamma(1) = 1$, we find $p(\delta) = 1 - (1-\delta)^n$. Finally, the following approximation is suggested,

$$\text{Pr(interval } (0,1) \text{ is covered after } n \text{ throws)} \simeq [1 - (1-\delta)^n]^{n-1}.$$

The function at the right-hand side starts at 0 when $n = 1$ and monotonically increases to 1 when $n \to \infty$. Note that this approximation is not exact because we silently assume that $X_{(i+1)} - X_{(i)}$ are independent. In fact, they are not.

(b) The R code with simulations is found in the file cov01.r. The following algorithm is used to determine whether n throws cover the interval: (i) Let there be N grid points and let X represent the center of the segment on the scale from 1 to N. (ii) Let INT=1:N be the array of indices and lefX and upX are the lower and the upper limit of the random interval. Then the first throw leaves in indices for which INT<lefX | INT>upX, where | is the or logical operator. (iii) If at the ith throw the length of INT is zero, then segments covered the interval completely and the throwing process is stopped (break); the number of throws required is saved in array ncov. The run with default values produced the following result: Empirical probability = 0.657 Approximation = 0.7763219

Problems

1. Prove that $\int_{-\infty}^{\infty} f_{\max}(x)dx = 1$ and $\int_{-\infty}^{\infty} f_{\min}(x)dx = 1$ using a change of variable.

2. Derive (3.57) from (3.56). [Hint: Apply the binomial probability.]

3. Find the density of the rth order statistic in the general case. Derive (3.58) to check your answer.

4. Define the range $R = X_{\max} - X_{\min}$. (a) Prove that the density of R is $f(r) = n(n-1)\int_{-\infty}^{\infty} f(u)f(u+r)[F(u+r) - F(u)]^{n-2}du$. (b) Write an R program to check this result via simulations using normally distributed random variables. (c) Derive the closed-form expression for $f(r)$ when observations are uniformly distributed on $[0,1]$.

5. Provide empirical evidence by simulation that $X_{(i+1)} - X_{(i)}$ are not independent: generate $X_1, X_2, X_3 \overset{\text{iid}}{\sim} \mathcal{R}(0,1)$ and plot $X_{(2)} - X_{(1)}$ versus $X_{(3)} - X_{(2)}$.

6. Plot the empirical probability of complete coverage versus n in Example 3.76, and superimpose with the theoretical approximation $[1 - (1-\delta)^n]^{n-1}$.

3.10 Multidimensional random vectors

This section requires familiarity with matrix algebra; see the Appendix.

So far, a bivariate 2×1 random vectors have been studied. The generalization to larger vectors case is fairly straightforward. A random vector of size m is typically understood as the column vector composed of m random variables:

$$\mathbf{X}^{m \times 1} = \begin{bmatrix} X_1 \\ \vdots \\ X_m \end{bmatrix}.$$

We use bold to indicate vectors and matrices throughout the book. The cdf and density are functions of $\mathbf{x}^{m \times 1} = (x_1, x_2, ..., x_m)'$ and are related via differentiation and integration as (here it is assumed that we deal with continuous random variables)

$$
\begin{aligned}
F(\mathbf{x}) &= \Pr(X_1 \leq x_1, X_2 \leq x_2, ..., X_m \leq x_m), \\
f(\mathbf{x}) &= \frac{\partial F}{\partial \mathbf{x}}, \quad F(\mathbf{x}) = \int_{-\infty}^{x_1} \int_{-\infty}^{x_2} \cdots \int_{-\infty}^{x_m} f(\mathbf{u}) d\mathbf{u}.
\end{aligned}
$$

Note that the lowercase is used since $\mathbf{x}$ is nonrandom. Many results and definitions of the bivariate distribution can be extended to multidimensional random vectors. For example, $X_1, X_2, ..., X_n$ are independent if and only if the joint cdf and density is the product of individual (marginal) cdfs and densities, respectively.

Although the term *random sample* and the abbreviation *iid* have been used many times before, below is the formal definition.

Definition 3.77 *We say that $X_1, X_2, ..., X_n$ are iid if they are iid with the same pdf/pmf f. Hence the joint density is the product of individual densities evaluated at x_i. In statistics, we refer to array $X_1, X_2, ..., X_n$ as a random sample of size n from a general population specified by pdf/pmf f.*

The concepts of iid random variables and a sample are centerpieces of statistics. Since $\{X_i\}$ are iid, the joint density is the product of individual densities:

$$f(x_1, x_2, ..., x_n) = \prod_{i=1}^{n} f(x_i),$$

where f is the density of the general population.

The expectation of $\mathbf{X}$ is the $m \times 1$ vector defined as the element-wise expectation vector,

$$\boldsymbol{\mu} = E(\mathbf{X}) = \begin{bmatrix} E(X_1) \\ \vdots \\ E(X_m) \end{bmatrix}.$$

The (i,j)th element of the $m \times m$ covariance matrix is defined as $\text{cov}(X_i, X_j)$ or in the matrix form as

$$\text{cov}(\mathbf{X}) = \begin{bmatrix} \text{cov}(X_1, X_1) & \cdots & \text{cov}(X_1, X_m) \\ \vdots & \ddots & \vdots \\ \text{cov}(X_m, X_1) & \cdots & \text{cov}(X_m, X_m) \end{bmatrix}$$

$$= \begin{bmatrix} \sigma_1^2 & \cdots & \sigma_{1m} \\ \vdots & \ddots & \vdots \\ \sigma_{1m} & \cdots & \sigma_m^2 \end{bmatrix} = \begin{bmatrix} \sigma_1^2 & \cdots & \rho_{1m}\sigma_1\sigma_m \\ \vdots & \ddots & \vdots \\ \rho_{1m}\sigma_1\sigma_m & \cdots & \sigma_m^2 \end{bmatrix}.$$

Alternatively, we express $\text{cov}(\mathbf{X}) = \{\rho_{ij}\sigma_i\sigma_j, i, j = 1, 2, ..., m\}$, where

$$\rho_{ij} = \frac{\sigma_{ij}}{\sigma_i\sigma_j}$$

is the correlation coefficient between X_i and X_j. Define

$$\mathbf{D} = \text{diag}(\sigma_1^2, \sigma_2^2, ..., \sigma_m^2) = \begin{bmatrix} \sigma_1^2 & 0 & 0 \\ \vdots & \ddots & \vdots \\ 0 & \cdots & \sigma_m^2 \end{bmatrix}$$

and $\mathbf{R}^{m\times m} = \{\rho_{ij}, i, j = 1, 2, ..., m\}$, the correlation matrix. Then we can compactly write

$$\text{cov}(\mathbf{X}) = \mathbf{D}^{1/2}\mathbf{R}\mathbf{D}^{1/2}$$

and reversely

$$\text{cor}(\mathbf{X}) = \mathbf{R} = \mathbf{D}^{-1/2}\text{cov}(\mathbf{X})\mathbf{D}^{-1/2}. \tag{3.61}$$

Sometimes, we need to consider the covariance matrix between two random vectors, $\mathbf{X}^{m\times 1}$ and $\mathbf{Y}^{n\times 1}$ as a $m \times n$ matrix of cross covariances,

$$\text{cov}(\mathbf{X}, \mathbf{Y}) = \{\text{cov}(X_i, Y_j), i = 1, ..., m; j = 1, ..., n\}.$$

It is obvious that $\text{cov}(\mathbf{X}) = \text{cov}(\mathbf{X}, \mathbf{X})$. In matrix notation,

$$\text{cov}(\mathbf{X}) = E[(\mathbf{X} - \boldsymbol{\mu})(\mathbf{X} - \boldsymbol{\mu})']$$

and

$$\text{cov}(\mathbf{X}, \mathbf{Y}) = E[(\mathbf{X} - \boldsymbol{\mu}_X)(\mathbf{Y} - \boldsymbol{\mu}_Y)'].$$

We say that vectors $\mathbf{X}$ and $\mathbf{Y}$ are uncorrelated if $\text{cov}(\mathbf{X}, \mathbf{Y}) = \mathbf{0}$. In this case, the correlation between each pair of the components of vectors $\mathbf{X}$ and $\mathbf{Y}$ is zero.

Theorem 3.78 *If* $\mathbf{X}$ *and* $\mathbf{Y}$ *are uncorrelated, then*

$$\text{cov}(\mathbf{X} + \mathbf{Y}) = \text{cov}(\mathbf{X}) + \text{cov}(\mathbf{Y}). \tag{3.62}$$

The proof follows from the component-wise examination. Basic properties of the mean and covariance matrix are presented below.

Theorem 3.79 *Let* $\mathbf{X}$, $\mathbf{Y}$, *and* $\mathbf{Z}$ *be random vectors,* $\mathbf{A}$ *and* $\mathbf{B}$ *be fixed matrices and a and b be scalars. Then, assuming the dimensions comply,*

$$E(a\mathbf{X} + b\mathbf{Y}) = aE(\mathbf{X}) + bE(\mathbf{Y}), \quad E(\mathbf{AX} + \mathbf{BY}) = \mathbf{A}E(\mathbf{X}) + \mathbf{B}E(\mathbf{Y}),$$
$$\text{cov}(\mathbf{X} + \mathbf{Y}, \mathbf{Z}) = \text{cov}(\mathbf{X}, \mathbf{Z}) + \text{cov}(\mathbf{Y}, \mathbf{Z}), \quad \text{cov}(\mathbf{AX}) = \mathbf{A}\text{cov}(\mathbf{X})\mathbf{A}'. \tag{3.63}$$

The proofs are straightforward. The last identity has an important corollary: if $\mathbf{A} = \mathbf{a}'$, the row vector, then

$$\text{var}(\mathbf{a}'\mathbf{X}) = \mathbf{a}'\text{cov}(\mathbf{X})\mathbf{a}. \tag{3.64}$$

The left-hand side is a linear combination of components of vector $\mathbf{X}$, and the right-hand side is a quadratic form (since $\mathbf{a}'\mathbf{X}$ is scalar, we use notation cov = var).

Example 3.80 *Variance of the sum of random variables. (3.64) to prove that the variance of the sum of pairwise uncorrelated random variables is the sum of variances.*

Solution. The sum of components of $\mathbf{X}$ can be written as $\mathbf{a}'\mathbf{X}$ where $\mathbf{a} = \mathbf{1}$. If the components of $\mathbf{X}$ are pairwise uncorrelated, then $\text{cov}(\mathbf{X}) = \text{diag}(\sigma_1^2, \sigma_2^2, ..., \sigma_m^2)$. Applying formula (3.64), one obtains $\text{var}(\mathbf{1}'\mathbf{X}) = \mathbf{1}'\text{cov}(\mathbf{X})\mathbf{1} = \sum_{i=1}^{m} \sigma_i^2$, the sum of variances.   □

A primary task of a statistician is to translate a practical problem into a probabilistic model. Often the translation is the most challenging task – after the problem is rigorously formulated the solution is easy. It is not uncommon that relevant assumptions must be made along the way. Below is an example.

Example 3.81 *Carpenter problem. A carpenter wants to cut 10 boards of the same length. He cuts one board and uses it as a meter for cutting the remaining nine boards. His partner suggests using a previously cut board as a meter. (a) Assuming that boards are cut imprecisely, is there a difference between the two methods, and if the difference exists, which method is better? (b) Express the length of the boards as random vectors.*

Solution. We need to express the problem in probabilistic terms by suggesting a rigorous model. Let us assume that the error of cutting is random with zero mean and variance σ^2 and the errors $\{\varepsilon_1, ..., \varepsilon_{10}\}$ are independent from board to board. (a) The actual length of the first board is $L + \varepsilon_1$, where L is the exact/desired length and ε_1 is the error of cutting. In the method suggested by the carpenter, the length of the ith board is the length of the first board plus an error, $(L + \varepsilon_1) + \varepsilon_i$, $i = 2, 3, ..., 10$. This yields that the variance of each cut after the first one is $\text{var}(\varepsilon_1 + \varepsilon_i) = 2\sigma^2$. In the method suggested by the partner, the length of the ith board is the length of the previous board plus ε_i, i.e. $(L + \varepsilon_{i-1}) + \varepsilon_i$, $i = 2, 3, ..., 10$. Hence the variance for the second board is $\text{var}(\varepsilon_1 + \varepsilon_2) = 2\sigma^2$, the variance of the third board is $3\sigma^2$, etc. and the variance of the 10th board is $10\sigma^2$. The first method of cutting is more precise. (b) Let $\boldsymbol{\varepsilon}$ be a 10×1 random vector with iid components having zero mean and variance σ^2. Introduce two 10×10 matrices $\mathbf{A}$ and $\mathbf{B}$ with $0/1$ entries: the first column of matrix $\mathbf{A}$ is $\mathbf{1}$ and $A_{ii} = 1$ for $i = 1, ..., 9$ and $B_{i,i+k} = 1$ for $k = 1, ..., i$ and 0 elsewhere. Then the length of boards in the first and second method are expressed as $L\mathbf{1} + \mathbf{A}\boldsymbol{\varepsilon}$ and $L\mathbf{1} + \mathbf{B}\boldsymbol{\varepsilon}$, respectively.

Example 3.82 *Covariance matrix.* *Prove that a covariance matrix is a nonnegative definite symmetric matrix.*

Solution. The symmetry follows from the fact that $\text{cov}(X_i, X_j) = \text{cov}(X_j, X_i)$. To prove that $\text{cov}(\mathbf{X}) = \boldsymbol{\Omega}$ is a nonnegative definite matrix, we need to prove that $\mathbf{a}'\boldsymbol{\Omega}\mathbf{a} \geq 0$ for every nonrandom vector $\mathbf{a}$. Consider a random variable as a linear combination, $Y = \mathbf{a}'\mathbf{X}$. From (3.64) we have $\text{var}(Y) = \mathbf{a}'\boldsymbol{\Omega}\mathbf{a} \geq 0$ because variance is nonnegative. Therefore a covariance matrix is a nonnegative definite matrix.

Theorem 3.83 *The following formulas hold*

$$\text{cov}(\mathbf{X}) = E(\mathbf{XX}') - \boldsymbol{\mu}\boldsymbol{\mu}', \quad \text{cov}(\mathbf{X}, \mathbf{Y}) = E(\mathbf{XY}') - \boldsymbol{\mu}_X\boldsymbol{\mu}_Y'$$
$$\text{cov}(\mathbf{X}) = E[(\mathbf{X} - \boldsymbol{\mu})\mathbf{X}'] = E[\mathbf{X}(\mathbf{X} - \boldsymbol{\mu})'],$$
$$\text{cov}(\mathbf{X}, \mathbf{Y}) = E[\mathbf{X}(\mathbf{Y} - \boldsymbol{\mu}_Y)'] = E[(\mathbf{X} - \boldsymbol{\mu}_X)\mathbf{Y}'].$$

Proof. For the first formula, using the previous theorem, we have

$$\text{cov}(\mathbf{X}) = E[(\mathbf{X} - \boldsymbol{\mu})\mathbf{X}'] - E[(\mathbf{X} - \boldsymbol{\mu})\boldsymbol{\mu}'] = E(\mathbf{XX}') - \boldsymbol{\mu}E(\mathbf{X}') - E(\mathbf{X})\boldsymbol{\mu}' + \boldsymbol{\mu}\boldsymbol{\mu}'$$
$$= E(\mathbf{XX}') - \boldsymbol{\mu}\boldsymbol{\mu}' - \boldsymbol{\mu}\boldsymbol{\mu}' + \boldsymbol{\mu}\boldsymbol{\mu}' = E(\mathbf{XX}') - \boldsymbol{\mu}\boldsymbol{\mu}',$$

which can be viewed as a generalization of $\text{var}(X) = E(X^2) - \mu^2$. Other formulas can be proven similarly.

Example 3.84 *Three-by-three covariance matrix.* *Let X, Y, and Z be independent (scalar) random variables. (a) Find the 3×3 covariance and correlation matrices for the random vector*

$$\mathbf{X} = \begin{bmatrix} X \\ Y - X \\ X + Y + Z \end{bmatrix}$$

by direct computation. (b) Find the covariance matrix of $\mathbf{X}$ *using identity (3.63).*

Solution. (a) To obtain the 3×3 correlation matrix, we need to have $(3 \times 4)/2 = 6$ quantities: three variances and three covariances. We repeatedly use the rule that the variance of the sum of independent random variables is equal to the sum of the variances. The variances of the second and the third component of vector $\mathbf{X}$ are $\text{var}(Y-X) = \sigma_X^2 + \sigma_Y^2$ and $\text{var}(X+Y+Z) = \sigma_X^2 + \sigma_Y^2 + \sigma_Z^2$, respectively. Compute covariances: $\text{cov}(X, Y-X) = -\text{var}(X) = -\sigma_X^2$, $\text{cov}(X, X+Y+W) = \sigma_X^2$, and $\text{cov}(Y-X, X+Y+W) = \sigma_Y^2 - \sigma_X^2$. These imply

$$\text{cov}(\mathbf{X}) = \begin{bmatrix} \sigma_X^2 & -\sigma_X^2 & \sigma_X^2 \\ -\sigma_X^2 & \sigma_X^2 + \sigma_Y^2 & \sigma_Y^2 - \sigma_X^2 \\ \sigma_X^2 & \sigma_Y^2 - \sigma_X^2 & \sigma_X^2 + \sigma_Y^2 + \sigma_Z^2 \end{bmatrix}.$$

Since the correlation coefficient between any W and V is $\rho = \dfrac{\text{cov}(W,V)}{\sqrt{\text{var}(W)\text{var}(V)}}$, the 3×3 correlation matrix for $(X, Y-X, X+Y+Z)$ is equal to

$$\begin{bmatrix} 1 & -\dfrac{\sigma_X}{\sqrt{\sigma_X^2+\sigma_Y^2}} & \dfrac{\sigma_X}{\sqrt{\sigma_X^2+\sigma_Y^2+\sigma_W^2}} \\ -\dfrac{\sigma_X}{\sqrt{\sigma_X^2+\sigma_Y^2}} & 1 & \dfrac{\sigma_Y^2-\sigma_X^2}{\sigma_X\sqrt{\sigma_X^2+\sigma_Y^2+\sigma_W^2}} \\ \dfrac{\sigma_X}{\sqrt{\sigma_X^2+\sigma_Y^2+\sigma_W^2}} & \dfrac{\sigma_Y^2-\sigma_X^2}{\sigma_X\sqrt{\sigma_X^2+\sigma_Y^2+\sigma_W^2}} & 1 \end{bmatrix}.$$

(b) We notice that $\mathbf{X} = \mathbf{AU}$, where

$$\mathbf{A} = \begin{bmatrix} 1 & 0 & 0 \\ -1 & 1 & 0 \\ 1 & 1 & 1 \end{bmatrix}, \quad \mathbf{U} = \begin{bmatrix} X \\ Y \\ Z \end{bmatrix}.$$

Therefore, from formula (3.63), we deduce

$$\text{cov}(\mathbf{X}) = \begin{bmatrix} 1 & 0 & 0 \\ -1 & 1 & 0 \\ 1 & 1 & 1 \end{bmatrix} \begin{bmatrix} \sigma_X^2 & 0 & 0 \\ 0 & \sigma_Y^2 & 0 \\ 0 & 0 & \sigma_Z^2 \end{bmatrix} \begin{bmatrix} 1 & 0 & 0 \\ -1 & 1 & 0 \\ 1 & 1 & 1 \end{bmatrix}'$$

$$= \begin{bmatrix} \sigma_X^2 & -\sigma_X^2 & \sigma_X^2 \\ -\sigma_X^2 & \sigma_X^2 + \sigma_Y^2 & \sigma_Y^2 - \sigma_X^2 \\ \sigma_X^2 & \sigma_Y^2 - \sigma_X^2 & \sigma_X^2 + \sigma_Y^2 + \sigma_Z^2 \end{bmatrix},$$

as in (a).

Example 3.85 *Special covariance matrix.* *Let random vector* $\mathbf{X}$ *and random variable* Y *be uncorrelated. Express* $\text{cov}(\mathbf{X}+Y\mathbf{1})$, *where* $\mathbf{1}$ *is the vector of ones, via the covariance matrix of* $\mathbf{X}$ *and variance of* Y.

Solution. Using Theorem 3.79, we obtain $\text{cov}(\mathbf{X}+Y\mathbf{1}) = \text{cov}(\mathbf{X}) + \text{cov}(Y\mathbf{1})$. In particular, we use formula (3.63) to find $\text{cov}(Y\mathbf{1})$ by letting $\mathbf{A} = \mathbf{1}$ and $\mathbf{X} = Y$, we have $\text{cov}(Y\mathbf{1}) = \mathbf{1}\text{var}(Y)\mathbf{1}' = \text{var}(Y)\mathbf{1}\mathbf{1}'$, the matrix with all entries equal $\text{var}(Y)$. Finally, $\text{cov}(\mathbf{X}+Y\mathbf{1}) = \text{cov}(\mathbf{X}) + \text{var}(Y)\mathbf{1}\mathbf{1}'$. □

The following example illustrates how to extend the LLN to a sequence of correlated random variables.

Example 3.86 *The correlated LLN.* *Consider the sequence of correlated random variables* $X_1, X_2, \ldots$ *with the same mean and variance,* $E(X_i) = \mu$ *and* $\text{var}(X_i) = \sigma^2$, *and correlation coefficient* $\rho(X_i, X_j) = \rho^{|i-j|}$ *where* $|\rho| < 1$. *Prove that* $\overline{X}_n = n^{-1}\sum_{i=1}^{n} X_i$ *quadratically converges to* μ: $E(\overline{X}_n - \mu)^2 \to 0$ *when* $n \to \infty$.

Solution. Since $E(\overline{X}_n) = \mu$, we have $E(\overline{X}_n - \mu)^2 = \text{var}(\overline{X}_n)$. First, find the variance of the average, $\text{var}(\overline{X}_n) = \frac{\sigma^2}{n^2}\sum_{i=1}^{n}\sum_{j=1}^{n}\rho^{|i-j|}$. Second, bound the variance from above to demonstrate that $\text{var}(\overline{X}_n) \to 0$. Indeed, for every i we obtain the upper bound as the sum of the geometric progression, $\sum_{j=1}^{n}\rho^{|i-j|} \leq \sum_{k=0}^{\infty}|\rho|^k = \frac{1}{1-|\rho|}$, and therefore, $\text{var}(\overline{X}_n) \leq \frac{\sigma^2}{n^2}\frac{n}{1-|\rho|} = \frac{\sigma^2}{n(1-|\rho|)} \to 0$. Interestingly, the order of the variance going to zero is the same as in the uncorrelated case, $O(n^{-1})$, but inflated by the coefficient $(1-|\rho|)^{-1} > 1$. □

It is easy to see that the correlation matrix in the above example originates from the autoregression of the first order, $X_i - \mu = \rho(X_{i-1} - \mu) + Z_i$, where Z_i and X_{i-1} do not correlate and $E(Z_i) = 0$ and $E(Z_i^2) = \sigma^2$.

3.10.1 Multivariate conditional distribution

The bivariate conditional density defined in Section 3.3 is easy to generalize to the multidimensional random vector as the ratio of the joint density of $\mathbf{Y}^{p\times1}$ and $\mathbf{X}^{q\times1}$ to the marginal density of $\mathbf{X}$:

$$f_{\mathbf{Y}|\mathbf{X}=\mathbf{x}}(\mathbf{y}) = \frac{f(\mathbf{x},\mathbf{y})}{f_{\mathbf{X}}(\mathbf{x})},$$

where f is the joint density and $f_{\mathbf{X}}$ is the marginal density. Similarly, we define the conditional mean as

$$E(\mathbf{Y}|\mathbf{X} = \mathbf{x}) = \int_{R^p} \mathbf{y}\, f_{\mathbf{Y}|\mathbf{X}=\mathbf{x}}(\mathbf{y})d\mathbf{y} = \int_{R^p} \mathbf{y}\,\frac{f(\mathbf{x},\mathbf{y})}{f_{\mathbf{X}}(\mathbf{x})}d\mathbf{y},$$

the p-valued function of the $q$dimensional argument.

Example 3.87 *Marginal and conditional independence are not equivalent. Use Example 3.85 to prove that marginal and conditional independence are not equivalent assuming that components of the $n \times 1$ random vector $\mathbf{X}$ are iid and $\mathbf{X}$ and Y are independent.*

Solution. Let $f(x)$ be the common density of n components of vector $\mathbf{X}$ and $g(y)$ be the density of Y. As follows from Section 3.6, the joint density of $\mathbf{Z} = \mathbf{X} + \mathbf{1}Y$ and Y, with some ambiguity of notation, takes the form

$$f(z_1, ..., z_n, y) = f(z_1 - y) \cdots f(z_n - y)g(y).$$

The marginal density of $\mathbf{Z}$ is derived by integrating out y:

$$f_{\mathbf{Z}}(z_1, ..., z_n) = \int_{-\infty}^{\infty} \prod_{i=1}^{n} f(z_i - y)g(y)dy.$$

The marginal density of Z_1 is derived by integrating out $y, z_2, ..., z_n$:

$$p_1(z_1) = \int_{-\infty}^{\infty} \cdots \int_{-\infty}^{\infty} f(z_1 - y) \prod_{i=2}^{n} f(z_i - y)g(y)dydz_2 \cdots dz_n$$

$$= \int_{-\infty}^{\infty} f(z_1 - y) \left(\prod_{i=2}^{n} \int_{-\infty}^{\infty} f(z_i - y)dz_i \right) g(y)dy = \int_{-\infty}^{\infty} f(z_1 - y)g(y)dy.$$

The joint density of Z_1 and Y is $f(z_1 - y)g(y)$, so that the conditional density can be written as

$$f_{Z_1|Y=u}(z_1) = \frac{f(z_1 - y)g(y)}{g(y)} = f(z_1 - y).$$

As follows from this derivation, marginal and conditional densities of Z_i are the same for every $i = 1, 2, .., n$. They are independent if and only if

$$\int_{-\infty}^{\infty} \prod_{i=1}^{n} f(z_i - y)g(y)dy = \prod_{i=1}^{n} \int_{-\infty}^{\infty} f(z_i - y)g(y)dy, \qquad (3.65)$$

which does not hold generally. Now we prove that $\{Z_i|Y = y, i = 1, 2, ..., n\}$ are independent. Indeed, it suffices to prove that

$$f_{\mathbf{Z}|Y=y}(z_1, ..., z_n) = \frac{\prod_{i=1}^{n} f(z_i - y)g(y)}{g(y)} = \prod_{i=1}^{n} f(z_i - y) = \prod_{i=1}^{n} f_{Z_i|Y=y}(z_i).$$

This is clearly true because $f_{Z_i|Y=y}(z_i) = f(z_i - y)$. In words, conditional and marginal independence are not equivalent. More insights into the difference between marginal and conditional independence for the multivariate normal distribution can be gained from Example 4.11.   □

The following example can be viewed as a generalization of the bivariate prediction problem formulated previously in Example 3.26.

Example 3.88 **Optimal prediction.** *What is an optimal prediction of* Y *given* $\mathbf{X}$*?*

Solution. This is a typical open-ended question where neither the probabilistic model nor the definition of "optimal" is formulated, and yet the question is fairly easy to comprehend. The task of a statistician is to translate a usually vague real-life question into a rigorous mathematical model using probabilistic language. First, we view Y and $\mathbf{X}$ as random variables. Second, Y and $\mathbf{X}$ have a known joint distribution. Third, we define the criterion for the quality of prediction as $E((Y - \mu(\mathbf{x}))^2 | \mathbf{X} = \mathbf{x})$, which is minimized when $\mu(\mathbf{x}) = E(\mathbf{Y}|\mathbf{X} = \mathbf{x})$ as is easy to see by differentiation with respect to μ since $\mathbf{x}$ is fixed. Thus, the optimal prediction of Y given $\mathbf{X} = \mathbf{x}$ is the regression/conditional mean of Y on $\mathbf{X}$.

3.10.2 Multivariate MGF

As a straightforward generalization of the univariate MGF (Section 2.5), we define the multivariate MGF of random vector $\mathbf{X}$ as $M(\mathbf{t}) = Ee^{\mathbf{t}'\mathbf{X}}$. Since $\partial M/\partial \mathbf{t} = E\left(\mathbf{X}e^{\mathbf{t}'\mathbf{X}}\right)$, we obtain

$$E(\mathbf{X}) = \left.\frac{\partial M}{\partial \mathbf{t}}\right|_{\mathbf{t}=\mathbf{0}}$$

as in the univariate case. Furthermore,

$$\frac{\partial^2 M}{\partial \mathbf{t}^2} = E\left(\mathbf{X}\mathbf{X}'e^{\mathbf{t}'\mathbf{X}}\right)$$

and

$$E(\mathbf{X}\mathbf{X}') = \left.\frac{\partial^2 M}{\partial \mathbf{t}^2}\right|_{\mathbf{t}=\mathbf{0}},$$

which is a generalization of formula (2.26) for $k = 2$. This means that once the multivariate MGF is determined, the moments are obtained through evaluation of the derivatives at $\mathbf{t} = \mathbf{0}$. Similarly, one defines the multivariate characteristic function as $Ee^{i\mathbf{t}'\mathbf{X}}$. Since there is a one-to-one correspondence between multivariate distributions and characteristic functions, the following important theoretical result holds.

Theorem 3.89 **Cramér–Wold.** *A multivariate distribution of* $\mathbf{X}$ *is uniquely specified by the set of univariate distributions of* $\mathbf{t}'\mathbf{X}$ *for all* $\mathbf{t}$.

Indeed, if the distributions of $\mathbf{t}'\mathbf{X}_1$ and $\mathbf{t}'\mathbf{X}_2$ are the same, then their characteristic functions are the same, and, therefore, their multivariate distributions coincide. For example, one concludes that $\mathbf{X}$ has a multivariate normal distribution if random variable $\mathbf{t}'\mathbf{X}$ is normally distributed for every $\mathbf{t}$. This fact is used in Example 6.47 to visually test if a 17×1 vector has a multivariate normal distribution (see function `normQQ`).

3.10.3   Multivariate delta method

The multivariate delta method is used to approximate the variance of a random variable that is a function of other random variables with known covariance matrix. The univariate delta method was discussed in Section 2.12.1. The goal of this section is to extend this method to several random variables.

Let $\mathbf{X}$ be an $m \times 1$ random vector with mean $\boldsymbol{\mu}$ and $\mathbf{Y} = \mathbf{f}(\mathbf{X})$ be another random vector, where $\mathbf{f} : R^m \to R^n$, so that $\mathbf{Y}$ is the $n \times 1$ random vector. The key to the delta method is the approximation of function $\mathbf{f}$ by a linear function using the Jacobian (see Appendix), $\mathbf{J} = \frac{\partial \mathbf{f}}{\partial \mathbf{x}}\big|_{\mathbf{x}=\boldsymbol{\mu}}$, a $n \times m$ matrix evaluated at $E(\mathbf{X}) = \boldsymbol{\mu}$. The Taylor series approximation of the first order (linear approximation) of $\mathbf{Y}$ at $\boldsymbol{\mu}=E(\mathbf{X})$ is $\mathbf{Y} \simeq \mathbf{f}(\boldsymbol{\mu}) + \mathbf{J}(\mathbf{x} - \boldsymbol{\mu})$. Then as follows from (3.64), the approximate covariance matrix of $\mathbf{Y}$ can be expressed via the covariance matrix of $\mathbf{X}$ as

$$\mathrm{cov}(\mathbf{Y}) \simeq \mathbf{J}\mathrm{cov}(\mathbf{X})\mathbf{J}'. \tag{3.66}$$

In a special case when components of vector $\mathbf{X}$ do not correlate and $Y = f(X_1, .., X_m)$ is a scalar random variable, we have

$$\mathrm{var}(Y) \simeq \sum_{i=1}^{m} \sigma_i^2 \left(\frac{\partial f}{\partial x_i} \right)^2,$$

where σ_i^2 is the variance of X_i and the ith derivative is evaluated at $x_i = \mu_i$. For example, the variances of the product and the ratio of two independent random variables can be approximated as

$$\mathrm{var}\,(UW) \simeq \sigma_U^2 \mu_W^2 + \sigma_W^2 \mu_U^2, \quad \mathrm{var}\left(\frac{U}{W}\right) \simeq \sigma_U^2 \frac{1}{\mu_W^2} + \sigma_W^2 \frac{\mu_U^2}{\mu_W^4}. \tag{3.67}$$

Note that the exact variance of the product is $\mathrm{var}\,(UW) = \sigma_U^2 \mu_W^2 + \sigma_W^2 \mu_U^2 + \sigma_U^2 \sigma_W^2$.

Example 3.90 *Multivariate delta method for ratio.* Let $Z_i \sim \mathcal{N}(0,1)$ and $Y_i \sim \mathcal{N}(0,1)$ be independent, $i = 1, 2, ..., n > 2$. The ratio

$$T = \frac{\sum_{i=1}^{n} Z_i/\sqrt{n}}{\sqrt{\sum_{i=1}^{n} Y_i^2/n}}$$

has t-distribution with n degrees of freedom with $\mathrm{var}(T) = n/(n-2)$. *Estimate the variance of T using the delta method.*

Solution. Express $T = U/\sqrt{W}$, where $U = \sum_{i=1}^{n} Z_i/\sqrt{n}$ and $W = \sum_{i=1}^{n} Y_i^2/n$. Using a derivation similar to (3.67), we obtain

$$\mathrm{var}\left(\frac{U}{\sqrt{W}}\right) \simeq \sigma_U^2 \frac{1}{\mu_W} + \sigma_W^2 \frac{\mu_U^2}{4\mu_W^3}. \tag{3.68}$$

But

$$\mu_U = 0, \sigma_U^2 = 1, \mu_W = 1, \sigma_W^2 = \frac{1}{n^2} \sum_{i=1}^{n} \text{var}(Y_i^2) = \frac{2}{n}.$$

Finally, plugging these values back into (3.68), the variance of the T statistic is approximated by the delta method as $\text{var}(T) \simeq 1$, which confirms the exact formula, $n/(n-2)$, for large n. $\qquad\square$

Sometimes, the function that defines the random variable is defined intrinsically, such as the solution of a nonlinear system of equations. Even if the function cannot be expressed in a closed form, the delta method applies. The derivative of the implicitly defined function as a part of the multivariable calculus is the key.

Example 3.91 *Implicit delta method. The $m \times 1$ random variable $\mathbf{Y}$ is defined intrinsically as the solution of the system of nonlinear equations $\mathbf{h}(\mathbf{X}, \mathbf{Y}) = \mathbf{0}$, where $\mathbf{X}$ is a $n \times 1$ random vector with mean $\boldsymbol{\mu}$ and covariance matrix $\boldsymbol{\Omega}$ and $\mathbf{h}$ is the $m \times 1$ vector function. Approximate the covariance matrix of $\mathbf{Y}$ via the delta method. Use linear function $\mathbf{h}$ to check the result.*

Solution. Let random vector $\mathbf{Y}$ be the solution of $\mathbf{h}(\mathbf{X}, \mathbf{Y}) = \mathbf{0}$ given $\mathbf{X}$. To apply the delta method we find the implicit Jacobian $\mathbf{J} = \partial \mathbf{Y}/\partial \mathbf{X}$ as the derivative of the implicitly defined function

$$\mathbf{J} = -\left(\left.\frac{\partial \mathbf{h}}{\partial \mathbf{Y}}\right|_{(\boldsymbol{\mu}, \boldsymbol{\mu}_Y)} \right)^{-1} \left(\left.\frac{\partial \mathbf{h}}{\partial \mathbf{X}}\right|_{(\boldsymbol{\mu}, \boldsymbol{\mu}_Y)} \right), \qquad (3.69)$$

where $\boldsymbol{\mu}_Y$ is the solution of $\mathbf{h}(\boldsymbol{\mu}, \boldsymbol{\mu}_Y) = \mathbf{0}$ and the derivative matrices are evaluated at $\mathbf{X} = \boldsymbol{\mu}$ and $\mathbf{Y} = \boldsymbol{\mu}_Y$. Hence the covariance matrix of $\mathbf{Y}$ is approximated by (3.66) with $\mathbf{J}$ as above. For a linear system $\mathbf{AY} + \mathbf{X} = \mathbf{0}$, we have $\partial \mathbf{h}/\partial \mathbf{Y} = \mathbf{A}$ and $\partial \mathbf{h}/\partial \mathbf{X} = \mathbf{I}$, so that $\mathbf{J} = -\mathbf{A}^{-1}$, and from (3.66), we obtain

$$\text{cov}(\mathbf{Y}) \simeq \mathbf{A}^{-1}\text{cov}(\mathbf{X})\mathbf{A}'^{-1}.$$

One obtains the same result by expressing $\mathbf{Y} = -\mathbf{A}^{-1}\mathbf{X}$ and using formula (3.63).

The following example applies the formula for the implicit Jacobian (3.69).

Example 3.92 *Random root. Random coefficients A and B of a cubic polynomial are independent and uniformly distributed on $[-1, 1]$. Approximate the variance of the root of the cubic equation $y^3 + Ay^2 + By + 1 = 0$. Use simulations to verify your answer.*

Solution. One could use the famous Cardano formula as a function of A and B to solve the cubic equation and apply approximation (3.66), but such approach would be very cumbersome. Instead, we obtain the needed derivatives with respect to A and B evaluated at the expected values $\mu_A = \mu_B = 0$ and

$\mu_y = -1$ because when $A = B = 0$ the polynomial turns into $y^3 + 1 = 0$. In our case, $n = 2$ and $m = 1$ with a 2×1 Jacobian and the covariance matrix

$$\boldsymbol{\Omega} = \begin{bmatrix} 1/3 & 0 \\ 0 & 1/3 \end{bmatrix}$$

since $\text{var}(A) = \text{var}(B) = 2^2/12 = 1/3$ because they are uniformly distributed. According to formula (3.69), we have

$$\frac{\partial \mathbf{h}}{\partial \mathbf{Y}}\bigg|_{(\boldsymbol{\mu}, \boldsymbol{\mu}_Y)} = \frac{\partial (y^3 + Ay^2 + By + 1)}{\partial y}\bigg|_{(y=-1, A=0, B=0)} = 3,$$

$$\frac{\partial \mathbf{h}}{\partial \mathbf{X}}\bigg|_{(\boldsymbol{\mu}, \boldsymbol{\mu}_Y)} = \begin{bmatrix} 1 \\ -1 \end{bmatrix}, \quad \mathbf{J} = -\frac{1}{3}\begin{bmatrix} 1 \\ -1 \end{bmatrix}.$$

Thus, if Y denotes the random root of the above cubic equation, we approximate $\text{var}(Y) \simeq 2/27 = 0.074$. See the R function `intrD3` with the printout: `Empirical variance = 0.07663353`. Simulations confirm the approximation.

Example 3.93 *The Kullback–Leibler entropy. Let the joint density of the m-dimensional random vector $\mathbf{X}$ be $f_\mathbf{X}$ and the marginal density of the ith component of $\mathbf{X}$ be f_i. Define the Kullback–Leibler entropy as*

$$\mathcal{E}_{\text{KL}} \stackrel{\text{def}}{=} -E \ln \frac{f_1(X_1) \cdots f_m(X_m)}{f_\mathbf{X}(X_1, ..., X_m)}.$$

(a) Using the concavity of $\log$ in combination with Jensen's inequality show that $\mathcal{E}_{\text{KL}} \geq 0$. (b) Show that if X_i are independent then $\mathcal{E}_{\text{KL}} = 0$.

Solution. (a) From Section 2.2, we know that, due to Jensen's inequality, $E(\ln U) \leq \ln E(U)$, where U is a positive random variable. Applying this inequality to $\mathcal{E}_{\text{KL}}$, we obtain

$$\mathcal{E}_{\text{KL}} \geq -\ln E \frac{f_1(X_1) \cdots f_m(X_m)}{f_\mathbf{X}(X_1, ..., X_m)}.$$

But

$$E \frac{f_1(X_1) \cdots f_m(X_m)}{f_\mathbf{X}(X_1, ..., X_m)} = \int_{R^m} \frac{f_1(x_1) \cdots f_m(x_m)}{f_\mathbf{X}(x_1, ..., x_m)} f_\mathbf{X}(x_1, ..., x_m) dx_1 ... dx_m$$

$$= \int_{R^m} f_1(x_1) \cdots f_m(x_m) dx_1 ... dx_m$$

$$= \left(\int_{R^1} f_1(x_1) dx_1\right) \times \cdots \times \left(\int_{R^1} f_m(x_m) dx_m\right) = 1.$$

This inequality implies $\mathcal{E}_{KL} \geq 0$. (b) When X_i are independent, the joint density equals the product of densities and the expression under ln turns into 1. Therefore $\mathcal{E}_{KL} = 0$. $\qquad\qquad\qquad\qquad\qquad\qquad\qquad\qquad\qquad\qquad$ □

One may view the Kullback–Leibler entropy as a measure of dependence among components of the random vector $\mathbf{X}$. The result $\mathcal{E}_{KL} \geq 0$ may be interpreted as saying that the joint distribution is more informative than the set of marginal distributions (recall that the Shannon entropy can be negative).

The above Kullback–Leibler entropy can be generalized to define the *relative entropy* or Kullback–Leibler divergence between the distributions of two random vectors $\mathbf{Y}$ and $\mathbf{X}$ defined by densities g and f, respectively, as

$$\mathcal{E}_{KL} = -E_f \left(\ln \frac{g(\mathbf{X})}{f(\mathbf{X})} \right), \qquad (3.70)$$

where the subindex f means that the expectation is over the random variable with density f. When $\mathbf{Y}$ and $\mathbf{X}$ are discrete random variables defined on the same set of values, the relative entropy is defined as $-\sum p_{Xi} \ln(p_{Yi}/p_{Xi})$. In short, (3.70) defines how different two distributions are. As we shall learn later in the statistics part of the book, the relative Kullback–Leibler entropy is closely related to the likelihood ratio test for statistical hypothesis testing – see Section 7.9.1.

3.10.4 Multinomial distribution

The multinomial distribution is a multivariate discrete distribution as a generalization of the binomial distribution to $m > 2$ categories. Note that the binomial distribution may be viewed as the distribution on $\{0,1\}^2$ with frequencies of two outcomes observed in n repeated experiments with the probability function $Cp_1^{x_1}p_2^{x_2}$, where $p_1 + p_2 = 1$ and $x_1 + x_2 = n$ (C is the normalizing coefficient). Now, assume that a random variable may take m different outcomes with probabilities $\{p_j, j = 1, 2, ..., m\}$ summing to 1. Denote X_1 as the number of possible outcomes in n repeated experiments from the first category, X_2 in the second category, etc. It can be proven that

$$\Pr(X_1 = x_1, X_2 = x_2, ..., X_m = x_m) = \frac{n!}{x_1! x_2! \cdots x_m!} p_1^{x_1} p_2^{x_2} \cdots p_m^{x_m}, \qquad (3.71)$$

where $x_1 + x_2 + \cdots + x_m = n$. The distribution specified by (3.71) is called multinomial. Obviously, $m = 2$ yields the binomial distribution. Some algebra proves that marginal distributions of the multinomial distribution are binomial: $E(X_j) = np_j$, $\mathrm{var}(X_j) = np_j(1 - p_j)$, and

$$E(X_j X_k) = n(n-1)p_i p_j, \quad \mathrm{cov}(X_j, X_k) = E(X_j X_k) - E(X_j)E(X_k) = -np_j p_k.$$

The pairs are negatively correlated. As in the case of the binomial distribution, one may think of multinomial distribution as the distribution of the sum of n multidimensional Bernoulli random variables.

Example 3.94 *The three-category multinomial distribution.* *Write R code to generate random numbers following a three-category multinomial distribution. Check the empirical probability with the theoretical (3.71).*

Solution. Without loss of generality we can label the three categories 1, 2, and 3. The user-defined probabilities are (p_1, p_2, p_3) where $p_3 = 1 - p_1 - p_2$. The number of simulated multinomial random variables is nSim, and the three-dimensional array x contains the frequencies of occurrence in each category, so that $n = x_1 + x_2 + x_3$. See the R function multB. After generating a large number of multinomial observations, we compute the proportion of simulations with frequency in each category equal to x[1], x[2], and x[3]. The sum of n iid Bernoulli random variables is used to generate the binomial distribution in each category. That is, a double loop is used, one a loop over n Bernoulli trials and another over the number of simulations. The output with default arguments indicates a good match: Simulated prob = 0.03782 Theoretical prob = 0.0378.

Problems

1. Let the components of the m-dimensional random vector $\mathbf{X}$ be iid with mean μ and variance σ^2. Express $E(\mathbf{X})$ and $\text{cov}(\mathbf{X})$ via $\mathbf{1} = (1, 1, ..., 1)'$ and the identity matrix $\mathbf{I}$.

2. Let the components of the m-dimensional random vector $\mathbf{X}$ be iid with a common variance σ^2 and $\mathbf{Y} = \mathbf{X} + Z\mathbf{1}$, where random variable Z has variance τ^2 and is independent from $\mathbf{X}$. Express $\text{cov}(\mathbf{Y})$ in the matrix form using $\mathbf{1}$ and $\mathbf{I}$.

3. Prove that if the m-dimensional random vector $\mathbf{X}$ has covariance matrix $\mathbf{\Omega}$, then $\text{var}(X_1 + X_2 + \cdots + X_n) = \mathbf{1}'\mathbf{\Omega}\mathbf{1}$.

4. Define the correlation matrix between the m-dimensional random vector $\mathbf{X}$ and the n-dimensional random vector $\mathbf{Y}$ using diagonal variance matrices $\mathbf{D_X}$ and $\mathbf{D_Y}$ similarly to (3.61).

5. Prove that $\text{cov}(\mathbf{X}, \mathbf{Y}) = (\text{cov}(\mathbf{Y}, \mathbf{X}))'$ where $'$ is the transposition sign.

6. Extend Theorem 3.78 to dependent $\mathbf{X}$ and $\mathbf{Y}$ using $\text{cov}(\mathbf{X}, \mathbf{Y})$.

7. Prove Theorem 3.79.

8. Use (3.64) to find the variance of the sum of the components of vector $\mathbf{Y}$ from Problem 2.

9. Using a proof similar to Example 3.82, prove that correlation matrix is symmetric and nonnegative definite.

10. The 3×1 random vector $\mathbf{X} = (Y - Z, Y + Z, X + Y - Z)$ where random variables X, Y, and Z are independent and have unit variance. Following Example 3.84, (a) find $\operatorname{cov}(\mathbf{X})$ using component-wise computation, (b) find $\operatorname{cov}(\mathbf{X})$ using identity (3.63), (c) find $\operatorname{cor}(\mathbf{X})$, and (d) find $\operatorname{var}(\mathbf{X}'\mathbf{1})$ using formula (3.64) and compare it with direct computation as the variance of the sum of independent random variables.

11. Random vector $\mathbf{X}$ has covariance matrix $\mathbf{\Omega}$. Express $\operatorname{cov}(\mathbf{a}'\mathbf{X}, \mathbf{b}'\mathbf{X})$ via $\mathbf{\Omega}$ where $\mathbf{a}$ and $\mathbf{b}$ are fixed vectors.

12. Use formula (3.63) to verify the variances in Example 3.81.

13. Under the conditions of Example (3.85), find $\operatorname{cov}(\mathbf{X} - Y\mathbf{1})$.

14. (a) Does Example 3.86 remain valid if $X_i = Z_i + U_i/i$? (b) Does Example 3.86 remain valid if $X_i = Z_i + U_i/i^p$ for any $p > 0$?

15. Let $\mathbf{X}$ and $\mathbf{Y}$ be two independent random vectors with the same mean. (a) Prove that $E \|\mathbf{X} - \mathbf{Y}\|^2 = \operatorname{tr}(\operatorname{cov}(\mathbf{X}) + \operatorname{cov}(\mathbf{Y}))$. (b) Apply this formula to find the expected squared distance between points randomly chosen in the unit cube and use simulations to verify your answer.

16. Prove that (3.65) does not hold when $f = g = \phi$.

17. (a) Prove that the MGF of a random vector with independent components is the product of individual MGFs. (b) Random variables Z and Y are independent. Express the MGF of the 2×1 random vector $\mathbf{X} = (X - Y, X + Y)$ via individual MGFs.

18. Prove that the characteristic function of $\mathcal{N}(\boldsymbol{\mu}, \mathbf{\Omega})$ is $\exp(it'\boldsymbol{\mu} - 0.5t'\mathbf{\Omega}t)$.

19. Verify how the accuracy of the delta method for the variance of the ratio of two independent random variables, U/W, assuming that U and W are uniformly distributed on $[0, a]$. Plot the exact and approximated variances as functions of a in the range from 0 to 2. [Hint: Derive the exact variance from the density of the ratio using the cdf of the ratio, $\Pr(U/W \leq z)$, by means of the bivariate uniform distribution on the square $[0, a]^2$.]

20. Approximate the variance of the roots of the quadratic equation $Ax^2 + 2x - B = 0$, where $A \sim \mathcal{R}(1/2, 3/2)$ and $B \sim \mathcal{R}(0, 1)$ are independent. Do the roots correlate? Use simulations to test the approximation.

21. Similarly to Example 3.92, approximate the variance of the root of the cubic equation $Dy^3 + Ay^2 + By + 1 = 0$, where $D \sim \mathcal{R}(0, 1)$ and all coefficients are independent. Compare with simulations.

22. Prove that $m = 2$ in distribution (3.71) yields a binomial distribution.

23. Find the correlation coefficient between X_j and X_k and prove that this coefficient lies in the interval $(-1, 0)$. [Hint: Let $p_i + p_j = c \leq 1$.]

24. Show that the multinomial distribution converges to the multivariate Poisson distribution when $n \to \infty$ and $p_j \to 0$ such that $np_j \to \lambda_j$.

25. Prove that the Shannon entropy in a sample of n iid observations is n times larger than in one observation.

26. (a) Derive a closed-form expression for $\mathcal{E}_{KL}$ for a bivariate normal distribution. (b) Show that $\mathcal{E}_{KL} = 0$ if and only if $\rho = 0$. (c) Show that $\mathcal{E}_{KL} > 0$ for $\rho \neq 0$ using the elementary inequality $\ln(1 + x) < x$ for $x \neq 0$. [Hint: Show that, without loss of generality, one can use the standard normal distribution.]

27. Modify function `multB` to generate multinomial random numbers with an arbitrary number of categories.

28. Develop a vectorized version of `multB` (no loop over simulations).

29. Find the correlation coefficient between X_j and X_k in the multinomial distribution.

30. (a) Prove that the Kullback–Leibler relative entropy defined by equation (3.70) is nonnegative. (b) Find how the uniform and normal distributions are different by computing the Kullback-Leibler relative entropy between $f = \mathcal{R}(-1, 1)$ and $g = \mathcal{N}(0, 1)$. (c) Derive a function of μ as $\mathcal{E}_{KL}$ computed for comparison of $f = \mathcal{N}(0, 1)$ with $g = \mathcal{N}(\mu, 1)$, and test your answer with `integrate` in R.

31. Derive the Kullback–Leibler entropy for the multinomial distribution.

32. Theorem 1.18 proves that the binomial distribution converges to the Poisson distribution when $n \to \infty$ and $p = \lambda/n$. Illustrate this fact by showing that $\mathcal{E}_{KL} \to 0$ when $n \to \infty$ where f is the probability mass function for the binomial and g is the pmf for the Poisson distribution.

Chapter 4

Four important distributions in statistics

In this chapter, we consider the four most important distributions in statistics: the multivariate normal distribution, the chi-square distribution (χ^2-distribution), the t-distribution, and the F-distribution. The multivariate normal distribution is the parent to all. These distributions are especially useful for statistical models under the most popular normal distribution assumption. Also, these distributions arise in asymptotic theory, when the sample size goes to infinity, even when the original distribution is not normal or even continuous.

It is especially convenient to work with multidimensional distributions using matrix algebra. The basics of matrix algebra are found in Section 10.2.

4.1 Multivariate normal distribution

The bivariate (two-dimensional, $m = 2$) normal distribution was introduced in Section 3.5. In this section, we study the multivariate normal distribution of an m-dimensional random vector

$$\mathbf{X} = \begin{bmatrix} X_1 \\ X_2 \\ \vdots \\ X_m \end{bmatrix}.$$

Thus, it is useful to review Section 3.10, where random vectors are studied regardless of their distribution. The fact that $\mathbf{X}^{m \times 1}$ has a multivariate normal

Advanced Statistics with Applications in R, First Edition. Eugene Demidenko.
© 2020 John Wiley & Sons, Inc. Published 2020 by John Wiley & Sons, Inc.

distribution will be written as

$$\mathbf{X} \sim \mathcal{N}(\boldsymbol{\mu}, \boldsymbol{\Omega}),$$

where $\boldsymbol{\mu}$ is the $m \times 1$ mean vector with the ith component μ_i and $\boldsymbol{\Omega}$ is the $m \times m$ covariance matrix defined in Section 3.10. In other notation,

$$E(\mathbf{X}) = \boldsymbol{\mu}, \quad \mathrm{cov}(\mathbf{X}) = E[(\mathbf{X} - \boldsymbol{\mu})(\mathbf{X} - \boldsymbol{\mu})'] = \boldsymbol{\Omega},$$

or in the coordinate form, $E(X_i) = \mu_i$ and $\mathrm{cov}(X_i, X_j) = \Omega_{ij}$ for $i, j = 1, 2, ..., m$. For $i = j$, we have $\Omega_{ii} = \mathrm{var}(X_i) = \sigma_i^2$. When $\boldsymbol{\mu} = \mathbf{0}$ and $\boldsymbol{\Omega} = \mathbf{I}$, we say that $\mathbf{X} = \mathbf{Z}$ is the standard normal vector and write $\mathbf{Z} \sim \mathcal{N}(\mathbf{0}, \mathbf{I})$.

The marginal distribution is also normal, $X_i \sim \mathcal{N}(\mu_i, \sigma_i^2)$. Any pair (X_i, X_j) has a marginal bivariate normal distribution $(i \neq j)$ or, symbolically,

$$\begin{bmatrix} X_i \\ X_j \end{bmatrix} \sim \mathcal{N} \left(\begin{bmatrix} \mu_i \\ \mu_j \end{bmatrix}, \begin{bmatrix} \sigma_i^2 & \rho_{ij}\sigma_i\sigma_j \\ \rho_{ij}\sigma_i\sigma_j & \sigma_j^2 \end{bmatrix} \right), \quad i, j = 1, 2, ..., m.$$

Note that matrix $\boldsymbol{\Omega}$, as a covariance matrix, is symmetric and nonnegative definite. This implies $\mathbf{a}'\boldsymbol{\Omega}\mathbf{a} \geq 0$ for every vector $\mathbf{a}$. Moreover, we shall assume that $\boldsymbol{\Omega}$ is a positive definite matrix, that is, $\mathbf{a}'\boldsymbol{\Omega}\mathbf{a} > 0$ for every $\mathbf{a} \neq \mathbf{0}$. Consequently, this matrix is nonsingular and the determinant is positive, $|\boldsymbol{\Omega}| > 0$. This means that a linear combination of $\mathbf{X}$ cannot be a fixed number.

The joint density of the multivariate normal vector $\mathbf{X}$ is given by

$$f(\mathbf{x}; \boldsymbol{\mu}, \boldsymbol{\Omega}) = (2\pi)^{-m/2} |\boldsymbol{\Omega}|^{-1/2} \, e^{-\frac{1}{2}(\mathbf{x}-\boldsymbol{\mu})'\boldsymbol{\Omega}^{-1}(\mathbf{x}-\boldsymbol{\mu})}, \quad \mathbf{x} \in R^m. \tag{4.1}$$

The argument of the exponential function is a quadratic form that can be represented in the coordinate form as

$$(\mathbf{x} - \boldsymbol{\mu})'\boldsymbol{\Omega}^{-1}(\mathbf{x} - \boldsymbol{\mu}) = \sum_{i=1}^{m}(x_i - \mu_i)^2\Omega_{ii}^{-1} + 2\sum_{i>j}^{m}(x_i - \mu_i)(x_j - \mu_j)\Omega_{ij}^{-1}. \tag{4.2}$$

The set of points $(\mathbf{x} - \boldsymbol{\mu})'\boldsymbol{\Omega}^{-1}(\mathbf{x} - \boldsymbol{\mu}) = \mathrm{const}$ defines an ellipsoid with the volume proportional to $|\boldsymbol{\Omega}|$. That is why the determinant of the covariance matrix is referred to as the *generalized variance*. It can be proven that if matrix $\boldsymbol{\Omega}$ is symmetric and positive definite, then $\boldsymbol{\Omega}^{-1}$ is symmetric and positive definite as well; see Chapter 10. This means that the above quadratic form takes positive values if $\mathbf{x} \neq \boldsymbol{\mu}$. It implies that the density (4.1) takes maximum at $\mathbf{x} = \boldsymbol{\mu}$ with the maximum value of the density equal to $(2\pi)^{-m/2}|\boldsymbol{\Omega}|^{-1/2}$. From the symmetry of $\boldsymbol{\Omega}$ it follows that the density (4.1) is symmetric about $\boldsymbol{\mu}$, i.e. $f(\mathbf{x} - \boldsymbol{\mu}; \boldsymbol{\mu}, \boldsymbol{\Omega}) = f(\boldsymbol{\mu} - \mathbf{x}; \boldsymbol{\mu}, \boldsymbol{\Omega})$.

Example 4.1 *Bivariate via multivariate.* *Show that the multivariate density (4.1) turns into the bivariate density (3.27) for $m = 2$.*

Solution. For $m = 2$, we have

$$\Omega = \begin{bmatrix} \sigma_1^2 & \sigma_{12} \\ \sigma_{12} & \sigma_2^2 \end{bmatrix}$$

with the determinant $|\Omega| = \sigma_1^2 \sigma_2^2 - \sigma_{12}^2 = (1 - \rho^2)\sigma_1^2\sigma_2^2$ and the inverse matrix

$$\Omega^{-1} = \frac{1}{(1-\rho^2)\sigma_1^2\sigma_2^2} \begin{bmatrix} \sigma_2^2 & -\sigma_1\sigma_2\rho \\ -\sigma_1\sigma_2\rho & \sigma_1^2 \end{bmatrix} = \frac{1}{1-\rho^2} \begin{bmatrix} 1/\sigma_1^2 & -\rho/(\sigma_1\sigma_2) \\ -\rho/(\sigma_1\sigma_2) & 1/\sigma_2^2 \end{bmatrix}.$$

Using expression (4.2) we obtain

$$(\mathbf{x} - \boldsymbol{\mu})'\Omega^{-1}(\mathbf{x} - \boldsymbol{\mu}) = \frac{\left(\frac{x_1-\mu_1}{\sigma_1}\right)^2 - 2\rho\left(\frac{x_1-\mu_1}{\sigma_1}\right)\left(\frac{x_2-\mu_2}{\sigma_2}\right) + \left(\frac{x_2-\mu_2}{\sigma_2}\right)^2}{1-\rho^2}.$$

Thus, for $m = 2$, density (4.1) turns into

$$f(x_1, x_2) = \frac{1}{(2\pi)\sigma_1\sigma_2\sqrt{1-\rho^2}} e^{-\frac{1}{2(1-\rho^2)}\left[\left(\frac{x_1-\mu_1}{\sigma_1}\right)^2 - 2\rho\left(\frac{x_1-\mu_1}{\sigma_1}\right)\left(\frac{x_2-\mu_2}{\sigma_2}\right) + \left(\frac{x_2-\mu_2}{\sigma_2}\right)^2\right]},$$

the same as the bivariate normal density given by (3.27).   □

The multivariate normal distribution with $m = 3$ is called the *trivariate* normal distribution. This distribution will be used when the partial correlation is introduced; see Example 4.13.

An important property of the multivariate normal distribution is that a linear transformation of a multivariate normal vector is again a multivariate normal vector.

Theorem 4.2 *If $\mathbf{X}$ is a $m \times 1$ random vector with a multivariate normal distribution $\mathbf{X} \sim \mathcal{N}(\boldsymbol{\mu}, \Omega)$ and $\mathbf{A}$ is $k \times m$ fixed matrix of full rank ($k \leq m$), then the $k \times 1$ random vector $\mathbf{Y} = \mathbf{AX}$ has a normal distribution, $\mathbf{Y} \sim \mathcal{N}(\mathbf{A}\boldsymbol{\mu}, \mathbf{A}\Omega\mathbf{A}')$.*

The condition that $\mathbf{A}$ has full rank implies that matrix $\mathbf{A}\Omega\mathbf{A}'$ is positive definite so that the m-variate normal distribution is well defined. In a special case when $\mathbf{A} = \mathbf{a}'$ we have $\mathbf{a}'\mathbf{X} \sim \mathcal{N}(\mathbf{a}'\boldsymbol{\mu}, \mathbf{a}'\Omega\mathbf{a})$. In words, a linear combination of the multivariate normal vector is a normally distributed random variable. In a very special case when $\mathbf{X} \sim \mathcal{N}(\mathbf{0}, \mathbf{I})$ and $\|\mathbf{a}\| = 1$, we have

$$\mathbf{a}'\mathbf{X} \sim \mathcal{N}(0, 1). \tag{4.3}$$

Example 4.3 *Joint distribution of the sum and difference. Random variables X and Y are independent and have the same normal distribution with mean μ and variance σ^2. Use Theorem 4.2 to find the joint distribution of the two-dimensional random vector with components as the sum and difference of X and Y, respectively. Prove that the components of the new vector are independent.*

Solution. Let X and Y be the original independent random variables with the same distribution, $\mathcal{N}(\mu, \sigma^2)$. We represent the 2×1 vector of interest as a linear transformation of the original 2×1 vector $\mathbf{X} = (X, Y)'$ as

$$\mathbf{Y} = \begin{bmatrix} X+Y \\ X-Y \end{bmatrix} = \begin{bmatrix} 1 & 1 \\ 1 & -1 \end{bmatrix} \begin{bmatrix} X \\ Y \end{bmatrix} = \mathbf{AX}, \quad \mathbf{A} = \begin{bmatrix} 1 & 1 \\ 1 & -1 \end{bmatrix}.$$

Find the mean and covariance matrix of $\mathbf{Y}$ as specified in Theorem 4.2:

$$E(\mathbf{Y}) = \mathbf{A}\boldsymbol{\mu} = \begin{bmatrix} 1 & 1 \\ 1 & -1 \end{bmatrix} \begin{bmatrix} \mu \\ \mu \end{bmatrix} = \begin{bmatrix} 2\mu \\ 0 \end{bmatrix} = \mu \begin{bmatrix} 2 \\ 0 \end{bmatrix},$$

$$\mathbf{A}\boldsymbol{\Omega}\mathbf{A}' = \begin{bmatrix} 1 & 1 \\ 1 & -1 \end{bmatrix} \begin{bmatrix} \sigma^2 & 0 \\ 0 & \sigma^2 \end{bmatrix} \begin{bmatrix} 1 & 1 \\ 1 & -1 \end{bmatrix} = 2\sigma^2 \begin{bmatrix} 1 & 0 \\ 0 & 1 \end{bmatrix}.$$

Therefore, $\mathbf{Y}$ has a bivariate normal distribution

$$\mathcal{N} \left(\mu \begin{bmatrix} 2 \\ 0 \end{bmatrix}, 2\sigma^2 \begin{bmatrix} 1 & 0 \\ 0 & 1 \end{bmatrix} \right).$$

Since the off-diagonal element is zero, the correlation between the first and the second components of vector $\mathbf{Y}$ is zero, so the components are independent. □

The following result is a necessary and sufficient condition for independence between two random vectors upon linear transformations.

Theorem 4.4 *If* $\mathbf{X}$ *is a* $m \times 1$ *random vector with a multivariate normal distribution* $\mathbf{X} \sim \mathcal{N}(\boldsymbol{\mu}, \boldsymbol{\Omega})$ *and* $\mathbf{A}$ *and* $\mathbf{B}$ *are* $k \times m$ *and* $p \times m$ *fixed matrices, then two random vectors* $\mathbf{AX}$ *and* $\mathbf{BX}$ *are independent if and only if* $\mathbf{A}\boldsymbol{\Omega}\mathbf{B}' = \mathbf{0}$.

Corollary 4.5 *If* $\mathbf{X}^{m \times 1} \sim \mathcal{N}(\boldsymbol{\mu}, \boldsymbol{\Omega})$ *and* $\mathbf{a}$ *and* $\mathbf{b}$ *are* $m \times 1$ *fixed vectors, then* $\mathbf{a}'\mathbf{X}$ *and* $\mathbf{b}'\mathbf{X}$ *are independent if and only if* $\mathbf{a}'\boldsymbol{\Omega}\mathbf{b} = 0$.

Example 4.6 ***Distribution of the sum and difference continued.*** *Use the above corollary to prove that the first and second components of the random vector* $\mathbf{Y}$ *in Example 4.3 are independent.*

Solution. We have

$$\mathbf{a} = \begin{bmatrix} 1 \\ 1 \end{bmatrix}, \quad \mathbf{b} = \begin{bmatrix} 1 \\ -1 \end{bmatrix}.$$

Therefore

$$
\mathbf{a'\Omega b} = \begin{bmatrix} 1 \\ 1 \end{bmatrix}' \begin{bmatrix} \sigma^2 & 0 \\ 0 & \sigma^2 \end{bmatrix} \begin{bmatrix} 1 \\ -1 \end{bmatrix} = \sigma^2 \begin{bmatrix} 1 \\ 1 \end{bmatrix}' \begin{bmatrix} 1 & 0 \\ 0 & 1 \end{bmatrix} \begin{bmatrix} 1 \\ -1 \end{bmatrix}
$$

$$
= \sigma^2 \begin{bmatrix} 1 \\ 1 \end{bmatrix}' \begin{bmatrix} 1 \\ -1 \end{bmatrix} = \sigma^2 (1 \times 1 - 1 \times 1) = 0.
$$

Hence $X+Y$ and $X-Y$ are independent. In fact, as we mentioned earlier, $X+Y$ and $X-Y$ are independent because the off-diagonal element of the covariance matrix is zero.

4.1.1 Generation of multivariate normal variables

Theorem 4.2 has an important corollary: any multivariate random vector can be linearly transformed into a standard normal vector $\mathbf{Z} \sim \mathcal{N}(\mathbf{0}, \mathbf{I})$, the random vector of independent normally distributed components with zero mean and unit variance. Indeed, let

$$
\mathbf{Z} = \mathbf{\Omega}^{-1/2}(\mathbf{X} - \boldsymbol{\mu}), \tag{4.4}
$$

where $\mathbf{\Omega}^{-1/2}$ is the *matrix square root* of $\mathbf{\Omega}^{-1}$; see Appendix. For example, the matrix square root can be computed as $\mathbf{\Omega}^{-1/2} = \mathbf{P}\mathbf{\Lambda}^{-1/2}\mathbf{P}'$, where $\mathbf{\Omega} = \mathbf{P}\mathbf{\Lambda}\mathbf{P}'$ is the spectral decomposition of matrix $\mathbf{\Omega}$ (column vectors of $\mathbf{P}$ are eigenvectors and the elements of the diagonal matrix $\mathbf{\Lambda}$ are eigenvalues of matrix $\mathbf{\Omega}$). Note that since $\mathbf{\Omega}$ is positive definite, all its eigenvalues, Λ_{ii}, are positive, so $1/\sqrt{\Lambda_{ii}}$ exists. The eigenvectors and eigenvalues are computed in R using function `eigen`; see Example 4.7. Result (4.4) can be interpreted as saying that any multidimensional normally distributed vector can be transformed into a vector with iid standard normally distributed components upon a linear transformation.

Reversely to (4.4), any multivariate random vector $\mathbf{X}$ can be expressed as a linear transform of $\mathbf{Z} \sim \mathcal{N}(\mathbf{0}, \mathbf{I})$, namely,

$$
\mathbf{X} = \boldsymbol{\mu} + \mathbf{\Omega}^{1/2}\mathbf{Z}. \tag{4.5}
$$

Instead of using the matrix square root one can use the *Cholesky decomposition*, $\mathbf{\Omega} = \mathbf{T}'\mathbf{T}$, where $\mathbf{T}$ is a nonsingular upper triangular matrix (elements below the diagonal are zero). Then, as follows from Theorem 4.2, we have

$$
\mathbf{X} = \boldsymbol{\mu} + \mathbf{T}'\mathbf{Z} \sim \mathcal{N}(\boldsymbol{\mu}, \mathbf{\Omega}). \tag{4.6}
$$

Indeed,

$$
\mathrm{cov}(\mathbf{X}) = \mathrm{cov}(\mathbf{T}'\mathbf{Z}) = \mathbf{T}'\mathrm{cov}(\mathbf{Z})\mathbf{T} = \mathbf{T}'\mathbf{T} = \mathbf{\Omega}.
$$

In short, a random vector with multivariate normal distribution can be expressed as a linear transformation of a random vector with independent standard normally distributed components upon a linear transformation. The Cholesky factor matrix, $\mathbf{T}$, is computed in R using function `chol`; see below.

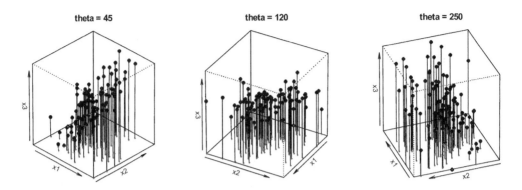

Figure 4.1: *One hundred normally distributed 3D points viewed at different* theta *angle. The depth of the points is achieved through projection of the points on the* (x, y) *plane. See the R function* mn3.

Example 4.7 ***Three-dimensional (3D) graphics in*** R. *Generate 100 3D multivariate normal points with*

$$\boldsymbol{\mu} = \begin{bmatrix} -1 \\ 2 \\ 3 \end{bmatrix}, \quad \boldsymbol{\Omega} = \begin{bmatrix} 3 & -1 & 1 \\ -1 & 2 & 1 \\ 1 & 1 & 2 \end{bmatrix}$$

using the matrix square root (4.5) and Cholesky decomposition (4.6). Write an R *program with arguments* nPoints, mu, *and* Omega *and plot the points using the* persp *function.*

Solution. The R program mn3 illustrates how to plot 3D points with different views/camera angles; see Figure 4.1. The argument sq controls the method of points generation: if sq=T (default), the matrix square root (4.5) is used and otherwise (4.6) is used. The internal R function p3 sets up the 3D plot using the persp command (actually, it plots nothing), and the points are plotted later using trans3d function. Three parameters/arguments specify the camera position: r specifies the distance of the camera from the center of the plot. If r is large, no perspective distortion is seen (for example, the vertical box lines look parallel). Argument theta specifies the angle on the (x, y) plane, and phi specifies the angle with z-axis (phi=0 would mean that the camera looks straight down from the top). Function trans3d transforms/projects an array of 3D vectors onto 2D. It returns a dataframe with two columns $x and $y. After this transformation, various R plot functions such as segments and lines can be used. To better understand the depth in 3D, we project the points onto the (x, y) plane using segments function.

We make several comments regarding point generation. We generate matrix **Z** with rows from the standard normal 3D distribution. To get the desired $\mathcal{N}(\boldsymbol{\mu}, \boldsymbol{\Omega})$, we multiply matrix $\mathbf{Z}$ at right by the transpose of $\mathbf{\Omega}^{1/2}$ or $\mathbf{T}'$. For the square root method, we do nothing because $\mathbf{\Omega}^{1/2}$ is symmetric. For the Cholesky decomposition, we multiply by $\mathbf{T}$. Finally, we check the point generation by computing the empirical mean and covariance matrix. For large nPoints, the empirical and true values must be close.

Alternatively, the empirical covariance matrix can be computed using a built-in function, var(X3). There will be a slight difference with ours because var uses nPoints-1, not nPoints. The trivariate normally distributed vectors are saved in matrix X3. If $\mathbf{z}_i$ and $\mathbf{x}_i$ are 1×3 vectors, as ith row of matrices $\mathbf{Z}$ and $\mathbf{X}$, respectively, then the line X3=Z%*%cOm+un%*%t(mu) means that $\mathbf{x}_i = \mathbf{z}_i \mathbf{T} + \boldsymbol{\mu}'$, where $\mathbf{T}$ is the Cholesky matrix factor. In terms of vector columns, we rewrite $\mathbf{x}_i' = \mathbf{T}'\mathbf{z}_i' + \boldsymbol{\mu}$.

4.1.2 Conditional distribution

The main result of this section is that the conditional multivariate normal distribution is also a normal distribution. Since the following is a generalization of the bivariate conditional distribution, we refer the reader to Section 3.5.1. Conditional density for multivariate random vectors was defined in Section 3.10.1.

Suppose that $\mathbf{X}^{m\times 1} \sim \mathcal{N}(\boldsymbol{\mu}, \mathbf{\Omega})$, where $\boldsymbol{\mu} = E(\mathbf{X})$ is the mean vector and $\mathbf{\Omega} = \text{cov}(\mathbf{X})$ is a positive definite matrix. Partition (split) this vector and the respective mean and covariance matrix as follows:

$$\mathbf{X} = \begin{bmatrix} \mathbf{X}_1 \\ \mathbf{X}_2 \end{bmatrix}, \quad \boldsymbol{\mu} = \begin{bmatrix} \boldsymbol{\mu}_1 \\ \boldsymbol{\mu}_2 \end{bmatrix}, \quad \mathbf{\Omega} = \begin{bmatrix} \mathbf{\Omega}_{11} & \mathbf{\Omega}_{12} \\ \mathbf{\Omega}_{12}' & \mathbf{\Omega}_{22} \end{bmatrix},$$

where $\mathbf{X}_1$ and $\mathbf{X}_2$ are $m_1 \times 1$ and $m_2 \times 1$ subvectors of $\mathbf{X}$, $m_1 + m_2 = m$ ($m_1 > 0$, $m_2 > 0$) and $\mathbf{\Omega}_{jk}$ are the respective $m_j \times m_k$ submatrices ($j, k = 1, 2$). Marginally, we have

$$\mathbf{X}_1 \sim \mathcal{N}(\boldsymbol{\mu}_1, \mathbf{\Omega}_{11}), \quad \mathbf{X}_2 \sim \mathcal{N}(\boldsymbol{\mu}_2, \mathbf{\Omega}_{22}).$$

Random vectors $\mathbf{X}_1$ and $\mathbf{X}_2$ are independent if and only if the off-diagonal submatrix is zero, $\mathbf{\Omega}_{12} = \mathbf{0}$.

The following theorem specifies the conditional distribution $\mathbf{X}_1 | \mathbf{X}_2 = \mathbf{x}$ (the proof is omitted).

Theorem 4.8 *Let the multivariate normal vector* $\mathbf{X}$ *be partitioned as above. Then the conditional distribution of* $\mathbf{X}_1$ *given* $\mathbf{X}_2 = \mathbf{x}$ *is multivariate normal. Namely,*

$$\mathbf{X}_1 | (\mathbf{X}_2 = \mathbf{x}) \sim \mathcal{N}\left(\boldsymbol{\mu}_1 + \mathbf{\Omega}_{12}\mathbf{\Omega}_{22}^{-1}(\mathbf{x} - \boldsymbol{\mu}_2), \mathbf{\Omega}_{11} - \mathbf{\Omega}_{12}\mathbf{\Omega}_{22}^{-1}\mathbf{\Omega}_{12}'\right). \quad (4.7)$$

The mean of the conditional distribution (conditional mean),

$$E(\mathbf{X}_1 | \mathbf{X}_2 = \mathbf{x}) = \boldsymbol{\mu}_1 + \mathbf{\Omega}_{12}\mathbf{\Omega}_{22}^{-1}(\mathbf{x} - \boldsymbol{\mu}_2) \quad (4.8)$$

is a linear function of $\mathbf{x}^{m_2 \times 1}$, sometimes referred to as the multidimensional multiple linear regression. The covariance matrix,

$$\operatorname{cov}(\mathbf{X}_1 | \mathbf{X}_2 = \mathbf{x}) = \boldsymbol{\Omega}_{11} - \boldsymbol{\Omega}_{12} \boldsymbol{\Omega}_{22}^{-1} \boldsymbol{\Omega}_{12}' \tag{4.9}$$

is a generalization of the conditional variance. Note that since the covariance matrix $\boldsymbol{\Omega}$ is positive definite, matrices $\boldsymbol{\Omega}_{11}$ and $\boldsymbol{\Omega}_{22}$ are positive definite as well as the main submatrices of $\boldsymbol{\Omega}$. Moreover, using matrix algebra, one can prove that the right-hand side of (4.9) is a positive definite matrix; see Section 10.4 and particularly Example 10.7.

We can interpret the decomposition of the covariance matrix in the same way we did the variance decomposition for the binormal distribution, as in Section 3.5.2:

Total covariance = Explained covariance + Unexplained covariance,

where the three matrices are defined as

$$
\begin{aligned}
\text{Total covariance} &= \quad \boldsymbol{\Omega}_{11}, \\
\text{Explained covariance} &= \quad \operatorname{cov}(\boldsymbol{\mu}_1 + \boldsymbol{\Omega}_{12} \boldsymbol{\Omega}_{22}^{-1} (\mathbf{X}_2 - \boldsymbol{\mu}_2)) = \boldsymbol{\Omega}_{12} \boldsymbol{\Omega}_{22}^{-1} \boldsymbol{\Omega}_{12}', \\
\text{Unexplained covariance} &= \quad \boldsymbol{\Omega}_{11} - \boldsymbol{\Omega}_{12} \boldsymbol{\Omega}_{22}^{-1} \boldsymbol{\Omega}_{12}'.
\end{aligned}
$$

The second line follows from the following derivation:

$$
\begin{aligned}
\operatorname{cov}(\boldsymbol{\mu}_1 + \boldsymbol{\Omega}_{12} \boldsymbol{\Omega}_{22}^{-1} (\mathbf{X}_2 - \boldsymbol{\mu}_2)) &= \operatorname{cov}(\boldsymbol{\Omega}_{12} \boldsymbol{\Omega}_{22}^{-1} (\mathbf{X}_2 - \boldsymbol{\mu}_2)) \\
&= E\left[(\boldsymbol{\Omega}_{12} \boldsymbol{\Omega}_{22}^{-1} (\mathbf{X}_2 - \boldsymbol{\mu}_2)(\boldsymbol{\Omega}_{12} \boldsymbol{\Omega}_{22}^{-1} (\mathbf{X}_2 - \boldsymbol{\mu}_2)' \right] \\
&= \boldsymbol{\Omega}_{12} \boldsymbol{\Omega}_{22}^{-1} \operatorname{cov}[(\mathbf{X}_2 - \boldsymbol{\mu}_2)(\mathbf{X}_2 - \boldsymbol{\mu}_2)'] \boldsymbol{\Omega}_{22}^{-1} \boldsymbol{\Omega}_{12}' \\
&= \boldsymbol{\Omega}_{12} \boldsymbol{\Omega}_{22}^{-1} \boldsymbol{\Omega}_{22} \boldsymbol{\Omega}_{22}^{-1} \boldsymbol{\Omega}_{12}' = \boldsymbol{\Omega}_{12} \boldsymbol{\Omega}_{22}^{-1} \boldsymbol{\Omega}_{12}'.
\end{aligned}
$$

The interpretation of the covariance decomposition in the multivariate case follows the same reasoning: When $\mathbf{X}_1$ and $\mathbf{X}_2$ are orthogonal, i.e. when $\boldsymbol{\Omega}_{12} = \mathbf{0}$, the explained covariance is zero. In the opposite situation, when $\mathbf{X}_1$ is a linear function of $\mathbf{X}_2$, we have $\boldsymbol{\Omega}_{11} - \boldsymbol{\Omega}_{12} \boldsymbol{\Omega}_{22}^{-1} \boldsymbol{\Omega}_{12}' = \mathbf{0}$.

In short, the conditional mean (4.8) is a linear function of $\mathbf{x}$ and conditional covariance is smaller than the total marginal covariance. By *smaller,* we mean that $\boldsymbol{\Omega}_{12} \boldsymbol{\Omega}_{22}^{-1} \boldsymbol{\Omega}_{12}'$ is a nonnegative definite matrix. This is equivalent to saying

$$\mathbf{u}' \boldsymbol{\Omega}_{12} \boldsymbol{\Omega}_{22}^{-1} \boldsymbol{\Omega}_{12}' \mathbf{u} \geq 0 \quad \forall \mathbf{u}. \tag{4.10}$$

Indeed, letting $\mathbf{v} = \boldsymbol{\Omega}_{12}' \mathbf{u}$, we can write $\mathbf{u}' \boldsymbol{\Omega}_{12} \boldsymbol{\Omega}_{22}^{-1} \boldsymbol{\Omega}_{12}' \mathbf{u} = \mathbf{v}' \boldsymbol{\Omega}_{22}^{-1} \mathbf{v}$. But $\boldsymbol{\Omega}_{22}^{-1}$ is positive definite since $\boldsymbol{\Omega}_{22}$ is positive definite. This implies that $\mathbf{v}' \boldsymbol{\Omega}_{22}^{-1} \mathbf{v} \geq 0$ and (4.10).

Sometimes we want to predict random variable Y given the $m \times 1$ random vector $\mathbf{X} = \mathbf{x}$. Rewrite (4.8) and (4.9) using Y and $\mathbf{X} = \mathbf{x}$ notation. Denote μ_y

and $\boldsymbol{\mu}_x$ as means of Y and $\mathbf{X}$, σ_y^2 and $\boldsymbol{\Omega}_x$ the variance of Y and the $m \times m$ covariance matrix of $\mathbf{X}$, and $\boldsymbol{\omega}_{yx}$ as the $m \times 1$ covariance vector between Y and $\mathbf{X}$. Then $Y|\mathbf{X} = \mathbf{x}$ has a normal distribution with the mean and variance given by

$$E(Y|\mathbf{X} = \mathbf{x}) = \mu_y + \boldsymbol{\omega}_{yx}'\boldsymbol{\Omega}_x^{-1}(\mathbf{x} - \boldsymbol{\mu}_x), \quad \text{var}(Y|\mathbf{X} = \mathbf{x}) = \sigma_y^2(1 - \rho^2), \quad (4.11)$$

where

$$\rho^2 = \sigma_y^{-2}\boldsymbol{\omega}_{yx}'\boldsymbol{\Omega}_x^{-1}\boldsymbol{\omega}_{yx} \quad (4.12)$$

is the coefficient of determination that tells the proportion of variance of Y explained by $\mathbf{X}$.

Example 4.9 *Bivariate as multivariate normal. Demonstrate that formula (4.7) gives the previously known formulas for conditional regression and variance for the bivariate normal distribution from Section 3.5.*

Solution. We need to show that (4.7) turns into (3.30). For the bivariate distribution we have $\mathbf{X}_1 = Y$ and $\mathbf{X}_2 = X$ and $\mathbf{x}_1 = y$ and $\mathbf{x}_2 = x$. Furthermore, $\boldsymbol{\mu}_1 = \mu_y$, $\boldsymbol{\mu}_2 = \mu_x$, $\boldsymbol{\Omega}_{12} = \sigma_{xy} = \rho\sigma_x\sigma_y$, $\boldsymbol{\Omega}_{11} = \sigma_y^2$, and $\boldsymbol{\Omega}_{22} = \sigma_x^2$. Hence the right hand side of (4.7) turns into $\mu_y + (\rho\sigma_x\sigma_y)\sigma_x^{-2}(x - \mu_x) = \mu_y + \rho\frac{\sigma_y}{\sigma_x}(x - \mu_x)$. Similarly, the right-hand side of (4.9) turns into $\sigma_y^2 - (\rho\sigma_x\sigma_y)\sigma_x^{-2}(\rho\sigma_x\sigma_y) = (1 - \rho^2)\sigma_y^2$. $\square$

The next example illustrates formulas (4.11) and (4.12) for time series prediction.

Example 4.10 *Time series prediction. The temperature in March (*`foold. csv`*) follows an autoregression of the first order, $T_{t+1} - \mu = \rho(T_t - \mu) + \varepsilon_t$, where $E(T_t) = \mu = 33° F$ is the average temperature in March, $E(\varepsilon_t) = 0$, $\varepsilon_t \overset{\text{iid}}{\sim} \mathcal{N}(0, \sigma^2)$ with $\sigma = 3° F$, and $\rho = 0.8$ (all variables have normal distribution), $t \le n = 31$ and T_{32} is the temperature on the April Fool's day (April 1). (a) Predict the temperature on the Fool's day using the March temperature data. (b) Find the probability that the temperature on the Fool's day will be below freezing.*

Solution. The sequence of random variables $\{T_t\}$ defined by the autoregression with $|\rho| < 1$ is an example of a stationary stochastic process; see Figure 4.2 produced by the R function `foold`. The mean and variance of such process are constant. To prove that $\text{var}(T_t) = \text{const}$, square both sides of the autoregression and take the expected value. Since T_t and ε_t do not correlate, we obtain $\sigma_T^2 = \rho^2\sigma_T^2 + \sigma^2$, that is, $\text{var}(T_t) \overset{\text{def}}{=} \sigma_T^2 = \sigma^2/(1 - \rho^2)$. Now we define the needed quantities in the notation of (4.11). We have $Y = T_{32}$ and $\mathbf{x} = (T_{31}, T_{30}, ..., T_1)$, so that marginal means are $\mu_Y = 33$ and $\boldsymbol{\mu}_x = 33 \times \mathbf{1}$ (note that the temperature vector is written in descending order starting from March 31). Derive matrix $\boldsymbol{\Omega}_y$ and vector $\boldsymbol{\omega}_{yx}$. First, we find the covariance between neighboring temperature observations,

$$\text{cov}(T_{t+1}, T_t) = E\left[\rho(T_t - \mu)^2 + (T_t - \mu)\varepsilon_t\right] = \rho \times \text{var}(T_t) = \rho\frac{\sigma^2}{1 - \rho^2},$$

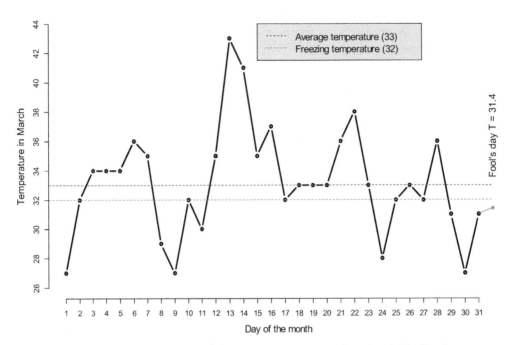

Figure 4.2: *Prediction of the temperature on the April Fool's day.*

and generally,

$$(\boldsymbol{\Omega}_y)_{t,t+\tau} = \text{cov}(T_{t+\tau}, T_t) = \rho^\tau \frac{\sigma^2}{1 - \rho^2}.$$

Thus, the (i, j)th element of matrix $\boldsymbol{\Omega}_y$ is $\sigma^2(1 - \rho^2)^{-1}\rho^{|i-j|}$ and the ith element of vector $\boldsymbol{\omega}_{yx}$ is $\sigma^2(1 - \rho^2)^{-1}\rho^i$. Matrix $\boldsymbol{\Omega}_y$ is a special matrix, called a Toeplitz matrix, because elements parallel to the main diagonal are the same. The inverse of such matrix is also a symmetric Toeplitz two-band matrix with nonzero elements on the main and adjacent diagonals, but other elements are zero. Skipping some algebra, we arrive at the prediction as conditional mean,

$$\widehat{T}_{FD} = (1 - \rho)\mu + \rho T_n. \tag{4.13}$$

This formula means that ρ serves as the coefficient at the linear combination between the average temperature, μ, and the present value of the time series, $T_n = 31$. In the particular case when there is a very strong correlation between present and past value ($\rho \simeq 1$), the best prediction for tomorrow is yesterday's temperature. On the other hand, if there is a very weak correlation ($\rho \simeq 0$), the best prediction is the average temperature. Since $\rho = 0.8$, $\mu = 33$ and $T_{31} = 31°F$ we have $(1 - 0.8) \times 33 + 0.8 \times 31 = 31.4$, just below freezing. The variance of $\widehat{T}_{FD}$ can be found either from formula (4.11) or directly from formula (4.13) as conditional variance,

$$\text{var}(\widehat{T}_{FD}) = \rho^2 \times \text{var}(T_n) = \sigma^2 \frac{\rho^2}{1 - \rho^2} = 16.$$

The probability that there will be a freezing temperature on the Fool's day is $\Phi((32-31.4)/4) = 0.56$. □

Sometimes conditional independence emerges in connection with a latent variable or confounder. Let the association between random variables X and Y be studied, and let there be another variable Z that is related to both X and Z but not of interest. Variables X and Y are *marginally independent* if the correlation coefficient between X and Y is zero. When the correlation coefficient between X and Y is zero in the conditional distribution given $Z = z$, we say that X and Y are *conditionally independent*. The following example provides more technical detail on the difference between marginal and conditional independence, and the following section introduces the partial correlation coefficient as a measure of conditional dependence.

Example 4.11 *Conditional independence for normally distributed vectors. Use Theorem 4.8 to argue that conditional and marginal independence are not equivalent, and demonstrate this by the vector $\mathbf{X}^{3\times 1}=(X+Z, Y+Z, Z)'$ where X, Y, X are iid standard normal random variables.*

Solution. To simplify, we shall assume that $\mathbf{X}_2 = X_2$ is a univariate random variable. Then $\mathbf{\Omega}_{12}$ is a $1 \times m_1$ vector and $\mathbf{\Omega}_{22} = \sigma_2^2$ is the variance of X_2. If the components of vector $\mathbf{X}_1$ are independent, then matrix $\mathbf{\Omega}_{11}$ is diagonal. For the components of vector $\mathbf{X}_1|(X_2 = x)$ to be independent, matrix $\mathbf{\Omega}_{11} - \sigma_2^{-2}\mathbf{\Omega}_{12}'\mathbf{\Omega}_{12}$ must be diagonal. In other words, marginal independence does not imply conditional independence and vice versa. The matrices cannot be both diagonal unless $\mathbf{\Omega}_{12} = \mathbf{0}$, i.e. components of $\mathbf{X}_1$ and X_2 are all independent.

The first two components of vector $\mathbf{X}$ are dependent with correlation coefficient $1/2$, but

$$\mathbf{\Omega}_{11} - \sigma_2^{-2}\mathbf{\Omega}_{12}'\mathbf{\Omega}_{12} = \begin{bmatrix} 2 & 1 \\ 1 & 2 \end{bmatrix} - \begin{bmatrix} 1 \\ 1 \end{bmatrix}\begin{bmatrix} 1 \\ 1 \end{bmatrix}' = \begin{bmatrix} 1 & 0 \\ 0 & 1 \end{bmatrix}.$$

Since the off-diagonal element is zero, the conditional distribution of $(X + Z)|Z$ and $(Y + Z)|Z$ is bivariate normal with zero correlation. That is, $X + Z$ and $Y + Z$ are conditionally independent. □

Sometimes, random variables X and Y are correlated because there is a third (maybe latent, invisible) variable Z that contributes to correlation between X and Y. Partial correlation is a new concept: it does not apply to the bivariate distribution because at least three random variables must be involved.

Definition 4.12 *A partial correlation coefficient is the correlation coefficient for conditional distribution, or*

$$\rho_{ij}|\mathbf{X}_2 = \frac{\sigma_{ij|\mathbf{X}_2}}{\sigma_{i|\mathbf{X}_2}\sigma_{j|\mathbf{X}_2}},$$

where $\sigma_{ij|\mathbf{X}_2}$, $\sigma_{i|\mathbf{X}_2}$, and $\sigma_{j|\mathbf{X}_2}$ are the $(i,j), (i,i)$, and (j,j) elements of matrix (4.9), $i, j = 1, ..., m_1$.

Partial correlation purifies the standard correlation by eliminating the influence of others. To distinguish between two types of correlations, we will refer to marginal (standard) and conditional (partial) correlation.

Example 4.13 Partial correlation with three RVs. *(a) Provide the closed-form formulas for partial correlation between X and Y conditional on Z for the trivariate normal distribution. (b) Prove that*

$$\rho_{XY|Z} = -\frac{\rho^{12}}{\sqrt{\rho^{11}\rho^{22}}},$$

where ρ^{jk} denotes the (j,k)th element of the inverse correlation matrix.

Solution. (a) Write the covariance matrix of the 3×1 vector $(X, Y, Z)'$ as

$$\Omega = \begin{bmatrix} \sigma_x^2 & \rho_{xy}\sigma_x\sigma_y & \rho_{xz}\sigma_x\sigma_z \\ \rho_{xy}\sigma_x\sigma_y & \sigma_y^2 & \rho_{yz}\sigma_y\sigma_z \\ \rho_{xz}\sigma_x\sigma_z & \rho_{yz}\sigma_y\sigma_z & \sigma_z^2 \end{bmatrix}.$$

The conditional covariance matrix, as follows from (4.9), takes the form

$$\begin{bmatrix} \sigma_x^2 & \rho_{xy}\sigma_x\sigma_y \\ \rho_{xy}\sigma_x\sigma_y & \sigma_y^2 \end{bmatrix} - \frac{1}{\sigma_z^2}\begin{bmatrix} \rho_{xz}\sigma_x\sigma_z \\ \rho_{yz}\sigma_y\sigma_z \end{bmatrix}\begin{bmatrix} \rho_{xz}\sigma_x\sigma_z \\ \rho_{yz}\sigma_y\sigma_z \end{bmatrix}'$$

$$= \begin{bmatrix} \sigma_x^2 - \sigma_z^{-2}(\rho_{xz}\sigma_x\sigma_z)^2 & \rho_{xy}\sigma_x\sigma_y - \sigma_z^{-2}(\rho_{xz}\sigma_x\sigma_z)\rho_{yz}\sigma_y\sigma_z \\ \rho_{xy}\sigma_x\sigma_y - \sigma_z^{-2}(\rho_{xz}\sigma_x\sigma_z)\rho_{yz}\sigma_y\sigma_z & \sigma_y^2 - \sigma_z^{-2}(\rho_{yz}\sigma_y\sigma_z)^2 \end{bmatrix}.$$

The partial correlation coefficient is the ratio of the $(1,2)$th diagonal element to the square root of the product of the diagonal elements:

$$\rho_{XY|Z} = \frac{\rho_{xy}\sigma_x\sigma_y - \sigma_z^{-2}(\rho_{xz}\sigma_x\sigma_z)\rho_{yz}\sigma_y\sigma_z}{\sqrt{\sigma_x^2 - \sigma_z^{-2}(\rho_{xz}\sigma_x\sigma_z)^2}\sqrt{\sigma_y^2 - \sigma_z^{-2}(\rho_{yz}\sigma_y\sigma_z)^2}} = \frac{\rho_{xy} - \rho_{xz}\rho_{yz}}{\sqrt{(1 - \rho_{xz}^2)(1 - \rho_{yz}^2)}}.$$

$$(4.14)$$

(b) The inverse correlation matrix is

$$\begin{bmatrix} 1 & \rho_{xy} & \rho_{xz} \\ \rho_{xy} & 1 & \rho_{yz} \\ \rho_{xz} & \rho_{yz} & 1 \end{bmatrix}^{-1} = \frac{1}{D}\begin{bmatrix} 1 - \rho_{yz}^2 & \rho_{xz}\rho_{yz} - \rho_{xy} & \rho_{xy}\rho_{yz} - \rho_{xz} \\ \rho_{xz}\rho_{yz} - \rho_{xy} & 1 - \rho_{xz}^2 & \rho_{xy}\rho_{xz} - \rho_{yz} \\ \rho_{xy}\rho_{yz} - \rho_{xz} & \rho_{xy}\rho_{xz} - \rho_{yz} & 1 - \rho_{xy}^2 \end{bmatrix},$$

where $D = 1 - \rho_{xy}^2 - \rho_{xz}^2 - \rho_{yz}^2 + 2\rho_{xy}\rho_{xz}\rho_{yz}$ is the determinant of the correlation matrix. Thus,

$$-\frac{\rho^{12}}{\sqrt{\rho^{11}\rho^{22}}} = -\frac{\rho_{xz}\rho_{yz} - \rho_{xy}}{\sqrt{(1 - \rho_{yz}^2)(1 - \rho_{xz}^2)}} = \rho_{XY|Z}.$$

$\square$

This rule for computation of the partial correlation coefficient holds in the general case, as stated below.

Theorem 4.14 *Denote $\rho_{ij|\cdot}$ the partial correlation coefficient between X_i and X_j conditional on the other components in the multivariate normal distribution and by $\mathbf{R}$ the correlation matrix. Then*

$$\rho_{ij|\cdot} = -\frac{\rho^{ij}}{\sqrt{\rho^{ii}\rho^{jj}}}$$

where ρ^{ij} is the (i, j)th element of matrix $\mathbf{R}^{-1}$.

An alternative way to compute a partial correlation coefficient is to follow the definition: (i) compute conditional means (4.8) for X_i and X_j using $\mathbf{X}_2^{(m-2) \times 1}$ as the remaining components and (ii) compute standard correlation coefficient between the residuals. This is how the partial correlation was computed in the above example.

Definition 4.15 X_i and X_j *are said to be conditionally independent (uncorrelated) if $\rho_{ij|\cdot} = 0$.*

As follows from (4.14), X and Y are conditionally independent of Z if $\rho_{xy} = \rho_{xz}\rho_{yz}$. Consequently, if X and Y are marginally uncorrelated but correlated with Z, they are conditionally correlated, or in general terms, $\rho_{ij|\cdot} = 0$ and $\rho_{ij} = 0$ are not equivalent.

Example 4.16 *Height of husband and wife. From a genetic standpoint, there is no explanation why the correlation between height of husband and wife is positive (the positive correlation was discovered by Francis Galton more than one hundred years ago). Use partial correlation to explain this phenomenon by referring to "attractiveness" as the latent variable.*

Solution. Consider the triple of random variables (H, W, A), where the components represent the height of husband and wife, and A represent the attractiveness between man and woman before getting married, a latent variable. We hypothesize that similarity of heights between partners is an attractive feature for both sexes. In other words, tall women like tall men and short men like short women, on average. Statistically, this means that $\rho_{HA} > 0$ and $\rho_{WA} > 0$. If couples married randomly without preference, then $\rho_{HW|A} = 0$, and, from formula (4.14), $\rho_{HW} = \rho_{HA} \times \rho_{WA} > 0$. In short, preference of height similarity at marriage leads to positive correlation although the partial correlation between man and woman height is zero.

4.1.3   Multivariate CLT

If $\mathbf{X}_i$ are iid random vectors with the mean vector $\boldsymbol{\mu}$ and covariance matrix $\boldsymbol{\Omega}$ and $\overline{\mathbf{X}}_n = \sum_{i=1}^n \mathbf{X}_i/n$ is the vector average, then the multivariate central limit theorem (CLT) says that $\sqrt{n}(\overline{\mathbf{X}}_n - \boldsymbol{\mu}) \simeq \mathcal{N}(\mathbf{0}, \boldsymbol{\Omega})$ when $n \to \infty$. The sign $\simeq$ means convergence in distribution. That is, the joint cdf of the vector random variable at left converges to the cdf of the multivariate normal distribution specified at right. Equivalently, one can rewrite this as $\boldsymbol{\Omega}^{-1/2}\mathbf{Z}_n \simeq \mathcal{N}(\mathbf{0}, \mathbf{I})$, where $\mathbf{Z}_n = \sqrt{n}(\overline{\mathbf{X}}_n - \boldsymbol{\mu})$ is the standardized random vector. In the spirit of the extended CLT, Theorem 2.58, we may have $\boldsymbol{\Omega}_n = \mathrm{cov}(\sqrt{n}\overline{\mathbf{X}}_n)$ and $\lim_{i \to \infty} \boldsymbol{\Omega}_n = \boldsymbol{\Omega}$, the limit covariance matrix.

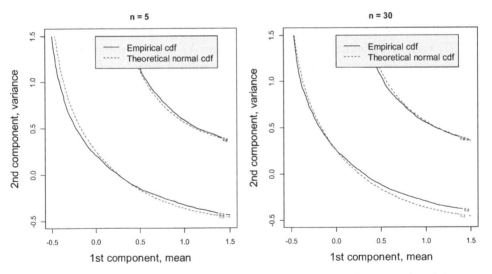

Figure 4.3: *The contours of the empirical and theoretical (approximate) bivariate normal cdf for the sample sizes $n = 5$ and $n = 30$. See the* R *function* multCLT.

Example 4.17 *Bivariate extended CLT.* *Random variables* $\{X_i, i = 1, ..., n\}$ *are independent and uniformly distributed on* $[0, 1]$. *Using simulations, demonstrate that the* 2×1 *vector*

$$\mathbf{U}_n^{2 \times 1} = \sqrt{n} \left[\begin{array}{c} (\overline{X}_n - 0.5)\sqrt{12} \\ 12 \left[\sum_{i=1}^n (X_i - \overline{X}_n)^2/n - (n-1)/(12n) \right] \end{array} \right]$$

has an approximate standard bivariate normal distribution with independent components when $n \to \infty$.

Solution. First we recall that the mean and variance of $\mathcal{R}(0,1)$ are 0.5 and $1/\sqrt{12}$, respectively; that explains why we center the components. Second, the multivariate CLT does not apply directly because terms $\{X_i - \overline{X}_n\}$ do correlate, so we apply the extended multivariate CLT. Now, we derive the limit covariance matrix $\mathbf{\Omega}^{2\times 2}$ (to simplify the derivation one may assume that X_i are distributed on the interval from $-1/2$ to $1/2$). (a) It is easy to confirm that the components of $\mathbf{U}_n$ have zero mean and components have zero covariance, $\Omega_{12} = 0$. (b) The variance of $\overline{X}_n - 0.5$ is $1/(12n)$ and the variance of $\sum_{i=1}^n (X_i - \overline{X}_n)^2/n$ is $1/(12^2 n)$. To graphically compare the empirical and approximate bivariate normal cdfs, we plot contours with the same levels (in our case 0.3 and 0.6). See R function `multCLT` and Figure 4.3, where the x-axis corresponds to the first component of $\mathbf{U}_n$ and the y-axis corresponds to the second component, the approximation is satisfactory even for small n.

Problems

1. Random variables X, Y, and Z are iid standard normal variables. Define a 3×1 random vector as $\mathbf{U} = (X+Y-Z, Y-X, X+Z)'$. Find the distribution of the random variable $H = \mathbf{U}'\mathbf{1}$ (a) using Theorem 4.2 by expressing $\mathbf{U}$ as a linear transformation of $(X, Y, Z)'$ and (b) directly. Show that both methods yield the same the distribution.

2. Prove that any pair of normally distributed random variables can be transformed into uncorrelated random variables upon plane rotation (derive the angle). [Hint: Consult the Appendix.]

3. Two vectors of length 3 are generated from the 5×1 normally distributed vector $\mathbf{X}$ using linear transformation $\mathbf{AX}$ and $\mathbf{BX}$ where $\mathbf{A}$ and $\mathbf{B}$ are matrices of full rank. Is it possible that the two vectors are independent? If your answer is positive, give an example; otherwise prove that it is impossible.

4. Prove that conditional distribution (4.7) turns into marginal distributions when $\mathbf{X}_1$ and $\mathbf{X}_2$ are independent.

5. Given random variables $X_i \overset{\text{iid}}{\sim} \mathcal{N}(\mu, \sigma^2)$, $i = 1, 2, ..., n$, find $E(X_1 | X_1 + X_2 + \cdots + X_n = x)$ and $\text{var}(X_1 | X_1 + X_2 + \cdots + X_n = x)$ using formulas (4.8) and (4.9).

6. Modify Example 4.10 to predict the temperature on April 2.

7. Generate 200 3D random points defined in Problem 1 and display them using function `mn3` for optimal visualization (find the best `theta`).

8. Prove that the multivariate moment generating function (MGF) of $\mathbf{X} \sim \mathcal{N}(\boldsymbol{\mu}, \boldsymbol{\Omega})$ is $M(\mathbf{t}) = \exp(0.5\mathbf{t}'\boldsymbol{\Omega}\mathbf{t} + \mathbf{t}'\boldsymbol{\mu})$. Show that the results of Section 3.10.2 produce $E(\mathbf{X}) = \boldsymbol{\mu}$ and $\text{cov}(\mathbf{X}) = \boldsymbol{\Omega}$.

9. The trivariate normal distribution is defined as $(U + \nu Z, V + \nu Z, Z)$, where U, V, and Z independent normally distributed random variables with zero mean and unit variance. (a) Compute the partial correlation coefficient between the first and second component as a function of ν on the interval $[-2, 2]$ using Example 4.13 and Theorem 4.14. (b) Plot the standard and partial correlation coefficients on the same graph using `matplot`. Explain the result.

10. The correlation matrix for a trivariate normal distribution of (X, Y, Z) is such that $\rho_{12} = 0$. (a) Show that X and Y may be conditionally dependent by showing that $\rho_{X,Y|Z} \neq 0$. (b) Show that X and Y are conditionally independent if, in addition, $\rho_{13} = 0$ or $\rho_{23} = 0$.

11. The triple (X, Y, Z) has a trivariate normal distribution. (a) Is it true that $E(Y|X, Z) = E(Y|X)$ if X and Z are independent. (b) Is it true that $E(Y|X, Z) = E(Y|X)$ if Y and Z are independent?

12. Apply the extended multivariate CLT in Example 4.17 by replacing $\boldsymbol{\Omega}$ with its estimate. [Hint: Use the R function `cov`.]

13. The triple (X, Y, Z) s normally distributed and independent. Prove that $U = X + Z$ and $V = Y + Z$ are conditionally independent using Theorem 4.8.

14. In Example 4.10, find the probability that on April 2 the temperature will be above freezing.

4.2   Chi-square distribution

The χ^2-distribution (or chi-square distribution) is the distribution of the sum of squared independent standard normal random variables. In mathematical terms, if $\{X_i, i = 1, 2, ..., n\}$ are iid random variables with distribution $\mathcal{N}(0, 1)$, then $\sum_{i=1}^{n} X_i^2$ has a χ^2-distribution with n degrees of freedom (df). We shall explain later why n is called *degrees of freedom* (df). To indicate the dependence on n, we use the notation $\chi^2(n)$. Symbolically,

$$\chi^2(n) = \sum_{i=1}^{n} \mathcal{N}^2(0, 1).$$

In a very special case, $n = 1$, we have $\mathcal{N}^2(0, 1) = \chi^2(1)$.

The distribution of $\chi^2(n)$ is a gamma distribution (Section 2.6) with parameters $\alpha = n/2$ and $\lambda = 1/2$. Therefore, the density given by

$$f(s; n) = \frac{1}{2^{n/2}\Gamma(n/2)} s^{n/2-1} e^{-s/2}, \quad s \geq 0. \qquad (4.15)$$

The cdf of the chi-square distributed random variable is $C(x; n) = \int_0^x f(s; n) ds$. The inverse cdf, the pth quantile, is denoted as $C^{-1}(p; n)$; all functions are available in R.

Example 4.18 *Chi-square distribution with df = 1. Derive the cdf and density of the chi-square distribution with df = 1 and check that the density coincides with (4.15).*

Solution. Let $s \geq 0$. The cdf is $F(s) = \Pr(Z^2 < s) = \Pr(|Z| < \sqrt{s}) = 1 - 2\Phi(-\sqrt{s})$, and the density is

$$\frac{dF}{ds} = \frac{1}{\sqrt{s}} \phi(-\sqrt{s}) = \frac{1}{\sqrt{2\pi s}} e^{-s/2}.$$

Using the fact that $\Gamma(1/2) = \sqrt{\pi}$ (Section 2.6) and (4.15), we obtain

$$f(s; 1) = \frac{1}{\sqrt{2}\Gamma(1/2)} s^{-1/2} e^{-s/2} = \frac{1}{\sqrt{2\pi s}} s^{-1/2} e^{-s/2},$$

the densities coincide, and $F(x) = C_1(x)$.   □

Three density functions are depicted in Figure 4.4. For $n = 1$, the density goes to ∞ when $s \to 0$; for $n = 2$ the chi-square distribution turns into the exponential distribution.

The mean and variance of $\chi^2(n)$ are elementary to find. Indeed, since $E(X_i^2) = 1$, we have

$$E(\chi^2(n)) = E\left(\sum_{i=1}^n X_i^2\right) = \sum_{i=1}^n E(X_i^2) = n.$$

Also, from formula (2.43), we derive $\text{var}(\chi^2(n)) = \sum_{i=1}^n \text{var}(X_i^2) = 2n$ since $\{X_i^2\}$ are independent.

It is instructive to find the mode of the χ^2-distribution. Since the maximum point of a positive function does not change after taking the log transformation, it suffices to solve the equation $(\ln f)' = [(n/2) - 1]/t - 1/2 = 0$, so

$$\text{mode} = n - 2.$$

This means that for $n \leq 2$ the density goes to $+\infty$ when s approaches zero. For $n > 2$ there exists a unique maximum of the density function. Note that the mode is less than the mean for the chi-square distribution. This distribution has a long right tail, especially for small df. We say that the chi-square distribution is skewed right.

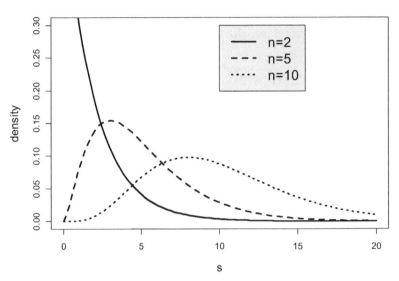

Figure 4.4: *Density functions of the chi-square distribution for $n = 2, 5$ and 10.*

Example 4.19 **MGF for the chi-square distribution.** *Find the MGF of the chi-square distribution (see Section 2.5). Find the first two moments using the MGF and compare with the mean and variance derived above.*

Solution. Using change of variable $u = (1 - 2t)s$ for $t < 1/2$, we obtain

$$
\begin{aligned}
M(t; n) &= \int_0^\infty e^{ts} f(s; n) dt = \frac{1}{2^{n/2} \Gamma(n/2)} \int_0^\infty e^{ts} s^{(n/2)-1} e^{-s/2} ds \\
&= \frac{1}{2^{n/2} \Gamma(n/2)} \int_0^\infty s^{(n/2)-1} e^{-(1-2t)s/2} ds \\
&= \frac{1}{2^{n/2} \Gamma(n/2)} \int_0^\infty \frac{1}{(1-2t)^{n/2}} u^{(n/2)-1} e^{-u/2} du \\
&= \frac{1}{(1-2t)^{n/2}} \frac{1}{2^{n/2} \Gamma(n/2)} \int_0^\infty u^{(n/2)-1} e^{-u/2} du = \frac{1}{(1-2t)^{n/2}}.
\end{aligned}
$$

Thus, the MGF is defined for $-\infty < t < 1/2$. To find the first and second moments, evaluate the first and second derivatives of MGF at $t = 0$:

$$
\begin{aligned}
E(\chi^2(n)) &= \left. \frac{dM(t; n)}{dt} \right|_{t=0} = n(1 - 2t)^{-n/2-1} \Big|_{t=0} = n, \\
E(\chi^2(n))^2 &= \left. \frac{d^2 M(t; n)}{dt^2} \right|_{t=0} = n(n+2)(1-2t)^{-n/2-2} \Big|_{t=0} = n(n+2).
\end{aligned}
$$

The variance is

$$
\text{var}(\chi^2(n)) = E(\chi^2(n))^2 - E^2(\chi^2(n)) = n(n+2) - n^2 = 2n,
$$

as above. □

The chi-square distribution has an important connection with the exponential distribution: the sum of n iid exponentially distributed random variables with the rate parameter λ has $(2\lambda)^{-1}\chi^2(n)$ distribution. More rigorously, the cdf of the sum of n iid exponentially distributed random variables with the rate parameter λ is equal $F(2\lambda x, n)$ where $F(\cdot, n)$ is the cdf of the chi-square distribution with $2n$ degrees of freedom. This fact will be used later to unbiasedly estimate λ and construct an exact confidence interval.

Example 4.20 *Standardized sum of squares.* Let $X_i \overset{iid}{\sim} N(\mu, \sigma^2)$. *Prove that the standardized sum of squares has a chi-square distribution, that is,*

$$\frac{1}{\sigma^2} \sum_{i=1}^{n} (X_i - \mu)^2 \sim \chi^2(n).$$

Solution. Represent $\sigma^{-2} \sum_{i=1}^{n} (X_i - \mu)^2 = \sum_{i=1}^{n} Z_i^2$, where $Z_i = (X_i - \mu)/\sigma \sim \mathcal{N}(0,1)$. Thus, $\sum_{i=1}^{n} Z_i^2 \sim \chi^2(n)$. □

Since $\{X_i^2\}$ are iid with mean 1 and variance 2 from the CLT, we infer that $(\chi^2(n) - n)/\sqrt{2n} \simeq \mathcal{N}(0,1)$, so, approximately

$$\chi^2(n) \simeq \mathcal{N}(n, 2n). \tag{4.16}$$

It immediately follows from the definition of the χ^2-distribution that

$$\chi^2(n_1) + \chi^2(n_2) \sim \chi^2(n_1 + n_2),$$

if the two chi-square distributions are independent. This statement is easy to generalize to an arbitrary sum of chi-square distributions.

R has four built-in functions for the chi-square distribution; see Table 4.1.

Table 4.1. Computation of the chi-square distribution in R

R function	Returns	Explanation
dchisq(x,df)	density (4.15)	df$= n$
pchisq(q,df)	cdf	cdf at q $= x$, $\mathcal{C}(x; n)$
qchisq(p,df)	inverse cdf	p=probability=χ^{-2}
rchisq(n,df)	random sample	$n =$ sample size

The following theorem is fundamental for statistical applications. A symmetric matrix $\mathbf{A}$ is called idempotent if $\mathbf{A}^2 = \mathbf{A}$. An idempotent matrix has eigenvalues 0 or 1; see Appendix.

Theorem 4.21 *Let* $\mathbf{A}$ *be a symmetric* $n \times n$ *idempotent matrix with* $tr(\mathbf{A}) = m$ *and* $\mathbf{Y} = (Y_1, Y_2, ..., Y_n)'$ *be a random vector with iid components distributed as* $\mathcal{N}(0, 1)$, *or equivalently in matrix notation,* $\mathbf{Y} \sim \mathcal{N}(\mathbf{0}, \mathbf{I})$. *Then*

$$\mathbf{Y}'\mathbf{A}\mathbf{Y} \sim \chi^2(m).$$

Proof. Since $\mathbf{A}$ is symmetric and idempotent matrix, it can be represented as $\mathbf{A} = \sum_{i=1}^{m} \mathbf{p}_i \mathbf{p}_i'$, where $\{\mathbf{p}_i, i = 1, ..., m\}$ are mutually orthogonal eigenvectors of unit length (see Appendix). Then

$$\mathbf{Y}'\mathbf{A}\mathbf{Y} = \sum_{i=1}^{m} (\mathbf{Y}'\mathbf{p}_i)(\mathbf{p}_i'\mathbf{Y}) = \sum_{i=1}^{m} (\mathbf{Y}'\mathbf{p}_i)^2 = \sum_{i=1}^{m} Z_i^2,$$

where $Z_i = \mathbf{Y}'\mathbf{p}_i$. But $\{Z_i, i = 1, 2, ..., m\}$ are independent and normally distributed $\mathcal{N}(0, 1)$. Indeed, they are normally distributed as linear combination of normally distributed random variables. Furthermore, $E(Z_i) = 0$ and since $\text{cov}(\mathbf{Y}) = \mathbf{I}$, due to Theorem 4.2, we have

$$\text{var}(Z_i) = \mathbf{p}_i'\text{cov}(\mathbf{Y})\mathbf{p}_i = \mathbf{p}_i'\mathbf{p}_i = \|\mathbf{p}_i\|^2 = 1.$$

They do not correlate and therefore independent because due to Theorem 4.4 for $i \neq j$ we have

$$\text{cov}(Z_i, Z_j) = E(\mathbf{Y}'\mathbf{p}_i\mathbf{Y}'\mathbf{p}_j) = \mathbf{p}_i'\text{cov}(\mathbf{Y})\mathbf{p}_j = \mathbf{p}_i'\mathbf{p}_j = 0.$$

Finally, $\mathbf{Y}'\mathbf{A}\mathbf{Y} = \sum_{i=1}^{m} Z_i^2 \sim \chi^2(m)$. $\square$

Below we apply this theorem to squared sum of the centered data.

Theorem 4.22 *Let* $X_i \overset{iid}{\sim} \mathcal{N}(\mu, \sigma^2)$. *Then*

$$\frac{1}{\sigma^2} \sum_{i=1}^{n} (X_i - \overline{X})^2 \sim \chi^2(n-1). \tag{4.17}$$

Proof. First, we express the left-hand side of (4.17) in terms of $Z_i \sim \mathcal{N}(0, 1)$:

$$\frac{1}{\sigma^2} \sum_{i=1}^{n} (X_i - \overline{X})^2 = \frac{1}{\sigma^2} \sum_{i=1}^{n} ((X_i - \mu) - (\overline{X} - \mu))^2$$

$$= \sum_{i=1}^{n} \left(\frac{X_i - \mu}{\sigma} - \frac{\overline{X} - \mu}{\sigma} \right)^2 = \sum_{i=1}^{n} (Z_i - \overline{Z})^2.$$

Second, we use matrix algebra to represent the quadratic form with $\mathbf{Z} = (Z_1, Z_2, ..., Z_n)'$. Then $\overline{Z} = n^{-1}\mathbf{1}'\mathbf{Z}$, where $\mathbf{1} = (1, 1, ..., 1)'$, so that in vector notation

$$\begin{bmatrix} Z_1 - \overline{Z} \\ Z_2 - \overline{Z} \\ \vdots \\ Z_n - \overline{Z} \end{bmatrix} = \begin{bmatrix} Z_1 \\ Z_2 \\ \vdots \\ Z_n \end{bmatrix} - \overline{Z}\begin{bmatrix} 1 \\ 1 \\ \vdots \\ 1 \end{bmatrix} = \mathbf{Z} - \overline{Z}\mathbf{1} = \mathbf{Z} - n^{-1}\mathbf{1}'\mathbf{Z}\mathbf{1}$$

$$= \mathbf{Z} - n^{-1}\mathbf{1}\mathbf{1}'\mathbf{Z} = (\mathbf{I} - n^{-1}\mathbf{1}\mathbf{1}')\mathbf{Z} = \mathbf{A}\mathbf{Z},$$

where $\mathbf{A} = \mathbf{I} - n^{-1}\mathbf{1}\mathbf{1}'$. This matrix is symmetric and idempotent. Indeed, since $\mathbf{1}'\mathbf{1} = n$, we have

$$\begin{aligned} \mathbf{A}^2 &= (\mathbf{I} - n^{-1}\mathbf{1}\mathbf{1}')(\mathbf{I} - n^{-1}\mathbf{1}\mathbf{1}') = \mathbf{I} - n^{-1}\mathbf{1}\mathbf{1}' - n^{-1}\mathbf{1}\mathbf{1}' + n^{-2}\mathbf{1}\mathbf{1}'\mathbf{1}\mathbf{1}' \\ &= \mathbf{I} - 2n^{-1}\mathbf{1}\mathbf{1}' + n^{-2}\mathbf{1}n\mathbf{1}' = \mathbf{I} - 2n^{-1}\mathbf{1}\mathbf{1}' + n^{-1}\mathbf{1}\mathbf{1}' = \mathbf{I} - n^{-1}\mathbf{1}\mathbf{1}' \end{aligned}$$

and

$$\text{tr}(\mathbf{A}) = \text{tr}(\mathbf{I} - n^{-1}\mathbf{1}\mathbf{1}') = n - n^{-1}\text{tr}(\mathbf{1}\mathbf{1}') = n - n^{-1}n = n - 1.$$

Thus,

$$\sum_{i=1}^{n}(Z_i - \overline{Z})^2 = (\mathbf{Z} - \overline{Z}\mathbf{1})'(\mathbf{Z} - \overline{Z}\mathbf{1}) = \mathbf{Z}'\mathbf{A}^2\mathbf{Z} = \mathbf{Z}'\mathbf{A}\mathbf{Z} \sim \chi^2(n-1)$$

by Theorem 4.21. $\qquad\qquad\qquad\qquad\qquad\qquad\qquad\qquad\qquad\qquad\quad\square$

This result explains why $n - 1$ is called degrees of freedom: for an idempotent matrix, $rank(\mathbf{A}) = \text{tr}(\mathbf{A}) = n - 1$. Thus degrees of freedom is the number of linearly independent columns of matrix $\mathbf{A}$.

Theorem 4.23 *Let* $\mathbf{Y}^{n \times 1} \sim \mathcal{N}(\mathbf{0}, \mathbf{I})$ *and* $\mathbf{A}$ *and* $\mathbf{B}$ *be two* $n \times n$ *symmetric idempotent matrices. If* $\mathbf{AB} = \mathbf{0}$, *then* $\mathbf{Y}'\mathbf{AY}$ *and* $\mathbf{Y}'\mathbf{BY}$ *are independent.*

Proof. The following fact of linear algebra, crucial for the proof, is sometimes referred to as simultaneous diagonalization: if two $n \times n$ symmetric matrices $\mathbf{H}$ and $\mathbf{G}$ commute, that is, $\mathbf{HG} = \mathbf{GH}$, then they can be simultaneously diagonalized by an orthogonal $n \times n$ matrix $\mathbf{P}$. That is, both matrices $\mathbf{PHP}'$ and $\mathbf{PGP}'$ are diagonal ($\mathbf{P}'\mathbf{P} = \mathbf{I}$). The symmetric matrices $\mathbf{A}$ and $\mathbf{B}$ commute because $\mathbf{AB} = \mathbf{0}$, implying that $(\mathbf{AB})' = \mathbf{0}'$. Since $(\mathbf{AB})' = \mathbf{B}'\mathbf{A}' = \mathbf{BA}$, we have $\mathbf{BA} = \mathbf{0}$, so $\mathbf{AB} = \mathbf{0} = \mathbf{BA}$. Following the proof of Theorem 4.21, let $\{\mathbf{p}_i, i = 1, 2, ..., \text{tr}(\mathbf{A})\}$ and $\{\mathbf{q}_j, j = 1, 2, ..., \text{tr}(\mathbf{B})\}$ be the column vectors of matrix $\mathbf{P}$ such that $\mathbf{A} = \sum \mathbf{p}_i\mathbf{p}_i'$ and $\mathbf{B} = \sum \mathbf{q}_j\mathbf{q}_j'$, where $\mathbf{p}_i'\mathbf{q}_j = 0$. This means that the quadratic forms $\mathbf{Y}'\mathbf{AY}$ and $\mathbf{Y}'\mathbf{BY}$ can be expressed as the sum of nonoverlapping sets of squared standard normal variables $\{U_i\}$ and $\{V_j\}$. Therefore, quadratic forms $\mathbf{Y}'\mathbf{AY}$ and $\mathbf{Y}'\mathbf{BY}$ are independent.

Corollary 4.24 *Let* $\mathbf{Y} \sim \mathcal{N}(\mathbf{0}, \mathbf{I})$ *and* $\mathbf{A}$ *and* $\mathbf{B}$ *be two* $n \times n$ *symmetric idempotent matrices such that* $\mathbf{AB} = \mathbf{0}$*. Then* $\mathbf{Y'AY} + \mathbf{Y'BY} \sim \chi^2(tr(\mathbf{A}) + tr(\mathbf{B}))$*.*

Example 4.25 *Weighted sum of squares.* *Let* $\mathbf{Y}^{n \times 1} \sim \mathcal{N}(\boldsymbol{\mu}, \boldsymbol{\Omega})$*. Prove that the weighted sum of squares has the chi-square distribution. Shortly we write* $(\mathbf{Y} - \boldsymbol{\mu})'\boldsymbol{\Omega}^{-1}(\mathbf{Y} - \boldsymbol{\mu}) \sim \chi^2(n)$*.*

Solution. Represent $\mathbf{Z} = \boldsymbol{\Omega}^{-1/2}(\mathbf{Y} - \boldsymbol{\mu})$. Then as follows from Section 4.1.1, $\mathbf{Z} \sim N(\mathbf{0}, \mathbf{I})$ and $(\mathbf{Y} - \boldsymbol{\mu})'\boldsymbol{\Omega}^{-1}(\mathbf{Y} - \boldsymbol{\mu}) = \|\mathbf{Z}\|^2 = \sum_{i=1}^{n} Z_i^2 \sim \chi^2(n)$.

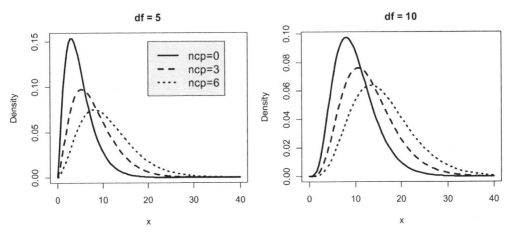

Figure 4.5: *Some densities of the noncentral chi-square distribution.*

4.2.1   Noncentral chi-square distribution

This distribution emerges in computation of power functions when testing statistical hypotheses.

Definition 4.26 *The noncentral chi-square distribution with* $df = n$ *and noncentrality parameter* $\delta = \sum_{i=1}^{n} \mu_i^2$ *is the distribution of*

$$\chi^2(\mathrm{df} = n, \mathrm{ncp} = \delta) = \sum_{i=1}^{n} \mathcal{N}^2(\mu_i, 1).$$

We use ncp to match the syntax used in R with default value ncp=0; see Table 4.1. For example, to compute the cdf of $\chi^2(n, \mathrm{ncp} = \delta)$, we write pchisq(q=x,ncp =delta). Figure 4.5 shows several densities of the noncentral chi-square distribution are shown. The noncentrality parameter moves the density to the right and makes it wider.

Example 4.27 *The cdf for the chi-square distribution with df = 1.* *Derive the cdf of* $\mathcal{N}^2(\mu, 1)$ *and confirm that it can be computed as* pchisq(q=x,df=1, ncp=mu^2)*.*

Solution. Similarly to Example 4.18, we have

$$F(s) = \Pr(X^2 < s) = \Pr(|X| < \sqrt{s}) = \Pr(X < \sqrt{s}) - \Pr(X < -\sqrt{s})$$
$$= \Phi(\sqrt{s} - \mu) - \Phi(-\sqrt{s} - \mu).$$

The R function is found in the file `chidf1.r`.   □

It is not difficult to compute the first two moments of the noncentral chi-square distribution. If $X \sim \chi^2(n, \mathrm{ncp} = \delta)$, then

$$E(X) = n + \delta, \quad \mathrm{var}(X) = 2n + 4\delta. \tag{4.18}$$

4.2.2   Expectations and variances of quadratic forms

Let $\mathbf{y} \sim \mathcal{N}(\boldsymbol{\mu}, \boldsymbol{\Omega})$ be a $n \times 1$ normally distributed random vector with mean vector $\boldsymbol{\mu}$ and $n \times n$ covariance matrix $\boldsymbol{\Omega}$, and let $\mathbf{A}$ and $\mathbf{B}$ be $n \times n$ fixed symmetric matrices. Then

$$E(\mathbf{y}'\mathbf{A}\mathbf{y}) = \mathrm{tr}(\mathbf{A}\boldsymbol{\Omega}) + \boldsymbol{\mu}'\mathbf{A}\boldsymbol{\mu},$$
$$\mathrm{var}(\mathbf{y}'\mathbf{A}\mathbf{y}) = 2tr(\mathbf{A}\boldsymbol{\Omega})^2 + 4\boldsymbol{\mu}'\mathbf{A}\boldsymbol{\Omega}\mathbf{A}\boldsymbol{\mu},$$
$$\mathrm{cov}(\mathbf{y}, \mathbf{y}'\mathbf{A}\mathbf{y}) = 2\boldsymbol{\Omega}\mathbf{A}\boldsymbol{\mu},$$
$$E((\mathbf{y} - \boldsymbol{\mu})'\mathbf{A}(\mathbf{y} - \boldsymbol{\mu})(\mathbf{y} - \boldsymbol{\mu})'\mathbf{B}(\mathbf{y} - \boldsymbol{\mu})) = \mathrm{tr}(\mathbf{A}\boldsymbol{\Omega})\mathrm{tr}(\mathbf{B}\boldsymbol{\Omega}) + 2tr(\mathbf{A}\boldsymbol{\Omega}\mathbf{B}\boldsymbol{\Omega}).$$

It is easier to prove these results starting from the assumption that $\boldsymbol{\mu} = \mathbf{0}$ and $\boldsymbol{\Omega} = \mathbf{I}$, and therefore $\mathbf{z} \sim \mathcal{N}(\mathbf{0}, \mathbf{I})$. Indeed, the first equation

$$E(\mathbf{z}'\mathbf{A}\mathbf{z}) = E\left(\sum_{i,j=1}^{n} A_{ij} z_i z_j\right) = \sum_{i=1}^{n} A_{ii} z_i^2 = \sum_{i=1}^{n} A_{ii} = \mathrm{tr}(\mathbf{A}).$$

Now we express $\mathbf{y} = \boldsymbol{\Omega}^{1/2}\mathbf{z} + \boldsymbol{\mu}$, and using the previous formula, we obtain

$$E(\mathbf{y}'\mathbf{A}\mathbf{y}) = E(\mathbf{z}'\boldsymbol{\Omega}^{1/2}\mathbf{A}\boldsymbol{\Omega}^{1/2}\mathbf{z}) + \boldsymbol{\mu}'\mathbf{A}\boldsymbol{\mu} = \mathrm{tr}(\boldsymbol{\Omega}^{1/2}\mathbf{A}\boldsymbol{\Omega}^{1/2}) = \mathrm{tr}(\mathbf{A}\boldsymbol{\Omega}) + \boldsymbol{\mu}'\mathbf{A}\boldsymbol{\mu}.$$

The second formula is a generalization of the well-known result $\mathrm{var}(y^2) = 2\sigma^4$, where y is a normally distributed random variable with variance σ^2. This formula was derived earlier (2.44) using the MGF.

Other three formulas are proved in a similar way.

4.2.3   Kronecker product and covariance matrix

In studying the statistical properties of estimators in statistics, we often must find variances or covariances of matrices of quadratic forms (some facts on variances of quadratic forms are presented in the proceeding section). Specifically, the problem is to find the covariance matrix of $\mathbf{X}\mathbf{X}'$, where $\mathbf{X}$ is the $m \times 1$ normally distributed vector, $\mathbf{X} \sim \mathcal{N}(\mathbf{0}, \boldsymbol{\Omega})$. To solve this problem, one has to define the covariance matrix of a random matrix. To answer this question, we to express the matrix as vector using the vec operator and then use its connection with the Kronecker matrix product; see Section 10.5. Following this idea, define the covariance matrix of $\mathbf{XX}'$ as the covariance of $\mathrm{vec}(\mathbf{XX}') = \mathbf{X} \otimes \mathbf{X}$. The following fact is crucial here: if $\mathbf{X} \sim \mathcal{N}(\mathbf{0}, \boldsymbol{\Omega})$, then

$$\mathrm{cov}(\mathbf{X} \otimes \mathbf{X}) = 2(\boldsymbol{\Omega} \otimes \boldsymbol{\Omega}). \tag{4.19}$$

Example 4.28 *Covariance of the Kronecker product. Check formula (4.19) for $m = 1$ and $m = 2$.*

Solution. For $m = 1$, we formally have $\boldsymbol{\Omega} = \sigma^2$ and formula (4.19) turns into $\mathrm{var}(Z^2) = 2\sigma^4$, where $Z \sim \mathcal{N}(0, \sigma^2)$. We knew this fact from Section 2.7. For $m = 2$, we check the formula in the simplest case when components of vector $\mathbf{X}$ are independent and have unit variance:

$$\mathbf{Z} \sim \mathcal{N}\left(\begin{bmatrix} 0 \\ 0 \end{bmatrix}, \begin{bmatrix} 1 & 0 \\ 1 & 0 \end{bmatrix} \right).$$

From the definition of the Kronecker product

$$\mathbf{Z} \otimes \mathbf{Z} = \begin{bmatrix} Z_1 \begin{bmatrix} Z_1 \\ Z_2 \end{bmatrix} \\ Z_2 \begin{bmatrix} Z_1 \\ Z_2 \end{bmatrix} \end{bmatrix} = \begin{bmatrix} Z_1^2 \\ Z_1 Z_2 \\ Z_2 Z_1 \\ Z_2^2 \end{bmatrix}.$$

Therefore (the matrix is symmetric so that we show only the upper part),

$$\mathrm{cov}(\mathbf{Z} \otimes \mathbf{Z}) = \begin{bmatrix} \mathrm{var}(Z_1^2) & \mathrm{cov}(Z_1^2, Z_1 Z_2) & \mathrm{cov}(Z_1^2, Z_2 Z_1) & \mathrm{cov}(Z_1^2, Z_2^2) \\ & \mathrm{var}(Z_1 Z_2) & \mathrm{cov}(Z_1 Z_2, Z_2 Z_1) & \mathrm{cov}(Z_1 Z_2, Z_2^2) \\ & & \mathrm{var}(Z_2 Z_1) & \mathrm{cov}(Z_2 Z_1, Z_2^2) \\ & & & \mathrm{var}(Z_2^2) \end{bmatrix}$$

$$= \begin{bmatrix} 2\sigma_1^4 & \mathrm{cov}(Z_1^2, Z_1 Z_2) & \mathrm{cov}(Z_1^2, Z_2 Z_1) & \mathrm{cov}(Z_1^2, Z_2^2) \\ & \mathrm{var}(Z_1 Z_2) & \mathrm{cov}(Z_1 Z_2, Z_2 Z_1) & \mathrm{cov}(Z_1 Z_2, Z_2^2) \\ & & \mathrm{var}(Z_2 Z_1) & \mathrm{cov}(Z_2 Z_1, Z_2^2) \\ & & & \mathrm{var}(Z_2^2) \end{bmatrix}.$$

The elementary element-by-element derivation relies on the fact that, since Z_1 and Z_2 are independent, we have $E(Z_1^2 Z_2^2) = E(Z_1^2)E(Z_2^2) = 1$ and $E(Z_1^3 Z_2) =$

$E(Z_1^3)E(Z_2) = 0$. Therefore,

$$\text{var}(Z_1^2) = 2, \quad \text{var}(Z_2^2) = 2,$$
$$\text{cov}(Z_1^2, Z_1 Z_2) = E[(Z_1^2 - 1)(Z_1 Z_2)] = E(Z_1^3 Z_2) - E(Z_1 Z_2) = 0$$
$$\text{cov}(Z_1^2, Z_2^2) = E[(Z_1^2 - 1)(Z_2^2 - 1)] = 1 - 1 - 1 + 1 = 0$$

Finally, we arrive at

$$\text{cov}(\mathbf{Z} \otimes \mathbf{Z}) = 2(\mathbf{I}_2 \otimes \mathbf{I}_2) = 2\mathbf{I}_4,$$

where the subindex indicates the dimension of the identity matrix. Now when $\mathbf{X} \sim \mathcal{N}(\mathbf{0}, \mathbf{\Omega})$ we express $\mathbf{X} = \mathbf{\Omega}^{1/2}\mathbf{Z}$, where $\mathbf{Z} \sim \mathcal{N}(\mathbf{0}, \mathbf{I})$. Then using the properties of the Kronecker product listed in Section 10.5, we obtain

$$\mathbf{X} \otimes \mathbf{X} = (\mathbf{\Omega}^{1/2}\mathbf{Z}) \otimes (\mathbf{\Omega}^{1/2}\mathbf{Z}) = (\mathbf{\Omega}^{1/2} \otimes \mathbf{\Omega}^{1/2})(\mathbf{Z} \otimes \mathbf{Z})$$

and therefore

$$\text{cov}(\mathbf{X} \otimes \mathbf{X}) = (\mathbf{\Omega}^{1/2} \otimes \mathbf{\Omega}^{1/2})\text{cov}(\mathbf{Z} \otimes \mathbf{Z})(\mathbf{\Omega}^{1/2} \otimes \mathbf{\Omega}^{1/2})$$
$$= 2(\mathbf{\Omega}^{1/2} \otimes \mathbf{\Omega}^{1/2})(\mathbf{I} \otimes \mathbf{I})(\mathbf{\Omega}^{1/2} \otimes \mathbf{\Omega}^{1/2}) = 2(\mathbf{\Omega} \otimes \mathbf{\Omega}).$$

Formula (4.19) is proved.

Problems

1. Investigate by simulations the performance of the normal approximation (4.16).

2. Find $\text{Pr}(\|\mathbf{y}\| \leq 1)$, where $\mathbf{y}^{n \times 1} \sim \mathcal{N}(\mathbf{0}, \mathbf{I})$.

3. Use simulations to test that $\sum_{i=1}^{n} Z_i^2 \sim \chi^2(n)$ where $Z_i \overset{iid}{\sim} \mathcal{N}(0, 1)$. Write an R program with n as an argument that plots the empirical and theoretical cdf (pchisq) on the same graph. Show the plot with $n = 3$.

4. Another way to approximate the cdf of $\chi^2(n)$ is $\Phi(\sqrt{2x} - \sqrt{2n - 1})$. Compare the performance of this approximation with (4.16).

5. Demonstrate by simulations that $-2\sum_{i=1}^{n} \ln U_i \sim \chi^2(2n)$, where $U_i \overset{iid}{\sim} \mathcal{R}(0, 1)$.

6. Let X and Y be two independent chi-square distributed random variables. Does $X/(X + Y)$ have a beta distribution? Use simulations to answer the question.

7. Demonstrate by simulations that the sum of n iid exponential distributions with the rate parameter λ has distribution $(2\lambda)^{-1}\chi^2(2n)$.

8. Connection to the Poisson distribution: using `ppois` and `pchisq`, verify that, for even n, we have `pchisq(q=x,df=n)=1-ppois(q=n/2-1,lambda=x/2)`. Express this identity in mathematical terms.

9. Demonstrate (4.17) with simulations. Write an `R` program with arguments `mu`, `sigma`, `n,` and `N` (the number of simulated values) using default values $-1, 2, 4$, and 10K, respectively, and plot the empirical cdf of the left-hand side and superimpose it with the theoretical chi-square cdf using different colors. Display the values of μ, σ, n, and N using the `title` command. Use vectorized computations through function `apply(X,1,var)` as in Section 1.5.

10. Are the average and SD independent for a nonnormal distribution? [Hint: Use simulations for samples from a chi-square distribution and plot average versus SD.]

11. Investigate how robust result (4.17) is to the normal distribution of X_i. Simulate X_i as lognormally distributed and superimpose the empirical with the theoretical cdf of the left-hand side. Show two plots when the match is satisfactory and poor using `par(mfrow=c(1,2))`. Specify other parameters as arguments of function `R` and display them on the plots.

12. Prove that the cdf of the sum of n iid exponentially distributed random variables with the rate parameter λ is equal to $\mathcal{C}(2\lambda x, n)$, where $\mathcal{C}(\cdot, n)$ is the cdf of the chi-square distribution with $2n$ degrees of freedom. [Hint: Start by proving that the exponential distribution with $\lambda = 1/2$ may be viewed as the chi-square distribution with 2 degrees of freedom, then use the fact that the sum of independent chi-square distributions is a chi-square distribution.]

13. Prove (4.18).

14. Prove the last three formulas at the beginning of Section 4.2.2.

15. Find the normal approximation to the noncentral chi-square distribution and test it via simulations using an increasing sequence of n.

16. As a continuation of Example 4.28, (a) express the 4×4 matrix $\text{cov}(\mathbf{X} \otimes \mathbf{X})$ where $\mathbf{X}$ is a bivariate normal variable with zero means, unit variances, and correlation coefficient ρ; (b) write an `R` program to test your formula via simulations.

4.3 t-distribution

The Student distribution, or t-distribution, is named after an anonymous author *Student* published in 1908 in the leading statistical journal *Biometrika*. In fact, it was William Sealy Gosset who published the paper, but this distribution does not bear his name. The *t*-distribution (sometimes called the *central t*-distribution) is the distribution of the ratio of a standard normal variable to the square root of the independent chi-square variable divided by its df, or symbolically,

$$t(n) = \frac{\mathcal{N}(0,1)}{\sqrt{\chi^2(n)/n}}.$$

The fact that a random variable X belongs to the *t*-distribution with df $= n$ is denoted as $X_n \sim \mathcal{T}(n)$. The density of the *t*-distribution is given by

$$f(x; n) = \frac{\Gamma\left(\frac{n+1}{2}\right)}{\Gamma\left(\frac{n}{2}\right)\sqrt{\pi n}} \frac{1}{(1 + x^2/n)^{(n+1)/2}}, \quad -\infty < x < \infty. \qquad (4.20)$$

The cdf of the *t*-distribution will be denoted as $\mathcal{T}$. The density is symmetric around 0 and, therefore, has the mean 0; the variance of the *t*-distribution is

$$\text{var}(X_n) = \frac{n}{n-2}.$$

The variance does not exist for $n \leq 2$.

Table 4.2. Computation of the *t*-distribution in R

R function	Returns	Explanation
dt(x,df)	density (4.20)	df$= n$
pt(q,df)	cdf	cdf at q $= x$
qt(p,df)	inverse cdf	p=probability
rt(n,df)	random sample	$n = $ sample size

Theorem 4.29 *When $n \to \infty$, the density (4.20) converges to the standard normal density.*

Proof. Use the famous limit

$$\lim_{n\to\infty} \left(1 + \frac{a}{n}\right)^n = e^a,$$

where a is a constant. Let $a = x^2$ and find the limit of the numerator in density (4.20) when $n \to \infty$:

$$\lim_{n\to\infty} (1 + x^2/n)^{(n+1)/2} = e^{x^2/2 \lim_{n\to\infty}(n+1)/n} = e^{x^2/2}.$$

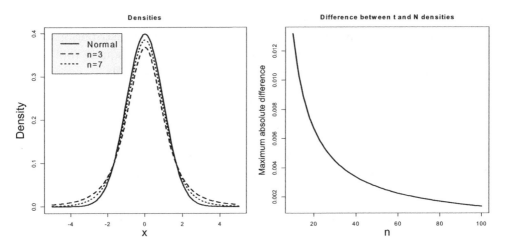

Figure 4.6: *Densities of the t-distribution and their normal approximation.*

Combine this result with the following mathematical fact

$$\lim_{n\to\infty} \frac{\Gamma\left(\frac{n+1}{2}\right)}{\Gamma\left(\frac{n}{2}\right)\sqrt{n}} = \frac{1}{\sqrt{2}},$$

and obtain that

$$\lim_{n\to\infty} f(x;n) = \frac{1}{\sqrt{2\pi}}e^{-x^2/2},$$

the standard normal density.   □

As follows from this theorem, the *t*-distribution approaches the standard normal distribution when the degrees of freedom increases to infinity. Some densities of the *t*-distribution are shown in Figure 4.6. Already $n = 20$ has a maximum absolute difference between densities around 0.007.

As follows from the theorem, the *t*-distribution converges to $\mathcal{N}(0,1)$: $\lim_{n\to\infty}$ $\Pr(X_n \le x) = \Phi(x)$. There exits an even better approximation of the cdf of the *t*-distribution:

$$\mathcal{T}(x) = \Pr(X_n \le x) \simeq \Phi\left(\frac{x(1 - 1/(4n))}{\sqrt{1 + x^2/(2n)}}\right). \qquad (4.21)$$

R has four functions associated with the *t*-distribution; see Table 4.2. The following theorem formulates a sufficient criterion for the *t*-distribution.

Theorem 4.30 *Let* $\mathbf{y}^{n\times 1}\sim\mathcal{N}(\mathbf{0},\mathbf{I})$*. Let vector* $\mathbf{a}$ *have unit length and the* $n \times n$ *matrix* $\mathbf{B}$ *be symmetric and idempotent with* $m =tr(\mathbf{B}) < n$*. If, in addition,*

$$\mathbf{Ba} = \mathbf{0}, \qquad (4.22)$$

then

$$\frac{\mathbf{a}'\mathbf{y}}{\sqrt{\mathbf{y}'\mathbf{By}/m}} \sim t(m). \qquad (4.23)$$

Proof. As follows from (4.3), $\mathbf{a}'\mathbf{y} \sim \mathcal{N}(0,1)$. The quadratic form in the denominator has a chi-square distribution, $\mathbf{y}'\mathbf{B}\mathbf{y} \sim \chi^2(m)$, as follows from Theorem 4.21. The fact that the linear and quadratic forms are independent follows from the following argumentation. Vector $\mathbf{a}$ is an eigenvector of matrix $\mathbf{B}$ corresponding to zero eigenvalue since $\mathbf{B}\mathbf{a} = \mathbf{0}$. Other eigenvectors can be chosen that are perpendicular to $\mathbf{a}$. This means that matrix $\mathbf{B}$ admits a representation $\mathbf{B} = \sum_{i=1}^{m} \mathbf{p}_i\mathbf{p}_i'$ where $\mathbf{p}_i \perp \mathbf{a}$ or algebraically, $\mathbf{a}'\mathbf{p}_i = 0$. This means that $X = \mathbf{a}'\mathbf{y}$ and $\mathbf{y}'\mathbf{B}\mathbf{y} = \sum_{i=1}^{m} Z_i^2$, where $Z_i = \mathbf{p}_i'\mathbf{y}$, are independent. Hence (4.23) is true. $\square$

The following application of the above theorem is an important statistical fact and gives rise to the *t*-test.

Theorem 4.31 *Let* $X_i \overset{\text{iid}}{\sim} \mathcal{N}(\mu, \sigma^2)$, $i = 1, 2, ..., n$. *Then*

$$\frac{\sqrt{n}(\overline{X} - \mu)}{\sqrt{\sum_{i=1}^{n}(X_i - \overline{X})^2/(n-1)}} \sim t(n-1). \tag{4.24}$$

Proof. We start by expressing the left-hand side of (4.24) in terms of $Z_i \overset{\text{iid}}{\sim} \mathcal{N}(0,1)$. Indeed, divide the numerator and denominator by σ. Then, for the numerator, we have

$$\sqrt{n}\frac{\overline{X} - \mu}{\sigma} = \sqrt{n}\frac{1}{\sigma n}\sum_{i=1}^{n}(X_i - \mu) = \frac{1}{\sqrt{n}}\sum_{i=1}^{n}\frac{X_i - \mu}{\sigma} = \frac{1}{\sqrt{n}}\sum_{i=1}^{n} Z_i = \mathbf{a}'\mathbf{Z},$$

where vector $\mathbf{a} = (1/\sqrt{n}, 1/\sqrt{n}, ..., 1/\sqrt{n})'$ has unit length and $\mathbf{Z} \sim \mathcal{N}(\mathbf{0}, \mathbf{I})$, so $\mathbf{a}'\mathbf{Z} \sim \mathcal{N}(0,1)$. For the denominator we have

$$\frac{1}{\sigma^2(n-1)}\sum_{i=1}^{n}(X_i - \overline{X})^2 = \frac{1}{\sigma^2(n-1)}\sum_{i=1}^{n}[(X_i - \mu) - (\overline{X} - \mu)]^2$$

$$= \frac{1}{n-1}\sum_{i=1}^{n}\left(\frac{X_i - \mu}{\sigma} - \frac{\overline{X} - \mu}{\sigma}\right)^2 = \frac{1}{n-1}\sum_{i=1}^{n}(Z_i - \overline{Z})^2.$$

From the solution to Example 4.22, we represent $\sum_{i=1}^{n}(Z_i - \overline{Z})^2 = \mathbf{Z}'\mathbf{B}\mathbf{Z}$, where $\mathbf{B} = \mathbf{I} - \frac{1}{n}\mathbf{1}\mathbf{1}'$. To apply Theorem 4.30, we must show that $\mathbf{B}\mathbf{a} = \mathbf{0}$. Indeed, since $\mathbf{1}'\mathbf{1} = n$, we obtain

$$\mathbf{B}\mathbf{a} = \left(\mathbf{I} - \frac{1}{n}\mathbf{1}\mathbf{1}'\right)\frac{1}{\sqrt{n}}\mathbf{1} = \frac{1}{\sqrt{n}}\left(\mathbf{1} - \frac{1}{n}\mathbf{1}\mathbf{1}'\mathbf{1}\right) = \frac{1}{\sqrt{n}}\left(\mathbf{1} - \frac{n}{n}\mathbf{1}\right) = \mathbf{0}.$$

Thus, all conditions of Theorem 4.30 are met and (4.24) is proved.

Corollary 4.32 *The average and sample variance are independent.*

As follows from the above example, if $X_i \overset{\text{iid}}{\sim} \mathcal{N}(\mu, \sigma^2)$, the average $\overline{X}$ and the sample variance $\sum_{i=1}^{n}(X_i - \overline{X})^2/(n-1)$ are independent because $\mathbf{Ba} = \mathbf{0}$.

Theorem 4.33 *Let two samples* $X_i \overset{\text{iid}}{\sim} \mathcal{N}(\mu, \sigma^2)$, $i = 1, 2, ..., n$ *and* $Y_i \overset{\text{iid}}{\sim} \mathcal{N}(\mu, \sigma^2)$, $i = 1, 2, ..., m$ *be independent. Then*

$$\frac{\overline{X} - \overline{Y}}{\widehat{\sigma}\sqrt{1/n + 1/m}} \sim t(n + m - 2),$$

where

$$\widehat{\sigma}^2 = \frac{1}{n + m - 2}\left(\sum_{i=1}^{n}(X_i - \overline{X})^2 + \sum_{i=1}^{m}(Y_i - \overline{Y})^2\right).$$

Proof. Divide numerator and denominator by σ and express

$$\sigma^{-1}(1/n + 1/m)^{-1/2}(\overline{X} - \overline{Y}) = \mathbf{1}'_n\mathbf{Z}_1 - \mathbf{1}'_m\mathbf{Z}_2,$$

where $\mathbf{Z}_1 \sim \mathcal{N}(\mathbf{0}, \mathbf{I}_n)$ and $\mathbf{Z}_2 \sim \mathcal{N}(\mathbf{0}, \mathbf{I}_m)$ are independent. Define a $(n + m) \times 1$ vector $\mathbf{u}$ such that the first n components are 1 and the remaining components are -1, and

$$\mathbf{Z} = \begin{bmatrix} \mathbf{Z}_1 \\ \mathbf{Z}_2 \end{bmatrix}, \quad \mathbf{A} = \begin{bmatrix} \mathbf{I}_n - n^{-1}\mathbf{1}_n\mathbf{1}'_n & \mathbf{0} \\ \mathbf{0} & \mathbf{I}_m - m^{-1}\mathbf{1}_m\mathbf{1}'_m \end{bmatrix}.$$

Then $\mathbf{Au} = \mathbf{0}$, matrix $\mathbf{A}$ is idempotent with $\text{tr}(\mathbf{A}) = n + m - 2$. Thus, Theorem 4.30 applies and the t-distribution with $n+m-2$ degrees of freedom is established.

4.3.1   Noncentral t-distribution

This distribution emerges in computing the power function of the t-test. Loosely speaking, the difference with the central distribution is that the numerator has a nonzero mean.

Definition 4.34 *The noncentral t-distribution is the distribution of the ratio of the normal random variable with mean* δ *and unit variance to the square root of the independent chi-square distribution divided by its degrees of freedom, or symbolically,*

$$t(n, \delta) = \frac{\mathcal{N}(\delta, 1)}{\sqrt{\chi^2(n)/n}},$$

where δ *is called the noncentrality parameter.*

The same four functions in R listed in Table 4.2 are used for the noncentral t-distribution but adding the parameter `ncp`, similarly to the noncentral chi-square distribution. For example, to compute the density with the noncentrality parameter δ, we use `dt(x,df,ncp=delta)`. Four densities of the noncentral t-distribution are shown in Figure 4.7.

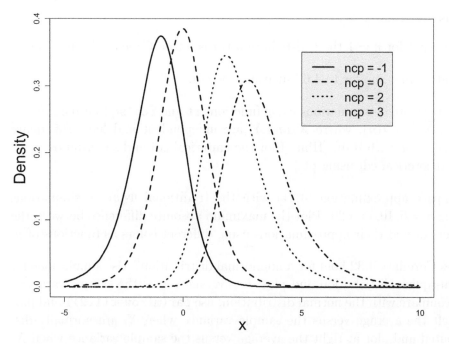

Figure 4.7: *Four densities of the noncentral t-distribution, n = 7.*

Example 4.35 *Noncentral t-distribution.* *Prove that*

$$\frac{\sqrt{n}\overline{X}}{\sqrt{\sum_{i=1}^{n}(X_i - \overline{X})^2/(n-1)}} \sim t(n-1, \delta = \mu\sqrt{n}), \qquad (4.25)$$

where $X_i \overset{\text{iid}}{\sim} \mathcal{N}(\mu, 1), \ i = 1, 2, ..., n.$

Solution. The numerator and denominator are still independent, but now $E(\sqrt{n}\overline{X}) = \mu\sqrt{n}$. This implies the result. $\square$

The noncentral *t*-distribution can be approximated with the normal distribution as follows: if X has a noncentral *t*-distribution with the noncentrality parameter δ and df $= n$, then

$$X \simeq \mathcal{N}(\delta, 1). \qquad (4.26)$$

There is a better approximation,

$$\frac{X(1 - 1/(4n)) - \delta}{\sqrt{1 + X^2/(2n)}} \simeq \mathcal{N}(0, 1), \qquad (4.27)$$

which implies that the cdf of the noncentral *t*-distribution can be approximated by Φ with the argument as the left-hand side of (4.27).

Problems

1. Show that for $n = 1$ the t-distribution turns into a Cauchy distribution.

2. Create a plot similar to 4.6 but using the cdf.

3. Demonstrate by simulations using functions `rt` and `rchisq` that $0.5\sqrt{n}(X - Y)/\sqrt{XY} \sim t(n)$, where X and Y are independent and have chi-square distribution with n df. [Hint: Plot the empirical cdf and superimpose with the theoretical cdf using `pt`.]

4. Compare approximation (4.21) with the traditional $\Phi(x)$ by simulations using $n = 5, 10$, and 20. Plot the maximum absolute difference between the exact cdf and their approximations using different colors as functions of n.

5. Does Corollary 4.32 hold for a nonnormal distribution? Use simulations for a chi-square distribution and plot the average versus the sample variance. To compare with the normal distribution, use `par(mfrow=c(1,2))` and plot at left the average versus the sample variance when X_i are normally distributed and plot at right the average versus the sample variance when X_i are chi-square distributed with degree of freedom m. Specify n and parameters of the distributions as arguments of the function. Explain why the plot at left indicates independence and the plot at right does not. [Hint: Implement vectorized simulations with `rowMeans` and `apply` for variance.]

6. Check distribution (4.25) through simulations for several n by matching the empirical and theoretical cdfs.

7. Check approximations (4.26) and (4.27) via cdfs. [Hint: Use arguments n and δ, plot the cdfs, and report the absolute difference from `pt`.]

8. Prove that the cdf of a noncentral t-distribution with the noncentrality parameter δ and df n evaluated at x can be written as $E_S\Phi(x\sqrt{S/n} - \delta)$, where S is the chi-square distributed random variable with df n. Use simulations to check this result. [Hint: Use the definition of the noncentral distribution and the repeated expectation Theorem 3.27.]

4.4 F-distribution

The F-distribution, after Ronald Fisher who discovered this distribution, is the distribution of the ratio of two independent chi-square distributions divided by the corresponding df,

$$F(m,n) = \frac{\chi^2(m)/m}{\chi^2(n)/n}, \quad m > 0, \ n > 0.$$

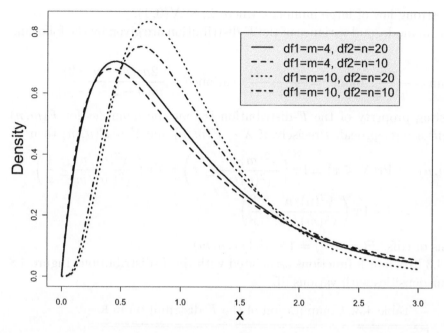

Figure 4.8: *Four densities of F-distribution with various degrees of freedom.*

It has the density

$$f(x; m, n) = \frac{\Gamma\left(\frac{m+n}{2}\right)\left(\frac{m}{2}\right)^{m/2} x^{m/2-1}}{\Gamma\left(\frac{m}{2}\right)\Gamma\left(\frac{n}{2}\right)\left(\frac{n}{2}\right)^{m/2}\left(1 + \frac{mx}{n}\right)^{m/2+n/2}}, \quad x \geq 0. \tag{4.28}$$

The F-distribution is related to the t-distribution,

$$F(1, n) = t^2(n). \tag{4.29}$$

In words, the F-distribution with the first df being one is the squared t-distribution. This can be seen from a simple fact $\mathcal{N}^2(0, 1) = \chi^2(1)$ or more rigorously by letting $m = 1$ in (4.28) and finding the density of X^2 where $X \sim t(n)$.

The F-distribution is related to the χ^2-distribution,

$$\lim_{n \to \infty} \mathcal{F}(x; m, n) = m\mathcal{C}(x; m) \tag{4.30}$$

for every x, where $\mathcal{F}$ denotes the cdf of the F-distribution and $\mathcal{C}$ denotes the cdf of the chi-square distribution. This can be shown rigorously by letting $n \to \infty$ in the density (4.28) and showing that it converges to $mf(x; m)$, where f is given by (4.20). Another way to see why (4.30) holds is to note that, with some ambiguity of notation,

$$\lim_{n \to \infty} \frac{1}{n}\chi^2(n) = \lim_{n \to \infty} \frac{1}{n}\sum_{i=1}^{n} Z_i^2 = 1 \quad \text{a.s.}$$

due to the strong law of large numbers, where $Z_i \sim \mathcal{N}(0, 1)$.

The mean, mode, and variance of the F-distribution are given by the following formulas:

$$\text{mean} = \frac{n}{n-2}, \quad \text{mode} = \frac{n(m-2)}{m(n+2)}, \quad \text{variance} = \frac{2n^2(m+n-2)}{m(n-2)^2(n-4)}.$$

An interesting property of the F-distribution is that the quantiles for $F(m, n)$ and $F(n, m)$ are reciprocal. Precisely, if $X \sim F(m, n)$ and $Y \sim F(n, m)$ then

$$\mathcal{F}(x; m, n) = \Pr(X \leq x) = \Pr\left(\frac{\chi^2(m)/m}{\chi^2(n)/n} \leq x\right) = \Pr\left(\frac{\chi^2(n)/n}{\chi^2(m)/m} \geq \frac{1}{x}\right)$$

$$= 1 - \Pr\left(\frac{\chi^2(n)/n}{\chi^2(m)/m} < \frac{1}{x}\right),$$

or, in terms of cdfs, $\mathcal{F}(x; m, n) = 1 - \mathcal{F}(1/x; n, m)$.

Table 4.3 lists four R functions associated with the F-distribution. Figure 4.8 depicts four densities with various df.

Table 4.3. Computation of the F-distribution in R

R function	Returns	Explanation
df(x,df1,df2)	density (4.28)	df1$= m$, df2$= n$
pf(q,df1,df2)	cdf	cdf at q $= x$
qf(p,df1,df2)	inverse cdf	p=probability
rf(n,df1,df2)	random sample	$n =$ sample size

The following result is frequently used to show that the ratio of two quadratic forms follows an F-distribution.

Theorem 4.36 *Let* $\mathbf{y}^{n \times 1} \sim \mathcal{N}(\mathbf{0}, \mathbf{I})$ *and let the* $n \times n$ *symmetric idempotent matrices* $\mathbf{A}$ *and* $\mathbf{B}$ *be such that*

$$\mathbf{AB} = \mathbf{0}.$$

Then

$$\frac{\mathbf{y}'\mathbf{A}\mathbf{y}/m}{\mathbf{y}'\mathbf{B}\mathbf{y}/n} \sim F(m, n),$$

where $m = tr(\mathbf{A})$ *and* $n = tr(\mathbf{B})$.

Proof. The key is that the quadratic forms in the numerator and denominator are independent due to Theorem 4.23. Since, as follows from Theorem 4.21, $\mathbf{y}'\mathbf{A}\mathbf{y} \sim \chi^2(m)$ and $\mathbf{y}'\mathbf{B}\mathbf{y} \sim \chi^2(n)$, the ratio adjusted by degree of freedom follows the F-distribution. $\qquad\square$

The following is an illustration of the theorem – this result is essential for testing the equality of variances from two independent samples. See Chapter 7.

Theorem 4.37 *Let* $X_i \overset{iid}{\sim} \mathcal{N}(\mu_X, \sigma^2)$ *and* $Y_j \overset{iid}{\sim} \mathcal{N}(\mu_Y, \sigma^2)$ *for* $i = 1, .., m$, $j = 1, ..., n$, *and, let* $\{X_i\}$ *and* $\{Y_j\}$ *be independent. Then*

$$\frac{\sum_{i=1}^{m}(X_i - \overline{X})^2/(m-1)}{\sum_{j=1}^{n}(Y_j - \overline{Y})^2/(n-1)} \sim F(m-1, n-1). \tag{4.31}$$

Proof. Since both variables have the same variance, σ^2, we can divide the numerator and denominator by σ^2 for normalization. Using the same technique as when proving (4.24), we can assume that $X_1, ..., X_m, Y_1, ..., Y_n \overset{iid}{\sim} \mathcal{N}(0, 1)$. Now we apply Theorem 4.36 by introducing the $(m+n) \times 1$ random vector $\mathbf{z}$ and two $(m+n) \times (m+n)$ matrices:

$$\mathbf{z} = \begin{bmatrix} \mathbf{x} \\ \mathbf{y} \end{bmatrix}, \ \mathbf{A} = \begin{bmatrix} \mathbf{I}_m - m^{-1}\mathbf{1}_m\mathbf{1}_m' & \mathbf{0} \\ \mathbf{0} & \mathbf{0} \end{bmatrix}, \ \mathbf{B} = \begin{bmatrix} \mathbf{0} & \mathbf{0} \\ \mathbf{0} & \mathbf{I}_n - n^{-1}\mathbf{1}_n\mathbf{1}_n' \end{bmatrix},$$

respectively. Then the numerator of (4.31) is $\mathbf{z}'\mathbf{A}\mathbf{z}/m$ and the denominator is $\mathbf{z}'\mathbf{B}\mathbf{z}/n$. But it is easy to see that $\mathbf{AB} = \mathbf{0}$ and also $\mathrm{tr}(\mathbf{A}) = m-1$ and $\mathrm{tr}(\mathbf{B}) = n-1$. Thus Theorem 4.36 applies and the distribution (4.31) is established. □

Sometimes we refer to the numerator and denominator in formula (4.31) as sample variances. In words, under the normal assumption, the ratio of sample variances has an F-distribution.

Definition 4.38 *The noncentral F-distribution with the noncentrality parameter* δ *is defined as*

$$F(m, n, \mathtt{ncp} = \delta) = \frac{\chi^2(m, \mathtt{ncp} = \delta)/m}{\chi^2(n)/n}.$$

The same R functions listed in Table 4.3 apply to the noncentral F-distribution using additional parameter $\mathtt{ncp}$. The noncentral F-distribution is used to derive the power function of the variance test.

Problems

1. Prove that the F-distribution is skewed to right. [Hint: Prove that mean > median > mode.]

2. Prove the formula for the mean and mode. [Hint: Express the mean integral in terms of another F-distribution. Take the log to find the maximum of the density.]

3. Prove the formula for the variance by computing $E(X^2)$ and then using $\mathrm{var}(X) = E(X^2) - E^2(X)$.

4. Demonstrate (4.31) by simulations. Write an R program letting $\mu_x = 1$, $\mu_y = 1$, $\sigma^2 = 2$, and m and n as arguments that plots the empirical distribution of the left-hand side and the theoretical cdf on the same graph. Display the plot for $m = 3$ and $n = 6$.

5. Derive the density of the left-hand side of (4.31) when variances for samples X and Y are different.

6. Find the limit of the F-distribution when $n \to \infty$.

7. Prove (4.29) by showing that the densities are the same.

8. Use simulations to verify how robust the result from Theorem 4.37 is to the normal distribution assumption. Simulate X_i and Y_i as chi-square distributed with df specified as the arguments of the R function with some default values. Superimpose the empirical cdf with the theoretical one under two scenarios when the match is satisfactory and when the match is poor using `par(mfrow=c(1,2))`. Implement vectorized simulations using `apply` for sample variances.

9. Plot the cdfs and densities of the noncentral F-distribution for several noncentrality parameters to see how they change starting from $\delta = 0$ using `par(mfrow=c(1,2))`. Use other parameters as arguments of the R function with some default values. Is it true that the cdf is an increasing function of δ? Display the values of the parameters of the distributions.

10. Prove that the cdf of the F-distribution with degrees of freedom m and n evaluated at x can be written as $E_S \mathcal{C}(xSm/n; n)$, where S is the chi-square distributed random variable with degrees of freedom n. Use simulations to check this result. [Hint: Use the definition of the F-distribution and the repeated expectation Theorem 3.27.]

Chapter 5

Preliminary data analysis and visualization

Data visualization is an important part of preliminary statistical analysis. Many data features, including skewness and heavy tails, the presence of outliers, clusters, and nonlinearity, can and should be seen and evaluated before starting statistical inference. The reason is that standard statistical models, such as the t-test and linear regression, typically assume a normal distribution, absence of clusters, and linearity. The best way to detect possible violations is to *see* the data. The saying "a picture is worth a thousand words" is very applicable here. We start this chapter by showing how the empirical cumulative distribution function (cdf) can be used to visualize and compare iid samples of observations. The receiver operator characteristic (ROC) curve naturally emerges by plotting one cdf against another. Other visualization techniques such as histogram, q-q and box plot, one- and two-dimensional kernel density estimation will be discussed as well. At the end of the chapter, we apply visualization techniques to spatial data for disease mapping.

5.1 Comparison of random variables using the cdf

If X and Y are numbers, it is simple to answer if $X < Y$. It is not that simple to answer if $X < Y$ when X and Y are random variables. Indeed, everybody knows that women are shorter than men. But what meaning do we put into *shorter*? Does it mean that a randomly chosen man is taller than a randomly chosen woman? Somebody may say that $X < Y$ if mean(X) < mean(Y). Does it mean that the probability that a randomly chosen woman is shorter than a randomly chosen man is greater than 1/2? Comparison of means (or medians) is meaningful but does not reflect the comparison on the *entire range* of population values (more

Advanced Statistics with Applications in R, First Edition. Eugene Demidenko.

discussion on when to use mean versus median is found in Section 1.4). In fact, many assertions in our life involve comparison of random variables; when we say that "prices in one grocery store are lower than in another" or "salaries in company A are higher than in company B," even if there are products in one store that cannot be bought in the other or if the companies are from different industries for employees with different titles and positions. The following definition of order for random variables applies to the entire range of values and uses the concept of the cdf.

Definition 5.1 *For two random variables X and Y we write $X \prec Y$ and say that X is stochastically smaller than Y, if $F_Y(u) < F_X(u)$ for every u, where F_X and F_Y are the cdfs of X and Y.*

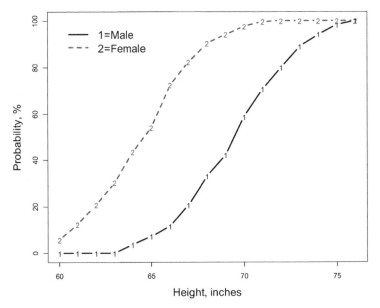

Figure 5.1: *Women are shorter than men because the cdf for men is below the cdf for women: the proportion of women shorter than x is greater than the proportion of men shorter than x (the height data are real).*

The above definition means that the probability that $Y < u$ is less than the probability $X < u$ for every u. It is easy to prove that if $X \prec Y$, then the mean and the median of X are less than the corresponding features of Y. The inequality for the median is obvious; the inequality for the mean follows from the following formula:

$$E(X) = \int_0^\infty (1 - F_X(x))dx - \int_{-\infty}^0 F_X(x)dx. \qquad (5.1)$$

Definition 5.1 is more stringent than inequality *on average* because the inequality $F_Y(u) < F_X(u)$ holds on the entire range of values.

In Figure 5.1, we show the cdfs of the heights of 17 women and 17 men (Dartmouth students, real data): the proportion of women who are shorter than x is larger than the proportion of men who is shorter than x for the entire range of height. We summarize this fact by saying that "men are taller than women." This statement is more stringent than just saying that the average height of men is greater than the average height of women.

An important property of Definition 5.1 is that relation $X \prec Y$ remains true upon any increasing transformation.

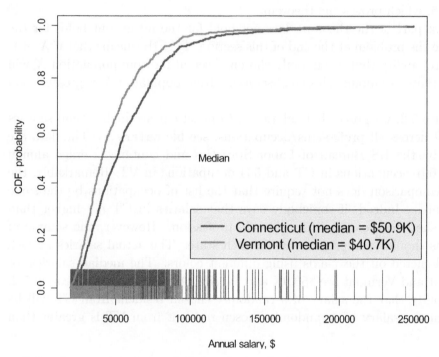

Figure 5.2: *An empirical proof that salaries in Connecticut are stochastically hiegher than salaries in VT: the proportion of employees in VT with salary less or equal x is greater than the proportion of employees in CT with salary less or equal x for any x.*

Theorem 5.2 *If $X \prec Y$, then (a) $g(X) \prec g(Y)$ for any strictly increasing function g, and (b) if X and Y are independent, then $\Pr(X < Y) > 0.5$.*

Proof. (a) If g is a strictly increasing function, then the inequality $g(X) \leq u$ is equivalent to the inequality $X \leq g^{-1}(u)$, where g^{-1} is the inverse function. Therefore denoting $v = g^{-1}(u)$ we obtain

$$F_{g(X)}(u) = \Pr(g(X) \leq u) = \Pr(X \leq v), \ F_{g(Y)}(u) = \Pr(g(Y) \leq u) = \Pr(Y \leq v).$$

Since $\Pr(X \leq v) < \Pr(Y \leq v)$, we have $F_{g(X)}(u) < F_{g(Y)}(u)$ for any u.

(b) Assuming that X and Y are continuous and independent random variables, we have

$$\Pr(X < Y) = \int_{-\infty}^{\infty} \left(\int_{x}^{\infty} f_Y(y) dy \right) f_X(x) dx$$

$$= \int_{-\infty}^{\infty} \left(1 - F_Y(x) \right) f_X(x) dx > 1 - \int_{-\infty}^{\infty} F_X(x) f_X(x) dx.$$

Integration by parts, $du = f_X(x) dx$ and $v = F_X(x)$ simplifies the last integral to $\int_0^1 t \, dt = 0.5$, which proves the theorem.   □

The first part of the theorem does not hold for the mean, but holds for the median; see the problem at the end of this section. Part (b) means that if $X \prec Y$ then the probability that a randomly chosen observation from population X will be smaller than a randomly chosen observation from population Y is greater than $1/2$.

In Figure 5.2, we *prove* that salaries in Connecticut are higher than salaries in Vermont across all professions/occupations; see file `salary.r`. The data are published by the US Bureau of Labor Statistics and contains average annual salaries of 691 occupations in CT and 534 occupations in VT. Remarkably, the cdf-based comparison does not require that the list of occupations be the same in both states. Indeed, if somebody says that salaries in CT are higher than in VT he/she does not refer to a specific profession. However, the sample of occupations should be representative in each state. The actual salaries in each state are depicted on the x-axis using different colors. The median salaries for Connecticut and Vermont are \$50.9K and \$40.7K. As follows from Theorem 5.2, the probability that the salary of a randomly chosen resident from CT will be higher than the salary of a randomly chosen resident from VT is greater than $1/2$.

5.1.1   ROC curve

Another statistical concept closely related to the cdf is the ROC curve, which was introduced during World War II to interpret radar images to determine the error of identification of an enemy airplane versus an artifact such as a bird. Nowadays, the ROC curve is commonly used in medicine, particularly radiology, to draw conclusions based on patient X-rays or to establish a diagnosis based on the outcome of a medical test.

Consider two random variables X and Y specified by cdfs F_X and F_Y associated with two groups. Assuming that the values of X and Y are mixed together, given observation U, one needs to identify what group the observation came from. Continuing the bird-versus-airplane classification example, let the radar reader identify the object in the sky based on its image size. In his previous experience he observed images that came from birds and airplanes, so that the distributions of sizes, such as the diameter of the image on the radar, in the form of cdfs F_X

and F_Y, airplanes and birds, respectively, are known. To classify the observation into one of two groups, a threshold, u, is needed: if the observed size of the radar image is smaller than u, then it is a bird and otherwise it is an airplane. The problem of such classification would be easy if sizes do not overlap but they do. For example, there may be a flock of birds looking large enough to be confused with an airplane. Reversely, at a certain angle, a small airplane may look like a large bird. An obvious assumption is that birds are smaller than airplanes, and hopefully $Y \prec X$, i.e. $F_X(u) < F_Y(u)$ for all u (note that, unlike in the previous section, where $X \prec Y$, we assume that $Y \prec X$ here); see Theorem 5.3.

The ROC curve is a parametrically defined on the unit square with the x-axis as $F_X(u)$ and the y-axis as $F_Y(u)$, where u runs from $-\infty$ to ∞. In short, the ROC curve is the plot of one cdf versus the other. In the ROC framework, $F_Y(u)$ is referred to as the sensitivity of the test that the radar image indicates a bird, $F_X(u)$ is referred to as the false positive (FP) rate or probability, and $1 - F_X(u)$ is referred to as the specificity of the test, or true negative (TN). In the radar example, a false positive decision is likely to make when the airplane is small. The false negative (FN) rate is complementary to sensitivity and defines the error of correct identification: the probability of the claim that the radar image represents a plane while in fact it is a bird.

In another example, mortgage defaulters are identified based on the applicant's family income. Let a bank collect data on an applicant's family income along with information on who did not default and who did. Respectively, let F_X be the cdf distribution of family income among nondefaulters and F_Y be the cdf distribution of family income among applicants who defaulted at some time in the mortgage loan (similarly, the problem may be formulated using the credit score when buying a car). Consider u as the income threshold: the rule is that if the applicant's income is smaller than u, the application is denied; otherwise, it is granted. Errors are inevitable: sensitivity means the proportion of applicants whose income was smaller than u and who defaulted – the probability of correct identification of defaulters, $F_Y(u)$ – and specificity identifies nondefaulters with low income.

Theorem 5.3 *Properties of the ROC curve. Let F_X and F_Y be the cdfs of two continuous random variables and the ROC curve be defined as a set of points on the unit square $\{u \in R^1 : (F_X(u), F_Y(u))\}$. Then the following holds:*

1. *It starts at origin $(0,0)$ and goes to $(1,1)$.*

2. *It is an increasing function.*

3. *It is invariant to an increasing transformation of X and Y.*

4. *The ROC curve is above the $45°$ line if and only if $Y \prec X$, i.e. $F_Y(u) > F_X(u)$; see Definition 5.1.*

5. *The minimum total error, defined as the sum of the false positive and false negative rates, is where the slope of the ROC curve is 1.*

6. *The area under the ROC curve (AUC) can be computed in three equivalent ways*

$$\text{AUC} = F_X(F_Y(X)) = 1 - E_Y(F_X(Y)) = \Pr(Y < X)$$

and is interpreted as the expected cdf of Y in the population X or the probability that $Y < X$.

7. *If $\{X_i, i = 1, ..., m\}$ and $\{Y_i, i = 1, ..., n\}$ are iid samples from populations X and Y, respectively, then the AUC can be unbiasedly estimated as $(mn)^{-1}\sum_{i,j} 1(Y_i < X_j)$, where 1 is the indicator function.*

Proof. Since X and Y are continuous random variables, $dF_X/du = f_X(u) > 0$, so that the inverse cdf exists, and $dF_Y/du = f_Y(u) > 0$ for any u. Note that the ROC curve can be regularly parametrized as a function of $x = F_X(u)$ as $R(x) = F_Y(F_X^{-1}(x))$ for $0 \le x \le 1$, where F_X^{-1} is the inverse cdf.

(i) This property holds because $F_X(-\infty) = F_Y(-\infty) = 0$ and $F_X(\infty) = F_Y(\infty) = 1$. (ii) We use the fact that the derivative of the inverse function is the reciprocal of the original function: $dR/dx = f_Y(F_X^{-1}(x))/f_X(x) > 0$, where f_Y and f_X are the density functions. (iii) Let g be an increasing transformation, $X' = g(X)$ and $Y' = g(Y)$. If g^{-1} is the inverse function, we have $F_{Y'}(y) = F_Y(g^{-1}(y))$ and $F_{X'}(x) = F_X(g^{-1}(x))$. Thus, the ROC curve as the set of points remains the same upon transformation g. (iv) Since F_Y is on the y-axis and F_X is on the x-axis, the curve is above the 45° line if and only if $F_Y(u) > F_X(u)$, Y is stochastically smaller than X. In our mortgage example, income of defaulters is stochastically smaller than income of nondefaulters. (v) The total error is $1-$ sensitivity + false positive rate, or in mathematical notation $1 - F_Y(u) + F_X(u)$. Since the ROC curve can be viewed as a function $y = y(x)$, where $x = F_X(u)$ and $y = F_Y(u)$, we can rewrite the total error as $1 - y(x) + x$. Differentiating with respect to x and setting it to zero we arrive at $dy/dx = 1$. This means that the minimum total error is where the tangent line has 45° slope (parallel to the main diagonal of the square). (vi) The AUC is defined as $\int_0^1 R(x)dx = \int_0^1 F_Y(F_X^{-1}(x))dx$. Upon the change of variable $z = F_X^{-1}(x)$, the AUC can be expressed as $\int_{-\infty}^{\infty} F_Y(z)f_X(z)dz = E_X(F_Y(X))$, the expected value of the cdf of Y over X. To obtain the second representation, apply integration by parts in the last integral by letting $u = F_Y(z)$ and $dv = f_X(z)dz$. For the last representation, we have $\Pr(Y < X) = \int_{-\infty}^{\infty} \int_{-\infty}^{x} f_X(x)f_Y(y)dxdy = \int_{-\infty}^{\infty} f_X(y)F_Y(y)dy$. (vii) The unbiasedness follows from the fact that $E(1(Y < X)) = \Pr(Y < X)$.

Remarks. (i) Many properties formulated above apply to discrete cdfs including empirical cdfs as estimates of the theoretical ones. (ii) Property 3 holds when the same transformation applies to X and Y. It also holds for random variables with support as an interval (finite or semi-infinite). For example, if X and

Y are positive continuous random variables, then the ROC curve for X and Y is the same as for $\ln X$ and $\ln Y$. In other terms, the ROC curve is invariant with respect to a monotonic transformation. (iii) Property 4 establishes the connection between stochastic inequality and the ROC curve. (iv) Property 5 defines the optimal discrimination threshold under the criterion of the sum of two errors. In a more general approach, we find an optimal threshold that minimizes a linear combination of errors that are associated with cost, see the mortgage example below. Another optimal threshold can be found from the condition that sensitivity equals specificity.

Binormal ROC curve

The simplest ROC curve, hereafter referred to as the binormal ROC curve, is when two independent random variables are normally distributed, $X \sim \mathcal{N}(\mu_X, \sigma_X^2)$ and $Y \sim \mathcal{N}(\mu_Y, \sigma_Y^2)$. Without loss of generality, we can assume that $\mu_Y < \mu_X$. Applied to our radar example, the last condition means that the average size of the bird image is smaller than the average size of the airplane image. The inequality of the means is necessary but not sufficient to claim that $Y \prec X$. Moreover, it is easy to prove that $Y \prec X$ if and only if $\mu_Y < \mu_X$ and $\sigma_X^2 = \sigma_Y^2$. The cdfs are easily expressed through the standard normal cdf, Φ as $F_X(u) = \Phi\left(\frac{u-\mu_X}{\sigma_X}\right)$ and $F_Y(u) = \Phi\left(\frac{u-\mu_Y}{\sigma_Y}\right)$, interpreted as the false positive ($1-$ specificity) and sensitivity of the test. Thus the binormal ROC curve can be derived (and plotted) as a parametrically defined curve with the x-coordinate $F_X(u)$ and the y-coordinate $F_Y(u)$ when u runs from $-\infty$ to ∞. Alternatively, the binormal ROC curve can be defined as the sensitivity R expressed directly through the false positive rate p as

$$R(p) = \Phi\left(\frac{\mu_X - \mu_Y + \sigma_X \Phi^{-1}(p)}{\sigma_Y}\right), \quad 0 < p < 1. \tag{5.2}$$

Now we find the area under the binormal ROC curve in closed form:

$$\text{AUC} = \int_0^1 R(p)dp = \int_0^1 \Phi\left(\frac{\mu_X - \mu_Y + \sigma_X \Phi^{-1}(p)}{\sigma_Y}\right) dp.$$

Make change of variable, $\Phi^{-1}(p) = u$ and obtain

$$\text{AUC} = \int_{-\infty}^{\infty} \Phi\left(\frac{\mu_X - \mu_Y + \sigma_X u}{\sigma_Y}\right) \phi(u)du = \Phi\left(\frac{\mu_X - \mu_Y}{\sqrt{\sigma_X^2 + \sigma_Y^2}}\right), \tag{5.3}$$

as follows from Theorem 2.35 of Section 2.7. Indeed, denoting $U = (\mu_X - \mu_Y + \sigma_X Z)/\sigma_Y \sim \mathcal{N}\left((\mu_X - \mu_Y)/\sigma_Y, \sigma_X^2/\sigma_Y^2\right)$, we obtain

$$E_{Z \sim \mathcal{N}(0,1)} \Phi\left(\frac{\mu_X - \mu_Y + \sigma_X Z}{\sigma_Y}\right) = E_U \Phi(U)$$

$$= \Phi\left(\frac{(\mu_X - \mu_Y)/\sigma_Y}{\sqrt{1 + \sigma_X^2/\sigma_Y^2}}\right) = \Phi\left(\frac{\mu_X - \mu_Y}{\sqrt{\sigma_X^2 + \sigma_Y^2}}\right).$$

Since $\mu_X > \mu_Y$, we have $0.5 <$ AUC <1. The greater the difference between the compared populations, the larger the AUC. The AUC under the normal distribution assumption has a nice interpretation following from the identity

$$\text{AUC} = \Pr(Y < X). \tag{5.4}$$

Indeed, $\Pr(Y < X) = \Pr(Y - X < 0) = \Pr(U < 0)$, where $U \sim \mathcal{N}(\mu_Y - \mu_X, \sigma_X^2 + \sigma_Y^2)$ because X and Y are independent. Therefore,

$$\text{AUC} = \Phi\left(\frac{\mu_X - \mu_Y}{\sqrt{\sigma_X^2 + \sigma_Y^2}}\right). \tag{5.5}$$

Note that identity (5.4) holds for any increasing transformation $g(X)$ and $g(Y)$ because $Y < X$ is equivalent to $g(Y) < g(X)$. In the radar example, if the size of radar images can be transformed to a normal distribution using an increasing transformation the AUC remains the probability that the image of a bird is smaller than an image of an airplane. In other words, the chance that the size of a random bird is smaller than the size of a random airplane on the radar is the AUC.

As follows from Property 4 of Theorem 5.3, the binormal ROC curve is above the main diagonal if

$$\mu_Y < \mu_X \text{ and } \sigma = \sigma_X = \sigma_Y. \tag{5.6}$$

This condition will be referred to as the equal-variance assumption. Under this favorable condition, the formulas above simplify:

$$R(p) = \Phi\left(\frac{\mu_X - \mu_Y}{\sigma} + \Phi^{-1}(p)\right), \quad \text{AUC} = \Phi\left(\frac{\mu_X - \mu_Y}{\sqrt{2}\sigma}\right). \tag{5.7}$$

First, we derive the threshold (specificity and sensitivity), where the total error (the sum of false negative and false positive rates), reaches its minimum under the equal-variance assumption. As follows from Property 5, it happens where $dR/dp = 1$, or

$$\frac{d}{dp}\Phi\left(\frac{\mu_X - \mu_Y}{\sigma} + \Phi^{-1}(p)\right) = \frac{\phi\left(\frac{\mu_X - \mu_Y}{\sigma} + \Phi^{-1}(p)\right)}{\phi(\Phi^{-1}(p))} = 1,$$

because $d\Phi^{-1}(p)/dp = 1/\phi(\Phi^{-1}(p))$. Thus, we conclude that the total error attains the minimum where the densities from the two groups intersect. The equality for densities implies a quadratic equation from which we infer that the optimal p is found from the equation $(\mu_X - \mu_Y)/\sigma + \Phi^{-1}(p) = -\Phi^{-1}(p)$, which finally yields $p_{\mathrm{opt}} = \Phi\left(-(\mu_X - \mu_Y)/(2\sigma)\right)$. The specificity and sensitivity probabilities are the same, $R_{\mathrm{opt}} = 1 - p_{\mathrm{opt}} = \Phi\left((\mu_X - \mu_Y)/(2\sigma)\right)$, and the optimal threshold is at the means' average, $u_{\mathrm{opt}} = (\mu_X + \mu_Y)/2$. Geometrically, this optimal discrimination rule corresponds to the intersection point of the density curves since they have the same width (SD). Summing up, for the binormal ROC curve under assumption (5.6), the optimal threshold u_{opt} produces the minimum total error and yields the same sensitivity and specificity, and the curve is symmetric around the cross diagonal. These properties are illustrated below in Example 5.5.

Second, we derive the optimal threshold when the variances are not the same. Differentiating (5.2) with respect to p and setting the derivative to zero, we conclude that as in the equal-variance case, the optimal discrimination threshold is where the densities intersect. By expressing $\Phi^{-1}(p) = (u - \mu_X)/\sigma_X$, we arrive at the quadratic equation for the optimal threshold,

$$(\mu_Y - u)^2/\sigma_Y^2 + \ln\sigma_Y^2 = (\mu_X - u)^2/\sigma_X^2 + \ln\sigma_X^2,$$

with the solution

$$u = \frac{A - \sqrt{A^2 - BC}}{B}$$

where

$$A = \mu_X\sigma_Y^2 - \mu_Y\sigma_X^2, \ B = \sigma_Y^2 - \sigma_X^2, \ C = \sigma_Y^2\mu_X^2 - \sigma_X^2\mu_Y^2 + \sigma_X^2\sigma_Y^2 \ln(\sigma_X^2/\sigma_Y^2).$$

For this threshold the tangent line to the ROC curve has slope equal to one confirming item 5 from Theorem 5.3. It is easy to see that the specificity and sensitivity probabilities are equal for $u = (\sigma_X\mu_Y + \sigma_Y\mu_X)/(\sigma_X + \sigma_Y)$. Geometrically, this is the point where the ROC curve intersects the cross diagonal of the unit square.

Linear discriminant analysis and the binormal ROC curve

Linear discriminant analysis developed by Ronald Fisher determines a plane that optimally separates two independent multivariate normal distributions with different means but the same covariance matrix. This separation naturally leads to a binormal ROC curve.

Theorem 5.4 *Linear discriminant analysis. Consider two independent multivariate normal distributions in R^m with different means but the same covariance matrix, $\mathcal{N}(\boldsymbol{\mu}_x, \boldsymbol{\Omega})$ and $\mathcal{N}(\boldsymbol{\mu}_y, \boldsymbol{\Omega})$. For given vectors $\mathbf{n}$ and $\mathbf{s}$, determine a linear discrimination rule such that if $(\mathbf{z} - \mathbf{s})'\mathbf{n} < 0$, then observation $\mathbf{z}$ belongs to the first distribution (the x distribution) and otherwise to the second distribution (the* *y distribution). Prove that the translation vector* $\mathbf{s} = (\boldsymbol{\mu}_x + \boldsymbol{\mu}_y)/2$ *and the normal vector* $\mathbf{n} = \boldsymbol{\Omega}^{-1}(\boldsymbol{\mu}_x - \boldsymbol{\mu}_y)$ *minimize the total misclassification error defined as*

$$E = \Pr((\mathbf{z} - \mathbf{s})'\mathbf{n} < 0 \,|\, \mathbf{z} \in \mathcal{N}(\boldsymbol{\mu}_y, \boldsymbol{\Omega})) + \Pr((\mathbf{z} - \mathbf{s})'\mathbf{n} > 0 \,|\, \mathbf{z} \in \mathcal{N}(\boldsymbol{\mu}_x, \boldsymbol{\Omega}))$$

with the total minimum classification error $2\Phi\left(-\delta/2\right),$ *where*

$$\delta = \sqrt{(\boldsymbol{\mu}_x - \boldsymbol{\mu}_y)'\boldsymbol{\Omega}^{-1}(\boldsymbol{\mu}_x - \boldsymbol{\mu}_y)}$$

is called the Mahalanobis distance.

Proof. Obviously, $(\mathbf{z} - \mathbf{s})'\mathbf{n} = 0$ defines a plane in R^m. This plane passes through vector $\mathbf{s}$, called the translation vector, and has the normal vector $\mathbf{n}$. Vectors on one side of the plane belong to the x-distribution and on the opposite side belong to the y-distribution. Note that the first term in E is the probability that vector $\mathbf{z}$ is ruled to belong to the x distribution but in fact it belongs the y distribution. A similar interpretation applies to the second term. To find optimal separation, we observe that, if $\mathbf{z}$ is a multivariate normally distributed vector, $(\mathbf{z} - \mathbf{s})'\mathbf{d}$ is a normally distributed random variable. Therefore, the total misclassification error can be expressed through the standard normal cdf as

$$E = \Phi\left(\frac{(\boldsymbol{\mu}_y - \mathbf{s})'\mathbf{n}}{\sqrt{\mathbf{n}'\boldsymbol{\Omega}\mathbf{n}}}\right) + 1 - \Phi\left(\frac{(\boldsymbol{\mu}_x - \mathbf{s})'\mathbf{n}}{\sqrt{\mathbf{n}'\boldsymbol{\Omega}\mathbf{n}}}\right).$$

Differentiating E with respect to $\mathbf{n}$ and setting the derivative to zero, one obtains

$$\frac{1}{\sqrt{\mathbf{n}'\boldsymbol{\Omega}\mathbf{n}}}\left[\phi\left(\frac{(\boldsymbol{\mu}_y - \mathbf{s})'\mathbf{n}}{\sqrt{\mathbf{n}'\boldsymbol{\Omega}\mathbf{n}}}\right) - \phi\left(\frac{(\boldsymbol{\mu}_x - \mathbf{s})'\mathbf{n}}{\sqrt{\mathbf{n}'\boldsymbol{\Omega}\mathbf{n}}}\right)\right]\mathbf{s} = \mathbf{0},$$

which implies $(\boldsymbol{\mu}_y - \mathbf{s})'\mathbf{n} = -(\boldsymbol{\mu}_x - \mathbf{s})'\mathbf{n}$, so that the optimal translation vector is $\mathbf{s} = (\boldsymbol{\mu}_x + \boldsymbol{\mu}_y)/2$. Now, differentiating E with respect to $\mathbf{n}$ and setting it to zero, we arrive at $\boldsymbol{\Omega}\mathbf{n} - (\boldsymbol{\mu}_y - \boldsymbol{\mu}_x) = \mathbf{0}$. Plugging in the optimal values for the translation and normal vectors into E, we obtain that both errors are the same and the total minimum classification error is $2\Phi\left(-\delta/2\right)$. □

According to the optimal discrimination rule, if $\mathbf{z}$ is an observation and $(\mathbf{z} - (\boldsymbol{\mu}_x + \boldsymbol{\mu}_y)/2)'\,\boldsymbol{\Omega}^{-1}(\boldsymbol{\mu}_x - \boldsymbol{\mu}_y) < 0$, we assign $\mathbf{z}$ to the x distribution and otherwise to the y distribution. This rule minimizes the total misclassification error. If errors have different consequences or one wants to know the interplay between specificity and sensitivity, the ROC curve can be constructed, where the two distributions are specified as

$$(\mathbf{z} - \mathbf{s})'\mathbf{n} \sim \begin{cases} \mathcal{N}(-\delta^2/2, \delta^2) \text{ if } \mathbf{z} \sim \mathcal{N}(\boldsymbol{\mu}_x, \boldsymbol{\Omega}) \\ \mathcal{N}(\delta^2/2, \delta^2) \text{ if } \mathbf{z} \sim \mathcal{N}(\boldsymbol{\mu}_y, \boldsymbol{\Omega}) \end{cases}.$$

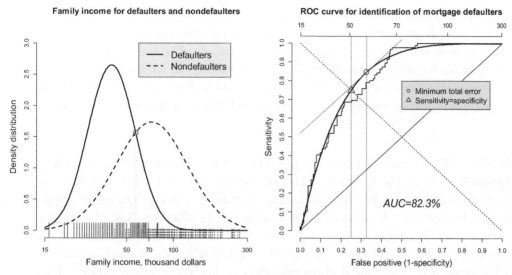

Figure 5.3: *Statistical identification of mortgage defaulters based on the family income of the applicant. An income threshold of $51K yields sensitivity equal to specificity equal to 75%. The income threshold of $57K yields the minimum combined false positive and false negative error.*

Since both distributions are normal, the ROC curve is binormal and can be derived through the threshold u by letting $F_Y(u) = \Phi\left(\frac{u - \delta^2/2}{\delta}\right)$ and $F_X(u) = \Phi\left(\frac{u + \delta^2/2}{\delta}\right)$. Geometrically, one can view the ROC curve by moving the plane $(\mathbf{z} - \mathbf{s})'\mathbf{n} = u$ when the threshold u runs from $-\infty$ to ∞, so that points for which $(\mathbf{z} - \mathbf{s})'\mathbf{n} < u$ are classified to the x distribution and points for which $(\mathbf{z} - \mathbf{s})'\mathbf{n} \geq u$ are classified to the y distribution. The AUC for this curve equals $\Phi\left(\delta/\sqrt{2}\right)$, and the error, the area above the ROC curve, is $\Phi\left(-\delta/\sqrt{2}\right)$.

Examples

Two examples apply the ROC curve to illustrate group discrimination. In both examples, the log transformation becomes very useful because the original data are positive and skewed to the right. Although the empirical ROC curve is invariant with respect to any monotonic transformation, we need to comply with the normal assumption to apply the binormal ROC curve.

The first example addresses the problem of identification of mortgage defaulters using the family income of the applicant. Obviously, the smaller the income, the greater the chance that the applicant will default – the ROC curve offers a rigorous analysis of overlooking a potential defaulter and the possibility of denying mortgage applications from nondefaulters.

Example 5.5 *Mortgage default ROC curve.* *File* mortgageROC.csv *contains the data on 375 mortgage applicants collected by a bank over the years. The*

first column Default *indicates that the applicant successfully paid off the mort-*
*gage (*no*) or defaulted (*yes*), and the second column records the family income*
(in thousands of dollars) at the time of the application. The bank wants to know
(i) how precise the decision based on family income can be made in terms of the
probability of the correct identification of a future defaulter, (ii) the probability of
denying applications of nondefaulters, and (iii) what the optimal income threshold
is.

Solution. The R function can be accessed by issuing source("c:\\StatBook\\
mortgageROC.r") in the R console. Before starting the ROC curve analysis, it
is advantageous to show the distribution of income in the two groups of applicants; see Figure 5.3. As follows from Property 3 of Theorem 5.3, although a
monotonic transformation does not affect the ROC curve, the log transformation
is advantageous to use the results of the binormal ROC curve. Indeed, as follows
from Example 5.16, the log10 transformation makes the original skewed data
normally distributed according to the q-q plot. The distribution of income on
the log10 scale is shown in the left plot using rug command. To distinguish
defaulters from nondefaulters, we use different height of the bars controlled by
the ticksize option and different style (lty). To facilitate the interpretation of
the log-transformed data, a nonuniform scale for the x-axis is used. To display
family income in thousands of dollars, we choose values on the original scale such
as 15, 50, 70, 100, and 300 used as labels in the axis command and display
them at positions after taking the log10 transform. The normal densities are
plotted using the empirical mean and sd functions in the two groups separately.
As seen from Figure 5.3 (the left plot), the mean income as well as SD for defaulters is smaller than for nondefaulters – thus the equal-variance assumption is
not appropriate for this example. The point where densities intersect corresponds
approximately to family income of \$60K.

Now we discuss the ROC curve; see the right plot. To plot the empirical ROC
curve, the option type="s" is used. The sensitivity (sens) and false positive
rates, as complementary to specificity (comp.spec), are computed as

    sens[i]=sum(XY<XY[i] & ind==1)/sum(ind==1)

    comp.spec[i]=sum(XY<XY[i] & ind==0)/sum(ind==0)

over the loop of 375 ordered values of income in the two groups combined (XY)
where the binary variable ind equals 1 for defaulters and 0 for nondefaulters.
Note that R translates the inequality condition such as XY<XY[i] & ind==1 to
1 if both conditions hold and 0 otherwise. The theoretical binormal ROC curve
is computed using the empirical means and SDs as in the densities from the left
plot. We use axis on the top of the plot (side=3) to display the income thresholds corresponding to the false positive rate on the x-axis. AUC is the area under
the empirical ROC curve computed as the sum of areas of individual rectangles
(Riemann sum), AUC=82.3%, is close to its theoretical counterpart, Binormal
AUC = 82.6%. As follows from (5.4), we interpret the AUC as the chance that a

randomly chosen mortgage defaulter has family income smaller than the income of a randomly chosen nondefaulter.

Two optimal thresholds are depicted. The triangle point on the ROC curve indicates where sensitivity equals specificity – the intersection of the curve with the cross diagonal (the dotted line that connects points (0,1) and (1,0)). If the threshold for the income is derived from equating sensitivity and specificity, we obtain $51K. An alternative threshold sets the sum of the errors to a minimum. As shown above, this is the point on the ROC curve where the slope is 1. This point is depicted with a circle. The intersection of densities in the plot at left and the point on the ROC curve with slope 1 produce the same threshold. ☐

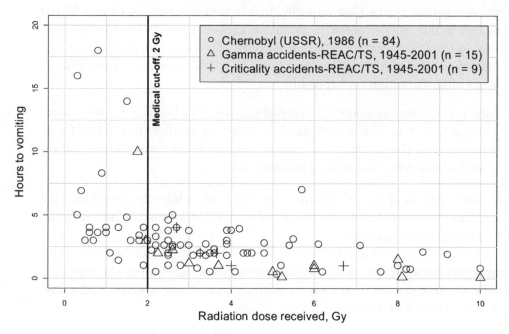

Figure 5.4: *Time to vomiting/emesis versus radiation exposure during three nuclear accidents (points represent victims). An individual requires medical attention if he/she receives radiation greater than 2 Gy. The distribution of time to vomiting is highly skewed and therefore the log transformation is needed.*

The next example is based on the work of Demidenko et al. [26] on using time to emesis/vomiting for rapidly deciding if an individual received a radiation dose greater than 2 Gy and as such prone to medical attention. As in the first example, the log transformation is applied, and the theory of the binormal ROC curve is used to compute the optimal threshold and to interpret the AUC.

Example 5.6 *Time to vomiting.* *One of the most prominent symptoms of exposure to radiation is vomiting (medical term "emesis"). The sooner vomiting begins, the more radiation the individual received. Figure 5.4 shows the data from*

108 radiation exposure accidents with a medical triage cut-off of 2 Gy. Using this data, the goal is to identify individuals who received radiation greater than 2 Gy and as such require medical attention.

Solution. The goal is to use the data presented in Figure 5.4 to understand the cons and pros of using threshold T for the time to vomiting to identify victims who received a radiation dose greater than 2 Gy. As follows from Figure 5.4, the sooner the victim starts vomiting, the greater the chance is that he/she received a dose greater than 2 Gy. See the R function vomit. Two types of mistakes can be made using time to vomiting:

1. False negative means that an individual is told not to seek care, whereas in fact he/she was exposed to radiation greater than 2 Gy. Complementary to the false negative rate is sensitivity, the correct identification of an individual who needs medical care because he/she vomited sooner than T hours after the radiation exposure.

2. False positive means that an individual was sent to a hospital for treatment, whereas in fact he/she was exposed to radiation smaller than 2 Gy. Complementary to the false positive rate is specificity, the correct identification of individuals who do not need medical care.

The consequences of the two mistakes are different: while the second mistake leads to unnecessary treatment and inconvenience, the first mistake increases the chance of death from exposure.

To compute the empirical ROC curve, for a given time threshold T, we count the proportion in the sample of who vomited less than T hours after the explosion and received a dose greater than 2 Gy (sensitivity), and those who received a dose smaller than 2 Gy (false positive):

$$\text{Sensitivity} = \frac{\#(\text{time to vomiting} < T \,\&\, \text{dose} > 2 \text{ Gy})}{\#(\text{dose} > 2 \text{ Gy})},$$

$$\text{False positive} = 1 - \text{Specificity} = \frac{\#(\text{time to vomiting} < T \,\&\, \text{dose} < 2 \text{ Gy})}{\#(\text{dose} < 2 \text{ Gy})}.$$

The empirical ROC curve is the plot of sensitivity (y-axis) versus false positive (x-axis); see Figure 5.5. Alternatively, one can depict the ROC curve by plotting the cdf of time to vomiting of those who received a dose greater than 2Gy versus time to vomiting of those who received dose less than 2Gy. To ensure that the cdfs have the same length, they should be computed on the ordered values of time to vomiting from the two groups.

Similarly to the previous example, we take the log transformation to normalize time to vomiting. Recall that this transformation does not affect the ROC curve, but it enables the application of the theory of the binormal ROC. The time-to-vomiting threshold is shown on the graph using axis(side=3,...) command.

The AUC is 83%, which means that the probability that a victim who received more than 2 Gy vomits sooner than a victim who received less than 2 Gy is 0.83. The point where the sum of errors is minimum is shown as an empty circle. The slope of the ROC curve at this point is 1 and the threshold is approximately two hours. This threshold yields 68% sensitivity and 84% specificity. The sensitivity is fairly low: 32% of victims who require medical attention will be overlooked. One may increase the threshold and say that the person should be given medical attention if he/she vomits within 3.5 hours, which improves the sensitivity to 80%, but the proportion who will be given the treatment in vain increases to 40%. One of the major lessons learned from the concept of the ROC curve is that there is an obvious trade-off between sensitivity and specificity: An increase in sensitivity leads to an increase in false positive rate. An important consideration to take into account when applied in real life is the capacity of healthcare facilities readily available to treat victims; see more discussion in the paper cited above.

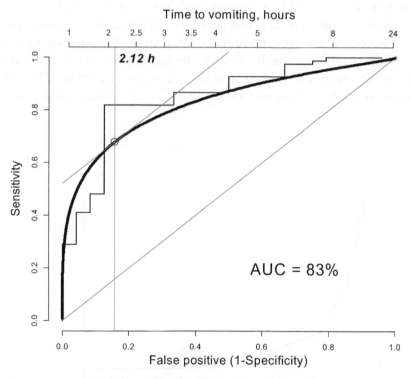

Figure 5.5: *Empirical and theoretical ROC curves to predict who received more than 2 Gy radiation based on the time to vomiting.*

5.1.2 Survival probability

From a mathematical standpoint, survival probability is complementary to the cdf, $S(x) = 1 - F(x)$. For example, if x counts days after a serious surgical operation, we are interested in the proportion who are alive by day x. If X denotes a random variable "time to death" the survival probability can be written as

$$S(x) = \Pr(X > x) = 1 - \Pr(X \leq x) = 1 - F(x),$$

where F is the cdf of X. By definition, X is a positive random variable. Survival probability is a decreasing function, $S(0) = 1$ and $S(\infty) = 0$.

More generally, we speak in terms of *time to event* where *event* may be necessarily be death. For example, in reliability theory, we speak of the time to device failure.

Survival analysis is a rich and important chapter of statistics with many clinical applications: the efficacy of a drug or a treatment is often measured by survival. Three types of survival models have been developed: (i) parametric, when the survival probability is defined up to a finite number of parameters, such as $S(x) = e^{-\alpha x}$; (ii) nonparametric, such as the Kaplan–Meier SC; and (iii) semiparametric, such as the Cox proportional hazard model. A big deal in survival data analysis is how to treat censored, or loss to follow-ups: patients may move from the area or do not show up and yet be alive, or live long enough, so that follow up is practically impossible. In fact, the Kaplan–Meier survival curve may be viewed as the empirical complementary cdf under censoring. We refer interested readers to books by Cox and Oakes [20] and Kalbfleisch and Prentice [62].

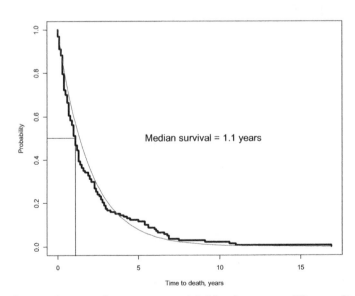

Figure 5.6: *Survival curve for patients with blood cancer. The median survival is 1.1 years.*

In Figure 5.6, we show an example of SC, the proportion of patients alive (time to death). These are blood-cancer patients after chemotherapy (the data are taken from a study by Rosenwald [84]. See the R function SCcancer. There are no censored observations in this data set because the analysis is restricted to dead people with complete follow up. As in the case of plotting the cdf, we order the observations (time to death) before plotting and use 1-(1:n)/n as the SC values on the y-axis. The median, low, and upper quartiles have clear interpretations: 50% of cancer patients die within approximately a year after treatment.

An important characteristic of survival is the *hazard function H* defined as

$$H(t) = \lim_{\Delta t \to 0} \frac{1}{S(t)} \frac{S(t) - S(t + \Delta t)}{\Delta t} = -\frac{S'(t)}{S(t)} = -\frac{d}{dt} \ln S(t).$$

In layman's language, the hazard function is the rate of dying per population alive. The only function with a constant hazard is the exponential survival function; see Section 2.4.

The ROC curve for survival

The fact that SC is complementary to the cdf allows us to use the theory of ROC curve for discrimination of survival time (or generally time to event) in two groups of patients. The following example illustrates how survival and ROC curves are connected to draw conclusions on the benefits of a new cancer drug. A new simulation technique, *bootstrap*, is used to discover the layman interpretation of the AUC. In previous simulations, the simulated data were drawn from a known theoretical distribution such as uniform or normal. In the bootstrap technique, observations are drawn from the empirical distribution, such as an empirical cdf constructed from the given data. An advantage of bootstrap is that no distribution assumption is made.

Example 5.7 *Cancer patient survival comparison. File* survcanc.csv *contains the results of a study on survival time in two groups of cancer patients: the first group (n = 103) used conventional chemotherapy (ind=0), and the second group (n = 59) used a new cancer drug. (a) Plot two SCs and the associated ROC curve to discriminate the two groups. (b) Compute the empirical AUC. (c) Use bootstrap simulations to confirm that the AUC can be interpreted as the proportion of patients who improve survival by using the new drug.*

Solution. The R function survROC produces Figure 5.7. (a) The empirical SC in each group is the proportion who lives longer than time t. Since SC is a stepwise function, we chose t as the ordered values of the survival times. From the left plot, we see that the proportion who died after t is greater in the new cancer drug group. It is not true that every patient in that group lived longer than every patient in the conventional chemotherapy group. To understand how groups can be discriminated in terms of their survival after the cancer treatment, the ROC curve is used – see the plot at right. The sensitivity is the proportion of 59 patients from the new cancer drug group and specificity is the proportion

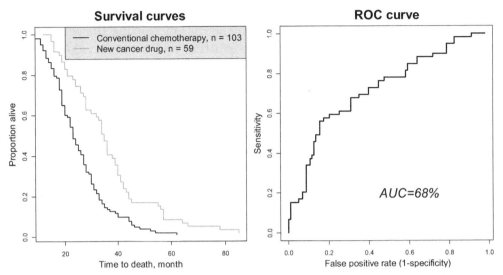

Figure 5.7: *Survival curves for two groups of cancer patients and the associated ROC curve. 68% of patients improve survival by taking new drug. In other words, the probability that a randomly chosen patient who used the new drug lives longer than a randomly chosen patients who used the conventional chemotherapy is 0.68.*

of 109 patients from the conventional chemotherapy group who lived longer than the threshold t. Again, since the empirical ROC curve is discrete, we use discrete values for the threshold, the ordered values of all $59 + 109 = 162$ survival times. Note that the rule of computation here is the same as for the regular ROC curve defined by the cdf, but now we use "greater than" to compute the proportions.

(b) The empirical AUC as the sum of the areas of small rectangles is 0.68. The interpretation is the same: based on survival time, we can discriminate patients into two groups with probability 0.68, or in layman's language, 68% of patients improve survival by taking the new drug.

(c) To verify this interpretation, the following bootstrap simulation is suggested: Imagine that many cancer patients are randomized to the two treatments: N patients receive the conventional chemotherapy and N matching patients take the new drug. The survival time of the patient from the first simulated group is a random number chosen from the sample of $n = 103$ patients (ind=0), and the survival time of the patient from the second simulated group is a random number chosen from the sample of $n = 59$ patients (ind=1). Then we compute the proportion of pairs for which the patient from the new drug group lived longer than one from the conventional chemotherapy group. In our function survROC, $N = 100,000$ was used, which yielded Percent patients who improve survival by taking new drug = 70.067, fairly close to the AUC = 68%.

Problems

1. Prove (5.1) using integration by parts and use it to show that, if $X \prec Y$, then the inequality holds for the mean.

2. Prove that $X \prec Y$ if the cdf of Y is the cdf of X shifted by a negative number.

3. Prove that Theorem 5.2 holds when g transforms $(-\infty, \infty)$ into a finite interval. Prove that part (b) of this theorem holds when X and Y are discrete random variables.

4. Does $\mu_X < \mu_Y$ for two binomial random variables with the same n imply $X \prec Y$? Is it true for the Poisson distribution?

5. Let X and Y be two independent normally distributed random variables with means μ_X and μ_Y and variances σ_X^2 and σ_Y^2, respectively. (a) Prove that a necessary condition for $X \prec Y$ is that $\mu_X < \mu_Y$. (b) Provide an example when $\mu_X < \mu_Y$ but $X \prec Y$ does not hold. (c) Prove that if $\sigma_X = \sigma_Y$, then $\mu_X < \mu_Y$ is a necessary and sufficient condition for $X \prec Y$.

6. Find an example of a pair of random variables such that $E(X) < E(Y)$ but $E(g(X)) > E(g(Y))$, where g is a strictly increasing function on the domain of two random variables. [Hint: Let X and Y be two discrete random variables that take values $(0, 1)$ and $(2/3, 3/4)$, respectively, with equal probability $1/2$. Consider function $g(x) = x^p$ where $p \to \infty$. Find p such that $E(X^p) > E(Y^p)$.]

7. (a) Prove that X is stochastically smaller than $X + Y$ where Y is a positive random variable. (b) Assuming that X and Y are continuous unimodal random variables with finite means, is it true that (a) the mean of $X + Y$ is greater than the mean of X and (b) the mode of $X + Y$ is greater than the mode of X? (c) Is it true that if X is stochastically smaller than $X + Y$, then Y is a positive random variable?

8. Prove that $E(X) = \int_0^\infty (1 - F(x))dx$ for a nonnegative random variable using (5.1).

9. Prove that $E(G) - E(X) = \int_{-\infty}^\infty (F_G(x) - F_X(x))dx$, where F_G and F_X are continuous cdfs of random variables G and X.

10. (a) Modify function salary to plot the data on the log10 scale. Display the actual salary data on the x-axis (see Figure 5.18). (b) Does the median change upon the log transformation? Does the mean change upon the log transformation? (c) Do you prefer the log scale over the original scale for this example? Explain why.

11. Modify function `salary` to compare the salaries using quantiles (order statistics). Do you prefer the quantile comparison?

12. What is the chance of living longer than five years for patients with blood cancer?

13. Reconstruct the plot in Figure 5.6. Find the lower and upper quartile as the 50% confidence interval of staying alive and display it on the plot.

14. Fit the survival data (function `SCcancer`) with the exponential cdf, and draw the line on the survival plot. [Hint: Replace θ with the survival mean; see Section 2.4.]

15. (a) Drug A has a treatment effect X and affects individuals according to $\mathcal{N}(\mu, \sigma^2)$. Drug B is stronger and affects individuals as $Y = X + \delta$, where $\delta > 0$ is an additive effect. Illustrate and interpret the difference between drugs using the cdf. (b) Do the same as above, but the effect is multiplicative: $\mu > 0$ and $Y = \delta X$, where $\delta > 1$ (use μ much larger than σ).

16. Use the definition of the hazard function as the limit to compute and plot the empirical hazard function for `DeathYears.csv` data in the R. Use `delta.t` $= \Delta t$ as an argument to the R function. Plot two empirical hazard functions with different `delta.t` to address the optimal choice of Δt.

17. (a) Sketch an ROC curve for uniformly distributed X and Y (nonoverlapping and overlapping). (b) Derive an ROC curve for exponentially distributed X and Y with rates λ_X and λ_Y, respectively. What condition on the rate parameters imply that the ROC curve is above 45° line? Derive the AUC using three formulas from Theorem 5.3. What threshold yields the maximum AUC?

18. (a) Prove that, if X and Y are positive continuous random variables, then the ROC curve for X and Y is the same as for $\ln X$ and $\ln Y$. (b) Prove that if X and Y are continuous random variables with support R^1, then the ROC curve for X and Y is the same as for $X' = g(X)$ and $Y' = g(Y)$, where $g(u) = e^u/(1+e^u)$. (c) Prove that the empirical ROC curve does not change upon an increasing transformation.

19. Prove that, if in the binormal ROC curve, $\mu_Y < \mu_X$ and $\sigma_X \neq \sigma_Y$, then the ROC curve is below the main diagonal at some point. [Hint: Use (5.2).]

20. Derive formula (5.3) using Property 5 from Theorem 5.3.

21. Under assumption (5.6), (a) express the maximum distance of the binormal ROC curve from the point $(1/2, 1/2)$ on the unit square as a function of $(\mu_X - \mu_Y)/\sigma$, and (b) find the minimum distance from the binormal ROC curve to the point $(0, 1)$.

22. Use data from Example 5.5 to plot the empirical total sum of errors and find the empirical optimal income threshold (in thousands of dollars).

23. How does one compute an average ROC curve given n samples of false positive and true positive rates from the same general population? Consider two methods: (a) combine the rates from all samples and plot x[order(x)] versus y[order(y)], and (b) compute n ROC curves on a common fine grid of false positive rates and take the average. Do the two methods produce the same ROC curve?

24. (a) Demonstrate by simulations that, if X and Y have lognormal distributions with densities specified by (2.64) with different means and variances but $\mu_Y < \mu_X$, then the proportion of $Y < X$ is well estimated by (5.5). (b) Assume that $X \sim \mathcal{N}(\mu_X, \sigma_X^2)$ and Y is lognormally distributed as in (a) and $\mu_Y < \mu_X$. Is the AUC formula still valid? Explain.

25. (a) Compute and plot the empirical sum of the two errors as a function of the threshold T in Example 5.6 and superimpose with the theoretical one. (b) Compute and plot the empirical sensitivity and specificity in Example 5.6 and superimpose with the theoretical ones. Find the threshold for which sensitivity equals specificity.

26. An empirical ROC curve is defined as a set of n pairs (f_i, t_i), where f_i and t_i are false positive and true positive rates. Approximate this curve with a binormal curve. [Hint: Use a quantile scale and simple linear regression.]

27. Reduce the problem of discriminating two samples with cdfs F_X and F_Y, respectively, with $\mu_X < \mu_Y$, to the ROC curve with the decision rule that an observation belongs to the second sample if $Y > u$. For example, in Example 5.5, we identify a nondefaulter if the applicant's family income is greater than u. (a) Modify the properties of Theorem 5.3 and reformulate the properties of the binormal ROC curve.

28. In Example 5.7, (a) write an R function with vectorized bootstrap simulations (no loop over i). (b) Find a matching theoretical lognormal distribution for each group of patients and plot the theoretical ROC curve with the AUC. [Hint: Use mean and sd to estimate μ_X, μ_Y, and the common σ.]

29. File usbflash.csv contains information on the time to failure of 179 existing USB flash drives and 97 new design USB flash drives (days). Following Example 5.7, conduct the analysis of time to failure and draw a probabilistic conclusion on the improvements of the new design using the AUC.

30. Body temperature is routinely used to identify whether a person is sick. Simulate 100 normal temperatures according to the normal distribution with mean 98.6 F° and SD = 1.5° F and 80 temperatures of people with the flu according to the normal distribution with mean 100.4° F and SD = 2° F. (a) Plot the empirical ROC curve and superimpose with the binormal ROC curve with and without the assumption that the variances are the same. (b) Compute the AUC and give an interpretation. (c) Derive an optimal temperature threshold to identify people with the flu.

31. A bank wants to develop a rule to grant or deny credit card applications. It has an access to 1996 credit card applicant data: some applicants failed a minimum payment and some did not, see R function `creditpr`. The first column indicates failed = 1, not failed = 0; the second column contains monthly paycheck; and the third column contains months at work. (a) Develop a rule to grant or not to grant credit card applications using linear discriminant analysis. (b) Plot the data points using different colors (use `legend`) and display the discriminant line. (c) Compute and display the empirical and theoretical total misclassification error.

5.2   Histogram

While it is easy to estimate the cdf, it is difficult, if possible at all, to estimate the probability density function (pdf) of a continuous random variable. One could suggest finding the empirical density as the derivative of the empirical distribution function, but it is not differentiable (it is not even continuous). The difficulty of reconstructing the density from the data can be compared with the difficulty of reconstructing the derivative of a function for which only a finite number of values are known.

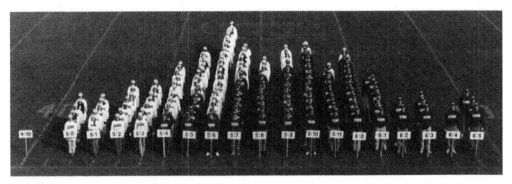

Figure 5.8: *143 students line up according to their height before a football game at Yale. Yale students wear black uniform.*

The traditional way to assess the pdf is through the *histogram*, a discretized density. The following three steps are involved in plotting the histogram:

1. Divide the range of n observed values into m equal length intervals/breaks/bins, $m < n$. The width of each bin (or bar) is $w = (x_{\max} - x_{\min})/m$ and is

called the *bandwidth*. The histogram need not be plotted on the actual range $[x_{\min}, x_{\max}]$; the range may be wider for easy presentation and interpretation.

2. Count the number of observations in each bin; these numbers are called *frequencies (or counts)*.

3. Plot m rectangles on each interval with the height equal to the number of observations from step #2. Such plot is called the frequency histogram.

It is crucial to select the right number of intervals, m. If m is small the histogram is flat and oversmooths the data; if m is large, the histogram is ragged (or undersmoothed).

A *living* histogram is depicted in Figure 5.8: 143 students line up according to their height with the natural bandwidth $w = 1$ in. The most frequent height is 5 feet 6 inches. As it happens most of the time, the counts are small at the ends with the peak somewhere in the middle. Yale students are taller following the definition of Section 5.1: the proportion of Yale students who are shorter than any h is smaller than the proportion of students from the other university.

R has a built-in function `hist`, which plots the histogram. Two options are available: the default option is `probability=F`, meaning that frequencies (counts) are plotted on the y-axis. The option `probability=T` (or `freq=F`) means that

$$\text{density} = \frac{\text{frequency}}{nw} \tag{5.8}$$

is used for the y-axis, where w is the bandwidth defined by the argument `bw` in R. Note that the right-hand side of (5.8) is not a probability but a density approximation (empirical density) as follows from (2.3) with the cdf estimated by (2.2). In particular, it may be greater than 1 unlike the option `pr=F`. Option `pr=T` is useful when the density is superimposed over the histogram. Option `breaks` controls either the number of bins/intervals or actual bin coordinates defined by `seq`.

Example 5.8 *Histogram of website hits. Plot a histogram of website hits. Use default `breaks` and your own to show the hour distribution. Show the actual data as short segments on the x-axis.*

Solution. The R code is in function `webhits.hist`, which produces Figure 5.9. The first call to `hist` uses the default bandwidth and breaks with frequencies on the y-axis. The second call uses probabilities with the breaks per hour. Note that we color the bins using `col` option. It is always informative to see the actual data – we show them using the `segments` function after `hist`. An alternative way to show the actual data on the x-axis is to use the R function `rug`.

The default number of breaks (12) oversmooths the distribution. For these data, it is natural to use one-hour breaks starting from 5 a.m., as is seen in the

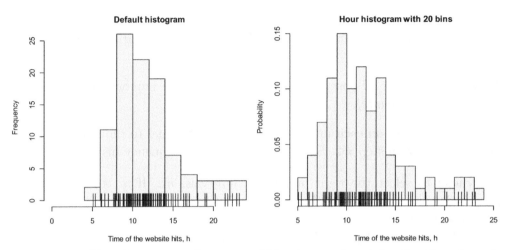

Figure 5.9: *Two histograms for comwebhits.dat produced by the R code web-hits.hist.*

right panel. As we can see, the maximum rate happens around 9–10 a.m., then there is a drop – people go for lunch. This example cautions that the default number of bins is just an option and may be subject to change. Several attempts may be taken before finding the right number of bins, breaks, and the range for the observations xlim. Note that bins are filled with the gray color gray95. The greater the number, the lighter the color.

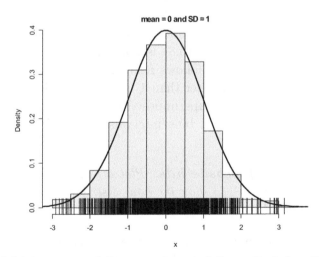

Figure 5.10: *A histogram and the superimposed theoretical density for a sample of 1,000 iid observations from $\mathcal{N}(0,1)$.*

Example 5.9 *Histogram of normal random variables.* *Generate 1,000 normally distributed random variables with mean and SD as the arguments to the function. Plot the histogram and superimpose with the theoretical density.*

Solution. The R function `histN` creates Figure 5.10. The option `probability=T` is used to plot the proportion of observations within the bin to match the density. Notice that the default limit on the x-axis does not cover the entire range of observation values. The option `breaks=seq(from=-4,to=4,by=1)` may be used instead.

Problems

1. Run `histN` with 10K observations and determine the appropriate number of bins. Specify guidelines for the choice of the *appropriate* number of bins in general terms.

2. Generate 1K iid observations from the F-distribution with $m = 3$ and $n = 10$ df. Use the default number of bins and your own most appropriate `breaks` as in Figure 5.9 (use `par(mfrow=c(1,2))`. Show the actual data on the x-axis as well).

3. Generate 100 iid observations from a Gaussian mixture distribution of $\mathcal{N}(1, 0.5^2)$ and $\mathcal{N}(2, 1)$ and $p = 0.3$ following Example 2.73. Then use a histogram for data visualization. Find the appropriate number of breaks and superimpose the histogram with the true density.

4. Count the number of football students behind each height sign from the two universities in Figure 5.8 and plot the two cdfs. Explain the statement "Yale students are taller."

5.3   Q-Q plot

A straightforward way to test whether a random sample $X_1, X_2, ..., X_n$ is drawn from a theoretical distribution with cdf F is to display the empirical and theoretical cdfs on the same plot to see how close they are. This method was used previously in simulation settings – when the number of simulations was large the two distributions were barely distinguishable. A more flexible way to test whether two distributions are the same is by visualization through quantiles. The q-q plot is justified by an obvious observation that two cdfs are the same if and only if their quantiles (inverse cdf) are the same. As follows from its definition (2.2) in Section 2.1, the empirical quantiles corresponding to probabilities $(1 - 0.5)/n, (2 - 0.5)/n, ..., (n - 0.5)/n$ are the order statistics (observations), $X_{(1)}, X_{(2)}, ..., X_{(n)}$. We subtract 0.5 to be able to compute theoretical quantiles for all observations, including maximum (otherwise, the quantile for cdf $= 1$ is not defined). Thus if the points with the x-coordinate as the theoretical quantile, $q_i = F^{-1}((i - 0.5)/n)$ and the y-coordinate as $X_{(i)}$ fall close to a straight line, there is a good indication that observations came from distribution F.

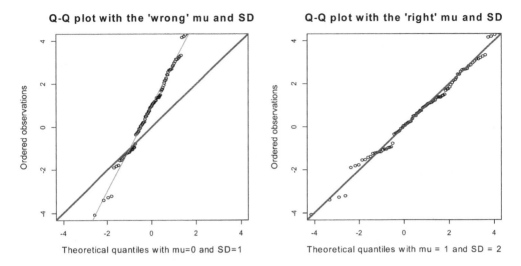

Figure 5.11: *Two q-q plots with thr "wrong" and "right" choice of μ and σ. Observations are drawn from $N(1, 2^2)$ distribution. In the plot at left, quantiles of the normal distribution with mu = 0 and SD = 1 are used and, in the plot at right, quantiles of the true mu = 1 and SD = 2 are used.*

An important advantage of the q-q plot over the cdf is that this method provides visual evidence not only for one distribution, but the entire family of distributions, up to location and scale. That is, if the q-q points fall on straight line, then X_i belong to a *family* of distributions:

$$X_i \sim F\left(\frac{x - \mu}{\sigma}\right), \tag{5.9}$$

where μ is called the location and σ the scale parameter. Importantly, parameters μ and σ are not specified, we simply assert that X_i belongs to a family of distributions specified by (5.9) with any μ and σ. For example, to verify that sample X_i has a normal distribution by matching the empirical and theoretical cdfs, we must specify the mean and variance, but those are often unknown. But if we plot $X_{(1)}, X_{(2)}, ..., X_{(n)}$ with $X_i \overset{\text{iid}}{\sim} \mathcal{N}(\mu, \sigma^2)$ versus theoretical quantiles of *any* normal distribution, such as $\mathcal{N}(0, 1)$, we always get a configuration of points close to a straight line – this follows from the identity (2.76). In short, if the order statistics plotted versus theoretical quantiles of the standard normal distribution look like a straight line, observations have a normal distribution. In fact, the intercept must be close to μ and the slope must be close to σ as follows again from (2.76). If the theoretical quantiles of the $\mathcal{N}(\mu, \sigma^2)$ distribution with exact μ and σ were used, the points should be close to the 45° line. It should be said that the q-q plot is not a rigorous. It is ad hoc technique useful for an eyeball assessment. The robust feature of the q-q plot visualization for testing the normal distribution assumption is illustrated in Figure 5.11, which is created by the R code `qqnill`.

In both plots, 100 observations are drawn from the normal distribution $\mathcal{N}(1, 2^2)$. In the left plot, quantiles of the standard normal distribution with "wrong" mu = 0 and SD = 1 are used, qnorm(x,mean=0,sd=1). Note that, even though μ and σ are not equal to the true values the points are close to a straight line, but not the 45° line. In the right plot the "right" mu=1 and SD=2 are used, qnorm(x,mean=1,sd=2). In both plots, points are close to a straight line; in the left plot points are close to the line with intercept=1 and slope=2 and in the plot at right points are close to the 45° line (intercept=0 and slope=1).

Alternatively, one can use a built-in function qqplot (or qqnorm to compare with the normal distribution); see the problem at the end of this section.

The eyeball assessment of the q-q plot works not only for the normal distribution; for example, this method is applied to the exponential distribution in the following example.

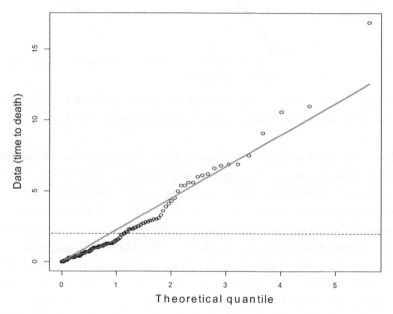

Figure 5.12: *The q-q plot for survival data: the exponential distribution is likely until time to death is one year. This suggests a possibility that factors of long-term survival are fundamentally different from those for short-term survival.*

Example 5.10 *Cancer survival.* *Use the q-q plot for data* DeathYears.csv *from Section 5.1.2 to test whether the survival time follows an exponential distribution.*

Solution. The cdf of the exponential distribution is $F(x) = 1 - e^{-\lambda x}$ with the ith quantile as the solution to equation $(i - 0.5)/n = 1 - e^{-\lambda x}$ that gives $x_i = -\log(1 - (i - 0.5)/n)/\lambda$. Hence if X would belong to an exponential distribution, the order statistics $X_{(i)}$ will be proportional to the theoretical quantiles, x_i. This means that points $(X_{(i)}, x_i)$ would fall on the straight line (no intercept). The R function `SCcancerQQ` creates Figure 5.12 upon the call `SCcancerQQ(job=1)`. As seen from this figure, the survival time follows an exponential distribution for a cohort of patients who die within a year. This analysis suggests that short and long survivals do not belong to the same distribution of patients. Note that the straight line should have a zero intercept because only the scale parameter specifies the exponential distribution ($\mu = 0$ in formula (5.9)). □

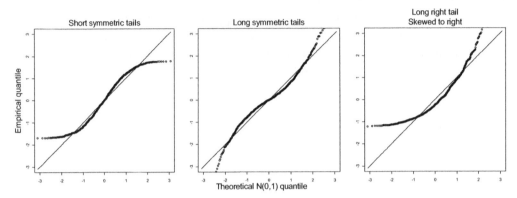

Figure 5.13: *Using q-q plots for comparison of the tails of the distribution with the normal tails. The most informative are the tails of the q-q plot.*

The q-q plot is convenient for comparison of the tails of the distribution with those of the normal distribution. Of a particular concern are observations with extreme large values, the outliers. Three situations are depicted in Figure 5.13 using simulated values: In the plot at left, the distribution had symmetric tails shorter than normal because the right empirical quantiles are smaller and the left quantiles are larger than normal. In the middle, the distribution also had symmetric tails, but they are longer/heavier than normal. In the plot at right, the distribution had a short left tail and a long right tail, the distribution is skewed to the right. Note that in all three cases the data were normalized by subtracting the mean and dividing by the standard deviation, so the 45° line is depicted for the benchmark comparison with the standard normal distribution.

In the following example, we use the q-q plot to test for the uniformity. This examples demonstrates that the q-q plot may be used to identify outliers.

Example 5.11 *National murder rate continued. Apply the q-q plot to the national murder rate from Example 1.4 to address the question whether the rates are uniformly distributed among 52 states.*

Solution. See the R function `QQmurder`. As follows from the q-q plot (not shown), the murder rates are uniformly distributed across states, except District of Columbia.

5.3.1 The q-q confidence bands

An obvious question is whether the q-q plot can be used to test the distribution assumption: if the q-q points $(q_i, X_{(i)})$ are close to a straight line, then the data X_i have the distribution F, as follows from (5.9). The q-q confidence band may be used for this purpose: we declare that data does not match the distribution if the q-q points are outside of the band. Here we explain how to compute the q-q band to test whether data come from a normal distribution. First, we normalize the data by subtracting the mean and dividing by SD (sometimes referred to as the Z-score), $Z_i = (X_i - \overline{X})/\text{SD}$ so that the working hypothesis is that $Z_i \sim \mathcal{N}(0, 1)$. Importantly, we assume that $\overline{X}$ and SD are satisfactory estimates of μ and σ, respectively. Second, we order Z and plot $Z_{(i)}$ versus theoretical quantiles $\Phi^{-1}((i - 0.5)/n)$. To compute the lower and upper bounds for the quantiles, we note that the variance of the empirical cdf, as the binomial probability, is $\Phi(q)(1-\Phi(q))/n$ so that an approximate $100\lambda\%$ confidence interval can be obtained as

$$\Phi(q) \pm Z_{(1+\lambda)/2}\sqrt{\frac{1}{n}\Phi(q)(1 - \Phi(q))}.$$

To obtain the confidence band for quantiles, transform the above confidence interval using the inverse cdf:

$$\Phi^{-1}\left(\Phi(q) \pm Z_{(1+\lambda)/2}\sqrt{\frac{1}{n}\Phi(q)(1 - \Phi(q))}\right). \tag{5.10}$$

Note that the argument of the inverse cdf may be outside $(0, 1)$ and there may be warning messages in R. This method is tested on the normal distribution in the following example.

Example 5.12 *Q-Q plot for normal distribution. Simulate $n = 20, 100$, and $1,000$ normally distributed random observations with zero mean and unit variance, do the q-q plots, and show the q-q confidence band with $\lambda = 0.95$.*

Solution. Since observations are normally distributed, the points should be within the band. The R code **qqband** creates Figure 5.14. Note that the width of the band is proportional to $1/\sqrt{n}$, that is, it is easier to detect a deviation from abnormality with large number of observations. As follows from these plots, the q-q band confirms the normal distribution. Not all points must be within the band, like the leftmost point on the right plot, because it is not 100% coverage. We expect to see about 5% of points outside the band. □

Abnormality can be detected only when the sample size is large enough: the following example illustrates this statement.

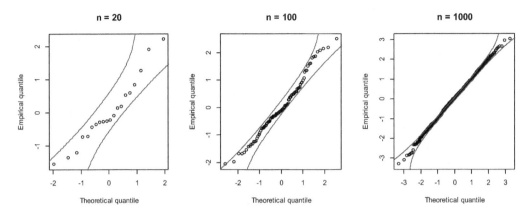

Figure 5.14: *Q-q confidence bands for testing the normal distribution assumption with different n. The confidence bands cover the majority of points: the distribution is indeed normal.*

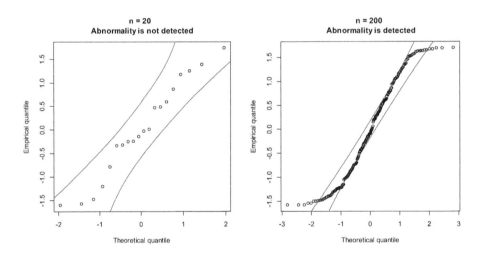

Figure 5.15: *Both samples are generated from the uniform distribution $\mathcal{R}(0,1)$. With $n = 20$, it is impossible to see the abnormality, but with $n = 200$, the abnormality is obvious. This tells us that the sample size should be large enough to see the abnormality.*

Example 5.13 *Q-Q plot for uniform distribution. Generate two samples from a uniform distribution on $(0,1)$ with $n = 20$ and $n = 200$ and use the q-q plot to detect abnormality.*

Solution. The R code is a slight modification of the R function `qqband` and is not presented here; see Figure 5.15. With $n = 20$ the abnormality is impossible to see, but with $n = 200$ the abnormality is obvious.

Example 5.14 *Q-Q plot for lognormal distribution. Simulate $n = 100$*

lognormally distributed random observations and plot the normal q-q plot with the band. Does this method detect the abnormality?

Solution. We simulate the lognormal distribution as `exp(rnorm(n=100))`; see Section 2.11. The R function is `qqLOGband` (the plot is not shown). Clearly, the points are outside of the band: the method works. □

The q-q confidence band can be generalized to other distributions. For example, consider the exponential distribution with the cdf $F(x) = 1-e^{-\lambda x}$. We start by normalizing the observations as $Z_i = X_i/\overline{X}$ so that Z_i's have an approximate cdf $F(z) = 1-e^{-z}$ ($\lambda = 1$). Since $F^{-1}(p) = -\ln(1-p)$, the q-q confidence band takes the form

$$-\ln\left(e^{-z} \pm q_{1-\alpha/2}\sqrt{e^{-z}(1-e^{-z})/n}\right), \quad z > 0, \tag{5.11}$$

where $q_{1-\alpha/2} = \Phi^{-1}(1-\alpha/2)$.

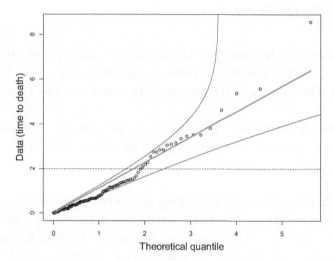

Figure 5.16: *The q-q confidence band (5.11) for cancer survival data from Example 5.15.*

Example 5.15 *Q-q plot for cancer survival.* *Use the q-q plot with the confidence band to see if the cancer survival data from Example 5.10 follow an exponential distribution.*

Solution. The call `SCcancerQQ(job=2)` creates Figure 5.16. The points are slightly outside the band around $z = 1$. However, the rejection of the exponential distribution is less obvious compared to Figure 5.12. We refer the reader to Example 9.20, where a two-component exponential model is used to fit the data. □

The next example continues the analysis of the family income as a predictor for mortgage default from Example 5.5. Recall that we suggested to taking

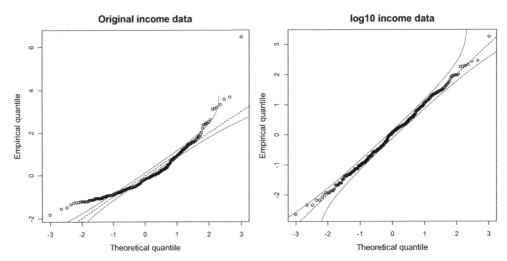

Figure 5.17: *The q-q plots for the original (thousand of dollars) and log-transformed data of the family income.*

the log of the income data to comply with the normal assumption. Below, we apply the q-q plot with the confidence band to illustrate that, after the log10 transformation, the distribution of the income data is indeed close to normal.

Example 5.16 *Family income. Apply the q-q plot with the 95% confidence band to family income data for mortgage defaulters and nondefaulters from file* mortgageROC.csv.

Solution. The caveat is that the defaulters and nondefaulters have different means and standard deviations. Therefore, we cannot apply the q-q plot to the whole sample. To circumvent this problem, we normalize the data for defaulters and nondefaulters separately and plot the combined Z-scores; see the R function mortgageQQ and Figure 5.17. In the plot at left, we show the q-q plot for the original income data, and we show log10 income in the plot at right. Obviously, the family income of mortgage applicants does not follow a normal distribution, but the log10-transformed data does.

Problems

1. Generate 100 iid observations from the uniform distribution on $[0, 1]$. Use the q-q plot to visually confirm that the distribution is uniform.

2. Generate 1,000 chi-square distributed random observations with df = 2, 4, and 100. Use the q-q plots to see how close they are to the normal distribution (use par(mfrow=c(1,3))). What conclusion, supported by CLT, can be drawn in relation to the df?

3. Use the `QQmurder` function from Example 5.11 to identify states for which murder rates follow a uniform distribution. Remove states with higher rates, recreate the q-q plot, and superimpose the straight line as in Figure 5.12.

4. Generate $n = 20$ and $n = 200$ uniformly distributed random observations from $(0, 1)$ and apply the q-q confidence band to test whether the samples are drawn from a normal distribution (use `mfrow=c(1,2)`). Draw a conclusion on the possibility of abnormality detection as a function of the sample size similarly to Figure 5.15.

5. Use the built-in function `qqnorm` in the function `qqnill`. What function does a better job?

6. Develop the q-q confidence band to test that observations are uniformly distributed. Test your band by generating uniformly distributed and normally distributed observations. [Hint: Normalize the data assuming that the support of the uniform distribution is $[\min X, \max X]$.]

7. Some claim that the square root transformation makes a Poisson distributed random variable close to normal. Use simulations and the q-q plot to test this claim.

8. File `toears.txt` contains toenail arsenic concentration data on 1057 New Hampshire residents (download the data using `x=scan("c:\\StatBook\\ toears.txt"`). Test whether `log10(x)` follows a normal distribution using the 95% q-q confidence band. More analysis of these data is found in Section 5.5.

9. File `c:\\StatBook\\Goldman.csv` contains several anatomical (osteometric) measurements of humans. (a) Use the q-q plot with the confidence band to see if the right (RHHD) and left (LHHD) humerus head diameter, and their difference are normally distributed (use `par(mfrow=c(1,3))`). (b) Does the log transformation of RHHD and LHHD make the distributions closer to normal? (c) Is `log(RHHD/LHHD)` is normally distributed? [Hint: Use `read.csv` to access the data.]

10. In Example 5.16, do not combine the Z-scores from defaulters and nondefaulters but plot them separately using different colors. Is family income normally distributed on the log scale for defaulters and nondefaulters considered separately?

11. Use the q-q plot as in Example 5.16 to provide graphical evidence that the time-to-vomiting data from Example 5.6 are normally distributed after taking the log transformation

5.4   Box plot

Box plot is used to visualize the distribution of data in a concise form. Typically, it is used when two or several samples are compared side by side.

We performed a comparison of salaries in Vermont and Connecticut earlier by means of cdf; see Figure 5.2. Now we use box plot; see function `salaryBAR` with the box plots for the two states on the original and log scales depicted in Figure 5.18.

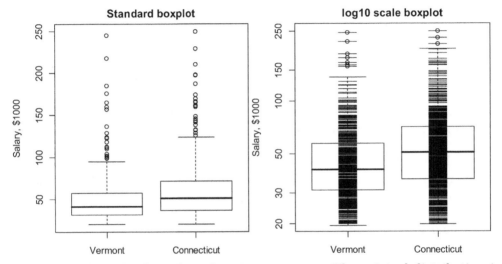

Figure 5.18: *Box plot for salary data in two states. The original distribution is not normal, but it becomes normal after taking the log (the y scale is not uniform).*

The data for the R function `boxplot` are supplied by `list` because vectors may have different length; the number of components dictates the number of box plots on the same graph, and the names are specified through the character vector `names`. The box/rectangle covers 50% of the data points: the bold line within the box is the median, the lower side of the box is the first quartile, and the upper side of the box is the third quartile (the difference is the interquartile range). The whiskers indicate either extreme values or the 1.5 interquartile range, whichever is less. Points (empty circles) outside of the box are considered as outliers. There are a few outliers with large salaries in both states. Obviously, the distribution of salary is not symmetric, so that a transformation is required to normalize the data for further statistical inference, such as testing whether the means of salaries are the same; see the plot at right. The difference between states is more visible if the data are plotted on the log scale. We make a couple of remarks on the R code when the data are plotted on the log scale: (i) we specify `xlim` to plot the horizontal segments to display the raw data and (ii) the y-axis is nonlinear but uses the actual $ scale as round amount specified in the variable `ysal`.

Problems

1. Plot a histogram from Figure 5.8 and count number of people behind each sign from $5:0$ to $6:5$. What is the median and what is the mode? Does mean have any sensible interpretation?

2. Add means to the left and right plots on Figure 5.18 and display them with different colors. Draw a conclusion on how well the mean represents the center of the distribution.

3. The file `c:\\StatBook\\Goldman.csv` contains several anatomical (osteo-metric) measurements of humans. Use `boxplot` to visualize RHHD and LHHD (use `par(mfrow=c(1,2))`). Is the log transformation useful here? [Hint: Use `read.csv()` to access the data.]

4. Download the cancer mortality rates from `https://www.cdc.gov/cancer/ npcr/uscs/download_data.htm`. Use `boxplot` to show the crude rate in all states.

5.5   Kernel density estimation

While estimation of the cdf is fairly easy, estimation of the density is not. Indeed, density estimation is somewhat similar to recovering the derivative of a function from the values known at a finite set of the argument. Not surprisingly, the density cannot be uniquely estimated without making some assumptions on the distribution family or the shape of the density.

There are two approaches to density estimation: *parametric* and *nonparametric*. In the parametric approach, the density is specified up to a finite number of unknown parameters. Then the density estimation reduces to estimation of those parameters, typically applying the method of maximum likelihood. This approach will be discussed later in the book. In the nonparametric approach, the dependence on parameters is not specified. The most popular nonparametric method of density estimation is based on the concept of the kernel density, or shortly, kernel. Silverman [95] is a classic book on nonparametric density estimation.

We start with the most popular kernel density, the standard normal density:

$$\phi(x) = \frac{1}{\sqrt{2\pi}} e^{-x^2/2}.$$

Let $\{X_i, i = 1, 2, ..., n\}$ be a random (iid) sample from a population with unknown density. Let h be a fixed positive parameter, called the *bandwidth*. We estimate the density as the average of n local kernel densities with center at X_i and SD $= h$, or more specifically as

$$f(x; h) = \frac{1}{nh} \sum_{i=1}^{n} \phi\left(\frac{x - X_i}{h}\right). \qquad (5.12)$$

Some authors call this density the Parzen density by the name of the statistician who suggested this method [79]. Formula (5.12) is merely the average of n normal densities with mean X_i and standard deviation h. Kernel density may be viewed as a smoothed histogram. Both require specification of the smoothing parameter. For histogram, one has to specify the number of bins (**breaks**) and for the kernel density one has to specify the bandwidth (**bw**).

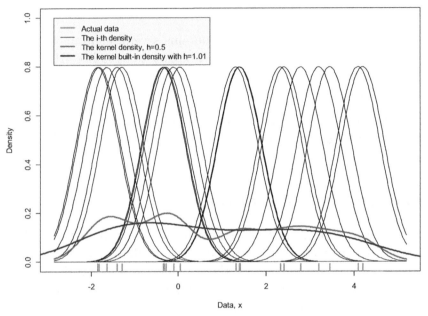

Figure 5.19: *Kernel density estimation is the average on n local densities. The ith local density has mean X_i and the same $SD = h$. The actual data are shown as bars on the x-axis (**rug** command).*

The kernel density is computed in R using the function **density**. Kernel density estimation is illustrated in Figure 5.19 for $n = 20$ randomly generated normally distributed observations with mean = 1 and SD = 2 shown as short green bars at the x-axis using the **rug** command (this figure is generated by the R function **kernavN**). The average of twenty individual densities $h^{-1}\phi\left(\frac{x-x_i}{h}\right)$, with $h = 0.5$, is shown in red; the normal kernel density computed with the built-in function **density** using an automatically computed bandwidth (default value) **bw** $= h = 1.01$ is shown in blue (run function **kernavN**). This figure illustrates that the choice of h is important: the larger the bandwidth, the smoother the density curve.

Other kernel densities can be used as well, but they should be normalized. They must have zero mean and unit variance. For example, a uniform kernel density given by $\phi(x) = 1/(2\sqrt{3})$ if $|x| < \sqrt{3}$ and 0 elsewhere may serve as a kernel. While the choice of the bandwidth is important (larger values of h oversmooth the density and smaller values make the density ragged), the choice

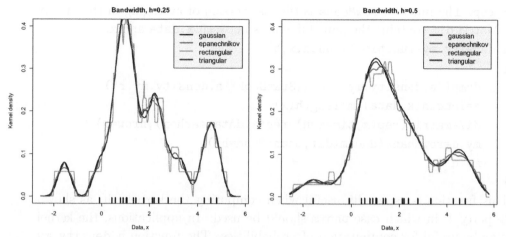

Figure 5.20: *Density estimation with four kernels. While the choice of h is important, the choice of the kernel is not.*

of the kernel is not very important, except the `rectangular` density option. Figure 5.20 illustrates this statement (see the R function `kernM`). We use the same bandwidth, h, for each kernel in each plot. The density estimates are close for each kernel, but the rectangles stand out.

Below, we list several properties of the kernel density (it is assumed that the kernel is normal):

1. Symbolically, we can express the kernel density as $n^{-1} \sum_{i=1}^{n} \mathcal{N}(X_i, h^2)$.

2. For any h and any kernel, (5.12) is a density because $h^{-1}\phi\left((x - X_i)/h\right)$ is a density.

3. When $h \to 0$, density $f(x; h)$ turns into spikes at x_i. When $h \to \infty$, the density vanishes, $f(x; h) \to 0$.

4. The kernel cdf is $F(x; h) = n^{-1} \sum_{i=1}^{n} \Phi\left((x - X_i)/h\right)$.

5. The expected value of the random variable with density $f(x; h)$, assuming that X_i are held fixed, is $\overline{X}$ with the variance h^2.

The second property is easy to prove. To prove 3, we show that $\lim_{h \to 0} h^{-1}$ $\phi\left(h^{-1}(x - X_i)\right) = 0$ for $x \neq x_i$. Express the function under the limit as $e^{g(h)}$, where $g(h) = -A/h^2 - \ln h$. Then, it is easy to see that $g(h)$ goes to $+\infty$ when $h \to 0$ or $h \to \infty$ (A and g are positive).

The kernel density can easily be computed directly from its definition; see function `n.density.my` below. The array `x.data` contains the data $\{X_i, i = 1, ..., n\}$, and `x` is the grid of values x at which the density is estimated. To avoid the loop over i, we use a matrix representation of $x - X_i$, which is computed using `rep`. The function `rowMeans` is the fast version of averaging in the matrix row. Matrix `dif` contains the pair differences between `x.data` and `x`.

```
n.density.my=function(x.data,x,h)
{
    dump("n.density.my","c:\\StatBook\\n.density.my.r")
    n=length(x.data);m=length(x)
    dif=matrix(rep(x,times=n)-rep(x.data,each=m),nrow=m)
    my.d=rowMeans(dnorm(dif,mean=0,sd=h))
    my.d
}
```

This function is especially useful when computing the kernel cdf, as shown in Property 4, in which case `pnorm` should be used. In applications, the kernel cdf may be useful for computation of probabilities. The function `n.density.my` is easy to modify for other kernels, then `dnorm` is replaced with another density function with zero mean and unit variance.

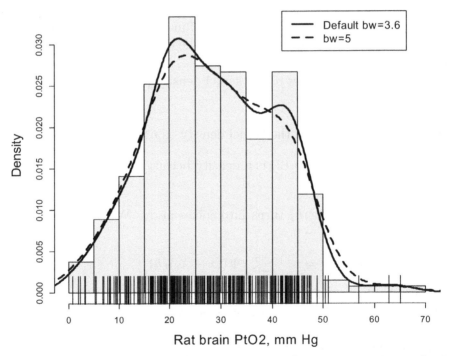

Figure 5.21: *The distribution of the rat brain pO_2 using 270 Eppendorf measurements. The density is estimated using the Gaussian kernel with two values of bandwidth. There is evidence that oxygen distribution in the brain is bimodal (two peaks).*

Example 5.17 *Rat brain oxygen distribution.* *The spatial distribution of oxygen in the brain is a fundamental biological problem (O'Hara et al. [77]).*

Understanding the distribution of the partial pressure of oxygen, pO_2, in living tissue, and in the brain particularly, is the key to understanding ischemia and stroke. One of the most popular methods of measuring pO_2 uses the Eppendorf polarographic microelectrode device. We analyze 270 Eppendorf measurements in the rat brain using histogram and kernel density.

Solution. The R function is `eppendorf`. It plots the histogram with the superimposed density – see Figure 5.21. We make several comments: (i) to see the actual data, we plot short segments at the bottom of the plot (alternatively, `rug(x)` can be used) and (ii) two Gaussian kernel densities are plotted: one uses the default bandwidth and the other uses `bw=5`. Both densities point to a bimodality of oxygen distribution.

This bimodality has an important biological interpretation: the peak at left is due to oxygen concentration in brain matter, and the second peak is due to blood vessels. Since oxygen is brought by blood, it is plausible that the density component with high values of pO_2, correspond to vessels and lower values correspond to brain matter. The saddle point, around 37 mmHg, can be used to discriminate pO_2 between the two parts of the brain.

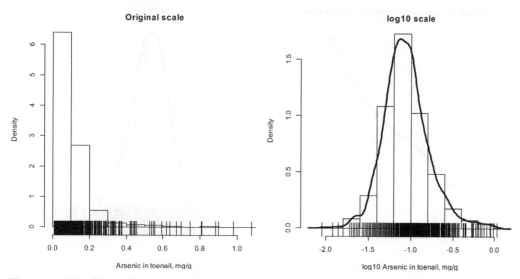

Figure 5.22: *Toenail arsenic distribution among 1,057 NH residents with default values. The log10 transformation normalizes the data.*

Example 5.18 *Arsenic toenail distribution in New Hampshire. Arsenic belongs to a group of toxic metals and may cause cancer. Several epidemiology papers relate the elevated concentration of arsenic in drinking water and the resulting excessive toenail arsenic concentration to bladder cancer. The distribution of arsenic in toenails may help environmental policymakers determine the threshold of the level of concentration above which the exposure to arsenic becomes* *dangerous for health [63]. Use histogram and kernel density for the original and log10 data on toenail arsenic distribution among 1,057 healthy New Hampshire residents. Identify the health-hazard toenail arsenic threshold.*

Solution. The default histograms on the original (mg/g) and log10 scale are shown in Figure 5.22 generated by `toears(job=1)`. Obviously, the original arsenic data are skewed, so the log transformation helps. We suggest using log10 transformation because it makes the interpretation easy, e.g. -1 means 10^{-1} mg/g arsenic. Figure 5.23 is generated by calling `toears(job=2)`. Note that we use the same limit for both axes (`xlim=c(-4,4)` and `ylim=c(-4,4)`), so that the $45°$ line connects the two corners. To identify a toenail arsenic threshold, we notice that, starting from 2, the empirical quantile is larger than the theoretical one. This means that the distribution of toenail arsenic deviates from normal after $q = 2$, which on the scale of mg/g, corresponds to 302 mg/g depicted by the vertical dotted line at the density plot at right. We may suggest that individuals with the toenail arsenic concentration above 302 mg/g have an abnormally/excessively high arsenic exposure and therefore may be considered as a health-hazard threshold.

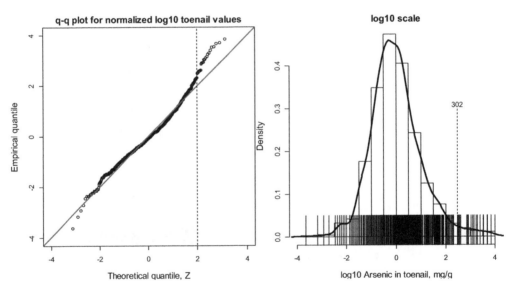

Figure 5.23: *The q-q plot to test the normality and the histogram with a larger number of bins.*

Example 5.19 *Asymmetric violin plot.* *Use kernel densities to plot the distribution of salary in VT and CT on the log10 scale side by side.*

Solution. When comparing two distributions, it is beneficial to see them side by side. The R function `asviol`, as a modification of function `salary` from Section 5.1, produces Figure 5.24. We draw attention to the nonlinear y-axis scale to display the dollar amount. Obviously, the distributions of salary in the two states are different. In particular, the proportion of VT residents with lower salary is higher than in CT.

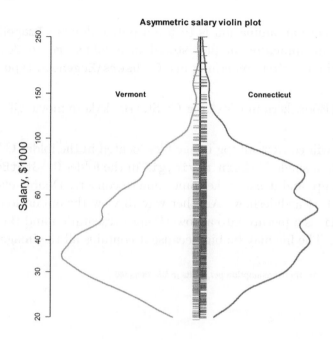

Figure 5.24: *The salary densities for two states drawn side by side.*

5.5.1 Density movie

Since the choice of bandwidth is important, it would be nice to see how density changes upon choice of bw before making the final choice on the bandwidth value. An obvious option is to plot density for an array of bw and view them in a moving fashion: we call this method the *density movie*. An example of a density movie is found in the R function kernM. The idea is simple: use several values of bandwidth (array hs) and save the graph in a file using jpeg format (other graphics formats, such as png, can be used as well). After a series of graphs have been stored, we can scroll them using keyboard arrows to see how bandwidth affects the density estimate. Alternatively, one can arrange the files in a movie using a slideshow or a special software such as XnView for PC, which makes a stand-alone executable program. The R function kern.movie generates n normally distributed observations, uses four kernels with bandwidth in the range from 0.1 to 0.5 with step 0.01, and saves the plots in the folder C:\StatBook\kern.movie\. Two typical plots are shown in Figure 5.20. We make several comments about this function: (i) the graph files should be organized as a sequence with increasing bw values;

for this purpose, we insert zeros to make the filenames of the same length and (ii) files are stored in the folder/directory `kern.movie` (this folder should be created before the function starts). Obviously, movies can be made up from any graphs in R.

Another way to create an animation is to use a public domain `ImageMagic` software. To create an animation of files stored in a folder, go to `Command Prompt`. In a black window after something like `C:\Users\Eugene>`, type

magick C:\StatBook\kern.movie*.jpg C:\StatBook\kern.movie.gif

The `magic` command will convert all jpg image files located in the folder C:\Stat Book\kern.movie\ into a single file `kern.movie.gif` in the folder `C:\StatBook\`. A black window pops up, and it may take some time to convert. Double-clicking on the `.gif` file will start a slideshow. Another way to view the animation is to insert `kern.movie.gif` as a picture into a PowerPoint presentation and then use Slide Show (press `F5`). The file may be big because it contains all the image files.

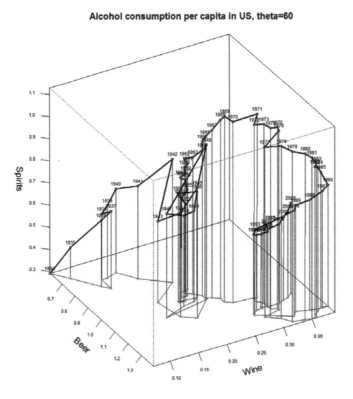

Figure 5.25: *3D scatterplot of alcohol consumption in USA over time (1934–2005) shown at the angle* `theta=60`. *Maximum spirit consumption was in the late 1960s to early 1970s.*

5.5.2 3D scatterplots

The 2D scatterplot of the sample $\{(X_i, Y_i), i = 1, 2, ..., n\}$ is essential for visualization of the relationship between Y and X. It can reveal outliers, clusters, patterns, or nonlinearity. When three variables are involved, the 3D plot is needed. The challenge here is to view the cloud of points at the right angle so that interesting patterns can be seen. To plot 3D points, the built-in R function `persp` is used in conjunction with `trans3d`. A scatterplot movie, when the 3D cloud is viewed in continuous motion at different angles, may be helpful to find an optimal angle. The following example illustrates this technique; see also Example 4.7.

Example 5.20 *Alcohol consumption in USA.* *The data on the three types of alcoholic beverages, namely, spirit, wine, and beer as per capita ethanol consumption in the United States, 1934– 2005, are obtained at http://www.niaaa.nih.gov /Resources/DatabaseResources/QuickFacts/AlcoholSales (see the dataset* `alcohol USA.csv`*). Plot the three types as a time series and view the plot at 360°, create a movie by saving the plots in the* `jpeg` *format.*

Solution. The R function is `alc3d`; it produces a 3D plot at `theta` $= 60°$ view shown in Figure 5.25. In this code, function `persp` does not plot anything; it just prepares the plot for function `trans3d` which computes the 2d coordinates of the points through projection. Showing the depths by projection on the (x, y) plane is important; the 2D coordinates of these points are computed using `trans3d` at the minimum z-values. There is a tendency of drinking more although the pattern is not straightforward.

All 360 plots are saved in the directory `c:\\StatBook\\alc3d\\`, and the movie can be played by double-clicking on `alc3d.gif` (about 23 MB). This movie was created by collecting the 360 files using the `ImageMagic` software by issuing

magick C:\StatBook\alc3d*.jpg C:\StatBook\alc3d.gif

at the command prompt, as in the previous section.

Problems

1. (a) Generate 100 random observations from $\mathcal{N}(-1, 2^2)$ and 200 random observations from $\mathcal{N}(4, 3^2)$. (b) Use the normal kernel density with the default `bw` for the combined 300 observations. (c) Use `rug` to display the original observations from the two populations using different colors. (d) Plot individual normal densities with the same color as `rug` and their liner combination with coefficients 1/3 and 2/3, respectively. (5) Find a better bandwidth by fitting a linear combination (define a criterion for the quality of the fit).

2. The same as above, but the second distribution is $\mathcal{N}(3, 1^2)$.

3. Check that n.density.my and the built-in density produce the same result using generated data.

4. (a) Generate 100 random observations from $\mathcal{R}(-2,2)$ and 200 random observations from $\mathcal{R}(1,3)$. (b) Use normal and uniform kernel densities with default bw for the combined 300 observations and superimpose with the theoretical rectangular density. (c) What kernel gives better results in terms of approximation to the true densities (provide and compute the criterion).

5. Repeat Problem 3 but instead of densities plot kernel cdf.

6. Prove properties 3 and 4 of the normal kernel density.

7. Do properties 2 – 5 hold for any kernel density (not necessarily normal)? Provide a proof or a counterexample.

8. (a) Suggest a kernel that is not listed among the options of the density function. (b) Generate a two-component sample using this kernel and smooth the density using the normal kernel (refer to Problem 1). (c) Assess how well the normal kernel density with the default bw reconstructs the true data distribution.

9. Describe the distribution of brain oxygen from Example 5.17 using a two-component normal density as the Gaussian mixture distribution $p\mathcal{N}(\mu_1, \sigma_1^2)$ $+(1-p)\mathcal{N}(\mu_2, \sigma_2^2)$. Suggest the means, SDs, and the component weight, p. Compare your mixture distribution model with the kernel density. Provide arguments in favor of the mixture over the kernel density.

10. Repeat the arsenic data analysis using a nonnormal kernel. Investigate how the health-hazard toenail arsenic threshold depends on the choice of the kernel. Which kernel is the most appropriate (provide justification)?

11. (a) Plot two estimated normal arsenic densities for theoretical quantiles $q < 2$ and $q > 2$ on the same graph as shown in Figure 5.23. (b) Use a confidence band (Section 5.3.1) in the q-q plot and derive the threshold where the points start deviating from normality and fall outside the confidence band.

12. Create an asymmetric violin density movie for the salary example using an array of bw values in the range from 0.01 to 0.1.

13. The file c:\\StatBook\\Goldman.csv contains several anatomical (osteometric) measurements of humans. (a) Use density to visualize the right (RHHD) and left (LHHD) humerus head diameter with different colors on the same plot (use par(mfrow=c(1,1)). (b) Use an asymmetric violin plot to see the differences between male and female for RHHD and LHHD (use par(mfrow=c(1,1)). Use read.csv to access the data.

14. The file c:\\StatBook\\autocrash.csv contains the data on automobile accidents from 2004 to 2016 in Allegheny County, Pennsylvania, centered at Pittsburgh (it is a subset of a larger data set found at https://data.wprdc. org/dataset/allegheny-county-crash-data). It contains 154,343 rows/ crash records with four columns CRASH_YEAR CRASH_MONTH DAY_ OF_WEEK TIME_OF_DAY. Investigate the intensity of crashes during the day of the week by plotting the kernel density of TIME_OF_DAY for all days and each day of the week using eight different colors, use legend to indicate the day of the week, and interpret the peaks.

15. Use the data on the time of crashes from the previous problem to plot the asymmetric violin plot. On the left-hand side, show weekday and, on the right-hand side, show weekend time data.

16. Report the time of crashes (see the problems above) over fifteen years 2004–2016 using par(mfrow=c(3,5)) to display the histogram with the superimposed kernel density as in Figure 5.22. Is the log transformation useful here?

5.6 Bivariate normal kernel density

Similarly to the univariate case, the goal is to estimate the joint density from n iid observation pairs (X_{1i}, X_{2i}), $i = 1, 2, ..., n$. The bivariate normal kernel density assumes that two random variables are correlated *locally* with the user-defined correlation coefficient ρ and standard deviations h_1 and h_2 for x_1 and x_2, respectively. Using the formula of the normal density from Section 3.5, the bivariate normal kernel density at point (u, v) around (x_{i1}, x_{i2}) is given by

$$f(u, v; h_1, h_2, \rho) = \frac{1}{C} \sum_{i=1}^{n} e^{-\frac{1}{2(1-\rho^2)} \left[\left(\frac{u-x_{1i}}{h_1} \right)^2 - 2\rho \left(\frac{u-x_{1i}}{h_1} \right) \left(\frac{v-x_{2i}}{h_2} \right) + \left(\frac{v-x_{2i}}{h_1} \right)^2 \right]}, \quad (5.13)$$

where $C = n(2\pi)h_1 h_2 \sqrt{1 - \rho^2}$ is the normalizing constant. It is easy to check that function f is a density as the average of densities. In the case of local independence, when $\rho = 0$, the normal kernel density simplifies considerably:

$$f(u, v; h_1, h_2) = \frac{1}{nh_1 h_2} \sum_{i=1}^{n} \phi \left(\frac{u - x_{1i}}{h_1} \right) \phi \left(\frac{v - x_{2i}}{h_2} \right). \quad (5.14)$$

Note that local independence ($\rho = 0$) does not imply the global independence (the sum of the products is not equal to the product of the sums).

The normal kernel density (5.13), viewed as the density of the pair of random variables (U, V), allows finding the conditional mean and variance by treating (x_{i1}, x_{i2}) as fixed, as stated below.

Theorem 5.21 *The conditional mean (regression) and the variance of the kernel density, assuming that (x_{1i}, x_{2i}) are fixed, are given by*

$$E(U|V = v) = \frac{\sum_{i=1}^{n}(x_{1i} + \rho\frac{h_1}{h_2}(v - x_{2i}))\phi\left(\frac{v - x_{2i}}{h_2}\right)}{\sum_{i=1}^{n}\phi\left(\frac{v - x_{2i}}{h_2}\right)}, \tag{5.15}$$

$$\text{var}(U|V = v) = (1 - \rho^2)h_1^2 + \frac{\sum_{i=1}^{n}[E(U|V = v) - x_{1i} - \rho\frac{h_1}{h_2}(v - x_{2i})]^2\phi\left(\frac{v - x_{2i}}{h_2}\right)}{\sum_{i=1}^{n}\phi\left(\frac{v - x_{2i}}{h_2}\right)}.$$

Proof. The bivariate kernel density (5.13) can be expressed as the average of n binormal normal distributions with means x_{1i}, x_{2i}, variances h_x^2, h_y^2 and correlation coefficient ρ, or symbolically as

$$\frac{1}{n}\sum_{i=1}^{n}\mathcal{N}\left(\begin{bmatrix} x_{1i} \\ x_{2i} \end{bmatrix}, \begin{bmatrix} h_1^2 & \rho h_1 h_2 \\ \rho h_1 h_2 & h_2^2 \end{bmatrix}\right).$$

According to Section 3.3, if (U, V) have density $f(u, v)$, the conditional mean of U given $V = v$ is

$$E(U|V = v) = \frac{\int_{-\infty}^{\infty} tf(t, v)dt}{f_2(v)},$$

where

$$f_2(v) = \int_{-\infty}^{\infty} f(t, v)dt = \frac{1}{nh_2}\sum_{i=1}^{n}\phi\left(\frac{v - x_{2i}}{h_2}\right)$$

is the marginal distribution of V. Now express the integral $E(U|V = v)$ as

$$\int_{-\infty}^{\infty} tf(t, v)dx_1 = \frac{1}{nh_2}\sum_{i=1}^{n}(x_{1i} + \rho\frac{h_1}{h_2}(v - x_{2i}))\phi\left(\frac{v - x_{2i}}{h_2}\right). \tag{5.16}$$

This proves (5.15). To prove the formula for the variance, we express

$$\text{var}(U|V = v) = E\left[(U - E(U|V = v))^2|V = v\right]$$

$$= E\left[(u - x_{1i} - \rho\frac{h_1}{h_2}(v - x_{2i}) + (x_{1i} + \rho\frac{h_1}{h_2}(v - x_{2i}) - E(U|V = v))^2|V = v\right]$$

$$= (1 - \rho^2)h_1^2 + E\left[E(U|V = v) - (x_{1i} + \rho\frac{h_1}{h_2}(v - x_{2i}))\right]^2.$$

Using the expression for $f_2(v)$ obtained above, we conclude the proof.   □

The regression of U on V given by formula (5.15) can be interpreted as the weighted local regression with the weights equal to $\phi\left((v - x_{2i})/h_2\right)$. The formula for $\text{var}(U|V = v)$ can be interpreted as the sum of the standard conditional variance of the binormal distribution and the variation of the nonlinear conditional mean around the standard linear regression.

The R function `bvn.density.my` computes the bivariate normal kernel density (5.13) given n observations on the plane passed by arrays `x.data` and `y.data`. Typically, density f is computed on a grid of values defined by `x` and `y` (they may have different length). Function `bvn.density.my` returns the matrix `Kxy` with number of rows equal to `length(x)` and the number of columns equal to `length(y)`. We make several comments on the `bvn.density.my` code. (i) Function f is computed using the loop/summation over i. (ii) The matrix `Kxy` is computed using vectors $\mathbf{u} = (\mathbf{x} - x_i)/h_x$ and $\mathbf{v} = (\mathbf{y} - y_i)/h_y$ based on matrix algebra: $\mathbf{u}^2 \mathbf{1}' - 2\rho\mathbf{u}\mathbf{v}' + \mathbf{1}(\mathbf{v}^2)'$. (iii) Specification of the local correlation coefficient, ρ, may be a problem; we set the local ρ equal to the global value computed using the data points as the default value.

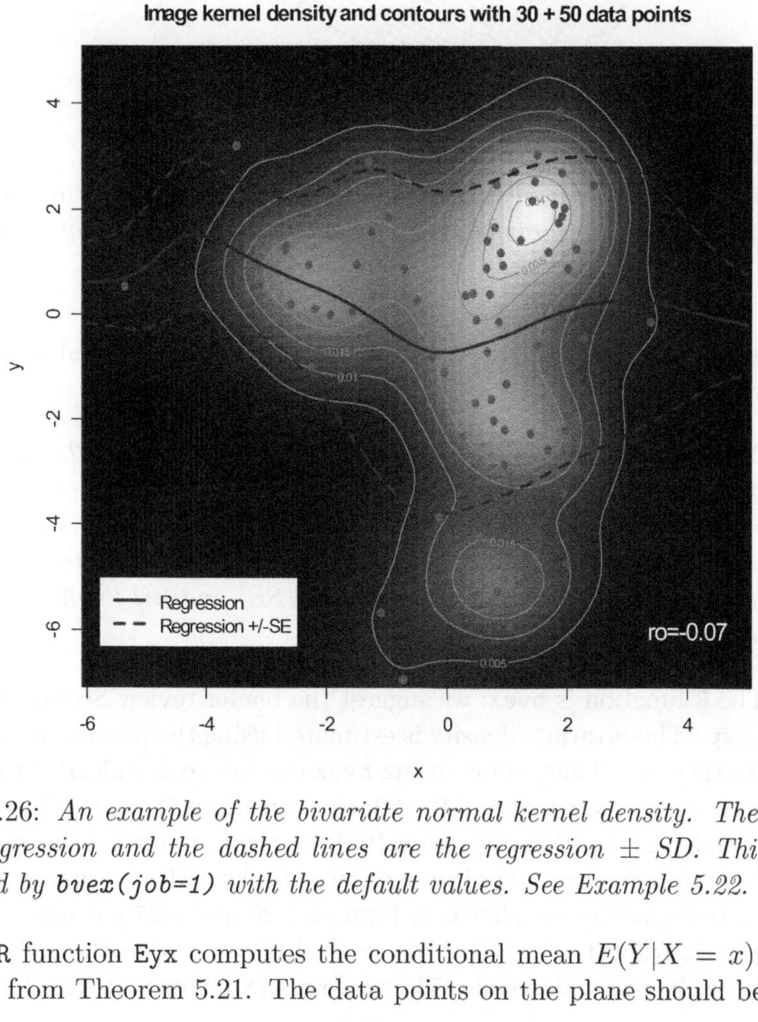

Image kernel density and contours with 30 + 50 data points

Figure 5.26: *An example of the bivariate normal kernel density. The solid line is the regression and the dashed lines are the regression $\pm$ SD. This graph is generated by* `bvex(job=1)` *with the default values. See Example 5.22.*

The R function `Eyx` computes the conditional mean $E(Y|X = x)$ using the formulas from Theorem 5.21. The data points on the plane should be provided

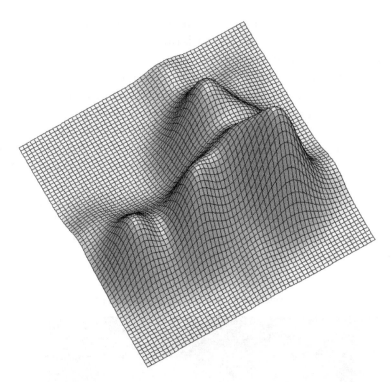

Figure 5.27: *The bivariate normal kernel density depicted using function* `persp` *with shade and illumination parameters. This image was created by issueing* `bvex(job=2).`

in the $n \times 2$ matrix `yx`. This function is used to display the conditional mean $\pm$ SE in the following example.

Example 5.22 *persp and* `image` *functions. Write an R program that generates n_1 iid normally distributed pairs from one bivariate distribution and n_2 iid normally distributed pairs from another bivariate distribution. Reconstruct the density using normal kernel. Display the density using (i) the R function* `image` *with added contours (show the conditional mean and SE) and (ii) the R function* `persp` *to plot the surface of the density.*

Solution. The R function is `bvex`; we suggest the reader review Section 4.1.1 on the use of `persp`. The bivariate density is estimated using the previously listed function `bvn.density.my`. Thus, before using `bvex` one has to download it to the R environment by using `source("c:\\StatBook\\bvn.density.my.r")`. The regression (`Exy`) and the conditional variance (`va`) are computed using the matrix algebra similarly to function `bvn.density.my`. Two graphs with contours and the bivariate density using `persp` are shown in Figures 5.26 and 5.27 corresponding to `job=1` and `job=2`, respectively. We make several comments on the code: (i) The image is shown in grayscale using 256 gray levels (8 bit image). (2) To add

contours to the image, the option add=T is used (otherwise it creates a new graph).
(3) The conditional mean and variance is computed in vectorized form (no for
loop over the grid of values x). (4) A fancy 3D plot is used when displaying the
bivariate kernel density using shade and illumination parameters (read more in
the help file by issuing ?persp).

job=3 creates and saves 360° view files of the surface, as in job=2, in the png
format. Before running this job a folder \bvex.movie\ should be created where
these files are saved. The filenames should have the same number of characters to
be saved in the order of the angle. For example, we insert two 0 for the filenames
with angles less than 10. To make a stand-alone movie run,

> convert C:\StatBook\bvex.movie*.png C:\StatBook\bvex3.gif

As discussed earlier in Section 5.5.1, ImageMagic is required to collect and convert
all png files from the folder \bvex.movie\ into one bvex3.gif file.

5.6.1   Bivariate kernel smoother for images

The kernel density can be used to smooth images. Indeed a grayscale image can
be viewed as a $n \times m$ matrix $\mathbf{M}$ with the (i, j)th element M_{ij} as the gray intensity.
The gray intensity can be measured on a different scale. In our particular example
below, we use the pgm image format, which represents a grayscale image M in
the 8-bit format as an integer from 0 (black) to 255 (white). Usually, image
data are stored as a one dimensional array, so as.matrix function is needed to
transform it into a matrix.

The reader should be aware that the numerical matrix representation and
its plotting using the image (or contour) command are different. The R code
matimage creates a random $n \times m$ matrix $\mathbf{M}$ with integer elements from 0 to
255 and plots this matrix using the image command to illustrate the difference.
Matrix values and the image are shown below. Note that the image is 90° coun-
terclockwise rotated relative to the matrix representation.

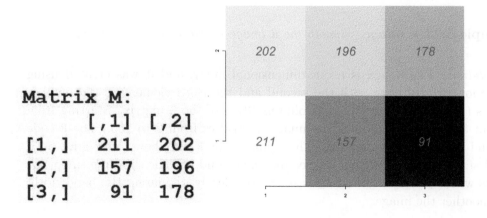

```
Matrix M:
        [,1]  [,2]
[1,]     211   202
[2,]     157   196
[3,]      91   178
```

To obtain the smoothed image, we apply the bivariate Gaussian kernel density with independent components using M_{ij} as weights in the linear combination of the normal density product:

$$K_{ij} = \sum_{k=1}^{n} \sum_{l=1}^{m} M_{kl} \phi\left(\frac{i-k}{h_x}\right) \phi\left(\frac{j-l}{h_x}\right), \quad i = 1, ..., n; j = 1, ..., m, \qquad (5.17)$$

where h_x and h_y are the bandwidths and, as such, dictate how smooth the resulting image will be. In this formula, the smoothed and the original matrix dimensions are the same, but it easy to generalize the smoothing procedure to the case when dimensions are different. If programmed in R, the above formula is very inefficient because it involves four loops over i, j, k, and l. Matrix algebra makes calculations much more efficient in R. Compute a $n \times m$ matrix $\mathbf{F}_x$ of values $\phi((i-k)/h_x)$ using the rep command with the option times for i values and each for k values, and do the same matrix computations for $\phi((j-l)/h_y)$ to create matrix $\mathbf{F}_y$. Then, to compute a matrix of smoothed values, use matrix multiplication $\mathbf{K} = \mathbf{F}_x \mathbf{M} \mathbf{F}_y'$. This method is applied to smooth the R letter in Figure 5.28.

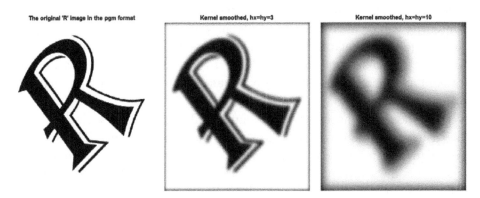

Figure 5.28: *Three smoothed images of R with different bandwidth values.*

Example 5.23 R *image. Smooth the R image provided in the file R.pgm.*

Solution. File R.pgm is a one-dimensional array, and it was created using PaintShop Pro software with the second and the third element as the number of rows (320) and the number of columns (311) of the R matrix. Starting from element 5, the array contains the image intensities from 0 to 255. The R code is found in function R.smooth. Three images of R are shown in Figure 5.28. The first image is the original image, and the second and the third are smoothed images with $h_x = h_y = 3$ and $h_x = h_y = 10$. Clearly, the larger the bandwidth, the smoother the image.

5.6.2   Smoothed scatterplot

A scatterplot depicts pairs of observations as a cloud of points. This cloud may reveal clusters, outliers, or a non-elliptical shape that may deny the bivariate normal distribution and, as such, a linear relationship. Thus, it would be informative to augment the plot with marginal and bivariate densities that may point out abnormality or nonlinear patterns. It is difficult to overemphasize the importance of visualization: "one picture's worth a thousand words." Real-world continuous data (with positive values) are skewed to right in most applications and therefore the log transformation may help; see Sections 2.10.2 and 2.11 for theoretical justification. However, when plotting observations on the log scale, we want to have axes on the original scale to ease the interpretation as illustrated in the following example.

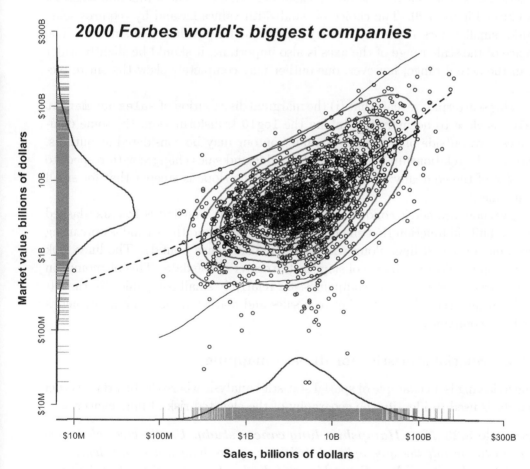

Figure 5.29: *The scatterplot market value versus sales in the World's 2000 biggest companies with superimposed one and two-dimensional kernel densities.*

Example 5.24 *Forbes market value. Use the data* `Forbes2000.csv` *to plot market value versus sales (millions of dollars) on the log scale among the World's 2000 biggest companies in 2004. Show individual distribution on the axes using the* **rug** *command and contour plots of the bivariate density to test the normal distribution assumption.*

Solution. The data are available in the package **HSAUR**. Also, it can be obtained as a part of a large data set in R and can be downloaded from `https://vincentare lbundock.github.io/Rdatasets/datasets.html`. When displaying the data on the log scale, it makes sense to use the base 10 log. We use the `rug` command and `density` to show the distribution of actual data on both axes. The two-dimensional kernel density function `bvn.density.my` is used to plot contours of the bivariate distribution of `sales` and `marketvalue` and the conditional mean function `Eyx` (they have to be downloaded beforehand). The R function `salmark` produces Figure 5.29. The choice of bandwidth values `hx` and `hy` requires some work: small values make the density jaggy and large values oversmooth. The choice of the scale range of the axes is also important, it should be slightly wider than the actual range; however, one outlier may completely skew the entire picture.

This scatterplot reveals that (i) the marginal distribution of `sales` and `market value` is close to normal after taking the `log10` transformation, (ii) some companies, say, outside of the smallest-value contour may be considered as outliers, and (iii) the relationship between market value and sales changes with respect to the size of the company: for large companies, the slope is steeper that for small companies.

Two nonparametric conditional means are shown. The first is `lowess` (dashed line), a built-in function that smooths the scatterplot, and the second is ours using `Eyx` function (solid line, along with the 95% confidence bound). The lines look similar and indicate a change of slope at the center of the data. This phenomenon may reflect heterogeneity of companies: for relatively small companies (to the left of the mean), the relationship between sales and market value is not as strong as for large companies.

5.6.3   Spatial statistics for disease mapping

The following is an example of spatial statistics analysis where the bivariate kernel density is used to identify the geography of the elevated risk of lung cancer.

Example 5.25 *New Hampshire lung cancer study. Use the bivariate kernel density to smooth the geographic distribution of the lung cancer incidence and population density in New Hampshire and display the ratio of the two densities to identify the geographic location of the elevated risk of lung cancer.*

Solution. The R function is `nhcancer`, which uses the following data files: (i) `NHtowns.csv` contains the (x, y) polygon coordinates of 259 towns in NH, (ii) xyNHcancer.csv contains the (x, y) coordinates of 10,439 lung cancers over a period of several years, and (iii) xyNHpopulation.csv contains the geographic location of 34,728 randomly chosen NH residents. The cancer and population distribution in NH is depicted in Figure 5.30 using nhcancer(job=1). Note that cancer and population are denser in southern NH. Both sets of points can be viewed as independent observations of pairs sampled from an unknown bivariate density. To estimate the density, we use function bvn.density.my. If this function is not in the current R session, you have to restore it by issuing source("C:\\StatBook\\ bvn.density. my.r"). Also to run job=3, the package jpeg must be installed (this package reads image and numerically represents it as a three-dimensional matrix corresponding to red, green, and blue color). You can download this package by issuing install.packages("jpeg").

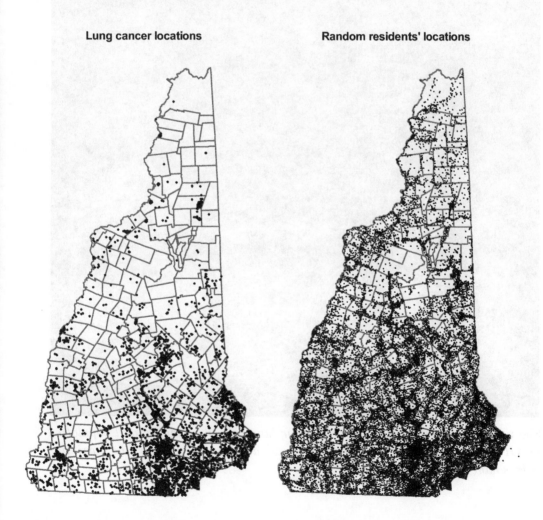

Figure 5.30: *Location of 10,439 lung cancers and 34,728 randomly chosen individuals in New Hampshire.*

job=2 plots the corresponding bivariate densities of the lung cancer incidence and NH population by contours. The maximum cancer density and population occurs at the same locations, which is not surprising (more people more cancer); see Figure 5.31. To find the location with the highest incidence per capita, we take the ratio of the two densities. The density ratio is justified by the following: let A be an area, say, a circle of radius r around a point (x_0, y_0) on the map.

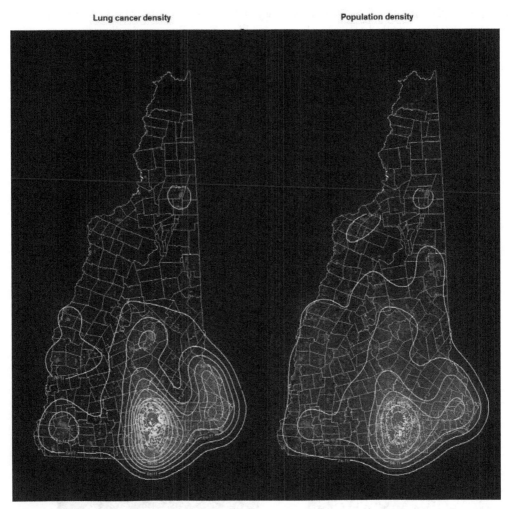

Figure 5.31: *Gaussian kernel densities for lung cancer incidence and population in NH shown via contours. The densities peak at the same location (the maximum is in southern NH).*

If $f_P(x, y)$ and $f_C(x, y)$ are densities of population and lung cancer incidence, then the rate of cancer in area A can be defined as the ratio of the number of cases in area A to the number of people living in the area A :

$$\frac{\int_A f_C(x,y)dxdy}{\int_A f_P(x,y)dxdy}.$$

If A shrinks to (x_0, y_0), then the ratio converges to

$$R(x_0, y_0) = \frac{f_C(x_0, y_0)}{f_P(x_0, y_0)},$$

the lung cancer rate. Although $R(x_0, y_0)$ is positive, it is not a density, but can be treated as density after normalization.

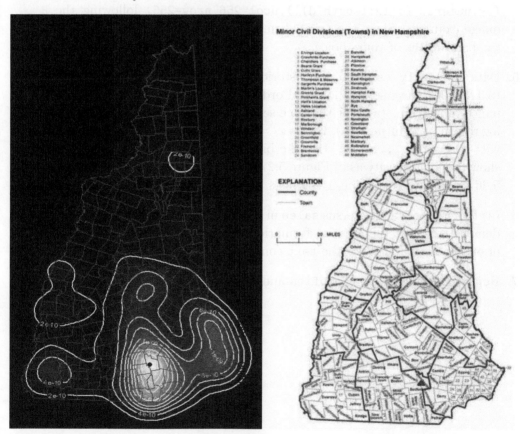

Figure 5.32: *Identification of the hot spot on the NH map, where the lung cancer rate is maximum. This graph is created by issuing* nhcancer(job=3).

job=3 identifies the hot spot on the map (triangle), where the rate of lung cancer is maximum; see Figure 5.32.

Problems

1. (a) Prove that the function of u and v defined by (5.14) is a bivariate density.
   (b) Find the bivariate cdf of the kernel density (5.14). (c) Define the kernel

density under the local independence assumption for other kernels, such as triangle or uniform, similarly to (5.14).

2. Is it true that local independence with the normal kernel implies that the regression defined in Theorem 5.21 is linear and parallel to the x-axis?

3. Generalize formula (5.17) to the case when the dimensions of the smoothed image are different from the original. Apply this formula to get a 200×200 smoothed R image.

4. Smooth the Lena image `d=scan("c:\\StatBook\\lena.pgm");` `d=matrix (as.numeric (d[12:length(d)]),ncol=256,nrow=256)` following the R image example. Show the original image and two smoothed images with the bandwidths of your choice.

5. Data `IBM_daily.csv` contains historical daily stock prices from 2/3/2016 to 1/2/1962 (reverse chronological order) downloaded from the website finance.yahoo.com. (a) Plot the time series from 1962 to 2016. (b) Plot the log return, $y_t = \ln(p_{t+1}/p_t)$, versus $x_t = \ln p_t$ using `type="l"` and `type="p"` with `ylim=c(-.1,.1)`. (c) Use the bivariate kernel density estimation to show the joint density as in Figure 5.27. (d) Use image density as in Figure 5.26 along with regression $\pm$ SD lines.

6. (a) Plot `marketvalue` versus `sales` in Example 5.24 on the original scale to demonstrate that the log transformation is needed. (b) Display the names of outlier companies (use the `text` command).

7. Repeat Example 5.24 for `profits` and `assets`.

Chapter 6

Parameter estimation

Fundamental to the statistical approach are the concepts of *general population* and *sample*. Statistics analyzes observations contained in the sample with the purpose of drawing conclusions about the general population. Collectively, the ensemble of statistical techniques is called *statistical inference*. For example, we may think of a general population of household incomes in a state or the entire country and conduct a survey that provides a sample of incomes. Statistics projects the knowledge gained from the sample to the general population. The simplest example of a statistical inference is a comparison of the average household income from the town's survey with the average income in the state, supposedly known. Importantly, we do not claim that the survey average equals the state family income. There is always an error of such estimation; the uncertainty associated with any statement regarding the comparison is unavoidable. However, one may expect that large sample size yields small uncertainty. Typically, the sample is random, so that the reported incomes $X_1, X_2, ..., X_n$ are independent and identically distributed (iid) observations drawn from a general population of all families with the probability density function (pdf), or shortly the density, f that can be succinctly written as $X_i \overset{\text{iid}}{\sim} f$.

Statistics can be parametric or nonparametric. In the former approach, f is known but defined up to an unknown parameter θ and we write $X_i \overset{\text{iid}}{\sim} f(x; \theta)$, where ; separates the argument of the density from the unknown true parameter θ. Sometimes, to reflect that θ is the parameter of the general population from which X_i are observed, we say that θ is the population parameter. If observations are continuous, f is the density as the derivative of cdf; if observations are discrete, f is the probability mass function (pmf). The unknown parameter may be multidimensional, $\boldsymbol{\theta} = (\theta_1, \theta_2, ..., \theta_m)$. Note that throughout the book we use upper case to denote random variables and lower case to denote arguments of functions, as in calculus. In nonparametric statistics, f is not defined through

Advanced Statistics with Applications in R, First Edition. Eugene Demidenko.
© 2020 John Wiley & Sons, Inc. Published 2020 by John Wiley & Sons, Inc.

parameters, and the only information given may be that the density is continuous with support on the entire line, or positive numbers. Sometimes we may use a semiparametric statistics by saying that X_i came from a distribution with the density $\sigma^{-1}f((x-\mu)/\sigma)$, where μ and σ are location and scale parameters and f is an unknown unimodal (maybe symmetric) density. In this book, we deal mostly with parametric statistics.

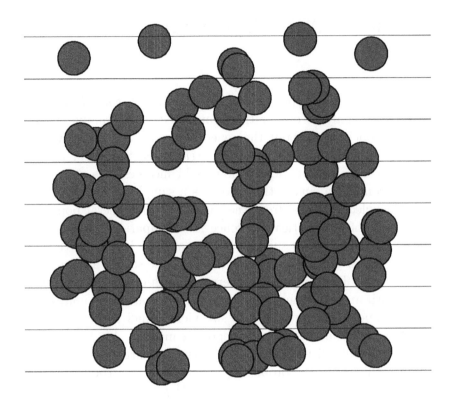

Figure 6.1: *Illustration to Example 6.1. Statistics is an inverse probability problem: what is the distance between lines if the diameter of the penny and the proportion of pennies that intersected a line are known?*

There are three major statistical tasks: (i) parameter (or point) estimation, (ii) interval estimation, (iii) statistical hypothesis testing. In a way, parameter estimation is essential because if a good estimator exists, it can be used to construct a confidence interval or treated as the test statistic for hypothesis testing. In this chapter, some fundamental methods for parameter estimation, such as the method of moments, least squares, and method of maximum likelihood, will be studied.

6.1   Statistics as inverse probability

Statistics solves an inverse problem of probability. In previous chapters, we covered probability problems where the distributions and their parameters were known. Starting from this chapter, the parameters are unknown – instead, we are given a sample from the distribution with the goal of estimating the unknown parameters. Formally, an *estimator* for θ is merely a function of data:

$$\widehat{\theta} = \widehat{\theta}(X_1, X_2, ..., X_n).$$

Sometimes a function of the data is called a *statistic*. We use "statistic" in hypothesis testing and "estimator" for parameter estimation. In statistics, we use hat to indicate that $\widehat{\theta}$ is an estimator of θ. Obviously, an estimator is a random variable since it depends on data. A value of an estimator is called *estimate*.

Basically, any probability problem can be inverted to a statistics problem.

Example 6.1 *Penny and lined paper.* *A penny of 0.75 inches in diameter is randomly dropped 100 times on a lined paper. It intersected a line in 80 throws; see Figure 6.1. Estimate the distance between the lines (d).*

Solution. Reduce the problem of tossing a coin on the lined paper to tossing a random interval $[X - 0.75/2, X + 0.75/2]$ on the real line with $X \sim \mathcal{R}(0, d)$, where X is the center of the coin between the lines. The coin falls between the lines if and only if $X - 0.75/2 > 0$ and $X + 0.75/2 < d$ or when $0.75/2 < X < d - 0.75/2$. Since X is uniformly distributed, the probability of falling between the lines is $(d - 0.75)/d$. Therefore, the probability of intersection a line in one toss is $0.75/d$. Thus, the expected number of intersections in 100 throws is $0.75/d \times 100$. This is the probabilistic answer. In statistics, we inverse this probability to get an estimate of d from the equation $0.75/d \times 100 = 80$, which gives

$$\widehat{d} = \frac{0.75 \times 100}{80} \simeq 0.94 \text{ in.}$$

Note that we put hat on d indicating that $\widehat{d}$ is an estimator of the unknown d. Importantly, $\widehat{d}$ itself is a random variable, hopefully close to the true d. In this particular estimation problem, it was easy to find an estimator for the unknown parameter using inverse probability; sometimes, this is not the case.

Example 6.2 *How much is in your bank account?* *Generate a random number from $0 to $10K and add money on your account. Guess how much money in the account.*

Solution. Let X be the random number distributed as $\mathcal{R}(0, 10000)$ and α be your account balance. Given $Z = X + \alpha$, how can we estimate α? Since $E(X) = 5000$, an estimator is $\widehat{\alpha} = Z - 5000$. Obviously, this estimator has a problem when $Z < 5000$.

Problems

1. Write an R program that simulates the "Penny and the lined paper" example. Use d as the argument of your function with the default value $d = 0.75$ and plot the d estimate as a function of throws from $N = 10$ to $N = 1000$ with step 10. Show the horizontal line with the height $= d$. Explain why $\hat{d}$ converges to d by appealing to the LLN.

2. Invert Buffon's needle problem from Example 3.68: given the number of random throws and the number of intersections of a needle of known length, L, estimate the distance between the lines. Write an R program with simulations with the default value $L = 7$ and the true unknown $D = 10$ (the needle is shorter than D). Display an estimate of D as a function of N from $N = 10$ to $N = 1000$; draw a horizontal line at the true D. [Hint: Modify the R function `buffon`.]

3. *Statistical experiment*: Take an empty clean can and scratch inside a diameter line at the bottom. The task is to estimate the diameter of the can by throwing a penny inside the can (you can shake the can for randomness) $n = 20$ times and observing the number of times it intersects the diameter. Develop an estimator for the diameter by finding the probability of intersection. Measure the diameter of the can and draw conclusions on the accuracy of your estimates. Write an R program that tests your analytical solution for the probability of intersection via simulations. [Hint: Assume that the center of the penny is randomly distributed in the circle of the diameter $d - 0.75$ where d is the diameter of the can.]

4. As a continuation of the previous problem, find what coin (diameter) is optimal for estimation of the diameter of the can. [Hint: If c is the diameter of the coin, find c such that the variance of the estimator $\hat{d}$ is minimum. Show that small- and large-diameter coins are not optimal.]

5. As in Problem 4, instead of scratching the diameter, make a hole (large enough) in the middle of the can and repeat all steps. Estimate the diameter of the can by counting the number of times the coin falls through.

6. Generalize Example 6.2 by knowing $Z_i = X_i + \alpha$, where $X_i \overset{iid}{\sim} \mathcal{R}(0, 10000)$, $i = 1, 2, ..., n$.

6.2   Method of moments

The method of moments (MM) is the most intuitively appealing method of parameter estimation. It is especially easy to implement when the theoretical mean can be expressed as a simple function of an unknown parameter. Indeed, let $X_1, ..., X_n$ be iid and the theoretical mean be written as $h(\theta)$, where h is a strictly monotonic function. Then the MM estimator is derived from the equation $h(\theta) = \overline{X}$, where $\overline{X}$ is the arithmetic average of observations. Symbolically, we may write

$$\widehat{\theta}_{MM} = h^{-1}(\overline{X}), \tag{6.1}$$

where h^{-1} is the inverse function (it uniquely exists because h is a strictly monotonic function).

Example 6.3 *MM estimator for λ in the exponential distribution.* *Given a random sample $X_1, X_2, ..., X_n$ find the MM estimator for the rate parameter λ in the exponential distribution.*

Solution. As follows from Section 2.4, the mean of the exponential distribution is $h(\lambda) = 1/\lambda$. Equating this theoretical mean to the observation average $\overline{X}$ yields $\widehat{\lambda}_{MM} = 1/\overline{X}$. ☐

An important decision in parametric statistics is the choice of the underlying distribution. In applied statistics, we not only solve the estimation problem, but also scrutinize the assumption on the distribution. The following example illustrates how one can address the choice of the distribution using the q-q plot from Section 5.3.

Table 6.1. Police arrival at the bank after alarm starts, min.

Date	Time	Date	Time	Date	Time	Date	Time	Date	Time
Sep 1	3.5*	Dec 1	4.1*	Feb 12	12.5	May 5	2.6*	July 28	1.9*
Sep 29	5.1	Dec 14	5.2	Feb16	3.9*	May 17	8.1	Jul 30	12.3
Oct 17	10.2	Dec 24	8.2	Feb 27	2.8*	May 30	9.9	Aug 3	9.4
Oct 30	3.5*	Dec 31	4.9*	Mar 23	1.7*	Jun 15	10.3	Aug 13	8.9
Nov 22	11.3	Jan 3	7.1	Apr 2	16.3	Jun 17	6.9	Aug 19	6.4
Nov 30	8.2	Jan 6	8.8	Apr 13	12.3	Jun 29	7.1	Aug 29	11.3

Example 6.4 *Police and bank robber.* *The robber knows that the alarm starts as soon as he breaks into the bank. The robber also knows that he needs 5 minutes to finish his business. (a) Assuming that the time of police arrival follows an exponential distribution, what is the probability that the robber will not be caught? (b) Use the q-q plot to test the exponential distribution assumption. (c) Estimate the probability directly from the data. See Table 6.1 (asterisk indicates that the robber was caught).*

Solution. (a) If the rate of the police arrival, λ, were known, the answer is easy to derive as complementary to the cdf (see Section 2.4):

$$\Pr(\text{robber will not be caught}) = \Pr(\text{police arrives after 5 minutes})$$
$$= 1 - (1 - e^{-5\lambda}) = e^{-5\lambda}.$$

However, as λ is unknown, we need to collect data on the time when police arrive at the bank after the alarm starts on 30 occasions, as is shown in Table 6.1. Since the average time is $\overline{X} = 7.49$ minutes, we obtain $\widehat{\lambda}_{MM} = 0.134$, and we estimate the probability that the robber will not be caught as $e^{-5 \times 0.134} = 0.51$. Obviously, the police should act faster to improve their performance.

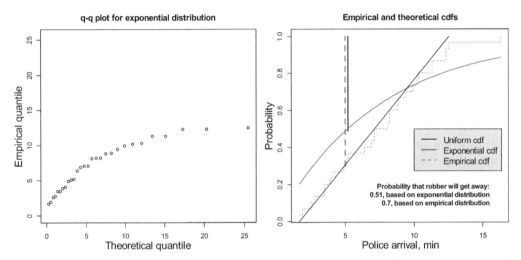

Figure 6.2: *The q-q plot and estimation of the probability that the robber will get away using parametric and nonparametric approaches. The q-q plot does not support an exponential distribution.*

(b) An important assumption in this solution is that police arrival follows an exponential distribution. We can test this assumption with a q-q plot. Plot the order statistics $X_{(1)}, X_{(2)}, ..., X_{(n)}$ on the y-axis and the theoretical quantiles of the exponential distribution on the x-axis, namely, $-\lambda^{-1} \ln(1 - (i - 0.5)/n)$ for $i = 1, 2, ..., n$. Remarkably, the choice of λ is not important for making a decision regarding the appropriateness of exponential distribution: the exponential assumption is correct if the configuration of points looks like a straight line regardless of the slope. We, however, use $\lambda = \widehat{\lambda}_{MM}$ to compute the theoretical quantiles (again, this choice is not important); see Figure 6.2. Note that the lower y-axis limit must be zero because the straight line of q-q points starts at zero. As can be seen from the left plot, exponential distribution is not a good choice to model the police arrival. Therefore, the implied probability 0.51 is questionable.

(c) For this particular example, we can estimate the probability without making an assumption on the distribution. In the spirit of nonparametric statistics, plot the empirical cdf, F, and estimate the probability that the robber gets away by evaluating $1 - F(5)$; see the right plot of Figure 6.2. Both the empirical and theoretical cdfs are shown; the lengths of the vertical segments at $x = 5$ above the cdfs are the probability estimates (they are plotted slightly apart for better visualization). The empirical probability indicates an even worse outcome: the robber gets away in 70% break-ins. This estimate is simply the proportion of times > 5 in Table 6.1, i.e. the proportion of no-asterisks. As follows from the cdf (stepwise dashed line), a better choice for the distribution of police arrival is a uniform distribution because the empirical cdf is close to straight line: it is plausible that police cars are located uniformly within the city at any specific time. The R code can be restored by issuing command `source("c:\\StatBook\\robpol.r")` in the R console.   □

The method of moments can be applied to discrete distributions as demonstrated below.

Example 6.5 *MM estimator for discrete distributions.* *Find the MM estimator for (a) the binomial probability, p, and (b) the rate of the Poisson distribution, λ.*

Solution. (a) Let $\{X_i, i = 1, 2, ..., n\}$ be the iid outcomes of a Bernoulli experiment with $\Pr(X_i = 1) = p$ and $\Pr(X_i = 0) = 1 - p$ where p is unknown; see Section 1.2. Since $E(X_i) = p$, the MM estimator is $\widehat{p}_{MM} = \sum_{i=1}^{n} X_i/n = m/n$, where m is the number of successes, $m = \sum_{i=1}^{n} X_i$. This is a well-known result: the probability is estimated as the proportion of successes in the series of Bernoulli trials. (b) Let $\{X_i\}$ be a random sample from a Poisson distribution with the rate parameter λ (see Section 1.7). Since $E(X_i) = \lambda$, we have $\widehat{\lambda}_{MM} = \overline{X}$. For both distributions, function h is linear and therefore estimators are unbiased.

6.2.1   Generalized method of moments

In the previous discussion, the goal was to estimate a single parameter. This section offers a generalization of the MM in two directions: (i) estimation of multiple parameters, and (ii) using other than the arithmetic mean functions of observations.

Suppose that several parameters, $\theta_1, \theta_2, ..., \theta_m$ are subject to estimation from a random sample $X_1, X_2, ..., X_n$, independently drawn from a general population with pdf $f(x; \theta_1, \theta_2, ..., \theta_m)$. Choose m estimating functions $M_1(x), ..., M_m(x)$ with expectations $E(M_1(X)), ..., E(M_m(X))$ expressed in a simple form, say, in a closed form, as a function of $\theta_1, \theta_2, ..., \theta_m$:

$$E(M_j(X)) = h_j(\theta_1, ..., \theta_m), \quad j = 1, 2, ..., m.$$

The generalized MM is the solution of m equations with m unknowns:

$$h_j(\theta_1, ..., \theta_m) = \frac{1}{n} \sum_{i=1}^{n} M_j(X_i), \quad j = 1, 2, ..., m.$$

Of course, not just any functions M_j qualify: they must be chosen such that the solution to the above system of equations exists and is unique. The choice of functions M_j is a big deal.

Example 6.6 *Police and bank robber continued.* *(a) Assuming that the time of police arrival is uniform, find the MM estimator for the lower and upper limit using the empirical mean and SD. (b) Apply the MM estimator to estimate the probability that the robber will get away.*

Solution. (a) Let $X_i \overset{\text{iid}}{\sim} \mathcal{R}(\alpha, \beta)$, $i = 1, 2, ..., n$. Use $\overline{X}$ and

$$\widehat{\sigma} = \sqrt{\frac{1}{n-1} \sum_{i=1}^{n} (X_i - \overline{X})}$$

for the MM estimators for α and β by solving two equations: $(\alpha + \beta)/2 = \overline{X}$ and $(\beta - \alpha)/\sqrt{12} = \widehat{\sigma}$. The MM estimators are $\widehat{\alpha}_{MM} = \overline{X} - \sqrt{3}\widehat{\sigma}$, $\widehat{\beta}_{MM} = \overline{X} + \sqrt{3}\widehat{\sigma}$. (b) For the data from Table 6.1, we obtain $\widehat{\alpha}_{MM} = 1.104075$ and $\widehat{\beta}_{MM} = 13.87592$. Therefore, the probability that the police arrive after five minutes (the robber gets away) $= (\widehat{\beta}_{MM} - 5)/(\widehat{\beta}_{MM} - \widehat{\alpha}_{MM}) = 0.69496$, very close to the nonparametric estimate.

Example 6.7 *Generalized MM estimator for λ in the exponential distribution.* *Use (a) $M(x) = e^{-x}$ and (b) $N(x) = x^k$ to derive generalized MM estimators for λ, where k is a given positive number.*

Solution. (a) It is easy to obtain the expected value,

$$\begin{aligned} E(M(X)) &= \lambda \int_0^\infty e^{-x} e^{-\lambda x} dx = \lambda \int_0^\infty e^{-(\lambda+1)x} dx \\ &= \frac{\lambda}{\lambda+1}(\lambda+1) \int_0^\infty e^{-(\lambda+1)x} dx = \frac{\lambda}{\lambda+1}, \end{aligned}$$

because $(\lambda + 1)e^{-(\lambda+1)x}$ is the density of the exponential distribution with the rate $\lambda + 1$. The solution of $\lambda/(\lambda + 1) = \overline{e^{-X}}$, where $\overline{e^{-X}} = \sum_{i=1}^n e^{-X_i}/n$, is $\widehat{\lambda}_{GMM} = \overline{e^{-X}}/(1 - \overline{e^{-X}})$. (b) Using the definition of the gamma distribution from Section 2.6, we obtain $E(N(X)) = \lambda \int_0^\infty x^k e^{-\lambda x} dx = \Gamma(k+1)/\lambda^k$. Therefore an alternative estimator of λ is

$$\widetilde{\lambda}_{GMM} = \left(\frac{1}{n\Gamma(k+1)} \sum_{i=1}^n X_i^k \right)^{-1/k}.$$

Example 6.8 *Generalized MM estimation of the gamma distribution.* *(a) Derive the MM estimator for the shape (α) and rate (λ) of the gamma distribution (Section 6.2.1). (b) Derive the generalized MM estimator for α and λ using two estimating functions $M_j(\alpha, \lambda) = X^{k_j} e^{-\theta_j X}$, where k_j and θ_j are fixed positive numbers ($j = 1, 2$). (c) Solve the nonlinear system of equations iteratively by Newton's algorithm. (d) Carry out simulations and display the mean values of two MM estimators as a function of the sample size.*

Solution. (a) From Section 6.2.1, we know that if X is gamma distributed with the shape parameter α and rate parameter λ, then $E(X) = \alpha/\lambda$ and $\mathrm{var}(X) = \alpha/\lambda^2$. We use this fact to find the MM estimators for α and λ by solving equations $\overline{X} = \alpha/\lambda$ and $\widehat{\sigma}^2 = \alpha/\lambda^2$, where $\overline{X}$ and $\widehat{\sigma}^2$ are sample mean and variance, $\widehat{\lambda}_{MM} = \overline{X}/\widehat{\sigma}^2$ and $\widehat{\alpha}_{MM} = \overline{X}^2/\widehat{\sigma}^2$.

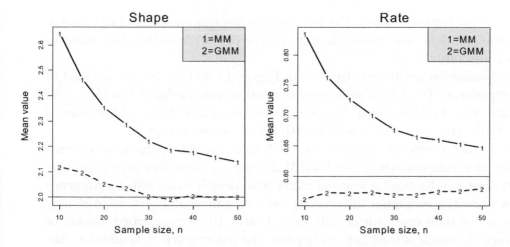

Figure 6.3: *Simulation results for MM and GMM for the shape (α) and the rate (λ) of the gamma distribution (the true $\alpha = 2$ and $\lambda = 0.5$). The GMM estimator has a smaller bias.*

(b) The suggested estimating functions admit a closed-form expression for the theoretical mean:

$$\frac{\lambda^\alpha}{\Gamma(\alpha)} \int_0^\infty x^{\alpha-1}e^{-\lambda x}x^k e^{-\theta x}dx = \frac{\lambda^\alpha}{\Gamma(\alpha)} \int_0^\infty x^{(\alpha+k)-1}e^{-(\lambda+\theta)x}dx$$

$$= \frac{\lambda^\alpha}{\Gamma(\alpha)} \times \frac{\Gamma(\alpha+k)}{(\lambda+\theta)^{\alpha+k}} = \frac{\lambda^\alpha \Gamma(k+\alpha)}{(\lambda+\theta)^{\alpha+k}\Gamma(\alpha)}.$$

Now let k_j and θ_j be any fixed positive numbers. Denote the empirical M values as $M_j = \sum_{i=1}^n X_i^{k_j}e^{-\theta_j X_i}/n$ for $j = 1, 2$. Then the generalized MM estimators for α and λ are the solutions to the following system of equations:

$$\alpha \ln \lambda - (\alpha+k_j)\ln(\lambda+\theta_j) + \ln\Gamma(\alpha+k_j) - \ln\Gamma(\alpha) - \ln M_j = 0, \quad j = 1, 2. \quad (6.2)$$

This system cannot be solved for α and λ in closed form – an iteration algorithm is required.

(c) We solve system (6.2) using Newton's algorithm; see Section 10.6.3. Let the 2×1 vector **g** denote the left-hand side of (6.2) and the 2×2 matrix **H** denote the matrix of their derivatives:

$$\mathbf{H} = \begin{bmatrix} \ln\lambda - \ln(\lambda+\theta_1) + \psi(\alpha+k_1) - \psi(\alpha) & \alpha/\lambda - (\alpha+k_1)/(\lambda+\theta_1) \\ \ln\lambda - \ln(\lambda+\theta_2) + \psi(\alpha+k_2) - \psi(\alpha) & \alpha/\lambda - (\alpha+k_2)/(\lambda+\theta_2) \end{bmatrix},$$

where $\psi = (\ln \Gamma)'$, the derivative of the log gamma function (`digamma` in R).
Newton's algorithm iterates as follows:

$$
\begin{bmatrix} \alpha_{s+1} \\ \lambda_{s+1} \end{bmatrix} = \begin{bmatrix} \alpha_s \\ \lambda_s \end{bmatrix} - \mathbf{H}_s^{-1} \mathbf{g}_s, \quad s = 0, 1, ...,
$$

where s is the iteration index. The iterative process continues until two consecutive iterations yield close values. It is reasonable to start from the MM estimates derived previously in (a).

(d) Simulation results are depicted in Figure 6.3 and the R code is found in file `gMMgamma.r`. The default number of simulations is `nSim=5000`, $k = (1, 2)$, and $\theta = (0.1, 0.5)$. As follows from Figure 6.3, the generalized MM yields estimates closer to the true value $\alpha = 2$ and $\lambda = 0.6$ for all sample sizes, n.

We make a couple of comments on the R code: (i) `gMM.gamma` is an internal function that computes the generalized MM estimates using Newton's iterations. Iterations stop if its number exceeds `maxIt` with default value 100. Convergence occurs when the maximum discrepancy between values at two sequential iterations is less than `eps` with default value 0.001. (ii) A regularized version of Newton's algorithm is used here to improve the convergence: 1 is added to the diagonal elements of matrix $\mathbf{H}$.   □

Method of moments is typically more robust to distribution specification, but less efficient compared with a more traditional method of maximum likelihood, however the distribution must be known; see Section 6.10 for more detail.

Problems

1. Explain why the *Penny and the lined paper* example from the previous section can be viewed as the method of moments.

2. Revisit Example 6.4. Assume that the distribution is uniform, $\mathcal{R}(0, \theta)$. Modify function `robpol`, replot Figure 6.2, and estimate the probability.

3. n iid observations are drawn from a uniform distribution, $\mathcal{R}(0, \theta)$. (a) Apply the generalized method of moments with $M(x) = x^k$ where $k > 0$. (b) Use simulations to plot the expected value of $\hat{\theta}$ versus n for $k = 1, 2$ and 3 with the true values in the interval $\theta \in (1/2, 2)$.

4. Estimation of α from Example 6.2: given n iid values of Z_i from $\mathcal{R}(\alpha, 10000 + \alpha)$, estimate how much in the account using MM and GMM with $M(x) = x^k$ where $k > 0$. Use simulations to address the problem of an optimal k.

5. n iid observations are drawn from a uniform distribution $\mathcal{R}(\tau, \theta)$. Use the first two moments to estimate τ and θ.

6. Find the method of moments estimator for λ in the Laplace distribution; see Section 2.4.1. [Hint: Use the fact that $\text{var}(X) = 2/\lambda^2$.]

7. Find the generalized MM estimator of λ in the Poisson distribution using $M(x) = e^{\nu x}$ where ν is fixed. [Hint: Use the MGF from Example 2.22.]

8. (a) Run function gMMgamma for larger values of n, say, n from 100 to 500. (b) Investigate how the choice of k and θ affects the quality of the GMM estimator. Draw conclusions on the optimal k.

9. A circle of unknown radius r is drawn at the center of the bottom of a jar of a known radius R, say, $R = 3$ in. Quarters (diameter $= 0.955$ in.) are dropped in the jar until the first overlap with the circle at the bottom. (a) Write simulations in R and plot the bottom of the jar with the dropped quarters similarly to Figure 6.1. (b) Estimate r. (c) Test your answer with simulations. [Hint: Consult Section 3.7.]

6.3 Method of quantiles

Instead of equating theoretical and empirical moments, we can equate theoretical and empirical quantiles of the distribution. This method of estimation, called the *method of quantiles* (MQ), is especially convenient when the population cdf is given in a form easy to invert. The empirical quantile, given probability p, is easy to obtain: if $X_1, X_2, ..., X_n$ is a random sample, the pth quantile is the (pn)th order statistic, $X_{(pn)}$, where $X_{(1)}, X_{(2)}, ..., X_{(n)}$ denote observations written in the ascending order (see Section 3.9). If pn is not an integer, we round it to the closest integer. If observations are in array X, we find the pth quantile as (X[order(X)])[p*n].

Example 6.9 *MQ estimator for the rate parameter.* *Find the MQ estimator of the rate parameter, λ, in the exponential distribution using the median.*

Solution. Since the theoretical 0.5 quantile is the solution to the equation $1 - e^{-\lambda M} = 0.5$, the MQ estimator of the rate parameter is $\widehat{\lambda}_{MQ} = \ln 2/M$, where M is the median.

Example 6.10 *MQ estimators for the scale and location parameters.* *(a) Find the MQ estimator for the scale and location parameters, θ and μ, based on an iid sample from the distribution defined by the cdf $F(x; \mu, \theta) = G((x-\mu)/\theta)$, where G is a known cdf, using two probabilities, p_1 and p_2. (b) Apply this method to estimate the mean, μ, and standard deviation, σ, using observations $X_i \overset{iid}{\sim} \mathcal{N}(\mu, \sigma^2)$, $i = 1, 2, ..., n$.*

Solution. (a) Define q_1 and q_2 as the respective quantiles of cdf G, $q_j = G^{-1}(p_j)$, $j = 1, 2$. The theoretical quantiles of cdf F are $\mu + \theta q_j$. The MQ estimators are found from solution of the equations $\mu + \theta q_1 = X_{(np_1)}$ and $\mu + \theta q_2 = X_{(np_2)}$:

$$\widehat{\theta}_{MQ} = \frac{X_{(np_2)} - X_{(np_1)}}{q_2 - q_1}, \quad \widehat{\mu}_{MQ} = \frac{X_{(np_1)} q_2 - X_{(np_2)} q_1}{q_2 - q_1}.$$

(b) Since the cdf of the normal distribution $\mathcal{N}(\mu, \sigma^2)$ is $\Phi((x - \mu)/\sigma)$, the MQ estimators can be expressed through inverse Φ as

$$\widehat{\sigma}_{MQ} = \frac{X_{(np_2)} - X_{(np_1)}}{\Phi^{-1}(p_2) - \Phi^{-1}(p_1)}, \quad \widehat{\mu}_{MQ} = \frac{X_{(np_1)} \Phi^{-1}(p_2) - X_{(np_2)} \Phi^{-1}(p_1)}{\Phi^{-1}(p_2) - \Phi^{-1}(p_1)}.$$

Of course, the choice of p_1 and p_2 is important and may be subject to optimization. Unfortunately, the optimal choice often depends on unknown μ and σ.

Problems

1. Find the MQ estimator of θ using n iid observations from $\mathcal{R}(0, \theta)$. Use theory and simulations to research the quality of estimation as a function of p dependent on θ. [Hint: Use the results of Section 3.9.]

2. Find the MQ estimators of a and b of the uniform distribution $\mathcal{R}(\alpha, \beta)$; see Section 2.3.

3. Find the MQ estimators for μ and λ in the double-exponential distribution; see Section 2.4.1.

4. The cdf of the power law distribution is $x^\alpha/(1 + x^\alpha)$, where $x > 0$ and α is a positive parameter. Estimate α using the MQ method.

5. Investigate the choice of p_1 and p_2 for optimal estimation of σ in Example 6.10(b) through simulations in R. [Hint: Define "optimal estimation."]

6.4 Statistical properties of an estimator

Not every function of data can be used as a reasonable estimator. In this section, we define some properties and criteria that can be used to select a good estimator. As a general, but vague guideline, an estimator $\widehat{\theta}$ is good if it is close to the true value of the population parameter θ. Three concepts are used to define how close an estimator is to the true value: unbiasedness, mean square error (MSE), and consistency. The first two are referred to as small-sample properties. The last is a large-sample property.

6.4.1 Unbiasedness

In previous examples, we informally judged on the quality of an estimator by how close its expected value was to the true parameter. Below, we give a formal definition.

Definition 6.11 *An estimator $\widehat{\theta}$ is called unbiased if its expected value equals the true parameter, $E(\widehat{\theta}) = \theta$.*

If an estimator $\widehat{\theta}$ is biased, we define the bias as $E(\widehat{\theta}) - \theta$. The bias is negative if $E(\widehat{\theta}) < \theta$ and the bias is positive if $E(\widehat{\theta}) > \theta$. In the former case we say that the estimator underestimates and overestimates θ in the latter.

Example 6.12 *Unbiasedness of the mean and sample variance.* Let $X_1, X_2, ..., X_n$ be iid observations from a distribution with unknown mean μ and variance σ^2. Then the mean (average) and sample variance,

$$\overline{X} = n^{-1} \sum_{i=1}^{n} X_i \text{ and } \widehat{\sigma}^2 = (n-1)^{-1} \sum_{i=1}^{n} (X_i - \overline{X})^2$$

are unbiased estimators for μ and σ^2, respectively.

Solution. The unbiasedness of the average follows from $E(\overline{X}) = \sum_{i=1}^{n} E(X_i)/n = \sum_{i=1}^{n} \mu/n = \mu$. To prove the unbiasedness of $\widehat{\sigma}^2$, we note that, without loss of generality, we can assume that X_i has zero mean because we can represent $X_i - \overline{X} = (X_i - \mu) - \overline{(X - \mu)}$. Now, expanding the square term in the variance, we obtain $\sum_{i=1}^{n}(X_i - \overline{X})^2 = \sum_{i=1}^{n} X_i^2 - 2\overline{X}\sum_{i=1}^{n} X_i + n\overline{X}^2$. Since $E(X_i) = 0$, we have $E(X_i^2) = \sigma^2$ and for the first term in the right-hand side of the above equation, we have $E\left(\sum_{i=1}^{n} X_i^2\right) = n\sigma^2$. For the second term, we have $E(X_iX_j) = 0$ if $i \neq j$, and $E(X_iX_j) = \sigma^2$ if $i = j$, due to independence. Therefore,

$$E\left(\overline{X}\sum_{i=1}^{n} X_i\right) = \frac{1}{n}E\left(\sum_{i=1}^{n}\sum_{j=1}^{n} X_iX_j\right) = \frac{1}{n}\sum_{i=1}^{n}\sum_{j=1}^{n} E(X_iX_j) = \frac{1}{n}n\sigma^2 = \sigma^2.$$

For the third term, we have

$$E(n\overline{X}^2) = \frac{1}{n}E\left(\sum_{i=1}^{n} X_i\right)^2 = \frac{1}{n}E\left(\sum_{i=1}^{n} X_i^2\right) + \frac{1}{n}E\left(\sum_{i\neq j}^{n} X_iX_j\right)$$

$$= \frac{1}{n}n\sigma^2 + \frac{1}{n}\sum_{i\neq j}^{n} E(X_iX_j) = \sigma^2.$$

Collect the terms as $E\left(\sum_{i=1}^{n}(X_i - \overline{X})^2\right) = n\sigma^2 - 2\sigma^2 + \sigma^2 = (n-1)\sigma^2$. Finally, $E(\widehat{\sigma}^2) = (n-1)\sigma^2/(n-1) = \sigma^2$, the estimator $\widehat{\sigma}^2$ is unbiased for σ^2. Importantly, these estimators remain unbiased regardless of the distribution. □

It is easy to see that $\overline{X}$ is unbiased even if X_i do correlate but have the same mean, and $\widehat{\sigma}^2$ remains unbiased if the observations are pairwise uncorrelated, $\text{cov}(y_i, y_j) = 0$ for $i \neq j$. The unbiased estimation of σ is considered later in the section.

An unbiased estimator may not exist. For example, there is no unbiased estimator of p^2 when only a single Bernoulli observation is available (the observation itself is an unbiased estimator of p). Indeed, let X be a binary outcome such that $\Pr(X = 1) = p$ and $\Pr(X = 0) = 1 - p$. If $f(X)$ is an unbiased estimator of p^2, then the following must hold: $E(f(X)) = pf(1) + (1 - p)f(0) = p^2$ for all $p \in (0, 1)$. But this is impossible because the left-hand side is a linear function and the right-hand side is a quadratic function.

In a more general case, when the distribution is continuous, for $\widehat{\theta} = g(X_1, ..., X_n)$ to be an unbiased estimator of θ, the following integral functional equation must hold.

$$\int_{R^n} g(x_1, ..., x_n) f(x_1; \theta) \cdots f(x_n; \theta) dx_1 \cdots dx_n = \theta, \quad \forall \theta$$

Only in rare cases does there exist function g such that this equation holds for all θ. Typically, unbiased estimators exist only for linear models and sometimes for the associated variances. For example, we do not know unbiased estimators for parameters of the gamma or beta distributions. The absence of an unbiased estimator significantly limits the classic theory of unbiased estimation and implied efficient estimation.

It is wrong to conclude that the MM estimator (6.1) is unbiased because $\overline{X}$ is unbiased. Generally, $\widehat{\theta}_{MM}$ is biased. Moreover, if function h^{-1} is convex, then the MM estimator has positive bias; if function h^{-1} is concave, then the estimator has negative bias. However, if function h is linear, then $\widehat{\theta}_{MM}$ is unbiased. The following is a continuation of Example 6.9.

**Example 6.13 *Estimation of the rate parameter of the exponential distribution.* **(a) Prove the method of moments estimator, $\widehat{\lambda}_{MM} = 1/\overline{X}$, is positively biased. (b) Prove that $\widehat{\lambda} = (n - 1)/\sum_{i=1}^{n} X_i$ is an unbiased estimator of λ by using the fact that $\sum_{i=1}^{n} X_i$ follows a gamma distribution with the shape parameter $\alpha = n$ and the rate parameter λ.*

Solution. (a) Recall that the method of moments estimator, $\widehat{\lambda}_{MM} = 1/\overline{X}$; see Example 6.3. This estimator is biased due to Jensen's inequality applied to a convex function $1/x$ (Section 2.4) and the fact that $E(X) = 1/\lambda$ for the exponential distribution (Section 2.4): $E(\widehat{\lambda}_{MM}) = E(1/\overline{X}) > 1/E(\overline{X}) = 1/(1/\lambda) = \lambda$.
(b) Find the expected value of $\left(\sum_{i=1}^{n} X_i\right)^{-1}$:

$$\frac{\lambda^n}{\Gamma(n)} \int_0^\infty \frac{1}{x} x^{n-1} e^{-\lambda x} dx = \frac{\lambda^n}{\Gamma(n)} \frac{\Gamma(n - 1)}{\lambda^{n-1}} \left(\frac{\lambda^{n-1}}{\Gamma(n - 1)} \int_0^\infty x^{n-2} e^{-\lambda x} dx\right) = \frac{\lambda}{n - 1}.$$

Hence $\widehat{\lambda} = (n - 1)/\sum_{i=1}^{n} X_i$ is an unbiased estimator of λ.

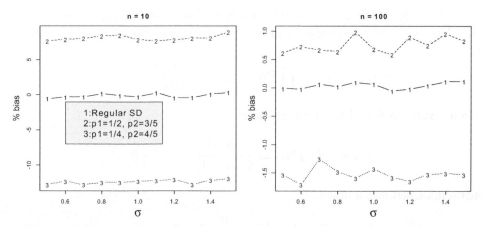

Figure 6.4: *The relative bias on the percent scale for normal σ of three estimators. The MQ estimators are more biased (nSim=100000).*

Example 6.14 *MQ estimator of normal SD.* *Use the normal distribution to compare the relative bias for the regular estimator of SD (sd) with the quantile estimator of SD using two sets of probabilities, $p_1 = 1/10, p_2 = 9/10$ and $p_1 = 3/5, p_2 = 1/2$, based on samples of size $n = 10$ and $n = 100$.*

Solution. The R function with simulations is found in file mqnorm.r; the results of simulations are depicted in Figure 6.4. This function uses the vectorized ordering of rows of matrix X as t(apply(t(X),MARGIN=2,FUN=sort)). This command returns the matrix for which elements in each row are reshuffled from minimum to maximum.

These simulations show that the MQ estimator of σ in a normal distribution is more biased than the traditional estimator. □

The following definition relies on the idea of the asymptotic property of an estimator: when the sample size, n, goes to infinity.

Definition 6.15 *An estimator $\widehat{\theta}_n$ is said to be asymptotically unbiased if its expectation converges to the true value when $n \to \infty$, or symbolically, $\lim_{n\to\infty} E(\widehat{\theta}_n) = \theta$.*

Sometimes we say that $\widehat{\theta}_n$ is unbiased in large sample.

Example 6.16 *Unbiased estimator in uniform distribution.* *Prove that $X_{\max} = \max X_i$ is asymptotically unbiased for θ, where $X_i \stackrel{iid}{\sim} \mathcal{R}(0,\theta)$. Find an unbiased estimator of θ in small sample (finite n).*

Solution. $X_{\max}$ is an obvious estimator of θ, but it has negative bias because $X_i < \theta$ for all $i = 1, 2, ..., n$, and therefore $E(X_{\max}) < \theta$. To find an adjustment that makes the estimator unbiased, we find the expected value of $X_{\max}$. Since

$X_{\max}$ is an nth order statistic, from Section 3.9, we know that $X_{\max}$ has density nx^{n-1}/θ^n. Alternatively, this density can be derived by differentiation of the cdf, which is

$$\Pr(X_{\max} < x) = \prod_{i=1}^{n} \Pr(X_i < x) = \left(\frac{x}{\theta}\right)^n.$$

The expected value is found from integration:

$$E(X_{\max}) = \frac{n}{\theta^n} \int_0^\theta x^n dx = \frac{n\theta^{n+1}}{\theta^n(n+1)} = \frac{n}{n+1}\theta.$$

Asymptotic unbiasedness follows:

$$\lim_{n\to\infty} E(X_{\max}) = \lim_{n\to\infty} \frac{n}{n+1}\theta = \theta.$$

Therefore, the unbiased estimator for every n is

$$\widehat{\theta} = \frac{n+1}{n} X_{\max}. \tag{6.3}$$

Example 6.17 *Smart soldier. A soldier saw enemy tank numbers 15, 45, and 38. Assuming that tanks are numbered sequentially 1,2,..., what is an unbiased estimate of the number of tanks in the enemy army?*

Solution. The tank numbers can be viewed as iid observations from a uniform distribution with unknown upper limit, θ. Thus, an unbiased estimator of tanks in the enemy army is $45 \times (4/3) = 60$. $\square$

The following result, due to Rao [82], says that an unbiased estimator $\widehat{\theta}$ should be a symmetric function of observations because otherwise it can be improved by averaging permutations.

Theorem 6.18 *Let $\widehat{\theta} = \widehat{\theta}(X_1, ..., X_n)$ be an unbiased estimator of θ where X_i are iid observations. Then the average of this estimator over all permutations of observations is unbiased and has the variance not smaller than $\widehat{\theta}$.*

Proof. The average of $\widehat{\theta}$ over all permutations is the estimator

$$\widetilde{\theta} = \widetilde{\theta}(X_1, ..., X_n) = \frac{1}{n!} \sum_{(i_1,...,i_n)\in\mathcal{P}} \widehat{\theta}(X_{i_1}, ..., X_{i_n}),$$

where $\mathcal{P}$ is the set of all permuted indices $(1, ,..., n)$. For example, for $n = 3$, set $\mathcal{P}$ contains $6 = 3!$ triplets: $(1, 2, 3)$, $(1, 3, 2)$, $(2, 1, 3)$, $(2, 3, 2)$, $(3, 1, 2)$, $(3, 2, 1)$. Because X_i are iid, the mean and variance do not change upon permutation of observations. In particular, this implies that $E(\widetilde{\theta}) = \theta$. Furthermore, if σ^2 is the variance of $\widehat{\theta}$, then $\text{cov}(\widehat{\theta}(X_{i_1}, ..., X_{i_n}), \widehat{\theta}(X_{j_1}, ..., X_{j_n})) \le \sigma^2$ for any two permutations $(i_1, .., i_n)$ and $(j_1, .., j_n)$. This follows from the elementary fact that for two random variables X and Y with the same variance σ^2 and correlation coefficient ρ, we have $\operatorname{cov}(X, Y) = \rho\sigma^2 \leq \sigma^2$. Now we find the variance of $\widetilde{\theta}$:

$$\operatorname{var}(\widetilde{\theta})$$
$$= \frac{1}{(n!)^2} \sum_{(i_1,...,i_n)} \operatorname{var}(\widehat{\theta}(X_{i_1}, ..., X_{i_n}))$$
$$+ \frac{1}{(n!)^2} \sum_{(i_1,...,i_n) \neq (j_1,...,j_n)} \operatorname{cov}(\widehat{\theta}(X_{i_1}, ..., X_{i_n}), \widehat{\theta}(X_{j_1}, ..., X_{j_n}))$$
$$\leq \frac{\sigma^2}{(n!)^2}n! + \frac{\sigma^2}{(n!)^2}((n!)^2 - n!) = \sigma^2,$$

where summation is over all $n!$ permutations. □

Clearly, if $\widehat{\theta}(X_1, ..., X_n)$ is a symmetric function (does not change upon permutation), then $\widetilde{\theta}$ coincides with $\widehat{\theta}$. Many unbiased estimators we know are symmetric functions of observations and therefore cannot be improved upon permutation.

Example 6.19 *Randomized response (Warner [102]). The randomized response design (RRD) was invented to estimate the population response rate to a sensitive question and yet protect the confidentiality of the respondent. The respondent is asked to answer a sensitive question truthfully if the answer to the a non-sensitive question is positive and its probability can be well estimated. Then the population response to the sensitive question can be unbiasedly estimated.*

Solution. An example of an RRD question is formulated as follows: If your mother was born between January and May, click *yes* and go to another question. If your mother was born after May answer truthfully *yes* or *no* to the question "Do you have a gun?" Obviously, even if the person answers yes, it does not mean that he/she has a gun because yes may mean that his/her mother was born between January and May. That is why the confidentiality of the respondent is protected. The trick is that we can safely assume that the probability that the mother was born between January and May is $b = 1/5$ and use this fact to estimate the proportion of people who own a gun. If p is the probability that a random person has a gun, we find the probability of yes as

$$\Pr(\text{yes}) = b + (1 - b)p.$$

Thus, if $\overline{Y}$ is the proportion of people who answered yes, then we derive an unbiased estimator for a random person to owning a gun as

$$\widehat{p} = \frac{\overline{Y} - b}{1 - b}.$$

Since $\overline{Y}$ is an unbiased estimator of yes, $\widehat{p}$ is an unbiased estimator of p as the proportion of the population who owns a gun. Obviously, when b is smaller the variance of $\widehat{p}$ is smaller, as is the confidentiality of the respondent.

Unbiased estimation of the standard deviation

Let $\{X_i, i = 1, 2, ..., n\}$ be iid observations with unknown mean μ and variance σ^2. A naive way to find an unbiased estimator for σ is to take the square root,

$$\widehat{\sigma} = \sqrt{\frac{\sum_{i=1}^{n}(X_i - \overline{X})^2}{n - 1}}, \tag{6.4}$$

where $n - 1$ is referred to as the degree of freedom (see Theorem 4.22). In fact, this is how we usually estimate σ. However, this estimator is negatively biased: $\widehat{\sigma}$ systematically underestimates σ. To show this we recall Jensen's inequality from Section 2.2: since $f(x) = \sqrt{x}$ is a concave function,

$$E(\widehat{\sigma}) < \sqrt{E\left(\frac{\sum_{i=1}^{n}(X_i - \overline{X})^2}{n - 1}\right)} = \sqrt{\sigma^2} = \sigma.$$

There is another elegant proof that $\widehat{\sigma}$ is biased by contradiction. Indeed, let us assume that $\widehat{\sigma}$ is unbiased for σ. Then

$$\text{var}(\widehat{\sigma}) = E(\widehat{\sigma}^2) - E^2(\widehat{\sigma}) = \sigma^2 - \sigma^2 = 0.$$

But $\widehat{\sigma}$ is not a fixed number, which implies $\text{var}(\widehat{\sigma}) > 0$, a contradiction.

Now we go back to the unbiased estimation of σ. An obvious way to find an unbiased estimator is to adjust the degrees of freedom,

$$\widetilde{\sigma} = \sqrt{\frac{\sum_{i=1}^{n}(X_i - \overline{X})^2}{n - v_n}},$$

where $v_n > 1$ is a constant dependent on the sample size, n. Then $n - v_n < n - 1$ and there is a hope that $E(\widetilde{\sigma}) = \sigma$. Unfortunately, there is no way to find v_n without specifying the distribution of X_i. Here, we suggest a solution when $X_i \sim \mathcal{N}(\mu, \sigma^2)$. From Section 4.2, we know that $\sigma^{-2} \sum_{i=1}^{n}(X_i - \overline{X})^2 \sim \chi^2(n-1)$, where the density of the chi-square distribution defined by (4.15). Using this density we can find the expected value of the square root in closed form:

$$E\sqrt{\sum_{i=1}^{n}(X_i - \overline{X})^2}$$

$$= \sigma \int_0^{\infty} \frac{\sqrt{s}}{2^{(n-1)/2}\Gamma((n-1)/2)} s^{(n-1)/2-1} e^{-s/2} ds$$

$$= \sigma \frac{2^{n/2}\Gamma(n/2)}{2^{(n-1)/2}\Gamma((n-1)/2)} \int_0^{\infty} \frac{1}{2^{n/2}\Gamma(n/2)} s^{n/2-1} e^{-s/2} ds = \sigma \frac{\sqrt{2}\Gamma(n/2)}{\Gamma((n-1)/2)}.$$

Thus, to make $\widehat{\sigma}_\nu$ unbiased, we have to use

$$v_n = n - \frac{2\Gamma^2(n/2)}{\Gamma^2((n-1)/2)}.$$

Using an asymptotic approximation of the gamma function, it is possible to prove that $\lim_{n \to \infty} \nu_n = 1.5$. Finally, we suggest

$$\widehat{\sigma} = \sqrt{\frac{\sum_{i=1}^{n}(X_i - \overline{X})^2}{n - 1.5}} \tag{6.5}$$

is almost unbiased for the normal SD.

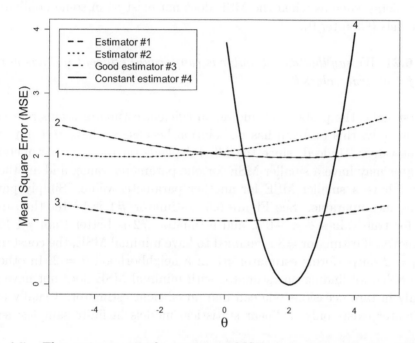

Figure 6.5: *The estimator with minimum MSE may not exist: the MSE as a function of the true parameter θ for four estimators. Estimator 3 is better than 1 and 2. There is no perfect estimator because the constant estimator $\widehat{\theta}_4 = 2$ with $MSE = (\theta - 2)^2$ is better than any other estimator in a neighborhood of 2.*

6.4.2 Mean Square Error

The best estimator should be close to the true value as much as possible. Since we deal with random data, $\widehat{\theta}$ cannot coincide with θ, but it can be within a proximity. In mathematical terms, we want to minimize the expected distance from $\widehat{\theta}$ to θ. This criterion gives rise to the following formalization, sometimes referred to as the loss function.

Definition 6.20 *The mean square error (MSE) is the expected squared distance between the estimator and the true value:*

$$\mathrm{MSE}(\theta) = E_\theta(\widehat{\theta} - \theta)^2, \tag{6.6}$$

where the subscript θ means that the expectation is taken assuming that the true value is θ.

Typically, the MSE is a nonconstant function of θ. This means that some parameter values are estimated more and some less precisely. One must realize that (i) the MSE is associated with quadratic loss and (ii) the MSE silently assumes a symmetric loss function. Both features may be inadequate for some statistical problems, such as when the MSE does not exist as in some nonlinear regression models (Chapter 9).

Definition 6.21 *We say that an estimator is (uniformly) efficient if it has minimum MSE for all true values θ.*

At the first sight, the problem of finding an efficient estimator seems reasonable. Unfortunately, this problem has no solution because we have to compare functions, not values. Indeed, since the MSE depends on the true parameter, θ, one estimator may have a smaller MSE for one parameter value, and another estimator may have a smaller MSE for another parameter value. Simply put, MSEs may be incomparable. See Figure 6.5: estimator #1 is better than estimator #2 for true values $\theta < -0.5$, and estimator #2 is better than #1 for $\theta > 0.5$. Moreover, if estimator #3 is claimed to have minimal MSE, the constant estimator $\widehat{\theta}_4 = 2$ outperforms estimator #3 in a neighborhood $\theta = 2$. In other words, the problem of finding an estimator with minimal MSE does not have a solution. Only in rare occasions one can find an efficient estimator. Usually an efficient estimator exists only in linear statistical models in finite samples; see Chapter 8.

To find an efficient estimator, we have to restrict the family of estimators under consideration. For example, we may find an efficient *unbiased* estimator among all unbiased estimators; in this regard see the Gauss–Markov theorem of Section 6.5.

The MSE can be decomposed into variance and bias squared:

$$\text{MSE} = \text{var} + Bias^2. \tag{6.7}$$

The proof is straightforward. Let $\mu = E(\widehat{\theta})$, and then subtracting and adding μ under the expectation sign, we obtain

$$\text{MSE} = E_\theta(\widehat{\theta} - \theta)^2 = E_\theta \left[(\widehat{\theta} - \mu) + (\mu - \theta) \right]^2.$$

Now we expand the square of the sum as

$$\left[(\widehat{\theta} - \mu) + (\mu - \theta) \right]^2 = (\widehat{\theta} - \mu)^2 + 2(\widehat{\theta} - \mu)(\mu - \theta) + (\mu - \theta)^2.$$

Taking the expectation, we obtain

$$\text{MSE} = E_\theta\left[(\widehat{\theta} - \mu)^2\right] + 2E_\theta\left[(\widehat{\theta} - \mu)(\mu - \theta)\right] + E_\theta\left[(\mu - \theta)^2\right].$$

The first term is the variance of $\widehat{\theta}$. The second term is zero because $(\mu - \theta)$ is a constant and $E_\theta(\widehat{\theta} - \mu) = 0$; the third term is the bias squared; formula (6.7) is proved.

We may say that an estimator with small MSE has small variance and is almost unbiased. Typically, variance and bias compete against each other in formula (6.7). It is a fallacy to say that an estimator has smaller MSE if the bias is zero because its variance may be larger.

In science and engineering, bias, sometimes called systematic bias, is referred to as *accuracy*, and the estimator SD = $\sqrt{\text{var}}$ is referred to as *precision*. In lay-language, the MSE is the sum of squared accuracy and precision. Sometimes, practitioners express the total uncertainty as $\sqrt{\text{var}} + |Bias|$. However, MSE expressed in formula (6.7) is preferable because it can be derived as the quadratic loss function, $E_\theta(\widehat{\theta} - \theta)^2$ and admits a closed-form expression for further analytical study. To obtain the uncertainty measured on the scale of measurements we use the root mean square error (RMSE), $\sqrt{\text{MSE}}$.

As follows from the decomposition formula (6.7), the MSE of an unbiased estimator equals its variance:

$$\text{MSE} = \text{var if bias=0}.$$

Of course other metrics can be used to judge how close the estimator and the true value are. For example, one may use the mean absolute error $E_\theta\left|\widehat{\theta} - \theta\right|$. Squared distance is easier to handle from a technical standpoint because of its relationship to the variance and the mean.

Example 6.22 *Comparison of estimators via the MSE.* *Compare the performance of the MM estimator with the estimator (6.3) using the MSE.*

Solution. The theoretical mean of $\mathcal{R}(0,\theta)$ is $\theta/2$ and the variance is $\theta^2/12$. From equating the theoretical mean to $\overline{X}$, we obtain the MM estimator, $\widehat{\theta}_1 = 2\overline{X}$. Since this estimator is unbiased, the variance is $\text{MSE}_1(\theta) = \text{var}(\widehat{\theta}_1) = 4\text{var}(\overline{X}) = (4/n)\text{var}(X) = \theta^2/(3n)$. Since $\widehat{\theta}_2 = (n+1)/nx_{\max}$ is unbiased, we find $\text{MSE}_2(\theta) = \text{var}(\widehat{\theta}_2) = E(\widehat{\theta}_2^2) - \theta^2$. The density of $X_{\max}$ is $f(x) = nx^{n-1}/\theta^n$, and therefore,

$$E(X_{\max}^2) = \int_0^\theta x^2 f(x)dx = (n/\theta^n)\int_0^\theta x^{n+1}dx = \frac{n\theta^2}{n+2}.$$

Finally,

$$\text{MSE}_2(\theta) = \frac{n}{n+2}\frac{(n+1)^2}{n^2}\theta^2 - \theta^2 = \frac{\theta^2}{n(n+2)}.$$

Compare the MSEs by taking the ratio, $\mathrm{MSE}_1(\theta)/\mathrm{MSE}_2(\theta) = (n+2)/3 > 1$ for $n > 1$. The answer is that estimator $\widehat{\theta}_2$ is better than $\widehat{\theta}_1$ because its expected squared distance to the true value (MSE) is smaller $(n > 1)$. Clearly, the two estimators are the same when $n = 1$.   □

Let $\widehat{\theta}_1$ and $\widehat{\theta}_2$ be two unbiased independent estimators of θ. There is a linear combination of estimators, $\widehat{\theta}_3 = \lambda\widehat{\theta}_1 + (1-\lambda)\widehat{\theta}_2$, that has a smaller variance than either of the two, assuming that the variances of both estimators are known. Indeed, first, we prove that $\widehat{\theta}_3$ is unbiased, $E(\widehat{\theta}_3) = E(\lambda\widehat{\theta}_1 + (1-\lambda)\widehat{\theta}_2) = \lambda\theta + (1-\lambda)\theta = \theta$. Second, for the unbiased estimator, MSE = variance, and so, due to independence, $\mathrm{var}(\widehat{\theta}_3) = \lambda^2\sigma_1^2 + (1-\lambda)^2\sigma_2^2$, where $\mathrm{var}(\widehat{\theta}_k) = \sigma_k^2$, $k = 1, 2$. Differentiating the above expression with respect to λ, find the optimal value $\lambda_{\mathrm{opt}} = \sigma_2^2/(\sigma_1^2+\sigma_2^2)$ with the minimum MSE $= \sigma_1^2\sigma_2^2/(\sigma_1^2+\sigma_2^2)$. This minimum is smaller than σ_1^2 and σ_2^2. For example, when $\sigma_1^2 = \sigma_2^2$, the variance of the optimal estimator is half as small.

Example 6.23 *Optimal estimator of the mean*. *The first random sample of size m produced an estimator of mean $\overline{X}_1$ and the second random sample of size n produced $\overline{X}_2$. Find an optimal estimator of the mean as a linear combination of $\overline{X}_1$ and $\overline{X}_2$ and prove that it is equal to the arithmetic average of the combined sample.*

Solution. Since $\mathrm{var}(\overline{X}_1) = \sigma^2/m$ and $\mathrm{var}(\overline{X}_2) = \sigma^2/n$, the optimal estimator of the mean, μ, is the linear combination with $\lambda_{\mathrm{opt}} = (1/n)/(1/m + 1/n) = m/(m+n)$. This optimal combination produces the following estimator:

$$\overline{X}_1\frac{m}{n+m} + \overline{X}_2\left(1 - \frac{m}{n+m}\right) = \overline{X}_1\frac{m}{n+m} + \overline{X}_2\frac{n}{n+m}$$

$$= \frac{1}{n+m}\left(\sum_{i=1}^{m}X_{1i} + \sum_{i=1}^{n}X_{2i}\right) = \overline{X}.$$

This is the average of the combined sample.   □

In the following example the unbiased and biased variance estimators of the normal sample are compared. Surprisingly, a biased estimator has a smaller MSE.

Example 6.24 *MSE of the variance estimator*. *Find and compare the MSEs of the unbiased and biased estimators of the variance of the normal distribution, σ^2:*

$$\sigma_{\mathrm{unbiased}}^2 = \frac{1}{n-1}\sum_{i=1}^{n}(X_i - \overline{X})^2, \quad \widehat{\sigma}_{\mathrm{biased}}^2 = \frac{1}{n}\sum_{i=1}^{n}(X_i - \overline{X})^2, \qquad (6.8)$$

where $X_i \overset{\mathrm{iid}}{\sim} \mathcal{N}(\mu, \sigma^2)$, $i = 1, 2, ..., n$.

Solution. Since X_i are normally distributed, as proven in Chapter 4, Example 4.22, we have $\sigma^{-2}\sum_{i=1}^{n}(X_i-\overline{X})^2 \sim \chi^2(n-1)$. This means that the following representation holds $\sigma^{-2}\sum_{i=1}^{n}(X_i-\overline{X})^2 = \sum_{i=1}^{n-1}Z_i^2$, where $Z_i \overset{\text{iid}}{\sim} \mathcal{N}(0,1)$. Therefore

$$\text{var}\left(\sum_{i=1}^{n}(X_i-\overline{X})^2\right) = \sigma^2 \text{var}\left(\sum_{i=1}^{n-1}Z_i^2\right) = \sigma^4(n-1)\text{var}(Z_1^2) = 2(n-1)\sigma^4.$$

Since the MSE of the unbiased estimator equals its variance, we have

$$\text{MSE}_{\text{unbiased}} = \text{var}\left(\frac{1}{n-1}\sum_{i=1}^{n}(X_i-\overline{X})^2\right) = 2\frac{n-1}{(n-1)^2}\sigma^4 = \frac{2\sigma^4}{n-1}.$$

The MSE of the biased variance is equal to its variance plus the bias squared:

$$\text{MSE}_{\text{biased}} = \text{var}\left(\frac{1}{n}\sum_{i=1}^{n}(X_i-\overline{X})^2\right) + \frac{1}{n^2}\sigma^4 = 2\frac{n-1}{n^2}\sigma^4 + \frac{1}{n^2}\sigma^4 = \frac{(2n-1)\sigma^4}{n^2}.$$

The difference between the two is proportional to

$$\frac{2}{n-1} - \frac{2n-1}{n^2} = \frac{2n^2 - (n-1)(2n-1)}{(n-1)n^2} = \frac{3n-1}{(n-1)n^2} > 0.$$

This means that the biased estimator, $\widehat{\sigma}^2_{\text{biased}}$ has a smaller MSE than the unbiased one, $\widehat{\sigma}^2_{\text{unbiased}}$. $\qquad\square$

This result gives rise to an optimal estimator of the normal variance.

Theorem 6.25 *Among all estimators of the normal variance of the type $\widetilde{\sigma}^2_v = (n-v)^{-1}\sum_{i=1}^{n}(X_i-\overline{X})^2$, the minimum MSE is attained at $v = -1$, i.e. the estimator*

$$\widetilde{\sigma}^2_{-1} = \frac{1}{n+1}\sum_{i=1}^{n}(X_i-\overline{X})^2$$

has minimum MSE.

Proof. As follows from the example above, $\text{var}(\widetilde{\sigma}^2_v) = 2(n-1)\sigma^4/(n-v)^2$. Since $E(\widetilde{\sigma}^2_v) = \sigma^2(n-1)/(n-\nu)$, the MSE, as a function of v, is

$$\text{MSE}(v) = \frac{2(n-1)+(1-v)^2}{(n-v)^2}.$$

Taking the derivative with respect to ν and equating it to zero, we obtain that the minimum MSE is attained at $v = -1$. Note that the normal assumption is crucial for $\widetilde{\sigma}^2_{-1}$.

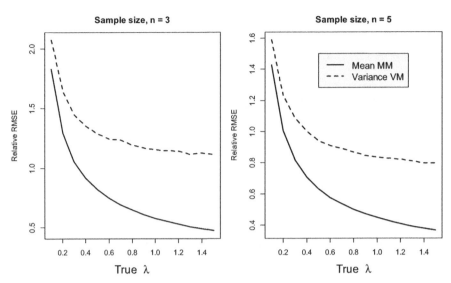

Figure 6.6: *Comparison of the mean and variance estimators of the Poisson rate λ using the relative RMSE ($N_{\text{sim}} = 100,000$). The Mean MM estimator has a smaller RMSE.*

MSE in simulations

Comparing estimators is a frequent statistical task. The MSE cannot always be computed analytically, so simulations may be required. If the true parameter is θ and the ith simulation produces estimator $\widehat{\theta}_i$, we compute an empirical MSE as

$$\text{MSE} = \frac{1}{N_{\text{sim}}} \sum_{i=1}^{N_{\text{sim}}} (\widehat{\theta}_i - \theta)^2. \tag{6.9}$$

Sometimes RMSE, $\sqrt{\text{MSE}}$, is used which, is on the same scale as θ. The relative RMSE, as the ratio of the RMSE to the true value of the positive parameter, has a clear interpretation when the RMSE increases with the parameter value. The following example illustrates how simulations can be used to compare two estimators of the Poisson rate.

Example 6.26 *Estimation of the Poisson rate. Use vectorized simulations to compare two estimators of the Poisson rate, λ: (i) the mean MM estimator, $\widehat{\lambda}_{MM} = \overline{X}$ (see Example 6.5), and (ii) the variance MM estimator, $\widehat{\lambda}_{VM} = \sum_{i=1}^{n} (X_i - \overline{X})^2 / (n-1)$. Plot the simulation-derived RMSE versus the true values of λ for two sample sizes, $n = 3$ and $n = 5$.*

Solution. The variance MM estimator is derived using the fact that the mean and variance of the Poisson distribution are the same. Since $\sum_{i=1}^{n} (X_i - \overline{X})^2 / (n-1)$ is an unbiased estimator of the variance, we estimate λ using $\widehat{\lambda}_{VM}$. The relative RMSE for the two estimators is shown in Figure 6.6 for $n = 3$ and $n = 5$; the

R code is found in file **pois2est.r**. For example, when $n = 3$ and $\lambda = 1$, the imprecision of the mean MM estimator is about 50%, and the imprecision of the variance MM estimator is about 120%. The former estimator uniformly outperforms the latter. We make a couple of comments on the code. (i) Matrix X contains all Poisson random numbers. The means for each sample are computed using the **rowMeans** command (the means in each row of matrix X). The variances for each sample are computed from **rowMeans(X^2)**. Alternatively, one could use **apply(X,1,var)** to compute **mm.var** but this method is slower. (ii) To display a mixture of regular and Greek letters, we use command **expression** for the x-label.

6.4.3 Multidimensional MSE

In the previous statistical estimation problems, only one parameter was unknown. Sometimes, the parameter vector $\boldsymbol{\theta}$ is multidimensional (therefore the boldface). Then the MSE should reflect the quality of estimation of all components of vector $\boldsymbol{\theta}$. We recommend the reader review the section on linear algebra and quadratic forms in the Appendix. Specifically, let the true unknown parameter vector, $\boldsymbol{\theta}$, be the $m \times 1$ vector, $\boldsymbol{\theta} = (\theta_1, \theta_2, ..., \theta_m)'$ and let $\widehat{\boldsymbol{\theta}} = (\widehat{\theta}_1, \widehat{\theta}_2, ..., \widehat{\theta}_m)'$ be the $m \times 1$ vector of estimators.

Definition 6.27 *The multidimensional MSE is the $m \times m$ matrix*

$$\mathrm{MSE}(\boldsymbol{\theta}) = E[(\widehat{\boldsymbol{\theta}} - \boldsymbol{\theta})(\widehat{\boldsymbol{\theta}} - \boldsymbol{\theta})'].$$

In particular, the sum of diagonal elements, $\mathrm{tr}(\mathrm{MSE}(\boldsymbol{\theta}))$, is called the total MSE.

The (i,j)th element of this matrix is $E[(\widehat{\theta}_i - \theta_i)(\widehat{\theta}_j - \theta_j)]$, and the (i,i)th diagonal element is the standard MSE of the ith component, $\mathrm{MSE}_i = E[(\widehat{\theta}_i - \theta_i)^2]$. For an unbiased estimator, $E(\widehat{\boldsymbol{\theta}}) = \boldsymbol{\theta}$, and therefore $\mathrm{MSE} = \mathrm{cov}(\widehat{\boldsymbol{\theta}})$, so the total MSE turns into total variance.

Definition 6.28 *We say that an estimator $\widehat{\boldsymbol{\theta}}_1$ is no less efficient than an estimator $\widehat{\boldsymbol{\theta}}_2$ if, for all true values of $\boldsymbol{\theta}$, we have*

$$\mathrm{MSE}_1(\boldsymbol{\theta}) \leq \mathrm{MSE}_2(\boldsymbol{\theta}).$$

That is, the difference between the right- and left-hand sides, is a nonnegative definite matrix (the eigenvalues of the difference are nonnegative).

In other words, the MSE of any linear combination of $\widehat{\boldsymbol{\theta}}_1$ is less that or equal to the MSE of the same linear combination of $\widehat{\boldsymbol{\theta}}_2$. In particular, an estimator $\widehat{\boldsymbol{\theta}}$ is an efficient estimator of $\boldsymbol{\theta}$ if the difference between its MSE and the MSE of another estimator is a nonnegative definite matrix. As follows from matrix algebra, the efficient estimator has the smallest individual component MSEs (the diagonal elements) and total MSE (the trace of the matrix).

Comparison of multidimensional estimators is not an obvious task because, for instance, the first estimator may have a smaller MSE for the first component, but the second estimator has smaller MSE for the second component. Only in rare statistical estimation problems, such as linear models to be studied in Chapter 8, there exists an efficient estimator in terms of Definition 6.28. In addition, comparison of estimators in terms of nonnegative definiteness may be too stringent in practice because a linear combination of parameters may have no sense, such as when the first component is the mean and the second is variance. From this perspective, the total MSE may be a better metric to judge estimation performance. The following example illustrates a comparison of multidimensional estimators.

Example 6.29 *Estimation of the lower and upper limits of the uniform distribution.* Let $\{X_i, i = 1, 2, ..., n\}$ be a random sample from $R(\theta_L, \theta_U)$. Compare three estimators of θ_L and θ_U using simulations (compute and compare three 2×2 MSE matrices): (i) the method of moments, (ii) minimum and maximum estimators, and (ii) minimum and maximum adjusted for bias.

Solution. The MM estimators for θ_L and θ_U are found from equating the empirical and theoretical mean and variance, $\overline{X} = (\theta_L + \theta_U)/2$ and $\widehat{\sigma}^2 = (\theta_U - \theta_L)^2/12$, respectively. Solving for θ_U and θ_L yields the MM estimators:

$$\widehat{\theta}_L = \overline{X} - \sqrt{3}\widehat{\sigma}, \quad \widehat{\theta}_U = \overline{X} + \sqrt{3}\widehat{\sigma}. \tag{6.10}$$

The obvious minimum and maximum estimators are

$$\widehat{\theta}_L = X_{\min}, \quad \widehat{\theta}_U = X_{\max}. \tag{6.11}$$

These estimators are positively and negatively biased, respectively. To eliminate the bias one can compute the theoretical expectation of $X_{\min}$ and $X_{\max}$ and find estimators by solving the respective equations as we did in the case when $\theta_L = 0$; see Example 6.22. Let us start with estimation of θ_U. The cdf and pdf of $X_{\max}$ are $(\theta_U - \theta_L)^{-n}(x - \theta_L)^n$ and $n(\theta_U - \theta_L)^{-n}(x - \theta_L)^{n-1}$, respectively. Thus,

$$E(X_{\max}) = n(\theta_U - \theta_L)^{-n} \int_0^{\theta_U - \theta_L} (x + \theta_L)x^{n-1}dx = \frac{\theta_U n - \theta_L}{n+1}.$$

For $X_{\min}$, the cdf is $1 - [1 - (\theta_U - \theta_L)^{-1}(x - \theta_L)]^n$ and the pdf is $n(\theta_U - \theta_L)^{-1}[1 - (\theta_U - \theta_L)^{-1}(x - \theta_L)]^{n-1}$. Thus

$$E(X_{\min}) = n(\theta_U - \theta_L)^{-1} \int_0^{\theta_U - \theta_L} (x + \theta_L)[1 - (\theta_U - \theta_L)^{-1}x]^{n-1}dx$$

$$= \frac{\theta_U^2 - \theta_L^2 n + \theta_L \theta_U (n-1)}{(\theta_U - \theta_L)(n+1)}.$$

The estimators for θ_L and θ_U are the solution to a system of equations:

$$\frac{n\theta_U - \theta_L}{n+1} = X_{\max}, \qquad \frac{\theta_U^2 + \theta_L\theta_U(n-1) - \theta_L^2 n}{(\theta_U - \theta_L)(n+1)} = X_{\min}.$$

Some elementary algebra gives unbiased estimators as a linear combination of $X_{\min}$ and $X_{\max}$:

$$\widehat{\theta}_L = \frac{n+1}{n^2+1}(nX_{\min} - X_{\max}) = X_{\min} - \Delta,$$

$$\widehat{\theta}_U = \frac{n+1}{n^2+1}(nX_{\max} + X_{\min}) = X_{\max} + \Delta, \qquad (6.12)$$

where $\Delta = (n-1)(X_{\max} - X_{\min})/(n^2+1)$. The R code that compares the three estimators defined in (6.10), (6.11), and (6.12) is found in the file luR.r. We make several comments: (i) The time of computation is printed out, since, for a large number of simulations, like Nsim=1000000, it may be time consuming, and (ii) computations are done in the vectorized fashion, use of the apply function for mean, SD, and min, and max is critical. Below are displayed three 2×2 MSE matrices for each method of estimation along with the total MSE (the sum on the diagonal).

```
        MSE1                        MSE2                       MSE3
 0.04765190 -0.02381576     0.035270484 -0.001912803    0.025361251 0.005115448
-0.02381576  0.04759992    -0.001912803  0.035187091    0.005115448 0.029528034
Total MSE = 0.09525182      Total MSE = 0.07045757      Total MSE = 0.05488929
```

According to the total MSE, the third method outperforms the other two. The minimum eigenvalues of the difference between matrices MSE1 and MSE3, and MSE2 and MSE3 are 0.0004 and -0.0088, respectively. While any linear combination of $\widehat{\theta}_L$ and $\widehat{\theta}_U$ has MSE smaller for the third than for the first method, it is not true when the second and the third methods are compared. If our concern is not estimation of any linear combination of θ_L and θ_U, the total MSE would be an adequate metric, and so estimators (6.12) are preferable.

6.4.4 Consistency of estimators

Consistency is an imaginary property of an estimator: a consistent estimator, $\widehat{\theta}_n$, approaches the true parameter value when the sample size, n, goes to infinity (notice that we use a subscript n to indicate that the estimator is a function of the sample size). Sometimes we say "large sample" to indicate that $n \to \infty$. We say "imaginary property" because in reality the sample size can never be infinity. Consistency is an asymptotic property of an estimator; another asymptotic property is the asymptotic unbiasedness, which was introduced in Definition 6.15. Many consistent estimators have a normal distribution in large sample.

Consistency is closely related to stochastic convergence of random variables discussed in Section 2.9; thus, we recommend reviewing that section. While in calculus convergence is uniquely defined through lim, there are several definitions of convergence in statistics depending on how one understands that a sequence of random variables approaches zero.

Three types of convergence/consistency are used when $n \to \infty$:

1. *Convergence in probability (weak convergence)*: The probability that an estimator differs from the true value θ by any positive ε approaches 0 when $n \to \infty$. Symbolically (large braces define a set of points),

$$\lim_{n \to \infty} \Pr \left\{ \left| \widehat{\theta}_n - \theta \right| > \varepsilon \right\} = 0 \quad \forall \varepsilon > 0. \tag{6.13}$$

   To indicate that estimator $\widehat{\theta}_n$ converges to θ in probability, we use notation $p\lim_{n \to \infty} \widehat{\theta}_n = \theta$.

2. *Convergence in quadratic sense*: The expected squared distance to the true value approaches zero, or MSE $\to 0$ when $n \to \infty$, which symbolically is written as

$$\lim_{n \to \infty} E(\widehat{\theta}_n - \theta)^2 = 0. \tag{6.14}$$

   This definition requires that the expectation at left exists for each n. Since the MSE is the sum of the variance and the squared bias, the variance and the asymptotic bias of a consistent estimator in the quadratic sense vanish. If estimator $\widehat{\theta}_n$ is unbiased, it is consistent if and only if its variance vanishes as n grows, or symbolically, $\lim_{n \to \infty} \mathrm{var}(\widehat{\theta}_n) = 0$. Limit (6.14) implies asymptotic unbiasedness, $\lim_{n \to \infty} E(\widehat{\theta}_n) = \theta$. Quadratic convergence is easy to generalize to convergency in the L^p norm as $\lim_{n \to \infty} E \left| \widehat{\theta}_n - \theta \right|^p = 0$, where $p \geq 1$.

3. *Convergence with probability 1, or almost surely (strong consistency)* or symbolically
$$\Pr \left(\lim_{n \to \infty} \widehat{\theta}_n = \theta \right) = 1.$$

   Sometimes a.s. is used to indicate the almost surely convergence and we write $\lim_{n \to \infty} \widehat{\theta}_n = \theta$ a.s. This means that the set of points for which the limit exists has the probability measure of 1. Equivalently, this definition can be expressed as $\lim_{n \to \infty} \Pr(\cup_{m=n}^{\infty} \left| \widehat{\theta}_n - \theta \right| > \varepsilon) = 0$ for every $\varepsilon > 0$.

These definitions are related as follows from the probability Section 2.9. For example, from Chebyshev's inequality, it follows that convergence in the quadratic sense implies convergence in probability and strong convergence implies weak convergence. Typically, the consistency of an estimator is an implication of the LLN with further application of the Slutsky theorem (see below).

Example 6.30 Sample mean and variance are consistent estimators.
Prove that $\overline{X}$ is a consistent estimator for μ. Moreover, prove that $\widehat{\sigma}^2_{\text{unbiased}}$ and $\widehat{\sigma}^2_{\text{biased}}$, as defined in Example 6.24, are consistent estimators for σ^2 in the quadratic sense if observations are normally distributed.

Proof. Since $\overline{X}$ is an unbiased estimator, it suffices to prove that its variance goes to zero when $n \to \infty$. But $\text{var}(\overline{X}) = \sigma^2/n \to 0$: the average is a consistent estimator. To prove the consistency of the variance estimators, we use the result of Example 6.8: if $n \to \infty$, then $\text{MSE}_{\text{unbiased}} = 2\sigma^4/(n-1) \to 0$ and $\text{MSE}_{\text{biased}} = (2n-1)\sigma^4/n^2 \to 0$. An alternative proof of consistency of $\widehat{\sigma}^2$ is found in Example 6.34 below. □

The following results are very convenient when proving the consistency of estimators.

Theorem 6.31 Slutsky. *Let $p\lim_{n\to\infty} \widehat{\tau}_n = \tau$ and $p\lim_{n\to\infty} \widehat{\theta}_n = \theta$. Then (a) $p\lim(\widehat{\tau}_n + \widehat{\theta}_n) = \tau + \theta$, (b) $p\lim(\widehat{\tau}_n\widehat{\theta}_n) = \tau\theta$, and (c) $p\lim(\widehat{\tau}_n/\widehat{\theta}_n) = \tau/\theta$ if $\theta \neq 0$.*

Proof. We prove (a). Let $\varepsilon > 0$. It is easy to see that

$$\left\{\left|(\widehat{\tau}_n - \tau) + (\widehat{\theta}_n - \theta)\right| > \varepsilon\right\} \subset \{|\widehat{\tau}_n - \tau| > \varepsilon/2\} \cup \left\{\left|\widehat{\theta}_n - \theta\right| > \varepsilon/2\right\}. \quad (6.15)$$

Use the following elementary inequality: $P(A \cup B) \leq \Pr(A) + \Pr(B)$. Letting $n \to \infty$ the right-hand side of (6.15) goes to zero, which implies (a). The proof of (b)–(c) is similar. □

The following is a generalization of the Slutsky theorem.

Theorem 6.32 *(a) Univariate case: let $p\lim_{n\to\infty} \widehat{\theta}_n = \theta$ and $f(\theta)$ be a continuous function of θ. Then $p\lim_{n\to\infty} f(\widehat{\theta}_n) = f(\theta)$. (b) Bivariate case: Let $p\lim_{n\to\infty} \widehat{\theta}_n = \theta$ and $p\lim_{n\to\infty} \widehat{\tau}_n = \tau$, and $f(\theta, \tau)$ be a continuous function of (θ, τ). Then $p\lim_{n\to\infty} f(\widehat{\theta}_n, \widehat{\tau}_n) = f(\theta, \tau)$.*

Proof. (a) We prove the consistency using the epsilon–delta definition of continuity: for every $\varepsilon > 0$, there is $\delta > 0$ such that, for every θ_*, where $|\theta_* - \theta| < \delta$, we have $|f(\theta_*) - f(\theta)| < \varepsilon$. Replacing θ_* with $\widehat{\theta}_n$, we note that the set $\{|\widehat{\theta}_n - \theta| < \delta\}$ is a subset of $\{|f(\widehat{\theta}_n) - f(\theta)| < \varepsilon\}$ since $|\theta_* - \theta| < \delta$ implies $|f(\theta_*) - f(\theta)| < \varepsilon$. For the complementary sets we can write

$$\left\{|f(\widehat{\theta}_n) - f(\theta)| \geq \varepsilon\right\} \subset \left\{\left|\widehat{\theta}_n - \theta\right| \geq \delta\right\}$$

and

$$\Pr\left\{|f(\widehat{\theta}_n) - f(\theta)| \geq \varepsilon\right\} \leq \Pr\left\{\left|\widehat{\theta}_n - \theta\right| \geq \delta\right\}.$$

Letting $n \to \infty$, we note that the right-hand side goes to zero, so the left-hand side goes to zero. That is, $p\lim_{n\to\infty} f(\widehat{\theta}_n) = f(\theta_0)$. The bivariate case is proved in a similar fashion. □

This proof can be extended to the case when f depends on n.

CHAPTER 6. PARAMETER ESTIMATION

Theorem 6.33 *Extended Slutsky.* Let $p\lim_{n\to\infty}\widehat{\theta}_n = \theta$ and continuous functions $f_n(\theta)$ converge to $f(\theta)$ such that derivatives of $f_n(\theta)$ are uniformly bounded in absolute value in a neighborhood of θ. Then $p\lim_{n\to\infty} f_n(\widehat{\theta}_n) = f(\theta)$.

Obviously, the above theorems can be generalized to functions with a vector argument.

Example 6.34 *Consistency of variance and SD.* Let X_i be iid with finite variance σ^2. Prove that $p\lim_{n\to\infty} n^{-1}\sum_{i=1}^n(X_i - \overline{X})^2 = \sigma^2$ and $p\lim_{n\to\infty}\widehat{\sigma} = \sigma$.

Solution. Note that one cannot apply the LLN directly to $n^{-1}\sum_{i=1}^n(X_i-\overline{X})^2$ because $\{(X_i - \overline{X})^2, i = 1, ..., n\}$ correlate. Here is how we work around this. Express

$$\frac{1}{n}\sum_{i=1}^n(X_i - \overline{X})^2 = \frac{1}{n}\sum_{i=1}^n X_i^2 - \left(\frac{1}{n}\sum_{i=1}^n X_i\right)^2$$

and let $\widehat{\tau}_n = \sum_{i=1}^n X_i^2/n$ and $\widehat{\theta}_n = \sum_{i=1}^n X_i/n$. Then, thanks to the Slutsky theorem,

$$p\lim_{n\to\infty} n^{-1}\sum_{i=1}^n(X_i - \overline{X})^2 = p\lim_{n\to\infty}\widehat{\tau}_n - \left(p\lim_{n\to\infty}\widehat{\theta}_n\right)^2 = (\mu^2 + \sigma^2) - \mu^2 = \sigma^2.$$

To prove the consistency of $\widehat{\sigma}$, we use Theorem 6.32 and the fact that the square root is a continuous function.

Problems

1. In Example 6.12, (a) prove that $\overline{x}$ is unbiased even if X_i do correlate but have the same mean, μ; (b) prove that $\widehat{\sigma}^2$ remains unbiased if the observations are pairwise uncorrelated, $\text{cov}(y_i, y_j) = 0$ for $i \neq j$.

2. Find an unbiased estimator of p^2 using $n > 1$ Bernoulli observations with $\Pr(X_i = 1) = p$ and derive its MSE. [Hint: Derive $E(\overline{X}^2)$ and use the fact that $E(\overline{X}) = p$.]

3. Prove that the variance of the asymptotically unbiased estimator of SD given by formula (6.5) can be approximated as $\sigma^2/(2n - 3)$. Carry out vectorized simulations to confirm this approximation for $n = 10, 15, 20$ and true σ^2 on the grid of values from 1 to 3 with the step 0.5.

4. The mean absolute error is defined as $\text{MAE} = E_\theta\left|\widehat{\theta} - \theta\right|$. (a) Is it true that $E_\theta|\widehat{\theta}_1 - \theta| \leq E_\theta|\widehat{\theta}_2 - \theta|$ implies $E_\theta(\widehat{\theta}_1 - \theta)^2 \leq E_\theta(\widehat{\theta}_2 - \theta)^2$? (b) Does the equality $E_\theta|\widehat{\theta} - \theta| = E_\theta|\widehat{\theta} - E(\theta)| + E|(\widehat{\theta}) - \theta|$ hold?

5. A statistician uses $\sum_{i=1}^{n} |\widehat{\theta}_i - \theta|/N$ to quantify the quality of an estimator, but gets a consistently smaller value than the theoretical RMSE. Explain. [Hint: Use Jensen's inequality.]

6. Let $\widehat{\theta}_1$ and $\widehat{\theta}_2$ be two unbiased independent estimators of θ with known variances σ_1^2 and σ_2^2, respectively. (a) Is it true that $(\widehat{\theta}_1 + \widehat{\theta}_2)/2$ has a smaller MSE than either of the two? (b) Find λ such that $\lambda\widehat{\theta}_1 + (1 - \lambda)\widehat{\theta}_2$ has minimum MSE.

7. Let $\overline{X}_1$ and $\overline{X}_2$ be two estimates of μ from two independent samples $\{X_{i1}, i = 1, ..., n_1\}$ and $\{X_{i2}, i = 1, ..., n_2\}$ from the same general population with mean μ and variance σ^2. Find λ such that $\widehat{\mu} = \lambda\overline{X}_1 + (1 - \lambda)\overline{X}_2$ has minimum MSE and compare it with the average over the combined sample.

8. Somebody wants to estimate the time to perform a specific task. Two methods are suggested: (i) measure the time of each of n tasks and take the average, and (ii) measure the total time required to perform all n tasks and divide by n. What method is better (assume that the variance of the measurement error of one task and the total time is the same)?

9. Let $\widehat{\sigma}_1^2$ and $\widehat{\sigma}_2^2$ be two sample variances from two independent normally distributed samples $\{X_{i1}, i = 1, ..., n_1\}$ and $\{X_{i2}, i = 1, ..., n_2\}$ from the same general population with mean μ and variance σ^2. Find λ such that $\widehat{\sigma}^2 = \lambda\widehat{\sigma}_1^2 + (1 - \lambda)\widehat{\sigma}_2^2$ has minimum MSE and compare it with the sample variance over the combined sample.

10. Let $\widehat{\theta}_1$ and $\widehat{\theta}_2$ be two unbiased estimators of θ with correlation coefficient ρ. (a) Is it true that $(\widehat{\theta}_1 + \widehat{\theta}_2)/2$ has a smaller MSE? (b) Find λ such that $\lambda\widehat{\theta}_1 + (1 - \lambda)\widehat{\theta}_2$ has minimum MSE. (c) For what ρ does the linear combination of estimators not reduce the MSE?

11. Use simulations to compute and display empirical relative MSE for two estimators of σ: the traditional (6.4) and $\widetilde{\sigma}$ with optimal ν_n. Use $n = 5$ and $n = 15$ with the sequence of true σ values from 0.1 to 1 with step 0.1.

12. (a) Derive analytically the MSE of the two estimators of λ in Example 6.26. (b) Plot the MSEs on the same graph as in Figure 6.6 to confirm the derivation. (c) Prove that the MSE of $\overline{X}$ is smaller.

13. (a) Analogously to Theorem 6.25, derive an optimal estimator of λ in Example 6.26. (b) Write an R function with simulations to test your derivation.

14. Demonstrate that computation of the vectorized variance using rowMeans (X^2) is faster than using apply(X,1,var) in Example 6.26.

15. (a) Prove that if $\widehat{\theta}$ is positive and unbiased for θ, then $\widehat{\theta}^{-1}$ is biased for θ^{-1}. Moreover, prove that $\widehat{\theta}^{-1}$ overestimates θ^{-1}. (b) Prove that if $\widehat{\theta}_1$ and $\widehat{\theta}_2$ are independent, where $\widehat{\theta}_1$ is unbiased for θ_1 and a positive $\widehat{\theta}_2$ is unbiased for positive θ_2, then $\widehat{\theta}_1/\widehat{\theta}_2$ is biased for θ_1/θ_2. [Hint: Use Jensen's inequality.]

16. Following Example 6.14 investigate the properties of the MM and MQ estimators of α and β of the uniform distribution $\mathcal{R}(\alpha, \beta)$ using simulations. [Hint: Plot percentage bias and RMSE for α and β for $n = 10$ and $n = 100$ on the grid of values of one parameter while holding the others fixed.]

17. Similarly to Example 6.22, compare two unbiased estimators of parameter θ in the uniform distribution $X_i \sim \mathcal{R}(\theta, 1)$, where $i = 1, 2, ..., n$ and $\theta < 1$. (a) Find an unbiased estimator based on $\overline{X}$. (b) Find an unbiased estimator based on $X_{\min}$. (c) Prove that the second estimator is better. (d) Plot the two MSEs for $n = 2$ with θ in the range from -1 to 1 and use simulations to confirm the theoretical MSEs for $\theta = -1$ and 0 (depict the points on the respective MSEs). [Hint: Use integrals $\int_\theta^1 x(1-x)^{n-1}dx = \frac{\theta n + 1}{(1+n)n}(1-\theta)^n$ and $\int_\theta^1 x^2(1-x)^{n-1}dx = \frac{\theta^2 n(n+1) + 2\theta n + 2}{(n^2+3n+2)n}(1-\theta)^n$.]

18. Let $\{X_i, i = 1, ..., n\}$ be a random sample of binomial random variables of size N. Two methods of estimating $\sigma^2 = \text{var}(X_1)$ are suggested: (a) using the formula $\sigma^2 = np(1-p)$ and (b) the general unbiased estimator $\widehat{\sigma}^2$. Which method is better? Use vectorized simulations with nSim=100000 and plot the RMSE as a function of n from $n = 5$ to $n = 20$ for $p = 0.1$ and $p = 0.5$ with $N = 30$. [Hint: Use par(mfrow=c(1,2)) and rowSums and rowMeans].

6.5 Linear estimation

In this section, we apply linear estimation theory to the simplest statistical model, the mean model. This model will be used to illustrate the fundamental task in statistics: find an optimal parameter estimator.

Specifically, let a set of n observations $Y_1, Y_2, ..., Y_n$ be randomly selected from a distribution with unknown mean α and variance σ^2. Shortly, we refer to this set as a random sample from the general population. Equivalently, this information can be represented as the *mean model*,

$$Y_i = \alpha + \varepsilon_i, \quad i = 1, 2, ..., n, \tag{6.16}$$

where ε_i and ε_j do not correlate $(i \neq j)$ and have zero mean and constant variance, $E(\varepsilon_i) = 0$ and $\text{var}(\varepsilon_i) = \sigma^2$ (ε_i are unobservable). Note that we use notation α for the mean (usually we use μ) because later α will be interpreted as the intercept of a linear model. In this section, we do not specify the distribution of ε (only the first two moments are involved); thus the distribution of Y_i is not necessarily normal. Also, we do not assume that ε_i are iid; the absence of correlation suffices at this point. Sometimes, the notation $Y_i \sim (\alpha, \sigma^2)$, $i = 1, 2, ..., n$ is used to indicate that $E(Y_i) = \alpha$ and $\mathrm{var}(Y_i) = \sigma^2$ without specifying the distribution (no letter in front of the parentheses).

As is easy to guess, the average, $\overline{Y}$, can be used as an estimator for α. The goal of this section is to prove some optimal properties of this estimator. As was mentioned above, the mean model may be viewed as the toy linear model. An elaborate coverage of the linear model is deferred to Chapter 8 because it involves statistical hypothesis testing and confidence intervals.

6.5.1 Estimation of the mean using linear estimator

We want to estimate mean α using a linear estimator $\widehat{\alpha} = \widehat{\alpha}(Y_1, Y_2, ..., Y_n) = \sum_{i=1}^{n} \lambda_i Y_i$, where coefficients $\lambda_1, \lambda_2, ..., \lambda_n$ are to be determined. Note that the arithmetic average, $\overline{Y} = \sum_{i=1}^{n} Y_i/n$ is a linear estimator with $\lambda_i = 1/n$.

Theorem 6.35 *Gauss–Markov for the mean.* *The arithmetic average estimator, $\overline{Y}$, has minimum variance among all unbiased linear estimators of α, or compactly BLUE (best linear unbiased estimator).*

Proof. Let $\widehat{\alpha}$ be a linear unbiased estimator, $E(\widehat{\alpha}) = \alpha$. This condition implies

$$E(\widehat{\alpha}) = E\left(\sum_{i=1}^{n} \lambda_i Y_i\right) = \sum_{i=1}^{n} \lambda_i E(Y_i) = \alpha \sum_{i=1}^{n} \lambda_i = \alpha \quad \forall \alpha.$$

Therefore the unbiasedness of $\widehat{\alpha}$ demands a constraint on the lambda coefficients:

$$\sum_{i=1}^{n} \lambda_i = 1. \tag{6.17}$$

However, since $\{Y_i\}$ are pairwise uncorrelated, the variance of the sum equals the sum of variances (see Section 3.10):

$$\mathrm{var}\left(\sum_{i=1}^{n} \lambda_i Y_i\right) = \sigma^2 \sum_{i=1}^{n} \lambda_i^2.$$

Since σ^2 is fixed, the problem of finding an optimal linear estimator reduces to an optimization problem

$$\min \sum_{i=1}^{n} \lambda_i^2 \tag{6.18}$$

under constraint (6.17). To solve this optimization problem, we use the Lagrange multiplier technique with Lagrange function (Section 10.6.4) given by

$$\mathcal{L}(\lambda_1, ..., \lambda_n; \omega) = \sum_{i=1}^{n} \lambda_i^2 - \omega\left(\sum_{i=1}^{n} \lambda_i - 1\right),$$

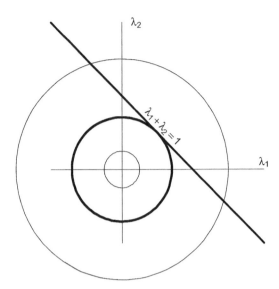

Figure 6.7: *The geometry of the Gauss–Markov theorem for $n = 2$. The circle with minimum radius (corresponding to minimum variance) touches the straight line $\lambda_1 + \lambda_2 = 1$ (unbiasedness condition) with the optimal solution $\lambda_1 = \lambda_2 = 1/2$. The large thin circle does not correspond to minimal variance, and the small thin circle does not meet the unbiasedness condition.*

where ω is the Lagrange multiplier. Differentiating with respect to λ_i, we obtain the necessary condition for the minimum,

$$\frac{\partial \mathcal{L}}{\partial \lambda_i} = 2\lambda_i - \omega = 0, \quad i = 1, 2, ..., n,$$

which means that all λ_i must be the same. Hence, from condition (6.17), we conclude that $\lambda_i = 1/n$: the optimal solution is the arithmetic average with the minimal variance σ^2/n. For the case when $n = 2$ the optimization problem (6.18) is illustrated in Figure 6.7. The straight line (bold) represents the unbiasedness constraint $\lambda_1 + \lambda_2 = 1$ and circles defined by $\lambda_1^2 + \lambda_2^2 = r^2$ correspond to the variance of the estimator. Among all circles, we need to find a circle with the minimum radius that intersects the line $\lambda_1 + \lambda_2 = 1$. This is the circle that touches the line (bold). Due to symmetry, the touching point lies on the $45°$ line, $\lambda_1 = \lambda_2 = 1/2$.

Example 6.36 *Slope regression model.* (a) *Find an unbiased linear estimator for β with the minimum variance in the slope regression model $Y_i = \beta x_i + \varepsilon_i$, where x_i are not all zero and fixed, and ε_i are the same as in the mean model above.* (b) *Prove that this estimator minimizes the sum of squares $\sum_{i=1}^{n}(Y_i - \beta x_i)^2$.*

Solution. (a) We seek an estimator of β as a linear function, $\widehat{\beta} = \sum_{i=1}^{n} \lambda_i Y_i$, where coefficients λ_i are subject to constrained optimization. For $\widehat{\beta}$ to be unbiased, we have to have

$$E(\widehat{\beta}) = \sum_{i=1}^{n} \lambda_i E(Y_i) = \beta \sum_{i=1}^{n} \lambda_i x_i = \beta \quad \forall \beta.$$

Thus, unbiasedness implies the following condition on the lambda coefficients

$$\sum_{i=1}^{n} \lambda_i x_i = 1. \tag{6.19}$$

Find the variance

$$\mathrm{var}(\widehat{\beta}) = \sum_{i=1}^{n} \lambda_i^2 \mathrm{var}(Y_i) = \sigma^2 \sum_{i=1}^{n} \lambda_i^2.$$

Thus, the optimal lambdas must minimize $\sum_{i=1}^{n} \lambda_i^2$ under constraint (6.19). Introduce the Lagrange function as in the mean model:

$$\mathcal{L}(\lambda_1, ... \lambda_n; \omega) = \sum_{i=1}^{n} \lambda_i^2 - \omega \left(\sum_{i=1}^{n} \lambda_i x_i - 1 \right),$$

where ω is the Lagrange multiplier. Differentiation of $\mathcal{L}$ with respect to λ_i gives $2\lambda_i = \omega x_i$ for $i = 1, 2, ..., n$. Plugging this solution back into (6.19) we arrive at $\lambda_i = x_i / \sum_{j=1}^{n} x_j^2$. The Gauss–Markov estimator, a linear unbiased estimator with minimum variance, is

$$\widehat{\beta} = \frac{\sum_{i=1}^{n} x_i Y_i}{\sum_{i=1}^{n} x_i^2}. \tag{6.20}$$

It has minimal variance $\mathrm{var}(\widehat{\beta}) = \sigma^2 / \sum_{i=1}^{n} x_i^2$ among all linear unbiased estimators. Note that, since not all x_i are zeros, the denominator is not zero and $\widehat{\beta}$ is well defined.

(b) Since the least squares estimator minimizes the sum of squares, the derivative at the solution must be zero,

$$\frac{d}{d\beta} \sum_{i=1}^{n} (Y_i - \beta x_i)^2 = -2 \sum_{i=1}^{n} (Y_i - \beta x_i) x_i = 0.$$

Solve for β and obtain that the optimal estimator $\widehat{\beta} = \sum_{i=1}^{n} x_i Y_i / \sum_{i=1}^{n} x_i^2$, the same as (6.20).

Example 6.37 *Two slope estimators. Consider two other linear estimators of β in the slope model, $\widetilde{\beta}_1 = n^{-1} \sum_{i=1}^{n} Y_i / x_i$ and $\widetilde{\beta}_2 = \sum_{i=1}^{n} Y_i / \sum_{i=1}^{n} x_i$ assuming that $x_i \neq 0$. (a) Prove that both estimators are unbiased. (b) Derive two algebraic inequalities from the optimality of estimator (6.20).*

Solution. (a) The unbiasedness follows from the fact that $\{x_i\}$ are fixed,

$$E(\widetilde{\beta}_1) = \frac{1}{n}\sum_{i=1}^{n} E\left(\frac{Y_i}{x_i}\right) = \frac{1}{n}\sum_{i=1}^{n}\frac{E(Y_i)}{x_i} = \frac{1}{n}\sum_{i=1}^{n}\frac{\beta x_i}{x_i} = \beta,$$

$$E(\widetilde{\beta}_2) = \frac{\sum_{i=1}^{n} E(Y_i)}{\sum_{i=1}^{n} x_i} = \beta\frac{\sum_{i=1}^{n} x_i}{\sum_{i=1}^{n} x_i} = \beta.$$

(b) Find the variances,

$$\mathrm{var}(\widetilde{\beta}_1) = \frac{1}{n^2}\sum_{i=1}^{n}\mathrm{var}\left(\frac{Y_i}{x_i}\right) = \sigma^2\frac{1}{n^2}\sum_{i=1}^{n}\frac{1}{x_i^2}, \quad \mathrm{var}(\widetilde{\beta}_2) = \sigma^2\frac{n}{\left(\sum_{i=1}^{n} x_i\right)^2}.$$

Both estimators $\widetilde{\beta}_1$ and $\widetilde{\beta}_2$ are unbiased and are linear functions of observations, so the Gauss–Markov theorem applies. Since estimator $\widehat{\beta}$ defined in equation (6.20) has minimal variance, the following algebraic inequalities hold:

$$\frac{1}{\sum_{i=1}^{n} x_i^2} \le \frac{1}{n^2}\sum_{i=1}^{n}\frac{1}{x_i^2}, \quad \frac{1}{\sum_{i=1}^{n} x_i^2} \le \frac{n}{\left(\sum_{i=1}^{n} x_i\right)^2}. \tag{6.21}$$

Note that these inequalities turn into equalities when $x_i = \mathrm{const}$. $\qquad\square$

The above example gives rise to student' grading and performance.

Example 6.38 *Student' grading. The instructor gives students a series of n assignments, say, homework, quizzes, etc. The maximum number of points in the ith assignment is x_i. Suppose that the ith student gains Y_i points in the ith assignment ($Y_i \le x_i$). To rank the students in the class, the instructor wants a metric for student' performance by finding the ratio of the number of points received to the maximum number of points. Two metrics are suggested: (a) Mean of the ratios, $\widetilde{\beta}_1 = n^{-1}\sum_{i=1}^{n} Y_i/x_i$, and (b) ratio of the means, $\widetilde{\beta}_2 = \sum_{i=1}^{n} Y_i/\sum_{i=1}^{n} x_i$. Suggest a better way to assess student' performance.*

Solution. To compare metrics/estimators a statistical model should be used. Here, we assume the slope model, $Y_i = \beta x_i + \varepsilon_i$ and treat the metrics as estimators of β for each student. According to the slope model student' performance is evaluated by the slope over the points gained during the course. As follows from Example 6.37, neither of the two metrics is optimal. A more precise way to evaluate student' performance is to use the Gauss–Markov estimator (6.20). The reader should understand the shortcomings of the slope model here due to the presence of the additive ε_i. For example, the number of gained points may be negative or larger than x_i which is impossible. A better model for student' grading is offered in Example 6.118. Interestingly, although that model is different, it leads to the same estimator $\widetilde{\beta}_2$. $\qquad\square$

Linear estimators are not always appropriate.

Example 6.39 *Linear estimator of the variance.* *Why is a linear estimator of σ^2 in the mean model is "bad"? More precisely, prove that a linear unbiased estimator of σ^2 does not exist unless the mean is proportional to the variance.*

Solution. The unbiasedness of estimator $\sum_{i=1}^{n} \lambda_i Y_i$ for σ^2 implies $\alpha \sum_{i=1}^{n} \lambda_i = \sigma^2$. Since λ_i are constants this equality cannot hold for all α and σ^2 unless α and σ^2 are proportionally related. Although a linear estimator of the variance does not make sense for the majority of distributions, it is appropriate for the Poisson distribution for which $\mu = \sigma^2$.

6.5.2   Vector representation

Vector representation substantially reduces the notation but most importantly employs powerful matrix algebra and calculus. Especially useful is vector/matrix representation for linear models. The goal of this section is to give a gentle introduction to vector representation for the simplest mean model and the respective Gauss–Markov theorem. The full advantage of linear algebra for a linear model can be appreciated from Chapter 8.

Introduce the following $n \times 1$ vectors:

$$\mathbf{Y} = \begin{bmatrix} Y_1 \\ Y_2 \\ \vdots \\ Y_n \end{bmatrix}, \quad \mathbf{1} = \begin{bmatrix} 1 \\ 1 \\ \vdots \\ 1 \end{bmatrix}, \quad \boldsymbol{\varepsilon} = \begin{bmatrix} \varepsilon_1 \\ \varepsilon_2 \\ \vdots \\ \varepsilon_n \end{bmatrix}.$$

Then the mean model can be written compactly as

$$\mathbf{Y} = \alpha \mathbf{1} + \boldsymbol{\varepsilon},$$

where $E(\boldsymbol{\varepsilon}) = \mathbf{0}$, $\mathrm{cov}(\boldsymbol{\varepsilon}) = \sigma^2 \mathbf{I}$, and $\mathbf{I}$ is the $n \times n$ identity matrix (we suggest the reader review Section 3.10). These equations imply that the components of vector $\boldsymbol{\varepsilon}$ have zero mean, have variance σ^2, and are uncorrelated. Since vector $\alpha \mathbf{1}$ is fixed we have $\mathrm{cov}(\mathbf{Y}) = \mathrm{cov}(\boldsymbol{\varepsilon}) = \sigma^2 \mathbf{I}$. Let $\boldsymbol{\lambda} = (\lambda_1, \lambda_2, ..., \lambda_n)'$ be the $n \times 1$ vector of fixed coefficients (note that we use the transposition sign $'$ for vector column to save space when introducing a column vector). Then a linear estimator of α can be expressed as $\hat{\alpha} = \boldsymbol{\lambda}' \mathbf{Y}$.

The following example emphasizes the advantage of the vector notation.

Example 6.40 *Vector notation for the slope model.* *(a) Use vector notation for the slope model in Example 6.36 and prove the Gauss–Markov theorem using a Lagrange function. (b) Prove that this estimator minimizes the sum of squares $\|\mathbf{Y} - \beta \mathbf{x}\|^2$. (c) Provide a geometrical interpretation to minimization of the sum of squares using Language projection.*

Solution. (a) Introduce the $n \times 1$ vector $\mathbf{x} = (x_1, x_2, ..., x_n)'$. Then the slope model can be written in the vector notation as $\mathbf{Y} = \beta\mathbf{x} + \boldsymbol{\varepsilon}$. Let $\boldsymbol{\lambda} = (\lambda_1, \lambda_2, ..., \lambda_n)'$ be the $n \times 1$ vector of fixed coefficients and $\widehat{\beta} = \boldsymbol{\lambda}'\mathbf{Y}$ be a linear estimator of the slope coefficient. We want to find vector $\boldsymbol{\lambda}$ such that $\widehat{\beta}$ is unbiased and has minimum variance/MSE. Using the results of Section 3.10, and particularly Theorem 3.79, we obtain

$$E(\widehat{\beta}) = \beta\boldsymbol{\lambda}'\mathbf{x}, \quad \mathrm{var}(\widehat{\beta}) = \boldsymbol{\lambda}'\mathrm{cov}(\mathbf{Y})\boldsymbol{\lambda} = \sigma^2 \|\boldsymbol{\lambda}\|^2.$$

From the first equation, we deduce that condition $\boldsymbol{\lambda}'\mathbf{x} = 1$ must hold to make $\widehat{\beta}$ unbiased (note that 1 is a number and $\mathbf{1}$ is a vector). The Lagrange function in the vector notation takes the form

$$\mathcal{L}(\boldsymbol{\lambda}; \omega) = \|\boldsymbol{\lambda}\|^2 - \omega\left(\boldsymbol{\lambda}'\mathbf{x} - 1\right),$$

where ω is the Lagrange multiplier. Differentiate with respect to vector $\boldsymbol{\lambda}$ to find the first-order condition for the minimum,

$$\frac{\partial\mathcal{L}}{\partial\boldsymbol{\lambda}} = 2\boldsymbol{\lambda} - \omega\mathbf{x} = \mathbf{0}.$$

This means that vector $\boldsymbol{\lambda}$ must be proportional to $\mathbf{x}$ with the coefficient of proportionality $\omega/2$. From the unbiasedness condition, $\boldsymbol{\lambda}'\mathbf{x} - 1 = 0$, we find the optimal $\boldsymbol{\lambda} = \|\mathbf{x}\|^{-2}\mathbf{x}$, so that the linear unbiased estimator of the slope with minimum variance can be written as

$$\widehat{\beta} = \frac{\mathbf{Y}'\mathbf{x}}{\|\mathbf{x}\|^2} \tag{6.22}$$

with the variance

$$\mathrm{var}(\widehat{\beta}) = \frac{\sigma^2}{\|\mathbf{x}\|^2}. \tag{6.23}$$

(b) Write the derivative of the sum of squares in vector form and equate it to zero,

$$\frac{d}{d\beta}\|\mathbf{Y} - \beta\mathbf{x}\|^2 = -2(\mathbf{Y} - \beta\mathbf{x})'\mathbf{x} = 0, \tag{6.24}$$

with the optimal solution $(\mathbf{Y}'\mathbf{x})/\|\mathbf{x}\|^2$, the same as (6.22). (c) The sum of squares may be viewed as the minimum squared Euclidean distance from point $\mathbf{Y} \in R^n$ to the straight line through the origin spanned by vector $\mathbf{x} \in R^n$. The condition of minimization of the sum of squares expressed by equation (6.24) means that vector $\mathbf{Y} - \beta\mathbf{x}$ is perpendicular to $\mathbf{x}$. In other words, the closest point from $\mathbf{Y}$ to the line is $\widehat{\mathbf{Y}} = \widehat{\beta}\mathbf{x}$.

Problems

1. Sketch the geometry of the least squares solution for the slope model when the number of observations is $n = 2$, similarly to Figure 6.7.

2. Define an estimator in the slope model as $\widetilde{\beta} = Y_k/x_k$, where $x_k = \max(x_i, i = 1, ..., n)$. (a) Is this estimator linear? (b) Is this estimator unbiased? (c) Prove that $\mathrm{var}(\widetilde{\beta}) \geq \mathrm{var}(\widehat{\beta})$, where $\widehat{\beta}$ is defined by (6.20). (d) Specify conditions under which $\mathrm{var}(\widetilde{\beta}) = \mathrm{var}(\widehat{\beta})$.

3. Define an estimator in the slope model as $\widetilde{\beta} = Y_k/x_k$, where $Y_k = \max(Y_i, i = 1, ..., n)$. Is this estimator a linear estimator? Prove analytically or show by simulations that $\widetilde{\beta}$ has positive bias.

4. Prove that $\widetilde{\beta} = (Y_n - Y_1)/(x_n - x_1)$ is unbiased for slope β but less efficient than the OLS estimator.

5. Is it true that estimator $\widetilde{\beta}_1$ is better than estimator $\widetilde{\beta}_2$ in Example 6.37? [Hint: Simulate $x_1, ..., x_n$ to test the inequality for variance.]

6. Provide a geometric illustration of the two estimators from Example 6.37.

7. Prove inequalities (6.21) algebraically. [Hint: Use Cauchy inequality (2.17) to prove the first inequality and the fact that the variance is nonnegative for the second inequality.]

8. Explain why a quadratic estimator in the form $\sum_{i=1}^{n} \lambda_i Y_i^2$ is generally "bad" for estimation of α in the mean model?

9. Express estimators $\widetilde{\beta}_1$ and $\widetilde{\beta}_2$ in Example 6.37 using vector notation.

10. Prove the unbiasedness of estimator (6.22) and derive variance (6.23) in vector language using the results of Section 3.10.

11. Sketch the OLS solution for the mean model with $\mathbf{Y} = (-1, 2)$, display the regression line, projection, and $\widehat{\mathbf{Y}}$. [Consult Example 6.40.]

12. Sketch the OLS solution for the slope model from Example 6.40 with $\mathbf{x} = (1, -2)$ and $\mathbf{Y} = (1, 2)$. Display the regression line, projection, and $\widehat{\mathbf{Y}}$. Compute and display the OLS projection point.

13. (a) Let $\widehat{\theta}_1$ and $\widehat{\theta}_2$ be two independent linear unbiased estimators of θ. Find a better linear unbiased estimator as a linear combination of these two estimators. When is a linear unbiased estimator not better than $\widehat{\theta}_1$ and $\widehat{\theta}_2$? (b) The same as above, but estimators may be dependent with the correlation coefficient ρ.

6.6 Estimation of variance and correlation coefficient

In the previous section, linear estimators have been used to estimate the mean. In this section, we turn our attention to *quadratic estimators* for the variance under the normal assumption. Then we discuss estimation of the correlation coefficient. Familiarity with matrix algebra and calculus is essential here, see the Appendix.

6.6.1   Quadratic estimation of the variance

Theorem 6.41 *Let $Y_1, Y_2, ..., Y_n$ be iid normally distributed observations with unknown mean and variance: $Y_i \overset{\text{iid}}{\sim} \mathcal{N}(\alpha, \sigma^2)$, $i = 1, 2, ..., n$. Then the estimator of the variance, sometimes referred to as the sample variance,*

$$\widehat{\sigma}^2 = \frac{1}{n-1} \sum_{i=1}^{n} (Y_i - \overline{Y})^2 \tag{6.25}$$

has minimum variance among all unbiased quadratic estimators, $\widetilde{\sigma}^2 = \mathbf{Y}'\mathbf{A}\mathbf{Y}$, where $\mathbf{Y} = (Y_1, Y_2, ..., Y_n)'$ is the observation vector and $\mathbf{A}$ is an $n \times n$ symmetric nonnegative definite matrix.

Proof. We start by showing that $\widehat{\sigma}^2$ can be expressed in the form $\mathbf{Y}'\mathbf{A}\mathbf{Y}$. Using simple linear algebra we write the sum of squares as the squared norm of the $n \times 1$ vector, $\sum_{i=1}^{n} (Y_i - \overline{Y})^2 = \left\| \mathbf{Y} - \mathbf{1}\overline{Y} \right\|^2$, where $\overline{Y} = n^{-1}\mathbf{Y}'\mathbf{1} = n^{-1}\mathbf{1}'\mathbf{Y}$ is the usual average. Express the deviation from the average in the matrix form,

$$\mathbf{Y} - \overline{Y}\mathbf{1} = \mathbf{Y} - \frac{1}{n}\mathbf{1}\mathbf{1}'\mathbf{Y} = \left(\mathbf{I} - \frac{1}{n}\mathbf{1}\mathbf{1}' \right) \mathbf{Y} = \mathbf{H}\mathbf{Y},$$

where $\mathbf{H} = \mathbf{I} - \frac{1}{n}\mathbf{1}\mathbf{1}'$ is a symmetric $n \times n$ matrix. Matrix $\mathbf{H}$ is nonnegative definite and idempotent: $\mathbf{H}^2 = \mathbf{H}$. Indeed, since $\mathbf{1}'\mathbf{1} = n$ we have

$$\begin{aligned} \mathbf{H}^2 &= \left(\mathbf{I} - \frac{1}{n}\mathbf{1}\mathbf{1}' \right) \left(\mathbf{I} - \frac{1}{n}\mathbf{1}\mathbf{1}' \right) = \mathbf{I} - \frac{1}{n}\mathbf{1}\mathbf{1}' - \frac{1}{n}\mathbf{1}\mathbf{1}' + \frac{1}{n}\mathbf{1}\mathbf{1}'\frac{1}{n}\mathbf{1}\mathbf{1}' \\ &= \mathbf{I} - \frac{1}{n}\mathbf{1}\mathbf{1}' - \frac{1}{n}\mathbf{1}\mathbf{1}' + \frac{1}{n}\mathbf{1}\mathbf{1}' = \mathbf{I} - \frac{1}{n}\mathbf{1}\mathbf{1}' = \mathbf{H}. \end{aligned}$$

Matrix $\mathbf{H}$ has $n-1$ zero eigenvalues and one nonzero eigenvalue, 1. Now we find the squared norm of $\mathbf{Y} - \overline{Y}\mathbf{1}$,

$$\left\| \mathbf{Y} - \mathbf{1}\overline{Y} \right\|^2 = \mathbf{Y}'\mathbf{H}^2\mathbf{Y} = \mathbf{Y}'\mathbf{H}\mathbf{Y} = \mathbf{Y}' \left(\mathbf{I} - \frac{1}{n}\mathbf{1}\mathbf{1}' \right) \mathbf{Y}.$$

Finally, $\widehat{\sigma}^2$ can be expressed as $\mathbf{Y}'\mathbf{A}\mathbf{Y}$, where

$$\mathbf{A} = \frac{1}{n-1} \left(\mathbf{I} - \frac{1}{n}\mathbf{1}\mathbf{1}' \right) \tag{6.26}$$

is a $n \times n$ symmetric nonnegative definite matrix.

Now let $\widetilde{\sigma}^2 = \mathbf{Y}'\mathbf{A}\mathbf{Y}$ be any quadratic unbiased estimator. To find its expected value and variance, use formulas $E(\mathbf{Y}'\mathbf{A}\mathbf{Y}) = \text{tr}(\mathbf{A}\mathbf{\Omega}) + \boldsymbol{\mu}'\mathbf{A}\boldsymbol{\mu}$ and $\text{var}(\mathbf{Y}'\mathbf{A}\mathbf{Y}) = 2\text{tr}(\mathbf{A}\mathbf{\Omega})^2 + 4\boldsymbol{\mu}'\mathbf{A}\mathbf{\Omega}\mathbf{A}\boldsymbol{\mu}$ from Section 4.2.2, where $\mathbf{Y} \sim \mathcal{N}(\boldsymbol{\mu}, \mathbf{\Omega})$. In our case $\mathbf{Y} \sim \mathcal{N}(\mu\mathbf{1}, \sigma^2\mathbf{I})$, so that we obtain

$$E(\widetilde{\sigma}^2) = \sigma^2\text{tr}(\mathbf{A}) + \mu^2\mathbf{1}'\mathbf{A}\mathbf{1}, \quad \text{var}(\widetilde{\sigma}^2) = 2\sigma^4\text{tr}(\mathbf{A}^2) + 4\mu^2\sigma^2\mathbf{1}'\mathbf{A}^2\mathbf{1}.$$

The unbiasedness dictates $\sigma^2 \mathrm{tr}(\mathbf{A}) + \mu^2 \mathbf{1}' \mathbf{A} \mathbf{1} = \sigma^2$ for all σ^2 and μ, or equivalently,

$$\mathbf{1}'\mathbf{A}\mathbf{1} = 0, \quad \mathrm{tr}(\mathbf{A}) = 1. \tag{6.27}$$

Since $\mathbf{A}$ is assumed to be nonnegative definite $\mathbf{1}'\mathbf{A}\mathbf{1} = 0$ implies $\mathbf{A}\mathbf{1} = \mathbf{0}$. Consequently, this equation implies that $\mathbf{A}^2\mathbf{1} = \mathbf{0}$ and $\mathbf{1}'\mathbf{A}^2\mathbf{1} = 0$, so $\mathrm{var}(\widetilde{\sigma}^2) = 2\sigma^4 \mathrm{tr}(\mathbf{A}^2)$. Therefore, to minimize the variance of $\widetilde{\sigma}^2$, one needs to solve the following matrix optimization problem: $\min \mathrm{tr}(\mathbf{A}^2)$ under conditions (6.27). Introduce the Lagrange function (we use coefficient $1/2$ to simplify the algebra after differentiation):

$$\mathcal{L}(\mathbf{A}; \nu, \lambda) = \frac{1}{2}\mathrm{tr}(\mathbf{A}^2) + \nu \mathbf{1}'\mathbf{A}\mathbf{1} - \lambda(\mathrm{tr}(\mathbf{A}) - 1),$$

where ν and λ and Lagrange multipliers. Using Section 10.5, we obtain

$$\frac{\partial \mathcal{L}}{\partial \mathbf{A}} = \frac{1}{2}\frac{\partial \mathrm{tr}(\mathbf{A}^2)}{\partial \mathbf{A}} + \nu \mathbf{1}\mathbf{1}' - \lambda \frac{\partial \mathrm{tr}(\mathbf{A})}{\partial \mathbf{A}} = \mathbf{A} + \nu \mathbf{1}\mathbf{1}' - \lambda \mathbf{I} = \mathbf{0},$$

which implies $\mathbf{A} = \lambda \mathbf{I} - \nu \mathbf{1}\mathbf{1}'$. Substituting this expression for $\mathbf{A}$ in the first condition of (6.27), we find $\lambda = n\nu$, so that $\mathbf{A} = \nu(n\mathbf{I} - \mathbf{1}\mathbf{1}')$. Substituting this expression for $\mathbf{A}$ into the second condition of (6.27), we find $n = 1/(n(n-1))$ and finally arrive at $\mathbf{A}$ given by formula (6.26). The theorem is proved.

Corollary 6.42 *As follows from the proof of the theorem,*

$$\mathrm{var}(\widehat{\sigma}^2) = \frac{2\sigma^4}{n-1}.$$

This formula implies that $\widehat{\sigma}^2$ is consistent in the quadratic sense (see Section 6.4.4) for σ^2 because $\widehat{\sigma}^2$ is unbiased and its variance vanishes with n.

From Section 4.2, we know that the normalized sum of squares has the chi-square distribution with $n-1$ degrees of freedom. Symbolically,

$$\frac{1}{\sigma^2}\sum_{i=1}^{n}(Y_i - \overline{Y})^2 \sim \chi^2(n-1).$$

One can rewrite this as $(n-1)\widehat{\sigma}^2/\sigma^2 \sim \chi^2(n-1)$, which will be used later in Section 7.5 to construct a confidence interval for σ^2.

It is worthwhile to note that the sample variance remains unbiased even when Y_i are not normally distributed as long as the third central moment is zero, $E(Y_i - \mu)^3 = 0$.

As a word of caution, the variance, as a measure of scatter, is most appropriate for symmetric distributions with relatively light tails, such as the normal distribution. Otherwise, a better, more robust measure of scatter is the nonparametric interquartile range or the parametric tight confidence range, as discussed earlier in the book; see Section 2.2.

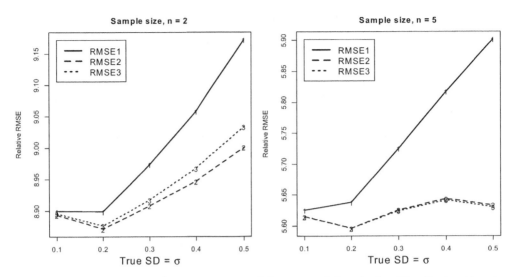

Figure 6.8: *Simulations for estimation of the area of the circle. The first estimator has a larger relative RMSE but the other two estimators are close. For small n the third estimator outperforms the first two.*

Example 6.43 *Unbiased estimation of the area of the circle.* *Two naive estimators for the area of the circle are suggested:* $\widehat{A}_1 = \pi \sum_{i=1}^{n} r_i^2 / n$ *and* $\widehat{A}_2 = \pi \bar{r}^2$, *where* r_i *are* n *iid unbiased measurements of the radius (the normal distribution is not assumed). (a) Prove that both estimators are positively biased. (b) Prove that* $\widehat{A}_3 = \widehat{A}_2 - \frac{\pi}{n}\widehat{\sigma}^2$ *is an unbiased estimator of the area, where* $\widehat{\sigma}^2$ *is the sample variance. (c) Use the relative root MSE (RMSE/σ) to compare the three estimators via simulations assuming that* r_i *are iid normally distributed.*

Solution. (a) Let $\{r_i\}$ be iid measurements of the true radius ρ with SD σ. Let $A = \pi \rho^2$ denote the true area. For the first estimator, since $E(r_i^2) = \operatorname{var}(r_i) + E^2(r_i)$, we have

$$E(\widehat{A}_1) = \frac{\pi}{n} \sum_{i=1}^{n} E(r_i^2) = \pi(\rho^2 + \sigma^2) = A + \pi \sigma^2.$$

This means that $\widehat{A}_1$ has bias $\pi \sigma^2$. Note that the bias does not decrease with n, so estimator $\widehat{A}_1$ has a systematic positive bias. The second estimator is also biased for finite n,

$$E(\widehat{A}_2) = \pi E(\bar{r}^2) = \pi(\operatorname{var}(\bar{r}) + E^2(\bar{r})) = \pi \sigma^2 / n + \pi r^2 = A + \pi \sigma^2 / n.$$

The bias goes to zero when n increases. We say that $\widehat{A}_2$ is consistent and asymptotically unbiased. (b) The expression of the expected value of the second estimator hints to how to make an unbiased adjustment. Indeed, since $\widehat{\sigma}^2$ unbiasedly estimates σ^2, an unbiased estimator of the area is $\widehat{A}_3 = \widehat{A}_2 - \pi \widehat{\sigma}^2 / n$. (c) The

R function `arMSE` generates the data and computes three area estimates in the vectorized fashion – the built-in functions `rowMeans` and `rowSums` are used for this purpose. Alternatively but slower, one may use `apply(X=r,MARGIN=1,FUN=var)` where `MARGIN=1` means that computations are done over the first dimension (rows) of matrix `X` (refer to Section 1.5.5). As follows from Figure 6.8 the third estimator outperforms the others.

6.6.2 Estimation of the covariance and correlation coefficient

We recommend reviewing Section 3.5.

Theorem 6.44 *Let (X_i, Y_i) be n iid observations from the bivariate normal distribution with unknown parameters. The sample covariance,*

$$\widehat{\sigma}_{xy} = \frac{1}{n-1} \sum_{i=1}^{n} (X_i - \overline{X})(Y_i - \overline{Y}), \qquad (6.28)$$

is an unbiased estimator of the population covariance σ_{xy}.

Solution. Denote $\mathrm{cov}(X_i, Y_i) = \sigma_{xy}$, $E(X_i) = \mu_x$ and $E(Y_i) = \mu_y$. Algebra simplifies if the covariance is redefined in terms of random variables with zero means,

$$\widehat{\sigma}_{xy} = \frac{1}{n-1} \sum_{i=1}^{n} (\varepsilon_i - \overline{\varepsilon})(\delta_i - \overline{\delta}),$$

where $\varepsilon_i = X_i - \mu_x$ and $\delta_i = Y_i - \mu_y$ have zero means and the same variances and covariance as X_i and Y_i. By expanding the product we obtain

$$E\left(\sum_{i=1}^{n}(\varepsilon_i - \overline{\varepsilon})(\delta_i - \overline{\delta})\right) = \sum_{i=1}^{n} E(\varepsilon_i \delta_i) - E\left(\overline{\varepsilon}\sum_{i=1}^{n}\delta_i\right) - E\left(\overline{\delta}\sum_{i=1}^{n}\varepsilon_i\right)$$

$$+ E\left(\sum_{i=1}^{n}\varepsilon_i \sum_{i=1}^{n}\delta_i\right) = n\sigma_{xy} - \frac{1}{n}n\sigma_{xy} - \frac{1}{n}n\sigma_{xy} + \frac{1}{n}n\sigma_{xy} = (n-1)\sigma_{xy},$$

which proves that $E(\widehat{\sigma}_{xy}) = \sigma_{xy}$. □

Now we turn our attention to estimation of the population correlation coefficient $\rho = \mathrm{cov}(X,Y)/\sqrt{\sigma_x^2 \sigma_y^2}$. Using unbiased estimators for variances and covariance, we arrive at the *sample correlation coefficient* (sometimes referred to as the Pearson correlation coefficient) :

$$r = \frac{\sum_{i=1}^{n}(X_i - \overline{X})(Y_i - \overline{Y})}{\sqrt{\sum_{i=1}^{n}(X_i - \overline{X})^2 \sum_{i=1}^{n}(Y_i - \overline{Y})^2}}. \qquad (6.29)$$

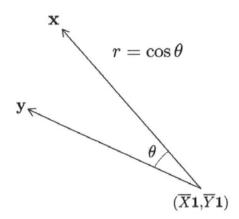

Figure 6.9: *The correlation coefficient is the cosine angle θ between vectors* $\mathbf{x} = (X_1 - \overline{X}, ..., X_n - \overline{X})$ *and* $\mathbf{y} = (Y_1 - \overline{Y}, ..., Y_n - \overline{Y})$. *Alternatively, we may think of the angle between* $(X_1, ..., X_n)$ *and* $(Y_1, ..., Y_n)$ *with the origin moved to* $(\overline{X}\mathbf{1}, \overline{Y}\mathbf{1})$.

Using elementary formulas

$$\sum_{i=1}^{n}(X_i - \overline{X})^2 = \sum_{i=1}^{n} X_i^2 - n\overline{X}^2, \quad \sum_{i=1}^{n}(Y_i - \overline{Y})^2 = \sum_{i=1}^{n} Y_i^2 - n\overline{Y}^2$$

$$\sum_{i=1}^{n}(X_i - \overline{X})(Y_i - \overline{Y}) = \sum_{i=1}^{n} X_i Y_i - n\overline{XY}, \tag{6.30}$$

the correlation coefficient can be alternatively expressed as

$$r = \frac{\sum_{i=1}^{n} X_i Y_i - n\overline{XY}}{\sqrt{\left(\sum_{i=1}^{n} X_i^2 - n\overline{X}^2\right)\left(\sum_{i=1}^{n} Y_i^2 - n\overline{Y}^2\right)}}. \tag{6.31}$$

In R the sample correlation coefficient is computed using function `cor`.

Geometric interpretation of the correlation coefficient

The correlation coefficient is the cosine of the angle between vectors $\mathbf{x} = (X_1 - \overline{X}, ..., X_n - \overline{X})$ and $\mathbf{y} = (Y_1 - \overline{Y}, ..., Y_n - \overline{Y})$ or, alternatively, between vectors $(X_1, ..., X_n)$ and $(Y_1, ..., Y_n)$ with the origin moved to the center of the vectors, $\overline{X}\mathbf{1}$ and $\overline{Y}\mathbf{1}$, where $\mathbf{1}$ is a vector of ones. To explain this interpretation, we express the right-hand side (6.29) in vector form. Indeed, the numerator can be viewed as the scalar product and the denominator can be viewed as the product of vector norms. Let $\mathbf{x}$ and $\mathbf{y}$ be $n \times 1$ vectors with the ith component X_i and Y_i, respectively. The correlation coefficient (6.29) can be written in vector form as

$$r = \frac{(\mathbf{x}-\overline{X}\mathbf{1})'(\mathbf{y}-\overline{Y}\mathbf{1})}{\|\mathbf{x}-\overline{X}\mathbf{1}\| \, \|\mathbf{y}-\overline{Y}\mathbf{1}\|}, \tag{6.32}$$

or

$$r = \cos\theta,$$

where $\theta \in [0, \pi]$ is the angle between $\mathbf{x}-\overline{X}\mathbf{1}$ and $\mathbf{y}-\overline{Y}\mathbf{1}$. See Figure 6.9 for a geometric illustration. Special cases when $\theta = 0$ or $\theta = \pi/2$ are important and will be discussed below.

Properties of the sample correlation coefficient

The sample correlation coefficient shares many properties with its theoretical counterpart, ρ, discussed in Section 3.5. One has to remember however that r, like ρ, has full justification for a random sample drawn from a bivariate normal distribution. In particular, this means that pairs (X_i, Y_i) must be iid. Consequently, $E(X_i) = \text{const}$ and $E(Y_i) = \text{const}$; see Section 6.7.4 where these conditions do not apply. Recall that the coefficient of determination (squared correlation coefficient); r^2 estimates the proportion of variance of Y explained by x. Although r can be always computed, even for discrete data, this interpretation of r^2, strictly speaking, is valid only for a bivariate normal distribution with iid observations. We shall discuss this in more detail in the chapter on linear model, where X_i are traditionally treated as fixed numbers.

The properties of r match the properties of the true correlation coefficient ρ from the bivariate normal distribution (Section 3.5).

1. $|r| \leq 1$. Geometrically, this property follows from the fact that r is the cosine of an angle. Algebraically, this property follows from the Cauchy inequality, $|\mathbf{u}'\mathbf{v}| \leq \|\mathbf{u}\| \, \|\mathbf{v}\|$, where $\mathbf{u} = \mathbf{x}-\overline{X}\mathbf{1}$ and $\mathbf{v} = \mathbf{y}-\overline{Y}\mathbf{1}$. The integral version of this inequality was discussed and proved in Section 2.2. The proof of the vector inequality uses the same idea: Introduce a quadratic function of β as $P(\beta) = \|\mathbf{u}-\beta\mathbf{v}\|^2 = \beta^2 \|\mathbf{v}\|^2 - 2\beta(\mathbf{u}'\mathbf{v}) + \|\mathbf{u}\|^2$. Since this quadratic function is nonnegative its discriminant must be nonpositive, i.e. $(\mathbf{u}'\mathbf{v})^2 - \|\mathbf{u}\|^2 \|\mathbf{v}\|^2 \leq 0$.

2. $r = \pm 1$ if and only if $\mathbf{y}-\overline{Y}\mathbf{1} = \beta(\mathbf{x}-\overline{X}\mathbf{1})$ or $\mathbf{y} = \alpha\mathbf{1}+\beta\mathbf{x}$ where $\alpha = \overline{Y}-\beta\overline{X}$. If $\beta > 0$, then $r = 1$ and, if $\beta < 0$, then $r = -1$. If $r = 1$, then $\theta = 0$ and, if $r = -1$, then $\theta = \pi$.

3. When $r = 0$, we say that X and Y are uncorrelated or vectors $\mathbf{x}-\overline{X}\mathbf{1}$ and $\mathbf{y}-\overline{Y}\mathbf{1}$ are orthogonal, $\theta = \pi/2$.

Theorem 6.45 *The distribution of r does not depend on $\mu_x, \mu_y, \sigma_x, \sigma_y$, but it does depend on ρ.*

Proof. First, we recognize that r does not depend on either μ_x or μ_y. Indeed, we can express $X_i - \overline{X} = (X_i - \mu_x) - (\overline{X} - \mu_x)$ and similarly for $Y_i - \overline{Y}$. Now, divide the numerator and denominator of (6.29) by $\sigma_x \sigma_y$,

$$
r = \frac{\sum_{i=1}^{n}(X_i/\sigma_x - \overline{X}/\sigma_x)(Y_i/\sigma_y - \overline{Y}/\sigma_y)}{\sqrt{\sum_{i=1}^{n}(X_i/\sigma_y - \overline{X}/\sigma_y)^2 \sum_{i=1}^{n}(Y_i/\sigma_y - \overline{Y}/\sigma_y)^2}}
$$

$$
= \frac{\sum_{i=1}^{n}(U_i - \overline{U})(V_i - \overline{V})}{\sqrt{\sum_{i=1}^{n}(U_i - \overline{U})^2 \sum_{i=1}^{n}(V_i - \overline{V})^2}}.
$$

This means that the distribution of r is the same as the distribution of r with normalized pairs (U_i, V_i) with $U_i \sim \mathcal{N}(0,1)$, $V_i \sim \mathcal{N}(0,1)$ and $\mathrm{cor}(U_i, V_i) = \rho$. $\square$

The fact that sample covariance and variances are unbiased may lead to a naive conclusion that the sample correlation coefficient is unbiased as well. Typically, an estimator as a nonlinear function is biased even though the arguments of the function are unbiased. For example, Jensen's inequality determines the direction of the bias if the function is convex (or concave). The following example takes advantage of Theorem 6.45 – only ρ need be specified when studying the distribution of r.

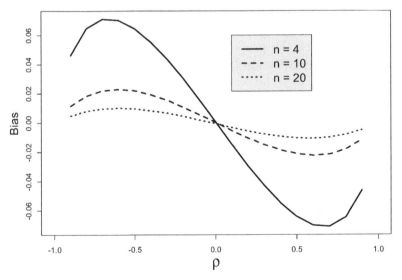

Figure 6.10: *Bias of the sample correlation coefficient, r, as a function of the true ρ with different sample sizes, n. The maximum bias is around $\rho = \pm 0.7$.*

Example 6.46 *Simulations for the correlation coefficient. Use vectorized simulations to show that r is biased for $n = 4, 10$, and 20. Find the true ρ for which the bias reaches its maximum.*

Solution. Since the distribution of r does not depend on the means and variances, it suffices to generate $X_i \sim \mathcal{N}(0,1)$ and $Y_i \sim \mathcal{N}(0,1)$ with correlation coefficient ρ. First, we generate $X_i \sim \mathcal{N}(0,1)$. Then, we generate $Y_i = \rho X_i + Z_i$, where $Z_i \sim \mathcal{N}(0, 1 - \rho^2)$ independent of X_i. Obviously, the pair (X_i, Y_i) satisfies the required conditions because

$$\mathrm{var}(Y_i) = \mathrm{var}(\rho X_i) + \mathrm{var}(Z_i) = \rho^2 + (1 - \rho^2) = 1,$$
$$\mathrm{cov}(X_i, Y_i) = \mathrm{cov}(X_i, \rho X_i + Z_i) = \rho \times \mathrm{var}(X_i) + \mathrm{cov}(X_i, Z_i) = \rho.$$

The R function is `robias`. It estimates the bias from simulations using vectorized computations based on the built-in functions `rowMeans` and `rowSums`. The result is shown in Figure 6.10. Note that we save computation time by generating X and Z before the loop over the `ro` values. Maximum bias decreases with n and peaks at about $\rho = \pm 0.7$.

Sample partial correlation

Partial correlation was introduced in Section 4.1.2. The partial correlation coefficient measures the correlation between X and Y *conditional* on other variables; thus, at least three variables must be involved. The sample partial correlation is computed as is the theoretical counterpart via inverse of the correlation matrix, as proved in Theorem 4.14,

$$r_{ij|\cdot} = -\frac{r^{ij}}{\sqrt{r^{ii} r^{jj}}},$$

where r^{ij} is the (i,j)th element of the $m \times m$ matrix $\mathbf{R}^{-1}$. Accordingly, r^{ii} and r^{jj} are the ith and jth diagonal elements of the inverse correlation matrix. The dot sign at the subindex indicates that the partial correlation is considered conditional on all other variables.

To compute the partial correlation, the sample size, n, must be greater than the number of variables, m; otherwise, matrix $\mathbf{R}$ is singular. To prove this we represent the data as the $n \times m$ matrix $\mathbf{X}$, where the jth column is a sample of size n of the jth variable. The $m \times 1$ average vector is $\bar{\mathbf{x}} = n^{-1}\mathbf{X}'\mathbf{1}$ (see the proof of Theorem 6.41). The centered data (means are subtracted) has the form $\mathbf{U} = \mathbf{X} - \mathbf{1}\bar{\mathbf{x}}'$. Finally, the $m \times m$ correlation matrix can be represented as the matrix product

$$\mathbf{R} = \mathbf{Z}'\mathbf{Z}, \tag{6.33}$$

where $\mathbf{Z} = \mathbf{U}\mathbf{D}^{-1/2}$ and $\mathbf{D} = \sqrt{\mathrm{diag}(\mathbf{U}'\mathbf{U})}$. In words, the average in each column of matrix $\mathbf{Z}$ is zero and the variance is 1. The correlation matrix in terms of the original data $\mathbf{X}$ is written as

$$\mathbf{R} = \mathbf{D}^{-1/2}(\mathbf{X} - \mathbf{1}\bar{\mathbf{x}}')'(\mathbf{X} - \mathbf{1}\bar{\mathbf{x}}')\mathbf{D}^{-1/2}. \tag{6.34}$$

From linear algebra (see Section 10.2 in the Appendix), we know that $\text{rank}(\mathbf{Z}'\mathbf{Z})$ $= \text{rank}(\mathbf{Z}) = \text{rank}(\mathbf{U})$. This means that, as follows from (6.33), if $m > n$, then the $m \times m$ matrix $\mathbf{R}$ has rank $n < m$ and therefore is singular and partial correlations do not exist. Simply put, the number of variables must be smaller than the number of observations.

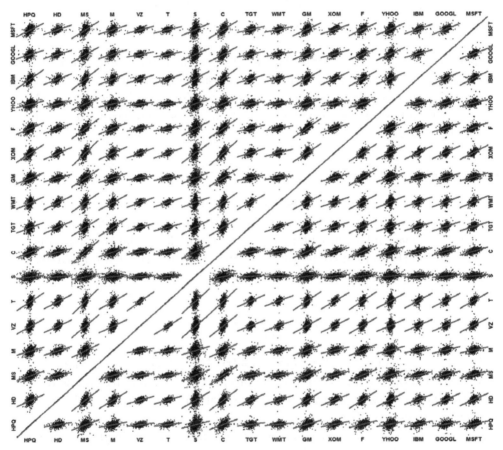

Figure 6.11: *Cross-correlation plots of returns of 17 stocks on the log scale (*`xlim` *and* `ylim` *are common to all plots).*

Example 6.47 *Correlation matrix heatmap of 17 major stock prices.*
The daily data on seventeen major stock prices is contained in the directory `c:\StatBook\stocks\` *in the csv format files (the data were downloaded from* `http://finance.yahoo.com`*) . Use the last column (Adj Close) to compute and display (a) the 17×17 heatmap correlation matrix for the stock prices on the log scale and (b) the color heatmap partial correlation matrix.*

Solution. The return of a stock at time t is computed as $r_t = (p_t - p_{t-1})/p_{t-1}$, where p_t is the price at the present time t and p_{t-1} is the price at the previous time, $t - 1$. The q-q plot (not shown) confirms that the ratio of prices on the log scale follows a normal distribution. Thus, we suggest showing and analyzing the data upon transformation $\ln(p_t/p_{t-1}) = \ln p_t - \ln p_{t-1}$. This transformation is convenient for modeling the stock prices on the log scale in the form of autoregression, $\ln p_t = \alpha + \beta \ln p_{t-1} + \varepsilon_t$. According to a well-known approximation, $\ln(1 + x) \simeq x$, we obtain

$$\ln \frac{p_t}{p_{t-1}} = \ln \left(1 + \frac{p_t - p_{t-1}}{p_{t-1}} \right) \simeq r_t.$$

Since the day-to-day variation of stock prices is relatively small, $\ln(p_t/p_{t-1})$ is close to the traditional definition of the stock price return, r_t. Two important features of the stock price data should be taken into account before computing the correlations: First, the time series on stock prices are presented from the most recent to oldest; we need to rearrange them in a regular order, from oldest to recent. Second, historic prices have different length for different stocks; we need to truncate time series to the shortest time domain. The R function is found in file `cimcorSP.r`.

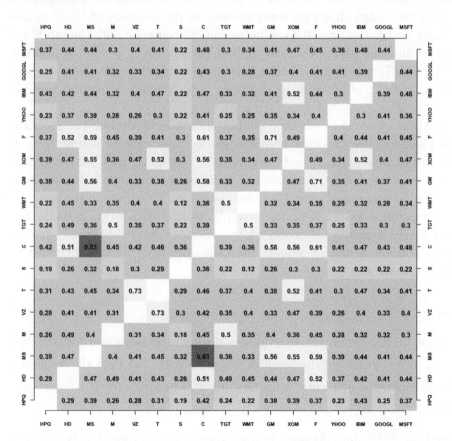

Figure 6.12: *The 17×17 heatmap correlation matrix for the stock prices.*

The R function `normQQ` investigates the normality of $\ln(p_t/p_{t-1})$ by means of the q-q plot applied to the standardized data, see the Cramér–Wold Theorem from Section 3.10.2 (the plot is not shown).

Figure 6.11 displays 17 returns on the log scale using the same `xlim` and `ylim` (the symbols of the stocks are shown at all four axes) produced by the R function `cimcorSP(job=1)`. Since the limits on the x and y axes are the same, we can judge the correlation based on the slope of the regression line (the steeper the slope, the greater the correlation). Interestingly, all stocks are positively correlated (all regression lines have positive slopes).

Figure 6.12 is produced by `cimcorSP(job=2)`, which plots the correlation matrix heatmap computed using the built-in function `cor`. This function computes the correlation coefficient between vectors `x` and `y`, but if `X` is a $n \times m$ matrix `cor(X)` returns an $m \times m$ correlation matrix between vector columns. Four colors are used to represent four ranges of the correlation coefficient: blue: $r \le 0.25$, green: $0.25 < r \le 0.5$, yellow: $0.5 < r \le 0.75$, and finally, red: $r > 0.75$. Note that "cold" colors are used to depict low correlation and "hot" colors are used to depict high correlation – that is why this plot is called a heatmap (other colors may be used from the file `Rcolor.pdf`). This heatmap representation is achieved with the `image` function using `breaks` as the intervals of the r values. The maximum correlation, $r = 0.83$, is between Citigroup (C) and Morgan Stanley (MS), two financial giants. Second highest correlation, $r = 0.73$, is between Verizon and AT&T, two giants of communication networks. The third highest correlation, $r = 0.71$, is between to giants in auto making, Ford and General Motors.

An important question is whether the correlation between any two stocks is due to their dependence on other common stocks. We recall a simple fact: If two stocks are expressed as $X = U + Z$ and $Y = V + Z$, they may correlate because they have a common component Z, such as the economic indicator, even if stock-unique components U and V do not correlate. To eliminate the common component (random variable Z in our example) partial correlation is used. The partial correlation heatmap is produced by calling `cimcorSP(job=3)`; partial correlation coefficient is usually smaller than the ordinary correlation coefficient (the graph is not shown).

Problems

1. (a) Find the minimum variance unbiased quadratic estimator for σ^2 when $Y_i \overset{iid}{\sim} \mathcal{N}(0, \sigma^2)$ following the proof of Theorem 6.41. (b) Find the minimum variance unbiased quadratic estimator for σ^2 when $Y_i \overset{iid}{\sim} \mathcal{N}(\mu, \sigma^2)$ and μ is known.

2. Prove that estimators $\widehat{A}_2$ and $\widehat{A}_3$ from Example 6.43 consistently estimate the unknown area of the circle. [Hint: Use Theorem 6.32.]

3. Formulate the problem of the unbiased quadratic estimation of the area of the circle in Example 6.43 similarly to variance. Find the optimal estimator by minimizing the trace of the squared matrix.

4. Consider two estimators of a ball's volume, $V = \frac{4}{3}\pi r^3$, as in Example 6.43, where $r_i \sim \mathcal{N}(\rho, \sigma^2)$ and neither ρ nor σ^2 is known. (a) Prove that $\widehat{V}_1 = \frac{4}{3}\pi \bar{r}^3$ overestimates the volume. (b) Prove that $\widehat{V}_2 = \frac{4}{3n}\pi \sum_{i=1}^{n} r_i^3$ overestimates the volume as well. (c) Consider $\widehat{V}_3 = \frac{4}{3}\pi \tilde{r}^3$, where $\tilde{r}$ is the root of the cubic equation $n^{-1}\sum_{i=1}^{n} r_i^3 = \tilde{r}^3 + 3\tilde{r}\tilde{\sigma}^2$. Is this estimator consistent? (d) Use simulations to compare the three estimators using the relative RMSE following Example 6.43.

5. Prove Corollary 6.42. [Hint: Use the fact that the variance of $\widehat{\sigma}^2$ is $2\sigma^4 \mathrm{tr}(\mathbf{A}^2)$, where $\mathbf{A}$ is defined in (6.26).]

6. Prove that the sample variance (6.25) is unbiased when $E(Y_i - \mu)^3 = 0$, instead assuming that Y_i are normally distributed.

7. Let X_i and Y_i be independent and drawn from a bivariate nonnormal distribution. (a) Does Theorem 6.44 still hold? (b) Can the correlation coefficient be defined? Does the cosine interpretation remain? (c) Does the interpretation of the coefficient of determination hold? (d) Which of the 1, 2 and 3 property of r hold? (e) Does Theorem 6.45 hold? [Hint: Consult Section 3.5.]

8. (a) Prove that r does not change upon linear transformations, $Y' = \alpha + \beta Y$ and $X' = \gamma + \tau X$ where $\beta\tau > 0$. (b) Does r change upon a nonlinear transformation? (c) Justify that, if X and Y are independent and follow a bivariate normal distribution, then the correlation coefficient upon any nonlinear transformation will be close to zero.

9. Derive the RMSE of the coefficient of determination via simulations as a function of the sample size. Using vectorized simulations (nSim=100000) from a bivariate normal distribution plot the RMSE versus n for $n = 10, 20, 50$, and 100 for $\rho^2 = 0$ and $\rho^2 = 0.5$ on the same graph using different colors.

10. (a) Simplify the vector representation (6.32) when $\mathbf{x}'\mathbf{1} = \mathbf{y}'\mathbf{1} = 0$. (b) Express (6.31) in vector form like (6.32).

11. Let $\mathbf{u}(a) = \mathbf{x} - a\mathbf{1}$ and $\mathbf{v}(b) = \mathbf{y} - b\mathbf{1}$ be vector functions of a and b, respectively. Denote $r(a, b) = (\mathbf{u}'(a)\mathbf{v}(b))/(\|\mathbf{u}(a)\| \, \|\mathbf{v}(b)\|)$. Prove that $r^2(a, b)$ reaches its minimum when $a = \overline{X}$ and $b = \overline{Y}$.

12. Modify function robias to evaluate the absolute maximum bias as a function of the sample size, n. Plot the bias versus sample size for $n = 4, 5, ..., 15$.

13. Use R to check formula (6.34): Generate a 10×5 matrix $\mathbf{X}$ of normally distributed iid observations and compute: (1) the 5×5 matrix (6.34) using vector computations; (2) 25 correlation coefficients, R_{ij}, in a `for` loop over $i, j = 1, 2, 3, 4, 5$; (3) using the built-in function `cor(X)`. Print out the three matrices and compare.

14. Using the data on 17 stocks modify function `cimcorSP` to display (a) time series of original stock prices p_t, (b) a time series of $\ln p_t$, (c) returns r_t, (d) returns on the log scale $\ln p_t - \ln p_{t-1}$.

15. Test the normality of returns for 17 stocks on the log scale by plotting q-q plots for $\ln p_t - \ln p_{t-1}$, display the 95% confidence bands as well (consult Section 5.3).

16. Modify function `cimcorSP` to display the correlation matrix heatmap with six breaks: $r \leq 0.2, 0.2 < r \leq 0.4, 0.4 \leq r < 0.6, 0.6 \leq r < 0.8$, and $r \geq 0.8$. Select colors from "cold" to "hot" of your choice from file `Rcolor.pdf`.

17. Display the partial correlation matrix heatmap with six breaks. Select colors of your choice.

6.7 Least squares for simple linear regression

Simple linear regression is a combination of the mean and slope models from Section 6.5,

$$Y_i = \alpha + \beta x_i + \varepsilon_i, \quad i = 1, 2, ..., n, \qquad (6.35)$$

where x_i is fixed, sometimes referred to as an independent variable or predictor (or covariate), α is the intercept and β is the slope, subject to estimation, and ε_i is an unobservable error term with zero mean and constant variance σ^2 (we say that error terms are homoscedastic). Note that the uppercase Y reflects that the dependent variable is random and the lowercase x reflects that the independent variable is fixed and known. In addition, it is assumed that $\{x_i\}$ take at least two distinct values, or equivalently

$$\sum_{i=1}^{n} (x_i - \overline{x})^2 \neq 0. \qquad (6.36)$$

As before, we assume that the error terms, ε_i and ε_j, do not correlate ($i \neq j$). Since $\alpha + \beta x_i$ are nonrandom, Y_i and Y_j do not correlate either. In another formulation, the simple linear regression model can be written as $E(Y_i) = \alpha + \beta x_i$. Note that we make no assumption on the distribution of ε_i at this point. The regression model (6.35) is a prototype of a general linear model to be studied in Chapter 8 in full detail. The reason to start with a simple statistical model is twofold:

(a) we can concentrate on statistical concepts by avoiding technical/matrix algebra detail, and (b) we can use this model for illustration of general statistical inference such as parameter estimation and hypothesis testing to follow.

The assumption that ε_i has zero mean is not very important: Indeed, if $E(\varepsilon_i) = \nu$, then we can combine α and ν and rewrite the model as $Y_i = \alpha' + \beta x_i + \varepsilon_i'$, where $\alpha' = \alpha + \nu$ and $\varepsilon_i' = \varepsilon_i - \nu$ with $E(\varepsilon_i') = 0$. However ν and the original intercept α cannot be identified simultaneously if ν is unknown, only the sum $\alpha + \nu$ can be identified.

Previously, we applied the least squares criterion to find the intercept and slope for a pair of random variables (X, Y) in the probabilistic terms – see Definition 3.47. Now, we treat x_i as fixed and, therefore, the integral criterion is replaced with the sum.

Definition 6.48 *The ordinary least squares (OLS) estimators of the intercept and slope in the simple linear regression minimize the residual sum of squares (RSS)*

$$S(\alpha, \beta) \stackrel{\text{def}}{=} \sum_{i=1}^{n}(Y_i - \alpha - \beta x_i)^2.$$

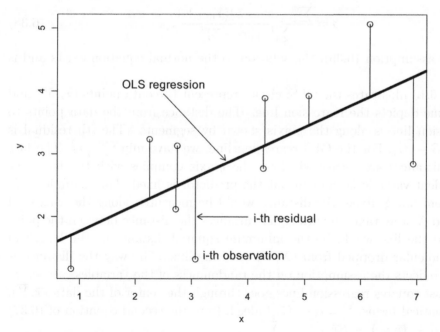

Figure 6.13: *Geometrical illustration of the OLS. The distance between observation and the regression line is along the y-axis.*

A remark regarding the terminology: we use "ordinary" to indicate that the sum of squares in not weighted due to homoscedastic errors. The distance between the ith observation and the regression line along the y-axis is called the i-th residual, $Y_i - \alpha - \beta x_i$; that is why we refer to S as the residual sum of squares. To find the least squares solution, we differentiate S with respect to α and β and equate the derivatives to zero,

$$\frac{\partial S}{\partial \alpha} = -2\sum_{i=1}^{n}(Y_i - \alpha - \beta x_i) = 0, \quad \frac{\partial S}{\partial \beta} = -2\sum_{i=1}^{n}(Y_i - \alpha - \beta x_i)x_i = 0.$$

These equations are the necessary conditions (later we shall prove that they are sufficient conditions as well) for the minimum of the residual sum of squares and are called *normal equations*. Rewrite the normal equations as a system of two linear equations

$$\alpha n + \beta \sum_{i=1}^{n} x_i = \sum_{i=1}^{n} Y_i, \quad \alpha \sum_{i=1}^{n} x_i + \beta \sum_{i=1}^{n} x_i^2 = \sum_{i=1}^{n} x_i Y_i.$$

One can apply the Cramér rule to solve this system for the unknown intercept and slope, which gives the solution for the OLS estimators:

$$\widehat{\beta} = \frac{\sum_{i=1}^{n} x_i Y_i - n\overline{x}\,\overline{Y}}{\sum_{i=1}^{n} x_i^2 - n\overline{x}^2}, \quad \widehat{\alpha} = \overline{Y} - \widehat{\beta}\overline{x}. \tag{6.37}$$

Equivalently, the slope coefficient can be computed as

$$\widehat{\beta} = \frac{\sum_{i=1}^{n}(x_i - \overline{x})(Y_i - \overline{Y})}{\sum_{i=1}^{n}(x_i - \overline{x})^2}. \tag{6.38}$$

Thanks to assumption (6.36), the solution to the normal equation exists and is unique.

Figure 6.13 illustrates the OLS: circles represent the data points (x_i, Y_i) and the bold line depicts the regression line. The distance from the data points to the regression line is along the y-axis shown by segments. The ith residual is $r_i = Y_i - \widehat{\alpha} - \widehat{\beta}x_i$. For the OLS regression line, we have min $\sum_{i=1}^{n} r_i^2$. The fact that the distances are measured along the y-axis complies with the fact that the dependent variable is random but the predictor is fixed. For example, if X were random but Y fixed, the distance would be measured along the x-axis. If both Y and X were random with equal variances the distance between the point (X, Y) and the line would be the minimum squared distance as the length of the perpendicular dropped from (X, Y) onto the line. The way the distance is computed reflects the assumption on the randomness of the variables.

The least squares regression lines goes through the center of the data $(\overline{x}, \overline{Y})$. In mathematical terms, $\overline{Y} = \widehat{\alpha} + \widehat{\beta}\overline{x}$. Indeed, from the second equation of (6.37) we obtain $\widehat{\alpha} + \widehat{\beta}\overline{x} = \overline{Y} - \widehat{\beta}\overline{x} + \widehat{\beta}\overline{x} = \overline{Y}$.

Function S is a quadratic, strictly convex function of α and β because the Hessian matrix,

$$\mathbf{H} = \begin{bmatrix} \frac{\partial^2 S}{\partial \alpha^2} & \frac{\partial^2 S}{\partial \alpha \partial \beta} \\ \frac{\partial^2 S}{\partial \beta \partial \alpha} & \frac{\partial^2 S}{\partial \beta^2} \end{bmatrix} = 2 \begin{bmatrix} n & \sum_{i=1}^{n} x_i \\ \sum_{i=1}^{n} x_i & \sum_{i=1}^{n} x_i^2 \end{bmatrix},$$

is positive definite (see Appendix). Indeed, this matrix is positive definite because the elements on the main diagonal are positive and the determinant is positive,

$$n \sum_{i=1}^{n} x_i^2 - \left(\sum_{i=1}^{n} x_i \right)^2 = n \sum_{i=1}^{n} (x_i - \overline{x})^2 > 0,$$

due to (6.36). Since function S is a strictly convex function the normal equations are sufficient conditions for the minimum and the minimum is attained at the unique point with coordinates given by (6.37). A typical RSS function is shown in Figure 6.14: when either of α or β goes to infinity, S goes to infinity (the surface rises).

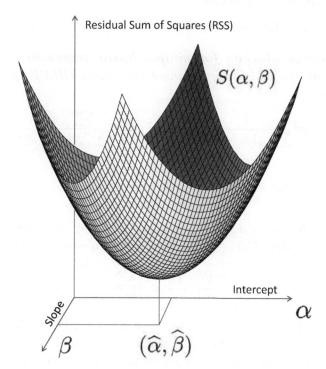

Figure 6.14: *Minimization of RSS as a function of α and β. Since $S(\alpha, \beta)$ is a strictly convex function the least squares solution exists and is unique.*

Example 6.49 *Two-group OLS. Prove that formulas (6.37) yield $\widehat{\alpha} = \overline{Y}_1$ and $\widehat{\beta} = \overline{Y}_2 - \overline{Y}_1$ if x_i takes only two distinct values 0 and 1, corresponding to Groups 1 and 2, respectively.*

Solution. Without loss of generality, we can rearrange the observations such that the first n_1 observations belong to Group 1, $x_i = 0$ for $i = 1, 2, .., n_1$. The last

n_2 observations belong to Group 2, $x_i = 1$ for $i = n_1 + 1, ..., n$, where $n_1 + n_2 = n$. Using elementary algebra, we obtain

$$\sum x_i Y_i = n_2, \quad \sum x_i^2 = n_2, \quad \overline{x} = n_2/n, \quad \overline{Y} = (n_1 \overline{Y}_1 + n_2 \overline{Y}_2)/(n_1 + n_2),$$

where $\overline{Y}_1$ and $\overline{Y}_2$ are the averages in Groups 1 and 2. Then, as follows from (6.37),

$$\widehat{\beta} = \frac{n_2 \overline{Y}_2 - n_2 \overline{Y}}{n_2 - n_2^2/n} = \frac{n}{n_1}(\overline{Y}_2 - \overline{Y}) = \overline{Y}_2 - \overline{Y}_1,$$

$$\widehat{\alpha} = \frac{n_1 \overline{Y}_1 + n_2 \overline{Y}_2}{n_1 + n_2} - (\overline{Y}_2 - \overline{Y}_1)\frac{n_2}{n_1 + n_2} = \overline{Y}_1.$$

6.7.1   Gauss–Markov theorem

The following is a generalization of Theorem 6.35 to a simple linear regression.

Theorem 6.50 *Gauss–Markov for simple linear regression.* *The OLS estimators (6.37) are the best linear unbiased estimators (BLUE) for β and α with minimal variances*

$$\mathrm{var}(\widehat{\beta}) = \sigma^2 \frac{1}{\sum_{i=1}^n (x_i - \overline{x})^2}, \quad \mathrm{var}(\widehat{\alpha}) = \sigma^2 \frac{\sum_{i=1}^n x_i^2/n}{\sum_{i=1}^n (x_i - \overline{x})^2}$$

and covariance

$$\mathrm{cov}(\widehat{\alpha}, \widehat{\beta}) = -\sigma^2 \frac{\overline{x}}{\sum_{i=1}^n (x_i - \overline{x})^2}.$$

Proof. First, we prove that $\widehat{\beta}$ and $\widehat{\alpha}$ are unbiased. Using representation (6.38) and the obvious result that $E(Y_i - \overline{Y}) = \alpha + \beta x_i - (\alpha + \beta \overline{x}) = \beta(x_i - \overline{x})$, we obtain

$$E(\widehat{\beta}) = \frac{\sum_{i=1}^n (x_i - \overline{x}) E(Y_i - \overline{Y})}{\sum_{i=1}^n (x_i - \overline{x})^2} = \beta \frac{\sum_{i=1}^n (x_i - \overline{x})^2}{\sum_{i=1}^n (x_i - \overline{x})^2} = \beta.$$

Since the slope estimator, $\widehat{\beta}$, is unbiased, we have $E(\widehat{\alpha}) = E(\overline{Y}) - \beta \overline{x} = \alpha + \beta \overline{x} - \beta \overline{x} = \alpha$, the OLS estimator of the intercept is unbiased as well. Second, we prove that $\widehat{\beta}$ has minimum variance in the family of unbiased linear estimators. Let $\widetilde{\beta} = \sum_{i=1}^n \lambda_i Y_i$ be a linear unbiased estimator (a linear function of observations with fixed coefficients). The unbiasedness implies $\sum \lambda_i(\alpha + \beta x_i) = \beta$, and so two linear equations (indices $i = 1, 2, ..., n$ are not shown for brevity) must hold:

$$\sum \lambda_i = 0, \quad \sum \lambda_i x_i = 1. \tag{6.39}$$

Since $\mathrm{var}(\widetilde{\beta}) = \sigma^2 \sum \lambda_i^2$, we arrive at the Lagrange function (the coefficient $1/2$ is chosen for convenience upon differentiation),

$$\mathcal{L}(\lambda_1, ...\lambda_n; \nu, \omega) = \frac{1}{2}\sum \lambda_i^2 - \nu \sum \lambda_i - \omega \left(\sum \lambda_i x_i - 1\right),$$

where ν and ω are Lagrange multipliers. Differentiation with respect to λ_i yields

$$\lambda_i = \nu + \omega x_i. \tag{6.40}$$

Restrictions (6.39) imply a pair of linear equations for ν and ω, namely, $\sum(\nu + \omega x_i) = 0$ and $\sum(\nu + \omega x_i)x_i = 1$. Expressing ν from the first equation and substituting into the second equation gives $\omega = (\sum(x_i - \overline{x})x_i)^{-1}$ and $\nu = -\overline{x}\omega$. Plugging these expressions back into (6.40) yields $\lambda_i = (x_i - \overline{x})\omega$, so the unbiased estimator of β with minimum variance is $\widetilde{\beta} = \sum(x_i - \overline{x})Y_i / \sum(x_i - \overline{x})x_i$. Now we must show that the numerator and denominator of $\widetilde{\beta}$ are the same as those of the least squares estimator (6.38):

$$\sum(x_i - \overline{x})(Y_i - \overline{Y}) = \sum(x_i - \overline{x})Y_i - \overline{Y}\sum(x_i - \overline{x}) = \sum(x_i - \overline{x})Y_i,$$

$$\sum(x_i - \overline{x})^2 = \sum(x_i - \overline{x})(x_i - \overline{x}) = \sum(x_i - \overline{x})x_i - \overline{x}\sum(x_i - \overline{x})$$

$$= \sum(x_i - \overline{x})x_i$$

because $\sum(x_i - \overline{x}) = 0$. The fact that $\widehat{\alpha} = \overline{Y} - \widehat{\beta}\overline{x}$ has minimum variance among unbiased linear estimators is proved as for $\widehat{\beta}$. The variance of $\widehat{\alpha}$ and $\widehat{\beta}$ and their covariance are easy to find by expressing $\widehat{\beta} - \beta = \sum \varepsilon_i d_i / \sum d_i^2$ and $\widehat{\alpha} - \alpha = \overline{\varepsilon} - \overline{x}\sum \varepsilon_i d_i / \sum d_i^2$, where $d_i = x_i - \overline{x}$ and the summation limits are omitted for brevity. Then

$$\text{var}(\widehat{\beta}) = \frac{\sum \text{var}(\varepsilon_i) d_i^2}{(\sum d_i^2)^2} = \frac{\sigma^2}{\sum d_i^2},$$

$$\text{var}(\widehat{\alpha}) = \text{var}(\overline{\varepsilon}) + \text{var}\left(\frac{\sum \varepsilon_i d_i}{\sum d_i^2}\overline{x}\right) - 2\text{cov}\left(\overline{\varepsilon}, \frac{\sum \varepsilon_i d_i}{\sum d_i^2}\overline{x}\right) = \sigma^2\left(\frac{1}{n} + \frac{\overline{x}^2}{\sum d_i^2}\right)$$

$$= \sigma^2 \frac{\sum_{i=1}^{n} x_i^2/n}{\sum_{i=1}^{n}(x_i - \overline{x})^2},$$

because

$$\text{cov}\left(\overline{\varepsilon}, \frac{\sum \varepsilon_i d_i}{\sum d_i^2}\overline{x}\right) = \sigma^2 \frac{\sum d_i}{n \sum d_i^2}\overline{x} = 0.$$

The formula for the covariance of $\widehat{\alpha}$ and $\widehat{\beta}$ is derived along the same lines. $\square$

The following example is the simplest application of the *optimal design of experiments* for linear models.

Example 6.51 *Where to nail? A stick of unit length is nailed to the wall at two places at given height and slope. Use the optimality of the OLS to justify that the best places to nail are the ends of the stick.*

Solution. The best places to nail give the most stable position of the stick; without loss of generality one can assume that the length of the stick is 1. The vertical position of the stick with the desired height α and slope β is expressed as $Y_i = \alpha + \beta x_i + \varepsilon_i$ for $i = 1, 2$, where $\mathrm{var}(\varepsilon_i) = \sigma^2$ represents the vertical imprecision of the stick location, and x_1 and x_2 are the places where the stick is nailed, $0 \le x_1 \le 1$ and $0 \le x_2 \le 1$. The stick instability can be associated with the variance of the slope or $\sigma^2/[(x_1 - \overline{x})^2 + (x_2 - \overline{x})^2]$, where $\overline{x} = (x_1 + x_2)/2$. Simple algebra gives $(x_1 - \overline{x})^2 + (x_2 - \overline{x})^2 = (x_1 - x_2)^2/2$. This means that stick will be most stable when $|x_1 - x_2| = \max$. But this is achieved when the nails are at the ends of the stick: $x_1 = 0$ and $x_2 = 1$. This example tells us that if we want to fix something in place, fix it at the ends to make it most stable.

Example 6.52 *Two-point slope estimator.* *The slope of the simple linear regression is estimated as the slope of a segment that connects two points on the plane, (x_i, Y_i) and (x_j, Y_j), assuming that $x_i \ne x_j$. That is, $\widetilde{\beta} = (Y_i - Y_j)/(x_i - x_j)$, where i and j are fixed, is called the two-point slope estimator. (a) Prove that the two-point slope is an unbiased estimator. (b) Find i and j that make this slope estimator the most precise. (c) Prove that $\mathrm{var}(\widetilde{\beta}) \ge \sigma^2/\sum_{i=1}^{n}(x_i - \overline{x})^2$ by referring to the Gauss–Markov theorem. (d) What algebraic inequality can be derived based on the previous point?*

Solution. (a) Prove the unbiasedness of the two-point slope estimator

$$E(\widetilde{\beta}) = \widetilde{\beta} = \frac{E(Y_i - Y_j)}{x_i - x_j} = \frac{E(Y_i) - E(Y_j)}{x_i - x_j} = \frac{(\alpha + \beta x_i) - (\alpha + \beta x_j)}{x_i - x_j} = \beta.$$

(b) To find optimal i and j, obtain the variance of the estimator

$$\mathrm{var}(\widetilde{\beta}) = \frac{\mathrm{var}(Y_i) + \mathrm{var}(Y_j)}{(x_i - x_j)^2} = \frac{2\sigma^2}{(x_i - x_j)^2}.$$

Therefore minimum variance occurs when $x_i - x_j = \max$ with solution $i = \arg\min x_i$ and $j = \arg\max x_i$. (c) Since $\widetilde{\beta}$ is a linear unbiased estimator from the Gauss–Markov theorem, we have $\mathrm{var}(\widetilde{\beta}) \ge \sigma^2/\sum_{i=1}^{n}(x_i - \overline{x})^2$. (d) As follows from the last inequality, we deduce that

$$\frac{1}{2}(x_{\max} - x_{\min})^2 \le \sum_{i=1}^{n}(x_i - \overline{x})^2 \qquad (6.41)$$

for any sequence of numbers $x_1, ..., x_n$.

6.7.2 Statistical properties of the OLS estimator under the normal assumption

The previous discussion of the OLS did not require an assumption on the distribution of errors ε_i. As we shall see later, the OLS becomes a superior method of estimation if errors are normally distributed. The following result is a landmark of linear models. It formulates the optimal properties of the OLS and major distributions for a simple linear regression. The reader is referred to Chapter 4 for a refresher on the t- and χ^2-distributions. More detailed discussion of the OLS will be given in the chapter on the multiple linear model.

Theorem 6.53 *Statistical properties of the OLS estimator.* If ε_i have normal distribution, then
(a) *The OLS estimators of intercept, $\widehat{\alpha}$, and slope, $\widehat{\beta}$, have minimum variances among all unbiased estimators (the improved Gauss–Markov theorem).*
(b) *$\widehat{\alpha}$ and $\widehat{\beta}$ have a bivariate normal distribution with the marginal distributions given by*

$$\widehat{\alpha} \sim \mathcal{N}\left(\alpha, \sigma^2 \frac{\sum_{i=1}^n x_i^2}{n \sum_{i=1}^n (x_i - \overline{x})^2}\right), \quad \widehat{\beta} \sim \mathcal{N}\left(\beta, \sigma^2 \frac{1}{\sum_{i=1}^n (x_i - \overline{x})^2}\right).$$

(c) *The estimator*

$$\widehat{\sigma}^2 = \frac{1}{n-2} \sum_{i=1}^n (Y_i - \widehat{\alpha} - \widehat{\beta} x_i)^2$$

is unbiased for σ^2 and the normalized sum of squares has a chi-square distribution:

$$(n-2)\frac{\widehat{\sigma}^2}{\sigma^2} \sim \chi^2(n-2).$$

(d) *$\widehat{\beta}$ minus its true value divided by its standard error has a t-distribution:*

$$\frac{\widehat{\beta} - \beta}{\widehat{\sigma}/\sqrt{\sum_{i=1}^n (x_i - \overline{x})^2}} \sim t(n-2).$$

The complete proof is deferred to later: item (a) is proved at the end of Section 6.9, and properties (c) and (d) will be proved in the general case of the multiple linear model. Property (b) follows from the fact that $\widehat{\alpha}$ and $\widehat{\beta}$ are linear functions of $Y_1, Y_2, ..., Y_n$, so that are normally distributed. Their unbiasedness and variances were derived in Theorem 6.50.

We emphasize the difference between BLUE Theorem 6.50 and property (a) of the above theorem. The former proves the optimality of the OLS among linear unbiased estimators, but the latter proves the optimality among all (linear and nonlinear) unbiased estimators of the slope and intercept. That is why we call this theorem "the improved Gauss–Markov theorem."

Example 6.54 *Simulations for simple linear regression.* Check by simulations the distributions in (b)–(d). Superimpose the empirical cdf with the theoretical one in a two-by-two plot (intercept, slope, chi-square, and t-distribution).

Solution. The R code is found in the file `olsim.r`. The default array x is the sequence of numbers from 1 to n. Note that x remains unchanged during simulations because it is fixed by the assumption. The plot is omitted.

6.7.3   The lm function and prediction by linear regression

Estimation of a linear regression by OLS is a frequent task in many statistical applications. In R, computation and relevant output of statistical information is provided by the built-in function lm. This function works for simple linear regression and multiple regression as well, with or without intercept. If y and x are arrays of the same length, we issue lm(y~x) to run a simple linear regression. The intercept is present by default. To estimate the slope model without intercept, as in Example 6.36, we write lm(y~x-1). Sometimes the dependent and independent variables are part of a dataframe. Then we refer to the variables by their names and specify the dataframe through the data option. For example, if the name of the dataframe is datfr and the names of the dependent and independent variables are ydep and xind, respectively, the call is lm(ydep~xdp,data=datfr).

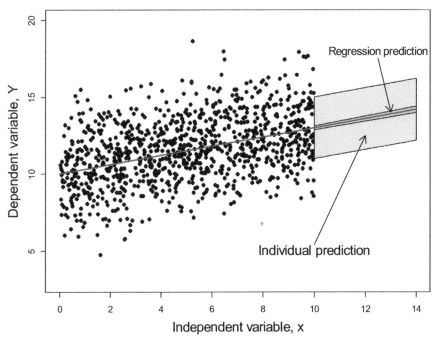

Figure 6.15: *The difference between individual and regression prediction. For individual prediction, $Y = \alpha + \beta x + \varepsilon$, but regression prediction predicts the point on the regression line, $\alpha + \beta x$. The value of the prediction is the same, but the former has the variance larger by σ^2.*

One of the most attractive features of linear regression is the ability to predict the dependent variable given the value of the independent variable. We distinguish two types of prediction: regression/mean prediction and individual prediction. Given a value of x, the former predicts the regression value $\alpha + \beta x$ and the latter predicts $\alpha + \beta x + \varepsilon$. In both cases the prediction is the same, $\widehat{\alpha} + \widehat{\beta}x$, but the variance is σ^2 larger for individual prediction. To emphasize the difference we denote $\widehat{Y} = \widehat{\alpha} + \widehat{\beta}x$ as the regression prediction and $\widetilde{Y} = \widehat{\alpha} + \widehat{\beta}x + \varepsilon$ as the individual prediction, where ε is independent of $\{Y_i, i = 1, ..., n\}$.

Example 6.55 *Variance of prediction.* *Derive the variance of the regression and individual prediction.*

Solution. The formulas derived in Theorem 6.50 are used here. Since $E(\widehat{Y}) = \alpha - \beta x$ and $\sum_{i=1}^{n} x_i^2/n = \sum_{i=1}^{n}(x_i - \overline{x})^2/n + \overline{x}^2$, we obtain

$$
\begin{aligned}
\text{var}(\widehat{Y}) &= E(\widehat{Y} - \alpha - \beta x)^2 = E[(\widehat{\alpha} - \alpha) + (\widehat{\beta} - \beta)x + \varepsilon]^2 \\
&= \text{var}(\widehat{\alpha}) + \text{var}(\widehat{\beta})x^2 + 2\text{cov}(\widehat{\alpha}, \widehat{\beta})x + \sigma^2 \\
&= \sigma^2 \left[\frac{\sum_{i=1}^{n} x_i^2/n}{\sum_{i=1}^{n}(x_i - \overline{x})^2} + \frac{x^2}{\sum_{i=1}^{n}(x_i - \overline{x})^2} - \frac{2\overline{x}x}{\sum_{i=1}^{n}(x_i - \overline{x})^2} \right] \\
&= \sigma^2 \left[\frac{1}{n} + \frac{\overline{x}^2 - x^2 - 2\overline{x}x}{\sum_{i=1}^{n}(x_i - \overline{x})^2} \right] = \sigma^2 \left[\frac{1}{n} + \frac{(\overline{x} - x)^2}{\sum_{i=1}^{n}(x_i - \overline{x})^2} \right].
\end{aligned}
$$

Since ε is independent of $\widehat{\alpha}$ and $\widehat{\beta}$, we have $\text{var}(\widetilde{Y}) = \text{var}(\widehat{Y}) + \sigma^2$. While the variance of the regression prediction decreases with n, the variance of the individual prediction is bounded by σ^2. Figure 6.15 illustrates the difference between the two types of prediction, where the shaded area depicts

$$
\widehat{\alpha} + \widehat{\beta}x \pm \sigma \sqrt{1 + \frac{1}{n} + \frac{(\overline{x} - x)^2}{\sum_{i=1}^{n}(x_i - \overline{x})^2}}.
$$

The difference between the mean and individual statistical inference will be further emphasized in Sections 7.10.4 and 8.5.

The R function `predict` can be used for prediction by linear regression estimated by `lm` with the somewhat confusing option `interval="confidence"` for regression prediction and `interval="prediction"` for individual prediction.

Theorem 6.56 *Statistical properties of prediction by regression.* We have

$$
\frac{\widehat{Y} - \alpha - \beta x}{\widehat{\sigma}\sqrt{q(x)}} \sim t(n - 2), \quad \frac{\widetilde{Y} - \alpha - \beta x}{\widehat{\sigma}\sqrt{1 + q(x)}} \sim t(n - 2), \tag{6.42}
$$

where $q(x) = 1/n + (\overline{x} - x)^2 / \sum(x_i - \overline{x})^2$.

The proof for more general multiple regression is found in Section 8.4.3. Distributions (6.42) make possible to construct an exact confidence interval for individual prediction:

$$
\text{Pr}\left(\widehat{Y} - t_{1-\alpha/2}(n - 2)\widehat{\sigma}\sqrt{1 + q(x)} < Y < \widehat{Y} + t_{1-\alpha/2}(n - 2)\widehat{\sigma}\sqrt{1 + q(x)} \right) = 1 - \alpha,
$$

where $t_{1-\alpha/2}(n - 2)$ is the $(1 - \alpha/2)$th quantile of the t-distribution with $n - 2$ degrees of freedom. We interpret this expression by saying that the double-sided interval $\widehat{Y} \pm t_{1-\alpha/2}(n-2)\widehat{\sigma}\sqrt{1+q(x)}$ covers future Y with probability $1-\alpha$. Sometimes, we are interested in a one-sided interval, say,

$$\Pr\left(Y > \widehat{Y} + t_\alpha(n-2)\widehat{\sigma}\sqrt{1+q(x)}\right) = 1 - \alpha.$$

Typically, $\alpha = 0.05$ is used, then we obtain the 95% confidence interval. More detail on confidence intervals can be learned from Section 7.8.

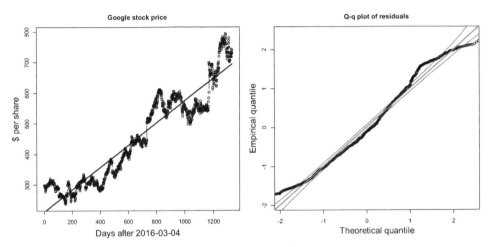

Figure 6.16: *Google stock price with the OLS fitted trend and the q-q plot of residuals.*

Example 6.57 *Prediction of the Google stock price.* *Use the data from Example 6.47 to (a) estimate the trend of Google daily stock price starting from August 19, 2004 until March 4, 2016, (b) do q-q plot to test whether residuals have normal distribution, and (c) predict stock prices from December 5, 2016 to January 3, 2017.*

Solution. (a) The R code is in function lm.trendSP. The summary of the lm output is shown below.

```
> lm.trendSP(job=1)
Call:
lm(formula = y ~ x)
Coefficients:
            Estimate Std. Error t value Pr(>|t|)
(Intercept) 2.118e+02  2.619e+00   80.85   <2e-16 ***
x           3.628e-01  3.407e-03  106.49   <2e-16 ***
Residual standard error: 47.75 on 1329 degrees of freedom
Multiple R-squared:  0.8951,     Adjusted R-squared:  0.895
F-statistic: 1.134e+04 on 1 and 1329 DF,  p-value: < 2.2e-16
```

The first line shows the call of the function. The second line gives summary statistics for residuals, $Y_i - \widehat{\alpha} - \widehat{\beta}x_i$: the minimum values, first quartile, median, the third quartile and maximum. The table of coefficients is the most important part of the output. It lists the OLS estimates of intercept and slope, their standard errors computed by formulas from Theorem 6.50 with σ replaced with its estimate $\sqrt{\hat{\sigma}^2}$, t-statistics as the ratio of the estimates to its standard error, and the p-values $\mathtt{Pr(>|t|)}$ to be explained later in Section 8.5. The residual standard error is $\hat{\sigma}$, degrees of freedom is n minus the number of coefficients estimated $(n-2)$. $\mathtt{Multiple\ R\text{-}squared}$ is computed by

$$R^2 = 1 - \frac{\sum_{i=1}^n r_i^2}{\sum_{i=1}^n (Y_i - \overline{Y})^2}, \tag{6.43}$$

and $\mathtt{Adjusted\ R\text{-}squared}$ is computed by the formula

$$R_{\text{adj}}^2 = 1 - \frac{\sum_{i=1}^n r_i^2/(n-2)}{\sum_{i=1}^n (Y_i - \overline{Y})^2/(n-1)}.$$

F-statistic and p-value will be discussed later. See the discussion and of the R-squared coefficient/coefficient of determination in the next section.

Figure 6.17: *Prediction of Google stock prices by a linear trend.*

(b) The price data and the fitted regression line along with the q-q plot are shown in Figure 6.16. We use option $\mathtt{type="b"}$ in the plot to show the points connected by segments. The 95% confidence band is shown in the q-q plot; see Section 5.3.1. The line in the middle is the 45° line. The points are outside of the confidence band on both sides: there is evidence that residuals do not follow a normal distribution. It is interesting to note that the empirical tails are shorter than the theoretical ones: short symmetric tails as labeled in Figure 5.13.

(c) The prediction of stock prices is depicted in Figure 6.17. Actual prices are shown at right using empty circles, so that they can be compared with prediction by regression. The short segments (up and down) represent standard errors of individual prediction. The predictions are amazingly accurate and all actual values are within $\pm \widehat{\sigma} \sqrt{1 + q(x)}$.

6.7.4 Misinterpretation of the coefficient of determination

Commonly, R^2 computed by formula (6.43), is misinterpreted as the proportion of variance explained by the independent variable/predictor. This interpretation is justified by a residual sum of squares (RSS) decomposition:

$$\sum_{i=1}^{n}(Y_i - \overline{Y})^2 = \sum_{i=1}^{n}(Y_i - \widehat{Y}_i)^2 + \sum_{i=1}^{n}(\widehat{Y}_i - \overline{Y})^2, \qquad (6.44)$$

where $\widehat{Y}_i = \widehat{\alpha} + \widehat{\beta}x_i$ are the predicted values. Expressing $Y_i - \overline{Y} = (Y_i - \widehat{Y}_i) + (\widehat{Y}_i - \overline{Y})$, it is obvious that the above equality holds if and only if $\sum(Y_i - \widehat{Y})(\widehat{Y}_i - \overline{Y}) = 0$. To prove that this sum vanishes, recall the first normal equation $\partial S/\partial \beta = 2\sum(Y_i - \widehat{Y})x_i = 0$. Indeed, since $\sum(Y_i - \widehat{Y}) = 0$, the normal equation can be equivalently written as $\sum(Y_i - \widehat{Y})(x_i - \overline{x}) = 0$. But since $\widehat{\alpha} = \overline{Y} - \widehat{\beta}\overline{x}$, we have $\widehat{Y}_i - \overline{Y} = \overline{Y} - \widehat{\beta}\overline{x} + \widehat{\beta}x_i - \overline{Y} = \widehat{\beta}(x_i - \overline{x})$, so

$$\sum(Y_i - \widehat{Y})(\widehat{Y}_i - \overline{Y}) = \widehat{\beta}\sum(Y_i - \widehat{Y})(x_i - \overline{x}) = 0.$$

The RSS decomposition (6.44) can be viewed as the variance decomposition by dividing both sides by n:

$$\frac{1}{n}\sum_{i=1}^{n}(Y_i - \overline{Y})^2 = \frac{1}{n}\sum_{i=1}^{n}(\widehat{Y}_i - \overline{Y})^2 + \frac{1}{n}\sum_{i=1}^{n}(Y_i - \widehat{Y}_i)^2,$$

which may be viewed as the empirical version of the variance decomposition in the bivariate normal distribution from Section 3.5.2: the left-hand side is an estimator of the variance of Y, the first term on the right-hand side is an estimator of the explained variance and the second term is the variance of the unexplained/residual variance. By representing

$$R^2 = 1 - \frac{\sum_{i=1}^{n} r_i^2/n}{\sum_{i=1}^{n}(Y_i - \overline{Y})^2/n},$$

one may say that the numerator $\sum_{i=1}^{n} r_i^2/n$ estimates the unexplained variance, the denominator $\sum_{i=1}^{n}(Y_i - \overline{Y})^2/n$ estimates the variance of Y, and therefore R^2 estimates the proportion of the variance of Y explained by x.

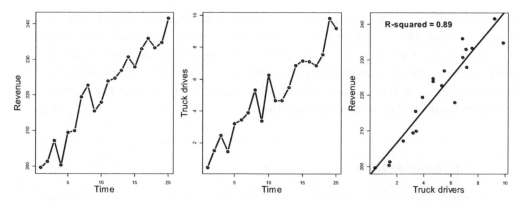

Figure 6.18: *Time series of a company's revenue and the number of truck drivers. As follows from the regression, about 90% of revenue can be explained by the number of truck drivers.*

The problem with the above argumentation is that Y_i are not identically distributed according to linear model (6.35): Y_i have different means, $E(Y_i) = \alpha + \beta x_i \neq$ const. There is a fundamental difference between regression in the bivariate normal distribution, where Y_i and X_i are iid, and the linear model (also called regression), where x_i are fixed numbers. Ignoring this difference leads to common mistakes such as expanding the above theorem on statistical properties of the OLS estimator to the case when independent variable x is random and observable together with Y or a misinterpretation of the coefficient of determination as the proportion of explained variance.

Unreasonably high values of R^2 can be expected when the dependent variable is a time series, as in Example 6.57, leading to spurious regression. If values of Y_i increase in time $\sum_{i=1}^{n}(Y_i - \overline{Y})^2/n$ is not an estimator of the variance of Y and $\overline{Y}$ is not an estimator of the mean of $\{Y_i, i = 1, 2, ..., n\}$. If Y_i are not iid, $\sum_{i=1}^{n}(Y_i - \overline{Y})^2/n$ becomes large compared with $\sum_{i=1}^{n} r_i^2/n$ and implies that R^2 is close to 1. Sometimes, interpretation of R^2 as the proportion of explained variance leads to paradoxes such as illustrated in the following example.

Example 6.58 *Want to improve business? Hire more truck drivers.* *Figure 6.18 depicts the time series of a company's revenue and the number of the number of truck drivers (the data are in file truckR.data.csv). Regress revenue on truck drivers and compute the coefficient of determination, R^2. Explain why R^2 cannot be interpreted as the proportion of variance of revenue explained by truck drivers and suggest an alternative, more sound, coefficient of determination.*

Solution. The R code truckR creates Figure 6.18. Following the traditional interpretation of R^2, almost 90% of variation in company's revenue is explained by the number of truck drivers – such regression is called *false regression*. This means that, to improve business, the CEO should hire more truck drivers, a obviously silly decision. The revenue is not iid, so the bivariate normal distribution of the squared correlation coefficient does not apply. To find a sound correlation coefficient, one must remove the trend from both time series and correlate the residuals (job=2). See Figure 6.19: the coefficient of correlation drops to 0.067. In Example 8.38, we will prove that this number can be derived using the concept of the generalized coefficient of determination.   □

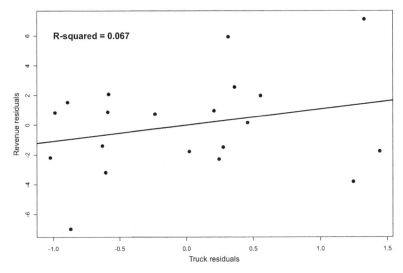

Figure 6.19: *The correlation between the residuals. Although there is a positive correlation between revenue and the number of truck drivers, it is negligible: now it makes sense.*

In Section 8.6.5, we introduce the generalized coefficient of determination to exclude unreasonably high value R^2 by comparing to a more sensible null model such as trend; see Example 8.37.

Problems

1. Find the OLS estimator of the slope if X were random and Y were fixed.

2. (a) Prove inequality (6.41) without referring to the Gauss–Markov theorem. (b) Find conditions under which the OLS and the two-point estimator coincide. [Hint: Let $u_k = x_{\max} - \overline{x}, u_l = x_{\min} - \overline{x}$, $u_i = x_i - \overline{x}$ and apply inequality $(u_k - u_l)^2/2 \le u_k^2 + u_l^2$.]

3. Find the intercept and slope in the two-group OLS example if x_i takes values -1 and 1.

4. Find the variance of the slope in the two-group OLS example. Write an R program to test your theoretical answer via simulations with default values

$n_1 = 10, n_2 = 7$, and $1 = \mu_1 = E(Y_i)$ for $i \leq n_1$ and $3 = \mu_2 = E(Y_i)$ for $i > n_1$ and $\sigma = 2$.

5. Explain why the problem of finding an optimal linear estimator without constraining the unbiasedness does not have a solution.

6. Prove the Gauss–Markov theorem using the variance of the intercept as the criterion. [Hint: Define $\tilde{\alpha} = \sum \lambda_i Y_i$ and prove that the unbiasedness leads to restrictions $\sum \lambda_i = 1$ and $\sum \lambda_i x_i = 1$.]

7. (a) Prove the formula for the covariance between $\hat{\alpha}$ and $\hat{\beta}$ in Theorem 6.50. (b) Derive the bivariate normal distribution for $(\hat{\alpha}, \hat{\beta})$. (c) Give the conditions under which $\hat{\alpha}$ and $\hat{\beta}$ are independent. (d) Derive the squared correlation coefficient between $\hat{\alpha}$ and $\hat{\beta}$. (e) Derive conditional means $E(\hat{\alpha}|\hat{\beta} = b)$ and $E(\hat{\beta}|\hat{\alpha} = a)$. [Hint: Consult Section 3.5.]

8. Prove that $\hat{\alpha}$ and $\hat{\beta}$ are independent of $\hat{\sigma}$. [Hint: Consult Section 4.3 and use Theorem 3.14.]

9. Consider function $S(\alpha, \beta)$ when condition (6.36) does not hold. (a) Is it true that S may go to infinity when $\alpha^2 + \beta^2 \to \infty$? (b) Find α and β such that $\alpha^2 + \beta^2 \to \infty$ but $S = \text{const}$. (c) Sketch $S(\alpha, \beta)$. (d) Show that the OLS solution exists, but is not unique.

10. Suppose that, in s simple linear regression, x_i take distinct values: $x_i \neq x_j$ for $i \neq j$. (a) Prove that the MSE of $\tilde{\beta} = (n(n-1))^{-1} \sum_{i \neq j}(Y_i - Y_j)/(x_i - x_j)$ cannot be smaller than $\sigma^2 / \sum_{i=1}^{n}(x_i - \bar{x})^2$. (b) Find conditions on x_i under which $\tilde{\beta}$ turns into the OLS estimator.

11. Suppose that the true slope in the simple linear regression is positive and $\{x_i\}$ take distinct values. Consider an estimator of the slope $\tilde{\beta} = (n(n-1))^{-1} \sum_{i \neq j} |Y_i - Y_j| / |x_i - x_j|$. (a) Is this estimator unbiased? (b) Can we refer to the Gauss–Markov theorem to claim that its variance cannot be smaller than $\sigma^2 / \sum_{i=1}^{n}(x_i - \bar{x})^2$? (c) Carry out simulations to compare the performance of $\tilde{\beta}$ and the OLS estimator.

12. (a) Derive the least squares estimators for the intercept and slope when the distance from (x_i, Y_i) to the straight line is measured along the x-axis. (b) Generate 1,000 pairs (x_i, Y_i) from a bivariate normal distribution with equal variances and correlation coefficient ρ as the argument of the R function with default value $\rho = 0.5$ and draw the two lines. (c) When do these estimates coincide with the regular OLS estimates?

13. (Continuation of the previous problem). (a) Derive estimates for the intercept and slope when the distance from (x_i, Y_i) to the straight line is measured as the minimal Euclidean distance to any point on the straight

line (the length of the perpendicular dropped from (x_i, Y_i) onto the line).
(b) Generate 1,000 pairs (x_i, Y_i) from a bivariate normal distribution with
equal variances and correlation coefficient ρ as the argument of the R function with default value $\rho = 0.5$ and draw the three lines (the ordinary OLS
line, the x-axis distance line from the previous problem, and the Euclidean
distance line). (c) When do the three estimates coincide? [Hint: Use the
Lagrange multiplier technique to find the minimal Euclidean distance from
the point to the line.]

14. Repeat Example 6.57 for IBM's stock. [Hint: Visit `finance.yahoo.com` to
    download the most recent IBM stock price, choose the time series period
    for a valid prediction.]

15. Prove that the "R-squared," as is computed by the `lm` function, is the
    squared correlation coefficient between Y_i and x_i.

16. The CEO of a company is making a decision on the expansion of the busi-
    ness. Over six years the profit was 10.3, 13.4, 12.9, 15.7, 14.9, and 15.8
    (millions of dollars), starting from year 1. To make the expansion success-
    ful, the profit of the company must be at least 20 million dollars in year 10.
    Estimate the probability of successful expansion.

17. The same as above, but to make the expansion successful, the cumulative
    profit of the company over years 7–10 must be at least 60 million dollars.
    Estimate the probability of successful expansion.

18. Consider prediction of Y at x by simple linear regression $Y_i = \alpha + \beta x_i + \varepsilon_i$,
    where, without loss of generality, we may assume that $x_1 \leq x_2 \leq \cdots \leq x_n$,
    and $x_1 < x_n < x$. The question is about prediction of Y using a subsample
    of $\{x_i\}$ of size $k < n$ (to simplify, it is assumed that σ^2 is known). (a) Is it
    true that any subsample yields unbiased prediction but with larger RMSE?
    (b) Consider prediction based on a subsample of length three: what is the
    optimal triple of values of x that gives the minimum RMSE? (c) Suppose
    that $\{x_i, i = 1, ..., n\}$ may be arbitrarily chosen on the interval $[a, b]$ where
    $b < x$. What is the optimal choice that minimizes the RMSE?

19. A university's annual endowments over six years were 3.78, 3.85, 4.01, 4.63,
    4.56, and 4.96 billion dollars. (i) Predict the endowment at year 10 and
    compute the 95% CI. Display the data, regression line, prediction, and CI
    with complete graph notations (use `par(mfrow=c(1,2))`). (ii) Using annual
    differences, plot and predict the annual difference at year 10 as in (1). (iii)
    Explain in statistical terms why the coefficients of determination are so
    different.

6.8 Sufficient statistics and the exponential family of distributions

Informally, sufficient statistics are functions of data that contain all information about the parameter. Sufficient statistics simplify statistical inference because the entire data are reduced to a few statistics without loss of statistical power. The formal definition is below.

Definition 6.59 *Let the n-dimensional vector of observations, $\mathbf{Y}$, have the pdf $f(\mathbf{y}; \boldsymbol{\theta})$, where $\boldsymbol{\theta}$ is an unknown m-dimensional parameter vector. The the m-dimensional statistic ($m \le n$), $\mathbf{U} = \mathbf{u}(\mathbf{Y})$, is sufficient for $\boldsymbol{\theta}$ if the conditional distribution of $\mathbf{Y}|\mathbf{U}$ does not depend on $\boldsymbol{\theta}$.*

Note the difference in notation: uppercase letters are used to indicate random variables, and the corresponding lowercase letters are used to indicate fixed values used as argument of the pdf. Recall that throughout the book bold means vector or matrix. The theorem below gives an easy and intuitively appealing criterion for sufficiency of $\mathbf{U}$.

Theorem 6.60 *Neyman factorization criterion. Under conditions of the previous definition, statistic $\mathbf{U} = \mathbf{u}(\mathbf{Y})$ is sufficient for $\boldsymbol{\theta}$ if the pdf can be factorized as*

$$f(\mathbf{y}; \boldsymbol{\theta}) = H(\mathbf{y})g(\mathbf{u}(\mathbf{y}); \boldsymbol{\theta}), \qquad (6.45)$$

where H and g are nonnegative functions.

In a nontrivial case, $m < n$, the original pdf dependent on the n-dimensional observation vector is reduced to the m-dimensional sufficient statistics $\mathbf{U}$. Sufficient statistics are for dimension reduction. The Neyman factorization criterion (sometimes attributed to Halmos and Savage [52]) works for continuous and discrete distributions. In the latter case, the pmf substitutes the pdf. Obviously, a one-to-one transformation yields a sufficient statistic. Sufficient statistics play an important role in the exponential family of distributions to be considered in the next section. A sufficient statistic may not optimally reduce the data, in which case the whole data $\mathbf{Y}$ is a sufficient statistic. However, the reader should understand limitations of the theory: in some practically important statistical problems, such as nonlinear regression, covered in Chapter 9, nontrivial sufficient statistics do not exist.

The concept of a *minimal sufficient statistic* has been developed such that any other sufficient statistic is a function of the minimal sufficient statistic. The interested reader can learn more about this concept in the textbook by Casella and Berger [17].

Example 6.61 *Sufficient statistic for the mean. Prove that $U = \overline{Y}$ is a sufficient statistic for μ when n independent observations $\{Y_i, i = 1, ..., n\}$ are normally distributed with mean μ and unit variance.*

Solution. The vector of observations, $\mathbf{Y} = (Y_1, ..., Y_n)$, has the joint density $f(\mathbf{y}; \mu) = (2\pi)^{-n/2} e^{-\frac{1}{2}\|\mathbf{y}-\mu\mathbf{1}\|^2}$ where $\mu = \theta$. Express the average in vector form as $\overline{Y} = \mathbf{Y}'\mathbf{1}/n$, where $\mathbf{1}$ is the $n \times 1$ vector of ones. Then, using familiar algebra, we obtain the expression for the argument of the exponential function ($\mathbf{y}$ is not random):

$$\|\mathbf{y}-\mu\mathbf{1}\|^2 = \|(\mathbf{y}-\overline{y}\mathbf{1})-(\mu-\overline{y})\mathbf{1}\|^2 = \|\mathbf{y}-\overline{y}\mathbf{1}\|^2 - 2(\mu-\overline{y})(\mathbf{y}-\overline{y}\mathbf{1})'\mathbf{1} + (\mu-\overline{y})^2 n.$$

But the middle term is zero because $(\mathbf{y}-\overline{y}\mathbf{1})'\mathbf{1} = \mathbf{y}'\mathbf{1}-\overline{y}n = \mathbf{y}'\mathbf{1} - \mathbf{y}'\mathbf{1} = 0$. Therefore, the original pdf can be rewritten as $f(\mathbf{y}; \mu) = (2\pi)^{-n/2} e^{-\frac{1}{2}\|(\mathbf{y}-\overline{y}\mathbf{1})\|^2} e^{-\frac{n}{2}(\overline{y}-\mu)^2}$. Letting $H(\mathbf{y}) = (2\pi)^{-n/2} e^{-\frac{1}{2}\|(\mathbf{y}-\overline{y}\mathbf{1})\|^2}$ and $g(\overline{y}; \mu) = e^{-\frac{n}{2}(\overline{y}-\mu)^2}$, we arrive at factorization (6.45). This proves that $U = \overline{Y}$ is a sufficient statistic for μ. Note that normal (not bold) font is used since U is scalar ($m = 1$).

Example 6.62 *Sufficient statistic for the Bernoulli distribution*. *Let Y_i be iid Bernoulli outcomes with probability p. Prove that the number of successes, $U = \sum_{i=1}^n Y_i$ is a sufficient statistic for p.*

Solution. The ith pmf is $p^{y_i}(1-p)^{1-y_i}$ and, due to independence, the joint pmf is the product,

$$f(y_1, ..., y_n; p) = \prod_{i=1}^n p^{y_i}(1-p)^{1-y_i} = p^u (1-p)^{n-u},$$

where $u = u(y_1, ..., y_n) = \sum y_i$ and $p = \theta$. Therefore, the joint pmf factorizes as (6.45), where $H(\mathbf{y}) = 1$ and $g(u; p) = p^u(1-p)^{n-u}$, which proves that $U = \sum_{i=1}^n Y_i$ is a sufficient statistic for p.

Example 6.63 *Sufficient statistic for the Poisson distribution*. *Let Y_i be n iid Poisson distributed observations with rate λ. Prove that $U = \sum_{i=1}^n Y_i$ is a sufficient statistic for λ.*

Solution. Since the ith observation has probability $(y_i!)^{-1}\lambda^{y_i}e^{-\lambda}$, the joint pmf takes the form

$$f(y_1, ..., y_n; \lambda) = \left(\prod_{i=1}^n y_i!\right)^{-1} e^{-\lambda}\lambda^{\sum_{i=1}^n y_i}.$$

Let $u = \sum_{i=1}^n y_i$ and introduce functions $H(\mathbf{y}) = 1/\prod_{i=1}^n y_i!$ and $g(u; \lambda) = e^{-\lambda}\lambda^u$. The joined pmf is expressed in the form (6.45). This proves that $U = \sum_{i=1}^n Y_i$ is a sufficient statistic. $\square$

In the above examples, the sufficient statistic was a scalar ($m = 1$). Below is an example of $m = 2$, a generalization of Example 6.61.

Example 6.64 *Sufficient statistics for the normal distribution.* *Prove that* $\mathbf{U}^{2\times1} = (\sum Y_i, \sum Y_i^2)$ *is a sufficient statistic for* $\boldsymbol{\theta} = (\mu, \sigma^2)$ *if the observations are iid normally distributed with unknown mean and variance.*

Solution. The joint pdf takes the form

$$f(y_1, ..., y_n; \mu, \sigma^2) = (2\pi)^{-n/2}(\sigma^2)^{-n/2}e^{-\frac{1}{2\sigma^2}\sum_{i=1}^{n}(y_i-\mu)^2}.$$

Similarly to the previous algebra, we have $\sum(y_i - \mu)^2 = \sum y_i^2 - 2\mu\sum y_i + n\mu^2$. This allows us to rewrite f as a function of $u_1 = \sum_{i=1}^{n} y_i$ and $u_2 = \sum_{i=1}^{n} y_i^2$:

$$f(y_1, ..., y_n; \mu, \sigma^2) = (2\pi)^{-n/2}(\sigma^2)^{-n/2}e^{-\frac{1}{2\sigma^2}u_2+\frac{\mu}{\sigma^2}u_1-\frac{n\mu^2}{2\sigma^2}}.$$

Hence, $\mathbf{U}$ is a sufficient statistic where $H(\mathbf{y}) = 1$. In words, one may say that the average and sample variance are sufficient statistics for the normal distribution $\mathcal{N}(\mu, \sigma^2)$. □

The following example defines sufficient statistics for a simple linear regression with a normally distributed dependent variable; see Section 6.7.2.

Example 6.65 *Sufficient statistics for linear regression.* *Prove that* $\mathbf{U}^{3\times1} = (\sum Y_i, \sum Y_i x_i, \sum Y_i^2)$ *defines sufficient statistics for* $\boldsymbol{\theta} = (\alpha, \beta, \sigma^2)$ *in a simple linear regression with a normally distributed dependent variable,* $Y_i \sim \mathcal{N}(\alpha + \beta x_i, \sigma^2)$, *where* x_i *are fixed numbers,* $i = 1, 2, ..., n$.

Solution. The joint pdf is

$$f(y_1, ..., y_n; \alpha, \beta, \sigma^2) = (2\pi)^{-n/2}(\sigma^2)^{-n/2}e^{-\frac{1}{2\sigma^2}\sum_{i=1}^{n}(y_i-\alpha-\beta x_i)^2}.$$

Express

$$\sum_{i=1}^{n}(y_i - \alpha - \beta x_i)^2 = \sum_{i=1}^{n} y_i^2 - 2\beta\sum_{i=1}^{n} y_i x_i - 2\alpha\sum_{i=1}^{n} y_i + G(\boldsymbol{\theta}),$$

where G is a function of unknown parameters. Therefore, the joint density can be expressed through $\sum_{i=1}^{n} y_i^2$, $\sum_{i=1}^{n} y_i x_i$, and $\sum_{i=1}^{n} y_i$. This proves that $\mathbf{U}$ is a sufficient statistic. □

The importance of sufficient statistics can be seen from the following theorem, which says that any unbiased estimator can be improved by conditioning on the sufficient statistic.

When a parameter is multidimensional, the MSE, as defined in Section 6.4.3, should be used to quantify how close the estimator comes to the true parameter. When the estimator is unbiased, the MSE turns into a covariance matrix. To compare covariance matrices, we say that matrix $\mathbf{A}$ is less or equal than matrix $\mathbf{B}$ (both matrices are symmetric), $\mathbf{A} \leq \mathbf{B}$, if matrix $\mathbf{B} - \mathbf{A}$ is nonnegative definite, i.e. $\mathbf{x}'(\mathbf{A} - \mathbf{B})\mathbf{x} \geq 0$ for every $\mathbf{x}$ (see Chapter 10). Now we are ready to formulate the theorem.

Theorem 6.66 *Rao–Blackwell. Let* $\mathbf{U} = \mathbf{u}(\mathbf{Y})$ *be a sufficient statistic and* $\mathbf{T}$ *be an unbiased estimator of* $\boldsymbol{\theta}$ *and let the conditional mean* $\mathbf{h}(\mathbf{u}) = E(\mathbf{T}|\mathbf{U} = \mathbf{u})$ *be independent of* $\boldsymbol{\theta}$. *Consider* $\mathbf{h}(\mathbf{U})$ *as a new estimator of* $\boldsymbol{\theta}$. *Then two properties hold: (a) this estimator is unbiased for* $\boldsymbol{\theta}$, *and (b) its covariance matrix is less or equal to the covariance matrix of* $\mathbf{T}$.

Proof. First, to minimize technical details, we prove the theorem for a scalar parameter θ. (a) The unbiasedness follows from the rule of repeated expectation from Section 3.3.1. For brevity, let $h = h(U) = E(T|U)$, so $E(h) = E_U[E(T|U)] = E(T) = \theta$ because T is an unbiased estimator. (b) To prove that the new estimator has a smaller variance, the rule of repeated expectation is applied to $(T - \theta)^2$:

$$\begin{aligned} \text{var}(T) &= E[(T - \theta)^2] = E_U[(T - \theta)^2|U] = E_U[((T - h) + (h - \theta))^2|U] \\ &= E_U[(T - h)^2|U] - 2E_U[(T - h)(h - \theta)|U] + E_U[(h - \theta)^2]. \end{aligned}$$

But the middle term vanishes because

$$E_U[(T - h)(h - \theta)|U] = E_U[(E(T|U) - h)(h - \theta)|U] = E_U[(h - h)(h - \theta)|U] = 0.$$

This means that $\text{var}(T) \geq E_U[(h - \theta)^2] = \text{var}(h)$. Second, to prove the theorem in the vector case, we use the repeated expectation rule in a vector form, $E(\mathbf{Y}) = E_{\mathbf{X}}[E(\mathbf{Y}|\mathbf{X})]$. The unbiasedness is obvious. To prove that the covariance matrix of the new estimator is smaller, use the following algebra.

$$\begin{aligned} \text{cov}(\mathbf{T}) &= E(\mathbf{T} - \boldsymbol{\theta})(\mathbf{T} - \boldsymbol{\theta})' = E_{\mathbf{U}}[((\mathbf{T} - \mathbf{h}) + (\mathbf{h} - \boldsymbol{\theta}))((\mathbf{T} - \mathbf{h}) + (\mathbf{h} - \boldsymbol{\theta}))'|\mathbf{U}] \\ &= E_{\mathbf{U}}[(\mathbf{T} - \mathbf{h})(\mathbf{T} - \mathbf{h})'|\mathbf{U}] - E_{\mathbf{U}}[(\mathbf{T} - \mathbf{h})(\mathbf{h} - \boldsymbol{\theta})'|\mathbf{U}] \\ &\quad - E_{\mathbf{U}}[(\mathbf{h} - \boldsymbol{\theta})(\mathbf{T} - \mathbf{h})'|\mathbf{U}] + E_{\mathbf{U}}[(\mathbf{h} - \boldsymbol{\theta})(\mathbf{T} - \mathbf{h})'] \end{aligned}$$

But

$$E_{\mathbf{U}}[(\mathbf{T} - \mathbf{h})(\mathbf{h} - \boldsymbol{\theta})'|\mathbf{U}] = E_{\mathbf{U}}[(\mathbf{h} - \boldsymbol{\theta})(\mathbf{T} - \mathbf{h})'|\mathbf{U}] = \mathbf{0},$$

as in the univariate case, so $\text{cov}(\mathbf{T}) \geq E_{\mathbf{U}}[(\mathbf{h} - \boldsymbol{\theta})(\mathbf{h} - \boldsymbol{\theta})'] = \text{cov}(\mathbf{h})$, which completes the proof. $\qquad\qquad\square$

The fact that the weighted average does not improve the arithmetic average for estimation of the mean in the iid case can be proved directly. Here, we apply the Rao–Blackwell theorem solely for the pedagogical purpose.

Example 6.67 *Weighted mean. Let* X_i *be iid observations from* $\mathcal{N}(\mu, \sigma^2)$, $i = 1, 2, ..., n$. *Let* $\widetilde{\mu}$ *be the weighted average,* $\widetilde{\mu} = \sum_{i=1}^n w_i X_i / \sum_{i=1}^n w_i$, *where* w_i *are fixed numbers with a nonzero sum. Prove that* $\widetilde{\mu}$ *is unbiased for* μ *and* $E(\widetilde{\mu}|\overline{X}) = \overline{X}$ *has a smaller variance than* $\widetilde{\mu}$ *if not all* w_i *are the same.*

Solution. The unbiasedness of $\widetilde{\mu}$ is easy to prove, $E(\widetilde{\mu}) = E\left(\sum_{i=1}^{n} w_i X_i\right)$ $/ \sum_{i=1}^{n} w_i = \sum_{i=1}^{n} w_i \mu / \sum_{i=1}^{n} w_i = \mu$. The joint distribution of $\widetilde{\mu}$ and $\overline{X}$ is bivariate normal (see Section 3.5):

$$\begin{bmatrix} \widetilde{\mu} \\ \overline{X} \end{bmatrix} \sim \mathcal{N}\left(\begin{bmatrix} \mu \\ \mu \end{bmatrix}, \sigma^2 \begin{bmatrix} \sum_{i=1}^{n} w_i^2/(\sum_{i=1}^{n} w_i)^2 & 1/n \\ 1/n & 1/n \end{bmatrix} \right).$$

Indeed,

$$\text{var}(\widetilde{\mu}) = \frac{\text{var}\left(\sum_{i=1}^{n} w_i X_i\right)}{(\sum_{i=1}^{n} w_i)^2} = \sigma^2 \frac{\sum_{i=1}^{n} w_i^2}{(\sum_{i=1}^{n} w_i)^2},$$

$$\text{cov}(\widetilde{\mu}, \overline{X}) = \frac{\text{cov}(\sum_{i=1}^{n} w_i X_i, \sum_{i=1}^{n} X_i)}{n \sum_{i=1}^{n} w_i} = \sigma^2 \frac{\sum_{i=1}^{n} w_i}{n \sum_{i=1}^{n} w_i} = \frac{\sigma^2}{n}.$$

As follows from the formula for conditional mean of the bivariate normal distribution (3.30) we have

$$E(\widetilde{\mu}|\overline{X}) = \mu + \frac{\text{cov}(\widetilde{\mu}, \overline{X})}{\text{var}(\overline{X})}(\overline{X} - \mu) = \mu + \frac{\sigma^2/n}{\sigma^2/n}(\overline{X} - \mu) = \overline{X}.$$

The Rao-Blackwell theorem applies, and therefore $E(\widetilde{\mu}|\overline{X})$ is unbiased with the variance equal or smaller than the variance of $\widetilde{\mu}$.

6.8.1 Uniformly minimum-variance unbiased estimator

Optimal/efficient parameter estimation is the central theme of theoretical statistics. In Section 6.4.2, we argued that the problem of efficient parameter estimation cannot be solved without restricting the family of estimators, such as considering only unbiased estimators. The Rao–Blackwell theorem tells how to improve an unbiased estimator through conditional mean, if a sufficient statistic is available. In this section, we learn that if sufficient statistic is complete, the resulting estimator is a uniformly minimum-variance unbiased estimator (UMVUE), or, in other words, efficient in the family of unbiased estimators. When a parameter is a vector, we use the term uniformly minimum-covariance unbiased estimator (UMCUE), where the minimum covariance (matrix) is understood as specified in the Rao–Blackwell theorem.

An alternative approach to prove that an unbiased estimator is UMCUE is to use the Cramér–Rao lower bound based on the concept of the Fisher information matrix; see Definition 6.96.

Definition 6.68 *We say that statistic T, whose distribution depends on parameter θ, is complete if, for every function g such that $E_\theta(g(T)) = 0$ for all θ, we have $g = 0$.*

In words, a statistic is complete if there is no nontrivial unbiased estimator of zero, as a function of T. This definition naturally extends to the case when $\mathbf{T}$ and $\boldsymbol{\theta}$ are multidimensional. We may speak of completeness of the entire distribution when $\mathbf{T}$ is the entire data.

Example 6.69 *Sufficiency of binomial proportion.* *Prove that binomial proportion is a complete sufficient statistic in a series of n iid Bernoulli trials.*

Solution. Here we have n iid Bernoulli random variables Y_i with binomial proportion $\hat{p} = m/n$, where $m = \sum Y_i$. As was shown in Example 6.62, m is sufficient; now we prove that m is complete. Let g be any function of m. Denote $g_k = g(k)$, $k = 0, 1, ..., n$ and assume that

$$E(g(m)) = \sum_{k=0}^{n} g_k \binom{n}{k} p^k (1-p)^{n-k} = 0$$

for all $0 < p < 1$. This can be viewed as a polynomial of the nth order of p. According to the main theorem of algebra, the number of real roots is no more than n, but this contradicts the assumption that the equality holds for all $0 < p < 1$. Therefore, $g_k = 0$ for all $k = 0, 1, ..., n$ which proves that $\hat{p}$ is a complete sufficient statistic.   □

Many statistics are complete. To motivate our statement, consider the following setup. Let the pdf of statistic T be $f(t; \theta)$ and parameter θ belong to $(-\infty, \infty)$. We need to prove that

$$\int_{-\infty}^{\infty} g(t) f(t; \theta) dt = 0 \quad \forall \theta \text{ implies } g = 0. \tag{6.46}$$

Consider the functional mapping $M : g \to h$ where $h = h(\theta)$ is $1/2$ times the left-hand side of (6.46). We want to show that M is a contraction mapping: $\sup |h_1(\theta) - h_2(\theta)| < \nu \times \sup |g_1(t) - g_2(t)|$ where $0 < \nu < 1$. Indeed, this inequality works for $\nu = 1/2$ because

$$|h_1(\theta) - h_2(\theta)| = \frac{1}{2} \left| \int_{-\infty}^{\infty} (g_1(t) - g_2(t)) f(t; \theta) dt \right| \leq \frac{1}{2} \int_{-\infty}^{\infty} |g_1(t) - g_2(t)| \, f(t; \theta) dt$$

$$\leq \frac{1}{2} \sup |g_1(t) - g_2(t)| \int_{-\infty}^{\infty} f(t; \theta) dt = \frac{1}{2} \sup |g_1(t) - g_2(t)|$$

since $\int_{-\infty}^{\infty} f(t; \theta) dt = 1$. Now we invoke the Banach fixed-point theorem (Kolmogorov and Fomin [66]): for a contraction mapping, there is at most one solution to equation $h(\theta) = 0$. That is, $g = 0$.

The following theorem is a milestone result of parameter estimation. It says that once an unbiased estimator and a complete sufficient statistic are available, an unbiased estimator with minimum variance can be obtained as the conditional mean. Another way to prove that an unbiased estimator has minimum variance is through the Fisher information considered in the next section. The pros and cons of the two approaches are discussed in Section 6.9.2.

Theorem 6.70 *Lehmann–Scheffé. An unbiased estimator that is a function of a complete sufficient statistic is UMVUE/UMCUE.*

Proof. Let $\widehat{\theta}$ be an unbiased estimator of θ as a function of complete sufficient statistic U: $\widehat{\theta} = \widehat{\theta}(U)$ and $\widetilde{\theta}$ be another unbiased estimator of θ. We want to prove that $\text{var}(\widehat{\theta}) \leq \text{var}(\widetilde{\theta})$. By use of the Rao–Blackwell theorem, $\widehat{\theta}_* = \widehat{\theta}_*(U) = E(\widetilde{\theta}|U)$, we can minimize the variance so that $\text{var}(\widehat{\theta}_*) \leq \text{var}(\widetilde{\theta})$. But, then, $\widehat{\theta}_* = \widehat{\theta}$ due to completeness of U. Indeed, by letting $g(U) = \widehat{\theta}(U) - \widehat{\theta}_*(U)$, we have $E(g(U)) = 0$ for all θ, so $\widehat{\theta}_* = \widehat{\theta}$. The proof of the multivariate version follows the same line. $\square$

The following, as a continuation of Example 6.13, is the first application of the Lehmann–Scheffé theorem.

Example 6.71 *The UMVUE for the exponential distribution. Prove that $\widehat{\theta} = \overline{X}$ and $\widehat{\lambda} = (n-1)/n\overline{X}$ are the UMVUEs for the scale and rate of the exponential distribution, respectively.*

Solution. Let $\{X_i, i = 1, ..., n\}$ be an iid sample from an exponential distribution with the pdf written as either $\lambda e^{-\lambda x}$, where λ is referred to as the rate, or $(1/\theta)e^{-x/\theta}$, where θ is referred to as the scale parameter (Section 2.4). Here we consider the scale parametrization: the proof for the rate parametrization is similar. The joint pdf takes the form $\lambda^n e^{-\lambda S}$, where $S = \sum_{i=1}^{n} X_i$. This means that the sum of observations is a sufficient statistic. This statistic is complete because S has a gamma distribution with shape n and rate λ and, as follows from criterion (6.46),

$$E(g(S)) = \frac{\lambda^n}{\Gamma(n)} \int_0^{\infty} g(u)u^{n-1}e^{-\lambda u}du = 0 \quad \forall \lambda > 0$$

implies that $g(u) = 0$ for all $u > 0$ because function $e^{-\lambda u}$ is continuously differentiable. Since $\widehat{\theta}$ is an unbiased estimator, sufficient and complete, this estimator is UMVUE.

Example 6.72 *The UMCUE the for normal distribution and simple linear regression. (a) Prove that (a) $\overline{Y}$ and the unbiased sample variance $\widehat{\sigma}^2$, where $Y_i \overset{iid}{\sim} \mathcal{N}(\mu, \sigma^2)$, are UMCUE and (b) the OLS estimators of intercept and slope in simple linear regression with normally distributed error terms are UMCUE.*

Solution. (a) $\overline{Y}$ and $\widehat{\sigma}^2$ are unbiased complete sufficient statistics for μ and σ^2 as follows from Example 6.64. Hence they are UMCUE. (b) The joint pdf was derived in Example 6.65. Express $\|\mathbf{y} - \alpha - \beta\mathbf{x}\|^2 = \|\mathbf{y}\|^2 - 2\beta\mathbf{y}'\mathbf{x} - 2\alpha\mathbf{y}'\mathbf{1} + \text{const}$, where const does not depend on $\mathbf{y}$. This means that the sufficient statistic is three-dimensional, $\mathbf{T} = (\|\mathbf{y}\|^2, \mathbf{y}'\mathbf{x}, \mathbf{y}'\mathbf{1})$. It is elementary to prove that the OLS estimators for β, α, and σ^2 are functions of $\mathbf{T}$ and, as was shown in Section 6.7, are unbiased. It is not difficult to prove that the joint density of $\mathbf{T}$ has derivatives with respect to α, β, and σ^2 up to any order, and therefore $\mathbf{T}$ is complete. Therefore, by the Lehmann–Scheffé theorem, the estimator $(\widehat{\alpha}, \widehat{\beta}, \widehat{\sigma}^2)$ has minimum covariance matrix, i.e. is UMCUE.   □

A somewhat paradoxical result, which follows from the Lehmann–Scheffé theorem, holds when there are no nontrivial sufficient statistics as stated below.

Theorem 6.73 *Let the distribution of $\mathbf{Y}$ have no nontrivial sufficient statistic, i.e. the only sufficient and complete statistic being $\mathbf{Y}$ itself. Then there is at most one unbiased estimator and this estimator is UMVUE.*

An example of a real statistical model that satisfies the conditions of the theorem is nonlinear regression: $Y_i = f_i(\theta) + \varepsilon_i$, $\varepsilon_i \overset{\text{iid}}{\sim} \mathcal{N}(0, \sigma^2)$ and f_i are nonlinear functions; see Chapter 9. Many nonlinear statistical models have no nontrivial sufficient statistics and unbiased estimators; see relevant discussion in Section 6.4.1. More applications of the Lehmann–Scheffé theorem, where the theory of sufficient statistics and UMVUE works, are found in the next section.

6.8.2   Exponential family of distributions

Many, but not all, distributions belong to the exponential family. This fact makes it easy to study those distributions under one umbrella of the exponential family with sufficient statistics easily expressed. We first consider distributions with a single parameter, and then extend the discussion to the multidimensional parameter space.

Definition 6.74 *We say that the pdf of a random variable Y dependent on unknown parameter θ belongs to the exponential family of distributions if it can be written as*

$$f(y;\theta) = e^{p(\theta)K(y) + S(y) - q(\theta)}, \qquad (6.47)$$

where $p(\theta)$, referred to as the link, is a strictly monotonic function and $K(y)$ is a sufficient statistic.

It is assumed that θ belongs to a known interval or the entire line and that the support of the distribution does not depend on θ. Function $e^{-q(\theta)}$ serves as the normalizing coefficient to ensure that the area under the density is 1. We note that function, $p(\theta)$, and the sufficient statistic, $K(y)$, enter as product. The fact that $K(y)$ is a sufficient statistic follows immediately from the Neyman factorization criterion where $H(y) = e^{S(y)}$, $u(y) = K(y)$, and $g(u(y), \theta) = e^{p(\theta)u(y) - q(\theta)}$.

This property makes the exponential family special and simplifies statistical inference. The fact that the pdf is expressed through the exponential function is not particularly important because any pdf can be written as an exponential function on its support.

Example 6.75 *Major distributions belong to exponential family. Show that the (a) exponential, (b) Bernoulli, (c) binomial, (c) Poisson, and (d) normal (σ^2 is known) distributions belong to the exponential family of distributions.*

Solution. See Table 6.2 with five the most popular distributions including pdf/pmf, p, q, and K. For example, the binomial distribution is written in the form (6.47)

$$\binom{n}{y}\pi^y(1-\pi)^{n-y} = (1-\pi)^n e^{y \ln \frac{\pi}{1-\pi} + \ln \binom{n}{y}}, \quad y = 0, 1, ..., n,$$

as follows from Section 1.6. □

As was mentioned above, not all distributions belong to exponential family. For example, the uniform distribution with unknown limit(s) does not belong to the exponential family because the distribution support depends on θ. Double-exponential (Laplace) distribution with unknown center does not belong to the exponential family either.

Table 6.2. Exponential family of distributions with single parameter.

Distribution	pdf	Support	Parameter	q	link, p	K	S
Exponential	$\lambda e^{-\lambda y}$	$y > 0$	$\lambda > 0$	$-\ln \lambda$	$-\lambda$	y	0
Bernoulli	$\pi^y(1-\pi)^{1-y}$	$y = 0, 1$	$0 < \pi < 1$	$-\ln(1-\pi)$	$\ln \frac{\pi}{1-\pi}$	y	0
Binomial	$\binom{n}{y}\pi^y(1-\pi)^{n-y}$	$y = 0, ..., n$	$0 < \pi < 1$	$-n\ln(1-\pi)$	$\ln \frac{\pi}{1-\pi}$	y	$\ln \binom{n}{y}$
Poisson	$\frac{1}{y!}\lambda^y e^{-\lambda}$	$y = 0, 1, ...$	$\lambda > 0$	λ	$\ln \lambda$	y	$-\ln y!$
Normal*	$\frac{1}{\sqrt{2\pi\sigma^2}}e^{-\frac{1}{2\sigma^2}(y-\mu)^2}$	$y \in R^1$	$\mu \in R^1$	$\frac{\mu^2+\sigma^2 \ln(2\pi\sigma^2)}{2\sigma^2}$	$\frac{\mu}{\sigma^2}$	y	$-\frac{y^2}{2\sigma^2}$

Note: * σ^2 is known

Definition 6.76 *The pdf from the exponential family is called canonical if the link function is identity, $p(\theta) = \theta$. If, in addition, the K function is identity, $K(y) = y$, the distribution is called double canonical.*

Obviously, any member of exponential family may be turned into double canonical after reparametrization $\tau = p(\theta)$. In particular, all distributions listed in Table 6.2 become double canonical after such reparametrization. For example, the binomial distribution turns into canonical when the unknown parameter is $\tau = \ln \frac{\pi}{1-\pi}$, not π; then the probability, in the form of equation (6.47), is written in a somewhat unusual way, $\Pr(Y = y; \tau) = e^{y\tau + \ln \binom{n}{y} - n\ln(1+e^\tau)}$. The canonical parametrization is convenient because the mean and variance are easy to express in terms of derivatives of function q as follows from the theorem and its corollary below.

Theorem 6.77 *Let the pdf of random variable Y belong to the exponential family defined by equation (6.47). Then*

$$E(K(Y)) = \frac{dq/d\theta}{dp/d\theta}, \quad \text{var}(K(Y)) = \frac{1}{(dp/d\theta)^2}\left[\frac{d^2q}{d\theta^2} - \frac{d^2p}{d\theta^2}\frac{dq/d\theta}{dp/d\theta}\right]. \tag{6.48}$$

Proof. Differentiating both sides of the equation $\int_{-\infty}^{\infty} e^{p(\theta)K(y)+S(y)-q(\theta)}dy = 1$ with respect to θ (if the distribution is discrete, use sum), we obtain

$$-\int_{-\infty}^{\infty}\frac{dq}{d\theta}e^{\theta K(y)+S(y)-q(\theta)}dy + \int_{-\infty}^{\infty}K(y)\frac{dp}{d\theta}e^{p(\theta)K(y)+S(y)-q(\theta)}dy$$

$$= -\frac{dq}{d\theta}\int_{-\infty}^{\infty}e^{p(\theta)K(y)+S(y)-q(\theta)}dy + \frac{dp}{d\theta}\int_{-\infty}^{\infty}K(y)e^{p(\theta)K(y)+S(y)-q(\theta)}dy$$

$$= -\frac{dq}{d\theta} + \frac{dp}{d\theta}E(K(Y)) = 0,$$

which implies the first equation of (6.48). Double differentiation of the integral over the density gives

$$-\frac{d^2q}{d\theta^2} + \left(\frac{dq}{d\theta}\right)^2 - 2\frac{dq}{d\theta}\frac{dp}{d\theta}E(K(Y)) + \frac{d^2p}{d\theta^2}E(K(Y)) + \left(\frac{dp}{d\theta}\right)^2 E(K^2(y)) = 0.$$

From this equation express expected value of $K^2(y)$ using the previously derived the expected value of $K(Y)$,

$$E(K^2(y)) = \frac{1}{(dp/d\theta)^2}\left[\frac{d^2q}{d\theta^2} - \left(\frac{dq}{d\theta}\right)^2 + 2\frac{dq}{d\theta}\frac{dp}{d\theta}E(K(Y)) - \frac{d^2p}{d\theta^2}E(K(Y))\right]$$

$$= \frac{1}{(dp/d\theta)^2}\left[\frac{d^2q}{d\theta^2} - \left(\frac{dq}{d\theta}\right)^2 + 2\frac{dq}{d\theta}\frac{dp}{d\theta}\frac{dq/d\theta}{dp/d\theta} - \frac{d^2p}{d\theta^2}\frac{dq/d\theta}{dp/d\theta}\right]$$

$$= \frac{1}{(dp/d\theta)^2}\left[\frac{d^2q}{d\theta^2} - \left(\frac{dq}{d\theta}\right)^2 + 2\left(\frac{dq}{d\theta}\right)^2 - \frac{d^2p}{d\theta^2}\frac{dq/d\theta}{dp/d\theta}\right]$$

$$= \frac{1}{(dp/d\theta)^2}\left[\frac{d^2q}{d\theta^2} + \left(\frac{dq}{d\theta}\right)^2 - \frac{d^2p}{d\theta^2}\frac{dq/d\theta}{dp/d\theta}\right]$$

and finally

$$\text{var}(K(Y)) = E(K^2(y)) - E^2(K(y)) = \frac{1}{(dp/d\theta)^2}\left[\frac{d^2q}{d\theta^2} + \left(\frac{dq}{d\theta}\right)^2 - \frac{d^2p}{d\theta^2}\frac{dq/d\theta}{dp/d\theta}\right]$$

$$-\frac{(dq/d\theta)^2}{(dp/d\theta)^2} = \frac{1}{(dp/d\theta)^2}\left[\frac{d^2q}{d\theta^2} - \frac{d^2p}{d\theta^2}\frac{dq/d\theta}{dp/d\theta}\right],$$

which implies the second equation of (6.48).

Corollary 6.78 *For the double canonical exponential family, we have $dp/d\theta = 1$ and $d^2p/d\theta^2 = 0$, which leads to*

$$E(K(Y)) = \frac{dq}{d\theta}, \quad \mathrm{var}(K(Y)) = \frac{d^2q}{d\theta^2}. \tag{6.49}$$

Moreover, for the double canonical exponential family, where $K(y) = y$, we have (a)

$$E(Y) = \frac{dq}{d\theta}, \quad \mathrm{var}(Y) = \frac{d^2q}{d\theta^2}, \tag{6.50}$$

and (b) the moment generating function (MGF) of Y is $M(t) = e^{q(\theta+t)-q(\theta)}$.

Proof. Equations (6.49) and (6.50) follow directly from (6.48). The derivation of the MGF for the double canonical exponential family is also straightforward:

$$M(t) = \int_{-\infty}^{\infty} e^{ty} e^{\theta y + S(y) - q(\theta)} dy = e^{q(t+\theta) - q(\theta)} \int_{-\infty}^{\infty} e^{(t+\theta)y + S(y) - q(t+\theta)} dy$$
$$= e^{q(t+\theta) - q(\theta)}$$

due to the fact that $e^{(t+\theta)y + S(y) - q(t+\theta)}$ is a density, with θ replaced with $t + \theta$. Recall (Section 2.5) that the MGF can be used to find the noncentral moments by evaluating derivatives at zero. For example, using the expression for $M(t)$ from the theorem, we get

$$E(Y) = \frac{dM}{dt}\bigg|_{t=0} = \frac{dq}{d\theta}, \quad E(Y^2) = \frac{d^2M}{dt^2}\bigg|_{t=0} = \frac{d^2q}{d\theta^2} + \left(\frac{dq}{d\theta}\right)^2.$$

□

This theorem will be used for estimation of generalized linear model in Section 8.8. Since variance is nonnegative, we assert that

$$\frac{d^2q}{d\theta^2} - \frac{d^2p}{d\theta^2}\frac{dq/d\theta}{dp/d\theta} \geq 0.$$

Previous discussion dealt with random variable Y; the extension to the random vector $\mathbf{Y}$ is straightforward.

Definition 6.79 *We say that pdf of a random vector $\mathbf{Y}$, dependent on unknown parameter θ, belongs to the exponential family if it can be written as*

$$f(\mathbf{y};\theta) = e^{p(\theta)K(\mathbf{y}) + S(\mathbf{y}) - q(\theta)}, \tag{6.51}$$

where $p(\theta)$, referred to as the link, is a strictly monotonic function, and $K(\mathbf{y})$ is a sufficient statistic.

The following result is trivial but comforting. In particular, this result tells us that sufficient statistics are just $K(\mathbf{Y})$.

Theorem 6.80 *If iid observations Y_i have the exponential family of distributions (6.47), then $\mathbf{Y} = (Y_1, Y_2, ..., Y_n)$ belongs to exponential family (6.51) and $K(\mathbf{y}) = \sum_{i=1}^{n} K(y_i)$ is a sufficient statistic. Moreover, if the original pdf is double canonical, then $\overline{Y}$ is a sufficient statistic and*

$$E(\overline{Y}) = \frac{dq}{d\theta}, \quad \text{var}(\overline{Y}) = \frac{1}{n}\frac{d^2q}{d\theta^2}. \tag{6.52}$$

Proof. Obviously, in the notation of Definition 6.79, $K(\mathbf{y}) = \sum_{i=1}^{n} K(y_i)$, $S(\mathbf{y}) = \sum_{i=1}^{n} S(y_i)$, and $q(\theta) = nq(\theta)$, with some ambiguity of notation. Sufficiency of $K(\mathbf{y})$ comes from the Neyman factorization criterion (6.45) with $H(\mathbf{y}) = e^{S(\mathbf{y})}$. To prove (6.52), we use formulas (6.50) with q replaced with nq. □

Now we discuss how sufficient statistics can be used for optimal unbiased estimation. The result below emphasizes the most attractive property of the exponential distribution.

Theorem 6.81 *Let Y_i be iid observations from the general population with distribution (6.47) where functions p and q are continuously differentiable. (a) If $(dq/d\theta)/(dp/d\theta) = \theta$, then $\overline{K} = \sum_{i=1}^{n} K(Y_i)/n$ is the UMVUE. (b) Let $(dq/d\theta)/(dp/d\theta) = w(\theta)$ be a one-to-one function of the unknown parameter. Then $\overline{K}$ is the UMVUE for $\theta_* = w(\theta)$.*

Proof. (a) The fact that upon $(dq/d\theta)/(dp/d\theta) = \theta$ statistic $\overline{K}$ is an unbiased estimator for θ follows from Theorem 6.77. Since functions p and q are continuously differentiable, statistic $\overline{K}$ is complete by criterion (6.46). Since $\overline{K}$ is sufficient, we finally conclude that it is UMVUE for θ. (b) The proof repeats after reparametrization $\theta_* = w(\theta)$.

Example 6.82 *UMVUE for major distributions.* *Use Theorem 6.81 to find UMVUEs for distributions listed in Table 6.2.*

Solution. First, we observe that all functions p and q from Table 6.2 have derivatives of every order and therefore $\overline{Y}$ is a complete sufficient statistic. Exponential: $q(\lambda) = -\ln\lambda$, $p(\lambda) = -\lambda$ and $(dq/d\lambda)/(dp/d\lambda) = 1/\lambda$. Therefore, the average, $\overline{Y}$, is UMVUE for $\theta = 1/\lambda$, the scale parameter (but not for λ). Bernoulli: $q(\pi) = -\ln(1-\pi)$, $p(\pi) = \ln[\pi/(1-\pi)]$ and $(dq/d\lambda)/(dp/d\lambda) = \pi$. This implies that $\overline{Y}$ is UMVUE for π. Poisson: $q(\lambda) = \lambda$, $p(\lambda) = \ln\lambda$ and $(dq/d\lambda)/(dp/d\lambda) = \lambda$. Thus $\overline{Y}$ is UMVUE for λ. Normal: $q(\mu) = (\mu^2 + \sigma^2\ln(2\pi\sigma^2))/\sigma^2$, $p(\mu) = \mu/\sigma^2$ and $(dq/d\mu)/(dp/d\mu) = \mu$. Thus, $\overline{Y}$ is UMVUE for μ. □

Finally, we define the exponential family for a multivariate random variable (random vector).

Definition 6.83 *We say that the pdf of a random vector* $\mathbf{Y}$, *dependent on unknown m-dimensional parameter vector* $\boldsymbol{\theta}$ *from an m-dimensional open set, belongs to the exponential family if it can be expressed as*

$$f(\mathbf{y}; \boldsymbol{\theta}) = e^{\mathbf{p}'(\boldsymbol{\theta})\mathbf{K}(\mathbf{y}) + S(\mathbf{y}) - q(\boldsymbol{\theta})}, \tag{6.53}$$

where $\mathbf{p}(\boldsymbol{\theta})$ *is a one-to-one* $m \times 1$ *vector function and* $\mathbf{K}(\mathbf{y})$ *is an* $m \times 1$ *sufficient statistic. If* $\mathbf{p}(\boldsymbol{\theta}) = \boldsymbol{\theta}$ *we call the exponential family canonical.*

Remark 6.84 *(i) Note that* $'$ *is the transposition sign and* $\mathbf{p}'(\boldsymbol{\theta})\mathbf{K}(\mathbf{y})$ *is the scalar product: in coordinate form, it can be written as* $\sum_{j=1}^{m} K_j(\mathbf{y})p_j(\boldsymbol{\theta})$. *(ii) The "one-to-one" means that* $\mathbf{p}(\boldsymbol{\theta}_1) = \mathbf{p}(\boldsymbol{\theta}_2)$ *implies* $\boldsymbol{\theta}_1 = \boldsymbol{\theta}_2$. *(iii) Typically we assume that functions* $\mathbf{p}$ *and* q *are continuously differentiable, which implies that statistic* $\mathbf{K}(\mathbf{Y})$ *is complete under mild assumptions. (iv) The joint distribution of a random sample from an exponential family belongs to the same exponential family; then the definition applies to* $\mathbf{Y}$ *as a collection of* n *iid observations whose distributions belong to the exponential family (6.47).*

The following is an extension of Theorem 6.77.

Theorem 6.85 *If the distribution of* $\mathbf{Y}$ *belongs to the canonical exponential family, then* $E(\mathbf{K}(\mathbf{Y})) = \partial q / \partial \boldsymbol{\theta}$ *and* $\mathrm{cov}(\mathbf{K}(\mathbf{Y})) = \partial^2 q / \partial \boldsymbol{\theta}^2$.

The proof is a straightforward generalization of Theorem 6.77 and its corollary is proved by vector differentiation.

Theorem 6.85 can be used to find a method of moments (MM) estimator for a canonical exponential family when $\mathbf{Y}$ is the combined vector of n iid observations $\mathbf{Y}_1, \mathbf{Y}_2, ..., \mathbf{Y}_n$. Define the MM estimator, $\widehat{\boldsymbol{\theta}}_{MM}$, as the solution to the equation

$$\frac{\partial q}{\partial \boldsymbol{\theta}} = \overline{\mathbf{K}}, \tag{6.54}$$

where $\overline{\mathbf{K}} = \sum_{i=1}^{n} \mathbf{K}(\mathbf{Y}_i)/n$. The covariance of $\widehat{\boldsymbol{\theta}}_{MM}$ can be approximated by the Hessian of function q,

$$\mathrm{cov}(\widehat{\boldsymbol{\theta}}_{MM}) \simeq \frac{1}{n} \left(\frac{\partial^2 q}{\partial \boldsymbol{\theta}^2} \bigg|_{\boldsymbol{\theta} = \widehat{\boldsymbol{\theta}}_{MM}} \right)^{-1}. \tag{6.55}$$

Indeed, express $\widehat{\boldsymbol{\theta}}_{MM} = \widehat{\boldsymbol{\theta}}_{MM}(\overline{\mathbf{K}})$ and approximate its covariance using the implicit delta method from Section 3.10.3

$$\mathrm{cov}(\widehat{\boldsymbol{\theta}}_{MM}) \simeq \left(\frac{\partial \widehat{\boldsymbol{\theta}}_{MM}}{\partial \overline{\mathbf{K}}} \right) \mathrm{cov}(\overline{\mathbf{K}}) \left(\frac{\partial \widehat{\boldsymbol{\theta}}_{MM}}{\partial \overline{\mathbf{K}}} \right)'.$$

To find the Jacobian, i.e. the derivative of $\widehat{\boldsymbol{\theta}}_{MM}$, as the solution of (6.54) with respect to $\overline{\mathbf{K}}$, differentiate both sides of equation (6.54) using the chain rule, $(\partial q / \partial \widehat{\boldsymbol{\theta}}_{MM})(\partial \widehat{\boldsymbol{\theta}}_{MM} / \partial \overline{\mathbf{K}}) = \mathbf{I}$, which implies

$$\frac{\partial \widehat{\boldsymbol{\theta}}_{MM}}{\partial \overline{\mathbf{K}}} = \left(\frac{\partial^2 q}{\partial \boldsymbol{\theta}^2} \bigg|_{\boldsymbol{\theta} = \widehat{\boldsymbol{\theta}}_{MM}} \right)^{-1}.$$

Finally,

$$\mathrm{cov}(\widehat{\boldsymbol{\theta}}_{MM}) \simeq \frac{1}{n} \left(\frac{\partial^2 q}{\partial \boldsymbol{\theta}^2} \bigg|_{\boldsymbol{\theta} = \widehat{\boldsymbol{\theta}}_{MM}} \right)^{-1} \left(\frac{\partial^2 q}{\partial \boldsymbol{\theta}^2} \bigg|_{\boldsymbol{\theta} = \widehat{\boldsymbol{\theta}}_{MM}} \right) \left(\frac{\partial^2 q}{\partial \boldsymbol{\theta}^2} \bigg|_{\boldsymbol{\theta} = \widehat{\boldsymbol{\theta}}_{MM}} \right)^{-1}$$

$$= \frac{1}{n} \left(\frac{\partial^2 q}{\partial \boldsymbol{\theta}^2} \bigg|_{\boldsymbol{\theta} = \widehat{\boldsymbol{\theta}}_{MM}} \right)^{-1},$$

as we intended to prove. In simulations, the Hessian is evaluated at the true parameter values. This method of estimation is applied to the beta distribution; see Section 2.14.

The following theorem establishes perhaps the most desirable property of the exponential family of distributions in connection with uniform minimum-covariance unbiased estimation.

Theorem 6.86 *Efficient estimation with the exponential family.* Let $\mathbf{Y}$ *belong to the exponential family with pdf (6.53). Then*

$$E(\mathbf{K}(\mathbf{Y})) = \left(\frac{\partial \mathbf{p}}{\partial \boldsymbol{\theta}} \right)'^{-1} \left(\frac{\partial q}{\partial \boldsymbol{\theta}} \right).$$

If the right-hand side equals $\boldsymbol{\theta}$, then $\mathbf{K}(\mathbf{Y})$ is a UMCUE for $\boldsymbol{\theta}$. If (6.53) is reparametrized in terms of $\boldsymbol{\theta}_*$, where $\boldsymbol{\theta}_*$ is the right-hand side, then $\mathbf{K}(\mathbf{Y})$ is a UMCUE for $\boldsymbol{\theta}_*$.

The proof that $\mathbf{K}(\mathbf{Y})$ is an unbiased estimator of $\boldsymbol{\theta}$ follows from differentiation of the integral over the density as in Theorem 6.77 and the proof that $\mathbf{K}(\mathbf{Y})$ has minimum covariance matrix is analogous to Theorem 6.81.

Example 6.87 *The normal distribution belongs to the exponential family.* *Confirm the joint density of $Y_i \overset{\text{iid}}{\sim} \mathcal{N}(\mu, \sigma^2)$, where μ and σ^2 are unknown, belongs to exponential family and that conditions of Theorem 6.86 apply.*

Solution. Let $\mathbf{y} = (y_1, ..., y_n)'$ and write the joint density as $f(\mathbf{y}; \mu, \sigma^2) = \exp(-0.5\sigma^{-2} \|\mathbf{y} - \mu \mathbf{1}\|^2 - 0.5n \ln(2\pi\sigma^2))$, where $\mathbf{1}$ is the $n \times 1$ vector of ones, and $\boldsymbol{\theta} = (\mu, \sigma^2)$. Function f takes the form (6.53), where $\mathbf{K}(\mathbf{y}) = (\mathbf{y}'\mathbf{1}, \|\mathbf{y}\|^2)'$ is a 2×1

vector, $\mathbf{p}(\mu, \sigma^2) = (\sigma^{-2}\mu, -0.5\sigma^{-2})'$ and $q(\mu, \sigma^2) = 0.5n\ln(2\pi\sigma^2) + 0.5n\sigma^{-2}\mu^2$. We have

$$\frac{\partial \mathbf{p}}{\partial \boldsymbol{\theta}} = \begin{bmatrix} \frac{\partial p_1}{\partial \mu} & \frac{\partial p_1}{\partial \sigma^2} \\ \frac{\partial p_2}{\partial \mu} & \frac{\partial p_2}{\partial \sigma^2} \end{bmatrix} = \begin{bmatrix} \sigma^{-2} & -\mu\sigma^{-4} \\ 0 & 0.5\sigma^{-4} \end{bmatrix} = 0.5\sigma^{-4} \begin{bmatrix} 2\sigma^2 & -2\mu \\ 0 & 1 \end{bmatrix},$$

$$\frac{\partial q}{\partial \boldsymbol{\theta}} = \begin{bmatrix} \frac{\partial q}{\partial \mu} \\ \frac{\partial q}{\partial \sigma^2} \end{bmatrix} = \begin{bmatrix} n\sigma^{-2}\mu \\ 0.5n\sigma^{-2} - 0.5n\sigma^{-4}\mu^2 \end{bmatrix} = 0.5n\sigma^{-4} \begin{bmatrix} 2\sigma^2\mu \\ \sigma^2 - \mu^2 \end{bmatrix}.$$

Thus,

$$\left(\frac{\partial \mathbf{p}}{\partial \boldsymbol{\theta}}\right)^{'-1} \left(\frac{\partial q}{\partial \boldsymbol{\theta}}\right) = 0.5n\sigma^{-2} \begin{bmatrix} 2\sigma^2\mu \\ 4\sigma^2\mu^2 + 2\sigma^4 - 2\sigma^2\mu^2 \end{bmatrix} = \begin{bmatrix} n\mu \\ n(\mu^2 + \sigma^2) \end{bmatrix}.$$

As follows from Theorem 6.86, UMVUEs of μ and $\mu^2 + \sigma^2$ can be obtained as $\mathbf{y}'\mathbf{1}/n$ and $\|\mathbf{y}\|^2/n$.

Example 6.88 *MM for beta distribution. Use equation (6.54) to find estimators for α and β having a sample $X_i \overset{\text{iid}}{\sim} \mathcal{B}(\alpha, \beta)$. (a) Prove that the beta distribution belongs to exponential family. (b) Find the MM estimator using Newton's algorithm for a series of simulated samples from a beta distribution. (c) Plot the simulation-based kernel densities and check formula (6.55).*

Solution. See Section 2.14 on the beta distribution.

(a) The pdf can be rewritten as

$$f(x; \alpha, \beta) = e^{(\alpha-1)\ln x + (\beta-1)\ln(1-x) - \ln B(\alpha,\beta)} = e^{\alpha \ln x + \beta \ln(1-x) - \ln x - \ln(1-x) - \ln B(\alpha,\beta)},$$

and therefore the beta distribution belongs to exponential family because it takes the form (6.53) with $\mathbf{K}(x) = (K_1(\mathbf{x}), K_2(\mathbf{x}))'$, $\mathbf{p}(\alpha, \beta) = (\alpha, \beta)'$ and $q = \ln B(\alpha, \beta)$, where $K_1 = K_1(\mathbf{X}) = \sum_{i=1}^{n} \ln X_i / n$ and $K_2 = K_2(\mathbf{X}) = \sum_{i=1}^{n} \ln(1 - X_i)/n$ are sufficient statistics. This implies that the beta distribution belongs to the canonical exponential family.

(b) Find derivatives of the q function with respect to α and β. Using the fact that $B(\alpha, \beta) = \Gamma(\alpha)\Gamma(\beta)/\Gamma(\alpha+\beta)$ and denoting $\psi(u) = d\ln\Gamma(u)/du$, the psigamma in R, we find $\partial q/\partial \alpha = \psi(\alpha) - \psi(\alpha + \beta)$ and $\partial q/\partial \beta = \psi(\beta) - \psi(\alpha + \beta)$. Thus, the MM estimators for α and β are found as solutions to a system of transcendent equations:

$$\psi(\alpha) - \psi(\alpha + \beta) = K_1, \quad \psi(\beta) - \psi(\alpha + \beta) = K_2.$$

This system is solved by Newton's algorithm

$$\begin{bmatrix} \alpha_{s+1} \\ \beta_{s+1} \end{bmatrix} = \begin{bmatrix} \alpha_s \\ \beta_s \end{bmatrix} - \begin{bmatrix} h_{11} & h_{12} \\ h_{21} & h_{22} \end{bmatrix}^{-1} \begin{bmatrix} \psi(\alpha_s) - \psi(\alpha_s + \beta_s) - K_1 \\ \psi(\beta_s) - \psi(\alpha_s + \beta_s) - K_2 \end{bmatrix},$$

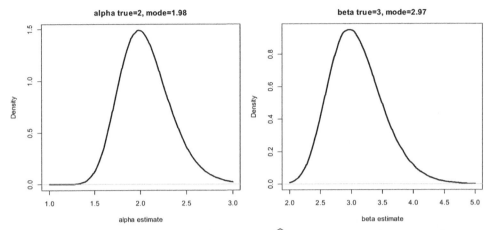

Figure 6.20: *Kernel densities of $\widehat{\alpha}_{MM}$ and $\widehat{\beta}_{MM}$ from 1 million simulations of the sample size $n = 100$ from the beta distribution with the true $\alpha = 2$ and $\beta = 3$ (it took four minutes for the R function* `betaMM.apply` *to complete).*

where

$$
\begin{bmatrix} h_{11} & h_{12} \\ h_{21} & h_{22} \end{bmatrix} = \begin{bmatrix} \psi_1(\alpha_s) - \psi_1(\alpha_s + \beta_s) & -\psi_1(\alpha_s + \beta_s) \\ -\psi_1(\alpha_s + \beta_s) & \psi_1(\beta_s) - \psi_1(\alpha_s + \beta_s) \end{bmatrix}
$$

is the 2×2 matrix of derivatives: h_{11} is the derivative of the first equation with respect to α, h_{12} is the derivative of the first equation with respect to β, and h_{21} and h_{22} are derivatives of the second equation with respect to α and β. Here ψ_1 is the derivative of the psigamma function, called `trigamma` in R. We start iterations from the method of moments estimates by equating the sample mean and variance, $\overline{X}$ and $\widehat{\sigma}^2$, to their theoretical counterparts from Section 2.14: $\alpha_0 = \overline{X}(\overline{X} - \overline{X}^2 - \widehat{\sigma}^2)/\widehat{\sigma}^2$ and $\beta_0 = (1 - \overline{X})(\overline{X} - \overline{X}^2 - \widehat{\sigma}^2)/\widehat{\sigma}^2$.

(c) The R function `betaMM.apply` carries out the simulations and plots the kernel densities of the estimates shown in Figure 6.20. The local function `betaMM` uses Newton's iterations and serves as the argument to function `apply` for vectorized computations. At the end of computations, the kernel densities of $\widehat{\alpha}$ and $\widehat{\beta}$ are depicted. The R function `rbeta` generates n iid beta-distributed observations (`shape1` $= \alpha$ and `shape2` $= \beta$). The estimates for α and β are stored in matrix `ab` and the diagonal elements of the inversed Hessian divided by n are stored in matrix `hv`. Although for most simulated values Newton's algorithm converges fairly quickly (<3 iterations), for some samples, especially when n is small, say, $n \leq 10$, the mode of the kernel density is a better summary, not the mean, because otherwise, it can yield erroneous results due to outliers. As follows from Figure 6.20, the modes of densities are close to the true values of α and β. That is, the MM estimation method is satisfactory. To check formula (6.55), we compute the variances of $\widehat{\alpha}_{MM}$ and $\widehat{\beta}_{MM}$ in three ways: (i) Compute empirical variance using nSim estimates from each simulation. (ii) Evaluate (6.55) at the true values of α and β. (iii) Evaluate (6.55) at each simulation using estimates $\widehat{\alpha}_{MM}$ and $\widehat{\beta}_{MM}$ and then take their means. The output of the function with default parameters is below.

```
> betaMM.apply(nSim=1000000)
[1] "Thu Dec 28 21:23:35 2017"
[1]  0.07882684 0.07068930 0.07656601
[1]  0.1927858 0.1724819 0.1871072
[1] "Thu Dec 28 21:27:24 2017"
```

The theoretical and simulation-based variances are in good agreement.

Example 6.89 *Multivariate normal distribution belongs to exponential family.* *(a) Prove that the distribution of the $k \times 1$ multivariate normal vector, $\mathbf{Y} \sim \mathcal{N}(\boldsymbol{\mu}, \boldsymbol{\Omega})$, with unknown mean and covariance matrix belongs to the exponential family, and (b) assuming that $\mathbf{Y}_i \overset{iid}{\sim} \mathcal{N}(\boldsymbol{\mu}, \boldsymbol{\Omega})$ find sufficient statistics for the joint distribution $(i = 1, 2, .., n)$.*

Solution. (a) The multivariate normal pdf is given by

$$f(\mathbf{y};\boldsymbol{\mu}, \boldsymbol{\Omega}) = (2\pi)^{-k/2} |\boldsymbol{\Omega}|^{-1/2} \times e^{-\frac{1}{2}(\mathbf{y}-\boldsymbol{\mu})'\boldsymbol{\Omega}^{-1}(\mathbf{y}-\boldsymbol{\mu})}.$$

To express the argument of the exponential function in the form (6.53) we employ the Kronecker matrix product (Appendix, Section 10.5) using the formula $\mathbf{x}'\mathbf{A}\mathbf{x} = \text{vec}'(\mathbf{A})(\mathbf{x} \otimes \mathbf{x})$. Then

$$-0.5(\mathbf{y} - \boldsymbol{\mu})'\boldsymbol{\Omega}^{-1}(\mathbf{y} - \boldsymbol{\mu}) = -0.5\mathbf{y}'\boldsymbol{\Omega}^{-1}\mathbf{y} + \boldsymbol{\mu}'\mathbf{y} - 0.5\boldsymbol{\mu}'\boldsymbol{\Omega}^{-1}\boldsymbol{\mu}$$

$$= -0.5\text{vec}'(\boldsymbol{\Omega}^{-1})(\mathbf{y} \otimes \mathbf{y}) + \boldsymbol{\mu}'\mathbf{y} - 0.5\boldsymbol{\mu}'\boldsymbol{\Omega}^{-1}\boldsymbol{\mu} = \mathbf{p}'(\boldsymbol{\theta})\mathbf{K}(\mathbf{y}) - 0.5\boldsymbol{\mu}'\boldsymbol{\Omega}^{-1}\boldsymbol{\mu},$$

where

$$\boldsymbol{\theta} = \begin{bmatrix} \text{vec}(\boldsymbol{\Omega}) \\ \boldsymbol{\mu} \end{bmatrix}, \ \mathbf{p}(\boldsymbol{\theta}) = \begin{bmatrix} -0.5\text{vec}(\boldsymbol{\Omega}^{-1}) \\ \boldsymbol{\mu} \end{bmatrix}, \ \mathbf{K}(\mathbf{y}) = \begin{bmatrix} \mathbf{y} \otimes \mathbf{y} \\ \mathbf{y} \end{bmatrix}$$

are $(k^2 + k) \times 1$ vectors. Finally, letting $Q(\boldsymbol{\theta}) = 0.5\boldsymbol{\mu}'\boldsymbol{\Omega}^{-1}\boldsymbol{\mu} + 0.5k\ln(2\pi)$, we express the normal pdf in the form (6.53).

(b) The joint density is

$$f(\mathbf{y}_1, ..., \mathbf{y}_n; \boldsymbol{\mu}, \boldsymbol{\Omega}) = (2\pi)^{-nk/2} |\boldsymbol{\Omega}|^{-n/2} e^{-\frac{1}{2}\sum_{i=1}^{n}(\mathbf{y}_i-\boldsymbol{\mu})'\boldsymbol{\Omega}^{-1}(\mathbf{y}_i-\boldsymbol{\mu})}.$$

As follows from the previous algebra, the sufficient statistics are $\sum_{i=1}^{n} \mathbf{Y}_i$ and $\sum_{i=1}^{n} \mathbf{Y}_i \otimes \mathbf{Y}_i$. It is easy to show that the latter statistic can be expressed through $\sum_{i=1}^{n} \mathbf{Y}_i \mathbf{Y}_i'$ because $\text{vec}(\mathbf{Y}_i \mathbf{Y}_i') = \mathbf{Y}_i \otimes \mathbf{Y}_i$; see the Appendix.

Problems

1. (a) Suggest how to improve two unbiased estimators by taking their linear combination. (b) Try to improve an estimator of μ by considering the standard and weighted means, $\overline{Y}$ and $\sum w_i Y_i / \sum w_i$, where Y_i are iid observations with mean μ and variance σ^2, and weights $w_i > 0$ are fixed.

2. Derive the result of Example 6.67 by obtaining the variance of the weighted average and using the Cauchy inequality.

3. Let $\widehat{\theta}_B$ be UMVUE and $\widehat{\theta}_A$ is another unbiased estimator. Prove that $\widehat{\theta}_B$ and $\widehat{\theta}_A$ are uncorrelated. [Hint: Consider a linear combination.]

4. Using the Neyman factorization criterion (a) find a sufficient statistic for $\mathbf{Y} = (Y_1, ..., Y_N)$, where Y_i are iid and have binomial distribution with parameters p and n, and (b) the same as before but $n_i \neq$ const.

5. Find a sufficient statistic for (a) the uniform distribution on $(0, \theta)$, (b) the gamma distribution, and (c) the beta distribution.

6. Let $Y_i \overset{\text{iid}}{\sim} \mathcal{N}(\mu, 1)$. Apply the Rao–Blackwell theorem for $T = \overline{Y}$ and $U = \sum_{i=1}^{k} Y_i / k$ where $k \leq n$. [Hint: Use the formula for the conditional mean of the bivariate normal distribution.]

7. Prove that, if $\mathbf{U}^{m \times 1}$ is a sufficient statistic and $\mathbf{V} = \mathbf{T}(\mathbf{U})$ is a nonlinear one-to-one transformation $R^m \to R^m$ such that $|\mathbf{T}(\mathbf{u})| \neq 0$ for every $\mathbf{u}$, then $\mathbf{V}$ is a sufficient statistic. [Hint: Use the formula for the joint density upon transformation from Section 3.6.]

8. Prove that the sample mean and variance are sufficient statistics in Example 6.64.

9. (a) Prove that $T = \sum_{i=1}^{m} Y_i / m$, where $0 < m < n$, is a sufficient statistic in Examples 6.62 and 6.63. (b) Using the Rao–Blackwell theorem prove that the conditional mean $E(T|U = u)$ is an unbiased estimator, where $U = \sum_{i=1}^{n} Y_i / n$.

10. (a) Prove that $\max Y_i$ is a complete statistic for θ, where $Y_i \overset{\text{iid}}{\sim} \mathcal{R}(0, \theta)$. (b) Prove that $\min Y_i$ is a complete statistic for θ, where $Y_i \overset{\text{iid}}{\sim} \mathcal{R}(\theta, 1)$, $\theta < 1$.

11. Prove that order statistics are sufficient statistics when observations are iid.

12. (a) Show that the exponential, chi-square, gamma, and beta distributions belong to exponential family. (b) Does the mixture normal distribution belong to the exponential family?

13. Prove that formulas (6.50) produce mean and variance previously derived for distributions listed in Table 6.2.

14. Derive the variance of Y from Theorem 6.50 using MGF.

15. Find the MGF for the double canonical exponential family with multiple parameters. [Hint: Use Definition 6.83 and Section 3.10.2.]

16. (a) Show that a simple linear regression with normally distributed error terms from Example 6.72 specifies a distribution that belongs to the exponential family. (b) Prove that the OLS estimator is UMCUE using Theorem 6.86. [Hint: Use $n = 1$.]

17. The pdf of the multivariate exponential distribution is defined as $f(\mathbf{y}; \boldsymbol{\theta}) = p(\boldsymbol{\theta})e^{-\boldsymbol{\theta}'\mathbf{y}}$, where $\mathbf{y} = (Y_1, ..., Y_m)'$ and $\boldsymbol{\theta} = (1/\theta_1, ..., 1/\theta_m)'$ have positive components, and $p(\boldsymbol{\theta}) = \prod \theta_j^{-1}$ is the normalizing coefficient. Show that this distribution belongs to the exponential family and find the UMVUE for θ_j.

18. Run the code `betaMM` with $n = 10$ and explain why the means of the `ab` columns are not good summaries of the proximity of the MM estimates to the true values.

19. Use $\|\mathbf{y} - \bar{y}\mathbf{1}\|^2$ as the statistic instead $\|\mathbf{y}\|^2$ in Example 6.87.

20. Using the result of Example 6.89, find sufficient statistics for n iid observations from the bivariate normal distribution with five unknown parameters.

21. Prove that the multinomial distribution from Section 3.10.4 belongs to the exponential family.

6.9 Fisher information and the Cramér–Rao bound

In the previous section, we discussed linear and quadratic unbiased estimation. Here we lay out the foundation for nonlinear unbiased parameter estimation assuming that the density of observations is known up to a finite number of parameters. We introduce Fisher information, the milestone of statistical inference, and derive the implied Cramér–Rao bound for the variance (covariance matrix in the case of a parameter vector). Clearly, the unbiased estimator whose variance attains this bound is called efficient. Importantly, the theory of the Cramér–Rao lower bound does not require that observations have the same distribution or be independent.

We work with random observations $Y_1, Y_2, ..., Y_n$ that can be combined in a random vector $\mathbf{Y}$. Observations may be dependent. but the joint pdf is known up an m-dimensional unknown parameter vector, giving the notation $f(y_1, y_2, ..., y_n; \boldsymbol{\theta})$. The theory below applies to discrete random variables as well; then f is understood as pmf and integration is substituted with summation. First, we present the theory for a single parameter, which is denoted as θ. We then extend to the multivariate case. In a special case when observations are drawn at random from the same general population with pdf $f(y; \boldsymbol{\theta})$, we have

$$f(y_1, y_2, ..., y_n; \boldsymbol{\theta}) = \prod_{i=1}^{n} f(y_i; \boldsymbol{\theta}) \qquad (6.56)$$

with certain abuse of notation (symbol f used for individual and joint pdf).

To shorten the notation, we use $f(\mathbf{y}; \boldsymbol{\theta})$ to denote the pdf of the vector of observations $\mathbf{Y} = (Y_1, Y_2, ..., Y_n)$. We comply with the rule: lowercase is used to indicate argument of functions and densities and uppercase is used to indicate random variables (boldface indicates vectors). Some technical conditions such as continuity of the second derivative of the pdf with respect to parameters and differentiation under the integral are assumed. In addition, we assume that the parameter space $\boldsymbol{\Theta}$ is open in R^m, so that differentiation is defined for every $\boldsymbol{\theta} \in \boldsymbol{\Theta}$. Another important assumption is that the support of the pdf, the set of $\mathbf{y}$ for which $f(\mathbf{y}; \boldsymbol{\theta}) > 0$, does not depend on $\boldsymbol{\theta}$. Clearly, this does not hold for a sample drawn from the uniform distribution $\mathcal{R}(0, \theta)$. A rather technical condition is differentiation under the integral sign if the random variable is continuous (differentiation of the integral is equivalent to differentiation of the integrand): all practically important densities comply with this condition. Specifically, to ensure that integration and differentiation are interchangeable, sometimes referred to as differentiation under the integral sign, we assume that $\ln f(\mathbf{y}; \boldsymbol{\theta})$ is differentiable with respect to parameters up to the third order, and, for every $\boldsymbol{\theta}$, there is $\delta > 0$ and a nonnegative function $G(\mathbf{y})$ such that

$$\left| \frac{\partial^3 \ln f(\mathbf{y}; \boldsymbol{\theta})}{\partial \boldsymbol{\theta}^3} \bigg|_{\boldsymbol{\theta}=\boldsymbol{\theta}'} \right| \leq G(\mathbf{y}),$$

$$\int_{R^n} G(\mathbf{y}) f(\mathbf{y}; \boldsymbol{\theta}) d\mathbf{y} < \infty \quad \text{for all } \boldsymbol{\theta}' \text{ such that } \|\boldsymbol{\theta} - \boldsymbol{\theta}'\| \leq \delta. \qquad (6.57)$$

More detail with examples can be found in Casella and Berger [17], Section 2.4. For a discrete distribution, integral is replaced with sum.

The distribution must be uniquely identified by the parameter: if $f(\mathbf{y}; \boldsymbol{\theta}_1) = f(\mathbf{y}; \boldsymbol{\theta}_2)$ for all $\mathbf{y}$ then $\boldsymbol{\theta}_1 = \boldsymbol{\theta}_2$. This condition will be referred to as the identifiability condition.

6.9.1   One parameter

In this section we assume that the unknown parameter, θ, is a scalar; the next section discusses the multiparameter case.

Definition 6.90 *Fisher information. Let $\mathbf{Y}$ have the joint pdf $f(\mathbf{y}; \theta)$, where θ is an unknown parameter. The Fisher information is defined as*

$$\mathcal{I}(\theta) = \int_{R^n} \left(\frac{d \ln f(\mathbf{y}; \theta)}{d\theta} \right)^2 f(\mathbf{y}; \theta) d\mathbf{y} = E \left(\frac{d \ln f(\mathbf{Y}; \theta)}{d\theta} \right)^2 .$$

If observations Y_i are iid with the common density $f(y;\theta)$, the information in the sample $\mathbf{Y} = (Y_1, ..., Y_n)$ is

$$\mathcal{I}(\theta) = n\mathcal{I}_1(\theta), \tag{6.58}$$

where

$$\mathcal{I}_1(\theta) = \int_{R^1} \left(\frac{d \ln f(y;\theta)}{d\theta} \right)^2 f(y;\theta) dy = E \left(\frac{d \ln f(Y_1;\theta)}{d\theta} \right)^2 \tag{6.59}$$

is the Fisher information in a single observation.

Remarks:

1. Here $\mathbf{y}$ denotes the argument of the pdf (nonrandom) and $\mathbf{Y}$ denotes the vector of observations (random vector). We follow this convention throughout the book.

2. It is silently assumed that integrals and the respective expectations exist. The first line of the definition says that information may be derived either through integration or, equivalently, through expectation.

3. Note that notation $\mathcal{I}$ is used for the information contained in the entire sample and $\mathcal{I}_1$ is used for the information contained in a single observation (it does not matter what the observation is because they are all iid, we use Y_1); (6.58) follows from the fact that, for iid observations, the log joint density equals the sum of log individual densities.

4. Fisher information depends on the true parameter, θ.

5. When $\mathbf{Y}$ is a discrete random variable, the integral will be replaced with summation.

The theorem below suggests an alternative way to compute the Fisher information as the variance of the derivative or the expected value of the second derivative.

Theorem 6.91 *(a) The expectation of the derivative of the log pdf is zero,*

$$E \left(\frac{d \ln f(\mathbf{Y};\theta)}{d\theta} \right) = 0. \tag{6.60}$$

(b) The Fisher information can be alternatively obtained as

$$\mathcal{I}(\theta) = \operatorname{var} \left(\frac{d \ln f(\mathbf{Y};\theta)}{d\theta} \right), \tag{6.61}$$

$$\mathcal{I}(\theta) = -E \left(\frac{d^2 \ln f(\mathbf{Y};\theta)}{d\theta^2} \right). \tag{6.62}$$

Proof. (a) Differentiating

$$1 = \int_{R^n} f(\mathbf{y};\theta) d\mathbf{y}$$

with respect to θ one obtains

$$0 = \int_{R^n} \frac{df(\mathbf{y};\theta)}{d\theta} d\mathbf{y} = \int_{R^n} \frac{d\ln f(\mathbf{y};\theta)}{d\theta} f(\mathbf{y};\theta) d\mathbf{y} = E\left(\frac{d\ln f(\mathbf{Y};\theta)}{d\theta}\right),$$

which proves (6.60). (b) To prove (6.61), we note that $E(X^2) = \mathrm{var}(X)$ if $E(X) = 0$, where X is thought of as $d\ln f(\mathbf{Y};\theta)/d\theta$. To prove (6.62) we use the previously derived identity (6.60). Differentiate

$$0 = \int_{R^n} \frac{d\ln f(\mathbf{y};\theta)}{d\theta} f(\mathbf{y};\theta) d\mathbf{y}$$

with respect to θ,

$$0 = \int_{R^n} \frac{d^2 \ln f(\mathbf{y};\theta)}{d\theta^2} f(\mathbf{y};\theta) d\mathbf{y} + \int_{R^n} \frac{d\ln f(\mathbf{y};\theta)}{d\theta} \frac{df(\mathbf{y};\theta)}{d\theta} d\mathbf{y}$$

$$= E\left(\frac{d^2 \ln f(\mathbf{Y};\theta)}{d\theta^2}\right) + \int_{R^n} \left(\frac{d\ln f(\mathbf{y};\theta)}{d\theta}\right)^2 f(\mathbf{y};\theta) d\mathbf{y} = E\left(\frac{d^2 \ln f(\mathbf{Y};\theta)}{d\theta^2}\right) + \mathcal{I}(\theta),$$

which implies (6.62).

Example 6.92 *Information for the normal distribution. Find the Fisher information for a random sample $Y_1, Y_2, ..., Y_n$ from $\mathcal{N}(\mu, \sigma^2)$ assuming that variance σ^2 is known. Demonstrate that all three formulas, (6.59), (6.61), and (6.62) yield the same answer.*

Solution. Here μ acts as θ. Since Y_i are iid, we compute the Fisher information in Y_1 and then multiply it by n as in (6.58). The log pdf is

$$\ln f(y; \mu) = -\frac{1}{2} \ln(2\pi\sigma^2) - \frac{1}{2\sigma^2}(y - \mu)^2$$

with the derivative $d\ln f(y;\mu)/d\mu = (y - \mu)/\sigma^2$. First, find the information in a single observation using the original formula (6.59) $\mathcal{I}_1(\mu) = E\left((Y_1 - \mu)/\sigma^2\right)^2 = 1/\sigma^2$. Second, use formula (6.61) $\mathcal{I}_1(\mu) = \mathrm{var}((Y_1 - \mu)/\sigma^2) = 1/\sigma^2$. Third, use formula (6.62), $d^2 \ln f(y;\mu)/d\mu^2 = -1/\sigma^2$. and therefore $\mathcal{I}_1(\mu) = 1/\sigma^2$. Thus, all three formulas produce the same answer. Finally, the information contained in the entire sample is $\mathcal{I}(\mu) = n/\sigma^2$. □

We mentioned that the Fisher information can be defined for a discrete distribution with pdf replaced with the pmf and integration replaced with sum – below is an example.

Example 6.93 *Information for a series of Bernoulli trials.* *Find the Fisher information in n iid Bernoulli outcomes $Y_1, Y_2, ..., Y_n$.*

Solution. Write the population pmf as $f(y; p) = p^y(1-p)^{1-y}$ where $y = 0, 1$. Then

$$\frac{d \ln f(y;p)}{dp} = \frac{d}{dp}\left(y \ln p + (1-y) \ln(1-p)\right) = \frac{y}{p} - \frac{1-y}{1-p} = \frac{y}{p(1-p)} - \frac{1}{1-p}.$$

Therefore, the Fisher information in one Bernoulli outcome is

$$\mathcal{I}_1(p) = \text{var}\left(\frac{Y}{p(1-p)} - \frac{1}{1-p}\right) = \text{var}\left(\frac{Y}{p(1-p)}\right) = \frac{\text{var}(Y)}{p^2(1-p)^2} = \frac{1}{p(1-p)}.$$

The Fisher information in a sample of size n is n times as large, $\mathcal{I}(p) = n\mathcal{I}_1(p) = n/[p(1-p)]$.

□

The following Cauchy inequality is instrumental to derive the Cramér–Rao lower bound for the variance of an unbiased estimator.

Lemma 6.94 *The Cauchy inequality in the integral/expectation form.* *Let $\mathbf{Y}$ be a random vector and $h = h(\mathbf{Y})$ and $g = g(\mathbf{Y})$ be random variables as functions of $\mathbf{Y}$. Then, assuming that the expectations exist,*

$$E^2(hg) \leq E(h^2)E(g^2). \tag{6.63}$$

The inequality turns into equality if and only if $h = \lambda g$, where λ is a constant.

Proof. Introduce a function of λ defined as $P(\lambda) = E(h - \lambda g)^2$. Represent

$$P(\lambda) = E(h - \lambda g)^2 = E(h^2) - 2\lambda E(hg) + \lambda^2 E(g^2),$$

a quadratic polynomial of λ if $E(g^2) \neq 0$. If $E(g^2) = 0$, inequality (6.63) turns into equality. Since $P(\lambda) \geq 0$ the discriminant must be nonpositive, $E^2(hg) - E(h^2)E(g^2) \leq 0$, which is equivalent to (6.63). If $h = \lambda g$, then the left-hand side equals the right-hand side. Reversely, if (6.63) turns into equality, then $h - \lambda g = 0$ for all $\mathbf{Y}$, or $h = \lambda g$. □

Now we are ready to formulate the main result.

Theorem 6.95 *The Cramér–Rao lower bound of the variance of the unbiased estimator.* *If $\widehat{\theta}(\mathbf{y})$ is an unbiased estimator of θ and $\mathcal{I}(\theta) \neq 0$, then*

$$\text{var}(\widehat{\theta}) \geq \frac{1}{\mathcal{I}(\theta)}, \tag{6.64}$$

or in words, the lower bound of the variance is the reciprocal of the Fisher information.

Proof. In the Cauchy inequality (6.63), letting

$$h(\mathbf{Y}) = (\widehat{\theta}(\mathbf{Y}) - \theta), \quad g(\mathbf{Y}) = \frac{d \ln f(\mathbf{Y}; \theta)}{d\theta},$$

we obtain

$$\left(\int_{R^n} (\widehat{\theta}(\mathbf{y}) - \theta) \frac{d \ln f(\mathbf{y}; \theta)}{d\theta} f(\mathbf{y}; \theta) d\mathbf{y} \right)^2$$

$$\leq \int_{R^n} (\widehat{\theta}(\mathbf{y}) - \theta)^2 f(\mathbf{y}; \theta) d\mathbf{y} \times \int_{R^n} \left(\frac{d \ln f(\mathbf{y}; \theta)}{d\theta} \right)^2 f(\mathbf{y}; \theta) d\mathbf{y}. \qquad (6.65)$$

Now prove that

$$\int_{R^n} (\widehat{\theta}(\mathbf{y}) - \theta) \frac{d \ln f(\mathbf{y}; \theta)}{d\theta} f(\mathbf{y}; \theta) d\mathbf{y} = 1.$$

Indeed, differentiating

$$\int_{R^n} \widehat{\theta}(\mathbf{y}) f(\mathbf{y}; \theta) d\mathbf{y} = \theta$$

with respect to θ, we obtain

$$\int_{R^n} \widehat{\theta}(\mathbf{y}) \frac{df(\mathbf{y}; \theta)}{d\theta} d\mathbf{y} = \int_{R^n} \widehat{\theta}(\mathbf{y}) \frac{d \ln f(\mathbf{y}; \theta)}{d\theta} f(\mathbf{y}; \theta) d\mathbf{y} = 1.$$

Looking back into formula (6.65), we notice that the left-hand side is 1, the first term in the right-hand side is $\mathrm{var}(\widehat{\theta})$ and the second term is $\mathcal{I}(\theta)$, the lower bound (6.64) is derived. □

The concept of UMVUE was introduced in the previous section. Fisher information provides another way to prove that an unbiased estimator is UMVUE, see more discussion following Example 6.107.

Definition 6.96 *An unbiased estimator $\widehat{\theta}$ with minimum variance among all unbiased estimators of θ will be referred to as the UMVUE in the Cramér–Rao sense.*

Be advised that the variance of the UMVUE, as defined in Section 6.8.1, may be greater than the Cramér–Rao lower bound. In other words, the bound may not be attainable. However, if the variance of an unbiased estimator is $1/\mathcal{I}(\theta)$, the estimator is UMVUE. When referring to UMVUE, we will silently assume in this section the Cramér–Rao sense.

Looking back into the proof of the theorem, we conclude that the equality in (6.64) occurs when there exists a function $h(\theta)$ such that $h(\theta)(\widehat{\theta}(\mathbf{y}) - \theta) = d \ln f(\mathbf{y}; \theta)/d\theta$. But this equation can be treated as the ordinary differential equation of the first order with the only solution $f(\mathbf{y}; \theta) = \exp(p(\theta)K(\mathbf{y}) + S(\mathbf{y}) - q(\theta))$, where $K(y) = \widehat{\theta}(\mathbf{y})$, $p(\theta) = \int h(\theta) d\theta$ and $q(\theta) = \int \theta h(\theta) d\theta$, which belongs to the exponential family (6.47). Thus, one concludes that the Cramér–Rao lower bound attains equality for the exponential family of distributions.

Theorem 6.97 *The arithmetic average is UMVUE for μ, p, λ, and θ for the following distributions: (a) normal with known variance, (b) Bernoulli, (c) Poisson and (d) exponential.*

Proof. (a) Example 6.92 derived that $\mathcal{I}(\mu) = n/\sigma^2$, so the variance of any unbiased estimator of μ cannot be smaller than σ^2/n. But $\mathrm{var}(\overline{Y}) = \sigma^2/n$ and therefore $\overline{Y}$ is UMVUE. (b) The Fisher information for the Bernoulli random sample $Y_1, Y_2, ..., Y_n$ was derived in Example 6.93, $\mathcal{I}(p) = np^{-1}(1-p)^{-1}$. Therefore the Cramér–Rao lower bound for the variance of an unbiased estimator is $p(1 - p)/n$. But $\mathrm{var}(\overline{Y}) = p(1 - p)/n$. Therefore, the arithmetic average is UMVUE. (c) The pmf for the Poisson distribution is $f(y; \lambda) = \lambda^y e^{-\lambda}/y!$ where $y = 0, 1, 2...$ (see Section 1.7), and so

$$\frac{d \ln f(y; \lambda)}{d\lambda} = \frac{y}{\lambda} - 1.$$

Since $\mathrm{var}(Y) = \lambda$ for $Y \sim \mathcal{P}(\lambda)$, we have $\mathcal{I}(\lambda) = n/\lambda$. But $\mathrm{var}(\overline{Y}) = \lambda/n$, so we deduce that $\overline{Y}$ is UMVUE. (d) If Y belongs to the exponential distribution with the scale parameter θ (Section 2.4) then $f(y; \theta) = \theta^{-1}e^{-y/\theta}$ and $\mathcal{I}(\theta) = n/\theta^2$. But $E(\overline{Y}) = \theta$ and $\mathrm{var}(\overline{Y}) = \theta^2/n$. Therefore, $\overline{Y}$ is UMVUE for the scale parameter θ. See Example 6.99 where the rate $\lambda = 1/\theta$ is estimated. ☐

When assumptions formulated at the beginning of the section are not satisfied, the Cramér–Rao lower bound may not work, as illustrated in an example below.

Example 6.98 ***When the Cramér–Rao bound does not hold.*** *Show that the Cramér–Rao lower bound does not hold for the uniform distribution, $X \sim \mathcal{R}(0, \theta)$.*

Solution. Let $Y_1, Y_2, ..., Y_n$ be a random sample from a uniform distribution on the interval $[0, \theta]$, where θ is an unknown positive parameter. Then $f(y; \theta) = 1/\theta$ for $0 \leq y \leq \theta$ and 0 elsewhere. Formally, $d \ln f(y; \theta)/d\theta = -1/\theta$, so that $\mathcal{I}(\theta) = n/\theta^2$. That is, the lower bound for the variance of an unbiased estimator of θ is θ^2/n. Now consider an unbiased estimator $\widehat{\theta} = 2\overline{Y}$ with the variance

$$\mathrm{var}(\widehat{\theta}) = \frac{4}{n}\mathrm{var}(Y) = \frac{4\theta^2}{12n} = \frac{\theta^2}{3n} < \frac{\theta^2}{n}.$$

This means that the variance of $\widehat{\theta}$ is smaller than the lower bound. The Cramér–Rao bound does not hold because the uniform distribution with unknown upper limit violates the assumption that the support of the distribution does not depend on the parameter. Formally, $d \ln f(y; \theta)/d\theta \neq -1/\theta$. Indeed, $d \ln f(y; \theta)/d\theta = 0$ for $y < 0$ or $y > \theta$ and $E(d \ln f(y; \theta)/d\theta) \neq 0$ when $0 \leq y \leq \theta$, which contradicts (6.60). ☐

Unbiased estimation of a nonlinear function of a parameter is not a trivial task because unbiasedness usually does not remain upon nonlinear transformation.

As follows from the previous theorem, the scale parameter of the exponential distribution is UMVUE, but finding an unbiased estimator for the rate parameter is a more difficult problem.

Example 6.99 *Efficient estimation of the rate parameter in the exponential distribution. Let Y_i be iid observations from the exponential distribution with unknown rate, λ. As was shown in Example 6.13 the estimator $\widehat{\lambda} = (n - 1)/(n\overline{Y})$ is unbiased, and as follows from Example 6.71 this estimator has minimum variance among all unbiased estimators. Prove that the variance of this estimator does not reach the Cramér–Rao lower bound.*

Solution. Following the derivation in Example 6.13, find

$$
E\left(\frac{1}{\overline{Y}}\right)^2 = n^2 E\left(\frac{1}{\sum_{i=1}^n Y_i}\right)^2 = 2\lambda n^2 \int_0^\infty \frac{1}{s^2}\frac{1}{2^n \Gamma(n)}(2\lambda s)^{n-1}e^{-\lambda s}ds
$$
$$
= \frac{n^2}{(n-1)(n-2)}\lambda^2.
$$

Therefore,

$$
\mathrm{var}(\widehat{\lambda}) = \left(\frac{n-1}{n}\right)^2 \mathrm{var}\left(\frac{1}{\overline{Y}}\right) = \frac{1}{n-2}\lambda^2.
$$

However, $\mathcal{I}(\lambda) = n/\lambda^2$, so the Cramér–Rao lower bound is $\lambda^2/n < \lambda^2/(n-2)$, the bound is not attained. This example illustrates that a UMVUE may not reach the Cramér–Rao lower bound.

6.9.2 Multiple parameters

The theory of the Cramér–Rao bound remains conceptually unchanged with multiple parameters – the only difference is that the variance is replaced with the covariance matrix. Specifically, let $\mathbf{Y}$ have the joint pdf $f(\mathbf{y}; \boldsymbol{\theta})$, where $\boldsymbol{\theta}$ is the unknown m-dimensional parameter vector. It is expected that all conditions formulated at the beginning of the section must be fulfilled: (1) The set of points $\mathbf{y}$ where $f(\mathbf{y}; \boldsymbol{\theta}) > 0$ (support) does not depend on $\boldsymbol{\theta}$, (2) the parameter space $\boldsymbol{\Theta}$ is open. (3) The identifiability condition holds. (4) Differentiation under the integral sign is permissible.

Definition 6.100 *The Fisher information is defined as the $m \times m$ matrix*

$$
\mathcal{I}(\boldsymbol{\theta}) = \int_{R^n}\left(\frac{\partial \ln f(\mathbf{y}; \boldsymbol{\theta})}{\partial \boldsymbol{\theta}}\right)\left(\frac{\partial \ln f(\mathbf{y}; \boldsymbol{\theta})}{\partial \boldsymbol{\theta}}\right)' f(\mathbf{y}; \boldsymbol{\theta})d\mathbf{y}
$$
$$
= E\left(\left(\frac{\partial \ln f(\mathbf{Y}; \boldsymbol{\theta})}{\partial \boldsymbol{\theta}}\right)\left(\frac{\partial \ln f(\mathbf{Y}; \boldsymbol{\theta})}{\partial \boldsymbol{\theta}}\right)'\right).
$$

If observations $\mathbf{Y}_i$ are iid with the population density $f(\mathbf{y}; \boldsymbol{\theta})$, then the information in the entire sample $\mathbf{Y} = (\mathbf{Y}_1, ..., \mathbf{Y}_n)$ is

$$\mathcal{I}(\boldsymbol{\theta}) = n\mathcal{I}_1(\boldsymbol{\theta}),$$

where

$$\mathcal{I}_1(\boldsymbol{\theta}) = \int_{R^n} \left(\frac{\partial \ln f(\mathbf{y}; \boldsymbol{\theta})}{\partial \boldsymbol{\theta}}\right) \left(\frac{\partial \ln f(\mathbf{y}; \boldsymbol{\theta})}{\partial \boldsymbol{\theta}}\right)' f(\mathbf{y}; \boldsymbol{\theta}) dy$$

$$= E\left(\left(\frac{\partial \ln f(\mathbf{Y}_1; \boldsymbol{\theta})}{\partial \boldsymbol{\theta}}\right) \left(\frac{\partial \ln f(\mathbf{Y}_1; \boldsymbol{\theta})}{\partial \boldsymbol{\theta}}\right)'\right)$$

is Fisher information in a single observation.

With some ambiguity in notation, f denotes the joint density of $\mathbf{Y}$ in the first integral and the same f denotes the density of $\mathbf{Y}_1$ in the second integral. This definition is illustrated below with the normal distribution. In Example 6.92, only μ was unknown. If variance σ^2 is also unknown, the Fisher information turns into a matrix, and so Definition 6.100 must apply.

Example 6.101 *Fisher information for the normal distribution.* *Derive the 2×2 Fisher information matrix for $\boldsymbol{\theta} = (\mu, \sigma^2)$ based on the sample of n iid observations $Y_i \sim \mathcal{N}(\mu, \sigma^2)$, $i = 1, 2, ..., n$.*

Solution. The log density is a function of two parameters (parameter σ^2 is treated as whole, not squared σ), $\ln f(y; \mu, \sigma^2) = -0.5 \ln(2\pi\sigma^2) - 0.5\sigma^{-2}(y - \mu)^2$, where the constant term is omitted, with the 2×1 derivative vector

$$\frac{\partial \ln f(y; \boldsymbol{\theta})}{\partial \boldsymbol{\theta}} = \begin{bmatrix} \frac{\partial \ln f(y; \mu, \sigma^2)}{\partial \mu} \\ \frac{\partial \ln f(y; \mu, \sigma^2)}{\partial \sigma^2} \end{bmatrix} = \begin{bmatrix} \frac{1}{\sigma^2}(y - \mu) \\ -\frac{1}{2\sigma^2} + \frac{1}{2\sigma^4}(y - \mu)^2 \end{bmatrix}.$$

Find the information matrix in a single observation (the subscript $_1$ is omitted for brevity)

$$\mathcal{I}_1(\boldsymbol{\theta}) = E \begin{bmatrix} \frac{1}{\sigma^2}(Y - \mu) \\ -\frac{1}{2\sigma^2} + \frac{1}{2\sigma^4}(Y - \mu)^2 \end{bmatrix} \begin{bmatrix} \frac{1}{\sigma^2}(Y - \mu) \\ -\frac{1}{2\sigma^2} + \frac{1}{2\sigma^4}(Y - \mu)^2 \end{bmatrix}'$$

$$= \frac{1}{\sigma^2} E \begin{bmatrix} \frac{1}{\sigma^2}(Y - \mu)^2 & \frac{1}{\sigma}(Y - \mu)\left(-\frac{1}{2\sigma} + \frac{1}{2\sigma^3}(Y - \mu)^2\right) \\ \frac{1}{\sigma}(Y - \mu)\left(-\frac{1}{2\sigma} + \frac{1}{2\sigma^3}(Y - \mu)^2\right) & \left(-\frac{1}{2\sigma} + \frac{1}{2\sigma^3}(Y - \mu)^2\right)^2 \end{bmatrix}.$$

In order to simplify the expectation, introduce the normalized normal variable, $Z = (Y - \mu)/\sigma \sim \mathcal{N}(0, 1)$. Using previously derived formulas $E(Z^2) = 1$, $E(Z^3) =$

0, and $E(Z^4) = 3$ (see Example 2.31) find the needed expectations,

$$E\left(\frac{1}{\sigma^2}(Y-\mu)^2\right) = E(Z^2) = 1,$$

$$E\left(\frac{1}{\sigma}(Y-\mu)\left(-\frac{1}{2\sigma}+\frac{1}{2\sigma^3}(Y-\mu)^2\right)\right) = E\left[Z\left(-\frac{1}{2\sigma}+\frac{Z^2}{2\sigma}\right)\right]$$

$$= -\frac{1}{2\sigma}E(Z)+\frac{1}{2\sigma}E(Z^3) = 0,$$

$$E\left(-\frac{1}{2\sigma}+\frac{1}{2\sigma^3}(Y-\mu)^2\right)^2 = E\left(-\frac{1}{2\sigma}+\frac{Z^2}{2\sigma}\right)^2$$

$$= \frac{1}{4\sigma^2}-\frac{E(Z^2)}{2\sigma^2}+\frac{E(Z^4)}{4\sigma^2} = \frac{1}{2\sigma^2}.$$

Finally, the 2×2 Fisher information matrix takes the form

$$\mathcal{I}(\boldsymbol{\theta}) = n\begin{bmatrix} \frac{1}{\sigma^2} & 0 \\ 0 & \frac{1}{2\sigma^4} \end{bmatrix}.$$

Note that the off-diagonal elements are zero. Obviously, this result may be viewed as the information matrix for the mean model (6.16) under the normal assumption. □

The following theorem is a straightforward generalization of Theorem 6.91 (note that we use bold $\boldsymbol{\theta}$ to denote a vector and therefore the symbol ∂ for partial differentiation now).

Theorem 6.102 (a) *The expectation of the log pdf derivative is zero,*

$$E\left(\frac{\partial \ln f(\mathbf{Y};\boldsymbol{\theta})}{\partial \boldsymbol{\theta}}\right) = \mathbf{0}. \tag{6.66}$$

(b) *The Fisher information can alternatively be derived as*

$$\mathcal{I}(\boldsymbol{\theta}) = \text{cov}\left(\frac{\partial \ln f(\mathbf{Y};\boldsymbol{\theta})}{\partial \boldsymbol{\theta}}\right), \tag{6.67}$$

$$\mathcal{I}(\boldsymbol{\theta}) = -E\left(\frac{\partial^2 \ln f(\mathbf{Y};\boldsymbol{\theta})}{\partial \boldsymbol{\theta}^2}\right). \tag{6.68}$$

Example 6.103 *Fisher information for the normal distribution via the second derivative.* Derive the Fisher information matrix in Example 6.101 using the double differentiation formula (6.68).

Solution. We have

$$\frac{\partial^2 \ln f(y;\boldsymbol{\theta})}{\partial \boldsymbol{\theta}^2} = \begin{bmatrix} \frac{\partial^2 \ln f}{\partial \mu^2} & \frac{\partial^2 \ln f}{\partial \mu \partial \sigma^2} \\ \frac{\partial^2 \ln f}{\partial \mu \partial \sigma^2} & \frac{\partial^2 \ln f}{\partial \sigma^4} \end{bmatrix} = \begin{bmatrix} -\frac{1}{\sigma^2} & -\frac{1}{\sigma^4}(y-\mu) \\ -\frac{1}{\sigma^4}(y-\mu) & \frac{1}{2\sigma^4}-\frac{1}{\sigma^6}(y-\mu)^2 \end{bmatrix}.$$

Using the Z-representation, the expectations simplify to

$$E\left(\frac{\partial^2 \ln f(Y;\boldsymbol{\theta})}{\partial \boldsymbol{\theta}^2}\right) = \begin{bmatrix} -\frac{1}{\sigma^2} & -\frac{1}{\sigma^3}E(Z) \\ -\frac{1}{\sigma^3}E(Z) & \frac{1}{2\sigma^4} - \frac{1}{\sigma^4}E\left(Z^2\right) \end{bmatrix} = -\begin{bmatrix} \frac{1}{\sigma^2} & 0 \\ 0 & -\frac{1}{2\sigma^4} \end{bmatrix},$$

and arrive at the same $\mathcal{I}(\boldsymbol{\theta})$. $\qquad\square$

The rest of the section uses matrix algebra heavily – we refer the reader to the Appendix. In particular, we note that $\mathbf{A} \geq \mathbf{0}$ means that matrix $\mathbf{A}$ is a nonnegative matrix, i.e. $\mathbf{x}'\mathbf{A}\mathbf{x} \geq 0$ for every vector $\mathbf{x}$.

Lemma 6.104 *Matrix Cauchy inequality in the expectation form.* *Let $\mathbf{Y}$ be a $n \times 1$ random vector, and $\mathbf{h} = \mathbf{h}(\mathbf{Y})$ and $\mathbf{g} = \mathbf{g}(\mathbf{Y})$ be $m \times 1$ random vectors. Suppose that the $m \times m$ matrix $E(\mathbf{gg}')$ is nonsingular. Then*

$$E\left(\mathbf{hh}'\right) - E\left(\mathbf{hg}'\right) E^{-1}(\mathbf{gg}')E\left(\mathbf{gh}'\right) \geq \mathbf{0}, \tag{6.69}$$

and the inequality turns into an equality if and only if $\mathbf{h} = \lambda\mathbf{g}$, where λ is a constant.

Proof. Define matrix $\mathbf{M} = E(\mathbf{hg}')E^{-1}(\mathbf{gg}')$ and consider the $m \times m$ nonnegative definite matrix

$$E\left[(\mathbf{h} - \mathbf{Mg})(\mathbf{h} - \mathbf{Mg})'\right] = E\left(\mathbf{hh}'\right) - \mathbf{M}E(\mathbf{gh}') - E(\mathbf{hg}')\mathbf{M}' + \mathbf{M}E(\mathbf{gg}')\mathbf{M}'.$$

This matrix is nonnegative definite because matrix $(\mathbf{h} - \mathbf{Mg})(\mathbf{h} - \mathbf{Mg})'$ is nonnegative definite. Using some matrix algebra, we obtain

$$\begin{aligned} \mathbf{M}E(\mathbf{gh}') &= E(\mathbf{hg}')E^{-1}(\mathbf{gg}')E(\mathbf{gh}'), \\ E(\mathbf{hg}')\mathbf{M}' &= E(\mathbf{hg}')E^{-1}(\mathbf{gg}')E(\mathbf{gh}'), \\ \mathbf{M}E(\mathbf{gg}')\mathbf{M}' &= E(\mathbf{hg}')E^{-1}(\mathbf{gg}')E(\mathbf{gg}')E^{-1}(\mathbf{gg}')E(\mathbf{gh}') \\ &= E(\mathbf{hg}')E^{-1}(\mathbf{gg}')E(\mathbf{gh}'). \end{aligned}$$

Collect the terms

$$E\left(\mathbf{hh}'\right) - E(\mathbf{hg}')E^{-1}(\mathbf{gg}')E(\mathbf{gh}') - E(\mathbf{hg}')E^{-1}(\mathbf{gg}')E(\mathbf{gh}')$$
$$+E(\mathbf{hg}')E^{-1}(\mathbf{gg}')E(\mathbf{gh}') = E\left(\mathbf{hh}'\right) - E(\mathbf{hg}')E^{-1}(\mathbf{gg}')E(\mathbf{gh}') \geq \mathbf{0}$$

and, finally, arrive at the inequality (6.69).

Remark 6.105 *In this lemma, we assumed that vectors $\mathbf{h}$ and $\mathbf{g}$ are functions of a random vector $\mathbf{Y}$. However, inequality (6.69) holds in the general case when $\mathbf{g}$ and $\mathbf{h}$ are any random vectors. The same comment is true for the univariate inequality (6.63).*

Theorem 6.106 *The Cramér–Rao lower bound for the covariance matrix of an unbiased estimator.* *If $\widehat{\boldsymbol{\theta}}(\mathbf{y})$ is an unbiased estimator of $\boldsymbol{\theta}$ and $\mathcal{I}(\boldsymbol{\theta})$ is nonsingular then*

$$\mathrm{cov}(\widehat{\boldsymbol{\theta}}) \geq \mathcal{I}^{-1}(\boldsymbol{\theta}), \tag{6.70}$$

or in words, the lower bound for the covariance matrix is the inverse Fisher information matrix.

Proof. Let in the Cauchy inequality (6.69) $\mathbf{h} = \widehat{\boldsymbol{\theta}}(\mathbf{y}) - \boldsymbol{\theta}$ and $\mathbf{g} = \frac{\partial \ln f(\mathbf{y};\boldsymbol{\theta})}{\partial \boldsymbol{\theta}}$. Then

$$
\begin{aligned}
E\left(\mathbf{h}\mathbf{h}'\right) &= \int_{R^n} (\widehat{\boldsymbol{\theta}}(\mathbf{y}) - \boldsymbol{\theta})(\widehat{\boldsymbol{\theta}}(\mathbf{y}) - \boldsymbol{\theta})' f(\mathbf{y};\boldsymbol{\theta}) d\mathbf{y} = \mathrm{cov}(\widehat{\boldsymbol{\theta}}), \\
E\left(\mathbf{h}\mathbf{g}'\right) &= \int_{R^n} (\widehat{\boldsymbol{\theta}}(\mathbf{y}) - \boldsymbol{\theta}) \left(\frac{\partial \ln f(\mathbf{y};\boldsymbol{\theta})}{\partial \boldsymbol{\theta}}\right)' f(\mathbf{y};\boldsymbol{\theta}) d\mathbf{y}, \\
E(\mathbf{g}\mathbf{g}') &= \int_{R^n} \left(\frac{\partial \ln f(\mathbf{y};\boldsymbol{\theta})}{\partial \boldsymbol{\theta}}\right) \left(\frac{\partial \ln f(\mathbf{y};\boldsymbol{\theta})}{\partial \boldsymbol{\theta}}\right)' f(\mathbf{y};\boldsymbol{\theta}) d\mathbf{y} = \mathcal{I}(\boldsymbol{\theta}).
\end{aligned}
$$

Prove that $E\left(\mathbf{h}\mathbf{g}'\right) = \mathbf{I}$, the identity $m \times m$ matrix. Indeed, differentiating the unbiasedness equation, $\int_{R^n} \widehat{\boldsymbol{\theta}}(\mathbf{y}) f(\mathbf{y};\boldsymbol{\theta}) d\mathbf{y} = \boldsymbol{\theta}$, with respect to $\boldsymbol{\theta}$ we obtain (refer to matrix calculus in the Appendix) $\int_{R^n} \widehat{\boldsymbol{\theta}}(\mathbf{y}) \left(\frac{\partial f(\mathbf{y};\boldsymbol{\theta})}{\partial \boldsymbol{\theta}}\right)' d\mathbf{y} = \mathbf{I}$. Using (6.66), represent the left-hand side as

$$\int_{R^n} \widehat{\boldsymbol{\theta}}(\mathbf{y}) \frac{\partial \ln f(\mathbf{y};\boldsymbol{\theta})}{\partial \boldsymbol{\theta}} f(\mathbf{y};\boldsymbol{\theta}) d\mathbf{y} = \int_{R^n} (\widehat{\boldsymbol{\theta}}(\mathbf{y}) - \boldsymbol{\theta}) \frac{\partial \ln f(\mathbf{y};\boldsymbol{\theta})}{\partial \boldsymbol{\theta}} f(\mathbf{y};\boldsymbol{\theta}) d\mathbf{y} = \mathbf{I}.$$

Plugging all these expressions in (6.69), we arrive at the Cramér–Rao inequality. □

The extension of the UMVUE to multidimensional parameter is straightforward: $\widehat{\boldsymbol{\theta}} = \widehat{\boldsymbol{\theta}}(\mathbf{y})$ is called the UMCUE in the Cramér–Rao sense if inequality (6.70) turns into equality.

Inequality (6.70) is equivalent to saying that the lower bound of the variance of the unbiased estimator $\mathbf{a}'\widehat{\boldsymbol{\theta}}$ is $\mathbf{a}'\mathcal{I}^{-1}(\boldsymbol{\theta})\mathbf{a}$ for any fixed $m \times 1$ vector $\mathbf{a}$. As a corollary, if, for example, one is interested in estimation of the jth component of the true parameter vector, other components are viewed as nuisance parameters, and $\widehat{\boldsymbol{\theta}}$ is UMCUE, then $\widehat{\theta}_j$ is UMVUE for θ_j. The following example illustrates this comment.

Example 6.107 *A UMVUE for the mean and variance.* (a) *Prove that the arithmetic average is a UMVUE of the mean in the normal distribution when σ^2 is unknown.* (b) *Prove that the unbiased variance estimator $\widehat{\sigma}_0^2 = n^{-1}\sum_{i=1}^{n}(Y_i - \mu)^2$ is a UMVUE in the Cramér–Rao sense for σ^2 when μ is known but $\widehat{\sigma}^2 = (n-1)^{-1}\sum_{i=1}^{n}(Y_i - \overline{Y})^2$ does not reach the lower bound when μ is unknown and yet is a UMVUE as follows from Example 6.72.*

Proof. (a) The 2×2 Fisher information matrix for a random sample from the normal distribution with unknown μ and σ^2 was derived in Example 6.101. Therefore, the covariance matrix of an unbiased estimator of μ and σ^2 is bounded from below by

$$\begin{bmatrix} \sigma^2/n & 0 \\ 0 & 2\sigma^4/n \end{bmatrix}.$$

Since $\overline{Y}$ is unbiased and $\mathrm{var}(\overline{Y}) = \sigma^2/n$, the arithmetic average is an UMVUE. Note that the zero off-diagonal element of the information matrix make $\overline{Y}$ UMVUE regardless of knowing or not knowing σ^2. (b) If μ is known, the lower bound for an unbiased estimator of σ^2 is $2\sigma^4/n$. But $\mathrm{var}(\widehat{\sigma}_0^2) = 2\sigma^4/n$, so this estimator is a UMVUE in the Cramér–Rao sense. However the unbiased estimator $\widehat{\sigma}^2$ does not reach the lower bound because $\mathrm{var}(\widehat{\sigma}^2) = 2\sigma^4/(n-1) > 2\sigma^4/n$. $\square$

This example highlights the difference between the Lehmann–Scheffé and Fisher information approaches to prove that an unbiased estimator has minimum variance: while $\widehat{\sigma}^2$ is a UMVUE, as follows from the Lehmann–Scheffé theorem, it does not reach the Cramér-Rao lower bound. Although the Fisher information criterion is more stringent, it allows one to approximate the variances of estimators needed for statistical hypothesis testing and confidence intervals (see the next chapter). Also, as we shall learn from the next section, the Cramér–Rao lower bound applies in large sample and, as such, becomes very instrumental in statistics.

The implication of the zero off-diagonal elements of the information matrix

Divide the $m \times 1$ parameter vector $\boldsymbol{\theta}$ into two parts. Suppose that the first part, the $m_1 \times 1$ vector $\boldsymbol{\theta}_1$ is the parameter of interest and the second part, $m_2 \times 1$ vector $\boldsymbol{\theta}_2$ is not of interest, a nuisance parameter. Partition the information matrix accordingly:

$$\mathcal{I}(\boldsymbol{\theta}) = \begin{bmatrix} \mathcal{I}_{11}(\boldsymbol{\theta}) & \mathcal{I}_{12}(\boldsymbol{\theta}) \\ \mathcal{I}_{12}'(\boldsymbol{\theta}) & \mathcal{I}_{22}(\boldsymbol{\theta}) \end{bmatrix}.$$

If the nuisance parameter vector were known, the information matrix for $\boldsymbol{\theta}_1$ would be $\mathcal{I}_{11}(\boldsymbol{\theta}_1) = \mathcal{I}_{11}(\boldsymbol{\theta})$ with the lower bound

$$\mathrm{cov}(\widehat{\boldsymbol{\theta}}_1) \geq \mathcal{I}_{11}^{-1}(\boldsymbol{\theta}).$$

If $\boldsymbol{\theta}_2$ is unknown and estimated, the lower covariance bound for $\boldsymbol{\theta}_1$ is the first $m_1 \times m_1$ block of the inverse matrix $\mathcal{I}(\boldsymbol{\theta})$. As follows from the block matrix inverse formula the $m_1 \times m_1$ block of the inverse matrix can be expressed through matrix blocks as follows (the argument $\boldsymbol{\theta}$ is omitted for brevity):

$$\mathcal{I}_{11}^{-1} + \mathcal{I}_{11}^{-1}\mathcal{I}_{12} \left(\mathcal{I}_{22} - \mathcal{I}_{12}'\mathcal{I}_{11}^{-1}\mathcal{I}_{12} \right)^{-1} \mathcal{I}_{12}'\mathcal{I}_{11}^{-1}.$$

Matrix $\mathcal{I}_{22} - \mathcal{I}_{12}'\mathcal{I}_{11}^{-1}\mathcal{I}_{12}$ is positive definite and matrix

$$\mathcal{I}_{11}^{-1}\mathcal{I}_{12}\left(\mathcal{I}_{22} - \mathcal{I}_{12}'\mathcal{I}_{11}^{-1}\mathcal{I}_{12}\right)^{-1} \times \mathcal{I}_{12}'\mathcal{I}_{11}^{-1}$$

is nonnegative definite. Two conclusions can be drawn from the above formula:
(1) Adding new parameters cannot improve estimation, or more precisely, adding
new parameters does not make the lower bound smaller. (2) If the cross infor-
mation is zero, $\mathcal{I}_{12} = \mathbf{0}$, the nuisance parameter does not affect the precision of
the parameter of interest. This can be also seen from the obvious block-diagonal
matrix inverse,

$$\mathcal{I}^{-1}(\boldsymbol{\theta}) = \begin{bmatrix} \mathcal{I}_{11}^{-1}(\boldsymbol{\theta}) & \mathbf{0} \\ \mathbf{0} & \mathcal{I}_{22}^{-1}(\boldsymbol{\theta}) \end{bmatrix}.$$

Example 6.108 *Cramér–Rao bound for regression.* *Derive the Cramér–
Rao lower bound for the regression parameters in the bivariate normal distribution.*

Solution. Since we are interested in the regression parameters, the bivariate
normal distribution is expressed as the conditional normal distribution of $Y|X =
x$ times the marginal distribution of X, or symbolically $Y|X = x \sim \mathcal{N}(\alpha+\beta x, \sigma^2)$
and $X \sim \mathcal{N}(\mu_x, \sigma_x^2)$; see Section 3.5.1. Then the log joint density, up to a constant
term, takes the form

$$\ln f(x, y; \alpha, \beta, \sigma^2, \mu_x, \sigma_x^2) = -\frac{1}{2}\ln \sigma^2 - \frac{1}{2\sigma^2}(y - \alpha - \beta x)^2 - \frac{1}{2}\ln \sigma_x^2 - \frac{1}{2\sigma_x^2}(x - \mu_x)^2.$$

Making simplifying notation $U = (Y - \alpha - \beta X)/\sigma$ and $V = (X - \mu_x)/\sigma_x$, where
U and V are independent standard normal random variables, write the derivative
of the log density as

$$\left.\frac{\partial \ln f}{\partial \boldsymbol{\theta}}\right|_{x=X, y=Y} = \begin{bmatrix} \frac{\partial \ln f}{\partial \alpha} \\ \frac{\partial \ln f}{\partial \beta} \\ \frac{\partial \ln f}{\partial \sigma^2} \\ \frac{\partial \ln f}{\partial \mu_x} \\ \frac{\partial \ln f}{\partial \sigma_x^2} \end{bmatrix} = \begin{bmatrix} -\frac{1}{\sigma^2}(Y - \alpha - \beta X) \\ -\frac{1}{\sigma^2}(Y - \alpha - \beta X)X \\ -\frac{1}{2\sigma^2} + \frac{1}{2\sigma^4}(Y - \alpha - \beta X)^2 \\ -\frac{1}{\sigma_x^2}(X - \mu_x) \\ \frac{1}{2\sigma_x^4}(X - \mu_x)^2 \end{bmatrix} = \begin{bmatrix} -\frac{U}{\sigma} \\ -\frac{U(\sigma_x V + \mu_x)}{\sigma} \\ -\frac{1}{2\sigma^2} + \frac{U^2}{2\sigma^2} \\ -\frac{V}{\sigma_x} \\ \frac{V^2}{2\sigma_x^2} \end{bmatrix}.$$

Make use of the following formulas

$$E(U^2) = E(V^2) = 1, \quad E(U^2 V) = E(U^3) = E(UV) = 0, \quad E(U^4) = 3.$$

Now the information matrix, as the covariance of the derivative vector, is easy to derive

$$
\mathcal{I}_1 =
\begin{bmatrix}
\sigma^{-2} & \mu_x \sigma^{-2} & 0 & 0 & 0 \\
\mu_x \sigma^{-2} & (\sigma_x^2 + \mu_x^2)\sigma^{-2} & 0 & 0 & 0 \\
0 & 0 & \sigma^{-4}/2 & 0 & 0 \\
0 & 0 & 0 & \sigma_x^{-2} & 0 \\
0 & 0 & 0 & 0 & \sigma_x^{-4}/2
\end{bmatrix}.
$$

Since the information matrix is block-diagonal, the Cramér–Rao lower bound given a sample (X_i, Y_i) from a bivariate normal distribution of size n is given by

$$
\mathrm{cov} \geq \frac{1}{n}
\begin{bmatrix}
\sigma^2 \sigma_x^{-2}(\sigma_x^2 + \mu_x^2) & -\mu_x \sigma^2/\sigma_x^2 & 0 & 0 & 0 \\
-\mu_x \sigma^2/\sigma_x^2 & \sigma^2/\sigma_x^2 & 0 & 0 & 0 \\
0 & 0 & 2\sigma^4 & 0 & 0 \\
0 & 0 & 0 & \sigma_x^2 & 0 \\
0 & 0 & 0 & 0 & 2\sigma_x^4
\end{bmatrix}.
$$

In particular, if $\widehat{\beta}$ is an unbiased estimator of β, we must have $\mathrm{var}(\widehat{\beta}) \geq \sigma^2/(n\sigma_x^2)$.

The lower bound for a biased estimator and the MSE

The Cramér–Rao lower bound for covariance matrix can be generalized to biased estimators.

Theorem 6.109 *The Cramér–Rao lower bound of the covariance matrix for a biased estimator.* *If $\widehat{\boldsymbol{\theta}}(\mathbf{y})$ is a biased estimator of $\boldsymbol{\theta}$ and $\mathcal{I}(\boldsymbol{\theta})$ is nonsingular, then*

$$
\mathrm{cov}(\widehat{\boldsymbol{\theta}}) \geq E\left(\frac{\partial}{\partial \boldsymbol{\theta}} E_{\boldsymbol{\theta}}(\widehat{\boldsymbol{\theta}})\right) \mathcal{I}^{-1}(\boldsymbol{\theta}) E\left(\frac{\partial}{\partial \boldsymbol{\theta}} E_{\boldsymbol{\theta}}(\widehat{\boldsymbol{\theta}})\right)'. \tag{6.71}
$$

Proof. Employ the matrix Cauchy inequality and follow the proof for an unbiased estimator. Differentiate an obvious equation

$$
\int_{R^n} \left(\widehat{\boldsymbol{\theta}}(\mathbf{y}) - E_{\boldsymbol{\theta}}(\widehat{\boldsymbol{\theta}})\right) f(\mathbf{y}; \boldsymbol{\theta}) d\mathbf{y} = \mathbf{0},
$$

where $E_{\boldsymbol{\theta}}(\widehat{\boldsymbol{\theta}}) = \int_{R^n} \widehat{\boldsymbol{\theta}}(\mathbf{y}) f(\mathbf{y}; \boldsymbol{\theta}) d\mathbf{y}$, with respect to $\boldsymbol{\theta}$ to obtain

$$
\int_{R^n} \left(\widehat{\boldsymbol{\theta}}(\mathbf{y}) - E_{\boldsymbol{\theta}}(\widehat{\boldsymbol{\theta}})\right) \left(\frac{\partial \ln f(\mathbf{y}; \boldsymbol{\theta})}{\partial \boldsymbol{\theta}}\right)' f(\mathbf{y}; \boldsymbol{\theta}) d\mathbf{y} = E\left(\frac{\partial}{\partial \boldsymbol{\theta}} E_{\boldsymbol{\theta}}(\widehat{\boldsymbol{\theta}})\right).
$$

Now let $\mathbf{h} = \widehat{\boldsymbol{\theta}}(\mathbf{y}) - E_{\boldsymbol{\theta}}(\widehat{\boldsymbol{\theta}})$ and $\mathbf{g} = \partial \ln f(\mathbf{y}; \boldsymbol{\theta})/\partial \boldsymbol{\theta}$ and apply the matrix Cauchy inequality to obtain $\operatorname{cov}(\widehat{\boldsymbol{\theta}}) - E\left(\frac{\partial}{\partial \boldsymbol{\theta}} E_{\boldsymbol{\theta}}(\widehat{\boldsymbol{\theta}})\right) \mathcal{I}^{-1}(\boldsymbol{\theta}) E\left(\frac{\partial}{\partial \boldsymbol{\theta}} E_{\boldsymbol{\theta}}(\widehat{\boldsymbol{\theta}})\right)' \geq 0$, which is equivalent to inequality (6.71). $\square$

Obviously, the unbiased version of the Cramér–Rao bound is a special case of (6.71) because, when $\widehat{\boldsymbol{\theta}}$ is unbiased, we have $\partial E_{\boldsymbol{\theta}}(\widehat{\boldsymbol{\theta}})/\partial \boldsymbol{\theta} = \partial \boldsymbol{\theta}/\partial \boldsymbol{\theta} = \mathbf{I}$. For a single parameter θ inequality (6.71) turns into $\operatorname{var}(\widehat{\theta}) \geq \left(\frac{d}{d\theta} E(\widehat{\theta})\right)^2 / \mathcal{I}(\theta)$. The inequality turns into equality when $\widehat{\theta} = \text{const}$.

Theorem 6.109 yields to obtain the lower bound for the MSE of any estimator

$$\text{MSE} \geq \left(E(\widehat{\boldsymbol{\theta}}) - \boldsymbol{\theta}\right)\left(E(\widehat{\boldsymbol{\theta}}) - \boldsymbol{\theta}\right)' + E\left(\frac{\partial}{\partial \boldsymbol{\theta}} E_{\boldsymbol{\theta}}(\widehat{\boldsymbol{\theta}})\right) \mathcal{I}^{-1}(\boldsymbol{\theta}) E\left(\frac{\partial}{\partial \boldsymbol{\theta}} E_{\boldsymbol{\theta}}(\widehat{\boldsymbol{\theta}})\right)'.$$

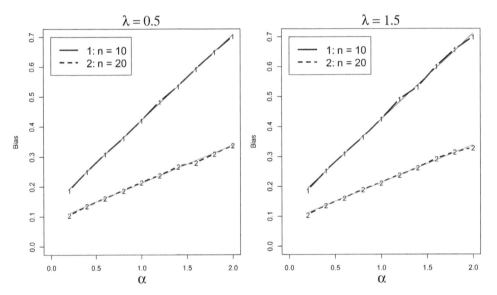

Figure 6.21: *Bias of the MM estimator $\widehat{\alpha}_{MM}$ as a function of the true α. The bias can be well fit with a linear function $a_n + b_n \alpha$.*

Example 6.110 *MM for gamma distribution.* *(1) Derive the 2×2 information matrix for the gamma distribution. (2) Derive the MM estimators $\widehat{\alpha}_{MM}$ and $\widehat{\beta}_{MM}$. (3) Use simulations to estimate the bias of $\widehat{\alpha}_{MM}$ for $n = 10$ and 20. (4) Derive the bias-adjusted estimator, $\widetilde{\alpha}_{MM}$. (5) Plot the empirical variance of the bias-adjusted $\widetilde{\alpha}_{MM}$ and its lower Cramér–Rao bound for $n = 10$ and 20.*

Solution. (1) As follows from Section 2.6, the log density function of the gamma distribution takes the form

$$\ln f(x; \alpha, \beta) = \alpha \ln \lambda - \ln \Gamma(\alpha) + (\alpha - 1) \ln x - \lambda x.$$

Find $\mathcal{I}_1$ as the expectation of the second derivatives

$$\frac{\partial^2 \ln f}{\partial \alpha^2} = -\frac{d^2}{d\alpha^2} \ln \Gamma(\alpha), \quad \frac{\partial^2 \ln f}{\partial \alpha \partial \lambda} = \frac{1}{\lambda}, \quad \frac{\partial^2 \ln f}{\partial \lambda^2} = -\frac{\alpha}{\lambda^2}.$$

Function $\psi_1(\alpha) = \frac{d^2}{d\alpha^2} \ln \Gamma(\alpha)$ is called the `trigamma` function and it is a built-in function in R (in the terminology of Abramowitz and Stegun [3], the `psigamma` function, hence the notation). These special functions were used in Example 6.88. The Fisher information matrix is expressed via ψ_1 as:

$$\mathcal{I}_1(\alpha, \beta) = \begin{bmatrix} \psi_1(\alpha) & -\frac{1}{\lambda} \\ -\frac{1}{\lambda} & \frac{\alpha}{\lambda^2} \end{bmatrix}.$$

(2) The MM estimator is found from equating the empirical and theoretical moments, $E(X) = \alpha/\lambda$ and $\operatorname{var}(X) = \alpha/\lambda^2$, with an obvious solution $\widehat{\alpha}_{MM} = \overline{X}^2/\widehat{\sigma}^2$ and $\widehat{\lambda}_{MM} = \overline{X}/\widehat{\sigma}^2$, where $\widehat{\sigma}^2$ is the unbiased sample variance. (3) The bias of $\widehat{\alpha}_{MM}$ is estimated using simulations (see `gammaInf` code, `job=1`) for the sample sizes $n = 10$ and $n = 20$ and $\lambda = 0.5$ and $\lambda = 1.5$, and presented in Figure 6.21. (4) The bias of the MM estimator can be well approximated with a linear function of α and λ with coefficients depending on n. If a_n and b_n are the intercept and slope from the linear fit, we set $E(\widehat{\alpha}_{MM}) - \alpha = a_n + b_n \alpha$ and so the bias-adjusted MM estimator takes the form $\widetilde{\alpha}_{MM} = (\widehat{\alpha}_{MM} - a_n)/(1 + b_n)$. As follows from our simulations, we can set $a_{10} = 0.13, b_{10} = 0.29$ and $a_{20} = 0.085, b_{20} = 0.126$.

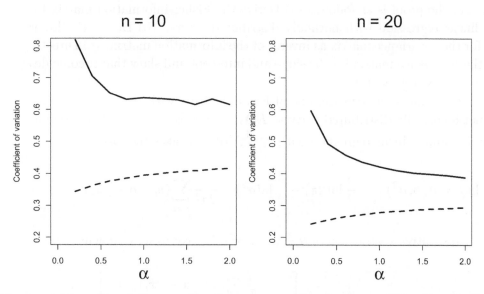

Figure 6.22: *Simulations with the gamma distribution: The Cramer-Rao bound and empirical variances for the MM estimator of α (generated by* `gammaInf(job=2)`*).*

(5) The Cramér–Rao lower bound for the variance of an unbiased estimator of α with n iid observations is n^{-1} times the $(1,1)$th element of the inverse matrix $\mathcal{I}_1(\alpha, \beta)$, that is,

$$\text{var}(\widetilde{\alpha}_{MM}) \geq \frac{1}{n} \frac{\alpha}{\alpha\psi(\alpha) - 1}.$$

Interestingly, the lower variance bound does not depend on λ. Simulations and the lower bound for $n = 10$ and $n = 20$ with $\lambda = 1$ are shown in Figure 6.22 using 50,000 simulations. Since the variances increase with α, we report the coefficient of variation, the ratio of SD to the true value α. For both values of the sample size, the lower bound for the coefficient of variation is smaller than the empirical one; with n increasing the difference gets smaller.

Gauss–Markov theorem for simple linear regression under normal assumption

Theorem 6.50 asserts that the OLS estimators are BLUE = best *linear* unbiased estimators of intercept and slope in the linear model (6.35) with fixed $x_1, x_2, ..., x_n$. Using the Fisher information and Cramér–Rao lower bound, we can strengthen this result: if, in addition, the residuals are normally distributed, the OLS estimators are the best in the family of all unbiased estimators (linear and nonlinear), or shortly

$$\text{OLS} = \text{UMVUE}. \tag{6.72}$$

The plan of the proof is as follows. (a) Derive the Fisher information matrix for simple linear regression with normally distributed errors. (b) Derive the lower bound for the covariance matrix as inverse of the information matrix. (c) Directly derive the covariance matrix for the slope and intercept and show that it coincides with the matrix from step (b).

(a) The derivation follows the line of Example 6.101 with the difference that Y_i are not identically distributed because $E(Y_i) = \alpha + \beta x_i \neq$ const. The log joint density for simple linear regression with $\varepsilon_i \overset{iid}{\sim} \mathcal{N}(0, \sigma^2)$ takes the form

$$\ln f(\mathbf{y}; \alpha, \beta, \sigma^2) = -\frac{n}{2}\ln(2\pi) - \frac{n}{2}\ln(\sigma^2) - \frac{1}{2\sigma^2}\sum_{i=1}^{n}(y_i - \alpha - \beta x_i)^2.$$

The 3×1 derivative vector, where $\boldsymbol{\theta} = (\alpha, \beta, \sigma^2)$, is easy to write down:

$$\frac{\partial \ln f(\mathbf{y}; \boldsymbol{\theta})}{\partial \boldsymbol{\theta}} = \begin{bmatrix} \frac{\partial \ln f}{\partial \alpha} \\ \frac{\partial \ln f}{\partial \beta} \\ \frac{\partial \ln f}{\partial \sigma^2} \end{bmatrix} = \begin{bmatrix} \frac{1}{\sigma^2}\sum(y_i - \alpha - \beta x_i) \\ \frac{1}{\sigma^2}\sum(y_i - \alpha - \beta x_i)x_i \\ -\frac{n}{2\sigma^2} + \frac{1}{2\sigma^4}\sum(y_i - \alpha - \beta x_i)^2 \end{bmatrix}.$$

Find the information matrix as the expectation of the covariance of this derivative

$$\mathcal{I}(\boldsymbol{\theta}) = E \begin{bmatrix} \frac{1}{\sigma^2} \sum (Y_i - \alpha - \beta x_i) \\ \frac{1}{\sigma^2} \sum (Y_i - \alpha - \beta x_i) x_i \\ -\frac{n}{2\sigma^2} + \frac{1}{2\sigma^4} \sum (Y_i - \alpha - \beta x_i)^2 \end{bmatrix} \begin{bmatrix} \frac{1}{\sigma^2} \sum (Y_i - \alpha - \beta x_i) \\ \frac{1}{\sigma^2} \sum (Y_i - \alpha - \beta x_i) x_i \\ -\frac{n}{2\sigma^2} + \frac{1}{2\sigma^4} \sum (Y_i - \alpha - \beta x_i)^2 \end{bmatrix}'.$$

Express the elements in terms of iid standard normal random variables $Z_i \sim \mathcal{N}(0,1)$ as in Example 6.101:

$$\frac{1}{\sigma} \sum (Y_i - \alpha - \beta x_i) = \frac{1}{\sigma} \sum \varepsilon_i = \sum Z_i,$$

$$\frac{1}{\sigma} \sum (Y_i - \alpha - \beta x_i) x_i = \frac{1}{\sigma} \sum \varepsilon_i x_i = \sum Z_i x_i$$

$$\frac{1}{\sigma^2} \sum (Y_i - \alpha - \beta x_i)^2 = \frac{1}{\sigma^2} \sum \varepsilon_i^2 = \sum Z_i^2.$$

The following expectations for covariances are straightforward to derive:

$$E\left(\sum Z_i\right) = 0, \quad E\left(\sum Z_i\right)^2 = E\left(\sum Z_i^2\right) = n,$$

$$\mathrm{cov}\left(\sum Z_i, \sum Z_i x_i\right) = E\left(\sum Z_i^2 x_i\right) = \sum x_i,$$

$$\mathrm{cov}\left(\sum Z_i, \sum Z_i^2\right) = \mathrm{cov}\left(\sum Z_i x_i, \sum Z_i^2\right) = 0$$

$$\mathrm{cov}\left(\sum Z_i x_i, \sum Z_i x_i\right) = E\left(\sum Z_i^2 x_i^2\right) = \sum x_i^2.$$

The information matrix for simple linear regression is given by

$$\mathcal{I}(\boldsymbol{\theta}) = \frac{1}{\sigma^2} \begin{bmatrix} n & \sum x_i & 0 \\ \sum x_i & \sum x_i^2 & 0 \\ 0 & 0 & n(2/\sigma^2) \end{bmatrix}.$$

Note that, as with the mean model from Example 6.101, the intercept and slope are orthogonal to the variance parameter.

(b) Since the 3×3 information matrix has a block-diagonal structure, to find the lower bound for the covariance matrix for α and β, it suffices to invert the 2×2 block. Thus, from the Cramér–Rao lower bound for α and β, we deduce that the covariance matrix of any unbiased estimator $\widetilde{\alpha}$ and $\widetilde{\beta}$ has a lower bound,

$$\mathrm{cov}\left(\begin{bmatrix} \widetilde{\alpha} \\ \widetilde{\beta} \end{bmatrix}\right) \geq \sigma^2 \begin{bmatrix} n & \sum x_i \\ \sum x_i & \sum x_i^2 \end{bmatrix}^{-1} = \begin{bmatrix} \frac{\sigma^2 \sum x_i^2}{n \sum x_i^2 - (\sum x_i)^2} & -\frac{\sigma^2 \sum x_i}{n \sum x_i^2 - (\sum x_i)^2} \\ -\frac{\sigma^2 \sum x_i}{n \sum x_i^2 - (\sum x_i)^2} & \frac{n\sigma^2}{n \sum x_i^2 - (\sum x_i)^2} \end{bmatrix}.$$

(c) The $(1,1)$th and the $(2,2)$th elements of the 2×2 covariance matrix (variances) were derived in Theorem 6.50 in a slightly different form. Using some algebra, from (6.30) it is obvious that the variances are the same as the diagonal elements of the lower bound matrix.

$$\operatorname{cov}(\widehat{\alpha}, \widehat{\beta}) = \operatorname{cov}(\overline{Y} - \widehat{\beta}\overline{x}, \widehat{\beta}) = \operatorname{cov}(\overline{Y}, \widehat{\beta}) - \overline{x} \times \operatorname{var}(\widehat{\beta}) = 0 - \overline{x}\frac{n\sigma^2}{n\sum x_i^2 - (\sum x_i)^2}$$

$$= -\frac{\sigma^2 \sum x_i}{n\sum x_i^2 - (\sum x_i)^2}.$$

To prove $\operatorname{cov}(\overline{Y}, \widehat{\beta}) = 0$, we express $\overline{Y} = \alpha + \beta\overline{x} + \overline{\varepsilon}$, which means that it suffices to prove that $E\left(\overline{\varepsilon}\sum(\varepsilon_i - \overline{\varepsilon})d_i\right) = 0$, where $d_i = x_i - \overline{x}$. We have

$$E\left(\overline{\varepsilon}\sum(\varepsilon_i - \overline{\varepsilon})d_i\right) = [E(\varepsilon_i^2)/n - E(\overline{\varepsilon}^2)]d_i = [\sigma^2/n - \sigma^2/n]d_i = 0.$$

Thus, we have proved that the Cramér–Rao lower bound is the same as the covariance matrix of the OLS estimators of slope and intercept, which implies that the OLS estimators are UMVUE.

Problems

1. Show that conditions on differentiation under the integral sign (6.57) hold for (a) the normal distribution $\mathcal{N}(\mu, 1)$ and (b) the Poisson distribution.

2. Does Fisher information change upon parameter reparametrization? Specifically, let $\tau = \tau(\theta)$ be a new parameter where τ is a one-to-one function. (a) Do sets of points/curves in R^2 defined as $(\theta, \mathcal{I}(\theta))$ and $(\theta, \mathcal{I}(\tau))$ coincide? (b) Use the Poisson distribution with $\tau = e^\lambda$ as an example (plot the two curves).

3. Find the Fisher information for parameter μ in the Laplace distribution with the density $f(y; \mu) = (\lambda/2)e^{-\lambda|y-\mu|}$ where $\lambda > 0$ is known. [Hint: $\lambda\operatorname{sign}(Y - \mu)$ is a discrete random variable which takes values $-\lambda$ or λ with probability $1/2$.]

4. Prove the information in the random sample is n times larger than from an individual sample expressed in (6.58).

5. In view of Remark 6.105, prove that $\operatorname{cov}'(Y, \mathbf{X})\operatorname{cov}^{-1}(\mathbf{X})\operatorname{cov}(Y, \mathbf{X}) \leq \operatorname{var}(Y)$, where Y is a random variable and $\mathbf{X}$ is a random vector and connect the result to the definition of the coefficient of determination (4.12), where Y and $\mathbf{X}$ have a multivariate normal distribution.

6. Find the Fisher information for λ in the Poisson distribution.

7. For the Poisson distribution, (a) check (6.60) directly, (b) show that formulas (6.61) and (6.62) produce the same result.

8. As a continuation of Example 6.99, prove that $\widetilde{\lambda} = ((n-2)/n)\,/\overline{Y}$ is biased but has a smaller MSE than $\widehat{\lambda}$.

9. Prove that, if a vector estimator is a UMCUE, then each component is a UMVUE.

10. Prove that if $\widehat{\boldsymbol{\theta}}$ is UMCUE for $\boldsymbol{\theta}$ then $\widehat{\theta}_k - \widehat{\theta}_j$ is UMVUE for $\theta_k - \theta_j$.

11. Use The generalized method of moments (Section 6.2.1) with $M_j(\alpha, \lambda) = X^{k_j}\,e^{-\theta_j X}$, $j = 1, 2$, to estimate parameters of the gamma distribution assuming that k_j and θ_j are given. [Hint: Use the fact that $\int_0^\infty \frac{\lambda^\alpha}{\Gamma(\alpha)} x^{\alpha-1} e^{-\lambda x} x^k e^{-\theta x} dx = (\lambda+\theta)^{-\alpha-k}\,\lambda^\alpha \frac{\Gamma(\alpha+k)}{\Gamma(\alpha)}.$]

12. (a) Prove that, if $\mathbf{X}$ and $\mathbf{Y}$ are independent random vectors with pdfs dependent on the common $\boldsymbol{\theta}$, then the Fisher information in $(\mathbf{X}, \mathbf{Y})$ is the sum of the information in $\mathbf{X}$ and $\mathbf{Y}$. (b) Does the information additivity hold for the sum, $\mathbf{X} + \mathbf{Y}$? [Hint: Consider $X \sim \mathcal{N}(\mu, \sigma^2)$ and $Y \sim \mathcal{N}(\mu, \tau^2)$ with σ^2 and τ^2 being known.]

6.10 Maximum likelihood

6.10.1 Basic definitions and examples

The method of maximum likelihood (ML) is the primary method of estimation in statistics: it is general and has optimal statistical properties when the sample is large. ML gives rise to interval estimation and hypothesis testing. ML was developed by Ronald Fisher, a British statistician, regarded as the founder of modern statistics.

We start with a fundamental question: What would be most *probable* value of parameter $\boldsymbol{\theta} \in R^m$ given observation $\mathbf{Y}$ drawn from the population with pdf $f(\mathbf{y}; \boldsymbol{\theta})$? A remark on the notation: as in the previous section, we use uppercase letters, such as $\mathbf{Y}$, to denote observations and corresponding lowercase letters, such as $\mathbf{y}$, to denote the argument of the pdf function.

To begin, consider estimation of unknown probability p given n independent Bernoulli outcomes $Y_1, Y_2, ..., Y_n$ with $\Pr(Y_i = 1) = p$. Here $\mathbf{Y}$ is the n-dimensional vector of outcomes, $\mathbf{Y} = (Y_1, Y_2, ..., Y_n)$. Due to independence, the joint probability is the product of individual probabilities,

$$L(p) = \prod_{i=1}^{n} p^{Y_i}(1-p)^{1-Y_i},$$

called the likelihood function. We draw attention to the notation: since the point of interest is p, we treat $Y_1, Y_2, ..., Y_n$ as fixed, so the joint likelihood is viewed as a function of p. A typical likelihood function, $L(p)$ as a function of $p \in (0, 1)$

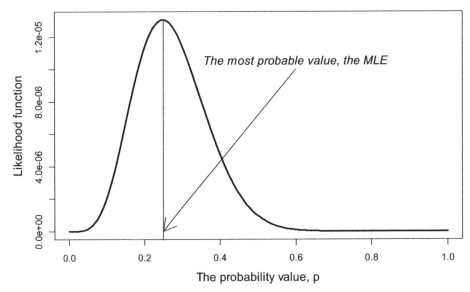

Figure 6.23: *The most probable value is the MLE, the point where the likelihood function reaches its maximum.*

is depicted in Figure 6.23. We define the most *probable* value of p as the point where $L(p)$ reaches its maximum. This point is called the *maximum likelihood estimator (MLE)*.

Definition 6.111 *Let the vector of observations* $\mathbf{Y}$ *have joint pdf* $f(\mathbf{y}; \boldsymbol{\theta})$. *Define the likelihood function as* $L(\boldsymbol{\theta}) = f(\mathbf{Y}; \boldsymbol{\theta})$ *and the log-likelihood function as* $l(\boldsymbol{\theta}) = \ln f(\mathbf{Y}; \boldsymbol{\theta})$. *The MLE,* $\widehat{\boldsymbol{\theta}}_{ML}$ *is the maximizer of* L *(or, equivalently, l),*

$$\widehat{\boldsymbol{\theta}}_{ML} = \arg\max L(\boldsymbol{\theta}) = \arg\max l(\boldsymbol{\theta}).$$

Obviously, different data yields different estimate. That is, the MLE is a function of data, $\widehat{\boldsymbol{\theta}}_{ML} = \widehat{\boldsymbol{\theta}}_{ML}(\mathbf{Y})$, and estimator. Recall that, while estimator is a function of data, the *estimate* is a value of that function. Note that the likelihood function is merely the joint pdf, the notation $L(\boldsymbol{\theta})$ is simply driven by the fact that the point of interest is $\boldsymbol{\theta}$ when $\mathbf{Y}$ are held fixed. ML applies to not necessarily iid data; however, its optimal asymptotic properties hold under the iid assumption. The reason the log-likelihood function is more convenient than the likelihood itself is that in the iid case, the product turns into a sum facilitating computation and study of statistical properties considerably. Differentiation of a sum is much easier than differentiation of a product.

Example 6.112 *The MLE for the Bernoulli probability.* *Find the MLE for* $p \in (0, 1)$ *from the series of n iid Bernoulli experiments,* $\Pr(Y_i = 1) = p$.

Solution. The likelihood function, $L(p)$, is given above and the log-likelihood function is given by $l(p) = \sum_{i=1}^{n} (Y_i \ln p + (1 - Y_i) \ln(1 - p))$. To find its maximum, differentiate and set the derivative to zero,

$$\frac{dl}{dp} = \sum_{i=1}^{n} \left(\frac{Y_i}{p} - \frac{1 - Y_i}{1 - p} \right) = 0.$$

It is easy to see that the MLE is the proportion of successes, called the Bernoulli proportion, $\widehat{p}_{ML} = \sum_{i=1}^{n} Y_i / n = \overline{Y}$. This value indeed returns the maximum of the log-likelihood because l is a concave function, that is, its second derivative is nonpositive,

$$\frac{d^2 l}{dp^2} = -\sum_{i=1}^{n} \left(\frac{Y_i}{1 - p} + \frac{1 - Y_i}{(1 - p)^2} \right) \leq 0$$

for every $Y_1, Y_2, ..., Y_n$.   □

It is more feasible to solve equations than find an extremum of a function: this leads to the notion of the score equation.

Definition 6.113 *The score equation is defined as $\partial l / \partial \boldsymbol{\theta} = \mathbf{0}$.*

For example, in the iid case when $\mathbf{Y}_i$ have common pdf $f(\mathbf{y}, \boldsymbol{\theta})$ the score equation takes the form

$$\sum_{i=1}^{n} \frac{\partial \ln f(\mathbf{Y}_i; \boldsymbol{\theta})}{\partial \boldsymbol{\theta}} = \mathbf{0}.$$

As we know (see Appendix), the derivative of the function being zero, sometimes referred to as the first-order condition, is a necessary, but not sufficient condition for the maximum. Indeed, the derivative is zero at the minimum and saddle point as well. When the log-likelihood function is concave, which is rarely the case, like in the example above, the first-order condition turns into a sufficient condition for the maximum. We must comfort the reader: major numerical methods of maximization of the log-likelihood that solve the score equation never yield a minimum or a saddle point but may find a *local maximum*, which may be not the global one (more detail to follow).

Maximum likelihood has a useful *invariance property*: if the point of interest is a function of a parameter, say, $\boldsymbol{\tau} = \boldsymbol{\tau}(\boldsymbol{\theta})$, where $\boldsymbol{\tau} : R^m \to R^p$ and $p \leq m$, then the MLE of $\boldsymbol{\tau}$ is $\boldsymbol{\tau}(\widehat{\boldsymbol{\theta}}_{ML})$, that is,

$$\widehat{\boldsymbol{\tau}}_{ML} = \boldsymbol{\tau}(\widehat{\boldsymbol{\theta}}_{ML}). \tag{6.73}$$

This property is illustrated by the following example.

Example 6.114 *Variance of the binomial proportion.* *Find the MLE of the variance of the binomial proportion.*

Solution. From the example above, $\widehat{p}_{ML} = \overline{Y}$, but $\mathrm{var}(\widehat{p}_{ML}) = \mathrm{var}(Y)/n = p(1-p)/n$. Therefore, we may view $\tau(p) = p(1-p)/n$. Thus, the MLE of the variance of the binomial proportion can be estimated as $\overline{Y}(1 - \overline{Y})/n$.

Example 6.115 *The Buffon's needle estimation problem. Estimate the ratio of the length of the needle to the distance between the lines from n random throws (see Example 3.68).*

Solution. We want to estimate $\rho = L/D$ given the number of intersections, m in n throws. Denote by $p(\rho)$ the theoretical probability of intersection given in Example 3.68. Let 1 denote the event if the needle intersects the line and 0 otherwise. Then the estimation problem reduces to recovering the Bernoulli probability $p = p(\rho)$ from m/n. That is, the MLE, $\widehat{p}_{ML}$ is the solution of the equation $p = p(\rho) = m/n$. Two cases should be distinguished: If $m/n < 2/\pi$ then the solution is easy, $\widehat{\rho}_{ML} = \pi m/(2n) < 1$. Otherwise, $\widehat{\rho}_{ML} > 1$ is the solution of the transcendent equation

$$\frac{2}{\pi} \arccos \frac{1}{\rho} + \frac{2\rho}{\pi}\left[1 - \sqrt{1 - \frac{1}{\rho^2}}\right] = \frac{m}{n}.$$

This equation can be solved by Newton's algorithm

$$\rho_{k+1} = \rho_k - \frac{\frac{2}{\pi}\arccos\frac{1}{\rho_k} + \frac{2\rho_k}{\pi}\left[1 - \sqrt{1 - \frac{1}{\rho_k^2}}\right] - \frac{m}{n}}{\frac{2}{\pi}\frac{\rho_k - \sqrt{\rho_k^2 - 1}}{\rho_k}}, \quad k = 0, 1, 2, ...$$

Iterations have sense: if for the current ρ_k, the difference between the left and right-hand side, is positive ρ_k decreases. Otherwise, it increases. The R code is found in file `bufprob.r`. The number of throws and the true ρ are the arguments of the function. First, we generate random throws (different `ss` values produce different Bernoulli outcomes and m). Then we compute the proportion and inverse the problem by estimating ρ given the proportion. The default run produces

```
bufprob()
[1] 0.800000 1.646926
```

The first number is the proportion of throws in which the needle intersects the lines (m/n), and the second number is the maximum likelihood estimator. It is estimated that the needle is 64% longer than the distance between the lines. See the simulation-based maximum likelihood subsection at the end of this section where L/D is estimated not from the formula for the probability, but based on simulations.

Example 6.116 *MM=ML. Prove that, for the canonical exponential family of distributions, the MM estimator defined by equation (6.54) coincides with the MLE.*

Solution. The joint density of the iid sample $\mathbf{Y}_i$ is given by

$$f(\mathbf{y}_1, ..., \mathbf{y}_n; \boldsymbol{\theta}) = e^{\boldsymbol{\theta}' \sum_{i=1}^n \mathbf{K}(\mathbf{y}_i) + S_*(\mathbf{y}_1, ..., \mathbf{y}_n) - nq(\boldsymbol{\theta})},$$

so the log-likelihood function, up to term S_* which does not contain the parameter, is given by $l(\boldsymbol{\theta}) = \boldsymbol{\theta}' \sum \mathbf{K}(\mathbf{y}_i) - nq(\boldsymbol{\theta})$. The first-order condition for the maximum is $\sum \mathbf{K}(\mathbf{y}_i) - n\frac{q(\boldsymbol{\theta})}{\partial \boldsymbol{\theta}} = \mathbf{0}$, which is equivalent to (6.54): MM=MLE.

Example 6.117 *Perch problem.* *A researcher wants to estimate the proportion of perch in a lake's fish population. He/she fishes on N days until the first perch is caught: on the ith day, he/she catches n_i fish none perch, and the $n_i + 1$ fish is a perch, where $n_i \geq 0$ (all fish are released back into the lake). Estimate the proportion of perch in the fish population.*

Solution. Let the proportion of perch in a fish population be p. The probability of catching first n_i other fish and the next being a perch is $p(1-p)^{n_i}$. Since the population of fish is not affected by fishing (fish are released) the likelihood function is the product of probabilities, $L(p) = \prod p(1-p)^{n_i}$. The log-likelihood is $l(p) = N \ln p + N_T \ln(1-p)$, where $N_T = \sum_{i=1}^N n_i$. Differentiate with respect to p and equate to zero, $N/p - N_T/(1-p) = 0$, and find the MLE $\widehat{p}_{ML} = N/(N + N_T)$. Check if this answer makes sense: (a) if there is only perch in the lake then $p = 1$. When fishing, the first fish caught will be always perch, i.e. $n_i = 0$ and $N_T = 0$ and the above formula gives $\widehat{p}_{ML} = 1$; (b) if there are only few perch in the lake, $p \simeq 0$ and there are many fish to catch before perch and n_i is large, which makes $\widehat{p}_{ML} \simeq 0$. □

We continue the discussion of a student's performance started in Example 6.38. The assumption is that the number of earned points in each assignment follows a binomial distribution with a probability that characterizes the student's performance.

Example 6.118 *Student' grading continued.* *Define the student's performance as the probability of earning a point in a series of n assignments with the maximum number of points x_i. Compare the quality of the two estimators of p considered in Example 6.38.*

Solution. The number of points earned, Y_i, in the assignment with the maximum number of points x_i is treated as the outcome of the binomial random variable in x_i Bernoulli experiments with probability p. Note that we use capital Y to denote a random variable, in this case, the number of points earned, and x to denote the maximum number of points, a fixed number. The probability of earning $Y_i = y_i$ points in the ith assignment is

$$\Pr(Y_i = y_i; p) = \binom{x_i}{y_i} p^{y_i}(1-p)^{x_i - y_i}, \quad i = 1, 2, ..., n.$$

The total log-likelihood function over a series of n assignments is given by

$$l(p) = \text{const} + \sum_{i=1}^{n} \left[Y_i \ln p + (x_i - Y_i) \ln(1-p) \right],$$

where const does not contain p. Differentiation with respect to p yields the MLE,

$$\widehat{p}_{ML} = \frac{\sum_{i=1}^{n} Y_i}{\sum_{i=1}^{n} x_i},$$

or in the notation of Example 6.38, $\widetilde{\beta}_2$. In words, the student's performance is measured by the proportion of the total number of points earned over the course. Now we compare the MLE with the naive estimator as a measure of the students' performance.

$$\widetilde{\beta}_1 = \frac{1}{n} \sum_{i=1}^{n} \frac{Y_i}{x_i}.$$

Both estimators are unbiased because $E(Y_i) = px_i$ but the MSE of $\widehat{p}_{ML}$ is smaller:

$$\text{var}(\widehat{p}_{ML}) = \frac{\sum_{i=1}^{n} \text{var}(Y_i)}{(\sum_{i=1}^{n} x_i)^2} = p(1-p) \frac{1}{\sum_{i=1}^{n} x_i},$$

$$\text{var}(\widetilde{\beta}_1) = \frac{1}{n^2} \sum_{i=1}^{n} \frac{\text{var}(Y_i)}{x_i^2} = p(1-p) \frac{1}{n^2} \sum_{i=1}^{n} \frac{1}{x_i}$$

because $(\sum_{i=1}^{n} x_i)^{-1} \le \sum_{i=1}^{n} x_i^{-1}/n^2$. This inequality follows from the Cauchy inequality $n^2 = (\sum x_i (1/x_i))^2 \le \sum x_i \sum 1/x_i$. The inequality turns into equality if $x_i = \text{const}$. Then, all assignments are of the same difficulty, the maximum number of points is the same. In that case, the two metrics of student's performance coincide. □

There are situations when the maximum of the likelihood is found directly without differentiation, as in the following example.

Example 6.119 *MLE for the uniform distribution. Observations $Y_1, ..., Y_n$ are drawn independently from a uniform distribution on the interval $[0, \theta]$, where $\theta > 0$ is unknown. Find the MLE.*

Solution. The density for the ith observation is θ^{-1} for $0 \le y \le \theta$ and zero elsewhere. Therefore, the likelihood function is the product

$$L(\theta) = \begin{cases} \theta^{-n}, & \text{if } \theta \ge \max Y_i \\ 0, & \text{elsewhere} \end{cases}.$$

Obviously, maximum of L occurs when $\widehat{\theta}_{ML} = \max Y_i$. □

Below is a continuation of Example 6.89 where the parameter to estimate is multidimensional. Then matrix calculus is invaluable; see Appendix.

Example 6.120 *The MLE of the multivariate normal distribution.* *Independent observation vectors* $\{\mathbf{Y}_i, i = 1, 2, ..., n\}$ *have a multivariate normal distribution* $\mathcal{N}(\boldsymbol{\mu}, \boldsymbol{\Omega})$, *where both* $\boldsymbol{\mu}$ *and* $\boldsymbol{\Omega}$ *are unknown. (a) Find the MLE for the mean* $\boldsymbol{\mu}$ *and covariance matrix* $\boldsymbol{\Omega}$. *(b) Derive the ML estimators for* μ *and* σ^2 *for univariate normal distribution in the special case* $k = 1$. *(c) Derive the ML estimators for variances, covariance and correlation coefficient,* ρ, *in the bivariate normal distribution* $(k = 2)$, *(d) Derive the MLE for intercept, slope, and conditional variance of the linear regression as conditional mean, and prove that OLS = ML for the intercept and slope.*

Solution. (a) As follows from Example 6.89, the likelihood function takes the form

$$L(\boldsymbol{\mu}, \boldsymbol{\Omega}) = (2\pi)^{-nk/2} |\boldsymbol{\Omega}|^{-n/2} e^{-\frac{1}{2} \sum_{i=1}^{n} (\mathbf{Y}_i - \boldsymbol{\mu})' \boldsymbol{\Omega}^{-1} (\mathbf{Y}_i - \boldsymbol{\mu})}.$$

Take ln to obtain the log-likelihood,

$$l(\boldsymbol{\mu}, \boldsymbol{\Omega}) = -\frac{nk}{2} \ln(2\pi) - \frac{n}{2} \ln |\boldsymbol{\Omega}| - \frac{1}{2} \sum_{i=1}^{n} (\mathbf{Y}_i - \boldsymbol{\mu})' \boldsymbol{\Omega}^{-1} (\mathbf{Y}_i - \boldsymbol{\mu}).$$

The first term contains neither $\boldsymbol{\mu}$ nor $\boldsymbol{\Omega}$ and, therefore, can be omitted when computing the maximum. To find the maximum differentiate l with respect to $\boldsymbol{\mu}$ and $\boldsymbol{\Omega}$ and set it to zero (consult the differentiation formulas from the Appendix). The score equation for $\boldsymbol{\mu}$ takes the form

$$\frac{\partial l}{\partial \boldsymbol{\mu}} = \sum_{i=1}^{n} \boldsymbol{\Omega}^{-1} (\mathbf{Y}_i - \boldsymbol{\mu}) = \boldsymbol{\Omega}^{-1} \sum_{i=1}^{n} (\mathbf{Y}_i - \boldsymbol{\mu}) = \mathbf{0}.$$

Since $\boldsymbol{\Omega}^{-1}$ is nonsingular this equation is equivalent to $\sum_{i=1}^{n} (\mathbf{Y}_i - \boldsymbol{\mu}) = \mathbf{0}$ with the solution

$$\widehat{\boldsymbol{\mu}}_{ML} = \overline{\mathbf{Y}} = \frac{1}{n} \sum_{i=1}^{n} \mathbf{Y}_i.$$

More advanced matrix calculus is required to get the MLE for $\boldsymbol{\Omega}$. Using formulas for differentiation with respect to a matrix from the Appendix, we obtain

$$\frac{\partial \ln |\boldsymbol{\Omega}|}{\partial \boldsymbol{\Omega}} = \boldsymbol{\Omega}^{-1}, \quad \frac{\partial}{\partial \boldsymbol{\Omega}} (\mathbf{Y}_i - \boldsymbol{\mu})' \boldsymbol{\Omega}^{-1} (\mathbf{Y}_i - \boldsymbol{\mu}) = -\boldsymbol{\Omega}^{-1} (\mathbf{Y}_i - \boldsymbol{\mu})(\mathbf{Y}_i - \boldsymbol{\mu})' \boldsymbol{\Omega}^{-1}.$$

Therefore,

$$\begin{aligned}
\frac{\partial l}{\partial \boldsymbol{\Omega}} &= -\frac{n}{2} \boldsymbol{\Omega}^{-1} + \frac{1}{2} \sum_{i=1}^{n} \boldsymbol{\Omega}^{-1} (\mathbf{Y}_i - \boldsymbol{\mu})(\mathbf{Y}_i - \boldsymbol{\mu})' \boldsymbol{\Omega}^{-1} \\
&= -\frac{n}{2} \boldsymbol{\Omega}^{-1} + \frac{1}{2} \boldsymbol{\Omega}^{-1} \left(\sum_{i=1}^{n} (\mathbf{Y}_i - \boldsymbol{\mu})(\mathbf{Y}_i - \boldsymbol{\mu})' \right) \boldsymbol{\Omega}^{-1} = \mathbf{0},
\end{aligned}$$

which yields

$$\widehat{\boldsymbol{\Omega}}_{ML} = \frac{1}{n} \sum_{i=1}^{n} (\mathbf{Y}_i - \overline{\mathbf{Y}})(\mathbf{Y}_i - \overline{\mathbf{Y}})'.$$

This estimator is referred to as the sample covariance matrix.

(b) Now we derive the ML estimators in a special case when $k = 1$, i.e. $Y_i \sim \mathcal{N}(\mu, \sigma^2)$, the univariate normal distribution using previously defined estimators. Clearly, $\widehat{\mu}_{ML} = \overline{Y}$ and $\widehat{\sigma}^2_{ML} = n^{-1} \sum_{i=1}^{n} (Y_i - \overline{Y})^2$.

(c) When $k = 2$, in the notation of Section 3.5.1 we have

$$\begin{bmatrix} Y_i \\ X_i \end{bmatrix} \overset{\text{iid}}{\sim} \mathcal{N}\left(\begin{bmatrix} \mu_y \\ \mu_x \end{bmatrix}, \begin{bmatrix} \sigma_y^2 & \sigma_{xy} \\ \sigma_{xy} & \sigma_x^2 \end{bmatrix} \right), \quad i = 1, 2, ..., n.$$

where $\sigma_{xy} = \rho \sigma_x \sigma_y$ is the covariance between Y and X. Using the previously derived estimates of $\boldsymbol{\mu}$ and $\boldsymbol{\Omega}$ with $k = 2$, we obtain familiar estimators,

$$\widehat{\mu}_y = \overline{Y}, \quad \widehat{\mu}_x = \overline{X}, \quad \widehat{\sigma}_y^2 = \frac{1}{n} \sum_{i=1}^{n} (Y_i - \overline{Y})^2, \quad \widehat{\sigma}_x^2 = \frac{1}{n} \sum_{i=1}^{n} (X_i - \overline{X})^2,$$

$$\widehat{\sigma}_{xy} = \frac{1}{n} \sum_{i=1}^{n} (Y_i - \overline{Y})(X_i - \overline{X}).$$

Note that unlike estimators derived in Section 6.6, MLE uses n, not $n-1$. Hence, the maximum likelihood produces biased estimators of the variance and covariance. In large sample, this bias vanishes. Now, using the invariance property (6.73) we obtain the ML estimator for the correlation coefficient,

$$r = \widehat{\rho} = \frac{\widehat{\sigma}_{12}}{\widehat{\sigma}_1 \widehat{\sigma}_2} = \frac{\sum_{i=1}^{n} (Y_i - \overline{Y})(X_i - \overline{X})}{\sqrt{\sum_{i=1}^{n} (Y_i - \overline{Y})^2 \times \sum_{i=1}^{n} (X_i - \overline{X})^2}},$$

which is previously called the "sample correlation coefficient" as follows from formula (6.29).

(d) Since the slope and intercept in linear regression as conditional mean of Y on X are expressed as $\beta = \sigma_{xy}/\sigma_x^2$ and $\alpha = \mu_y - \beta \mu_x$, using the invariance property of the maximum likelihood, we obtain the MLE for the slope and intercept by plugging in the previously derived MLE,

$$\widehat{\beta} = \frac{\widehat{\sigma}_{xy}}{\widehat{\sigma}_x^2} = \frac{\sum_{i=1}^{n} (Y_i - \overline{Y})(X_i - \overline{X})}{\sum_{i=1}^{n} (X_i - \overline{X})^2}, \quad \widehat{\alpha} = \overline{Y} - \widehat{\beta}\overline{X}.$$

Comparison with the ordinary least squares estimators of Section 6.7 concludes that OLS = MLE.

We know that when $x_1, x_2, ..., x_n$ are fixed as in the linear model (6.35), the OLS estimator is UMVUE, symbolically indicated in equation (6.72), because

(a) it is unbiased and (b) its variance reaches the Cramér–Rao lower bound. A natural question arises: Does the OLS estimator remain UMVUE in the Cramér–Rao sense when x are random, or more precisely, when (Y_i, X_i) are iid observations from the bivariate normal distribution? The answer is negative.

Example 6.121 *OLS does not reach Cramér–Rao bound. Prove that, if (Y_i, X_i) are iid observations from the bivariate normal distribution, then (a) the OLS/ML estimator of the slope, $\widehat{\beta}$, is unbiased for β, (b) the distribution of $\sqrt{n}\sigma_x(\widehat{\beta} - \beta)/\sigma$ is the t-distribution with $n - 1$ degrees of freedom, so (c) the OLS estimator does not reach the Cramér–Rao lower bound.*

Solution. (a) The unbiasedness of $\widehat{\beta}$ follows from the rule of repeated expectation (3.13). Applying this rule, we express

$$E(\widehat{\beta}) = E_{\mathbf{X}}(E(\widehat{\beta}|\mathbf{X} = \mathbf{x})), \tag{6.74}$$

where $\mathbf{X} = (X_1, ..., X_n)$. Since, for fixed $\mathbf{X} = \mathbf{x}$, we have $E(\widehat{\beta}|\mathbf{X} = \mathbf{x}) = \beta$, as shown in Theorem 6.50, we deduce $E(\widehat{\beta}) = \beta$.

(b) The density of $\widehat{\beta}$ is found from the combination of the following two distributions

$$\widehat{\beta}|\mathbf{X} = \mathbf{x} \sim \mathcal{N}\left(\beta, \frac{\sigma^2}{\sum_{i=1}^{n}(x_i - \overline{x})^2}\right), \quad \frac{1}{\sigma_x^2}\sum_{i=1}^{n}(X_i - \overline{X})^2 \sim \chi^2(n - 1).$$

The conditional distribution of $\widehat{\beta}$ with $\mathbf{X}$ fixed at $\mathbf{x}$ follows from Theorem 6.50, and the second distribution was derived in Example 4.22. We find the marginal density of $\widehat{\beta}$ by integrating out $s = \sum_{i=1}^{n}(X_i - \overline{X})^2$ from the joint density as the product of the two. More technically, the density of $\widehat{\beta}$ is found as the integral

$$p_{\widehat{\beta}}(b) = \int_0^\infty \frac{\sigma_x\sqrt{s}}{\sigma\sqrt{2\pi}} e^{-\frac{s\sigma_x^2}{2\sigma^2}(b-\beta)^2} f(s)\,ds$$

where

$$f(s) = \frac{1}{2^{(n-1)/2}\Gamma((n-1)/2)} s^{((n-1)/2)-1} e^{-s/2}$$

is the density of $\chi^2(n - 1)$ as presented in Section 4.2. The trick is that the integrand, after some algebra, can be expressed as another chi-square density. More precisely,

$$
\begin{aligned}
p_{\widehat{\beta}}(b) &= \frac{\sigma_x}{\sigma\sqrt{2\pi}} \frac{1}{2^{(n-1)/2}\Gamma((n-1)/2)} \int_0^\infty e^{-\frac{s\sigma_x^2}{2\sigma^2}(b-\beta)^2} s^{n/2-1} e^{-s/2} ds \\
&= \frac{\sigma_x}{\sigma\sqrt{2\pi}} \frac{1}{2^{(n-1)/2}\Gamma((n-1)/2)} \int_0^\infty s^{n/2-1} e^{-s\left(1/2 + \frac{\sigma_x^2}{2\sigma^2}(b-\beta)^2\right)} ds \\
&= \frac{\sigma_x}{\sigma\sqrt{2\pi}} \frac{1}{2^{(n-1)/2}\Gamma((n-1)/2)} \frac{\Gamma(n/2)}{\left[\frac{1}{2}\left(1 + \frac{\sigma_x^2}{\sigma^2}(b-\beta)^2\right)\right]^{n/2}} \\
&= \frac{\sigma_x}{\sigma\sqrt{\pi}} \frac{\Gamma\left(\frac{1}{2}n\right)}{\Gamma\left(\frac{1}{2}n - \frac{1}{2}\right)} \times \frac{1}{\left[1 + \frac{\sigma_x^2}{\sigma^2}(b-\beta)^2\right]^{n/2}}.
\end{aligned}
$$

The last function is the density of the t-distribution with $n-1$ degrees of freedom, as follows from Section 4.3. Succinctly,

$$
\sqrt{n}\frac{\sigma_x}{\sigma}(\widehat{\beta} - \beta) \sim t(n-1).
$$

(d) Again, referring to Section 4.3, the variance of the t-distributed random variable with $n - 1$ degrees of freedom is $(n-1)/(n-3)$, so

$$
\text{var}(\widehat{\beta}) = \frac{1}{n}\frac{\sigma^2}{\sigma_x^2}\frac{n-1}{n-3} > \frac{1}{n}\frac{\sigma^2}{\sigma_x^2}.
$$

The variance of the OLS estimator in the bivariate normal distribution is greater than the Cramér–Rao lower bound. However, although the lower bound for the variance is not achieved, we will show in Section 8.4.5 that the OLS estimator has minimum variance in the family of all unbiased estimators as follows from the Lehmann–Scheffé theorem.   □

The method of finding the density of the slope coefficient can be applied to any distribution of X when the distribution of $\sum_{i=1}^n (X_i - \overline{X})^2$ is known, continuous or discrete.

Example 6.122 *Linear regression with random* X. *Observations* (Y_i, X_i) *are iid drawn from a bivariate normal distribution specified as the conditional distribution* $Y|X = x \sim \mathcal{N}(\alpha + \beta x, \sigma^2)$ *and the marginal distribution* $X \sim f(\cdot; \boldsymbol{\gamma})$, *where neither* f *or* $\boldsymbol{\gamma}$ *is known. (a) Prove that the distribution of* X *does not affect estimators* α, β, *and* σ^2: *OLS = MLE. (b) Prove that* $\widehat{\alpha}_{dataset}$ *and* $\widehat{\beta}_{dataset}$ *remain unbiased for* α *and* β. *(c) Express the exact density of* $\widehat{\beta}_{dataset}$ *as an integral when* X *is continuous and as a sum when* X *is discrete. (d) Write* R *code with vectorized simulations to estimate the cdfs of* $\widehat{\beta}_{dataset}$ *under normal, uniform, and discrete distributions of* X.

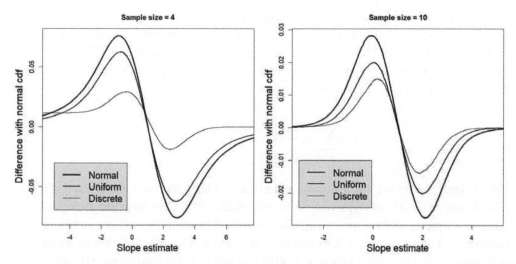

Figure 6.24: *The difference of cdfs for three distributions of X with the same mean and variance: normal, uniform, and the two-value discrete distributions for $n = 4$ and $n = 10$.*

Solution. (a) The joint distribution of (Y, X) is the product of the conditional distribution of Y given $X = x$ and the marginal distribution of X; see Section 3.3. Hence, if (X_i, Y_i) are n iid observations from the bivariate distribution, the log-likelihood function, up to a constant term, takes the form

$$l(\alpha, \beta, \sigma^2, \gamma) = -\frac{n}{2} \ln \sigma^2 - \frac{1}{\sigma^2} \sum_{i=1}^{n} (Y_i - \alpha - \beta X_i)^2 + \sum_{i=1}^{n} \ln f(X_i; \gamma).$$

Taking the derivative with respect to α and β, we arrive at the familiar OLS = MLE of intercept and slope

$$\widehat{\beta}_{ML} = \widehat{\beta}_{dataset} = \frac{\sum_{i=1}^{n}(Y_i - \overline{Y})(X_i - \overline{X})}{\sum_{i=1}^{n}(X_i - \overline{X})^2}, \quad \widehat{\alpha}_{ML} = \widehat{\alpha}_{dataset} = \overline{Y} - \widehat{\beta}_{ML}\overline{X}.$$

Note the difference in the notation compared to Section 6.7 where x was fixed and known: here X is random, so we use uppercase. (b) The unbiasedness holds even when X is not fixed or not normally distributed because (6.74) holds regardless of of distribution of X. (c) The derivation of the exact distribution of $\widehat{\beta}_{dataset}$ in the previous example is easy to generalize to the case when X has any distribution. Indeed, let $f_S(s)$ be the cdf of $S = \sum_{i=1}^{n}(X_i - \overline{X})^2$. Conditional of X_i, the distribution of $\widehat{\beta}_{dataset}$ is normal with mean β and variance $\operatorname{var}(\widehat{\beta}_{dataset}) = \sigma^2/S$. Therefore, the exact density of $\widehat{\beta}_{dataset}$ can be expressed as an integral,

$$f_{\widehat{\beta}}(b) = \int_0^\infty \frac{\sigma}{s} \phi\left(\frac{b - \beta}{\sigma}s\right) f_S(s) ds. \tag{6.75}$$

In the previous example, $f_S(s)$ was a chi-square distribution, up to a constant. When X has a discrete distribution, $\sum_{i=1}^{n}(X_i - \overline{X})^2$ has a discrete distribution as well and the integral in (6.75) is replaced with a sum. For example, if $\sum_{i=1}^{n}(X_i - \overline{X})^2$ takes values $s_1, s_2, ..., s_K$ with probabilities $p_1, p_2, ..., p_K$, respectively, then the distribution of $\widehat{\beta}_{dataset}$ takes the form

$$f_{\widehat{\beta}}(b) = \sum_{k=1}^{K} p_k \frac{\sigma}{s_k} \phi \left(\frac{b - \beta}{\sigma} s_k \right). \tag{6.76}$$

(d) The R code is found in function `regrD.r`. The difference between cdfs for $n = 4$ and $n = 10$ is depicted in Figure 6.24. (1) The internal function `ols.slope` computes nSim slopes using the vectorized built-in functions `rowMeans` and `rowSums`. For example, if X is nSim by n, then `rowMeans(X)` returns nSim means in each row. (2) Three types of X with the same mean (mux) and SD (sdx) are simulated: normal, uniform, and discrete (takes only two different values). (3) All cdfs are plotted as the difference from the normal cdf `pnorm((b-beta)/se.b)`, where to `se.b` is computed by the formula $\sigma/(\sqrt{n}\sigma_x)$ to see the differences. The difference between cdfs cannot be seen if plotted as the regular cdfs in the range from 0 to 1.

As follows from Figure 6.24 the difference in cdfs quickly reduces as the sample size, n, grows. The distribution of the slope estimates is symmetric around β due to the symmetry of the distribution of X around its mean. The smallest differences are for the discrete distribution, then the uniform and finally the normal distribution of X. $\qquad\square$

In the following example, the exact distribution of the OLS estimator of the slope is derived under the assumption that X is a Bernoulli random variable and uses the previously obtained general formula (6.76).

Example 6.123 *Regression with a Bernoulli regressor.* *The Bernoulli random variable X takes value 1 with probability p and 0 with probability $1 - p$. The conditional distribution of Y given $X = x$ is normal with mean $\alpha + \beta x$ and variance σ^2. Given a random sample (Y_i, X_i) of size n (a) derive the exact distribution of the OLS estimator $\widehat{\beta}$ as a finite linear combination of normal distributions, and (b) check this distribution through simulations.*

Solution. (a) First we find the marginal distribution of $\sum_{i=1}^{n}(X_i - \overline{X})^2 = \sum_{i=1}^{n} X_i^2 - \left(\sum_{i=1}^{n} X_i\right)^2/n$. Since X_i takes either 0 or 1, we have $\sum_{i=1}^{n} X_i^2 = \sum_{i=1}^{n} X_i$, and therefore, $\sum_{i=1}^{n}(X_i - \overline{X})^2 = S - S^2/n$, where $S = \sum_{i=1}^{n} X_i$ with possible values $0, 1, ..., n$. We have to exclude possibilities $S = 0$ and $S = n$ because the OLS estimator is not defined in those cases. Therefore, we derive the distribution of the slope under the assumption that $\sum_{i=1}^{n}(X_i - \overline{X})^2 > 0$, which excludes $S = 0$ and $S = n$. Hence, we conclude that $\sum_{i=1}^{n}(X_i - \overline{X})^2$ takes values

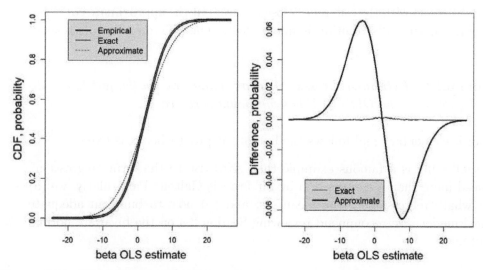

Figure 6.25: *The empiirical, exact, and approximate cdfs for the regression slope with binary X ($p = 0.1$, $n = 6$).*

$m - m^2/n$ with probability

$$p_m = \frac{\binom{n}{m}p^m(1-p)^{n-m} - (1-p)^n - p^n}{1 - (1-p)^n - p^n}, \quad m = 1, ..., n-1.$$

Finally, the exact cdf of the OLS estimator takes the form

$$\Pr(\widehat{\beta} \leq b) = \sum_{m=1}^{n-1} p_m \Phi\left(\frac{b-\beta}{\sigma}\sqrt{\frac{m(n-m)}{n}}\right).$$

(b) The simulations are carried out in the R code `regdic` and the cdfs are depicted in Figure 6.25. The exact cdf is compared with the empirical cdf and the single normal approximation

$$F_0(b) = \Phi\left(\frac{b-\beta}{\sigma}\sqrt{np(1-p)}\right).$$

We make several remarks about the code. (1) Vectorized computation is used: values for X and Y are generated once and are stored as `nSim` by `n` matrices. (2) `nSim` values of $\sum_{i=1}^{n}(X_i - \overline{X})^2$ are computed using the `rowSums` function; note that `rowSums(X^2)` returns `nSim` values $\sum_{i=1}^{n} X_i^2$ for each simulation. (3) Simulations for which $\sum_{i=1}^{n}(X_i - \overline{X})^2 = 0$ are removed because the OLS is not defined. (4) To compute the probability distribution of $\sum_{i=1}^{n}(X_i - \overline{X})^2$, we must combined repeated values $\{m - m^2/n, m = 1, 2, ..., n-1\}$ and sum the respective probabilities. (5) Since the cdfs are very close, we show the difference with the empirical cdf in the right plot. The empirical and exact theoretical cdf practically coincide (left plot) with the difference close to zero, but the approximate cdf based on the single normal distribution is not perfect. The difference in cdfs reaches 0.06.

Theorem 6.124 *Extended Gauss–Markov Theorem. In the family of estimators $E(\widetilde{\beta}|\mathbf{x}) = \beta$, the OLS/MLE has minimum variance.*

Proof is elementary and follows the line of the proof when $\mathbf{x}$ is fixed.

The following is a famous example that gives rise to the term "regression" introduced more than a century ago by Sir Francis Galton. Particularly, we (a) explain what "regression to the mean" is, and (b) how to build an adequate statistical model. We recommend reviewing Section 3.5 on the bivariate normal distribution.

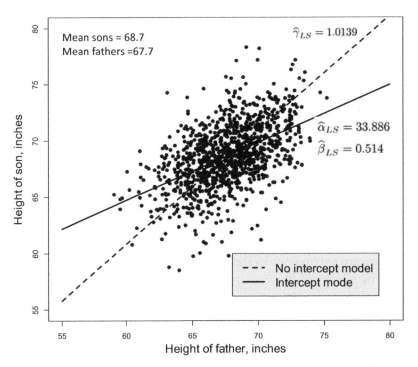

Figure 6.26: *Galton height data: the height of sons versus height of fathers (n = 1078) is fitted with intercept (solid line) and without intercept (dotted line). This phenomenon is called "regression to the mean."*

Example 6.125 *Regression to the mean: heights of fathers and sons. Sir Francis Galton (1822–1911) was a British scientist who studied the relationship between the heights of fathers and sons in his famous paper "Regression Towards Mediocrity in Hereditary Stature" published in 1885. File* `father.son.csv` *contains 1,078 data points. (a) Depict the heights and compute the regressions* *with and without intercept. (b) Decide which model, the slope (no intercept) model or a linear regression model with intercept, is appropriate to relate the heights.*

Solution. (a) The height data, along with the two regressions, are depicted in Figure 6.26. The cloud of points has an elliptical shape indicating that the data (F_i, S_i) where F_i and S_i are the height of the ith father and son, respectively, $i = 1, 2, ..., n = 1,078$, and follow a bivariate normal distribution. We want to predict the height of the son by the height of the father. Two models can be suggested:

1. The slope model S_i, or the model without intercept, $S_i = \gamma F_i + \varepsilon_i$.

2. The traditional linear model with the intercept $S_i = \alpha + \beta F_i + \varepsilon_i$.

The error term, ε_i, has zero mean and constant variance. Both models can be justified. The first model relates the heights following a simple proportional rule: if $\gamma > 1$ the sons are taller then fathers and if $\gamma < 1$ the reverse is true, on average. The second model is justified from the standpoint of the bivariate normal distribution as the expectation of the height of sons conditional on the height of fathers. The LS estimates with and without intercept are shown in Figure 6.26. The fact that the slope estimate, $\widehat{\gamma}_{LS} = 1.039 > 1$, reflects the fact that sons are taller than fathers by one inch. According to this model, sons are taller than fathers by 1.4%, on average.

In the intercept model, the answer to who is taller depends on how tall the father is. Since $E(S) = \alpha + \beta F$, the son is taller than his father, again on average, if $33.886 + 0.514F > F$. This means that if the father's height is less than $33.886/(1 - 0.514) = 69.724$ inches then the son is taller. Reversely, if the father's height is greater than 69.724 inches then the son is shorter. We interpret it as saying that taller fathers have shorter sons and vice versa, or as Sir Galton put in "Towards Mediocrity." Today, this phenomenon is known as "regression to the mean."

Which model, no intercept or intercept is better? Let us take a look from the perspective of human-race dynamics. Figure 6.27 depicts simulations according to both models from generation to generation over 15 generations. At Generation 1, we have the distribution of heights of 1,078 grandsons according to each model using standard deviations estimated from Galton's data. For example, in the no intercept model (left plot), the distribution of the grandson heights is derived as 1,078 random normally generated data with means $1.039 S_i$ and SD $= 2.796$. In the intercept model (right plot), random heights are generated with means $33.886 + 0.514 S_i$ and SD $= 2.437$. At Generation 2, grandsons become fathers and the process of data generation continues. The left plot reveals that since the gamma estimate is greater than one, according to the no intercept model, the human race diverges in height. Indeed, looking at the relationship $S_i = \gamma F_i$ as a recursive equation, $y_{i+1} = \gamma y_i$, we derive $y_i = y_0 \gamma^i \to \infty$, where the generation index, $i \to \infty$. Conversely, according to the intercept model, the human race is stable. Indeed, now we have $y_{i+1} = \alpha + \beta y_i$ and $y_i = a \sum_{j=0}^{i} \beta^j + y_0 \beta^i$. But, since $0 < \beta < 1$, we have $\sum_{j=0}^{i} \beta^j \to (1 - \beta)^{-1}$ and finally, we deduce that human height is stable over generations with the limit

$$\lim_{i \to \infty} y_i = \frac{\alpha}{1 - \beta} = \text{const.}$$

In fact, since using the estimates of the intercept and slope from Galton's data we obtain the human-race height is around $33.886/(1 - 0.514) = 69.7$ inches (my height is 70 inches). The intercept model is biologically plausible unlike the no intercept model and therefore is more adequate to describe the father–son relationship.

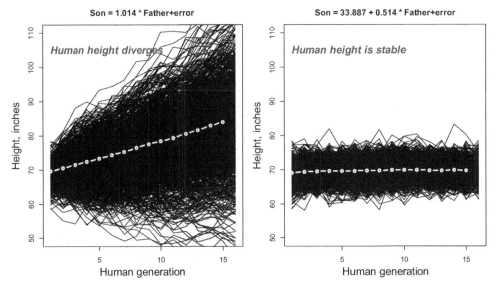

Figure 6.27: *Simulation of 1,078 heights over 15 human generations.*

Somebody may argue that people's height increases. However, this increase is not due to heredity but better nutrition and living conditions. This hypothesis is supported by the fact that the height of animals remains stable over centuries. □

The following is a continuation of Example 3.36, where it was proven that the average of the ratio of the circumference to the diameter of the tire does not converge to π even when infinitely many tires are measured. Here, we suggest two other estimators of π, least squares and the MLE. The latter requires specification of the distribution of the measurement error – we assume that measurements follow a normal distribution.

Example 6.126 *Who said π continued.* *Assume that measurements of the tire circumference and diameter follow a normal distribution, $\mathcal{N}(0, \sigma^2)$, where* *the variance $\sigma^2 > 0$ is unknown. Derive the MLE for π based on n independent measurements and compare four estimators of π via simulations: (1) average of ratios (naive), (2) ratio of averages, (3) OLS estimator, and (4) MLE.*

Solution. The measured diameter and circumference of the ith tire are $D_i = d_i + \delta_i$ and $C_i = \pi d_i + \varepsilon_i$, respectively, where d_i is the true diameter of the ith tire, and δ_i and ε_i are independent measurement errors with positive variances. It was proven in Example 3.36 that the naive estimator, $\widehat{\pi}_N$ is inconsistent (systematically biased), but the ratio of averages, $\widehat{\pi}_R$ is consistent. $C_i = \pi d_i + \varepsilon_i$ can be treated as the slope model from Example 6.36 with the OLS estimator

$$\widehat{\pi}_{LS} = \frac{\sum_{i=1}^n C_i D_i}{\sum_{i=1}^n D_i^2},$$

where, in the absence of d_i, its measurement, D_i, is used. This estimator is inconsistent under assumption that $\lim_{n\to\infty} n^{-1} \sum_{i=1}^n d_i^2 = A > 0$. Indeed, from the LLN,

$$p \lim_{n\to\infty} \frac{1}{n} \sum_{i=1}^n C_i D_i = \lim_{n\to\infty} \frac{1}{n} \sum_{i=1}^n \pi d_i^2 = \pi A,$$

$$p \lim_{n\to\infty} \frac{1}{n} \sum_{i=1}^n D_i^2 = \lim_{n\to\infty} \frac{1}{n} \sum_{i=1}^n d_i^2 + p \lim_{n\to\infty} \frac{1}{n} \sum_{i=1}^n \delta_i^2 = A + \sigma^2,$$

where $\sigma^2 = \mathrm{var}(\delta_i) > 0$. Therefore, from Section 6.4.4, we conclude that

$$p \lim_{n\to\infty} \widehat{\pi}_{LS} = \frac{\pi}{1 + \sigma^2/A} < \pi.$$

Note that the bias is small in practice because σ^2/A is small. So far, we have not used a specific distribution for the measurement error. Now we find the MLE under the assumption that both circumference and the diameter are measured with normally distributed errors, $\varepsilon_i, \delta_i \overset{\text{iid}}{\sim} \mathcal{N}(0, \sigma^2)$. Since measurements for the ith tire are independent, the joint pdf of (C_i, D_i) takes the form

$$f(C_i, D_i; \pi, \sigma^2, d_1, ..., d_n) = \frac{1}{2\pi\sigma^2} e^{-\frac{1}{2\sigma^2}\left[(C_i - \pi d_i)^2 + (D_i - d_i)^2\right]},$$

where $\{d_1, ..., d_n\}$ are unknown and treated as nuisance parameters. The total log-likelihood function takes the form

$$l(\pi, \sigma^2, d_1, ..., d_n) = -n \ln(2\pi) - n \ln \sigma^2 - \frac{1}{2\sigma^2} \sum_{i=1}^n \left[(C_i - \pi d_i)^2 + (D_i - d_i)^2\right].$$

Note that this log-likelihood violates a condition of the maximum likelihood theory because the number of unknown parameters increases with n. Differentiation with respect to σ^2 gives $\sigma^2 = (2n)^{-1} \sum_{i=1}^{n} \left[(C_i - \pi d_i)^2 + (D_i - d_i)^2 \right]$. Plugging this solution back into the log-likelihood l, the problem reduces to

$$\min_{\pi, d_1, \ldots, d_n} \sum_{i=1}^{n} \left[(C_i - \pi d_i)^2 + (D_i - d_i)^2 \right].$$

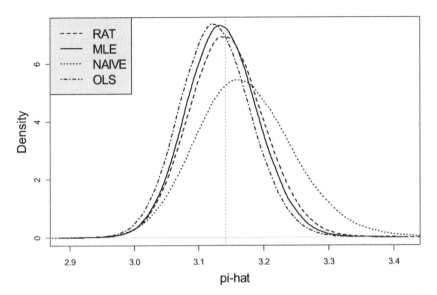

Figure 6.28: *The Gaussian kernel densities for four estimators of π based on* nSim=500000 *simulations. The naive estimator, $\widehat{\pi}_N$ (dotted line) overestimates π, the ratio of the sums, $\widehat{\pi}_R$ is unbiased but the density is wider, The LS estimator $\widehat{\pi}_{LS}$ slightly underestimates, π, but is otherwise close to the MLE.*

Differentiation with respect to d_i gives $-2\pi(C_i - \pi d_i) - 2(D_i - d_i) = 0$ and $d_i = (D_i + \pi C_i)/(1 + \pi^2)$. After some algebra, we obtain $(C_i - \pi d_i)^2 + (D_i - d_i)^2 = (C_i - \pi D_i)^2/(1 + \pi^2)$. After unknown d_i have been eliminated maximization of the log-likelihood reduces to minimization of

$$H(\pi) = \frac{1}{1 + \pi^2} \sum_{i=1}^{n} (C_i - \pi D_i)^2.$$

The first-order condition for the minimum takes the form

$$2\pi \sum_{i=1}^{n} (C_i - \pi D_i)^2 + 2(1 + \pi)^2 \sum_{i=1}^{n} (C_i - \pi D_i) D_i = 0$$

which simplifies to a quadratic equation $F\pi^2 - H\pi - E = 0$, where

$$F = n^{-1} \sum_{i=1}^{n} (C_i + 2D_i) D_i, \; H = n^{-1} \sum_{i=1}^{n} (C_i^2 - D_i^2 + 2C_i D_i), \; E = n^{-1} \sum_{i=1}^{n} C_i D_i,$$

so that, finally,

$$\widehat{\pi}_{ML} = \frac{H + \sqrt{H^2 + 4EF}}{2F}.$$

It is possible to prove that $\widehat{\pi}_{ML}$ is a consistent estimator of π (see the problem at the end of the section) despite the fact that the number of unknown parameters increases with n.

The R code with simulations is `piest`. The default number of measured tires is n=20 assuming that the exact diameter varies from 5 to 20 length units, centimeters or inches. To generate the true diameters the uniform distribution is used (matrix d contains the true values the same in each row). The default measurement error standard deviation is `sigma=1`. The results on accuracy and precision of the four estimators are returned in the tabular form and the densities are depicted in Figure 6.28.

```
> piest(n=20,nSim=500000)
                Bias         RMSE  Pr(abs(Est-pi)<0.01)
RAT    -2.630037e-05 0.05763228              0.138186
MLE    -6.097152e-03 0.05504837              0.143810
NAIVE   2.633785e-02 0.08102537              0.104438
OLS    -1.647930e-02 0.05664590              0.137862
```

Three summary statistics characterize the quality of the π estimation: (1) Bias as the difference $E(\widehat{\pi}) - \pi$, (2) empirical MSE, and (3) the proportion of estimates that fall within ± 0.01 interval around π. As expected from the theory, $\widehat{\pi}_R$ and $\widehat{\pi}_{ML}$ have small empirical bias (it is not zero due to the finite number of simulations), the naive estimator has a significant positive bias, and the least square estimator has a negative bias. MLE and OLS have the smallest MSE. The bias and MSE may be not adequate measures of the quality of some estimators, such as $\widehat{\pi}_N$ because the ratio C_i/D_i has a Cauchy distribution with infinite mean and variance. A better metric is the probability that the estimator falls within a fixed-width interval, say $|\widehat{\pi} - \pi| < 0.01$. This empirical quantity is reported in the last column; 14.4% of all MLE values are within close proximity while the naive estimator yields 10.4%. The OLS and RAT estimators are very close in this respect. In summary, under the normal assumption, the MLE outperforms other estimators.

6.10.2   Circular statistics and the von Mises distribution

Circular, sometimes called directional, statistics analyses angular data. Several excellent books can be recommended for further reading, such as Mardia and Jupp [71] and Fisher [49]. Periodic data events that occur during the day or the year also can be studied using methods of circular statistics, such as time of 911 police calls (period is 24 hours), auto crashes during the day, time of website clicks around the clock as in Example 2.1, time of precipitation at a certain geographic location during the year, etc. Due to periodicity, the traditional distributions

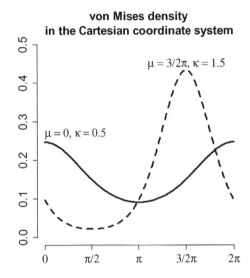

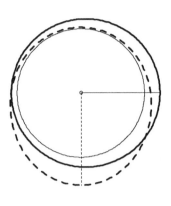

Figure 6.29: *Two von Mises densities plotted as functions of angle θ in the Cartesian and polar coordinate systems. The angle between minimum and maximum is $\pi/2$.*

do not work. The following von Mises distribution on the circle with the radian angle θ has the probability density function

$$f(\theta; \mu, \kappa) = \frac{1}{2\pi I_0(\kappa)} e^{\kappa \cos(\theta - \mu)}, \quad 0 \le \theta < 2\pi, \tag{6.77}$$

where μ is the center of the distribution and belongs to the interval $[0, 2\pi)$, and $\kappa > 0$ is the precision parameter similar to $1/\sigma$ in the normal distribution (see below). $I_0(\kappa)$ is the zero-order modified Bessel function of the first kind. Since the area under the density is 1, we have the following integral representation

$$I_0(\kappa) = \frac{1}{2\pi} \int_0^{2\pi} e^{\kappa \cos(\theta - \mu)} d\theta = \frac{1}{2\pi} \int_0^{2\pi} e^{\kappa \cos \theta} d\theta.$$

The Bessel function is an example of special functions and frequently emerges as the solution of ordinary differential equations; more information can be gained from an authoritative book by Abramowitz and Stegun [3].

The density (6.77) is an analog of the normal density on the circle. It is unimodal and symmetric around μ. To find its connection with the normal density approximate the argument of the exponential function around μ as follows

$$\kappa \cos(\theta - \mu) \simeq \text{const} - \frac{1}{2(1/\sqrt{2\kappa})^2}(\theta - \mu)^2.$$

Example 6.127 *von Mises density. Plot von Mises densities as functions of angle θ in the Cartesian and polar (wrapped around the circle) coordinate systems for $\mu = 0, \kappa = 0.5$ and $\mu = 3/2\pi, \kappa = 1.5$.*

Solution. See Figure 6.29. The first density (solid line) reaches maximum at $\theta = 0$ and the second density (dashed line) reaches its maximum at $\theta = 3/2\pi$. Correspondingly, in the polar coordinate system, the density with $\mu = 0, \kappa = 0.5$ has maximum deviation from the unit circle at the angle $\theta = 0$ and minimum deviation at $\theta = \pi$. The second density has maximum deviation at the angle $\theta = 3/2\pi$ and minimum deviation at $\theta = \pi/2$. $\qquad\qquad\square$

Now we discuss the maximum likelihood estimation of μ and κ. Given iid angle observations $\theta_1, ..., \theta_n$ distributed according to the von Mises density, the log-likelihood function takes the form

$$l(\mu, \kappa) = -n \ln J_0(\kappa) + \kappa \sum_{i=1}^{n} \cos(\theta_i - \mu).$$

Rewrite the sum of cosines in a form suitable for maximization using the trigonometric identity $\cos(\alpha - \beta) = \cos\alpha\cos\beta + \sin\alpha\sin\beta$,

$$\sum \cos(\theta_i - \mu) = \sum (\cos\theta_i \cos\mu + \sin\theta_i \sin\mu)$$
$$= \cos\mu \sum \cos\theta_i + \sin\mu \sum \sin\theta_i = A\cos\mu + B\sin\mu,$$

where $A = \sum \cos\theta_i$ and $B = \sum \sin\theta_i$ (index i is omitted for brevity). Combine the terms using the same trigonometric formula and arrive at $\sum \cos(\theta_i - \mu) = n\overline{R}\cos(\overline{\theta} - \mu)$, where

$$\overline{R} = \frac{1}{n}\sqrt{\left(\sum\cos\theta_i\right)^2 + \left(\sum\sin\theta_i\right)^2}, \quad \cos\overline{\theta} = \frac{\sum\cos\theta_i}{\sqrt{\sum(\cos\theta_i)^2 + (\sum\sin\theta_i)^2}}. \tag{6.78}$$

This implies that the MLE for the center of the distribution is $\widehat{\mu}_{ML} = \overline{\theta}$ and the MLE for the precision parameter κ is the solution of the equation $I_0'(\kappa) - \overline{R}I_0(\kappa) = 0$. Note that $\widehat{\mu}_{ML}$ is not the average of angles but arccos function of the right-hand side of $\cos\overline{\theta}$ in (6.78). Newton's algorithm can be used to solve the equation for κ as follows:

$$\kappa_{s+1} = \kappa_s - \frac{I_0'(\kappa_s) - \overline{R}I_0(\kappa_s)}{I_0''(\kappa_s) - \overline{R}I_0'(\kappa_s)}, \tag{6.79}$$

where $'$ and $''$ are the first and second derivatives of the Bessel function I. For $\overline{R} < 1$ one can use the starting value $\kappa_0 = 0.5(1 - \overline{R})^{-1}$. The next example illustrates the maximum likelihood estimation of a von Mises distribution. We refer the reader to books cited above to read more about circular statistics and to the R package `circular`.

Example 6.128 *When students go to bed.* *On the question "When did you last go to bed?" nine students answered 23.1, 0.3, 22.6, 21.5, 23.7, 1.2, 0.7, 22.1, 23.3 (minutes were transformed to decimals as min/60). Use a von Mises distribution to (a) estimate what time students go to bed on average, and (b) estimate the proportion of students who go to bed after midnight.*

Solution. (a) The naive choice to use average of times does not work because, although 23.1 and 0.3, are close in time their average does not make sense. To apply the von Mises distribution, we must convert the hour data to polar angles θ (radians) using the formula

$$\text{radian} = \frac{30 - \text{hour}}{12}\pi \bmod 2\pi.$$

According to this formula midnight translates into $\theta = \pi/2$. The mod operator in R is represented as the double percent sign %%. See the R code gotobed. The application of the above algorithm yields mu.ML=1.730645 and R.bar=0.955461. Starting from 11.226119, it took four iterations of (6.79) to converge to kappa.ML = 11.495083030. To convert radians to hours, we use the back transformation $\bar{h} = 30 - 12\hat{\mu}_{ML}/\pi$. We conclude that students go to bed on average at hgotobed=23.389423 (11:20 p.m.).

(b) Estimating the proportion of students who go to bed after midnight seems straightforward but is a tricky question. From the data itself, three out of nine students go to bed after midnight, that is, the proportion is about 1/3, a nonparametric estimate. Now we integrate the density but the problem with the integral evaluation is that the lower limit is unclear. We may suggest to assuming that the latest hour students go to bed is 6 a.m. ($\theta = 0$) that implies that the proportion of students who go to bed after midnight ($\theta = \pi/2$) can be estimated as

$$\frac{1}{I_0(\kappa)}\left(\frac{1}{2\pi}\int_0^{\pi/2} e^{\kappa\cos(\theta-\bar{\theta})}d\theta\right) = 0.2962471,$$

fairly close to the nonparametric estimate of 1/3.

Example 6.129 *Auto crashes around the clock.* *Generalize the Parzen density estimation for circular data using a von Mises density and apply this density to display the distribution of auto crash for the days of the week from file autocrash.csv (see Section 5.5 and the problem at the end of that section).*

Solution. Following the idea of the kernel density estimation from Section 5.5, given angular data $\theta_1, \theta_2, ..., \theta_n$, we estimate the density at θ as the average of n densities with the center at the observed angle θ_i and fixed κ (similar to the bandwidth parameter),

$$f(\theta) = \frac{1}{n}\sum_{i=1}^{n}\frac{1}{2\pi I_0(\kappa)}e^{\kappa\cos(\theta-\theta_i)}, \quad 0 \leq \theta < 2\pi.$$

The R function autocrash carries out computations and produces Figure 6.30. The week of the day is in the column DAY_OF_WEEK and the time of the crash in the column TIME_OF_DAY. The first two digits of the time represent the hour and the last two digits represent minutes. To transfer this format to decimal compute hours as hf=floor(hour) and then convert minutes into decimal hour=hf+(hour-hf)*10/6 where hour is the array TIME_OF_DAY. Then the decimal time data are transformed into angular format by the formula from the previous example, theta=(pi*(30-hid)/12) %%(2*pi) where hid is the hour data for the specific day of the week. The estimated circular densities for each day of the week are depicted in Figure 6.30. The familiar morning and afterwork rush hours are around 8 a.m. and 5:30 p.m. Surprisingly, there is considerable risk of driving late Saturday and Sunday nights, actually early morning Sunday and Monday, around 2 a.m.

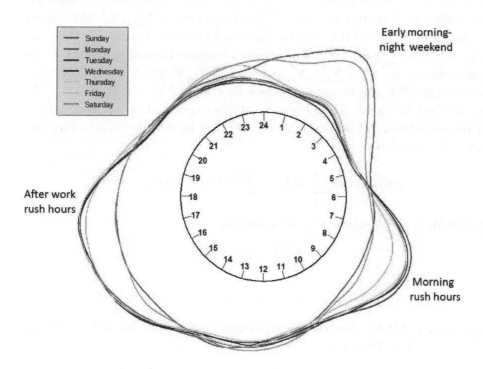

Figure 6.30: *Parzen circular densities of auto crashes around the clock from Example 6.129.*

6.10.3 Maximum likelihood, sufficient statistics and the exponential family

The MLE is a function of sufficient statistics. Indeed, if $\mathbf{u}(\mathbf{Y})$ is a sufficient statistic for $\mathbf{Y}$, as follows from the Neyman factorization criterion (Section 6.8), the density can be expressed as $f(\mathbf{y}; \boldsymbol{\theta}) = H(\mathbf{y})g(\mathbf{u}(\mathbf{y}); \boldsymbol{\theta})$ and, therefore, the maximizer/MLE of the log likelihood $l(\boldsymbol{\theta}) = \ln H(\mathbf{Y}) + \ln g(\mathbf{u}(\mathbf{Y}); \boldsymbol{\theta})$ is expressed through the function of data $\mathbf{u}$ as

$$\widehat{\boldsymbol{\theta}}_{ML} = \mathbf{Q}(\mathbf{u}(\mathbf{Y})). \tag{6.80}$$

In words, the MLE is a function of sufficient statistics.

Example 6.130 *MLEs are functions of sufficient statistics. Demonstrate that MLEs for Bernoulli, bivariate, and multivariate normal distributions are functions of sufficient statistics.*

Solution. As follows from Example 6.62, the sufficient statistic of n independent Bernoulli trials is $U = \sum_{i=1}^{n} Y_i$. From Example 6.112, we know that $\widehat{p}_{ML} = \overline{Y} = U/n$, meaning that $Q = 1/n$. For the bivariate normal distribution, the sufficient statistics are $S_1 = \sum_{i=1}^{n} Y_i$ and $S_2 = \sum_{i=1}^{n} Y_i^2$. As follows from Example 6.120, the MLE can be expressed through S_1 and S_2, thus, we set $\sum (Y_i - \overline{Y})^2 = \sum Y_i^2 - n\overline{Y}^2$. For the multivariate normal distribution the sufficient statistics are $\mathbf{S}_1 = \sum_{i=1}^{n} \mathbf{Y}_i$ and $\mathbf{S}_2 = \sum_{i=1}^{n} \mathbf{Y}_i \mathbf{Y}_i'$. As in the bivariate distribution, the MLE can be expressed as a functions of $\mathbf{S}_1$ and $\mathbf{S}_2$. $\qquad \square$

Now we find the MLE for the exponentially family. Let us start with a single parameter family of iid observations specified by density (6.47). The log-likelihood takes the form

$$l(\theta) = p(\theta) \sum_{i=1}^{n} K(Y_i) + \sum_{i=1}^{n} S(Y_i) - nq(\theta).$$

Clearly, the MLE is the solution to the equation

$$\frac{q'(\theta)}{p'(\theta)} = \overline{K}$$

where $\overline{K} = \sum_{i=1}^{n} K(Y_i)/n$, the average of the K values.

Example 6.131 *MLEs for major distributions. Find the MLE of θ for distributions listed in Table 6.2.*

Solution. First, we notice that for all distributions $\overline{K} = \overline{Y}$. The q and p functions are listed below (for the binomial distribution $Y_1, Y_2, ..., Y_N$ is the number of successes in n Bernoulli events). MLEs are estimators used before.

	Exponential	Bernoulli	Binomial	Poisson	Normal
q'	$-\lambda^{-1}$	$(1-\pi)^{-1}$	$n(1-\pi)^{-1}$	1	$\mu\sigma^{-2}$
p'	-1	$\pi^{-1}(1-\pi)^{-1}$	$\pi^{-1}(1-\pi)^{-1}$	λ^{-1}	σ^{-2}
MLE	$\widehat{\lambda}_{ML} = \overline{Y}^{-1}$	$\widehat{\pi}_{ML} = \overline{Y}$	$\widehat{\pi}_{ML} = \overline{Y}/n$	$\widehat{\lambda}_{ML} = \overline{Y}$	$\widehat{\mu}_{ML} = \overline{Y}$

□

For the multidimensional exponential family the MLE of $\boldsymbol{\theta}^{m\times 1}$ is obtained as the solution to the system of m equations

$$\left(\frac{\partial \mathbf{p}}{\partial \boldsymbol{\theta}}\right)\overline{\mathbf{K}} - \frac{\partial q}{\partial \boldsymbol{\theta}} = \mathbf{0}$$

where $\overline{\mathbf{K}} = n^{-1}\sum_{i=1}^{n}\mathbf{Y}_i$ is an $m\times 1$ vector and $\partial \mathbf{p}/\partial\boldsymbol{\theta}$ and $\partial q/\partial\boldsymbol{\theta}$ are a $m\times m$ matrix and $m\times 1$ vector, respectively.

6.10.4 Asymptotic properties of ML

The MLE has optimal asymptotic (large sample) properties: the estimator is consistent, normally distributed, and efficient. In this section, we formulate the assumptions and the main theorem on the properties when the sample size, n, goes to infinity.

ML assumptions

A *IID assumption.* Observations $\mathbf{Y}_1, \mathbf{Y}_2, ..., \mathbf{Y}_n$ are independent and identically distributed with the common pdf $f(\mathbf{y};\boldsymbol{\theta})$, where the vector of unknown parameters, $\boldsymbol{\theta} \in R^m$ and m is known and fixed (does not depend on n).

B *The parameter space $\boldsymbol{\theta} \in \Theta$ is an open set in R^m.*

C *Common support.* The support of f, the set $\{\mathbf{y} : f(\mathbf{y};\boldsymbol{\theta}) > 0\}$, does not depend on the parameter.

D *Regularity conditions.* For each $\mathbf{y}$ from the support set, function $\ln f(\mathbf{y};\boldsymbol{\theta})$ has derivatives and finite expectation up to the third order so that (6.57) holds and differentiation with respect to $\boldsymbol{\theta}$ under the integral sign is permissible and interchangeable.

E *Identifiability conditions.* (E1) $f(\mathbf{y};\boldsymbol{\theta}_1) = f(\mathbf{y};\boldsymbol{\theta}_2)$ for all $\mathbf{y}$ implies $\boldsymbol{\theta}_1 = \boldsymbol{\theta}_2$. (E2) The score equation has unique root in the sense that

$$E_\theta\left(\left.\frac{\partial \ln f(\mathbf{y};\boldsymbol{\theta})}{\partial \boldsymbol{\theta}}\right|_{\boldsymbol{\theta}=\boldsymbol{\theta}_*}\right) = \mathbf{0} \tag{6.81}$$

implies $\boldsymbol{\theta} = \boldsymbol{\theta}_*$. (E3) The information matrix is nonsingular for all $\boldsymbol{\theta} \in \Theta$.

Example 6.132 *Conditions for the Bernoulli distribution.* *Check the identifiability conditions E for the Bernoulli distribution.*

Solution. In the previous notation, for the Bernoulli distribution, $\boldsymbol{\theta} = p$, and $\ln f(\mathbf{y};\boldsymbol{\theta}) = y\ln p + (1-y)\ln(1-p)$ with the parameter space being $p \in (0,1) = \Theta$.

To check (E1), we must prove that, if equation $y \ln p_1 + (1 - y) \ln(1 - p_1) = y \ln p_2 + (1 - y) \ln(1 - p_2)$ holds for $y = 1$ and $y = 0$, then $p_1 = p_2$. Indeed, if $y = 1$ we have $\ln p_1 = \ln p_2$ and therefore $p_1 = p_2$. To check (E2), we must show that equation

$$E_{p_1} \left(\frac{y}{p_2} - \frac{1 - y}{1 - p_2} \right) = \frac{p_1}{p_2} - \frac{1 - p_1}{1 - p_2} = 0$$

implies $p_1 = p_2$. Indeed, keeping p_2 fixed and solving for p_1, we obtain $p_1 = p_2$. For (E3), the information matrix is $(p(1 - p))^{-1} \neq 0$ for $0 < p < 1$.

Example 6.133 *Multidimensional canonical exponential family.* *Prove that condition E holds for the multidimensional canonical exponential family (6.53) if and only if*

$$\left. \frac{\partial q}{\partial \boldsymbol{\theta}} \right|_{\boldsymbol{\theta} = \boldsymbol{\theta}} = \left. \frac{\partial q}{\partial \boldsymbol{\theta}} \right|_{\boldsymbol{\theta} = \boldsymbol{\theta}_*}$$

implies $\boldsymbol{\theta} = \boldsymbol{\theta}_*$.

Solution. For the canonical exponential family, the pdf is given by $f(\mathbf{y}; \boldsymbol{\theta}) = e^{\boldsymbol{\theta}' \mathbf{K}(\mathbf{y}) + S(\mathbf{y}) - q(\boldsymbol{\theta})}$, so by Theorem 6.85, we have

$$E_{\boldsymbol{\theta}} \left(\left. \frac{\partial \ln f(\mathbf{y}; \boldsymbol{\theta})}{\partial \boldsymbol{\theta}} \right|_{\boldsymbol{\theta} = \boldsymbol{\theta}_*} \right) = E_{\boldsymbol{\theta}}(\mathbf{K}(\mathbf{Y})) - \left. \frac{\partial q}{\partial \boldsymbol{\theta}} \right|_{\boldsymbol{\theta} = \boldsymbol{\theta}_*} = \left. \frac{\partial q}{\partial \boldsymbol{\theta}} \right|_{\boldsymbol{\theta} = \boldsymbol{\theta}} - \left. \frac{\partial q}{\partial \boldsymbol{\theta}} \right|_{\boldsymbol{\theta} = \boldsymbol{\theta}_*}.$$

In words, $\partial q / \partial \boldsymbol{\theta}$ must be a one-to-one transformation of the parameter space to guarantee that the multidimensional canonical exponential family satisfies the identifiability conditions.   □

The following result is fundamental for asymptotic statistical inference with the Fisher information matrix as the central concept. Recall, we use notation $\mathcal{I}_1(\boldsymbol{\theta})$ to denote the information matrix in a single observation. The information in the entire sample of n iid observations is $\mathcal{I}(\boldsymbol{\theta}) = n\mathcal{I}_1(\boldsymbol{\theta})$; see Section 6.9.

Theorem 6.134 *Asymptotic properties of the MLE.* *Under the ML assumptions A – E, when the sample size, n, increases to infinity, the MLE, $\widehat{\boldsymbol{\theta}}_{ML}$, has the following properties.*

1. *Consistency. With probability 1, the MLE converges to the true value,*

$$\Pr \left(\lim_{n \to \infty} \widehat{\boldsymbol{\theta}}_{ML} = \boldsymbol{\theta} \right) = 1. \tag{6.82}$$

2. *Asymptotic normality. In the limit, the MLE has a multivariate normal distribution with inverse of the information matrix as the covariance matrix:*

$$\sqrt{n}(\widehat{\boldsymbol{\theta}}_{ML} - \boldsymbol{\theta}) \simeq \mathcal{N} \left(\mathbf{0}, \mathcal{I}_1^{-1}(\boldsymbol{\theta}) \right)$$

3. *Asymptotic efficiency. In the limit, the covariance matrix of any other consistent M-estimator $\widetilde{\boldsymbol{\theta}}$ is larger than of the MLE. That is, if $\sqrt{n}(\widetilde{\boldsymbol{\theta}} - \boldsymbol{\theta}) \simeq \mathcal{N}\left(\mathbf{0}, \mathbf{A}(\boldsymbol{\theta})\right)$, then $\mathbf{A}(\boldsymbol{\theta}) \geq \mathcal{I}_1^{-1}(\boldsymbol{\theta})$. We say that MLE is asymptotically efficient.*

Proof. The results of the previous section are instrumental. The definition of the M-estimator and the proof of asymptotic efficiency is deferred to Section 6.11.

(1) Consistency. The MLE, $\widehat{\boldsymbol{\theta}}_{ML} = \widehat{\boldsymbol{\theta}}_n$, solves the score equation

$$\frac{1}{n} \sum_{i=1}^{n} \frac{\partial \ln f(\mathbf{Y}_i; \boldsymbol{\theta})}{\partial \boldsymbol{\theta}} = \mathbf{0}. \tag{6.83}$$

Consider a limiting point of the sequence $\widehat{\boldsymbol{\theta}}_n$ when $n \to \infty$. In other words, let a subsequence $\widehat{\boldsymbol{\theta}}_{n_k}$ converge to $\boldsymbol{\theta}_*$ when $k \to \infty$ and $n_k \to \infty$. By the strong law of large numbers

$$\Pr\left(\lim_{k \to \infty} \frac{1}{n} \sum_{i=1}^{n} \frac{\partial \ln f(\mathbf{Y}_i; \boldsymbol{\theta})}{\partial \boldsymbol{\theta}} \bigg|_{\boldsymbol{\theta} = \boldsymbol{\theta}_{n_k}} = E_{\boldsymbol{\theta}}\left(\frac{\partial \ln f(\mathbf{Y}; \boldsymbol{\theta})}{\partial \boldsymbol{\theta}} \bigg|_{\boldsymbol{\theta} = \boldsymbol{\theta}_*} \right) \right) = 1.$$

But the left-hand side of the expression under the probability sign is zero due to (6.83), so

$$E_{\boldsymbol{\theta}}\left(\frac{\partial \ln f(\mathbf{Y}; \boldsymbol{\theta})}{\partial \boldsymbol{\theta}} \bigg|_{\boldsymbol{\theta} = \boldsymbol{\theta}_*} \right) = \mathbf{0},$$

which implies that $\boldsymbol{\theta} = \boldsymbol{\theta}_*$ due to the identifiability condition E. This means that any limiting point of the sequence $\widehat{\boldsymbol{\theta}}_n$ is $\boldsymbol{\theta}$ and therefore (6.82) holds.

(2) Asymptotic normality. The central point of the proof is the fact that

$$\frac{1}{\sqrt{n}} \sum_{i=1}^{n} \frac{\partial \ln f(\mathbf{Y}_i; \boldsymbol{\theta})}{\partial \boldsymbol{\theta}} \simeq \mathcal{N}\left(\mathbf{0}, \mathcal{I}_1(\boldsymbol{\theta})\right). \tag{6.84}$$

Expand the left-hand side of the $m \times 1$ score equation into a first-order Taylor series around the true $\boldsymbol{\theta}$ as

$$\frac{1}{\sqrt{n}} \sum_{i=1}^{n} \frac{\partial \ln f(\mathbf{Y}_i; \boldsymbol{\theta})}{\partial \boldsymbol{\theta}} \bigg|_{\boldsymbol{\theta} = \boldsymbol{\theta}_*} = \frac{1}{\sqrt{n}} \sum_{i=1}^{n} \frac{\partial \ln f(\mathbf{Y}_i; \boldsymbol{\theta})}{\partial \boldsymbol{\theta}}$$

$$+ \frac{1}{\sqrt{n}} \sum_{i=1}^{n} \frac{\partial^2 \ln f(\mathbf{Y}_i; \boldsymbol{\theta})}{\partial \boldsymbol{\theta}^2} (\boldsymbol{\theta}_* - \boldsymbol{\theta}) + o(\|\boldsymbol{\theta}_* - \boldsymbol{\theta}\|).$$

Substituting $\boldsymbol{\theta}_*$ with the MLE $\widehat{\boldsymbol{\theta}}_{ML}$ we obtain

$$\mathbf{0} = \frac{1}{\sqrt{n}} \sum_{i=1}^{n} \frac{\partial \ln f(\mathbf{Y}_i; \boldsymbol{\theta})}{\partial \boldsymbol{\theta}} + \left(\frac{1}{n} \sum_{i=1}^{n} \frac{\partial^2 \ln f(\mathbf{Y}_i; \boldsymbol{\theta})}{\partial \boldsymbol{\theta}^2} \right)$$

$$\times \left(\sqrt{n}(\widehat{\boldsymbol{\theta}}_{ML} - \boldsymbol{\theta}) \right) + o(\|\widehat{\boldsymbol{\theta}}_{ML} - \boldsymbol{\theta}\|). \tag{6.85}$$

As was proven before, the first $m \times 1$ vector term of the right-hand side of (6.85) converges in distribution to a multivariate normal (6.84). Also, from the LLN,

$$\lim_{n \to \infty} \frac{1}{n} \sum_{i=1}^{n} \frac{\partial^2 \ln f(\mathbf{Y}_i; \boldsymbol{\theta})}{\partial \boldsymbol{\theta}^2} = E\left(\frac{\partial^2 \ln f(\mathbf{Y}; \boldsymbol{\theta})}{\partial \boldsymbol{\theta}^2}\right) = -\mathcal{I}_1(\boldsymbol{\theta})$$

because $\mathbf{Y}_i$ are iid and, therefore, the $m \times m$ matrices $\partial^2 \ln f(\mathbf{y}_i; \boldsymbol{\theta})/\partial \boldsymbol{\theta}^2$ are iid as well (the right-hand side was established in Theorem 6.102). The last term in (6.85) converges to zero with probability 1 due to the strong consistency of $\widehat{\boldsymbol{\theta}}_{ML}$. Now we use Theorem 4.2, which says that, if $\mathbf{X} \sim \mathcal{N}(\boldsymbol{\mu}, \boldsymbol{\Omega})$, then $\mathbf{AX} \sim \mathcal{N}(\mathbf{A}\boldsymbol{\mu}, \mathbf{A}\boldsymbol{\Omega}\mathbf{A}')$. Letting $\mathbf{X} \sim \mathcal{N}(\mathbf{0}, \boldsymbol{\Omega})$, where $\boldsymbol{\Omega} = \mathcal{I}_1(\boldsymbol{\theta})$ and $\mathbf{A} = \mathcal{I}_1^{-1}(\boldsymbol{\theta})$, we express the limiting distribution of $\sqrt{n}(\widehat{\boldsymbol{\theta}}_{ML} - \boldsymbol{\theta})$ as $\mathbf{AX}$, which has zero mean and covariance matrix $\mathcal{I}_1^{-1}(\boldsymbol{\theta})\mathcal{I}_1(\boldsymbol{\theta})\mathcal{I}_1^{-1}(\boldsymbol{\theta}) = \mathcal{I}_1^{-1}(\boldsymbol{\theta})$. Finally, we obtain the desired result $\sqrt{n}(\widehat{\boldsymbol{\theta}}_{ML} - \boldsymbol{\theta}) \simeq \mathcal{N}(\mathbf{0}, \mathcal{I}_1^{-1}(\boldsymbol{\theta}))$ when $n \to \infty$.

(3) *Asymptotic efficiency* is proven in Theorem 6.159 by showing that, among M-estimators, the MLE has the minimal asymptotic covariance matrix.   □

This theorem is at the center of statistical inference to derive estimators for parameters and their approximate distributions, confidence intervals and hypothesis testing. Of course, one never gets an infinitely large sample size, but, as an approximation, this theorem is the most powerful tool, especially in applied statistics. In short, one can say that the MLE is asymptotically normally distributed with the mean as a true parameter and covariance matrix $\mathcal{I}^{-1}(\boldsymbol{\theta}) = n^{-1}\mathcal{I}_1^{-1}(\boldsymbol{\theta})$. In practice, we approximate the distribution of the MLE as

$$\widehat{\boldsymbol{\theta}}_{ML} \simeq \mathcal{N}\left(\boldsymbol{\theta}, \mathcal{I}^{-1}(\boldsymbol{\theta})\right). \tag{6.86}$$

Typically, the information matrix is a continuous function of parameters; then, as follows from Section 6.4.4, we assert that the distribution of the MLE is approximately normal with the covariance matrix estimated at the MLE,

$$\widehat{\boldsymbol{\theta}}_{ML} \simeq \mathcal{N}\left(\boldsymbol{\theta}, \mathcal{I}^{-1}(\widehat{\boldsymbol{\theta}}_{ML})\right).$$

Once the Fisher information is derived, the asymptotic distribution is easy to write, as illustrated in the following example.

Example 6.135 *Large-sample distribution of the slope coefficient.* Derive the asymptotic distribution of the OLS/MLE slope estimator in the linear regression if $n \to \infty$ iid observations are drawn from the bivariate normal distribution.

Solution. As follows from Example 6.120, the OLS = MLE of the slope coefficient,

$$\widehat{\beta}_{ML} = \widehat{\beta}_{OLS} = \frac{\sum_{i=1}^{n}(X_i - \overline{X})(Y_i - \overline{Y})}{\sum_{i=1}^{n}(X_i - \overline{X})^2}.$$

Now we use Example 6.108, where the information matrix was derived, particularly the fact that the $(2,2)$th element of the inverse information matrix corresponding to the slope coefficient β is σ^2/σ_x^2. From Theorem 6.134, it follows that, for large n,

$$\sqrt{n}\left(\widehat{\beta}_{ML} - \beta\right) \simeq \mathcal{N}\left(0, \frac{\sigma^2}{\sigma_x^2}\right).$$

Since neither σ^2 nor σ_x^2 is known in practice, we approximate

$$\widehat{\beta}_{ML} \simeq \mathcal{N}\left(\beta, \frac{\widehat{\sigma}_{ML}^2}{\sum_{i=1}^{n}(X_i - \overline{X})^2}\right),$$

where

$$\widehat{\sigma}_{ML}^2 = \frac{1}{n}\sum_{i=1}^{n}(Y_i - \widehat{\alpha}_{ML} - \widehat{\beta}_{ML}X_i)^2.$$

Example 6.136 *MLE of the multinomial distribution. Derive the MLE for parameters of the multinomial distribution and specify their asymptotic properties.*

Solution. As follows from Section 3.10.4, the log-likelihood function, up to a term which contains observations only, takes a simple form

$$l(p_1, p_2, ..., p_m) = \sum_{j=1}^{m} X_j \ln p_j,$$

where X_j is the observed frequency of category j in n experiments and $\{p_1, ..., p_m\}$ are subject to estimation under the restriction that $\sum_{j=1}^{n} p_j = 1$. The Lagrangian multiplier technique is employed here to maximize l under restriction,

$$\mathcal{L}(p_1, p_2, ..., p_m; \lambda) = \sum_{j=1}^{m} X_j \ln p_j - \lambda\left(\sum_{j=1}^{m} p_j - 1\right).$$

Differentiation with respect to p_j gives

$$\frac{\partial \mathcal{L}}{\partial p_j} = \frac{X_j}{p_j} - \lambda = 0, \quad j = 1, 2, ..., m$$

which means that p_j are proportional to X_j. Since $\{X_j\}$ are independent and $\partial^2\mathcal{L}/\partial p_j^2 = -X_j/p_j^2$, $E(X_j) = np_j$, the information matrix is $\mathcal{I} = n \times diag(1/p_1, 1/p_2, ..., 1/p_m)$. Now we find the ML estimators for p_j given an iid sample of size n. Using the restriction, find λ

$$1 = \sum_{j=1}^{m} p_j = \frac{1}{\lambda}\sum_{j=1}^{m} X_j = \frac{n}{\lambda},$$

and finally, $\widehat{p}_j = X_j/n$. The ML estimator is unbiased since $E(X_j) = np_j$ and the $m \times m$ covariance matrix can be derived directly using formulas presented in Section 3.10.4,

$$C_{jk} = \frac{1}{n} \times \begin{cases} p_j(1-p_j) \text{ if } j = k \\ \\ -p_j p_k \text{ if } j \neq k \end{cases}.$$

A peculiar property of this covariance matrix is that it singular. Indeed, introduce the $m \times 1$ vector of probabilities $\mathbf{p} = (p_1, p_2, ..., p_m)'$ and $\mathbf{P}$, the diagonal matrix composed of this vector, or, symbolically, $\mathbf{P} = diag(\mathbf{p})$. Then it is easy to check that the $m \times m$ covariance matrix of the ML estimates can be compactly written as

$$\mathbf{C} = \frac{1}{n}(\mathbf{P} - \mathbf{pp}').$$

We prove that elements in each row (or column) add to 0. Indeed, let $\mathbf{1}$ be the $m \times 1$ vector of ones. It is equivalent to prove that $\mathbf{C1} = \mathbf{0}$. We have

$$\mathbf{C1} = \frac{1}{n}(\mathbf{P} - \mathbf{pp}')\mathbf{1} = \frac{1}{n}(\mathbf{P1} - \mathbf{pp}'\mathbf{1}).$$

But $\mathbf{P1} = \mathbf{p}$ and $\mathbf{pp}'\mathbf{1} = \mathbf{p}(\mathbf{p}'\mathbf{1}) = \mathbf{p}$ because $\mathbf{p}'\mathbf{1} = \sum_{j=1}^{m} p_j = 1$. In matrix notation we write $\mathbf{C1} = \mathbf{0}$, which means that matrix $\mathbf{C}$ is singular. In a special case when the multinomial distribution becomes binomial, we have $m = 2$ and $p = p_1 = 1 - p_2$:

$$\mathbf{C} = \frac{1}{n}\begin{bmatrix} p(1-p) & -p(1-p) \\ -p(1-p) & p(1-p) \end{bmatrix}$$

with zero determinant. Matrix $\mathbf{C}$ degenerates due to restrictions of the probability parameters, $\sum_{j=1}^{m} p_j = 1$. The same is true for their ML estimates because $\sum_{j=1}^{m} X_j = m$. Although ML estimates for p are consistent and asymptotically normal, the limiting multivariate normal distribution $\sqrt{n}(\widehat{\mathbf{p}} - \mathbf{p}) \simeq \mathcal{N}(\mathbf{0}, \mathbf{P} - \mathbf{pp}')$ degenerates because the covariance matrix is singular. More insights into maximum likelihood estimation with restricted parameters can be learned from Example 6.148.   □

The following is a continuation of Example 6.120. Matrix algebra and matrix calculus is instrumental for this and the following example; see Appendix. Also we use the result on the covariance matrix of the vector Kronecker product from Section 4.2.3.

Example 6.137 *The MLE for the multivariate normal distribution.* *Formulate the asymptotic properties of the MLE based on a sample from a multivariate normal distribution,* $\mathbf{Y}_i \overset{\text{iid}}{\sim} \mathcal{N}(\boldsymbol{\mu}, \boldsymbol{\Omega})$, $i = 1, 2, ..., n$.

Solution. The information matrix is derived as the covariance matrix of the first derivatives with respect to vector $\boldsymbol{\mu}$ and matrix $\boldsymbol{\Omega}$. Since $\partial \ln f / \partial \boldsymbol{\mu} = \boldsymbol{\Omega}^{-1}(\mathbf{Y} - \boldsymbol{\mu})$, we obtain the information block corresponding to $\boldsymbol{\mu}$ as

$$\mathrm{cov}\left(\frac{\partial \ln f}{\partial \boldsymbol{\mu}}\right) = E\left[\boldsymbol{\Omega}^{-1}(\mathbf{Y} - \boldsymbol{\mu})\right]\left[\boldsymbol{\Omega}^{-1}(\mathbf{Y} - \boldsymbol{\mu})\right]'$$

$$= \boldsymbol{\Omega}^{-1} E\left[(\mathbf{Y} - \boldsymbol{\mu})(\mathbf{Y} - \boldsymbol{\mu})'\right] \boldsymbol{\Omega}^{-1} = \boldsymbol{\Omega}^{-1}\boldsymbol{\Omega}\boldsymbol{\Omega}^{-1} = \boldsymbol{\Omega}^{-1}.$$

To obtain the information block corresponding to matrix $\boldsymbol{\Omega}$, we write the derivative as

$$\frac{\partial \ln f}{\partial \boldsymbol{\Omega}} = -\frac{1}{2}\boldsymbol{\Omega}^{-1} + \frac{1}{2}\boldsymbol{\Omega}^{-1}(\mathbf{Y} - \boldsymbol{\mu})(\mathbf{Y} - \boldsymbol{\mu})'\boldsymbol{\Omega}^{-1}.$$

The covariance of the matrix is found as the covariance of the stacked vector columns of matrix $\partial \ln f / \partial \boldsymbol{\Omega}$ using formulas from the Appendix,

$$\mathrm{vec}\left(\frac{\partial \ln f}{\partial \boldsymbol{\Omega}}\right) = -\frac{1}{2}\mathrm{vec}(\boldsymbol{\Omega}^{-1}) + \frac{1}{2}\mathrm{vec}[\boldsymbol{\Omega}^{-1}(\mathbf{Y} - \boldsymbol{\mu})(\mathbf{Y} - \boldsymbol{\mu})'\boldsymbol{\Omega}^{-1}]$$

and, therefore,

$$\mathrm{vec}[\boldsymbol{\Omega}^{-1}(\mathbf{Y} - \boldsymbol{\mu})(\mathbf{Y} - \boldsymbol{\mu})'\boldsymbol{\Omega}^{-1}] = \boldsymbol{\Omega}^{-1} \otimes \boldsymbol{\Omega}^{-1}[(\mathbf{Y} - \boldsymbol{\mu}) \otimes (\mathbf{Y} - \boldsymbol{\mu})].$$

Now find the covariance of the vector using the formula from Section 4.2.3 $\mathrm{cov}[(\mathbf{Y} - \boldsymbol{\mu}) \otimes (\mathbf{Y} - \boldsymbol{\mu})'] = 2(\boldsymbol{\Omega} \otimes \boldsymbol{\Omega})$,

$$\mathrm{cov}\left(\boldsymbol{\Omega}^{-1} \otimes \boldsymbol{\Omega}^{-1}[(\mathbf{Y} - \boldsymbol{\mu}) \otimes (\mathbf{Y} - \boldsymbol{\mu})]\right)$$

$$= (\boldsymbol{\Omega}^{-1} \otimes \boldsymbol{\Omega}^{-1})\mathrm{cov}[(\mathbf{Y} - \boldsymbol{\mu}) \otimes (\mathbf{Y} - \boldsymbol{\mu})'](\boldsymbol{\Omega}^{-1} \otimes \boldsymbol{\Omega}^{-1})$$

$$= (\boldsymbol{\Omega}^{-1} \otimes \boldsymbol{\Omega}^{-1})(\boldsymbol{\Omega} \otimes \boldsymbol{\Omega})(\boldsymbol{\Omega}^{-1} \otimes \boldsymbol{\Omega}^{-1}) = 2(\boldsymbol{\Omega}^{-1} \otimes \boldsymbol{\Omega}^{-1}),$$

so

$$\mathrm{cov}\left[\mathrm{vec}\left(\frac{\partial \ln f}{\partial \boldsymbol{\Omega}}\right)\right] = 2(\boldsymbol{\Omega}^{-1} \otimes \boldsymbol{\Omega}^{-1}).$$

Now we prove that the covariance between the derivative with respect to the mean and the covariance matrix is zero,

$$\mathrm{cov}\left(\frac{\partial \ln f}{\partial \boldsymbol{\Omega}}, \mathrm{vec}\left(\frac{\partial \ln f}{\partial \boldsymbol{\Omega}}\right)\right) = \mathbf{0}. \tag{6.87}$$

Indeed, the left-hand side of (6.87) can be rewritten as

$$E\left[\boldsymbol{\Omega}^{-1}(\mathbf{Y} - \boldsymbol{\mu})\left(\boldsymbol{\Omega}^{-1} \otimes \boldsymbol{\Omega}^{-1}[(\mathbf{Y} - \boldsymbol{\mu}) \otimes (\mathbf{Y} - \boldsymbol{\mu})]\right)'\right]$$

$$= \boldsymbol{\Omega}^{-1}E\left[(\mathbf{Y} - \boldsymbol{\mu})(\mathbf{Y} - \boldsymbol{\mu})' \otimes (\mathbf{Y} - \boldsymbol{\mu})'\right]\boldsymbol{\Omega}^{-1} \otimes \boldsymbol{\Omega}^{-1}.$$

But each element of the matrix under the expectation sign can be expressed as the sum of $Z_k Z_l Z_q$, where the Zs are iid standard normal random variables, so

(6.87) holds. Finally, the asymptotic distribution is multivariate normal. More specifically,

$$
\sqrt{n}
\begin{bmatrix}
\overline{\mathbf{Y}} - \mu \\
\mathrm{vec}(\widehat{\mathbf{\Omega}}_{ML}) - \mathrm{vec}(\mathbf{\Omega})
\end{bmatrix}
=
\begin{bmatrix}
\mathbf{\Omega} & \mathbf{0} \\
\mathbf{0} & 0.5(\mathbf{\Omega} \otimes \mathbf{\Omega})
\end{bmatrix},
\tag{6.88}
$$

where $\overline{\mathbf{Y}} = \widehat{\mu}_{ML}$ and $\widehat{\mathbf{\Omega}}_{ML} = n^{-1} \sum_{i=1}^{n} (\mathbf{Y}_i - \overline{\mathbf{Y}})(\mathbf{Y}_i - \overline{\mathbf{Y}})'$. The MLE for $\mathbf{\Omega}$ follows from $\partial \ln |\mathbf{\Omega}| / \partial \mathbf{\Omega} = \mathbf{\Omega}^{-1}$.

The fact below, sometimes referred to as the *asymptotic delta method*, is useful if the vector of parameters of interest, τ, is a function of the original vector of parameters, θ. Maximum likelihood is invariant in this respect as expressed in equation (6.73). Moreover, it is not difficult to prove that under the conditions of the previous theorem

$$
\sqrt{n}(\widehat{\tau}_{ML} - \tau) \simeq \mathcal{N}\left(0, \left(\frac{\partial \tau}{\partial \theta}\right) \mathcal{I}_1^{-1}(\theta) \left(\frac{\partial \tau}{\partial \theta}\right)' \right), \quad n \to \infty.
\tag{6.89}
$$

Essentially, the covariance matrix of $\widehat{\tau}_{ML} = \tau(\widehat{\theta}_{ML})$ is derived via the delta method using formula (3.66) from Section 3.10.3. If τ is the p-dimensional parameter, then $\partial \tau / \partial \theta$ is the $p \times q$ matrix and the resulting covariance matrix is $p \times p$.

We apply this result to find the distribution of the correlation coefficient in large sample, a classic result of mathematical statistics.

Example 6.138 *Asymptotic distribution of the correlation coefficient.*
Prove that, when $n \to \infty$, the distribution of the sample correlation coefficient in the bivariate normal distribution converges to normal as

$$
\sqrt{n}(r - \rho) \simeq \mathcal{N}\left(0, \frac{1}{(1 - \rho^2)^2} \right).
\tag{6.90}
$$

Solution. We use asymptotic normality (6.89) with the 2×2 covariance matrix

$$
\mathbf{\Omega} =
\begin{bmatrix}
\sigma_x^2 & \sigma_{xy} \\
\sigma_{xy} & \sigma_y^2
\end{bmatrix}
$$

and the correlation coefficient $\rho = \sigma_{xy}(\sigma_x^2)^{-1/2}(\sigma_y^2)^{-1/2}$ as a function of σ_x^2, σ_{xy}, and σ_y^2, corresponding to the one-dimensional function τ in the formula (6.89). To derive the asymptotic distribution for the sample correlation coefficient r, as follows from the previous example, we must obtain the 4×4 Kronecker matrix

$$
\mathbf{\Omega} \otimes \mathbf{\Omega} =
\begin{bmatrix}
\sigma_x^4 & \sigma_x^2 \sigma_{xy} & \sigma_{xy} \sigma_x^2 & \sigma_{xy}^2 \\
\sigma_x^2 \sigma_{xy} & \sigma_x^2 \sigma_y^2 & \sigma_{xy}^2 & \sigma_{xy} \sigma_y^2 \\
\sigma_{xy} \sigma_x^2 & \sigma_{xy}^2 & \sigma_x^2 \sigma_y^2 & \sigma_y^2 \sigma_{xy} \\
\sigma_{xy}^2 & \sigma_{xy} \sigma_y^2 & \sigma_y^2 \sigma_{xy} & \sigma_y^4
\end{bmatrix}
$$

and the 4×1 derivative vector

$$
\frac{\partial \rho}{\partial \boldsymbol{\theta}} =
\begin{bmatrix}
\frac{\partial \rho}{\partial \sigma_x^2} \\[4pt]
\frac{1}{2} \frac{\partial \rho}{\partial \sigma_{xy}} \\[4pt]
\frac{1}{2} \frac{\partial \rho}{\partial \sigma_{xy}} \\[4pt]
\frac{\partial \rho}{\partial \sigma_y^2}
\end{bmatrix}
=
\begin{bmatrix}
-\frac{1}{2}\sigma_{xy}(\sigma_x^2)^{-3/2}(\sigma_y^2)^{-1/2} \\[4pt]
\frac{1}{2}(\sigma_x^2)^{-1/2}(\sigma_y^2)^{-1/2} \\[4pt]
\frac{1}{2}(\sigma_x^2)^{-1/2}(\sigma_y^2)^{-1/2} \\[4pt]
-\frac{1}{2}\sigma_{xy}(\sigma_x^2)^{-1/2}(\sigma_y^2)^{-3/2}
\end{bmatrix},
$$

where $\boldsymbol{\theta} = (\sigma_x^2, \sigma_{xy}, \sigma_{xy}, \sigma_y^2)$. The element σ_{xy} repeats to reflect its repetition in the vector representation of $\boldsymbol{\Omega}$. The rest of the proof involves some tedious algebra (omitted) that leads to

$$
\begin{bmatrix}
-\frac{1}{2}\sigma_{xy}(\sigma_x^2)^{-3/2}(\sigma_y^2)^{-1/2} \\[4pt]
\frac{1}{2}(\sigma_x^2)^{-1/2}(\sigma_y^2)^{-1/2} \\[4pt]
\frac{1}{2}(\sigma_x^2)^{-1/2}(\sigma_y^2)^{-1/2} \\[4pt]
-\frac{1}{2}\sigma_{xy}(\sigma_x^2)^{-1/2}(\sigma_y^2)^{-3/2}
\end{bmatrix}^{T}
\begin{bmatrix}
\sigma_x^4 & \sigma_x^2\sigma_{xy} & \sigma_{xy}\sigma_x^2 & \sigma_{xy}^2 \\[4pt]
\sigma_x^2\sigma_{xy} & \sigma_x^2\sigma_y^2 & \sigma_{xy}^2 & \sigma_{xy}\sigma_y^2 \\[4pt]
\sigma_{xy}\sigma_x^2 & \sigma_{xy}^2 & \sigma_x^2\sigma_y^2 & \sigma_y^2\sigma_{xy} \\[4pt]
\sigma_{xy}^2 & \sigma_{xy}\sigma_y^2 & \sigma_y^2\sigma_{xy} & \sigma_y^4
\end{bmatrix}
$$

$$
\times
\begin{bmatrix}
-\frac{1}{2}\sigma_{xy}(\sigma_x^2)^{-3/2}(\sigma_y^2)^{-1/2} \\[4pt]
\frac{1}{2}(\sigma_x^2)^{-1/2}(\sigma_y^2)^{-1/2} \\[4pt]
\frac{1}{2}(\sigma_x^2)^{-1/2}(\sigma_y^2)^{-1/2} \\[4pt]
-\frac{1}{2}\sigma_{xy}(\sigma_x^2)^{-1/2}(\sigma_y^2)^{-3/2}
\end{bmatrix}
= \frac{1}{2}\left(1 - \rho^2\right)^2.
$$

Thus, formula (6.89) gives the result (6.90).

6.10.5   When maximum likelihood breaks down

The goal of this section is to illustrate that the ML assumptions A–E listed above are important for asymptotic properties of the MLE. When they do not hold, the maximum likelihood may break down.

The identifiability condition does not hold

A necessary condition for the ML is the identifiability condition E: the pdf is defined uniquely by the parameters: $f(\mathbf{y}; \boldsymbol{\theta}_1) = f(\mathbf{y}; \boldsymbol{\theta}_2)$ for all $\mathbf{y}$ implies that $\boldsymbol{\theta}_1 = \boldsymbol{\theta}_2$. When this condition does not hold, we say that parameters are nonidentifiable and the model is overspecified. Obviously, if, for some $\boldsymbol{\theta}_1 \neq \boldsymbol{\theta}_2$, we have $f(\mathbf{y}; \boldsymbol{\theta}_1) = f(\mathbf{y}; \boldsymbol{\theta}_2)$ for all $\mathbf{y}$, the ML estimator may converge to either $\boldsymbol{\theta}_1$ or $\boldsymbol{\theta}_2$. Or, does not converge at all simply because, from the probabilistic standpoint, parameters $\boldsymbol{\theta}_1$ and $\boldsymbol{\theta}_2$ are indistinguishable. Below is a concrete example of an overspecified model for which the information matrix becomes deficient because condition E is not satisfied.

Example 6.139 *Three-mean overspecified model.* *Two iid samples have normal distribution with same known variance* σ^2 *but different group-specific means. Apply maximum likelihood to estimate the group-specific means using the model* $Y_{ij} = \mu + \delta_j + \varepsilon_{ij}$, *where* $\varepsilon_{ij} \overset{iid}{\sim} \mathcal{N}(0, \sigma^2)$ *and the variance* σ^2 *is known.*

Solution. As follows from this model, the first group of observations has mean $\mu_1 = \mu + \delta_1$ and the second group has the mean $\mu_2 = \mu + \delta_2$. That is, δ_1 and δ_2 are deviations of the group-specific means from the gross mean, μ. The log-likelihood function, up to a constant term, takes the form

$$l(\mu, \mu_1, \mu_2) = -\frac{1}{2\sigma^2} \sum_{i=1}^{n} \left[(Y_{i1} - \mu - \mu_1)^2 + (Y_{i2} - \mu - \mu_2)^2 \right].$$

The second derivatives are

$$\frac{\partial^2 l}{\partial \mu^2} = -\frac{2n}{\sigma^2}, \quad \frac{\partial^2 l}{\partial \mu_1^2} = -\frac{n}{\sigma^2}, \quad \frac{\partial^2 l}{\partial \mu_2^2} = -\frac{n}{\sigma^2},$$

$$\frac{\partial^2 l}{\partial \mu \partial \mu_1} = -\frac{n}{\sigma^2}, \quad \frac{\partial^2 l}{\partial \mu \partial \mu_2} = -\frac{n}{\sigma^2}, \quad \frac{\partial^2 l}{\partial \mu_1 \partial \mu_2} = 0,$$

so the 3×3 Fisher information matrix

$$\mathcal{I} = \frac{n}{\sigma^2} \begin{bmatrix} 2 & 1 & 1 \\ 1 & 1 & 0 \\ 1 & 0 & 1 \end{bmatrix}.$$

This matrix is deficient (singular) because the first column is the sum of the second and the third. Therefore, $\mathcal{I}$ does not have full rank. Consequently, asymptotic normality of the MLE cannot be claimed. We note that the score equation for the means leads to a system of three equations

$$\overline{Y}_1 - \mu - \mu_1 = 0, \quad \overline{Y}_2 - \mu - \mu_2 = 0, \quad \overline{Y}_1 + \overline{Y}_2 - 2\mu - \mu_2 = 0.$$

Obviously, the third equation is the sum of the first two: the MLE cannot be derived uniquely. In short, the gross mean and group-specific means cannot be identified simultaneously – the model is overspecified. To correct this model, assume that $\delta_1 = -\delta_2$. Alternatively, we may write the correct model as $Y_{ij} = \mu_j + \varepsilon_{ij}$. Note here that we assumed that σ^2 is known – matrix $\mathcal{I}$ remains deficient when σ^2 is unknown as well.

MLE may be inconsistent

It is commonly believed that the maximum likelihood estimator is always consistent. The following examples illustrates that, when condition A is not satisfied, the MLE may be inconsistent.

Example 6.140 *Decay model.* *Consider the slope model* $Y_t = \beta e^{-\gamma t} + \varepsilon_t$, *where the decay rate* $\gamma > 0$ *is known and* $\varepsilon_t \overset{iid}{\sim} \mathcal{N}(0, \sigma^2)$, $t = 1, 2, ..., n$. *To simplify, it is assumed that the variance of observations,* σ^2, *is known as well. This model is used to estimate* β, *the mean of* Y *at the start of the decay process* $(t = 0)$. *Show that* $\widehat{\beta}_{ML}$ *is inconsistent when* $n \to \infty$.

Solution. For this example, condition A does not hold because Y_t are not identically distributed, $E(Y_t) = \beta e^{-\gamma t} \neq$ const. The log-likelihood function, up to a constant term, takes the form

$$l(\beta) = -\frac{n}{2} \ln \sigma^2 - \frac{1}{2\sigma^2} \sum_{t=1}^{n} (Y_t - \beta e^{-\gamma t})^2.$$

Taking the derivative with respect to β and setting it to zero leads to the MLE expressed in the closed form as

$$\widehat{\beta}_{ML} = \frac{\sum_{t=1}^{n} Y_t e^{-\gamma t}}{\sum_{t=1}^{n} e^{-2\gamma t}}.$$

One may interpret the decay model with a known rate as the slope model from Example 6.36 where $x_t = e^{-\gamma t}$ and therefore

$$\mathrm{var}(\widehat{\beta}_{ML}) = \frac{\sigma^2}{\sum_{t=1}^{n} e^{-2\gamma t}}.$$

To show that the MLE is not consistent, we prove that the variance of $\widehat{\beta}_{ML}$ does not vanish when the sample size increases. Indeed, since $e^{-2\gamma t}$ may be viewed as the geometric progression, we have

$$\lim_{n \to \infty} \sum_{t=1}^{n} e^{-2\gamma t} = \sum_{t=1}^{\infty} e^{-2\gamma t} = \frac{e^{-2\gamma}}{1 - e^{-2\gamma}} > 0,$$

and therefore $\mathrm{var}(\widehat{\beta}_{ML})$ does not go to zero. This means that the MLE is not consistent. $\qquad\square$

Below, we consider a famous example where the dimension of the unknown parameter vector increases with the sample size – assumption A does not hold and the MLE is not consistent.

Example 6.141 *Neyman and Scott problem.* nk *independent normally distributed observations are presented in tabular form as* $\{Y_{ij}, i = 1, ..., n; j = 1, ..., k\}$. *It is assumed that all observations have the same variance* σ^2, *but observations from group i have different mean,* μ_i. *Compactly, we write* $Y_{ij} \sim \mathcal{N}(\mu_i, \sigma^2)$. *Prove that when the sample size, n, goes to infinity, but k is fixed, the MLE of neither* μ_i *nor* σ^2 *is consistent.*

CHAPTER 6. PARAMETER ESTIMATION

Solution. The $(n+1)$-dimensional parameter vector is $\boldsymbol{\theta} = (\mu_1, ..., \mu_n, \sigma^2)$ and its dimension increases with the sample size, an obvious violation of the maximum likelihood theory assumption. The log-likelihood function is

$$l(\mu_1, ..., \mu_n, \sigma^2) = -\frac{1}{2}\left[nk \ln \sigma^2 + \frac{1}{\sigma^2} \sum_{i=1}^{n} \sum_{j=1}^{k} (Y_{ij} - \mu_i)^2 \right]$$

with the ML estimates

$$\widehat{\mu}_i = \overline{Y}_i = \frac{1}{k} \sum_{j=1}^{k} Y_{ij}, \quad i = 1, 2, ..., n, \quad \widehat{\sigma}^2 = \frac{1}{nk} \sum_{i=1}^{n} \sum_{j=1}^{k} (Y_{ij} - \overline{Y}_i)^2.$$

The fact that $\widehat{\mu}_i$ do not converge to μ_i when $n \to \infty$ is easy to understand because $\text{var}(\widehat{\mu}_i) = \sigma^2/k = \text{const}$ and does not vanish with n. But $\widehat{\sigma}^2$ does not converge to σ^2 either. In fact, $\widehat{\sigma}^2$ has a systematic bias because as follows from Section 4.2

$$E \sum_{j=1}^{k} (Y_{ij} - \overline{Y}_i)^2 = \sum_{j=1}^{k} E(Y_{ij} - \overline{Y}_i)^2 = (k-1)\sigma^2,$$

so that $E(\widehat{\sigma}^2) = \sigma^2 - \sigma^2/k$. Thus, we conclude that the MLE for σ^2 is not consistent because it has a systematic negative bias σ^2/k. $\qquad\square$

The bandwidth parameter in the kernel density estimation is critical, see Section 5.5. The next example shows that this parameter cannot be estimated from the data.

Example 6.142 *Bandwidth cannot be estimated.* *The bandwidth parameter h of the Gaussian kernel density (5.12) cannot be estimated by maximum likelihood.*

Solution. Let $X_1, X_2, .., X_n$ be iid distributed with the Parzen density

$$f(x; h) = \frac{1}{nh} \sum_{j=1}^{n} \phi\left(\frac{x - X_j}{h}\right),$$

where $\phi(s) = (2\pi)^{-1/2} e^{-\frac{1}{2}s^2}$ is the normal kernel. The log-likelihood function, up to a constant term, takes the form

$$l(h) = -n \ln h + \sum_{i=1}^{n} \ln \left[\sum_{j=1}^{n} \phi\left(\frac{X_i - X_j}{h}\right) \right]. \tag{6.91}$$

We will show that this function goes to $+\infty$ when $h \to 0$, the maximum likelihood estimate does nor exist. Indeed,

$$\sum_{i=1}^{n} \ln \left[\sum_{j=1}^{n} \phi \left(\frac{X_i - X_j}{\sigma} \right) \right] = \sum_{i=1}^{n} \ln \left[\phi(0) + \sum_{j \neq i} \phi \left(\frac{X_i - X_j}{\sigma} \right) \right]$$

$$\geq \sum_{i=1}^{n} \ln \phi(0) = -0.5n \ln(2\pi),$$

which implies $l(h) \geq -n \ln h - 0.5n \ln(2\pi)$. Since $l(\sigma) \to +\infty$ when $\sigma \to 0$, the MLE does not exist. The problem is that the kernel density involves a number of parameters increasing with n, so that condition A does not hold. Indeed, the correct log-likelihood function should include the means of each term in the Parzen density,

$$f(x; h, \mu_1, ..., \mu_n) = \frac{1}{nh} \sum_{j=1}^{n} \phi \left(\frac{x - \mu_j}{h} \right)$$

and therefore the log-likelihood takes the form

$$l(h, \mu_1, ..., \mu_n) = -n \ln h + \sum_{i=1}^{n} \ln \left[\sum_{j=1}^{n} \phi \left(\frac{X_i - \mu_j}{h} \right) \right],$$

which takes the form (6.91) after replacing μ_j with their ML estimates $\widehat{\mu}_j = X_j$. Thus, implicitly, kernel density estimation involves a number of parameters μ_j increasing with n. □

The MLE is not always inconsistent when the number of parameters increases with the sample size. Here is an example.

Example 6.143 *MLE may be consistent when the number of parameters increases with* n. *Observations have the same tabular form as in Example 6.141, but now* $Y_{ij}|(x_1, ..., x_k) \sim \mathcal{N}(\mu_i + \beta x_j, \sigma^2)$, *where* $x_1, x_2, ..., x_k$ *are iid with a possibly unknown density dependent on parameter* τ *independent of* μ_i *and* β. *Prove that the MLE for* β *is consistent when* $n \to \infty$ *and* k *is fixed.*

Solution. As with the above example, the log-likelihood takes the form

$$l(\mu_1, ..., \mu_n, \sigma^2) = -\frac{1}{2} \left[nk \ln \sigma^2 + \frac{1}{\sigma^2} \sum_{i=1}^{n} \sum_{j=1}^{k} (Y_{ij} - \mu_i - \beta x_j)^2 \right] + \sum_{j=1}^{k} \ln f(x_j).$$

Differentiate with respect to μ_i, set to 0, and express μ_i through β,

$$\mu_i = \frac{1}{k} \sum_{j=1}^{k} (Y_{ij} - \beta x_j) = \overline{Y}_i - \beta \overline{x}, \quad i = 1, 2, ..., n.$$

Substituting these values back into l, we arrive at the slope model of Section 6.5,

$$\widehat{\beta} = \frac{\sum_{j=1}^{k}(\overline{Y}_j - \overline{\overline{Y}})(x_j - \overline{x})}{\sum_{j=1}^{k}(x_j - \overline{x})^2}.$$

On the other hand, regression with subject-specific intercepts is consistent because

$$\text{var}(\widehat{\beta}) = \frac{1}{nk}\frac{\sigma^2}{\sum_{j=1}^{k}(x_j - \overline{x})^2} \to 0 \text{ when } n \to \infty.$$

$\square$

In the example below, Assumption A does not hold because observations correlate. In fact, the coefficient of correlation between any pair of observations is the same – this model was studied earlier in Example 2.46 of Section 2.9. We show that the MLE is inconsistent then.

Example 6.144 *Observations are correlated.* *Consider the following statistical model:*

$$Y_i = \mu + \eta + \varepsilon_i, \ i = 1, 2, ..., n,$$

where μ is the parameter of interest, and

$$\eta \sim \mathcal{N}(0, \sigma_\eta^2), \quad \varepsilon_i \sim \mathcal{N}(0, \sigma_\varepsilon^2)$$

are unobservable independent random variables (variances σ_η^2 and σ_ε^2 are assumed known for simplicity).

Solution. Since ε_i and η are independent $\text{var}(Y_i) = \sigma_\varepsilon^2 + \sigma_\eta^2$ and $\text{cov}(Y_i, Y_j) = \sigma_\eta^2$ for $i \neq j$. The model can be succinctly written as multivariate normal in matrix notation

$$\mathbf{Y} \sim \mathcal{N}(\mu\mathbf{1}, \sigma_\varepsilon^2(\mathbf{I}+\rho\mathbf{1}\mathbf{1}')),$$

where $\mathbf{Y} = (Y_1, ..., Y_n)'$, $\mathbf{1}$ is the $n \times 1$ vector of ones and $\rho = \sigma_\eta^2/\sigma_\varepsilon^2$ (we recommend reviewing Section 4.1). This model can be viewed as the mean model of Section 6.5. Find the MLE for μ by maximizing the log-likelihood function,

$$l(\mu) = -\frac{n}{2}\ln(2\pi) - \frac{1}{2}\ln|\sigma_\varepsilon^2(\mathbf{I}+\rho\mathbf{1}\mathbf{1}'))| - \frac{1}{2\sigma_\varepsilon^2}(\mathbf{Y}-\mu\mathbf{1})'(\mathbf{I}+\rho\mathbf{1}\mathbf{1}')^{-1}(\mathbf{Y}-\mu\mathbf{1}).$$

Note that the first two terms do not depend on μ, so the maximum over μ is equivalent to the minimum of the quadratic function

$$\min_{\mu}(\mathbf{Y}-\mu\mathbf{1})'(\mathbf{I}+\rho\mathbf{1}\mathbf{1}')^{-1}(\mathbf{Y}-\mu\mathbf{1}).$$

Differentiating with respect to μ we obtain the first-order condition for the minimum, $\mathbf{1}'(\mathbf{I}+\rho\mathbf{1}\mathbf{1}')^{-1}(\mathbf{Y}-\mu\mathbf{1}) = 0$, which gives

$$\widehat{\mu}_{ML} = \frac{\mathbf{1}'(\mathbf{I}+\rho\mathbf{1}\mathbf{1}')^{-1}\mathbf{y}}{\mathbf{1}'(\mathbf{I}+\rho\mathbf{1}\mathbf{1}')^{-1}\mathbf{1}}.$$

Find its variance using some matrix algebra as

$$\mathrm{var}(\widehat{\mu}_{ML}) = \sigma_\varepsilon^2 \frac{\mathbf{1}'(\mathbf{I}+\rho\mathbf{1}\mathbf{1}')^{-1}(\mathbf{I}+\rho\mathbf{1}\mathbf{1}')(\mathbf{I}+\rho\mathbf{1}\mathbf{1}')^{-1}\mathbf{1}}{(\mathbf{1}'(\mathbf{I}+\rho\mathbf{1}\mathbf{1}')^{-1}\mathbf{1})^2} = \frac{\sigma_\varepsilon^2}{\mathbf{1}'(\mathbf{I}+\rho\mathbf{1}\mathbf{1}')^{-1}\mathbf{1}}.$$

The inverse of matrix $\mathbf{I}+\rho\mathbf{1}\mathbf{1}'$ can be written in closed form as

$$(\mathbf{I}+\rho\mathbf{1}\mathbf{1}')^{-1} = \mathbf{I} - \frac{\rho}{1+n\rho}\mathbf{1}\mathbf{1}',$$

see formula (10.1) of the Appendix. Further,

$$\mathbf{1}'(\mathbf{I}+\rho\mathbf{1}\mathbf{1}')^{-1}\mathbf{1} = n - \frac{\rho}{1+n\rho}n^2 = \frac{n}{1+n\rho} \to \frac{1}{\rho}$$

when $n \to \infty$. Thus, the MLE is not consistent because the variance of $\widehat{\mu}_{ML}$ does not go to zero. $\qquad\square$

In general the MLE is consistent whether the parameter space, Θ, is open or not. However, Θ must be open to claim asymptotic normality. In the next example, the parameter space is not open (condition B does not hold) and we show that the limiting distribution is not normal. There may be several situations when a parameter is nonnegative: It may be nonnegative due to common sense, say, height or weight, or it may be the variance. The naive solution of assuming that the parameter is positive (the parameter space is open) leads to another problem: the ML estimate will not exist with positive probability and, moreover, this probability may be up to 50% when the true parameter is close to zero. Specifically, if the parameter belongs to the boundary of Θ (when Θ is not open), then the limiting distribution is a mixture of probability mass and normal distribution, as shown in an example below.

Example 6.145 *Parameter belongs to the boundary of the parameter space. Show that the distribution of the MLE of μ in the standard normal distribution with known σ^2, when $\mu \geq 0$, is not normal.*

Solution. Observations $Y_i \overset{\text{iid}}{\sim} \mathcal{N}(\mu, \sigma^2)$, where $\mu \geq 0$ and σ^2 is known. The ML estimator takes the form

$$\widehat{\mu}_{ML} = \max(0, \overline{Y}) = \begin{cases} 0 \text{ if } \overline{Y} < 0 \\ \overline{Y} \text{ if } \overline{Y} \geq 0 \end{cases}.$$

The cdf of $\widehat{\mu}_{ML}$ is

$$\Pr(\widehat{\mu}_{ML} \leq x) = \begin{cases} 0 \text{ if } x < 0 \\ \Phi\left(-\frac{\mu}{\sigma}\sqrt{n}\right) \text{ if } x = 0 \\ \Phi\left(-\frac{x-\mu}{\sigma}\sqrt{n}\right) \text{ if } x > 0 \end{cases},$$

so the cdf of $\sqrt{n}(\widehat{\mu}_{ML} - \mu)$ is

$$\Pr\left(\sqrt{n}(\widehat{\mu}_{ML} - \mu) \leq x\right) = \begin{cases} 0 \text{ if } x < -\sqrt{n}\mu \\ \Phi\left(-\frac{\mu}{\sigma}\sqrt{n}\right) \text{ if } x = -\sqrt{n}\mu \\ \Phi\left(\frac{x}{\sigma}\right) \text{ if } x > -\sqrt{n}\mu \end{cases} .$$

If the true $\mu > 0$, then $\lim_{n\to\infty} \Pr\left(\sqrt{n}(\widehat{\mu}_{ML} - \mu) \leq x\right) = \Phi\left(x/\sigma\right)$. That is, $\sqrt{n}(\widehat{\mu}_{ML} - \mu) \simeq \mathcal{N}(0, \sigma^2)$, the same as the theorem of asymptotic normality of the MLE. However, when $\mu = 0$, the cdf of $\sqrt{n}(\widehat{\mu}_{ML} - \mu)$ is not normal,

$$\Pr\left(\sqrt{n}(\widehat{\mu}_{ML} - \mu) \leq x\right) = \begin{cases} 0 \text{ if } x < 0 \\ 1/2 \text{ if } x = 0 \\ \Phi\left(\frac{x}{\sigma}\right) \text{ if } x > 0 \end{cases} .$$

This is a combination of the mass probability $1/2$ at $x = 0$ and normal distribution for $x > 0$. One may suggest complying with Condition B by assuming that $\mu > 0$. But then we face another problem: with probability close to $1/2$, the MLE does not exist for small μ. Reparametrization like $\mu = e^{-\nu}$ does not help either because $\widehat{\nu}_{ML} \to \infty$ when $\overline{Y} \leq 0$. $\qquad\qquad\qquad\square$

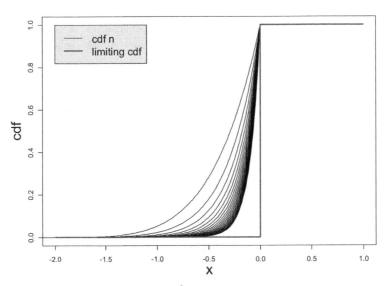

Figure 6.31: *The limiting cdf of $\sqrt{n}(\widehat{\theta}_{ML} - \theta)$ for uniform distribution is a step function when $n \to \infty$.*

Example 6.146 *Density support is parameter-dependent.* *Let $Y_i \overset{\text{iid}}{\sim} \mathcal{R}(0, \theta)$. The MLE for $\theta > 0$ is consistent, but the limiting distribution is not normal.*

Solution. As follows from Example 6.119, the cdf of the MLE $\widehat{\theta}_{ML} = \max Y_i$ is

$$\Pr\left(\widehat{\theta}_{ML} \leq x\right) = \begin{cases} 0 \text{ if } x < 0 \\ (x/\theta)^n \text{ if } 0 \leq x \leq \theta \\ 1 \text{ if } x > \theta \end{cases}.$$

First, using this expression, we show that $\widehat{\theta}_{ML}$ is consistent by showing that, when $n \to \infty$, we have $\Pr\left(\left|\widehat{\theta}_{ML} - \theta\right| \leq \varepsilon\right) \to 1$ for any small $\varepsilon > 0$. Indeed,

$$\Pr\left(\left|\widehat{\theta}_{ML} - \theta\right| \leq \varepsilon\right) = \Pr(\widehat{\theta}_{ML} < \theta + \varepsilon) - \Pr(\widehat{\theta}_{ML} < \theta - \varepsilon) = 1 - \Pr(\widehat{\theta}_{ML} < \theta - \varepsilon)$$

because $\theta + \varepsilon > \theta$. But $\Pr(\widehat{\theta}_{ML} < \theta - \varepsilon) = ((\theta - \varepsilon)/\theta)^n \to 0$ when $n \to \infty$, so $\Pr(|\widehat{\theta}_{ML} - \theta| \leq \varepsilon) \to 1$. Second, we derive the cdf of the normalized MLE as

$$\Pr\left(\sqrt{n}(\widehat{\theta}_{ML} - \theta) \leq x\right) = \begin{cases} 0 \text{ if } x < -\sqrt{n}\theta \\ (1 + (x/\theta)n^{-1/2})^n \text{ if } -\sqrt{n}\theta \leq x \leq 0 \\ 1 \text{ if } x > 0 \end{cases}.$$

These cdfs do not converge to $\Phi(x)$ when $n \to \infty$ simply because they all equal 1 for $x > 0$. Moreover, they converge to a step function at 0 as illustrated in Figure 6.31.   □

There may be multiple solutions to the score equation, however, under conditions A–E they all converge to the true parameter value when the sample size increases.

Example 6.147 *Multiple solutions in the quadratic model.* *Consider n iid observations (Y_i, X_i) specified by the quadratic model $Y_i | X_i = x_i \sim \mathcal{N}(\theta + \theta^2 x_i, \sigma^2)$, where σ^2 is known and the marginal distribution of X_i does not depend on θ.*

Solution. The log-likelihood, up to a constant, takes the form

$$l(\theta) = -\frac{n}{2}\ln\sigma^2 - \frac{1}{2\sigma^2}\sum_{i=1}^{n}(Y_i - \theta - \theta^2 X_i)^2 + \sum_{i=1}^{n}\ln f_X(X_i),$$

where f_X is the marginal density of X_i. The solution to the score equation turns into a cubic equation for θ,

$$\sum_{i=1}^{n}(\theta + \theta^2 X_i - Y_i)(1 + 2\theta X_i) = 0,$$

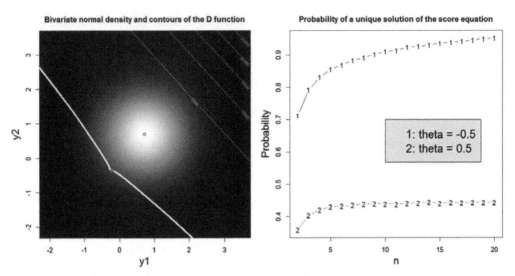

Figure 6.32: *ML estimation of the quadratic model. Left: the image of the con-*
ditional on (X_1, X_2) *bivariate normal distribution of* (Y_1, Y_2) *for* $\theta = -.5$ *and*
$\sigma = 1$ $(n = 2)$. *Right: the probability that the score (cubic) equation has a unique*
root for $\theta = -.5$ *and* $\theta = .5$ *as a function of* n.

or
$$a\theta^3 + b\theta^2 + c\theta + d = 0,$$

where

$$a = 2\sum_{i=1}^{n} X_i^2, \quad b = 3\sum_{i=1}^{n} X_i, \quad c = n - 2\sum_{i=1}^{n} X_i Y_i, \quad d = -\sum_{i=1}^{n} Y_i.$$

The cubic equation may have up to three real roots. It has a unique real root if
and only if the discriminant is not positive,

$$D(\mathbf{Y}) = 18abcd - 4b^3 d + b^2 c^2 - 4ac^3 - 27a^2 d^2 \leq 0.$$

In the left plot of Figure 6.32, the bivariate normal density of (Y_1, Y_2) conditional on $(X_1 = x_1, X_2 = x_2)$ is shown as an image $(n = 2)$ with $\theta = .5$
and $\sigma = 1$ (the R code is cubMLE). The point in the middle of the light region is $(\theta + \theta^2 X_1, \theta + \theta^2 X_2)$. Yellow curves depict contour levels of function
$D = D(Y_1, Y_2)$; the bold line corresponds to level $D = 0$. For these particu-
lar values of x_1 and x_2, the mean gives a negative D-value: a unique solution to
the score equation. The right plot depicts the probability that the score equation
has a unique solution/MLE with $\sigma = 1$ and $\theta = -.5$ and $\theta = .5$. Interestingly,
while the probability of multiple solutions goes to zero for $\theta = -.5$ with n increasing, it remains about 50% for $\theta = .5$ regardless of the sample size. This is
explained below.

Check the identifiability condition (6.81): the quadratic model is identifiable if

$$E_\theta \left(\frac{\partial \ln f(Y,X;\theta)}{\partial \theta} \bigg|_{\theta=\theta_*} \right) = E_\theta \left[(\theta_* + \theta_*^2 X - Y)(1 + 2\theta_* X) \right]$$

$$= 2\theta_* + 3\theta_*^2 \mu_X + 2\theta_*^3 (\mu_X^2 + \sigma_X^2) - (\theta + \theta^2 \mu_X) - 2\theta_* (\theta \mu_X + \theta^2 (\mu_X^2 + \sigma_X^2)) = 0$$

implies $\theta_* = \theta$. Let θ_* be any value. Then $(\theta + \theta^2 \mu_X) - 2\theta_* (\theta \mu_X + \theta^2 (\mu_X^2 + \sigma_X^2)) = A_*$, where A_* is the first term of the last expression. Since the quadratic equation for θ may have two solutions, the identifiability condition E does not hold. This explains why the probability of a unique solution of the score equation does not approach 1 even when $n \to \infty$. □

The following example is a continuation of Example 6.136 with the multinomial distribution, here we illustrate what happens when parameters obey a restriction, so Assumption B does not hold. When parameters are linearly restricted, the information matrix becomes singular and the asymptotic covariance matrix cannot be derived via matrix inverse. To avoid the matrix singularity, reduce the dimension of the parameter space using a restriction.

Example 6.148 *Information for the multinomial distribution.* *Avoid information matrix singularity in the multinomial distribution by eliminating the last probability. Demonstrate that the covariance matrix for the other parameters does not change.*

Solution. The sum of probabilities in the multinomial distribution is 1 so that the parameter space $\Theta = \{\mathbf{p} \in R^m, p_j > 0, \sum p_j = 1\}$ is not an open set in R^m. Therefore, Assumption B is violated. Recall that the covariance matrix of the MLE was previously derived directly using the formulas provided in Section 3.10.4. Can we derive this matrix as the inverse of the information matrix? Using the fact that $\partial l / \partial p_j = X_j / p_j$, we get the (j, k)th element of the information matrix,

$$\mathcal{I}_{jk} = n \times \begin{cases} (1 - p_j)/p_j \text{ if } j = k \\ -1 \text{ if } j \neq k \end{cases}.$$

The information matrix is singular. Indeed, using notation $\mathbf{P}$, represent this matrix as $\mathcal{I} = n \left(\mathbf{P}^{-1} - \mathbf{11}' \right)$ where $\mathbf{P}^{-1}$ is the diagonal matrix with $1/p_j$ on the diagonal. Prove that $\mathcal{I}\mathbf{p} = \mathbf{0}$. Indeed,

$$n \left(\mathbf{P}^{-1} - \mathbf{11}' \right) \mathbf{p} = n \left(\mathbf{P}^{-1}\mathbf{p} - \mathbf{11}'\mathbf{p} \right) = n (\mathbf{1} - \mathbf{1}) = \mathbf{0}$$

because $\mathbf{P}^{-1}\mathbf{p} = \mathbf{1}$. Since matrix $\mathcal{I}$ is singular, the asymptotic covariance matrix cannot be derived as the inverse of $\mathcal{I}$. One might suggest using the generalized matrix inverse, but this inverse does not produce the covariance matrix as desired, see problems at the end of this section.

To fix the problem with restrained parameters, reduce the dimension of the parameter space. In the case of the multinomial distribution, let $\boldsymbol{\theta} = (p_1, ..., p_{m-1}) \in R^{m-1}$, so that $p_m = 1 - \sum_{j=1}^{m-1} p_j$ and the log-likelihood function, up to a constant term, takes the form

$$l(p_1, ..., p_{m-1}) = \sum_{j=1}^{m-1} X_j \ln p_j + X_m \ln \left(1 - \sum_{j=1}^{m-1} p_j \right)$$

with the first derivatives

$$\frac{\partial l}{\partial p_j} = \frac{X_j}{p_j} - \frac{X_m}{1 - \sum_{j=1}^{m-1} p_j}.$$

It is easy to confirm that the MLE are the same as in the unrestricted parametrization because the score equations imply $X_j/p_j = \text{const.}$ The (j,k)th element of the $(m-1) \times (m-1)$ information matrix takes the form

$$\mathcal{I}_{jk} = n \times \begin{cases} 1/p_j + 1/p_m \text{ if } j = k \\ 1/p_m \text{ if } j \neq k \end{cases}.$$

We want to be sure that the inverse of this matrix produces the covariance matrix of the MLE derived previously in Example 6.136 without restriction on the probabilities. For this purpose, we use formula (10.6) of the Appendix by observing that the $(m-1) \times (m-1)$ information matrix can be written as $\mathbf{P}^{-1} + (1/p_m)\mathbf{1}\mathbf{1}'$, where $\mathbf{P}$ is a diagonal matrix with p_j on the diagonal. Letting $\mathbf{A} = \mathbf{P}^{-1}$, $\mathbf{a} = p_m^{-1}\mathbf{1}$ and $\mathbf{b} = \mathbf{1}$, we obtain

$$n\mathcal{I}^{-1} = \frac{1}{n}\mathbf{P} - \frac{1/p_m}{1 + 1/p_m \sum_{i=1}^{m-1} p_j}\mathbf{p}\mathbf{p}' = \mathbf{P} - \frac{1}{p_m + \sum_{i=1}^{m-1} p_j}\mathbf{p}\mathbf{p}' = \mathbf{P} - \mathbf{p}\mathbf{p}'.$$

This is exactly the $(m-1) \times (m-1)$ covariance matrix of $\hat{\mathbf{p}}$ derived in Example 6.136: the two approaches are equivalent. In words, the dimension reduction by elimination of p_m is equivalent to cutting out the $(m-1) \times (m-1)$ block of the original $m \times m$ covariance matrix. $\square$

In the following example, all conditions hold, but the MLE may not be robust to presence of outliers.

Example 6.149 *MLE of the maximum unemployment rate.* *A sample $\{X_i\}$ of unemployment rates in $n = 200$ randomly selected areas of the country is shown in Figure 6.33. Estimate the maximum unemployment rate, θ, using maximum likelihood based on the assumption that the rates are iid and follow a uniform distribution with the upper bound θ.*

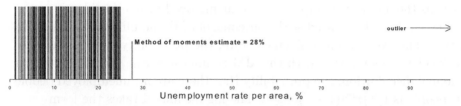

Figure 6.33: *The unemployment rate in a sample of 200 areas. One outlier (100% unemployment rate) spoils the MLE. On the other hand, the MM estimate, $\widehat{\theta}_{MM} = 2\overline{X} = 28\%$ is less efficient but robust to outliers.*

Solution. The problem with this data is that there is one area where all residents are unemployed (rate = 100%), that is, formally, the maximum unemployment rate is 100%. It may happen that only 1–2 unemployed people live in that area. Therefore, one may suggest that this particular area should be excluded. A robust (and unbiased) rate estimator may be computed using the method of moments (MM), $\widehat{\theta}_{MM} = 2\overline{X}$, see Example 6.22. This MM estimate, $\widehat{\theta}_{MM} = 28\%$, seems more reasonable: it has larger variance when the model is correct, but it is robust to outliers (see function mleUNE). Ironically, the ML becomes more and more sensitive to outliers when the sample size, n, increases because in practical applications the likelihood of an outlier increases with n. In contrast, the method of moments depends less on the distribution specification because it produces the same estimate for all densities f symmetric around $\theta/2$. Indeed, $f(\theta/2 - x) = f(\theta/2 + x)$ that implies $E(X) = \theta/2$ and, therefore, $\widehat{\theta}_{MM} = 2\overline{X}$ is unbiased. This comment can be generalized: when observations obey the distribution (no outliers), the method of maximum likelihood is optimal, at least when the sample size is large enough. However, when some observations do not follow the distribution, ML deteriorates quickly. □

The above example raises a question: Is the maximum unemployment rate valid? A better formulated question is, "What is the average unemployment rate?"

Example 6.150 *MLE of the average unemployment rate.* *As in the previous example, we want to estimate the average of the unemployment rate using the data on the employed (k_i) and unemployed (m_i) people in the area.*

Solution. The unemployment rate in the ith area is $X_i = m_i/(m_i + k_i)$, $i = 1, 2, ..., n$ with the average rate $\overline{X}$. However, this estimator silently assumes that (a) X_i have the same variance and (b) X_i are normally distributed because the average is efficient when observations are normally distributed. None of these assumptions hold, so $\overline{X}$ is not an optimal estimator. In particular, this estimator is heavily influenced by small areas as in Figure 6.33. A better average rate can be computed using the probabilistic model by treating the unemployment rate as the probability that a randomly chosen person is unemployed. Denote

$n_i = m_i + k_i$ as the number of employed and unemployed people (residents) in the ith area. Then m_i may be treated as the outcome of n_i Bernoulli experiments with probability r. The average unemployment rate, the probability that a randomly chosen resident is unemployed, with the MM estimator is $\widehat{r}_{MM} = \overline{X}$. According to the binomial distribution, the probability that there are m_i unemployed people among n_i people is $\binom{n_i}{m_i} r^{m_i}(1-r)^{n_i - m_i}$. The log-likelihood takes the form

$$l(r) = \sum_{i=1}^{n}\left[\ln\binom{n_i}{m_i} + m_i \ln r + (n_i - m_i)\ln(1-r)\right].$$

Taking the derivative and setting it to zero gives the ML estimator, $\widehat{r}_{ML} = \sum m_i / \sum n_i$. Both estimates of the unemployment rate, $\widehat{r}_{MM}$ and $\widehat{r}_{ML}$ are unbiased because $E(m_i) = rn_i$, so

$$E(\widehat{r}_{MM}) = \frac{1}{n}\sum_{i=1}^{n}E\left(\frac{m_i}{n_i}\right) = \frac{1}{n}\sum_{i=1}^{n}\frac{rn_i}{n_i} = r, \quad E(\widehat{r}_{ML}) = \frac{\sum_{i=1}^{n} rn_i}{\sum_{i=1}^{n} n_i} = r.$$

Now find the variances,

$$\mathrm{var}(\widehat{r}_{MM}) = \frac{1}{n^2}\sum_{i=1}^{n}\mathrm{var}\left(\frac{m_i}{n_i}\right) = \frac{1}{n^2}\sum_{i=1}^{n}\frac{r(1-r)n_i}{n_i^2} = r(1-r)\frac{1}{n^2}\sum_{i=1}^{n}\frac{1}{n_i},$$

$$\mathrm{var}(\widehat{r}_{ML}) = r(1-r)\frac{\sum_{i=1}^{n} n_i}{\left(\sum_{i=1}^{n} n_i\right)^2} = r(1-r)\frac{1}{\sum_{i=1}^{n} n_i}.$$

Recall the inequality $n^{-2}\sum_{i=1}^{n} n_i^{-1} \geq \left(\sum_{i=1}^{n} n_i\right)^{-1}$. It turns into equality if and only if $n_i = \mathrm{const}$. This inequality follows from the Cauchy inequality (see the Appendix, Section 10.2),

$$1 = \left(\frac{1}{n}\sum_{i=1}^{n}\frac{1}{\sqrt{n_i}}\sqrt{n_i}\right)^2 \leq \frac{1}{n^2}\sum_{i=1}^{n}\frac{1}{n_i}\sum_{i=1}^{n} n_i.$$

This inequality implies that $\mathrm{var}(\widehat{r}_{ML}) \leq \mathrm{var}(\widehat{r}_{MM})$ with the equality if and only if an equal number of people live in each area. The variance of $\widehat{r}_{MM}$ will be large if some areas are sparsely populated.

6.10.6   Algorithms for log-likelihood function maximization

Only in rare cases such as linear models, discussed in Chapter 8, can the maximum of the likelihood function be found in closed form. Typically, its maximization requires iterations. Three algorithms may be used to maximize a general log-likelihood function: Newton–Raphson (NR), Fisher scoring (FS), and empirical Fisher scoring (EFS). Here, we do not consider the expectation-maximization (EM) algorithm because it applies to specific distributions whereas the above three algorithms can be applied to any distribution. Plus, the EM algorithm does not produce the covariance matrix required for further statistical inference.

All three algorithms use iterations in the form

$$\boldsymbol{\theta}_{k+1} = \boldsymbol{\theta}_k + \lambda_k \mathbf{H}_k^{-1} \left(\left. \frac{\partial l}{\partial \boldsymbol{\theta}} \right|_{\boldsymbol{\theta}=\boldsymbol{\theta}_k} \right), \quad k = 0, 1, 2, ..., \tag{6.92}$$

where $\boldsymbol{\theta}_k$ and $\boldsymbol{\theta}_{k+1}$ are the values of the parameter vector at iteration k and $k+1$, respectively, $\lambda_k > 0$ is the step length (typically, $\lambda_k = 1$), and $\mathbf{H}_k$ is a positive definite matrix. In multivariate calculus, the derivative $\partial l / \partial \boldsymbol{\theta}$ is referred to the gradient. Since matrix $\mathbf{H}_k$ is positive definite, matrix $\mathbf{H}_k^{-1}$ is positive definite as well. Therefore, the angle between the direction of maximization in (6.92) and the gradient is positive. In a special case, when $\mathbf{H}_k$ is the identity matrix, algorithm (6.92) is called the steepest ascent algorithm.

There are two important properties of the general algorithm (6.92). First, at convergence, the limiting point of successful iterations is the stationary point of the log-likelihood function, the solution of the score equation. Indeed, if $\lim_{k\to\infty} \boldsymbol{\theta}_k = \lim_{k\to\infty} \boldsymbol{\theta}_{k+1} = \boldsymbol{\theta}_*$ and $\lim_{k\to\infty} \lambda_k = \lambda_* > 0$ and $\lim_{k\to\infty} \mathbf{H}_k = \mathbf{H}_*$ is a positive definite matrix then equation (6.92) turns into

$$\boldsymbol{\theta}_* = \boldsymbol{\theta}_* + \lambda_* \mathbf{H}_*^{-1} \left(\left. \frac{\partial l}{\partial \boldsymbol{\theta}} \right|_{\boldsymbol{\theta}=\boldsymbol{\theta}_*} \right),$$

which yields

$$\left. \frac{\partial l}{\partial \boldsymbol{\theta}} \right|_{\boldsymbol{\theta}=\boldsymbol{\theta}_*} = \mathbf{0}.$$

This means that any limiting point is the solution of the score equation. Second, since matrix $\mathbf{H}_k$ is positive definite $\mathbf{H}_k^{-1}$ is positive definite as well. Therefore, if

$$\left. \frac{\partial l}{\partial \boldsymbol{\theta}} \right|_{\boldsymbol{\theta}=\boldsymbol{\theta}_k} \neq \mathbf{0},$$

there exists a positive step length λ_k such that the value of the log-likelihood function at iteration $k+1$ is greater than at iteration k. Sometimes, $\lambda_k = 1$ may not produce a larger value of l at each iteration or $\boldsymbol{\theta}_k$ diverges. Then the step should be shrunken, say, by half until the log-likelihood takes a larger value.

The three algorithms mentioned above differ by how matrix $\mathbf{H}$ is computed. Below we assume that the total log-likelihood function $l(\boldsymbol{\theta})$ is the sum of n individual log-likelihoods corresponding to observations $i = 1, 2, ..., n$:

$$l(\boldsymbol{\theta}) = \sum_{i=1}^{n} l_i(\boldsymbol{\theta}).$$

Observations are independent but may not necessarily be identically distributed.

1. Newton-Raphson (NR) algorithm:

$$\mathbf{H} = -\frac{\partial^2 l}{\partial \boldsymbol{\theta}^2},$$

the Hessian.

2. Fisher-scoring (FS) algorithm:

$$\mathbf{H} = -E\left(\frac{\partial^2 l}{\partial \boldsymbol{\theta}^2}\right) = \mathcal{I},$$

the expected Hessian, or the Fisher information matrix.

3. Empirical Fisher-scoring (EFS) algorithm:

$$\mathbf{H} = \sum_{i=1}^{n} \left(\frac{\partial l_i}{\partial \boldsymbol{\theta}}\right)\left(\frac{\partial l_i}{\partial \boldsymbol{\theta}}\right)'.$$

Remarks: (a) NR and FS require computation of second derivatives, but EFS deals only with the first derivatives of l. (b) While the Hessian may be not positive definite the expected Hessian is positive definite if the identifiability Condition E holds. (c) The justification for EFS is the fact that, if the observations are iid, then

$$p\lim_{n\to\infty} \frac{1}{n}\sum_{i=1}^{n}\left(\frac{\partial l_i}{\partial \boldsymbol{\theta}}\right)\left(\frac{\partial l_i}{\partial \boldsymbol{\theta}}\right)' = E\left[\left(\frac{\partial l_1}{\partial \boldsymbol{\theta}}\right)\left(\frac{\partial l_1}{\partial \boldsymbol{\theta}}\right)'\right] \qquad (6.93)$$

due to the LLN since $\partial l_i/\partial \boldsymbol{\theta}$ are iid. Therefore,

$$\sum_{i=1}^{n}\left(\frac{\partial l_i}{\partial \boldsymbol{\theta}}\right)\left(\frac{\partial l_i}{\partial \boldsymbol{\theta}}\right)' \simeq \mathcal{I}.$$

(d) The same justification can be used to claim that, with iid observations and n large enough,

$$-\frac{\partial^2 l}{\partial \boldsymbol{\theta}^2} = -\sum_{i=1}^{n}\frac{\partial^2 l_i}{\partial \boldsymbol{\theta}^2} \simeq \mathcal{I}$$

in view of Theorem 6.102.

In short, (a) all three algorithms should be close for large n if observations are iid, (b) the FS algorithm is preferable because the inverse to the expected Hessian estimates the covariance matrix of the parameters needed for further statistical inference.

Example 6.151 *MLE for Cauchy distribution.* *Estimate the center of the Cauchy distribution, θ, using the three algorithms.*

Solution. The density of the Cauchy distribution with center θ takes the form (Section 2.13.1) $f(y; \theta) = [\pi(1 + (y - \theta)^2)]^{-1}$. If Y_i are iid observations, the log-likelihood function, up to a constant, is written as

$$l(\theta) = -\sum_{i=1}^{n} \ln(1 + (Y_i - \theta)^2)$$

with the score equation

$$\frac{dl}{d\theta} = -2 \sum_{i=1}^{n} \frac{Y_i - \theta}{1 + (Y_i - \theta)^2} = 0.$$

This equation does not admit a solution in closed form – iterations are required:
(1) The second derivative is required to apply the NR algorithm,

$$\frac{d^2 l}{d\theta^2} = 2 \sum_{i=1}^{n} \frac{(Y_i - \theta)^2 - 1}{(1 + (Y_i - \theta)^2)^2}.$$

The iterations take the form

$$\theta_{k+1} = \theta_k - \frac{\sum_{i=1}^{n} \frac{Y_i - \theta_k}{1 + (Y_i - \theta_k)^2}}{\sum_{i=1}^{n} \frac{(Y_i - \theta_k)^2 - 1}{(1 + (Y_i - \theta_k)^2)^2}}, \quad k = 0, 1, \ldots$$

starting from $\theta_0 = $ median.

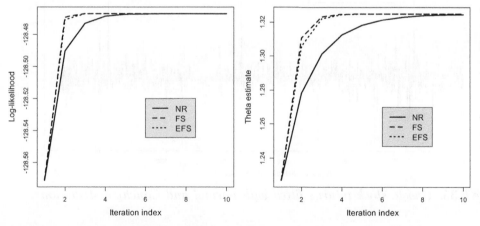

Figure 6.34: *Three algorithms for θ estimation in the Cauchy distribution. The true $\theta = 1$, and the sample size is $n = 50$. The FS and EFS algorithms are close but the convergence of the NR is slow.*

(2) To derive the FS algorithm, we find the Fisher information:

$$\mathcal{I} = -E\left(\frac{d^2 l}{d\theta^2}\right) = -\frac{2}{\pi} \int_{-\infty}^{\infty} \frac{(y - \theta)^2 - 1}{(1 + (y - \theta)^2)^3} dy = -\frac{2}{\pi} \int_{-\infty}^{\infty} \frac{y^2 - 1}{(1 + y^2)^3} dy = \frac{1}{2}.$$

Thus, the FS algorithm has a simpler form,

$$\theta_{k+1} = \theta_k + \frac{4}{n} \sum_{i=1}^{n} \frac{Y_i - \theta_k}{1 + (Y_i - \theta_k)^2}, \quad k = 0, 1, \ldots$$

(3) The EFS algorithm uses the squared first derivatives to estimate the expected second derivative. Thus, the second derivative is not required:

$$\theta_{k+1} = \theta_k + \frac{\sum_{i=1}^{n} \frac{Y_i - \theta_k}{1 + (Y_i - \theta_k)^2}}{2 \sum_{i=1}^{n} \left(\frac{Y_i - \theta_k}{1 + (Y_i - \theta_k)^2} \right)^2}, \quad k = 0, 1, \ldots$$

Convergence of the three algorithms to the MLE is depicted in Figure 6.34 using a sample $Y_1, \ldots, Y_n$ generated by the built-in function `rcauchy` with $n = 50$. Alternatively, `atan` could be used to generate observations using inverse cdf; see Section 2.13.1. Note that the MLE is close to the median. The NR converges slower compared to the other two algorithms. The R code is in function `cauchy.theta`. Typically, the FS algorithm is the fastest. □

The next example is a continuation of the previous example with the addition of the scale parameter, λ.

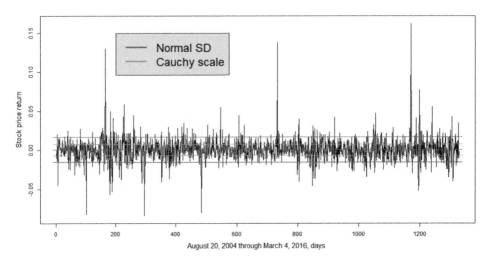

Figure 6.35: *Google stock returns fitted with normal and Cauchy distributions.*

Example 6.152 *MLE for the Cauchy distribution with an unknown scale parameter.* *(a) Develop the maximum likelihood estimation algorithm for the Cauchy distribution with two unknown parameters, the mode (θ) and the scale parameter (λ), using the Fisher scoring algorithm. (b) Test the algorithm via vectorized simulations. (c) Apply the FS algorithm to estimate Google's stock price returns assuming that they follow Cauchy distribution. (d) Compute the probability of negative returns for normal and Cauchy distributions.*

Solution. (a) The density of the Cauchy distribution with two parameters takes the form

$$f(y; \theta, \lambda) = \frac{1}{\pi\lambda} \frac{1}{1 + ((y - \theta)/\lambda)^2}.$$

If Y_i are iid observations, the log-likelihood, up to a constant, is written as

$$l(\theta, \lambda) = -n \ln \lambda - \sum_{i=1}^{n} \ln(1 + ((Y_i - \theta)/\lambda)^2).$$

Differentiate

$$\frac{\partial l}{\partial \theta} = -\frac{2}{\lambda} \sum_{i=1}^{n} \frac{(Y_i - \theta)/\lambda}{1 + ((Y_i - \theta)/\lambda)^2}, \quad \frac{\partial l}{\partial \lambda} = -\frac{n}{\lambda} + \frac{2}{\lambda} \sum_{i=1}^{n} \frac{((Y_i - \theta)/\lambda)^2}{1 + ((Y_i - \theta)/\lambda)^2}.$$

Since $Z = (Y - \theta)/\lambda$ has the ordinary Cauchy distribution, we obtain

$$\mathcal{I}_{11} = \frac{4}{\lambda^2} E\left(\frac{Z^2}{(1 + Z^2)^2}\right) = \frac{1}{2\lambda^2}, \quad \mathcal{I}_{12} = -\frac{4}{\lambda^2} E\left(\frac{Z^3}{(1 + Z^2)^2}\right) = 0,$$

$$\mathcal{I}_{22} = \frac{1}{\lambda^2} E\left(\frac{1 - Z^2}{1 + Z^2}\right)^2 = \frac{1}{2\lambda^2}.$$

The off-diagonal element of the information matrix is zero. Therefore, the FS iterations, split for θ and λ, are

$$\theta_{k+1} = \theta_k + \frac{4\lambda_k}{n} \sum_{i=1}^{n} \frac{(Y_i - \theta_k)/\lambda_k}{1 + ((Y_i - \theta_k)/\lambda_k)^2},$$

$$\lambda_{k+1} = \lambda_k + \frac{2\lambda_k}{n} \left(2 \sum_{i=1}^{n} \frac{((Y_i - \theta_k)/\lambda_k)^2}{1 + ((Y_i - \theta_k)/\lambda_k)^2} - n\right).$$

Obtain the asymptotic variances as the reciprocal of the Fisher information, $\text{var}(\widehat{\theta}_{ML}) \simeq 2\lambda^2/n$ and $\text{var}(\widehat{\lambda}_{ML}) \simeq 2\lambda^2/n$. Interestingly, the asymptotic variances for the center and scale parameters are the same.

(b) The R code is `cauchy.google`. Vectorized simulations look remarkably fast and elegant: it takes less than a second to run `nSim=10000` FS iterations. The key to vectorized simulations is to generate all Cauchy distributed random variables at once and store them in a matrix `nSim` by n. Then we use `rowSums` to compute `nSim` derivatives with respect to θ and λ. Below is the output of `cauchy.google` with default arguments; the mean values for θ and λ are close to the theoretical values. Empirical and theoretical variances are close as well.

```
> cauchy.google(job=1)
[1] -0.9999028 2.0056966
[1] 0.1712636 0.1694950 0.1600000
```

(c) The Google stock prices as a time series $Y_1, Y_2, ..., Y_n$ were analyzed in Example 6.57. Here we analyze the distribution of returns, $r_i = (Y_i - Y_{i-1})/Y_{i-1}$.

Returns, not stock prices, themselves are major quantities of interest for investors. As can be seen from Figure 6.35, returns exhibit large variation and spikes (positive and negative) are not seldom – the normal distribution may be not adequate, the q-q plot confirms abnormality (not shown). The Cauchy distribution of returns may be a satisfactory alternative to the normal distribution. The FS iterations produce $\widehat{\theta}_{ML} = 0.0006029807$ and $\widehat{\lambda}_{ML} = 0.0073432496$.

(d) The chance to get negative returns is an important risk – benefit analysis of stock volatility: while positive returns are desirable negative returns may devastate the investor. The crucial task of risk management is to estimate the probability of negative returns. Under assumption that returns are normally distributed this probability is estimated as $\Phi(r/\widehat{\sigma})$. If returns follow a Cauchy distribution they are estimated using the Cauchy cdf. As can be seen from Figure 6.36, the chances of large negative returns, not surprisingly, are considerably higher under the Cauchy model.

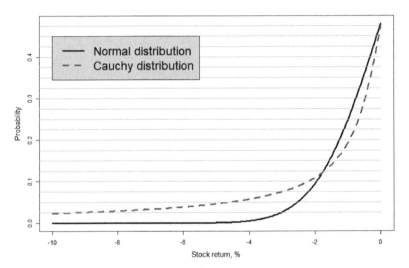

Figure 6.36: *Probability of obtaining a negative return with two distribution assumptions.*

Example 6.153 *FS algorithm for the exponential family.* *Derive the FS iterations for the exponential family with the m-dimensional parameter specified by density (6.53).*

Solution. If $\mathbf{Y}_1, \mathbf{Y}_2, ..., \mathbf{Y}_n$ are iid random vectors with the density specified by (6.53), the log-likelihood function takes the form

$$l(\boldsymbol{\theta}) = \mathbf{p}'(\boldsymbol{\theta}) \sum_{i=1}^{n} \mathbf{K}(\mathbf{Y}_i) + \sum_{i=1}^{n} S(\mathbf{Y}_i) - nq(\boldsymbol{\theta}).$$

The estimation problem simplifies after reparametrization, $\boldsymbol{\tau} = \mathbf{p}(\boldsymbol{\theta})$ since $\mathbf{p}$ is a one-to-one mapping. Denote $\mathbf{p}^{-1}$ as the inverse transformation. Then q can be expressed in terms of new parameters as $Q(\boldsymbol{\tau}) = q(\boldsymbol{\tau}^{-1}(\boldsymbol{\theta}))$ and the log-likelihood, after omitting the term $\sum_{i=1}^{n} S(\mathbf{Y}_i)$ which does not affect maximization, becomes $l(\boldsymbol{\tau}) = \boldsymbol{\tau}'\mathbf{K} - nQ(\boldsymbol{\tau})$. Thus, the MLE reduces to the solution of the score equation $\partial l/\partial \boldsymbol{\tau} = \mathbf{K} - n(\partial Q/\partial \boldsymbol{\tau}) = \mathbf{0}$. Since the Hessian of the log-likelihood, $\partial^2 l/\partial \boldsymbol{\tau}^2 = -n\partial^2 Q/\partial \boldsymbol{\tau}^2$, is not random, the NR and FS algorithms coincide and the iterations take the form

$$\boldsymbol{\tau}_{k+1} = \boldsymbol{\tau}_k + \left(\left.\frac{\partial^2 \mathbf{Q}}{\partial \boldsymbol{\tau}^2}\right|_{\boldsymbol{\tau}=\boldsymbol{\tau}_k} \right)^{-1} \left(\left.\frac{\partial \mathbf{Q}}{\partial \boldsymbol{\tau}}\right|_{\boldsymbol{\tau}=\boldsymbol{\tau}_k} - \overline{\mathbf{K}} \right), \quad k = 0, 1, ...,$$

where $\overline{\mathbf{K}} = \mathbf{K}/n$. Due to the invariance property of the ML, we obtain $\widehat{\boldsymbol{\theta}}_{ML} = \mathbf{p}^{-1}(\widehat{\boldsymbol{\tau}}_{ML})$, where $\widehat{\boldsymbol{\tau}}_{ML}$ is the limiting vector of the iterations above, or, practically speaking, the vector from the final iteration upon convergence.

In the following example, there are two parameters, and the partial derivatives and matrices should be used.

Example 6.154 *MLE for the gamma distribution. Write R code for the maximum likelihood estimation of parameters of the gamma distribution using the NR, FS, and EFS algorithms and compare their performance in terms of convergence to the ML estimate using simulated data.*

Solution. Recall that the method of moments was applied to estimate parameters of the gamma distribution in Examples 6.8 and 6.110. Now, we use ML with the function to maximize

$$l(\alpha, \lambda) = n\alpha \ln \lambda - n \ln \Gamma(\alpha) + (\alpha - 1)A_0 - \lambda A_1,$$

where $A_0 = \sum_{i=1}^{n} \ln X_i$ and $A_1 = \sum_{i=1}^{n} X_i$ are sufficient statistics. As in Example 6.8, the following terminology is used with regard to the gamma function. The first and second derivative of the log gamma function,

$$\psi_0(x) = \frac{d}{dx}\ln\Gamma(x), \quad \psi_1(x) = \frac{d^2}{dx^2}\ln\Gamma(x),$$

are called the `digamma` and `trigamma` functions, respectively (they are built-in functions in R). With the help of function ψ, the first derivatives of the log-likelihood function are

$$\frac{\partial l}{\partial \alpha} = n \ln \lambda - n\psi_0(\alpha) + A_0, \quad \frac{\partial l}{\partial \lambda} = \frac{n\alpha}{\lambda} - A_1,$$

Matrices $\mathbf{H}$ for NR, FS and EFS are

$$n\begin{bmatrix} \psi_1(\alpha) & -1/\lambda \\ -1/\lambda & \alpha/\lambda^2 \end{bmatrix}, \quad n\begin{bmatrix} \psi_1(\alpha) & -1/\lambda \\ -1/\lambda & \alpha/\lambda^2 \end{bmatrix},$$

$$\sum_{i=1}^{n} \begin{bmatrix} \ln \lambda - \psi_0(\alpha) + \ln X_i \\ \alpha/\lambda - X_i \end{bmatrix} \begin{bmatrix} \ln \lambda - \psi_0(\alpha) + \ln X_i \\ \alpha/\lambda - X_i \end{bmatrix}',$$

respectively. Note that HR and FS coincide because the Hessian is not random (it does not contain random variables). See the R code `mle.gamma.OPT`.

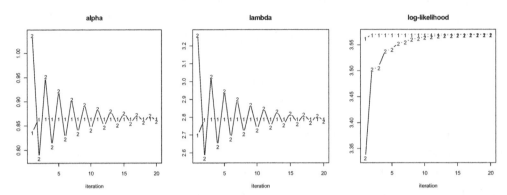

Figure 6.37: *Convergence of two algorithms for maximization of the log-likelihood function with gamma-distributed simulated data: 1=NR/FS, 2=EFS.*

We make a few comments on the code: (a) Argument `ss` controls simulation values with the default seeding number `ss=3` (without `set.seed`, every run will generate different data and debugging will be difficult). (b) ML iterations start with the MM estimates for α and λ, `aMM` and `lMM`, respectively, as derived in the examples above. (c) The two-dimensional arrays `all1` and `all2` store values of the MLE approximations, the likelihood values and the sum of absolute values of the score equations at each iteration for NR/FS and EFS, respectively. (d) Matrix $\mathbf{H}$ for the EFS algorithm is computed as $\mathbf{D'D}$, where $\mathbf{D}$ is the $n \times 2$ matrix of derivatives (variable `der2`), not as a sum over $i = 1, 2, ..., n$.

There is no guarantee that any of these algorithms converge. For example, the EFS algorithm may not converge for small n. However, for large n, say, $n > 100$, the difference between algorithms is small due to (6.93). A typical convergence of two algorithms is shown in Figure 6.37: $n = 20$ gamma-distributed observations were generated using the `rgamma` command with $\alpha = 1$ and $\lambda = 2$ (the default values). The iterations for both algorithms started from the MM estimate as derived in Example 6.110. The NR/FS algorithm converges quickly but convergence of the EFS algorithm is much slower.

Example 6.155 *Poverty in CT*. *Using the salary data analyzed in Section 5.1 (the R code* `salary`*), estimate the percentage of people in poverty in CT by fitting the gamma distribution.*

Solution. The FS algorithm is used to estimate α and λ by ML. The call `mle.gamma.CT()` produces the following output:

```
[1] "alphaML= 4.619 , lambdaML= 8.11e-05"
[1] "Minimum salary = 19870"
```

The percent poverty is computed by evaluating the gamma cdf (the R function `pgamma`) at the poverty salary with the default value $11,000.

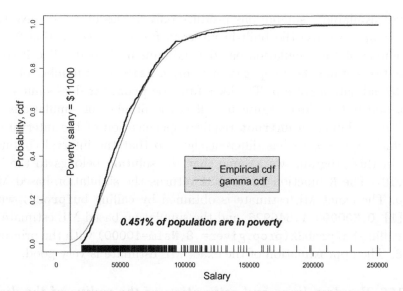

Figure 6.38: *Empirical and gamma-distribution cdfs of salaries in CT. The percent poverty is estimated as 0.45%*

The cdf-based analysis of poverty in CT is depicted in Figure 6.38. First, as follows from this figure, the gamma distribution fits the empirical salary distribution well. Second, the data itself cannot be used to identify the portion of people with salary less than \$11,000 because the minimum salary, \$19,870, is greater than the poverty level, \$11,000, so some model for salary distribution is required to estimate the probability of the left tail. Third, the difference between the empirical and gamma cdf at left is not indicative of a poor fit but simply due to the fact that low income was not recorded. Indeed, assuming that the minimum wage is \$10/h, the minimum salary is $52 \times 40 \times 10 = \$20,800$, which hints that salaries of only full time workers were recorded. But people may work part time and earn less than \$11,000. That is, there are people below the poverty level. The proportion of those people is 0.45% according to our analysis.

Simulation-based maximum likelihood

Some statistical models are so complicated that the density is intractable, but simulations are still available. Can we use simulations instead to apply the maximum likelihood? The answer is positive. Consider "The Buffoon's needle estimation problem" in Example 6.115. The maximum likelihood estimator of $\rho = L/D$ is the solution to the equation $p(\rho) = m/n$, where p is the probability of needle intersection in a single random throw. Imagine that function $p(\rho)$ is not available. Since simulations are in vectorized form and fairly fast, we can replace p with $\widehat{p}$, the estimate of p from simulations, and solve the equation $\widehat{p}(\rho) = m/n$. The problem is that Newton's algorithm is not applicable anymore because the

derivative is not available – simulations produce only values of p. To solve the
equation $\widehat{p}(\rho) = m/n$ we use the built-in `uniroot` function (see Section 2.13).
This is an example of the simulation-based maximum likelihood. The R code
`bufprobSA` has two arguments: `prop.given` = m/n, `Nsim` is the number of simulations used to estimate p given ρ. The local function `prop.est` does simulation
to estimate p, and it has three arguments: `N` is the number of simulations; it
returns $\widehat{p}(\rho) - m/n$. Function `uniroot` requires specification of the interval on
the ends of which $\widehat{p}(\rho) - m/n$ has different signs so that the interval contains
the solution. In this program, we assume that the solution belongs to the interval $(1/100, 10)$. The R function `bufprobSA` returns the simulation-based ML
estimate for ρ. The exact ML estimate is obtained by calling `bufprob()` with
the printout `[1] 0.800000 1.646926`, and the simulation-based ML estimate is
computed by calling `bufprobSA(prop.given=.8,Nsim=10000)` with the printout
`[1] 1.643168`. The approximation of the exact ML estimate is very good.

Example 6.156 *Random lines and estimation of the radius of the disk.*
n random lines intersected a disk of unknown radius r on m occasions (see Ex-
ample 3.69). Estimate r by simulation-based maximum likelihood.

Solution. Let $p(r)$ be the probability that a random line on a ± 1 square
intersects the disk of radius $r < 1$. Since p is unknown, replace it with $\widehat{p}$ from
vectorized simulations. The R code `ranlRest` is similar to the previous one, but
simulations are different. Arguments are: `Nsim` is the number of simulations
to estimate the proportion of intersections from simulations, `prop` is the given
proportion of times random lines intersected the disk. We use the proportion
from Example 3.69 with the true radius 0.25. The call `ranlRest()` gives the `[1]`
`0.2210333`   □

Simulation-based ML estimation can be applied to more general ML estimation problems with a multidimensional parameter. Then, R built-in optimization
functions, such as `nlm`, `nlminb` or `optim` without gradient specification, could be
used.

Problems

1. Using the chain rule show that the score equation in the tau parametrization
   holds whenever the score equation holds for the theta parametrization in
   connection to equation (6.73).

2. Is it true that, if the log-likelihood function is a concave function, then
   the likelihood function is concave too? First consider a single-parameter
   problem and then the multidimensional case. [Hint: Refer to the Appendix.]

3. Interpret the result of Example 6.112 in view of the result of Example 6.116.

4. Prove that $\widehat{\pi}_{ML}$ from Example 6.126 is consistent for π. [Hint: Find the limits of F, H, and E.]

5. In Example 6.126, prove that $\widehat{\pi}_{ML} < \widehat{\pi}_{LS}$ by showing that the derivative of $H(\pi)$ evaluated at $\widehat{\pi}_{LS}$ is negative.

6. (a) Derive the MLE for π from Example 6.126 under assumption that measurements δ_i and ε_i have different variances. (b) Prove that the MLE is consistent. (c) Modify function piest to produce a figure similar to (6.28).

7. Use function autocrash from Example 6.129 to plot Parzen densities for each year using the par(mfrow=c(3,4)) layout.

8. Given n iid observations derive the MM and ML estimators for all distributions listed in Table 6.2 and demonstrate that they coincide.

9. Prove that $\widehat{p}_{ML}$ in the *Perch problem* example returns the maximum of the likelihood function. [Hint: Prove that l is a concave function.]

10. Extend the *Perch problem* example to the case when the researcher stops fishing after two perch are caught.

11. Correct the three-mean overspecified model by reducing the problem of parameter estimation to two means. Derive the distribution of the MLE and show that the asymptotic normality approximation is exact.

12. Derive the MLE for μ, σ^2 and ρ given the sample of normally distributed observations with mean μ, variance σ^2 and correlation coefficient $\rho > 0$ between any two pairs, or shortly $\mathbf{Y} \sim \mathcal{N}(\mu\mathbf{1}, \sigma^2((1-\rho)\mathbf{I} + \rho\mathbf{11}'))$.

13. Prove that the MLE is inconsistent in Example 6.140 when σ^2 is unknown.

14. Are μ_i and σ^2 consistent in Example 6.141 when $k \to \infty$?

15. Can the bandwidth be consistently estimated using another kernel density?

16. The scatterplot weight versus height among Korean individuals was presented in Figure 3.13. (a) Estimate the two regressions by ML and display them as in Figure 3.13. (b) Display contours of the bivariate normal density using option add=T. [Use function heightweight.]

17. Using simulated vector **p** in Example 6.148, show that the generalized matrix inverse of the information matrix generally does not produce the correct covariance matrix. Prove that, in a special case when all probabilities are the same and equal to $1/m$, the covariance matrix is correct.

18. Redo Example 6.146 using the triangular density $f(y; \theta) = 2\theta^{-2}(\theta - y)$ for $0 \leq y \leq \theta$. (a) Derive the score equation and develop the FS algorithm. (b) Carry out simulations to test the asymptotic normality of the MLE.

19. In Example 6.151 compare $\widehat{\theta}_{ML}$ with the median in terms of the probability $\Pr\left(\left|\widehat{\theta} - \theta\right| < \varepsilon\right)$, where ε is a given number. Use simulations and depict the probabilities for $\widehat{\theta}_{ML}$ and the median as functions of the sample size, n.

20. Extend Example 6.151 to simultaneously estimate the center and the scale parameter using the density $f(y; \theta, \tau) = (\pi\tau)^{-1}1/[1 + ((y - \theta)/\tau)^2]$. Derive the three algorithms and compare their performance.

21. Estimate the length of the random segment in Example 3.70 using the simulation-based MLE.

22. The Weibull distribution is often used in survival analysis and engineering applications, including lifetime, waiting time, etc. Its cdf is defined as $F(x; \lambda, k) = 1 - e^{-(x/\lambda)^k}$ for $x \geq 0$. (a) Develop an iterative algorithm to estimate parameters λ and k. (b) Test the performance of the algorithm by simulations similarly to Example 6.154.

23. The R code `rws` is a random-walk game on the $n \times n$ square with user-defined probabilities `probx` and `proby` defining the probabilities of moving in the right and up direction, respectively, until the moving point reaches the destination point `Tx` and `Ty`. Estimate `probx` and `proby` given the number of steps required to reach the destination point. [Hint: Start with solving the problem when `probx` and `proby` are the same.]

6.11   Estimating equations and the M-estimator

As we know from Section 6.10, if the distribution is known, ML is the best method of parameter estimation, at least with a large sample. Obviously, this method requires knowledge of the distribution/density function. What if we do not know the distribution completely, but, for example, know how the expected value of the observation is expressed through parameters? Then the estimating equation (EE) approach can be applied. The EE approach can be viewed as a generalization of ML where estimating function is the score function (see below).

Definition 6.157 *EE and the estimating function. Let $\mathbf{Y}_1, \mathbf{Y}_2, ..., \mathbf{Y}_n$ be iid random vectors with the common pdf $f(\mathbf{y}; \boldsymbol{\theta})$ dependent on the m-dimensional parameter $\boldsymbol{\theta}$. An $m \times 1$ continuously differentiable vector function $\boldsymbol{\Psi} = \boldsymbol{\Psi}(\mathbf{y}; \boldsymbol{\theta})$ is called an estimating function if its expectation is zero (hereafter referred to as the zero expectation condition),*

$$E_{\boldsymbol{\theta}}\boldsymbol{\Psi}(\mathbf{Y}, \boldsymbol{\theta}) = \mathbf{0}.$$

Moreover, we assume that

$$E_{\boldsymbol{\theta}}\boldsymbol{\Psi}(\mathbf{Y}; \boldsymbol{\theta}_*) = \mathbf{0} \tag{6.94}$$

implies $\boldsymbol{\theta} = \boldsymbol{\theta}_*$, and the $m \times m$ matrix

$$\mathbf{H}(\boldsymbol{\theta}) = E_{\boldsymbol{\theta}} \left(\frac{\partial \boldsymbol{\Psi}(\mathbf{Y}; \boldsymbol{\theta})}{\partial \boldsymbol{\theta}} \right) \tag{6.95}$$

is nonsingular for every $\boldsymbol{\theta}$. Then the M-estimator, $\widehat{\boldsymbol{\theta}}_M$ is defined as the solution of m estimating equations

$$\sum_{i=1}^{n} \boldsymbol{\Psi}(\mathbf{Y}_i; \boldsymbol{\theta}) = \mathbf{0}.$$

The M-estimator turns into the ML estimator and the EE turns into the score equation in a special case when the estimating function is the derivative of the log pdf, the score function,

$$\boldsymbol{\Psi}(\mathbf{y}, \boldsymbol{\theta}) = \frac{\partial \ln f(\mathbf{y}; \boldsymbol{\theta})}{\partial \boldsymbol{\theta}}. \tag{6.96}$$

Note that the zero expectation condition holds from (6.66) of Theorem 6.102.

Sometimes, it is more appealing to find an estimator that minimizes a function rather than a solution to a system of equations.

Definition 6.158 *Minimum contrast estimator.* *Let $\boldsymbol{\Psi}(\mathbf{y}; \boldsymbol{\theta})$ be an estimating function under the terms of the previous definition. The quasi-log-likelihood function is defined as*

$$\boldsymbol{\Upsilon}(\mathbf{y}; \boldsymbol{\theta}) = \int_{-\infty}^{\mathbf{y}} \boldsymbol{\Psi}(\mathbf{u}; \boldsymbol{\theta}) d\mathbf{u}$$

provided that the integral exists and matrix $\mathbf{H}(\boldsymbol{\theta})$ defined in (6.95) is positive definite. The minimum contrast estimator is defined as the minimizer of the quasi-log-likelihood function for the sample $\mathbf{Y}_1, \mathbf{Y}_2, ..., \mathbf{Y}_n$, or symbolically,

$$\widehat{\boldsymbol{\theta}}_{MC} = \arg\min \sum_{i=1}^{n} \boldsymbol{\Upsilon}(\mathbf{Y}_i; \boldsymbol{\theta}).$$

In most estimation problems the M-estimator and the minimum contrast estimator coincide, $\widehat{\boldsymbol{\theta}}_M = \widehat{\boldsymbol{\theta}}_{MC}$, due to the Fundamental Theorem of Calculus,

$$\frac{\partial}{\partial \mathbf{y}} \boldsymbol{\Upsilon}(\mathbf{y}; \boldsymbol{\theta}) = \boldsymbol{\Psi}(\mathbf{y}; \boldsymbol{\theta}),$$

because the first-order condition for the minimum of function $\boldsymbol{\Upsilon}$ turns into the EE. The condition on positive definiteness for matrix $\mathbf{H}$ guarantees that, at least in the neighborhood of the true parameter, the Hessian is positive and minimization is valid. If matrix $\mathbf{H}$ is negative definite, we can negate function $\boldsymbol{\Psi}$ and make matrix $\mathbf{H}$ a positive definite matrix. An advantage to using the quasi-log-likelihood instead of the estimating function is that sometimes the latter is discrete but the former is continuous, see examples below.

The following fact about the asymptotic properties of the M-estimator is a generalization of Theorem 6.134.

Theorem 6.159 *Asymptotic properties of the M-estimator.* *Under the A–E assumptions with (6.81) replaced by (6.94), the M-estimator is (1) consistent and (2) asymptotically normal,*

$$\sqrt{n}(\widehat{\boldsymbol{\theta}}_M - \boldsymbol{\theta}) \simeq \mathcal{N}\left(\mathbf{0}, \mathbf{H}^{-1}(\boldsymbol{\theta})\mathbf{Q}(\boldsymbol{\theta})\mathbf{H}'^{-1}(\boldsymbol{\theta})\right), \quad n \to \infty, \tag{6.97}$$

where the $m \times m$ matrix $\mathbf{Q}$ is defined as

$$\mathbf{Q}(\boldsymbol{\theta}) = E_{\boldsymbol{\theta}}\left(\boldsymbol{\Psi}(\mathbf{Y};\boldsymbol{\theta})\boldsymbol{\Psi}'(\mathbf{Y};\boldsymbol{\theta})\right).$$

(3) The optimal choice of the estimating function is the score function (6.96):

$$\mathcal{I}^{-1}(\boldsymbol{\theta}) \le \mathbf{H}^{-1}(\boldsymbol{\theta})\mathbf{Q}(\boldsymbol{\theta})\mathbf{H}'^{-1}(\boldsymbol{\theta}). \tag{6.98}$$

In words, the M-estimator has a larger covariance matrix than the MLE, which proves the asymptotic efficiency of the maximum likelihood.

Proof. (1) As with the proof of Theorem 6.134, consider the average of the estimating equations (EE) $n^{-1}\sum_{i=1}^{n}\boldsymbol{\Psi}(\mathbf{Y}_i;\boldsymbol{\theta})$ when $n \to \infty$. Since vectors $\boldsymbol{\Psi}(\mathbf{Y}_i;\boldsymbol{\theta})$ are iid, the average of EE converges to its mean with probability 1 due to the strong LLN, $n^{-1}\sum_{i=1}^{n}\boldsymbol{\Psi}(\mathbf{Y}_i;\boldsymbol{\theta}) \to E_{\boldsymbol{\theta}}\boldsymbol{\Psi}(\mathbf{Y};\boldsymbol{\theta}) = \mathbf{0}$. Let a subsequence of M-estimators $\widehat{\boldsymbol{\theta}}_{n_k}$ converge to some $\boldsymbol{\theta}_*$. Then $E_{\boldsymbol{\theta}}\boldsymbol{\Psi}(\mathbf{Y};\boldsymbol{\theta}_*) = \mathbf{0}$ which implies $\boldsymbol{\theta} = \boldsymbol{\theta}_*$ due to condition (6.94). Since any subsequence of M-estimators converges to the true $\boldsymbol{\theta}$ the M-estimator, $\widehat{\boldsymbol{\theta}}_M$, is strongly consistent; see Section 6.4.4.

(2) Similarly to the proof of asymptotic normality of the MLE, we prove that the normalized EE have the multivariate normal distribution. We start by stating that

$$\frac{1}{\sqrt{n}}\sum_{i=1}^{n}\boldsymbol{\Psi}(\mathbf{Y}_i;\boldsymbol{\theta}) \simeq \mathcal{N}\left(\mathbf{0}, \mathbf{Q}(\boldsymbol{\theta})\right) \tag{6.99}$$

due to the multivariate CLT of Section 4.1.3, where we let $\mathbf{X}_i = \boldsymbol{\Psi}(\mathbf{Y}_i;\boldsymbol{\theta})$ and $\text{cov}(\boldsymbol{\Psi}(\mathbf{Y}_i;\boldsymbol{\theta})) = E_{\boldsymbol{\theta}}\left(\boldsymbol{\Psi}(\mathbf{Y};\boldsymbol{\theta})\boldsymbol{\Psi}'(\mathbf{Y};\boldsymbol{\theta})\right)$. Expand the normalized EE (6.99) up to the first order for large n

$$\mathbf{0} = \frac{1}{\sqrt{n}}\sum_{i=1}^{n}\boldsymbol{\Psi}(\mathbf{Y}_i;\boldsymbol{\theta}) + \left(\frac{1}{n}\sum_{i=1}^{n}\frac{\partial\boldsymbol{\Psi}(\mathbf{Y}_i;\boldsymbol{\theta})}{\partial\boldsymbol{\theta}}\right)\left(\sqrt{n}(\widehat{\boldsymbol{\theta}}_M - \boldsymbol{\theta})\right) + o(\left\|\widehat{\boldsymbol{\theta}}_M - \boldsymbol{\theta}\right\|).$$

But

$$\frac{1}{n}\sum_{i=1}^{n}\frac{\partial\boldsymbol{\Psi}(\mathbf{Y}_i;\boldsymbol{\theta})}{\partial\boldsymbol{\theta}} \to E_{\boldsymbol{\theta}}\left(\frac{\partial\boldsymbol{\Psi}(\mathbf{Y};\boldsymbol{\theta})}{\partial\boldsymbol{\theta}}\right) = \mathbf{H}(\boldsymbol{\theta})$$

with probability 1, so the asymptotic normality (6.97) is established.

(3) Because the expectation of the estimating function is zero, we write in the integral form $\int \boldsymbol{\Psi}(\mathbf{y};\boldsymbol{\theta})f(\mathbf{y};\boldsymbol{\theta})d\mathbf{y} = \mathbf{0}$. Differentiating this equation with respect

to $\boldsymbol{\theta}$ and using the product rule, we obtain

$$\mathbf{0} = \int \frac{\partial \boldsymbol{\Psi}(\mathbf{y}; \boldsymbol{\theta})}{\partial \boldsymbol{\theta}} f(\mathbf{y}; \boldsymbol{\theta}) d\mathbf{y} + \int \boldsymbol{\Psi}(\mathbf{y}; \boldsymbol{\theta}) \left(\frac{\partial f(\mathbf{y}; \boldsymbol{\theta})}{\partial \boldsymbol{\theta}} \right)' d\mathbf{y}$$

$$= \mathbf{H}(\boldsymbol{\theta}) + \int \boldsymbol{\Psi}(\mathbf{y}; \boldsymbol{\theta}) \left(\frac{\partial \ln f(\mathbf{y}; \boldsymbol{\theta})}{\partial \boldsymbol{\theta}} \right)' f(\mathbf{y}; \boldsymbol{\theta}) d\mathbf{y}.$$

Apply the Cauchy inequality (6.69) to the second term by letting

$$\mathbf{h} = \boldsymbol{\Psi}(\mathbf{y}; \boldsymbol{\theta}), \quad \mathbf{g} = \frac{\partial \ln f(\mathbf{y}; \boldsymbol{\theta})}{\partial \boldsymbol{\theta}}.$$

Since $E_{\boldsymbol{\theta}} \left[\mathbf{hg}' \right] = -\mathbf{H}(\boldsymbol{\theta})$ and $E_{\boldsymbol{\theta}} \left[\mathbf{gg}' \right] = \mathcal{I}(\boldsymbol{\theta})$, the Cauchy inequality implies $\mathbf{Q}(\boldsymbol{\theta}) - \mathbf{H}(\boldsymbol{\theta}) \mathcal{I}^{-1}(\boldsymbol{\theta}) \mathbf{H}'(\boldsymbol{\theta}) \geq \mathbf{0}$. Since matrix $\mathbf{H}(\boldsymbol{\theta})$ is nonsingular the multiplication by $\mathbf{H}^{-1}(\boldsymbol{\theta})$ at left and $\mathbf{H}'^{-1}(\boldsymbol{\theta})$ at right leads to inequality (6.98). $\square$

Sometimes, estimation involves parameters or interest and nuisance parameters. The following result, as a consequence of the proved theorem, tells that if parameters are orthogonal in the sense that the expectation of the cross-derivative is zero then use of any consistent estimator of nuisance parameter does not affect the efficiency of parameters of interest as discussed in Section 6.9.2 when parameters are estimated by ML.

Corollary 6.160 *Let, under conditions of the above theorem, the vector of parameters $\boldsymbol{\theta}$ split into two subvectors, the vector of parameters of interest, $\boldsymbol{\gamma}$, and the vector of nuisance parameters, $\boldsymbol{\nu}$. Let the parameter of interest be estimated by the method of maximum likelihood, but the nuisance parameter be estimated via the EE with the system of estimating equations taking the form*

$$\boldsymbol{\Psi}_1(\mathbf{Y}; \boldsymbol{\gamma}, \boldsymbol{\nu}) = \mathbf{0}, \quad \boldsymbol{\Psi}_2(\mathbf{Y}; \boldsymbol{\nu}) = \mathbf{0},$$

where $\boldsymbol{\Psi}_1(\mathbf{Y}; \boldsymbol{\gamma}, \boldsymbol{\nu}) = \partial \ln f(\mathbf{Y}; \boldsymbol{\gamma}, \boldsymbol{\nu}) / \partial \boldsymbol{\gamma}$ is the score equation and $\boldsymbol{\Psi}_2(\mathbf{Y}; \boldsymbol{\nu})$ is an estimating function for $\boldsymbol{\nu}$. If the expectation of the cross-derivative is zero, i.e.

$$E_{\boldsymbol{\theta}} \left(\frac{\partial \boldsymbol{\Psi}_1(\mathbf{Y}; \boldsymbol{\gamma}, \boldsymbol{\nu})}{\partial \boldsymbol{\nu}} \right) = \mathbf{0}, \tag{6.100}$$

then the asymptotic covariance matrix of the parameter of interest does not depend on the estimating function $\boldsymbol{\Psi}_2$. In other terms, the estimator of $\boldsymbol{\gamma}$ is asymptotically efficient.

Proof. As follows from (6.100) matrix $\mathbf{H}$ is block-diagonal, or more specially (the arguments of the vector functions are omitted for brevity)

$$\mathbf{H} = \begin{bmatrix} E_{\boldsymbol{\theta}} \left(\frac{\partial \boldsymbol{\Psi}_1}{\partial \boldsymbol{\gamma}} \right) & \mathbf{0} \\ \mathbf{0} & E_{\boldsymbol{\theta}} \left(\frac{\partial \boldsymbol{\Psi}_2}{\partial \boldsymbol{\nu}} \right) \end{bmatrix}, \quad \mathbf{Q} = \begin{bmatrix} E_{\boldsymbol{\theta}} \left(\boldsymbol{\Psi}_1 \boldsymbol{\Psi}_1' \right) & E_{\boldsymbol{\theta}} \left(\boldsymbol{\Psi}_1 \boldsymbol{\Psi}_2' \right) \\ E_{\boldsymbol{\theta}} \left(\boldsymbol{\Psi}_2 \boldsymbol{\Psi}_1' \right) & E_{\boldsymbol{\theta}} \left(\boldsymbol{\Psi}_2 \boldsymbol{\Psi}_2' \right) \end{bmatrix}.$$

Therefore the asymptotic covariance matrix for the M-estimator, $\widehat{\boldsymbol{\theta}}$, is (since only the first block is of interest we mark other three blocks by three asterisks)

$$\mathbf{H}^{-1}\mathbf{Q}\mathbf{H}'^{-1} = \begin{bmatrix} E_{\boldsymbol{\theta}}^{-1}\left(\frac{\partial\boldsymbol{\Psi}_1}{\partial\boldsymbol{\gamma}}\right)E_{\boldsymbol{\theta}}\left(\boldsymbol{\Psi}_1\boldsymbol{\Psi}_1'\right)E_{\boldsymbol{\theta}}^{-1}\left(\frac{\partial\boldsymbol{\Psi}_1}{\partial\boldsymbol{\gamma}}\right)' & *** \\ *** & *** \end{bmatrix},$$

where the superscript -1 at the expectation sign is a shortcut notation for the expectation of the inverse matrix. The first block is the asymptotic covariance matrix of the MLE for the parameter of interest, and it does not depend on the estimating equation for the nuisance parameter. □

Simply put, if the orthogonality condition (6.100) holds, we can use any consistent estimator of the nuisance parameter to estimate the parameter of interest without loss of efficiency. This result is especially useful when the joint distribution can be expressed as the product of conditional and marginal distributions that depends on the nuisance parameter only. Then we can achieve full efficiency even if the marginal distribution is not known, below is an example.

An attractive feature of the M-estimator is that its covariance matrix can be consistently estimated without knowing the true pdf,

$$\text{cov}(\widehat{\boldsymbol{\theta}}_M)\simeq\left(\frac{1}{n}\sum_{i=1}^{n}\frac{\partial\boldsymbol{\Psi}_i}{\partial\boldsymbol{\theta}}\right)^{-1}\left(\frac{1}{n}\sum_{i=1}^{n}\boldsymbol{\Psi}_i\boldsymbol{\Psi}_i'\right)\left(\frac{1}{n}\sum_{i=1}^{n}\frac{\partial\boldsymbol{\Psi}_i'}{\partial\boldsymbol{\theta}}\right)^{-1},$$

where function $\boldsymbol{\Psi}$ and its derivative are evaluated at $\widehat{\boldsymbol{\theta}}_M$. Sometimes, this formula is referred to as the *sandwich formula*.

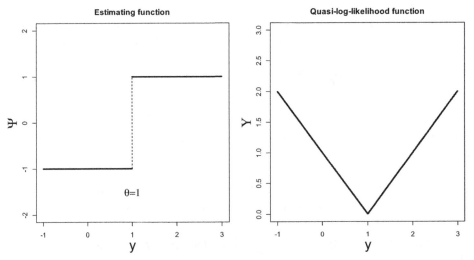

Figure 6.39: *The estimating and quasi-log-likelihood functions for the median.*

From introductory statistics, we know that the average is not robust to outliers but the median is. In the example below, we rigorously study the robust property of the median.

Example 6.161 *Comparison of average and median. Prove that the average of iid observations from the Laplace distribution is consistent for the center but has an asymptotic variance twice that of the median. Write R code to confirm this via vectorized simulations.*

Solution. We assume that Y_i are iid and have the Laplace (double exponential) distribution with pdf $f(y; \theta) = (\lambda/2)e^{-\lambda|y-\theta|}$ with a known rate parameter $\lambda > 0$; see Section 2.4.1. The MLE of the location parameter θ, with observations following the Laplace distribution, is the median. Indeed, the log-likelihood is $l(\theta) = 0.5n\lambda - \lambda\sum_{i=1}^{n}|Y_i - \theta|$, which attains its maximum where $\sum_{i=1}^{n}|Y_i - \theta|$ takes minimum. Differentiation of this function gives the condition for the MLE, $\sum_{i=1}^{n}\text{sign}(Y_i - \theta) = 0$. This equation means that $\widehat{\theta}_{ML}$ is such that the number of observations to left of $\widehat{\theta}_{ML}$ is equal to the number of observations to right of $\widehat{\theta}_{ML} = M$, that is, MLE = median for the Laplace distribution. The median may be viewed as an M-estimator with the estimating function

$$\Psi(y; \theta) = \begin{cases} -1 \text{ if } y < \theta \\ 1 \text{ if } y \geq \theta \end{cases} \qquad (6.101)$$

and the quasi-log-likelihood function $\Upsilon(y; \theta) = |y - \theta|$; see Figure 6.39.

Since MLE = median, we find its asymptotic distribution from Theorem 6.134. The Fisher information from an individual observation is

$$\mathcal{I}_1(\theta) = \lambda^2 \text{var}(\text{sign}(Y_1 - \theta)) = \lambda^2 E(\text{sign}^2(Y_1 - \theta)) = \lambda^2.$$

Thus, for large n, the median has normal distribution $\sqrt{n}(M - \theta) \simeq \mathcal{N}(0, 1/\lambda^2)$. The average, $\overline{Y}$, is another M-estimator with the estimating function $\Psi(y; \theta) = y - \theta$ and estimating equation $\sum_{i=1}^{n}(Y_i - \theta) = 0$ which produces $\widehat{\theta} = \overline{Y}$, the mean. Since $E(\Psi(y; \theta)) = E(y - \theta) = 0$, one infers that the average is a consistent estimator of the location parameter. Find its limiting distribution using the theory of the estimating equation (6.97):

$$H(\theta) = E_\theta \frac{d\Psi(Y, \theta)}{d\theta} = E_\theta(-1) = -1, \quad Q(\theta) = E_\theta(Y - \theta)^2 = \text{var}(Y) = 2/\lambda^2.$$

Therefore, for large sample size, $\sqrt{n}(\overline{Y} - \theta) \simeq \mathcal{N}(0, 2/\lambda^2)$. From comparison of the two limiting normal distributions, we deduce that the average applied to the Laplace distribution has the asymptotic variance twice that of the median. Vaguely, we say that the average is not a robust estimator of the center of the distribution with heavy tails, but the median is.

We use the method of generation for random variables following the Laplace distribution described in Section 2.4.1. The R code is found the file `meanmed.r`. We use function `apply` to compute the median and the mean in a vectorized fashion. Below is a typical output with default sample size $n = 10$.

```
meanmed()
[1] "Variance of median= 0.145392977656977"
[1] "Variance of mean= 0.199901757015213"
[1] "Ratio= 1.37490654800974"
```

When $n = 1,000$, the ratio of the variances is close to 1.9. Interestingly, the advantages of using the median over the mean are slim for small n, say, $n = 3$, as follows from the run below.

```
meanmed(n=3)
[1] "Variance of median= 0.647246433875679"
[1] "Variance of mean= 0.670009421378491"
[1] "Ratio= 1.03516896549975"
```

The mean is almost as efficient as the median. A heuristic explanation is that for small n, the chance of getting a large-value observation is small, so the two estimates of the center are close.

6.11.1   Robust statistics

The aim of robust statistics is to find estimators that perform well even when the original distribution assumption is violated. For example, we want to find an estimator of the center (location) of the distribution in the presence of outliers. We use Example 6.161, where it was shown that the arithmetic average is efficient when the distribution is normal, but not robust to heavy tails (the possibility of outliers). For such distributions, the median is a robust estimator of the center: when the distribution is normal, the arithmetic average is more efficient, but the presence of symmetric outliers (heavy distribution tails) the median gives smaller variance as an estimator of the center (location parameter). Several robust estimators have been suggested to estimate the location parameter of a symmetric distribution. In fact, the M-estimator was originally developed in connection to robust statistics; see Huber [58]. The goal of this section is to demonstrate how the M-estimator can be used for robust statistics.

Let Y_i be n iid observations drawn from a distribution with pdf $f(y - \theta)$, where f is an unknown density with symmetrically heavy tails (heavier than normal), so that outliers are possible, $f(-y) = f(y)$, and θ is the center/location parameter. The median is a robust estimator of θ with the estimating function (6.101). Several other estimating functions and respective estimators have been suggested to reduce the impact of extremely large or extremely small observations. For example, the Huber estimating function, sometimes referred to as the Huber

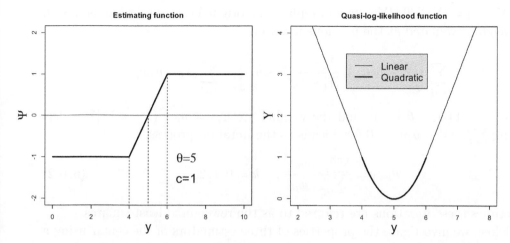

Figure 6.40: *Estimating and quasi-log-likelihood functions for the Huber proposal.*

proposal, takes the form

$$
\Psi(y;\theta) =
\begin{cases}
y - \theta \text{ if } |y - \theta| < c \\
c \text{ if } y - \theta \geq c \\
-c \text{ if } y - \theta < -c
\end{cases}
.
$$

This function complies with the zero expectation condition, $E\Psi(Y - \theta) = 0$, because Y_i are symmetrically distributed around the center θ. Elementary integration gives the quasi-log-likelihood function

$$
\Upsilon(y;\theta) =
\begin{cases}
(y - \theta)^2 \text{ if } |y - \theta| < c \\
2c\,|y - \theta| - c^2 \text{ if } |y - \theta| \geq c
\end{cases}
.
$$

Within the interval $(-c, c)$, this function is quadratic, but outside is linear. The quasi-log-likelihood function (Υ) and its derivative (Ψ) are continuous functions, where c is a given (user-defined) parameter; see Figure 6.40.

Parameter c defines the interval around θ within which the data are normally distributed. The Huber estimating function can be viewed as a compromise between the mean and the median: in the interval $(-c, c)$ it behaves like the estimating function for the mean and outside as the estimating function for the median. The estimating equation takes the form

$$
\sum_{i=1}^{n}[(Y_i - \theta)u_i + \text{sign}(Y_i - \theta)(1 - u_i)] = 0,
$$

where $u_i = 1\,(|Y_i - \theta| < c)$ and 1 is the indicator function. To find the solution of the estimating equation (the counterpart of the score equation), we represent

$\text{sign}(Y_i - \theta) = (Y_i - \theta)/|Y_i - \theta|$ and replace the original estimating function with the weights evaluated at the previous iteration,

$$\sum_{i=1}^{n} \left[(Y_i - \theta)u_i + \frac{Y_i - \theta}{|Y_i - \theta|}(1 - u_i) \right] = \sum_{i=1}^{n} (Y_i - \theta)w_i,$$

where $u_i = 1\left(|Y_i - \theta_k| < c\right)$ and the weights are $w_{ik} = u_i + (1 - u_i)/|Y_i - \theta_k|$. Solving $\sum_{i=1}^{n}(Y_i - \theta)w_i = 0$ for θ leads to the iteration process:

$$\theta_{k+1} = \frac{\sum_{i=1}^{n} w_{ik}Y_i}{\sum_{i=1}^{n} w_{ik}}, \quad k = 0, 1, 2, \dots \qquad (6.102)$$

Sometimes these iterations are referred to as the reweighted least squares.

Below, we investigate the properties of three estimators of the center using a two-component Gaussian mixture distribution via vectorized simulations.

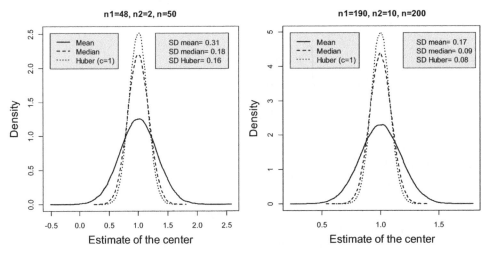

Figure 6.41: *Densities of three estimators of the center in the two-component Gaussian mixture (number of simulations, nSim=100,000). In simulations at left, the proportion of outliers is $\delta \simeq 2/50$ and in simulations at right the proportion is $\delta \simeq 10/200$. In both cases the median is about twice as efficient as the mean.*

Example 6.162 *Estimation of the noisy distribution. Let Y_i be iid with the density specified by the two-component normal/Gaussian mixture, $Y_i \sim (1 - \delta)\mathcal{N}(\theta, \sigma^2) + \delta\mathcal{N}(\theta, \sigma_N^2)$, where $\sigma_N^2 >> \sigma^2$ and δ is small. Plot the kernel densities and compare SDs of (a) mean, (b) median, and (c) Huber estimator using vectorized simulations.*

Solution. The parameters of the Gaussian mixture are interpreted as follows: (a) $\mathcal{N}(\theta, \sigma^2)$ specifies the distribution of the "correct" general population and $\mathcal{N}(\theta, \sigma_N^2)$ specifies the distribution of the noise; they have the same center

θ but the variance of the noise is much larger; (b) δ specifies the proportion of noisy observations in the sample. For example, if n is the total sample size, the proportion of "correct" observations is $(1-\delta)n$ and the proportion of "noisy" observations (outliers) is δn. This comment is the basis for generating observations from the two-component model: given n, we generate δn observations from $\mathcal{N}(\theta, \sigma^2)$ and $(1-\delta)n$ observations from $\mathcal{N}(\theta, \sigma_N^2)$; the same method was used earlier in Example 2.73. The R code with vectorized computations is found in file robloc.r. We make a few comments: (a) The mean is computed using rowMeans command which returns nSim means across columns in the nSim by n observation matrix. (b) The median is computed using apply where MARGIN=1 and FUN=median means that nSim medians across columns of matrix Y are returned. (c) nSim Huber estimates of θ with $c=1$ (we use the name cc because c is taken by "combine values" in R) used as default are computed using the local function HubEstF. This is where vectorized computations are used by processing $\mathbf{Y}$ as a matrix in the iterative formula (6.102), not a $n \times 1$ vector, as the traditional loop would do. To comply with dimensions, we use the $n \times 1$ vector of ones (un), so that the nSim by n matrix $\boldsymbol{\theta}\mathbf{1}'$ is computed as hub.est%*%t(un). This code generates Figure 6.41. All three estimators are unbiased and have symmetric distribution around θ, but the density of the mean estimator is wider than the median and Huber estimators. The number of noisy observations with $n=50$ is $n_2 = 2$ and the number of correct observations is $n_1 = 48$. The Huber estimator has a slightly smaller SD than the median estimator of θ. However, the Huber proposal requires the specification of parameter c which makes it problematic in practical applications. $\square$

The following is an application of the noise model from the previous example to study the distribution of the Google stock returns from Example 6.152. As noted in that example, stock returns typically exhibit spikes (atypically high positive or negative return values) and Google stock price is no exception. The Cauchy distribution was applied in Example 6.152; now we use the Gaussian mixture from the previous example. This model enables us to answer important questions such as how frequently one may except spikes and what is the magnitude of the spikes measured on the scale of the standard deviation.

Example 6.163 *Estimation of Google stock prices. (a) Develop the maximum likelihood estimation of the Gaussian mixture with the same mean $Y_i \sim (1-\delta)\mathcal{N}(\theta, \sigma^2) + \delta\mathcal{N}(\theta, \sigma_N^2)$ using the EFS algorithm. (b) Apply this model to Google stock returns to estimate the frequency of spikes and their SD.*

Solution. The density of the two-component Gaussian mixture with a single center is

$$f(y; \theta, \delta, \sigma, \sigma_N) = \frac{1-\delta}{\sigma}\phi\left(\frac{y-\theta}{\sigma}\right) + \frac{\delta}{\sigma_N}\phi\left(\frac{y-\theta}{\sigma_N}\right).$$

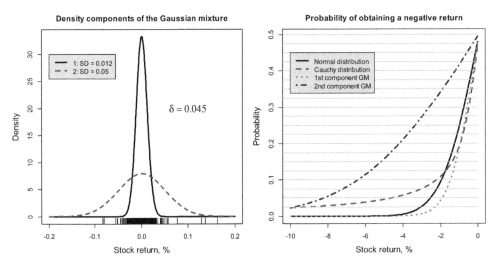

Figure 6.42: *The two-component densities and the associated probabilites of observing negative returns. Spikes are expected in 4.5% of the daily sample; one spike each three weeks. The magnitude of spikes is five times larger than the regular variation of returns.*

Using $\phi'(x) = -x\phi(x)$, we obtain the needed derivatives:

$$\frac{\partial f}{\partial \theta} = \left[\frac{1-\delta}{\sigma^3}\phi\left(\frac{y-\theta}{\sigma}\right) + \frac{\delta}{\sigma_N^3}\phi\left(\frac{y-\theta}{\sigma_N}\right)\right](y-\theta),$$

$$\frac{\partial f}{\partial \delta} = -\frac{1}{\sigma}\phi\left(\frac{y-\theta}{\sigma}\right) + \frac{1}{\sigma_N}\phi\left(\frac{y-\theta}{\sigma_N}\right), \quad \frac{\partial f}{\partial \sigma} = \frac{(1-\delta)((y-\theta)^2 - \sigma^2)}{\sigma^4}\phi\left(\frac{y-\theta}{\sigma}\right),$$

$$\frac{\partial f}{\partial \sigma_N} = \frac{\delta}{\sigma_N^4}\phi\left(\frac{y-\theta}{\sigma_N}\right)((y-\theta)^2 - \sigma_N^2).$$

The log-likelihood function is

$$l(\theta, \delta, \sigma, \sigma_N) = \sum_{i=1}^{n} \ln\left[\frac{1-\delta}{\sigma}\phi\left(\frac{Y_i-\theta}{\sigma}\right) + \frac{\delta}{\sigma_N}\phi\left(\frac{Y_i-\theta}{\sigma_N}\right)\right].$$

Let $\mathbf{d}_1$ be a $n \times 1$ vector with the ith component as the derivative $\partial \ln f/\partial \theta = (\partial f/\partial \theta)/f$ evaluated at Y_i at the kth iteration. Similarly, define the vector of derivatives with respect to the other three parameters $\mathbf{d}_2, \mathbf{d}_3$, and $\mathbf{d}_4$. Let $\mathbf{D}$ be the $n \times 4$ matrix with the respective vector columns. Then the score equations can be written in the matrix form as $\mathbf{D'1}$ where $\mathbf{1}$ is the $n \times 1$ vector of ones, and the estimated FS algorithm takes the form

$$\mathbf{p}_{k+1} = \mathbf{p}_k + (\mathbf{D}_k'\mathbf{D}_k)^{-1}\mathbf{D}_k'\mathbf{1}, \quad k = 0, 1, 2, ...,$$

where the subindex k indicates that the involved terms are evaluated at the kth iteration, and $\mathbf{p} = (\theta, \delta, \sigma, \sigma_N)'$ is the 4×1 vector of unknown parameters.

The R code is found in function gng. The iterations are shown below: it takes seven iterations to converge with the tolerance eps=.00001. The first column is the iteration index. The second column is the value of the log-likelihood function. The next four columns are parameters θ, δ, σ, and σ_N. The last column is the gradient as the norm of the derivative vector.

```
[1] 1.000000e+00 3.793050e+03 0.000000e+00 3.458647e-02 1.170470e-02 5.927901e-02 9.681352e+03
[1] 2.000000e+00 3.793895e+03 6.261431e-04 4.240726e-02 1.198556e-02 5.086973e-02 6.009189e+02
[1] 3.000000e+00 3.793939e+03 6.251297e-04 4.479996e-02 1.199631e-02 5.055888e-02 9.648902e+00
[1] 4.000000e+00 3.793939e+03 6.247740e-04 4.490459e-02 1.199502e-02 5.044998e-02 1.144107e+01
[1] 5.000000e+00 3.793939e+03 6.247344e-04 4.480053e-02 1.199682e-02 5.043734e-02 3.944097e+00
[1] 6.000000e+00 3.793939e+03 6.245870e-04 4.482522e-02 1.199626e-02 5.042718e-02 2.253615e+00
[1] 7.000000e+00 3.793939e+03 6.246227e-04 4.480720e-02 1.199661e-02 5.042785e-02 9.780251e-01
```

The starting values for the four parameters are derived as follows: the mean parameter θ is set to zero; to find a rough estimate for σ, σ_N, and δ, we first compute the SD of the entire sample of Google returns and then set σ as the SD of the subsample with the values within the interval ($-2\times$SD, $2\times$SD). An estimator for δ_N is obtained in a similar way, but for a subsample with values outside of this interval. The proportion of returns that follow the second component is computed as the proportion of the number of values outside of above the interval. The summary of estimated parameters is shown below. The first column is the ML estimate at the final iteration of the EFS algorithm. The second column is its standard error computed as the square root of the diagonal of matrix $(\mathbf{D}_k'\mathbf{D}_k)^{-1}$ at the final iteration. The Z-values, as the ratio of the first to the second column, are shown in the third column.

```
[1,] 0.000624596 0.0003508764  1.780102
[2,] 0.044814192 0.0095375723  4.698700
[3,] 0.011996459 0.0002811400 42.670765
[4,] 0.050426440 0.0041934680 12.024997
```

Figure 6.42 shows the results of the two-component Gaussian mixture estimation. The plot at left depicts the normal densities of the first (regular variation of returns) and second component (spikes). Bars at the x-axis represent actual Google return data using the R function rug. The SD of the spike component (SD = 0.05) is almost five times larger than the SD (SD = 0.012). According to our analysis, there are 4.5% spikes in the data. A spike occurs about every three weeks on average. The right plot of the figure depicts the probability of negative return computed as $\Phi(r/\text{SD})$. The first two probability curves are copied from Figure 6.36. Regular variation of returns have small probability of occurrence, but the probability of spikes is very large – the good news is that they happen infrequently, only once every three weeks.

Problems

1. Prove that the sandwich formula yields a consistent estimate of the covariance matrix of the M-estimator. [Hint: Use results of Section 6.4.4.]

2. (a) Modify function meanmed from Example 6.161 to investigate the advantage of using the median over the mean for $n = 3, 10, 100, 1K, 10K$ with

fixed λ by plotting the ratio of variances versus n on the log scale. (b) Investigate how λ affects the ratio of variances. (c) Replace `var` with the mean absolute value criterion `mean(abs(..))`. Do conclusions withstand the change in the criterion? Explain.

3. (a) Define the L_p estimator of the center of the distribution, θ, as the minimizer of the quasi-log-likelihood function $\Upsilon(y;\theta) = |y - \theta|^p$, where $p > 0$. (b) Show that this family of estimators covers the MLE for normal and double Laplace distributions. (c) Develop an algorithm for estimation of θ in the form of the reweighted least squares (6.102). (d) Modify function `robloc` from Example 6.162 to test this estimator using the Gaussian two-component model.

4. (a) Define Andrews' estimator of the center of the distribution using the estimating function $\Upsilon(y;\theta) = c(1 - \cos(y/c))$ for $|y| \leq \pi c$ and $\Upsilon(y;\theta) = 2c$ for $|y| > \pi c$ where $c > 0$. (b) Plot this function for $c = 1, 2, 3$.(c) Develop an algorithm for estimation of θ in the form (6.102). (d) Define the "optimal" c as the inflection point (the point where convexity turns into concavity) and prove that $c = 6/\pi$. (e) Apply this function to Example 6.162 and compare the results.

5. Use simulations to test how biased are the SE of parameters in the two-component Gaussian mixture. Run `gng` with simulations and compute the empirical variance of the estimates and the mean of the diagonal elements of matrix $(\mathbf{D}_k'\mathbf{D}_k)^{-1}$ at the final iteration.

6. The logistic regression model for iid pairs (Y_i, X_i), where Y_i is binary and $\mathbf{X}_i$ is normally distributed, is defined by the conditional distribution $\Pr(Y = 0|\mathbf{X} = x) = (1 + e^{\alpha+\beta x})^{-1}$ and the marginal normal density for $\mathbf{X}$ with mean μ and variance σ^2, see Section 8.8.2 for more detail where $\mathbf{X}_i$ is fixed. (a) Develop the maximum likelihood estimation for α and β and derive the asymptotic covariance matrix for the ML estimator. (b) Derive the M-estimator by minimizing the sum of squares $\sum(Y_i - e^{\alpha+\beta x}(1+e^{\alpha+\beta x})^{-1})^2$, and show that this estimator is consistent but less efficient than the ML estimator. (c) Use simulations to verify your analytic solution. [Hint: Use the rule of repeated expectation (3.27) and `integrate` to compute covariance matrices.]

7. Run function `meanmed` using increasing n to demonstrate that the ratio of the variances indeed approaches 2 when $n \to \infty$.

8. (a) Run an analysis of the Hewlett-Packard stock return similarly to Example 6.163. (b) File `AMZN.csv` contains Amazon stock prices from 2015 to 2019. Run the analysis similarly to Example 6.163.

Chapter 7

Hypothesis testing and confidence intervals

Together with parameter estimation, hypothesis testing and confidence intervals are the backbones of statistical inference. In a way, parameter estimation is essential because once a satisfactory estimator exists, one can use it to test hypotheses or construct confidence intervals. Properties of statistical tests and confidence intervals are studied from two perspectives: (i) small-sample properties, when the sample size, n, is finite, and (ii) asymptotic properties, when $n \to \infty$. From a practical standpoint, the latter never takes place – we treat asymptotic results as an approximation. Hypothesis testing and confidence intervals are intrinsically related. We discuss how to test statistical hypotheses, and then turn our attention to interval estimation.

Special attention is given to explanation of major statistical concepts, such as the p-value, in layman's terms. Often, the work of statisticians is used by nonstatisticians. Therefore, the ability to explain statistics in lay-language is of paramount importance – this is how our work is judged, and ultimately, paid.

7.1 Fundamentals of statistical testing

We give an intuitive introduction to hypothesis testing using the concept of the *normal* (or expected) *range*. More theoretical discussion starts in the next section. Plenty of details on statistical hypothesis testing can found in the authoritative book by Lehmann and Romano [69].

Let $Y_1, Y_2, ..., Y_n$ be a random sample from a general population with the probability density function $f(y; \theta)$ dependent on unknown parameter θ. For example, Y_i may be income reported by the ith household in a town survey. If surveys are conducted on a randomly chosen subpopulation of town families, we may safely

Advanced Statistics with Applications in R, First Edition. Eugene Demidenko.
© 2020 John Wiley & Sons, Inc. Published 2020 by John Wiley & Sons, Inc.

assume that Y_i are independent and identically distributed (iid). Under the most comfortable distribution assumption, Y_i are normally distributed with unknown mean and variance, $Y_i \overset{\text{iid}}{\sim} \mathcal{N}(\mu, \sigma^2)$. However, to concentrate on conceptual points we shall assume in this section that σ^2 is known. The town officials want to know if the town's household income is equal to the state's household income, μ_0, supposedly known. However, since the town's average, $\overline{Y}$, is a continuous random variable, it will never be exactly equal μ_0, precisely, $\Pr(\overline{Y} = \mu_0) = 0$. Thus, the problem of statistical testing $\mu = \mu_0$ is not trivial. In vague language, we want to show that $\overline{Y}$ is close to μ_0. "How close?" is a paramount question.

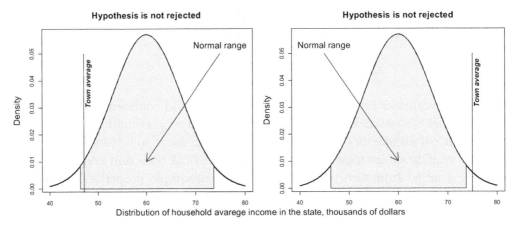

Figure 7.1: *Illustration of hypothesis testing via the normal range of the household income averaged over n families in the state. The outcome of the test depends on whether the observed town average belongs to the normal range.*

We formulate the hypothesis as

$$H_0 : \mu = \mu_0, \qquad (7.1)$$

which will be referred to as the *null hypothesis*. μ_0 will be referred to as the *null value*. The idea of testing reduces to providing some evidence that the data confirm the null hypothesis. Clearly, the null hypothesis will hold if $\overline{Y}$ is close to μ_0 and $|\overline{Y} - \mu_0|$ is small. Note that the above "hard" condition $\overline{Y} = \mu_0$ is replaced with the "soft" condition $|\overline{Y} - \mu_0|$ is small. The crucial task is to define what is small. To answer this question, we find the *normal range* of household incomes averaged over n families in a typical state town, i.e. the town with the household income equal the null value, μ_0. Traditionally, we use the normal range as the 95% interval interval

$$\mu_0 \pm q_{1-\alpha/2} \frac{\sigma}{\sqrt{n}}, \qquad (7.2)$$

where $\alpha = 0.05$ and $q_{1-\alpha/2} = \Phi^{-1}(1-0.05/2)$ is the 0.975 quantile of the standard normal distribution, where α is referred to as the significance level. The range

(7.2) is the expected range: it is expected that $(1 - \alpha)100\%$ observations will be within this interval. Now the hypothesis testing follows a simple rule: if $\overline{Y}$ belongs to the normal range specified by (7.2) the null hypothesis that the town's household income is the same as in a typical state town is not rejected, and if $\overline{Y}$ is outside of (7.2) it is rejected. Equivalently, we may formulate the test statistic as

$$Z_{\text{obs}} = \frac{\overline{Y} - \mu_0}{\sigma/\sqrt{n}}, \tag{7.3}$$

the observed Z-value. Under the null hypothesis (7.1), $Y_i \overset{\text{iid}}{\sim} \mathcal{N}(\mu_0, \sigma^2)$ and the test statistic Z has standard normal distribution $\mathcal{N}(0,1)$, and therefore hypothesis H_0 is rejected if $|Z_{\text{obs}}| > q_{1-\alpha/2}$. It is important to remember that $\alpha = 0.05$, referred to as the significance level, was chosen arbitrarily, out of tradition. Consequently, 5% of average incomes will fall outside interval (7.2) even though the survey is conducted in a typical state town.

Figure 7.1 illustrates the above hypothesis testing. Two scenarios are depicted. In the left plot, the hypothesis (7.1) is not rejected because the town average is within the normal range. In the right plot, the hypothesis is rejected because the town average is outside the normal range. The density depicts the distribution of $\overline{Y}$ under the null hypothesis, $\overline{Y} \sim \mathcal{N}(\mu_0, \sigma^2/n)$, with the state's household income null value $\mu_0 = \$60K$. The shaded area depicts the 95% normal range of income averaged over n families. Note that this is not the 95% range of family income but the 95% range of the *average* income, due to the $\sqrt{n}$ in the denominator. The rejection area is the unshaded/white area at the tails. The hypothesis that the town's household income is the same as in the state is accepted if the observed value $\overline{Y}$ belongs to the shaded area. Otherwise, it is rejected.

7.1.1 The p-value and its interpretation

Instead of reporting *yes* or *no* on rejecting the null hypothesis, we often compute

$$p\text{-value} = \Pr\left(|Z_{\text{obs}}| > Z\right) = 2\Phi(-|Z_{\text{obs}}|) = 2(1 - \Phi(|Z_{\text{obs}}|)),$$

where Z_{obs} is treated as a fixed value and $Z \sim \mathcal{N}(0,1)$. In Figure 7.1, the p-value is twice the area under the density to the left of *Town average* in the left plot and to the right of *Town average* in the right plot. The p-value is more informative because it tells how strongly the data support the null hypothesis, or how far the average is from the null value (in our example 60). The p-value is the probability to get a Z-value larger than $|Z_{\text{obs}}|$ in a typical state town, i.e. in the town with the household income μ_0. According to statistical dogma, if p-value < 0.05, we say that the difference between the household income in a town and the state is *statistically significant*. The larger the observed value of the test statistic, the smaller p-value, the greater the evidence that the data do not support the null hypothesis. However, in order this claim to be true, the power of the test must be monotonic on the both sides of the null value. In terms of the cumulative distribution function (cdf) $F_Z(z; \mu_1) > F_Z(z; \mu_2)$ for $\mu_0 < \mu_1 < \mu_2$ and $F_Z(z; \mu_1) > F_Z(z; \mu_2)$ for $\mu_1 < \mu_2 < \mu_0$, where F_Z is the cdf of Z defined in the right-hand side of (7.3). See more detail in Section 5.1. In layman's language, the p-value is the chance of getting a test statistic greater than that observed if the null hypothesis is true.

The example below illustrates the interpretation of the p-value through simulations. More discussion of the p-value and its shortcomings is found in Section 7.10.

Example 7.1 *Simulations for p-value.* *A town survey of $n = 100$ family incomes compared to the state income yields the observed Z-statistic $= 1.9$ and the p-value $= 0.0588$. (a) Illustrate by simulations that this p-value equals the proportion of Z-statistics for which $|Z| > 1.9$ or p-values < 0.0588 if the Z-test were conducted on many other imaginary surveys of 100 families from a typical state town. (b) Prove that the result does not depend on μ_0, σ, and n, because under the null hypothesis, the p-values have a uniform distribution on $(0, 1)$.*

Solution. (a) If $\{Y_1, Y_2, ..., Y_{100}\}$ is a sample of incomes in town, the observed p-value $= 0.0588$ is computed in R as `2*(1-pnorm(abs(Z)))` where `Z=(av-60)/SD*sqrt(n)` and `av=mean(Y)`. Given the p-value, we compute the absolute value of Z as `Z.obs=qnorm(pvalue.obs/2,lower. tail=F)`. The code below repeats this computation on many imaginable surveys conducted in a typical town with the average household income \$60K. In this code, we set $\sigma = SD = 10$ as the default value, but it does not affect the answer as can be verified by simulations or theoretically as follows from (b).

```
function(incA=60,pvalue.obs=0.0588,SD=10,n=100,nSurv=10000)
{
#incA=state official family income, the null hypothesis value
#Z.obs=observed Z-statistic
#pvalue.obs=observed p-value based on the survey of 100 families
#SD=standard deviation on family income in town and state, known
#n=town survey sample size
#nSurv=imaginary survey sample size in a typical state town
dump("pvsim","c:\\StatBook\\pvsim.r")
Z.obs=qnorm(pvalue.obs/2,lower.tail=F)
z.nSurv=pv.nSurv=rep(NA,nSurv)
for(i in 1:nSurv) #imaginary nSurv surveys
{
  inc=rnorm(n,mean=incA,sd=SD) # n random incomes in typical town
  av=mean(inc)
Z=(av-incA)/SD*sqrt(n)
    z.nSurv[i]=Z
```

```
pv.nSurv[i]=2*(1-pnorm(abs(Z)))
}
cat("Observed Z = Z.obs =",Z.obs,"\nProportion of Z-values in a
typical state town for which abs(Z)>=Z.obs =",
        mean(abs(z.nSurv)>=Z.obs),"\n")
cat("Observed p-value = pvalue.obs =",pvalue.obs,"\nProportion of

    p-values in a typical state town < pvalue.obs =",
    mean(pv.nSurv<pvalue.obs),"\n")
}
```

A typical run produces the following output.

```
> pvsim()
Observed Z = Z.obs = 1.889686
Proportion of Z-values in a typical state town for which
        abs(Z)>=Z.obs=0.0579
Observed p-value = pvalue.obs = 0.0588
Proportion of p-values in a typical state town < pvalue.obs
        0.0579
```

The difference between the observed p-value and the simulation-based p-value is due to the finite number of imaginary surveys (simulations), nSurv=10000. The reader can repeat simulations with large nSurv to ensure that the two quantities are basically the same. Moreover, the result does not depend upon the choice of the default values of the function arguments. (b) The proof that p-value $\sim \mathcal{R}(0,1)$ under the null hypothesis is similar to that of the proof of Theorem 2.71 of Section 2.13. Let x be in $(0,1)$. Then, under the null hypothesis, $Z \sim \mathcal{N}(0,1)$ and

$$\Pr(p\text{-value} \leq x) = \Pr(2\Phi(-|Z|) \leq x) = \Pr(|Z| \geq -\Phi^{-1}(x/2))$$
$$= 2\Phi(\Phi^{-1}(x/2)) = x.$$

This means that the cdf of p-value is x and, therefore, it has a uniform distribution on $(0,1)$. Now, based on the uniformity of the p-values, we prove that the output of function pvsim does not depend on the arguments, but does depend on nSurv, which defines how close the simulation-based p-value is to the observed. Indeed, it is sufficient to prove that under the null hypothesis

$$\Pr(p\text{-value} \leq \text{observed } p\text{-value}) = \text{observed } p\text{-value},$$

which follows from the uniformity of p-value. Since the proof does not depend on μ_0, σ, and n, the simulation result does not depend of the choice of the program arguments. □

In vague terms, the observed p-value indicates how strongly the data support the null hypothesis (the smaller, the stronger) or the likelihood of getting a p-value smaller than the observed one if the null hypothesis holds. In different words, the observed p-value is the proportion of p-values smaller than the observed in many other imaginary samples for which the null hypothesis holds. See more relevant discussion in Section 7.10.

7.1.2   Ad hoc statistical testing

In this section, we provide two other ad hoc examples where hypothesis testing is conducted using the concept of the normal range.

The multiple-choice test is a common exam format: We want to test the hypothesis that the student simply guessed when answering questions. Surprisingly, one may score pretty well in a two-choice test by guessing at random.

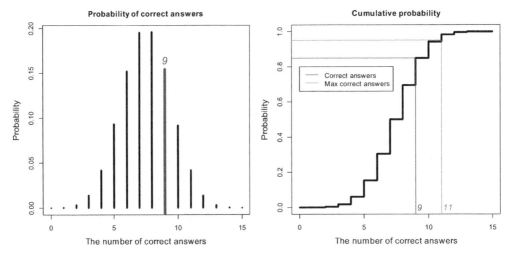

Figure 7.2: *Illustration of the two-choice test hypothesis testing. A student may get up to 11 correct answers with the probability 0.95. Since the student answered 9 questions correctly the hypothesis that he randomly picked the answer is not rejected.*

Example 7.2   *Two-choice test.* *A student answered correctly on 9 out of 15 yes/no questions. (a) Test the hypothesis that he/she picked answers at random (simple guess) by computing the maximum number of points under the assumption of random choice using the significance level $\alpha = 0.05$. (b) Use vectorized simulations to check your answer.*

Solution. (a) Since the answer to each question is binary, we can model the number of correct answers as a binomial probability. Thus, if X_i is the Bernoulli random variable which takes 1 if the answer to the ith question is correct and 0 otherwise, the total number of points gained is $\sum_{i=1}^{15} X_i$. The maximum number of points on the test is 15 (when all answers are correct) and the minimum number of points is 0 (no single answer is correct). To test the hypothesis H_0 :

$p = 1/2$ we need to find the normal range as the maximum number of points if the student guessed. In mathematical terms, we need to find $r_{\max}$ such that $\Pr\left(\sum_{i=1}^{15} X_i \geq r_{\max}\right) = 0.05$. Then, if $9 > r_{\max}$ we reject the hypothesis that the answers were random. Hence, $r_{\max}$ may be interpreted as a threshold beyond which the number of points with answers chosen at random is improbable. In R, $r_{\max}$ can be computed as `rmax=qbinom(0.95,size=15,prob=0.5)` which gives `rmax=11`. Since $9 < 11$ we cannot reject the hypothesis that answers were chosen at random with error 5%. This test is called one-sided because the decision is made based on one inequality: the hypothesis is rejected if the total number of points is greater than $r_{\max}$. We underscore that, although the value $\alpha = 0.05$ is traditional is statistics, there is nothing special about this choice: to a certain extent it is arbitrary and yet crucial for the outcome of hypothesis testing. For example, if $\alpha = 0.2$ the critical value would be $r_{\max} = 9$ and the hypothesis that the student guessed should be rejected. (b) The R code `multcht` produces Figure 7.2. We make several comments. (a) Matrix X represents random yes/no answers coded 1/0, respectively; the number of rows can be interpreted as the number of students who took the test and the jth column codes 1 if the answer is correct and 0 otherwise. Variable `m.correct` is the number of correct answers for each student, an array of length `nSim`. (b) Array `pr` contains probabilities to m points from $m = 0$ (all answers are wrong) to $m = n = 15$ (all answers are correct), again under the assumption that answers are random. (c) The left plot in Figure 7.2 depicts the distribution of the points on the test. These probabilities are computed based on simulations; the same probabilities could be computed as `dbinom(0:15,size=15,prob=0.5)`. The right plot depicts the cdf, which is needed to compute the maximum number of points that could be gained if answers were random. Since $\alpha = 0.05$ we invert the cdf and obtain `rmax = 11` (green line). □

The reader should not forget that the outcome of hypothesis testing depends on the choice of α, and there is no particular reason why $\alpha = 0.05$ is chosen – it is just a tradition. Roughly speaking, the problem of hypothesis testing is substituted with the problem of choosing the significance level. More discussion on the choice of significance level can be found in Section 7.10.

The next example demonstrates that statistical hypothesis testing is not the only way to make arguments on the plausibility of events. Sometimes a mere report on the chance of the event under normal circumstances may serve as a convincing answer instead of reporting the p-value.

Example 7.3 *Statistics in court: insider trading.* *On Monday May 24, 2016 at 3:37:21 the presidents of USA and Vietnam announced a deal in the amount of $11.3 billion to buy wide-body 747-8 Booing VietJet airplanes to boost tourism to Vietnam. As a result, Boeing shares rose 0.2% to $127.58. Mr. John Smith is accused of insider trading by receiving a call about the deal before it was announced and bought Boeing shares at 3:37:24 after the announcement in the* *amount of $10 million. As such, he instantly earned $20K. Estimate the chance that he legitimately bought stocks using a sample of the 19 fastest trading times of other stock dealers in similar announcements (the time data is found in the file* `boengtr.csv`*).*

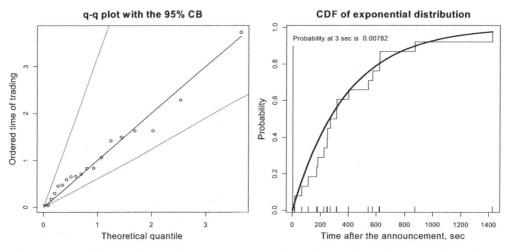

Figure 7.3: *The analysis of* `boengtr.csv` *data from Example 7.3. As follows from the q-q plot, the time of trading follows an exponential distribution. As follows from the cdf analysis, the chance that Mr. John Smith could have bought stocks 3 seconds after the announcement without insider trading is* 0.00782.

Solution. The time of trading stocks after a major company announcement is likely to follow an exponential distribution (traders want to react as quickly as possible which makes the density decreasing) with the cdf $F(x; \lambda) = 1 - e^{-\lambda x}$. We use the q-q plot to visually test this hypothesis for the 19 time data points from file `boengtr.csv`. See Figure 7.3. In the left plot, we use the q-q plot with the 95% confidence band (CB) to compare the empirical quantiles, which are the ordered time versus theoretical quantile under the assumption that the time follows an exponential distribution – we recommend reviewing Section 5.3 and particularly equation (5.11). In the plot at right, we show the empirical and theoretical cdfs with $\widehat{\lambda}_{ML} = 1/\overline{X}$. The history data have minimum time of trading 15 seconds after the major deal announcement. Although 3 seconds is considerably smaller the lawyer for the defendant argues that if more data were collected the minimum time could be much smaller and therefore 3 seconds is not beyond a reasonable doubt. What is the chance that a stock dealer may trade within 3 seconds? Since the exponential distribution is a reasonable approximation of the time of trading, we estimate this probability as $1 - e^{-\widehat{\lambda}_{ML} \times 3} = 0.00782$. In other words, if Mr. John Smith acted without the tip, the chances that he could buy Boeing stocks only 3 seconds after the announcement is eight out of one thousand. We will revisit this problem in Example 7.32 using rigorous derivations.

Problems

1. (a) Is it true that any null hypothesis can be rejected by choosing α small enough? (b) Explain in lay-language why a smaller σ in the town income example increases the chance of rejecting the null hypothesis.

2. Is it true that the p-value is the probability of the test statistic falling within rejection area under the null hypothesis?

3. Prove that the Z-test matches the monotonicity of the cdf on the both sides of the null μ_0, that is, the larger the value of $|Z_{\text{obs}}|$, the more evidence that the null hypothesis does not hold.

4. Repeat simulations by function `pvsim` in Example 7.1 to demonstrate that the result does not change upon the choice of the default values and explain why.

5. Generalize Example 7.2 to a three-choice test with the hypothesis that the student guessed to pick the answer on three-choice questions if he/she gained 9 out of 15 points. Use simulations to check the answer.

6. Generalize Example 7.2 to an arbitrary multiple-choice test as described in the previous problem (use the number of choices, k, as an argument in the R code).

7. Generate 100K trading times from the exponential distribution with $\lambda = \widehat{\lambda}_{ML}$ and estimate the proportion of times that are less than or equal 3 seconds in Example 7.3.

7.2 Simple hypothesis

Basic hypothesis testing was illustrated intuitively in the previous section. Here we use rigorous language.

A statistical hypothesis is called a two-sided simple hypothesis if the distribution depends on a single parameter with the null hypothesis $H_0 : \theta = \theta_0$ against the alternative $H_A : \theta \neq \theta_0$. Note that the alternative is *two-sided* because in the case when the null hypothesis does not hold we may have either $\theta > \theta_0$ or $\theta < \theta_0$.

The procedure for hypothesis testing involves four steps:

1. Find a "good" estimator of θ and define the distance, the difference between the estimator and the null value, called the *test statistic*, T.

2. Find the distribution of the test statistic under the null hypothesis $\theta = \theta_0$. This distribution is called the null distribution of the test statistic.

3. Given positive and small significance level, α, such as $\alpha = 0.05$, find the critical value, T_0, such that

$$\Pr(|T| > T_0) = \alpha \qquad (7.4)$$

under assumption that the null hypothesis is true.

4. Carry out the test: if the observed value of T is outside the interval $(-T_0, T_0)$, then the null hypothesis is rejected.

The reasoning behind hypothesis testing is that if the observed value is outside the test statistic normal range, computed under the assumption that the null hypothesis is true, then the hypothesis should be rejected.

Using the previous language, the interval $(-T_0, T_0)$ is the normal range of T. The union of intervals $T > T_0$ and $T < -T_0$ is called the *rejection region* and α is called the *size* of the test or *significance level*. As follows from the definition of the test expressed in equation (7.4), α is the probability of rejection of the null hypothesis when it is true, and sometimes referred to as the *type I error*. All three terms for α are interchangeable. In our town example, 5% of towns will be declared atypical (the null hypothesis is rejected) even though additional surveys may confirm that the income average falls within the "normal" range.

Equation (7.4) may have no solution for T_0 when the distribution is discrete. Then we pick T_0 such that the difference between the left, and the right-hand side is minimum in absolute value but $\Pr(|T| > T_0) \leq \alpha$.

A statistical hypothesis is call *left-sided* if the alternative is $H_A : \theta < \theta_0$, and a statistical hypothesis is call *right-sided* if the alternative is $H_A : \theta > \theta_0$. Such hypotheses are called *one-sided*. An example of a one-sided hypothesis is the above two-choice example. The null hypothesis is that the answers were picked at random, $H_0 : p = 1/2$ against the alternative $H_A : p > 1/2$. Two-sided alternative requires two-tail test and one-sided alternative requires one-sided test. One-sided alternative is appealing in practice but two-sided hypothesis are more frequently studied in the theory of hypothesis testing; more discussion can be found later in the chapter.

Below we provide several illustrative examples on testing double-sided and one-sided hypotheses.

There is a rich body of literature on the distribution of sexes. Many studies report a slight prevalence of newborn boys over girls. It is well documented that the ratio of boys to girls varies from country to country – see the article "Human sex ratio" in Wikipedia. The question on the distribution of sexes reduces to hypothesis testing with $p = 0.5$.

Example 7.4 *Boy or girl?* *The file NHBirths2003_2009.csv contains birth information on 33,666 New Hampshire newborns from 2003 to 2009. Use this data to test the hypotheses that (a) the probabilities of giving birth to a boy and a* *girl are the same and (b) the chance of giving birth to a girl is smaller than for a boy.*

Solution. To read the data use the command `da=read.csv("c:\\StatBook\\ NHBirths 2003_2009.csv",stringsAsFactors=F)`. To compute the probability of a girl and the number of girls in the data set, issue `c(mean(da$CHILD_SEX=="F") ,sum(da$CHILD_SEX=="F"))` which returns `4.914454e-01 1.654500e+04`. If p is the probability that the newborn is a boy the null hypothesis is $H_0 : p = 0.5$. To test the hypothesis, derive the distribution of the number of newborn girls, m, in $n = 33,666$ births. Since the births are independent, the probability that m girls are born, as follows from Section 1.6, can be expressed as the binomial probability $\Pr(m) = \binom{n}{m} 2^{-n}$.

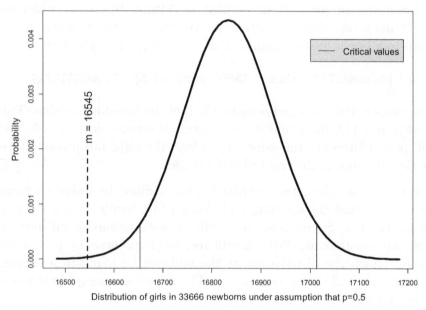

Distribution of girls in 33666 newborns under assumption that p=0.5

Figure 7.4: *The distribution of girls under the null hypothesis that the newborn is a girl, $p = 0.5$. The normal range is between the vertical bars. Since the observed number of girls does not belong to the normal range the null hypothesis should be rejected.*

(a) The alternative hypothesis is $H_A : p \neq 0.5$. The null hypothesis will be rejected if either too many boys or too many girls were born – the test must have two tails. Find the critical values of the test as `qbinom(.025,size=33666,prob=0.5) =16653` and `qbinom(1-.025,size=33666,prob=0.5)=17013`. In Figure 7.4, we depict the binomial distribution with $n = 33,666$ and $p = 0.5$. The critical values are shown as a segment symmetric around $n/2 = 16833$. Since the observed number of girls is outside the interval, the hypothesis on equal probabilities should be rejected with the type I error 5%. Compute the p-value as $2 \times \Pr(m \leq m_{\text{obs}})$, where m is the binomial random variable with $p = 0.5$ and `size` $= 33666$, and

$m_{\mathrm{obs}} = 16545,$

$$p\text{-value} = 2 * \texttt{pbinom}(16545, \texttt{size} = 33666, \texttt{prob} = 0.5) = 0.001725148.$$

We say that the difference of probabilities of birth for the two genders is statistically significant. The double-sided alternative reveals are differences in gender probabilities, but does not test which probability is smaller (or larger).

(b) That more boys are being born is an established fact. The one-sided alternative, $H_A : p > 1/2$, is more informative. To answer whether the NH data supports the null hypothesis, we find the maximum number of girls to be born among 33666 newborn babies if $p = 1/2$ using the 95%th quantile of the binomial distribution, `qbinom(1-.05,size=33666,prob=0.5)=16984`. If the number of boys in the file is greater than 16984 the null hypothesis is rejected. Indeed, the number of boys in the sample is $33666 - 16545 = 17121 > 16984$. We reject the null hypothesis and accept the alternative: the NH data confirms that the chance of giving birth to a boy is greater than that of giving birth to a girl with

$$p\text{-value} = \texttt{1-pbinom}(17121, \texttt{size} = 33666, \texttt{prob} = 0.5) = 0.0008310724.$$

As the reader can see, the one-sided p-value is half of the two-sided p-value. This is because, when $p = 1/2$, the binomial distribution is symmetrical the left- and the right-tail probabilities are the same. Therefore, the right tail probability is half smaller that the sum of the two tail probabilities.   □

In the previous discussion, the distribution was defined by unknown parameters. We may say that the distribution belonged to a family of distributions specified by parameters. Sometimes, the family of distributions is unknown or is very broad, say, continuous. We can still test the hypothesis by rephrasing the problem for which the distribution at the null can be found in a simpler form. Such tests are called *nonparametric*. The following example reduces a nonparametric test to the binomial test studied in Example 7.4.

Example 7.5 *The sign test.* *Observations Y_i are drawn from an unknown distribution. (a) Test the hypothesis that the median of the distribution is m_0 by reducing the problem to the binomial distribution. (b) A house buyer wants to know whether the median house price in the area where he wants to buy is the same as where he lives now, $350K$. The real estate agent supplied him with the price list of 78 houses recently sold (see file houseprice.txt). Test the hypothesis that the median price is $350K$.*

Solution. (a) The key to finding the test statistic is the fact that, if Y_i are iid from a distribution with median μ_0, we must have $\Pr(Y_i < \mu_0) = 0.5$. As a word of caution, the mean does not work here. Moreover, since Y_i are independent, one can view $Y_i < \mu_0$ as a Bernoulli event with probability 0.5. Thus, we let the test statistic be the number of observations for which $Y_i < \mu_0$. The command sum(scan("c:\\StatBook\\houseprice.txt")<350) produces $m = 38$. Now the test proceeds as in Example 7.4. (b) Here, we have $m = 38$ and $n = 78$. Compute the 2.5% and 0.975% quantiles as qbinom(.025,size=78,prob=0.5)=30 and qbinom(.975,size=78,prob=0.5)=48. Since $30 < m < 48$ we cannot reject the hypothesis that the median house price in the area is \$350K. Compute the p-value as 2*binom(38,size=78,prob=0.5)=0.9099465. This is called the sign test. Section 7.4.5 compares this test to a parametric t-test.

Example 7.6 *Poverty test.* *The country's poverty threshold is \$15K annual income per family with 10% of families in poverty. The file "familyincome.csv" contains a random sample of 1,000 family incomes collected in a state. Test the hypothesis that the portion of families in poverty in the state is the same as in the country without assuming income distribution.*

Solution. We use a nonparametric test similar to Example 7.5, but with $p = 1/10$. Let $Y_i < 15$ be a binary event and the number of families with income less then \$15K be m. If the statewide poverty percent is the same as in the country, then m should be distributed as a binomial random variable with $p = 1/10$ and $n = 1000$. The R code is found in **familyincome.r** with the output

```
familyincome()
Read 1000 items
Proportion < $15K = 0.135 , #families < $15K = m = 135
m1 = 82 m2 = 119
```

Since $m = 135$ is outside $(82, 119)$, we reject the hypothesis that the poverty rate in the state is the same as in the country.

Problems

1. (a) Repeat Example 7.4 with the null hypothesis that the probability of a newborn boy is 0.5. First, use the data set, and second, test the hypothesis without reading the data set by using the data on girls presented in the Example. (b) Test the null hypothesis using $\alpha = 0.01$. (c) Prove that the p-value does not change whether the null hypothesis is formulated for boys or girls.

2. In Example 7.5, test the null hypothesis with the alternative that the median house price where he lives now is greater that \$350K (compute the p-value).

3. Suppose in Example 7.5 that the house price follows a lognormal distribution. Use simulations to report how frequently the Z-test applied to the data on the log scale and the binomial test simultaneously reject the null hypothesis.

4. (a) Compute the p-value in Example 7.6. (b) There are 5% rich families in the country with income \$300K and above. Test the hypothesis that the percent rich families in the state is the same as in the country.

7.3 The power function of the Z-test

The goal of this section is to introduce the main concepts of statistical testing using the simplest Z-test. This test will serve as a prototype with minimal technical details. Moreover, this test, in connection with the Wald test, emerges as one of the most common asymptotic tests where the distribution of the test statistic is approximated with the normal distribution, as we shall learn later in the chapter.

7.3.1 Type II error and the power function

The type I error, or the size of the test (it is often referred to as the significance level in applied statistics), is the probability of rejecting the null hypothesis $H_0 : \theta = \theta_0$ when it is actually true. Since notation α is typically used, one calls this error the alpha error. There exists another error, accepting the null when the alternative $\theta \neq \theta_0$ is true. This error is called the type II error and is sometimes referred to as the beta error. In short,

1. The alpha error (type I, α) is the error of rejecting the null hypothesis when it is actually true.

2. The beta error (type II, β) is the error of accepting the null hypothesis when the alternative hypothesis is actually true.

These errors apply not only to statistical hypothesis testing, but to every decision made based on some belief or *a priori* knowledge formulated by means of the null hypothesis. Unfortunately, it is impossible to simultaneously minimize both errors even when the alternative is known. Some trade-off is expected. The following example amplifies that the consequences of the two errors may be strikingly different.

Example 7.7 *Biopsy: to do or not to do?* *Illustrate the consequences of the two types of errors on breast cancer detection using a mammogram under the null hypothesis that the breast is normal.*

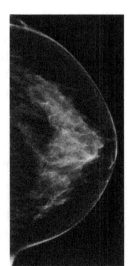

Solution. Mammography is the gold standard X-ray technique for breast cancer detection. A typical mammogram is

shown at right. If the mammogram looks somewhat abnormal, the patient is sent for a biopsy (a tissue sample is
extracted using a needle). Suppose that the null hypothesis is that the breast is
normal versus the alternative that there is a cancer tumor. Errors are inevitable
and consequences vary. The type I error means that an unaffected woman is
encouraged to undergo a biopsy. In this case, the woman experiences anxiety,
pain, and other unwanted and unnecessary displeasures. The type II error means
the doctor does not send the woman with cancer for a biopsy. Such a decision
implies overlooking breast cancer and may lead to death. In short, while the
type I error is associated with inconvenience, the type II error is associated with
possible death. Consequently, in this example, the preference should be given to
minimizing the beta error while keeping the alpha error at a reasonable level of
discomfort.

Definition 7.8 *The power function of a test is the probability of rejecting the
null hypothesis under the alternative, the complementary probability of the beta
error.*

The power function depends on the alternative value – one can expect that,
for alternatives far from the null, the power function will be close to 1 (the beta
error is close to zero). Power functions are used to compare tests. If two tests
have the same type I error, we choose the test with a larger power. From this
perspective, the power function is as important for comparing tests as the MSE is
for parameter estimation. In practical applications, one uses the power function
to determine the sample size.

Computation of the power function is illustrated below with the Z-test, the
simplest statistical test for the mean of the normal distribution with known variance. Despite this seemingly unpractical assumption, this test is very important
as an asymptotic or approximation test applied to general hypothesis testing by
replacing the true variance with its estimate.

The Z-test with the test statistic (7.3) applies to testing

$$H_0 : \mu = \mu_0 \text{ versus } H_A : \mu \neq \mu_0$$

given n iid observations drawn from the normal distribution $\mathcal{N}(\mu, \sigma^2)$, where μ
is unknown and σ^2 is known.

Example 7.9 *The power of the Z-test. Derive the power of the Z-test.*

Solution. The null hypothesis is rejected if $\left| \overline{Y} - \mu_0 \right|$ is large, specifically, when
$\left| \overline{Y} - \mu_0 \right| > s$. Let α be the chosen type I error (the size of the test), $\alpha = 0.05$. We
want to find $s > 0$ from this condition. Since the type I error is the probability
of rejecting the null when it is true, we deduce that

$$\Pr(\left| \overline{Y} - \mu_0 \right| > s \,|\, Y_i \overset{\text{iid}}{\sim} \mathcal{N}(\mu_0, \sigma^2)) = \alpha.$$

Since $\overline{Y} \sim \mathcal{N}(\mu_0, \sigma^2/n)$, we obtain that the above condition can be rewritten as $2\Phi\left(-s/\sigma\sqrt{n}\right) = \alpha$ or $s = \sigma\Phi^{-1}(1 - \alpha/2)$. Finally, the hypothesis $H_0 : \mu = \mu_0$ is rejected when

$$\left|\overline{Y} - \mu_0\right| > \sigma n^{-1/2} q_{1-\alpha/2}, \tag{7.5}$$

where $q_{1-\alpha/2} = \Phi^{-1}(1 - \alpha/2)$ is the $(1 - \alpha/2)$th quantile of the standard normal cdf. For example, when $\alpha = 0.05$, we have $q_{1-\alpha/2} = 1.96$. This test leads to rejection of the null hypothesis when it is actually is true in 1 out of 20 cases. In Example 7.7, it means that 1 out 20 women will be sent for unnecessary biopsy.

The power of test (7.5) is the probability of rejecting $\mu = \mu_0$ when in fact the opposite is true,

$$P(\mu) = \Pr\left(\left|\overline{Y} - \mu_0\right| > \sigma n^{-1/2} q_{1-\alpha/2} | Y_i \overset{\text{iid}}{\sim} \mathcal{N}(\mu, \sigma^2)\right).$$

One immediately observes that the power function at the null equals the size of the test, α. Under alternative, $\overline{Y} \sim \mathcal{N}(\mu, \sigma^2/n)$, and therefore, $(\overline{Y} - \mu)/\sqrt{\sigma^2/n} \sim \mathcal{N}(0, 1)$. Using this fact, we represent the power function, $P(\mu)$, as

$$1 - \Pr\left(\left|\overline{Y} - \mu_0\right| \leq \sigma n^{-1/2} q_{1-\alpha/2}\right)$$

$$= 1 - \Pr\left(\overline{Y} - \mu_0 \leq \sigma n^{-1/2} q_{1-\alpha/2}\right) + \Pr\left(\overline{Y} - \mu_0 \leq -\sigma n^{-1/2} q_{1-\alpha/2}\right)$$

$$= 1 - \Pr\left(\frac{\overline{Y} - \mu}{\sigma n^{-1/2}} \leq \frac{\mu_0 - \mu}{\sigma n^{-1/2}} + q_{1-\alpha/2}\right) + \Pr\left(\frac{\overline{Y} - \mu}{\sigma n^{-1/2}} \leq \frac{\mu_0 - \mu}{\sigma n^{-1/2}} - q_{1-\alpha/2}\right).$$

Finally, the power of the Z-test is given by

$$P(\mu) = 1 - \Phi\left(\sqrt{n}\frac{\mu_0 - \mu}{\sigma} + q_{1-\alpha/2}\right) + \Phi\left(\sqrt{n}\frac{\mu_0 - \mu}{\sigma} - q_{1-\alpha/2}\right). \tag{7.6}$$

Several power functions for $\mu_0 = 0$ with $\alpha = 0.05$ and shown in Figure 7.6. Note that a smaller σ a higher the power of the test. Also, the power increases with n. See Example 7.12 below, which discusses other properties of the test. □

Expected power and p-value

It is tempting to compute the expected (or ad-hoc) power using $\overline{Y}$ as an estimate of μ, namely, $\widehat{P} = 1 - \Phi\left(s + q_{1-\alpha/2}\right) + \Phi\left(s - q_{1-\alpha/2}\right)$, where $s = \sqrt{n}(\mu_0 - \overline{Y})/\sigma$. In the notation of Example 7.9, p-value is computed as $p = 2\Phi\left(-|s|\right)$. Expressing s through p we obtain the relationship between expected power and the p-value in explicit form:

$$\widehat{P}(p) = 1 - \Phi\left(-\Phi^{-1}(p/2) + q_{1-\alpha/2}\right) + \Phi\left(-\Phi^{-1}(p/2) - q_{1-\alpha/2}\right). \tag{7.7}$$

It is possible to prove $\widehat{P}$ is a decreasing function of p, and $\widehat{P}(0) = 1$ and $\widehat{P}(1) = 0$; see Figure 7.5. Interestingly, the relationship between the observed power and the p-value in the Z-test does not depend on either the observed $\overline{Y}$, or n, μ_0 or σ.

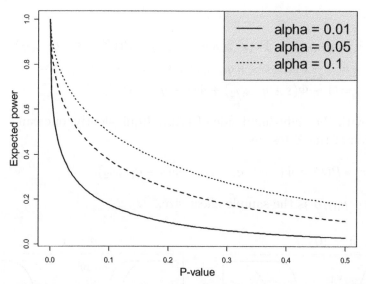

Figure 7.5: *Expected power and p-value in the Z-test for three significance levels.*

Effect size

The previous example suggests that the power of the test can be expressed through the quantity $(\mu_0 - \mu)/\sigma$. This quantity is so important that deserves a definition.

Definition 7.10 *Effect size.* *The difference between the null and alternative in the units of the population SD, $c = (\mu - \mu_0)/\sigma$, is called the **individual effect size** (or just effect size). The quantity $s = \sqrt{n}c$ is called the **sample effect size**.*

The effect size was introduced by Cohen [19] and is sometimes referred to as the Cohen effect size. The effect size compares the difference between the null and the alternative hypothesis for variables measured on different scales. From a practical standpoint, the alternative hypothesis expressed via effect size is sometimes more appealing than the absolute difference. For example, consider testing of the treatment effect of a drug intended to lower blood pressure. If σ is the standard deviation of the blood pressure among people, a treatment effect of size $c = 0.1$ means the drug is expected to lower the blood pressure by 10%. We shall explain later why we prefer calling c the *individual* effect size. The sample effect size is $\sqrt{n}$ times the individual sample size because it is measured on the scale of the standard deviation of $\overline{Y}$ as follows from the elementary observation that $\mathrm{SD}(\overline{Y}) = \sigma/\sqrt{n}$. The effect size has practical appeal as the relative deviation of the mean; in particular, it has a nice interpretation as the coefficient of variation with log-normal distribution, see more detail in Example 7.25.

The power function (7.6) can easily be expressed through the sample effect size, s, as $P(s) = \Pr(|Z - s| > q_{1-\alpha/2})$, where Z is the standard normal random variable. Indeed,

$$\Pr(|Z - s| > q_{1-\alpha/2}) = \Pr(Z > s + q_{1-\alpha/2}) + \Pr(Z < s - q_{1-\alpha/2})$$
$$= 1 - \Pr(Z < s + q_{1-\alpha/2}) + \Pr(Z < s - q_{1-\alpha/2}) =$$
$$= 1 - \Phi(s + q_{1-\alpha/2}) + \Phi(s - q_{1-\alpha/2}),$$

which coincides with the right-hand side of (7.6). Equivalently, one can rewrite the power function of the Z-test as

$$P(s) = \Phi(-s - q_{1-\alpha/2}) + \Phi(s - q_{1-\alpha/2}), \tag{7.8}$$

where $s = \sqrt{n}(\mu - \mu_0)/\sigma$ is the sample effect size.

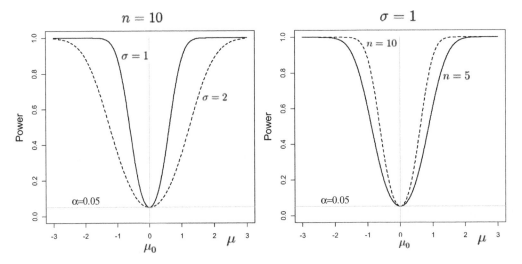

Figure 7.6: *The power function of the Z-test with different population SD, σ, and the sample size, n. The power function is symmetric around the null value, μ_0. Larger sample sizes and smaller population SDs increase the power of the test.*

The following describes properties of a *good* statistical test.

Definition 7.11 *(a) The test is called monotonic if its power function increases when the alternative moves away from the null. (b) The test is called symmetric if its power function is symmetric around the null value. (c) The test is unbiased if the power function at the alterative is greater then at the null. (d) The test is locally unbiased if the first derivative of the power function at the null is zero and the second derivative is positive. (e) The test is called consistent if the power function approaches 1 at any alternative when n goes to infinity.*

These properties make sense and are expected to hold: In lay-language, monotonicity of the test means that it is easier to reject the null when the alternative moves away from the null; tests that do not comply with this requirement are called aberrant. The symmetry of the test is not required but is reasonable. Test unbiasedness is related to monotonicity. Indeed, it is unlikely that the error of rejecting the null at the alternative would be smaller than at the null value. Local unbiasedness is self-explanatory: it simply means that the test is unbiased in a neighborhood of the null value. Local unbiasedness is instrumental and fairly easy to check (one may refer to unbiasedness as global unbiasedness). Finally, the consistency of the test means that, in large samples, we can discriminate the alternative from the null with probability 1.

The following example illustrates the above properties.

Example 7.12 *Power of the Z-test. Show that the Z-test in Example 7.9 holds all properties (a)–(e).*

Solution. See Figure 7.6 for illustration. (a) The monotonicity follows from the simple fact that $f(x) = \Pr(|Z - x| > q_{1-\alpha/2})$ is an increasing function of x for $x > 0$ and a decreasing function of x for $x < 0$. To prove this express $f(x) = 1 - \Phi(x + q_{1-\alpha/2}) + \Phi(x - q_{1-\alpha/2})$ and find the derivative, $f'(x) = -\phi(x + q_{1-\alpha/2}) + \phi(x - q_{1-\alpha/2})$. If $x > 0$ we have $\phi(x + q_{1-\alpha/2}) - \phi(x - q_{1-\alpha/2}) < 0$ because $(x + q_{1-\alpha/2})^2 - (x - q_{1-\alpha/2})^2 = 4xq_{1-\alpha/2} > 0$ and therefore $f'(x) > 0$. If $x < 0$ we have $\phi(x + q_{1-\alpha/2}) - \phi(x - q_{1-\alpha/2}) > 0$ and therefore $f'(x) < 0$. (b) The power function (7.6) is symmetric around μ_0. This is easy to see if the power function is written as a function of the sample effect size,

$$P(\mu) = \Phi\left(-\sqrt{n}\frac{\mu_0 - \mu}{\sigma} - q_{1-\alpha/2}\right) + \Phi\left(\sqrt{n}\frac{\mu_0 - \mu}{\sigma} - q_{1-\alpha/2}\right) \qquad (7.9)$$

since $1 - \Phi(z) = \Phi(-z)$. (c) The unbiasedness of the test follows from its monotonicity. (d) We have

$$\left.\frac{dP}{d\mu}\right|_{\mu=\mu_0} = \left.\frac{\sqrt{n}}{\sigma}\phi\left(-\sqrt{n}\frac{\mu_0 - \mu}{\sigma} - q_{1-\alpha/2}\right) - \frac{\sqrt{n}}{\sigma}\phi\left(\sqrt{n}\frac{\mu_0 - \mu}{\sigma} - q_{1-\alpha/2}\right)\right|_{\mu=\mu_0}$$

$$= \frac{\sqrt{n}}{\sigma}\phi\left(-q_{1-\alpha/2}\right) - \frac{\sqrt{n}}{\sigma}\phi\left(-q_{1-\alpha/2}\right) = 0.$$

To find the second derivative, use $\phi'(x) = -x\phi(x)$, so that

$$\frac{d^2P}{d\mu^2} = \frac{n}{\sigma^2}\left(\sqrt{n}\frac{\mu_0 - \mu}{\sigma} + q_{1-\alpha/2}\right)\phi\left(-\sqrt{n}\frac{\mu_0 - \mu}{\sigma} - q_{1-\alpha/2}\right)$$

$$-\frac{n}{\sigma^2}\left(\sqrt{n}\frac{\mu_0 - \mu}{\sigma} - q_{1-\alpha/2}\right)\phi\left(\sqrt{n}\frac{\mu_0 - \mu}{\sigma} - q_{1-\alpha/2}\right)$$

and, therefore,

$$\left.\frac{d^2P}{d\mu^2}\right|_{\mu=\mu_0} = \frac{n}{\sigma^2}q_{1-\alpha/2}\phi\left(-q_{1-\alpha/2}\right) + \frac{n}{\sigma^2}q_{1-\alpha/2}\phi\left(-q_{1-\alpha/2}\right)$$

$$= 2\frac{n}{\sigma^2}q_{1-\alpha/2}\phi\left(-q_{1-\alpha/2}\right) > 0.$$

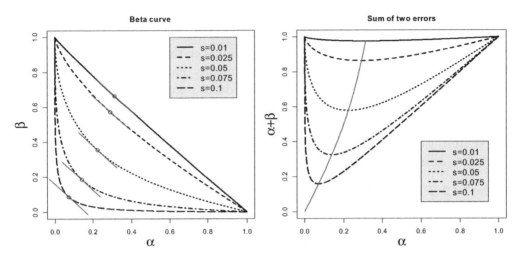

Figure 7.7: *The beta curve and the optimal α that minimizes the sum of two type errors in the Z-test. The tangent segments at the optimal α have the slope -1.*

Note that the local power increases with n and decreases with σ^2 getting larger. (e) The Z-test is consistent because, when $n \to \infty$ and $\mu > \mu_0$, the first term in (7.9) approaches 1. If $\mu < \mu_0$ the second term approaches 1.

7.3.2   Optimal significance level and the ROC curve

Typically, the significance level (the type I error) is $\alpha = 0.05$. This level of statistical significance is rarely defended in applications yet is crucial for the outcome of hypothesis testing. In this section, we explore a possibility of deriving an optimal value of α that minimizes the sum of the two type errors.

An obvious desire is to simultaneously minimize alpha and beta. However, decreasing the alpha error implies increasing the beta error and vice versa, as illustrated in the following example. There is a trade-off. Both errors cannot be minimized simultaneously. Below is a continuation of Example 7.12 where α minimizes the sum of the two errors.

Example 7.13 *Optimal type I error. (a) Plot the beta curve as a function of α for several given alternatives and fixed σ and n. (b) Prove that the beta error is a decreasing function of α. (c) Find the optimal α that minimizes the sum of the two type errors given the alternative and prove that the optimal α is where the slope of the beta curve is -1. (d) Investigate the properties of the optimal α as a function of the sample effect size.*

Solution. (a) The beta error is a complementary probability to the power function given by (7.6). Therefore, in the notation of the sample effect size (see Definition 7.10), $\beta(\alpha) = \Phi\left(s + q_{1-\alpha/2}\right) - \Phi\left(s - q_{1-\alpha/2}\right)$. See the left side of Figure 7.7, where several beta curves are shown for the sample effect size

$s = 0.01, 0.025, 0.05, 0.075$, and 0.1. Note how steeply $\beta(\alpha)$ drops at zero as the alternative moves away from the null. (b) To prove that $\beta(\alpha)$ is a decreasing function, use the calculus formula for the derivative of the inverse function, $d\Phi^{-1}(z)/dz = 1/\phi(\Phi^{-1}(z))$, which implies

$$\frac{d\beta}{d\alpha} = -\frac{\phi(s + q_{1-\alpha/2}) + \phi(s - q_{1-\alpha/2})}{2\phi(q_{1-\alpha/2})} < 0.$$

(c) The sum of two errors, $C(\alpha) \overset{\text{def}}{=} \alpha + \beta(\alpha)$, takes a minimum when $dC/d\alpha = 0$

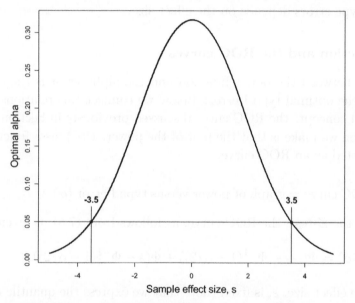

Figure 7.8: *The optimal α as a function of the sample effect size. The traditional $\alpha = 0.05$ is optimal when the sample effect size is equal ± 3.5.*

which implies $d\beta/d\alpha = -1$. The slope $-45°$ is shown in the left plot of Figure 7.7 by the tangent. From the above expression for $d\beta/d\alpha$, obtain the condition $\phi(s + q_{1-\alpha/2}) + \phi(s - q_{1-\alpha/2}) = 2(q_{1-\alpha/2})$, or equivalently, $e^{-\frac{1}{2}(s+q_{1-\alpha/2})^2} + e^{-\frac{1}{2}(s-q_{1-\alpha/2})^2} = 2e^{-\frac{1}{2}q_{1-\alpha/2}}$, which after some algebra is reduced to a quadratic equation for $e^{sq_{1-\alpha/2}}$, namely, $e^{-sq_{1-\alpha/2}} + e^{sq_{1-\alpha/2}} = 2e^{\frac{1}{2}s^2}$, with the optimal solution

$$\alpha_{\text{opt}} = 2\Phi\left(\frac{1}{|s|}\ln\left(e^{\frac{1}{2}s^2} - \sqrt{e^{s^2} - 1}\right)\right). \tag{7.10}$$

The right plot in Figure 7.7 illustrates the optimal α and the tangent with $-45°$ slope. (d) Let $\alpha_{\text{opt}} = \alpha_{\text{opt}}(s)$ be defined by formula (7.10). By Taylor series expansion, $\ln\left(e^{\frac{1}{2}s^2} - \sqrt{e^{s^2} - 1}\right) = -s - s^3/12 + O\left(s^4\right)$, we have $\alpha_{\text{opt}}(0) = 2\Phi(-1) \simeq 0.317$. By L'Hôpital's rule for the argument of the Φ function, we have

$$\lim_{s \to \infty} \frac{\ln\left(e^{\frac{1}{2}s^2} - \sqrt{e^{s^2} - 1}\right)}{s} = -\lim_{s \to \infty} s\frac{e^{\frac{1}{2}s^2}}{\sqrt{e^{s^2} - 1}} = -\infty.$$

Figure 7.8 depicts the graph of α_{opt} as a function of s. The traditional $\alpha = 0.05$ minimizes the sum of type I and II errors when the sample effect size is about ± 3.5. For example, with the sample size $n = 12$ the traditional $\alpha = 0.05$ corresponds to the effect size equal to one. $\qquad\square$

We learned from this example that the significance level that minimizes the sum of the two type errors depends on the effect size.

The power function and the ROC curves

The relationship between the power function and the alpha error is the central point of finding the optimal type I error. Below we connect this relationship to a well-established concept: the ROC curve discussed previously in Section 5.1.1.

The connection we make is that the plot of the power, $P(\alpha)$, as a function of α can be interpreted as an ROC curve,

ROC curve = graph of power versus type I error (α).

For example, for the Z-test, the ROC curve, as follows from (7.8), is given by

$$\text{ROC}(\alpha) = \Phi(-s - \Phi^{-1}(1 - \alpha/2)) + \Phi(s - \Phi^{-1}(1 - \alpha/2))$$

where the sample effect size, s, is fixed (note that we express the quantile $q_{1-\alpha/2}$ through $\Phi^{-1}(1 - \alpha/2)$ explicitly). In the table below, we connect the terminology of the ROC curve with that of statistical hypothesis testing.

ROC term	Statistics term/notation
Sensitivity	Power, $1 - \beta$
False negative	beta error, β
Specificity	$1 - \alpha$
False positive	alpha error, α

Thus, area under the ROC curve (AUC) as the area under the power function can be interpreted as the overall power of the test:

$$\text{AUC} = \int_0^1 \text{ROC}(\alpha)d\alpha = \int_0^1 \left[\Phi\left(-s - q_{1-\alpha/2}\right)d\alpha + \Phi\left(s - q_{1-\alpha/2}\right)\right]d\alpha$$

A change of variable, $q_{1-\alpha/2} = -z$, gives an alternative representation:

$$\text{AUC} = 2 \int_{-\infty}^{0} \left[\Phi\left(z - s\right) + \Phi\left(z + s\right) \right] \phi(z) dz.$$

It is especially easy to establish the connection between the ROC curve and AUC for one-sided hypotheses.

7.3.3 One-sided hypothesis

The two-sided hypothesis may be not practical. Consider testing a new drug compared to a conventional treatment. In the case when the two-sided hypothesis is rejected, we conclude that the new drug does not have the same treatment effect as the conventional drug. But the doctors want to know if the new drug is better than the old one. This question leads to a one-sided hypothesis:

$$H_0 : \theta \leq \theta_0 \text{ versus } H_A : \theta > \theta_0.$$

The difference with the one-sided hypothesis is that now the null values belong to an interval, not a point. Such hypotheses are called *composite*.

Conceptually, testing a one-sided hypothesis is similar to the two-sided hypothesis. As before one needs to have a test statistic such that large values of T are expected for large θ. For instance, in the drug-testing example, T may be survival after the disease diagnosis: the better the drug the longer survival. Then the null hypothesis is rejected if the observed value T is larger the normal threshold, or the critical value, T_0:

$$\Pr(T > T_0) = \alpha. \tag{7.11}$$

The critical value T_0 can be interpreted as the maximum value of the test statistic under the null hypothesis. If the observed value of T is greater than T_0, we reject the null hypothesis. A similar justification applies to testing $H_0 : \theta \geq \theta_0$ versus $H_A : \theta < \theta_0$.

For illustration, we continue the Poverty Example 7.6. The standard answer with the two-sided test "the hypothesis that the poverty rate in the state is the same as in the country is rejected" is somewhat unsatisfactory because it does not distinguish between being below and above the poverty rate. We are concerned that the poverty rate in the state is higher, i.e. the number of families with income less than \$15K is larger than 10%.

Example 7.14 *Poverty rate continued.* *Test the hypothesis that more than 10% of families are in poverty.*

Solution. The only difference is that we now use a one critical value for the number of families m whose income is less than \$15K.

```
> qbinom(.95,size=1000,prob=1/10)
[1] 116
```

We interpret 116 as the maximum number of families among 1,000 in a typical state whose income is less than \$15K. If the number of families is large than 116 we do not reject the hypothesis that the state is in poverty. Indeed, since the observed number of families with income less than \$15K is 135, which is greater than 116, we conclude that the state's poverty is higher than average. However, we must add that this statement is subject to a 5% error.   □

Below is the examination of the one-sided Z-test as the counterpart of the two-sided test from Example 7.12.

Example 7.15 *The one-sided Z-test.* *(a) Test the one-sided null hypothesis $H_0 : \mu \leq \mu_0$ with the alternative $H_A : \mu > \mu_0$ using n observations $Y_i \overset{\text{iid}}{\sim} \mathcal{N}(\mu, \sigma^2)$ where σ^2 is known. Explain the test in lay-language. (b) Derive and plot the power of the test. (c) Check the properties of the test listed in (7.11). (d) Find the optimal α that minimizes the sum of two errors and prove that for this α the type I and II errors are the same.*

Solution. (a) Large values of $\overline{Y}$ would indicate that the null hypothesis should be rejected. The threshold/critical value is found from the equation $\Pr(\overline{Y} - \mu_0 > c | \mu = \mu_0) = \alpha$, where α is the chosen significance level, say $\alpha = 0.05$. Since $Z = \sqrt{n}(\overline{Y} - \mu_0)/\sigma \sim \mathcal{N}(0, 1)$, we obtain that the critical value is $\mu_0 + \sigma/\sqrt{n} Z_{1-\alpha}$, where $Z_{1-\alpha} = \Phi^{-1}(1 - \alpha)$, the $(1 - \alpha)$th quantile. Thus, the hypothesis $H_0 : \mu \leq \mu_0$ is rejected if $\overline{Y} > \mu_0 + \sigma/\sqrt{n} Z_{1-\alpha}$. In lay-language, if $\mu \leq \mu_0$, the observed value of $\overline{Y}$ should be not much larger than μ_0. Specifically, if $\mu = \mu_0$ the average, $\overline{Y}$ belongs to the semi-infinite interval $(-\infty, \mu_0 + \sigma/\sqrt{n} Z_{1-\alpha})$ with probability $1 - \alpha$. Thus, if the observed value $\overline{Y}$ is greater than $\mu_0 + \sigma/\sqrt{n} Z_{1-\alpha}$ the null hypothesis $\mu = \mu_0$ should be rejected. The error of this decision is α.

(b) The power of the test is the probability of rejecting the null when it is not true,

$$P(\mu) = \Pr\left(\overline{Y} - \mu_0 > \sigma n^{-1/2} q_{1-\alpha/2} | \mu\right) = \Phi\left(\sqrt{n}\frac{\mu - \mu_0}{\sigma} - Z_{1-\alpha}\right), \quad \mu \geq \mu_0. \tag{7.12}$$

Note that the power is the probability of rejection interval $(\mu_0 + \sigma/\sqrt{n} Z_{1-\alpha}, \infty)$ evaluated at the alternative, μ, indicated by the bar $|$. By definition, the power equals α when $\mu = \mu_0$. The power functions with different population SDs are depicted in Figure 7.9.

(c) The test is monotonic because $P(\mu)$ is a monotonic function due to monotonicity of Φ. Symmetry does not apply because it is a one-sided test. Unbiasedness follows from the monotonicity. However, unlike the two-sided test, the derivative at μ_0 is not zero,

$$\left.\frac{dP}{d\mu}\right|_{\mu=\mu_0} = \left.\frac{\sqrt{n}}{\sigma}\phi\left(\sqrt{n}\frac{\mu - \mu_0}{\sigma} - q_{1-\alpha/2}\right)\right|_{\mu=\mu_0} = \frac{\sqrt{n}}{\sigma}\phi\left(q_{1-\alpha/2}\right).$$

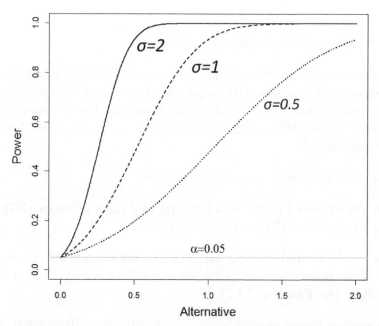

Figure 7.9: *Three power functions for one-sided hypothesis testing. A smaller SD yiekds a higher power.*

The test is consistent because the power goes to 1 when n approaches infinity.

(d) As follows from (7.12), the sum of alpha and beta errors is $C(\alpha) = \alpha + \Phi\left(\Phi^{-1}(1-\alpha) - s\right)$, where $s = \sqrt{n}(\mu - \mu_0)/\sigma$ is the sample effect size. Differentiating with respect to α and equating to zero, we arrive at the equation $\phi(Z_{1-\alpha} - s) = \phi(Z_{1-\alpha})$, which yields $Z_{1-\alpha} = s/2$ and $\alpha_{\text{opt}} = \Phi(-s/2)$. This type I error yields the beta error $\beta = \Phi\left(\Phi^{-1}(1-\alpha) - s\right) = \Phi(s/2 - s) = \Phi(-s/2) = \alpha_{\text{opt}}$. That is, α_{opt} yields the same type I and type II errors.

The ROC curve and AUC

The ROC curve is given by $\text{ROC}(\alpha) = \Phi\left(s - \Phi^{-1}(1-\alpha)\right)$, where s is the sample effect size. Now we find the area under the ROC curve similarly to derivation for the binormal ROC curve from Section Section 5.1.1,

$$\text{Area under the power function} = \text{AUC}$$

$$= \int_0^1 \text{ROC}(\alpha)d\alpha = \int_{-\infty}^{\infty} \Phi(s-z)\,\phi(z)dz = E_{Z \sim \mathcal{N}(0,1)}\Phi(Z-s) = \Phi\left(\frac{s}{\sqrt{2}}\right).$$

The one-sided Z-test can be interpreted as the binormal ROC curve with $\overline{X} \sim \mathcal{N}(\mu_0, \sigma^2/n)$ and $\overline{Y} \sim \mathcal{N}(\mu, \sigma^2/n)$ where $\mu_0 < \mu$. This explains the appearance of $\sqrt{2}$ in the denominator.

Problems

1. Write the R code which reproduces Figure 7.6.

2. (a) Prove that a smaller σ yields a higher the power of the Z-test in Example 7.9. (b) Write an R function with arguments μ_0, n, and α that plots the power functions for three $\sigma = 1,2,3$ on the same plot (use different colors and legend). [Hint: Differentiate with respect to σ and show that the derivative is positive.]

3. Prove that unbiased test is also locally unbiased.

4. Prove that the expected power as a function of p-value given by (7.7) is a decreasing function and $\widehat{P}(0) = 1$ and $\widehat{P}(1) = 0$.

5. Somebody wants to find an optimal $\alpha > 0$ that maximizes the second derivative of the power function at the null hypothesis value. Does such an α exist? [Hint: Use Example 7.12.]

6. The state median family income is $\mu_0 = \$60K$ with $\sigma = SD = \$30K$. Plot the power of the Z-test for the null hypothesis $H_0 : \mu = \mu_0$ based on $n = 100$ town family reports as a function of (a) alternative, μ, (b) effect size, c, and (c) sample effect size, s, using par(mfrow=c(1,3)). Plot two curves with $\alpha = 0.01$ and $\alpha = 0.05$ using different colors. Use legend to label the curves with different α. Label the axes, and use title for each plot.

7. Days between consecutive earthquakes in China and California was measured: (452, 732, 583, 651) and (841, 1302, 983, 1193, 903, 1278), respectively; see Example 2.19. (a) Assuming that earthquakes follow an exponential distribution use the Z-score to test if the scale parameter θ is the same in the two regions. Provide two versions of the Z-test depending on how variance of $\widehat{\theta}$ is estimated: (i) use the variance of the exponential distribution, and (ii) use the general formula for an unbiased variance estimate. (b) Compare the two versions by means of the power function using simulations. Compare the powers for different sample sizes.

8. Using the same μ_0 and σ as in the previous problem, plot the power for the alternative $\mu = \$65K$ as a function of n. How many families must the researcher ask about their household income to achieve power $= 0.9$ with $\alpha = 0.05$?

9. Using the same μ_0, μ, and σ as above, and $n = 200, 300$, (a) plot the sum of two errors as in the right plot of Figure 7.7 for two ns using different colors, (b) compute optimal α and β, and (c) show the optimal α and the minimum sum of errors on the graph.

10. Plot the AUC in the two-sided Z-test as a function of the sample effect size using `integrate`. [Hint: Define a local function with arguments z and s that computes the integrand, and consult Section 2.9.2.]

11. Use the Z-test to test $H_0 : \mu = \mu_0$ given a sample of normally distributed observations with mean μ, variance σ^2 and correlation coefficient $\rho > 0$ between any two pairs: $\mathbf{Y} \sim \mathcal{N}(\mu \mathbf{1}, \sigma^2((1-\rho)\mathbf{I} + \rho \mathbf{1}\mathbf{1}'))$. Use simulations to plot the empirical type I error as a function of ρ for $n = 5$ and $n = 30$. [Hint: Use values for μ_0 and σ as arguments of the R function.]

12. Derive the power of the Z-test for testing the $A = \mathrm{AUC}$ (see Section 5.1.1) using the expression for the variance of the empirical AUC given by Bamber [6] $[A(1 - A) + (n - 1)(Q_1 - A^2) + (m - 1)(Q_2 - A^2)]/(nm)$, where $Q_1 = A/(2 - A)$ and $Q_2 = 2A^2/(1 + A)$, and n and m are the sample sizes. Derive optimal n and m by maximizing the power over n and m under the restriction that $n + m = N$. Use simulations from the binormal ROC curve to test your derivations.

7.4 The t-test for the means

The Z-test served as a toy test to introduce the principals of statistical hypothesis testing. A critical assumption of the test was that SD $= \sigma$ was known. Starting from this section, SD is unknown and estimated, implying that the test statistic is t-distributed. Testing the means of the normal distribution is the most frequent task in applied statistics.

7.4.1 One-sample t-test

Assume that n observations, Y_i, are drawn from the normal distribution with unknown mean μ and unknown variance σ^2. Shortly, we write $Y_i \overset{\text{iid}}{\sim} \mathcal{N}(\mu, \sigma^2)$, $i = 1, 2, ..., n$. We aim to test that the mean is μ_0 with the null hypothesis is $H_0 : \mu = \mu_0$ and the double-sided alternative $H_A : \mu \neq \mu_0$. The test statistic is

$$T = \frac{\sqrt{n}(\overline{Y} - \mu_0)}{\widehat{\sigma}}, \tag{7.13}$$

where $\overline{Y} = n^{-1} \sum_{i=1}^{n} Y_i$ is the average and $\widehat{\sigma}^2 = (n-1)^{-1} \sum_{i=1}^{n} (Y_i - \overline{Y})^2$ is the unbiased estimator of σ^2 (see Section 6.6). As follows from Theorem 4.31, under the null hypothesis ($\mu = \mu_0$), statistic T has a t-distribution with $n - 1$ degrees of freedom. It is critical that its distribution does not depend on unknown parameters. If α is a given significance level and $t_{1-\alpha/2}$ is the $(1 - \alpha/2)$th quantile, the null hypothesis H_0 is rejected if the observed value T lies outside of the interval $(-t_{1-\alpha/2}, t_{1-\alpha/2})$, or simply written as $|T| > t_{1-\alpha/2}$. In R, the p-value of the test is computed as `2*(1-pt(abs(T),df=n-1))`. The coefficient 2 is because the test is double-sided and therefore two tails must be accounted for. A computationally more accurate p-value is computed as `2*pt(abs(T),df=n-1,lower.tail=F)`, especially when $|T|$ is large. Alternatively, the t-test may be performed using the built-in function `t.test`.

Example 7.16 *College student living costs. One of the largest expenses of college education is the living cost (housing, meals, etc.). It can even surpass tuition. College official say that the average living cost is $17K. To verify this estimate, a survey of 23 students has been conducted resulted in an average cost of $21K with the standard deviation of $8K. (a) Test the hypothesis that the living cost is as the officials say, (b) compute the p-value and explain what it means through simulations of imaginable surveys.*

Solution. We have to make several assumptions to proceed: (1) The living cost of all students (general population) follows a normal distribution. (b) The student survey asks students at random. (c) The responses are independent: no roommates. These assumptions ensure that the reported costs $\{Y_i, i = 1, .., 23\}$ are iid and follow distribution $\mathcal{N}(\mu, \sigma^2)$. Based on these assumptions we can rigorously answer the questions: (a) In the language of the t-test we have $\mu_0 = 17$ with the alternative $H_A : \mu \neq 17$. The observed T-statistic is

$$T_{\text{obs}} = \frac{\sqrt{n}(\overline{Y} - \mu_0)}{\widehat{\sigma}} = \frac{\sqrt{23}(21 - 17)}{8} = 2.398.$$

Using the significance level (type I error) $\alpha = 5\%$, the normal range for the T-statistic is $(-t_{1-\alpha/2}, t_{1-\alpha/2})$, where $t_{1-\alpha/2}$ is the $(1 - \alpha/2)$th quantile of the t-distribution with $23 - 1 = 22$ degrees of freedom. In R, we compute the quantile as `qt(1-.05/2,df=22)` that gives 2.073873. This means that if the officials were correct the T-statistic would take values within the interval ± 2.073873 with probability 95%. Since the observed value of the T-statistic is outside of this range we reject the hypothesis that living costs are $17K. (b) The p-value, computed as `2*pt(2.398,df=23-1,lower.tail=F)`, gives 0.025. To understand what this value means, we appeal to simulations: Imagine that (a) the college officials are correct, i.e. the living costs are $\mu_0 = \$17K$, and the true σ is any value, (b) we conduct many other student surveys with $n = 23$ students and for every survey compute the average $\overline{Y}$, $\widehat{\sigma}$ and the T-statistic, as we did above. Then, in 2.5% of surveys, the absolute value of the T-statistic would be larger than $T_{\text{obs}} = 2.398$. The code which realizes this imaginable task is given below.

```
pvalcost=function(N=1000,n=23,sigma=12)
{
#N=the number of imaginable surveys with n students
dump("pvalcost","c:\\StatBook\\pvalcost.r")
mu0=17 # the null value
Tstat=rep(NA,N)
```

```
for(i in 1:N)
{
  Y=rnorm(n,mean=17,sd=sigma)
  Y.bar=mean(Y);sigma.hat=sd(Y)
  Tstat[i]=sqrt(n)*(Y.bar-17)/sigma.hat
}
cat("\nProportion of abs(Tstat)>2.398 =",
           mean(abs(Tstat)>2.398),"\n")
}
```

A few comments on the code: (1) The default number of surveys N=1000 and the default SD living costs among students is `sigma=12`. (2) `mu0=17` means that the official is correct. In words, if the null hypothesis were true, the p-value is the proportion of future surveys for which the absolute value of the T-statistic will be greater than the absolute value of the observed T-statistic derived from the actual data.

Power function

The power function of the t-test uses the noncentral t-distribution discussed earlier in Section 4.3.1. By definition, the power function is the probability of rejecting the null under the alternative. If $Y_i \overset{\text{iid}}{\sim} \mathcal{N}(\mu, \sigma^2)$ with μ not necessarily equal to μ_0 following this definition we derive the power function as follows (vertical bar means that the probability is computed at the alternative value μ):

$$P(\delta) = \Pr\left(|T| > t_{1-\alpha/2}|\mu\right) = \Pr\left(\left|\frac{\sqrt{n}(\overline{Y} - \mu_0)}{\widehat{\sigma}}\right| > t_{1-\alpha/2}|\mu\right)$$

$$= \Pr\left(\frac{\sqrt{n}(\overline{Y} - \mu_0)/\sigma}{\widehat{\sigma}/\sigma} < -t_{1-\alpha/2}|\mu\right) + 1 - \Pr\left(\frac{\sqrt{n}(\overline{Y} - \mu_0)/\sigma}{\widehat{\sigma}/\sigma} < t_{1-\alpha/2}|\mu\right)$$

$$= \mathcal{T}(-t_{1-\alpha/2}; n-1, \delta) + 1 - \mathcal{T}(t_{1-\alpha/2}; n-1, \delta) \qquad (7.14)$$

where $\mathcal{T}$ denotes the cdf of the noncentral t-distribution (`pt`) with the associated degrees of freedom (df) $n-1$ and the noncentrality parameter $\delta = \sqrt{n}(\mu - \mu_0)/\sigma$. We emphasize that δ is the sample effect size as it appeared earlier in the power of the Z-test, see Definition 7.10. The noncentral t-distribution is a built-in function in R; it has the same call as the traditional (central) t-distribution but with the noncentrality parameter specified in the `ncp` option.

Example 7.17 *Properties of the one-sample test.* (a) *Confirm the power function (7.14) using vectorized simulations. Prove that the power function is (b) an increasing function for $\mu \geq \mu_0$ and decreasing for $\mu \leq \mu_0$, (c) a symmetric function around μ_0, and (d) takes value α at $\mu = \mu_0$. (e) Prove that the t-test is unbiased.*

Solution. (a) The R code is found in file `powsim.r` (the plot is not shown). `job=1` performs simulations for the one-sample test (this example) and `job=2` performs simulations for the two-sample test (Example 7.18). The default call of `pt` produces a warning `full precision may not have been achieved` for large arguments. To avoid this the option, `lower.tail=F` is used with `1-pt(ta,df=n-1, ncp=NCP)` replaced with `pt(ta,df=n-1,ncp=NCP,lower.tail=F)`. The arguments are self-explanatory: `NP` is the number of points for plotting the theoretical power function and `NPSIM` is the number of points (empty circles) to simulate. Vectorized simulations generate `nSim*n` random numbers at once and store them in the `nSim` by `n` matrix `Y`. The mean and variance arrays of length `nSim` are computed using vectorized functions `rowMeans` and `rowSums`. The command `mean(abs(Ti)>ta)` returns the proportion of simulations for which condition `abs(Ti)>ta` holds, the empirical power of the t-test. (b) Using the definition of the noncentral t-distribution from Section 4.3 and the rule of repeated expectation formulated in Theorem 3.27 (see the problem at the end of that section) the power function of the test can be written as $P(\delta) = E_U \Phi(U - \delta) + 1 - E_U \Phi(U + \delta)$, where $U = t_{1-\alpha/2}\sqrt{S/(n - 1)}$ and S is the chi-square distributed random variable with $n - 1$ df, and δ is the sample effect size. Using this expression, it is easy to derive other properties of the t-test. Indeed, using differentiation with respect to μ under the expectation sign, we derive that function P is monotonic on both sides of μ_0. (c) The symmetry is obvious. (d) At $\mu = \mu_0$, the noncentrality parameter turns to zero and the power function equals α. (e) The unbiasedness of the test follows from properties *(b)–(d)*.

7.4.2   Two-sample t-test

In the two-sample t-test, two independent random samples from $X \sim \mathcal{N}(\mu_X, \sigma^2)$ and $X \sim \mathcal{N}(\mu_Y, \sigma^2)$ of size m and n with averages $\overline{X}$ and $\overline{Y}$, respectively, are available and the null hypothesis is that the population means are the same, $H_0 : \mu_X = \mu_Y$. Sometimes, this test is called the unpaired t-test to distinguish from the paired t-test to be discussed in Section 7.4.4. First, we present the test under the assumption that the samples are drawn from populations with the same variance, which is usually referred to as the *equal-variance assumption*. The test statistic is

$$T = \frac{\overline{X} - \overline{Y}}{\widehat{\sigma}\sqrt{1/m + 1/n}}, \tag{7.15}$$

where $\widehat{\sigma}^2$ is the pooled variance estimator,

$$\widehat{\sigma}^2 = \frac{1}{n + m - 2}\left(\sum_{i=1}^{m}(X_i - \overline{X})^2 + \sum_{i=1}^{n}(Y_i - \overline{Y})^2\right). \tag{7.16}$$

As follows from Theorem 4.33, under the null hypothesis, $T \sim t(n + m - 2)$. The test procedure is similar to the previous test: (a) Given two samples, compute the observed value of the T statistic, T_{obs}. (b) If $|T_{\text{obs}}|$ is outside the interval $\pm t_{1-\alpha/2}$, where $t_{1-\alpha/2}$ is the $(1-\alpha/2)$th quantile of the t-distribution with $n+m-2$ degrees of freedom, the hypothesis is rejected. Alternatively, the built-in function $\texttt{t.test}$ can be used; see Example 7.20 below.

The power function of the two-sample t-test takes the form

$$P(\delta) = \mathcal{T}(-t_{1-\alpha/2}; n+m-2, \delta) + 1 - \mathcal{T}(t_{1-\alpha/2}; n+m-2, \delta), \qquad (7.17)$$

where $\mathcal{T}$ is the cdf of the noncentral t-distribution and $\delta = \Delta(1/n+1/m)^{-1/2}/\sigma$ is the noncentrality parameter, the sample effect size, where $\Delta = \mu_X - \mu_Y$ is the difference of the population means. The effect size, $c = (\mu_X - \mu_Y)/\sigma$, is the distance between the alternative and the null on the scale of the population SD. The two-sample test has properties similar to the one-sample test studied earlier in Example 7.17.

Example 7.18 *Properties of the two-sample test.* *(a) Confirm that the power function of the t-test is given by (7.17) via vectorized simulations. Prove that the power function (b) is an increasing function for $\delta \geq 0$ and decreasing for $\delta \leq 0$, (c) is symmetric around zero (even function), and (d) takes value α at $\delta = 0$, (e) that the two-sample t-test is unbiased.*

Solution. The R function is found in the file $\texttt{powsim.r(job=2)}$, the plot is not shown here. To prove the properties of the power function (7.17), we note that it is similar to the power (7.14) except that the degree of freedom and noncentrality parameters are different. □

The following example illustrates how the power function can be used to design studies before data collection.

Example 7.19 *Optimal design for the two-sample t-test.* *A researcher wants to detect the difference in cholesterol in two groups of patients using the t-test. The cost of measurement per patient is fixed and known. (a) Prove that, to maximize the statistical power of detection, one must have the same number of patients in the two groups. (b) What is the minimum detectable effect size with 100 patients in both groups that yields the power 80% with the significance level $\alpha = 0.05$?*

Solution. We seek n and m such that $n+m = N$, where N is the total number of patients in the two groups. Let the type I error, α, and the effect size $c = \Delta/\sigma$ be given. (a) Rewrite the power of detection as

$$\mathcal{T}(-t_{1-\alpha/2}; N-2, c(1/n+1/m)^{-1/2}) + 1 - \mathcal{T}(t_{1-\alpha/2}; N-2, c(1/n+1/m)^{-1/2}), \qquad (7.18)$$

where $t_{1-\alpha/2}$ depends on N and therefore is fixed. As follows from Example 7.18, the power function in an increasing function of Δ. Since the cost per patient is fixed, to maximize the power, we need to maximize $(1/n + 1/m)^{-1/2}$. That is, we must minimize $1/n + 1/m$. The Lagrange function takes the form $\mathcal{L}(n, m, \lambda) = 1/n + 1/m - \lambda(N - n - m)$. Differentiation with respect to n and m implies that $n = m$. The power is maximized with the same number of patients in each group. (b) The optimal number of patients in each group is $n = m = 50$. Compute quantile $t_{1-\alpha/2}$ as `qt(1-0.05/2,df=100-2)` and then compute the power (7.18) on a fine grid of values c where $(1/n + 1/m)^{-1/2} = 1/\sqrt{2/50} = 5$. Since the power is an even function of c, it is sufficient to use positive values. Finally, find c that gives the power closest to 0.8. The output of `powsim(job=2)` gives `Minimum detectable effect size, c.min = 0.5675676`.

The invariance of the two-sample t-test to transformation

The normal distribution assumption is crucial for the t-test. The idea of using a normalizing transformation, the transformation which makes a not normally distributed random variable normally distributed, is justified by the following two facts: (1) if the distributions of X and Y are the same, then the distributions of $f(X)$ and $f(Y)$ are the same, and (2) as follows from Theorem 2.74, the normalizing transformation f which makes $f(X)$ and $f(Y)$ normally distributed exists. This means that under the equal-variance assumption, if the null hypothesis is true, $\mu_X = \mu_Y$, the distribution of X and Y is the same and therefore $E(f(X)) = E(f(Y))$. In other words, the null hypothesis on the equality of the means can be equivalently tested on the transformed data. We note that the equal-variance assumption is important because, otherwise, common transformation is not applicable.

The following is a continuation of Example 7.16 about studying the living costs of college students. This example emphasizes that the log transformation replaces the null hypothesis on the equality of the means with the equality of the medians, which may be especially attractive to comply with normal distribution in the presence of heavy tails such as cost. Note that the interpretation of the null hypothesis will change upon transformation.

Example 7.20 *Living costs for freshmen and sophomores. The living costs for freshmen and sophomores are compared: 45 freshman and 23 sophomore students were asked about their living costs (file stucost.csv). Test the hypothesis that the living costs are the same. (a) Use the traditional t-test. (b) Use the q-q plot to visually test if costs and log costs follow the normal distribution. (c) Apply the log transformation and reformulate the comparison in the median language.*

Solution. All computations are done by the R function presented in the file `stucost.r`. The mean costs for freshmen is \$24.1K and for sophomores \$29.5K. (a) The R function `t.test` with option `var.equal=T` is used to compute the p-value, `0.03`. (b) It is suspected that the cost data are skewed to right as many

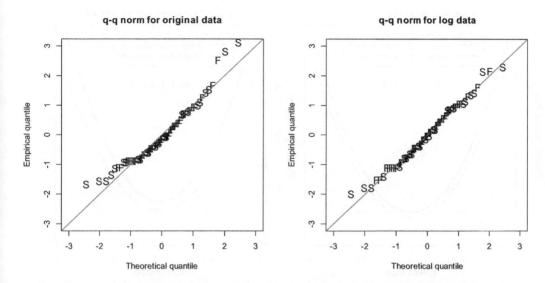

Figure 7.10: *The q-q plots for the original centered living costs data and the log transformed. Distribution of the data on the log scale looks closer to a normal distribution. Symbols F and S indicate freshman and sophomore, respectively.*

positive data. (b) The q-q plot reveals that indeed the costs data are skewed and that the distribution of the log transformed costs are closer to normal. We would like to make the following comment: it would be wrong to use the q-q plot for the original data because the means of two samples may be not the same (the null hypothesis does not hold). Therefore, before combining the samples for the q-q plot, we subtract the respective means, referred to as *centered data*. The q-q plots for the original centered and log transformed data are shown in Figure 7.10. The log transformation makes the distribution closer to normal as required by the t-test. (c) If the cost data follow the lognormal distribution it makes more sense to compare the medians, not the means, because they may be affected by the heavy right tail and possibility of outliers. Indeed, in Figure 7.10, we see three outliers in the left plot. As follows from Section 2.11, the median of the lognormal distribution transforms into the mean upon log transformation. Therefore, instead of the null hypothesis $H_0 : \mu_X = \mu_Y$, where X and Y are costs for freshmen and sophomores, we formulate the null hypothesis in terms of the median as $H_0 : \text{median}_X = \text{median}_Y$. The key point is that the median for the log normal distribution is e^μ and upon the log transformation turns into the mean, μ. The call `t.test(lX,lY)`, where `lX` and `lY` are the log costs for freshmen and sophomores, returns `p-value = 0.06943`. Accepting the standard significance level 0.05, the median difference between costs cannot claimed to be statistically significant.

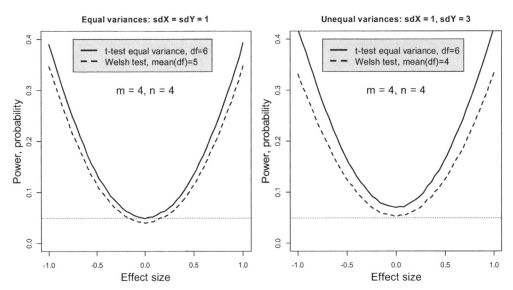

Figure 7.11: *The power functions of the two tests estimated by vectorized simulations (number of simulations is 100,000, $\alpha = 0.05$), see Example 7.21.*

t-test with unequal variances

If the variances in two samples are not the same, the t-distribution does not work – the problem of finding an optimal test is referred to as the Behrens–Fisher problem. An obvious modification of the t-test is to estimate the variance of $\overline{X} - \overline{Y}$ as

$$\widetilde{\sigma}^2 = \frac{1}{m}\widehat{\sigma}_X^2 + \frac{1}{n}\widehat{\sigma}_Y^2,$$

where $\widehat{\sigma}_X^2$ and $\widehat{\sigma}_Y^2$ are unbiased variance estimators, with the test statistic $\widetilde{T} = (\overline{X} - \overline{Y})/\widetilde{\sigma}$. However, $\widetilde{T}$ does not have a t-distribution with $m + n - 2$ df. The built-in function t.test uses the Welsh test, which computes the approximate degree of freedom as

$$\nu = \frac{\widetilde{\sigma}^4}{\widehat{\sigma}_X^4/(m^2(m-1)) + \widehat{\sigma}_Y^4/(n^2(n-1))}.$$

Obviously, ν is not integer but the t-distribution can be defined even when ν is not an integer (ν is an estimator of df). Simple derivation implies

$$\min(m, n) - 1 < \nu < \max(m, n) - 1. \tag{7.19}$$

This means that Welsh's df is smaller than $m + n - 2$ used in the equal variance t-test. The larger $\widehat{\sigma}_X^2$ compared to $\widehat{\sigma}_Y^2$, the closer ν is to $m - 1$. This test is not exact: the size of the test is not equal α, especially when σ_X^2 and σ_Y^2 are close to each other. Although, for large n, this shortcoming quickly disappears. Below, we present simulations that support this statement.

Example 7.21 *Welsh and t-test. Compare the power of the t-test with df =*
$m + n - 2$ and the Welsh test with df = ν using simulations.

Solution. The R function is found in file `ttest2pow.r` which produces Figure
7.11. Note that when m and n become large, the t-test converges to the Z-test
discussed in detail in the previous section. Indeed, as follows from the graph at
left (the true variances are the same), one may conclude that the equal variance
test has greater power, but this is because the size of the Walsh test is smaller
(the value of the power function at zero). When the true variances are not equal,
the size of the Walsh is close to nominal, $\alpha = 0.05$, but the traditional t-test has
an inflated size. When m and n are large, the both tests are close.   □
   Till now no test yields the nominal type I error for testing the means of
normally distributed samples with different unknown variances. R uses the Walsh
test by default.

7.4.3   One-sided t-test

As we have argued earlier, the one-sided test is practically more interesting. Indeed, in Example 7.19 with a new drug, the question we want to ask is whether
it lowers cholesterol, not whether it is different from the conventional drug. Consequently, the null hypothesis is $H_0 : \Delta \geq 0$, and the alternative is $H_A : \Delta < 0$.
To test this hypothesis under the equal-variance assumption, we compute the
same T-statistic (7.15) and reject the null if its value is smaller than t_α, the αth
quantile of the t-distribution with $m + n - 2$ df. This test is called the left-sided
tow-sample t-test. The p-value in the one-sided test is half of that of the two-
sided test. The power of this test is $\mathcal{T}(t_\alpha; n + m - 2, c(1/n + 1/m)^{-1/2})$, where
$c = \Delta/\sigma$, the effect size ($\Delta \leq 0$). The power function looks like the left branch of
the power function of the traditional two-sided test but starts at α, not $1 - \alpha/2$.

Example 7.22 *Sample size one- versus two-sided test. Is it true that*
under the same conditions the total sample size for a one-sided test is half of that
of a two-sided test (the sample size in the two groups is the same)?

Solution. If p is the desired power, the total sample size, N, in the two-sided
test is the solution to the equation

$$\mathcal{T}(-t_{1-\alpha/2}; N - 2, 2c/\sqrt{N}) + 1 - F(t_{1-\alpha/2}; N - 2, 2c/\sqrt{N}) = p$$

and the sample size in the one-sided test is $\mathcal{T}(t_\alpha; N - 2, 2c/\sqrt{N}) = p$. The R function is found in the file `sampt2.r`. For example, default values `eff.size=-1,alpha`
`=.02,p=.8` require `N1=180` observations (90 observations per group) for the one-
sample test and `N2=211` observations for the two-sample test. Although it is
always true that the two-sample test requires a larger sample size, the answer in
negative.

7.4.4 Paired versus unpaired t-test

In the paired t-test, observations from the two samples are paired or matched $\{(X_i, Y_i), i = 1, 2, .., n\}$. Matching means that the pairing observations are collected under the same conditions with the values dependent on the same common factors. In Example 7.23 below, the employees in each pair hold the same positions, years at work, or collectively depend on common factors, yet have different salaries. The statistical model for a paired t-test takes the form

$$X_i = \mu_i + \varepsilon_{Xi}, \quad Y_i = \mu_i + \Delta + \varepsilon_{Yi}, \quad i = 1, 2, ..., n,$$

where n pairs $(\varepsilon_{Xi}, \varepsilon_{Yi})$ have zero means, are independent and have bivariate normal distributions,

$$\begin{bmatrix} \varepsilon_{Xi} \\ \varepsilon_{Yi} \end{bmatrix} \sim \mathcal{N}\left(\begin{bmatrix} 0 \\ 0 \end{bmatrix}, \begin{bmatrix} \sigma_1^2 & \sigma_{12} \\ \sigma_{12} & \sigma_2^2 \end{bmatrix} \right).$$

Two assumptions on common factors, μ_i, can be made: They are either (a) iid normally distributed and independent of ε_{Xi} and ε_{Yi}, with a certain mean and variance, say, η^2, or (b) fixed but unknown. The quantity of interest is Δ which is a fixed (nonrandom) unknown parameter. The null hypothesis is $H_0 : \Delta = 0$. We draw the reader's attention to how flexible and general the model for the paired t-test is because μ_i are not specified. The beauty of the paired t-test is that it is independent of μ_i because they are eliminated by taking the difference.

The paired t-test suggests to use the one-sample t-test applied to the difference of pairs. Indeed, since

$$D_i = Y_i - X_i \overset{\text{iid}}{\sim} \mathcal{N}(\Delta, \tau^2),$$

we simply use the one sample t-test with the null hypothesis $H_0 : \Delta = 0$. Specifically, $\tau^2 = \sigma_1^2 - 2\sigma_{12} + \sigma_2^2$, but we do not need to know these sigmas. The key of the paired t-test is that μ_i are eliminated by taking the difference. The paired t-test typically produces a smaller p-value than the traditional unpaired t-test. To understand the reason, we write the test statistics for the two tests we make some simplifying assumptions: (1) the variances are known, (2) $\sigma_1^2 = \sigma_2^2 = \sigma^2$ and $\sigma_{12} = 0$. The last assumption implies that D_i has variance $2\sigma^2$, and X_i and Y_i have the same variance, $\sigma^2 + \eta^2$, and therefore, the equal-variance assumption holds. The test statistics for the paired and unpaired t-tests are

$$\frac{\overline{Y} - \overline{X}}{\sqrt{2\sigma^2/n}}, \quad \frac{\overline{Y} - \overline{X}}{\sqrt{2(\sigma^2 + \eta^2)/n}}, \tag{7.20}$$

respectively. In fact, the t-test turns into the Z-test because variances are known, but it makes no difference to our explanation. The tests have the same numerator but different denominators: the numerator of the unpaired t-test is larger for $\eta^2 > 0$. This means that the value of the test statistic shrinks to zero and,

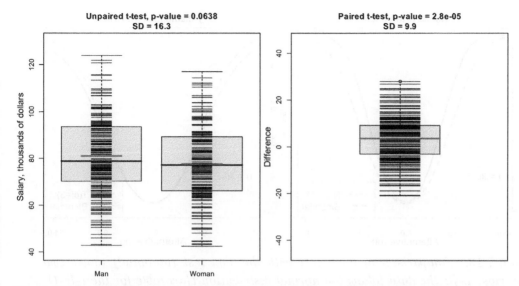

Figure 7.12: *Unpaired versus paired t-test for the gender-specific salary comparison. The horizontal short segments depict the means and the wide segments depict the medians. Note, SD in the unpaired t-test is more than 50% higher than in the paired t-test.*

therefore, the p-value is larger for the unpaired t-test. In words, the variance in the unpaired t-test contains the baseline variance but in the paired t-test this variance is eliminated by taking the difference. The following example illustrates the advantage of the paired t-test through matching.

Example 7.23 *Salaries for men and women*. *A big corporation is being accused of unequal pay for men and women. To address this issue, 158 men and pair-matched women have been randomly chosen and their salary reported, see file salaryMW_paired.csv. The matched pairs had the same position, years at work, etc. Test the hypothesis that men and women are compensated equally.*

Solution. The R code is found in file salaryMW.r and can be restored by issuing source("C:\\StatBook\\salaryMW.r") at the R console. The two tests are visualized in Figure 7.12. The actual salaries are depicted as segments and the bold segment (red) depicts the means (the wide bold segment depicts the median). The mean salary of male and female employees is about $81K and $78K, respectively. The unpaired t-test with equal variance yields $p = 0.0638$ with the SD of the combined data about 16.3. This SD reflects not only the variation of salaries per se, but the variation due to employee qualifications such as position, experience, years at work, etc. A better comparison uses the fact that the pairs have been matched – see the plot at right in Figure 7.12. The difference eliminates factors irrelevant to gender and SD is reduced to 9.9: the paired t-test yields $p = 2.8 \times 10^{-5}$.

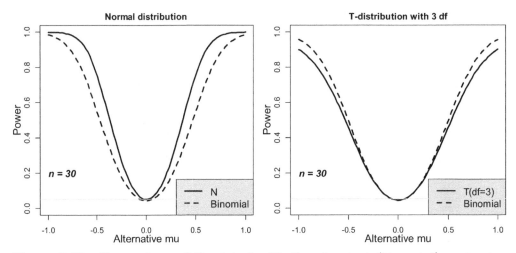

Figure 7.13: *Comparison of the t-test with the sign test (binomial) under two scenarios. Left: the data follow the normal distribution (favorable for the t-test). Right: the data follow the t-distribution with df = 3 (unfavorable for the t-test).*

7.4.5   Parametric versus nonparametric tests

In Example 7.5 a nonparametric sign test is suggested to test the hypothesis $H_0 : \mu = \mu_0$ versus $H_A : \mu \neq \mu_0$ when the distribution was unknown. The goal of this section is to compare the sign test with the t-test using a power function estimated via simulations. It is expected that if the distribution is normal the t-test will have larger power; otherwise, the sign test will outperform.

To investigate how distribution affects the power, a density of the form $f((x - \mu)/\sigma)$ is suggested, where μ is referred to as the location and σ as the scale. Function f is assumed to be symmetric around zero and have unit variance, but is not specified otherwise. Two situations are presented. In the first case, $f = \phi$, the standard normal density. In the second case, f is the density of a t-distribution with low degrees of freedom, say, $df = 3$ (heavy tails). When generating random observations, we divide the t-distributed random variables by $\sqrt{n/(n-2)}$, see Section 4.3, to comply with the assumption that f has unit variance. The R code is found in file `nonpN.r` and the results are presented graphically in Figure 7.13. As expected, the t-test is more powerful then the sign test, as seen in the left plot, because dichotomization reduces the information in the sample. When observations have heavy tails (the right plot), the binomial test is preferable because its power function is higher.

Problems

1. Prove that one- and two-sample t-tests converge to respective Z-tests when $n \to \infty$ and $m \to \infty$. [Hint: Show that the distribution of the T statistic under the null and the alternative converges to a normal distribution.]

2. The acceptance region of the test is composed of observations which lead to acceptance of the null hypothesis. (a) Display the acceptance region in the observation space on the plane for the one-sample t-test $H_0 : \mu = 1$ versus $H_A : \mu \neq 1$ when $n = 2$ and $\lambda = 0.05$. (b) Generate 10000 samples under the null hypothesis and plot the points in red if they belong to the acceptance region and in green if they belong to the rejection region (use `pch="."`). Compute and show the proportion of samples that reject the null hypothesis. [Hint: Use the `contour` function.]

3. Define the acceptance region in R^n for the one-sample t-test $H_0 : \mu = 1$ versus $H_A : \mu \neq 1$ for an arbitrary n and $\lambda = 0.05$. [Hint: Define the acceptance region as an ellipsoid $\{\mathbf{x} \in R^n : (\mathbf{x} - \mathbf{b})' \mathbf{A} \, (\mathbf{x} - \mathbf{b}) \leq c\}$.]

4. The test statistic for the two-sample t-test under the equal-variance assumption is computed as $(\overline{X} - \overline{Y})/\tilde{\sigma}$, where $\tilde{\sigma}$ is the standard error of the two samples combined. (a) Does this test statistic follow a t-distribution? (b) Develop the test for the null hypothesis that the means are the same. (c) Use simulations to compare its power function with (7.17).

5. (a) Use simulations to support the claim that $\widetilde{T} = (\overline{X} - \overline{Y})/\tilde{\sigma}$ in the two-sample t-test where $\{X_i \sim \mathcal{N}(\mu_X, \sigma_X^2), i = 1, ..., m\}$ and $\{Y_i \sim \mathcal{N}(\mu_Y, \sigma_Y^2), i = 1, ..., n\}$ do not have a t-distribution with $m + n - 2$ df when $\sigma_X \neq \sigma_Y$. (b) Given σ_X and σ_Y, find the optimal df k by fitting the empirical cdf with the theoretical one on the grid of values of k from $m + n - 2 - \Delta$ to $m + n - 2 + \Delta$ and compare it with Welsh's ν. [Hint: Use $\max |F_k(T) - \widehat{F}(T)|$ as the criterion where F_k and $\widehat{F}$ are the theoretical cdf of the t-distribution with k df and empirical from simulations. Modify function `ttest2pow`.]

6. (a) Prove inequality (7.19). (b) Express ν as a function of $\rho = \widehat{\sigma}_Y^2/\widehat{\sigma}_X^2$ and show that this function is monotonic. Prove that if $m < n$ then ν converges to $m - 1$ when $\rho \to 0$ and converges to $n - 1$ when $\rho \to \infty$.

7. (a) Run function `ttest2pow` from Example 7.21 for various m, n, and σ_X^2, σ_Y^2 to investigate how close the tests are. (b) Use an alternative estimator of the degrees of freedom, $\kappa = 2(\widehat{\sigma}_X^2(m - 1) + \widehat{\sigma}_Y^2(n - 1))/(\widehat{\sigma}_X^2 + \widehat{\sigma}_Y^2)$ and compare this test with the Walsh test using simulations.

8. Prove that the p-value in the one-sided test is half that in the two-sided test.

9. Does the one-sided test in Example 7.20 change the conclusion on living costs?

10. Prove that under the conditions of Example 7.22, the two-sided test requires more observations than the one-sided test.

11. Modify function `sampt2` from Example 7.22 to compare the minimum detectable effect size in one- and two-sided tests.

12. Use simulations to show that the unpaired t-test produces p-values larger than the paired t-test. Simulate X_i and Y_i according to the statistical model in Section 7.4.4 and compute p-values (variances unknown and unequal). Return the proportion of simulations when the p-value from the unpaired t-test is smaller than the p-value from the paired t-test. [Hint: Make n, $\eta^2, \sigma_1^2, \sigma_2^2$ and σ_{12} the arguments of the R function.]

13. The same as above, but compare the tests using the power function.

14. To test a new anticholesterol drug, 16 patients have been recruited. The LDL cholesterol level at the baseline was 163, 179, 168, 189, 173, 179, 197, 173, 185, 201, 182, 169, 186, 169, 199, 188 mg/dL. Two months after the drug administration cholesterol was measured again in the same patients: 161, 181, 158, 177, 171, 180, 175, 171, 182, 198, 172, 163, 182, 166, 187, 179. Use box plots to display the data, display the raw data on the boxes as segments (`par(mfrow =c(1,2)`) as in Example 7.23. (a) First, apply the unpaired/regular t-test under the equal-variance assumption and then the paired t-test. Display the p-value in the title. (b) Demonstrate that the function `t.test` and your own computations produce the same p-values. (c) Explain the difference in the p-values between unpaired and paired tests. Which test is more appropriate?

7.5   Variance test

Sometimes we are less interested in the mean than the variance of the normal distribution. In the example below we compare the volatility of returns of two stocks – a typical investment task. In this section, we discuss the test about the variance in one sample that employs the chi-square distribution. The test with two variances is deferred to the next section, as a special case of the general inverse-cdf test.

7.5.1   Two-sided variance test

The null hypothesis of the one-sample two-sided variance test is written as $H_0 : \sigma^2 = \sigma_0^2$ versus the alternative $H_A : \sigma^2 \neq \sigma_0^2$ given a random sample $X_1, X_2, ..., X_n$ from the normal distribution $\mathcal{N}(\mu, \sigma^2)$. The basis for the test is the fact that

$$\frac{1}{\sigma^2} \sum_{i=1}^{n} (X_i - \overline{X})^2 \sim \chi^2(n-1),$$

as has been proven in Theorem 4.22 from Section 4.2. Therefore, the test statistic takes the form

$$\frac{1}{\sigma_0^2}\sum_{i=1}^{n}(X_i - \overline{X})^2. \tag{7.21}$$

If the observed value of this statistic is outside the *normal* range, the range of values (7.21) under the null hypothesis, then we reject the null hypothesis. Traditionally, equal-tail probabilities are used to determine the normal range: given significance level (type I error) α, compute quantiles $q_1 = C_{n-1}^{-1}(\alpha/2)$ and $q_2 = C_{n-1}^{-1}(1 - \alpha/2)$, where C_{n-1} is the cdf of the chi-square distribution (pchisq) with $n - 1$ df, and C_{n-1}^{-1} is its inverse (qchisq). If (7.21) is outside the interval (q_1, q_2) the null hypothesis is rejected. However, the equal-tail test is biased. Indeed, since the hypothesis is rejected when $\sum_{i=1}^{n}(X_i - \overline{X})^2 > \sigma_0^2 q_2$ or $\sum_{i=1}^{n}(X_i - \overline{X})^2 < \sigma_0^2 q_1$ the power of the equal-tail test is

$$P(\sigma^2; \sigma_0^2) = 1 + C_{n-1}\left(\sigma_0^2/\sigma^2 q_1\right) - C_{n-1}\left(\sigma_0^2/\sigma^2 q_2\right). \tag{7.22}$$

To prove that the test is biased, we need to evaluate the derivative of P at $\sigma^2 = \sigma_0^2$ using the density from Section 4.2,

$$\left.\frac{dP}{d\sigma^2}\right|_{\sigma^2=\sigma_0^2} = \frac{1}{2^{(n-1)/2}\Gamma((n-1)/2)}\left[q_1^{(n-3)/2}e^{-q_1/2} - q_2^{(n-3)/2}e^{-q_2/2}\right]. \tag{7.23}$$

This quantity is not zero for any α and n, and therefore the test is biased (see Example 7.24 for more detail). To derive an unbiased test, the tail probabilities must be unequal to make (7.23) zero. When the probabilities of the tails are not the same the expression for the power function remains the same, (7.22), but now the tail quantiles (critical values) q_1 and q_2 are chosen such that $C_{n-1}(q_2) - C_{n-1}(q_1) = 1 - \alpha$ to make the derivative (7.23) zero. The problem of finding $q_1 < q_2$ reduces to a system of two equations,

$$(n-3)\left(\ln q_2 - \ln q_1\right) - (q_2 - q_1) = 0, \quad C_{n-1}\left(q_1\right) - C_{n-1}\left(q_2\right) + (1 - \alpha) = 0. \tag{7.24}$$

To solve this system we use the Newton's algorithm with the 2×1 adjustment vector computed at each iteration:

$$\begin{bmatrix} (n-3)/q_1 - 1 & -(n-3)/q_2 + 1 \\ -f(q_1) & f(q_2) \end{bmatrix}^{-1} \begin{bmatrix} (n-3)\left(\ln q_2 - \ln q_1\right) - (q_2 - q_1) \\ C_{n-1}\left(q_1\right) - C_{n-1}\left(q_2\right) + (1 - \alpha) \end{bmatrix},$$

starting from the equal-tail quantiles. In R, the above f, C and C^{-1} are computed using the built-in functions dchisq, pchisq and qchisq, respectively, with df=n-1. Newton's algorithm is realized in the R function varq. The algorithm is fast. Typically, it requires 3–4 iterations to converge. The following example illustrates the power function and application of the variance test with unequal-tail probabilities.

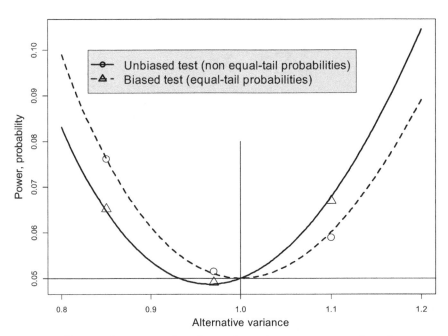

Figure 7.14: *Power functions for the unbiased variance test based on non equal tail-probabilities and the biased test based on equal tail-probabilities. The points are derived from vectorized simulations,* `nSim=1000000`*.*

Example 7.24 *Variance test.* *(a) Plot the power function for equal- and unequal-tail variance test for $n = 20$ and $\sigma_0^2 = 1$ with the size $\alpha = 0.05$. (b) To check the analytical power, add to the plot the power values obtained from simulations for three alternative values $\sigma^2 = 0.85, 0.97,$ and 1.1. (c) A new technology aims at improving the precision of production measured on the scale of millimeters. The old technology had precision $SD = 0.1$ mm and twelve products produced by the second-generation technology had measurements 138.508, 138.368, 138.372, 138.305, 138.383, 138.391, 138.424, 138.349, 138.428, 138.344, 138.335, 138.341. Test the hypothesis that the precision of the new technology is the same as the old one.*

Solution. (a) The R code is found in function `vartest` and the plot is shown in Figure 7.14. The traditional equal tail-probabilities test is biased because for some variances $\sigma^2 < \sigma_0^2 = 0.01$ the chance of rejecting the null hypothesis is larger than the error of rejecting the null when it is true, $\alpha = 0.05$, which makes no sense. The unbiased test guarantees that the derivative of $P(\sigma^2)$ at the null value is zero and the minimum power is attained at $\sigma^2 = \sigma_0^2$. (b) Simulations confirm the power function for both tests (empty circle for the unbiased test and triangle for the biased test). With vectorized simulations, we generate `nSim` by n matrix X of random numbers distributed as $\mathcal{N}(\mu, \sigma^2)$ and then use `apply(X,1,var)` to obtain variances for each row. (c) Formulate the rejection criterion in terms of $SD = \widehat{\sigma}$:

the null hypothesis is rejected if $\widehat{\sigma} > \sigma_0\sqrt{q_2/(n-1)}$ or $\widehat{\sigma} < \sigma_0\sqrt{q_1/(n-1)}$, which specify the normal SD range. Newton's algorithm realized in function `vartest` quickly converges to $q_1 = 4.099444$ and $q_2 = 23.134794$. Thus we have $\widehat{\sigma} = 0.05425361$ and $\sigma_0\sqrt{q_1/(n-1)} = 0.06104725$ and $\sigma_0\sqrt{q_2/(n-1)} = 0.14502286$. Since $\widehat{\sigma}$ is outside of the normal SD range, the null hypothesis is rejected.

7.5.2 One-sided variance test

From the practical perspective, the one-sided test is more appealing. Indeed, let us revisit the variance test on the new technology in Example 7.24. The question of interest is whether the new technology improves precision, that is, leads to a smaller variance, but when the two-sided hypothesis is rejected we claim that the two variances are not equal. In reality, nobody would try to use statistical hypothesis testing if new measurements produced SD greater than 0.1 mm.

The null hypothesis in the one-sided test is $H_0 : \sigma^2 \geq \sigma_0^2$ versus the alternative is $H_A : \sigma^2 < \sigma_0^2$. The one-sided test is simpler than the two-sided test because the issue of equal-tail probabilities does not emerge. The null hypothesis is rejected if $\widehat{\sigma} < \sigma_0\sqrt{q/(n-1)}$, where q is the αth quantile of the chi-square distribution with $n-1$ degrees of freedom with the power function $P(\sigma^2) = \mathcal{C}_{n-1}\left(\sigma_0^2/\sigma^2 q\right)$, where $\sigma^2 \leq \sigma_0^2$ and $q = \mathcal{C}_{n-1}^{-1}(\alpha)$ is the αth quantile. Another advantage of the one-sided test is that the p-value is easy to compute as $\mathcal{C}_{n-1}\left((n-1)\widehat{\sigma}^2/\sigma_0^2\right)$. In Example 7.24 we compute `pchisq((20-1)*0.05425361^2/0.1^2,df=20-1)=0.001270186`.

Example 7.25 *Coefficient of variation. A researcher wants to know how many repeated measurements to make to test that the relative precision is smaller than κ_0, a given positive number. (a) Use the lognormal distribution for measurements and connect the standard deviation on the log scale to the coefficient of variation. (b) Five measurements with precision at least 1% have been taken. What is the maximum detectable relative precision (CV) with the power 80% and significance level 5%?*

Solution. The lognormal distribution was advocated in the probability part of the book as a natural distribution for positive random variables just as the normal distribution was advocated for values on the entire real line–see Section 2.10.2 and Section 2.11. A peculiar property of the lognormal distribution is that the standard deviation, σ, of the log-transformed variable is close to the coefficient of variation ($\kappa = \text{CV}$) of the original random variable, $\sigma \simeq \kappa$. Thus, if the relative precision of measurements is understood as CV, we rephrase the question to ask if the standard deviation of the log-transformed measurements (σ) is greater than κ_0 versus the alternative $H_A : \sigma < \kappa_0$. As follows from the above discussion, if n log measurements produce $\widehat{\sigma} < \kappa_0$ and moreover, $\widehat{\sigma} < \kappa_0\sqrt{q/(n-1)}$ where $q = \mathcal{C}_{n-1}^{-1}(\alpha)$, we reject the null with the type I error α. (b) Since the power function is $P(\sigma^2) = \mathcal{C}_4\left(\kappa_0^2/\sigma^2 q\right)$, the maximum detectable CV is $0.01/\sqrt{\mathcal{C}^{-1}(0.05)/\mathcal{C}^{-1}(0.8)}100\% = 0.34\%$, where 4 is degrees of freedom, omitted for brevity. In other words, alternatives $\kappa < 0.0034$ yield power less than 0.8 with $n = 5$ measurements and $\alpha = 5\%$.

Problems

1. Prove that if $\alpha < 0.5$ the one-sided variance hypothesis is rejected if $\hat{\sigma} < \sigma_0$. [Hint: prove that $\mathcal{C}_u(u)$ is a decreasing function of $u \geq 2$ and $\lim_{u \to \infty} \mathcal{C}_u(u) = 0.5$.]

2. (a) Display the acceptance region in the observation space on the plane for the one-sided variance test $H_0 : \sigma_0^2 \leq 1$ versus $H_A : \sigma_0^2 > 1$ when $n = 2$ and $\lambda = 0.05$. (b) Generate 10,000 samples under the null hypothesis and plot the points in red if they belong to the acceptance region and in green if they belong to the rejection region (use `pch="."`). Compute and show the proportion of samples that reject the null hypothesis.

3. Develop the one- and two-sided variance tests when μ is known.

4. Use `vartest` from Example 7.24 to investigate how the sample size affects the difference between the biased and unbiased tests. Plot the maximum absolute difference between the two powers as a function of n (use the σ_0^2 value as an argument to the R function).

5. Use `vartest` from Example 7.24 to investigate how n and σ_0^2 affect the bias of the equal-tail test measured as the derivative of the power function at the null. Plot the bias as a function of n for three different values of σ_0^2.

6. Repeat Example 7.24 to test the one-sided hypothesis $H_0 : \sigma_0 \leq 0.1$ versus $H_A : \sigma > 0.1$. Plot the power function, check with simulations and compute the p-value. Does the one-sided test have more practical interest?

7. Use the test statistic (7.15) to test that two independent samples from normal distributions have the same mean and variance. (a) Prove that this statistic has a t-distribution with $m + n - 2$ df. (b) Estimate the power function using vectorized simulations on the grid of values for the difference of the means of the ratios of the variances (set the variance from the first sample to 1) and plot contours. (c) Is the test locally unbiased?

7.6   Inverse-cdf test

In this section, we discuss a general statistical hypothesis test, called the *inverse-cdf test*, which gives rise to many practically important tests, such as the variance test, tests on binomial proportion, Poisson rate, and the correlation coefficient. This test is discussed in Casella and Berger [17] who refer to Mood et al. [75] as the originators of this method. A point of special attention is the derivation of the power function. This function will be used to compare tests and develop optimal study designs such as sample size determination. We underscore that the inverse-cdf test is exact for continuous random variables meaning that it yields the desired type I error α if the cdf is a continuous function. When the random variable is discrete the nominal type I error is not guaranteed. In contrast, tests that rely on asymptotic properties do not guarantee the desired level of statistical significance even for a continuous distribution.

7.6.1 General formulation

Definition 7.26 *A pivotal quantity, P, is a function of the data and unknown parameters, $P = P(X_1, ..., X_n; \theta)$ such that its distribution does not depend on parameters.*

An example of a pivotal quantity is the t-test statistic, $P = \sqrt{n}(\overline{Y} - \mu)/\widehat{\sigma}$ which depends on μ but its distribution depends on neither μ nor σ^2. The idea of using a pivotal quantity for testing $H_0 : \theta = \theta_0$ against the two-sided alternative $H_A : \theta \neq \theta_0$ is as follows. Since the distribution is known, we can compute the normal range as the interval between two quantiles. Then if $P(X_1, ..., X_n; \theta_0)$ is outside of the normal range, the hypothesis is rejected. Alternatively, we can compute $\widehat{\theta}_L$ and $\widehat{\theta}_U$ from the two quantiles and reject the null hypothesis if θ_0 is outside $(\widehat{\theta}_L, \widehat{\theta}_U)$. The pivotal quantity is especially convenient when there is a statistic which cumulative distribution function (cdf) is known up to parameter θ. It is easy to see that this cdf is a pivotal quantity because its distribution is uniform on $(0, 1)$ as follows from Theorem 2.71.

Example 7.27 *Pivotal quantity for uniform distribution.* The iid observations $\{X_i, i = 1, ..., n\}$ belong to a uniform distribution $\mathcal{R}(0, \theta)$. (a) Test the one-sided hypothesis $H_0 : \theta \leq \theta_0$ against $H_A : \theta > \theta_0$ using the pivotal quantity $X_{\max}/\theta$. (b) Ten iid observations produced $X_{\max} = 1.9$. Test the hypothesis $\theta_0 \leq 2$.

Solution. It is easy to see that $X_i/\theta \sim \mathcal{R}(0, 1)$ and therefore $X_{\max}/\theta$ is indeed a pivotal quantity because $X_{\max}/\theta$ has a cdf independent of θ, namely, x^n, see Example 6.22. (a) To find the normal range when $\theta = \theta_0$, we need to chose a significance level α and then find q such that $q^n = 1 - \alpha$, that is, $q = \sqrt[n]{1 - \alpha}$. Thus, we reject the hypothesis that $\theta \leq \theta_0$ if $X_{\max} > \theta_0 \sqrt[n]{1 - \alpha}$. (b) Let $\alpha = 0.05$. Since $2\sqrt[10]{1 - 0.05} = 1.99 > 1.9$ we cannot reject the hypothesis that $\theta_0 \leq 2$.

Definition 7.28 *Inverse-cdf test.* Let T be a statistic with continuous cdf $F(t; \theta)$ where F is an decreasing and differentiable function of θ for every t. Let α_1 and α_2 be any positive (nonrandom) numbers such that $\alpha_1 + \alpha_2 = \alpha$, the significance level. The inverse-cdf test for the null hypothesis $H_0 : \theta = \theta_0$ against $H_A : \theta \neq \theta_0$ rejects the null if either $F(T_{\text{obs}}; \theta_0) < \alpha_2$ or $F(T_{\text{obs}}; \theta_0) > 1 - \alpha_1$.

Example 7.29 *Inverse-cdf test for uniform distribution.* *Show that the inverse-cdf test applied to the one-sided hypothesis test from Example 7.27 leads to the same rejection rule.*

Solution. The test statistic is $X_{\max}$ and $F(t; \theta) = (t/\theta)^n$ for $0 < t < \theta$. According to the one-sided inverse-cdf test, the hypothesis $\theta \leq \theta_0$ is rejected if $(X_{\max}/\theta_0)^n > 1 - \alpha$ which leads to the same rejection rule as in Example 7.27. The pivotal quantity and the inverse-cdf tests are equivalent because they are based on the same statistic, $X_{\max}$.

Example 7.30 *Inverse-cdf test for variance.* *Show that the two-sided variance test discussed in the previous section can be recast as the inverse-cdf test.*

Solution. In the framework of the variance test, the statistic is $S = \sum_{i=1}^{n}(X_i - \overline{X})^2$. Since $S/\sigma^2 \sim \chi^2(n-1)$, the cdf of S is given by $F(s; \sigma^2) = \mathcal{C}(s/\sigma^2; n-1)$. Clearly, $F(s; \sigma^2)$ is a decreasing function of σ^2 when s is held fixed. To simplify, we assume equal-tail probabilities: let α be given and $q_1 = \mathcal{C}(\alpha/2; n-1)$ and $q_1 = \mathcal{C}(1 - \alpha/2; n-1)$ be like in the equal tail-variance test. If S_{obs} is the observed $\sum_{i=1}^{n}(X_i - \overline{X})^2$, the null hypothesis $H_0 : \sigma^2 = \sigma_0^2$ is rejected if either $\mathcal{C}(S_{\mathrm{obs}}/\sigma_0^2; n-1) > 1 - \alpha/2$ or $\mathcal{C}(S_{\mathrm{obs}}/\sigma_0^2; n-1) < \alpha/2$, or equivalently if $S_{\mathrm{obs}} > q_2\sigma_0^2$ or $S_{\mathrm{obs}} < q_1\sigma_0^2$. This is the same rule as has been suggested earlier.   □

The following result justifies why the inverse-cdf test is very attractive: (a) The power function is readily available. (b) The unbiased test with unequal probability tails reduces to a simple numerical algorithm.

Theorem 7.31 *Locally unbiased inverse-cdf test.* *Let T be a test statistic with continuous cdf $F(t; \theta)$, where F is a differentiable and decreasing function of θ for every t. According to the inverse-cdf test, the null hypothesis $H_0 : \theta = \theta_0$ is rejected if either $F(T_{\mathrm{obs}}; \theta_0) > F(q_2; \theta_0)$ or $F(T_{\mathrm{obs}}; \theta_0) < F(q_1; \theta_0)$. The power function of the two-sided inverse-cdf test is $P(\theta) = 1 + F(q_1; \theta) - F(q_2; \theta)$, where $q_1 = q_1(\theta_0)$ and $q_2 = q_2(\theta_0)$ are the quantiles as solutions of the equations $F(q_1; \theta_0) = \alpha_2$ and $F(q_2; \theta_0) = 1 - \alpha_1$, respectively ($\alpha_1 + \alpha_2 = \alpha$). The test is locally unbiased if $q_1 < q_2$ satisfy the following system of equations*

$$F(q_2; \theta_0) - F(q_1; \theta_0) = 1 - \alpha, \quad F'_\theta(q_1; \theta_0) - F'_\theta(q_2; \theta_0) = 0, \qquad (7.25)$$

where F'_θ means the derivative of the cdf with respect to θ.

The proof is elementary, the local unbiasedness equation follows from differentiation of the power function and equating it to zero evaluated at $\theta = \theta_0$. Obviously, the value of the power function is α at $\theta = \theta_0$. This theorem will be applied to practically important tests below. The optimal tail probabilities that make the inverse-cdf test unbiased require a numerical iterative solution using Newton's algorithm. Note that the equal-tail probabilities usually lead to biased tests. The locally unbiased test turns into a globally unbiased test if the power function is increasing for $\theta > \theta_0$ and deceasing for $\theta < \theta_0$.

The problem of insider trading was solved *ad hoc* in Example 7.3; the next example solves this problem rigorously by computing the p-value.

Example 7.32 *Testing λ in exponential distribution. (a) Apply the inverse-cdf test to test the rate parameter, λ, in the exponential distribution for one- and two-sided hypotheses using the fact that the sum of m iid exponentially distributed random variables has cdf $\mathcal{C}(2\lambda x, 2n)$, where $\mathcal{C}$ is the cdf of the chi-square distribution. (b) Apply the inverse-cdf test to test that the two rate parameters from the two independent exponentially distributed samples are the same. Derive the one- and two-sided tests. Use simulations to test the analytic powers. (c) Using the data from Example 7.3 compute the p-value to test the hypothesis that Mr. Smith's rate of trade is the same as or smaller than the regular rate of trade.*

Solution. The test statistic is $S = \sum_{i=1}^{m} X_i \sim \mathcal{C}(2\lambda x, 2m)$. (a) In the case of the one-sided hypothesis the null is $H_0 : \lambda \leq \lambda_0$ with the alternative $H_A : \lambda > \lambda_0$. If α is the significance level the hypothesis is rejected if $S_{\text{obs}} < (2\lambda_0)^{-1}q$, where $q = \mathcal{C}^{-1}(\alpha, 2m)$ is the αth quantile of the chi-square distribution with m df (larger λ yields smaller S). The power function is $P(\lambda) = \mathcal{C}(\lambda q/\lambda_0, m)$ for $\lambda \geq \lambda_0$. For the two-sided hypothesis $H_0 : \lambda = \lambda_0$ versus $H_A : \lambda \neq \lambda_0$ we find quantiles q_1 and q_2 such that

$$\mathcal{C}(2\lambda_0 q_2, 2m) - \mathcal{C}(2\lambda_0 q_1, 2m) = 1 - \alpha.$$

If the tail probabilities are the same, we set $q_1 = \mathcal{C}^{-1}(\alpha/2, 2m)$ and $q_2 = \mathcal{C}^{-1}(1 - \alpha/2, 2m)$. To make the test unbiased we solve a system of equations as in the variance test. (b) Now we have $(2\lambda_X)^{-1}S_X \sim \mathcal{C}(\cdot, 2m)$ and $(2\lambda_Y)^{-1}S_Y \sim \mathcal{C}(\cdot, 2n)$ where $S_X = \sum_{i=1}^{m} X_i$ and $S_Y = \sum_{i=1}^{n} Y_i$ are independent. Consider the test statistic as the ratio of averages, $(S_X/m)/(S_Y/n)$. This ratio, as the ratio of chi-square distributions adjusted for degrees of freedom, has an F-distribution with cdf $\mathcal{F}((\lambda_X/\lambda_Y)x, \text{df1}= 2m, \text{df2}= 2n)$. The one-sided test is obvious (see the next item), the two-sided test follows the line of the F-test for variance described below. (c) The null hypothesis is $H_0 : \lambda_X/\lambda_Y \leq 1$ versus $H_A : \lambda_X/\lambda_Y > 1$, where λ_X is the Mr. Smith's rate of trade ($m = 1$) and λ_Y is that of other traders ($n = 19$). The observed value of the test statistic is $(S_X/m)/(S_Y/n) = (3/1)/(7259/19) = 0.00785$. The p-value is the cdf of the F-distribution evaluated at the observed value of the test statistic: p = `pf(0.00785,df1=2,df2=2*19)=0.00782`. The value of the cdf close to the argument follows from the fact that, for $m = 2$, the density at zero is well approximated by function x as follows from formula (4.28). This p-value is very close to the nonparametric p-value from Example 7.3.

7.6.2 The F-test for variances

First, we derive the test as the ratio of two variances. Second, we apply Theorem 7.31 to obtain the power function and use it to develop an unbiased test.

We want to test the equality of variances from two independent normally distributed samples $X_i \overset{\text{iid}}{\sim} \mathcal{N}(\mu_1, \sigma_1^2)$ and $Y_j \overset{\text{iid}}{\sim} \mathcal{N}(\mu_2, \sigma_2^2)$, where $i = 1, 2, ..., n_1$ and $j = 1, 2, ..., n_2$. The null hypothesis is two-sided: $H_0 : \sigma_1^2 = \sigma_2^2$ against $H_A : \sigma_1^2 \neq \sigma_2^2$. If $\widehat{\sigma}_1^2$ and $\widehat{\sigma}_2^2$ denote the unbiased estimators of the two variances (sample variances), the ratio of the variances under the null, as follows from Theorem 4.37 of Section 4.4, follows the F-distribution, $\widehat{\sigma}_1^2/\widehat{\sigma}_2^2 \sim F(n_1-1, n_2-1)$. Hence, according to the F-test, the null hypothesis is rejected when the ratio of sample variances is either small or large. Specifically, let the size of the test (significance level) be α, and positive tail probabilities α_1 and α_2 be such that $\alpha_1 + \alpha_2 = \alpha$. Let $q_1 < q_2$ be the α_1th and $(1 - \alpha_2)$th quantiles, respectively, so that the random variable with F-distribution with $n_1 - 1$ and $n_2 - 1$ degrees of freedom falls within interval (q_1, q_2) with probability $1 - \alpha$. The null hypothesis is rejected if the ratio of variances, $\widehat{\sigma}_1^2/\widehat{\sigma}_2^2$, is either smaller than q_1 or greater than q_2. In the equal-tail F-tes,t we chose $\alpha_1 = \alpha_2 = \alpha/2$. Question: Is such a test unbiased and, if not, how can we make it unbiased?

To answer this question, we write $\widehat{\sigma}_1^2/\widehat{\sigma}_2^2 = \nu(\widehat{\sigma}_1^2/\sigma_1^2)/(\widehat{\sigma}_2^2/\sigma_2^2)$, where $\nu = \sigma_1^2/\sigma_2^2$, and rephrase the null and the alternative hypothesis as $H_0 : \nu = 1$ against $H_A : \nu \neq 1$. This reformulation allows us to express the cdf of the test statistic (the ratio of variances) as $\Pr(\widehat{\sigma}_1^2/\widehat{\sigma}_2^2 \leq x) = \mathcal{F}(x/\nu; n_1 - 1, n_2 - 1)$. Now, it is obvious that the above F-test takes the form of the inverse-cdf test. The following theorem is a straightforward reformulation of Theorem 7.31 applied to the F-test. Interestingly, the equal-tail probability test is unbiased when two samples have the same size.

Theorem 7.33 *Properties of the F-test. The power function of the F-test is given by*

$$P(\nu) = 1 - \mathcal{F}(q_2/\nu; n_1 - 1, n_2 - 1) + \mathcal{F}(q_1/\nu; n_1 - 1, n_2 - 1),$$

where $\nu = \sigma_1^2/\sigma_2^2$ is the alternative variances ratio. The test with equal-tail probabilities is unbiased when $n_1 = n_2$. To make the F-test unbiased when $n_1 \neq n_2$ the quantiles q_1 and q_2 should be found from the solutions of the following equations:

$$\mathcal{F}(q_2; n_1 - 1, n_2 - 1) - \mathcal{F}(q_1; n_1 - 1, n_2 - 1) - (1 - \alpha) = 0,$$

$$k_1 \ln \frac{q_2}{q_1} - (k_1 + k_2) \ln \frac{k_2 + k_1 q_2}{k_2 + k_1 q_1} = 0,$$

where $k_1 = (n_1 - 1)/2$ and $k_2 = (n_2 - 1)/2$. Newton's iterations are given below.

Proof. The power function follows from Theorem 7.31. The tail probabilities are $\alpha_1 = \mathcal{F}(q_1; n_1 - 1, n_2 - 1)$ and $\alpha_2 = 1 - \mathcal{F}(q_2; n_1 - 1, n_2 - 1)$. When $n_1 = n_2$ the test is unbiased because

$$\mathcal{F}^{-1}_{\alpha/2}(n_1 - 1, n_2 - 1) = \frac{1}{\mathcal{F}^{-1}_{1-\alpha/2}(n_2 - 1, n_1 - 1)},$$

$$\mathcal{F}^{-1}_{1-\alpha/2}(n_1 - 1, n_2 - 1) = \frac{1}{\mathcal{F}^{-1}_{\alpha/2}(n_2 - 1, n_1 - 1)},$$

as indicated in Section 4.4. Now we discuss Newton's algorithm to find the quantiles when $n_1 \neq n_2$. The density of the F-distribution is proportional to $x^{k_1-1}/(k_2 + k_1 x)^{k_1+k_2}$, where $k_j = (n_j - 1)/2$ for $j = 1, 2$. It is easy to see the unbiasedness equation after taking the log transformation is equivalent to the second equation of the system to solve. The adjustment 2×1 vector for $(q_1, q_2)'$ in Newton's algorithm at each iteration is as follows:

$$
\begin{bmatrix} -k_1/q_1 + k_1(k_1 + k_2)/(k_2 + k_1 q_1) & k_1/q_2 - k_1(k_1 + k_2)/(k_2 + k_1 q_2) \\ -f(q_1) & f(q_2) \end{bmatrix}^{-1}
$$

$$
\times \begin{bmatrix} k_1 \ln q_2 - (k_1 + k_2)\ln(k_2 + k_1 q_2) - k_1 \ln q_1 + (k_1 + k_2)\ln(k_2 + k_1 q_1) \\ \mathcal{F}(q_2) - \mathcal{F}(q_1) - (1 - \alpha) \end{bmatrix},
$$

where f is the density and $\mathcal{F}$ is the cdf of the F-distribution, computed as `df` and `pf`, respectively (df is omitted for brevity). We take equal-tail quantiles as initial values for iterations. Typically, it requires $2 - 3$ iterations to converge. Note that these quantiles are independent of the alternative $\nu = \sigma_1^2/\sigma_2^2$. $\qquad\square$

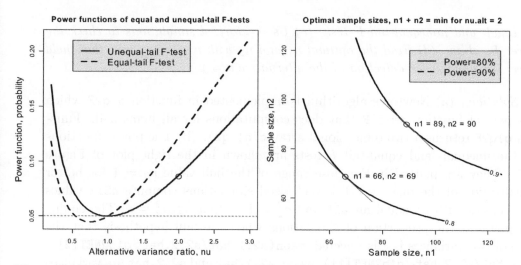

Figure 7.15: *The graphical output from the R function* smallF *in Example 7.34. Left: The equal-tail F-test is biased, but the unequal-tail F-test is unbiased. Right: The contours of the power function of the unequal F-test to detect the ratio of variances $\nu = 2$. The circle on the curve is where power is 0.8 or 0.9 with minimum $n_1 + n_2$.*

The following example illustrates how the power function can be used to design optimal studies and achieve the desired power with the minimum total number of measurements.

Example 7.34 *Comparison of equal and unequal F-tests for variances.* (a) *Write an R function for Newton's algorithm.* (b) *Plot two power functions*

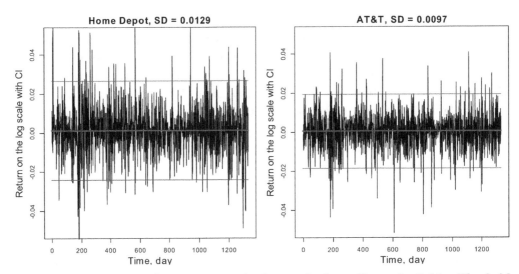

Figure 7.16: *Two stock returns on the log scale from Example 7.35. The bold horizontal lines depict the mean and the two thin lines depict the 95% confidence interval. This figure is produced by calling* `vartestSP()`.

for equal- and nonequal-tail F-test. (c) Use vectorized simulations to check the power function. (d) Find the optimal n_1 and n_2 with $n_1 + n_2 = \min$ that yield power 0.8 and 0.9 for detection of the alternative $\nu = 1.5$.

Solution. (a) Newton's algorithm is implemented in function `q1q2F` which is a part of function `smallF` that does computations for all items a–d. Function `q1q2F` returns a two-dimensional array `c(q1,q2)`. (b) The power functions for the unequal- and equal-tail F-tests are shown in the right plot of Figure 7.15. They are plotted in the close range of the null value $\nu_0 = 1$ for better visualization of the bias. (c) The vectorized simulations compute `nSim` values of variances with σ_1^2 equal `nu.alt` and $\sigma_2^2 = 1$ so that $\sigma_1^2/\sigma_2^2 = \nu$. The empirical power is the proportion of simulations when the null hypothesis that the ratio of two variances is 1 is rejected, `mean(s2.1.hat/s2.2.hat>q1q2NET[2] | s2.1.hat/s2.2.hat<q1q2NET[1])`, where `s2.1.hat` and `s2.2.hat` are variances from simulations and `q1q2NET` is the two-dimensional array returned by `q1q2F` with specified `n1` and `n2`. In Figure 7.15, the empirical power is indicated by a little circle – the match with the theoretical power is very good. (d) The right plot in Figure 7.15 depicts two contours with levels 0.8 and 0.9. To find n_1 and n_2 on the contour curve such that $n_1 + n_2 = \min$, we need to find the point where the tangent slope is -1. For power 0.8, we need to take a few more measurements for the second variance, but for power 0.9 the number of measurements should be almost the same.

Example 7.35 *Testing the volatility of two stocks. Test if a pair of stocks from Example 6.47 have the same variance.*

Solution. Volatility of stocks is an important factor of investment risk. For example, if variances of two stocks are the same, the optimal proportion in the investment portfolio is 50/50, as follows from Section 3.8. The R function `vartestSP` displays two daily stock prices analyzed previously in function `cimcorSP`, shown in Figure 7.16. The default arguments are `stock1=2` and `stock2=6`, Home Depot and AT&T, respectively. The call `vartestSP()` produced the following output:

```
lower nu=0.8980113, upper nu=1.113572,
nu=var1/var2=1.761471, p-value = 1.183452e-24.
```

The normal range of the ratio of variances, Home Depo over AT&T $(\widehat{\nu})$, is from 0.898 to 1.114. Since the observed value is $\widehat{\nu}_{\text{obs}} = 1.761$ we reject the null hypothesis that the variances are the same. In this example, the sample size in the first and second samples are the same and, therefore, as follows from Theorem 7.33, the probability tails at $(\alpha/2)$ and $(1 - \alpha/2)$ are the same. The p-value is computed as the probability of the F-distributed random variable to be greater than 1.761 or smaller than 1/1.761.

Example 7.36 *Variance test for regressions. Two independent simple linear regressions with sample sizes $n_1 = 13$ and $n_2 = 9$ produced regression variance estimates $\sigma_1^2 = 1.9$ and $\sigma_2^2 = 2.6$, respectively. Test the null hypothesis that $\sigma_1^2 > \sigma_2^2$ with $\lambda = 0.05$ and compute the p-value.*

Solution. We make a couple of remarks: (i) Typically, the null hypothesis includes the equality sign but in our formulation we have a strong inequality, (ii) the alternative hypothesis is not mentioned, so we specify that the alternative hypothesis as $H_A : \sigma_1^2 \leq \sigma_2^2$. Since $(n_1-2)\widehat{\sigma}_1^2/\sigma_1^2 \sim \chi^2(n_1-2)$ and $(n_2-2)\widehat{\sigma}_2^2/\sigma_2^2 \sim \chi^2(n_2 - 2)$ as follows from Theorem 6.53 and $\widehat{\sigma}_1^2$ and $\widehat{\sigma}_2^2$ are independent, $\widehat{\sigma}_1^2/\widehat{\sigma}_2^2$ follows F-distribution with $n_1 - 2$ and $n_2 - 2$ df if $\sigma_1^2 = \sigma_2^2$. To test the null hypothesis we appeal to the concept of the normal range: if the null hypothesis were true, $\widehat{\sigma}_1^2/\widehat{\sigma}_2^2$ would be large, so find minimum value ν_L such that $\widehat{\sigma}_1^2/\widehat{\sigma}_2^2 > \nu_L$ for all $\sigma_1^2 > \sigma_2^2$ with probability $1-\alpha$. The interval (ν_L, ∞) is declared the natural range and if the observed value $\widehat{\sigma}_1^2/\widehat{\sigma}_2^2 \leq \nu_L$ the null hypothesis is rejected. Since ν_L is computed as `qf(0.05,13-2,9-2)=0.3319689` and $\widehat{\sigma}_1^2/\widehat{\sigma}_2^2 = 1.9/2.6 = 0.731$, we cannot reject the null hypothesis that $\sigma_1^2 > \sigma_2^2$ with type I error 5%. The p-value is computed as `pf(1.9/2.6,13-2,9-2)=0.3079748`.

7.6.3 Binomial proportion

Probability testing with Bernoulli trials is one of the oldest and most common statistical tasks. As before, we may be interested in one- or two-sided test having one or two samples. Two fundamentally different approaches are used: (i) asymptotic tests (the sample size is large) are reduced to a normal distribution and (ii) inverse-cdf test, such as the Clopper-Pearson test, is supposed to work for finite sample size. However neither of the tests yields the nominal type I error because of the distribution discreteness – see Theorem 2.71. There is a fundamental problem with discrete distributions – for a given α, such as $\alpha = 0.05$, no test of that size necessarily exists.

One-sample test

We want to test the two-sided null hypothesis $H_0 : p = p_0$ versus $H_A : p \neq p_0$ from n iid Bernoulli trials, where p is the probability of individual success, $\Pr(X_i = 1) = p$. The maximum likelihood estimator is the binomial proportion, $\widehat{p} = \sum_{i=1}^{n} X_i/n$ and, therefore, as follows from the maximum likelihood theory under the null hypothesis, we can approximate the distribution as $\widehat{p} \simeq \mathcal{N}(p_0, p_0(1 - p_0)/n)$. Thus, the null hypothesis is rejected if $|\widehat{p} - p_0| > q_{1-\alpha/2}\sqrt{p_0(1 - p_0)/n}$. This test will be referred to as the *null Wald test* because the variance of the binomial proportion is estimated under the null hypothesis, $\text{var}(\widehat{p}) \simeq p_0(1-p_0)/n$. Clearly, this test can be viewed as the Z-test because the distribution is assumed normal and the variance is fixed and known as well. In the *estimated Wald test*, the binomial proportion itself is used to estimate the variance. Therefore, we write $\widehat{p} \simeq \mathcal{N}(p_0, \widehat{p}(1 - \widehat{p})/n)$ which leads to the rejection rule $|\widehat{p} - p_0| > q_{1-\alpha/2}\sqrt{\widehat{p}(1 - \widehat{p})/n}$.

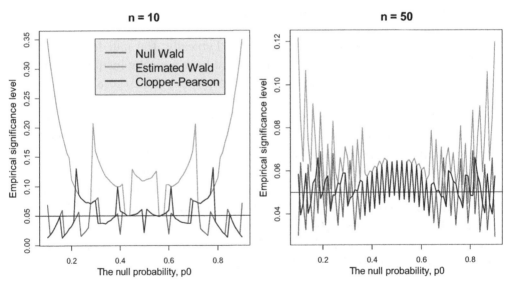

Figure 7.17: *Example 7.37: The type I error from simulations of the null and estimated Wald tests. These simulations illustrate that even an "exact" test has the type I error different from the theoretical α.*

Example 7.37 *Simulations for the Wald test.* *Use simulations to investigate how well the empirical significance level of the Wald test matches the nominal significance level, α.*

Solution. The R function is found in file `binprop1.r`. The true p_0 takes values on the grid from 0.01 to 0.09. Statistic $\sum_{i=1}^{n} X_i$ is generated as the number of successes in n Bernoulli outcomes using `rbinom` function (alternatively, but slower, one could generate a matrix of Bernoulli random variables with the number of rows equal to the number of simulations and the number of columns equal n, and then use `rowSums`). Two sample sizes, $n = 10$ and $n = 50$, are taken to illustrate the discrepancy between the nominal and empirical type I errors. We intentionally use a fairly small sample size because, for large n, the discrepancy will be negligible, see Figure 7.17. Neither of the tests produces the nominal $\alpha = 0.05$. The match is especially poor for the estimated Wald test at the ends of the interval $(0, 1)$.   □

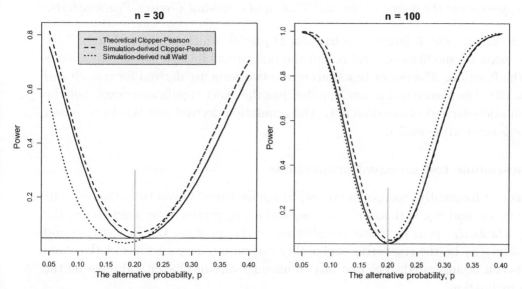

Figure 7.18: *The power functions for binomial proportion in the Wald and Clopper-Pearson tests ($\alpha = 0.05$ and $p_0 = 0.2$).*

The original *Clopper-Pearson test* for binomial proportion can be viewed as the inverse-cdf test and uses the fact that the cdf of the number of successes in n Bernoulli trials is the cdf of the beta distribution, see Section 2.14. For our purpose, we will approximate the cdf of the binomial proportion, $\widehat{p}$, as

$$\Pr(\widehat{p} \leq x) \simeq 1 - \mathcal{B}(p; xn + 0.5, n(1 - x) + 0.5),$$

where 0.5 is used as a continuity correction. Unlike the binomial cdf the beta distribution is continuous which is convenient from a numerical standpoint such as application of Newton's algorithm. However, such "smoothing" does not eliminate the fundamental problem of finding the test with exact type I error.

The one-sided test is easy to apply: if $H_0 : p \leq p_0$ and $H_A : p > p_0$, according to Theorem 7.31, the null hypothesis is rejected if $\mathcal{B}(p_0; \widehat{p}_{\text{obs}} n + 0.5, n(1 - \widehat{p}_{\text{obs}}) +$

$0.5) < \alpha$, where the left-hand side is the p-value. The following is an illustration of this test applied to Example 7.4.

Example 7.38 *Boy or girl continued. Compute the p-value of the Clopper-Pearson test with the null hypothesis $H_0 : p \leq 0.5$ versus the alternative $H_A : p > 0.5$, where p is the probability that a newborn is a boy.*

Solution. Since $\widehat{p}_{\text{obs}} = 1 - 0.491445 = 0.50855$ and $n = 33666$ we compute the p-value as `pbeta(0.5,0.50855*33666+0.5,33666*(1-0.50855)+0.5)=0.00085`, very close to the 0.000831 derived in Example 7.4.

Example 7.39 *Simulations for the Clopper-Person test. Use simulations to approximate the power of the null Wald and equal-tail Clopper-Pearson tests.*

Solution. The R function is `binprop1(job=2)` which produces Figure 7.18. The nominal significance level α and the null probability p_0 are the arguments of the function. The power functions of the two tests are derived for $n = 30$ and $n = 100$. The theoretical power function has the exact significance level, but the simulation-derived power does not. The simulation-derived null Wald test power deteriorates with small n.

Two-sample test on equal proportions

This test frequently emerges in statistical applications. Given two binomial outcomes, m_1 and m_2, with sample sizes n_1 and n_2, respectively, we want to test the hypothesis $H_0 : p_1 = p_2$ versus the alternative $H_0 : p_1 \neq p_2$. Let $\widehat{p}_1 = m_1/n_1$ and $\widehat{p}_2 = m_2/n_2$ be the respective probability estimates. Two versions of the Wald tests can be suggested. The first has an unequal variance flavor and relies on the approximation

$$\frac{\widehat{p}_1 - \widehat{p}_2}{\sqrt{\widehat{p}_1(1 - \widehat{p}_1)/n_1 + \widehat{p}_2(1 - \widehat{p}_2)/n_2}} \sim \mathcal{N}(0, 1),$$

and the second has an equal-variance favor and relies on the approximation

$$\frac{\widehat{p}_1 - \widehat{p}_2}{\sqrt{\widehat{p}(1 - \widehat{p})(1/n_1 + 1/n_2)}} \sim \mathcal{N}(0, 1),$$

where $\widehat{p} = (m_1 + m_2)/(n_1 + n_2)$ is the common probability estimator. In both cases, the hypothesis is rejected with the type I error α if the observed test statistic (the left-hand side) falls outside the interval $(-q_{1-\alpha/2}, q_{1-\alpha/2})$. The one-sided test is obvious.

Example 7.40 *Equal- versus unequal-variance test. Use simulations to estimate (a) the empirical the type I error and (b) the two-dimensional power functions for the unequal and equal-variance approximation.*

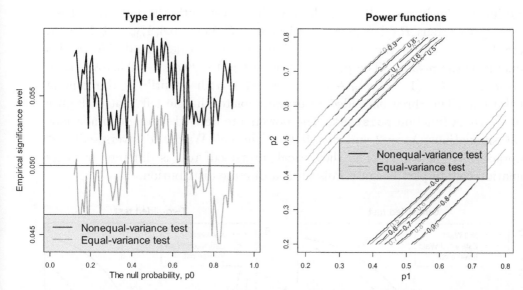

Figure 7.19: *The simulation derived type I error (left) (nominal $\alpha = 0.05$) and the level sets of the power as a function of $n_1 = 50$ and $n_2 = 40$ for the nonequal and equal Wald tests (right).*

Solution. Figure 7.19 is produced by the R function `binprop2` with the sample sizes $n_1 = 50$ and $n_2 = 40$ as default arguments. The type I error of the equal-variance Wald test is closer to the nominal level and the power functions have similar values.

7.6.4 Poisson rate

Given n iid Poisson-distributed observations, we want to test the rate parameter, λ. The Wald test uses the fact that $Y = \sum_{i=1}^{n} Y_i$ follows a Poisson distribution with parameter λn. Therefore, under the null hypothesis $\lambda = \lambda_0$ for the CLT, we have

$$\frac{Y - \lambda_0 n}{\sqrt{\lambda_0 n}} \simeq \mathcal{N}(0, 1), \quad n \to \infty. \tag{7.26}$$

Similarly to the binomial proportion, we can refer to this test as the null Wald test because the variance is evaluated at the null value. In the estimated Wald test $\lambda_0 n$ in the denominator is replaced with Y as an estimator of $\lambda_0 n$. As in the previous problem of binomial proportion, this test can be viewed as the Z-test because the distribution is assumed to be normal and the variance $\sqrt{\lambda_0 n}$ is fixed and known.

The inverse-cdf test uses the fact that, for finite n, the cdf of the Poison distribution can be better approximated by the gamma distribution (2.34) than by the normal distribution as indicated in (7.26). A continuity adjustment may be applied – then λ in the gamma cdf is replaced with $\lambda + 0.5$. For example, according to the inverse-cdf test, the one-sided test with the null hypothesis $H_0 : \lambda \le \lambda_0$ versus $H_A : \lambda > \lambda_0$ is rejected if $F(\lambda_0 n + 0.5; Y_{\mathrm{obs}} + 1, 1) < \alpha$, where F is the cdf of the gamma distribution and the left-hand side of this inequality is the p-value. The power of this test is $1 - F(q_{1-\alpha} + 0.5, n\lambda + 1, 1)$, where $q_{1-\alpha} = F^{-1}(1 - \alpha, n\lambda_0 + 1, 1)$ is the $(1 - \alpha)$th quantile of the gamma distribution. The three one-sided tests are depicted in the left plot of Figure 7.20; see the R function `poistest`. The powers are derived through simulations (100K simulations, $\lambda_0 = 2$ and $\alpha = 0.05$). The null Wald and inverse-cdf test are practically the same. The theoretical inverse-cdf test is different from its simulation counterpart because Poisson is a discrete distribution.

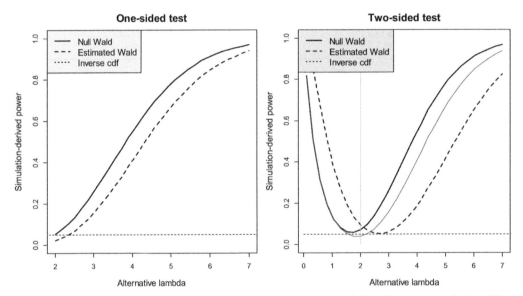

Figure 7.20: *The simulation-derived power functions of the three tests derived by function* `poistest` *with default arguments* $n = 2$, $\lambda_0 = 2$, *and* $\alpha = 0.05$.

Now we discuss the two-sided test $H_0 : \lambda = \lambda_0$ versus $H_A : \lambda \ne \lambda_0$. As follows from Theorem 7.31, the null hypothesis is rejected if either $F(\lambda_0 n + 0.5; Y_{\mathrm{obs}} + 1, 1) < \alpha_1$ or $F(\lambda_0 n + 0.5; Y_{\mathrm{obs}} + 1, 1) > 1 - \alpha_2$, where F is the cdf of the gamma distribution and positive α_1 and α_2 are such that $\alpha_1 + \alpha_2 = \alpha$, the nominal significance level. The theoretical power of the test is $1 - F(n\lambda, q_1 + 1, 1) + F(n\lambda, q_2 + 1, 1)$. For example, in the equal-tail probability test, we set $\alpha_1 = \alpha_2 = \alpha/2$. The unbiased inverse-cdf test requires the solution of the following system of equations for the gamma quantiles $q_1 < q_2$:

$$F(n\lambda_0 + .5, q_1 + 1, 1) - F(n\lambda_0 + .5, q_2 + 1, 1) = 1 - \alpha,$$
$$(q_1 - q_2)\ln(n\lambda_0 + .5) - (\ln \Gamma(q_1 + 1) - \ln \Gamma(q_2 + 1)) = 0. \qquad (7.27)$$

To solve this system, we employ the `nls` function to minimize the squared difference between the left- and right-hand sides of the above equations. More detail on this function can be found in Chapter 9 – see function `poistest`. Note that, for the Poisson distribution, the power depends on $n\lambda_0$. It can be explained by the fact that the sum of n iid Poisson-distributed random variables with the rate λ is a Poisson-distributed random variable with the rate $\nu\lambda$. The right plot of Figure 7.20 depicts three powers approximated by simulations and the theoretical power of the inverse-cdf test. As in the one-sided test, the null and inverse-cdf tests are practically the same. The theoretical power function may be used for sample size determination as illustrated in the following example.

Example 7.41 *Poisson sample size determination. A European country is concerned with low birthrate with $\lambda = 1.3$ measured as the average number of children per family. To boost the birthrate, government offered financial support to new families. Two years after the initiative, a family survey is conducted. Assuming that the number of children per family follows a Poisson distribution, what should be the minimum number of families to survey, n, to detect the difference in the birthrate as low as 0.1 with the type I error $\alpha = 0.05$ and the power of detection 90%. Use the null Wald and the inverse cdf two-sided tests to find n.*

Solution. The null Wald test can be treated as the Z-test with the power function, similar to (7.8), given by $P_w(\lambda, n) = \Phi(-s(\lambda) - q_{1-\alpha/2}) + \Phi(s(\lambda) - q_{1-\alpha/2})$ where $s(\lambda) = \sqrt{n}(\lambda - \lambda_0)/\sqrt{\lambda_0}$ is the sample effect size. The required sample size is found from solution of the equation $P_w(1.4, n) = 0.9$. The sample size based on the unbiased inverse-cdf test is found from the equation $P_{icdf}(\lambda, n) = 0.9$, where $P_{icdf}(\lambda, n) = 1 - F(1.4n, q_1+1, 1) + F(1.4n, q_2+1, 1)$ and q_1 and q_2 are the solution of system (7.27). Our R code is found in file `poissamn.r`. This function has the arguments with the default values `lambda0=1.2`, `mindet=.1`, `alpha=.05`, `pow=.9`. The two power functions are plotted against n in the range from 20 to 2000 (not shown). The required sample sizes for the null Wald and the unbiased inverse-cdf test are $n = 1261$ and $n = 1313$, respectively. Formally, since the double-side test is used, we must run the power analysis with the minimum detectable difference `mindet=-.1`. However, the theoretical power function P_w is symmetric around λ_0 and the power function P_{icdf} is almost symmetric due to large $n\lambda_0$. Therefore the sample sizes do not change. We emphasize that the required sample size was computed based on the theoretical powers. A practically more reliable method is based on simulations although it would require considerable computation time for a large number of simulations.

Problems

1. Derive the power function of the test in Example 7.29 and carry out simulations to test it.

2. The one-sided hypothesis $H_0 : \lambda_0 < 2$ versus $H_A : \lambda \geq 2$ is rejected if $Y \geq 3$ where Y is a single Poisson-distributed observation. Find the type I error

and the power function. Plot the power functions for $\lambda \geq 2$ using `ppois` and based on the gamma approximation.

3. (a) Modify function `poissamn` from Example 7.41 to compute the required sample size for the one-sided alternative $H_A : \lambda > 1.4$. (b) Prove that the one-sided alternative yields a smaller sample size than the two-sided alternative.

4. Modify function `poissamn` from Example 7.41 to plot the required sample size as a function of the minimum detectable difference of rates for the two tests with one- and two-sided alternatives (use `mfrow=c(1,2)`).

5. Use simulations in Example 7.41 to test how theoretical and simulation-derived sample sizes are different by plotting the two power functions on the same graph. [Hint: Modify functions `poistest` and `poissamn`.]

6. (a) Use function `vartestSP` from Example 7.35 to test the variance equality of another pair of stocks. Compute the normal range and justify computation of the p-value. (b) Test the hypothesis on the equality of two regression variances of two stocks from regressing the log return on time as in Example 7.36.

7. The R code `vartestSP2` plots recent stock prices for GOOGLE and AMAZON. Modify this code to test the hypothesis that the variance of stock returns on the log scale is the same.

7.7 Testing for correlation coefficient

The Pearson correlation coefficient, r, is the major statistical parameter that characterizes the relationship between two random variables given n iid observation pairs (X_i, Y_i). Although the correlation coefficient can be computed for any paired data the original interpretation of the squared correlation coefficient/coefficient of determination, as the proportion of variance of Y explained by X, is valid only for bivariate normal distributions. We recommend the reader to review Section 3.5 and particularly Section 3.5.2 to refresh the notion of the population correlation coefficient, ρ, in the framework of the bivariate normal distribution and its sample counterpart, r, discussed in Section 6.6.

Ronald Fisher suggested a simple transformation of r that makes the distribution close to normal:

$$\frac{1}{2} \ln \frac{1+r}{1-r} \simeq \mathcal{N}\left(\frac{1}{2} \ln \frac{1+\rho}{1-\rho}, \frac{1}{n-3}\right), \quad n \to \infty. \tag{7.28}$$

The Fisher Z-transformation $\vartheta(u) = 0.5 \ln[(1+u)/(1-u)]$ produces an approximation accurate even for moderate n. Testing of the one-sided or two-sided

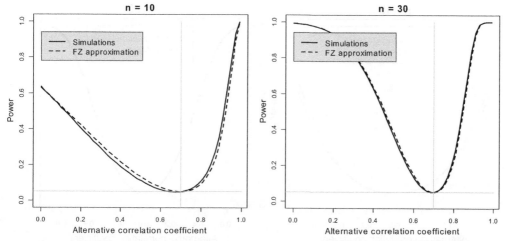

Figure 7.21: *The simulation-derived power and function (7.29) for testing H_0 : $\rho = 0.7$ against $H_A : \rho \neq 0.7$. This plot is generated by the R function* corn0.

hypothesis with the help of approximation (7.28) is straightforward (this test will be called the FZ-test). For example, if Y denotes the observed left-hand side of (7.28) then the p-value for testing the hypothesis $H_0 : \rho = \rho_0$ versus $H_A : \rho \neq \rho_0$ is computed as $2\Phi(-\sqrt{n-3}\,|Y - \vartheta(\rho_0)|)$. Clearly, there is a parallel between this test and the Z-test studied in detail in Section 7.3 because the variance, $1/(n-3)$, is known. For example, the approximate power of the FZ-test takes the form

$$P_{FZ}(\rho; \rho_0) = 1 - \Phi\left(\sqrt{n-3}(\vartheta(\theta) - \vartheta(\theta_0)) + q_{1-\alpha/2}\right)$$
$$+ \Phi\left(\sqrt{n-3}(\vartheta(\theta) - \vartheta(\theta_0)) - q_{1-\alpha/2}\right), \qquad (7.29)$$

where $q_{1-\alpha/2}$ is the $(1 - \alpha/2)$th quantile of the standard normal cdf. Applying the back transformation, we infer that the zero-correlation hypothesis ($\rho_0 = 0$) is not rejected with the type I error α if

$$\frac{e^{-2q_{1-\alpha/2}/\sqrt{n-3}} - 1}{e^{-2q_{1-\alpha/2}/\sqrt{n-3}} + 1} < r < \frac{e^{2q_{1-\alpha/2}/\sqrt{n-3}} - 1}{e^{2q_{1-\alpha/2}/\sqrt{n-3}} + 1}. \qquad (7.30)$$

Example 7.42 *Simulations for the FZ-test. Write an R function that compares the simulation-derived power for the FZ-test with its approximation given by (7.29) for $n = 10$ and $n = 30$ with $H_0 : \rho = 0.7$ against $H_A : \rho \neq 0.7$.*

Solution. The code is found in file corn0.r which produces Figure 7.21. Vectorized simulations are used to compute 100K simulated correlation coefficients as a vector. The simulation-derived power and its FZ approximation power given by (7.29) are very close. $\qquad \square$

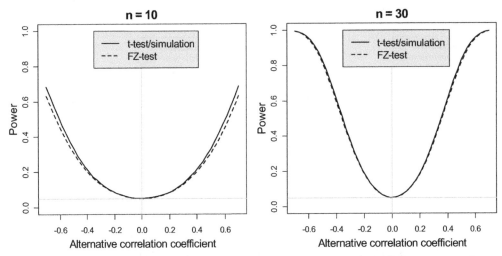

Figure 7.22: *The simulation-derived power and function (7.29) for testing the no correlation hypothesis $H_0 : \rho = 0$ against $H_A : \rho \neq 0$. This plot is generated by the R function corO from Example 7.43.*

For the zero-correlation hypothesis, $H_0 : \rho = 0$ an exact test exists based on fact that under the null ($\rho = 0$):

$$\frac{r\sqrt{n-2}}{\sqrt{1-r^2}} \sim t_{n-2}. \tag{7.31}$$

Using this fact, we reject the zero-correlation hypothesis if $|r|\,\sqrt{n-2}/\sqrt{1-r^2} > \mathcal{T}^{-1}(1-\alpha/2; n-2)$, where $\mathcal{T}$ is the cdf of the central t-distribution. Moreover, as follows from Section 4.3.1, the distribution of the left-hand side of (7.31) can be well approximated by the noncentral t-distribution:

$$\frac{r\sqrt{n-2}}{\sqrt{1-r^2}} \sim t_{n-2}\left(n-2, \text{ncp} = \frac{\rho\sqrt{n-2}}{\sqrt{1-\rho^2}}\right). \tag{7.32}$$

Thus the power of the zero-correlation test is approximated as

$$P_t(\rho_0; \rho) = 1 - \mathcal{T}(q_{1-\alpha/2}; n-2, \text{ncp} = \nu(\rho)) + \mathcal{T}(-q_{1-\alpha/2}; n-2, \text{ncp} = \nu(\rho)), \tag{7.33}$$

where $\nu(\rho) = \rho\sqrt{n-2}/\sqrt{1-\rho^2}$ is the noncentrality parameter.

Example 7.43 ***Power for the FZ-test.*** *Write an R function which computes and graphs the power functions for the FZ- and t-test given by (7.33) for testing the zero-correlation hypothesis and superimpose with the simulation-derived power with $n = 10$ and $n = 30$.*

Solution. The code is found in file corO.r which produces Figure 7.22. Vectorized simulations are used to compute 100K simulated correlation coefficients as a vector. The FZ-test and the exact t-test are very close.

Problems

1. Plot the bounds of statistical significance of the correlation coefficient given by (7.30) versus $n = 5, 6, ..., 20$ using α as an argument of the R function with the default value 0.05. Use `polygon` with `col="gray90"` to fill the bounds.

2. Derive the density of r using (7.31). Check you derivation by plotting the theoretical cdf versus simulation-based cdf on the same graph for $n = 10$ and $n = 30$. [Hint: Consult Section 4.3 and use `integrate` to compute the theoretical cdf of r when $\rho = 0$; see Section 2.9.2.]

3. (a) Compute $E(r)$ using function `integrate` based on approximation (7.28) on a grid of ρ values as in Figure 7.22. (b) Plot the theoretical bias and the bias obtained from simulations as in Example 6.46 for $n = 10$ and $n = 30$.

4. (a) Prove that the zero-correlation test based on the t-distribution is unbiased. (b) Use simulations to demonstrate that the test of $H_0 : \rho = \rho_0$, where $\rho_0 \neq 0$ with the power function given by (7.33), is biased. [Hint: Modify function `corn0` and plot the simulation-derived power in close proximity to ρ_0.]

5. What sample size is required to reject the null hypothesis $H_0 : \rho = 0.25$ versus the double-sided alternative with type I and II errors 0.05 and 0.1, respectively? Use simulations to check your answer. [Hint: Modify function `corn0`.]

6. Plot the three power functions to test $H_0 : \rho = 0.5$ versus $H_A : \rho \neq 0.5$ with $n = 8$.

7. Similarly to Example 7.42, derive and plot the power functions for testing the null hypothesis $\rho \geq \rho_0$ versus $\rho < \rho_0$, as in Figure 7.21.

8. (a) Similarly to Example 7.42, derive and plot the power function for testing the null hypothesis regarding the coefficient of determination, r^2. (b) Use the fact that $t^2 \sim F$ to derive the distribution of r^2 under the zero-correlation hypothesis.

7.8 Confidence interval

Conceptually speaking, there is a problem with point estimation. Indeed, let for example, the average annual income in Vermont be $43,324.73 (either the arithmetic average or median – it does not matter at this point). Does it mean that the income of a random person in Vermont is $43,324.73? No. Moreover, the probability that a randomly asked person has annual income $43,324.73 is zero. It may turn out that nobody's income in Vermont is \$43,324.73. What is \$43,324.73, and how does one interpret it?

To obtain a more sensible estimate of income, one has to speak in terms of the *confidence interval (CI)* that covers the true unknown μ with given probability. For example, one may find I_L and I_U such that interval (I_L, I_U) covers μ with probability 0.95. Thus, we come to the idea of the *interval estimation*. The lower I_U and the upper I_U limits of the CI are random variables because they are functions of data, that is, statistics. We emphasize that in this section we derive the confidence interval for a parameter. The confidence range, introduced in Section 2.2.2, defines the interval for the random variable from the general population. Roughly speaking, the confidence range is $2\sqrt{n}$ wider than the confidence interval.

Before construction of a CI, one should specify the probability or the *confidence level*, λ. For example, if I is the true annual income in Vermont a $100\lambda\%$ confidence interval means that I_L and I_U contain I with probability λ, or in mathematical terms

$$\Pr(I_L \leq I \leq I_U) = \lambda.$$

Now, it is clear why we add the word "confidence" to interval estimation. Although many people understand that a range of values (confidence interval) better describes the distribution than just one value, like mean or median, the confidence level is rarely given in popular literature when reporting the range of the distribution. Unfortunately, without specifying the confidence level the range does not mean much. Sometimes, the mean $\pm$ standard error is reported, and sometimes the mean $\pm$ SD. In this sense, the CI is more precise because it tells the coverage probability. Now we are ready for a formal definition of CI.

Definition 7.44 *Let $X_1, ..., X_n$ be a random sample from a general population with unknown parameter θ. We say that $(\widehat{\theta}_L, \widehat{\theta}_U)$ is a $100\lambda\%$ double-sided confidence interval (CI) for the true unknown θ if*

$$\Pr(\widehat{\theta}_L \leq \theta \leq \widehat{\theta}_U) = \lambda, \tag{7.34}$$

where λ is called the confidence level and the probability is computed at the true parameter θ. Lower and upper one-sided confidence intervals are defined as $\Pr(\theta \leq \widehat{\theta}_L) = \lambda$ and $\Pr(\theta \geq \widehat{\theta}_U) = \lambda$, respectively.

Sometimes, we use $\alpha = 1 - \lambda$ as the significance level, or the probability that the CI does not contain the true value θ. To underscore that the coverage probability is exactly λ we say that the CI is *exact*. In other words, $\widehat{\theta}_L$ and $\widehat{\theta}_U$ are referred to as the lower and upper confidence bounds for θ.

There is a close relationship between hypothesis testing and confidence intervals. For example, in Section 7.6, we used the pivotal quantity and its variant, inverse cdf, for hypothesis testing. The same method applies to construct CI: Let $P = P(X_1, ..., X_n; \theta)$ be a pivotal quantity. That is, its cdf, F, does not depend on θ. The following method can be used to construct a CI given λ.

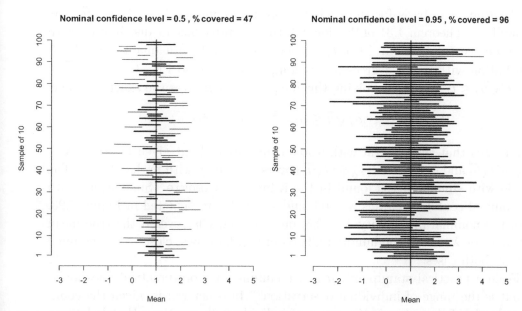

Figure 7.23: *Simulations with the 95% CI for μ. This plot is generated by the R code* CImovie. *The bold segments are used when the CI contains the true μ.*

1. Find quantiles q_L and q_U such that $F(q_U) - F(q_L) = \lambda$.

2. Solve equations $P(X_1, ..., X_n; \theta) = q_L$ and $P(X_1, ..., X_n; \theta) = q_U$ to find $\widehat{\theta}_L$ and $\widehat{\theta}_U$ that imply (7.34), provided that P is an increasing function of θ for any $X_1, ..., X_n$ (if P is a decreasing function exchange $\widehat{\theta}_L$ and $\widehat{\theta}_U$).

Obviously this method can be applied to construct a one-sided CI as well, as illustrated in the following example.

Example 7.45 *CI for uniform distribution.* *Find the lower CI for θ based on n iid observations from $\mathcal{R}(0, \theta)$. Write R code to check the coverage probability through simulations.*

Solution. Let λ be a given confidence level, $0 < \lambda < 1$. The pivotal quantity is $X_{\max}/\theta$ which has cdf x^n. Therefore, in Step 1, we have $\Pr(X_{\max}/\theta \leq \sqrt[n]{\lambda}) = \lambda$. In Step 2, we solve for θ to obtain the lower CI $\theta \geq X_{\max}/\sqrt[n]{\lambda}$. The CI $(X_{\max}/\sqrt[n]{\lambda}, \infty)$ covers the true population parameter with probability λ. The R code is found in file ciumax.r. The R function has the following default arguments: theta=1,n=5,lambda=.95,nSim=1000000. The call ciumax() returns 0.949886.

Example 7.46 *CI for the mean of a normal distribution.* *Given $X_i \overset{\text{iid}}{\sim} \mathcal{N}(\mu, \sigma^2)$, construct the $100\lambda\%$ CI for the mean. Write R code to simulate the coverage of μ.*

Solution. The basis for the CI is the pivotal quantity $(\overline{X} - \mu)/(\widehat{\sigma}/\sqrt{n})$. As follows from Theorem 4.31 of Section 4.3, this quantity has a t-distribution with $n - 1$ degrees of freedom. Let $t_{1-\alpha/2}$ be the $0.5(1 + \lambda)$th quantile of the t-distribution with $n - 1$ degrees of freedom, $\alpha = 1 - \lambda$, so $\Pr(-t_{1-\alpha/2} \leq (\overline{X} - \mu)/(\widehat{\sigma}/\sqrt{n}) \leq t_{1-\alpha/2}) = \lambda$. Solving this inequality for μ gives the exact coverage:

$$\Pr(\overline{X} - t_{1-\alpha/2}\widehat{\sigma}/\sqrt{n} \leq \mu \leq \overline{X} + t_{1-\alpha/2}\widehat{\sigma}/\sqrt{n}) = \lambda.$$

This means that the interval with a lower and upper limit given by $\widehat{\theta}_L = \overline{X} - t_{1-\alpha/2}\widehat{\sigma}/\sqrt{n}$ and $\widehat{\theta}_U = \overline{X} + t_{1-\alpha/2}\widehat{\sigma}/\sqrt{n}$ covers the true μ with probability λ. The R code with simulations is found in the R function `CImovie`. Simulations with the nominal confidence levels $\lambda = 0.5$ and $\lambda = 0.95$ are depicted in Figure 7.23. With the nominal probability $\lambda = 0.5$, one hundred CIs covered the true mean 47 times, for $\lambda = 0.95$ one hundred CIs covered μ 96 times, close to the nominal value in both cases. $\qquad\square$

From a practical standpoint, μ is a phantom. A more practical question is "What is the range of individual observation?" In other words, given the confidence level $100\lambda\%$, what is the range of individual observations that belong to this range with probability λ. The problem of the tight confidence range was introduced in Section 2.2.2. For symmetric distributions, such as normal, the tight confidence range leads to a symmetric interval around $\overline{X}$.

Example 7.47 *Confidence range for a normal distribution.* *Construct the $100\lambda\%$ confidence range given a random sample from a normal distribution with unknown mean and variance.*

Solution. We want to find X_L and X_U such that $\Pr(X_L < X < X_U) = \lambda$ where $\{X_i, i = 1, ..., n\}$ and X are iid from $\mathcal{N}(\mu, \sigma^2)$. We prove that

$$X_L = \overline{X} - t_{(1+\lambda)/2}\widehat{\sigma}\sqrt{1 + 1/n}, \quad X_U = \overline{X} + t_{(1+\lambda)/2}\widehat{\sigma}\sqrt{1 + 1/n}.$$

Indeed, using the suggested formulas, $\Pr(X_L < X < X_U)$ can be written as

$$\Pr(\overline{X} - t_{(1+\lambda)/2}\widehat{\sigma}\sqrt{1 + 1/n} < X < \overline{X} + t_{(1+\lambda)/2}\widehat{\sigma}\sqrt{1 + 1/n})$$
$$= \Pr\left(-t_{(1+\lambda)/2} < \frac{\sigma^{-1}(X - \overline{X})/\sqrt{1 + 1/n}}{\sigma^{-1}\widehat{\sigma}} < t_{(1+\lambda)/2}\right) = \lambda$$

because $\sigma^{-1}(X - \overline{X})/\sqrt{1 + 1/n} \sim \mathcal{N}(0, 1)$ and $(\sigma^{-1}\widehat{\sigma})^2 \sim \chi^2(n-1)$ and, therefore, the random variable in the inequality bracket is t-distributed with $n-1$ degrees of freedom. Roughly, the confidence range is $2\sqrt{n}$ wider than CI for μ, and one may say that the 95% confidence range is $\pm$ twice standard deviation. We emphasize the difference between the confidence range and classic CI in Example 7.46: the former is constructed to cover the individual value of an observation, but the latter is constructed to cover an unknown mean. This illustrates the difference between parameter- and individual-based statistical inference.

Example 7.48 *CI for the difference of the means.* *Let* $X_i \overset{\text{iid}}{\sim} \mathcal{N}(\mu_1, \sigma^2)$ *for* $i = 1, 2, ..., m$ *and* $Y_j \overset{\text{iid}}{\sim} \mathcal{N}(\mu_2, \sigma^2)$ *for* $j = 1, 2, ..., n$ *be independent random samples (variance in both samples is the same). Construct the exact* $100\lambda\%$ *CI for* $\mu_1 - \mu_2$.

Solution. The pivotal quantity is

$$\frac{(\overline{X} - \overline{Y}) - (\mu_1 - \mu_2)}{\widehat{\sigma}\sqrt{1/m + 1/n}} \sim t(m + n - 2),$$

where $\widehat{\sigma}^2$ is the pooled variance (7.16). Therefore the $100\lambda\%$ CI for $\mu_1 - \mu_2$ is $\overline{X} - \overline{Y} \pm \widehat{\sigma}\sqrt{1/m + 1/n} \times t_{1-\alpha/2}$, where $t_{1-\alpha/2}$ is the $0.5(1 + \lambda)$th quantile of the t-distribution with $m + n - 2$ degrees of freedom. $\square$

The following example illustrates how to find the CI when the cdf can be approximated by a normal distribution. See Definition 7.10 for the effect size.

Example 7.49 *CI for sample effect size.* *Find the CI for the sample effect size,* $s = \sqrt{n}\mu/\sigma$, *using observations* $X_i \overset{\text{iid}}{\sim} \mathcal{N}(\mu, \sigma^2)$, $i = 1, 2, ..., n$ *based on approximation (4.27).*

Solution. The test statistic $T = \sqrt{n}\overline{X}/\widehat{\sigma}$ seems to be a natural choice and is distributed as

$$T = \frac{\sqrt{n}(\overline{X} - \mu)/\sigma + \theta}{\widehat{\sigma}/\sigma} \sim t(n - 1, \text{ncp} = s),$$

the noncentral t-distribution with $n - 1$ df and noncentrality parameter s. In fact, Lehmann and Romano [69], p. 224, proved that this noncentral t-distribution leads to the uniformly most powerful (UMP) invariant test for μ/σ. From approximation (4.27), we have

$$\frac{T(1 - 1/(4(n - 1))) - s}{\sqrt{1 + T^2/(2(n - 1))}} \simeq \mathcal{N}(0, 1).$$

To simplify the CI, denote $Q = T(1 - 1/(4(n-1)))$ and $v = \sqrt{1 + T^2/(2(n-1))}$. Then, approximate $Q \simeq \mathcal{N}(s, v^2)$ and, therefore, an approximate $100\lambda\%$th confidence level CI for s is $Q \pm Z_{(1+\lambda)/2}v$. $\square$

The following example (a) illustrates how to derive a CI for a function of a parameter and (b) emphasizes the difference between the CI for the parameter and the confidence range for the individual value. More discussion on the difference between parameter and individual statistical inference is found in Section 7.10.

Example 7.50 *House price on the log scale.* *A sample of six recent selling houses in town is $348K, $297K, $734K, $298K, $503K, and $465K. Derive a 50% confidence interval for the median and a 50% confidence range for an individual house price in town. Write an R program to check if the coverage is 50% through vectorized simulations.*

Solution. It is reasonable to assume that the distribution of prices is lognormal: $\ln X_i \sim \mathcal{N}(\mu, \sigma^2)$, $i = 1, 2, .., 6$. An advantage of the lognormal distribution, besides addressing the heavy right tail, is that the median is e^μ; see Section 2.11. This means that we can estimate the median of the prices as the geometric mean $(\Pi X_i)^{1/n} = e^{\overline{\ln X}}$. The 50% CI for μ is $\overline{\ln X} \pm t_{1-\alpha/2}\widehat{\sigma}/\sqrt{n}$, where $n = 6$, $\alpha = 0.5$, $t_{1-\alpha/2}$ is the $(1 - \alpha/2)$th quantile of the t-distribution with $n - 1$ df, and $\widehat{\sigma}$ is the SD estimate of $\ln X_i$. Obviously, any increasing transformation, such as exp, maintains the confidence level and, therefore, the 50% CI for the median is $e^{\overline{\ln X} \pm t_{1-\alpha/2}\widehat{\sigma}/\sqrt{n}}$. The call `houprCI()` produces the following result: `The 50% CI for the median: (376,464)`. The 50% confidence range for the individual house price is $e^{\overline{\ln X} \pm t_{1-\alpha/2}\widehat{\sigma}}$ with the output `The 50% confidence range for the individual house price: (323,540)`. This means that if a buyer sees a house he/she likes there is a 50% chance that the price of the house is between $323K and $540K. The confidence interval for the median may be of interest to town or state officials, but the confidence range is more informative to an individual. Vectorized simulations confirm the 50% probability.

7.8.1 Unbiased CI and its connection to hypothesis testing

The unbiased CI is defined similarly to the unbiased test.

Definition 7.51 *Unbiased CI.* *The* $100\lambda\%$ *confidence interval* $(\widehat{\theta}_L, \widehat{\theta}_U)$ *is unbiased if*

$$\Pr(\widehat{\theta}_L \leq \theta \leq \widehat{\theta}_U | \theta_*) \leq \lambda. \tag{7.35}$$

We say that the CI is locally unbiased if

$$\frac{d}{d\theta} \Pr(\widehat{\theta}_L \leq \theta \leq \widehat{\theta}_U | \theta_*) \bigg|_{\theta=\theta_*} = 0. \tag{7.36}$$

Notation $|\theta_*$ means that the probability is evaluated at θ_* while the true parameter is θ. The interval $(\widehat{\theta}_L, \widehat{\theta}_U)$ is supposed to cover the true parameter θ but the coverage probability is computed at a different value of the parameter, θ_*. Of course, $\widehat{\theta}_L$ and $\widehat{\theta}_U$ do not depend on the true parameter and cover θ with probability λ, that is, $\Pr(\widehat{\theta}_L \leq \theta \leq \widehat{\theta}_U | \theta) = \lambda$. In words, inequality (7.35) says that for an unbiased CI, the probability of covering a wrong parameter value is smaller than that of covering the true parameter value. Obviously, if CI is unbiased, it is locally unbiased as follows from the simple fact that the derivative of the function is zero at the local or global minimum. Consequently, if CI is locally biased it cannot be unbiased. Local unbiasedness is fairly easy to check and instrumental because it reduces to examining the derivative of the coverage probability. Typically, if a CI is locally unbiased it is (globally) unbiased.

Example 7.52 *Unbiasedness of the CI for the mean.* *Prove that the CI for μ from Example 7.46 is (a) locally unbiased and (b) (globally) unbiased.*

Solution. Express probability (7.35), where θ and θ_* are replaced with μ and μ_*, as a function of $\delta = \sqrt{n}(\mu_* - \mu)/\sigma$, as we did for the power function (7.14),

$$C(\delta) = \Pr\left(\left|\frac{\sqrt{n}(\overline{Y} - \mu)}{\widehat{\sigma}}\right| < t_{1-\alpha/2}|\mu_*\right) = 1 - P(\delta) = E\Phi(U+\delta) - E\Phi(-U+\delta),$$

and take the derivative, $dC/d\delta = E\phi(U+\delta) - E\phi(-U+\delta)$. (a) $dC/d\delta$ evaluated at $\delta = 0$ is zero because $\phi(-U) = \phi(U)$, that is, the CI is locally unbiased. (b) Rewrite

$$\frac{dC}{d\delta} = \frac{1}{\sqrt{2\pi}} E\left[e^{-\frac{1}{2}(U+\delta)^2}\left(1 - e^{2U\delta}\right)\right].$$

Since U is a positive random variable $dC/d\delta < 0$ for $\delta > 0$ and $dC/d\delta > 0$ for $\delta < 0$. This means that inequality (7.35) holds for any μ_* and μ.

Theorem 7.53 *Duality between hypothesis testing and CI*. *Let the null hypothesis $H_0 : \theta = \theta_0$ versus the double-sided alternative $H_A : \theta \neq \theta_0$ be accepted with the type I error α if $T_L \leq \theta_0 \leq T_U$ for all θ_0, where T_L and T_U are random and do not depend on θ_0. Then (T_L, T_U) may serve as a CI and covers the true θ with probability $1 - \alpha$. Conversely, if $(\widehat{\theta}_L, \widehat{\theta}_U)$ is the CI with the confidence level $\lambda = 1 - \alpha$ then the null hypothesis $H_0 : \theta = \theta_0$ is rejected with the type I error α if θ_0 is outside the CI. Moreover if the hypothesis test is unbiased then the respective CI is unbiased as well.*

Proof. If the null hypothesis is accepted with the type I error α, we have $\Pr(T_L \leq \theta_0 \leq T_U|\theta_0) = \lambda$ for all θ_0. But this is exactly the definition of the CI for the true θ_0 with confidence level λ. Conversely, if $\Pr(\widehat{\theta}_L \leq \theta_0 \leq \widehat{\theta}_U|\theta_0) = \lambda$, we reject the hypothesis if θ_0 is outside $(\widehat{\theta}_L, \widehat{\theta}_U)$ at the type I error of α. The unbiasedness of the test with the acceptance condition $T_L \leq \theta_0 \leq T_U$ means that $\Pr(\widehat{\theta}_L \leq \theta_0 \leq \widehat{\theta}_U|\theta_*) \leq \Pr(\widehat{\theta}_L \leq \theta_0 \leq \widehat{\theta}_U|\theta_0) = \lambda$ for every θ_* and θ_0, which is exactly the definition of the unbiased CI. $\square$

An important condition is that the limits T_L and T_U do not depend on θ. An immediate application of Theorem 7.53 is that CIs for the mean and difference of the means derived in Examples 7.46 and 7.48 are unbiased because as it was shown in Examples 7.17 and 7.18, the power functions reach a minimum at the null hypothesis value.

7.8.2 Inverse cdf CI

This method may be used in the framework of pivotal quantity, but it is easier to derive the CI directly from the cdf similarly to hypothesis testing discussed in Section 7.6. Let statistic T have a cdf $F(t; \theta)$, where θ is a one-dimensional parameter. Suppose that $F(t; \theta)$ is continuous in both arguments and is a strictly decreasing function of θ for each t, the same assumptions we made for the inverse-cdf test. First, consider an equal-tail CI. Let $\widehat{\theta}_L$ and $\widehat{\theta}_U$ be the solutions of the equations $F(T; \theta) = 1 - \alpha/2$ and $F(T; \theta) = \alpha/2$, where T is the observed statistic T and $\lambda = 1 - \alpha$ is the desired confidence level. By convention, we let $\widehat{\theta}_L = -\infty$ and $\widehat{\theta}_U = +\infty$ if the respective equations do not have a solution. If θ belongs to a finite interval, we choose the lower and the upper bound, respectively. The interval $(\widehat{\theta}_L, \widehat{\theta}_U)$ covers the true value θ with probability λ. The proof is straightforward: since F is decreasing in θ, we have

$$\Pr\left(\alpha/2 \le F(T; \theta) \le 1 - \alpha/2\right) = \Pr\left(F(T; \widehat{\theta}_U) \le F(T; \theta) \le F(T; \widehat{\theta}_L)\right) = \lambda.$$

Since F is a strictly decreasing function of θ the above probability is equivalent to (7.34). This method of interval estimation is known for a long time and was used by David [22] who tabulated the exact confidence interval for the Pearson correlation coefficient long before the computer era.

Below is a direct application of the inverse cdf CI, see Zou [109].

Example 7.54 *CI for effect size.* *Apply the inverse cdf CI to the individual effect size in the one-sample t-test framework using approximation (4.27).*

Solution. In the one-sample t-test, we have $Y_i \overset{\text{iid}}{\sim} \mathcal{N}(\mu, \sigma^2)$, $i = 1, 2, ..., n$ with the null hypothesis $H_0 : \mu = \mu_0$. The individual effect size is defined as $c = (\mu - \mu_0)/\sigma$, see Definition 7.10. Since $T = \sqrt{n}(\overline{Y} - \mu_0)/\widehat{\sigma}$ follows a noncentral t-distribution with $\text{df} = n - 1$ and noncentrality parameter c, we can find the exact CI for c by solving equations $\mathcal{T}(T, \text{df} = n - 1, \sqrt{n}\widehat{c}_L) = 1 - \alpha/2$ and $\mathcal{T}(T, \text{df} = n - 1, \sqrt{n}\widehat{c}_U) = \alpha/2$. These equations may be solved numerically, but approximation (4.27) allows us to write the CI in closed form as $c_1 \pm c_2 q_{1-\alpha/2}$ where $c_1 = T\left(1 - 1/(4(n-1))\right)/\sqrt{n}$ and $c_2 = \sqrt{1 + T^2/(2(n-1))}q_{1-\alpha/2}/\sqrt{n}$. $\square$

Clearly, the coverage probability does not change when unequal-tail probabilities are used: $F(T; \theta) = 1 - \alpha_1$ and $F(T; \theta) = \alpha_2$, provided that $\alpha_1 + \alpha_2 = \alpha$. This observation gives rise to the minimum width CI as the solution to the following optimization problem: find $\theta_L < \theta_U$ such that

$$\min(\theta_U - \theta_L) \tag{7.37}$$

under restriction

$$F(T; \theta_U) - F(T; \theta_L) + \lambda = 0. \tag{7.38}$$

Optimization problem (7.37) can be reduced to a nonlinear equation using the Lagrange function

$$\mathcal{L}(\theta_L, \theta_U, \nu) = \theta_U - \theta_L - \nu\left[F(T; \theta_U) - F(T; \theta_L) + \lambda\right],$$

where ν is the Lagrange multiplier. Taking the derivatives with respect to θ_L and θ_U, we obtain

$$\frac{\partial \mathcal{L}}{\partial \theta_L} = -1 + \nu F_\theta'(T; \theta_L) = 0, \quad \frac{\partial \mathcal{L}}{\partial \theta_U} = 1 - \nu F_\theta'(T; \theta_U) = 0,$$

where F'_θ denotes the derivative of the cdf with respect to the parameter. The above conditions imply

$$F'_\theta(T;\theta_U) - F'_\theta(T;\theta_L) = 0, \tag{7.39}$$

as a replacement of the original optimization equation (7.37). The confidence interval $(\widehat{\theta}_L, \widehat{\theta}_U)$ with the lower and upper limits as solutions to equations (7.38) and (7.39) is referred to as the *shortest CI*. This interval is narrower than the equal-tail inverse cdf CI. For a simple type of cdf, the shortest CI has the exact confidence level and is unbiased as stated below.

Theorem 7.55 *Let the cdf of statistic T be expressed in the form $F(h(t)+\theta r(t))$, where $F = F(x)$ is a known cdf and r is a positive function. Then the inverse cdf shortest CI for θ has lower and upper CI limits given by $\widehat{\theta}_L = (q_L - h(T))/r(T)$ and $\widehat{\theta}_U = (q_U - h(T))/r(T)$, respectively. It has exact confidence level λ, where $q_L < q_U$ are the solutions to the following equations: $f(q_L) - f(q_U) = 0$ and $F(q_U) - F(q_L) + \lambda = 0$, where $f = dF/dx$ is the density.*

Proof. Let α_1 and α_2 be positive such that $\alpha_1 + \alpha_2 = \alpha = 1 - \lambda$. Then equations for the lower and the upper limits are $r(t)\theta_L + h(t) = q_L$ and $r(t)\theta_U + h(t) = q_U$, where q_L and q_U are respective quantiles such that $F(q_L) = 1 - \alpha_1$ and $F(q_U) = \alpha_2$. These equations allow one to rewrite the width of the CI as $\theta_U - \theta_L = (q_U - q_L)/r(T)$. This means that the problem of the shortest CI turns into an the optimization problem $\min(q_U - q_L)$ under restriction $F(q_U) - F(q_L) + \lambda = 0$, which does not depend on statistic T. Consequently, the coverage probability of this CI is λ. $\qquad\square$

Positiveness of function r is not critical: when r is a negative, quantiles q_L and q_U interchange. In another version of this theorem, we may have $F(h(x)+r(x)/\theta)$; then q_L and q_U interchange as well. The minimum-width CI applies below to the variance and SD of the normal distribution.

7.8.3 CI for the normal variance and SD

The CI for the normal variance closely follows the derivation of the inverse-cdf test from Section 7.5: if statistic $S = \sum_{i=1}^n (X_i - \overline{X})^2$ the equal-tail $(1-\alpha)100\%$ CI for σ^2 is $S/q_2 \leq \sigma^2 \leq S/q_1$, where $q_1 = \mathcal{C}_{n-1}(\alpha/2)$ and $q_2 = \mathcal{C}_{n-1}(1-\alpha/2)$ are the $\alpha/2$ and $1-\alpha/2$ quantiles of the chi-square distribution with $n-1$ df. The shortest CI for the normal variance was developed by Tate and Klett [98] based on the observation that optimal quantiles do not yield equal probability tails. Guenther (1972) reduced the CI for normal variance to the incomplete gamma function, however the computation algorithm was presented. Here, we derive the shortest CI for σ^2 following Lagrange multiplier technique from the previous section. Since the width of the interval $S/q_2 \leq \sigma^2 \leq S/q_1$ is proportional to $1/q_1 - 1/q_2$, we are looking for quantiles q_1 and q_2 such that $1/q_1 - 1/q_2$ takes minimum under the restriction $\mathcal{C}_{n-1}(q_2) - \mathcal{C}_{n-1}(q_1) = \lambda$ or, in the terms of Lagrange function,

$$\mathcal{L}(q_1, q_2, \nu) = \frac{1}{q_1} - \frac{1}{q_2} - \nu\left[\mathcal{C}_{n-1}(q_2) - \mathcal{C}_{n-1}(q_1) - \lambda\right],$$

where ν is the Lagrange multiplier. Differentiating $\mathcal{L}$ with respect to the unknowns we obtain

$$\frac{\partial \mathcal{L}}{\partial q_1} = -\frac{1}{q_1^2} + \nu c_{n-1}(q_1) = 0, \quad \frac{\partial \mathcal{L}}{\partial q_2} = \frac{1}{q_2^2} - \nu c_{n-1}(q_2) = 0,$$

where c_{n-1} denotes the density of the chi-square distribution, $c_{n-1} = \mathcal{C}'_{n-1}$. The system of equations is the same as (7.24) with the first equation replaced by $(n+1)(\ln q_2 - \ln q_1) - (q_2 - q_1) = 0$. Thus Newton's algorithm presented in Section 7.5 is easy to modify.

Example 7.56 *Shortest CI*. *Write an R program that computes the shortest CI quantiles using Newton's algorithm and compares the width of the interval with the equal tail CI.*

Solution. The width of the CI with equal-tail probabilities is proportional to $1/\mathcal{C}_{n-1}(\alpha/2) - 1/\mathcal{C}_{n-1}(1 - \alpha/2)$ and the shortest width is $1/q_1 - 1/q_2$, where q_1 and q_2 are the solutions of the system of equations solved by Newton's algorithm. The R code is found in file `civar.r` and the plot is shown in Figure 7.24. Since the width of the CI is proportional to the observed value of S, we show $1/q_1 - 1/q_2$ on the y-axis in the left plot. When n is small, say, $n < 6$ there is up to 30% width reduction when the optimal quantiles are used.

Example 7.57 *Shortest CI for SD*. *Modify the previous example to construct the shortest CI for the standard deviation, σ.*

Solution. Since the CI for σ is sought in the form $\sqrt{S/q_2} \leq \sigma \leq \sqrt{S/q_1}$, we want to minimize $1/\sqrt{q_1} - 1/\sqrt{q_2}$ under the same restriction $\mathcal{C}_{n-1}(q_2) - \mathcal{C}_{n-1}(q_1) = \lambda$. Simple algebra proves that equation $n(\ln q_2 - \ln q_1) - (q_2 - q_1) = 0$ should be used to optimize the width of the CI for σ. See `civar(job=2)`, the plot is not shown here. The shortest CI for σ reduces the width compared to the equal-tail CI by 20% for $n = 4$ and 5% for $n = 15$.

7.8.4   CI for other major statistical parameters

The confidence interval for binomial probability is built using the same normal approximation used for hypothesis testing and is derived by converting the test to the interval for the unknown probability, see Section 7.6.3. The estimated Wald test is easy to convert: if $\widehat{p}$ is the estimated probability, we convert the acceptance rule $|\widehat{p} - p| \leq q_{1-\alpha/2}\sqrt{\widehat{p}(1-\widehat{p})/n}$ to a CI for p as $\widehat{p} \pm q_{1-\alpha/2}\sqrt{\widehat{p}(1-\widehat{p})/n}$.

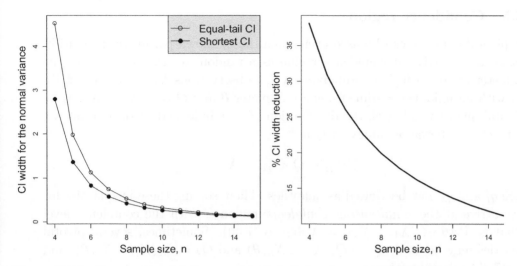

Figure 7.24: *Comparison of the equal-tail and shortest width 95% CI for normal variance, see Example 7.56. This figure is generated by* `civar(job=1)`.

A quadratic equation must be solved to convert the null Wald test with the acceptance rule $|\widehat{p} - p| \le q_{1-\alpha/2}\sqrt{p(1-p)/n}$ to an asymmetric CI

$$\widehat{p} - \frac{1}{2}\frac{q_{1-\alpha/2}}{Z^2_{1-\alpha/2} + n}\left(2\widehat{p}q_{1-\alpha/2} - q_{1-\alpha/2} \pm \sqrt{D}\right),$$

where $D = q_{1-\alpha/2} - 4\widehat{p}^2 n + 4\widehat{p}n$ is the discriminant. See the R function `ci.binpr` which does simulations for coverage probability of the two methods with default arguments `n=30`, `lambda=.95`, `nSim=1000000`. The coverage probability of the null Wald CI is consistently closer to the nominal value than that of the estimated Wald CI.

The CI for the Poisson rate follows the same idea: the estimated Wald CI is given by $\widehat{\lambda} \pm q_{1-\alpha/2}\sqrt{\widehat{\lambda}/n}$ and the null Wald CI (sometimes called the Wilson CI) is

$$\widehat{\lambda} - \frac{1}{2}\frac{q_{1-\alpha/2}}{n}\left(-q_{1-\alpha/2} \pm \sqrt{q_{1-\alpha/2} + 4n\widehat{\lambda}}\right).$$

The R function `ci.pois` investigates the coverage probability of the Poisson rate similarly to the previous R function for binomial probability. As before, the null Wald (Wilson) CI is preferable.

Finally, we quickly discuss the CI for correlation coefficient using the Fisher transformation. Converting the acceptance rule into a CI, we obtain that the true correlation coefficient, ρ, belongs to the interval $(AB - 1)/(AB + 1)$, where $A = (1 + r)/(1 - r)$ and $B = \exp(\pm 2q_{1-\alpha/2}/\sqrt{n - 3})$ approximately with probability $1 - \alpha$. Function `ci.cor` studies coverage probability through simulations: the above CI covers ρ with probability close to the nominal.

7.8.5 Confidence region

The pivotal quantity can be used even when the unknown parameter is multidimensional; we arrive at confidence region as a random set that covers the true parameter vector with given probability λ. Let observations $X_1, ..., X_n$ be distributed with an unknown m-dimensional parameter $\boldsymbol{\theta}$ and $Q = Q(X_1, .., X_n; \boldsymbol{\theta})$ be a pivotal quantity. That is, the distribution of Q is independent of $\boldsymbol{\theta}$. First, we construct an interval of values for Q as

$$\Pr(q_1 < Q < q_2) = \lambda, \tag{7.40}$$

where q_1 and q_2 may be viewed as quantiles. Then, solving these inequalities for $\boldsymbol{\theta}$, we arrive at the *simultaneous confidence region* C with the confidence level λ, that is, $\Pr(\boldsymbol{\theta} \in C(X_1, .., X_n)) = \lambda$ due to (7.40). Sometimes, several pivotal quantities may be used, $Q_1 = Q_1(X_1, .., X_n; \boldsymbol{\theta})$ and $Q_2 = Q_2(X_1, .., X_n; \boldsymbol{\theta})$, and then (7.40) takes the form

$$\Pr(q_{11} < Q_1 < q_{12} \cap q_{21} < Q_2 < q_{12}) = \lambda,$$

where q_{jk} are functions of λ. Again, inverting these inequalities for $\boldsymbol{\theta}$, we arrive at the confidence region C with the confidence level λ. This method is especially attractive when Q_1 and Q_2 are independent. An alternative approach is to construct a confidence region as a contour of the joint pdf. These two approaches are illustrated in the example below.

Example 7.58 *Simultaneous confidence region for mean and standard deviation.* (a) Given n iid observations, $X_i \overset{\text{iid}}{\sim} \mathcal{N}(\mu, \sigma^2)$, construct a trapezoid confidence region for (σ, μ) using separate confidence intervals for σ and μ, and then as a level set of the joint pdf of Q_1 and Q_2. (b) Apply this confidence region to test the hypothesis $H_0 : \mu = \mu_0$, $\sigma = \sigma_0$ and plot the simulation-derived power as a function of (σ, μ).

Solution. (a) We construct a confidence region based on two pivotal quantities:

$$Q_1 = \frac{n-1}{\sigma^2}\hat{\sigma}^2 \sim \chi^2(n-1), \quad Q_2 = \sqrt{n}\frac{\overline{X} - \mu}{\sigma} \sim \mathcal{N}(0, 1).$$

We start by constructing the $100\lambda_1\%$ and $100\lambda_2\%$ intervals Q_1 and Q_2 such that $\lambda_1\lambda_2 = \lambda$. For example, we may let $\lambda_1 = \lambda_2 = \sqrt{\lambda}$ then define $\alpha = 1 - \sqrt{\lambda}$. Since Q_1 and Q_2 are independent, the rectangle with sides as individual intervals has the joint coverage probability λ: $\Pr(c_{\alpha/2} < Q_1 < c_{1-\alpha/2} \cap q_{\alpha/2} < Q_2 < q_{1-\alpha/2}) = \lambda$, where $q_{\alpha/2} = -q_{1-\alpha/2}$, $c_{\alpha/2}$ and $c_{1-\alpha/2}$ are the $(1 - \alpha/2)$th quantiles of the standard normal distribution and the $(\alpha/2)$th and the $(1-\alpha/2)$th quantile of the chi-square distribution with $n - 1$ degrees of freedom, respectively. Solving

$$c_{\alpha/2} < \frac{n-1}{\sigma^2}\hat{\sigma}^2 < c_{1-\alpha/2}, \quad -q_{1-\alpha/2} < \sqrt{n}\frac{\overline{X} - \mu}{\sigma} < q_{1-\alpha/2}$$

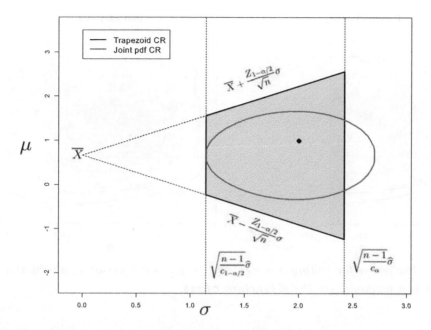

Figure 7.25: *Two exact* $100\lambda\%$ *simultaneous confidence regions for* (σ, μ) *using a sample from a normal distribution with mean* μ *and variance* σ^2 *(the dot depicts the true value).*

for μ and σ^2, we obtain the joint confidence region,

$$\sqrt{\frac{n-1}{c_{1-\alpha/2}}}\widehat{\sigma} < \sigma < \sqrt{\frac{n-1}{c_{\alpha/2}}}\widehat{\sigma}, \quad \overline{X} - \frac{q_{1-\alpha/2}}{\sqrt{n}}\sigma < \mu < \overline{X} + \frac{q_{1-\alpha/2}}{\sqrt{n}}\sigma,$$

which covers (σ, μ) with probability λ. This system defines a trapezoid in the plane with the x-axis as σ and the y-axis as μ. See Figure 7.25 and the R code `cfmus.r` with default values $n = 20$, $\mu = 1$ and $\sigma = 2$.

Alternatively, we construct the confidence region for (σ, μ) using the joint pdf of Q_1 and Q_2 as the product of the chi-square pdf with $n-1$ degrees of freedom from Section 4.2 and the standard normal pdf, $f(q_1, q_2) = A^{-1}e^{(n-3)/2 \ln q_1 - q_1/2 - q_2^2/2}$, where $A = 2^{n/2}\sqrt{\pi}\Gamma((n-1)/2)$. Choose $c > 0$ such that $\int_{f(q_1,q_2) \geq c} f(q_1, q_2)dq_1dq_2 = \lambda$ and then define the region in the plane (σ, μ) such that $f(Q_1, Q_2) = c$. To solve the above equation for c we use vectorized integration as described in Example 1.12 of Section 1.5. Several methods can be used. Here we use the `expand.grid` command. Once the left-hand side of the above integral equation is evaluated, the R function `uniroot` is used to find c. To plot the upper and lower parts of the curve, $f(Q_1, Q_2) = c$, we take advantage of the fact that μ can be found from σ using the square root function,

$$\mu_{1,2} = \overline{X} \pm \frac{\sigma}{\sqrt{n}}\sqrt{(n-3)\ln q_1 - q_1 - 2\log(Ac)}.$$

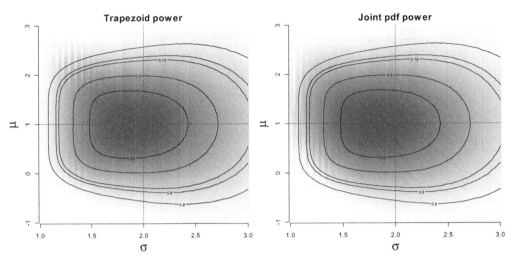

Figure 7.26: *The power for testing $\sigma = \sigma_0$ and $\mu = \mu_0$ as a function of alternative σ and μ for two methods, see the R function* cfmus.

The confidence region based on the joint pdf is tighter and yet has the same coverage probability. The empirical probabilities are derived in cfmus(job=2).

(b) In view of the duality Theorem 7.53, the trapezoid confidence region can be used to test the null hypothesis that $\mu = \mu_0$ and $\sigma = \sigma_0$: the hypothesis is accepted if (μ_0, σ_0) belongs to the trapezoid or if (μ_0, σ_0) falls within the contour $f(Q_1, Q_2) = c$. First, we check if $\widehat{\sigma}$ is within the limits on the x-axis and second if $\overline{X}$ is between the $\pm n^{-1/2} q_{1-\alpha/2} \widehat{\sigma}$ slope wedge. Since the hypothesis is rejected if either $\sigma = \sigma_0$ or $\mu = \mu_0$ is outside of the region, we set $\alpha = 1 - \lambda$ to guarantee that the power function equals $1 - \alpha$ at (σ_0, μ_0). The power functions for both methods on the grid of values for σ and μ is computed in job=3 and returned as a list. Figure 7.26 depicts two power functions as images with contour added. the contour plot is derived in job=4. To speed up vectorized simulation in the loop over alternatives σ and μ, we derive $\overline{X}$ and $\widehat{\sigma}$ for standard normal random variables and then adjust them as $\sigma\widehat{\sigma}$ and $\overline{X} + \mu$. Note that even for a moderate sample size, $n = 20$, the confidence regions are rather large.

Problems

1. Construct the two-sided CI for the upper limit of the uniform distribution from Example 7.45 and check its coverage probability with simulations.

2. Does the CI for θ in Example 7.45 change if instead $X_{\max}$ an unbiased estimator is used, $X_{\max}(n + 1)/n$?

3. (a) Construct the two-sided CI for μ Example 7.46 using unequal-tail probabilities. That is, given λ and α define the quantiles of the t-distribution t_α and $t_{1-\alpha-\lambda}$, where $0 < \lambda < 1$ and $0 < \alpha \leq (1 - \lambda)/2$. (b) Prove that

the expected length of the interval with $\alpha \neq (1 - \lambda)/2$ is greater than with $\alpha = (1 - \lambda)/2$. (c) Write an R program with simulations to display the average length of the to CIs as a function of α.

4. Modify the R function `CImovie` from Example 7.46 to show the coverage of the true μ by the one-sided CI as in Figure 7.23. Use the lower- or upper-sided CI specified by the argument of the function.

5. Modify the R function `CImovie` from Example 7.46 to show the coverage probability of the difference of the means in Example 7.48.

6. Find the shortest CI for λ in exponential distribution using the result of Example 7.32.

7. Use function `nonctp` to test the coverage probability of the sample effect size in Example 7.49 through simulations. Plot the coverage probability versus s for $n = 10$ and $n = 20$.

8. Develop a CI for the area of the circle $\pi \rho^2$ given n iid measurements $r_i \sim \mathcal{N}(\rho, \sigma^2)$ in the form $\left(\pi(\overline{r} + t_\alpha \widehat{\sigma}/\sqrt{n})^2 \leq \pi \rho^2 \leq \pi(\overline{r} + t_{1-\lambda-\alpha}\widehat{\sigma}/\sqrt{n})^2\right)$, where σ is relatively small compared to ρ. Find α that minimizes $t_{1-\lambda-\alpha}^2 - t_\alpha^2$.

9. Generalize Example 7.54 to the two-sample t-test. Use vectorized simulations to test the coverage probability.

10. Plot recent stock prices and log returns for GOOGLE and AMAZON and compute the 95% CI for the correlation coefficients. [Hint: Modify function `vartestSP2`, use `par(mfrow=c(1,2))`.]

7.9 Three asymptotic tests and confidence intervals

The previous discussion of hypothesis testing and confidence interval estimation is rather limited because (a) only one parameter is involved and (b) the pivotal quantity or a statistic's cdf is supposed to be known. Neither of these two conditions hold for more complicated statistical models. Then we must rely on the asymptotic distribution when $n \to \infty$. Not surprisingly, the asymptotic tests are built based on maximum likelihood theory. Hence the results of Chapter 6, in particular those discussed in Section 6.10 and Theorem 6.134, are fundamental for this section.

Let $\mathbf{Y}_i$ be n iid random vectors with pdf $f(\mathbf{y}; \boldsymbol{\theta})$, where $\boldsymbol{\theta}$ is an unknown m-dimensional parameter vector. It is assumed that $\boldsymbol{\theta}$ belongs to an m-dimensional open set in R^m. Other requirements listed in Section 6.10 apply as well.

In this section, we partition $\boldsymbol{\theta}$ into p-dimensional $\boldsymbol{\beta}$ and k-dimensional $\boldsymbol{\gamma}$, where $m = p+k$. The beta parameters are referred to as the parameters of interest and the gamma parameters are referred to as nuisance parameters. Although hypothesis testing and confidence intervals are stated in terms of $\boldsymbol{\beta}$, nuisance parameters may be important as well because they equally affect the properties of asymptotic methods. Denote $\widehat{\boldsymbol{\theta}} = (\widehat{\boldsymbol{\beta}}, \widehat{\boldsymbol{\gamma}})$ the maximum likelihood estimator as the solution of m score equations $\partial l/\partial \boldsymbol{\theta} = \mathbf{0}$, where

$$l(\boldsymbol{\beta}, \boldsymbol{\gamma}) = \sum_{i=1}^{n} \ln f(\mathbf{Y}_i; \boldsymbol{\beta}, \boldsymbol{\gamma})$$

is the log-likelihood function. The null hypothesis is stated as $H_0 : \boldsymbol{\beta} = \boldsymbol{\beta}_0$ against the alternative $H_A : \boldsymbol{\beta} \neq \boldsymbol{\beta}_0$, where the nuisance parameter $\boldsymbol{\gamma}$ is not specified. Three asymptotic tests, Wald, likelihood ratio (LR), and score tests have been developed to test this hypothesis. Their test statistics with the respective asymptotic distributions under assumption that $\boldsymbol{\beta} = \boldsymbol{\beta}_0$ are listed below. Note that for $p \geq 1$, we use the chi-square, but for $p = 1$ we use the normal distribution (this distribution is helpful for one-sided tests).

Table 7.1. Three asymptotic test statistics and their distributions under the null hypothesis.

Test	$\simeq \chi^2(p)$, general case $p \geq 1$	$\simeq \mathcal{N}(0,1)$, single parameter $p = 1$
Wald	$(\widehat{\boldsymbol{\beta}} - \boldsymbol{\beta}_0)'\widehat{\boldsymbol{\Omega}}_{\beta\beta}^{-1}(\widehat{\boldsymbol{\beta}} - \boldsymbol{\beta}_0)$	$(\widehat{\beta} - \beta_0)/\sqrt{\widehat{\Omega}_{\beta\beta}}$
LR	$2\left[l(\widehat{\boldsymbol{\beta}}, \widehat{\boldsymbol{\gamma}}) - l(\boldsymbol{\beta}_0, \widehat{\boldsymbol{\gamma}}_0)\right]$	$\text{sign}(\widehat{\beta} - \beta_0)\sqrt{2\left[l(\widehat{\beta}, \widehat{\gamma}) - l(\beta_0, \widehat{\gamma}_0)\right]}$
score	$\mathbf{d}_0'\widehat{\boldsymbol{\Omega}}_{\beta\beta}\mathbf{d}_0$	$d_0\sqrt{\widehat{\Omega}_{\beta\beta}}$

In the score test,

$$\mathbf{d}_0 = \left.\frac{\partial l}{\partial \boldsymbol{\beta}}\right|_{\boldsymbol{\beta}=\boldsymbol{\beta}_0, \boldsymbol{\gamma}=\widehat{\boldsymbol{\gamma}}_0}$$

is the derivative of the log-likelihood function evaluated at the null hypothesis $\boldsymbol{\beta} = \boldsymbol{\beta}_0$ and at $\boldsymbol{\gamma} = \widehat{\boldsymbol{\gamma}}_0$ that maximizes l when $\boldsymbol{\beta} = \boldsymbol{\beta}_0$; see more detail below.

The tests apply similarly to how we used them before. For example, if the observed Wald statistic is greater than the $(1 - \alpha)$th quantile of the chi-square distribution with p df the null hypothesis is rejected with the approximate type I error α. These tests are supposed to yield the exact type I error, α, and the respective p-values for large n but these tests may be different for a small n. It is very important to find out which test is most appropriate for a specific statistical model. Usually, we rely on simulations because small-sample properties are intractable. We refer the reader to Chapter 9 where these tests are compared for nonlinear regression.

1. Here $\widehat{\boldsymbol{\Omega}}_{\beta\beta}^{-1}$ is the inverse of the $p \times p$ submatrix of the $m \times m$ matrix $\widehat{\boldsymbol{\Omega}}$ as the asymptotic covariance matrix for $\widehat{\boldsymbol{\theta}}$ defined through the $m \times m$ Fisher

information matrix,

$$\boldsymbol{\Omega} = \mathcal{I}^{-1} = - \left[\sum_{i=1}^{n} E\left(\frac{\partial^2 l}{\partial \boldsymbol{\theta}^2} \right) \right]^{-1} = \left[\sum_{i=1}^{n} E\left(\left(\frac{\partial l}{\partial \boldsymbol{\theta}} \right) \left(\frac{\partial l}{\partial \boldsymbol{\theta}} \right)' \right) \right]^{-1}.$$

Two methods may be used to get $\widehat{\boldsymbol{\Omega}}$ as an estimator of $\boldsymbol{\Omega}$: (i) Use $\boldsymbol{\beta} = \boldsymbol{\beta}_0$ and $\boldsymbol{\gamma} = \widehat{\boldsymbol{\gamma}}_0$; this method will be referred to as the null Wald (score) test because the beta parameters are taken at the null hypothesis value. (ii) Use MLEs $\boldsymbol{\beta} = \widehat{\boldsymbol{\beta}}$ and $\boldsymbol{\gamma} = \widehat{\boldsymbol{\gamma}}$; this method will be referred to as the estimated Wald (score) test. There is no difference between the two methods for large n, but there may be a difference for small n. We have seen the difference in Section 7.6.3. Note that $\boldsymbol{\Omega}_{\beta\beta}^{-1}$ is not the $p \times p$ submatrix of the information matrix unless it is block-diagonal or there are no nuisance parameters; see Section 6.9.2.

2. Vector $\widehat{\boldsymbol{\gamma}}_0$ in the formulation of LR and score tests is defined as the solution to the equation

$$\left. \frac{\partial l}{\partial \boldsymbol{\gamma}} \right|_{\boldsymbol{\beta}=\boldsymbol{\beta}_0} = \mathbf{0}. \tag{7.41}$$

Equivalently, under the assumption of existence and uniqueness of this solution, $\widehat{\boldsymbol{\gamma}}_0$ is the maximizer of the log-likelihood function when $\boldsymbol{\beta}$ is held fixed at $\boldsymbol{\beta} = \boldsymbol{\beta}_0$, so the LR test statistic is rewritten as $2\left[\max_{\boldsymbol{\theta}} l(\boldsymbol{\theta}) - \max_{\boldsymbol{\gamma}} l(\boldsymbol{\beta}_0, \boldsymbol{\gamma})\right]$.

3. Due to the iid assumption, one may write $\boldsymbol{\Omega}^{-1} = n\mathcal{I}_1$ where $\mathcal{I}_1$ is the Fisher information matrix in one observation. It should be said that the approximations listed in the table may work even when the distribution of $\mathbf{Y}_i$ varies with i as in linear regression with fixed x_i. Then some assumptions must be made on convergence of f_i. See more discussion in Section 9.4.1 applied to nonlinear regression.

Application of the tests to hypothesis testing is obvious: if the observed test statistic is greater than the $(1 - \alpha)$th quantile of the chi-square distribution with p degrees of freedom the null hypothesis is rejected. These tests can be applied to a one-sided test using the normal approximation when $p = 1$. The tests can be inverted to derive the confidence interval by solving for β_0. It is especially easy to invert the Wald test: the $(1 - \alpha)$th CI is $\widehat{\beta} \pm q_{1-\alpha/2}\sqrt{\Omega_{\beta\beta}}$. Of course, we cannot state that this interval contains the true parameter for finite n with probability $1 - \alpha$, but this probability should converge to $1 - \alpha$ when $n \to \infty$. More examples on application of these tests and their statistical properties with finite n are found in Chapter 9.

The tests simplify when there are no nuisance parameters ($k = 0$), as a continuation of discussion from Section 7.6.3.

Example 7.59 *Asymptotic tests for binomial proportion.* *(a) Apply the three asymptotic tests to test the binomial proportion from Section 7.6.3. (b) Examine how close the true and asymptotic distributions are through simulations by plotting the empirical cdf superimposed with the standard normal cdf.*

Solution. (a) The Fisher information in the sample of n Bernoulli experiments is $n/[p(1-p)]$ and the asymptotic variance $\text{var}(\widehat{p}) = p(1-p)/n$. Therefore, the Wald test statistic turns into $n(\widehat{p}-p_0)^2/[p(1-p)] \simeq \chi^2(1)$, where $p = p_0$ yields the null Wald test and $p = \widehat{p}$ yields the estimated Wald test because $\chi^2(1) = \mathcal{N}^2(0,1)$. In the LR test, we turn our attention to the log likelihood value. If m is the number of successes, the log-likelihood function takes the form $l(p) = m\ln p + (n-m)\ln(1-p)$, and therefore $2(l(\widehat{p}) - l(p_0)) \simeq \chi^2(1)$. The score test distribution takes the form $(m/p_0 - (n-m)/(1-p_0))^2 p(1-p)/n$, where $p = p_0$ leads to the null and $p = \widehat{p}$ the estimated score test. (b) Simulations and plot of empirical cdfs with superimposed theoretical cdf are performed by the R function `thastest` with arguments n, p0, and nSim; see Figure 7.27. The empirical cdf has a stepwise shape due to discreteness of the binomial proportion discussed earlier in Section 7.6.3. The normal approximation improves with n.

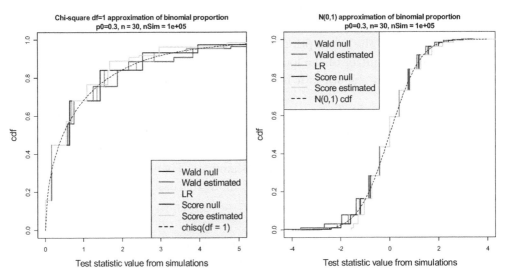

Figure 7.27: *Empirical and chi-square (df = 1) cdf for five test statistics for testing the binomial probability, Example 7.59. This graph is produced by function* `thastest`.

Example 7.60 *Linear regression with fixed predictor.* *Derive the test statistics for the three asymptotic tests for the null hypothesis $H_0 : \beta = \beta_0$ in simple linear regression.*

Solution. We use notation and results from Sections 6.7.2 and 6.9.2. For the null Wald test, we have $\widehat{\Omega}_{\beta\beta} = \widehat{\sigma}_0^2/\sum(x_i - \overline{x})^2$, where $\widehat{\sigma}_0^2 = n^{-1}\sum((Y_i - \overline{Y}) - \beta_0(x-$

$\overline{x}))^2$ maximizes the log-likelihood function, l, under restriction $\beta = \beta_0$. For the LR test, we have $\boldsymbol{\gamma} = (\alpha, \sigma^2)$ and $\widehat{\alpha}_0 = \overline{Y} - \beta_0\overline{x}$ where $\widehat{\sigma}_0^2$ is as in the Wald test. After some elementary algebra we obtain $2\left[l(\widehat{\beta}, \widehat{\boldsymbol{\gamma}}) - l(\beta_0, \widehat{\boldsymbol{\gamma}}_0)\right] = n \ln \widehat{\sigma}_0^2/\widehat{\sigma}_{ML}^2$, where $\widehat{\sigma}_{ML}^2 = n^{-1}\sum((Y_i - \overline{Y}) - \widehat{\beta}(x - \overline{x}))^2$ is the (unrestricted) minimum of the residual sum of squares. Using approximation $\ln(1 + x) \simeq x$, this test statistic can be approximated as

$$n\frac{\widehat{\sigma}_0^2 - \widehat{\sigma}_{ML}^2}{n\widehat{\sigma}_{ML}^2} = \frac{(\widehat{\beta} - \beta_0)^2\sum(x_i - \overline{x})^2}{\widehat{\sigma}_{ML}^2},$$

the same as the null Wald test statistic but with $\widehat{\sigma}_0^2$ replaced by $\widehat{\sigma}_{ML}^2$. The score test statistic takes the form

$$\frac{\left[\sum((Y_i - \overline{Y}) - \beta_0(x_i - \overline{x}))x_i\right]^2}{\widehat{\sigma}_0^2\sum(x_i - \overline{x})^2} = \frac{(\widehat{\beta} - \beta_0)^2}{\widehat{\sigma}_0^2/\sum(x_i - \overline{x})^2},$$

the same as the Wald test. Finally, we conclude that the Wald and score tests are the same, but the LR test is different and close to the first two when the unrestricted and restricted sums of squares are close – we expect the power of the tests to be close in a neighborhood of the null. □

The following example continues studying the statistical properties of linear regression in the bivariate normal distribution (x is random). See Examples 6.108 and 6.121.

Example 7.61 *Linear regression with random predictor.* *(a) Derive the normal approximation for the slope coefficient in three tests in the linear regression of the multivariate normal distribution. (b) Verify the distribution of the test statistic through simulations by plotting the empirical cdf superimposed with the theoretical one. (c) Plot vectorized simulation-based power functions for the three tests. (d) Improve the match between empirical and nominal significance level by using the t-distribution.*

Solution. (a) The triple $(Y, X, \mathbf{U})$ has a multivariate normal distribution with linear regression specified as conditional mean, $E(Y|X = x, \mathbf{U} = \mathbf{u}) = \beta x + \boldsymbol{\tau}'\mathbf{u}$, and conditional variance $\mathrm{var}(Y|X = x, \mathbf{U} = \mathbf{u}) = \sigma^2$, where $\boldsymbol{\gamma} = (\boldsymbol{\tau}, \sigma^2)$ in the previous notation. The joint marginal distribution of X and $\mathbf{U}$ is multivariate normal with $\mathrm{var}(X) = \sigma_x^2$, $\mathrm{cov}(X, \mathbf{U}) = \boldsymbol{\sigma}_{xu}$, and $\mathrm{cov}(\mathbf{U}) = \boldsymbol{\Omega}_u$, where it is assumed that all parameters of this distribution are known, to simplify the derivation . The log density, up to a constant, is given by $\ln f = -(y - \beta x - \boldsymbol{\tau}'\mathbf{u})^2/(2\sigma^2) - 0.5\ln\sigma^2$ and simple derivation implies the information matrix takes the form

$$\mathcal{I}_1 = \frac{1}{\sigma^2}\begin{bmatrix} \sigma_x^2 & \boldsymbol{\sigma}_{xu}' & 0 \\ \boldsymbol{\sigma}_{xu} & \boldsymbol{\Omega}_u & 0 \\ 0 & 0 & \frac{1}{2\sigma^2} \end{bmatrix}.$$

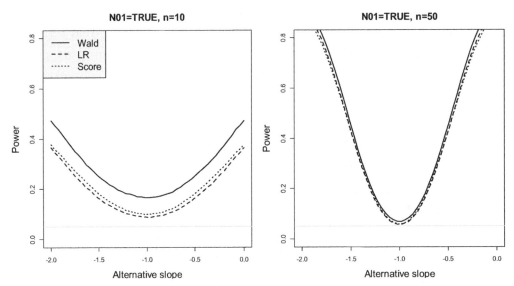

Figure 7.28: *Simulation-derived power functions for the three asymptotic tests from Example 7.61 (function* powlinmod*) with normal distribution, N01=TRUE.*

As follows from Example 6.120, given n iid vector observations $(Y_i, X_i, \mathbf{U}_i)$, the MLEs of the slope coefficients are $\widehat{\boldsymbol{\theta}} = \sum (\mathbf{Z}_i \mathbf{Z}_i')^{-1} \mathbf{Z}_i Y_i$, where $\mathbf{Z}_i = (X_i, \mathbf{U}_i')'$, and $\widehat{\sigma}^2 = S_{\min}/n$, where $S_{\min} = \sum (Y_i - \widehat{\beta} X_i - \widehat{\boldsymbol{\tau}}' \mathbf{U}_i)^2$ is the minimum sum of squared residuals. The MLE of the slope of interest, $\widehat{\beta}$, is the first component of vector $\widehat{\boldsymbol{\theta}}$ with the asymptotic variance as the $(1,1)$the element of the inverse information matrix, $\Omega_{\beta\beta} = \sigma^2/(nv^2)$, where $v^2 = \sigma_x^2 - \boldsymbol{\sigma}_{xu}' \boldsymbol{\Omega}_u^{-1} \boldsymbol{\sigma}_{xu}$. The estimator of the asymptotic variance $\widehat{\Omega}_{\beta\beta}$ is $\Omega_{\beta\beta}$ evaluated at $\sigma^2 = \widehat{\sigma}^2$. The key to the LR approximation is to find $\boldsymbol{\gamma}_0 = \boldsymbol{\gamma}_0(\beta_0)$, which is defined from equation (7.41) or, equivalently, which maximizes the log-likelihood when β is held fixed at the null value, β_0. Following the derivation of the unconstrained maximization we obtain $\widehat{\boldsymbol{\tau}}_0 = \sum (\mathbf{U}_i \mathbf{U}_i')^{-1} \mathbf{U}_i (Y_i - \beta_0 X_i)$ and $\widehat{\sigma}_0^2 = S_0/n$, where $S_0 = \sum (Y_i - \beta_0 X_i - \widehat{\boldsymbol{\tau}}_0' \mathbf{U}_i)^2$ is the minimum sum of squared residuals under constrained $\beta = \beta_0$. Finally, the LR and score approximations take the form $\text{sign}(\widehat{\beta} - \beta_0) \sqrt{n \ln(S_0/S_{\min})} \simeq \mathcal{N}(0, 1)$. The derivation of the score approximation follows the same line of justification $\sum (Y_i - \beta_0 X_i - \widehat{\boldsymbol{\tau}}_0' \mathbf{U}_i) X_i / (\widehat{\sigma} v \sqrt{n}) \simeq \mathcal{N}(0, 1)$.

(b) The vectorized simulations and the plot of the cdfs are carried out by the R function thlinmod (it takes less than a second to carry out 100K simulations). The arguments of this function with default values are as follows: n=10, b0=-1, tau=.6, s2=2, sdx=1, sdu=.5, roxu=.7, nSim=100000 (the plot of cdfs is not shown). For small n the approximations are slightly different from standard normal, however, the distributions of the LR and score statistics are consistently closer to normal compared to the Wald test.

(c) The powers for the three asymptotic tests are derived in function powlinmod and displayed in Figure 7.28. The regression model takes the form $E(Y|X =$

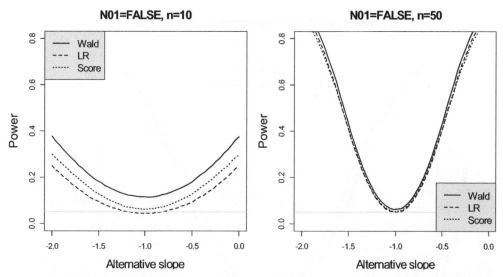

Figure 7.29: *Simulation-derived power functions for the three asymptotic tests from Example 7.61 (function* `powlinmod`*) with the t-distribution distribution, N01=FALSE.*

$x, U = u) = \beta x + \tau u$, where (Y, X, U) have the three dimensional normal distribution (no intercept). The parameters of the model are specified by the arguments of the function, the default values are as follows: N01=TRUE, alpha=0.05, b0=-1, tau=.6, s2=2, sdx=1, sdu=.5, roxu=.7, nSim=100000. The argument N01 defines what distribution to use: if TRUE (default) the standard normal distribution is used to compute the critical value of the test statistic, as given in the right column of Table 7.1, and otherwise the t-distribution with $n - 2$ df (used in d below). The null value for the slope is $\beta_0 = -1$. Due to large amount of computation the run may take a while. As follows from Figure 7.28, when the sample size is large, like $n = 50$ (the plot at right); the power functions of the three tests are close and the empirical type I error is close to the nominal, $\alpha = 0.05$. However, for small n (the plot at left), like $n = 10$, the empirical significance level does not match $\alpha = 0.05$, especially for the Wald test. The significance level is the closest for the LR test.

(d) To make the empirical significance level closer to the nominal t-distribution with $n-2$ df is used here (N01=FALSE). See Figure 7.29. Indeed, the t-distribution is very helpful – the score test, again, yields the closest match. In general, these simulations confirm that for a small sample size, n, the actual size of the test may be different from the nominal size (in this case $\alpha = 0.05$). When n increases, asymptotic tests behave as predicted.   □

As follows from the previous example, all three tests fail to produce the nominal type I error $\alpha = 0.05$. This situation is typical for the tests based on approximations, including asymptotic tests. Unequal type I errors (the value of the power functions at the null hypothesis) make it impossible to compare tests.

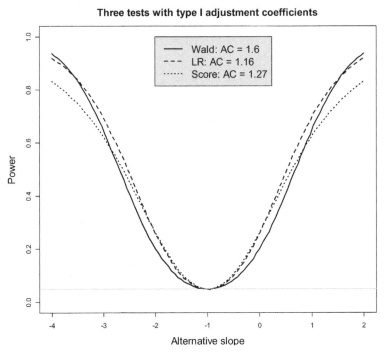

Figure 7.30: *The type I error adjusted power functions for the three tests. There is a slight superiority of the likelihood ratio test.*

Below, we illustrate how to achieve the nominal type I error using simulations at the null hypothesis. The idea is to shrink the rejection area if the power of the test at the null hypothesis is greater than α, and expand it otherwise. For example, if the rejection area has the form $|T_{\text{stat}}| > T_{\text{crit}}$ and this test yields the power at the null greater than α, we find the coefficient c such that the probability of $|T_{\text{stat}}| > T_{\text{crit}} \times c$ at the null is approximately α, where T_{stat} is the test statistic at the null value and T_{crit} is the critical value. This method is illustrated in the following example.

Example 7.62 *Tests with the type I adjustment coefficients.* *Find type I adjustment coefficients (AC) for each test from Example 7.61 and plot the respective power functions.*

Solution. The R computations are found in the function `powlinmodC`. Since all three tests produce the type I error greater than $\alpha = 0.05$, we are seeking $c > 1$ making $\Pr(|T_{\text{stat}}| > T_{\text{crit}} \times c) \simeq \alpha$. To find the type I error AC, we generate T_{stat} assuming that the slope coefficient equals the null value (the default value is -1) and then increase c starting from 1 until $\Pr(|T_{\text{stat}}| > T_{\text{crit}} \times c) \leq \alpha$. The adjusted power functions with respective AC are depicted in Figure 7.30. Note that the AC for the Wald test is the greatest because its power at the null is greater that the other two. Now, all three tests have the type I error close to 0.05, and making the tests comparable. The likelihood-ratio test is slightly superior.

7.9.1 Pearson chi-square test

This section discusses application of the asymptotic tests to the multinomial distribution, see Section 3.10.4 and Example 6.136. Given m frequencies, X_j, we want to test the null hypothesis $H_0 : p_1 = p_{10}, ..., p_m = p_{m0}$ versus the alternative that $p_j \neq p_{j0}$ for at least one j. Note that since p_j and p_{j0} add to 1 the actual number of the null hypotheses and the degrees of freedom for the chi-square distribution is $m - 1$.

First, we apply the LR test. Since $l = \sum_{j=1}^{m} X_j \ln p_j$, the MLE is $\widehat{p}_j = X_j/n$, and the difference of twice log-likelihoods takes the form

$$2n \left[\sum_{j=1}^{m} \widehat{p}_j \ln \widehat{p}_j - \sum_{j=1}^{m} \widehat{p}_j \ln p_{j0} \right] = 2n \sum_{j=1}^{m} \widehat{p}_j \ln \frac{\widehat{p}_j}{p_{j0}} \simeq \chi^2(m-1), \qquad (7.42)$$

if the null hypothesis holds. The left-hand side is equivalent to the relative Kullback-Leibler entropy discussed in the probability part of the book (Section 3.10.3).

Second, since the information matrix is $\mathcal{I} = n \times diag(1/p_1, 1/p_2, ..., 1/p_n)$ the null Wald test (the information matrix is evaluated at the null) is written as

$$n \sum_{j=1}^{m} \frac{(\widehat{p}_j - p_{j0})^2}{p_{j0}} \simeq \chi^2(m-1), \qquad (7.43)$$

the famous Pearson chi-square test. Hence, we refer to this test as the Pearson/Wald test.

Example 7.63 Pearson/Wald test. *Show that the Pearson chi-square test with $m = 2$ is equivalent to the one-sample null Wald test for a binomial proportion from Section 7.6.3.*

Solution. Recall that the null Wald test from Section 7.6.3 rejects the null $H_0 : p = p_0$ if $|\widehat{p} - p_0| > q_{1-\alpha/2}\sqrt{p_0(1-p_0)/n}$. The rejection rule for test statistic (7.43) takes the form

$$n \left[\frac{(\widehat{p} - p_0)^2}{p_0} + \frac{((1 - \widehat{p}) - (1 - p_0))^2}{1 - p_0} \right] > \chi_{1-\alpha}^{-2}(m - 1).$$

After some algebra on the left-hand side, one obtains

$$n \left[\frac{(\widehat{p} - p_0)^2}{p_0} + \frac{(\widehat{p} - p_0)^2}{1 - p_0} \right] = \frac{(\widehat{p} - p_0)^2}{p_0(1 - p_0)/n}.$$

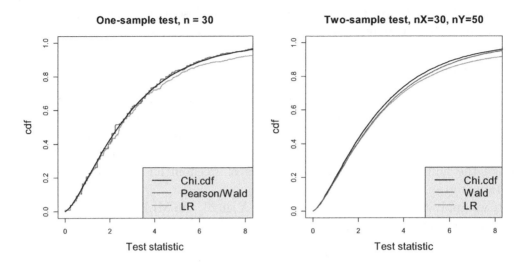

Figure 7.31: *Simulation-based verification of how close the test statistics are to the chi-square distribution. The Wald test outperforms the LR test, see Example 7.65.*

But the chi-square distribution is the distribution of the squared standard normal variable, or symbolically, $\chi^2(1) = \mathcal{N}^2(0,1)$ and, therefore, $\chi^{-2}_{1-\alpha}(m-1) = q_{1-\alpha/2}$, which proves the equivalence of the tests.   □

Asymptotic equivalence of the Wald and LR tests follows from the general theory of maximum likelihood and was discussed at the beginning of this section. Below we demonstrate this equivalence by elementary calculus.

Example 7.64 Pearson = Wald. *Use calculus to demonstrate that the Pearson/ Wald and LR test statistics are equivalent when $\widehat{p}_j \simeq p_{j0}$, $j = 1, 2, ..., m$.*

Solution. We want to show that the left-hand sides of (7.42) and (7.43) are close when $\widehat{p}_j$ and p_{j0} are close. Introduce function $f(x) = 2(x+p_0) \ln((x+p_0)/p_0)$ and apply Taylor's series expansion up to the second order around $x = 0$,

$$ f(x) = f(0) + xf'(0) + \frac{1}{2}x^2 f''(0) + O(x^3) = 2x + \frac{x^2}{p_0} + O(x^3). $$

Therefore, letting $x = \widehat{p}_j - p_{j0}$, the Pearson/Wald test statistic can be approximated as

$$ 2n \sum_{j=1}^{m} \widehat{p}_j \ln \frac{\widehat{p}_j}{p_{j0}} \simeq 4n \sum_{j=1}^{m} (\widehat{p}_j - p_{j0}) + n \sum_{j=1}^{m} \frac{(\widehat{p}_j - p_{j0})^2}{p_{j0}} = n \sum_{j=1}^{m} \frac{(\widehat{p}_j - p_{j0})^2}{p_{j0}} $$

because $\sum_{j=1}^{m} (\widehat{p}_j - p_{j0}) = \sum_{j=1}^{m} \widehat{p}_j - \sum_{j=1}^{m} p_{j0} = 1 - 1 = 0$.   □

The Pearson test is a one-sample test. In a two-sample test, frequencies X_j and Y_j from two samples of sizes n_X and n_Y are compared with the null hypothesis $H_0 : p_{1X} = p_{1Y}, ..., p_{mX} = p_{mY}$ the test statistic of the LR test is

$$2 \left[n_X \sum_{j=1}^m \widehat{p}_{jX} \ln \widehat{p}_{jX} + n_Y \sum_{j=1}^m \widehat{p}_{jY} \ln \widehat{p}_{jY} - (n_X + n_Y) \sum_{j=1}^m \widehat{p}_j \ln \widehat{p}_j \right] \simeq \chi^2(m-1),$$

where $\widehat{p}_j = (X_j + Y_j)/(n_X + n_Y)$, the probability estimates under the null (the probabilities are the same). The Wald test takes the form

$$\frac{1}{1/n_X + 1/n_X} \sum_{j=1}^m \frac{(\widehat{p}_{jX} - \widehat{p}_{jY})^2}{\widehat{p}_j} \simeq \chi^2(m-1). \tag{7.44}$$

Example 7.65 *Simulations for Wald and LR tests.* *Use simulations to compare how close the test statistics in the one- and two-sample tests are to the chi-square distributions for Wald and LR tests.*

Solution. The R function `chismult` performs simulations and produces the plot portrayed in Figure 7.31. The default call is `chismult(p=c(.1,.2,.5,.2),` `n=30,nY=50,nSim=100000))` where array p specifies the probabilities (the default is $m = 4$), n=30 and nY=50 are the sample sizes of the first and second sample, respectively, and nSim is the number of simulations. As follows from our analysis, the Wald test outperforms the LR test. The distribution of the test statistics is closer to the chi-square distribution for one- and two-sample tests. □

The following is a continuation of Example 1.14 from Chapter 1.

Example 7.66 *Pearson test for text analysis.* (a) *Apply the one-sample Pearson/Wald test to test if the distribution of English letters in Jack London's novel* Call of the Wild *is the same as in a typical English text.* (b) *Apply the two-sample Wald test to test the difference in distributions of letters in Jack London's novel and Mark Twain's novel* The Adventure of Tom Sawyer.

Solution. The R function `fr1JL` computes the p-values for the one- and two-sample Wald tests under assumption that letter frequencies follow a multinomial distribution. (a) To test that the distribution of letters in Jack London's novel follows the distribution of a typical English text the Pearson/Wald test chi-square statistic (7.43) is computed. The observed value is 856.2, which yields $p = 1.61 \times 10^{-164}$. (b) The observed value of the two-sample chi-square (7.44) is 413.9 which yields $p = 4.38 \times 10^{-72}$. See Figure 7.32. Both tests indicate that the distributions of letters is statistically significant. As will be discussed in the following section, such small p-values are driven by a large number of letters in the novels. Discussion of how n affects the p-value is found in Section 7.10.

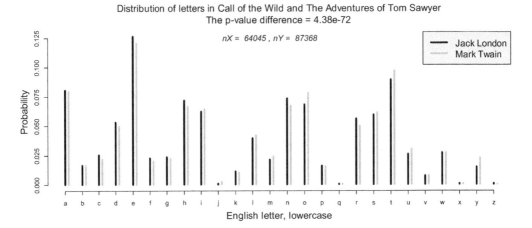

Figure 7.32: *Distribution of letters in Jack London's and Mark Twain's novels; the numbers of letters are 64,045 and 87,368, respectively (see Example 7.66). The p-values are computed based on the test statistics (7.44).*

7.9.2   Handwritten digit recognition

Handwritten digit recognition is one of the most famous computer science classification problems. The goal of this section is to illustrate how a computer can predict what digit is written. The Pearson/Wald test statistics (7.43) is the criterion by which the digit is predicted. The problem of handwritten digit recognition emerged in computer recognition of the five digit zip code from scanned envelopes by the U.S. Postal Service, as one of many other computer vision problems. Many methods with low misclassification error have been suggested over the years including neural networks and random forest. The interested reader is referred to the papers by LeCun, et al. [68]. We will be using a widely popular MNIST data set freely available at https://pjreddie.com/projects/mnist-in-csv/. This site contains two data sets: train set and test set as 28 × 28 pixel 8-bit grayscale images of handwritten digits. That is, the intensity of each pixel varies from 0 to 255 where 0 codes the black color and 255 codes the white color (the background is black). The train data set mnist_train.csv contains 59,999 grayscale images of handwritten digits as a matrix 59,999 by 785; the first column is the digit and the remaining $28^2 = 784$ columns are the intensities at each pixel (a number from 0 to 255). The test data set mnist_test.csv has the same format but the number of digits/rows is 1,000. That is, it is a matrix 1,000 by 785. Note that we do not test the hypothesis that a particular handwritten image is an image of a specific digit but use test statistics (7.43) to find the digit for which the value of this test statistic criterion is minimum. As the authors who prepared this data set said, "The test set is notoriously difficult, and a 2.5% error rate is excellent."

We start by plotting ten handwritten digits from the train data set. To save

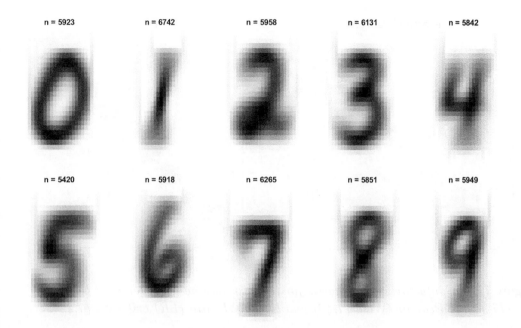

Figure 7.33: *Composite images from 59,999 handwritten digits from the train dataset* mnist_train.csv. *The intensity at each pixel of the digit is the average of n intensities. This figure is created by calling* dig.mnist(job=1).

ink, we plot the digits on the white background by computing $255-$ gray intensity value, see Figure 7.33. This figure is created by calling dig.mnist(job=1), which uses digit images from mnist_train.csv data set. The number of instances of each digit in the set is shown on the top. The intensity of each pixel is the average of 59,999 intensities from the train data set. Calling dig.mnist(job=2) plots a similar figure for digits from mnist_test.csv data set (not shown). Now we describe how to predict the digit given its handwritten image; the formula 255-gray intensity is used to exclude zero probability values. We treat average intensity at each pixel of a digit from the train data set mnist_train.csv as p_{j0} of the multinomial distribution with 784 categories. Hence each row in the test set mnist_test.csv is treated as a frequency of multinomial distribution with the null hypothesis that the probabilities are those obtained by averaging from the train data set, p_{j0}. dig.mnist(job=3) computes the chi-square criterion (7.43) and the LR criterion (7.42) for each of 10,000 rows with $\widehat{p}_j$ computed from the frequencies in the row. Interestingly, sometimes in the literature, criterion (7.42) is used for classification and referred to as the entropy without mentioning its connection to the LR test. Array achisq contains values of the chi-square and array lr contains values of the LR criterion for each digit. The predicted digit has the minimum criterion value. At the end of the job, 10,000 true digits and their predicted digits are saved in a 10,000 by 3 matrix idmin. This job may take several hours.

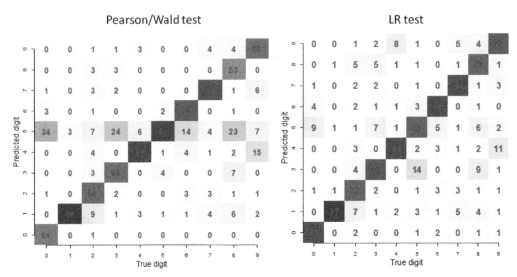

Figure 7.34: *Confusion image/matrix for recognition a handwritten digit based on the chi-square Pearson/Wald (7.43) and likelihood ratio (LR) criterion (7.42).*

The next job, `dig.mnist(job=4)` computes the confusion matrix as the probability of classification for two criteria and plots Figure 7.34 (the darker the cell the higher the probability). The (i, j)th element of the confusion matrix is

Pr(image is recognized as digit $(j-1)$ | the true digit is $(i-1)$), $i, j = 1, 2, ..., 10.$

The (i, i)th element of the confusion matrix is the proportion of digits recognized correctly, the (i, j)th element is the proportion of digits predicted as $(j-1)$ while the true digit was $(i-1)$. For example, as follows from Figure 7.34, using the Pearson test statistic, 34% of digit images of '0' were wrongly recognized as '5'. Indeed, as follows from this figure '0' is the least recognizable handwritten digit, and '5' is the most incorrect. The LR test outperforms the Pearson test in this example.

Problems

1. Use function `thastest` from Example 7.59 to plot the maximum absolute difference between approximate and the standard normal cdfs for five methods as a function of n. Use `mfrow(par=c(1,2))` to plot the difference for $p_0 = 0.1$ and $p_0 = 0.5$.

2. Derive the LR and score tests for testing the null hypothesis $H_0 : \mu = \mu_0$ versus $H_A : \mu \neq \mu_0$ using n iid observations $Y_i \sim \mathcal{N}(\mu, \sigma^2)$ where σ^2 is unknown and treated as the nuisance parameter along the lines of Example 7.60.

3. In Example 7.60, (a) derive the estimated Wald test, (b) adjust the estimated Wald test and the score test to make them exact (the type I error equals the nominal, α).

4. (a) Derive the three asymptotic tests for the null hypothesis $H_0 : \lambda = \lambda_0$ versus $H_A : \lambda \neq \lambda_0$ using n iid observations Y_i from exponential distribution. (b) Derive the respective confidence intervals. (c) Use vectorized simulations to plot the coverage probability as a function of λ for given n.

5. A pharmaceutical company is seeking a treatment effect of a new drug to treat epileptic patients. The time between two consecutive epileptic seizures in $n = 7$ patients who took the drug was $10, 5, 7, 2, 2, 6, 4$ and in $m = 11$ who took placebo was $2, 4, 3, 1, 7, 5, 2, 3, 1, 3, 4$. Assuming that the time between seizures follows an exponential distribution, test the hypothesis that the new drug does not improve the seizure rate. Use the Wald and LR tests.

6. Based on results in Example 7.59, (a) derive the LR and score CIs for binomial proportion by inverting the tests and (b) use simulations to study their coverage probability as in Example 7.37.

7. The same as above but for the Poisson rate, λ.

8. Modify function thlinmod from Example 7.61 to investigate how the maximum absolute difference between the asymptotic cdfs and standard normal cdf depends on six parameters b0=-1, tau=.6, s2=2, sdx=1, sdu=.5, roxu=.7 with fixed n. Use par(mfrow=c(2,3)).

9. The R function wtext computes the frequencies of English letters in the novel by Jane Austen *Pride and Prejudice*. This text has been downloaded as a part of the Gutenberg project from the website http://www.gutenberg .org. (a) Add the probabilities to the plot from Example 7.66. (b) Test the hypothesis that the distribution of letters is the same as in a typical English text. (c) Test the hypotheses that the distribution of letters is the same as in Jack London's and Mark Twain's novels.

10. Develop three asymptotic tests to test that two students from (a) the same and (b) different classes have the same performance (probability of earning the point) in Example 6.118.

11. Develop three asymptotic tests to test the hypothesis that $\mu_i = \text{const}$ in Example 6.141.

12. In Example 6.141, (a) develop three asymptotic tests for testing $H_0 : \mu = \mu_0$, (b) construct the respective confidence intervals for μ, and (c) use simulations to test the performance of the tests and CIs.

13. Apply the LR test for testing $p_1 = p_2$ from Example 1.15 for all four cases depicted in Figure 1.7.

14. Apply the LR test for testing $\delta_j = 0$ in Example 6.139 and use simulations to derive the power of the test.

15. (a) Prove that the test statistic for the LR test in Example 6.126 can written as $T = n \ln(H(\pi)/H(\widehat{\pi}_{ML})$ where $\pi = 3.141593$. (b) Use vectorized simulations to show that $T \simeq \chi^2(1)$ for fairly large n.

16. Construct three asymptotic confidence intervals for α in Example 6.141 and test their coverage probability using simulations.

17. Construct a confidence region for (α, β) in Example 6.141 using the three tests, generate a random sample of size n and plot the three regions using different colors.

18. Run function `dig.mnist(job=3)` with the original black background, i.e. when 255 is not subtracted. Does the white background help?

19. (a) Switch `mnist_test.csv` and `mnist_train.csv` data set and repeat computations. (b) Combine `mnist_test.csv` and `mnist_train.csv` data sets and repeat computations. Compare the results.

7.10 Limitations of classical hypothesis testing and the d-value

First of all, we would like to emphasize that there is nothing wrong with statistical hypothesis testing and the p-value. Mathematics is correct. Much confusion and misunderstanding comes from when the results of the null hypothesis statistical testing (NHST) are interpreted and layman statements are made: "lost in translation" very much applies here. How do we interpret statistical significance, and why we need this concept in the first place? To answer these questions, we have to ask ourselves how the results of statistical hypothesis testing are going to be used. Why is 0.05 a good choice for the study at hand? The common mistake people make is *overinterpreting* the p-value and statistical hypothesis testing in general when statistical significance is substituted with practical significance, Gelman and Stern [35]. For example, a typical mistake is to think of the p-value as the probability that the null hypothesis is false or to associate small value with the strength of the treatment effect.

Interpretation of statistical hypothesis testing and the associated p-value has been a topic of discussion for more than fifty years. See a collection of papers entitled "What If There Were No Significance Tests?" [53]. For example, Cohen [19] wrote "...null hypothesis significance testing ... does not tell us what we want to know..." and Schmidt and Hunter [88] used even a stronger language: "Statistical significance testing retards the growth of scientific knowledge; it never makes a positive contribution." See also a recent paper by Smith [96] for a short bibliography list. The latest burst of discussion was fueled by an editorial published in *Basic and Applied Social Psychology* journal by Trafimow and Marks [100] who announced that the null hypothesis significance testing procedure is invalid and therefore their journal is going to ban the p-value < 0.05 as a proof of a scientific finding. Not long after the American Statistical Association published an unprecedented statement on statistical significance and the p-value which could be compared to a major science quake (Wasserstein and Lazar [103]). The goal of this section is to discuss the limitations of statistical hypothesis testing and the p-value from the practical standpoint using rigorous/mathematical argumentation and to illustrate when the p-value is an adequate feature and when it is not.

7.10.1   What the p-value means?

We discussed the interpretation of the p-value in Example 7.1 of Section 7.1.1 and Example 7.16 of Section 7.4. Recall that if the null hypothesis were true, values of the test statistic are expected to be contained in a close interval is called the *normal range*. Thus, if the observed value of the test statistic, T_{obs}, is outside of the normal range, the null hypothesis is rejected. The p-value reflects how far T_{obs} is from its expected value (usually zero) on the probability scale. Loosely speaking, the p-value is the chance of getting a test statistic larger than the observed if the null hypothesis were true. Smaller p-values provide more evidence that the alternative holds provided the distribution of the test statistic (or the power of the test) is monotonic on both sides of the null value (for the double-sided test). Many other interpretations of the p-value such as those mentioned at the beginning of the section or given at the website `http://rpsychologist.com/d3/NHST/` are frequent overstatements. The p-value merely tells how close the estimator (or a summary statistic) is to the null value on the probability scale, if the null hypothesis were true. Although the interpretation of the p-value was discussed several times, we provide another example due to the importance of a clear interpretation of this value in layman's terms understandable by nonstatisticians.

Example 7.67 *What is the p-value?* *Imagine that your concern is an elevated concentration of a carcinogen in urine of residents from town X. The average concentration in the country is 0.1 mg/L. Researchers collected $n = 100$ urine samples from the town with the average $\overline{X} = 0.12$ mg/L and SD $= \widehat{\sigma} = 0.075$ mg/L. Under the normal distribution assumption, the test statistic of the t-test that the carcinogen concentration in town X is the same as in the country was*

$$T = \frac{\overline{X} - 0.1}{\widehat{\sigma}}\sqrt{n} = \frac{0.12 - 0.1}{0.075}\sqrt{100} = 2.67$$

with a two-sided p-value = 2*pt(-2.67,df=100-1)=0.0089. Explain what the *p-value is through imaginable sample collection in typical towns, i.e. towns where the carcinogen concentration is 0.1 mg/L.*

Solution. Imagine that you collected 100 samples in many other typical towns in the country ("typical" means that the average concentration of the carcinogen in each town equals the national average $= 0.1$ mg/L). For each town's data, you repeated the above hypothesis testing and computed the test statistic T and the p-value with town-specific $\overline{X}$ and $\widehat{\sigma}$. Then the proportion of test statistics larger in absolute value then 2.67 must be 0.89%. Equivalently, the proportion of p-values smaller than 0.0089 must be 0.89%. The statement about the p-value follows from the fact that, under the null hypothesis, the p-values are uniformly distributed on the interval $(0, 1)$.

7.10.2   Why $\alpha = 0.05$?

Statistical dogma suggests the significance level $\alpha = 0.05$. Why 0.05? This question is rarely discussed in statistics and its applications. Ronald Fisher [48] proposed $\alpha = 0.05$ as the probability to be outside ± 1.96 sigma range of the normal distribution. He declared: "Deviations exceeding twice the standard deviation are thus formally regarded as significant." Today, many prominent statisticians rise up against "statistical significance" implied by the rule $p < 0.05$ (Amrhein et al. [5], Wasserstein et al. [104]). The problem with the term "statistical significance" is that, in many cases, stakeholders and decision makers, such as juries, do not understand what the p-value is and interpret "statistical significance" as the proof that the null hypothesis is wrong. To understand the consequences of such interpretation consider the following example.

Example 7.68 *Statistical significance and forensic science.* *Estimate the number of innocent people on the death row if the p-value less than 0.05 were used as evidence of capital crime.*

Solution. Imagine that court is looking for a supporting evidence from the t-test (or any other statistical test). For example, it may be a blood stains on the shirt of the defendant who claims that this is his own blood, not the blood of the victim. Since the stains are old and faded the result of the t-test is prone to uncertainty. The null hypothesis is that the blood on the shirt belongs to the defendant (presumption of innocence) and $\alpha = 0.05$ means that the probability that the defendant will be accused of murder, when innocent, is 0.05. If their capital crime were proven with statistical verdict "statistical significant" there would be a good chance of the jury finding the suspect guilty. In 2019 alone, there were about $2,500$ death row inmates and the rule $\alpha = 0.05$ would lead to execution of 125 innocent people.   □

Any particular threshold is difficult to justify, but this does not mean that statisticians should avoid discussion of the subject or at least do not understand the consequences of choosing $\alpha = 0.05$. The choice of α should involve an analysis of interplay between type I and type II errors. Since the latter is a function of the alternative, this analysis should involve a range of values. In the following example, we illustrate how significance level can be chosen to minimize the total expected cost due to errors in breast cancer detection considered previously in Example 7.7. The analysis of the type I and II errors is further used for computation of the optimal significance level for testing whether computer-aided diagnosis (CAD) in a sample of n digital mammograms is statistically distinct from the traditional visual radiologist evaluation. We remind the reader that other relevant details on the trade-off between type I and II errors and its connection with ROC curve have been discussed earlier in Section 7.3.

Example 7.69 *Optimal α. Assume that a priori probability of a woman coming for mammography has breast cancer is 1 out of 1000. Let the cost of biopsy be \$1K and the total cost of the breast cancer treatment at the later stage, when tumor is overlooked in the mammogram, be \$1 million. (a) Prove that, to minimize the total cost of errors, the significance level should be chosen such the sum of type I and II errors, $\alpha + \beta$, is minimum. (b) The CAD and visual/traditional radiology analysis are compared based on the one-sample two-sided t-test of the tumor-like (white spot) volume in n breasts with the target expressed as the effect size, $c = (\mu - \mu_0)/\sigma$, where μ_0 is the average tumor-like volume detected by the traditional technique (known) and σ is the standard deviation of white-spot volumes. Find an optimal α that minimizes the total error.*

Solution. (a) Let n be the number of women who take the mammography exam. The cost associated with sending normal women to do the biopsy (type I error) is $1000 \times n \times 999/1000 \times \alpha$. The cost associated with overlooking breast cancer (type II error) and the implied cancer treatment at the later stage is $1000000 \times n \times 1/1000 \times \beta$. The total cost is

$$\frac{1}{1000}1000 \times n \times 999 \times \alpha + \frac{1}{1000}1000000 \times n \times \beta = 1000n\left(\frac{999}{1000}\alpha + \beta\right) \propto \alpha + \beta,$$

where $\propto$ is the proportionality sign. (b) Given a sample of n white-spot volumes in the CAD system, under the null hypothesis is $H_0 : \mu = \mu_0$, the sum of two type of errors (total error) is given by

$$\alpha + \mathcal{T}(t_{1-\alpha/2}; n - 1, \delta = \sqrt{n}c) - \mathcal{T}(-t_{1-\alpha/2}; n - 1, \delta = \sqrt{n}c), \tag{7.45}$$

where $\mathcal{T}$ is the cdf of the noncentral t-distribution with $n-1$ df and noncentrality parameter $\delta = \sqrt{n}c$, as follows from (7.14). Recall that the total error as a function of α for the Z-test was derived in Example 7.13 and plotted in Figure 7.7. The total error as a function of the optimal significance level, α, defined by (7.45) is depicted in Figure 7.35. The larger the sample size, the smaller the optimal α, and the larger the effect size the smaller α. $\square$

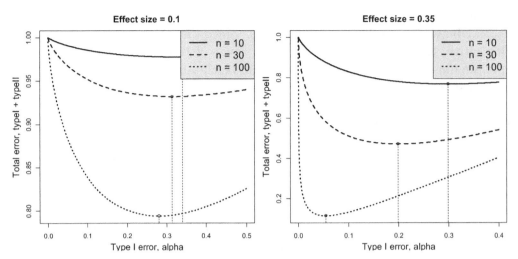

Figure 7.35: *Derivation of the optimal α that minimizes the total error for two effect sizes and three sample sizes. The traditional $\alpha = 0.05$ is optimal when effect size equals 0.35 and $n = 100$. Optimal α increases with small sample size and small effect size.*

The above example is discussed solely for illustrative purpose. Nevertheless, it underscores consideration of two pieces of information required to compute the optimal statistical significance level: (a) quantitative assessment of the consequences of type I and II errors, such as cost, and (b) specification of the effect size. Statistical dogma "5% test size fits all" does not work. The statistical significance level, α, must be justified, or at least discussed, every time because the consequences of rejecting the null hypothesis when it is actually true are problem dependent.

7.10.3   The null hypothesis is always rejected with a large enough sample size

Most statistical tests are consistent. That is, the power of the test converges to 1 when the sample size goes to infinity. Perhaps because of that statisticians avoid saying "accept," preferring the term "do not reject" the null hypothesis. Indeed, collection of enough new data will always lead to rejection of the null hypothesis even if there is a negligible difference from the null value. Consistency of the test has a devastating implication on the p-value: it vanishes with increasing n even if the difference from the alternative is as small as 10^{-10}. To illustrate, consider the simplest Z-test from Section 7.1.1 with

$$p\text{-value} = 2\Phi\left(-\frac{|\overline{Y} - \mu_0|}{\sigma/\sqrt{n}}\right).$$

If the alternative is true $Y_i \sim \mathcal{N}(\mu, \sigma^2)$, where $\mu \neq \mu_0$, and for every $|\mu - \mu_0| = \delta > 0$ from the law of large numbers of Section 2.9 and generalized Slutsky theorem of Section 6.4.4, we have

$$p \lim_{n \to \infty} p\text{-value} = p \lim_{n \to \infty} 2\Phi\left(-\frac{|\overline{Y} - \mu_0|}{\sigma/\sqrt{n}}\right) = 2\Phi\left(-p \lim_{n \to \infty} \frac{|\overline{Y} - \mu_0|}{\sigma/\sqrt{n}}\right)$$

$$= 2\Phi\left(-p \lim_{n \to \infty} \frac{|\mu - \mu_0|}{\sigma/\sqrt{n}}\right) = 2\Phi\left(-p \lim_{n \to \infty} \frac{\delta}{\sigma}\sqrt{n}\right) = 0.$$

The fact that everything is statistically significant for large data sets is well known to scientists who deal with big data such as from Medicare or genomics. The p-value merely tells us how close is $\overline{Y}$ to μ_0 on the probability scale. Not surprisingly, when $n \to \infty$, the average approaches the true mean making p-values smaller and smaller.

The following example illustrates the problem with statistical significance when used for decision making: Traditional hypothesis testing contradicts common sense.

Example 7.70 *Two jobs. A statistician wants to change his job and considers two companies, small and large. The industry salary average at the level of his education and experience is $90K. A sample of 10 salaries from the small company has average $87K with SD=$7K. A sample of 500 salaries from the large company has average $89K and the same SD. What company offers the better package?*

Solution. Although the average in the small company is smaller than the industry average, this may happen due to natural sample variation. Test the null hypothesis that the salary in the small company is the same as in the industry by computing the two-sided p-value = 2*pnorm((87-90)/7*sqrt(10)) =0.1753341. According to statistical dogma, since the p-value > 0.05, the null hypothesis cannot be rejected and statistician is inclined to apply for a small company. The same computation with the large company yields the p-value = 2*pnorm((89-90)/7*sqrt(500)) =0.001401302. The one-sided p-values yield similar results: the difference in salary in the small company from the industry average is not statistically significant but is statistically significant for the large company. A paradox: according to statistical dogma, the statistician should apply for the company with low salaries. We will return to this problem when considering the d-value as an alternative to the p-value. □

Sometimes, the confidence interval is advocated to support the null hypothesis. However, the confidence, like the p-value, shrinks to zero as the sample size gets larger. The p-value, as an indicator of scientific finding, is useless: if p-value > 0.05 one may just increase n and eventually get the desired "statistically significant" result.

7.10.4   Parameter-based inference

The reason the p-value drops to zero with large n is because the null hypothesis is formulated in terms of the population-based *parameter* such as mean, μ. No wonder that the null hypothesis is rejected when a tiny, practically unimportant difference exists between the null and the alternative because the test statistic/estimator converges to the alternative when $n \to \infty$.

Are we really interested in testing the population parameter? In some cases we are, and then the fact that the p-value goes to zero with n should not be disturbing. The following examples illustrate when the p-value makes sense.

Example 7.71 *Who's the next president?* *Candidates A and B are running for president. In the recent poll of 10,000 people 5,122 voted for candidate A and 4,878 for candidate B. Use statistical hypothesis testing to predict who will be the next president (we assume that the president is elected by the maximum number of population votes).*

Solution. Let p denote the true unknown probability that a random person will vote for candidate A. Thus, if $p < 0.5$ then candidate B will win the election and otherwise candidate A. This is an example when the double-sided null hypothesis $H_0 : p = 0.5$ does not make sense because it does not answer the question who will be the new president. The single-sided null hypothesis $H_0 : p \leq 0.5$ with alternative $H_A : p > 0.5$ is more relevant. This is an example where our testing concerns the population-based parameter p because it answers our question. Various testing methods can be used to compute the p-value. The null Wald test p-value is computed as `pnorm((.4878-0.5)/sqrt(0.5*(1-0.5)/1000))=0.00734`. The exact p-value is computed using the binomial distribution as `pbinom(4878, size=10000,prob=0.5)=0.007547342`. The fact that the p-value vanishes, even if the alternative p is just slightly greater than 0.5, when the number of respondents increases is not uncomfortable. However, using a threshold, like $\alpha = 0.05$ to claim "statistical significance" does not help here because it is a matter of the sample size.

Example 7.72 *Will an asteroid hit the Earth?* *An asteroid is flying toward the Earth and there is considerable fear that there will be a devastating contact. n independent astronomical measurements/methods have been taken to project the asteroid trajectory resulted in the mean shortest distance to the Earth $\overline{D}$ with SD estimate $= \widehat{\sigma}$. The asteroid will hit the Earth if it flies closer than D_0 distance from the Earth surface (this value is derived from theoretical considerations and is known). Will the asteroid hit the Earth (measurements of the shortest distance are assumed to be iid normally distributed)?*

Solution. The estimated probability that the asteroid hits the Earth is

$$\widehat{p} = \mathcal{T}\left(\frac{\overline{D} - D_0}{\widehat{\sigma}/\sqrt{n}}, \mathrm{df} = n - 1\right),$$

where $\mathcal{T}$ is the cdf of the central t-distribution with $n-1$ df. This probability is the p-value of the one-sample one-sided t-test with the null hypothesis H_0 : $D_{\text{true}} \geq D_0$ versus alternative $H_A : D_{\text{true}} < D_0$, where D_{true} is the true shortest distance (see Section 7.4). In fact, the p-value approaches 1 or 0. More precisely,

$$p \lim_{n \to \infty} p\text{-value} = \begin{cases} 1 \text{ if } D_{\text{true}} < D_0 \\ 0 \text{ if } D_{\text{true}} \geq D_0 \end{cases}.$$

The fact that the p-value approaches zero when $D_{\text{true}} \geq D_0$ is not uncomfortable: it means that if $D_{\text{true}} \geq D_0$ lots of measurements lead to the conclusion that the asteroid will not hit the Earth. Three lessons can be learned: (i) The p-value makes perfect sense as the probability that the asteroid will hit the Earth. (ii) The fact that the p-value converges to either 0 or 1 depending of the inequality $D_{\text{true}} \geq D_0$ is not uncomfortable. (iii) Statistical hypothesis testing and establishing "statistical significance" does not help here because it requires specification of the threshold type I error. Reporting the p-value is sufficient.   □

The parameter-based inference and NHST takes place if the fact that the p-value vanishes when the sample size increase to infinity is not disturbing. That is, the null hypothesis will be sooner or later be rejected, even if the alternative is negligibly close to null. The above two examples describe exactly this: even one extra vote decides the president, or even $D_0 - D_{\text{true}} \leq 1$ m, life on Earth will be exterminated.

7.10.5   The d-value for individual inference

In many applications, we are interested in *individual* inference. The difference between population-based and individual inference has been discussed earlier in the context of prediction by regression from Section 6.7.3 (Figure 6.15). If the dependent variable is the revenue of the company, the CEO wants to know if they can afford an expansion in future years based on the actual revenue values, not the regression mean (population parameter). Although both approaches yield the same predicted value, individual prediction accounts for uncertainty of the revenue from year to year and consequently the confidence interval will be wider and the probability for a successful company expansion smaller. Similarly, in contrast to the classic confidence interval for the mean, Example 7.47 finds the confidence range for the individual observation. We are mostly interested in knowing the range of income of an individual rather then an abstract value μ.

The goal of this section is to introduce the d-value as the metric for population comparison on the individual scale based on a recent paper by the author [29]. Variants of this metric stemming from reporting $\Pr(X < Y)$ have been proposed previously by several authors including Wolfe and Hogg [106], McGraw and Wong [72] (who used the term *common effect size*), Schuemie et al. [90], and Browne [15]. Here we establish its connection to the p-value and statistical hypothesis testing and other central concepts of statistics such as the ROC curve, effect size and the Mann-Whitney nonparametric test by contrasting parameter-based inference with individual comparison.

To avoid technical detail and concentrate on the concept, we suggest a two-sample comparison problem under the equal-variance assumption having independent observations $X_i \overset{\text{ind}}{\sim} \mathcal{N}(\mu_X, \sigma^2)$ and $Y_j \overset{\text{ind}}{\sim} \mathcal{N}(\mu_Y, \sigma^2)$ from population X and Y, respectively. Moreover, we start with the assumption that σ is known. In many practical problems, we want to know if one population is smaller than another. Following the classical hypothesis testing paradigm, order on populations is expressed in terms of the *population* mean with the null hypothesis $H_0 : \mu_X \geq \mu_Y$ and the alternative hypothesis $H_A : \mu_X < \mu_Y$. Hence we refer to this approach as parameter-based inference and the associated p-value takes a familiar form

$$p\text{-value} = \Phi\left(\frac{\overline{X} - \overline{Y}}{\sqrt{2}\sigma}\sqrt{n}\right) \simeq \Pr(\overline{X} < \overline{Y}),$$

where $\simeq$ means estimation. In words, the p-value in the two-sample test estimates the probability that $\overline{X} < \overline{Y}$. Clearly, the p-value approaches zero when $n \to \infty$ if μ_X is slightly smaller than μ_Y.

In *individual* comparison we are interested in the probability that a random observation from population X is smaller than a random observation from population Y or

$$\Pr(X < Y) \simeq \Phi\left(\frac{\overline{X} - \overline{Y}}{\sqrt{2}\sigma}\right) \overset{\text{def}}{=} d\text{-value}. \tag{7.46}$$

The d-value estimates the probability

$$\Pr(X_i < Y_j) = \Phi\left(\frac{\mu_X - \mu_Y}{\sqrt{2}\sigma}\right), \quad i, j = 1, 2, ..., n,$$

or writing side-by-side

$$p \simeq \Pr(\overline{X} < \overline{Y}), \quad d \simeq \Pr(X_i < Y_j). \tag{7.47}$$

From the computational standpoint, the difference between the two metrics is the presence of $\sqrt{n}$ in the p-value formula. Thus, we can say that the d-value is the n-of-1 p-value. The d-value has a simple interpretation as the chance that a random observation from population X is smaller than a random observation from population Y. Unlike the p-value, the d-value does not vanish with n and converges to the population difference expressed via probability as

$$p \lim_{n\to\infty} d = \Phi\left(\frac{\mu_X - \mu_Y}{\sqrt{2}\sigma}\right).$$

The difference between p- and d-value is illustrated in cartoon Figure 7.36: the larger the sample size the smaller the p-value. From the individual perspective, statistical significance is not relevant. The d-value for comparison of man versus woman is 0.498 – the probability that the IQ of a random woman is smaller than the IQ of a random man is 0.502, so the difference is 0.2%.

Figure 7.36: *Who is the smartest in the family? Husband: "According to the latest, the most extensive research study that involved one million men and one million women, the IQ of a man is higher than the IQ of a woman with the p-value=10^{-6}." Wife: "I don't know what you are talking about: The chance that your IQ is higher than mine is only 0.2%, practically zero."*

The d-value is connected to other statistical concepts, such as effect size, ROC curve and the Mann-Whitney test:

1. The d-value may be viewed as the effect size on the probability scale; see Definition 7.10. Effectively, we convert the sample effect size to the individual effect size by dividing by $\sqrt{n}$. Unlike effect size, probability is more appealing in practical applications when weighing the chances as illustrated in examples below. Many traditional two-sided hypothesis tests and the associated p-values can be turned into d-value by replacing n with 1. For example, the d-value can be expressed via the p-value for the null hypothesis $H_0 : \mu = \mu_0$ versus $H_A : \mu \neq \mu_0$ in the two-sided Z-test of Section 7.3 as

$$d = \Phi\left(\frac{1}{\sqrt{n}}\Phi^{-1}(0.5 \times p)\right). \tag{7.48}$$

2. The d-value is the area above the ROC curve in the discrimination of population X versus population Y (thus the d),

$$d \simeq \Pr(X < Y) = 1 - \text{AUC}.$$

See formula (5.7) from Section 5.1.1. To report the benefit value, we use the complementary probability,

$$b = 1 - d = \text{AUC}.$$

For example, if the first group of patients (X) receives no treatment and the second group (Y) receives a treatment, supposed to lower cholesterol, the b-value computed as $\Phi((\overline{Y} - \overline{X})/\sigma)$ estimates the proportion of patients who benefit from the treatment (we assume that $\overline{X} < \overline{Y}$).

3. Due to (7.47), the d-value is unbiasedly estimated as

$$d \simeq \frac{1}{mn} \sum_{i=1}^{m} \sum_{j=1}^{n} 1(X_i < Y_j), \qquad (7.49)$$

where 1 is an indicator function (takes 1 if the inequality holds and 0 if not). Details and the proof are provided in Theorem 5.3. This formula does not require normal distribution and is referred to as the nonparametric d-value. The double sum, $U = \sum_{i=1}^{m} \sum_{j=1}^{n} 1(X_i < Y_j)$ is called the Mann-Whitney U test statistic, see more detail in Hollander and Wolfe [56].

The following two examples illustrate application of the d-value for decision making on the individual basis. Solutions in both examples address a specific question, not just computing the p-value and claiming (or not claiming) "statistical significance."

Example 7.73 *Two jobs continued.* *Use the d-value to decide what company to apply to.*

Solution. The paradoxical decision to apply for the company with low salaries in Example 7.70 was driven by the difference in the sample size when computing the p-value. Instead, we appeal to estimating the probability that a randomly chosen employee has a salary greater than in the industry and decide what company to apply to, which corresponds to the highest probability. An estimator of this probability is the d-value $\simeq \Pr(\overline{Z} > 90)$ where $\overline{Z}$ is the average salary in the company. For the small company, we have `pnorm((87-90)/7)=0.3341176` and for the large company we have d-value = `pnorm((89-90)/7)=0.4432015`. Note that we do not have $\sqrt{2}$ in the d-value like in formula (7.46) because the industry average is known (one-group comparison). Since the chance of earning more in the large company is greater the statistician should apply to the large company – it makes sense now.

The next example illustrates how the d-value can be used when several treatments are compared. Certainly, the treatments can be compared using the effect size, but the probabilistic scale has more appeal for practitioners especially when weighing options.

Example 7.74 *Drug or not to drug?* *A pharmaceutical company developed a new drug to lower cholesterol. In a phase II clinical trial, $m = 600$ patients were given a placebo (`treatm=0`) and $n = 500$ patients were given the drug (`treatm=1`). The percent difference of the cholesterol level from the baseline after completion the course of the treatment was recorded; see file `saldisc.csv`. The results of the trial were submitted to the Food and Drug Administration. Give the drug a go or no go?*

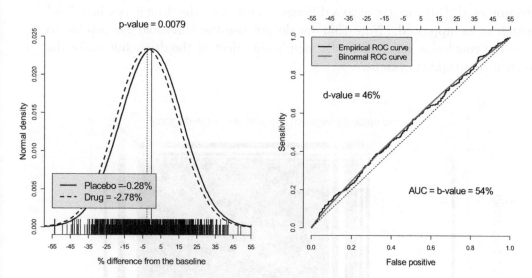

Figure 7.37: *Cholesterol distributions in two groups of patients, placebo ($m = 600$) and new drug ($n = 500$). Although the drug effect is statistically significant the chance that an individual will benefit from taking the dug is 54%. If an individual switches to a low-fat diet, the chance of lowering cholesterol is 60%. Therefore the drug effect is not substantial enough to be approved.*

Solution. The R function is `saldisc` and the call `saldisc(job=1)` produces Figure 7.37. The mean % difference for placebo (X) and drug (Y) groups are -0.028% (the so-called placebo effect) and -2.78%, respectively. The one-sided t-test under the assumption that the variances in the two groups are the same produces the p-value $= 0.0079 < 0.05$. According to statistical dogma, the drug is declared statistically significant and as such can be approved. The trouble is that the drug improves cholesterol level only by $2.78 - 0.28 = 2.5\%$ or in the scale of the effect size $2.5/17.07 = 0.15$. The low p-value reflects the difference in the means over 600 placebo and 500 drug patients, but the drug is supposed to lower the cholesterol level of the individual! A better way to communicate the treatment effect is to estimate the probability that a randomly chosen individual from the placebo group will have lower cholesterol level than a random individual from the drug group estimated by the d-value $= \Pr(X < Y) \simeq$

pnorm$((0.028-2.78)/17.07)$=0.459 with the b-value $= 1 - 0.459 = 0.541$. This means that only 54% of individuals benefit from taking the drug. Moreover, this probability can be used to compare the drug effect with an alternative treatment such as lowering cholesterol by switching to a low-fat diet which includes oats, garlic, nuts, etc. It is important to realize that the p-values with different treatments cannot be compared because different numbers of individuals are involved. Even if the same number of individuals are involved, the p-value comparison will be equivalent to the d-value comparison, but the latter has a better interpretation. Assuming that the low-fat diet improves cholesterol level in 60% of individuals (Brinton et al. [14]), we come to a different conclusion: the drug is not beneficial enough to be approved. Note that we do not use the threshold for p-value to make the conclusion on statistically significant effect of the drug, but make the decision by weighing options.

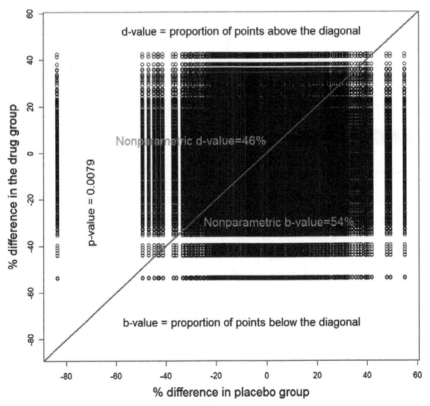

Figure 7.38: *300,000 points (X_i, Y_j) for the nonparametric d- and b-value in the spirit of the Mann-Whitney test. The parametric and nonparametric values practically coincide. The d-value is the proportion of points above the 45° line.*

The plot at right in Figure 7.37 depicts the empirical and binormal ROC curves for discrimination of placebo against drug groups, see Section 5.1.1. The `side=3` axis shows the respective threshold for discrimination.

The (X_i, Y_j) points for computing the nonparametric d- and b-values are shown in Figure 7.38. This figure is obtained by calling `saldisc(job=2)`. Individuals with higher cholesterol level are below the 45° line. Note that the distribution of points on the vertical lines is the same (the same is true for points on the horizontal lines). An individual in the placebo group improved his cholesterol level by more than 80% (leftmost set of points) and can be regarded as an outlier. The nonparametric d-value is the proportion of points above the diagonal. Equivalently, this proportion can be computed as the area under the empirical ROC curve in Figure 7.37 as the Riemann sum, see Theorem 5.3. The nonparametric and parametric d-values are practically the same. □

As with any population parameter, the d-value may be subject to statistical inference, such as computing a confidence interval (the same method can be applied to construction of the CI for the p-value, see Boos and Stefanski [11]).

Example 7.75 *CI for the d-value.* *Adopt Example 7.54 to compute the $(1 - \alpha)100\%$ CI for the d-value for the one- and two-sample t-tests.*

Solution. In the case of the one-sample one-sided t-test, the theoretical d-value takes the form $\Phi\left((\mu - \mu_0)/\sigma\right)$ estimated as $\Phi\left((\overline{Y} - \mu_0)/\widehat{\sigma}\right)$, where $\mu \leq \mu_0$. The lower and upper limits of the exact CI for the d-value are computed as $\Phi(c_L)$ and $\Phi(c_U)$, where c_L and c_U are the confidence limits for the effect size $c = (\mu - \mu_0)/\sigma$, computed as specified in Example 7.54. These limits can be computed exactly from the noncentral t-distribution or approximately. In the case of the two-sample one-sided t-test, we want to find a CI for $\Phi((\mu_X - \mu_Y)/\sqrt{2\sigma^2})$ using statistic $(\overline{X} - \overline{Y})/\widehat{\sigma}(1/m + 1/n)^{-1/2}$, which has a noncentral distribution with $m + n - 2$ df and the noncentrality parameter $(\mu_X - \mu_Y)/\sigma(1/m + 1/n)^{-1/2}$. □

An extension of the d-value to linear models is found in Section 8.5 and its application in Example 8.32 and Section 8.6.

Problems

1. (a) Provide the p-value interpretation for the one-sided t-test in Example 7.67. (b) Following the line of this example explain the p-value using the gun ownership study based on the binomial proportion test from Section 7.6.3.

2. Repeat Example 7.69 under the assumption that the total cost of the breast cancer treatment at the later stage is \$500K. Plot the graphs as in Figure 7.35.

3. Demonstrate graphically and by simulations that the p-value vanishes when n goes to infinity but the d-value converges to its theoretical counterpart.

4. (a) Prove that the p-value in the one-sample one-sided t-test vanishes with n when $\mu \neq \mu_0$ analogously to the proof from Section 7.10.3. (b) Prove

that the p-value in the two-sample one-sided t-test vanishes when $\mu_X \neq \mu_Y$ if $\min(m, n) \to \infty$. [Hint: Use the results of Section 6.4.4.]

5. Derive the limits of the p-value as in the previous problem for one-sided t-tests.

6. Demonstrate by computation using formula (7.48) that the p- and d-value in the caption of Figure 7.36 match each other.

7. Rephrase Example 7.4 to answer the question on the individual prediction and compute the d-value.

8. In Example 7.70, replace the two-sided test with the one-sided test. Does the conclusion on applying to a company with lower salary remain?

9. Compute the d-value in Example 7.1 based on the one-sided test. Describe a practical situation when the d-value is more relevant than the p-value.

10. Is probability 0.00782 in Example 7.3 associated with the p-value or d-value?

11. Modify function dvalREG(job=1) to investigate by simulations which is a better estimator of the theoretical d-value, pnorm or pt.

12. Based on allegations that a worldwide company practices race discrimination when compensating its employees, the salaries of $n = 15$ pair-matched white versus nonwhite employees were surveyed to produce the p-value $= 0.08$. The plaintiff lawyers insisted in conducting an expanded survey with $n = 100$ which produced the p-value 3×10^{-5}. How would you explain the difference in the courtroom referring to formula (7.48)? How would you solve the problem of the race discrimination assuming that the data on the industry difference between compensation of whites and nonwhites is available?

13. Compute the d-value in Example 7.5 to predict that a specific house is priced under \$350K. Use (a) normal assumption of prices from the file houseprice.txt. (b) Assume the lognormal distribution of prices. (c) Compute the nonparametric estimate of the d-value.

14. Compute and provide an interpretation of the d-value in Example 7.16.

15. In view of Example 7.75, (a) develop the exact CI for the d-value using the uniroot function and (b) use simulations based on function cic to report on the coverage probability of the d-value of the exact and approximate CIs.

16. The same as above but apply to the p-value.

Chapter 8

Linear model and its extensions

Linear model is arguably the most popular statistical model. It combines tractable theory with flexibility – many data analyses can be reduced to a linear model. The simple linear regression model was studied earlier in Section 6.7 and we strongly recommend the reader review the material. This chapter discusses the *multiple linear model*, sometimes called *multivariate* or *linear regression* with multiple predictors, sometimes called independent or explanatory variables. Matrix formulation is a primary method of statistical treatment in this chapter – we recommend the reader consult the Appendix, as well as Section 3.10.

The chapter is organized in the following way: First, we formulate general properties of linear model and then illustrate its applications with real-life examples. Next, we extend the linear model to study tabular data using the ANOVA model, and finally discuss the generalized linear model when the dependent variable is binary (logistic regression) or count (Poisson regression).

8.1 Basic definitions and linear least squares

Multiple linear model (regression) relates the dependent variable, Y, to several independent variables (or predictors) $x_1, x_2, ..., x_{m-1}$ in a linear fashion as

$$Y_i = \beta_1 + \beta_2 x_{i1} + ... + \beta_m x_{i,m-1} + \varepsilon_i, \quad i = 1, 2, ..., n, \tag{8.1}$$

where i codes the ith observation. The matter of concern is estimation and testing the beta coefficients. Unobservable error terms $\varepsilon_1, \varepsilon_2, ..., \varepsilon_n$ have zero mean, constant variance σ^2, and are mutually independent (or uncorrelated with each other). Parameter β_1 is the intercept and formally can be expressed as the coefficient at the independent variable that takes value 1. The presence of the

---

Advanced Statistics with Applications in R, First Edition. Eugene Demidenko.

intercept term is not required, but it is typical. Coefficients at the predictors, β_j, $j \geq 2$ are referred to as the slope coefficients, and their statistical treatment is the major focus of this chapter. It is important to remember that, in the classic linear model, Y_i and ε_i are random variables but x_{ij} are fixed/nonrandom. Although this contradicts many real-life examples where x_{ij} are observed (not designed), it simplifies the theory dramatically – see more discussion in Section 8.4.5.

Now we introduce the following vectors and matrices:

$$
\mathbf{y}^{n \times 1} = \begin{bmatrix} Y_1 \\ Y_2 \\ \vdots \\ Y_n \end{bmatrix}, \quad
\mathbf{X}^{n \times m} = \begin{bmatrix} 1 & x_{11} & \cdots & x_{1,m-1} \\ 1 & x_{21} & \cdots & x_{2,m-1} \\ \vdots & \vdots & \vdots & \vdots \\ 1 & x_{n1} & \cdots & x_{n,m-1} \end{bmatrix}, \quad
\boldsymbol{\beta}^{m \times 1} = \begin{bmatrix} \beta_1 \\ \beta_2 \\ \vdots \\ \beta_m \end{bmatrix}, \quad
\boldsymbol{\varepsilon}^{n \times 1} = \begin{bmatrix} \varepsilon_1 \\ \varepsilon_2 \\ \vdots \\ \varepsilon_n \end{bmatrix}.
$$

Here, we use lowercase bold to denote the random vector $\mathbf{y}$ instead of the uppercase to distinguish it from matrix. Sometimes, since x_{ij} are nonrandom, we call $\mathbf{X}$ a *design matrix*. Using the above notation, the linear model can be compactly written in vector/matrix form as

$$
\mathbf{y} = \mathbf{X}\boldsymbol{\beta} + \boldsymbol{\varepsilon}. \tag{8.2}
$$

In order to estimate $\boldsymbol{\beta}$ uniquely, we assume the *identifiability condition*:

$$
\text{rank}(\mathbf{X}) = m < n. \tag{8.3}
$$

We say that matrix $\mathbf{X}$ has full rank; $m < n$ means that the number of observations (sample size) is greater than the number of the beta parameters to estimate. Condition (8.3) means that the column vectors of matrix $\mathbf{X}$ are linearly independent: no column can be expressed as a linear combination of others. If condition (8.3) does not hold, the coefficients cannot be uniquely estimated.

Linear model (8.2) can be succinctly written as

$$
\mathbf{y} \sim (\mathbf{X}\boldsymbol{\beta}, \sigma^2 \mathbf{I}) \tag{8.4}
$$

meaning that $E(\mathbf{y}) = \mathbf{X}\boldsymbol{\beta}$ and $\text{cov}(\mathbf{y}) = \sigma^2 \mathbf{I}$. Note that (8.4) specifies only the first two moments of $\mathbf{y}$ without specifying the distribution. In a sense, (8.4) may be referred to as a semiparametric statistical model. If $\mathbf{y}$ has a multivariate normal distribution, we write

$$
\mathbf{y} \sim \mathcal{N}(\mathbf{X}\boldsymbol{\beta}, \sigma^2 \mathbf{I}). \tag{8.5}
$$

According to model (8.4), Y_i and Y_k do not correlate for $i \neq k$, and, according to model (8.5), they are mutually independent. Since matrix $\mathbf{X}$ is fixed, $\text{cov}(\boldsymbol{\varepsilon}) = \text{cov}(\mathbf{y})$. The errors have the same variance (we call them homoscedastic), and under model (8.5), they are normally distributed.

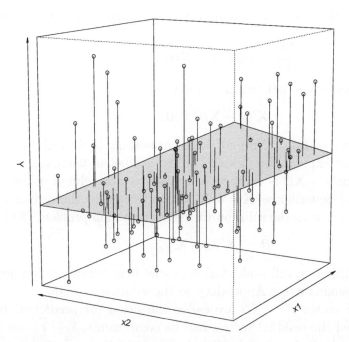

Figure 8.1: *Observation-space geometry of the OLS ($m = 2$). The least squares plane minimizes the sum of squared vertical distances between observation Y_i and the projected point (fitted value). Function* olsg() *saves 360 *.jpg files in the folder* olsg *for viewing at all angles.*

Linear model can be expressed in terms of vector columns of matrix $\mathbf{X}$ as

$$\mathbf{y} = \beta_1 \mathbf{x}_1 + \beta_2 \mathbf{x}_2 + \ldots + \beta_m \mathbf{x}_m + \varepsilon,$$

where $\mathbf{X}$ is composed of the n-dimensional vectors $\{\mathbf{x}_j, j = 1, \ldots, m\}$, or in other notation $\mathbf{X} = [\mathbf{x}_1, \mathbf{x}_2, \ldots, \mathbf{x}_m]$ and, in the presence of the intercept, $\mathbf{x}_1$ is a vector of ones.

Similarly to simple linear regression from Section 6.7, the least squares is used to estimate the beta coefficients as the solution of an optimization problem:

$$\min_{\beta} \|\mathbf{y} - \mathbf{X}\beta\|^2. \tag{8.6}$$

The ordinary least squares (OLS) defined by this criterion finds the plane in R^{m+1} that minimizes the sum of squared vertical distances (along the y-axis); see Figure 8.1. This geometric illustration will be referred to as the *observation-space geometry*.

The function to minimize is quadratic in β and therefore admits a closed-form solution. Define the *residual sum of squares (RSS)* as a function of β:

$$S(\beta) = \|\mathbf{y} - \mathbf{X}\beta\|^2, \quad \beta \in R^m. \tag{8.7}$$

The necessary condition for the minimum is that the derivative at the solution is zero,

$$\frac{\partial S}{\partial \boldsymbol{\beta}} = -2\mathbf{X}'(\mathbf{y} - \mathbf{X}\boldsymbol{\beta}) = \mathbf{0}.$$

The system of m equations, written in vector form as

$$\mathbf{X}'(\mathbf{y} - \mathbf{X}\boldsymbol{\beta}) = \mathbf{0}, \tag{8.8}$$

is called the *normal equation*. "Normal" here has nothing to do with the normal distribution, see the explanation below. In the geometric language, we write $\mathbf{y} - \mathbf{X}\boldsymbol{\beta} \perp \mathbf{X}$: vector $\mathbf{y} - \mathbf{X}\boldsymbol{\beta}$ is orthogonal to each vector column of matrix $\mathbf{X}$. Rewrite the normal equation as $\mathbf{X}'\mathbf{y} = \mathbf{X}'\mathbf{X}\boldsymbol{\beta}$ and come to the ordinary least squares (OLS) estimator as the solution to the optimization problem (8.6)

$$\widehat{\boldsymbol{\beta}} = (\mathbf{X}'\mathbf{X})^{-1}\mathbf{X}'\mathbf{y}. \tag{8.9}$$

Note that the condition on full rank (8.3) is crucial here because it implies that matrix $\mathbf{X}'\mathbf{X}$ is nonsingular (see Appendix), so the solution is unique.

Components of vector $\widehat{\mathbf{y}} = \mathbf{X}\widehat{\boldsymbol{\beta}}$ are called the *fitted (or predicted) values*. Vector $\mathbf{y} - \widehat{\mathbf{y}}$ is called the residual vector, and its components, $Y_i - \widehat{Y}_i$, are called least squares *residuals*. In Figure 8.1, $\widehat{Y}_i$ is the height of the projected point and the residual, $Y_i - \widehat{Y}_i$, is the difference between the observation and the fitted value.

In matrix form, the fitted values can be expressed as the projection of $\mathbf{y}$ onto the linear space generated by vector columns of matrix $\mathbf{X}$, or algebraically, $\widehat{\mathbf{y}} = \mathbf{P}\mathbf{y}$, where $\mathbf{P}$ is the $n \times n$ matrix, called the *hat (or projection) matrix* because it "puts a hat on $\mathbf{y}$,"

$$\mathbf{P} = \mathbf{X}(\mathbf{X}'\mathbf{X})^{-1}\mathbf{X}'. \tag{8.10}$$

This matrix is symmetric and idempotent. The minimum sum of squares can be expressed in matrix form using matrix

$$\mathbf{H} = \mathbf{I} - \mathbf{P} = \mathbf{I} - \mathbf{X}(\mathbf{X}'\mathbf{X})^{-1}\mathbf{X}',$$

sometimes called the *annihilator matrix*. Geometrically, as follows from the normal equation, $\mathbf{y} - \widehat{\mathbf{y}} \perp \mathbf{X}$, or algebraically, $\mathbf{X}'(\mathbf{y} - \widehat{\mathbf{y}}) = \mathbf{0}$. This explains why the fitted values are sometimes called the projected values. This follows directly from the normal equation (8.8),

$$\mathbf{X}'(\mathbf{y} - \mathbf{P}\mathbf{y}) = \mathbf{X}'\mathbf{H}\mathbf{y} = (\mathbf{X}' - \mathbf{X}'\mathbf{X}(\mathbf{X}'\mathbf{X})^{-1}\mathbf{X}')\mathbf{y} = (\mathbf{X}' - \mathbf{X}')\mathbf{y} = \mathbf{0}.$$

Matrix $\mathbf{H}$ is also symmetric and idempotent, $\mathbf{H}^2 = \mathbf{H}$. Now we prove that

$$\min_{\boldsymbol{\beta}} \|\mathbf{y} - \mathbf{X}\boldsymbol{\beta}\|^2 = \mathbf{y}'\mathbf{H}\mathbf{y}.$$

Indeed, the minimum RSS can be expressed as

$$\left\|\mathbf{y} - \mathbf{X}\widehat{\boldsymbol{\beta}}\right\|^2 = \left\|\mathbf{y} - \mathbf{X}(\mathbf{X}'\mathbf{X})^{-1}\mathbf{X}'\mathbf{y}\right\|^2 = \mathbf{y}'\mathbf{H}'\mathbf{H}\mathbf{y} = \mathbf{y}'\mathbf{H}\mathbf{y}.$$

Example 8.1 *Residuals and fitted values.* *Show that (a) the sum of residuals in a linear model with no intercept, and (b) the average of original observations equals the average of the fitted values.*

Solution. (a) Consider the first equation/component of the normal equation (8.8). Since $\mathbf{1}$ is the first column the first equation reads $\sum r_i = 0$ where $r_i = Y_i - \widehat{Y}_i$ is the ith residual. (b) Rewrite the previously derived equation as $\sum Y_i/n = \sum \widehat{Y}_i/n$, which means $\overline{Y} = \overline{\widehat{Y}}$.

Example 8.2 *The OLS for the simple linear regression.* *Using general formula (8.9), derive the OLS estimators for (i) the mean, (ii) slope, and (ii) and simple linear regression model.*

Solution. (i) The mean model, or the intercept model from Section 6.5, has no predictors, $Y_i = \beta_1 + \varepsilon_i$. In vector notation it can be written as $\mathbf{y} = \beta_1 \mathbf{1} + \boldsymbol{\varepsilon}$. For this model, $\mathbf{X'X} = \mathbf{1'1} = n$, $(\mathbf{X'X})^{-1} = 1/n$, $\mathbf{X'y} = \mathbf{1'y} = \sum Y_i$, and the OLS estimator (8.9) turns into the usual mean, $\widehat{\beta}_1 = \sum Y_i/n = \overline{Y}$.

(ii) The slope model from Section 6.7 has no intercept and a single predictor, $Y_i = \beta_1 x_i + \varepsilon_i$. In vector form, $\mathbf{y} = \beta_1 \mathbf{x} + \boldsymbol{\varepsilon}$. Then, $\mathbf{X'X} = \|\mathbf{x}\|^2 = \sum x_i^2$, $(\mathbf{X'X})^{-1} = 1/\|\mathbf{x}\|^2$, $\mathbf{X'y} = \mathbf{x'y} = \sum x_i Y_i$. The OLS estimator (8.9) turns into $\widehat{\beta}_1 = \mathbf{x'y}/\|\mathbf{x}\|^2 = \sum x_i Y_i / \sum x_i^2$, as derived in Example 6.36.

(iii) The simple linear regression is a combination of the two models and can be written as $\mathbf{y} = \beta_1 \mathbf{1} + \beta_2 \mathbf{x} + \boldsymbol{\varepsilon}$. In formulation (8.2), $\mathbf{X} = [\mathbf{1}, \mathbf{x}]$ and $\boldsymbol{\beta}^{2\times1} = (\beta_1, \beta_2)'$, and we have

$$\mathbf{X'X} = \begin{bmatrix} \mathbf{1'1} & \mathbf{x'1} \\ \mathbf{1'x} & \mathbf{x'x} \end{bmatrix}, \quad (\mathbf{X'X})^{-1} = \frac{1}{n\sum x_i^2 - (\sum x_i)^2} \begin{bmatrix} \sum x_i^2 & -\sum x_i \\ -\sum x_i & n \end{bmatrix},$$

$$\mathbf{X'y} = \begin{bmatrix} \mathbf{1'y} \\ \mathbf{x'y} \end{bmatrix} = \begin{bmatrix} \sum_{i=1}^{n} Y_i \\ \sum_{i=1}^{n} x_i Y_i \end{bmatrix}$$

Therefore,

$$\widehat{\boldsymbol{\beta}} = \frac{1}{n\sum x_i^2 - (\sum x_i)^2} \begin{bmatrix} \sum x_i^2 & -\sum x_i \\ -\sum x_i & n \end{bmatrix} \begin{bmatrix} \sum Y_i \\ \sum x_i Y_i \end{bmatrix} = \begin{bmatrix} \frac{\sum x_i^2 \sum Y_i - \sum x_i \sum x_i Y_i}{n\sum x_i^2 - (\sum x_i)^2} \\ \frac{n\sum x_i Y_i - (\sum x_i)(\sum Y_i)}{n\sum x_i^2 - (\sum x_i)^2} \end{bmatrix}.$$

The slope and intercept, after some algebra, take the familiar form presented in Section 6.7,

$$\widehat{\beta}_2 = \frac{\sum x_i Y_i - n\left(\sum x_i/n\right)\left(\sum Y_i/n\right)}{\sum x_i^2 - n\left(\sum x_i/n\right)^2} = \frac{\sum (x_i - \overline{x})(Y_i - \overline{Y})}{\sum (x_i - \overline{x})^2},$$

$$\widehat{\beta}_1 = \frac{\sum x_i^2 \sum Y_i/n - \sum x_i/n \sum x_i Y_i}{\sum x_i^2 - n\left(\sum x_i/n\right)^2} = \overline{Y} - \widehat{\beta}_2 \overline{x}.$$

Example 8.3 *Transformed OLS.* *The design matrix,* $\mathbf{X}$, *is linearly transformed by the rule:* $\mathbf{X} \rightarrow \mathbf{X}_T \overset{\text{def}}{=} \mathbf{XT}$, *where* $\mathbf{T}$ *is a square* $m \times m$ *nonsingular matrix. How is the OLS estimate transformed?*

Solution. First of all, we observe that the transformed design matrix has full rank because matrix $\mathbf{T}$ is nonsingular. Therefore the identifiability condition holds. Second, by the definition, the OLS estimator for the transformed design matrix takes the form

$$\widehat{\boldsymbol{\beta}}_T = \left(\mathbf{X}_T'\mathbf{X}_T\right)^{-1}\mathbf{X}_T'\mathbf{y} = \left(\mathbf{T}'\mathbf{X}'\mathbf{XT}\right)^{-1}\mathbf{T}'\mathbf{X}'\mathbf{y} = \mathbf{T}^{-1}\left(\mathbf{X}'\mathbf{X}\right)^{-1}\mathbf{T}'^{-1}\mathbf{T}'\mathbf{X}'\mathbf{y}$$
$$= \mathbf{T}^{-1}\left(\mathbf{X}'\mathbf{X}\right)^{-1}\mathbf{X}'\mathbf{y} = \mathbf{T}^{-1}\widehat{\boldsymbol{\beta}}.$$

In words, to find the OLS estimate for the transformed data, one must conduct the back transformation of the original OLS estimate. For example, if one of the predictors was measured in pounds and one needs to know its slope estimate on the scale of kilograms, multiply the estimate by 0.4536. The linear transformation of the beta coefficients upon linear transformation of the design matrix is a characteristic property of the linear model. Alternatively, to demonstrate the relationship between $\widehat{\boldsymbol{\beta}}_T$ and $\widehat{\boldsymbol{\beta}}$, one could appeal to the least square criterion $\|\mathbf{y} - \mathbf{X}\boldsymbol{\beta}\|^2 = \|\mathbf{y} - \mathbf{XTT}^{-1}\boldsymbol{\beta}\|^2 = \|\mathbf{y} - \mathbf{X}_T\boldsymbol{\beta}_T\|^2$, which means that the beta coefficients after transformation can be computed as $\widehat{\boldsymbol{\beta}}_T = \mathbf{T}^{-1}\widehat{\boldsymbol{\beta}}$.

8.1.1   Linear model with the intercept term

The goal of this section is to prove that the intercept term can be eliminated by averaging out $\mathbf{y}$ and $\mathbf{X}$.

Theorem 8.4 *The slope estimates of the linear model with the intercept term can be obtained as slope estimates of a linear model without intercept term by averaging out the data, i.e. by replacing* Y_i *and* x_{ij} *with* $Y_i - \overline{Y}$ *and* $x_{ij} - \overline{x}_j$, *respectively.*

Proof. Let $\mathbf{X} = [\mathbf{1}, \mathbf{X}_s]$ and $\widehat{\boldsymbol{\beta}} = (\beta_1, \boldsymbol{\beta}_s')'$ where the vector columns of the $n \times (m-1)$ matrix $\mathbf{X}_s$ are $\{\mathbf{x}_1, ..., \mathbf{x}_{m-1}\}$ and $\boldsymbol{\beta}_s$ is the $(m-1) \times 1$ vector of the slope coefficients. Introduce the $n \times 1$ vector and the $n \times (m-1)$ matrix defined as $\overset{\circ}{\mathbf{y}} = \mathbf{H}_1\mathbf{y}$ and $\overset{\circ}{\mathbf{X}} = \mathbf{H}_1\mathbf{X}_s$ where $\mathbf{H}_1 = \mathbf{I} - \mathbf{11}'/n$ is a $n \times n$ idempotent matrix, $\mathbf{H}_1^2 = \mathbf{H}_1$ (we use $_1$ in the subscript because $\mathbf{H}_1$ is a special case of the annihilator matrix $\mathbf{H}$ with $\mathbf{X} = \mathbf{1}$). It is easy to see that the ith element of vector $\overset{\circ}{\mathbf{y}}$ is $Y_i - \overline{Y}$ and the (i, j)th element of matrix $\overset{\circ}{\mathbf{X}}$ is $x_{ij} - \overline{x}_j$. One may call matrix $\mathbf{H}_1$ an *average-correcting* matrix. In this notation, the OLS estimate (8.9) can be rewritten as

$$\begin{bmatrix} \widehat{\beta}_1 \\ \widehat{\boldsymbol{\beta}}_s \end{bmatrix} = \begin{bmatrix} n & \mathbf{1}'\mathbf{X}_s \\ \mathbf{X}_s'\mathbf{1} & \mathbf{X}_s'\mathbf{X}_s \end{bmatrix}^{-1} \begin{bmatrix} \mathbf{1}'\mathbf{y} \\ \mathbf{X}_s'\mathbf{y} \end{bmatrix}.$$

Now apply formula (10.12) for the partitioned matrix inverse (we are interested in the second row of the inverse matrix). In the notation of the Appendix, we have

$$\mathbf{E} \stackrel{\mathrm{def}}{=} \mathbf{X}_2'\mathbf{X}_2 - \frac{1}{n}(\mathbf{X}_2'\mathbf{1})(\mathbf{X}_2'\mathbf{1})' = (\mathbf{H}_1\mathbf{X}_2)'(\mathbf{H}_1\mathbf{X}_2) = \overset{\circ}{\mathbf{X}}{}'\overset{\circ}{\mathbf{X}},$$

so that the covariance matrix of the slope coefficients is $\sigma^2 \left(\overset{\circ}{\mathbf{X}}{}'\overset{\circ}{\mathbf{X}} \right)^{-1}$, see Theorem 8.8. In particular, the $(m-1) \times (m-1)$th block of matrix $(\mathbf{X}'\mathbf{X})^{-1}$ is $\left(\overset{\circ}{\mathbf{X}}{}'\overset{\circ}{\mathbf{X}} \right)^{-1}$. Finally the OLS slopes can be computed as

$$\widehat{\beta}_s = -\frac{1}{n}\mathbf{E}^{-1}\mathbf{X}_2'\mathbf{1}\mathbf{1}'\mathbf{y} + \mathbf{E}^{-1}\mathbf{X}_2'\mathbf{y} = \mathbf{E}^{-1}(\mathbf{H}_1\mathbf{X}_2)'\mathbf{H}_1\mathbf{y} = \left(\overset{\circ}{\mathbf{X}}{}'\overset{\circ}{\mathbf{X}} \right)^{-1} \overset{\circ}{\mathbf{X}}{}'\overset{\circ}{\mathbf{y}},$$

as we intended to show. □

This theorem reduces the dimension of the inverse matrix from $m \times m$ to $(m-1) \times (m-1)$. As a corollary, one can assert that the estimates of slope coefficients do not change if a constant is added to some or all predictors. For example, the OLS slope coefficient in the simple linear regression is equal to the OLS coefficient in the slope model upon averaging out.

8.1.2 The vector-space geometry of least squares

As was mentioned above, typically, a linear model contains an intercept term. Since we are mostly interested in the slope coefficients, the intercept term can be eliminated through averaging as indicated in Theorem 8.4. Thus in this section, when referring to $\mathbf{y}$ or $\mathbf{x}_j$ we mean $\mathbf{y} - \overline{Y}\mathbf{1}$ and $\mathbf{x}_j - \overline{x}_j\mathbf{1}$. After subtracting the averages, the vectors are orthogonal if and only if they have zero correlation. Note that correlation refers here to fixed $\mathbf{x}_j$ in the algebraic sense (see more discussion in Section 8.4.5).

In the *vector-space geometry,* observations on the dependent and independent variables are combined in vectors – see Figure 8.2 where $n = 3$ and $m = 2$. That is, the linear model takes the form $\mathbf{y} = \beta_1\mathbf{x}_1 + \beta_2\mathbf{x}_2 + \boldsymbol{\varepsilon}$. The plane spanned by $\mathbf{x}_1$ and $\mathbf{x}_2$ is shown as a gray parallelogram: $\widehat{\mathbf{y}} = \widehat{\beta}_1\mathbf{x}_1 + \widehat{\beta}_2\mathbf{x}_2$ is the closest to $\mathbf{y}$ as the solution to the optimization problem (8.6). As follows from the normal equation (8.8), vector $\mathbf{y} - \widehat{\mathbf{y}}$ is orthogonal to the plane and, as such, is orthogonal to every vector on the plane, including $\mathbf{x}_1$ and $\mathbf{x}_2$. Symbolically,

$$\mathbf{y} - \widehat{\mathbf{y}} \perp \mathbf{x}_1, \quad \mathbf{y} - \widehat{\mathbf{y}} \perp \mathbf{x}_2. \tag{8.11}$$

In statistical language, we interpret this by saying that the residuals do not correlate with the predictors.

In Figure 8.2, the square angle is 90° in compliance with the normal equation. Geometrically, $\widehat{\mathbf{y}}$ is the base of the perpendicular dropped from $\mathbf{y}$ onto the plane spanned by $\mathbf{x}_1$ and $\mathbf{x}_2$. From the parallelogram rule, we have $\widehat{\mathbf{y}} = \widehat{\beta}_1 \mathbf{x}_1 + \widehat{\beta}_2 \mathbf{x}_2$, where $\widehat{\beta}_1$ and $\widehat{\beta}_2$ are the OLS estimates. The orthogonality between residual vectors and $\mathbf{x}_1$ and $\mathbf{x}_2$ written in (8.11) follows from the normal equation (8.8), which can be rewritten as scalar products $(\mathbf{y}-\widehat{\mathbf{y}})'\mathbf{x}_j = 0$, $j = 1,2,...,m$. Therefore, in Figure 8.2, the angle between $\mathbf{y}-\widehat{\beta}_1\mathbf{x}_1$ and $\mathbf{x}_1$ is $90°$; the angle between $\mathbf{y}-\widehat{\beta}_2\mathbf{x}_2$ and $\mathbf{x}_2$ is also $90°$. We call equation (8.8) normal because $\mathbf{y} - \widehat{\mathbf{y}}$ is normal (orthogonal) to the plane spanned by the vector columns of matrix $\mathbf{X}$. The squared cosine angle θ, the angle between $\mathbf{y}$ and the plane spanned by the vector columns of matrix $\mathbf{X}$, is the coefficient of determination; see more discussion in Section 8.1.3. A smaller the angle (greater R^2) indicates a the better regression.

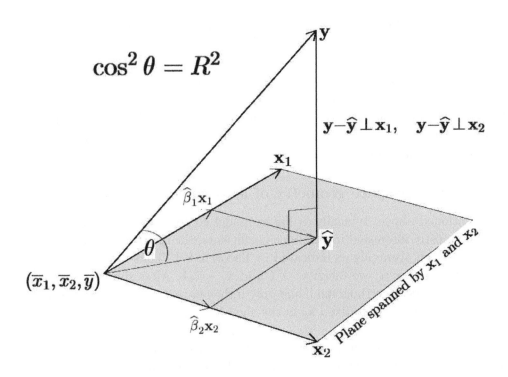

Figure 8.2: *The vector-space geometry of the OLS. The least squares solution finds the point on the plane closest to* $\mathbf{y}$. *This point,* $\widehat{\mathbf{y}}$, *is the base of the perpendicular (projection) dropped from* $\mathbf{y}$ *onto the plane. A smaller angle θ indicates a better regression.*

Theorem 8.5 *Pythagorean theorem of regression analysis. The following decomposition of the sum of squares holds.*

$$\|\mathbf{y}\|^2 = \|\mathbf{y}-\widehat{\mathbf{y}}\|^2 + \|\widehat{\mathbf{y}}\|^2 \tag{8.12}$$

Proof. Rewrite $\|\mathbf{y}\|^2 = \|(\mathbf{y}-\widehat{\mathbf{y}}) + \widehat{\mathbf{y}}\|^2 = \|\mathbf{y}-\widehat{\mathbf{y}}\|^2 + 2(\mathbf{y}-\widehat{\mathbf{y}})'\widehat{\mathbf{y}} + \|\widehat{\mathbf{y}}\|^2$. But the middle term is zero, $(\mathbf{y}-\widehat{\mathbf{y}})'\widehat{\mathbf{y}} = [(\mathbf{y} - \mathbf{X}\widehat{\beta})'\mathbf{X}]\widehat{\beta} = 0$, due to normal equation

(8.8). Note that decomposition (8.12) holds regardless of the presence of the intercept. The sum of squares decomposition can be viewed as the Pythagorean theorem applied to the triangle in Figure 8.2 with the $\mathbf{y}$ as the hypotenuse, and $\mathbf{y}-\widehat{\mathbf{y}}$ and $\widehat{\mathbf{y}}$ as the legs of the right-angled triangle. The decomposition (8.12) can be read as the squared length of the hypotenuse equals the sum of the squared length of the catheti. Since by convention all variables were centered, and as follows from Example 8.1, the average of residuals, $\mathbf{y}-\widehat{\mathbf{y}}$ is zero, we can rewrite decomposition (8.12) as

$$\left\|\mathbf{y}-\overline{Y}\mathbf{1}\right\|^2 = \left\|\mathbf{y}-\widehat{\mathbf{y}}\right\|^2 + \left\|\widehat{\mathbf{y}} - \overline{Y}\mathbf{1}\right\|^2, \tag{8.13}$$

where we use the fact $\overline{Y} = \overline{\widehat{Y}}$ proved earlier in Example 8.1.

Example 8.6 Uncorrelated predictors. *Assuming that predictors in the linear model (8.1) do not correlate, (a) show that the OLS estimate of slope coefficients can be reduced to $m - 1$ simple linear regressions and (d) provide a geometric illustration for $m = 3$.*

Solution. (a) Since predictors do not correlate we have $\sum_{i=1}^n (x_{ij} - \overline{x}_j)(x_{ik} - \overline{x}_k) = 0$ or geometrically $\mathbf{x}_j - \mathbf{1}\overline{x}_j \perp \mathbf{x}_k - \mathbf{1}\overline{x}_k$, $j \neq k$. In the notation of Theorem 8.4, matrix $\overset{\circ}{\mathbf{X}}{}'\overset{\circ}{\mathbf{X}}$ is diagonal,

$$\overset{\circ}{\mathbf{X}}{}'\overset{\circ}{\mathbf{X}} = \begin{bmatrix} \sum(x_{i1} - \overline{x}_1)^2 & 0 & \cdots & 0 \\ 0 & \sum(x_{i2} - \overline{x}_2)^2 & 0 & 0 \\ 0 & 0 & \ddots & 0 \\ 0 & 0 & 0 & \sum(x_{i,m-1} - \overline{x}_{m-1})^2 \end{bmatrix},$$

and therefore, as follows from that theorem, the OLS vector of slopes can be written as

$$\widehat{\boldsymbol{\beta}}_s = \left(\overset{\circ}{\mathbf{X}}{}'\overset{\circ}{\mathbf{X}}\right)^{-1}\overset{\circ}{\mathbf{X}}{}'\overset{\circ}{\mathbf{y}} = \begin{bmatrix} \frac{\sum(x_{i1}-\overline{x}_1)(Y_i-\overline{Y})}{\sum(x_{i1}-\overline{x}_1)^2} \\ \frac{\sum(x_{i2}-\overline{x}_2)(Y_i-\overline{Y})}{\sum(x_{i2}-\overline{x}_2)^2} \\ \vdots \\ \frac{\sum(x_{i,m-1}-\overline{x}_{m-1})(Y_i-\overline{Y})}{\sum(x_{i,m-1}-\overline{x}_{m-1})^2} \end{bmatrix},$$

a stack of slopes from $m - 1$ simple linear regressions, see Section 6.7.

(b) The geometry of the OLS when $\mathbf{x}_1$ and $\mathbf{x}_2$ do not correlate is illustrated in Figure 8.3 where the origin is shifted to the average point in R^3. Here, $\mathbf{x}_1 \perp \mathbf{x}_2$ means that vectors $\mathbf{x}_1 - \mathbf{1}\overline{x}_1$ and $\mathbf{x}_2 - \mathbf{1}\overline{x}_2$ are orthogonal. Marked angles are $90°$. In the bivariate model, vector $\mathbf{y}-\widehat{\mathbf{y}}$ is orthogonal to the plane spanned by $\mathbf{x}_1$ and $\mathbf{x}_2$, so it is orthogonal to vector $\widehat{\mathbf{y}} = \widehat{\beta}_1\mathbf{x}_1 + \widehat{\beta}_2\mathbf{x}_2$. Denote $\widehat{\mathbf{y}}_1$ as the

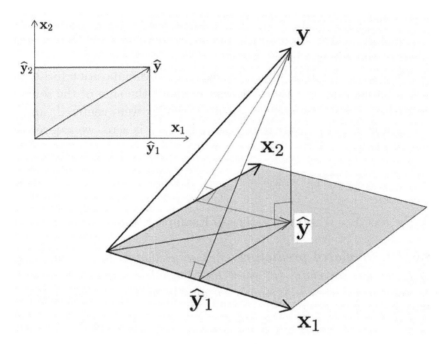

Figure 8.3: *Geometric illustartion when $\mathbf{x}_1 \perp \mathbf{x}_2$: the bivariate regression reduces to two separate simple linear regressions. Marked angles are $90°$ angles.*

projection of vector $\mathbf{y}$ onto the line spanned by vector $\mathbf{x}_1$. That is, $\widehat{\mathbf{y}}_1$ is the vector of fitted values in simple linear regression of Y on $\mathbf{x}_1$. We prove geometrically that $\widehat{\mathbf{y}}_1 = \widehat{\beta}_1 \mathbf{x}_1$ due to the condition that the angle between $\mathbf{x}_1$ and $\mathbf{x}_2$ is $90°$. It suffices to prove that $\widehat{\mathbf{y}} - \widehat{\beta}_1 \mathbf{x}_1 \perp \mathbf{x}_1$. Indeed, since $\widehat{\mathbf{y}} - \widehat{\beta}_1 \mathbf{x}_1$ is parallel to $\mathbf{x}_2$ and $\mathbf{x}_1 \perp \mathbf{x}_2$ we have $\widehat{\mathbf{y}} - \widehat{\beta}_1 \mathbf{x}_1 \perp \mathbf{x}_1$. This follows from elementary geometry: If a vector is orthogonal to the plane, it is orthogonal to every line on the plane. $\square$

The sum of squares decomposition (8.13) is a fundamental result of regression analysis and gives rise to the coefficient of determination, the most popular goodness-of-fit measure.

8.1.3 Coefficient of determination

Coefficient of determination (sometimes referred to as the multiple coefficient of determination) is commonly used to characterize the quality of the regression fit (we assume that regression has the intercept term). In the coordinate form, equation (8.13) is written as

$$\sum_{i=1}^{n}(Y_i - \overline{Y})^2 = \sum_{i=1}^{n}(Y_i - \widehat{Y}_i)^2 + \sum_{i=1}^{n}(\widehat{Y}_i - \overline{Y})^2$$

and can be interpreted as the sum of squares (SS) decomposition

$$\text{Total SS} = \text{Residual SS} + \text{Fitted SS}.$$

The coefficient of determination for the multiple linear model is defined in the way it was defined for the simple linear regression from Section 6.7.4,

$$R^2 = 1 - \frac{\|\mathbf{y}-\widehat{\mathbf{y}}\|^2}{\|\mathbf{y}-\overline{y}\mathbf{1}\|^2}. \tag{8.14}$$

The coefficient of determination has a nice geometric interpretation as the cosine squared of the angle between the vector of the dependent variable and the plane spanned by the independent variables/predictors – see Figure 8.2. Consider the right triangle with vertices $(\overline{x}_1, \overline{x}_2, \overline{y})$, $\widehat{\mathbf{y}}$, and $\mathbf{y}$ with the sine of angle θ defined as

$$\sin \theta = \frac{\|\mathbf{y}-\widehat{\mathbf{y}}\|}{\|\mathbf{y}-\overline{y}\mathbf{1}\|}.$$

Using the identity $\cos^2 \theta = 1 - \sin^2 \theta$, we derive

$$\cos^2 \theta = 1 - \frac{\|\mathbf{y}-\widehat{\mathbf{y}}\|^2}{\|\mathbf{y}-\overline{y}\mathbf{1}\|^2} = R^2.$$

Coefficient of determination is invariant with respect to multiplication of the dependent variable by a constant and invariant with respect to linear transformation of the independent variables; see Example 8.3.

Example 8.7 *Prove that coefficient of determination stays the same or increases with addition of a new predictor.*

Solution. As follows from (8.14), it is sufficient to prove that the RSS does not decrease upon addition of a new predictor. The following general fact is employed: the minimum of a function cannot be smaller if the domain of the minimization expands. Indeed, addition of a new predictor can be viewed as expanding from R^m to R^{m+1}. Therefore, $\|\mathbf{y}-\widehat{\mathbf{y}}\|^2$ cannot decrease. More discussion on this topic is found in Section 8.6.4.   □

Coefficient of determination is routinely interpreted as the proportion of variance of the dependent variable explained by independent variables. In Section 6.7.4, we cautioned that such interpretation may not be valid when Y_i do not have the same mean. We refer the reader to Section 8.4.4, where the generalized coefficient of determination is introduced to address this shortcoming of the traditional R^2.

Problems

1. (a) In a linear model without intercept, $\{\varepsilon_i\}$ do not correlate and have a constant variance, but the mean is not zero, $E(\varepsilon_i) = \eta$. Reduce this model to a linear model with an intercept. Is η estimable? (b) The same question, but the original model has an intercept.

2. Explain why if condition (8.3) does not hold, the beta coefficients cannot be estimated uniquely. [Hint: Use the least squares solution.]

3. (a) Suppose that in a linear model with an intercept, the sum of x_{ij} in each row is zero. Is the model identifiable? Does the model become identifiable after dropping the intercept term? (b) Suppose that, in the linear model (8.2), $\mathbf{Xp} = \mathbf{0}$, where $\mathbf{p}$ is a nonzero vector. Is the model identifiable? (c) The same as above, but the sum in each row of matrix $\mathbf{X}$ is zero. Is the model identifiable?

4. A statistical model has the form $Y_{ij} = \mu + \beta_i + \varepsilon_{ij}$, where $\{\varepsilon_{ij}\}$ are iid with zero mean, $i = 1, 2$ and $j = 1, 2, ..., n$. (a) Express this model as a linear model with a $(2n) \times 3$ design matrix. (b) Is this model identifiable? (c) The same as above, but the model takes the form $Y_{ij} = \beta_i + \varepsilon_{ij}$. Is the model identifiable?

5. Suppose that matrix $\mathbf{X}$ does not have full rank. (a) Does the RSS function $S(\boldsymbol{\beta})$ attain its minimum? (b) Does the minimum exist? (c) Find $\mathbf{p} \neq \mathbf{0}$ such that $S(\boldsymbol{\beta}_0 + \lambda \mathbf{p})$ converges to a finite number when $\lambda \to \infty$ for any vector $\boldsymbol{\beta}_0$.

6. Prove that the OLS slope estimates in a linear mode with an intercept term do not change if one or more predictors change by a constant. Provide two proofs: (a) considering the RSS, and (b) using formula (8.9).

7. Express the mean of the dependent variable as a linear combination of the predictors means in the linear model without intercept term. Use this result to explain why the intercept term is typically present in many linear models.

8. (a) Prove that the projection matrix (8.10) is a symmetric idempotent matrix. (b) Let $\mathbf{u} \in R^n$ and $\widehat{\mathbf{u}}$ is the projection of $\mathbf{u}$ onto the subspace spanned by $\mathbf{X}$. Prove that the projection of $\widehat{\mathbf{u}}$ onto the subspace spanned by $\mathbf{X}$ is the same as $\widehat{\mathbf{u}}$.

9. Referring to Figure 8.2, (a) what are roughly the OLS estimates for β_1 and β_2, (b) can the angle θ be obtuse?

10. (a) Prove inequality $\left\| \mathbf{y} - \widehat{\beta}_1 \mathbf{x}_1 \right\| \geq \left\| \mathbf{y} - \widehat{\beta}_1 \mathbf{x}_1 - \widehat{\beta}_2 \mathbf{x}_2 \right\|$, where $\widehat{\boldsymbol{\beta}} = (\widehat{\beta}_1, \widehat{\beta}_2)'$ is the OLS estimate in the regression of $\mathbf{y}$ on $\mathbf{x}_1$ and $\mathbf{x}_2$, using the result of Example 8.34. (b) Provide a geometric illustration by referring to Figure 8.2.

11. Write R code with n, m, and σ^2 as arguments that first generates random $\boldsymbol{\beta}$ and $\mathbf{X}$ and then $\mathbf{y}$ using a normal distribution. Check decomposition (8.12) numerically.

12. Prove that the coefficient of determination is invariant with respect to multiplication of the dependent variable by a constant and invariant with respect to linear transformation of independent variables.

13. Prove that the multiple coefficient of determination is greater than or equal to the individual coefficient of determination between the dependent variable and each independent variable, and provide a geometric illustration.

14. Prove that if predictors do not correlate as in Example 8.6, the multiple coefficient of determination is equal to the sum of squared correlation coefficients.

8.2   The Gauss–Markov theorem

The Gauss–Markov theorem for a simple linear regression was proved in Section 6.7.1. Here, we extend this result to multiple linear regression. First, we prove that the OLS estimator is unbiased and find its covariance matrix. Then prove that OLS is BLUE (best linear unbiased estimator). Note that we do not specify the distribution of the error term here. That is, model (8.4) is assumed.

Theorem 8.8 *The OLS estimator is unbiased with the covariance matrix*

$$\text{cov}(\widehat{\boldsymbol{\beta}}) = \sigma^2 (\mathbf{X'X})^{-1}. \tag{8.15}$$

Proof. Substituting $\mathbf{y} = \mathbf{X}\boldsymbol{\beta} + \boldsymbol{\varepsilon}$ into formula (8.9), we obtain

$$
\begin{aligned}
\widehat{\boldsymbol{\beta}} &= (\mathbf{X'X})^{-1}\mathbf{X'y} = (\mathbf{X'X})^{-1}\mathbf{X'}(\mathbf{X'}\boldsymbol{\beta} + \boldsymbol{\varepsilon}) \\
&= (\mathbf{X'X})^{-1}\mathbf{X'X}\boldsymbol{\beta} + (\mathbf{X'X})^{-1}\mathbf{X'}\boldsymbol{\varepsilon} = \boldsymbol{\beta} + (\mathbf{X'X})^{-1}\mathbf{X'}\boldsymbol{\varepsilon}.
\end{aligned}
$$

Since $E(\boldsymbol{\varepsilon}) = \mathbf{0}$, the unbiasedness of the OLS estimator follows,

$$E(\widehat{\boldsymbol{\beta}}) = \boldsymbol{\beta} + (\mathbf{X'X})^{-1}\mathbf{X'}E(\boldsymbol{\varepsilon}) = \boldsymbol{\beta}.$$

Now we derive the covariance matrix using the representation from above, $\widehat{\boldsymbol{\beta}} - \boldsymbol{\beta} = (\mathbf{X'X})^{-1}\mathbf{X'}\boldsymbol{\varepsilon}$, and the fact that $\text{cov}(\boldsymbol{\varepsilon}) = \sigma^2\mathbf{I}$,

$$
\begin{aligned}
\text{cov}(\widehat{\boldsymbol{\beta}}) &= E(\widehat{\boldsymbol{\beta}} - \boldsymbol{\beta})(\widehat{\boldsymbol{\beta}} - \boldsymbol{\beta})' = E\left[(\mathbf{X'X})^{-1}\mathbf{X'}\boldsymbol{\varepsilon}\boldsymbol{\varepsilon}'\mathbf{X}(\mathbf{X'X})^{-1}\right] \\
&= (\mathbf{X'X})^{-1}\mathbf{X'}E(\boldsymbol{\varepsilon}\boldsymbol{\varepsilon}')\mathbf{X}(\mathbf{X'X})^{-1} = \sigma^2(\mathbf{X'X})^{-1}\mathbf{X'X}(\mathbf{X'X})^{-1} = \sigma^2(\mathbf{X'X})^{-1}.
\end{aligned}
$$

This completes the proof.   □

Note that the intercept may be eliminated by averaging and, as follows from the proof of Theorem 8.4, the covariance matrix of the slope coefficients is expressed as $\sigma^2 \left(\overset{\circ}{\mathbf{X}}' \overset{\circ}{\mathbf{X}} \right)^{-1}$.

Now we are ready to formulate and prove the Gauss–Markov theorem as a generalization of Theorem 6.50. A critical point of the proof is formulation of the best estimator. For a one-dimensional parameter, the answer is the mean square error (MSE). Since $\boldsymbol{\beta}$ is multidimensional, the concept of the multidimensional MSE from Section 6.4.3 is employed. Because we are dealing with unbiased estimators of $\boldsymbol{\beta}$, we say that an unbiased estimator, $\widehat{\boldsymbol{\beta}}$, is more efficient, or better, than another unbiased estimator, $\widetilde{\boldsymbol{\beta}}$, if

$$\mathrm{cov}(\widehat{\boldsymbol{\beta}}) \leq \mathrm{cov}(\widetilde{\boldsymbol{\beta}}), \qquad (8.16)$$

meaning that the difference between the right- and the left-hand side is a non-negative definite matrix (see the Appendix). Equivalently, the above inequality holds if $\mathbf{p}'\mathrm{cov}(\widehat{\boldsymbol{\beta}})\mathbf{p} \leq \mathbf{p}'\mathrm{cov}(\widetilde{\boldsymbol{\beta}})\mathbf{p}$ for every vector $\mathbf{p}$. Since $\mathbf{p}'\mathrm{cov}(\widehat{\boldsymbol{\beta}})\mathbf{p} = \mathrm{var}(\mathbf{p}'\widehat{\boldsymbol{\beta}})$, we may equivalently assert that $\widehat{\boldsymbol{\beta}}$ is more efficient than $\widetilde{\boldsymbol{\beta}}$ if the variance of any linear combination of vector $\widehat{\boldsymbol{\beta}}$ is smaller than the variance of that linear combination of vector $\widetilde{\boldsymbol{\beta}}$.

Theorem 8.9 *Gauss–Markov. The OLS estimator is BLUE.*

Proof. The unbiasedness has been proven in Theorem 8.8. Now we prove that $\widehat{\boldsymbol{\beta}}$ is the best in the family of linear unbiased estimators. We say that an estimator $\widetilde{\boldsymbol{\beta}}$ is linear if it is expressed as a linear function of $\mathbf{y}$ meaning that the estimator can be written as $\widetilde{\boldsymbol{\beta}} = \mathbf{L}\mathbf{y}$, where $\mathbf{L}$ is a $m \times n$ fixed matrix. It is easy to see that the OLS estimator is linear with $\mathbf{L} = (\mathbf{X}'\mathbf{X})^{-1}\mathbf{X}'$. To find the condition on $\mathbf{L}$ implied by unbiasedness of $\widetilde{\boldsymbol{\beta}}$, we have $E(\widetilde{\boldsymbol{\beta}}) = \mathbf{L}E(\mathbf{y}) = \mathbf{L}\mathbf{X}\boldsymbol{\beta} = \boldsymbol{\beta}$ for all $\boldsymbol{\beta}$ which means that matrix $\mathbf{L}$ must satisfy m^2 equations $\mathbf{L}\mathbf{X} = \mathbf{I}$, where $\mathbf{I}$ is the $m \times m$ identity matrix. Since $\mathrm{cov}(\widehat{\boldsymbol{\beta}}) = \sigma^2(\mathbf{X}'\mathbf{X})^{-1}$ and $\mathrm{cov}(\widetilde{\boldsymbol{\beta}}) = \mathbf{L}\mathrm{cov}(\mathbf{y})\mathbf{L}' = \sigma^2\mathbf{L}\mathbf{L}'$, in view of definition (8.16), the OLS estimator is more efficient than $\widetilde{\boldsymbol{\beta}}$ if and only if

$$\mathbf{p}'\mathbf{L}\mathbf{L}'\mathbf{p} \geq \mathbf{p}'(\mathbf{X}'\mathbf{X})^{-1}\mathbf{p} \quad \forall \mathbf{p} \qquad (8.17)$$

under constraint

$$\mathbf{L}\mathbf{X} = \mathbf{I}, \qquad (8.18)$$

where, without loss of generality, we can assume that $\|\mathbf{p}\| = 1$. To prove (8.17), introduce a quadratic function of scalar λ

$$\left\| \mathbf{X}(\mathbf{X}'\mathbf{X})^{-1}\mathbf{p} - \lambda\mathbf{L}'\mathbf{p} \right\|^2$$
$$= \lambda^2 \mathbf{p}'\mathbf{L}\mathbf{L}'\mathbf{p} - 2\mathbf{p}'(\mathbf{X}'\mathbf{X})^{-1}\mathbf{X}'\mathbf{L}'\mathbf{p} + \mathbf{p}'(\mathbf{X}'\mathbf{X})^{-1}\mathbf{X}'\mathbf{X}(\mathbf{X}'\mathbf{X})^{-1}\mathbf{p}.$$

Since this function is nonnegative, the discriminant must be nonpositive,

$$(\mathbf{p}'(\mathbf{X}'\mathbf{X})^{-1}\mathbf{X}'\mathbf{L}'\mathbf{p})^2 \leq \mathbf{p}'\mathbf{L}\mathbf{L}'\mathbf{p} \times \mathbf{p}'(\mathbf{X}'\mathbf{X})^{-1}\mathbf{p}.$$

But, due to constraint (8.18), we have $\mathbf{X}'\mathbf{L}' = \mathbf{I}$, so

$$(\mathbf{p}'(\mathbf{X}'\mathbf{X})^{-1}\mathbf{p})^2 \leq \mathbf{p}'\mathbf{L}\mathbf{L}'\mathbf{p} \times \mathbf{p}'(\mathbf{X}'\mathbf{X})^{-1}\mathbf{p}$$

and, finally, $\mathbf{p}'(\mathbf{X}'\mathbf{X})^{-1}\mathbf{p} \leq \mathbf{p}'\mathbf{L}\mathbf{L}'\mathbf{p}$, which means that for any linear unbiased estimator $\widehat{\beta}$ the inequality (8.16) holds. The theorem is proved.

Example 8.10 *Minimum variance slope estimator.* *Somebody is interested in the first component of vector β and treats other components as nuisance parameters. Find an unbiased linear estimator of β_1 with minimum variance.*

Solution. Let $\widehat{\boldsymbol{\beta}}$ be the OLS estimator and $\widehat{\beta}_1$ its first component. We claim that $\widehat{\beta}_1$ is an unbiased linear estimator with minimum variance. Indeed, $\widehat{\beta}_1$ is unbiased because all components are unbiased; $\widehat{\beta}_1$ is a linear estimator because $\widehat{\boldsymbol{\beta}}$ is a linear estimator. The inequality (8.16) means that, for $\mathbf{p} = (1, 0, ..., 0)'$,

$$\mathrm{var}(\widehat{\beta}_1) = \mathbf{p}'\mathrm{cov}(\widehat{\boldsymbol{\beta}})\mathbf{p} \leq \mathbf{p}'\mathrm{cov}(\widetilde{\boldsymbol{\beta}})\mathbf{p} = \mathrm{var}(\widetilde{\beta}_1)$$

for any linear unbiased estimator $\widetilde{\beta}_1$.

8.2.1 Estimation of regression variance

Besides regression coefficients, we have to estimate the regression variance, σ^2. An obvious choice is to use a quadratic function of regression residuals, $\mathbf{y}-\widehat{\mathbf{y}}$. The goal of this section is to prove that

$$\widehat{\sigma}^2 = \frac{1}{n-m}\|\mathbf{y}-\widehat{\mathbf{y}}\|^2 \tag{8.19}$$

is an unbiased estimator of σ^2 referred to as the mean square error, where $\widehat{\sigma}$ is referred to as the residual standard error. The denominator is referred to as the degree of freedom: n is the number of observations and m is the number of estimated coefficients including the intercept, if present. Recall that the unbiasedness of this estimator in a special case of regression without predictors (intercept only) was proven earlier in Theorem 6.41 with the degree of freedom $n-1$.

The formula for the expected value of a quadratic form below, with a normally distributed, $\mathbf{y}$ has been derived in Section 4.2.2. Here, we do not use the normality of $\mathbf{y}$.

Lemma 8.11 *Let $\mathbf{y} = \boldsymbol{\mu} + \boldsymbol{\varepsilon}$, where $E(\boldsymbol{\varepsilon}) = \mathbf{0}$ and $\mathrm{cov}(\boldsymbol{\varepsilon}) = \sigma^2\mathbf{I}$ and $\boldsymbol{\mu}$ is a fixed vector. Let $\mathbf{A}$ be a fixed symmetric matrix. Then $E(\mathbf{y}'\mathbf{A}\mathbf{y}) = \sigma^2\mathrm{tr}(\mathbf{A}) + \boldsymbol{\mu}'\mathbf{A}\boldsymbol{\mu}$.*

Proof. Let Σ denote the summation over indices i and j. Then taking the expected value of

$$\mathbf{y}'\mathbf{A}\mathbf{y} = \Sigma\varepsilon_i\varepsilon_j A_{ij} + \Sigma\mu_i\varepsilon_j A_{ij} + \Sigma\varepsilon_i\mu_j A_{ij} + \Sigma\mu_i\mu_j A_{ij}$$

one obtains the desired result:

$$E(\mathbf{y}'\mathbf{A}\mathbf{y}) = \Sigma_{i=j}E\left(\varepsilon_i^2\right)A_{ij} + \boldsymbol{\mu}'\mathbf{A}\boldsymbol{\mu} = \sigma^2\mathrm{tr}(\mathbf{A}) + \boldsymbol{\mu}'\mathbf{A}\boldsymbol{\mu},$$

because $E(\varepsilon_i) = E(\varepsilon_j) = 0$ and $E(\varepsilon_i\varepsilon_j) = 0$ if $i \neq j$ and $E(\varepsilon_i^2) = \sigma^2$.

Theorem 8.12 *The estimator (8.19) is unbiased for σ^2.*

Proof. Express $\widehat{\mathbf{y}} = \mathbf{X}\widehat{\boldsymbol{\beta}} = \mathbf{X}(\mathbf{X}'\mathbf{X})^{-1}\mathbf{X}'\mathbf{y}$, so that, $\mathbf{y}-\widehat{\mathbf{y}} = \mathbf{y} - \mathbf{X}(\mathbf{X}'\mathbf{X})^{-1}\mathbf{X}'\mathbf{y}$
$= \mathbf{H}\mathbf{y}$, where $\mathbf{H}$ is a symmetric and idempotent annihilator matrix, so $\|\mathbf{y}-\widehat{\mathbf{y}}\|^2 = \mathbf{y}'\mathbf{H}\mathbf{y}$. Moreover, it is easy to see that

$$\mathbf{H}\mathbf{X} = \left[\mathbf{I} - \mathbf{X}(\mathbf{X}'\mathbf{X})^{-1}\mathbf{X}'\right]\mathbf{X} = \mathbf{X}-\mathbf{X}(\mathbf{X}'\mathbf{X})^{-1}\mathbf{X}'\mathbf{X} = \mathbf{0}.$$

Now using the formula from the above lemma where $\mathbf{y} = \boldsymbol{\mu} + \boldsymbol{\varepsilon}$ and $\boldsymbol{\mu} = \mathbf{X}\boldsymbol{\beta}$, we note that $\mathbf{H}\boldsymbol{\mu} = \mathbf{0}$, and we finally obtain

$$\begin{aligned} E\left(\|\mathbf{y}-\widehat{\mathbf{y}}\|^2\right) &= \sigma^2 \mathrm{tr}(\mathbf{A}) = \sigma^2 \mathrm{tr}[\mathbf{I} - \mathbf{X}(\mathbf{X}'\mathbf{X})^{-1}\mathbf{X}'] \\ &= n\sigma^2 - \sigma^2[\mathrm{tr}(\mathbf{I})-\mathrm{tr}(\mathbf{X}'\mathbf{X})^{-1}(\mathbf{X}'\mathbf{X})] = (n - m)\sigma^2, \end{aligned}$$

which implies the unbiasedness of $\widehat{\sigma}^2$.

Problems

1. The variance of each component of the first unbiased multidimensional estimator is smaller than the variance of the corresponding component of the second unbiased estimator. Does this imply that the first estimator is more efficient than the second one? [Hint: Consider a two-dimensional estimator when providing an example.]

2. Using the results from Example 8.2, prove that the covariance matrix (8.15) yields the variance of the slope estimate in the simple linear regression given in Theorem 6.50.

3. A linear model takes the form $Y_i = \alpha x_i - \beta z_i + \varepsilon_i$, where x_i and z_i take values 0 or 1 and $x_i z_i = 0$. Find a linear unbiased estimator of $\alpha - \beta$ referring to the Gauss–Markov theorem.

4. Provide an alternative proof of the Gauss–Markov theorem using the Lagrange multiplier technique (see Appendix). Derive the solution of the optimization problem $\mathbf{p}'\mathbf{L}\mathbf{L}'\mathbf{p} = \|\mathbf{L}'\mathbf{p}\|^2$ under restriction (8.18), where $\mathbf{p}$ is fixed. [Hint: use $\mathbf{l} = \mathrm{vec}(\mathbf{L}')$ and rewrite (8.18) as $(\mathbf{I} \otimes \mathbf{X})\mathbf{l} = \mathrm{vec}(\mathbf{I})$ and express $\mathbf{L}'\mathbf{p} = (\mathbf{p}'\otimes\mathbf{I})\mathbf{l}$.]

5. Prove if $\mathbf{L} \neq (\mathbf{X}'\mathbf{X})^{-1}\mathbf{X}'$ in the Gauss–Markov theorem, then there exists $\mathbf{p}$ such that $\mathbf{p}'\mathbf{L}\mathbf{L}'\mathbf{p} > \mathbf{p}'(\mathbf{X}'\mathbf{X})^{-1}\mathbf{p}$.

6. Somebody is interested in the difference between two slope coefficients in a multiple linear regression. Prove that the best linear estimator is the difference of the corresponding components of the OLS estimator.

7. In the context of Example 8.10, somebody is interested in estimating the squared coefficient, β_1^2. Does the result of the example hold?

8. Apply Lemma 8.11 to $y \sim \mathcal{R}(0,1)$ where $\mathbf{A} = a$ is a number and then check the equality by direct computation using the uniform distribution (see Section 2.3).

9. Use Lemma 8.11 to prove that the sample variance (6.25) is unbiased in the model $\mathbf{y} = \mu\mathbf{1} + \varepsilon$.

10. Prove that matrix $\mathbf{A}$ in the proof of Theorem 8.12 is symmetric.

11. Prove that $\widehat{\sigma}^2$ remains unbiased even if the identifiability condition (8.3) does not hold. [Hint: Split matrix $\mathbf{X} = [\mathbf{X}_1, \mathbf{X}_2]$ where $\mathbf{X}_2$ is a linear combination of matrix $\mathbf{X}_1$, i.e. $\mathbf{X}_2 = \mathbf{X}_1\mathbf{Q}$.]

8.3   Properties of OLS estimators under the normal assumption

Properties of the OLS studied in the previous section did not require specification of the error distribution. However, under assumption that the errors follow a normal distribution the linear model becomes especially attractive because the key statistics have known distributions such as chi-square or t-distribution. Moreover, in the linear model under the normal assumption (8.5), the Gauss–Markov theorem can be strengthened and the OLS estimator turns into an efficient estimator. The proofs are simplified by of the ground work done in previous chapters, particularly Chapters 4 and 6.

Theorem 8.13 *The following properties hold under the normal assumption:*

1. *The OLS and maximum likelihood (ML) estimators of the beta coefficients coincide, OLS = ML.*

2. *$\widehat{\beta}$ has a multivariate normal distribution with the mean as the true parameters β and the covariance matrix (8.15):*
$$\widehat{\beta} \sim \mathcal{N}(\beta, \sigma^2(\mathbf{X}'\mathbf{X})^{-1}).$$

3. *The OLS estimator is efficient among all unbiased estimators: if $\widetilde{\beta}$ is another, maybe nonlinear, unbiased estimator of β, then $\mathrm{cov}(\widehat{\beta}) \leq \mathrm{cov}(\widetilde{\beta})$.*

4. *The estimator of the variance (8.19), upon normalization, has a chi-square distribution with $n - m$ degrees of freedom,*
$$\frac{n-m}{\sigma^2}\widehat{\sigma}^2 \sim \chi^2(n-m).$$

5. *Estimators $\widehat{\boldsymbol{\beta}}$ and $\widehat{\sigma}^2$ are independent, and the ratio of the deviation of the jth OLS coefficient from its true value to its estimate of the standard error, called the t-statistic, has a t-distribution with $n - m$ degrees of freedom,*

$$\frac{\widehat{\beta}_j - \beta_j}{\widehat{\sigma}\sqrt{(\mathbf{X}'\mathbf{X})^{-1}_{jj}}} \sim t(n - m), \quad j = 1, 2, ..., m. \tag{8.20}$$

Proof. 1. Since $\mathbf{y} \sim \mathcal{N}(\mathbf{X}\boldsymbol{\beta}, \sigma^2\mathbf{I})$, as follows from expression (4.1), where $\boldsymbol{\mu} = \mathbf{X}\boldsymbol{\beta}$ and $\boldsymbol{\Omega} = \sigma^2\mathbf{I}$, its joint probability density function (pdf) takes the form

$$f(\mathbf{y}; \boldsymbol{\beta}, \sigma^2) = (2\pi)^{-n/2}\sigma^{-n}e^{-\frac{1}{2\sigma^2}\|\mathbf{y} - \mathbf{X}\boldsymbol{\beta}\|^2}$$

with the log-likelihood, up to a constant term,

$$l(\boldsymbol{\beta}, \sigma^2) = -\frac{1}{2}\left[n\ln\sigma^2 + \frac{1}{\sigma^2}\|\mathbf{y} - \mathbf{X}\boldsymbol{\beta}\|^2\right].$$

The maximum of l over $\boldsymbol{\beta}$ is attained where $\|\mathbf{y} - \mathbf{X}\boldsymbol{\beta}\|^2 = \min$. Therefore OLS = ML.

2. The fact that $\widehat{\boldsymbol{\beta}}$ has a multivariate normal distribution with mean $\boldsymbol{\mu}$ and covariance matrix $\sigma^2(\mathbf{X}'\mathbf{X})^{-1}$ follows from Theorem 8.8 and the fact that $\widehat{\boldsymbol{\beta}}$ has the normal distribution because it is a linear function of $\mathbf{y}$, see Theorem 4.2.

3. To prove that the OLS estimator is efficient, it suffices to prove that its covariance matrix reaches the Cramér–Rao lower bound similarly to Example 6.137. The Fisher information matrix for the linear model is derived using the covariance matrix of the first derivative of the log pdf with respect to $\boldsymbol{\beta}$ and σ^2,

$$\begin{bmatrix} \frac{\partial \ln f}{\partial \boldsymbol{\beta}} \\ \frac{\partial \ln f}{\partial \sigma^2} \end{bmatrix} = \frac{1}{\sigma^2}\begin{bmatrix} \mathbf{X}'(\mathbf{y} - \mathbf{X}\boldsymbol{\beta}) \\ -0.5n + 0.5\sigma^{-2}\|\mathbf{y} - \mathbf{X}\boldsymbol{\beta}\|^2 \end{bmatrix}.$$

We have

$$\text{cov}\left(\frac{\partial \ln f}{\partial \boldsymbol{\beta}}\right) = \frac{1}{\sigma^4}\mathbf{X}'\text{cov}(\boldsymbol{\varepsilon})\mathbf{X} = \frac{1}{\sigma^2}\mathbf{X}'\mathbf{X},$$

$$\text{cov}\left(\frac{\partial \ln f}{\partial \boldsymbol{\beta}}, \frac{\partial \ln f}{\partial \sigma^2}\right) = \frac{0.5}{\sigma^6}\text{cov}\left(\mathbf{X}'\boldsymbol{\varepsilon}, \|\boldsymbol{\varepsilon}\|^2\right) = \mathbf{0},$$

$$\text{cov}\left(\frac{\partial \ln f}{\partial \sigma^2}\right) = \frac{1}{4\sigma^8}\text{var}(\|\boldsymbol{\varepsilon}\|^2) = \frac{1}{4\sigma^8}2n\sigma^4 = \frac{n}{2\sigma^4}.$$

The middle result follows from the fact that $E(\varepsilon_i\varepsilon_j^2) = 0$ and the third result follows from the fact that $\text{var}(\sum \varepsilon_i^2) = \sum \text{var}(\varepsilon_i^2) = 2n\sigma^4$ because $\text{var}(\varepsilon_i^2) = 2\sigma^4$.

See details in Example 6.101. Finally, the Fisher information matrix for the linear model takes the block-diagonal form

$$
\mathcal{I}(\boldsymbol{\beta},\sigma^2) = \begin{bmatrix} \frac{1}{\sigma^2}\mathbf{X'X} & \mathbf{0} \\ \mathbf{0'} & \frac{n}{2\sigma^4} \end{bmatrix}.
$$

Its inverse yields the Cramér–Rao lower bound,

$$
\mathcal{I}^{-1}(\boldsymbol{\beta},\sigma^2) = \begin{bmatrix} \sigma^2(\mathbf{X'X})^{-1} & \mathbf{0} \\ \mathbf{0'} & \frac{2\sigma^4}{n} \end{bmatrix}.
$$

Since $\text{cov}(\widehat{\boldsymbol{\beta}}) = \sigma^2(\mathbf{X'X})^{-1}$, we conclude that the covariance matrix of the OLS estimator reaches the Cramér–Rao lower bound and as such $\widehat{\boldsymbol{\beta}}$ is efficient.

4. Theorem 4.21 and the representation from Theorem 8.12 are employed, $\|\mathbf{y}-\widehat{\mathbf{y}}\|^2 = \mathbf{y'Ay}$, where matrix $\mathbf{A}$ is defined in (8.15). Moreover, since it was shown that $\mathbf{AX} = \mathbf{0}$, we can further express

$$
\|\mathbf{y}-\widehat{\mathbf{y}}\|^2 = (\boldsymbol{\varepsilon} + \mathbf{X}\boldsymbol{\beta})'\mathbf{A}(\boldsymbol{\varepsilon} + \mathbf{X}\boldsymbol{\beta}) = \boldsymbol{\varepsilon}'\mathbf{A}\boldsymbol{\varepsilon} + 2\boldsymbol{\varepsilon}'\mathbf{AX}\boldsymbol{\beta} + \boldsymbol{\beta}'\mathbf{X'AX}\boldsymbol{\beta} = \boldsymbol{\varepsilon}'\mathbf{A}\boldsymbol{\varepsilon}.
$$

As was shown, matrix $\mathbf{A}$ is idempotent. Find its trace (we indicate the dimension of the identity matrices),

$$
\text{tr}(\mathbf{A}) = \text{tr}(\mathbf{I}_n) - \text{tr}\left(\mathbf{X}(\mathbf{X'X})^{-1}\mathbf{X'}\right) = n - \text{tr}\left((\mathbf{X'X})^{-1}\mathbf{X'X}\right) = n - \text{tr}(\mathbf{I}_m) = n - m.
$$

Hence, $\|\mathbf{y}-\widehat{\mathbf{y}}\|^2/\sigma^2 = \mathbf{Z'AZ} \sim \chi^2(n-m)$, where $\mathbf{Z} = \boldsymbol{\varepsilon}/\sigma \sim \mathcal{N}(\mathbf{0},\mathbf{I})$ proves the property.

5. Theorem 4.30 is employed for the proof. Instead of considering the jth component, we prove a more general result for any linear combination of the OLS estimator. Let $\mathbf{p}$ be any nonzero $m \times 1$ vector. We want to prove that

$$
\frac{\mathbf{p}'(\widehat{\boldsymbol{\beta}} - \boldsymbol{\beta})}{\widehat{\sigma}\sqrt{\mathbf{p}'(\mathbf{X'X})^{-1}\mathbf{p}}} \sim t(n-m).
$$

Express the numerator and denominator of the left-hand side as functions of $\mathbf{Z}$ by dividing the numerator and denominator by σ. As follows from the earlier proof of this theorem, we can express $\widehat{\sigma}/\sigma = \sqrt{\mathbf{Z'AZ}/(n-m)}$. Now we express

$$
\frac{1}{\sigma}\frac{\mathbf{p}'(\widehat{\boldsymbol{\beta}} - \boldsymbol{\beta})}{\sqrt{\mathbf{p}'(\mathbf{X'X})^{-1}\mathbf{p}}} = \frac{\mathbf{p}'(\mathbf{X'X})^{-1}\mathbf{X'Z}}{\sqrt{\mathbf{p}'(\mathbf{X'X})^{-1}\mathbf{p}}} = \mathbf{a'Z},
$$

where $\mathbf{a} = \mathbf{X}(\mathbf{X'X})^{-1}\mathbf{p}/\sqrt{\mathbf{p}'(\mathbf{X'X})^{-1}\mathbf{p}}$. Show that the norm of this vector is 1,

$$
\mathbf{a'a} = \frac{1}{\mathbf{p}'(\mathbf{X'X})^{-1}\mathbf{p}}\mathbf{p}'(\mathbf{X'X})^{-1}\mathbf{X'X}(\mathbf{X'X})^{-1}\mathbf{p} = \frac{1}{\mathbf{p}'(\mathbf{X'X})^{-1}\mathbf{p}}\mathbf{p}'(\mathbf{X'X})^{-1}\mathbf{p} = 1.
$$

It is left to demonstrate that $\mathbf{Aa} = \mathbf{0}$, which follows from the previous proof of $\mathbf{AX} = \mathbf{0}$. This fact implies independence of $\widehat{\boldsymbol{\beta}}$ and $\widehat{\sigma}^2$. To prove the 5th statement for various j we set $\mathbf{p}$ to be a vector with the jth component 1 and zero elsewhere.   □

The third property means that even if the error term is not necessarily normally distributed the OLS estimator is the best in the family of linear unbiased estimators. Under the normal assumption, it is the best in the family of all (linear and nonlinear) unbiased estimators.

Especially useful is the fifth property: it gives rise to testing the significance of the jth predictor which is equivalent to $\beta_j = 0$. This method is the basis for model building and variable selection.

Theorem 8.14 *Gauss–Markov theorem for regression variance.* *Under the normal assumption, $\widehat{\sigma}^2$ has minimum variance among all unbiased estimators as quadratic functions of* $\mathbf{y}$.

The proof is found in Demidenko [28], p. 158. This theorem asserts that the variance of any unbiased quadratic estimator $\widetilde{\sigma}^2 = \mathbf{y}'\mathbf{Ay}$ is not smaller than the variance of the estimator (8.19) given by

$$\mathrm{var}(\widehat{\sigma}^2) = \frac{2\sigma^4}{n-m}.$$

This formula corresponds to the previously derived formula for the iid sample presented in Corollary 6.42 for the case when $m = 1$.

8.3.1   The sensitivity of statistical inference to violation of the normal assumption

Statistical inference for the linear model assumes that the error term is normally distributed. The goal of this section is to understand how sensitive/robust the inference is to violating this assumption, and particularly how the properties specified in Theorem 8.13 are affected. Analytical investigation is very complicated when the distribution is not normal – instead, we appeal to simulations. Two types of nonnormal distributions are considered: (1) the t-distribution with a small degree of freedom, which is symmetric but has heavy tails with the possibility of outliers, and (2) the chi-square distribution with a small degree of freedom, which is asymmetric and also has a right heavy tail. Two examples will be offered to demonstrate the consequences of violating the normal assumption. In the first example, we look at the distribution of the slope coefficient, regression variance and t-statistic; in the second example, we look at statistical significance testing of the slope coefficient using the power function.

Below, we describe the first example and its R code, which generates errors according to the t-distribution with k degrees of freedom. Using the t-distribution is convenient because its degree of freedom controls how close the distribution is to the normal distribution. If k is large, say, $k > 30$, the t-distribution is very close to normal (Section 4.3), so one may expect the statements of Theorem 8.13 to hold. On the other hand, a small k implies a distribution with heavy tails (in the extreme case, $k = 1$ yields the Cauchy distribution). Heavy tails lead to the possibility of outliers and less statistical efficiency.

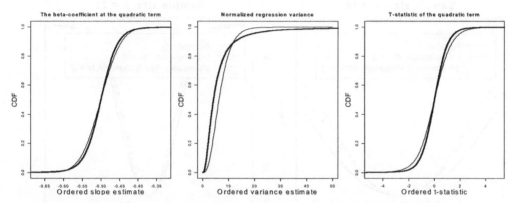

Figure 8.4: *The empirical (bold) and theoretical cdfs for properties 2, 4, and 5 specified in Theorem 8.13 when the error term follows the t-distribution with $k = 3$ degrees of freedom ($n = 10$, $\sigma = 1$). See Example 8.15 for detail.*

Example 8.15 *Regression with a nonnormal distribution. The linear model is a quadratic function of x given by $Y_i = \beta_1 + \beta_2 x_i + \beta_3 x_i^2 + \varepsilon_i$, where $x_i = i$ for $i = 1, 2, ..., n$, $E(\varepsilon_i) = 0$ and $\mathrm{var}(\varepsilon_i) = \sigma^2$. Use simulations to study how sensitive properties 2,4, and 5 are to violation of the normal assumption by modeling ε_i through the t-distribution with k degrees of freedom.*

Solution. The R code `simLM` has the default call `simLM(n=10,sigma=1,beta=c(-1,1,-.5),nSim=10000,k=3)`. In this code, we study the properties of the OLS coefficient at the quadratic term. To compare the empirical distribution of the OLS estimator $\widehat{\beta}_3$ (obtained from simulations) with the theoretical one we plot the empirical and theoretical cdfs on the same graph. The same method is used to assess the difference in cdfs for $\widehat{\sigma}^2$ when errors have normal and t-distributions – see Figure 8.4. A few remarks on the code: (1) Since matrix $\mathbf{X}$ does not change during simulations, we compute $(\mathbf{X}'\mathbf{X})^{-1}\mathbf{X}'$ beforehand (matrix LX). (2) To make the error term have variance σ^2, the simulated values obtained from `rt` are multiplied by $\sigma/\sqrt{k/(k-2)}$ because the variance of the t-distributed random variable is $k/(k-2)$. As follows from the figure, the distributions of $\widehat{\beta}_3$ and the t-statistic are fairly close to normal and the t-distribution, respectively. In both cases the empirical cdfs are wider, which means that the theoretical variance of the OLS estimate is larger than its theoretical counterpart derived under the assumption that the errors have a normal distribution. On the other hand, the distribution of $\widehat{\sigma}^2$ is more sensitive (less robust) to deviation from the normal distribution, as is typical. The sample size is an important parameter of robustness is the sample size: for large n the distribution of the OLS estimator and the t-statistics are close to theoretical. With large n, we are less concerned with abnormal ε_i, thanks to the CLT(see Section 2.10).   □

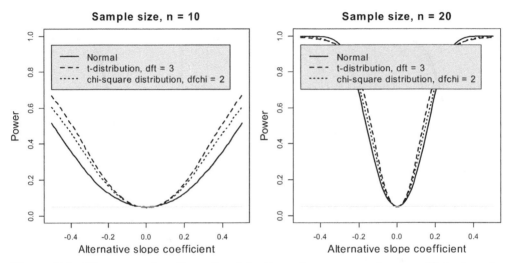

Figure 8.5: *The power functions for detection of a nonzero slope coefficient under different assumptions of the error term, see Example 8.16 for detail.*

In the second example, we investigate how t- or chi-square distribution of the error term affects statistical testing of the slope coefficient using the power function, which can be estimated as the proportion of simulations in which the p-value is smaller than α, the specified significance level, say, $\alpha = 0.05$.

Example 8.16 *Violation of the normal assumption.* *Using simulations, investigate how violation of the normal assumption affects the power of statistical significance testing of the slope coefficient in the simple linear regression. Use three distributions: normal, t-distribution and chi-square distribution for two sample sizes.*

Solution. The R code is found in `roblinreg2` which produces Figure 8.5, where as in the previous example, we use $x_i = i$ for $i = 1, 2, ..., n$ (it takes about 3 minutes to do the simulations and plot the graph). The two sample sizes, σ, degrees of freedom for the distributions, and number of simulations (the default `nSim=100000`) are specified as the arguments of the function. For the t-distribution, we normalize the simulated errors to make $\text{var}(\varepsilon_i) = \sigma^2$ as shown before. For the chi-square distribution we normalize as $\sigma(\varepsilon - \text{dfchi})/\sqrt{2 \times \text{dfchi}}$ to ensure that $\text{var}(\varepsilon_i) = \sigma^2$. We use vectorized simulations to approximate the power by generating a `nSim` by `n` matrix of error terms and computing $\widehat{\beta}$ and $\widehat{\sigma}$ using `rowSums` and `rowMeans`. An alternative method of vectorized simulations using the `apply` function is found in `roblinreg`. The local function `simlr` uses `lm` to get the p-value row-wise. This method may be used for a multiple regression but it is very slow. Surprisingly, as follows from Figure 8.5, the t- and chi-distributed errors have little effect on the power of significance testing of the slope coefficient: The type I error is just slightly smaller than the nominal $\alpha = 0.05$ when $n = 10$ but the power of a nonzero detection is even higher than when errors are normally distributed! The power of detection with nonnormal distributions is also higher when $n = 20$ and $|\beta| < 0.3$.

There is an opinion that statistical testing is invalid when errors do not follow a normal distribution. These two examples do not confirm this claim starting from a moderate n.

Problems

1. Prove that an estimator $\mathbf{p}'\widehat{\boldsymbol{\beta}}$ is an unbiased efficient estimator of $\mathbf{p}'\boldsymbol{\beta}$ where $\mathbf{p}$ is fixed.

2. Find an unbiased estimator for $\beta_1 + \sigma^2$. Does this estimator have minimum variance?

3. Show that estimator $\widehat{\sigma}^2$ does not reach the Cramér–Rao lower bound.

4. (a) Prove that the choice of the true beta coefficients and `sigma` in the R code `simLM` from Example 8.15 do not affect the robustness of the OLS. (b) Investigate how n affects that robustness.

5. Modify function `simLM` to investigate distributions using the q-q plot instead of the cdf (see Section 5.3) .

6. Modify function `simLM` to investigate properties of the slope coefficient.

7. Modify function `simLM` to investigate how robust OLS to an asymmetric distribution, such as the chi-square distribution with k degrees of freedom. [Hint: Normalize the chi-square distribution to have zero mean and variance σ^2.]

8. Suggest an unbiased estimator of β_j^2. Is it an unbiased estimator with minimum variance? Does the Markov-Gauss theorem apply?

9. Following Example 8.15, explore how k affects the properties of the OLS estimation in connection with n. Plot the maximum absolute deviation of cdfs as a function of n for $k = 3, 4, 7$.

10. Modify function `roblinreg2` from Example 8.16 to investigate how the degree of freedom in the t-distribution affects the power function: Plot the power at the alternative slope coefficient 0.2 as a function of `dft`.

8.4   Statistical inference with linear models

The linear model under consideration is compactly written as (8.5). We will use the results of Chapter 4 extensively for the proofs in this section. Theorem 8.13 is the foundation for statistical inference based on linear model including confidence intervals and hypothesis testing. In this section, we derive the F-test for the general linear hypothesis and its power function, lay out the foundation for prediction by linear regression, and correctly interpret the coefficient of determination as a popular goodness-of-fit measure.

8.4.1   Confidence interval and region

The $(1-\alpha)100\%$ confidence interval for the jth regression coefficient immediately follows from Property 5 of Theorem 8.13,

$$\widehat{\beta}_j \pm t_{1-\alpha/2}^{-1}\widehat{\sigma}\sqrt{(\mathbf{X}'\mathbf{X})_{jj}^{-1}}, \quad j = 1, 2, ..., m, \qquad (8.21)$$

where $t_{1-\alpha/2}^{-1}$ is the $(1-\alpha/2)$th quantile of the t-distribution with $n-m$ df.

Sometimes, it is advantageous to show the *confidence region* for a pair of parameters of interest on the plane, which has the shape of an ellipse. The result below provides a theoretical foundation for the construction of such elliptical regions. This region gives us a clue how correlation among predictors affects correlation among parameters.

Theorem 8.17 *Let s be a subset from $\{1, 2, ..., m\}$ containing k indices, and $\boldsymbol{\beta}_s$, $\widehat{\boldsymbol{\beta}}_s$, and $(\mathbf{X}'\mathbf{X})_s^{-1}$ be the corresponding true subvector, the $k \times 1$ OLS estimator, and the $k \times k$ submatrix of $(\mathbf{X}'\mathbf{X})^{-1}$. Then,*

$$\frac{1}{k\widehat{\sigma}^2}(\widehat{\boldsymbol{\beta}}_s - \boldsymbol{\beta}_s)' \left[(\mathbf{X}'\mathbf{X})_s^{-1}\right]^{-1} (\widehat{\boldsymbol{\beta}}_s - \boldsymbol{\beta}_s) \sim F(k, n-m). \qquad (8.22)$$

Proof. The central point of the proof is an obvious observation that $\widehat{\boldsymbol{\beta}}_s - \boldsymbol{\beta}_s \sim \mathcal{N}\left(\mathbf{0}, \sigma^2 (\mathbf{X}'\mathbf{X})_s^{-1}\right)$ since a subvector of a random vector with multivariate normal distribution has a multivariate normal distribution – see Section 4.1. Then as follows from Section 4.2, we have

$$\frac{1}{\sigma^2}(\widehat{\boldsymbol{\beta}}_s - \boldsymbol{\beta}_s)' \left[(\mathbf{X}'\mathbf{X})_s^{-1}\right]^{-1} (\widehat{\boldsymbol{\beta}}_s - \boldsymbol{\beta}_s) \sim \chi^2(k).$$

Divide the numerator and denominator of (8.22) by σ^2,

$$\frac{(\widehat{\boldsymbol{\beta}}_s - \boldsymbol{\beta}_s)' \left[(\mathbf{X}'\mathbf{X})_s^{-1}\right]^{-1} (\widehat{\boldsymbol{\beta}}_s - \boldsymbol{\beta}_s)/(k\sigma^2)}{\widehat{\sigma}^2/\sigma^2}.$$

The numerator has a chi-square distribution divided by its degrees of freedom and the denominator has a chi-square distribution divided by its degrees of freedom, see the proof of Property 5 in Theorem 8.13. Moreover, the numerator and denominator are independent because matrix $\mathbf{X}_s$, as a submatrix of $\mathbf{X}$, is orthogonal to $\mathbf{I} - \mathbf{X}(\mathbf{X}'\mathbf{X})^{-1}\mathbf{X}$. From Theorem 4.36 of Section 4.4 this implies that the left-hand side of (8.22) has the F-distribution with k and $n - m$ degrees of freedom. □

Based on result (8.22), define the $(1 - \alpha)100\%$ confidence region for $\boldsymbol{\beta}_s$ as the set of points

$$\left\{ \boldsymbol{\beta}_s \in R^k : \frac{1}{k\widehat{\sigma}^2}(\widehat{\boldsymbol{\beta}}_s - \boldsymbol{\beta}_s)' \left[(\mathbf{X}'\mathbf{X})_s^{-1} \right]^{-1} (\widehat{\boldsymbol{\beta}}_s - \boldsymbol{\beta}_s) \leq F_{1-\alpha}^{-1}(k, n - m) \right\}. \quad (8.23)$$

Clearly, this region comprises points within the ellipse with the center at the OLS estimate $\widehat{\boldsymbol{\beta}}_s$.

Example 8.18 F and t test. *Demonstrate that for $k = 1$ region (8.23) turns into a CI (8.21).*

Solution. Let the index set, s, contain a single j. Then $\left[(\mathbf{X}'\mathbf{X})_s^{-1} \right]^{-1} = 1/(\mathbf{X}'\mathbf{X})_{jj}^{-1}$ and (8.22) turns into

$$\frac{(\widehat{\beta}_j - \beta_j)^2}{\widehat{\sigma}^2(\mathbf{X}'\mathbf{X})_{jj}^{-1}} \sim F(1, n - m) = t^2(n - m),$$

because of Property 5 from Theorem 8.13. The equality at the right-hand side should be interpreted as saying that the squared random variable with a t-distribution has the F-distribution with 1 degree of freedom – see Section 4.4. □

Computation of the confidence region under the multiplicative error scheme is illustrated in the following example.

Cobb–Douglas production function

Production function is a central concept of mathematical economics and relates the output (production) Y to the capital K and labor L of a manufacturing business. In fact, any two (not necessarily capital and labor), or even multiple resources can be used to model the output. In the popular Cobb–Douglas production function, the output is related to the cost of capital and labor (all variables are in dollar amounts) as

$$Y = AK^\alpha L^\beta. \quad (8.24)$$

Parameters α and β are called *elasticities* because, as is easy seen, $\alpha = \frac{\partial Y/Y}{\partial K/K}$ and $\beta = \frac{\partial Y/Y}{\partial L/L}$. This is interpreted as saying that a 1% increase of capital, when labor is held fixed, leads to an $\alpha\%$ increase of output. Similarly, a 1% increase of labor, when capital is held fixed, leads to a $\beta\%$ increase of output. The language of elasticity is popular in econometrics.

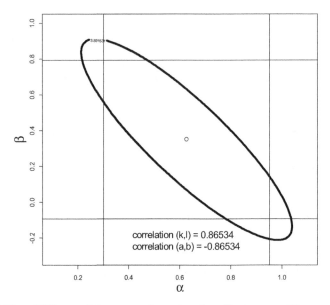

Figure 8.6: *The 95% confidence region for (α, β) is an ellipse with the center at the OLS estimate. The vertical and horizontal lines depict the lower and the upper individual 95% confidence limits. This graph was created by calling* CDpf(job=1). *The correlation coefficient between k and l is the same in absolute value as between the OLS estimates a and b.*

Statistical models such as the power function (8.24) are conveniently estimated under the *multiplicative error scheme* (versus the traditional additive error scheme),

$$Y_i = A K_i^\alpha L_i^\beta e^{\varepsilon_i}, \quad i = 1, 2, ..., n,$$

where $\varepsilon_i \overset{iid}{\sim} \mathcal{N}(0, \sigma^2)$. This model implies that Y_i has a lognormal distribution with the mean $A' K_i^\alpha L_i^\beta$, where $A' = A e^{\sigma^2/2}$, as follows from Section 2.11. The standard deviation of Y_i is proportional to the mean. Therefore, the deviation on the relative scale is constant – more discussion about the multiplicative and the additive error is found in Example 9.1. Besides having a clear interpretation, the multiplicative error has an important practical advantage; after taking log the original nonlinear model turns into a linear model:

$$y_i = \ln A + \alpha k_i + \beta l_i + \varepsilon_i, \tag{8.25}$$

where the lowercase indicates the log value. Many real-life regression problems with heavy tail distributions improve upon taking the log transformation. Making predictors normally distributed has another statistical advantage as follows from

Section 8.4.5 – the OLS estimator has minimum variance even if the predictors, in our case K_i and L_i, are random.

Estimation of the Cobb–Douglas production function is illustrated given annual data on Y, K, and L found in file CDpf.csv. The R code with computations is in file CDpf.r. The output of this code with job=1 is presented below. As follows from this printout, the OLS estimates are $\widehat{\alpha} = 0.6266$ and $\widehat{\beta} = 0.3502$ with the 95% CIs computed by formula (8.21), where $t_{1-0.05/2}^{-1}(17)$ is computed as qt(1-.05/2,df=17) = 2.109816. The explanation of the F-statistic and Multiple R-squared is found in Section 8.4.4.

```
> CDpf(job=1)
            Estimate Std. Error t value Pr(>|t|)
(Intercept)   0.5657     0.3163   1.788 0.091570 .
k             0.6266     0.1542   4.063 0.000809 ***
l             0.3502     0.2096   1.671 0.113072
Residual standard error: 0.2432 on 17 degrees of freedom
Multiple R-squared:  0.8791,    Adjusted R-squared:  0.8649
F-statistic:  61.8 on 2 and 17 DF,  p-value: 1.587e-08
[1] "95% CI:"
      lower limit upper limit
alpha  0.30122939   0.9519826
beta  -0.09202123   0.7924037
```

The 95% confidence region for (α, β) computed by (8.23) is shown in Figure 8.6. We make a couple of comments on the code: (1) The 2×2 matrix $\widehat{\sigma}^2 (\mathbf{X'X})_s^{-1}$, where $s = (2,3)$, is extracted from the covariance matrix for the three parameters of the model (the first component is the intercept) using the vcov function. (2) The ellipse is shown as the contour of the quadratic function computed on the grid of values for α and β (we recommend reviewing the bivariate normal distribution in Section 3.5). Here we use vectorized computations based on the rep function; more detail is found in Example 1.12. It is possible to prove that the correlation coefficient between the estimators of elasticities, a and b, are the same as the correlation between k and l but have opposite signs. The vertical and horizontal lines depict the lower and the upper individual confidence limits of the parameters.

8.4.2 Linear hypothesis testing and the F-test

There are two types of hypothesis testing for multiple regression: (1) testing statistical significance of a regression coefficient, which addresses the fundamental question of whether the presence of a predictor in the model is supported by the data, and (2) testing the general linear hypothesis that involves several regression coefficients.

Testing statistical significance of regression coefficient

The simple double-sided hypothesis test for the jth regression coefficient $H_0 : \beta = \beta_{0j}$ with the alternative $H_A : \beta \neq \beta_{0j}$ follows immediately from Property 5 of Theorem 8.13: the null hypothesis is rejected if

$$\left| \frac{\widehat{\beta}_j - \beta_{0j}}{\widehat{\sigma}\sqrt{(\mathbf{X}'\mathbf{X})_{jj}^{-1}}} \right| > t_{1-\alpha/2}^{-1}(n-m), \tag{8.26}$$

where $t_{1-\alpha/2}^{-1}(n-m)$ is the $(1-\alpha/2)$th quantile of the t-distribution with $n-m$ degrees of freedom. The p-value is the probability of the t-statistic being larger than the observed absolute value, under the null hypothesis, and computed in R as 2*(1-pt(LHS,df=n-m)). The most frequent test uses $\beta_{0j} = 0$ as the null hypothesis which is equivalent to testing whether the predictor has zero effect on the dependent variable. The corresponding t-statistic is given by

$$t_j = \frac{\widehat{\beta}_j}{\widehat{\sigma}\sqrt{(\mathbf{X}'\mathbf{X})_{jj}^{-1}}}, \quad j = 1, 2, ..., m,$$

along with the p-value shown in the table for regression coefficients of the lm function. Since, for large degree of freedom, the t-distribution is close to the normal distribution (see Section 4.3), we can roughly say that the predictor is statistically significant if the absolute value of the t-statistic is greater than 2, with significance level 5%.

Testing the general linear hypothesis

The general linear hypothesis is formulated in matrix language as

$$H_0 : \mathbf{C}\boldsymbol{\beta} = \mathbf{a}, \tag{8.27}$$

where the $k \times m$ constant matrix $\mathbf{C}$ has full rank $k < m$ and $\mathbf{a}$ is a constant $k \times 1$ vector. This hypothesis is tested using the F-distribution and the respective test is referred to as the F-test. This test requires derivation of the minimum RSS (8.7) under the restriction, $S_0 = \min_{\mathbf{C}\boldsymbol{\beta}=\mathbf{a}} S(\boldsymbol{\beta})$. Before formulating the F-test, we discuss how the restricted RSS can be computed.

Introduce the Lagrange function (see Appendix),

$$\mathcal{L}(\boldsymbol{\beta}, \boldsymbol{\lambda}) = \|\mathbf{y} - \mathbf{X}\boldsymbol{\beta}\|^2 - \boldsymbol{\lambda}'(\mathbf{C}\boldsymbol{\beta} - \mathbf{a}),$$

where $\boldsymbol{\lambda}$ is the $k \times 1$ vector of Lagrange multipliers. Differentiate with respect to $\boldsymbol{\beta}$ and express the solution through Lagrange multipliers, $\boldsymbol{\beta} = \widehat{\boldsymbol{\beta}} + 0.5(\mathbf{X}'\mathbf{X})^{-1}\mathbf{C}'\boldsymbol{\lambda}$. Plug this solution back into the restriction equation to obtain

$$\boldsymbol{\lambda} = 2\left(\mathbf{C}(\mathbf{X}'\mathbf{X})^{-1}\mathbf{C}'\right)^{-1}(\mathbf{a} - \mathbf{C}\widehat{\boldsymbol{\beta}}),$$

and finally find the beta-vector, which yields the minimum RSS under the restriction $\mathbf{C}\boldsymbol{\beta} = \mathbf{a}$,

$$\widehat{\boldsymbol{\beta}}_R = \widehat{\boldsymbol{\beta}} + (\mathbf{X}'\mathbf{X})^{-1}\mathbf{C}'\left(\mathbf{C}(\mathbf{X}'\mathbf{X})^{-1}\mathbf{C}'\right)^{-1}(\mathbf{a} - \mathbf{C}\widehat{\boldsymbol{\beta}}).$$

Note that matrix $\mathbf{C}(\mathbf{X}'\mathbf{X})^{-1}\mathbf{C}'$ is nonsingular because $\mathbf{C}$ has full rank. The restricted RSS can be computed in a straightforward manner as $S_0 = \left\|\mathbf{y} - \mathbf{X}\widehat{\beta}_R\right\|^2$.

In the theorem to follow, we operate with the difference, $S_0 - S_{\min}$, where $S_{\min}$ is the global minimum derived by means of the OLS. Using the restricted solution, we can find this difference explicitly as follows. Let $\mathbf{r} = \mathbf{y} - \mathbf{X}\widehat{\beta}$ be the OLS residual vector, so that $S_{\min} = \|\mathbf{r}\|^2$. Then we express

$$S_0 = \left\| \mathbf{r} - \mathbf{X}(\mathbf{X}'\mathbf{X})^{-1}\mathbf{C}' \left(\mathbf{C}(\mathbf{X}'\mathbf{X})^{-1}\mathbf{C}'\right)^{-1} (\mathbf{a} - \mathbf{C}\widehat{\beta}) \right\|^2.$$

Since $\mathbf{X}'\mathbf{r} = \mathbf{0}$, after some elementary simplification, we obtain

$$S_0 = \|\mathbf{r}\|^2 + (\mathbf{a} - \mathbf{C}\widehat{\beta})' \left(\mathbf{C}(\mathbf{X}'\mathbf{X})^{-1}\mathbf{C}'\right)^{-1} (\mathbf{a} - \mathbf{C}\widehat{\beta}),$$

and, therefore,

$$S_0 - S_{\min} = (\mathbf{a} - \mathbf{C}\widehat{\beta})' \left(\mathbf{C}(\mathbf{X}'\mathbf{X})^{-1}\mathbf{C}'\right)^{-1} (\mathbf{a} - \mathbf{C}\widehat{\beta}). \tag{8.28}$$

Now we are ready to formulate the F-test applied to a general linear hypothesis (8.27). This test is the omnibus for many tests concerning parameters of the linear model including t-tests and ANOVA.

Theorem 8.19 *Under the null hypothesis (8.27),*

$$\frac{(S_0 - S_{\min})/k}{S_{\min}/(n - m)} \sim F(k, n - m). \tag{8.29}$$

Proof. Theorem 4.36 of Section 4.4 is used. Divide numerator and denominator of (8.29) by σ^2. Then as was shown previously, the distribution of the denominator can be viewed as $\chi^2(n - m)/(n - m)$. Show that the numerator can be expressed as $\chi^2(k)/k$ using representation (8.28). Indeed, since $\mathbf{a} - \mathbf{C}\widehat{\beta} = \mathbf{a} - \mathbf{C}(\beta + (\mathbf{X}'\mathbf{X})^{-1}\mathbf{X}'\varepsilon) = -\mathbf{C}(\mathbf{X}'\mathbf{X})^{-1}\mathbf{X}'\varepsilon$ under the null, we rewrite $(S_0 - S_{\min})/\sigma^2 = \mathbf{Z}'\mathbf{A}\mathbf{Z}$, where $\mathbf{A} = \mathbf{X}(\mathbf{X}'\mathbf{X})^{-1}\mathbf{C}' \left(\mathbf{C}(\mathbf{X}'\mathbf{X})^{-1}\mathbf{C}'\right)^{-1} \mathbf{C}(\mathbf{X}'\mathbf{X})^{-1}\mathbf{X}'$, and $\mathbf{Z} = \varepsilon/\sigma \sim \mathcal{N}(\mathbf{0}, \mathbf{I})$. Matrix $\mathbf{A}$ is orthogonal to $\mathbf{B} = \mathbf{I} - \mathbf{X}(\mathbf{X}'\mathbf{X})^{-1}\mathbf{X}'$ from the denominator since $S_{\min}/\sigma^2 = \mathbf{Z}'\mathbf{B}\mathbf{X}$. Now show that matrix $\mathbf{A}$ is an idempotent matrix:

$$\mathbf{A}^2 = \mathbf{X}(\mathbf{X}'\mathbf{X})^{-1}\mathbf{C}' \left(\mathbf{C}(\mathbf{X}'\mathbf{X})^{-1}\mathbf{C}'\right)^{-1} \mathbf{C}(\mathbf{X}'\mathbf{X})^{-1}\mathbf{C}' \left(\mathbf{C}(\mathbf{X}'\mathbf{X})^{-1}\mathbf{C}'\right)^{-1}$$

$$\times \mathbf{C}(\mathbf{X}'\mathbf{X})^{-1}\mathbf{X}' = \mathbf{X}(\mathbf{X}'\mathbf{X})^{-1}\mathbf{C}' \left(\mathbf{C}(\mathbf{X}'\mathbf{X})^{-1}\mathbf{C}'\right)^{-1} \mathbf{C}(\mathbf{X}'\mathbf{X})^{-1}\mathbf{X}' = \mathbf{A}.$$

Lastly, we need to show that $\mathrm{tr}(\mathbf{A}) = k$,

$$
\begin{aligned}
\mathrm{tr}(\mathbf{A}) &= \mathrm{tr}\left(\mathbf{X}(\mathbf{X}'\mathbf{X})^{-1}\mathbf{C}'\left(\mathbf{C}(\mathbf{X}'\mathbf{X})^{-1}\mathbf{C}'\right)^{-1}\mathbf{C}(\mathbf{X}'\mathbf{X})^{-1}\mathbf{X}'\right) \\
&= \mathrm{tr}\left(\left(\mathbf{C}(\mathbf{X}'\mathbf{X})^{-1}\mathbf{C}'\right)^{-1}\mathbf{C}(\mathbf{X}'\mathbf{X})^{-1}\mathbf{X}'\mathbf{X}(\mathbf{X}'\mathbf{X})^{-1}\mathbf{C}'\right) \\
&= \mathrm{tr}\left(\left(\mathbf{C}(\mathbf{X}'\mathbf{X})^{-1}\mathbf{C}'\right)^{-1}\mathbf{C}(\mathbf{X}'\mathbf{X})^{-1}\mathbf{C}'\right) = \mathrm{tr}(\mathbf{I}_k) = k.
\end{aligned}
$$

All conditions of Theorem 4.21 from Section 4.2 are fulfilled, so we deduce that $(S_0 - S_{\min})/\sigma^2 \sim \chi^2(k)/k$. Therefore, (8.29) follows from Theorem 4.36 of Section 4.4.

Example 8.20 *F turns into t.* *Show that the F-test for (8.27) can be reduced to the t-test (8.26) when $k = 1$.*

Solution. When $k = 1$, write (8.27) as $\mathbf{c}'\boldsymbol{\beta} = a$, where $\mathbf{c}$ is a $m \times 1$ vector and a is a number. The numerator in (8.29), as follows from (8.28), takes the form $(a - \mathbf{c}'\widehat{\boldsymbol{\beta}})^2 / \mathbf{c}'(\mathbf{X}'\mathbf{X})^{-1}\mathbf{c}$ and the F-statistic becomes

$$
\frac{(a - \mathbf{c}'\widehat{\boldsymbol{\beta}})^2}{\widehat{\sigma}^2 \mathbf{c}'(\mathbf{X}'\mathbf{X})^{-1}\mathbf{c}} = \left(\frac{a - \mathbf{c}'\widehat{\boldsymbol{\beta}}}{\sqrt{\widehat{\sigma}^2 \mathbf{c}'(\mathbf{X}'\mathbf{X})^{-1}\mathbf{c}}}\right)^2.
$$

Since

$$
\frac{a - \mathbf{c}'\widehat{\boldsymbol{\beta}}}{\widehat{\sigma}\sqrt{\mathbf{c}'(\mathbf{X}'\mathbf{X})^{-1}\mathbf{c}}} \sim t(n - m)
$$

and $t^2(n - m) = F(1, n - m)$, the F-test (8.29) is equivalent to the t-test (8.26) when $a = \beta_{0j}$ and $\mathbf{c}$ is a vector with zero components but the jth component is 1. $\qquad\square$

The following is a continuation of the Cobb–Douglas production function example. This example illustrates how to run the F-test (8.29) in R: five different ways are discussed including function anova.

Example 8.21 *Testing the economy of scale.* *(a) Test the hypothesis that there is no economy of scale. That is, the proportional expansion of capital and labor yields a proportional increase of the output. (b) Derive the optimality condition for capital and labor that maximize the profit and show if this condition holds over the years.*

Solution. (a) We say that there is no economy of scale (also known as constant return to scale) if changing capital and labor to ρK and ρL yields the output ρY. If the output follows the Cobb–Douglas production function, this change

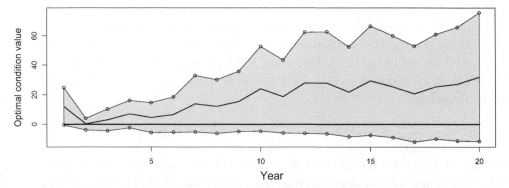

Figure 8.7: *The 95% confidence region for the optimality condition.*

yields $A(\rho K)^{\alpha}(\rho L)^{\beta} = A\rho^{\alpha+\beta}K^{\alpha}L^{\beta}$. Therefore, there is no economy of scale if $\alpha + \beta = 1$. Under the economy of scale, the Cobb–Douglas production function takes the form $Y = AK^{\alpha}L^{1-\alpha}$ and, therefore, can be expressed in a simple form as $(Y/L) = A(K/L)^{\alpha}$, where Y/L and K/L are referred to productivity and capital-to-labor ratio. Testing the economy of scale is an important step before making decisions on expanding the business. We aim to test the hypothesis $\alpha + \beta = 1$ for the data `CDpf.csv` used above. Several methods, implemented in `CDpf(job=2)`, are used to test this hypothesis. However, different methods produce the same p-value. First, rewrite model (8.25) so that $\alpha + \beta - 1$ becomes a new regression coefficient,

$$\alpha k_i + \beta l_i = (\alpha + \beta - 1)l_i - (\alpha + \beta - 1)l_i + \alpha k_i + \beta l_i = (\alpha + \beta - 1)l_i + \alpha(k_i - l_i) + l_i.$$

Therefore, the p-value for testing the null hypothesis $\alpha + \beta - 1 = 0$ can be read from the output of the `lm` call with dependent variable $y_i - l_i$ and predictors l_i and $k_i - l_i$. Second, the t-test from Example 8.20 can be used where $a = 1$ and $\mathbf{c} = (0, 1, 1)$. Third, the F-test (8.29) can be used with the numerator computed by formula (8.28), which gives $(a - \mathbf{c}'\widehat{\boldsymbol{\beta}})^2/(\widehat{\sigma}^2\mathbf{c}'(\mathbf{X}'\mathbf{X})^{-1}\mathbf{c})$. Fourth, S_0 can be computed as the RSS under the assumption that $\alpha + \beta = 1$, which can be obtained from the `lm` call with dependent variable $y_i - l_i$ and a single predictor $k_i - l_i$. Fifth, the comparison between two linear models, the full model and the restricted model under the null hypothesis, using the F-test can be accomplished based on function `anova`. Note, the two models must have the same dependent variable. In our case, the full model is estimated as `suFULL=lm(y~1+k)` and the restricted model is estimated as `suRESTR=lm(y~offset(l)+x2)` where `x2=k-l` with `offset(l)=l`, the coefficient is not estimated. Note that these runs have the same dependent variables. Now, to compare the models, we run `anova(suFULL,suRESTR)`. All five methods produce the same result, p-value $= 0.83$. In summary, there is strong evidence that there is no economy of scale: the hypothesis $\alpha + \beta = 1$ cannot be rejected.

(b) Since $Y = AK^{\alpha}L^{1-\alpha}$ defines the revenue, one can find the optimal capital

– labor allocation given the total cost $C = K + L$ that maximizes the profit $Y - C$. Since C is fixed, the optimization problem reduces to maximization of Y. By taking the log transformation, the optimization problem is written as $\max_{K+L=C}(\alpha \ln K + (1 - \alpha) \ln L)$. Introduce the Lagrange function (see the Appendix), $\mathcal{L}(K, L, \lambda) = \alpha \ln K + (1 - \alpha) \ln L - \lambda(K + L - C)$. Differentiate with respect to K and L

$$\frac{\partial \mathcal{L}}{\partial K} = \frac{\alpha}{K} - \lambda = 0, \quad \frac{\partial \mathcal{L}}{\partial L} = \frac{1 - \alpha}{L} - \lambda = 0$$

to yield the optimality condition $(1-\alpha)K - \alpha L = 0$. To test whether this condition holds over the years, we compute and plot the optimal condition $(1 - \widehat{\alpha})K_i - \widehat{\alpha}L_i$ along with the 95% confidence region

$$(1 - \widehat{\alpha})K_i - \widehat{\alpha}L_i \pm t_{.975}^{-1}(n - 2)(K_i + L_i)SE(\widehat{\alpha}), \quad i = 1, 2, ..., n.$$

The condition fails where the line is outside of the confidence region. Figure 8.7 shows evidence that the resource allocation was near optimum.

Power function and sample size determination

The goal of this section is to derive the power function of the F-test expressed through the noncentral F-distribution introduced in Section 4.4.

Theorem 8.22 *The power function of the F-test (8.29) for testing the null hypothesis (8.29) versus the alternative hypothesis $H_A : \delta \stackrel{\text{def}}{=} \mathbf{C}\boldsymbol{\beta} - \mathbf{a} \neq \mathbf{0}$ in linear model (8.5) is given by*

$$P(\lambda) = 1 - \mathcal{F}(F_{1-\alpha}^{-1}, \mathrm{df1} = k, \mathrm{df2} = n - m, \mathrm{ncp} = \lambda), \tag{8.30}$$

where $\mathcal{F}$ is the cdf of the noncentral F-distribution, $F_{1-\alpha}^{-1}$ is the $(1 - \alpha)th$ quantile of the central F-distribution with k and $n - m$ degrees of freedom, and

$$\lambda = \frac{1}{\sigma^2}\boldsymbol{\delta}'\left(\mathbf{C}(\mathbf{X}'\mathbf{X})^{-1}\mathbf{C}'\right)^{-1}\boldsymbol{\delta}$$

is the noncentrality parameter.

Proof. The result follows from expression (8.28) and from

$$\frac{1}{\sigma}(\mathbf{a} - \mathbf{C}\widehat{\boldsymbol{\beta}}) = -(\boldsymbol{\delta}/\sigma) - \mathbf{C}(\mathbf{X}'\mathbf{X})^{-1}\mathbf{X}\mathbf{Z},$$

where $\mathbf{Z} \sim \mathcal{N}(\mathbf{0}, \mathbf{I})$. This representation implies that

$$\frac{1}{\sigma^2}(\mathbf{a} - \mathbf{C}\widehat{\boldsymbol{\beta}})'\left(\mathbf{C}(\mathbf{X}'\mathbf{X})^{-1}\mathbf{C}'\right)^{-1}(\mathbf{a} - \mathbf{C}\widehat{\boldsymbol{\beta}}) \sim \chi^2(\mathrm{df} = k, \mathrm{ncp} = \lambda),$$

which, finally, leads to (8.30). See Sections 4.2 and 4.4 for detail. □

Below is an example where the power function (8.30) is used to find the required n; this task is usually referred to as the *sample size determination*.

Example 8.23 *Power of the F-test. (a) Derive the power of the test for $k = 1$ and particularly when the alternative hypothesis is that the jth slope coefficient is $a \neq 0$. (b) For how long one should observe a time series to detect the coefficient 0.01 at the quadratic term in the trend with the power 0.8 and the type I error 0.05 assuming that $\sigma = 2, 3, 4$?*

Solution. (a) As in Example 8.20, we write $H_A : \mathbf{c}'\boldsymbol{\beta} = a$. Then the power function becomes

$$P(\lambda) = 1 - \mathcal{F}(F_{1-\alpha}^{-1}, \mathrm{df1} = 1, \mathrm{df2} = n - m, \mathrm{ncp} = \lambda)$$

and the noncentrality parameter simplifies to

$$\lambda = \frac{1}{\sigma^2} \frac{a^2}{\mathbf{c}'(\mathbf{X}'\mathbf{X})^{-1}\mathbf{c}}.$$

In a special case when the alternative hypothesis is $H_A : \beta_j = a$, the noncentrality parameter turns into

$$\lambda_j = \frac{a^2}{\sigma^2 (\mathbf{X}'\mathbf{X})_{jj}^{-1}}.$$

One may say that the noncentrality parameter is the squared alternative on the scale of the parameter standard error.

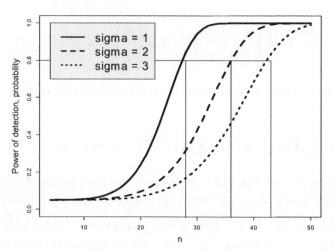

Figure 8.8: *The sample size determination in Example 8.23 to detect the coefficient at the quadratic term $\beta_3 = a = 0.01$ with power 0.8.*

(b) The linear model takes the form $Y_t = \beta_1 + \beta_2 t + \beta_3 t^2$, where time $t = 1, 2, ..., n$. The null hypothesis is $H_0 : \beta_3 = 0$ and the alternative hypothesis is $H_A : \beta_3 = 0.01$. Matrix $\mathbf{X}'\mathbf{X}$ can be computed in R or be given explicitly as below,

$$\mathbf{X'X} = \begin{bmatrix} n & n(n+1)/2 & n(2n^2+3n+1)/6 \\ n(n+1)/2 & n(2n^2+3n+1)/6 & n^2(n+1)^2/4 \\ n(2n^2+3n+1)/6 & n^2(n+1)^2/4 & n(6n^4+15n^3+10n^2-1)/30 \end{bmatrix}.$$

The result of sample determination is depicted in Figure 8.8 and the R code is found in file qtrpow.r. Note that n does not yield the exact power 0.8 because of discreteness. The required period of observation should be $n = 28, 36$ and 43 under scenarios $\sigma = 1, 2$, and 3.   □

In the above theorem, vector $\boldsymbol{\delta}$ reflects how different the null and the alternative are, but what is interesting is that the power function solely depends on the noncentrality parameter, λ. When $\boldsymbol{\delta}$ is close to zero, λ is also close to zero and the power is close to the type I error, α. When $\lambda \to \infty$, the power function converges to 1. Also, one can see that larger σ decreases λ and the power.

Example 8.24 *Simulations for the F-test.* *Use simulations with quadratic regression* $y_i = \beta_1 + \beta_2 x_i + \beta_3 x_i^2 + \varepsilon_i$ *to estimate the empirical power of the F-test for testing linear hypotheses* $\beta_2 = \beta_3$ *and* $\beta_1 = 1$. *Plot the empirical and theoretical powers as functions of* λ *on the same graph to ensure that the curves are appear identical.*

Solution. Matrix $\mathbf{C}$ and vector $\mathbf{a}$ are given as

$$\mathbf{C} = \begin{bmatrix} 0 & 1 & -1 \\ 1 & 0 & 0 \end{bmatrix}, \quad \mathbf{a} = \begin{bmatrix} 0 \\ 1 \end{bmatrix}.$$

To generate y_i, we must have an alternative 3×1 beta-coefficient vector, $\boldsymbol{\beta}_A$, that given λ, satisfies

$$\lambda = \frac{1}{\sigma^2}(\mathbf{C}\boldsymbol{\beta}_A - \mathbf{a})' \left(\mathbf{C}(\mathbf{X'X})^{-1}\mathbf{C}'\right)^{-1} (\mathbf{C}\boldsymbol{\beta}_A - \mathbf{a}).$$

Let $\boldsymbol{\beta}$ satisfy the null hypothesis $\mathbf{C}\boldsymbol{\beta} = \mathbf{a}$. Define the alternative $\boldsymbol{\beta}_A = \boldsymbol{\beta} + \nu\mathbf{d}$, where $\mathbf{d}$ is any 3×1 nonzero vector and ν is a scalar such that $\lambda = \nu^2 D/\sigma^2$, where $D = \mathbf{d}'\mathbf{C}' \left(\mathbf{C}(\mathbf{X'X})^{-1}\mathbf{C}'\right)^{-1} \mathbf{C}\mathbf{d}$, which implies $\nu = \pm\sigma\sqrt{\lambda/D}$. The R code is found in function linpower.r. The call is linpower(n=10,m=3,k=2 ,NL=10,beta=c(1,.5,.5),sigma=2,alpha=.05,nSim=10000,st=3) where n is the sample size, m is the number of estimated parameters, k=2 is the number of hypotheses, NL is the number of lambda points to simulate with, beta is the true vector of parameters, sigma is the regression SD, alpha is the type I error, nSim is the number of simulations, and st is the simulation seed number. Vector $\mathbf{a}$ is not defined but computed as $\mathbf{C}\boldsymbol{\beta}$ to make the code more general. We make several remarks on the code. (1) Since matrices $\mathbf{X}$ and $\mathbf{C}$ are fixed, we compute several matrices including $(\mathbf{X}'\mathbf{X})^{-1}$ and $\left(\mathbf{C}(\mathbf{X}'\mathbf{X})^{-1}\mathbf{C}'\right)^{-1}$ beforehand. (2) The $m \times 1$ direction vector $\mathbf{d}$ is taken randomly with uniformly distributed components. This choice should not affect the power function which depends solely on λ. However, this vector is used to compute the alternative beta coefficients (`beta.alt`) as described above. (3) The lambda values, which control how far the alternative is from the null, are in array `lambda` of length `NL`. They should start from 0 with an upper limit that produces a power close to 1. (4) The observed values of the F-statistic are saved in array `Fobs`. The empirical power is the proportion of simulations for which `Fobs` is greater than the critical F-distribution level computed as `qff=qf(1-alpha,df1=k,df2=n-m)`. The plot generated by the above `linpower` call is not presented here, but the empirical and theoretical power curves match each other.

8.4.3 Prediction by linear regression and simultaneous confidence band

Prediction by a simple linear regression was discussed in Section 6.7.3. Given predictor vector $\mathbf{x}$, the value of the dependent variable is $Y = \mathbf{x}'\boldsymbol{\beta}+\varepsilon$ predicted as $\mathbf{x}'\widehat{\boldsymbol{\beta}}$ with the prediction error variance $\sigma^2 + \mathbf{x}'\mathrm{cov}(\widehat{\boldsymbol{\beta}})\mathbf{x}$. As discussed in Section 6.7.3, we refer to this prediction as an *individual prediction* (the term ε is present) in contrast with the traditional regression prediction of $\mathbf{x}'\boldsymbol{\beta}$. See also Example 7.47. The basis for statistical inference of individual prediction is given below.

Theorem 8.25 *Individual prediction by regression. Let $\mathbf{x}$ be an $m \times 1$ vector and $y \sim \mathcal{N}(\mathbf{x}'\boldsymbol{\beta},\sigma^2)$ be independent of the vector of observations $\mathbf{y}$. Then*

$$\frac{Y - \mathbf{x}'\widehat{\boldsymbol{\beta}}}{\widehat{\sigma}\sqrt{1 + \mathbf{x}'(\mathbf{X}'\mathbf{X})^{-1}\mathbf{x}}} \sim t(n - m). \tag{8.31}$$

Proof. We use Property 5 of Theorem 8.13 by expressing the numerator as

$$Y - \mathbf{x}'\widehat{\boldsymbol{\beta}} = \mathbf{x}'(\boldsymbol{\beta}-\widehat{\boldsymbol{\beta}}) + \varepsilon \sim \mathcal{N}\left(0, \sigma^2\sqrt{1 + \mathbf{x}'(\mathbf{X}'\mathbf{X})^{-1}\mathbf{x}}\right).$$

The rest of the proof follows the proof of Theorem 8.13. □

Using the distribution (8.31), we obtain the $(1 - \alpha)100\%$ CI for y as

$$\mathbf{x}'\widehat{\boldsymbol{\beta}} \pm t_{1-\alpha/2}^{-1}(n - m)\widehat{\sigma}\sqrt{1 + \mathbf{x}'(\mathbf{X}'\mathbf{X})^{-1}\mathbf{x}}.$$

The $(1 - \alpha)100\%$ CI for the mean/regression value, $\mathbf{x}'\boldsymbol{\beta}$ is the same as above without 1 under the square root:

$$\mathbf{x}'\widehat{\boldsymbol{\beta}} \pm t_{1-\alpha/2}^{-1}(n - m)\widehat{\sigma}\sqrt{\mathbf{x}'(\mathbf{X}'\mathbf{X})^{-1}\mathbf{x}}. \tag{8.32}$$

The difference between the individual and mean/regression prediction is driven by the presence of ε in the former – see Figure 6.15 from Section 6.7.3 for an illustration. Sometimes, we refer to this prediction as the pointwise CI prediction in contrast to simultaneous confidence band prediction discussed below.

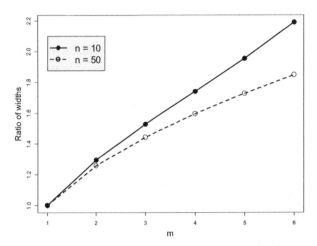

Figure 8.9: *The ratio of widths in the simultaneous (8.33) versus the pointwise prediction (8.32) for $\alpha = 0.05$. A smaller degree of freedom generates a bigger difference.*

Simultaneous confidence band

The CI for prediction by regression specified by (8.32) covers the regression value $\boldsymbol{\beta}'\mathbf{x}$ for a specific vector of predictors $\mathbf{x}$. It is interesting to construct a confidence region for the entire regression plane in R^m, such a confidence region is referred to as the simultaneous confidence band (CB) or Scheffé's simultaneous confidence intervals by the name of the statistician who introduced and solved this problem. The theoretical basis for the CB is the following result, which follows immediately from (8.22) when $k = m$,

$$\frac{1}{m\widehat{\sigma}^2}(\widehat{\boldsymbol{\beta}} - \boldsymbol{\beta})'(\mathbf{X}'\mathbf{X})(\widehat{\boldsymbol{\beta}} - \boldsymbol{\beta}) \sim F(m, n - m),$$

because, when $s = \{1, 2, ..., m\}$, we have $(\mathbf{X}'\mathbf{X})_s = (\mathbf{X}'\mathbf{X})$. Therefore, the ellipse around the OLS estimate in R^m defined as

$$\mathcal{B} = \left\{ \boldsymbol{\beta} \in R^m : \frac{1}{m\widehat{\sigma}^2}(\widehat{\boldsymbol{\beta}} - \boldsymbol{\beta})'(\mathbf{X}'\mathbf{X})(\widehat{\boldsymbol{\beta}} - \boldsymbol{\beta}) \leq F_{1-\alpha}^{-1}(m, n - m) \right\}$$

is the $(1 - \alpha)$th confidence region for $\boldsymbol{\beta}$. with probability $1 - \alpha$.

Now we construct the simultaneous confidence band. Let $\mathbf{x}$ be any vector from R^m. Define the lower and the upper limits of the band from two optimization problems,

$$\min_{\beta \in B} \mathbf{x}'\beta, \quad \max_{\beta \in B} \mathbf{x}'\beta,$$

respectively. To find the two extrema, introduce the Lagrange function

$$\mathcal{L}(\beta,\lambda) = \mathbf{x}'\beta - \frac{1}{2}\lambda\left((\hat{\beta}-\beta)'(\mathbf{X}'\mathbf{X})(\hat{\beta}-\beta) - m\hat{\sigma}^2 F_{1-\alpha}^{-1}(m, n-m)\right),$$

which yields the solution $\tilde{\beta} = \hat{\beta} - \lambda(\mathbf{X}'\mathbf{X})^{-1}\mathbf{x}$ and the Lagrange multiplier

$$\lambda = \pm \frac{\hat{\sigma}\sqrt{m}\sqrt{F_{1-\alpha}^{-1}(m, n-m)}}{\sqrt{\mathbf{x}'(\mathbf{X}'\mathbf{X})^{-1}\mathbf{x}}}.$$

Thus, the simultaneous lower and upper limits are

$$\mathbf{x}'\hat{\beta} \pm \hat{\sigma}\sqrt{m F_{1-\alpha}^{-1}(m, n-m)\mathbf{x}'(\mathbf{X}'\mathbf{X})^{-1}\mathbf{x}}. \tag{8.33}$$

Note that the CB is similar to the CI but wider by a factor $\sqrt{m F_{1-\alpha}^{-1}(m, n-m)}$ $/t_{1-\alpha/2}(n-m)$. The ratio of these widths for $\alpha = 0.05$ with two sample sizes is shown in Figure 8.9. Until $m = 3$, the difference between pointwise and simultaneous prediction is not considerable, but it then quickly increases.

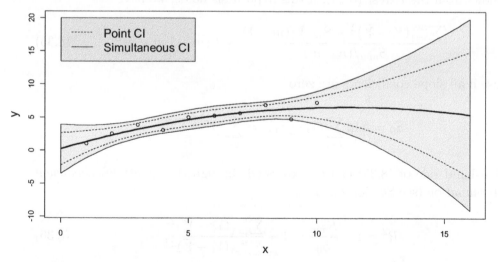

Figure 8.10: *The simultaneous 95% confidence band for quadratic regression. The band quickly widens outside of the observation domain – a typical feature of polynomial regression.*

Example 8.26 Simultaneous CB. *Compute and display the simultaneous confidence band for the quadratic regression from Example 8.24.*

Solution. The R code is in file `simCB.r` and the plot is in Figure 8.10. The call to the function is `simCB(job=1,n=10,beta=c(1,1,-.05),sigma=1,NX=100, alpha=.05,sk=3,nSim=10000,xmin=-20,xmax=30)`, where NX specifies the number of grid values in `x=seq(from=xmin,to=xmax,length=NX)` used in `job=3` for testing that the regression curve belongs to the CB in simulations with probability $1 - \alpha$. The simulations are similar to the R code `linpower` discussed above. Since the point CI computed by formula (8.32) has a smaller width, the dotted lines are within the CB. Recall that the ratio of the widths is depicted in Figure 8.9. We draw the readers's attention to a large widening of the CB outside of the observation range 1–10. Such widening is a characteristic property of polynomial regression: while fitting the (x_i, y_i) data with a polynomial could be used for interpolation, polynomial prediction suffers from an increasing imprecision when applied to extrapolation.

8.4.4   Testing the null hypothesis and the coefficient of determination

In this section, we assume that the multiple regression has the form (8.1): it contains the intercept term. The null hypothesis is that all slope coefficients are zero,

$$H_0 : \beta_2 = 0, ..., \beta_m = 0. \tag{8.34}$$

As follows from the F-test (8.29), if this hypothesis holds, we have

$$\frac{\left(\sum_{i=1}^n (Y_i - \overline{Y})^2 - S_{\min}\right)/(m-1)}{S_{\min}/(n-m)} \sim F(m-1, n-m), \tag{8.35}$$

because if all slope coefficients are zero

$$S_0 = \min_{\beta_1} \sum_{i=1}^n (Y_i - \beta_1)^2 = \sum_{i=1}^n (Y_i - \overline{Y})^2.$$

The left-hand side of (8.35) can be expressed through the (multiple) *coefficient of determination* (see Section 8.1.3) as

$$R^2 = 1 - \frac{S_{\min}}{S_0} = 1 - \frac{\sum_{i=1}^n (Y_i - \widehat{Y}_i)^2}{\sum_{i=1}^n (Y_i - \overline{Y})^2}, \tag{8.36}$$

so (8.35) can be rewritten as

$$\frac{R^2}{1 - R^2} \times \frac{n-m}{m-1} \sim F(m-1, n-m), \tag{8.37}$$

under the null hypothesis (8.34) or equivalently that the true coefficient of determination is zero.

The coefficient of determination (`multiple R-squared`) is a part of the `summary` of the `lm` call. To illustrate, we use the output of `CDpf(job=1)` for estimation of the Cobb–Douglas production function. The reported `F-statistic` is the left-hand side of test (8.35), or equivalently (8.37), computed as $\frac{0.8791}{1-0.8791} \times \frac{20-3}{3-1} = 61.806$, with the corresponding p-value `pf(61.806,df1= 3-1,df2=20-3, lower.tail=F)=1.587178e-08`. Strong data-supported evidence suggests that capital and labor are good predictors of the output.

`Adjusted R-squared` is a modification of R^2 to account for degrees of freedom of the sum of squares in the numerator and denominator of (8.36),

$$R_{\text{adj}}^2 = 1 - \frac{\sum_{i=1}^{n}(Y_i - \widehat{Y}_i)^2/(n - m)}{\sum_{i=1}^{n}(Y_i - \overline{Y})^2/(n - 1)},$$

Unlike the traditional coefficient of determination R_{adj}^2 may be negative when R^2 and $n - m$ are small. A positive feature is that an addition of a new predictor does not necessarily increase R_{adj}^2, so it may be used for selection of predictors.

The coefficient of determination is a popular goodness-of-fit measure to convey the regression strength. It is frequently interpreted as the proportion of variance of the dependent variable explained by the predictors. We warned in Section 6.7.4 that this interpretation may be wrong especially for time series data with a positive trend. Succinctly, the problem of interpretation of R^2 as the proportion of the variance of y explained by predictors is that $\sum_{i=1}^{n}(Y_i - \overline{Y})^2/(n-1)$ is not an estimator of the variance in the framework of linear model. A generalization of the coefficient of determination is deferred to Section 8.6.5.

8.4.5 Is X fixed or random?

This question aims at the Achilles heel of linear model. Linear model theory is built on the assumption that matrix $\mathbf{X}$ is fixed. However, in most applications, $\mathbf{X}$ is observed along with $\mathbf{y}$. For example, capital and labor in the above Cobb–Douglas production function are random as is the output. In all other linear model examples considered in the next section, matrix $\mathbf{X}$ is random as well. Only in rare applications are the regressors fixed such as when matrix $\mathbf{X}$ is chosen by the experimentalist or is a function of time as in time series analysis.

Two analyses of the properties of the OLS estimator with random $\mathbf{X}$ are offered here: First, we discuss the efficiency, as an extension of the Gauss–Markov theorem, and then we discuss statistical properties as outlined in Theorem 8.13.

Note that the asymptotic properties of the OLS estimator can be studied in both cases when $\mathbf{X}$ is fixed or random – see Section 9.4.1 in the more general setting of nonlinear regression. In general, OLS is consistent and asymptotically normally distributed under mild conditions when $n \to \infty$, even when ε_i are not normally distributed. In vague language, these properties follow from the LLN and CLT (Sections 2.9 and 2.10). However, if, in addition, ε_i are normally

distributed, the OLS estimator is asymptotically efficient as follows from the
general theory of maximum likelihood covered in Section 6.10.

Efficiency of the OLS estimator with random X

The Gauss–Markov theorem remains true under some special circumstances that
will be discussed below. In short, unbiasedness is replaced with conditional unbiasedness, see Shaffer [92] for more detail.

Definition 8.27 *Conditional unbiasedness. Let, in the linear model (8.2),
matrix $\mathbf{X}$ be random and independent of $\boldsymbol{\varepsilon}$. We say that an estimator $\widetilde{\boldsymbol{\beta}}$ is con-
ditionally unbiased for $\boldsymbol{\beta}$ if $E(\widetilde{\boldsymbol{\beta}}|\mathbf{X}) = \boldsymbol{\beta}$.*

Conditional unbiasedness implies traditional unbiasedness due to the rule of
repeated expectation from Section 3.3.1,

$$E(\widetilde{\boldsymbol{\beta}}) = E_{\mathbf{X}}E(\widetilde{\boldsymbol{\beta}}|\mathbf{X}) = E_{\mathbf{X}}(\boldsymbol{\beta}) = \boldsymbol{\beta}.$$

Theorem 8.28 *Gauss–Markov with random $\mathbf{X}$. Let, in the linear model
(8.2), matrix $\mathbf{X}$ be random and independent of $\boldsymbol{\varepsilon}$ (as before, the components of
the error vector have zero mean, constant variance, and mutually independent).
Then the OLS estimator $\widehat{\boldsymbol{\beta}}$ is efficient in the family of conditionally unbiased and
linear estimators of $\mathbf{y}$.*

Proof. Let $\widetilde{\boldsymbol{\beta}}$ be a conditionally unbiased linear estimator; $\widetilde{\boldsymbol{\beta}} = \mathbf{L}(\mathbf{X})\mathbf{y}$ where
$\mathbf{L}$ is an $m \times n$ function of $\mathbf{X}$ such that $E(\widetilde{\boldsymbol{\beta}}|\mathbf{X}) = \boldsymbol{\beta}$. Since $\mathbf{X}$ and $\boldsymbol{\varepsilon}$ are independent,
we have

$$\begin{aligned} E(\widetilde{\boldsymbol{\beta}}|\mathbf{X}) &= E(\mathbf{L}(\mathbf{X})\mathbf{y}|\mathbf{X}) = E(\mathbf{L}(\mathbf{X})\mathbf{X}\boldsymbol{\beta} + \mathbf{L}(\mathbf{X})\boldsymbol{\varepsilon}|\mathbf{X}) \\ &= \mathbf{L}(\mathbf{X})\mathbf{X}\boldsymbol{\beta} + \mathbf{L}(\mathbf{X})E(\boldsymbol{\varepsilon}|\mathbf{X}) = \mathbf{L}(\mathbf{X})\mathbf{X}\boldsymbol{\beta} = \boldsymbol{\beta} \end{aligned}$$

because $E(\boldsymbol{\varepsilon}|\mathbf{X}) = \mathbf{0}$. This implies $\mathbf{L}(\mathbf{X})\mathbf{X} = \mathbf{I}$. Following the proof of Theorem
8.9, we derive that $E(\text{cov}(\widehat{\boldsymbol{\beta}})|\mathbf{X}) \leq E(\text{cov}(\widetilde{\boldsymbol{\beta}})|\mathbf{X})$ where $\widehat{\boldsymbol{\beta}}$ is the OLS estimator.
Taking the expectation over $\mathbf{X}$ and applying the rule of repeated expectation, we
obtain $\text{cov}(\widehat{\boldsymbol{\beta}}) \leq \text{cov}(\widetilde{\boldsymbol{\beta}})$. The OLS estimator is efficient in the family of conditionally unbiased linear estimators.   □

Now we turn our attention to the efficiency of the OLS estimator assum-
ing normal distribution of $\boldsymbol{\varepsilon}$. As follows from Theorem 8.13, the OLS estimator
has the smallest variance when the distribution of $\mathbf{Y}$ is normal and $\mathbf{X}$ is fixed.
Under what conditions does the OLS estimator remain efficient when $\mathbf{X}$ is random? Loosely speaking, the answer is positive when the rows of matrix $\mathbf{X}$ are iid
normally distributed, see Section 6.8.1.

Theorem 8.29 *Let $(Y_i, \mathbf{X}_i)$ be iid observations from a multivariate normal dis-
tribution with conditional mean $E(Y_i|\mathbf{X}_i = \mathbf{x}_i) = \beta_0 + \boldsymbol{\beta}'\mathbf{x}_i$, and marginally
$\mathbf{X}_i \overset{\text{iid}}{\sim} \mathcal{N}(\boldsymbol{\mu}, \boldsymbol{\Omega})$, $i = 1, 2, ..., n$. Then the OLS the estimator for regression parameters $(\beta_0, \boldsymbol{\beta})$ is a uniformly unbiased minimum-variance estimator (UMVUE).*

Proof. This result follows from the fact that the multivariate normal distribution is complete and $\sum Y_i \mathbf{x}_i$ and $\sum \mathbf{x}_i \mathbf{x}_i'$ are sufficient statistics. The combination of these two facts allows application of the Lehmann–Scheffé Theorem 6.8.1. □

Note that the proofs of efficiency for fixed and random $\mathbf{X}$ are different: when $\mathbf{X}$ is fixed, we prove the efficiency by showing that the Cramér–Rao lower bound for the information matrix is attained. When $\mathbf{X}$ is random we use the concepts of completeness and sufficiency. We could not prove efficiency using the information matrix because

$$\left(E \sum_{i=1}^{n} \mathbf{X}_i \mathbf{X}_i' \right)^{-1} < E \left(\sum_{i=1}^{n} \mathbf{X}_i \mathbf{X}_i' \right)^{-1} . \tag{8.38}$$

More precisely, if $g(\mathbf{x})$ is the density of $\mathbf{X}_i$ independent of $\boldsymbol{\beta}$, the log joint density for the ith observation takes the form $-(Y_i - \beta_0 - \boldsymbol{\beta}' \mathbf{X}_i)^2/(2\sigma^2) - \ln g(\mathbf{X}_i)$. As follows from the proof of Item 3 of Theorem 8.13, an unbiased estimator would have minimum covariance matrix equal the left-hand side of (8.38), but the covariance of the OLS estimator is the right-hand side. Therefore, the OLS estimator does not achieve the Cramér–Rao lower bound to be UMVUE if g is a normal density.

An important practical conclusion can be drawn from Theorem 8.29: To efficiently estimate regression coefficients, continuous predictors, in addition to the dependent variable, should have a normal distribution. This means that normalizing transformations, such as the log transformation of positive predictors, are encouraged.

Now we turn our attention on the distribution of the OLS estimator when $\mathbf{X}$ is random and derive the exact distribution in a special case. Since, conditionally on $\mathbf{X}'\mathbf{X}$, the distribution of the OLS estimator is multivariate normal, $\widehat{\boldsymbol{\beta}} \sim \mathcal{N}(\boldsymbol{\beta}, \sigma^2(\mathbf{X}'\mathbf{X})^{-1})$, we can derive the marginal distribution of $\widehat{\boldsymbol{\beta}}$ by integrating (or summing) over the distribution of $\mathbf{X}'\mathbf{X}$. This method can be applied to either density or cdf. This method was employed in Example 6.121 to derive the distribution for the OLS slope coefficient in the simple linear regression with normally distributed x_i as in Theorem 8.29, namely, $\sqrt{n}\sigma_x(\widehat{\beta} - \beta)/\sigma \sim t$-distribution with $n-1$ degrees of freedom. The example below continuous Example 6.123 and illustrates how the distribution can be derived when X_i is a Bernoulli random variable.

Example 8.30 *Exact distribution of the OLS slope estimator with random predictors*. *(a) Derive the distribution of the OLS estimator of the slope when $\{(Y_i, X_i), i = 1, .., n\}$ are iid observations following the slope model $Y|X = x \sim \mathcal{N}(\beta x, \sigma^2)$, where X is the Bernoulli random variable with $\Pr(X = 1) = p$. (b) Use vectorized simulations to confirm the distributions by plotting the empirical and theoretical cdfs on the same graph.*

Solution. (a) The critical point is the distribution of $\sum_{i=1}^{n} X_i^2$, that follows a binomial distribution with $\Pr(\sum X_i^2 = k) = \binom{n}{k} p^k (1 - p)^{n-k}$. Therefore, the

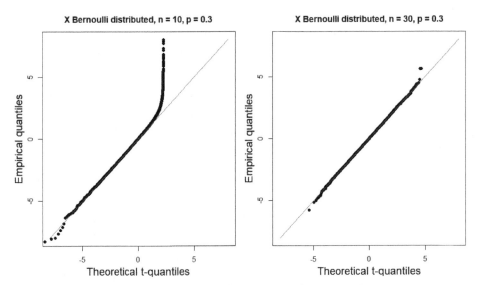

Figure 8.11: *The q-q plot of the OLS slope coefficient in linear regression with the intercept and Bernoulli predictors from Example 8.31. For small $n = 10$, the t-statistic (8.39) does not have the t-distribution with $n - 2$ degrees of freedom. This graph was produced by function* `olsnormT(job=2)`.

marginal cdf of the OLS slope estimator is

$$\frac{1}{1-(1-p)^n} \sum_{m=1}^{n} \Phi\left(\frac{b-\beta}{\sigma}\sqrt{m}\right) \binom{n}{m} p^m (1-p)^{n-m}.$$

Note that this derivation requires that $\sum_{i=1}^{n} X_i^2 > 0$. Therefore, the summation starts from $k = 1$ and therefore we divide by the probability that at least one X_i equals 1. (b) The simulations are found in the R function `olsbern.r`. The default arguments are `p=0.3`, `n=10`, `alpha=1`, `beta=.5`, `sigma=.5`, `nSim=500000`. The cdfs practically overlap (the graph is not shown).

Statistical properties of the OLS estimator

While the OLS estimator with random predictors remains efficient under certain circumstances, the distributions listed in Theorem 8.13 do not hold even for normally distributed $\mathbf{X}_i$ as in Theorem 8.29. We investigated the power of the test in the two-variable regression in the trivariate normal distribution in Example 7.61 and found that the likelihood-ratio (LR) test gave the best result. In this section, we are concerned with the distribution of the t-statistic (8.20) – the backbone of the theory of linear models. When $(Y_i, \mathbf{X}_i)$ are iid as in Theorem 8.29, the asymptotic theory of maximum likelihood applies and we can claim that when $n \to \infty$,

$$\frac{\widehat{\beta}_j - \beta_j}{\widehat{\sigma}\sqrt{(\mathbf{X}'\mathbf{X})_{jj}^{-1}}} \simeq \mathcal{N}(0, 1), \quad j = 1, 2, ..., m \tag{8.39}$$

under the null hypothesis that the true slope coefficient is β_j. However, for small n the left-hand side of (8.20) does not have the t-distribution, as illustrated in the following example.

Example 8.31 *T-statistic for regression with random predictors. Use vectorized simulations and q-q plot to visually test whether the t-statistics for the OLS slope coefficient have the t-distribution with $n - 2$ df when (1) X_i have the normal distribution, and (2) X_i have the Bernoulli distribution.*

Solution. The R function is `olsnormT` with `job=1` corresponding to normal and `job=2` to Bernoulli distribution. For both jobs, the q-q plots are depicted for $n = 10$ and $n = 30$. When $X_i \sim \mathcal{N}(\mu, \sigma_X^2)$, the t-statistic computed as $(\widehat{\beta} - \beta)/\widehat{\sigma}\sqrt{\sum(X_i - \overline{X})^2}$ has a distribution close to t-distribution with $n - 2$ df (the plot is not shown). However, when X_i has the Bernoulli distribution, the t-statistic is very skewed at right for small $n = 10$, see Figure 8.11. The t-statistic for the normally distributed predictor closely follows a t-distribution with $n - 2$ df. This observation is in a good agreement with the power function derived in Example 7.61 (d) when the t-distribution was used to compute the critical value of the test statistic.

Problems

1. Prove that matrix $\mathbf{X}_s$ is orthogonal to matrix $\mathbf{I} - \mathbf{X}(\mathbf{X}'\mathbf{X})^{-1}\mathbf{X}$ in the proof of Theorem 8.17.

2. Demonstrate by simulations that the CI defined by (8.21) covers the true value of the slope coefficient with the nominal probability. Modify function `CDpf` for this purpose (the data on K and L do not change). Print out the empirical coverage probability for the two elasticity parameters. [Hint: Use other parameters for simulations from the printout.]

3. Demonstrate that, under the multiplicative error, the coefficient of variation of the dependent variable is constant (consult Section 2.11). Provide a layman's explanation why in most practical statistical problems, including the production function, the multiplicative error model is more appropriate than the additive/traditional error model.

4. Develop an algorithm to compute the p-value for testing the hypothesis $H_0 : \mathbf{c}'\boldsymbol{\beta} = a$ based on the F-test (8.29) using function `lm`.

5. As follows from the output of `CDpf(job=1)` the coefficient at 1 is not statistically significant (the p-value $= 0.113072 > 0.05$). Does this mean that one can omit this variable and consider the output only as a function of capital?

6. Figure 8.6 shows the elliptical and rectangular confidence regions for (α, β). (a) Does the rectangular confidence region cover the true α and β with probability 0.95?

7. Prove that in regression with two predictors $\mathbf{x}_1$ and $\mathbf{x}_2$ the correlation coefficient between $\mathbf{x}_1$ and $\mathbf{x}_2$ is the same in the absolute value as the correlation coefficient between the OLS estimators $\widehat{\beta}_1$ and $\widehat{\beta}_2$ but opposite in sign.

8. File `pfx123.csv` contains the annual data on the revenue of a manufacturing company and three production components (all in million dollars). (1) Provide an interpretation of the regression coefficients. (2) Plot the 95% confidence regions for the first two slope coefficients similarly to Figure 8.6. (3) Following Example 8.21, test the hypothesis of no economy of scale using three methods. (4) Using the distribution (8.31), compute the probability that $c1 = 110$, $c2 = 18$, $c3 = 10$ yield the output greater than 10 billion dollars.

9. Does the rectangle defined by the individual confidence intervals for α and β in Figure 8.6 cover the true (α, β) with probability 0.95? Do you expect to have coverage probability greater or smaller than 0.95? Write R code to answer the question.

10. Explain why the plot from the `linpower` function does not change upon the choice of `st` argument despite the fact that $\boldsymbol{\beta}_A$ changes.

11. Modify function `simCB` to construct and display CB for a cubic regression, similarly to Figure 8.10 in Example 8.26.

12. Check distribution (8.37) by simulation with quadratic regression. Specify the coefficient values and σ as the arguments of your R function and use `summary (lm())$r.squared` to get R^2 from the `lm` fit. Plot the empirical and theoretical cdf of the left-hand side of (8.37) on the same graph.

13. Run function `olsbern` to research how the distribution is affected by other parameters, α, β, and σ, and n. Is it true that the distribution is not affected by α, β, and σ and it approaches the normal distribution with large n even when $p \neq 0.5$?

14. Derive the exact cdf and density of the OLS slope estimator in simple linear regression with a Poisson distributed predictor. Use simulations to check your analytic derivation. [Hint: The sum of n iid Poisson distributed random variables with λ has a Poisson distribution with mean λn.]

15. Express the exact cdf and density of the OLS slope estimator in simple linear regression with chi-square distributed predictor in terms of the integral over $(0, \infty)$. Use simulations to check your analytic derivation. [Hint: The sum of n iid chi-square distributed random variables with df k has a chi-square distribution with df kn.]

8.5 The one-sided p- and d-value for regression coefficients

The traditional interpretation of the regression coefficient as the mean change in the dependent variable for a unit increase of the predictor value suffers from ignoring the uncertainty in the coefficient estimate. In this section, we measure the strength of the predictor on the probability scale using the one-sided p-value and a generalization of the d-value introduced in Section 7.10.5.

The one-sided test is used to provide a meaningful probabilistic interpretation. The one-sided test, in general, is appealing from the practical standpoint and was promoted previously in Section 7.4. For example, if one is interested in a drug treatment effect, which is supposed to lower the cholesterol level, we want to see the evidence that, among people who took the drug, the level is lower (the one-sided test), not just that the treatment effect is not zero, as the traditional two-sided test suggests.

Table 8.1. Interpretation of the jth regression coefficient on the probability scale

Concept	Estimate	R
Population level		
One-sided p-value	$\Pr(\widehat{Y}_0 > \widehat{Y}_0)$	`pt(-abs(bj)/sej,df=n-m)=ct[j,4]/2`
Individual level		
d-value	$\Pr(\widehat{Y}_0 + \varepsilon_0 > \widehat{Y}_1 + \varepsilon_1)$	`pt(-abs(bj)/sqrt(2*sig^2+sej^2),df=n-m)`
b-value	$\Pr(\widehat{Y}_0 + \varepsilon_0 < \widehat{Y}_1 + \varepsilon_1)$	`pt(abs(bj)/sqrt(2*sig^2+sej^2),df=n-m)`

`su=summary(lm());ct=su$coefficients;sig=su$sigma;bj=ct[j,1];sej=ct[j,2]`

Two types of interpretations of the regression coefficient and the associated strength of the regressor on the probability scale are recognized in this section:

- *Population level*: one unit increase of the predictor value increases the mean of the dependent variable (no error term), quantified by the one-sided p-value.

- *Individual level*: one unit increase of the predictor value increases the value of the dependent variable (with the error term), quantified by the d- or b-value.

The interpretation at the population or individual level is determined by the application of the regression analysis: The former should be used when prediction of the mean regression value is the goal and the latter when prediction on the individual level is of concern. The recognition of the difference between the two levels of the application statistical of analysis has been the topic of discussion in the previous parts of the book starting from Section 1.4.1 and later in Section 7.10. For example, in medical applications, the d-value is associated with harm to an individual and the complementary b-value is associated with the individual treatment effect.

8.5.1 The one-sided p-value for interpretation on the population level

Without loss of generality, let the coefficient of interest be the coefficient at the first regressor so that the multiple linear regression with m coefficients and n observations is written as $Y = \beta x + \gamma' z + \varepsilon$. If $\widehat{\beta}$ and $\widehat{\gamma}$ are the regression coefficient estimates, the traditional interpretation is based on the elementary representation $\widehat{Y}_1 - \widehat{Y}_0 = \widehat{\beta}$, where $\widehat{Y}_1 = \widehat{Y}|(x+1) = \widehat{\beta}(x+1) + \widehat{\gamma}' z$ and $\widehat{Y}_0 = \widehat{Y}|x = \widehat{\beta}x + \widehat{\gamma}' z$. Thus, we interpret the beta-coefficient as an increase of the mean of the dependent variable due to an increase of predictor x by 1 under the assumption that other predictors, z, are held fixed. Such an interpretation treats regression as a linear function and $\widehat{\beta}$ as a known coefficient.

To account for the uncertainty of $\widehat{\beta}$, we consider the increase of the dependent variable on the probability scale assuming that $\widehat{\beta} > 0$. Then

$$\Pr(\widehat{Y}_0 > \widehat{Y}_1) = \Pr(\widehat{\beta} < 0) \simeq \mathcal{T}(-|\widehat{\beta}|/SE_\beta, \mathrm{df} = n - m) \stackrel{\mathrm{def}}{=} \text{one-sided } p\text{-value},$$
$$(8.40)$$

where the sign $\simeq$ means "estimate" and $\mathcal{T}$ is the cdf of the t-distribution; see Theorem 8.13. The one-sided p-value can be viewed as the p-value of the one-sided test $H_0 : \beta \le 0$ versus the alternative $H_A : \beta > 0$. As follows from (8.40), the one-sided p-value estimates the probability that the regression mean value of the dependent variable evaluated at x is greater than the mean value evaluated at $x + 1$. The larger the t-statistic $= |\widehat{\beta}|/SE_\beta$, the smaller the probability. See Table 8.1 on how to compute this value in R where j stands for the jth regression coefficient. Note that the one-sided p-value is half of the traditional two-sided p-value which is extracted from the summary of the lm object as ct[j,4]/2. In the cholesterol example, x is the group indicator variable: $x_i = 1$ if individual i took the drug and $x_i = 0$ if the individual took a placebo with expectation that $\widehat{\beta} < 0$. Then the one-sided p-value equal to $\mathcal{T}(-|\widehat{\beta}|/SE_\beta, \mathrm{df} = n - m)$ estimates the probability that the mean cholesterol level in the treatment group is higher than in the placebo group (the harm), assuming that the groups have the same values of covariates z. The complementary probability, $1 - p$, estimates the benefit value of taking the drug on the population/mean level.

8.5.2 The d-value for interpretation on the individual level

As criticized in Section 7.10, the p-value of any test approaches zero when the sample sizes increases – "statistical significance" can be always achieved with large n even when the difference between the null and the alternative is practically negligible. This property of the p-value holds for the regression analysis as well. To show this, we assume that $\mathbf{x}_i$ are random iid vectors with covariance matrix $\mathbf{\Omega_x}$. Then as follows from the LLN the n^{-1} times covariance matrix for the slope coefficients converges to $\sigma^2 \mathbf{\Omega_x}^{-1}$,

$$ p \lim_{n\to\infty} \frac{1}{n} \mathrm{cov}(\widehat{\boldsymbol{\beta}}) = \sigma^2 p \lim_{n\to\infty} \frac{1}{n} \left(\overset{\circ}{\mathbf{X}}{}' \overset{\circ}{\mathbf{X}} \right)^{-1} = \sigma^2 \mathbf{\Omega_x}^{-1}. $$

Since the standard error of the OLS estimator goes to zero with the order of $1/\sqrt{n}$, the p-value approaches zero with n. In other words, statistical significance can be achieved with n large enough whatever the threshold for the type I error is: the regression mean values can be perfectly separated even for a tiny nonzero regression coefficient.

In many applications, we are interested in the effect of the predictor on the individual level like in the cholesterol example. As doctors say "We treat an individual not a group of individuals." See more examples of this kind in Section 7.10.5. In fact, the treatment effect can be always statistically significant on the population level in terms of regression mean even if it is practically negligible. Now we ask: what is the probability that an individual will benefit from taking a new drug? To answer this question, we follow the same idea of comparison of populations corresponding to $x + 1$ and x as we did above. However, instead of comparing the means, we look at the individual level and, therefore, the error term is present. That is, the populations to compare are $\widehat{Y}_1 + \varepsilon_1$ and $\widehat{Y}_0 + \varepsilon_0$, where ε_1 and ε_0 are the independent error terms with zero mean and regression variance σ^2. Following the same idea as in (8.40), we estimate the probability as

$$ \mathrm{Pr}(\widehat{Y}_0 + \varepsilon_0 > \widehat{Y}_1 + \varepsilon_1) = \mathrm{Pr}(\widehat{\beta} + \varepsilon_0 - \varepsilon_1 < 0) $$

$$ \simeq \mathcal{T} \left(-\frac{|\widehat{\beta}|}{\sqrt{2\widehat{\sigma}^2 + SE_{\beta}^2}}, \mathrm{df} = n - m \right) \overset{\mathrm{def}}{=} d-\mathrm{value}. $$

Compared to (8.40), the term $2\widehat{\sigma}^2$ is added to the variance in the numerator under the square root; see Table 8.1. Continuing the previous example, the d-value is interpreted as the probability that the cholesterol level of a randomly chosen person who took the drug will be higher (the harm effect) than the cholesterol level of a randomly chosen person from the control group with the same predictors/characteristics specified by vector $\mathbf{z}$. For large df, say, df > 20 one may use the normal approximation. The benefit of the drug is measured by the complementary probability,

$$b-\text{value} \stackrel{\text{def}}{=} \mathcal{T}\left(\frac{|\widehat{\beta}|}{\sqrt{2\widehat{\sigma}^2 + SE_\beta^2}}, \text{df} = n - m\right),$$

as the probability that the cholesterol level of a randomly chosen person who took the drug will be lower (beneficial effect) than the cholesterol level of a randomly chosen person from the control group. When n is large, the variance of the regression coefficient, SE_β^2 is close to zero. However, due to $2\widehat{\sigma}^2$, the d-value does not approach zero because of the variation from individual to individual, which is always an important factor when treating individuals. A similar probability assessment can be carried out in a more general situation when individual responses are compared with a combination of predictors. The next example illustrates the interpretation in more detail. More applications of the d- and b-values for interpretation of regression coefficients on the probabilistic scale are found in Section 8.6.1 and Examples 8.40 and 8.48.

The d-value for personalized medicine

Hypertension is one of the world's most costly health conditions and contributes to risk of strokes and heart attacks. It is well known that the risk of high blood pressure increases with age, but also depends on other factors, such as smoking, family history, BMI, and lack of physical exercise. Many pharmaceutical companies work on developing drugs that lower the blood pressure. Much cost is involved not only at the research stage, but also when conducting a clinical trial for testing the drug efficacy. How to test the drug and what metric to use is a question of paramount importance. The following example emphasizes the difference between measuring the drug effect on the population/group level and from the individual perspective of an individual.

Example 8.32 *D-value for personalized medicine. In a clinical trial to test a new drug to lower the blood pressure, one thousand participants were given the drug and one thousand participants were given a placebo. Six months after administration, the difference of the systolic blood pressure from the baseline was measured. Other relevant factors, such as gender, smoking status, age, family history, BMI, and time exercised per week were recorded as well; see file dvalPMED.csv. (a) Explain the p-value for* **drug***. (b) Compute and explain the d- and b-value in layman's language. (c) Estimate the probability that a smoker with low exercise activity will experience improved blood pressure if he/she takes the drug, quits smoking and increase exercise by two hours a week.*

Solution. Below is the `summary(lm)` output (with some omissions).

```
> dvalPMED()
Coefficients:
             Estimate Std. Error t value Pr(>|t|)
(Intercept) -2.45551    3.39175  -0.724  0.46917
drug        -5.46558    0.89063  -6.137 1.01e-09 ***
smoking      3.20301    1.23333   2.597  0.00947 **
BMI          0.14421    0.04710   3.062  0.00223 **
man          2.28628    0.88916   2.571  0.01020 *
age          0.14046    0.04283   3.279  0.00106 **
famhist      3.13778    0.95343   3.291  0.00102 **
exer        -2.03759    0.14237 -14.312  < 2e-16 ***

Residual standard error: 19.85 on 1992 degrees of freedom
Multiple R-squared:  0.1319,     Adjusted R-squared:  0.1288
F-statistic: 43.22 on 7 and 1992 DF,  p-value: < 2.2e-16

B-value of the drug alone = 58%
B-value of the drug + quitting smoke + increasing exersize = 67%
```

(a) All predictors are statistically significant according to statistical 5% dogma. On average, patients who take the drug lower their systolic blood pressure by 5.46558 mmHg, and the p-value is impressively small. The one-sided b-value $= 1 - 1.01 \times 10^{-9}/2$ is practically 1. This means that the average of blood pressure over 2000 patients will be reduced with probability close to 1. First, it is important to remember that this probability is computed for the average of $n = 2000$ patients, for a smaller n their probability would decrease considerably. Second, this probability does not take into account the patient-to-patient variability of response to the drug. Indeed, this variability estimated by the residual standard error $\hat{\sigma} = 19.85$ is quite substantial. Moreover, `multiple R-squared` (coefficient of determination) indicates that only 13% of variation in blood pressure is explained by the predictors. (b) To account for the patient-to-patient response to the drug, the b-value must be employed. Using the last row command of Table 8.1, we compute the d-value as 42%. This means that, in 42% of patients who took the drug, the blood pressure will increase (the drug harm). Complementarily, the b-value $= 58\%$ meaning that 58% will benefit from the drug (treatment effect). From the perspective of personalized medicine, the chance that the blood pressure of a randomly chosen patient who took the drug is smaller than the blood pressure of a randomly chosen patient with the same smoking status, BMI, age, family history, and the exercise activity, who did not take the drug is 0.58. (c) Now we compare the patient who does not take the drug, smokes, and does not exercise with a patient who takes the drug, quits smoking, and increases exercising by two hours a week. The expected difference in blood pressure is `num` $= -5.46558 - 3.20301 - 2 \times 2.03759 = -12.744$ mmHg. However, this average does not tell the whole story: how does one lower the blood pressure? To answer this question, we need to compute the b-value. First, we need to compute the variance of the sum as `varde=t(cc)%*%COV%*%cc` where `cc=c(0,1,-1,0,0,0,0,2)` and `COV=vcov(o)` is the 8×8 covariance matrix of all regression coefficients. Then the b-value is computed as `pt(-num/sqrt(2*so$sigma^2+varde),df=o$df.residual)` is 67%. This means that the chance that an individual who takes the drug, quits smoking, and exercises will lower his/her blood pressure is 67% compared those who do not take the drug, smokes, and does not exercise.

Problems

1. Prove that the regression d-value turns into the two-sample d-value from Section 7.10.5 when there are no predictors, just an intercept term.

2. Is it true that the ratio between the one-sided p-value and the complementary one-sided b-value converges to 1 when $n \to \infty$?

3. Is it true that the difference between the one-sided p- and d-value is greater when the coefficient of determination in smaller? [Hint: Use simple linear regression as an example.]

4. In Example 8.32, (a) interpret the d-value, (b) compute and interpret the d- and the b-value for a person who does not take the drug but instead who quit smoking, reduces her/his BMI by 5 kg/m^2 and increases exercising time by 5 hours/day. (c) Estimate the probability that men have a higher cholesterol level than women assuming that all other conditions are the same. Rephrase the answer in terms of comparing a random man with a random woman.

5. Compute and interpret the d- and b-value for the β elasticity coefficient (labor) in the Cobb–Douglas production function (`CDpf`).

8.6 Examples and pitfalls

The goal of this section is to provide some examples of multiple regression analysis and discuss possible pitfalls. In particular, we discuss how to interpret regression coefficients and why regression of Y on X is different from X on Y. In addition, we provide a geometric illustration of the addition of a new predictor and introduce the generalized coefficient of determination to avoid false discovery.

8.6.1 Kids drinking and alcohol movie watching

Underage drinking in early adolescence is not only a cause of immediate life problems, including reckless driving and school dropout, but is also associated with higher risk of developing alcoholism in adulthood. Dartmouth researchers led by Sargent et al. [86] found a positive correlation between watching alcohol scenes in movies and initiation of drinking. This finding may have a potential outcome on the movie industry by restricting the audience and consequently reducing industry profits. In the following example, we use the data on the age of

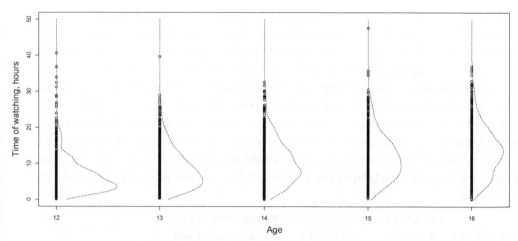

Figure 8.12: *Time of watching alcohol related scenes in movies over age. The data are considerably skewed – a log transformation may help to reduce the unwanted effect of outliers.*

drinking initiation and alcohol scene movie watching analyzed by Demidenko et al. [27]. The dependent variable is adolescents' exposure to alcohol scenes in popular movies derived through telephone interviews (see details in the paper cited). The data file `kidsdrink.csv` contains information on 3,805 kids of aged from 12 to 16. The first column (`alcm`) is the cumulative time in hours the kid watched alcohol related scenes over a two year period. The second column (`drink`) is a binary variable which indicates that he/she already tried alcohol at the time of interview. Other covariates include: `sex` (boy = 1, girl = 0), `race` (white = 1, otherwise = 0), having any alcohol branded merchandise (like a *Corona* T-shirt) `alcbr` (have an alcohol item = 1, otherwise = 0), parents' education `pared` (high school or higher education = 1, otherwise = 0), family income `inc` (income < $20K = 1 , otherwise 0), and performance in school `grade` (below average = 1, otherwise 0). The R code is in file `kidsdrink.r` and can be downloaded using the `source()` command. It does two jobs; in particular, `kidsdrink(job=1)` produces Figure 8.12 and `kidsdrink(job=2)` produces Figure 8.13.

To understand the distribution at a specific time, we use the Gaussian kernel density; see Section 5.5. The kernel densities are 90° rotated and plotted at each age. Not surprisingly, the time is highly skewed. The majority of kids are exposed to alcohol scenes about 15 minutes a month, but some kids watch alcohol movies more than 20 hours a month.

To "unskew" the data and reduce the unwanted effect of outliers (kids who many alcohol movies) the log transformation should be used. The problem is that some observations have zero values (no alcohol scenes watched). There are at least three methods to handle zero observations. The easiest method is to omit observations/cases entirely. This may be applied when the proportion of zero data is negligible. The second method is to add 1, or any other positive constant, before using the log transformation, $\log(1 + y)$. The advantage of this transformation is that, for small y, we have $\log(1 + y) \simeq y$ and, for large y, we have $\log(1 + y) \simeq \log(y)$. Since the time of alcohol scenes in each movie was originally measured in seconds, we take the transformation $\log(1/60^2 + y)$. The third method is to treat zeros as measurements below the detectable level. Since the time record starts at 1 second it makes sense to replace zeros with $1/60^2$ and then take the natural log transformation, $\ln(y)$. The last two approaches yield very similar results, below we use the last one. The `summary` output of the `lm(logalcm ~drink+age+boy+race+alcbr+pared+inc+grade, data=evc.dat)` function is shown below.

```
               Estimate Std. Error t value Pr(>|t|)
(Intercept) -0.09386    0.13303    -0.706 0.480526
drink        0.43270    0.02834    15.270 < 2e-16   ***
age          0.13707    0.00946    14.490 < 2e-16   ***
boy          0.04831    0.02430     1.988 0.046877  *
race         0.26678    0.04582     5.822 6.29e-09  ***
alcbr        0.26749    0.04017     6.658 3.17e-11  ***
pared       -0.02216    0.02703    -0.820 0.412186
inc          0.08445    0.03793     2.226 0.026059  *
grade       -0.09075    0.02598    -3.493 0.000483  ***
Residual standard error: 0.7417 on 3796 degrees of freedom
Multiple R-squared:  0.1969,    Adjusted R-squared:  0.1952
F-statistic: 116.3 on 8 and 3796 DF,  p-value: < 2.2e-16
```

All variables are statistically significant, according to statistical dogma threshold $\alpha = 0.05$, except for parents' education (`pared`). The most important variable contributing to watching alcohol related movies is drinking status with the regression coefficient 0.4328. To interpret this coefficient on the log scale we denote the expected time of watching of a kid who does not drink as Y_n and who drinks as Y_d with the difference on the log scale as $\ln Y_d - \ln Y_n = 0.4328$. Now using the Taylor series approximation $\ln(1 + x) \simeq x$ for small x we write

$$0.4328 = \ln Y_d - \ln Y_n = \ln\left(1 + \frac{Y_d - Y_n}{Y_n}\right) \simeq \frac{Y_d - Y_n}{Y_n}.$$

This means that adolescents who drink watch alcohol movies approximately 43% more than adolescents who do not drink alcohol. Also, from the regression output, we learn that children watch movies at an increasing rate as they get older. Following the same line of argumentation, we infer that kids watch more movies with the rate 13.7% a year.

In Figure 8.13, we display the regression results graphically with the emphasis on how drinking and having an alcohol related item affect alcohol movie watching. Since the model uses the dependent variable on the log scale, some caution should be taken when displaying the fitting results to make the presentation convincing.

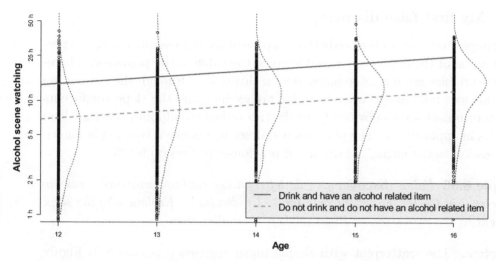

Figure 8.13: *Distribution of alcohol scene movie watching and the expected time in two cohorts of kids. Drinking and owing an alcohol related item considerably contribute to alcohol movie watching.*

To do this, we chose some hour numbers in the range from 1 to 50, like those used in Figure 8.13 but display them at the vertical position on the ln scale (a similar method was used earlier in the book such as in Section 5.4). This method of display makes values on the y-axis interpretable because they are shown on the original scale and at the same time the problem of skewness is solved. Indeed, the kernel densities look symmetric and the distribution is much closer to normal. To amplify the effect of drinking and having an alcohol related item, we superimpose two lines (run `kidsdrink` to see colors). The first line (red solid) displays the trend of alcohol movie watching for a "bad" kid, who initiated drinking alcohol and owns an alcohol related item. The second line (green dotted) displays a "good" kid who did not initiate drinking alcohol and does not own an alcohol related item. Both kids are white boys (`boy=1, race=1`), have educated and not poor parents (`pared=1, inc=1`) and are good students (`grade=1`). The combination of two factors, alcohol drinking and having an alcohol related item, considerably contribute to alcohol movie watching: the difference is about 15 hours.

Now we assess alcohol movie watching by comparing two kids, one who drinks and another who does not using the concept of the d-value by computing `pt(-0.4327/sqrt(2*0.7417^2+0.02834^2),df=3796)=0.34`. This means that the chance that a kid who does not drink watches more alcohol movies than the kid who does drink is 34%, assuming that all other factors are the same. It is important that this statement does not depend on the transformation of the dependent variable, the log transformation in this case.

8.6.2   My first false discovery

This example underscores the predictive purpose of the regression, namely, to predict the value of the dependent variable given the value of the predictor. Therefore, the variables are not interchangeable in this case. Although the correlation coefficient does not change when y and x are exchanging, the slope coefficients do. This comment seems obvious from the theoretical standpoint, but may lead to puzzles in applications. The phenomenon described below is commonly known as "regression to the mean," as discussed previously in Section 6.125.

Example 8.33 *False discovery. File leftright.csv contains measurements (in centimeters) of the left and right arm for 174 individuals. Explain why the slope coefficient in the regression of Right on Left is smaller than 1.*

Solution. The scatterplot with simple linear regression is shown in Figure 8.14. The regression output of Right on Left is below (see function `leftright`).

```
              Estimate Std. Error t value Pr(>|t|)
(Intercept)   3.56897    1.16306    3.069   0.0025  **
left          0.92225    0.02533   36.403   <2e-16 ***
```

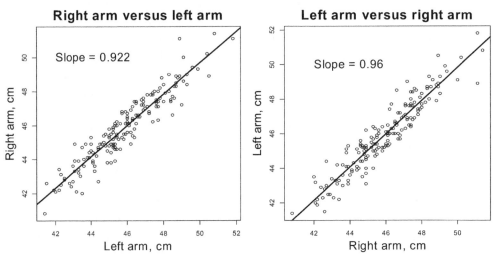

Figure 8.14: *Left: Since the slope estimate < 1 it follows that the right arm is shorter than the left one. Right: the left arm is shorter. A paradox.*

As follows from this regression, the slope coefficient is less than 1. Moreover, the null hypothesis that the slope equals 1 is rejected with the p-value `2*pt((1-.92225)/0.02533,df=174 -2, lower.tail=F)=0.002492323`. Note that we use the option `lower.tail=F` since the value is close to zero. Does this mean that the right arm is shorter? However, when regressing Left or Right an opposite result is obtained.

	Estimate	Std. Error	t value	Pr(>|t|)
(Intercept)	1.81420	1.21118	1.498	0.136
right	0.96038	0.02638	36.403	<2e-16 ***

As follows from this output, we obtain a reverse result: the left arm is shorter. How to solve this puzzle?

One has to remember that *regression is used for prediction* of the dependent variable Y given the predictor value $X = x$ because the OLS minimizes the distance between the observation points and the regression line along the y-axis. For example, as discussed in Section 3.5 the scatterplot of points drawn from a bivariate normal distribution has an elliptical shape. Moreover, as was depicted in Figures 3.12 and 3.13, regressions of Y on X and X on Y are different even though X and Y are drawn from the same population. If X and Y had equal roles, the main principal axis would be a better representation of their relationship.

To demonstrate this phenomenon theoretically consider the following statistical model: let A denote the arm length, regardless of left or right, in the population of people assuming that $A \sim \mathcal{N}(\mu_A, \sigma^2)$. In other words, theoretically, the lengths of left and right arms are the same for each individual but differs across individuals with average μ_A and variance σ^2. The data we have are results of measurements $L = A + u$ and $R = A + v$, where u and v are random variables representing independent normally distributed measurement errors with zero mean and variance τ^2. According to Section 3.5.1, the pair (L, R) has a bivariate normal distribution

$$\begin{bmatrix} L \\ R \end{bmatrix} \sim \mathcal{N}\left(\begin{bmatrix} \mu_A \\ \mu_A \end{bmatrix}, \begin{bmatrix} \sigma^2 + \tau^2 & \sigma^2 \\ \sigma^2 & \sigma^2 + \tau^2 \end{bmatrix} \right)$$

with regressions as conditional mean

$$E(L|R=r) = \mu_A + \frac{\sigma^2}{\sigma^2 + \tau^2}(r - \mu_A), \quad E(R|L=l) = \mu_A + \frac{\sigma^2}{\sigma^2 + \tau^2}(l - \mu_A).$$

If measurement errors were zero ($\tau = 0$), the two regressions would have slope 1 and coincide. Otherwise, the slope in the two regressions is smaller than 1. We say that the slope is attenuated.

To compare the arms, a paired t-test should be used. t.test(x=right,y=left, paired =T) that gives the p-value = 0.4733: The two arms have the same length on average.

8.6.3 Height, foot, and nose regression

Naive thinking is that if the dependent variable, Y, has positive slope estimates in simple linear regressions on both predictors x_1 and x_2, then, in the bivariate regression of Y on x_1 and x_2, the slope coefficients must be positive as well. However, this is not always the case. The following is an example.

File `HeightFootNose.csv` contains measurements of 497 people's height and length of foot and nose (in inches); the R code is found in file `hfn.r`. The scatterplots `Height` versus `Foot` and `Nose` are shown in Figure 8.15. As expected, people with longer feet and noses are taller, although the correlation between `Height` and `Nose` is only $r = 0.13$, so only 1.7% of height variation can be explained by nose length ($0.13^2 = 0.0169$). The correlation between `Foot` and `Nose` is higher, $r^2 = 0.36^2 = 0.13$ proportion of variation of `Nose` can be explained by `Foot`.

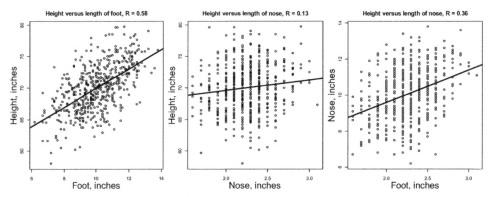

Figure 8.15: *Although the regressions of `Height` on `Foot` and `Nose` have positive slopes (the first two plots) the slope at `Nose` in the birvariate regression is negative because `Foot` and `Nose` correlate (the third plot).*

The output of the linear model with both predictors is below. The interpretation of the coefficient follows: If a person's foot is 10 inches and his nose is 2.5 inches, we expect that his height is `56.63+1.614*10-1.25*2.5=69.6` inches. This model explains 35% of height variation among males. Note that the coefficient at `Nose` is negative with a fairly small p-value. First of all, the negative sign does not contradict the simple linear regression and the plot at right in Figure 8.15. From regression of `Height` on `Nose`, one concludes that people with a longer nose tend to be taller. From the multiple regression one concludes that people with a longer nose are shorter among the subpopulation of people with the same foot size. Since the groups are different, we cannot expect the same correlation from the simple and multiple models.

```
lm(formula = Height ~ Foot + Nose, data = da)
            Estimate Std. Error t value Pr(>|t|)
(Intercept)  56.6307     1.2436  45.537   <2e-16 ***
Foot          1.6140     0.1017  15.874   <2e-16 ***
Nose         -1.2499     0.5125  -2.439   0.0151 *
Residual standard error: 3.048 on 494 degrees of freedom
Multiple R-squared:  0.3486,    Adjusted R-squared:  0.346
F-statistic: 132.2 on 2 and 494 DF,  p-value: < 2.2e-16
```

Now we investigate under what conditions the slope is positive in a simple linear regression but negative in the bivariate linear regression. For this purpose, we employ the formula for the conditional expectation of a multivariate normal distribution from Section 4.1.2 given in Theorem 4.8. Thus, we assume that Y, X_1, and X_2 follow a trivariate normal distribution with conditional expectation/linear regression given by $E(Y|X_1 = x_1, X_2 = x_2) = \beta_0 + \beta_1 x_1 + \beta_2 x_2$, where, to be concrete, Y is height, X_1 is foot, and X_2 is nose measurements. In the notation of Theorem 4.8, the vector of slope coefficients is $\Omega_{22}^{-1}\Omega_{21}$. To simplify the analysis, we shall assume that Y, X_1, and X_2 have unit variance, so, for the trivariate normal distribution, we obtain

$$\Omega_{22} = \begin{bmatrix} 1 & \rho \\ \rho & 1 \end{bmatrix}, \quad \Omega_{21} = \begin{bmatrix} \rho_1 \\ \rho_2 \end{bmatrix},$$

where ρ is the correlation coefficient between X_1 and X_2, and ρ_1 and ρ_2 are correlation coefficients between Y and X_1, and Y and X_2, respectively. Note that since variances are equal to 1 the slopes in the simple linear regressions of Y on X_1 and Y on X_2 are equal to ρ_1 and ρ_2, respectively. The vector of the slope coefficients in the bivariate regression is given by

$$\begin{bmatrix} \beta_1 \\ \beta_2 \end{bmatrix} = \begin{bmatrix} 1 & \rho \\ \rho & 1 \end{bmatrix}^{-1} \begin{bmatrix} \rho_1 \\ \rho_2 \end{bmatrix} = \frac{1}{1-\rho^2} \begin{bmatrix} \rho_1 - \rho\rho_2 \\ \rho_2 - \rho\rho_1 \end{bmatrix}.$$

As follows from the right-hand side, even though ρ_1 and ρ_2 are positive the second component (the slope coefficient at Nose) may be negative if the following three conditions hold: (i) $\rho > 0$, (ii) ρ_2 is small, and (iii) ρ_1 is large such that $\rho_2 - \rho\rho_1 < 0$. This is exactly what happens in our example: (i) Foot and Nose do positively correlate with $\rho = 0.36$, (ii) Height and Nose have low correlation, $\rho_2 = 0.13$, and (iii) Height and Foot have high positive correlation $\rho_1 = 0.58$, which implies $0.13 - 0.58 \times 0.36 = -0.079 < 0$. More insights into when the slopes from simple and multiple regressions have different signs can be gained from the geometric illustration in Example 8.35 to follow.

Interpretation of regression coefficients

Traditionally, we interpret the slope coefficient β_j in linear model (8.1) as the expected increase of variable Y when predictor x_j increases by 1 assuming that other predictors do not change. This interpretation cannot be used to compare the strength of predictors measured on different scales. For example, a one inch increase of foot has a greater impact on height than a one inch increase of nose. Other shortcomings and its connection to the p- and d-value are found in Section 8.5.

A better comparison would be on the scale of standard deviation of predictors. This interpretation is valid for observational studies when predictors are sampled from a general population. For example, we may compare the effect on height using 10% SD increase for foot and nose. Since $\mathrm{SD}_{\mathrm{foot}} = 1.44$ and $\mathrm{SD}_{\mathrm{nose}} = 0.29$ we interpret the regression as follows.

- In the population of people with the same nose length, an increase of foot by $10\%\mathrm{SD}_{\mathrm{foot}}$ leads to increase of height by $1.614 \times 0.144 = 0.23$ inch.

- In the population of people with the same foot size an increase of nose by $10\%\mathrm{SD}_{\mathrm{nose}}$ leads to decrease of height only by $1.25 \times 0.029 = 0.036$ inch.

Now the increase of the two predictors is comparable because they are expressed on the scale of their variation, so they are scale independent. Obviously, measurements of nose are not as important for prediction of height as those of foot.

Now we make a comment on the interpretation of the intercept. As follows from model (8.1), the intercept is the expected value of the dependent variable when all predictors are zero. In many real-life regressions zero predictors makes no sense. Consequently, the p-value of the intercept often has no sense and is usually close to zero. For example, as follows from the above regression, we interpret the intercept $= 56.6$ as the height of a person with no feet and no nose! A much more sensible value to interpret is the expected value of the dependent variable when predictors take their average values with the estimate $\widehat{\beta}_2\bar{x}_1 + \widehat{\beta}_3\bar{x}_2 + ... + \widehat{\beta}_m\bar{x}_{m-1}$. One can either compute this estimate and obtain the standard error to conduct further statistical hypothesis testing regarding the expected average value of the dependent variable or, alternatively, subtract averages from the predictors. Below we show the output of the lm with averages for Foot and Nose subtracted (f and n, respectively).

```
               Estimate Std. Error t value Pr(>|t|)
(Intercept)    70.0435      0.1367 512.274   <2e-16 ***
f               1.6140      0.1017  15.874   <2e-16 ***
n              -1.2499      0.5125  -2.439   0.0151 *
Residual standard error: 3.048 on 494 degrees of freedom
Multiple R-squared:  0.3486,    Adjusted R-squared:  0.346
F-statistic: 132.2 on 2 and 494 DF,  p-value: < 2.2e-16
```

Compared to the output above, nothing changed but the row corresponding to Intercept. Now intercept makes perfect sense: the height of the person with average foot and nose measurements.

8.6.4   A geometric interpretation of adding a new predictor

This section discusses issues related to the addition of a new predictor using geometrical arguments. In particular, in examples to follow, we emphasize that neither RSS nor R^2 is useful to identify the optimal number of predictors in the linear model because they decrease with increasing m. We also illustrate when these goodness-of-fit measures do not change upon addition of a new predictor.

(a) Addition of a new predictor

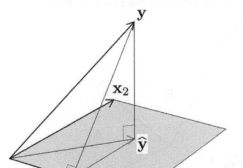

(b) New predictor does not change RSS

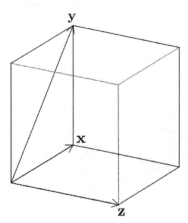

Figure 8.16: *Geometrical illustration to Example 8.34.*

Example 8.34 *When RSS does not decrease.* (a) Prove that the addition of a new predictor to a linear model leads to the same or smaller RSS (higher coefficient of determination). (b) Provide geometric argumentation. (c) Prove that the addition of a new predictor does not change the OLS estimate and the RSS if the new predictor is orthogonal to the residual vector from the original regression.

Solution. (a) Let the linear model be (8.2) and the new predictor be $\mathbf{z}^{n\times 1}$ with coefficient γ. Let RSS_{new} and RSS_{old} denote the minimum RSS for the extended and the original linear model. We have

$$RSS_{\text{new}} = \min_{\beta,\gamma} \|\mathbf{y} - \mathbf{X}\beta - \gamma\mathbf{z}\|^2 \leq \min_{\beta,\gamma=0} \|\mathbf{y} - \mathbf{X}\beta - \gamma\mathbf{z}\|^2$$
$$= \min_{\beta} \|\mathbf{y} - \mathbf{X}\beta\|^2 = RSS_{\text{old}}.$$

We note that the domain of minimization at the left-hand side contains the domain of minimization at the right-hand side. Obviously, the minimum of any function, including RSS, on a domain set is equal to or smaller than that on its subset. The inequality between RSSs can be immediately translated into an inequality between coefficients of determination.

(b) As in Figure 8.2, we assume that the original linear model is a slope model (Section 6.5.2): it contains only $\mathbf{x}_1$; see the left plot in Figure 8.16. The minimum RSS in the slope model is the squared length of the perpendicular dropped from $\mathbf{y}$ onto $\mathbf{x}_1$, or, equivalently, the squared distance between $\mathbf{y}$ and the basis of the projection $\widehat{\mathbf{y}}_1$ (note that $\mathbf{y} - \widehat{\mathbf{y}}_1 \perp \mathbf{x}_1$). Consider the right triangle with vertices $\mathbf{y}$, $\widehat{\mathbf{y}}$, and $\widehat{\mathbf{y}}_1$. In this triangle, $\mathbf{y} - \widehat{\mathbf{y}}$ is the cathetus and $\mathbf{y} - \widehat{\mathbf{y}}_1$ is the hypotenuse. As we know, the length of the hypotenuse cannot be smaller that the

Even if both individual slopes are positive
one of the bivariate slopes may be negative

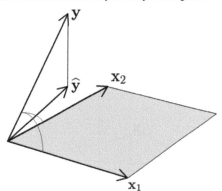

Both bivariate slopes cannot be negative
if both individual slopes are positive

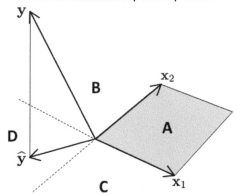

Figure 8.17: *Geometric illustration of the bivariate regression from Example 8.35. New predictor does not change RSS (and the coefficient of determination) if it is orthogonal to residuals.*

length of the cathetus, $\|\mathbf{y}-\widehat{\mathbf{y}}_1\| \leq \|\mathbf{y}-\widehat{\mathbf{y}}\|$, which means that $RSS_{\text{new}} \leq RSS_{\text{old}}$. (c) Let $\mathbf{z}$ be orthogonal to the residual vector or algebraically $\mathbf{z}'(\mathbf{y} - \mathbf{X}\widehat{\boldsymbol{\beta}}) = \mathbf{0}$. Since the OLS estimate is unique it suffices to prove that, after the addition of $\mathbf{z}$, the beta coefficients do not change and the coefficient at $\mathbf{z}$ is zero. The normal equations for the extended matrix $[\mathbf{X}, \mathbf{z}]$ can be written as a pair of equations $\mathbf{X}'(\mathbf{y} - \mathbf{X}\boldsymbol{\beta}-\gamma\mathbf{z}) = \mathbf{0}$ and $\mathbf{z}'(\mathbf{y} - \mathbf{X}\boldsymbol{\beta}-\gamma\mathbf{z}) = 0$. Substituting $\boldsymbol{\beta} = \widehat{\boldsymbol{\beta}}$ and $\gamma = 0$, we get $\mathbf{X}'(\mathbf{y} - \mathbf{X}\widehat{\boldsymbol{\beta}}) = \mathbf{0}$ and $\mathbf{z}'(\mathbf{y} - \mathbf{X}\widehat{\boldsymbol{\beta}}) = 0$. This situation is depicted in the left plot of Figure 8.16: the regression is the slope model, $\mathbf{y} = \beta\mathbf{x} + \boldsymbol{\varepsilon}$ and $\mathbf{z}$ is an additional predictor. The three vectors are vertices of a cube. The projection of $\mathbf{y}$ on $\mathbf{x}$ is the same as projection of $\mathbf{y}$ on the plane spanned by $\mathbf{x}$ and $\mathbf{z}$. In this particular case $\widehat{\beta} = 1$ and $\mathbf{z}$ is orthogonal to both $\mathbf{y}$ and $\mathbf{x}$, and therefore $\mathbf{z}$ is orthogonal to the residual vector, $\mathbf{y} - \mathbf{x}$. $\square$

In the height, foot and nose example, the slopes from two simple regressions were positive, but, in the bivariate regression, one of the coefficients was negative. The following example tackles this puzzle from a geometrical perspective. Specifically, we pose the following question. Let the dependent variable have positive correlations with two predictors. Under what conditions can one of the OLS coefficients in the bivariate regression be negative?

Example 8.35 *Positive slopes.* *The OLS estimate of the slope coefficients in simple individual regressions of Y on x_1 and Y on x_2 are positive. (a) Is it possible for one of the slope coefficient in the bivariate regression of Y on x_1 and x_2 to be negative? (b) Is it possible for both slope coefficients to be negative? Provide a geometric illustration.*

Solution. See Figure 8.17. It is assumed that the origin is shifted to the average values, so that the cosine angle between vectors is the correlation coefficient. We use a simple fact that the correlation coefficient is positive if and only if the slope coefficient in simple linear regression is positive. (a) Yes, it is possible; see the left plot. Consider a triple of vectors $\mathbf{x}_1$, $\mathbf{x}_2$, and $\mathbf{y}$ such that the angles between $\mathbf{x}_1$ and $\mathbf{y}$, and between $\mathbf{x}_2$ and $\mathbf{y}$ are positive (meaning that the correlation is positive and the slopes of individual regressions are positive, the angles depicted by the arcs are acute). If the correlation between $\mathbf{x}_1$ and $\mathbf{x}_2$ is high (the angle is small), the projection of $\mathbf{y}$ onto the plane spanned by $\mathbf{x}_1$ and $\mathbf{x}_2$, the basis of the perpendicular, $\widehat{\mathbf{y}}$, may be outside the parallelogram built on these vectors (shaded area). This means that the coefficient $\widehat{\beta}_1$ in the bivariate regression $\widehat{\mathbf{y}} = \widehat{\beta}_1\mathbf{x}_1 + \widehat{\beta}_2\mathbf{x}_2$ is negative. (b) No, it is impossible. Without loss of generality, one can assume that $\mathbf{x}_1$ and $\mathbf{x}_2$ have unit length. By condition, $\mathbf{y}'\mathbf{x}_1$ and $\mathbf{y}'\mathbf{x}_2$ are positive. The slope coefficients from the bivariate regression are

$$\widehat{\beta}_1 = \frac{1}{\Delta}[\mathbf{y}'\mathbf{x}_1 - (\mathbf{x}_1'\mathbf{x}_2)(\mathbf{y}'\mathbf{x}_2)], \ \ \beta_2 = \frac{1}{\Delta}[\mathbf{y}'\mathbf{x}_2 - (\mathbf{x}_1'\mathbf{x}_2)(\mathbf{y}'\mathbf{x}_1)],$$

where $\Delta = 1 - (\mathbf{x}_1'\mathbf{x}_2)^2 > 0$. We prove the impossibility by contradiction: let $\widehat{\beta}_1$ and $\widehat{\beta}_2$ both be negative. Then we have $(\mathbf{x}_1'\mathbf{x}_2) > (\mathbf{y}'\mathbf{x}_1)/(\mathbf{y}'\mathbf{x}_2)$ and $(\mathbf{x}_1'\mathbf{x}_2) > (\mathbf{y}'\mathbf{x}_2)/(\mathbf{y}'\mathbf{x}_1)$. Multiplying these inequalities gives $(\mathbf{x}_1'\mathbf{x}_2) > 1$, which contradicts $\Delta > 0$. The plot at right shows four areas for $\widehat{\mathbf{y}}$. (1) If $\widehat{\mathbf{y}}$ belongs to area **A,** both $\widehat{\beta}_1$ and $\widehat{\beta}_2$ are positive. (2) If $\widehat{\mathbf{y}}$ belongs to area **B,** coefficient $\widehat{\beta}_1$ is negative and $\widehat{\beta}_2$ is positive. (3) If $\widehat{\mathbf{y}}$ belongs to area **C,** coefficient $\widehat{\beta}_1$ is positive and $\widehat{\beta}_2$ is negative. (4) If $\widehat{\mathbf{y}}$ belongs to area **D** both $\widehat{\beta}_1$ and $\widehat{\beta}_2$ are negative. If the latter is the case, as shown, at least one individual slope is negative, so they cannot both be positive.

8.6.5 Contrast coefficient of determination against spurious regression

The coefficient of determination is commonly used in regression analysis as a goodness-of-fit measure. In Section 6.7.4, we saw that the coefficient of determination can be misinterpreted as the variance of the dependent variable explained by predictors (independent variables), especially when y has a positive trend as in time series in economic applications. For example, as discussed in Example 6.58, following a widespread interpretation, 90% of the revenue of a company can be explained by the number of truck drivers. This is an example of a spurious regression. The goal of this section is to emphasize the limitations of the traditional interpretation of R^2 and suggest a solution.

The problem with interpretation of R^2 as the proportion of the variance explained by predictors is that the variance of the dependent variable with increasing mean does not make sense. To suggest an alternative, we look at R^2 from the perspective of statistical hypothesis testing of Section 8.4.4. In fact, R^2 tells the strength of the alternative to the null hypothesis (8.34). In other words, the coefficient of determination reflects the quality of the fit compared to the constant null model, $E(y_i) = \text{const}$. Now it becomes clear why R^2 is close to 1 for time series data with a positive (or negative) trend. The null hypothesis $E(y_i) = \text{const}$ makes no sense. To make the comparison sensible, one has to contrast the regression with a more reasonable null model. Below, we offer a generalization of the coefficient of determination in the case when the null hypothesis is more realistic. As such, it is more complicated than the constant-expectation model.

We distinguish two types of predictors in a linear regression model. The first type of predictors $\mathbf{x}_{11}, ..., \mathbf{x}_{1p}$, which constitute matrix $\mathbf{X}_1$, will be referred to as null predictors. For example, $\mathbf{x}_{11}$ may be a vector of ones that corresponds to the intercept term, the second predictor, $\mathbf{x}_1$ may be a vector of time $(1, 2, ..., n)'$, as in time series analysis. The purpose of the null predictors is to catch an obvious pattern or trend of the dependent variable. The second type of predictors, $\mathbf{x}_{21}, ..., \mathbf{x}_{2q}$, which constitute matrix $\mathbf{X}_2$, are actual predictors that explain variation of the dependent variable. That is, the following partition takes place: $\mathbf{X} = [\mathbf{X}_1, \mathbf{X}_2]$. The linear regression model takes the form (full model)

$$\mathbf{y} = \mathbf{X}_1\boldsymbol{\beta}_1 + \mathbf{X}_2\boldsymbol{\beta}_2 + \boldsymbol{\varepsilon}, \tag{8.41}$$

with the null hypothesis

$$H_0 : \boldsymbol{\beta}_2 = \mathbf{0}. \tag{8.42}$$

The difference of the contrast (generalized) versus traditional coefficient of determination is that the set of null predictors is expanded and therefore may include others, not only a constant term. The formal definition is given below.

Definition 8.36 *The contrast coefficient of determination is defined as*

$$R_C^2 = 1 - \frac{\min_{\beta_1,\beta_2} \|\mathbf{y} - \mathbf{X}_1\boldsymbol{\beta}_1 - \mathbf{X}_2\boldsymbol{\beta}_2\|^2}{\min_{\beta_1} \|\mathbf{y} - \mathbf{X}_1\boldsymbol{\beta}_1\|^2}. \tag{8.43}$$

We call this the contrast coefficient contrast because it contrasts two linear models. The contrast coefficient of determination turns into the traditional coefficient when $\mathbf{X}_1$ consists of just a vector of ones. Its connection to testing the null hypothesis (8.42) is similar to (8.37),

$$\frac{R_C^2}{1 - R_C^2} \times \frac{n - p - q}{q} \sim F(q, n - p - q).$$

To compute R_C^2 in R, we run two models: the full model (8.41) and the model without actual predictors (the null predictors only). The traditional coefficient of determination for the full and null models, denoted as R^2 and R_0^2, respectively, can be extracted using `summary(lm())$r.squared` (we reserve notation R^2 for the standard coefficient of determination when the null model is constant). Then the contrast coefficient of determination can be computed as

$$R_C^2 = 1 - \frac{1 - R^2}{1 - R_0^2} = \frac{R^2 - R_0^2}{1 - R_0^2}.$$

Simple algebra yields $R_C^2 \leq R^2$. Figure 8.19 provides a geometric interpretation and more discussion is found at the end of this section.

An alternative way to compute R_C^2 is to regress the dependent variable and all actual predictors on the null predictors as shown in the example below.

Example 8.37 *Contrast coefficient of determination. Show that the contrast coefficient of determination is equal to the traditional coefficient of determination in regression of $\mathbf{r}$ on $\mathbf{R}_2$, where $\mathbf{r}$ and each vector column of matrix $\mathbf{R}_2$ are residual vectors of regression $\mathbf{y}$ and $\mathbf{X}_2$ on $\mathbf{X}_1$.*

Solution. The following discussion uses the definition of the annihilator matrix from Section 8.1. We assume that the first column of $\mathbf{X}_1$ is the vector of ones (regression with intercept). First, we note that $\min_{\boldsymbol{\beta}_1} \|\mathbf{y} - \mathbf{X}_1\boldsymbol{\beta}_1\|^2 = \|\mathbf{r}\|^2$ and $\mathbf{r}'\mathbf{1} = 0$ where $\mathbf{r}$ is the residual vector in regression of $\mathbf{y}$ on $\mathbf{X}_1$. Second, we note that the matrix of residual vectors of $\mathbf{X}_2$ on $\mathbf{X}_1$ can be written as $\mathbf{R}_2 = \mathbf{H}_1\mathbf{X}_2$, where $\mathbf{H}_1 = \mathbf{I} - \mathbf{X}_1(\mathbf{X}_1'\mathbf{X}_1)^{-1}\mathbf{X}_1'$ is the annihilator matrix. This means that, as follows from definition (8.43), it suffices to prove that

$$\min_{\boldsymbol{\beta}_1, \boldsymbol{\beta}_2} \|\mathbf{y} - \mathbf{X}_1\boldsymbol{\beta}_1 - \mathbf{X}_2\boldsymbol{\beta}_2\|^2 = \min_{\boldsymbol{\beta}_2} \|\mathbf{r} - \mathbf{R}_2\boldsymbol{\beta}_2\|^2.$$

But the left-hand side can be expressed as

$$\min_{\boldsymbol{\beta}_1} \|(\mathbf{y} - \mathbf{X}_2\boldsymbol{\beta}_2) - \mathbf{X}_1\boldsymbol{\beta}_1\|^2 = (\mathbf{y} - \mathbf{X}_2\boldsymbol{\beta}_2)'\mathbf{H}_1(\mathbf{y} - \mathbf{X}_2\boldsymbol{\beta}_2)$$
$$= (\mathbf{H}_1\mathbf{y} - \mathbf{H}_1\mathbf{X}_2\boldsymbol{\beta}_2)'(\mathbf{H}_1\mathbf{y} - \mathbf{H}_1\mathbf{X}_2\boldsymbol{\beta}_2) = \|\mathbf{r} - \mathbf{R}_2\boldsymbol{\beta}_2\|^2.$$

Despite the fact that neither $\mathbf{y}$ nor $\mathbf{X}$ follows a multivariate normal distribution, this result may be interpreted through the concept of partial correlation of Section 4.1.2: the contrast coefficient of determination is the squared partial correlation coefficient between $\mathbf{y}$ and $\mathbf{X}_2$ conditional on $\mathbf{X}_1$. □

The contrast coefficient of determination has the same interpretation as the proportion of variance of the dependent variable explained by actual predictors but it may make more sense because the null variables exclude the presence of an obvious pattern that has little to do with the actual predictors.

Example 8.38 *Truck drivers. Compute the contrast coefficient of determination in the truck driver Example 6.58.*

Solution. According to the traditional coefficient of determination almost 90% of the revenue of , Y, could be explained by the number of truck drivers d. We argued that the constant model, $Y_t = $ const, makes no sense as the null hypothesis model because revenue increases steadily in time making R^2 close to 1 because $\sum_{t=1}^{n}(Y_t - \overline{Y})^2$ is large. Instead a more reasonable null model is the trend model, $E(Y_t) = \gamma_1 + \gamma_2 t$. This modification dictates the following model for revenue: $E(Y_t) = \beta_1 + \beta_2 d_t + \beta_3 t$. To compute R_C^2, we run two regressions, one as displayed above, and the other is the trend model. The R code is in function truckR(job=3) and the output is listed below. The contrast coefficient of determination is

$$R_C^2 = \frac{R^2 - R_0^2}{1 - R_0^2} = \frac{0.9399 - 0.9356}{1 - 0.9356} = 0.067.$$

```
truckR(job=3)
lm(formula = revenue ~ truc.dr + ti, data = da)
            Estimate Std. Error t value Pr(>|t|)
(Intercept) 198.1421     1.6662 118.918  < 2e-16 ***
truc.dr       1.0807     0.9792   1.104  0.28515
ti            1.6380     0.4281   3.826  0.00135 **
Residual standard error: 3.311 on 17 degrees of freedom
Multiple R-squared:  0.9399,    Adjusted R-squared:  0.9328
F-statistic: 132.9 on 2 and 17 DF,  p-value: 4.182e-11

lm(formula = revenue ~ ti, data = da)
            Estimate Std. Error t value Pr(>|t|)
(Intercept) 198.8487     1.5475  128.50  < 2e-16 ***
ti            2.0887     0.1292   16.17 3.65e-12 ***
Residual standard error: 3.331 on 18 degrees of freedom
Multiple R-squared:  0.9356,    Adjusted R-squared:  0.932
F-statistic: 261.4 on 1 and 18 DF,  p-value: 3.652e-12

[1] "R2G = 0.0668508039113752"
```

The paradox is resolved: Truck drivers have a negligible effect on the revenue. The value $R_C^2 = 0.067$ is equals to that obtained in Example 6.58 as the squared correlation coefficient between residuals from regressions on time t as follows from Example 8.37. $\qquad\square$

Autoregression is a popular statistical model of time series analysis. Recall that this model was used in Example 4.10 in the probability part of the book to predict the temperature on the Fool's Day as an illustration of the multivariate normal distribution where all parameters must be known. In statistics, we use data to estimate those parameters. The coefficients of autoregression can be viewed as a linear model with predictors that are the previous values of the time series. Clearly, the assumption on a fixed matrix $\mathbf{X}$ is violated here. Nevertheless, the least squares theory is applicable, but should be treated just as an approximation.

The goal of the following example is twofold: (1) to illustrate how autoregression can be used for prediction and (2) to convince the reader why the contrast coefficient of determination is a more appropriate measure of goodness-of-fit than the traditional coefficient of determination.

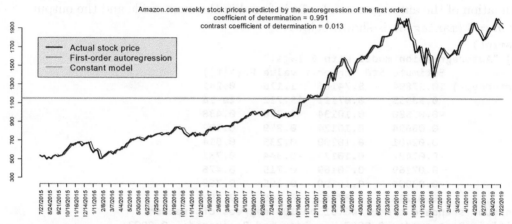

Figure 8.18: *Amazon.com weekly stock prices (Adj Close) and fitted/predicted values from the autoregression of the first order. The traditional coefficient of determination is highly inflated because the null hypothesis is that prices vary around the mean.*

Example 8.39 *Autoregression for prediction of Amazon stock prices.*
Historical stock prices can be downloaded from finance.yahoo.com. *File AMZN_weekly.csv contains weekly stock prices from the week of 7/27/2015 to the week of 7/29/2019 (211 data points). Use these data to estimate the autoregression with the specified* maxlag *order and predict Amazon.com prices one week ahead. Compute the contrast coefficient of determination using the first-order autoregression as the null model.*

Solution. The autoregression of the mth order with the time series being $Y_1, Y_2, ..., Y_n$ is the regression of Y_t on its past values $Y_{t-1}, Y_{t-2}, ..., Y_{t-m}$. Specifically, the vector of the dependent variable and the matrix of m predictors, following the format of linear model, is as follows:

$$\mathbf{y}^{(n-m)\times 1} = \begin{bmatrix} Y_{m+1} \\ Y_{m+2} \\ \vdots \\ Y_n \end{bmatrix}, \quad \mathbf{X}^{(n-m)\times(m+1)} = \begin{bmatrix} 1 & Y_m & Y_{m-1} & \cdots & Y_1 \\ 1 & Y_{m+1} & Y_m & & Y_2 \\ \vdots & \vdots & \vdots & \ddots & \vdots \\ 1 & Y_{n-1} & Y_{n-2} & \cdots & Y_{n-m} \end{bmatrix}.$$

The second column of matrix $\mathbf{X}$ is the time series shifted by 1 (lag=1), the third column is shifted by 2 (lag=2), etc. Thus the autoregression takes the form of a linear model $Y_t = \beta_1 + \beta_2 Y_{t-1} + \beta_3 Y_{t-2} + \ldots + \beta_{m+1} Y_{t-m} + \varepsilon_t$, $t = m+1, m+2, \ldots, n$. Although an assumption of the linear model is violated, because predictors are random variables and $\{Y_t\}$ are dependent even if $\varepsilon_t \overset{\text{iid}}{\sim} \mathcal{N}(0, \sigma^2)$, we estimate the beta coefficients by least squares using `lm` in R. The R code for estimation of the autoregression by `lm` is found in the `amzn.r` file and the output for $m = 8$ (`maxlag=8`) is shown below.

```
> amzn()
[1] "Autoregression model with 8 lags:"
            Estimate Std. Error t value Pr(>|t|)
(Intercept) 10.27894    8.74166   1.176   0.241
X1           0.98955    0.07211  13.722  <2e-16 ***
X2          -0.07880    0.10134  -0.778   0.438
X3           0.08694    0.10124   0.859   0.392
X4           0.02401    0.10200   0.235   0.814
X5          -0.03497    0.10172  -0.344   0.731
X6           0.07269    0.10169   0.715   0.476
X7          -0.03004    0.10190  -0.295   0.768
X8          -0.03244    0.07283  -0.445   0.657
Residual standard error: 46.62 on 193 degrees of freedom
Multiple R-squared:  0.9907,    Adjusted R-squared:  0.9903
F-statistic:  2575 on 8 and 193 DF,  p-value: < 2.2e-16
```

Two important observations can be made: (1) all lags but the first one are statistically insignificant. (2) The coefficient of determination $R^2 = 0.9903$ makes an impression that the autoregression almost perfectly predicts the next week's stock price. One can get a parsimonious model by eliminating lags greater than 1, which leads to the autoregression of the first order; see below.

```
            Estimate Std. Error t value Pr(>|t|)
(Intercept) 10.275402   8.641197   1.189    0.236
X[, 1]       0.997013   0.006869 145.140   <2e-16 ***
Residual standard error: 46.1 on 200 degrees of freedom
Multiple R-squared:  0.9906,    Adjusted R-squared:  0.9905
F-statistic: 2.107e+04 on 1 and 200 DF,  p-value: < 2.2e-16
```

Based on this model, if Y_t is the present week's value of the stock price the next week's price is predicted (fitted values) as $10.275 + 0.997 Y_t$. The fitted values and the actual prices Y_t are depicted in Figure 8.18.

Does $R^2 = 0.9907$ reflect the quality of prediction, i.e. is it true that 99% variation of Amazon.com stock price next week can be predicted by the price in the present week? The question depends on what we mean by "variation." If Y_t varied around the mean, $\overline{Y} = \sum_{t=2}^{n} Y_t / (n-1)$, then the traditional coefficient determination would reflect the quality of prediction. But $\{Y_t\}$ do not vary around the mean because they have an obvious trend. As in the previous example with truck drivers, the null hypothesis that $\{Y_t\}$ vary around the mean does not make sense, so $\sum_{t=2}^{n} (Y_t - \overline{Y})^2$ is a big number that inflates R^2 almost to 1. A more reasonable null hypothesis is to assume that the prices next week can be well predicted by the last week prices, or formally, in the language of the first-order autoregression, $H_0 : \beta_2 = 0, \ldots, \beta_8 = 0$ which gives to the contrast coefficient of determination

$$R_C^2 = \frac{R^2 - R_0^2}{1 - R_0^2} = 0.013,$$

where $R^2 = 0.9907$ and $R_0^2 = 0.9906$. In layman's words, the improvement of the eight-lag autoregression compared to the first-order autoregression is only 1.3%.

Example 8.43 in the next section illustrates how to use the contrast coefficient in observational studies (not necessarily time series).

Geometric interpretation of the contrast coefficient of determination

We use the following representation of the contrast coefficient of determination,

$$R_C^2 = 1 - \frac{\|\mathbf{y} - \widehat{\mathbf{y}}\|^2}{\|\mathbf{y} - \widehat{\mathbf{y}}_1\|^2}, \tag{8.44}$$

where $\widehat{\mathbf{y}}$ is the vector of fitted values from regression on $\mathbf{X}_1$ and $\mathbf{X}_2$ and $\widehat{\mathbf{y}}_1$ is the vector of fitted values from regression on $\mathbf{X}_1$. To concentrate on conceptual issues, $\mathbf{X}_1$ and $\mathbf{X}_2$ are assumed below to be predictor vectors $\mathbf{x}_1$ and $\mathbf{x}_2$.

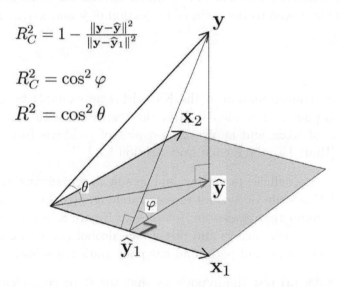

Figure 8.19: *Geometry of the contrast coefficient of determination (90° angles are marked by corners).*

The contrast coefficient of determination has a nice geometric interpretation, similar to the traditional coefficient of determination discussed above. As before, we consider a bivariate linear model with two predictors $\mathbf{x}_1$ and $\mathbf{x}_2$. To eliminate the intercept term, the averages are subtracted from observations which in geometric language means that the origin is moved to the point $(\bar{\mathbf{x}}_1 \mathbf{1}, \bar{\mathbf{x}}_2 \mathbf{1}, \bar{\mathbf{y}} \mathbf{1}) \in R^n$. Referring to Figure 8.19, the traditional coefficient of determination equals the cosine squared angle θ between $\mathbf{y}$ and the plane spanned by $\mathbf{x}_1$ and $\mathbf{x}_2$, where angle θ is between $\mathbf{y}$ and $\widehat{\mathbf{y}}$, the vector of the predicted values from the bivariate regression. When reporting the quality of fit, such as when using the coefficient of determination, we distinguish between the predictor of interest, $\mathbf{x}_2$ and predictor $\mathbf{x}_1$ viewed as the adjustment or baseline predictor. Thus, the null model is $E(\mathbf{y}) = \beta_{10}\mathbf{x}_1$ and we want to express the quality of fit by the bivariate model $E(\mathbf{y}) = \beta_1\mathbf{x}_1 + \beta_2\mathbf{x}_2$ compared to the null model using the contrast coefficient of determination computed by formula (8.44), where $\widehat{\mathbf{y}}_1$ is the projection of $\mathbf{y}$ onto $\mathbf{x}_1$. Consider the right triangle with vertices $\mathbf{y}$, $\widehat{\mathbf{y}}$, and $\widehat{\mathbf{y}}_1$. From elementary geometry $R_C^2 = \cos^2 \varphi$ because $\sin \varphi = \|\mathbf{y}-\widehat{\mathbf{y}}\| \,/\, \|\mathbf{y}-\widehat{\mathbf{y}}_1\|$. Thus we interpret R_C^2 similarly to the traditional R^2 as the cosine squared angle with the origin moved to the projection of $\mathbf{y}$ onto $\mathbf{x}_1$. Now we prove that $R_C^2 \leq R^2$. It suffices to prove that $\tan \varphi \geq \tan \theta$. Consider the respective right triangles with angles θ and φ; note that they share the same cathetus $[\mathbf{y},\widehat{\mathbf{y}}]$. Vector $\widehat{\mathbf{y}}-\widehat{\mathbf{y}}_1$ is orthogonal to $\mathbf{x}_1$ because $\mathbf{x}_1$ is orthogonal to $\mathbf{y}-\widehat{\mathbf{y}}$ and $\mathbf{y}-\widehat{\mathbf{y}}_1$ and therefore is orthogonal to any line that belongs to the plane spanned by these vectors including $\widehat{\mathbf{y}} - \widehat{\mathbf{y}}_1$. Since $\|\widehat{\mathbf{y}} - \widehat{\mathbf{y}}_1\|$ is the cathetus and $\|\widehat{\mathbf{y}}\|$ is the hypotenuse length, we have $\|\widehat{\mathbf{y}} - \widehat{\mathbf{y}}_1\| \leq \|\widehat{\mathbf{y}}\|$ which proves that $\tan \varphi \geq \tan \theta$.

In general, the contrast coefficient of determination has the same interpretation, but the origin is moved to the point of projection of $\mathbf{y}$ and $\mathbf{x}_{11}, ..., \mathbf{x}_{1p}$ onto space of $\mathbf{x}_{21}, ..., \mathbf{x}_{2q}$.

Problems

1. Justify the log transformation in the Kids drinking example by means of the q-q plot applied to residuals: Run the regression with and without log transform of `alcm` and apply the q-q plot of residuals from the two regressions. [Hint: Use the R code from Section 5.3.]

2. (a) Interpret the coefficients at `boy` and `grade` in `kidsdrink` regression using the relative scale of alcohol movie watching. (b) Omit the `pared` predictor and rerun the regression. (c) Add to Figure 8.13 the regression line for a black girl who drinks and possess an alcohol related item, whose parents are uneducated and poor, and has poor grades at school.

3. In Example 8.33, (a) test the hypothesis that the slope coefficient is 1 in regression of `right` on `left`, (b) run regression of `right` on `left` and `left` on `right` without intercept, and (c) test the hypothesis that the slope is 1. Explain why the slopes are close to 1.

4. To illustrate slope attenuation from Example 8.33 simulate samples Y and x of sufficient sample size from the same normal distribution with variance σ^2 and u with variance σ_u^2. Plot $(x_i + u_i, Y_i + u_i)$ and the regression of $Y_i + u_i$ on $x_i + u_i$ (display the slope coefficient and variance) and run the code with increasing `sigma2`. Draw conclusions and explain.

5. Compute the 3×3 correlation matrix using data from `HeightFootNose.csv` and compute the partial correlation using two methods: by inverse of the correlation matrix and as the correlation between residuals referring to Section 4.1.2. Use this result to explain why the slope at `Nose` is negative.

6. Use data from `HeightFootNose.csv` to predict `Nose` by `Height` and `Foot`. (a) Is one of the slopes negative? (b) Is it generally true that if in regression of Y on x_1 and x_2 one slope will be positive and one slope is negative then in regression of x_1 on Y and x_2 one slope is positive and another slope will be negative?

7. The jth predictor in the multiple regression is rescaled as γx_j. What changes: the OLS estimate, SE, t-statistic, p-value, or R^2? Justify by formulas.

8. Is the interpretation of the slope coefficients on the scale of standard deviation in simple linear regressions with the same dependent variable and different predictors equivalent to comparison of t-statistics, i.e. slope$_1$ < slope$_2$ is equivalent to $t_1 < t_2$? Does it hold for multiple regression as well?

9. A new independent variable z is added to a regression with m predictors. It is known that the dependent variable Y and z are uncorrelated. Is it true that the OLS coefficient at z is zero? Specify conditions under which the coefficient will be zero and provide a geometric illustration.

10. Derive the condition expressed in terms of correlations that positive slopes in simple linear regression of Y on x_1 and Y on x_2 imply positive slopes in the bivariate regression of Y on x_1 and x_2. [Hint: Use a trivariate normal distribution with unit variances.]

11. The slopes in simple linear regression of Y on x_1 and Y on x_2 are negative. Is it possible that the slopes in the bivariate regression of Y on x_1 and x_2 are both positive?

12. (a) Prove that $0 \le R_C^2 \le 1$. Describe in geometric terms when $R_C^2 = 1$ and when $R_C^2 = 0$. (b) Provide a geometric interpretation of the inequality $R_C^2 \le R_0^2$. (c) Provide a geometrical interpretation of the case when $R_C^2 = R_0^2$.

13. Generate a triple of correlated normally distributed random variables, compute the squared partial correlation coefficient and R_C^2 and demonstrate that they give the same result. [Hint: Use methods of Section 4.1.1.]

14. Provide a geometric illustration of $R_C^2 = 0$ referring to Figure 8.19.

8.7   Dummy variable approach and ANOVA

Dummy variables are used in regression to code categories (groups). It is a simple and yet very powerful technique to account for differences in regressions with combined data. The dummy variable approach may be used to recast the *t*-test or ANOVA as a linear model that may be beneficial with repeated measurements or when testing a relationship adjusted by other variables (sometimes called confounders). Special attention is given to the interpretation of regression coefficients, especially on the log scale.

8.7.1   Dummy variables for categories

Sometimes data have groups or categories. To account for group differences in regression analysis, the concept of a dummy (or indicator) variable becomes very handy. We set a dummy variable $d_i = 1$ if the ith observation belongs to group k and zero otherwise. The dummy variable takes either 0 or 1 when coding just two categories. Dummy variables may be used to detect differences between groups in terms of either the intercept or slope. The basic idea is to combine the data from different groups, associate them with dummy variable(s) and run one multiple regression. It is important to remember that, under this approach, the regression variance (σ^2) is silently assumed to be the same in all groups.

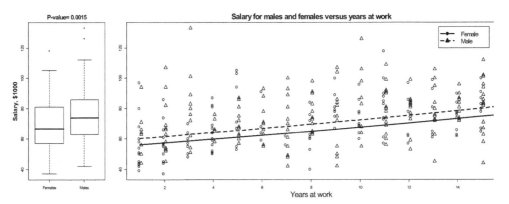

Figure 8.20: *Left: The boxplot of salaries for Females and Males with the crude p-value = 0.0015. Right: Salary as a function of Years at work with the adjusted p-value = 0.00748. Years at work in an important variable for studying the gender difference in salary.*

Example 8.40 *Gender difference in salary. There is a concern about gender discrimination of salary at a big company – a lawsuit is on the table and it is up to a statistician to decide if the difference in salaries is statistically significant. The file Salary.csv contains the salaries of 300 randomly chosen employees. The first column (dptm) indicates the department of work (Sales, IT, Finance, Production), the second column (sexC) indicates the gender (0 = female, 1 = male),*

*the third column (year) is years at work, and, finally, the fourth column (y) is
annual salary in thousands of dollars. Is there a statistically significant difference
in salary between men and women?*

Solution. A naive answer on the gender difference is based on the t-test.
See the left plot in Figure 8.20 and the R code salMW.r. This code uses function layout to display plots of different sizes (recall that par(mfrow=c(1,2))
allocates equal areas for two plots). The rectangular area occupied by plots
must be specified by a matrix. In function salMW, the layout is defined as
m=matrix(c(1,1,2,2,2,2,2,2,2,2),ncol=5); layout(m). This means that
the area of the left plot is a quarter the size of the right plot. This technique is
useful when one needs to show the main plot and use others as supporting plots.

The box plot is the traditional way to show the data summary for a two-group
comparison (see Section 5.4). The p-value $= 0.0015$ indicates strong evidence
toward difference in salary. Somebody may argue that experience measured as
years at work is not taken into account. Maybe women have less experience, and
that is why their salary is smaller. To address this concern we run the regression
with sexC and year as predictors. Then we refer to the standard t-test p-value as
the crude p-value. Inclusion of years at work as a predictor hints to a possibility
of using the log transformation of salary because annual salary normally increases
on the percent scale to catch up with inflation. Indeed, the q-q plot reveals that
the log transformation of y is well justified (not shown). Note that, strictly
speaking, the q-q plot may not be applicable because y is not iid but because
salaries were observed for employees in a random fashion this plot is useful. We
start with the linear model $\ln(y) = \beta_0 + \beta_1 \times \text{sexC} + \beta_2 \times \text{year} + \varepsilon$ with the output
of lm(LY~sexC+year) shown below (LY=log(y)).

	Estimate	Std. Error	t value	Pr(>|t|)
(Intercept)	4.054014	0.027950	145.045	< 2e-16
sexC	0.067429	0.025038	2.693	0.00748
year	0.020660	0.002773	7.451	1.02e-12

The interpretation of the regression coefficients is as follows.

1. Each year salary increases by 2% regardless of the gender because if the
   salary for female is modeled as $Y = e^{a+bt}$, where t is in years and $a = 4.054014$ and $b = 0.02066$. The instantaneous increase is computed as the
   derivative of $\ln(y)$ with respect to t:

$$\frac{1}{y}\frac{d}{dt}(e^{a+bt}) = b.$$

   If t is in years then, formally speaking, the relative salary increase is
   $(e^{a+b(t+1)} - e^{a+bt})/e^{a+bt} = e^b - 1$. But, since $b \simeq 0$, we have $e^b - 1 \simeq b$
   as follows from the Taylor series expansion. The same answer is for males
   where $a = 4.054014 + 0.067429$.

2. Males earn by 7% more than females even if they have the same experience (years at work) because following the same argumentation one obtains

$$\frac{e^{4.054014+0.067429+0.02066t} - e^{4.054014+0.02066t}}{e^{4.054014+0.02066t}}$$

$$= e^{0.067429} - 1 = 0.06975 \simeq 0.067429.$$

Note, for small coefficient at `sexC`, as in this model, we have $e^{0.067429} - 1 \simeq 0.067429$, thanks to the Taylor series expansion.

3. The salary of a male with five years' experience is approximately

$$e^{4.05+0.067+0.02\times5} = \$67.830.$$

This model can be used to compute a confidence interval for the salary given years at work and test the hypothesis if she or he is underpaid.

The regression coefficient at `sexC` is statistically significant but the employer may disagree with the result of the gender difference because the average salary differs from department to department. Is it possible that the difference becomes statistically insignificant after adjusting for where the employee works? To address the difference across departments, we introduce four dummy variables. Each dummy variable takes value either 0 or 1: if the employee works in the Sales department `d1=1`; otherwise, `d1=0`. If the employee works in the IT department `d2=1`; otherwise, `d2=0`, and finally, for the Finance department. Regression `lm(LY~sexC+year+d1+d2+d3+d4)` produces a `NA` result for one of the dummy variable. The reason is that by default the regression has an intercept term but since `d1+d2+d3+d4=1` (every employee belongs to one of the departments) the regression becomes overspecified because the columns of matrix $\mathbf{X}$ are linearly dependent.

```
lm(formula = LY ~ sexC + year + dptm)
                Estimate Std. Error t value Pr(>|t|)
(Intercept)     4.092871   0.034264 119.452  < 2e-16 ***
sexC            0.069478   0.024953   2.784  0.00571 **
year            0.020398   0.002763   7.382 1.61e-12 ***
dptmIT         -0.037626   0.034301  -1.097  0.27357
dptmProduction -0.083143   0.034610  -2.402  0.01692 *
dptmSales      -0.038584   0.033576  -1.149  0.25144
Residual standard error: 0.2122 on 294 degrees of freedom
Multiple R-squared:  0.2021,    Adjusted R-squared:  0.1885
F-statistic: 14.89 on 5 and 294 DF,  p-value: 4.955e-13

B-value of sexC = 59%
```

Two strategies may be pursued to avoid the overspecification: (i) Run regression without intercept `lm(LY~sexC+year+d1+d2+d3+d4-1)`. (ii) Run the regression omitting the fourth dummy variable `lm(LY~sexC+year+d1+d2+d3)`. In the latter strategy the intercept represents the log salary in the Finance department – we say that Finance is a reference category. The difference in average salary between the Sales and IT from the Finance department (intercept) is not statistically significant. However, the average salary in the Production department is about 8% lower than in the Finance department. To make the model parsimonious, we could omit d1 and d2 and leave in only d3 but this would not change the detected gender difference. In short, even after adjusting for differences in the departments there is strong evidence that men earn 7% or more than women.

This assessment of the difference in the salary does not take into account the imprecision of the beta-coefficient at sexC and variation of salaries across individuals. A probabilistic assessment involves the d- or b-value. Using the simple computations shown in Table 8.1 of Section 8.5, we obtain b-value $= 0.59$, which means that under equal conditions the chance that a women has smaller salary than a man is about 60%. The traditional and the b-value interpretations of the regression coefficient compliment each other: the former reflects the quantitative assessment of the salary difference, and the latter reflects the chance of a woman being underpaid. Alternatively, one may avoid computation of dummy variables and simply use dptm in the lm call. R automatically creates four categories, adds the intercept and removes one category (in this case Finance) to avoid linear dependence among predictors. Compared to the previous output, the coefficients at sexC and year do not change, but now the intercept has a slightly different interpretation. Since Finance department is the reference category, a female who just started working in that department is expected to have salary $e^{4.0092871} = \$55,080$. The coefficients at dptmIT, dptmProduction and dptmSales also change because they reflect the difference in salary compared to Finance. Although convenient, the automatic specification of categories is less flexible. For example, we may want to use Production as a reference category or combine several categories if the difference is not statistically significant. The user-defined dummies are more more flexible for model building. □

The following example illustrates how to use dummy variables when a change point is of concern: Does the regression coefficient change after a certain point in time? One can look at the change of the regression intercept, slope or both. Below, we discuss the problem of studying the Nile river flow in connection with building the High Dam at Aswan, Egypt in late sixties. Many regarded the construction of the dam pivotal to the country's industrialization and economic growth.

Example 8.41 *Nile river flow. Cobb [18] presented and analyzed the data on the annual volume of the Nile River discharged at Aswan, Egypt, 10^8 m^3. The data are also available in a data repository at https://vin centarelbundock.git hub.io/Rdatasets/datasets.html. Use the dummy variable approach to model the flow of Nile.*

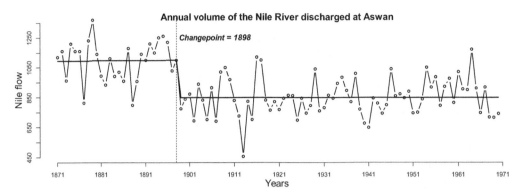

Figure 8.21: *Nile flow from 1871 to 1971 with the changepoint at 1898. The regression model has a slightly positive slope from 1871 to 1898 and flats afterwards.*

Solution. The data is found in the file `NileFlow.csv` and the R program in the file `nile.r`; see Figure 8.21. There is an obvious change of flow around 1898. Define the dummy variable `d=rep(0,n);d[year<=yc]=1` with the default value yc=1898. The predictor is `I(d*year)` where operator `I` means that R treats `d*year` as a numeric vector. This model will be referred to as the trend model. The output is shown below.

```
lm(formula = flow ~ I(d * year))
                Estimate Std. Error  t value  Pr(>|t|)
(Intercept) 849.95838    15.04397   56.498   < 2e-16 ***
I(d * year)   0.13151     0.01509    8.717  7.32e-14 ***
Residual standard error: 127.7 on 98 degrees of freedom
Multiple R-squared:  0.4367,    Adjusted R-squared:  0.431
F-statistic: 75.99 on 1 and 98 DF,  p-value: 7.319e-14
```

In words, this model means that before year yc=1899 the river flow follows a linear trend 849.95838+0.13151*year and after 1898 the water flow is constant 849.95838. We notice that the trend is positive and statistically significant, but perhaps can be ignored from the practical standpoint because from 1871 to 1898 it increases only by 0.13151*(1898-1871)=3.55077. This is just 2.8% of the residual standard error, the variation from year to year. Thus, the model with predictor d will be a good choice, we refer to this model as the step model.

```
lm(formula = flow ~ d)
            Estimate Std. Error  t value  Pr(>|t|)
(Intercept)   849.97     15.05   56.490   < 2e-16 ***
d             247.78     28.44    8.714  7.44e-14 ***
Residual standard error: 127.7 on 98 degrees of freedom
Multiple R-squared:  0.4366,    Adjusted R-squared:  0.4308
F-statistic: 75.93 on 1 and 98 DF,  p-value: 7.439e-14
```

The advantage of the step model is that the estimate of the difference between river flows at the change point is readily available as the coefficient at d: the volume of water dropped in 1898 by 247.78×10^8 m^3 per year. In layman's words,

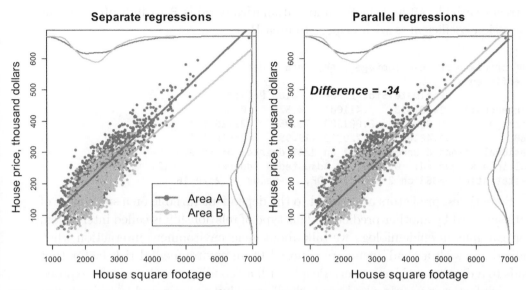

Figure 8.22: *House price as a function of square footage in two areas. The difference between house prices with the same footage in two areas is 34 thousands of dollars.*

the Nile river flow was 30% higher prior 1898. There is strong evidence that the estimated difference is statistically significant. □

The dummy variable approach can be used to test if two regressions are parallel or have the same slope. The following example illustrates this technique.

Example 8.42 *Are two regressions parallel?* *House prices are compared in two areas A and B (column AB is a character). Since price depends on the size of the house (square footage) one needs to compare not average prices but regressions of the price on the house square footage. Use file **houseprice.csv** on 2,244 house prices and their square footage to test if the two regressions are parallel: have the same slope but significantly different intercepts. Find the difference in house prices in the two areas.*

Solution. The R function found in file housepr.r produces Figure 8.22. The scatterplot at left depicts houses' square footage and price and their marginal distribution in the two areas as Gaussian kernel densities; see Example 5.24. We notice that area A has a wider distribution of price and footage than area B and both are skewed to the right. If regressions are computed separately, they have different intercepts and slopes. We hypothesize that while the relationship between price and footage is the same in two areas, prices in area A are greater for houses of the same footage. The following regression confirms that there is a very strong dependence of price on footage and that the areas are different meaning that the same house in area B is cheaper by $34 thousand. Note that

R treats variable AB as a factor and automatically adds B to indicate that the
regression coefficient is with respect to area B.

```
Call:
lm(formula = price ~ footage + AB, data = dat)
Coefficients:
              Estimate Std. Error t value Pr(>|t|)
(Intercept)   1.791618   3.411004   0.525    0.599
footage       0.099188   0.001289  76.954   <2e-16 ***
ABB         -34.401665   1.720752 -19.992   <2e-16 ***
Residual standard error: 40.07 on 2241 degrees of freedom
Multiple R-squared:  0.7515,    Adjusted R-squared:  0.7512
F-statistic:  3388 on 2 and 2241 DF,  p-value: < 2.2e-16
```

Sometimes, predictors are related to the dependent variable on a subsample of
data specified by another predictor. This type of relationship is called interaction.
For example, in epidemiology we talk about gene–environment interaction when
individuals with a "bad" gene are exposed to an environmental risk factor that
leads to a decease such as cancer. People with a "good" gene who are not exposed
to a risk factor or people who have a "bad" gene but not exposed to a risk factor
do not develop cancer. An example concerning obesity is presented below to
illustrate the point.

Example 8.43 *BMI-Gene interaction. Researchers have identified two genes
that increase the chance of developing obesity. The data file obesegene.csv contains data on 748 people (rows) with six types of information: G1 indicates the
first gene expression (G1 = 1: overexpressed, G1 = 0: underexpressed), G2 indicates the second gene expression (G2 = 1: overexpressed, G2 = 0: underexpressed), age in years, sex (sex = 0: woman, sex = 1: man), BMIP is the average
BMI of parents and BMI is the BMI of the person in the row. Use regression
analysis to identify risk factors for obesity expressed through BMI.*

Solution. To understand the distribution of the body mass index (BMI) in
parents and children, we may start with plotting kernel densities on the same
graph (see Section 5.5), see Figure 8.23 and the R code obesegene.r. Notice that
the plots have different sizes. This is achieved by commands m=matrix(c(1,1,1,1,2,
2,2,2,2,2),ncol=5);layout(m). This means that the second graph is wider
that the first one. According to medical standard a person with BMI > 30 is
called obese. We plot the raw BMI of children and parents at the bottom and the
top, respectively. Two important observations can be made: (1) BMI is skewed to
the right, and (2) children have a lighter right tail. The first observation indicates
that the log transformation will be helpful to comply with the normal distribution
assumption. The regression output with log10(BMI) as the dependent variable
is shown below (LBMI=log10(BMI) and LBMIP=log10(BMIP)).

Although both genes are statistically significant, the second gene has a much
larger impact on the weight of the individual: individuals of the same gender and
parents with the same BMI who carry G2 have BMI greater by $(e^{0.4} - 1)100\% =$

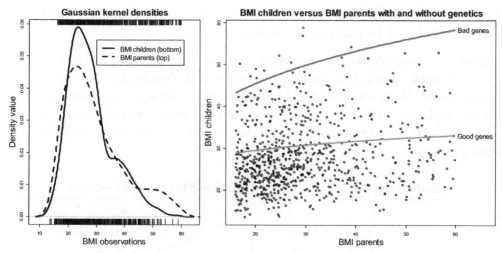

Figure 8.23: *Regression analysis of the BMI–Gene interaction. The* R *function* layout *is used to display plots of different sizes.*

50% compared to those who do not carry this gene. To check that these genes have an interaction with obese parents, we run the following regression:

```
lm(formula = LBMI ~ G1 + age + sex + LBMIP + I(G2 * LBMIP))
                 Estimate Std. Error t value Pr(>|t|)
(Intercept)    2.4803329  0.0476247  52.081  < 2e-16 ***
G1             0.0652123  0.0091949   7.092 3.09e-12 ***
age            0.0098928  0.0003863  25.612  < 2e-16 ***
sex           -0.1964888  0.0091605 -21.450  < 2e-16 ***
LBMIP          0.1043810  0.0134943   7.735 3.38e-14 ***
I(G2 * LBMIP)  0.1224113  0.0029318  41.753  < 2e-16 ***
Residual standard error: 0.1226 on 740 degrees of freedom
Multiple R-squared:  0.8007,    Adjusted R-squared:  0.7993
F-statistic: 594.5 on 5 and 740 DF,  p-value: < 2.2e-16
```

Here I(G2*LBMIP) is the interaction term. Note that the term G2, commonly referred to as the main effect, was omitted from this regression, is not required. The coefficient of determination $R^2 = 0.8$ indicates that 80% of variation of BMI can be explained by the predictors. To quantify the contribution of the genetic factors, we compute the contrast coefficient of determination:

$$R_C^2 = \frac{R_F^2 - R_0^2}{1 - R_0^2} = \frac{0.8007 - 0.3196}{1 - 0.3196} = 0.7.$$

This means that out of 80% BMI variation 70% is explained by genetic factors. The difference between BMI of two cohorts of women of age 60 with bad genes (G1 = 1 and G2 = 1) and good genes (G1 = 0 and G2 = 0) is displayed in Figure 8.23. Children who have good genes even with obese parents are on the border of obesity. However, having bad genes and obese parents leads to severe obesity as follows from the following `lm` run. Note that we use the `I()` command to tell R that `G2*LBMIP` should be treated as a product. □

Overlooking clusters/groups in regression analysis may lead to poor results; below is an example. For a few clusters, regression with dummy variables is an appropriate model. However, when the number of clusters is large, say, more than five, a more sophisticated statistical model, called the *mixed model*, should be applied. We refer the reader to books by Fitzmaurice et al. [50] and Demidenko [28], among others.

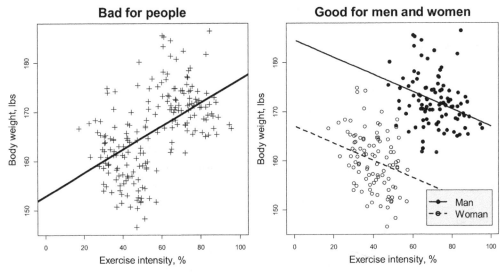

Figure 8.24: *Simpson paradox: good for men and women and bad for people. Overlooking clusters may lead to erroneous results. If gender is not taken into account exercise increases body weight (left). If gender is represented by a dummy variable exercising decreases body weight (right).*

Example 8.44 *Simpson's paradox: Good for men and women, bad for people. The relationship between body weight and exercise intensity is studied. File* `simpson.csv` *contains the data on exercise for 200 people. The first column is* Exint, *the second column is* BodyW, *and the third column is* Sex, *with self-explanatory meaning. Plot the data and fit a simple linear regression. Then fit a regression with* Sex *as a dummy variable. Explain the difference in results.*

Solution. The R code is found in file `simpson.r` which produces Figure 8.24. In the plot at left there is no distinction between the sexes – the more people exercise the heavier they are: exercising is bad. In the plot at right, we admit that (a) men are heavier and (b) men exercise more intensely. These hypotheses lead to a regression with the dummy variable `Sex` to reflect the gender difference in weight.

```
Call:
lm(formula = BodyW ~ ExInt + Sex, data = d)
Coefficients:
             Estimate Std. Error t value Pr(>|t|)
(Intercept) 184.56727    2.91376  63.343  < 2e-16 ***
ExInt        -0.17535    0.04028  -4.353 2.15e-05 ***
SexW        -17.46252    1.45850 -11.973  < 2e-16 ***

Residual standard error: 5.325 on 197 degrees of freedom
Multiple R-squared:  0.5817,     Adjusted R-squared:  0.5774
F-statistic: 136.9 on 2 and 197 DF,  p-value: < 2.2e-16
```

A few comments on the `lm` call: (a) Instead of using arrays computed locally in the function, we use the data frame `d=read.csv("c:\\StatBook\\Simpson.csv"`, `header=T)`; then the variables are referred to by the names in the data frame. (b) In previous examples, we used a numeric dummy variable with values 0 and 1. Here, we directly use the `Sex` variable which takes either `W` or `M`. R automatically recognizes the two categories and creates variable `SexW` which is equivalent to coding 1 for `W` and 0 for `M` (note that we usually apply a reverse coding when done manually).

According to the model with a dummy variable, exercise is good for men and women because it reduces weight. Needless to say, clustering may be difficult to detect in multiple regression when plotting the dependent variable versus a specific predictor if it correlates with other predictors.

8.7.2 Unpaired and paired t-test

The goal of this section is to show that the t-test can be cast into a linear model using the dummy variable approach. This representation allows for two-sample testing adjusted for other variables/confounders.

Unpaired t-test

We recommend the reader review Section 7.4. Recall that, in the two-sample test, we have two independent normally distributed groups of observations, X and Y, having different means and the same variance σ^2 with the null hypothesis that the two means, μ_X and μ_Y, are the same. To express the t-test as a linear model, combine the observations from the two groups into one sample, and introduce a dummy variable: $d_i = 1$ if the ith observation comes from the first group and $d_i = 0$ otherwise. We can write this model compactly as

$$Y_i \sim \mathcal{N}(\mu_Y + d_i(\mu_X - \mu_Y), \sigma^2), i = 1, 2, ..., n,$$

where n is the total number of observations in the two groups. This model is equivalent to the simple linear regression model $Y_i = \beta_1 + \beta_2 d_i + \varepsilon_i$. We refer the reader to Example 6.49, where we showed that testing the significance of the regression slope is equivalent to the t-test. Note that the equal-variance assumption is crucial here. However, an advantage of reducing the t-test to a linear model becomes evident when there is a *confounder*, another variable, that may impact the differences in the groups and, therefore, should be added to the regression as in the following example.

Example 8.45 *Movie rating. A survey is conducted to test for gender difference in the rating of a recently released movie. File* movrat.csv *contains the data on viewers' score (in the range from 0 to 100), gender, and age. Apply the unpaired t-test and then adjust for age to see if the difference can be explained by age, not gender.*

Solution. The R code in file movrat.r produces Figure 8.25. In the plot at left the standard t-test is applied to compare the scores from men and women with the p-value $= 0.028$. The p-value can be computed using either function t.test or lm with the dummy variable which codes the gender of the viewer. We challenge this analysis and test if age is a confounder and may explain or at least alter the rating differences. The reason for possible confounding is that the average age among men in the survey was 53 and the average age of women was 32. The high score may just reflect the age difference. To test if age makes a difference in scoring, we run a bivariate regression of score on sex and age.

```
lm(formula = score ~ sex)
               Estimate Std. Error  t value  Pr(>|t|)
(Intercept)     47.380      1.111    42.630    <2e-16 ***
sex              3.447      1.569     2.197    0.0283 *
Residual standard error: 23.37 on 885 degrees of freedom
Multiple R-squared:  0.005423,  Adjusted R-squared:  0.004299
F-statistic: 4.825 on 1 and 885 DF,  p-value: 0.0283


lm(formula = score ~ sex + age)
               Estimate Std. Error  t value  Pr(>|t|)
(Intercept)    41.10054    2.39557   17.157    <2e-16 ***
sex            -0.55430    2.06727   -0.268    0.7887
age             0.19571    0.06622    2.956    0.0032 **
Residual standard error: 23.26 on 884 degrees of freedom
Multiple R-squared:  0.01515,   Adjusted R-squared:  0.01293
F-statistic: 6.801 on 2 and 884 DF,  p-value: 0.001171
```

From the regression of score on sex (the first run), we conclude that men like the movie better than women and rate it 3.4 points above the women's average. But from the regression of score on sex and age (second run), we surprisingly infer that the difference in movie rating depends mainly not on gender, but the age of viewers. For example, a person older by 10 years rates the movie by 2 points more. In the plot at right, we apply boxplot to scores adjusted for age: the first box corresponds to score[sex==0]-0.19571*age[sex==0] and the second box corresponds to score[sex==1] -0.19571*age[sex==1]. The difference between scores vanishes. This example illuminates a possibility that group differences may be explained by other variables and therefore more information should be collected every time one conducts a survey. Then a linear model should be used instead the t-test because it allows one to incorporate other relevant variables.

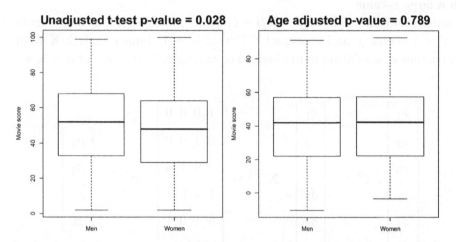

Figure 8.25: *Unadjusted and adjusted p-values for gender difference in movie rating. The rating difference is explained by age, not gender. This phenomenon happens when an important predictor is missing in the regression.*

Paired t-test

The paired t-test is a clever way to eliminate confounders by matching the pairs of observations to increase statistical power and decreases the p-value compared to the unpaired/standard t-test; see Section 7.4.4. To be specific, let us consider the problem of testing a new drug which is supposed to lower cholesterol. In the paired t-test design, the cholesterol is measured in n people before the drug was given $\{x_i, i = 1, ..., n\}$ and after the drug was administered $\{z_i, i = 1, ..., n\}$. Then the difference is computed $d_i = z_i - x_i$ and the one-sample t-test is applied to test the hypothesis that the mean of d_i is zero. We can rewrite the paired t-test as a linear model by expressing

$$x_i = \mu_i + \varepsilon_i, \quad z_i = \mu_i + \Delta + \delta_i, \quad i = 1, 2, ..., n, \quad (8.45)$$

where ε_i and δ_i are iid normally distributed random variables with zero mean and variance σ^2, μ_i is a fixed unknown parameter which represents the baseline cholesterol level of the ith individual, and Δ is the drug effect. The advantage of the paired t-test is that the baseline cholesterol, expressed through μ_i, is eliminated. As was shown in Section 7.4.4, both t-test statistics have the same numerator, $\bar{z} - \bar{x}$, but different denominator. For the unpaired t-test, in addition to the estimate of σ^2, the variance of $\{\mu_i\}$ is added, which makes the absolute value of the t-statistic smaller and the p-value larger. In words, the paired t-test eliminates the population variation and, as such, is more powerful compared to the unpaired/standard t-test. In contrast, in the traditional unpaired t-test, the difference between groups is buried under individual variation, and as such, results in a large p-value.

To rewrite (8.45) in the form $\mathbf{y} = \mathbf{X}\boldsymbol{\beta} + \boldsymbol{\varepsilon}$, combine cholesterol measurements into a $(2n) \times 1$ vector $\mathbf{y}$ and construct a $(2n) \times (n+1)$ binary matrix $\mathbf{X}$ with dummy variables representing individuals. For example, in the case of $n = 3$, we have

$$
\mathbf{y}^{6\times 1} = \begin{bmatrix} x_1 \\ x_2 \\ x_3 \\ z_1 \\ z_2 \\ z_3 \end{bmatrix}, \; \boldsymbol{\varepsilon}^{6\times 1} = \begin{bmatrix} \varepsilon_1 \\ \varepsilon_2 \\ \varepsilon_3 \\ \delta_1 \\ \delta_2 \\ \delta_3 \end{bmatrix}, \; \mathbf{X}^{6\times 1} = \begin{bmatrix} 1 & 0 & 0 & 0 \\ 0 & 1 & 0 & 0 \\ 0 & 0 & 1 & 0 \\ 1 & 0 & 0 & 1 \\ 0 & 1 & 0 & 1 \\ 0 & 0 & 1 & 1 \end{bmatrix}, \; \boldsymbol{\beta}^{4\times 1} = \begin{bmatrix} \mu_1 \\ \mu_2 \\ \mu_3 \\ \Delta \end{bmatrix}.
$$

It is easy to check by examination that the model (8.45) can be expressed as a linear model $\mathbf{y} = \mathbf{X}\boldsymbol{\beta} + \boldsymbol{\varepsilon}$. In general, matrix $\mathbf{X}$ is composed of two stacked identity matrices $\mathbf{I}_n$ and the $(2n) \times 1$ dummy variable with the first n components 0 and the last n components 1. It is possible to show that the OLS estimate from this model, as follows from Theorem 8.13 is $\widehat{\Delta} = \bar{z} - \bar{x}$ and the test statistic for the null hypothesis is $\sqrt{n}(\bar{z} - \bar{x})/\widehat{\sigma}$, where $\widehat{\sigma}^2 = \sum_{i=1}^{n}(d_i - \bar{d})^2/(n-1)$ and $d_i = z_i - x_i$.

8.7.3   Modeling longitudinal data

The dummy variable approach may not work if it leads to an overspecified model (too many parameters), below is an example.

Example 8.46 *Blood pressure treatment. To test whether a new drug helps to reduce the systolic blood pressure (BP, mm Hg) 30 patient volunteers have been hired; the data are found in file BPdata.csv. For each volunteer, the blood pressure was measured three times: at the baseline $(t = 0)$ and at two consecutive months $t = 1, 2$ after the drug administration. (a) Show that the linear model with a patient-specific intercept/baseline $Y_{it} = \alpha_i + \gamma t + \beta_1 M_i + \beta_2 A_i + \varepsilon_{it}$ is not identifiable, where Y_{it} is the BP of ith person at time t, $M_i = 1$ codes male and 0 female, A_i is the age of the patient, and ε_{it} are iid normally distributed with zero mean and variance σ^2. (b) Estimate a parsimonious model with the dependent variable as the difference of BP from the baseline and test whether the drug reduces BP. (c) Provide an explanation for why the difference model leads* to a more statistically significant result (the p-value is smaller) by referring to the paired t-test.

Solution. The data are depicted in Figure 8.26 and the R code is found in function `BPlong.r`. Some people call this a spaghetti plot. (a) As follows from the plot at left, there is a substantial variation of BP at the baseline ($t = 0$). The standard linear regression,

$$Y_{it} = \alpha + \gamma t + \beta_1 M_i + \beta_2 A_i + \varepsilon_{it}, \quad t = 0, 1, 2, \quad (8.46)$$

yields the $p = 0.071$ at the month variable (t). One could improve this model by using the patient-specific intercept, leading to a linear model with 30 dummy variables representing 30 intercept terms/baselines. The problem with this model is that it is not identifiable because matrix $\mathbf{X}$ does not have full rank. One can confirm this computationally by constructing a 90×33 matrix $\mathbf{X}$ and trying to invert $\mathbf{X}'\mathbf{X}$ using `solve(t(X)%*%X)`. Hence, the patient-specific intercept model is overspecified. (b) To reduce the number of parameters while accounting for a high scatter baseline BP a *difference model* can be used,

$$d_{it} = \gamma t + \varepsilon_{it}, \quad t = 0, 1, \quad (8.47)$$

where $d_{it} = Y_{it} - Y_{i0}$ is the difference from the baseline. The `lm` outputs for both models are show below. Note that neither intercept gender (M) nor age (A) is included in the model because they are eliminated by taking the difference.

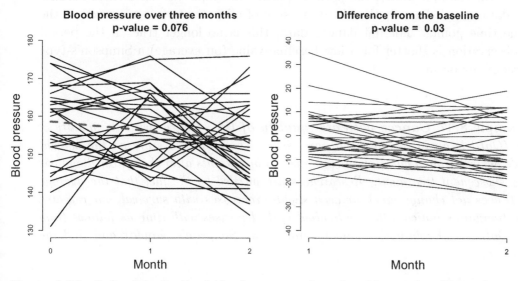

Figure 8.26: *Left: Longitudinal blood pressure data for 30 patients (the reduction of BP is not statistically significant); the slope is estimated by model (8.46). Right: The difference from the baseline (the reduction of BP is statistically significant); the slope is estimated by model (8.47).*

```
lm(formula = Y ~ mo + M + A)
Coefficients:
            Estimate Std. Error t value Pr(>|t|)
(Intercept) 153.42298    4.67473  32.820   <2e-16 ***
mo           -2.43333    1.35634  -1.794   0.0763 .
M            -0.94590    2.30292  -0.411   0.6823
A             0.10782    0.08176   1.319   0.1908
Residual standard error: 10.51 on 86 degrees of freedom
Multiple R-squared:  0.05706,   Adjusted R-squared:  0.02417
F-statistic: 1.735 on 3 and 86 DF,  p-value: 0.1659
```

```
lm(formula = diff ~ mo - 1)
Coefficients:
   Estimate Std. Error t value Pr(>|t|)
mo   -4.867      2.195  -2.218   0.0304 *
Residual standard error: 12.02 on 59 degrees of freedom
Multiple R-squared:  0.07694,   Adjusted R-squared:  0.06129
F-statistic: 4.918 on 1 and 59 DF,  p-value: 0.03045
```

As follows from this parsimonious model, the statistical significance of the slope is improved. (c) The reason for the improvement is that the baseline variation was reduced by taking the difference. The same method was used in the paired t–test. Moreover, the difference makes the patient-specific factors are unneccesary. This example teaches us that complex models do not necessarily lead to better results. □

In the example above, the patient-specific intercept was not critical because the data were balanced: the blood pressure of all patients was measured at the same time points. For unbalanced data, this is no longer true: if the period of observation is shorter for a low baseline value (on average) a Simpson's-type paradox my occur.

Example 8.47 *Cancer patients' quality of life. Cancer patients' quality of life (QoL) is an important healthcare issue. The QoL is a score on the percent scale derived from a questionnaire completed by patients after the cancer diagnosis and subsequent treatment. Researchers are puzzled by the fact that the average QoL does not change much or even slightly increases with survival, but repeated questionnaires indicate that individual QoL decreases with time as follows from file QoL.csv. Explain this phenomenon using Simpson's paradox and find the solution.*

Solution. The spaghetti plot for 50 cancer patients (each line represents a patient) over the span of 20 years is shown in Figure 8.27 and `summary(lm)` output is shown below.

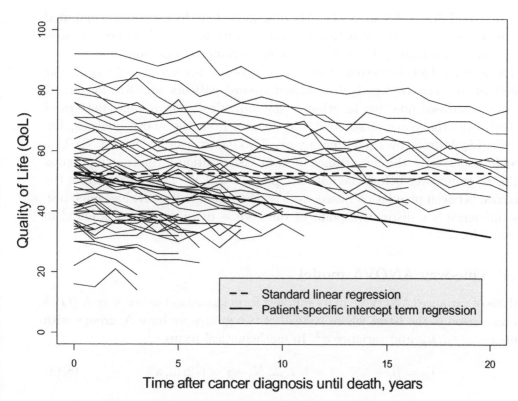

Figure 8.27: *Cancer patients' quality of life (QoL) versus years of survival after cancer diagnosis (baseline). Simpson's paradox: Although individual QoLs gradually drop the average increases.*

```
lm(formula = QOL ~ ttL, data = d)
            Estimate Std. Error t value Pr(>|t|)
(Intercept) 52.45533    0.74282  70.617   <2e-16 ***
ttL          0.01696    0.04831   0.351    0.726
Residual standard error: 13.85 on 699 degrees of freedom
Multiple R-squared:  0.0001762, Adjusted R-squared:  -0.001254
F-statistic: 0.1232 on 1 and 699 DF,  p-value: 0.7257
```

Indeed, although not statistically significant, the slope is slightly positive (the dashed line in Figure 8.27): does the quality of life improve in cancer patients?

```
lm(formula = Q ~ D + tt - 1)
     Estimate Std. Error t value Pr(>|t|)
tt  -1.021340   0.009683 -105.47   <2e-16 ***
Residual standard error: 2.016 on 650 degrees of freedom
Multiple R-squared:  0.9987,    Adjusted R-squared:  0.9986
F-statistic: 1e+04 on 51 and 650 DF,  p-value: < 2.2e-16
```

This is what the researchers may puzzle with: repeated questions indicate that for each individual the quality of life drops but on average it is just the opposite (solid line in Figure 8.27). The reason is that the average is computed for a heterogeneous group of patients: patients with low baseline QoL die early and patients with high QoL live long, leading to Simpson's paradox.

The average QoL is meaningless: our concern is not an average QoL, but an individual. The average QoL is biased toward patients who live longer. A correctly rephrased question is: what is the individual, not average, pattern of the QoL over the years of survival. As follows from the above analysis, the individual QoL drops by 1 point each year. Two lessons can be learned: (i) This example underscores the importance of data visualization – Simpson's paradox is difficult to uncover before the data are plotted. (ii) We need to recognize the difference between population-based, such as the mean, and individual inference. This difference is a distinctive point of the book – more discussion is found in Section 7.10.

8.7.4   One-way ANOVA model

Analysis of variance (ANOVA), in its simplest form known as the one-way ANOVA, is an extension of the t-test where instead of two groups we have N groups with means $\mu_1, \mu_2, ..., \mu_N$ and variance σ^2. In mathematical terms,

$$Y_{ij} = \mu_i + \varepsilon_{ij}, \quad i = 1, 2, ..., N, \quad j = 1, 2, ..., n_i, \tag{8.48}$$

where $\varepsilon_{ij} \overset{\text{iid}}{\sim} N(0, \sigma^2)$. Here i codes the group and j denotes the jth observation in the ith group. This model is called the unbalanced one-way ANOVA model because n_i may vary across i; the model is called balanced if $n_i = \text{const}$. Although many design studies intend to collect an equal number of observations per group missing values are not uncommon – by ignoring missing observations one arrives at the unbalanced model (see the Influenza incidence example below).

The null hypothesis is that the group means are the same,

$$H_0 : \mu_1 = \mu_2 = ... = \mu_N. \tag{8.49}$$

The key to testing this hypothesis is (a) recasting the ANOVA model into a multiple regression with N dummy variables, and (b) applying the F-test of Theorem 8.19.

(a) Combine the n_i observations from the ith group into a $n_i \times 1$ vector $\mathbf{y}_i$. Further, combine these N vectors $\mathbf{y}_i$ into a $K \times 1$ vector $\mathbf{y}$, where $K = \sum_{i=1}^{N} n_i$ is the total number of observations, and do the same with error terms ε_{ij}. Denote the $N \times 1$ vector of means $\boldsymbol{\mu} = (\mu_1, \mu_2, ..., \mu_N)'$, and find the $K \times N$ matrix $\mathbf{X}$ such that (8.48) can be written as $\mathbf{y} = \mathbf{X}\boldsymbol{\mu} + \boldsymbol{\varepsilon}$, making the ANOVA model a linear model. Since all groups have the same variance, the components of vector $\boldsymbol{\varepsilon}$ have zero mean, independent and normally distributed, i.e. all conditions of the linear model hold. Now we construct matrix $\mathbf{X}$ with entries either 0 or 1, such that the ANOVA model (8.48) can be rewritten as a linear model without intercept:

Let the first dummy variable take 1 if the observation belongs to the first group and 0 otherwise, the second dummy variable take 1 if the observation belongs to the second group and 0 otherwise, etc. Combine the N dummy variables into a matrix $\mathbf{X}$ with the $n_i \times 1$ vector of ones $\mathbf{1}_i$ on the diagonal as follows:

$$\mathbf{X}^{K \times N} = \begin{bmatrix} \mathbf{1}_1 & 0 & 0 & 0 \\ 0 & \mathbf{1}_2 & 0 & 0 \\ 0 & 0 & \ddots & 0 \\ 0 & 0 & 0 & \mathbf{1}_N \end{bmatrix}.$$

It is easy to verify by a row-by-row examination that the linear model $\mathbf{y} = \mathbf{X}\boldsymbol{\mu} + \boldsymbol{\varepsilon}$ is equivalent to the ANOVA model (8.48). Certainly, having $\mathbf{y}$ and $\mathbf{X}$ one could run `lm` to obtain the OLS estimates of μ_i, but it easy to find the closed-form expression for $\widehat{\mu}_i$ as follows. The residual sum of squares to minimize takes the form $\sum_{i=1}^{N} \sum_{j=1}^{n_i} (Y_{ij} - \mu_i)^2$ with the OLS estimates

$$\widehat{\mu}_i = \overline{Y}_i = \frac{1}{n_i} \sum_{j=1}^{n} Y_{ij}, \quad i = 1, 2, .., N$$

and $\widehat{\sigma}^2 = \sum_{i=1}^{N} \sum_{j=1}^{n_i} (Y_{ij} - \overline{Y}_i)^2 / (K - N)$, where $K - N$ is the df.

(b) In the notation of Theorem 8.19, we have

$$S_{\min} = \sum_{i=1}^{N} \sum_{j=1}^{n_i} (Y_{ij} - \overline{Y}_i)^2, \quad S_0 = \sum_{i=1}^{N} \sum_{j=1}^{n_i} (Y_{ij} - \overline{Y})^2,$$

where $\overline{Y} = K^{-1} \sum_{i=1}^{N} \sum_{j=1}^{n_i} Y_{ij}$ is the gross average. Thus, the F-test (8.29) for the null hypothesis (8.49) uses the test statistic

$$\frac{\sum_{i=1}^{N} \sum_{j=1}^{n_i} (Y_{ij} - \overline{Y})^2 - \sum_{i=1}^{N} \sum_{j=1}^{n_i} (Y_{ij} - \overline{Y}_i)^2}{\sum_{i=1}^{N} \sum_{j=1}^{n_i} (Y_{ij} - \overline{Y}_i)^2} \times \frac{K - N}{N - 1} \sim F(N - 1, K - N).$$

$$(8.50)$$

The F-test gave rise to the name of this model, "analysis of variance." Indeed, divided by its degrees of freedom, the numerator is the estimate of the explained variance and the denominator is the estimate of the residual variance as was discussed earlier in Section 6.7.4.

The example below illustrates the application of the F-test; most importantly, it shows an extension of ANOVA when some additional information in the form of another dummy variable or predictor is available. Then, recasting the ANOVA model into a linear model $\mathbf{y} = \mathbf{X}\boldsymbol{\beta} + \boldsymbol{\varepsilon}$ becomes very beneficial.

Example 8.48 *Influenza incidence. The table below shows the incidence of influenza in an area per one thousand people by age group over seven years.* (a)

Test the hypothesis that the incidence is the same across all age groups. (b) In 2000, influenza vaccination was administered to everybody in the area. Test if the effect of vaccination reduces the flu incidence by introducing a dummy variable, and compute the b-value to quantify the benefit of vaccination.

Solution. We assume that (i) the incidence rate has a normal distribution with the same variance across years and groups, (ii) incidence is independent from year to year and across age groups, (iii) the means may be different for different age groups, but they stay constant over the years. These assumptions allow us to apply the one-way ANOVA model with $N = 8$ and n_i varying from 6 to 7. Note that NA in the table means "not available" and must be treated as missing, that is, $n_1 = 6$. The R code is found in the file `flu.r` and the rates of influenza in eight age groups over five years are shown in Figure 8.28. The standard `boxplot` command was superimposed with the raw data (empty circles), and the average in each age group shown as the filled circle (see Section 5.4).

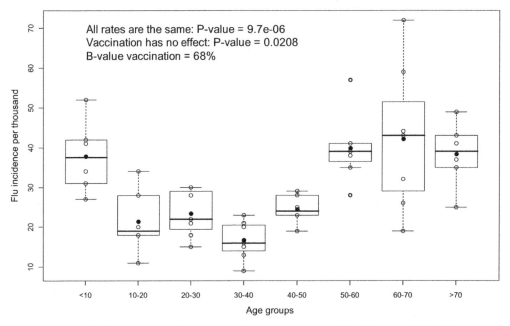

Figure 8.28: *Influenza incidence in eight age groups. The F-test ANOVA rejects the null hypothesis that the rates are the same across age groups with the $p=9.7 \times 10^{-6}$.*

(a) The p-value for the null hypothesis that all age rates are the same is computed in two ways: computation of the left-hand side in (8.50) and construction of vector $\mathbf{y}$ (missing values are skipped) and matrix $\mathbf{X}$ as above by running `lm(y~X)`. Note that it makes the last regression coefficient not estimable because the intercept term is present by default, the F-test produces the right p-value because the null hypothesis is that the rates are the same (alternatively, one could remove the last column of matrix $\mathbf{X}$). Both methods give the same $p = 9.67 \times 10^{-6}$. The hypothesis that the rates are the same in different age groups is rejected.

Influenza incidence by age group per 1,000

Age	1998	1999	2000	2001	2002	2003	2004
<10	41	NA	52	34	27	31	42
10-20	20	34	28	18	NA	11	18
20-30	30	28	18	22	15	21	30
30-40	9	16	21	15	20	13	23
40-50	29	19	28	25	23	NA	23
50-60	39	38	41	28	57	35	41
60-70	59	26	72	19	43	32	44
>70	41	35	49	NA	43	25	37

(b) Representation of the ANOVA model as a linear model is crucial here: create a dummy variable (vac), which takes 0 for years 1998, 1999, and 2000 and 1 for years starting from 2001, and add this variable to the regression. Then the effect of the vaccine follows regular coefficient significance testing. See the output below.

```
lm(formula = y ~ X + vac - 1)
Coefficients:
      Estimate Std. Error t value Pr(>|t|)
X1      41.926      4.092  10.246 4.11e-13 ***
X2      24.569      3.933   6.246 1.60e-07 ***
X3      26.936      3.741   7.200 6.61e-09 ***
X4      20.222      3.741   5.405 2.66e-06 ***
X5      27.569      3.933   7.009 1.25e-08 ***
X6      43.365      3.741  11.592 8.11e-15 ***
X7      45.651      3.741  12.203 1.47e-15 ***
X8      41.403      3.933  10.526 1.78e-13 ***
vac     -6.139      2.558  -2.400   0.0208 *
Residual standard error: 9.111 on 43 degrees of freedom
Multiple R-squared:  0.9376,    Adjusted R-squared:  0.9245
F-statistic: 71.75 on 9 and 43 DF,  p-value: < 2.2e-16
```

The p-value for the null hypothesis that the vaccination has no effect is a part of the standard summary(lm()), $p = 0.0208$. The regression analysis implies that vaccination reduces the rate by 6.139. Now we quantify the effect of vaccination on the probabilistic scale by comparing flu incidence across the groups of people who got vaccinated and who did not. Using the R command from Table 8.1, we compute the b-value $= 0.68$. This means that the chance that flu incidence is smaller in people who get vaccinated is 68%.

Example 8.49 *Linear regression with repeated measurements.* *To understand a gender-specific difference in school performance across the country, SAT math scores were collected in N schools for n_i students in each school, $i = 1, 2, ..., N$. (a) Extend the one-way ANOVA model to account for school differences and reduce the problem to a slope regression problem. (b) Use simulated data to confirm numerically that the* lm *function and statistical inference based on the slope model yield the same results.*

Solution. (a) Let individual SAT scores and gender be represented as $\{Y_{ij}, i = 1, ..., N, j = 1, ..., n_i\}$ and $\{x_{ij}, i = 1, ..., N, j = 1, ..., n_i\}$, respectively ($x_{ij} = 1$ if a boy and $x_{ij} = 0$ if a girl). While school performances may vary, it is believed that the effect of gender, if any, is constant. This assumption leads to a linear regression with repeated measurements as an extension of the ANOVA model,

$$Y_{ij} = \mu_i + \beta x_{ij} + \varepsilon_{ij} \qquad (8.51)$$

where $\varepsilon_{ij} \overset{\text{iid}}{\sim} \mathcal{N}(0, \sigma^2)$. This model implies that the difference of the SAT score between boys and girls is β. This model can be reduced to a linear model $\mathbf{y} = \mathbf{X}\boldsymbol{\beta} + \boldsymbol{\varepsilon}$ but we will find a simplified solution as follows. The sum of squares to minimize is

$$\min_{\mu_1,...,\mu_N,\beta} \sum_{i=1}^{N} \sum_{j=1}^{n_i} (Y_{ij} - \mu_i - \beta x_{ij})^2 = \min_{\beta} \sum_{i=1}^{N} \left[\min_{\mu_i} \sum_{j=1}^{n_i} [(Y_{ij} - \beta x_{ij}) - \mu_i]^2 \right].$$

We can view minimization over μ_i independently of i by replacing $z_{ij} = Y_{ij} - \beta x_{ij}$, where β is held fixed. Letting $\overline{Y}_i = \sum Y_{ij}/n_i$ and $\overline{x}_i = \sum x_{ij}/n_i$, being average SAT scores and gender in the ith school, we express

$$\widehat{\mu}_i = \frac{1}{n_i} \sum_{j=1}^{n_i} z_{ij} = \frac{1}{n_i} \sum_{j=1}^{n_i} (Y_{ij} - \beta x_{ij}) = \overline{Y}_i - \beta \overline{x}_i.$$

Plugging this solution back into the original sum of squares, we reduce the problem with $N + 1$ slope coefficients to a single slope coefficient (see Example 6.36 of Section 6.5),

$$\min_{\beta} \sum_{i=1}^{N} \sum_{j=1}^{n_i} (\widetilde{Y}_{ij} - \beta \widetilde{x}_{ij})^2,$$

where $\widetilde{Y}_{ij} = Y_{ij} - \overline{Y}_i$ and $\widetilde{x}_{ij} = x_{ij} - \overline{x}_i$. The OLS estimate of the parameter of interest takes the familiar form

$$\widehat{\beta} = \frac{\sum_{i=1}^{N} \sum_{j=1}^{n_i} \widetilde{Y}_{ij} \widetilde{x}_{ij}}{\sum_{i=1}^{N} \sum_{j=1}^{n_i} \widetilde{x}_{ij}^2}.$$

(b) The slope model yields the same statistical inference as the full linear model with $\sum_{i=1}^{N} n_i - 1$ degrees of freedom. The R function linrep generates Y_{ij} and

x_{ij} and estimates the difference of the math SAT score between boys and girls using the full model in the form $\mathbf{y} = \mathbf{X}\boldsymbol{\beta} + \boldsymbol{\varepsilon}$ and the slope model. A truncated output is shown below.

```
> linrep()
Call:
lm(formula = Y ~ X - 1)

X31    29.918        1.117   26.77   <2e-16 ***

Residual standard error: 19.61 on 1221 degrees of freedom
Multiple R-squared:  0.9986,    Adjusted R-squared:  0.9985
F-statistic: 2.768e+04 on 31 and 1221 DF,  p-value: < 2.2e-16

beta.hat= 29.91803
Res. st.error= 19.61318
t value= 26.77348
Pr(>|t|)= 1.285404e-124
```

All critical values from the models, such as the beta coefficient, residual standard error, t-value and p-value coincide.

Example 8.50 *Synergy of drugs in pharmacology.* *Sometimes new cancer treatments are developed as a combination of known drugs. In vague terms, we say that a combination of drugs is synergistic if its effect is stronger than each of the drugs used separately. Below, we define synergy of drugs for cancer cell culture experiments where "effect" is associated with cell survival, and we show that testing for synergy reduces to the one-way ANOVA on the log scale.*

Solution. Let treatments A and B lead to immediate cancer cell kill with surviving fraction S_A and S_B, respectively. Let the combination of treatments lead to a surviving fraction S_{AB}. According to Bliss ([10]), we say that there is a synergy of treatments if $S_A S_B > S_{AB}$. We say that drugs act independently if $S_A S_B = S_{AB}$, and drugs are antagonistic if $S_A S_B < S_{AB}$. This definition can be explained by a simple observation that if treatment A is applied right after treatment B it acts only on the surviving population of cancer cells: After drug A the proportion of surviving cells is S_A; since the drug acts only on living/surviving cells the proportion of surviving cells after consequent administration of drug B will be $S_A S_B$. Therefore, in the absence of synergy the proportion of surviving cells after the treatments (the order does not matter) is the product, $S_A S_B$.

This definition of synergy has a nice probabilistic interpretation: associate surviving fractions S_A and S_B with probabilities of surviving due to the respective drugs. If the events of surviving are denoted as A and B, their probabilities are $\Pr(A) = S_A$ and $\Pr(B) = S_B$. If drugs act independently, the probability of surviving, when two drugs administered simultaneously, is

$$S_{AB} = \Pr(A \cap B) = \Pr(A)\Pr(B) = S_A S_B.$$

In the above definition, "immediate" means that cells die immediately after the treatment, that is, the proportion of surviving cells is measured at the time of the treatment. In pharmacology, drugs do not imply immediate cell death and it takes some time before a cancer cell dies. Moreover, since time is involved surviving cells may proliferate. To account for time-related cell growth or natural death, we have to have a control group C with no treatment and the associated fraction S_C. Again, appealing to probability theory, we now treat the observed surviving fraction S_A as the proportion of cells not killed by drug A and dying naturally, or in mathematical terms, $S_A = \Pr(A \cap C)$. Then the probability of survival can be viewed as a conditional probability: $\Pr(A|C) = \Pr(A \cap C)/\Pr(C) = S_A/S_C$. A similar probability of survival can be obtained for population of cell, treated with drug B, and simultaneous treatment A and B,

$$\Pr(B|C) = \frac{\Pr(B \cap C)}{\Pr(C)} = \frac{S_B}{S_C}, \quad \Pr(AB|C) = \frac{\Pr(AB \cap C)}{\Pr(C)} = \frac{S_{AB}}{S_C}.$$

Now, applying the above definition of synergy, we arrive at a more general definition when the control's group population reduces due to natural cell death,

$$\frac{S_A}{S_C} \times \frac{S_B}{S_C} > \frac{S_{AB}}{S_C}.$$

or equivalently $S_A S_B > S_C S_{AB}$. From the computational point of view, it is easier to rewrite the synergy condition on the log scale,

$$\ln S_A + \ln S_B - \ln S_C - \ln S_{AB} > 0. \tag{8.52}$$

We say that drugs act independently if the inequality turns into equality.

Now we apply the definition of synergy (8.52) to experimental data that involves four cell survival experiments with replicates: each cell culture experiment is repeated n times and the surviving fraction is measured; C_j = control, A_j = drug A, B_j = drug B, and D_j = combination of drugs A and B, $j = 1, 2, ..., n$ (n may be treatment group specific). Since survival fraction is positive, it makes sense to model the observed quantities on the log scale using a multiplicative error scheme, as we previously used in the production function example,

$$C_j = e^{\mu_0 + \varepsilon_{0j}}, \quad A_j = e^{\mu_1 + \varepsilon_{1j}}, \quad B_j = e^{\mu_2 + \varepsilon_{2j}}, \quad D_j = e^{\mu_3 + \varepsilon_{3j}},$$

where the μ's are the true survival fractions on the log scale and the ε's are iid normally distributed random deviations with zero mean and variance σ^2. Take the log and denote survival fractions on the log scale as

$$Y_{0j} = \ln C_j, \quad Y_{1j} = \ln A_j, \quad Y_{2j} = \ln B_j, \quad Y_{3j} = \ln D_j.$$

This system can be rewritten as a one-way ANOVA model

$$Y_{ij} = \mu_i + \varepsilon_{ij}, \quad i = 0, 1, 2, 3, \quad j = 1, ..., n_i.$$

According to the synergy definition (8.52), we want to test the null hypothesis that the drugs act independently,

$$H_0 : \mu_1 + \mu_2 - \mu_3 - \mu_0 = 0. \tag{8.53}$$

Since $\overline{Y}_i$ is an estimate of μ_i, we say that synergy is observed if

$$\overline{Y}_1 + \overline{Y}_2 - \overline{Y}_3 - \overline{Y}_0 > 0.$$

which does not imply that the true synergy, or more specifically, independence action expressed in equation (8.53) is true. To test the null hypothesis H_0, the following t-test is used:

$$\frac{\overline{Y}_1 + \overline{Y}_2 - \overline{Y}_3 - \overline{Y}_0}{\widehat{\sigma}} \sim t\left(\sum_{i=0}^{3} n_i - 4\right), \tag{8.54}$$

where

$$\widehat{\sigma}^2 = \frac{1}{\left(\sum_{i=0}^{3} n_i - 4\right)\sum_{i=0}^{3} n_i^{-1}}\left(\sum_{i=0}^{3}\sum_{j=1}^{n_i}(Y_{ij} - \overline{Y}_i)^2\right)$$

is an estimate of σ^2.

Survival fraction of ZR75 cancer cells

Treatment	Rep 1	Rep 2	Rep 3
Control	0.82927	1.0662	1.10105
BYL	0.75958	0.79443	0.79791
GSK	033798	0.53659	0.52613
BYL+GSK	0.16028	0.30662	0.21254

ZR75 is the name of the breast cancer ductal carcinoma cancer cell line. Both BYL and GSK (GlaxoSmithKline) are novel pharmacological drugs for breast cancer treatment (Hosford et al. [57]). Is there a synergy between the two drugs? Test the independence action using the survival fraction data in the table above. In this experiment, $n = 3$ and

$$\overline{Y}_1 + \overline{Y}_2 - \overline{Y}_3 - \overline{Y}_0 = 0.50269082 > 0,$$

which hints toward synergy of drugs BYL and GSK. The observed test statistic, the left-hand side of expression (8.54), is equal to 1.95207032 and, therefore, the p-value of the null hypothesis of drug independence is computed as pt(1.95207032,df=3*4-4,lower.tail=F)=0.04335717. According to statistical dogma, the synergy is statistically significant.

8.7.5   Two-way ANOVA

As above, we have N groups, but now we assume that the means across rows are not the same, or more specifically,

$$Y_{ij} = \mu_i + \nu_j + \varepsilon_{ij}, \quad i = 1, 2, ..., N, \ j = 1, 2, ..., n, \tag{8.55}$$

where $\{\mu_i, i = 1, ..., N\}$ and $\{\nu_j, j = 1, ..., n\}$ are subject to estimation under the constraint

$$\sum_{j=1}^{n} \nu_j = 0. \tag{8.56}$$

This constraint makes the model identifiable. The meaning of μ and ν parameters will be clarified later in example to follow. As before, it is assumed that $\varepsilon_{ij} \overset{iid}{\sim} \mathcal{N}(0, \sigma^2)$ and the number of observations in each group is the same. To find the OLS estimates for μ and ν parameters we need to minimize the residual sum of squares (RSS),

$$\min_{\mu_i, \nu_j} \sum_{i=1}^{N} \sum_{j=1}^{n} (Y_{ij} - \mu_i - \nu_j)^2.$$

Holding ν_j fixed, find the OLS estimate for μ_i,

$$\widehat{\mu}_i = \frac{1}{n} \sum_{j=1}^{n} (Y_{ij} - \nu_j) = \overline{Y}_{i\cdot},$$

where $\overline{Y}_{i\cdot} = \sum_{j=1}^{n} Y_{ij}/n$ is the average in the ith row, due to constraint $\sum_{j=1}^{n} \nu_j = 0$. After plugging these averages back into the RSS, we obtain the OLS estimate for the ν parameters,

$$\widehat{\nu}_j = \frac{1}{N} \sum_{i=1}^{N} (Y_{ij} - \overline{Y}_{i\cdot}) = \overline{Y}_{\cdot j} - \overline{Y},$$

where $\overline{Y}_{\cdot j} = \sum_{i=1}^{N} Y_{ij}/N$ is the average in the jth column, and

$$\overline{Y} = \frac{1}{N} \sum_{i=1}^{N} \overline{Y}_{i\cdot} = \frac{1}{Nn} \sum_{i=1}^{N} \sum_{j=1}^{n} Y_{ij}$$

is the gross average. It is easy to see that the sum of $\widehat{\nu}_j$ is zero – the constraint (8.56) holds. A remark about the dot-notation: the dot indicates "averaging out" meaning that $\overline{Y}_{i\cdot}$ is the average of Y_{ij} over j and $\overline{Y}_{\cdot j}$ is the average over i. The minimum of the RSS is instrumental for various hypothesis tests,

$$S_{\min} = \sum_{i=1}^{N} \sum_{j=1}^{n} (Y_{ij} + \overline{Y} - \overline{Y}_{i\cdot} - \overline{Y}_{\cdot j})^2.$$

Various linear hypotheses regarding the μ and ν parameters may be tested. The general procedure involves finding the restricted RSS and applying the F-test (8.29). For example, to test that all μ parameters are the same ($\mu_i = \mu$), we find the restricted RSS as

$$S_0 = \min_{\mu,\nu_j} \sum_{i=1}^{N} \sum_{j=1}^{n} (Y_{ij} - \mu - \nu_j)^2 = \sum_{i=1}^{N} \sum_{j=1}^{n} (Y_{ij} + \overline{Y} - \overline{Y}_{\cdot j})^2$$

with the first degree of freedom $N - 1$ and the same second degree of freedom $Nn - N - (n-1)$. To test that all ν parameters are zero ($\nu_j = 0$) we apply (8.29), where

$$S_0 = \min_{\mu_i} \sum_{i=1}^{N} \sum_{j=1}^{n} (Y_{ij} - \mu_i)^2 = \sum_{i=1}^{N} \sum_{j=1}^{n} (Y_{ij} - \overline{Y}_{i\cdot})^2$$

with the first degree of freedom $n - 1$ and the same second degree of freedom.

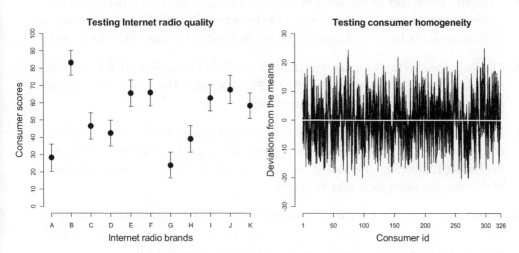

Figure 8.29: *Internet radio analysis. The left plot depicts average scores +/- SD for 11 radio brands and the right plot depicts consumer scores as the deviation from the average radio brand score.*

Example 8.51 *Internet radio. File* consIR.csv *contains 326 consumer satisfaction scores on 11 Internet radio brands. (a) Test the hypothesis that the brands are equivalent in terms of consumer satisfaction. (b) Test the hypothesis that consumers' opinions are homogeneous.*

Solution. The R code found in the file consIR.r produces Figure 8.29. Satisfaction scores are modeled as a two-way ANOVA model, where $\{\mu_i, i = 1, ..., 11\}$ represent satisfaction with internet brands and $\{\nu_j, j = 1, ..., 326\}$ represent consumer-specific differences in scoring. (a) To visualize consumers' scores we plot the average score for each brand +/- SD. As follows from this plot Internet radios vary considerably in their quality – it is not surprising that the formal F-test rejects the null hypothesis that the scores are the same. (b) The null hypothesis is that ν_j are close to zero, that is, consumers' opinion is homogeneous. To visually assess this hypothesis we plot $\widehat{\nu}_j = \overline{Y}_{.j} - \overline{Y}$. Obviously there is sufficient variation in $\widehat{\nu}_j$ that the F-test rejects the null as well. □

One can rewrite the two-way ANOVA model (8.55) as a linear model in the form $\mathbf{y} = \mathbf{X}\boldsymbol{\beta} + \boldsymbol{\varepsilon}$ as we did for the one-way ANOVA. This method is more powerful because it is easy to (1) account for different numbers of observations per group ($n_i \neq$ const) and (2) adjust for predictors.

Problems

1. In Example 8.40, explain why the coefficient at d3 and dptmProduction have the same absolute value but different sign.

2. Compare the trend model used in the Nile Example 8.41 with (a) the model where the predictor is computed as I(d*(year-yc)) and (b) the step model. Which model is better? Argue based on statistical testing with the F-test.

3. Challenge the default choice yc=1898 in the Nile Example 8.41 for the step model: run the models for yc=1880:1960 and plot the residual standard error versus yc. What is the optimal yc?

4. File CollegeSalary.csv contains salary information on 36 college employees, including the department where they work, gender, and years at work. Following Example 8.40; test if there is a statistically significant difference of salaries between men and women.

5. Show that the one-way ANOVA model (8.48) can be equivalently rewritten as $Y_{ij} = \mu + \alpha_i + \varepsilon_{ij}$ where $\sum_{i=1}^{n} \alpha_i = 0$. Give an interpretation for α_i and find the OLS estimates for μ and α_i in the new formulation.

6. Demonstrate by simulations that the random variable at the left-hand side has the distribution specified in the right-hand side of (8.50) by matching the empirical distribution with theoretical one. [Hint: Simulate N and n_i as random integers and $\mu_i = \mu$ and σ^2 as continuous random variables.]

7. Build a parsimonious model by reducing the number of age groups in Example 8.48 to two: the first age group is 20–50 and the second age group includes all others. Estimate the model with the vac variable and test if the parsimonious model is statistically different from the full model.

8. In Example 8.48, compare the advantage of vaccination among young people ages 10 to 20 against unvaccinated older people of age > 70 using the b-value.

9. (a) Express the two-way ANOVA model (8.55) as a linear model. Generate Y_{ij}, compute S_0 and $S_{\min}$, and compute these quantities using the output of lm. (b) Express the two-way ANOVA model as a linear model with different n_i and check your representation through lm.

10. There is a hypothesis that the price of the Internet radio in Example 8.51 drives the quality and the differences between radio brands. The prices are: 126, 236, 167, 178, 211, 195, 169, 125, 201, 189, 122. Incorporate these data into the two-way ANOVA using the linear model representation (as in the previous problem) to test this hypothesis.

11. Extend model (8.51) to several independent variables, $Y_{ij} = \mu_i + \boldsymbol{\beta}' \mathbf{x}_{ij} + \varepsilon_{ij}$ and derive the OLS estimate for $\boldsymbol{\beta}$ expressed through differences.

8.8 Generalized linear model

The generalized linear model (GLM) is a generalization of the linear model. It handles discrete (Bernoulli or binomial), count (Poisson) and continuous (exponential, normal) distributions of the dependent variable under one methodological umbrella of the exponential family of distributions; see Section 6.8.2. First, we introduce a general idea how any distribution can be turned into a regression when the mean is expressed as a linear combination of predictors. Second, we discuss a more restrictive, but yet very flexible approach based on the exponential family of distributions.

The idea of modeling the dependent variable, discrete or continuous, defined by the probability density function (pdf) is as follows. Let $Y_1, Y_2, ..., Y_n$ be independent observations with means $\mu_1, \mu_2, ..., \mu_n$, respectively, and common parameter $\boldsymbol{\theta}$ with the pdf defined as $f(\cdot; \mu_i, \boldsymbol{\theta})$, $i = 1, 2, ..., n$. Let $\mathbf{x}_1, \mathbf{x}_2, ..., \mathbf{x}_n$ be the respective vectors of predictors that are related to the means as $g(\mu_i) = \boldsymbol{\beta}' \mathbf{x}_i$, where g is an increasing function, known as the *link*. The key assumption is that the distribution of Y_i is defined as a linear combination of predictors with unknown coefficients. Then the model for the dependent variable takes the form $Y_i \sim f(\cdot; g^{-1}(\boldsymbol{\beta}' \mathbf{x}_i), \boldsymbol{\theta})$, where g^{-1} is the inverse function, with the log-likelihood function given by (see Section 6.10)

$$l(\boldsymbol{\beta}, \boldsymbol{\theta}) = \sum_{i=1}^{n} \ln f(Y_i; g^{-1}(\boldsymbol{\beta}' \mathbf{x}_i), \boldsymbol{\theta}).$$

This model is called *linear,* because upon a transformation, the means of Y_i are expressed as a linear combination of predictors, and it is called *generalized* because the distribution f may be other than normal; however, when $f = \phi$, g is the identity function, and $\boldsymbol{\theta} = \sigma^2$ we arrive at the classic linear model. Obviously, this approach works if any other parameter, such as median, can be expressed as a linear combination of predictors. The following example illustrates this approach.

Example 8.52 *Black Friday shopping. File* `blackfriday.csv` *contains the data on 548 shoppers at Walmart on Black Friday. The first column is the time the shopper walks in after the opening at 6 a.m., the second column is the age, and the third column is the gender of the shopper (1 = male, 0 = female). (a) Use the exponential distribution for the time of shopping to estimate the time of shopping adjusted for* **age** *and* **gender** *and estimate the model by maximum likelihood using the Fisher scoring algorithm. (b) Plot the average time to walk in after 6 a.m. for a female as a function of age along with the 95% confidence band. (c) Plot the proportion of female shoppers as a function of time of age 20 and 50. (d) Compare males and females by plotting the proportion as a function of time.*

Solution. The R function is `walD`. To begin, we assume that age and gender make no difference on the time when people come to Walmart. Then the density of the distribution is $\lambda e^{-\lambda y}$, where Y is the time (hour) after 6 a.m., with the log-likelihood function $l(\lambda) = n \ln \lambda - \lambda \sum_{i=1}^{n} Y_i$. Differentiating with respect to λ and equating the derivative to zero estimates λ by the maximum likelihood, $\widehat{\lambda}_{ML} = 1/\overline{Y} = 0.1851733$. For example, the average time after 6 a.m. when people come to Walmart is estimated as $1/\widehat{\lambda}_{ML} = 1/0.1851733 = 5.4$ hours, i.e. the largest crowd is expected around 11:20 a.m. The proportion of people who show up at Walmart by noon is estimated as $1 - e^{-6 \times 0.1851733} = 0.67$ (we suggest reviewing the exponential distribution in Section 2.4).

(a) Now we want to incorporate the information on the shopper, age and gender. Since $1/\lambda$ is the mean of the exponential distribution, λ will vary from shopper to shopper as a function of age and gender. Two points should be addressed: (a) Since $\lambda > 0$ the function must be positive. (b) It is plausible that the dependence on age is nonlinear, so that a quadratic function of age may be used. Taking this into account, we suggest specifying the shopper-specific rate λ_i as $\lambda_i = e^{\beta_0 + \beta_1 a_i + \beta_2 a_i^2 + \beta_3 g_i}$, $i = 1, 2, ..., n$, where the coefficients, as in linear model, are subject to estimation. Clearly, this specification leads to a GLM with the link $g = -\ln$, since $E(Y_i) = 1/\lambda_i$ and $\mathbf{x}_i = (1, a_i, a_i^2 g_i)'$, $\boldsymbol{\beta} = (\beta_0, \beta_1, \beta_2, \beta_3)'$. The log-likelihood function is similar to the case above and is given by

$$l(\boldsymbol{\beta}) = \sum_{i=1}^{n} \ln \lambda_i - \sum_{i=1}^{n} Y_i \lambda_i = \boldsymbol{\beta}' \mathbf{r} - \sum_{i=1}^{n} Y_i e^{\boldsymbol{\beta}' \mathbf{x}_i}, \qquad (8.57)$$

where $\mathbf{r} = \sum_{i=1}^{n} \mathbf{x}_i$ is a 4×1 vector. The Fisher scoring (FS) algorithm is used to find the maximum of l. See Section 6.10.6. Differentiate l with respect to $\boldsymbol{\beta}$ once and twice to find the score equation and the Hessian,

$$\frac{\partial l}{\partial \boldsymbol{\beta}} = \mathbf{r} - \sum_{i=1}^{n} Y_i e^{\boldsymbol{\beta}' \mathbf{x}_i} \mathbf{x}_i = \mathbf{0}, \quad \frac{\partial^2 l}{\partial \boldsymbol{\beta}^2} = -\sum_{i=1}^{n} Y_i e^{\boldsymbol{\beta}' \mathbf{x}_i} \mathbf{x}_i \mathbf{x}_i'.$$

The negative expected Hessian or the Fisher information matrix is given by

$$-E\left(\frac{\partial^2 l}{\partial \beta^2}\right) = \sum_{i=1}^{n} E(Y_i) e^{\beta' \mathbf{x}_i} \mathbf{x}_i \mathbf{x}_i' = \sum_{i=1}^{n} \frac{1}{e^{\beta' \mathbf{x}_i}} e^{\beta' \mathbf{x}_i} \mathbf{x}_i \mathbf{x}_i' = \sum_{i=1}^{n} \mathbf{x}_i \mathbf{x}_i'.$$

Now the FS algorithm takes the form

$$\beta_{k+1} = \beta_k + \left(\sum_{i=1}^{n} \mathbf{x}_i \mathbf{x}_i'\right)^{-1} \left(\mathbf{r} - \sum_{i=1}^{n} Y_i e^{\beta_k' \mathbf{x}_i} \mathbf{x}_i\right), \quad k = 0, 1, \ldots \quad (8.58)$$

To simplify the notation introduce ($m = 4$)

$$\mathbf{X}^{n \times m} = \begin{bmatrix} \mathbf{x}_1' \\ \mathbf{x}_2' \\ \vdots \\ \mathbf{x}_n' \end{bmatrix}, \quad \mathbf{D}^{n \times n} = \mathrm{diag}(e^{\beta_k' \mathbf{x}_1}, \ldots, e^{\beta_k' \mathbf{x}_n}), \quad \mathbf{y} = \begin{bmatrix} Y_1 \\ Y_2 \\ \vdots \\ Y_n \end{bmatrix}.$$

Then $\sum_{i=1}^{n} Y_i e^{\beta_k' \mathbf{x}_i} \mathbf{x}_i = \mathbf{X}' \mathbf{D} \mathbf{y}$ and finally the FS algorithm can be succinctly written as

$$\beta_{k+1} = \beta_k + \left(\mathbf{X}'\mathbf{X}\right)^{-1} \mathbf{X}' \mathbf{D} \mathbf{x}.$$

The iterations converge to the maximum likelihood estimate, $\lim_{k \to \infty} \beta_k = \widehat{\beta}_{ML}$ and asymptotically (see Section 6.10),

$$\widehat{\beta}_{ML} \simeq \mathcal{N}\left(\beta, \left(\mathbf{X}'\mathbf{X}\right)^{-1}\right) \text{ when } n \text{ is large.} \quad (8.59)$$

The distribution is strikingly similar to the distribution of the OLS estimator (σ^2 is absent) – this is a common feature for all GLMs.

The FS iterations are shown below.

```
> walD(job=2)
[1] "Lambda from exp. distribution with no predictors:"
[1] 0.1851733
[1] "FS iteration for MLE: iter, loglik, beta0, beta1, beta2, beta3"
[1]  1.000000e+00 -1.552776e+03 -3.322407e-01 -5.352632e-02  8.932895e-05 -9.316698e-02
[1]  2.000000e+00 -1.478375e+03  6.256251e-01 -1.126365e-01  1.019182e-03 -8.965306e-02
[1]  3.000000e+00 -1.462000e+03  8.967239e-01 -1.300855e-01  1.296030e-03 -8.794725e-02
[1]  4.000000e+00 -1.461127e+03  8.830362e-01 -1.289025e-01  1.275408e-03 -8.748002e-02
[1]  5.000000e+00 -1.461124e+03  8.851275e-01 -1.290472e-01  1.277617e-03 -8.749330e-02
[1]  6.000000e+00 -1.461124e+03  8.848069e-01 -1.290225e-01  1.277193e-03 -8.748909e-02
      ML betas          SE         P-value
1   0.884806939 0.2265491926 9.399803e-05
2  -0.129022535 0.0183885322 2.275513e-12
3   0.001277193 0.0003603908 3.942369e-04
4  -0.087489089 0.0937682291 3.508016e-01
```

To start iterations, we set $\beta_0 = \ln 0.1851733$ and the other three coefficients to zero. It takes only six iterations to converge. For example, the expected time of a female shopper of age a to show up at Walmart (reciprocal of lambda) is

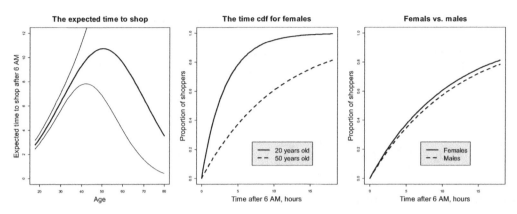

Figure 8.30: *Graphical output of* `walD(job=3)` *for the Black Friday Walmart shopping example.*

estimated as $e^{-0.8845+0.129a-0.001277a^2}$. The maximum value is for a female of age $0.129/(2 \times 0.001277) \simeq 51$ years old. One can say that the age of a late shopper (female and male) is around 51. Conversely, young adults and elderly people tend to come to Walmart earlier.

(b) Figure 8.30 is created by calling `walD(job=3)`. After the beta coefficients are estimated, we plot $e^{-0.8845+0.129a-0.001277a^2}$ versus a on the range from 20 to 80 (see the left plot). Since the coefficient at a^2 is negative the branches of the curve are down with the maximum at $a_{\max} = 51$ as computed above. Now we explain how to compute the 95% confidence band for females. We compute the variance of $\widehat{\beta}_1 + \widehat{\beta}_2 a + \widehat{\beta}_3 a^2$, where a is female age using the first 3×3 block (`cov13`) of the 4×4 matrix $(\mathbf{X}'\mathbf{X})^{-1}$. To do this, we construct a 61×3 matrix `X13=cbind(rep(1,length(age)),age,age^2)`, where 63 is the number of elements in the array `18:80`. We then compute the variances of the linear predictor as

$$\texttt{var.age} = \texttt{diag}(\texttt{X13\%} * \%\texttt{cov13\%} * \%\texttt{t(X13)}).$$

Finally, the lower 95% confidence limit is computed as

$$\texttt{low.age} = -\texttt{bet[1]} - \texttt{bet[2]} * \texttt{age} - \texttt{bet[3]} * \texttt{age\^{}2} - \texttt{Z1a} * \texttt{sqrt(var.age)}$$

Surprisingly, the confidence band is very wide, especially for older shoppers. This can be explained by the large contribution of the quadratic term, a^2. (c) To plot the proportion of shoppers at specific time t, we use the cdf, $1-e^{-\lambda_i t}$ where λ_i is the shopper-specific rate as a function of age and gender (the middle plot). Since 51 year olds are the latest shoppers, the cdf for a 20-year-old female is above the 50-year-old female. (d) The plot at right depicts the differences between females and males, again, using the cdf. The difference is not substantial.

Similar estimation results can be obtained using the R function `glm`, see below.

8.8.1 MLE estimation of GLM

The goal of this subsection is to provide a general discussion of the maximum likelihood estimation (MLE) for the generalized linear model (GLM) under the umbrella of the exponential family of distributions introduced in Section 6.8.2. This family includes the exponential model considered in the above example, logistic, probit, and Poisson regression models, among others. The most popular generalized linear models are listed in Table 8.2. Specifically, following the scheme outlined above, it is assumed that independently distributed observations, Y_i, have the marginal probability density function defined as

$$f(y_i; \boldsymbol{\beta}) = e^{p(\boldsymbol{\beta}'\mathbf{x}_i)K(y_i)+S(y_i)-q(\boldsymbol{\beta}'\mathbf{x}_i)}, \quad i = 1, 2, ..., n. \tag{8.60}$$

Table 8.2. GLM regressions ($K(y_i) = y_i$)

Regression	support	parameter	$p(\boldsymbol{\beta}'\mathbf{x}_i)$	$q(\boldsymbol{\beta}'\mathbf{x}_i)$	mean
Exponential	$Y_i > 0$	$\lambda_i = e^{\boldsymbol{\beta}'\mathbf{x}_i}$	$e^{\boldsymbol{\beta}'\mathbf{x}_i}$	$-\boldsymbol{\beta}'\mathbf{x}_i$	$e^{-(\boldsymbol{\beta}'\mathbf{x}_i)}$
Logistic	$Y_i = 0, 1$	$\pi_i = \frac{e^{\boldsymbol{\beta}'\mathbf{x}_i}}{1+e^{\boldsymbol{\beta}'\mathbf{x}_i}}$	$\boldsymbol{\beta}'\mathbf{x}_i$	$-\ln(1+e^{\boldsymbol{\beta}'\mathbf{x}_i})$	$\frac{e^{\boldsymbol{\beta}'\mathbf{x}_i}}{1+e^{\boldsymbol{\beta}'\mathbf{x}_i}}$
Probit	$Y_i = 0, 1$	$\pi_i = \Phi(\boldsymbol{\beta}'\mathbf{x}_i)$	$\ln\frac{\Phi(\boldsymbol{\beta}'\mathbf{x}_i)}{1-\Phi(\boldsymbol{\beta}'\mathbf{x}_i)}$	$\ln(1-\Phi(\boldsymbol{\beta}'\mathbf{x}_i))$	$\Phi(\boldsymbol{\beta}'\mathbf{x}_i)$
Binomial	$Y_i = 0, ..., n_i$	$\pi_i = \frac{e^{\boldsymbol{\beta}'\mathbf{x}_i}}{1+e^{\boldsymbol{\beta}'\mathbf{x}_i}}$	$n_i(\boldsymbol{\beta}'\mathbf{x}_i)$	$-n_i\ln(1+e^{\boldsymbol{\beta}'\mathbf{x}_i})$	$n_i\frac{e^{\boldsymbol{\beta}'\mathbf{x}_i}}{1+e^{\boldsymbol{\beta}'\mathbf{x}_i}}$
Poisson	$Y_i = 0, 1, ...$	$\lambda_i = e^{\boldsymbol{\beta}'\mathbf{x}_i}$	$e^{\boldsymbol{\beta}'\mathbf{x}_i}$	$\boldsymbol{\beta}'\mathbf{x}_i$	$e^{\boldsymbol{\beta}'\mathbf{x}_i}$

The log-likelihood function, up to a constant term $\sum S(Y_i)$, takes the form

$$l(\boldsymbol{\beta}) = \sum_{i=1}^{n}(p(\boldsymbol{\beta}'\mathbf{x}_i)K_i - q(\boldsymbol{\beta}'\mathbf{x}_i)), \tag{8.61}$$

where $K_i = K(Y_i)$. The first and second derivatives of the log-likelihood function are

$$\frac{\partial l}{\partial \boldsymbol{\beta}} = \sum_{i=1}^{n}(\dot{p}_i K_i - \dot{q}_i)\mathbf{x}_i, \quad \frac{\partial^2 l}{\partial \boldsymbol{\beta}^2} = \sum_{i=1}^{n}\left(\ddot{p}_i K_i - \ddot{q}_i\right)\mathbf{x}_i\mathbf{x}_i',$$

where the dot sign indicates the derivation, and the following notation is applied.

$$\dot{p}_i = \dot{p}(\boldsymbol{\beta}'\mathbf{x}_i), \quad \dot{q}_i = \dot{q}(\boldsymbol{\beta}'\mathbf{x}_i), \quad \ddot{q}_i = \ddot{q}(\boldsymbol{\beta}'\mathbf{x}_i).$$

Theorem 6.77 is used to find the Fisher information matrix

$$\mathcal{I}(\boldsymbol{\beta}) = -E\left(\frac{\partial^2 l}{\partial \boldsymbol{\beta}^2}\right) = \sum_{i=1}^{n}\left(\ddot{q}_i - \frac{\ddot{p}_i \dot{q}_i}{\dot{p}_i}\right)\mathbf{x}_i\mathbf{x}_i'.$$

Hence, in general form, the Fisher scoring (FS) algorithm is written as

$$\boldsymbol{\beta}_{k+1} = \boldsymbol{\beta}_k + \mathcal{I}^{-1}(\boldsymbol{\beta}_k) \sum_{i=1}^{n} (\dot{p}_i K_i - \dot{q}_i)\mathbf{x}_i, \quad k = 0, 1, 2, \ldots, \tag{8.62}$$

where derivatives $\dot{p}_i$ and $\dot{q}_i$ are evaluated at the kth iteration.

We use Example 8.52 to illustrate that (a) the general log-likelihood function (8.61) turns into (8.57), and (b) the general FS algorithm (8.62) turns into (8.58). (a) Indeed, the exponential distribution of Y_i used in Example 8.52 can be written as $e^{-y_i e^{\boldsymbol{\beta}'\mathbf{x}_i} + \boldsymbol{\beta}'\mathbf{x}_i}$, so we have $K(y_i) = y_i$, $p(u) = -e^u$, and $q(u) = -u$. Thus, function (8.61) turns into $\sum(-e^{\boldsymbol{\beta}'\mathbf{x}_i}Y_i + \boldsymbol{\beta}'\mathbf{x}_i) = \mathbf{r}'\boldsymbol{\beta} - \sum Y_i e^{\boldsymbol{\beta}'\mathbf{x}_i}$. (b) For the information matrix, we have $\ddot{q}_i = 0$ and, therefore, $\mathcal{I} = \mathbf{X}'\mathbf{X}$, so that (8.62) takes the form (8.58).

Three other types of GLM regressions with the density (8.60) will be studied in this section, the list with specific function of p and q is presented in Table 8.2 complementary to Table 6.2 from Section 6.8.2. Although the listed regressions can be estimated using the general framework outlined in this section, it is worth studying them separately. They are members of the same family, but they are quite different from the application standpoint.

8.8.2   Logistic and probit regressions for binary outcome

Logistic and probit regressions are used to model the dependent variable which takes binary (0 or 1), or binomial (count) values. First, we will be dealing with the most popular binary logistic regression, $Y_i \in \{0, 1\}$ with the probability defined as

$$\Pr(Y_i = 1) = \frac{e^{\boldsymbol{\beta}'\mathbf{x}_i}}{1 + e^{\boldsymbol{\beta}'\mathbf{x}_i}}. \tag{8.63}$$

The log-likelihood function may be derived from (8.61) with the help of Table 8.2 or directly since the outcome, $Y_i = y$, can be written compactly as

$$\Pr(Y_i = y) = \left(\frac{e^{\boldsymbol{\beta}'\mathbf{x}_i}}{1 + e^{\boldsymbol{\beta}'\mathbf{x}_i}} \right)^y \left(\frac{1}{1 + e^{\boldsymbol{\beta}'\mathbf{x}_i}} \right)^{1-y},$$

yielding

$$l(\boldsymbol{\beta}) = \mathbf{r}'\boldsymbol{\beta} - \sum_{i=1}^{n} \ln(1 + e^{\boldsymbol{\beta}'\mathbf{x}_i}),$$

where $\mathbf{r} = \sum_{i=1}^{n} Y_i \mathbf{x}_i = \sum_{Y_i=1} \mathbf{x}_i$ is considered fixed when maximizing the log-likelihood function. The Fisher information matrix and the asymptotic covariance matrix are given by

$$\mathcal{I}(\boldsymbol{\beta}) = \sum_{i=1}^{n} \frac{e^{\boldsymbol{\beta}'\mathbf{x}_i}}{(1 + e^{\boldsymbol{\beta}'\mathbf{x}_i})^2} \mathbf{x}_i \mathbf{x}_i', \quad \mathrm{cov}(\widehat{\boldsymbol{\beta}}) = \mathcal{I}^{-1},$$

respectively.

For the *probit regression* the probability of the outcome is derived through the standard normal cdf,

$$\Pr(Y_i = 1) = \Phi(\boldsymbol{\beta}'\mathbf{x}_i).$$

The probit regression is popular in toxicology studies (Finney [47], Demidenko [28], [30]) and can be justified by the following. Imagine that the harm of a toxin is measured on the continuous scale as variable H that can be modeled as a linear regression $H_i = \beta_{*0} + \boldsymbol{\beta}_*'\mathbf{x}_i + \varepsilon_i$, where ε_i are iid normally distributed with zero mean and variance σ^2. However H_i is not observable (latent variable). Instead, we know that if H_i exceeds a threshold c, the organism dies of the toxin. If Y_i is the observed death event we have $Y_i = 0$ if $H_i < c$ and $Y_i = 1$ if $H_i \geq c$. Then, the probability of death is $\Pr(Y_i = 1) = \Phi(\beta_0 + \boldsymbol{\beta}'\mathbf{x}_i)$ where $\beta_0 = (\beta_* - c)/\sigma$ and $\boldsymbol{\beta} = \boldsymbol{\beta}_*/\sigma$, as in the probit regression. The probit and logit are closely related: the absolute difference between $e^s/(1+e^s)$ and $\Phi(s/1.7)$ is less than 0.01. In other words, if the MLE coefficients in the logistic regression were $\widehat{\boldsymbol{\beta}}$, the coefficients in the probit regression will be roughly $\widehat{\boldsymbol{\beta}}/1.7$.

Example 8.53 *Constant logistic regression*. *Derive the ML estimator for the intercept in the constant logistic regression model.*

Solution. We have $\mathbf{x}_i = 1$ and $\Pr(Y_i) = e^{\beta_0}/(1+e^{\beta_0})$ with the log-likelihood function $l(\beta_0) = n_1\beta_0 - n\ln(1+e^{\beta_0})$, where n_i is the number of positive outcomes, $\sum_{i=1}^n Y_i$. Differentiating with respect to β_0 and equating to zero, we obtain $n_1 - ne^{\beta_0}/(1+e^{\beta_0}) = 0$ and finally, the MLE is $\widehat{\beta}_0 = \ln(n_1/(n - n_1))$. For example, if the number of positive and negative outcomes is the same ($n_1 = n - n_1$) the intercept is zero. The ML estimator makes perfect sense because it can be derived from equating n_1/n to the theoretical probability, following the idea of the method of moments (see Section 6.2).

Logistic regression with a binary predictor

The following is a generalization of the previous example which gives rise to the notion of the odds ratio, a popular interpretation of the regression coefficient with binary x_i. Note that $\mathbf{x}_i$ may be treated as a binary random variable and the probability (8.63) can be understood as the conditional probability, $\Pr(Y_i = 1 | \mathbf{X}_i = \mathbf{x}_i)$. See more discussion in Section 9.4.1, where two schemes are suggested: deterministic and stochastic. Importantly, the MLE does not change if the distribution of $\mathbf{x}_i$ is beta-independent. When both outcome and predictor/covariate are binary, the data can be treated as a two-by-two contingency table.

Example 8.54 *MLE for a two-by-two contingency table*. *(a) Find the ML estimator when x_i is binary. (b) Derive the 2×2 Fisher information matrix and the asymptotic variance of the slope estimate.*

Solution. Predictor x_i takes values 0 and 1. Define frequency $n_{1Y} = \sum_{i=1}^{n} Y_i$ as well as $n_{1x} = \sum_{i=1}^{n} x_i$ and $n_{0x} = \sum_{i=1}^{n}(1 - x_i)$ as the number of observations with $x_i = 0$ and $x_i = 1$, respectively. In addition, denote

$$n_{00} = \sum_{i=1}^{n}(1-Y_i)(1-x_i), \; n_{01} = \sum_{i=1}^{n}(1-Y_i)x_i, \; n_{10} = \sum_{i=1}^{n} Y_i(1-x_i), \; n_{11} = \sum_{i=1}^{n} Y_i x_i.$$

The probability model with a single predictor is given by $\Pr(Y_i = 1) = e^{\beta_0 + \beta_1 x_i} / (1 + e^{\beta_0 + \beta_1 x_i})$. The log-likelihood function with binary x is

$$l(\beta_0, \beta_1) = \beta_0 n_{1Y} + \beta_1 n_{11} - n_{0x} \ln(1 + e^{\beta_0}) - n_{1x} \ln(1 + e^{\beta_0 + \beta_1}).$$

(a) Differentiating with respect to β_0 and β_1 and setting to zero, we arrive at the system of equations to be solved for β_0 and β_1:

$$n_{1Y} - \frac{n_0}{1 + e^{\beta_0}} - \frac{n_{1x}}{1 + e^{\beta_0 + \beta_1}} = 0, \quad n_{11} - \frac{n_{1x}}{1 + e^{\beta_0 + \beta_1}} = 0.$$

The solution simplifies if expressed through variables $B_0 = e^{\beta_0}$ and $B_{01} = e^{\beta_0 + \beta_1}$:

$$n_{1Y} - \frac{n_0}{1 + B_0} - \frac{n_{1x}}{1 + B_{01}} = 0, \quad n_{11} - \frac{n_{1x}}{1 + B_{01}} = 0.$$

From the second equation derive $B_{01} = (n_{1x} - n_{11})/n_{11}$ and substitute into the first equation to finally obtain the ML estimators for the intercept and slope as

$$\widehat{\beta}_0 = \ln \frac{n_{1Y} - n_{11}}{n_{0x} + n_{11} - n_{1Y}}, \quad \widehat{\beta}_1 = \ln \frac{(n_{0x} + n_{11} - n_{1Y})n_{11}}{(n_{1Y} - n_{11})(n_{1x} - n_{11})}.$$

It is easy to see that the quantities under the ln sign are nonnegative. However, they may be zero. That is, the ML estimator may not exist with a positive probability. Rewrite the ML estimator for the slope coefficient as $\widehat{\beta}_1 = \ln \text{OR}$, where

$$\text{OR} = \frac{n_{00} n_{11}}{n_{01} n_{10}} \tag{8.64}$$

is called the odds ratio (OR). More explanation is found below.

(b) The 2×2 Fisher information matrix takes the form

$$\mathcal{I}(\beta_0, \beta_1) = \begin{bmatrix} a + b & b \\ b & b \end{bmatrix},$$

where $a = n_{0x} e^{\beta_0}/(1 + e^{\beta_0})^2$, $b = n_{1x} e^{\beta_0 + \beta_1}/(1 + e^{\beta_0 + \beta_1})^2$. Elementary algebra implies

$$\text{var}(\widehat{\beta}_1) = \frac{(1 + e^{\beta_0})^2}{n_{0x} e^{\beta_0}} + \frac{(1 + e^{\beta_0 + \beta_1})^2}{n_{1x} e^{\beta_0 + \beta_1}}.$$

Using expressions for the MLE, $\widehat{\beta}_0$ and $\widehat{\beta}_1$, we arrive at the following useful estimate of the slope variance

$$\text{var}(\widehat{\beta}_1) \simeq 1/n_{00} + 1/n_{01} + 1/n_{10} + 1/n_{11}.$$

Definition of OR in the language of the 2×2 contingency table

The logistic regression with binary predictor can be expressed as the following 2×2 contingency table.

	$Y = 0$	$Y = 1$
$x = 0$	$\widehat{p}_{00} = n_{00}/n$	$\widehat{p}_{01} = n_{01}/n$
$x = 1$	$\widehat{p}_{10} = n_{10}/n$	$\widehat{p}_{11} = n_{11}/n$

This table specifies probabilities of the joint distribution of two binary random variables as $p_{ij} = \Pr(Y = i, x = j)$ for $i, j \in \{0, 1\}$, and its estimates are expressed through the frequencies n_{ij} defined above. Define the odds as the ratio of the conditional probabilities in the two groups corresponding to $x = 1$ and $x = 0$ as

$$\text{Odd}_1 = \frac{\Pr(Y = 1 | x = 1)}{\Pr(Y = 0 | x = 1)} = \frac{\Pr(Y = 1, x = 1)/\Pr(x = 1)}{\Pr(Y = 0, x = 1)/\Pr(x = 1)} = \frac{p_{11}}{p_{10}},$$

$$\text{Odd}_0 = \frac{\Pr(Y = 1 | x = 0)}{\Pr(Y = 0 | x = 0)} = \frac{\Pr(Y = 1, x = 0)/\Pr(x = 0)}{\Pr(Y = 0, x = 0)/\Pr(x = 0)} = \frac{p_{01}}{p_{00}}.$$

The theoretical odds ratio is defined as the ratio of the odds,

$$\text{OR} = \frac{\text{Odd}_1}{\text{Odd}_0} = \frac{p_{00}p_{11}}{p_{01}p_{10}} \simeq \frac{n_{00}n_{11}}{n_{01}n_{10}}.$$

This is the OR formula (8.64) defined in the previous example on the basis of the ML logistic regression. This means that OR from the two-by-two table can be estimated by running the logistic regression and taking the exponent of the regression slope.

The language of OR is routinely used in epidemiology to measure the risk of developing a disease, such as lung cancer, associated with an exposure, such as smoking. This association was not clear until the fifties, mostly due to the work of Ronald Fisher who denied the harm of smoking, now an established fact. His opinion "was flawed by an unwillingness to examine the entire body of data available and prematurely drawn conclusions. His views may also have been influenced by personal and professional conflicts, by his work as a consultant to the tobacco industry, and by the fact that he was himself a smoker" as Stolley [97] put it. The following example illustrates the connection between developing lung cancer and smoking via the 2×2 table and OR.

Example 8.55 *Smoking and lung cancer. 770 lung cancer patients and 15,900 cancer-free people were asked if they smoke. Among the total 16,670 people, there were 4,500 smokers and 12,170 nonsmokers. The frequency data are represented below as a two-by-two table. Compute and interpret the odds ratio between smoking and developing lung cancer.*

	Cancer-free	Cancer	Sum
Nonsmoker	12,048	122	12,170
Smoker	3,852	648	4,500
Sum	15,900	770	16,670

Solution. First, consider smokers:

$$\text{Chance of getting cancer if you smoke} = p_{11} = \frac{648}{648 + 3852} = \frac{648}{4500},$$

$$\text{Chance of being cancer-free if you smoke} = p_{10} = \frac{3852}{648 + 3852} = \frac{3852}{4500}$$

with the odds as the ratio $\text{Odd}_{\text{smoker}} = \frac{p_{11}}{p_{10}} = \frac{648}{3852}$.

Now consider nonsmokers:

$$\text{Chance of getting cancer if you do not smoke} = p_{01} = \frac{122}{122 + 12048} = \frac{122}{12170},$$

$$\text{Chance of being cancer-free if you do not smoke} = p_{00} = \frac{12048}{122 + 12048} = \frac{12048}{12170}$$

with $\text{Odd}_{\text{nonsmoker}} = \frac{p_{01}}{p_{00}} = \frac{122}{12048}$.

Finally, the odds ratio of association between lung cancer and smoking is

$$\text{OR} = \frac{\text{Odd}_{\text{smoker}}}{\text{Odd}_{\text{nonsmoker}}} = \frac{p_{11}p_{00}}{p_{10}p_{01}} = \frac{648 \times 12048}{3852 \times 122} = 16.6.$$

A public message: the chance of developing lung cancer is almost 17 times higher if you smoke.

The glm function in R

The R function glm estimates GLM by maximum likelihood using the FS algorithm. The syntax and the output of glm is similar to lm with the only difference that it requires specification of the GLM family (see Table 8.1). Recall that statistical inference for linear model is exact. For GLM, we rely on the asymptotic theory. Consequently, the standard error of the coefficient is just an approximation; for small sample size, such approximation may be poor. The default family is gaussian (linear regression); family=binomial means logistic regression. For example, one may estimate the exponential regression from Example 8.52 using family=Gamma(link='log'). However, this call will fail because some shopping times are zero. One could replace them with small values, y[y==0]=0.0001 but the choice of *small* is questionable. Note that our coefficients have the opposite sign because the expected value of Y is the reciprocal of λ.

To run a probit regression, one has to use family=binomial(link="probit"). The dependent variable in the traditional logistic regression takes values either 0 or 1. Function `glm` can also be used when the dependent variable is the binomial outcome. Then the variable before ~ must have two columns: the first column must specify the number of positive outcomes (successes) and the second column must specify the number of negative outcomes (failures) following the language of the binomial distribution in Section 1.6. In the following example, we run `glm` to double check the OR derived from the contingency table using standard logistic (very inefficient) and binomial regression (very efficient).

Example 8.56 *Two-by-two contingency table. (a) Recast the 2×2 contingency table from Example 8.55 as (1) a logistic regression and (2) a binomial regression to confirm that OR = 16.6. (b) Test the null hypothesis that smoking has no effect on lung cancer.*

Solution. See the R function `lungsm`. (a) To run `glm` for logistic regression, we must represent the frequencies as two binary arrays for the dependent variable Y and predictor x of length 16670 (the sample size = 16670, very inefficient). For each cell, compose a binary matrix with two columns and the number of rows equal to the frequency of the cell. For example, for the first cell, we have yx11=cbind(rep(1,648),rep(1,648)) and for the second cell we have yx10=cbind(rep(1,122),rep(0,122)). Finally we create matrix yx of dimension 16670 by 2 using the `rbind` command. The output of the `glm` call is shown below.

```
lungsm()
Call:
glm(formula = yx[, 1] ~ yx[, 2], family = binomial)
Coefficients:
            Estimate Std. Error z value Pr(>|z|)
(Intercept) -4.59263    0.09099  -50.47   <2e-16 ***
yx[, 2]      2.81018    0.10041   27.99   <2e-16 ***
    Null deviance: 6239.3  on 16669  degrees of freedom
Residual deviance: 5075.3  on 16668  degrees of freedom
AIC: 5079.3
Number of Fisher Scoring iterations: 7
```

The format of the MLE coefficients is the same as for the linear model `lm`. The OR is computed as exp(2.81018)=16.61291. The deviance in the framework of GLM is twice the negative log-likelihood function suitable for hypothesis testing. The `Null deviance` is twice the negative log-likelihood function for the intercept model and `Residual deviance` is twice the negative log-likelihood function for the model in the call. The binomial regression is efficient because we specify only frequencies (the sample size = 2, very efficient): suc=c(648,122); fail=c(3852,12048) and x=c(1,0). The call is glm(cbind(suc,fail)~x,family =binomial) and yields the same result.

(b) To test the null hypothesis that all slope coefficients of logistic regression are zero, we take the difference between the deviances and use the fact that according to the likelihood ratio (LR) test this difference has a chi-square distribution with degree of freedom equal to the number of slope coefficients. In our case the null hypothesis is that the coefficient of the smoking variable is zero. The p-value is computed as `pchisq(6239.3-5075.3,df=1,lower.tail=F)` which results in `4.066332e-255`. Note that the F-test from Theorem 8.19 for linear regression is exact for any n while the LR test is asymptotic and has the type I error close to the nominal value for large n. The Akaiki information criterion (AIC) computed as the residual deviance plus twice the number of parameters in GLM; in our model AIC $= 5075.3 + 2 \times 2 = 5079.3$. The AIC is used to compare the quality of fit of different statistical models with different number of parameters penalized by the number of parameters. A model with small AIC has satisfactory fit (low deviance) and small number of parameters (parsimonious). $\qquad\square$

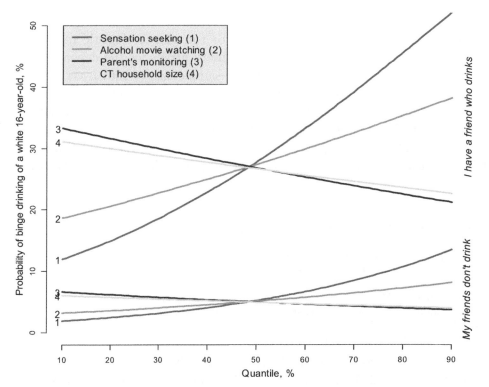

Figure 8.31: *Probability of binge drinking of a 16-year-old white individual as a function of four continuous predictors on the quantile scale.*

The next example studies the problem of adolescent drinking and identification of factors contributing to the initiation of drinking discussed earlier in Section 8.6.1. We use this example to illustrate how to interpret, graphically display, and compare the importance of continuous variables/predictors measured on different scales. We underscore that for linear models the regression coefficient, or the beta-coefficient, can be used to compare the strength of the association across variables. GLM is a nonlinear model and therefore graphical representation is more valuable.

Example 8.57 *Binge drinking among adolescents.* *File* kidsdrinkDAT.csv *contains data on binge drinking among 2,951 adolescents aged 13–19 along with the following variables/predictors: an indicator whether a kid's friend is a drinker, race (black = 1, white = 0), sensation seeking on the scale from 5 to 20, alcohol movie watching as described in Section 8.6.1, parent's monitoring on the scale from 0 to 15, and the census-tract average household size. (a) Run logistic regression, compute OR for binary predictors and provide an interpretation. (b) Display the effect of continuous variables on the probability of binge drinking using the quantile scale. (c) Plot the expected probability of binge drinking as a function of alcohol movie watching for an average white kid 16 years old and plot a 95% confidence band.*

Solution. (a) The R code in the file geodrink.r produces the output shown below. Since frndalc is binary, the OR is computed as $\exp(1.94614) = 7.0016$. It has the same interpretation as in the above example with a single binary predictor. In words, if a kid has a friend who drinks, the chance that he/she starts binge drinking increases by a factor of seven compared to the kid whose friends do not drink, assuming that other variables/predictor values are the same, specifically, age, race, sensation seeking, etc. As follows from this regression, the proportion of binge drinkers among black kids is smaller: $\exp(-1.87944) = 0.15268$, almost one seventh. We can compute the OR for age as $\exp(0.49956) = 1.648$ and interpret it as the increase chance of binge drinking per year.

```
geodrink()
Call:
glm(formula = bingedr ~ frndalc + age + black + senseek + alcmovie +
    parentmont + cthhsz, family = binomial, data = dat)
Coefficients:
             Estimate Std. Error z value Pr(>|z|)
(Intercept) -12.22016    0.87215 -14.012  < 2e-16 ***
frndalc       1.94614    0.26863   7.245 4.34e-13 ***
age           0.49956    0.03978  12.557  < 2e-16 ***
black        -1.87944    0.29272  -6.421 1.36e-10 ***
senseek       0.26059    0.01971  13.219  < 2e-16 ***
alcmovie      1.88290    0.25855   7.283 3.28e-13 ***
parentmont   -0.12338    0.02298  -5.369 7.91e-08 ***
cthhsz       -0.44289    0.11992  -3.693 0.000222 ***
    Null deviance: 3404.4  on 2950  degrees of freedom
Residual deviance: 2461.1  on 2943  degrees of freedom
  (702 observations deleted due to missingness)
```

(b) The interpretation of OR for continuous predictors is less clear: although we still can interpret the exponent of the regression coefficient as OR when comparing groups with x and $x + 1$, we cannot use this feature to compare the strength of association across predictors when they are measured on different scales. Simply put, an increase by 1 unit of measurement in one variable may be incomparable to an increase by 1 unit of measurement in another variable. To compare the effect of different continuous variables, we suggest displaying the probability of the event, such as binge drinking, on the quantile scale; see Figure 8.31. To display the probability on the quantile scale, we specify the race (`black=0`) and age (`age=16`). Two groups are shown: the lower group of kids has no friends who drink (`frndalc=0`) and the upper group does have friends who are drinkers (`frndalc=1`). Not surprisingly, the groups are very different because the OR $= 7$. For each continuous variable, we compute the 10th and 90th percentiles, say, $q_{.1}$ and $q_{.9}$ and plot the probability curve $e^{c+\widehat{\beta}x}/(1+e^{c+\widehat{\beta}x})$, where $q_{.1} \leq x \leq q_{.9}$, $\widehat{\beta}$ is the coefficient of the respective continuous variable, and c is the constant. For example, for the lower group (`frndalc=0`), when x is the `senseek` variable, we have $c = -12.22016 + 0.49956 \times 16 + 1.8829 \times \overline{alcmovie} - 0.12338 \times \overline{parentmont} - 0.44289 \times \overline{cthhsz}$. As follows from Figure 8.31, sensation seeking is the most influential variable: when an adolescent has a friend who drinks and is in the top 10% of sensation seekers the chance that he/she is a binge drinker is 50%. The second most important variable is alcohol movie watching: the chance of being a binge drinker for a adolescent who watches many alcohol movies (90th quantile) is about 10% higher than an average kid. Parent's monitoring and the number of people in the family are protective against binge drinking.

(c) See `geodrink(job=2)`. To compute the 95% confidence band for the probability, we compute the variance of $\widehat{\beta}'\mathbf{x}$ as $\mathbf{x}'\text{cov}(\widehat{\beta})\mathbf{x}$, where $\mathbf{x}$ is the 8×1 vector of predictors and $\text{cov}(\widehat{\beta})$ is extracted from the `glm` call as `vcov(o)` where o is the output (the plot is not presented here).

8.8.3 Poisson regression

Poisson regression models count data with the mean as the exponential function of a linear combination of predictors (see Section 1.7 for refreshing on the Poisson distribution). Specifically, the Poisson regression model specifies the probability that count Y_i takes value $0, 1, \dots$ with the rate parameter $\lambda_i = e^{\beta'\mathbf{x}_i}$, implying $E(Y_i) = e^{\beta'\mathbf{x}_i}$. The regression may or may not have an intercept term, but the $n \times m$ matrix $\mathbf{X}$, composed of vectors $\mathbf{x}_i$, must have full rank be able to identify parameters β uniquely. Since Y_i has a Poisson distribution $\text{var}(Y_i) = e^{\beta'\mathbf{x}_i}$. The log-likelihood function, up to a constant term, is $l(\beta) = \beta'\mathbf{r} - \sum_{i=1}^{n} e^{\beta'\mathbf{x}_i}$, where vector $\mathbf{r} = \sum_{i=1}^{n} Y_i\mathbf{x}_i$ may be treated as fixed when l is maximized. Typically, the FS algorithm requires only a few iterations to converge, with the Fisher information matrix estimated as $\mathbf{X}'\mathbf{E}\mathbf{X}$ where $\mathbf{E}$ is the $n \times n$ diagonal matrix with the ith diagonal element $e^{\widehat{\beta}'\mathbf{x}_i}$ and $\widehat{\beta}$ is the ML estimate. Statistical inference is based on the fact that $\widehat{\beta}_{ML} \simeq \mathcal{N}(\beta, (\mathbf{X}'\mathbf{E}\mathbf{X})^{-1})$. A larger sample size yields a better approximation. In R, Poisson regression is estimated using the `glm` function with the option `family=poisson`. The following example applies Poisson regression to traffic violations (tickets).

Example 8.58 *Poisson regression for traffic violations.* *An auto insurance company is interested in identifying demographic factors contributing to*

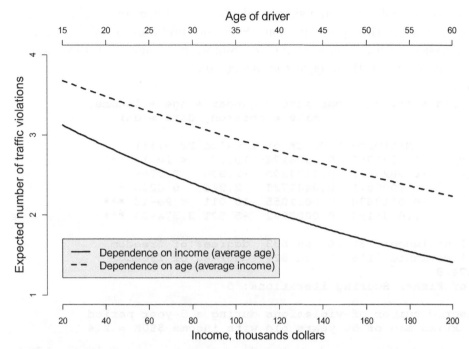

Figure 8.32: *The expected number of traffic violations as a function of income and age for a married male driver.*

traffic violations. File `Traffic.Viol.csv` *contains information on traffic violations (variable* `trviol`*) of 835 drivers over five years along with other socio-demographic variables including marriage status (1 if married, 0 otherwise), gender (male = 1, female = 0), age, and income (thousand dollars). (a) Run the Poisson regression and interpret the regression coefficient at* `trviol` *on the percent scale. (b) Estimate the expected number of violations during a five-year period for a married man of 50 years with income \$80K. (c) Display how age and income affect the expected frequency of traffic violations of a married man by plotting the expected number of violations using* `axis` *at sides 1 and 3.*

Solution. The R code is found in the file `poisR.r` and the Poisson regression output upon the call `poisR()` shown below. (a) Marriage status (slightly), age, and income protect from violations while males get more tickets than females on average. To interpret how gender affects the number of violations, we write the expected number of tickets for males and females as $e^{c+.1}$ and e^{c}, respectively, where c is the constant, common to both sexes representing shared predictors, such as being married, age and the same income (the same idea was used in Example 8.40 for the interpretation of the coefficient at the variable on the log scale). Then the percent violations for male compared to female is

$$\frac{e^{c+.1} - e^c}{e^c} \times 100\% = (e^{.1} - 1) \times 100\% = 10.517\% \simeq 10\%.$$

This approximation follows from a simple calculus, $e^x - 1 \simeq x$ for small x. Note that c does not affect the approximation. Such interpretation is valid for any Poisson regression with a binary predictor. Simply put, males get 10% more violations than females with other factors being equal.

```
> poisR()
glm(formula = trviol ~ marriage + gender + age + income,
                     family = poisson, data = da)
Coefficients:
              Estimate Std. Error  z value Pr(>|z|)
(Intercept)  1.7577394  0.0906176   19.397  < 2e-16 ***
marriage    -0.0620409  0.0474395   -1.308   0.1909
gender       0.1000811  0.0441733    2.266   0.0235 *
age         -0.0110670  0.0011055  -10.011  < 2e-16 ***
income      -0.0044124  0.0007992   -5.521 3.37e-08 ***

    Null deviance: 1309.6  on 833  degrees of freedom
Residual deviance: 1169.7  on 829  degrees of freedom
AIC: 3174.8
Number of Fisher Scoring iterations: 5

The expected number of violations during a 5-year period
for a married man of 50 years old with income $80K = 2.4
```

(b) To estimate the expected number of violations for a married man of 50 years and income $80K, we compute $\exp(1.7577394 - 0.0620409 + 0.1000811 - 50 \times 0.011067 - 80 \times 0.0044124) = 2.4338$. (c) The plot of the expected number of violations as a function of income and age of a married man is portrayed in Figure 8.32. When computing the dependence on income, we take the drivers' average age (51.5) and when computing the dependence on age, we take the average income ($74.3K). See function poisR to learn how to use axis with option side=1 and side=3. □

There is an interesting connection between Poisson regression and complementary log-log binary regression with the probability of a binary event Y_i defined as $\Pr(Y_i = 1) = 1 - e^{-e^{\beta' x_i}}$. This regression can be estimated by glm using the option family=binomial(link="cloglog") and is illustrated below as a continuation of Example 8.58.

```
> cloglogV()
glm(formula = trviol ~ marriage + gender + age + income,
            family = binomial(link = "cloglog"), data = da0)
Coefficients:
             Estimate Std. Error  z value Pr(>|z|)
(Intercept)  1.826734   0.199200    9.170  < 2e-16 ***
marriage    -0.071897   0.099119   -0.725  0.46823
gender       0.071417   0.091391    0.781  0.43454
age         -0.014941   0.002347   -6.365 1.95e-10 ***
income      -0.004590   0.001517   -3.025  0.00249 **
    Null deviance: 697.75  on 833  degrees of freedom
Residual deviance: 641.98  on 829  degrees of freedom
```

Example 8.59 *No traffic violations records. Assume that in the file Traffic.Viol.csv instead of the number of traffic violations the responders provided a binary answer if they had at least one traffic violation. Run binary regression with the link "cloglog" to see how close the results are to Poisson regression.*

Solution. The same data set is used as in the previous example but with column `trviol` replaced with `da0$trviol[da0$trviol>0]=1`, see the output above (the R code is in the file `cloglogV.r`). The coefficients are close to the Poisson regression from Example 8.58 although the *p*-value of the coefficient at `gender` is above the traditional threshold 0.05. Not surprisingly, there is some loss of information if the number of traffic violations is reduced to a binary variable.

Problems

1. Prove that the log-likelihood function $l(\beta)$ in Example 8.52 is concave, assuming that $Y_i > 0$ and matrix $\mathbf{X}$ has full rank. [Hint: Prove that the Hessian matrix is negative definite, see the Optimization section of the Appendix.]

2. Compute and display the 95% confidence band for the proportion of female shoppers (the right-hand plot in Figure 8.30).

3. Use the Wald and likelihood-ratio tests in Example 8.52 to test whether gender and age are statistically significant, $H_0 : \lambda_i = \lambda$ (compute the *p*-values).

4. Estimate the time of shopping from Example 8.52 using `glm` with an appropriate `family`. Explain why the results are different.

5. Use the FS algorithm (8.62) to estimate the logistic regression coefficients and their standard errors from Example 8.55 and compare your results with the `glm` output.

6. Sometimes epidemiologists use the relative risk (RR) ratio defined as $\Pr(Y = 1|x = 1)/\Pr(Y = 1|x = 0)$ in contrast to OR for easy interpretation. (a) Compute and provide an interpretation of the RR for Example 8.55. (b) Express RR as a function of OR, β_0 and $p = \Pr(x = 1)$, and plot RR versus OR for several β_0 letting $p = 1/2$.

7. Apply the probit regression to the two-by-two table from Example 8.55 using `glm` and compare the results using the conversion coefficient 1.7 for the ML estimates and their standard errors.

8. The Weibull distribution is a generalization of the exponential distribution and is often used in survival analysis and engineering applications. Its cdf is $F(x; \lambda, k) = 1 - e^{-(\lambda x)^k}$, where λ and k are positive parameters, called *scale*

and *shape*, respectively. When $k = 1$, the Weibull distribution reduces to an exponential distribution, but for $k > 1$, its mode is positive unlike the exponential distribution. The moment generating function of the Weibull distribution is given by $E(e^{t \ln X}) = \lambda^t \Gamma(t/k + 1)$ with the expected value $\lambda \Gamma(1/k+1)$. Apply ML to fit the shopping time from Example 8.52 using (a) an unadjusted model, (b) a model adjusted for age and gender, assuming that the time of shopping follows a Weibull distribution with unknown but fixed shape. [Hint: Assume that λ_i depends on age and gender as in Example 8.52.]

9. Derive the ML estimator and its asymptotic variance for the intercept in closed form in the probit model as in Example 8.53.

10. The file `amazshop.csv` contains the following information on 1000 Amazon.com shoppers: `shop`=1 indicates the internal definition of active shoppers and 0 otherwise, `age, sex`=1 codes male and 0 female, `total` contains the dollar amount spent during the year, and `npurch` indicates the number of items sold. (a) Run logistic regression on the four variables and develop a parsimonious model by selecting only statistically significant variables. (b) Test the hypothesis on the validity of the logistic regression using the likelihood ratio test by testing that all slope coefficients are zero. (c) Plot the probability to be qualified as an active shopper versus total amount for the shopper of age 20 and age 60. (d) Compute and display the 95% confidence interval for the probability of a person who spends $2000 per year and is 60 years old.

11. Derive the MLE for Poisson regression with a single binary predictor as in Example 8.54.

12. In Example 8.58, (a) is it true that people older by 10 years are getting 11% fewer traffic violations? (b) exclude variable `marriage`, rerun the regression and redo the figure, (c) test the null hypothesis that all regression coefficients are zero, (d) compute the 95% confidence interval for the expected number of traffic violations of a single 30-year-old woman with income of $50K.

Chapter 9

Nonlinear regression

Nonlinear regression is a powerful statistical tool rarely covered in traditional statistics textbooks. Nonlinear regression is not only a practically important technique, but also an important example of a real-life statistical model where classical theory of unbiased estimation and sufficient statistics do not work. Undoubtedly, the linear model is the champion among statistical techniques when it comes to modeling relationships between variables. However, sometimes the association is not linear such as when the response has a sigmoid shape – then nonlinear regression must be applied. Unlike linear regression, nonlinear regression is a complex statistical model where small sample properties are difficult to study – here, we rely on asymptotic properties. The major method of estimation is the nonlinear least squares, which, unlike linear least squares, requires iterations. Various numerical issues arise in the nonlinear regression model: (1) finding satisfactory starting value, (2) existence of the solution of the nonlinear least squares, (3) multiple local minima for the residual sum of squares. This chapter covers major concepts of nonlinear regression, illustrated with various examples, and its implementation in R. The chapter ends with the design of experiments emerging in engineering or chemical sciences where the values of the independent variable may be chosen by the experimentalist.

9.1 Definition and motivating examples

Nonlinear regression is an obvious generalization of linear regression (linear model). The main difference is that, in nonlinear regression, the expected value of the dependent variable is a nonlinear function of parameters. For example, quadratic regression

$$Y_i = \beta_0 + \beta_1 x_i + \beta_2 x_i^2 + \varepsilon_i$$

Advanced Statistics with Applications in R, First Edition. Eugene Demidenko.
© 2020 John Wiley & Sons, Inc. Published 2020 by John Wiley & Sons, Inc.

is linear with respect to parameters and therefore can be estimated by the linear least squares applied to the linear model after a change of variable, $z_i = x_i^2$. Regression models that can be reduced to a linear model by introducing new variables are called *curvilinear models*. A model

$$Y_i = g(\beta_0 + \beta_1 x_i + \beta_2 z_i) + \varepsilon_i, \tag{9.1}$$

where g is a monotonic function is called *quasilinear*, and cannot be reduced to a linear model; see Example 9.1 below.

Intrinsically linear models are nonlinear regression models that can be reduced to a linear model upon reparametrization:

$$Y_i = g_1(\beta_1, \beta_2) x_i + g_1(\beta_1, \beta_2) z_i + \varepsilon_i, \tag{9.2}$$

where g_1 and g_2 are one-to-one functions. Indeed, letting $\gamma_1 = g_1(\beta_1, \beta_2)$ and $\gamma_2 = g_2(\beta_1, \beta_2)$ we reduce the model to linear. *Intrinsically nonlinear models* are those that cannot be reduced to a linear model by reparametrization.

If $\mathbf{x}_i$ denotes a vector of independent/explanatory variables/predictors, nonlinear regression is defined as

$$Y_i = f(\mathbf{x}_i; \boldsymbol{\beta}) + \varepsilon_i, \quad i = 1, 2, ..., n, \tag{9.3}$$

where $\{\varepsilon_i\}$ are (unobservable) uncorrelated random terms with zero mean and constant variance σ^2. Function f is called the *regression function*, and sometimes the *response function*. Linear regression is a special case of (9.3) with $f(\mathbf{x}_i; \boldsymbol{\beta}) = \boldsymbol{\beta}'\mathbf{x}_i$, however, in nonlinear regression we do not even require the dimensions of $\boldsymbol{\beta}$ and $\mathbf{x}_i$ to match. Moreover, we usually use the notation

$$f_i(\boldsymbol{\beta}) = f(\mathbf{x}_i; \boldsymbol{\beta}) \tag{9.4}$$

to shorten the formulas; then we write

$$Y_i = f_i(\boldsymbol{\beta}) + \varepsilon_i.$$

This notation means that nonlinear regression functions f_i may be any and are not necessary associated with predictors. Usually, $f_i(\boldsymbol{\beta})$ are continuous and differentiable functions, but not always, as we shall see later. We assume that the m-dimensional parameter vector $\boldsymbol{\beta}$ belongs to the entire R^m or a convex m-dimensional parameter set, $\boldsymbol{\Theta} \subset R^m$. To estimate parameters uniquely, we need to assume that the nonlinear regression is *identifiable:* $f_i(\boldsymbol{\beta}_1) = f_i(\boldsymbol{\beta}_2)$ for all $i = 1, 2, ..., n$ implies $\boldsymbol{\beta}_1 = \boldsymbol{\beta}_2$.

Sometimes the regression function is nonlinear, as in the case of quasilinear regression (9.1), but can be reduced to a linear function by taking transformation g^{-1} of both sides of the regression equation. The example below underscores how the statistical model changes upon such transformation. Consequently, linear least squares applied to $g^{-1}(Y_i)$ does not produce the same estimates produced by the original nonlinear regression model.

Example 9.1 *Estimation of an exponential model.* *The profit of a company grows following an exponential trend $y = e^{\alpha + \beta t}$ over the years, with some variation. Compare two regression models: (a) the error term is multiplicative, $\ln Y_t = \alpha + \beta t + \delta_t$, and (b) the error term is additive, leading to the quasilinear regression, $Y_t = e^{\alpha + \beta t} + \varepsilon_i$, where both error terms are iid normally distributed with zero mean.*

Solution. Model (a) implies that Y_i has a lognormal distribution with $E(Y_t) = e^{\alpha + \beta t + \sigma^2/2}$ and $\text{var}(Y_t) = (e^{\sigma^2} - 1)e^{2(\alpha + \beta t) + \sigma^2}$, which implies that the ratio of the SD to the mean is constant. It reduces to a linear model after taking the log transformation. On the other hand, model (b) assumes a constant SD because the error is additive – see Figure 9.1. This model is quasilinear with $g = \exp$. The multiplicative error model implies that the variation is proportional to the trend value, whereas the additive error model has constant variation (SD). The choice is not obvious but it is well confirmed that real-life data often follows the lognormal distribution (Section 2.11) and therefore the multiplicative error model is more sound. □

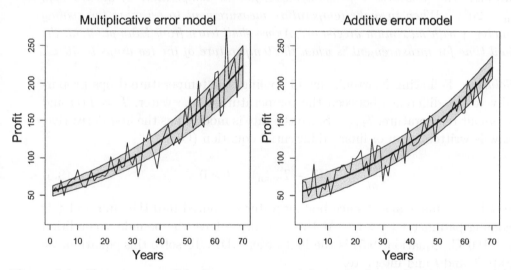

Figure 9.1: *Two error models for an exponential trend: the multiplicative model implies that variation of profit is proportional to the trend but the additive model has a constant variation of profit around the trend.*

When dealing with nonlinear regression, the choice of f is always a big deal. Typically, nonlinear regression emerges in science and engineering where physical laws dictate the relationship. In those instances the regression function is the solution to the underlying differential equation. Below are two motivating examples.

Example 9.2 *Temperature in a tea cup.* *Derive the model of the temperature drop in the tea cup after pouring in boiled water. Provide an interpretation of* *the parameters and find the time of comfortable drinking when the temperature of the tea is 50°C. Set up the nonlinear regression for parameter estimation.*

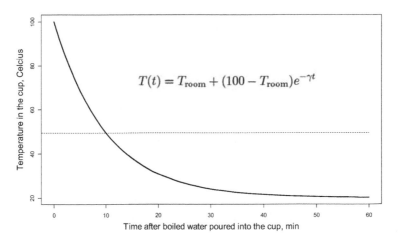

Figure 9.2: *The nonlinear regression model for the temperature of the tea cup, $T_{\text{room}} = 20° C$. When to make temperature measurement to estimare the cooling parameter γ with maximum precision? As we shall learn from Example 9.33, the optimal time for measurement is when the temperature of the tea drops to 70°C.*

Solution. Following Newton's law of cooling, the temperature drops proportionally to the difference between the temperature of the water, $T = T(t)$ and the room air temperature, T_{room}. Since the rate is modeled as the first derivative, this law is written as an ordinary differential equation (ODE),

$$\frac{dT}{dt} = -\gamma(T - T_{\text{room}}), \quad t \geq 0, \tag{9.5}$$

where t is the time elapsed since boiling water is poured into the cup, and γ is the cooling parameter ($\gamma > 0$). At time $t = 0$, we have the boundary condition $T(0) = 100°C$. Equation (9.5) is the first-order ODE. To solve this equation, we separate T and t and integrate,

$$\int \frac{dT}{T - T_{\text{room}}} = -\gamma \int dt,$$

yielding $T(t) = T_{\text{room}} + e^{-\gamma t + C}$, where C is a constant. Taking into account the boundary condition, we find $e^C = 100 - T_{\text{room}}$, that is, that the final model for the temperature of the water in the cup takes the form

$$T(t) = T_{\text{room}} + (100 - T_{\text{room}})e^{-\gamma t}, \tag{9.6}$$

where parameter γ is subject to estimation; see Figure 9.2.

Of course, this model is a simplification. The volume, the shape of the cup and the material it is made of can make a difference, but we ignore those details.

Now we can give a more precise interpretation of parameter γ as the rate (the relative decrease) at which the temperature drops in the cup,

$$\gamma = -\frac{dT/dt}{T - T_{\text{room}}}.$$

Model (9.6) can answer the question, when the tea is comfortable to drink? Say the comfortable temperature (not too hot and not too cold) is $T_{\text{comfort}} = 50°C$; then the optimal time is

$$t_{\text{opt}} = \frac{1}{\gamma} \ln \frac{100 - T_{\text{room}}}{T_{\text{comfort}} - T_{\text{room}}}.$$

Now we set up the nonlinear regression. Let a temperature sensor be available (it is important that this sensor does not affect the temperature of the water in the cup) and let n measurements of the water at times $t_1, t_2, ..., t_n$ be taken. Due to unavoidable measurement error, the model takes the form of a nonlinear regression model

$$T_i = T_{\text{room}} + (100 - T_{\text{room}})e^{-\gamma t_i} + \varepsilon_i, \qquad (9.7)$$

where $E(\varepsilon_i) = 0$ and $\text{var}(\varepsilon_i) = \sigma^2$. If the temperature in the room is not known, we have to estimate two parameters, T_{room} and γ, so that the regression function is

$$f_i(T_{\text{room}}, \gamma) = T_{\text{room}} + (100 - T_{\text{room}})e^{-\gamma t_i}.$$

It is a great deal of interest to find time measurements $\{t_i, i = 1, 2, ..., n\}$ that imply the most efficient estimates of the parameters – this is a topic of the optimal design of experiments to be discussed later; see Example 9.33.

Example 9.3 *Change-point regression: The DNA repair threshold. Exposure of the cellular DNA to ionizing radiation inflicts various types of damage resulting in reduced cell survival. Cells can repair DNA at a low radiation dose, but fail beyond a threshold. To estimate the threshold, cell experiments have been conducted with increasing doses of radiation and measured survival in vitro, see Figure 9.3. Suggest a nonlinear regression model for cell survival as a function of given radiation.*

Solution. The survival time (time to death) is skewed to the right and, in accordance with the exponential survival function (Section 5.1.2) can be modeled on the log scale. The most convenient is the log base 10 for ease of reading: 2 is interpreted as 100, -1 as 0.1, etc. Let L be the survival time on the $\log_{10}$ scale of a typical cell exposed to a single dose of radiation r at time 0. If radiation is low, it is expected that $L = \alpha - \beta r$, where α is the life expectancy (hours on the $\log_{10}$ scale) of an exposed cell able to withstand the radiation, and $\beta > 0$ is the survival rate. It is known that beyond a threshold r_T the survival rate drops by

$\delta > 0$. That is, the change-point relationship between survival time and radiation takes the form

$$L = L(r; \alpha, \beta, \delta, r_T) = \begin{cases} \alpha - \beta r \text{ if } r < r_T \\ \alpha - \beta r - \delta(r - r_T) \text{ if } r \geq r_T \end{cases},$$

where neither α, β, δ, or r_T is known.

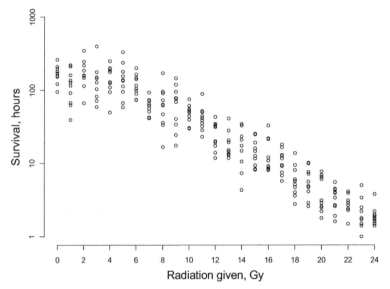

Figure 9.3: *Change-point regression: the cell survival time as a function of a single radiation exposure. The DNA damage can be repaired at low dose. Can the threshold (change-point) be identified?*

Obviously, the L function can be represented more compactly as

$$L = \alpha - \beta r - \delta \max(r - r_T, 0).$$

Question: given a series of *in vitro* experiments $\{(r_i, Y_i), i = 1, 2, ..., n\}$, where Y_i is the measured survival time, how do we estimate the involved parameters α, β, δ and particularly r_T as the threshold (change-point)? Since survival time is a subject of measurement error and, most importantly, the cell-specific response to radiation exposure, there should be inevitable deviation of measurements from the model leading to a nonlinear regression

$$Y_i = \alpha - \beta r_i - \delta \max(r_i - r_T, 0) + \varepsilon_i, \quad i = 1, 2, ..., n, \tag{9.8}$$

where $\{\varepsilon_i, i = 1, 2, ..., n\}$ are random uncorrelated errors with zero mean and constant variance. The change-point parameter r_T is central to the estimation problem and is intrinsically nonlinear. Indeed, if r_T were known, the regression could be reduced to a linear regression with three coefficients,

$$Y_i = \alpha - \beta r_i - \delta M_i + \varepsilon_i$$

where $M_i = \max(r_i - r_T, 0)$. Model (9.8) is a particular nonlinear regression model for which the standard `nls` function does not work. More details are found in Example 9.13. $\qquad\square$

The ancient Greek philosopher Aristotle (384–322 BC) asserted that the velocity of a free-falling object is proportional to its mass. It took almost two thousand years to put the Aristotelian theory to the test. The great Italian scientist, Galileo Galilei (1564–1642) demonstrated, contrary to the Aristotelian physics, that two heavy objects of different masses dropped from the Leaning Tower of Pisa struck the ground almost at the same time. He continued the experiments with balls rolling down ramps and meticulously measured "a full hundred times" until he had achieved "an accuracy such that the deviation between two observations never exceeded one-tenth of a pulse beat." By itself, that was an amazing achievement because there was no clock at the time. Those experiments led him to conclude that "free fall was uniformly accelerated motion." To many, Galileo Galilei was the first scientist; to me, he was the first statistician because he collected the data, analyze them and drew a conclusion. Born the same year Galileo died, Isaac Newton (1642–1726) suggested a formula for the distance covered by a free-falling body,

$$S(t) = \frac{1}{2}gt^2, \quad t \geq 0, \tag{9.9}$$

where g is a constant close to 9.8 m/s^2, as a part of his three-law theory of mechanics. Although his theory was first confirmed by an orbiting planet, formula (9.9) contradicts common sense: according to this formula feather and metal ball fall at the same speed. That is why the Aristotelian teaching dominated humankind's thinking for over two thousand years! Only later experiments in vacuum tube and, finally, a video-recorded experiment on the moon by the Commander astronaut David Scott in 1971, who dropped a hammer and a feather from the same height and they hit the ground at the same time (see Wikipedia) confirmed Newton's formula.

The reason Newton's formula does not work on Earth is because of the air resistance. Several models for free-falling to adjust for the air resistance have been suggested over the years. One of those, used in the following example, leads to a nonlinear regression.

Example 9.4 *Free-fall equation: Aristotle or Galilei?* Allmaraz et al. [4] *suggested the model, hereafter referred to as the C-model,*

$$S(t; C) = \frac{1}{C} \ln \frac{e^{t\sqrt{gC}} + e^{-t\sqrt{gC}}}{2}, \quad t \geq 0 \tag{9.10}$$

to describe the distance covered by a free-falling object, where $C \geq 0$ is an object-specific constant closely related to the ratio of its volume to mass, and $g = 9.7935$ m/s^2 is Earth's gravity (acceleration) constant. Prove that this model combines Aristotle and Newton's law, and set up a nonlinear regression to estimate C.

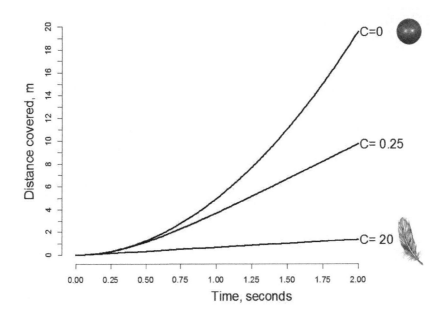

Figure 9.4: *Distance covered by free-falling objects. The lead ball has $C = 0$ and the distance covered follows the Newtonian gravitation law, $0.5gt^2$. The feather has a large $C = 20$ and covers only about a meter after 2 seconds of free fall while a regular object with $C = 0.25$ covers about 9 meters.*

Solution. We intentionally indicate in (9.10) that S depends on parameter C because it will a subject of estimation. First, we show that when $C \to 0$ (there is no air resistance or the mass of the object is infinitely large) (9.10) converges to the Newton's law (9.9). Apply the L'Hôpital's rule,

$$\lim_{C \to 0} \frac{1}{C} \ln \frac{e^{t\sqrt{gC}} + e^{-t\sqrt{gC}}}{2} = t\sqrt{g} \lim_{C \to 0} \frac{1}{2\sqrt{C}} \frac{e^{t\sqrt{gC}} - e^{-t\sqrt{gC}}}{e^{t\sqrt{gC}} + e^{-t\sqrt{gC}}}.$$

Now expand $e^{t\sqrt{gC}}$ and $e^{-t\sqrt{gC}}$ into Taylor series,

$$e^{t\sqrt{gC}} = 1 + t\sqrt{g}\sqrt{C} + \frac{1}{2}t^2 gC + o(C), \quad e^{-t\sqrt{gC}} = 1 - t\sqrt{g}\sqrt{C} + \frac{1}{2}t^2 gC + o(C),$$

and, finally, obtain

$$\lim_{C \to 0} \frac{1}{C} \ln \frac{e^{t\sqrt{gC}} + e^{-t\sqrt{gC}}}{2} = t\sqrt{g} \lim_{C \to 0} \frac{1}{2\sqrt{C}} \frac{e^{t\sqrt{gC}} - e^{-t\sqrt{gC}}}{e^{t\sqrt{gC}} + e^{-t\sqrt{gC}}}$$

$$= t\sqrt{g} \lim_{C \to 0} \frac{2t\sqrt{g}\sqrt{C}}{4\sqrt{C}} = \frac{1}{2}gt^2.$$

Second, we investigate how the object with $C > 0$ flies after some period of time, or more specifically find the speed of the object when $t \to \infty$,

$$\lim_{t \to \infty} \frac{S(t)}{t} = \lim_{t \to \infty} \frac{dS(t)}{dt} = \frac{\sqrt{gC}}{C} \lim_{t \to \infty} \frac{e^{t\sqrt{gC}} - e^{-t\sqrt{gC}}}{e^{t\sqrt{gC}} + e^{-t\sqrt{gC}}} = \sqrt{\frac{g}{C}}.$$

This means that, on Earth, a free-falling object reaches its maximum speed, $\sqrt{g/C}$ and does not accelerate any more – this is what Aristotle taught. Figure 9.4 illustrates formula (9.10) and the two limiting cases. The feather has a very large C and, therefore, after dropping, quickly reaches its linear trajectory and parachutes down slowly. On the other hand, a lead ball has C close to zero and drops according to the Newton's law.

Now investigate another extreme case, when $t \simeq 0$. Find the acceleration of free-falling objects on Earth as the second derivative of the distance (9.10) evaluated at $t = 0$,

$$\frac{d^2 S}{dt^2}\bigg|_{t=0} = \frac{4ge^{-2t\sqrt{gC}}}{\left(1 + e^{-2t\sqrt{gC}}\right)^2}\bigg|_{t=0} = g.$$

This means that when the fall starts, it's kinetics follows Newton's law (9.9) regardless of the geometry of the object. In short, when it comes to free-falling objects on Earth, Aristotle was right at infinity and Galileo was right at zero.

For any free-fall data as measurements of time t_i and the distance covered by this time, S_i, we have to estimate an object-specific parameter C in the nonlinear regression

$$S_i = \frac{1}{C} \ln \frac{e^{t_i\sqrt{gC}} + e^{-t_i\sqrt{gC}}}{2} + \varepsilon_i. \tag{9.11}$$

The problem of estimating the C-model is considered later in Example 9.18.

Problems

1. The Cobb–Douglas production function relates an output, such as revenue, from inputs, such as labor and capital, in the form $Y = AK^\alpha L^{1-\alpha}$. Set up a nonlinear regression and repeat Example 9.1 by reproducing Figure 9.1.

2. Consider parameter estimation in Example 9.2. (a) Suppose that all n measurements of the temperature in the cup are done at the same time.

Suggest an estimator of parameter γ when T_{room} is known. (b) Suppose that n measurements of the temperature in the cup are done at time t_* and $2t_*$. Suggest estimators of parameters γ and T_{room}.

3. The data for Example 9.3 is in the file dnaRAD.csv. Suggest and implement a method for estimation of the threshold by fitting two linear regressions. [Hint: The fitting model is not continuous.]

4. Write an R program to reproduce Figure 9.4.

9.2   Nonlinear least squares

Nonlinear least squares (NLS) is an obvious generalization of linear least squares and minimizes the residual sum of squares (RSS),

$$S(\boldsymbol{\beta}) = \sum_{i=1}^{n} (Y_i - f_i(\boldsymbol{\beta}))^2. \tag{9.12}$$

The minimizer, $\widehat{\boldsymbol{\beta}} \in \boldsymbol{\Theta}$ is called the nonlinear least squares estimate of $\boldsymbol{\beta}$. We can express the sum of squares in vector form through

$$\mathbf{y} = \begin{bmatrix} Y_1 \\ Y_2 \\ \vdots \\ Y_n \end{bmatrix}, \ \mathbf{f}(\boldsymbol{\beta}) = \begin{bmatrix} f_1(\boldsymbol{\beta}) \\ f_2(\boldsymbol{\beta}) \\ \vdots \\ f_n(\boldsymbol{\beta}) \end{bmatrix},$$

so that the regression function may be viewed as a map from $\boldsymbol{\Theta}$ to R^n, $\mathbf{f} : \boldsymbol{\Theta} \to R^n$. Since we assume that the nonlinear regression is identifiable, the mapping $\mathbf{f}$ is one-to-one: different points from $\boldsymbol{\Theta} \subset R^m$ are mapped into different points in R^n.

In the vector notation, the nonlinear regression can be compactly written as

$$\mathbf{y} = \mathbf{f}(\boldsymbol{\beta}) + \boldsymbol{\varepsilon}$$

with the RSS given by

$$S(\boldsymbol{\beta}) = \|\mathbf{y} - \mathbf{f}(\boldsymbol{\beta})\|^2, \tag{9.13}$$

the squared Euclidean norm. In geometric language, $S(\boldsymbol{\beta})$ is the squared distance between the data vector $\mathbf{y}$ and regression (or response) surface $\mathbf{f}(\boldsymbol{\beta})$. In Figure 9.5, we illustrate the nonlinear least squares geometrically, assuming that the number of observations, $n = 3$, and the number of unknown parameters (the dimension of vector $\boldsymbol{\beta}$), $m = 2$. The regression surface is defined as $\mathbf{f} : R^2 \to R^3$ and parametrized as $\mathbf{f} = \mathbf{f}(\boldsymbol{\beta})$. The nonlinear least squares solution is the point

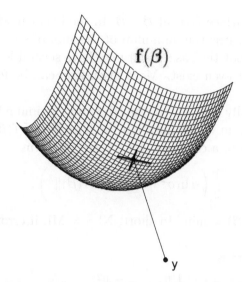

f(β)

y

Figure 9.5: *The geometry of the nonlinear least squares. The NLS finds a point* $\mathbf{f}(\widehat{\beta})$ *on the m-dimensional surface* $\mathbf{f}(\beta)$ *closest to the data point,* $\mathbf{y}$. *The residual vector* $\mathbf{y} - \mathbf{f}(\widehat{\beta})$ *is orthogonal to the tangent plane (cross) as follows from the normal equation (9.14).*

on the surface $\mathbf{f}$ that is the closest to data vector $\mathbf{y}$. Recall that for the linear model surface $\mathbf{f}(\beta)$ will be a plane spanned by predictors $\mathbf{x}_1, ..., \mathbf{x}_m$. When $m = 1$ the regression surface turns into a curve parametrized by $\beta_1 = \beta$.

The RSS attains the minimum where the vector of first derivatives $\partial S(\beta)/\partial\beta$ turns zero. Thus, we arrive at the definition of the *normal equation*, which can be written in matrix form as

$$\mathbf{J}'(\beta)(\mathbf{y} - \mathbf{f}(\beta)) = \mathbf{0}, \tag{9.14}$$

where $\mathbf{J}$ is the *Jacobian*, the $n \times m$ matrix of the derivatives of the regression function,

$$\mathbf{J} = \frac{\partial \mathbf{f}}{\partial \beta} = \begin{bmatrix} \frac{\partial f_1}{\partial \beta_1} & \frac{\partial f_1}{\partial \beta_1} & \cdots & \frac{\partial f_1}{\partial \beta_m} \\ \frac{\partial f_2}{\partial \beta_1} & \frac{\partial f_2}{\partial \beta_2} & \cdots & \frac{\partial f_2}{\partial \beta_m} \\ \vdots & \vdots & \ddots & \vdots \\ \frac{\partial f_n}{\partial \beta_1} & \frac{\partial f_n}{\partial \beta_2} & \cdots & \frac{\partial f_n}{\partial \beta_m} \end{bmatrix} = \begin{bmatrix} \frac{\partial \mathbf{f}}{\partial \beta_1}, & \frac{\partial \mathbf{f}}{\partial \beta_2}, & \cdots, & \frac{\partial \mathbf{f}}{\partial \beta_m} \end{bmatrix},$$

where $\partial \mathbf{f}/\partial \beta_j$ is the jth column-vector of the Jacobian ($j = 1, ..., m$). We will assume that $\mathbf{J}$ is a matrix of full rank (the vector columns are linearly independent). For linear model, $\mathbf{J} = \mathbf{X}$ and does not depend on β. The normal equation (9.14) is a necessary first-order condition for the minimum of the RSS. It means that, if $\widehat{\beta}$ is the NLS estimate, then the residual vector $\mathbf{y} - \mathbf{f}(\widehat{\beta})$ is orthogonal to

the tangent plane of the surface $\mathbf{f}(\boldsymbol{\beta})$ at $\boldsymbol{\beta} = \widehat{\boldsymbol{\beta}}$. In the linear model, the solution
to the normal equation returns the minimum of the sum of squares. For a non-
linear model this may be not the case: $S(\boldsymbol{\beta})$ may have several local minima and
the NLS estimate may not even exist. More discussion can be found in Section
9.7.

When errors are normally distributed, the NLS and maximum likelihood (ML)
estimates coincide. Indeed, in vector notation, we have $\mathbf{y} \sim \mathcal{N}(\mathbf{f}(\boldsymbol{\beta}), \sigma^2 \mathbf{I})$, and the
log-likelihood function, up to a constant term, takes the form

$$l(\boldsymbol{\beta}) = -\frac{1}{2}\left(n \ln \sigma^2 + \frac{1}{\sigma^2} \|\mathbf{y} - \mathbf{f}(\boldsymbol{\beta})\|^2 \right). \tag{9.15}$$

Clearly, $l = \max$ when $S(\boldsymbol{\beta}) = \min$. In short, NLS = ML if errors are normally
distributed.

The ML estimator of σ^2 is

$$\widehat{\sigma}^2 = \frac{1}{n}\left\| \mathbf{y} - \mathbf{f}(\widehat{\boldsymbol{\beta}}) \right\|^2.$$

One may use $n - m$ in the denominator, but, unlike the linear model, it does not
make the $\widehat{\sigma}^2$ unbiased.

Even if Y_i are independent and normally distributed, the NLS estimator is
not efficient for finite n, unlike linear model. Studying small sample properties
for nonlinear regression is a difficult task; see Section 9.4.

Minimization of the sum of squares (9.12) does not admit a closed-form solution. Instead, iterative algorithms should be used. Sometimes, when models are
complicated, algorithms do not converge – the starting values of the parameter
vector are important.

Problems

1. The circle nonlinear regression ($n = 2$) is defined as $f_1(\theta) = \cos\theta$ and
   $f_2(\theta) = \sin\theta$. (a) Find the NLS estimate of θ given $Y = (Y_1, Y_2)$ directly
   without differentiation by letting $\tau = a\cos(Y_1/\sqrt{Y_1^2 + Y_2^2})$. (b) Plot the
   RSS for two random $\mathbf{y}$ on the same graph and prove that the RSS is bimodal:
   it has one local minimum and maximum. (c) Find the true and false NLS
   estimate as solutions of the normal equation.

2. The ocean shoreline is specified by a parametrically defined curve $x(\tau) =
   3 - \tau$, $Y(\tau) = -0.1\tau^2 + 0.5\tau + 3$. Express the squared distance to the
   shore from a point in the ocean $(50, 2)$ as a nonlinear regression and find
   the shortest distance using NLS graphically by plotting the distance as a
   function of τ.

3. (a) Reduce the nonlinear least squares estimation of regression $Y_i = \alpha(1 +
   \beta x_i) + \varepsilon_i$, where both α and β are unknown parameters, to linear least
   squares. (b) Formulate the condition of identifiability.

4. Reduce a sinusoidal regression problem, $Y_i = \alpha \sin(x_i + \beta) + \varepsilon_i$, where both α and β are unknown parameters, to linear least squares.

5. Can the NLS problem in regression (9.1) with $g' > 0$ be reduced to linear least squares by applying g^{-1} to both sides? Do the linear and nonlinear least squares solutions coincide? Provide an example.

9.3 Gauss–Newton algorithm

The Gauss–Newton (GN) algorithm is a popular and effective way to find the minimum of the sum of squares in nonlinear regression. The idea of the algorithm is based on repetitive approximation of the nonlinear regression function with a linear function (linear model) and use of the formula for linear least squares to obtain the next approximation for the parameter vector. Indeed, if $\mathbf{b}_k$ is the parameter vector at iteration k, we approximate the regression function as

$$\mathbf{f}(\boldsymbol{\beta}) \simeq \mathbf{f}(\mathbf{b}_k) + \mathbf{J}_k(\boldsymbol{\beta} - \mathbf{b}_k), \tag{9.16}$$

where $\mathbf{J}_k$ is the Jacobian, or the $n \times m$ matrix of the derivatives of the regression function evaluated at $\mathbf{b}_k$,

$$\mathbf{J}_k = \left. \frac{\partial \mathbf{f}}{\partial \boldsymbol{\beta}} \right|_{\boldsymbol{\beta}=\mathbf{b}_k}.$$

The linear approximation (9.16) is used for computational purposes here as well as for approximate statistical inference in the following sections. Using approximation (9.16), we replace the original nonlinear sum of squares (9.12) with the linear sum of squares

$$\|\mathbf{y} - \mathbf{f}(\mathbf{b}_k) - \mathbf{J}_k(\boldsymbol{\beta} - \mathbf{b}_k)\|^2,$$

which leads to the solution

$$\boldsymbol{\beta} - \mathbf{b}_k = (\mathbf{J}_k'\mathbf{J}_k)^{-1}\mathbf{J}_k'(\mathbf{y} - \mathbf{f}(\mathbf{b}_k)).$$

The GN iterations take the form

$$\mathbf{b}_{k+1} = \mathbf{b}_k + (\mathbf{J}_k'\mathbf{J}_k)^{-1}\mathbf{J}_k'(\mathbf{y} - \mathbf{f}(\mathbf{b}_k)), \quad k = 0, 1, 2, \dots. \tag{9.17}$$

To start iterations, one needs to have a starting value, $\mathbf{b}_0$.

Theorem 9.5 *The Gauss–Newton algorithm is equivalent to the Fisher scoring algorithm (Section 6.10.6) when errors have a normal distribution.*

Proof. The log-likelihood function for nonlinear regression with normally distributed errors is given by (9.15). The derivatives, expressed through the Jacobian $\mathbf{J}$ and $\boldsymbol{\varepsilon} = \mathbf{y} - \mathbf{f}(\boldsymbol{\beta})$, can be written as

$$\frac{\partial l}{\partial \boldsymbol{\beta}} = \frac{1}{\sigma^2}\mathbf{J}'\boldsymbol{\varepsilon}, \quad \frac{\partial l}{\partial \sigma^2} = -\frac{n}{\sigma^2} + \frac{\|\boldsymbol{\varepsilon}\|^2}{2\sigma^4},$$

where $\varepsilon \sim \mathcal{N}(\mathbf{0}, \sigma^2 \mathbf{I})$. Therefore,

$$\text{cov}\left(\frac{\partial l}{\partial \boldsymbol{\beta}}\right) = \frac{1}{\sigma^4}\mathbf{J}'\text{cov}(\varepsilon)\mathbf{J} = \frac{1}{\sigma^2}(\mathbf{J}'\mathbf{J}), \quad \text{cov}\left(\frac{\partial l}{\partial \boldsymbol{\beta}}, \frac{\partial l}{\partial \sigma^2}\right) = \mathbf{0},$$

$$\text{var}\left(\frac{\partial l}{\partial \sigma^2}\right) = \frac{1}{4\sigma^8}\text{var}(\|\varepsilon\|^2) = \frac{2n\sigma^4}{4\sigma^8} = \frac{n}{2\sigma^4}$$

because $E(\varepsilon_i \varepsilon_j \varepsilon_k) = 0$ and $\text{var}(\varepsilon_i) = 2\sigma^4$ for any i, j, and k, as in the linear regression model. Finally, the $(m+1) \times (m+1)$ Fisher information matrix for $\mathbf{y}$ is block-diagonal and looks the same as for the linear model,

$$\mathcal{I} = \begin{bmatrix} \frac{1}{\sigma^2}(\mathbf{J}'\mathbf{J}) & \mathbf{0} \\ \mathbf{0}' & n/(2\sigma^4) \end{bmatrix}, \tag{9.18}$$

with the difference that, for linear regression, matrix $\mathbf{J} = \mathbf{X}$ is constant, but, for nonlinear regression, $\mathbf{J}$ is a function of $\boldsymbol{\beta}$. The inverse matrix,

$$\mathcal{I}^{-1} = \begin{bmatrix} \sigma^2(\mathbf{J}'\mathbf{J})^{-1} & \mathbf{0} \\ \mathbf{0}' & 2\sigma^4/n \end{bmatrix},$$

gives the lower Cramér–Rao bound for unbiased estimators. As follows from Section 6.10.6, we interpret $\sigma^2(\mathbf{J}'\mathbf{J})^{-1}$ as the inverse of the expected Hessian, so the Fisher scoring algorithm takes the form

$$\mathbf{b}_{k+1} = \mathbf{b}_k + \sigma^2(\mathbf{J}_k'\mathbf{J}_k)^{-1}\left(\left.\frac{\partial l}{\partial \boldsymbol{\beta}}\right|_{\boldsymbol{\beta}=\mathbf{b}_k}\right) = \mathbf{b}_k + \sigma^2(\mathbf{J}_k'\mathbf{J}_k)^{-1}\frac{1}{\sigma^2}\mathbf{J}_k'(\mathbf{y} - \mathbf{f}(\mathbf{b}_k))$$

$$= \mathbf{b}_k + (\mathbf{J}_k'\mathbf{J}_k)^{-1}\mathbf{J}_k'(\mathbf{y} - \mathbf{f}(\mathbf{b}_k)),$$

the same as the GN iterations. $\square$

More discussion on studying the small-sample properties of the NLS estimator is found in Section 9.4.3.

Theorem 9.6 *For a linear model, the Gauss–Newton iterations converge after the first iteration regardless of the starting point.*

Proof. For a linear model we have $\mathbf{f}(\mathbf{b}) = \mathbf{X}\mathbf{b}$, and the Jacobian is a fixed matrix, $\mathbf{J} = \mathbf{X}$. If $\mathbf{b}_0$ is a starting point, at the first iteration, we obtain

$$\mathbf{b}_k = \mathbf{b}_0 + (\mathbf{X}'\mathbf{X})^{-1}\mathbf{X}'(\mathbf{y} - \mathbf{X}\mathbf{b}_0) = \mathbf{b}_0 + (\mathbf{X}'\mathbf{X})^{-1}\mathbf{X}'\mathbf{y} - (\mathbf{X}'\mathbf{X})^{-1}(\mathbf{X}'\mathbf{X})\mathbf{b}_0$$

$$= (\mathbf{X}'\mathbf{X})^{-1}\mathbf{X}'\mathbf{y} = \widehat{\mathbf{b}},$$

the linear least squares estimate. $\square$

To carry out iterations (9.17), the Jacobian should have full rank, i.e. matrix $\mathbf{J}'_k \mathbf{J}_k$ should be nonsingular at each iteration. Thus, we extend the identifiability condition by assuming that the Jacobian is a continuous function on $\boldsymbol{\Theta}$ and has full rank, or equivalently,

$$|\mathbf{J}'(\boldsymbol{\beta})\mathbf{J}(\boldsymbol{\beta})| \neq 0 \quad \forall \boldsymbol{\beta} \in \boldsymbol{\Theta}. \tag{9.19}$$

There is no guarantee that the GN iterations converge for every nonlinear regression starting at any point. However, we can prove that, if iterations converge, then the gradient of the sum of squares at the limit point is zero.

Theorem 9.7 *If iterations (9.17) converge,* $\lim_{k \to \infty} \mathbf{b}_k = \mathbf{b}_*$, *then the gradient of S at $\mathbf{b}_*$ is zero,*

$$\left.\frac{\partial S}{\partial \boldsymbol{\beta}}\right|_{\beta=\mathbf{b}_*} = \mathbf{0}.$$

Proof. Letting $k \to \infty$ in both sides of (9.17), we obtain

$$\mathbf{b}_* = \mathbf{b}_* + (\mathbf{J}'_* \mathbf{J}_*)^{-1} \mathbf{J}'_*(\mathbf{y} - \mathbf{f}(\mathbf{b}_*))$$

due to continuity of $\mathbf{J}$. This implies that

$$\left.\frac{\partial S}{\partial \boldsymbol{\beta}}\right|_{\beta=\mathbf{b}_*} = -2\mathbf{J}'_*(\mathbf{y} - \mathbf{f}(\mathbf{b}_*)) = \mathbf{0},$$

which proves the theorem. $\qquad\qquad\qquad\qquad\qquad\qquad\qquad\qquad\qquad\qquad\quad\square$

An indicative characteristic of convergence is that the value of the RSS decreases from iteration to iteration but the GN algorithm does not guarantee such a decrease. To ensure a decrease, a fraction of a step in the direction of the GN should be used,

$$\mathbf{b}_{k+1} = \mathbf{b}_k + \rho_k (\mathbf{J}'_k \mathbf{J}_k)^{-1} \mathbf{J}'_k(\mathbf{y} - \mathbf{f}(\mathbf{b}_k)), \tag{9.20}$$

where $0 < \rho_k \leq 1$ such that

$$S(\mathbf{b}_{k+1}) < S(\mathbf{b}_k) \text{ if } \left.\frac{\partial S}{\partial \boldsymbol{\beta}}\right|_{\beta=\mathbf{b}_k} \neq \mathbf{0}.$$

Then, one can prove that, under mild conditions, the sequence $\{\mathbf{b}_k, k = 0, 1, ...\}$ converges to the point of local minimum of function $S(\boldsymbol{\beta})$. A general criterion of optimization theory to ensure convergence is $\sum_{k=1}^{\infty} \rho_k = \infty$. Often, the choice $\rho_k = 1$ leads to convergence if the starting point is fairly close to the solution. More expensive strategies can be employed to ensure that S decreases from iteration to iteration. For example, ρ_k may be chosen by halving until the RSS becomes smaller, or golden-section search can be used (Dennis and Schnabel [33]).

Several (nonequivalent) convergency criteria may be employed: (a) stop iterations when

$$\frac{|b_{k+1,j} - b_{k,j}|}{1 + |b_{k,j}|} < \varepsilon, \quad j = 1, ..., m, \tag{9.21}$$

where ε is a small number, say, $\varepsilon = 10^{-5}$, or (b) when the total angular criterion [28] is satisfied,

$$\frac{(\mathbf{y} - \widehat{\mathbf{f}}_k)' \mathbf{J}_k (\mathbf{J}_k' \mathbf{J}_k)^{-1} \mathbf{J}_k' (\mathbf{y} - \widehat{\mathbf{f}}_k)}{\sqrt{m} \left\| \mathbf{y} - \widehat{\mathbf{f}}_k \right\|^2} < \varepsilon. \tag{9.22}$$

Several other modifications of the GN algorithm have been developed over the years including the Levenberg–Marquardt algorithm. Matrix $\mathbf{J}_k' \mathbf{J}_k$ can become close to singular when parameters of regression are difficult to identify even though condition (9.19) is fulfilled. Levenberg suggested using $\mathbf{J}_k' \mathbf{J}_k + \lambda_k \mathbf{I}$ where $\lambda_k > 0$ and setting $\lambda_k \to 0$ when $k \to \infty$ (λ_k is referred to as the regularization parameter). The Levenberg–Marquardt algorithm is popular in engineering applications when solving ill-posed inverse problems. However, none of existing iterative algorithms guarantee the convergence to a global minimum because S may have several local minima. A good practice is to start iterations from several starting values to confirm that they converge to the same point. More detail on numerical complications and the definition of a successful starting value, along with respective criteria can be found in Section 9.7.

Problems

1. (a) Generate $n = 30$ random observations $\{Y_i, i = 1, 2, ..., n\}$ from the model $Y_i = \alpha(1 + \beta x_i) + \varepsilon_i$ where $\alpha = -1$, $\beta = 0.1$, $x_i = i$, and $\varepsilon_i \sim \mathcal{N}(0, 1)$. (b) Apply the GN algorithm in the form (9.17) to find the NLS estimates for α and β and make sure that they converge to the values obtained from the linear least squares (print out and stop iterations when criterion (9.21) or (9.22) holds). Start iterations from $\alpha_0 = \overline{Y}$ and $\beta_0 = 0$. (c) Prove that the normal equations of the NLS estimation (9.14) and the linear least squares are equivalent.

2. (a) Reduce the problem of estimation of the nonlinear regression $Y_i = \beta(1 + \beta x_i) + \varepsilon_i$ to the solution of a cubic equation with help of the `polyroot` function in R. (b) Generate $n = 100$ random values $\{x_i, i = 1, ..., n\}$ and $Y_i = \beta(1 + \beta x_i) + \varepsilon_i$ where $\varepsilon_i \sim \mathcal{N}(0, 1)$ with $\beta = -0.02$. Estimate β by GN iterations and compare with the cubic root (use one of the stopping criteria).

3. (a) May Levenberg algorithm converge if λ_k are bounded from zero (i.e. there exists such positive δ that $\lambda_k > \delta$ for all iterations k)? (b) To avoid computation and inverse of matrix $\mathbf{J}_k' \mathbf{J}_k$ at each iteration, somebody suggests using $\mathbf{A} = \mathbf{J}_0' \mathbf{J}_0$ for all $k > 0$. Do Theorems 9.6 and 9.7 hold?

4. Show that the left-hand side of criterion (9.22) converges to zero when the GN iterations converge.

9.4 Statistical properties of the NLS estimator

Statistical properties of any estimator, including the NLS estimator $\widehat{\boldsymbol{\beta}}$, can be studied from two perspectives depending on the sample size: (a) the large sample (or asymptotic) properties (Ivanov [59]), when the sample size goes to infinity, $n \to \infty$, and (b) the small sample properties, when n is finite and fixed. Large n leads to normal distribution while the distribution of $\widehat{\boldsymbol{\beta}}$ for finite n is complicated. Moreover, the NLS estimator may have infinite mean and variance for finite n. Simulations are usually used to study small sample properties of the NLS estimator, see Section 9.6. However, simulations have limitations because the gained knowledge is pertinent to the choice of parameter values.

9.4.1 Large sample properties

The large sample properties of the NLS estimator can be studied under two schemes. In the deterministic scheme, $\{\mathbf{x}_i\}$ are treated as fixed and the conditions upon which asymptotic properties hold are expressed in terms of $f_i = f(\mathbf{x}_i; \boldsymbol{\beta})$. Under this scheme, we assume that ε_i are iid, but not necessarily normally distributed. Clearly, in this scheme, Y_i are not identically distributed because $E(Y_i) = f_i \neq \text{const}$, so the asymptotic theory of ML is not applicable. Recall that we had the same situation in the linear model. Under the stochastic scheme, $\{\mathbf{x}_i\}$ are assumed random iid and ε_i are iid normally distributed, so that the nonlinear regression is viewed as a conditional mean (see Section 3.3.1). These assumptions imply that $(Y_i, \mathbf{x}_i)$ are iid, so the classic ML theory applies (NLS = ML). In general, under the iid assumption, the NLS estimator is consistent and asymptotically normal. In addition, if ε_i are normally distributed, the NLS is asymptotically efficient. This statement applies to linear models as well as a special case of nonlinear regression.

Deterministic scheme

Several conditions for consistency and asymptotic normality of the NLS estimator have been suggested when $\mathbf{x}_i$ are fixed in (9.4). It was proven that if the limit

$$Q(\boldsymbol{\beta}, \boldsymbol{\beta}_0) = \lim_{n \to \infty} \frac{1}{n} \sum_{i=1}^{n} (f_i(\boldsymbol{\beta}) - f_i(\boldsymbol{\beta}_0))^2 \quad \forall \boldsymbol{\beta} \neq \boldsymbol{\beta}_0 \tag{9.23}$$

exists, such that $Q(\boldsymbol{\beta}, \boldsymbol{\beta}_0) = 0$ if and only if $\boldsymbol{\beta} = \boldsymbol{\beta}_0$, then the NLS estimator $\widehat{\boldsymbol{\beta}}$ is consistent. Wu [107] generalized the above criterion: the NLS estimator is consistent if and only if

$$\lim_{n\to\infty} \sum_{i=1}^{n} (f_i(\boldsymbol{\beta}) - f_i(\boldsymbol{\beta}_0))^2 = \infty \quad \forall \boldsymbol{\beta} \neq \boldsymbol{\beta}_0. \tag{9.24}$$

Clearly, condition (9.23) implies (9.24).

Example 9.8 OLS consistency. *Derive the necessary and sufficient condition for consistency of the linear least squares by applying Wu's condition (9.24).*

Solution. For the linear model, we have $\mathbf{f}(\beta) = \mathbf{X}\beta$, so

$$\|\mathbf{X}\boldsymbol{\beta} - \mathbf{X}\boldsymbol{\beta}_0\|^2 = (\boldsymbol{\beta} - \boldsymbol{\beta}_0)'(\mathbf{X}'\mathbf{X})(\boldsymbol{\beta} - \boldsymbol{\beta}_0) \geq \|\boldsymbol{\beta} - \boldsymbol{\beta}_0\|^2 \lambda_{\min}(\mathbf{X}'\mathbf{X}) \to \infty$$

as $n \to \infty$ when $\boldsymbol{\beta} \neq \boldsymbol{\beta}_0$. Thus, the necessary and sufficient condition for the OLS consistency is that the minimum eigenvalue of matrix $\mathbf{X}'\mathbf{X}$ goes to infinity when the sample size increases. □

The following example illustrates that the NLS estimator is not consistent for a sigmoid exponential model when observations are time series on infinite time domain.

Example 9.9 NLS inconsistency. *Using condition (9.24), prove that the parameters of the sigmoid exponential model $f_i = \alpha_1 + \alpha_2 e^{-\alpha_3 t_i}$, where $t_1 < t_2 < t_3 \cdots \to \infty$ (infinite time domain) as $n \to \infty$, cannot be consistently estimated (all parameters are positive).*

Solution. Let $\boldsymbol{\beta} = (\alpha_1, \alpha_2, \alpha_3)$ and $\boldsymbol{\beta}_0 = (\alpha_1, \alpha_2, \alpha_{30})$. The left-hand side of (9.24) takes the form

$$\alpha_2^2 \sum_{i=1}^{n} \left(e^{-\alpha_3 t_i} - e^{-\alpha_{30} t_i} \right)^2 < \infty,$$

when $t_i \to \infty$. Hence, Wu's criterion does not hold and the three parameters cannot be consistently estimated. On the other hand, the asymptote parameter, α_1, can be consistently estimated because $p\lim_{n\to\infty} \overline{Y}_n = \alpha_1$. This example illustrates that, if Wu's condition does not hold, then at least one parameter in inconsistent and other parameters may be consistent. □

Now we turn our attention to asymptotic normality. Unlike linear least squares, nonlinear least squares may lead to multiple solutions of the normal equation (9.14). It is very difficult to provide conditions on the uniqueness of the NLS estimate for a finite sample (Demidenko [23]). At least, we need to be sure that the normal equation has a unique solution in the asymptotic sense, that is, the following limit exists:

$$\mathbf{G}(\boldsymbol{\beta}, \boldsymbol{\beta}_0) = \lim_{n\to\infty} \frac{1}{n} \sum_{i=1}^{n} (f_i(\boldsymbol{\beta}) - f_i(\boldsymbol{\beta}_0)) \left. \frac{\partial f_i}{\partial \boldsymbol{\beta}} \right|_{\boldsymbol{\beta} = \boldsymbol{\beta}_0}, \tag{9.25}$$

and $\mathbf{G}(\boldsymbol{\beta}, \boldsymbol{\beta}_0) = \mathbf{0}$ if and only if $\boldsymbol{\beta} = \boldsymbol{\beta}_0$. Further, we refer to this condition as the *asymptotic uniqueness condition*. Suppose that, in addition to existence of limits (9.23) and (9.25), remotely similar to the Lindeberg–Feller condition of Theorem 2.58, the derivatives are uniformly bounded: for every $\delta > 0$ and every $\boldsymbol{\beta}_0$, there is such K that uniformly,

$$\left\| \frac{\partial f_i}{\partial \boldsymbol{\beta}} \right\| \leq K, \quad i = 1, 2, \dots \tag{9.26}$$

for all $\boldsymbol{\beta}$ such that $\|\boldsymbol{\beta} - \boldsymbol{\beta}_0\| \leq \delta$. Finally, we assume that the limit

$$\mathbf{A}(\boldsymbol{\beta}) = \lim_{n \to \infty} \frac{1}{n} \sum_{i=1}^{n} \left(\frac{\partial f_i}{\partial \boldsymbol{\beta}} \right) \left(\frac{\partial f_i}{\partial \boldsymbol{\beta}} \right)' \quad \forall \boldsymbol{\beta},$$

exists, and, moreover, matrix $\mathbf{A}(\boldsymbol{\beta})$ is nonsingular for every $\boldsymbol{\beta}$. Then the NLS estimator is normally distributed in large sample (recall that ε_i are not necessarily normally distributed):

$$\sqrt{n}(\widehat{\boldsymbol{\beta}} - \boldsymbol{\beta}) \simeq \mathcal{N}\left(0, \sigma^2 \mathbf{A}^{-1}(\boldsymbol{\beta})\right), \quad n \to \infty, \tag{9.27}$$

where $\boldsymbol{\beta}$ is the true parameter vector. This result implies that the covariance matrix of $\widehat{\boldsymbol{\beta}}$ can be consistently estimated as

$$\operatorname{cov}(\widehat{\boldsymbol{\beta}}) = \widehat{\sigma}^2 (\widehat{\mathbf{J}}'\widehat{\mathbf{J}})^{-1}, \tag{9.28}$$

where

$$\widehat{\sigma}^2 = \frac{1}{n-m} \sum_{i=1}^{n} (Y_i - f_i(\widehat{\boldsymbol{\beta}}))^2 \tag{9.29}$$

is the estimated error variance, and

$$\widehat{\mathbf{J}} = \left. \frac{\partial \mathbf{f}}{\partial \boldsymbol{\beta}} \right|_{\boldsymbol{\beta} = \widehat{\boldsymbol{\beta}}}$$

is the Jacobian evaluated at the NLS estimate (computed at the last iteration of the Gauss–Newton algorithm). Practically, the distribution of the NLS estimator is approximately normal for large n, which can be written, with some ambiguity, as

$$\widehat{\boldsymbol{\beta}} \simeq \mathcal{N}\left(\boldsymbol{\beta}, \widehat{\sigma}^2 (\widehat{\mathbf{J}}'\widehat{\mathbf{J}})^{-1}\right). \tag{9.30}$$

Unfortunately, unlike linear model, we cannot claim unbiasedness of the NLS estimator or prove that the ratio of the NLS estimator minus the true value to its standard error has the t-distribution. Also, we cannot claim that the NLS estimator reaches the Cramér–Rao lower bound and, therefore, is efficient, as in linear model, simply because $\widehat{\boldsymbol{\beta}}$ is biased in the majority of nonlinear regressions.

Stochastic scheme

Under this scheme, we treat the nonlinear regression as the conditional mean, so

$$E(Y_i|\mathbf{X} = \mathbf{x}_i) = f(\mathbf{x}_i; \beta), \quad \mathbf{X}_i \overset{\text{iid}}{\sim} p(\mathbf{x}; \theta),$$

where p is the marginal density of $\mathbf{X}$ with some unknown, nuisance parameter θ. For example, when Y_i is normally distributed conditional on $\mathbf{X}_i$, we may think of a general population of $(Y, \mathbf{X})$ with conditional density

$$Y|\mathbf{X} = \mathbf{x} \sim \mathcal{N}(f(\mathbf{x}; \beta), \sigma^2)$$

and marginal density $\mathbf{X} \sim p(\mathbf{x}; \theta)$. Then the classic ML theory of Section 6.10 applies with the log-likelihood function, up to a constant term, given by

$$
\begin{aligned}
l(\beta, \theta) &= \sum_{i=1}^{n} \left(-\frac{1}{2} \ln \sigma^2 - \frac{1}{2\sigma^2}(Y_i - f(\mathbf{x}_i; \beta))^2 + \ln p(\mathbf{x}_i; \theta) \right) \\
&= -\frac{n}{2} \ln \sigma^2 - \frac{1}{2\sigma^2} \sum_{i=1}^{n}(Y_i - f(\mathbf{x}_i; \beta))^2 + \sum_{i=1}^{n} \ln p(\mathbf{x}_i; \theta).
\end{aligned}
$$

The critical condition here is that vectors β and θ do not have common components and therefore maximization of l with respect to β is not affected by θ. This means that Y_i are iid and NLS = ML, and the Fisher information is block-diagonal, that is, estimation of θ does not affect estimation of β; see Section 6.9. Following the conditions of the ML theory, and particularly condition (6.81), under assumption

$$G(\beta, \beta_0) = \mathbf{0} \text{ iff } \beta = \beta_0,$$

where

$$G(\beta, \beta_0) = E_{\mathbf{X}} \left((f(\mathbf{X}; \beta) - f(\mathbf{X}; \beta_0)) \left. \frac{\partial f(\mathbf{X}; \beta)}{\partial \beta} \right|_{\beta = \beta_0} \right), \qquad (9.31)$$

we deduce that the NLS estimator is asymptotically efficient and has an asymptotic normal distribution

$$\sqrt{n}(\widehat{\beta} - \beta) \simeq \mathcal{N}\left(\mathbf{0}, \sigma^2 \mathbf{B}^{-1}(\beta)\right), \quad n \to \infty, \qquad (9.32)$$

where

$$\mathbf{B}(\beta) = E_{\mathbf{X}} \left[\left(\frac{\partial f(\mathbf{X}; \beta)}{\partial \beta} \right) \left(\frac{\partial f(\mathbf{X}; \beta)}{\partial \beta} \right)' \right]$$

is a positive definite matrix.

In fact, the statements on the asymptotic normality of the NLS in the deterministic (9.27) and stochastic approaches (9.32) look very similar with the only difference that the former uses the empirical average and the latter the expectation. Due to the law of large numbers and the Extended Slutsky Theorem from Section 6.4.4, we conclude that, in practice, both approaches come to the same approximation of the covariance matrix of the NLS estimator given by (9.28), and the distribution approximation (9.30) is valid under both schemes.

Example 9.10 *Asymptotic properties of quadratic regression.* *Consider the asymptotic properties of the quadratic (in parameter sense) regression $Y_i = \beta(1 + \beta x_i) + \varepsilon_i$ under deterministic and stochastic schemes. Formulate the assumptions and state the asymptotic distribution of the NLS estimator.*

Solution. First, consider the deterministic scheme. Suppose that the following limits exist:

$$\lim_{n\to\infty} \frac{1}{n} \sum_{i=1}^{n} x_i = m_x, \quad \lim_{n\to\infty} \frac{1}{n} \sum_{i=1}^{n} x_i^2 = s_x^2 < m_x^2.$$

It is easy to prove that, in general, $s_x^2 \geq m_x^2$, and $s_x^2 = m_x^2$ if and only if $\{x_i\}$ converge, meaning that there exists x_0 such that $\lim_{i\to\infty} x_i = x_0$. To avoid this case, we assume that $s_x^2 > m_x^2$. We start with proving that condition (9.23) holds,

$$
\begin{aligned}
Q(\beta, \beta_0) &= (\beta - \beta_0)^2 \lim_{n\to\infty} \frac{1}{n} \sum_{i=1}^{n} (1 + (\beta + \beta_0)x_i)^2 \\
&= (\beta - \beta_0)^2 \left(1 + 2(\beta + \beta_0)m_x + (\beta + \beta_0)^2 s_x^2\right),
\end{aligned}
$$

where $\beta \neq \beta_0$. The second factor at the right-hand side, as a quadratic polynomial of $\beta + \beta_0$, is positive because the discriminant, $m_x^2 - s_x^2 < 0$. This means that $Q > 0$ if and only if $\beta \neq \beta_0$. Now we investigate the limit (9.25),

$$
\begin{aligned}
G(\beta, \beta_0) &= (\beta - \beta_0) \lim_{n\to\infty} \frac{1}{n} \sum_{i=1}^{n} (1 + (\beta + \beta_0)x_i)(1 + 2\beta_0 x_i) \\
&= (\beta - \beta_0) \left[1 + (\beta + 3\beta_0)m_x + 2\beta_0(\beta + \beta_0)s_x^2\right].
\end{aligned}
$$

If the domain of β and β_0 is the entire real line $G(\beta, \beta_0) = 0$ for some $\beta \neq \beta_0$. Thus to comply with the asymptotic uniqueness condition, we must restrict the parameter set. For example, we can assume that β and β_0 are both positive along with m_x. The fact that the asymptotic uniqueness condition does not hold for all β links with Example 9.30, to be considered later, because both regressions are quadratic functions of β. To comply with condition (9.26), we assume that the sequence $\{x_i\}$ is bounded. Now we find the expression for $A(\beta)$ to formulate asymptotic normality using (9.27),

$$A(\beta) = \lim_{n\to\infty} \frac{1}{n} \sum_{i=1}^{n} (1 + 2\beta x_i)^2 = 1 + 2\beta m_x + 4\beta^2 s_x^2.$$

Again, $A(\beta) > 0$ because the quadratic polynomial is positive since the discriminant is negative. Finally, the asymptotic distribution of the NLS estimator under the deterministic scheme takes the form

$$\sqrt{n}(\widehat{\beta} - \beta) \simeq \mathcal{N}\left(0, \frac{\sigma^2}{1 + 2\beta m_x + 4\beta^2 s_x^2}\right).$$

Second, we consider the stochastic scheme under assumption that $Y|X = x \sim \mathcal{N}(\beta(1 + \beta x), \sigma^2)$, where X is a random variable with unspecified marginal distribution, $E(X) = \mu_x$, and $E(X^2) = \sigma_x^2$. Simple algebra yields

$$G(\beta, \beta_0) = (\beta - \beta_0)E_X\left[(1 + (\beta + \beta_0)X)(1 + 2\beta_0 X)\right].$$

As with the deterministic scheme, function G may reach zero for some $\beta \neq \beta_0$ if no restrictions are put on the parameter set. However, as before, if β, β_0 and μ_x are positive $G \neq 0$ for $\beta \neq \beta_0$. Since

$$B(\beta) = E_X(1 + 2\beta X)^2 = 1 + 2\beta\mu_x + 4\beta^2\sigma_x^2$$

under the stochastic scheme, we have

$$\sqrt{n}(\widehat{\beta} - \beta) \simeq \mathcal{N}\left(0, \frac{\sigma^2}{1 + 2\beta\mu_x + 4\beta^2\sigma_x^2}\right).$$

The asymptotic distribution of the NLS estimator in both schemes is the same with the only difference is that under the deterministic scheme the sample values/estimates are used for the true μ_x and σ_x^2. Moreover, if μ_x and σ_x^2 are unknown, the two schemes lead to the same result when μ_x and σ_x^2 are replaced with their sample estimates.

9.4.2   Small sample properties

The small sample (n is fixed) properties of the NLS estimator are difficult to study analytically hampered by numerical complications such as nonexistence and nonuniqueness of the solution to the normal equation (see Section 9.7). Then theory of the lower Cramér–Rao bound does not work here simply because unbiased estimators do not exist for many nonlinear statistical models including nonlinear regression. The mentioned difficulties are illustrated with a simple nonlinear regression where the NLS estimator does not have finite mean and variance.

Example 9.11 *Fieller regression.* *Consider a nonlinear regression model $Y_i = \nu(1 + \tau x_i) + \varepsilon_i$, where x_i are fixed and ε_i are iid normally distributed random variables with zero mean $(i = 1, 2, ..., n)$. Show that the NLS estimator for τ has the Cauchy distribution and therefore has infinite mean and variance.*

Solution. The NLS estimators for ν and τ minimize the RSS function $\sum_{i=1}^{n}(Y_i - \nu(1 + \tau x_i))^2$. Using an obvious reparametrization, we reduce this problem to linear least squares: $\beta_1 = \nu$ and $\beta_2 = \nu\tau$. Therefore, if $\widehat{\beta}_1$ and $\widehat{\beta}_2$ are the least squares estimates of the intercept and slope in the simple linear regression of Y on x, the NLS estimate of τ is found as the ratio, $\widehat{\tau} = \widehat{\beta}_2/\widehat{\beta}_1$. Since $\widehat{\beta}_1$ and $\widehat{\beta}_2$ have normal distribution, the ratio has a Cauchy distribution (see Example 3.64), and as such has infinite mean and variance as shown by Fieller almost one hundred years ago. $\square$

The exact, quite complicated, density of the NLS estimator for a one-parameter nonlinear regression with iid normally distributed errors, $\mathbf{y} = \mathbf{f}(\beta) + \boldsymbol{\varepsilon}$, under assumption that the normal equation has a unique solution, was derived by Demidenko [31]. Denote

$$A = A(b) = \left\| \dot{\mathbf{f}}(b) \right\|^2, \quad B = B(b;\beta) = \ddot{\mathbf{f}}'(b)(\mathbf{f}(b) - \mathbf{f}(\beta)),$$

$$C = C(b) = \dot{\mathbf{f}}'(b)\ddot{\mathbf{f}}(b), \quad D = D(b;\beta) = \dot{\mathbf{f}}'(b)(\mathbf{f}(b) - \mathbf{f}(\beta)),$$

$$E = E(b) = \left\| \ddot{\mathbf{f}}(b) \right\|^2, \quad Q = Q(b;\beta,\sigma^2) = \frac{A^2 + AB - CD}{\sqrt{\sigma^2 A(AE - C^2)}},$$

where one dot over $\mathbf{f}$ denotes the $n \times 1$ derivative vector and two dots denotes the $n \times 1$ second derivative vector. The density of the NLS estimator is given by

$$p_{EX}(b;\beta,\sigma^2) = \frac{\left\| \dot{\mathbf{f}}(b) \right\|}{\sqrt{2\pi\sigma^2}} \exp\left[-\frac{1}{2\sigma^2} \frac{\left(\dot{\mathbf{f}}'(b)(\mathbf{f}(b) - \mathbf{f}(\beta)) \right)^2}{\left\| \dot{\mathbf{f}}(b) \right\|^2} \right] \times a_{EX}(b;\beta,\sigma)$$

where

$$a_{EX}(b;\beta,\sigma) = \left(1 + \frac{AB - CD}{A^2} \right)(2\Phi(Q) - 1) + \frac{2\sigma}{A}\sqrt{\frac{AE - C^2}{A}}\phi(Q),$$

where Φ and $\phi = \Phi'$ are the standard normal cdf and density.

Some generalizations to the nonlinear regression and estimating equation approach can be found in the cited paper. Although it is quite remarkable that the exact distribution of the NLS estimator depends on the first two derivatives of the nonlinear regression function, the practical applications of this result are limited due to its complexity. Some satisfactory approximations are offered below.

9.4.3 Asymptotic confidence intervals and hypothesis testing

Since the finite sample properties of nonlinear regression are complicated, we rely on asymptotic properties for hypothesis testing and confidence intervals (CI). Under the normal distribution assumption, the theory of the ML applies and, in fact, methods described below can be deduced from the asymptotic normality of the NLS = ML estimator. Three types of approximations are discussed here: Wald, likelihood ratio, and score/near exact tests. Their comparison through simulations for an exponential regression is found in Section 9.6.

Wald approximation

The Wald CI and the associated hypothesis testing with the asymptotic normal approximation (n is large) is straightforward: if (9.28) is the estimated covariance matrix, the $(1-\alpha)100\%$ symmetric CI for β_j is $\widehat{\beta}_j \pm q_{1-\alpha/2}SE_j$, where $q_{1-\alpha/2} = \Phi^{-1}(1-\alpha/2)$ and SE_j is the estimate of the standard error of the NLS estimate computed as the square root of the (j,j)th diagonal element of matrix (9.28),

$$SE_j = \widehat{\sigma}\sqrt{(\widehat{\mathbf{J}'\mathbf{J}})^{jj}}, \quad j = 1, 2, ..., m,$$

where $\widehat{\sigma}^2$ is the estimate of σ^2 as the mean squared residuals (9.29), and the superscript indicates the diagonal element of the inverse matrix. Respectively, if the null hypothesis is $H_0 : \beta_j = \beta_{j0}$, as follows from (9.30), the hypothesis is rejected if $|\widehat{\beta}_j - \beta_{j0}|/SE_j > q_{1-\alpha/2}$ with the p–value $= 2\Phi(-|\widehat{\beta}_j - \beta_{j0}|/SE_j)$. The built-in R function nls (see the next section) reports the standard error SE_j, the t-statistic $\widehat{\beta}_j/SE_j$, and the p-value under the null hypothesis $\beta_{j0} = 0$.

An advantage of the Wald approach is that the covariance matrix is estimated once at the last Gauss–Newton iteration and then used to compute the CIs and hypothesis tests for all m parameters. The Wald test is easy to modify when a linear function of parameters is tested, say, $\mathbf{a}'\boldsymbol{\beta} = c$, where $\mathbf{a}$ is a fixed $m \times 1$ vector and c is a constant. Then the Z-score takes the form

$$Z = \frac{\mathbf{a}'\widehat{\boldsymbol{\beta}} - c}{\widehat{\sigma}\sqrt{\mathbf{a}'(\widehat{\mathbf{J}'\mathbf{J}})^{-1}\mathbf{a}}} \simeq \mathcal{N}(0, 1), \quad (9.33)$$

with the p-value computed as $2\Phi(-|Z|)$.

The approximate $(1-\alpha)$ confidence region for parameters follows from approximation (9.30) and has an elliptical shape with the center at the NLS estimate, $\widehat{\boldsymbol{\beta}}$,

$$\left\{ \boldsymbol{\beta} \in R^m : (\boldsymbol{\beta} - \widehat{\boldsymbol{\beta}})'(\widehat{\mathbf{J}'\mathbf{J}})(\boldsymbol{\beta} - \widehat{\boldsymbol{\beta}}) \leq \widehat{\sigma}^2 \chi^{-2}(1-\alpha, m) \right\}, \quad (9.34)$$

where $\chi^{-2}(1-\alpha, m)$ is the $(1-\alpha)$ quantile of the chi-square distribution with m degrees of freedom. We expect that the confidence region covers the true vector $\boldsymbol{\beta}$ with probability $1-\alpha$ at least for sufficiently large n. Some authors suggest using the F-distribution for the right-hand side inequality in (9.34), which means that the chi-quantile is replaced with m times the F-quantile, $mF^{-1}(1-\alpha, m, n-m)$, to mimic the linear model confidence region. Note that, for large n, the two sets will be close because when the second degree of freedom is getting large the chi- and F-distributions converge; see equation (4.30) in Section 4.4.

Likelihood ratio approximation

A more sophisticated likelihood ratio (LR) test involves estimation of an additional nonlinear regression to test $\beta_j = \beta_{j0}$ for each fixed $j = 1, 2, ..., m$. Recall that, for the general LR test

$$-2\left(\max l_{\texttt{restricted}} - \max l_{\texttt{unrestricted}}\right) \simeq \chi^2(k),$$

where l is the log-likelihood function, $\max l_{\text{unrestricted}}$ is the ML over unrestricted $\boldsymbol{\beta}$ values, $\max l_{\text{restricted}}$ is the ML over restricted $\boldsymbol{\beta}$ values, and k is the number of restrictions/constraints. We apply this general test to testing $\beta_j = \beta_{j0}$. Let $\widehat{S}_{j0}$ and $\widehat{S}$ denote the minimal values of the RSS under restriction $\beta_j = \beta_{j0}$ and no restriction,

$$\widehat{S}_{j0} = \min_{\boldsymbol{\beta}:\beta_j=\beta_{j0}} \|\mathbf{y} - \mathbf{f}(\boldsymbol{\beta})\|^2 , \quad \widehat{S} = \min_{\boldsymbol{\beta}} \|\mathbf{y} - \mathbf{f}(\boldsymbol{\beta})\|^2 , \quad (9.35)$$

respectively. Then, after some algebra using expression (9.15), we obtain

$$-2(\max l_{\text{restricted}} - \max l_{\text{unrestricted}}) = n \ln(\widehat{S}_{j0}/\widehat{S}),$$

where $\widehat{S}_{j0}$ is computed from the nonlinear regression with the jth parameter fixed at $\beta_j = \beta_{j0}$. Therefore, for large n, we derive that, under the null hypothesis $\beta_j = \beta_{j0}$,

$$n \ln(\widehat{S}_{j0}/\widehat{S}) \simeq \chi^2(1) \quad (9.36)$$

with the p-value computed as 1-pchisq where pchisq is evaluated at $n \ln(\widehat{S}_j/\widehat{S})$ with one degree of freedom. A numerical disadvantage of the LR test, compared to the Wald test, is that testing a linear combination of parameters $\mathbf{a}'\boldsymbol{\beta} = c$ requires minimization of the sum of squares under the linear restriction.

Since $Z \sim \mathcal{N}(0,1)$ implies $Z^2 \sim \chi^2(1)$, we can rewrite approximation (9.36) as

$$\text{sign}(\widehat{\beta}_j - \beta_{j0})\sqrt{n \ln(\widehat{S}_{j0}/\widehat{S})} \simeq \mathcal{N}(0,1), \quad (9.37)$$

where $\widehat{\beta}_j$ is the jth component of the NLS estimate $\widehat{\boldsymbol{\beta}}$. Approximation (9.37) will be referred to as the *signed likelihood ratio* approximation. Note that the number under the square root is nonnegative because $\widehat{S}_{j0} \geq \widehat{S}$ since $\widehat{S}_{j0}$ is the RSS under restriction.

The LR test gives rise to a CI with the lower and the upper limits for parameter β_j as solutions of

$$n \ln(\widehat{S}_{j0}(\beta_j)) - n \ln \widehat{S} - \chi^{-2}(1 - \alpha, 1) = 0, \quad (9.38)$$

where $\widehat{S}_{j0}(\beta_j)$ is the residual sum of squares when the jth parameter is held fixed at β_j, and $\chi^{-2}(1 - \alpha, 1)$ is the $(1 - \alpha)$ quantile of the chi-square distribution with one degree of freedom. This method will be referred to as the LR-profile CI. One can plot the left-hand side as a function of β_j to compute where the curve intersects the x-axis. Alternatively, the R function uniroot can be used to solve this equation; see Section 2.13. Examples of the LR-profile CI are found later in Section 9.6. The method of profiling, where the residual sum of squares is expressed as a function of a single parameter β_j, is also used in the score approximation.

The confidence region for regression parameters can be derived from the above chi-square approximation $(k = m)$,

$$\left\{ \boldsymbol{\beta} \in R^m : \ln \| \mathbf{y} - \mathbf{f}(\boldsymbol{\beta}) \|^2 - \ln \widehat{S} \leq \frac{1}{n} \chi^{-2}(1 - \alpha, m) \right\}. \tag{9.39}$$

One can derive an alternative confidence region by mimicking the linear model,

$$\left\{ \boldsymbol{\beta} \in R^m : \| \mathbf{y} - \mathbf{f}(\boldsymbol{\beta}) \|^2 - \widehat{S} \leq \widehat{\sigma}^2 \chi^{-2}(1 - \alpha, m) \right\}. \tag{9.40}$$

A slightly larger confidence region will be obtained by replacing χ^{-2} with mF^{-1}, as in the Wald approach above, to account for the finite degrees of freedom. More details on hypothesis testing and CI comparison via simulations, are found in Section 9.6.

The score approximation

The major result of the ML theory is that, under the iid assumption, the distribution of the derivative of the log-likelihood function evaluated at the true parameter value converges to the multivariate normal distribution with zero mean and covariance matrix equal to the Fisher information matrix. This fact gives rise to the score test. Originally, the score test was developed by a famous Indian statistician C.R. Rao [82]. In econometrics, it is called the Lagrange multiplier test. Since, from the proof of Theorem 9.5 the derivative of the log-likelihood function (9.15) is $\sigma^{-2} \mathbf{J}'(\mathbf{y} - \mathbf{f}(\boldsymbol{\beta}))$ and the Fisher information matrix is $\sigma^{-2}(\mathbf{J}'\mathbf{J})$ and since $\widehat{\boldsymbol{\beta}} \to \boldsymbol{\beta}$ with probability 1, we infer that, for large n, in the spirit of the score test,

$$(\widehat{\mathbf{J}}'\widehat{\mathbf{J}})^{-1/2} \widehat{\mathbf{J}}'(\mathbf{y} - \mathbf{f}) \simeq \mathcal{N}(\mathbf{0}, \sigma^2 \mathbf{I}),$$

where $\widehat{\mathbf{f}} = \mathbf{f}(\widehat{\boldsymbol{\beta}})$ and $\widehat{\mathbf{J}} = \mathbf{J}(\widehat{\boldsymbol{\beta}})$. This will be referred to as the *score approximation*. Moreover, since the NLS solution obeys $\widehat{\mathbf{J}}'(\mathbf{y} - \widehat{\mathbf{f}}) = \mathbf{0}$, the above approximation can be equivalently written as

$$(\widehat{\mathbf{J}}'\widehat{\mathbf{J}})^{-1/2} \widehat{\mathbf{J}}'(\widehat{\mathbf{f}} - \mathbf{f}) \simeq \mathcal{N}(\mathbf{0}, \sigma^2 \mathbf{I}). \tag{9.41}$$

This approximation, in the framework of nonlinear regression, was termed by Pazman [80] the near exact approximation. The score approximation (9.41) gives rise to construction of a confidence region with an approximate $(1 - \alpha)100\%$ confidence level

$$\left\{ \boldsymbol{\beta} \in R^m : (\widehat{\mathbf{f}} - \mathbf{f}(\boldsymbol{\beta}))' \widehat{\mathbf{J}} (\widehat{\mathbf{J}}'\widehat{\mathbf{J}})^{-1} \widehat{\mathbf{J}}' (\widehat{\mathbf{f}} - \mathbf{f}(\boldsymbol{\beta})) \leq \widehat{\sigma}^2 \chi^{-2}(1 - \alpha, m) \right\}. \tag{9.42}$$

Approximation (9.41) can be used to test the null hypothesis about the jth parameter $H_0 : \beta_j = \beta_{j0}$. Let $\widehat{\boldsymbol{\beta}}_j$ denote the NLS estimate that minimizes $\widehat{S}_{j0}$ in (9.35), where β_j is held fixed (restricted). Compute the restricted fitted value

$\widehat{\mathbf{f}}_j = \mathbf{f}(\widehat{\boldsymbol{\beta}}_j)$, the variance estimate $\widehat{\sigma}_j^2 = \left\| \mathbf{y} - \widehat{\mathbf{f}}_j \right\|^2 / (n - m + 1)$, and the restricted Jacobian $\widehat{\mathbf{J}}_j = \mathbf{J}(\widehat{\boldsymbol{\beta}}_j)$. Compute the test statistic

$$\widehat{\sigma}_j^{-2} (\widehat{\mathbf{f}} - \widehat{\mathbf{f}}_j)' \widehat{\mathbf{J}}_j (\widehat{\mathbf{J}}_j' \widehat{\mathbf{J}}_j)^{-1} \widehat{\mathbf{J}}_j' (\widehat{\mathbf{f}} - \widehat{\mathbf{f}}_j), \tag{9.43}$$

where $\widehat{\mathbf{f}} = \mathbf{f}(\widehat{\boldsymbol{\beta}})$ be the unrestricted fitted value as above. Then, if the observed value of (9.43) is greater than the quantile of the chi-square distribution with 1 degree of freedom, then we reject the null $H_0 : \beta_j = \beta_{j0}$.

To find the *score* CI for β_j, we use the profile idea by expressing other parameters through β_j as in the likelihood ratio test ($1 \le j \le m$ is fixed). In a way, the original regression with m parameters is reduced to a single parameter β_j. See Bates and Watts [7] and Brazzale et al. [16] for more detail. The algorithm for the score CI for parameter β_j is proceeds as follows (the above notation applies).

1. Define the test statistic as a function of β_j,

$$T_j(\beta_j) = \widehat{\sigma}_j^{-2} (\widehat{\mathbf{f}} - \mathbf{f}(\widetilde{\boldsymbol{\beta}}_j))' \widehat{\mathbf{J}}_j (\widehat{\mathbf{J}}_j' \widehat{\mathbf{J}}_j)^{-1} \widehat{\mathbf{J}}_j' (\widehat{\mathbf{f}} - \mathbf{f}(\widetilde{\boldsymbol{\beta}}_j)). \tag{9.44}$$

2. The $(1 - \alpha)$ score CI for β_j is the solution to the equation

$$T_j(\beta_j) - \chi^{-2}(1 - \alpha, 1) = 0. \tag{9.45}$$

As in the LR CI, the `uniroot` function is used to solve equation (9.45). See Example 9.15 and Section 9.6 for detail.

Prediction by nonlinear regression

An important application of regression is prediction of the dependent variable $Y = f(\mathbf{x}; \boldsymbol{\beta}) + \varepsilon$ given the predictor value $\mathbf{x}$. The prediction is obvious, $\widehat{Y} = f(\mathbf{x}; \widehat{\boldsymbol{\beta}})$. We use the delta method of Section 3.10.3 to approximate the standard error of prediction as

$$SE(\widehat{Y}) \simeq \widehat{\sigma} \sqrt{1 + \mathbf{j}'(\widehat{\mathbf{J}}'\widehat{\mathbf{J}})^{-1}\mathbf{j}}, \tag{9.46}$$

where $\widehat{\mathbf{J}} = \mathbf{J}(\widehat{\boldsymbol{\beta}})$ is the $n \times m$ Jacobian and $\mathbf{j} = \partial f(\mathbf{x}; \boldsymbol{\beta})/\partial \boldsymbol{\beta}$ is the $m \times 1$ derivative at prediction $\mathbf{x}$, both evaluated at the NLS estimate, $\widehat{\boldsymbol{\beta}}$. Consequently, we obtain an approximate $(1 - \alpha)$th symmetric confidence interval (CI) for Y as

$$f(\mathbf{x}; \widehat{\boldsymbol{\beta}}) \pm \Phi^{-1}(1 - \alpha/2)SE(\widehat{Y}).$$

Note that this CI covers future observation Y, that is, individual prediction. To obtain the CI for regression/mean prediction, 1 should be removed from the square root in formula (9.46). Recall that we discussed the difference between individual and mean prediction in Section 6.7.3 in the framework of linear regression.

9.4.4   Three methods of statistical inference in large sample

The asymptotic properties of the Wald, LR and score tests under iid assumption in large sample under the umbrella of the ML theory discussed in Section 7.9. The goal of this section is to demonstrate that we may expect similar statistical properties from a numerical point of view by referring to confidence regions produced by three asymptotic methods. Although this analysis is somewhat ambiguous it offers a big picture.

1. *Equivalence between (9.34) and (9.40).* Expand $\|\mathbf{y} - \mathbf{f}(\boldsymbol{\beta})\|^2$ into Taylor series around the NLS estimate up to the second term. Since the derivative is zero at $\widehat{\boldsymbol{\beta}}$, we have

$$\|\mathbf{y} - \mathbf{f}(\boldsymbol{\beta})\|^2 - \left\|\mathbf{y} - \mathbf{f}(\widehat{\boldsymbol{\beta}})\right\|^2 \;\simeq\; \frac{1}{2}(\boldsymbol{\beta}-\widehat{\boldsymbol{\beta}})' \left.\frac{\partial^2 \|\mathbf{y} - \mathbf{f}(\boldsymbol{\beta})\|^2}{\partial \boldsymbol{\beta}}\right|_{\boldsymbol{\beta}=\widehat{\boldsymbol{\beta}}} (\boldsymbol{\beta}-\widehat{\boldsymbol{\beta}})$$

$$\simeq (\boldsymbol{\beta}-\widehat{\boldsymbol{\beta}})'(\widehat{\mathbf{J}}'\widehat{\mathbf{J}})(\boldsymbol{\beta}-\widehat{\boldsymbol{\beta}}).$$

   The second approximation is due to the fact that the expected Hessian is $\widehat{\mathbf{J}}'\widehat{\mathbf{J}}$, which supports the claim that the Gauss–Newton and Fisher scoring algorithms are equivalent for nonlinear regression.

2. *Equivalence between (9.34) and (9.39).* Using an elementary calculus approximation, $\ln(1 + x) \simeq x$ for small x, we represent

$$\ln \|\mathbf{y} - \mathbf{f}(\boldsymbol{\beta})\|^2 - \ln \widehat{S} = \ln\left(1 + \frac{\|\mathbf{y} - \mathbf{f}(\boldsymbol{\beta})\|^2 - \widehat{S}}{\widehat{S}}\right)$$

$$\simeq \frac{\|\mathbf{y} - \mathbf{f}(\boldsymbol{\beta})\|^2 - \widehat{S}}{\widehat{S}} \simeq \frac{\|\mathbf{y} - \mathbf{f}(\boldsymbol{\beta})\|^2 - \widehat{S}}{\widehat{\sigma}^2},$$

   which implies the claimed equivalence. Note that we assume that $\widehat{S}$ and $\|\mathbf{y} - \mathbf{f}(\boldsymbol{\beta})\|^2$ are close, which is true when $\boldsymbol{\beta}$ and $\widehat{\boldsymbol{\beta}}$ are close.

3. *Equivalence between (9.34) and (9.42).* Approximate the nonlinear regression function around $\widehat{\boldsymbol{\beta}}$ with a linear function as in (9.16), $\mathbf{f}(\boldsymbol{\beta}) \simeq \widehat{\mathbf{f}} + \widehat{\mathbf{J}}(\boldsymbol{\beta}-\widehat{\boldsymbol{\beta}})$ where $\widehat{\mathbf{f}} = \mathbf{f}(\widehat{\boldsymbol{\beta}})$. Then, substituting this into the left-hand side of expression (9.42) we obtain

$$(\boldsymbol{\beta}-\widehat{\boldsymbol{\beta}})'(\widehat{\mathbf{J}}'\widehat{\mathbf{J}})(\widehat{\mathbf{J}}'\widehat{\mathbf{J}})^{-1}(\widehat{\mathbf{J}}'\widehat{\mathbf{J}}) = (\boldsymbol{\beta}-\widehat{\boldsymbol{\beta}})'(\widehat{\mathbf{J}}'\widehat{\mathbf{J}})(\boldsymbol{\beta}-\widehat{\boldsymbol{\beta}}),$$

   as in (9.34). Clearly, the difference may be considerable for small n or large σ^2.

An important observation is that all the approximations rely on proximity of $\boldsymbol{\beta}$ and $\widehat{\boldsymbol{\beta}}$. These means that the power functions of the respective tests are supposed to look similar in the neighborhood of the null value, but may diverge when the distance between the null and alternative values increases. This claim is supported by simulations in Section 9.6.

Problems

1. Does the NLS estimator of β in regression $Y_i = \beta/(1+\beta)+\varepsilon_i$ with normally distributed errors, conditional on $\overline{Y} > 0$, have a finite mean? [Hint: The integration is over positive values.]

2. Formulate conditions on the sequence $\{x_i, i = 1, 2, ...\}$ such that the NLS estimator in the Fieller regression is consistent under the deterministic approach using function Q.

3. Prove that the Wu condition (9.24) implies that $Q(\boldsymbol{\beta}, \boldsymbol{\beta}_0) = 0$ if and only if $\boldsymbol{\beta} = \boldsymbol{\beta}_0$, assuming that limit (9.23) exists.

4. Referring to Example 9.8, show that the Wu condition (9.24) holds for the quasilinear regression model $Y_i = g(\boldsymbol{\beta}'\mathbf{x}_i)+\varepsilon_i$, where g is a strictly increasing function such that $g' \geq \delta > 0$, and $\lambda_{\min}(\mathbf{X}'\mathbf{X}) \to \infty$ when $n \to \infty$.

5. (a) Prove that $p\lim_{n\to\infty} \overline{Y}_n = \alpha_1$ in Example 9.9. (b) Can parameter α_2 be consistently estimated?

6. Prove that, under the stochastic scheme, matrix $\mathbf{B}(\boldsymbol{\beta})$ is equal to matrix $\mathbf{A}(\boldsymbol{\beta})$ where, instead of lim, we use $p\lim$.

7. Formulate asymptotic properties of the LS estimator for parameters of linear regression $Y_i = \mathbf{x}_i'\boldsymbol{\beta}+\varepsilon_i$ in the stochastic approach where $\mathbf{x}_i \overset{\text{iid}}{\sim} \mathcal{N}(\boldsymbol{\mu}, \boldsymbol{\Omega})$ and $\boldsymbol{\Omega}$ is a positive definite matrix. Prove that $G(\boldsymbol{\beta}, \boldsymbol{\beta}_0) = \mathbf{0}$ iff $\boldsymbol{\beta} = \boldsymbol{\beta}_0$, and express matrix $\mathbf{B}(\boldsymbol{\beta})$ via $\boldsymbol{\mu}$ and $\boldsymbol{\Omega}$.

8. Following Example 9.10, formulate the asymptotic distribution for the exponential regression $Y_i = e^{\beta x_i} + \varepsilon_i$ where, under the stochastic scheme, $X \sim \mathcal{N}(0, \sigma_x^2)$.

9. Consider the intrinsically linear regression model (9.2) under the stochastic scheme. Prove that the linear least squares for g_1 and g_2 and the consequent back transformation for β_1 and β_2 have the same asymptotic covariance matrix as the NLS estimator for β_1 and β_2.

10. Formulate asymptotic properties of the NLS estimator for parameters of the Fieller regression in the stochastic approach assuming $x_i \overset{\text{iid}}{\sim} \mathcal{N}(\mu_x, \sigma_x^2)$ and $\mu_x \neq 0$, $\sigma_x^2 > 0$.

11. Demonstrate by simulations that p_{EX} is the exact density for parameter β in the nonlinear regression $Y_i \overset{\text{iid}}{\sim} \mathcal{N}(i\beta^3, \sigma^2)$, $i = 1, 2, ..., n$. Derive density p_{EX} and superimpose the empirical cdf with the theoretical cdf derived from integration of p_{EX} (use `integrate`).

12. Generalize the previous problem to the nonlinear regression model $\mathbf{y} \sim \mathcal{N}(\beta\mathbf{x}+e^{\beta\mathbf{z}}, \sigma^2\mathbf{I})$, where $n \times 1$ vectors $\mathbf{x}$ and $\mathbf{z}$ are not collinear. [Hint: Use `uniroot` to find the NLS estimate.]

13. Prove that the score approximation (9.41) is exact for a linear model with known σ^2.

14. Derive the density of the NLS estimator in a one-parameter nonlinear regression from approximation (9.41) and show that it yields density p_{EX} from Section 9.4.2 with $a_{EX} = 1+(AB-CD)/A^2$ (see more detail in [31]).

15. Derive (9.36) from the general LR test based on the $\chi^2(k)$ approximation.

9.5   The nls function and examples

In this section, we introduce the R function `nls` and show by examples how to use it. To run this function, the user has to specify the formula for nonlinear regression and starting values. We use the marathon world record data from 1908 to 2013 from the website http://en.wikipedia.org/wiki /Marathon_world_record_ progression for illustration; see Figure 9.6. As the reader can see, the best running time may not necessarily decrease from year to year (marathon length is 26 miles 385 yards). It is interesting to know if there is a limit, the absolute World record.

Example 9.12 *Estimation of the absolute marathon world record.* (a) *Use the exponential function with asymptote to estimate the absolute marathon world record. (b) Test the hypothesis that the absolute world record is two hours using the Wald and LR tests.*

Solution. The data are in file `marathonWR2.txt` and the R code is in function `marathon` with the `summary` output listed below. Note that the data time format is `55:18.4` with the number 2 (hours) omitted. Thus we need to cut out the minute data and transform characters to `numeric` type to obtain array `ti`.

The marathon record time t_i in year Y_i is modeled as an exponential function with an asymptote

$$t_i = \alpha_1 + \alpha_2 e^{-\alpha_3(Y_i-1900)} + \varepsilon_i, \quad i = 1, 2, ..., n.$$

According to this model, in year $Y = 1900$, the expected record time is $\alpha_1 + \alpha_2$ and the predicted absolute marathon world record is α_1. Parameter α_3 specifies the annual rate of the running time decrease. In words, this parameter tells us how fast the year record approaches the absolute world record. The call to the built-in function `nls` is

`nls(ti~a1+a2*exp(-a3*(year-1900)),start=list(a1=130,a2=10,a3=.01)).`

As in the case of linear model, function `lm`, the variable before ~ specifies the name of the dependent variable and the expression after ~ specifies the nonlinear regression formula. When the regression model is more complex than just a formula `function` should be used. For example, we may define the function before the `nls` call

```
mwr=function(a1,a2,a3,time_year) a1+a2*exp(-a3*(time_year-1900))
```

and then use

```
out=nls(ti~mwr(a1,a2,a3,time_year=year),
        start=list(a1=a1.0,a2=a2.0,a3=a3.0))
```

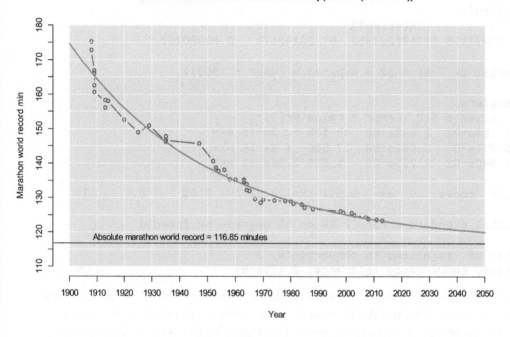

Figure 9.6: *Fit of the marathon world record by* `nls` *using the exponential function with the estimated asymptote as the absolute world record.*

Parameter names may be any and their starting values must be given using the `start` option (this is different from `lm`). When starting values are far from the NLS estimate the minimization algorithm may not converge. There is no general method to find satisfactory starting values. However, in the case with only one intrinsically nonlinear parameter one may use a loop over a range of values and estimate other parameters using the `lm` function. Indeed, in the case of the exponential function only parameter `a3` is intrinsically nonlinear and parameters `a1` and `a2` can be estimated by the `lm` function – see the output and the R code `marathon`. The output below is generated by the command `summary(out)` where out is the output of function `nls`. As discussed in the previous section, Std.Error of the jth parameter is SE_j and the t value is the ratio of the Estimate to Std.Error. Pr($>|$t$|$) is the p-value computed based on the t-distribution with 48 - 3 = 45 degrees of freedom under the null hypothesis that the true parameter is zero. Note that unlike linear regression neither standard error nor p-value are exact for nonlinear regression. Residual standard error is $\hat{\sigma}$ as the square root of the minimum residual sum of squares divided by degrees of freedom. If out is the output of function nls, the Parameters table is accessed as summary(out)\$coefficients or coef(summary(out)), and $\hat{\sigma}$ is accessed as summary(out)\$sigma. Parameter estimates can be accessed as coef(out).

Below is the printout of the marathon function.

```
marathon()
            (Intercept)               x
 2.63473315 117.71237352  57.26242173     0.02020202


Formula: ti ~ a1 + a2 * exp(-a3 * (year - 1900))

Parameters:
   Estimate Std. Error t value Pr(>|t|)
a1 1.168e+02  2.648e+00  44.130  < 2e-16 ***
a2 5.771e+01  2.092e+00  27.586  < 2e-16 ***
a3 1.940e-02  2.265e-03   8.566 5.26e-11 ***
---
Signif. codes:  0 '***' 0.001 '**' 0.01 '*' 0.05 '.' 0.1 ' ' 1

Residual standard error: 2.66 on 45 degrees of freedom

Number of iterations to convergence: 3
Achieved convergence tolerance: 2.426e-07
```

According to this model, we estimate that each year the record time decreases by 2% and the absolute World record is 117 minutes. Recall that p-value tests that the respective parameter is zero. For example, the p-value for parameter a1 does not make much sense because the null hypothesis is the absolute world record is zero, the p-value for a2 does not make much sense either. The p-value for parameter a3 tests if marathon time decreases from year to year on average. All these hypotheses obviously have little practical interest and it is not surprising that the p-values are so small.

(b) Now we illustrate how to test a hypothesis with nonlinear regression as discussed in the previous section. Let us test if the absolute marathon world record is exactly two hours, $H_0 : \alpha_1 = 120$. The Z-score of the Wald test is

$$(\hat{\alpha}_1 - 120)/SE_1 = (116.8 - 120)/2.648 = -1.2085 > -1.96.$$

Following the traditional language, the null hypothesis cannot be rejected with the p-value computed as 2*(1-pnorm(1.2086)) =0.226855.

To apply the likelihood ratio test we need to run regression $(t_i - 120) = \alpha_2 e^{-\alpha_3 (Y_i - 1900)} + \varepsilon_i$. The following lines of codes have been added to compute the LR test p-value:

```
#LRT H0:a1=120
ti120=ti-120
out120=nls(ti120~a2*exp(-a3*(year-1900)),
        start=list(a2=a2.0,a3=a3.0))
n=length(ti)
S.hat=summary(out)$sigma^2*(n-3)
S.hat120=summary(out120)$sigma^2*(n-2)
LHS=n*log(S.hat120/S.hat)
pvLRT=1-pchisq(LHS,df=1)
paste("LRT p-value =",pvLRT)
```

with the output

```
[1] "LRT p-value = 0.176937930208949"
```

The two methods give the same result: with type I error 0.05, the hypothesis that the absolute world record is two hours cannot be rejected.   □

We make a few caution comments on the interpretation of the `nls` output:

1. The reason that the ratio of the parameter estimate to its standard error is called the t-statistic in the `nls` software, not Z-statistic, is because the t-distribution with $n - m$ degrees of freedom is used to compute the p-value to mimic the linear model. However, while in linear model $\widehat{\beta}_j / SE_j$ follows the t-distribution, this ratio does not follow the t-distribution in nonlinear regression.

2. It is worthwhile to remember that the p-value reported by `nls` is associated with the null hypothesis that the coefficient (more accurately 'parameter') is zero. In linear model, the zero-coefficient ($\beta_j = 0$) hypothesis tests the presence of the independent variable (predictor) in regression, while in nonlinear regression may have no predictor associated with β_j. Moreover, the null hypothesis $\beta_j = 0$ may have no sense. Consequently, small p-values may not necessarily point to a practically important result. For example, in the marathon example the p-value $< 2 \times 10^{-16}$ for parameter `a1` is associated with the null hypothesis that a human can cover 26 miles in no time.

Example 9.13 *Change-point regression continued. The data on the change-point regression are found in file dnaRAD.csv (see Figure 9.3). (a) Find a starting value and use the Gauss–Newton iterations to estimate four parameters. (b) Find the 95% CI for the change-point using the Wald and LR profile CIs.*

Solution. (a) As follows from Example 9.3 the only intrinsic nonlinear parameter is r_T, other parameters are linear. Therefore to get good starting values we can run linear regressions with $r_T = 1, 2, ..., 24$ and choose with minimum RSS. Below is summary of `lm` output with minimum RSS corresponding to $r_T = 5$.

```
Coefficients:
             Estimate Std. Error t value Pr(>|t|)
(Intercept)  3.16123    0.04431  71.346  < 2e-16 ***
x1           0.01117    0.01113   1.003   0.317
x2           0.08716    0.01253   6.958 3.09e-11 ***
---
Signif. codes:  0 '***' 0.001 '**' 0.01 '*' 0.05 '.' 0.1 ' ' 1

Residual standard error: 0.2032 on 247 degrees of freedom
Multiple R-squared:  0.9089,    Adjusted R-squared:  0.9082
F-statistic:  1232 on 2 and 247 DF,  p-value: < 2.2e-16

[1] 5
```

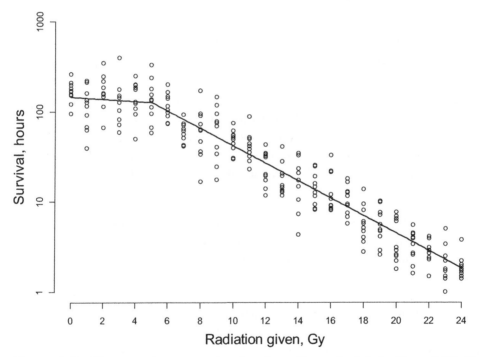

Figure 9.7: *Change-point regression for the cell survival data, $\hat{r}_T = 5.05$ Gy.*

However, the `nls` function starting from these values does not work for this model and produces an error (see the R code `dnaRAD`):

```
Error in nls(log10Surv~a1-a2*rad-a3*pmax(rad-a4,0),
    start=list(a1=b[1],:  step factor 0.000488281 reduced below
    'minFactor' of 0.000976562
```

The reason is that the regression function is not differentiable with respect to r_T where $r_i = r_T$. To work around this problem, apply Gauss–Newton iterations by constructing the Jacobian matrix

$$J = \text{cbind}(\text{rep}(1, n), -\text{rad}, x2, b[3] * (\text{rad} > b[4]))$$

where the array b has elements corresponding to α, β, δ and r_T. The last column is the derivative of the L function with respect to r_T. The GN iterations (9.17) produce the following result (see the code).

```
beta-est              SE         P-value
b1 3.16261344  0.04670953  0.000000e+00
b2 0.01220410  0.01542765  4.289127e-01
b3 0.08478215  0.01566269  6.197688e-08
b4 5.05302712  0.66594786  3.264056e-14
```

(b) The Wald CI for the change point (parameter b4) is found as $5.053 \pm 1.96 \times 0.666 = (3.75, 6.36)$. Computation of the LR profile 95% CI is found in function dnaRAD1 and gives a nearly identical CI: $(3.7, 6.34)$. □

Nonlinear regression models are often used in pharmacokinetics to model drug concentration in the body. Below is a typical example concerning imaging.

Several imaging techniques, such as magnetic resonance or PET, require injection of a contrast agent. When should imaging begin before the agent starts diffusing and washing out? The combination of two processes, agent diffusion and washing out give rise to a two-compartment pharmacokinetics model.

Example 9.14 *Two-compartment pharmacokinetics model. The injected contrast agent reaches the vital body organ according to the exponential model*

$$V_D(t) = V_0(1 - e^{-\alpha t}),$$

where $\alpha > 0$ is the diffusion rate and V_0 is the maximum drug concentration in the organ measured as mg/l. On the other hand, the drug washes out of the body following a decay model $V_D(t)e^{-\beta t}$ resulting in the three-parameter drug concentration model

$$V(t) = V_0(1 - e^{-\alpha t})e^{-\beta t}, \quad t \geq 0.$$

Using the data in file twocph.csv, *(a) find good starting values on the grid for V_0, α and β values using vectorized computations; (b) estimate parameters V_0, α, and β by the NLS; (c) compute and display the 95% CI for fitted value; and (d) estimate the optimal time of imaging, its standard error, and the 95% CI.*

Solution. Figure 9.8 displays the data and the fitted model. The R code is found in function twopch. Here we use a local function VtM to compute the two-compartment model and its derivatives/gradient/Jacobian with respect to three parameters.

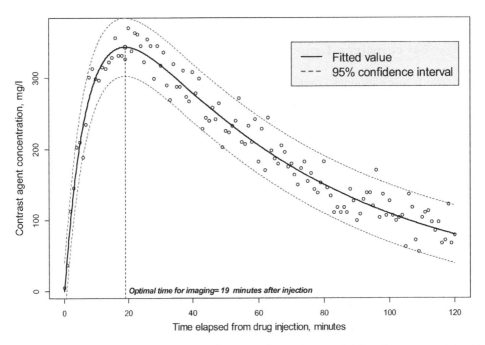

Figure 9.8: *The two-compartment pharmacokinetics model for drug concentration after injection. Roughly speaking, imaging should be done within 18–20 minute interval.*

```
VtM=function(V0,alpha,beta,x) # Two-compartment function
{
    e1=exp(-alpha*x);e2=exp(-beta*x)
    Vt.ret=V0*(1-e1)*e2
    dVdV0=(1-e1)*e2
    dVda=V0*e1*e2*x
    dVdb=-Vt.ret*x
    attr(Vt.ret,"gradient")=cbind(dVdV0,dVda,dVdb)
    return(Vt.ret)
}
```

Note that the gradient is attached to the return value as an attribute. For example, if `Vall` is the return value of `VtM`, we can extract the Jacobian as `attr(Vall,"gradient")`. We need the Jacobian to compute the CI for the fitted value. The call to the `nls()` function is `out=nls(Vt~VtM(V0,alpha, beta,x=min),start=list(V0=V00,alpha=a0,beta=b0))`.

(a) A straightforward way to get starting values for the three parameters is to define the grid of values for α and β using a double loop and then find the optimal V_0 that minimizes the RSS, $\sum_{t=1}^{n}(V_t - V_0 E_t)^2$, where $E_i = (1-e^{-\alpha t})e^{-\beta t}$, as in the slope model. It is easy to see that the minimum of RSS is equivalent to the maximum of $(\sum_{t=1}^{n} V_t E_i)^2 / \sum_{t=1}^{n} E_i^2$. Using vectorized computations, we create the double grid for the two parameters using either `expand.grid` or `rep` with the `times` (default) and `each` options (see Example 1.12) and use the RSS from the slope model. Once the double (extended) grid for α and β is constructed, we compute a matrix of $V(t)$ values with the number of rows equal the length of the double grid times the length of `min`. This can be done via matrix multiplication as `alphs%*%t(min)` and `betas%*%t(min)`, where `alphs` and `betas` are double grid arrays. Another way to find good starting values is to use one grid for α and estimate V_0 and β from linear regression $\ln(V_t/(1-e^{-\alpha t})) = \ln V_0 - \beta t + \eta$.

(b) Once starting values are determined, use of the of `nls` function is straight-forward (see function `twocph`). (c) The standard error for predicted value $Y = f(\mathbf{x}; \widehat{\boldsymbol{\beta}}) + \varepsilon$ is computed by formula (9.46).

(d) The optimal time of imaging is when the volume of the contrast agent is maximum. Take the derivative $dV(t)/dt = V_0 \left[\alpha e^{-\alpha t} - \beta(1 - e^{-\alpha t}) \right] e^{-\beta t}$ and equate to zero, which gives $t_{\mathrm{opt}} = \alpha^{-1} \ln(1 + \alpha/\beta)$. Matrix $(\mathbf{J'J})^{-1}$ evaluated at the iteration can be retrieved from the `nls` output using `summary(out.nls)$cov.unscaled` so that the 3×3 covariance matrix for the NLS estimate, $\mathbf{C}$, is esti-mated as `summary(out.nls)$sigma^2*summary(out. nls)$cov.unscaled`. Al-ternatively, one could obtain this matrix by `vcov(out.nls)`. The optimal time estimated as $\widehat{t}_{\mathrm{opt}} = \widehat{\alpha} \ln\left(1 + \widehat{\alpha}/\widehat{\beta}\right)$ is a nonlinear function of $\widehat{\alpha}$ and $\widehat{\beta}$, so the delta method may be used to estimate the variance of $\widehat{t}_{\mathrm{opt}}$ as

$$
\mathrm{var}(\widehat{t}_{\mathrm{opt}}) \simeq
\begin{bmatrix}
\left. \dfrac{\partial t_{\mathrm{opt}}}{\partial \alpha} \right|_{\alpha=\widehat{\alpha}, \beta=\widehat{\beta}} \\[2ex]
\left. \dfrac{\partial t_{\mathrm{opt}}}{\partial \beta} \right|_{\alpha=\widehat{\alpha}, \beta=\widehat{\beta}}
\end{bmatrix}'
\mathbf{C}
\begin{bmatrix}
\left. \dfrac{\partial t_{\mathrm{opt}}}{\partial \alpha} \right|_{\alpha=\widehat{\alpha}, \beta=\widehat{\beta}} \\[2ex]
\left. \dfrac{\partial t_{\mathrm{opt}}}{\partial \beta} \right|_{\alpha=\widehat{\alpha}, \beta=\widehat{\beta}}
\end{bmatrix},
\tag{9.47}
$$

where

$$
\frac{\partial t_{\mathrm{opt}}}{\partial \alpha} = \frac{(\alpha + \beta)^{-1}\alpha - (\ln(\alpha + \beta) - \ln \beta)}{\alpha^2}, \quad
\frac{\partial t_{\mathrm{opt}}}{\partial \beta} = \frac{(\alpha + \beta)^{-1} - \beta^{-1}}{\alpha}.
$$

```
> twocph()
Starting values:  752.1721 0.06154545 0.02118182
Formula:  Vt ~VO * (1 - exp(-alpha * min)) * exp(-beta * min)
Parameters:
 Estimate Std.  Error t value Pr(>|t|)
VO 5.324e+02 1.715e+01 31.04 <2e-16 ***
alpha 1.084e-01 7.220e-03 15.01 <2e-16 ***
beta 1.587e-02 5.291e-04 29.99 <2e-16 ***
Signif.  codes:  0 '***' 0.001 '**' 0.01 '*' 0.05 '.'  0.1 ' ' 1
Residual standard error:  20.54 on 118 degrees of freedom
Number of iterations to convergence:  5
Achieved convergence tolerance:  7.238e-06
```

```
topt = 18.99124 , SE topt = 0.5476612
```

The optimal time of imaging is 19 minutes after injection with the standard error 0.55 with the 95% CI $19 \pm 1.96 \times 0.55 = (18, 20)$. Roughly speaking, imaging should be done within 18–20 minutes.

Example 9.15 *Asymptotic CI. Compute three CIs, Wald, LR, and score, for parameter α in the two-compartment pharmacokinetics model from Example 9.14 with the confidence level 95% and 99%..*

Solution. The three methods for confidence interval estimation have been discussed in Section 9.4.3. The Wald CI is easy to derive as the NLS estimate $+/-$ SE times the critical value of the Z-statistic $= \Phi^{-1}(1 - \alpha/2)$. The likelihood ratio and score CIs involve solving equations (9.38) and (9.45), respectively, using the `uniroot` function. The code is in function `twocphCI`. We make several comments regarding the code: Three internal functions are part of the code: `VtM` computes the regression function with the Jacobian as an attribute. The other two functions compute the left-hand sides of (9.38) and (9.45) to be solved by the `uniroot` function. Function `uni3` computes equation (9.38) for the LR and function `unNE.prof` computes (9.45) for the score CI. Note that both functions call `nls` with the regression specified explicitly by the formula. Four calls to `uniroot` is required for the lower and upper limit in each method. The `interval` must be chosen such that the equation to solve has opposite signs at the ends of the interval. For the CI lower limit, we choose the left end as the NLS estimate minus four SE and for the right end as the estimate itself. For the CI upper limit, we choose the left end as the NLS estimate and the right end as the NLS estimate plus four SE. The two sets of CIs are shown below.

```
> twocphCI(alpha=.05)          > twocphCI(alpha=.01)
          Lower      Upper             Lower      Upper
Wald   0.09420460 0.1225047    Wald   0.08975833 0.1269510
LR     0.09498390 0.1231466    LR     0.09090318 0.1282504
Score  0.09468905 0.1235045    Score  0.09044946 0.1289123
```

The CIs are tight and are very similar due to the high precision of the alpha estimate.

Example 9.16 *Forensic science. A person died at 3 p.m. after a heart attack. A criminal investigator suspected a murder because an empty ampoule 20 mg of liquid poison has been found at the crime scene. Assuming that the injection of the poison invokes a heart attack when its cumulative concentration (area under the curve) reaches 22 mg/L and the pharmacokinetics estimates for α and β along with their covariance matrix, $\mathbf{C}$, are derived in Example 9.14, estimate the time when the injection took place and compute its 95% CI.*

Solution. The time to death after injection satisfies the equation $20 \int_0^T (1 - e^{-\alpha t}) e^{-\beta t} dt = 22$. This integral admits a closed-form expression, which leads to an equation for the time to death, T:

$$\beta + \alpha e^{-(\alpha+\beta)T} - (\alpha+\beta)e^{-\alpha T} = C, \qquad (9.48)$$

where $C = (22/20)\alpha(\alpha+\beta)$. To find T, we plug in the estimates for α and β from Example 9.14 and solve equation (9.48) by Newton's algorithm

$$T_{k+1} = T_k + \frac{\beta + \alpha e^{-(\alpha+\beta)T} - (\alpha+\beta)e^{-\alpha T} - C}{\alpha(\alpha+\beta)\left(e^{-(\alpha+\beta)T} - e^{-\alpha T}\right)}, \quad k = 0, 1, 2, \ldots$$

starting from T_0 found from equation $\beta - (\alpha+\beta)e^{-\alpha T} - C = 0$ (see function twocph). This starting value is greater than the solution and guarantees convergence of Newton's algorithm (see Section 10.6.3). A few iterations yield $T = 38$. Thus, we conclude that the estimated time when injection took place is 2:22 p.m. To estimate the 95% CI, we find the variance from the delta method with the covariance matrix for $(\widehat{\alpha}, \widehat{\beta})$ derived in Example 9.14. The derivatives $\partial T/\partial\alpha$ and $\partial T/\partial\beta$ of the intrinsic function T are obtained from derivation of the left-hand side of (9.48) with respect to α and β. These derivatives could be obtained by hand, but here we apply an R built-in symbolic differentiation using the D function in conjunction with expression command.

```
>D(expression(a3+a2*exp(-T*(a2+a3))-(a2+a3)*exp(-a2*T)),'a2')
exp(-T*(a2+a3))-a2*(exp(-T*(a2+a3))*T)-(exp(-a2*T)-(a2+a3)*exp(-a2*T)*T))
>D(expression(a3+a2*exp(-T*(a2+a3))-(a2+a3)*exp(-a2*T)),'a3')
1-a2*(exp(-T*(a2+a3))*T)-exp(-a2*T)
>D(expression(a3+a2*exp(-T*(a2+a3))-(a2+a3)*exp(-a2*T)),'T')-
(a2*(exp(-T*(a2+a3))*(a2+a3))-(a2+a3)*(exp(-a2*T)*a2))
```

These derivatives are obtained in the console and copied and pasted into the R code. We conclude that with 95% confidence probability the injection took place somewhere from 2:15 to 2:29 p.m. □

The following is an example of the likelihood ratio test when the Wald test is not applicable. The Cobb–Douglas production function, under the assumption of no economy of scale, relates the two resources, capital (K) and labor (L) to the output as $Y = AK^\alpha L^{1-\alpha}$. A popular generalization of this function is the constant elasticity (CES) production function, a popular model in econometrics defined by

$$Y = A(\alpha K^\rho + (1-\alpha)L^\rho)^{1/\rho},$$

where ρ is the elasticity parameter (it can be positive and negative). See more detail in Greene [43]. When $\rho \to 0$ the CES function converges to the Cobb-Douglas production function. To show this, apply L'Hôpital's rule to the log CES function,

$$\lim_{\rho\to 0} \ln Y = \ln A + \lim_{\rho\to 0} \frac{\ln(\alpha K^\rho + (1-\alpha)L^\rho)}{\rho}$$

$$= \ln A + \frac{\alpha \ln K + (1-\alpha)\ln L}{\alpha + (1-\alpha)} = \ln A + \alpha \ln K + (1-\alpha)\ln L.$$

Therefore, after exponentiating, we obtain the desired result: $\lim_{\rho \to 0} Y = AK^\alpha L^{1-\alpha}$.

Example 9.17 *CES production function. File* `ces.csv` *contains 50 data points for K, L, and Y. (a) Estimate parameters of the CES production function, plot the estimated surface along with the actual data and values. (b) Test the hypothesis that the CES and the Cobb–Douglas functions are statistically indistinguishable.*

Solution. The R code is found in file `ces.r`. The nonlinear regression for estimation of the CES function takes the form

$$Y_i = A(\alpha K_i^\rho + (1 - \alpha)L_i^\rho)^{1/\rho} + \varepsilon_i, \quad i = 1, 2..., n,$$

where $\varepsilon_i \overset{\text{iid}}{\sim} \mathcal{N}(0, \sigma^2)$. (a) The NLS fitting for this function requires fairly accurate starting values. Since parameter A is linear, it suffices to use the grid over values α_l and ρ_j. To find the optimal values, we use a double loop and compute vector $(\alpha_l K^{\rho_j} + (1 - \alpha_l)L^{\rho_j})^{1/\rho_j}$. Then we find the optimal A as the slope coefficient from linear least squares.

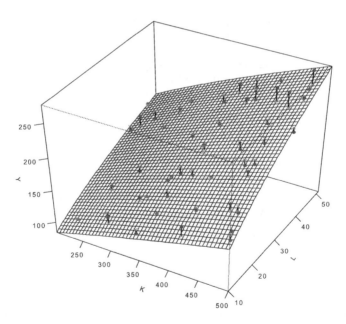

Figure 9.9: *The fitted surface for the CES production function with actual and fitted values as projections.*

The output of the `ces` function is below.

```
ces()
[1] "Starting values:  0.484813 0.445555 0.243343"
Formula:  Y ~A * (alpha * K^ro + (1 - alpha) * L^ro)^(1/ro)
```

```
Parameters:
 Estimate Std.  Error t value Pr(>|t|)
A 1.6783 0.2517 6.668 2.61e-08 ***
alpha 0.4414 0.1167 3.784 0.000437 ***
ro 0.2492 0.1824 1.366 0.178355
Signif.  codes:  0 '***' 0.001 '**' 0.01 '*' 0.05 '.'  0.1 ' ' 1
Residual standard error:  8.962 on 47 degrees of freedom
Number of iterations to convergence:  3
Achieved convergence tolerance:  2e-06
Formula:  Y ~A * K^alpha * L^(1 - alpha)
Parameters:
 Estimate Std.  Error t value Pr(>|t|)
A 1.37692 0.05149 26.74 <2e-16 ***
alpha 0.60054 0.01478 40.63 <2e-16 ***
Signif.  codes:  0 '***' 0.001 '**' 0.01 '*' 0.05 '.'  0.1 ' ' 1
Residual standard error:  9.043 on 48 degrees of freedom
Number of iterations to convergence:  4
Achieved convergence tolerance:  9.615e-06
[1] "LRT p-value= 0.162944425087874"
```

Since the p-value corresponds to the null hypothesis $H_0 : \rho = 0$, the Wald p-value can be obtained directly from the readout, 0.178355.

(b) To compute the p-value for the LR test, we have to estimate the Cobb–Douglas production function from the nonlinear regression $Y_i = AK_i^\alpha L_i^{1-\alpha} + \varepsilon_i$. We underscore that using linear regression after taking the log transformation is not appropriate here because it silently assumes the multiplicative error; we need to stay with the additive error to compare the RSS from the two models. The LR test follows the lines of testing in Example 9.12. Figure 9.9 depicts the estimated surface of the CES production function with the actual triple data (K_i, L_i, Y_i) and fitted values as projections onto the surface along the z-axis. As we can see from the plot, it fits is fairly well.

Example 9.18 *Falling hat.* *The data on the distance covered by a free falling hat is found in file FallingHat.csv. Compare fitting the data with the C-model discussed in Example 9.4 with Newton's formula $S(t) = 0.5gt^2$ and this formula adjusted by a linear term.*

Solution. The data and three models are depicted in Figure 9.10. Fitting with the C-model requires calling nls; function ff(time,C.est) computes the distance covered using the C-model given by formula (9.10). As follows from the output below, $\widehat{C} = 0.3318$. A competing model is the Newton gravitation formula $S(t) = 0.5gt^2$ where $g = 9.7935$. A naive idea to account for air resistance is to augment Newton's formula with a linear term, $S(t) = 0.5gt^2 + at$, referred to as "Linear" in Figure 9.10. Indeed as follows from the output below $\widehat{a} = -0.02925 <$

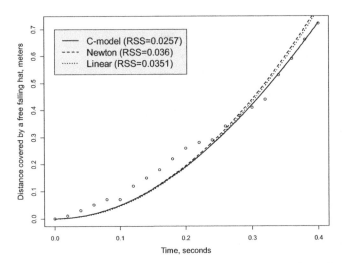

Figure 9.10: *Falling hat example: residual sum of squares (RSS) from the three fitting models.*

0, but statistically insignificant, indicates that the fall is slower than predicted by Newton's formula with no air resistance.

```
Formula: St ~ ff(time, C.est)
Parameters:
      Estimate Std. Error t value Pr(>|t|)
C.est    0.3318      0.1312    2.53    0.0199 *
Residual standard error: 0.03584 on 20 degrees of freedom
Number of iterations to convergence: 3
Achieved convergence tolerance: 3.674e-06

lm(formula = Stg ~ da$time - 1)
Coefficients:
        Estimate Std. Error t value Pr(>|t|)
da$time -0.02925    0.03907  -0.748    0.463
Residual standard error: 0.04186 on 20 degrees of freedom
Multiple R-squared: 0.02725,   Adjusted R-squared:  -0.02139
F-statistic: 0.5602 on 1 and 20 DF,  p-value: 0.4629
```

All three models poorly predict the hat fall at the time of release perhaps because the hat went down sideways. The C-model is preferable.

9.5.1 NLS-cdf estimator

In this section, we describe how nonlinear regression can be used to estimate parameters of distribution defined by a cumulative distribution function (cdf). Let Y_i be n iid observations drawn from a general population with cdf $F(y; \boldsymbol{\theta})$ dependent on unknown vector parameter $\boldsymbol{\theta}$. Recall that the empirical cdf can be unbiasedly estimated as the proportion of observations to the left of y,

$$F_i = \frac{\sum_{j=1}^n 1(Y_j \leq Y_i)}{n}.$$

See Section 2.1.1. The proportion F_i can be treated as the dependent variable in a nonlinear regression of F_i on $F(Y_i; \boldsymbol{\theta})$. That is, the parameter vector $\boldsymbol{\theta}$ can be estimated by the nonlinear least squares of the difference between empirical and theoretical cdfs,

$$\min_{\boldsymbol{\theta}} \sum_{i=1}^n (F_i - F(Y_i; \boldsymbol{\theta}))^2.$$

If observations are reindexed in ascending order (see Section 3.9), one easy computes $F_i^* = (i - 1/2)/n$. The $1/2$ correction avoids 0 and 1 values for the empirical cdf which may be convenient. The nonlinear least squares problem is rewritten as

$$\min_{\boldsymbol{\theta}} \sum_{i=1}^n (F_i^* - F(Y_{(i)}; \boldsymbol{\theta}))^2,$$

where $Y_{(i)}$ is the order statistic. This method will be referred to as the *NLS-cdf* method of estimation. An attractive feature of this method is that no special software is required because the `nls` function suffices.

The variance of F_i can be approximated either empirically as $F_i(1 - F_i)$ or theoretically as $F(Y_i; \boldsymbol{\theta})(1 - F(Y_i; \boldsymbol{\theta}))$. This suggests the *weighted NLS-cdf* estimator

$$\min_{\boldsymbol{\theta}} \sum_{i=1}^n (F_i - F(Y_i; \boldsymbol{\theta}))^2 w_i,$$

where

$$w_i = \frac{1}{F_i(1 - F_i)} \text{ or } w_i = \frac{1}{F(Y_i; \widehat{\boldsymbol{\theta}})(1 - F(Y_i; \widehat{\boldsymbol{\theta}}))}$$

and $\widehat{\boldsymbol{\theta}}$ is an estimate of $\boldsymbol{\theta}$. The `weights` is an option in `nls`, so this method is easy to implement. The NLS-cdf estimate may provide a reasonable starting value for ML algorithm.

Example 9.19 *NLS for the gamma distribution.* *Use* `nls` *to estimate the parameters of the gamma distribution. Compare the method of moments (MM) with the NLS-cdf method of estimation via simulations by reporting the bias and the relative MSE.*

Solution. If $Y_1, ..., Y_n$ is a random sample from a gamma distribution with parameters α and λ, the MM estimators are

$$\widehat{\lambda}_{MM} = \frac{\overline{Y}}{\widehat{\sigma}^2}, \quad \widehat{\alpha}_{MM} = \frac{\overline{Y}^2}{\widehat{\sigma}^2}.$$

See Example 6.8 from Section 2.6. The NLS-cdf estimate minimizes

$$\sum_{i=1}^{n} \left(F_i^* - G(Y_{(i)}; \alpha, \lambda) \right)^2,$$

where G is the cdf of the gamma distribution (pgamma). The R code is found in the file gammaNLS.r with default values $\alpha = 1$ and $\lambda = 2$. The output is shown below.

	Bias alpha	Bias lambda	RMSE alpha	RMSE lambda
MM	0.04708847	0.11169929	0.2014257	0.4647877
NLS-cdf	1.07268411	0.07268411	0.4253850	0.4253850
NLS-cdf weighted	1.13400573	0.13400573	0.4028067	0.4028067

Interestingly, the MM produces less bias and a smaller RMSE (see Section 6.8) for the alpha-parameter, but the NLS-cdf estimator for the lambda-parameter has better characteristics.   □

Especially attractive is the NLS-cdf estimator for mixture distributions; see Section 3.3.2. For example, parameters of the Gaussian mixture distribution can be estimated by the NLS-cdf method by minimizing

$$\sum_{i=1}^{n} \left[F_i - p\Phi\left(\frac{Y_i - \mu_1}{\sigma_1}\right) - (1-p)\Phi\left(\frac{Y_i - \mu_2}{\sigma_2}\right) \right]^2. \tag{9.49}$$

Again, this minimization can be accomplished by running the nls function.

In the example below, this method is applied to the mixture exponential distributions to identify short and long survivors.

Example 9.20 *Mixture exponential distribution. The cancer survival data were analyzed in Section 5.1.2. Use the mixture exponential distribution to identify two groups of survivors who die early and live long.*

Solution. The cdf of the exponential distribution is $1 - e^{-\lambda x}$. If the survival data represents a mixture of two groups of patients with different λ, we can model the common survival as

$$F(x; p, \lambda_1, \lambda_2) = p(1 - e^{-\lambda_1 x}) + (1 - p)1 - e^{-\lambda_2 x},$$

where p is the proportion of patients with the hazard rate λ_1 and $(1 - p)$ is the proportion of patients with the hazard rate λ_2. We obtain estimates of these parameters from nls. The R code is found in file SCcancerQQ2, the printout is below (the figure is not shown).

```
Formula:  y~p*(1-exp(-lambda1*x))+(1-p)*(1-exp(-lambda2*x))
Parameters:
 Estimate Std.  Error t value Pr(>|t|)
 p 0.70601 0.05833 12.103 < 2e-16 ***
```

```
lambda1 0.89129 0.05090 17.511 < 2e-16 ***
lambda2 0.23195 0.03684 6.297 3.97e-09 ***
Signif. codes: 0 '***' 0.001 '**' 0.01 '*' 0.05 '.' 0.1 ' ' 1
Residual standard error: 0.01878 on 135 degrees of freedom
Number of iterations to convergence: 15
Achieved convergence tolerance: 5.535e-06
```

As follows from this output, all parameters have small relative error (t-value) and the empirical cdf is fitted with the residual standard error 0.0188. The proportion of short-lived cancer patients is 0.7 with the rate $\lambda_1 = 0.89$. This means that 50% of those patients die within $\ln 2/0.89 = 0.8$ year, the median survival. The long-lived survivors constitute 0.3 of the sample with rate 0.23 and median survival $\ln 2/0.23 = 2.2$.

Problems

1. Use the theory from the previous sections to reproduce standard errors of the parameters, t- and p-values, and the residual standard error of the output in Example 9.12.

2. Use a hyperbolic function $f(Y_i; \beta_1, \beta_2, \beta_3) = \beta_1 + \beta_2/((Y_i - 1900) + \beta_3)$ to fit the marathon world record data in Example 9.12. Find starting values using linear least squares (lm). Superimpose the fitted hyperbolic curve on the plot in Figure 9.6. Make your judgment on what fitting model is better.

3. Investigate the consistency of the NLS estimator of the change-point in Example 9.13 using simulation under the stochastic scheme by assuming that "Radiation given" has a uniform marginal distribution on $[0, 24]$. Use $n = 100$, 500, $1,000$, and $10,000$ and compute the average of the b4 parameter as a function of n and show that it converges to the true value. [Hint: Use the b-values in the dnaRAD output as the true values for data simulation.]

4. Apply the multiplicative error model to estimate three parameters in Example 9.14. Discuss the advantages of the multiplicative versus additive model.

5. In Example 9.16, $V_0 = 20$ is not fixed, but estimated with SD $= 0.5$. Research how it affects the time estimate T and its CI.

6. Prove that the NLS-cdf estimation produces consistent estimates using the fact that the empirical cdf converges to the theoretical one by referring to Section 6.11.

7. Apply the NLS-cdf estimation to parameters μ and σ using observations $Y_i \overset{iid}{\sim} \mathcal{N}(\mu, \sigma^2)$, $i = 1, 2, ..., n$. (a) Write R code to compare the empirical

standard error from simulations with the average standard error from the
`nls` function. (b) Compare bias and RMSE of the NLS-cdf estimator with
the standard estimators $\bar{y}$ and $\hat{\sigma}$.

8. Apply the NLS-cdf to estimate parameters of the two-mixture distribution
   by minimizing (9.49). Write an R program to research the unbiasedness of
   parameters and their standard errors with increasing n.

9. A nonlinear regression model is defined as $\mathbf{Y} = \beta\mathbf{u} + g(\beta)\mathbf{z} + \boldsymbol{\varepsilon}$, where $\mathbf{x}$
   and $\mathbf{z}$ are $n \times 1$ nonzero vectors and g is a function of β. (a) Is the least
   squares estimate of β from the linear slope regression of $\mathbf{Y}$ on $\mathbf{u}$ unbiased
   (see Example 6.40)? (b) Formulate conditions when the estimate from (a)
   is unbiased. (c) Is the NLS estimate of β is unbiased. (d) Prove that an
   unbiased estimate of β can be obtained from the slope model by regressing
   $\mathbf{Y}$ on $\mathbf{x} = \mathbf{u} - (\mathbf{u}'\mathbf{z})/\|\mathbf{z}\|^2$ even when function g is unknown. (d) Carry out
   vectorized simulations to compare the MSE of the NLS estimator and the
   estimator from (c) using $\mathbf{x} = (1, 2, ..., n)'$, $\mathbf{z} = \mathbf{x}^2$ and $g(\beta) = e^{-\beta}$. Use n
   and σ as arguments to the R function. Plot MSEs on the range of β values
   from -1 to 1.

10. In Example 9.18, the hat was falling from a height of two meters. Using
    the C-model predict when the hat hits the ground. Compute the standard
    error of prediction. [Hint: Use the delta method.]

11. Consider estimating parameters $\boldsymbol{\beta}$ of a logistic regression from Section 8.8.2
    by the NLS. (a) Is the NLS estimator consistent? (b) Derive the asymptotic
    distribution of the NLS estimator in a special case when $\boldsymbol{\beta}'\mathbf{x}_i = \beta_1 + \beta_2 x_i$.
    (c) Use simulations to compare NLS and MLE.

9.6 Studying small sample properties through simulations

Small sample properties of the NLS estimator are difficult to study analytically
– instead simulations are usually used. The theory of sufficient statistics and
uniform minimum-variance unbiased estimation (UMVUE) does not work for
nonlinear regression because unbiased estimators typically do not exist (see Theorem 6.73). However, simulations have their own limitation because they are
pertinent to specific parameter values. Indeed, unlike linear model, statistical
properties of the NLS estimator may change depending on the true parameter
value, regression variance, and the sample size. A good practice is to conduct
simulations on a range of values to see how robust the simulation results are. In
general, when dealing with nonlinear models, one has to be careful in making
statements on the effectiveness of statistical inference, such as hypothesis testing

or CI. It may happen that one method outperforms for one regression model but underperforms for another.

Four types of statistical inference for nonlinear regression with finite n are illustrated through simulations in this section: (1) comparison of approximations for the distribution of parameter estimators, (2) comparison of three asymptotic statistical tests by means of the power function, (2) comparison of confidence region for parameters, and (4) evaluating the performance of several methods for CIs. A common problem is that during simulations the `nls()` function may produce an error that will stop the program – to continue computation the `try(nls())` command should be used.

9.6.1 Normal distribution approximation

As was mentioned before, the distribution of the NLS estimator with small n is very complicated even when the error terms are iid normally distributed. Typically, the Wald approximation is used as specified in equation (9.33), or more specifically,

$$Z_j = \frac{\widehat{\beta}_j - \beta_j}{\widehat{\sigma}\sqrt{(\widehat{\mathbf{J}'\mathbf{J}})_{jj}^{-1}}} \simeq \mathcal{N}(0, 1), \quad j = 1, 2, ..., m, \tag{9.50}$$

where $\widehat{\beta}_j$ is the NLS estimate of the true regression parameter β_j, $\widehat{\sigma}$ is the regression estimate of σ, and $\widehat{\mathbf{J}}$ is the $n \times m$ Jacobian matrix evaluated at $\widehat{\beta}$. Recall that, in `nls`, the Z-value is called the `t-value`. The normal approximation is supposed to work for large n, but, for small sample size (or large σ) it is a matter of research through simulations to assess the quality of this approximation. More sophisticated approximation has been suggested in Section 9.4.3: near exact (9.41) and LR (9.37). The quality of these approximations will be evaluated in the following example by comparing the three simulation-derived cdfs of Z-values for each parameter with the standard normal cdf (`pnorm`). When testing the quality of the distribution approximation, the cdf is preferable over the density. Clearly, the density is more informative, but reconstructing the kernel density from empirical values involves the choice of the bandwidth (see Section 5.5). In contrast, the empirical cdf unbiasedly estimates the true cdf.

Example 9.21 *Distribution of the NLS estimates. Compare the three distribution approximations for the NLS estimates in the exponential regression $y_i = \alpha_1 - \alpha_2 e^{-\alpha_3 x_i} + \varepsilon_i$ using simulations.*

Solution. The LR and score approximations were defined earlier by equations (9.37) and (9.41). The R code with simulations is in file `nen.r`. The approximations are tested and compared using the cdf for the three parameters, α_1, α_2, and α_3. The default true alpha parameters are `a1.true=10,a2.true=2,a3.true=.01`. Besides the regression parameters, we have to specify the sample size, n (the default value n=20), and the regression SD, σ (the default value `sigma=0.2`); x_i are defined as equidistant values from 1 to `nMAX` (the default value `nMAX=300`). When specifying simulations, the parameters should be carefully chosen so that not too many `nls` calls fail and at the same time the difference between the methods is not negligible, so a conclusion could be drawn. Moreover, it is instructive to show a few simulated samples of $\{y_i\}$ with the fitted curve before starting simulations (`job=0`).

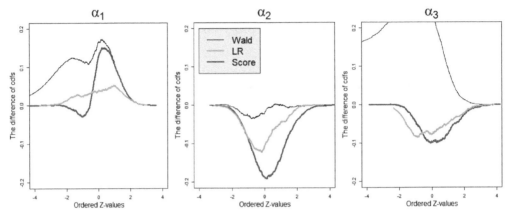

Figure 9.11: *The difference from the empirical cdf for three approximations: Wald, score and LR. The ordered Z-statistics, computed from 5,000 simulations, are shown at the x-axis and the cdf differnce at the y-axis. The Wald approximation of the NLS estimate $\widehat{\alpha}_3$ is not satisfactory.*

It is convenient to separate the time consuming simulations from their further analysis and plotting: `job=1` carries out simulations and returns the parameter values, and `job=2` and `job=3` do the plotting. We recommend saving the results of simulations as an R object. For example, by issuing `nen.out1=nen(job=1)` the NLS estimates will be saved in the R object `nen.out1`. Every time `nen` is running, this object will be attached to the code itself due to the command line `dump(c("nen","nen.out1"),"c:\\StatBook\\nen.r")`. This means that next time the user continues work on the project the results of simulations will be restored along with the `nen` function. While the Z-values for Wald and score approximations can be derived for all three parameters using one `nls` call, we need to run separate regressions for each parameter to compute Z-values for the LR approximation.

`job=2` plots the four cdfs for each parameter. Instead of plotting four cdfs a better visual is to plot the difference between the empirical cdf of Z-values and `pnorm`, as depicted in Figure 9.11. As follows from these simulations, the LR approximation is the most precise. While the popular Wald approximation (9.50) is satisfactory for linear parameters α_1 and α_2, it is unsatisfactory for the nonlinear parameter α_3. As we shall see later, this conclusion remains valid for other methods of statistical inference.

9.6.2 Statistical tests

In Section 9.4.3, we saw how three approximations gave rise to three statistical tests: Wald, LR, and score. The goal of this section is to compare these tests using a concrete nonlinear regression through simulations. The power function is routinely used to compare statistical tests. Recall that the power function is the probability of rejecting the null when the alternative is true – see Chapter 7 and particularly Section 7.3 for more detail. The power function for testing the null hypothesis $H_0 : \theta = \theta_0$ has two branches rising at θ_0 and it is supposed to reach 1 when the alternative goes away from θ_0. Ideally, the power function takes value α, the specified type I error (significance level or the size of the test), at $\theta = \theta_0$ and is above α at both sides. However, in real nonlinear statistical problems, including nonlinear regression, some unwanted/aberrant behavior may occur: the power may not reach 1, may not increase, and the value of the power function at θ_0 may not equal α. This kind of misbehavior is likely with small sample size, n.

The following example continues studying the nonlinear regression from Example 9.21 for testing the null hypothesis on the rate parameter, α_3. We deliberately have chosen this parameter because, for the linear parameters (α_1 and α_2), it is expected that the difference between tests will be less noticeable.

Example 9.22 *Power function.* *Use simulations to compare the Wald, LR, and score tests for testing the null hypothesis $H_0 : \alpha_3 = 0.01$ in the exponential regression model from Example 9.21.*

Solution. See the R function `power.nls`. The parameters chosen are the same as in Example 9.21, but here we use $\sigma = 0.2$. `job=1` performs simulations and `job=1.1` does the plotting of powers depicted in Figure 9.12 as functions of the alternative value α_3. The default number of simulations is 500, but to get sufficiently accurate results, `nSim=5000` would be a minimum. As in the previous example, when doing time consuming simulations, it is convenient to split the task into two jobs: first performing and saving simulations and then the analysis and power plotting. In the `power.nls` function, `job=1` carries out the simulation and saves the power values for the three tests. The fourth column is the number of simulations when `nls` failed. For each alternative value α_3 defined by `seq(from=.01-.008,to=.01+.012,length=n.a3)` and each simulation `isim`, we compute the p-value either using `pnorm` (Wald) or `pchisq` (LR and score). Then, the power is computed as the proportion of simulations where the p-value is less than `alpha`. Since some simulations fail, we have to remove those before processing by using `[!is.na(pW)]`. The result of `job=1` is saved as an R object `power.nls_1.out` by issuing `power.nls_1.out=power.nls(job=1,nSim=5000)` which may take up to half an hour, of course, depending on the machine.

We make several comments on the implementation of simulations. First, to compute the restricted RSS in the LR test, we run the `nls` function with

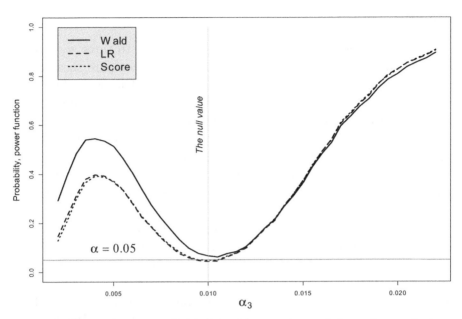

Figure 9.12: *Power function of the three tests estimated through simulations (the number of simulations nSim = 10K). The null hypothesis is that the rate parameter $\alpha_3 = 0.01$. Note that the power function of the Wald test is high because its value at the null value is greater than the nominal $\alpha = 0.05$.*

the a3 parameter fixed at the alternative (the `nls` output is `onlsLR`). Second, the exponential regression function is specified as a local function `fr` with the Jacobian matrix (**J**) returned as an attribute `attr(va,"grad")`. The Jacobian is needed for the score test. Third, when the alternative value of the rate parameter, α_3 is close to zero, matrix $\mathbf{J}'\mathbf{J}$ becomes close to deficient and the `solve` function fails, forcing simulations to stop. To avoid singularity of this matrix and, more importantly, to continue simulations, the generalized matrix inverse is used – this is accomplished in the local function `geninvese`. This function uses the singular value decomposition and treats eigenvalue with absolute value less that 10^{-10} as zero; the resulting matrix is called the Moore–Penrose generalized inverse of a symmetric matrix.

As follows from Figure 9.12 the type I error of the LR test is close to the nominal $\alpha = 0.05$, but the power function of the Wald test is greater than $\alpha = 0.05$, so the functions are not comparable. The aberrant behavior of the tests to the left in the neighborhood of $\alpha_3 = 0.004$ is obvious. Generally, we expect that the power increases when the alternative moves away from the null, but not in our case. The reason is that when the rate approaches zero the model is poorly identifiable for small α_3 and so the power of the tests drops. In conclusion, the LR and score tests are very close and the fact that the power of the Wald test is greater to the left of the null value is due to noncompliance with the condition that the power must be 5% at the null value.

9.6.3 Confidence region

In this section, we demonstrate how a confidence region can be plotted and assessed using simulations for a two-parameter regression function.

Example 9.23 *Confidence region.* *Use simulations for the two-parameter nonlinear regression* $f_i(A, \lambda) = A(1 - e^{-\lambda x_i})$ *under the multiplicative error scheme to construct the* $(1 - \alpha)$ *confidence region based on the normal (9.34) and (9.42) score approximations.*

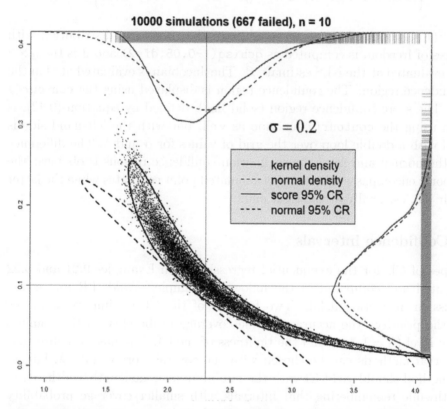

Figure 9.13: *Graphical representation of statistical simulations in Example 9.23. The red vertical and horizontal lines depict the true value of α and λ. The elliptical confidence region (dashed line) based on the linear approximation is unsatisfactory.*

Solution. After taking the log transformation, the nonlinear regression takes the form $y_i = \alpha + \log(1 - e^{-\lambda x_i}) + \varepsilon_i$ where $\alpha = \ln A$. See the R code `expgr`. The default parameters are `A.true=10`, `lambda.true=0.1`, `x=1:10`, `sigma=.2`. As in the example above, `job=0` displays four simulated samples and depicts the true and fitted curve. `job=1` performs simulations and produces the very informative Figure 9.13 with `nSim=10000` points as the NLS estimates for α and λ (in fact, 667 simulation points were not depicted because `nls` did not converge in 10,00 simulations); 9,333 values $\widehat{\alpha}$ and $\widehat{\lambda}$ are shown using the `rug()` command on `side=3` and `side=4`, respectively. The two corresponding densities are shown: the dashed line is the asymptotic normal density with the true parameters and the solid line is the Gaussian kernel density (see Section 5.5) using the NLS estimates. Interestingly, while the density of $\widehat{\lambda}$ is close to normal, the distribution of $\widehat{\alpha}$ is skewed with a heavy right tail. The normal approximation 95% confidence region (dashed ellipse) is

$$\left\{ \boldsymbol{\beta} = (\alpha, \lambda) : (\boldsymbol{\beta} - \widehat{\boldsymbol{\beta}})'(\widehat{\mathbf{J}}'\widehat{\mathbf{J}})^{-1}(\boldsymbol{\beta} - \widehat{\boldsymbol{\beta}}) = \widehat{\sigma}\chi^{-2}(1 - \alpha, 2) \right\},$$

where $\chi^{-2}(1 - \alpha, 2)$ is the $(1 - \alpha)$ quantile of the chi-square distribution with two degrees of freedom is computed as `qchisq(1-0.05,df=2)` and $\mathbf{J}$ is the $n \times 2$ Jacobian evaluated at the NLS estimate $\widehat{\boldsymbol{\beta}}$. The Jacobian is evaluated at $\boldsymbol{\beta}$ in the score confidence region. The confidence region is displayed using the `contour()` function. The score confidence region (solid line), defined by equation (9.42), is computed using the `contour()` function as well, but with the left-hand side is computed with a double loop over the grid of values for α and λ. The difference between the normal and score approximation confidence regions is obvious: the former poorly encompasses the cloud of simulated point estimates while the latter does a fair job, especially at the right end.

9.6.4 Confidence intervals

Three types of CIs for the exponential regression from Examples 9.21 and 9.22 are computed and compared in the following example: Wald, LR and score, per discussion in Section 9.4.3. Two features of the CI are important in the order of the priority: the accuracy of the coverage probability to the nominal confidence level ($\lambda = 1 - \alpha$) and the tightness of the CI. One has to realize that many interval methods can be chosen with the coverage probability λ, but an optimal method should yield CI with the minimum average length/width. Also, it is worthwhile remembering that intervals with smaller coverage probability typically have smaller width. In other words, there is an interplay between the coverage probability and the width of CI. This relationship is similar to what we observed above when the power of the tests were analyzed .

Example 9.24 *Confidence interval.* *Test the coverage probability and the width of three CIs for α_3 in the exponential regression from Example 9.22.*

Solution. The R code is found in `ci.nls`; it may take several minutes to run because the simulated data are attached. This function passes two arguments:

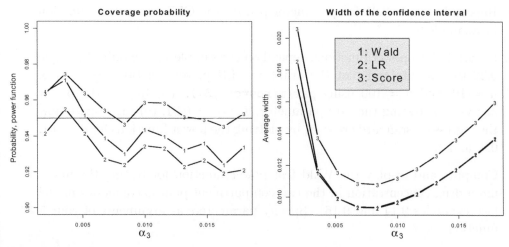

Figure 9.14: *Coverage probability and the average width for three types of CI as functions of the true parameter α_3 in the exponential regression.*

nSim is the number of simulations and `alpha` is the significance level (the confidence level, $\lambda = 1 - \alpha$). Similar to `power.nls` from Example 9.22, job=1 does simulations and saves them in an R object upon completion of `ci.nls_1.out= ci.nls(job=1,nSim=500)`. job=1.1 processes the lower and the upper limits and makes plots as shown in Figure 9.14. The default number of simulations is 500 although at least `nSim=5000` is needed to ensure satisfactory estimation of the coverage probability and the average width. Simulations here use the same parameters as in Example 9.22. The three CIs were computed in Example 9.15. The Wald CI is easy, but LR and score require solving an equation – in our code the `uniroot` function is used. Function `uni3` specifies this equation for the LR CI and function `unNE.prof` specifies this equation for the score CI.

As follows from Figure 9.14, the coverage probability of the score CI is the closest to the nominal $\lambda = 95\%$ level. Surprisingly, the coverage probability of the Wald test is slightly closer to 0.95 than the LR CI. The score CIs are slightly wider than for Wald and LR CIs, but the latter have the coverage probability closer to the nominal value. Interestingly, the average width increases when the true parameter value approaches zero. This a reflection of the same phenomenon discovered in testing of the power functions explained by the fact that the model becomes poorly identifiable for small α_3.

Problems

1. Use function `nen` from Example 9.21 to confirm that large σ affects the distribution approximation in a similar way as small n. Specifically, derive the distribution approximation for $\widehat{\alpha}_3$ on a grid of values for increasing σ and decreasing n and compare their performance.

2. Repeat Example 9.21 for parameter ρ from the CES production function in Example 9.17.

3. (a) Study how power functions depend on the sample size: modify Example 9.22 to compute and plot the Wald and LR power functions for $n = 5$, $n = 10$, and $n = 100$ using `par(mfrow=c(1,3))`. (b) Repeat the power analysis for testing the rate parameter $H_0 : \alpha_3 = 0.01$. (c) Investigate how the regression standard error, σ, by plotting the power functions for $\sigma = 0.1$, 0.3, and 1.

4. Compute and display the Wald test power function for testing the maximum drug concentration in the two-compartment pharmacokinetics model (see examples 9.14 and 9.16). Specify parameters as arguments of your R function.

9.7   Numerical complications of the nonlinear least squares

As we know, the least squares solution for the linear regression model is expressed in closed form and requires a single matrix inverse. The solution is unique if the design matrix has full rank. Nonlinear regression leads to a much more difficult computational problem. First, it requires iterations that may not converge. Second, the least squares solution may not exist. Third, the solution may be not unique because the residual sum of squares may have several local minima. We recommend reviewing the section on optimization in the Appendix. This section does not intend to provide a comprehensive coverage of the topic, but just familiarize the reader with pitfalls and possible solutions.

To illustrate the possibility of nonexistence of the NLS estimate, we consider the following example of the exponential decay regression function.

Example 9.25 *Decay model.* *The exponential decay model has the form of nonlinear regression*

$$Y_i = e^{-\beta i} + \varepsilon_i, \quad i = 1, .., n,$$

where β is an unknown decay rate subject to estimation by nonlinear least squares. Illustrate the possibility of nonexistence of the NLS estimate and possible divergence of the RSS minimization algorithm.

Solution. If the variance of errors is large, observations y_i may be negative especially for large i and positive β. Although we expect the decay parameter be positive, we do not exclude the possibility that the NLS estimate is negative, that is, the parameter space is the whole real line. When $\beta \to -\infty$ the residual sum of squares (RSS), $S(\beta) \to \infty$. But when $\beta \to \infty$, we have

$$\lim_{\beta \to \infty} S(\beta) = \lim_{\beta \to \infty} \sum_{i=1}^{n} \left(y_i - e^{-\beta i} \right)^2 = \sum_{i=1}^{n} Y_i^2.$$

In other words, the RSS has a left asymptote. Now we turn our attention to the derivative of the RSS,

$$\frac{dS}{d\beta} = 2 \sum_{i=1}^{n} \left(Y_i - e^{-\beta i} \right) e^{-\beta i}.$$

For large β, the sign of the derivative is dominated by the term $i = 1$, namely, $Y_1 e^{-\beta}$. This means that if the first observation is negative, which is unlikely, but possible, then $dS/d\beta < 0$ for large β. This means that $S(\beta)$ approaches its left asymptote from above and the least squares may not exist or may have a false local minimum. Practically, if the starting value is large, the minimization algorithm may diverge.   □

It is worthwhile to mention that under the multiplicative error model (see Example 9.1), written as $Y_i = e^{-\beta i + \varepsilon_i}$, numerical complications disappear and the estimation is easy because it reduces to a linear regression upon taking the log transformation, $z_i = -\beta i + \varepsilon_i$ where $z_i = \ln Y_i$. This model implies that observations are positive, so negative observations are impossible.

It is desirable to have a criterion for existence of the NLS estimate before starting the minimization algorithm. It is easy to develop an existence criterion for a one-dimensional nonlinear regression with $\beta \in (-\infty, \infty)$. Indeed, assume that the left and the right asymptotes exist, that is, $S_- = \lim_{\beta \to -\infty} S(\beta)$ and $S_+ = \lim_{\beta \to \infty} S(\beta)$. If there exists β_0 such that $S(\beta_0) < \min(S_-, S_+)$, then the least squares estimate exists. The next section generalizes the criteria for existence to multiple parameters.

9.7.1 Criteria for existence

The parameter space $\beta \in \Theta \subset R^m$ is assumed convex and $\partial \Theta$ denotes the boundary of Θ, infinite points included. For example, if $\beta > 0$ we have $\partial \Theta = \{0, \infty\}$ and notation $\beta_k \to \partial \Theta$ means that either $\beta_k \to 0$ or $\beta_k \to \infty$.

Definition 9.26 *Existence level.* *The existence level for the nonlinear least squares problem is defined as*

$$\overline{S}_E = \inf_{\beta_k \to \partial \Theta} \sum_{i=1}^{n} (Y_i - f_i(\boldsymbol{\beta}_k))^2.$$

In words, the existence level is the minimum of the minimum sum of squares on the boundary of the parameter set.

Example 9.27 *Existence level.* *Find the existence level for the decay model from Example 9.25.*

Solution. Since $\partial\Theta = \{-\infty, \infty\}$, we seek

$$\lim_{\beta\to\infty}\sum_{i=1}^{n}(Y_i - e^{-\beta i})^2 = \sum_{i=1}^{n}Y_i^2, \quad \lim_{\beta\to-\infty}\sum_{i=1}^{n}(Y_i - e^{-\beta i})^2 = \infty.$$

Therefore, the existence level is $\overline{S}_E = \sum_{i=1}^{n}Y_i^2$.   □

The main result on the existence level is formulated below (Demidenko [23], [25]).

Theorem 9.28 *If $\beta_0 \in \Theta$ exists such that $\sum_{i=1}^{n}(Y_i - f_i(\beta_0))^2 < \overline{S}_E$, then the NLS estimate exists. Moreover, the set of $\beta \in \Theta$ such that $\sum_{i=1}^{n}(Y_i - f_i(\beta))^2 \leq \sum_{i=1}^{n}(Y_i - f_i(\beta_0))^2$ is compact and GN iterations (9.20) starting from β_0 converge to a local minimum, provided the RSS decreases from iteration to iteration.*

The condition formulated in this theorem not only guarantees convergence, but also determines the rule for a successful starting value.

Example 9.29 *Successful starting value.* *As a continuation of Example 9.27, find a successful starting value using Theorem 9.28 assuming that all observations Y_i are positive.*

Solution. Let $\beta_0 = -\max_{i=1,\dots,n}(\ln Y_i)/i$ and show that it satisfies the conditions of Theorem 9.28. Let $\beta_0 = -\ln Y_{i_0}/i_0$, where i_0 is among $i = 1,\dots,n$ that gives the minimum. Then, by definition, $-\beta_0 \leq \ln a_i/i$ and $e^{-\beta_0 i} \leq Y_i$ for all i and $e^{-\beta_0 i_0} = Y_{i_0}$. This implies

$$S(\beta_0) = \sum_{i=1}^{n}(Y_i - e^{-\beta_0 i})^2 = (Y_{i_0} - e^{-\beta_0 i_0})^2 + \sum_{i\neq i_0}(Y_i - e^{-\beta_0 i})^2 \leq \sum_{i\neq i_0}Y_i^2 < \overline{S}_E,$$

where $\overline{S}_E = \sum_{i=1}^{n}Y_i^2$ is the existence level derived in Example 9.27. The conditions of Theorem 9.28 are fulfilled and the set $S(\beta) \leq S(\beta_0)$ is compact.   □

Existence criteria for other nonlinear regressions can be found in the literature cited above.

9.7.2   Criteria for uniqueness

As we mentioned earlier, the RSS function in nonlinear regression, unlike linear regression, may have several local minima. For example, as was shown by Demidenko [23], the probability that $S(\beta)$ has two local minima is positive for any nonlinear regression with a single parameter if it has infinite tails, i.e. $\|\mathbf{f}(\beta)\| \to \infty$ when $|\beta| \to \infty$. The following example taken from Demidenko [31] geometrically illustrates possible multimodality for $n = 2$.

Example 9.30 *Probability of local minima. Define the region on the plane where the RSS for the quadratic regression with $f_1(\beta) = \beta$ and $f_2(\beta) = \beta^2$ has two local minima ($n = 2$), and compute the probability that the RSS has two local minima.*

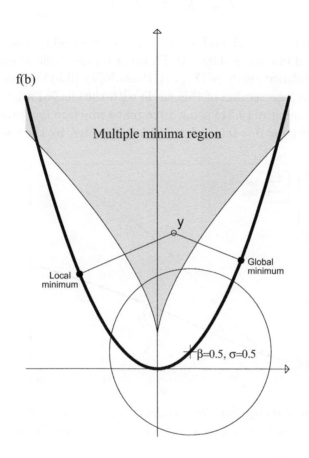

f(b)

Multiple minima region

y

Global minimum

Local minimum

$\beta=0.5, \sigma=0.5$

Figure 9.15: *Quadratic regression on the plane ($n = 2$). If the data point $\mathbf{y}$ belongs to the shaded region, the sum of squares, $(Y_1 - \beta)^2 + (Y_2 - \beta^2)^2$, has two local minima.*

Solution. The RSS function for the quadratic regression on the plane, $S(\beta) = (Y_1 - \beta)^2 + (Y_2 - \beta^2)^2$ is a polynomial of the fourth order and the normal equation may have up to three real roots, two of them correspond to local minima (one of them being global). Geometrically (see Section 9.2), nonlinear least squares turns into a problem of finding the closest point on the parabola β^2 from the data point $\mathbf{y} = (Y_1, Y_2) \in R^2$; see Figure 9.15. The normal equation reduces to a cubic equation which has three real roots if $Y_2 > 3/4^{1/3} |Y_1|^{2/3} + 1/2$. In the figure, the data point $\mathbf{y}$ leads to two local minima as the distance to the

parabola; the positive value is the true NLS estimate (the global minimizer) and
the negative value is the false NLS estimate. For given β and σ, one can compute
the probability of two local minima as the integral

$$\text{Pr(two local minima)} = \sigma^{-1} \int_{-\infty}^{\infty} \phi \left(\frac{y_1 - \beta}{\sigma} \right) \Phi \left(\frac{\beta^2 - \frac{3}{4^{1/3}} |y_1|^{2/3} - 0.5}{\sigma} \right) dy_1.$$

$$(9.51)$$

For example, for $\beta = 0.5$ and $\sigma = 0.5$, the probability that the sum of
squares has two local minima is 0.026. In Figure 9.15, the circle around $(0.5, 0.5^2)$
shows the 95% confidence region of (Y_1, Y_2). Probability (9.51) equals the density-
weighted area of the intersection of this circle with the shaded area. The R function `q2pr` evaluates integral (9.51) using `integrate` function and produces Figure
9.16. The probability for $\beta = 0.5$ and $\sigma = 0.5$ is depicted by the dashed line.

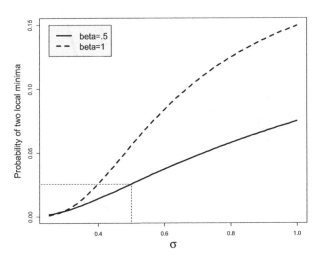

Figure 9.16: *The probability that the quadratic regression on the plane has two
local minima of the residual sum of squares computed as the integral (9.51).*

Development of criteria for uniqueness of the NLS estimate is a difficult problem, but they have a simple rule: if $\boldsymbol{\beta}_*$ is the point of local minimum of the RSS
with the value less than $\overline{S}_U$, the uniqueness level, then $\boldsymbol{\beta}_*$ is the point of the
global minimum (the NLS estimate). As a rule of thumb, a smaller residual sum
of squares yields a better chance that the found local minimum is the global one.
The uniqueness criteria for some simple nonlinear regressions are developed by
Demidenko [24], [25].

Problems

   1. Illustrate the possibility of numerical complications in Example 9.25. (a)
   Generate Y_i for $n = 5$ and positive β and run `nls`. (b) Plot $S(\beta)$ and

display the NLS solution as the point on the RSS plot. (c) When $Y_1 < 0$, use a large starting value to make `nls` fail.

2. (a) Reduce the NLS problem for the decay model to the solution of a polynomial equation with $x = e^{-\beta}$ using the `polyroot` function in R. (b) Use $n = 3$ and simulate Y_i with $\sigma = 1$, estimate β using the `nls` function and then using the `polyroot` function (select only positive real roots). (c) Using simulations, estimate the probability that the NLS estimate does not exist (no positive real roots). [Hint: Use `Re` and `Im` functions.]

3. Prove that $\overline{S}_E = \infty$ for a nonlinear regression model $Y_i = e^{\beta x_i} + \varepsilon_i$, where $-\infty < \beta < \infty$ and $x_1 < x_2 < \cdots < x_n$ are such that $x_1 x_n < 0$.

4. Find $\overline{S}_E$ for a nonlinear regression model $Y_i = e^{\beta x_i} + \varepsilon_i$, where $-\infty < \beta < \infty$ and $x_i > 0$.

5. Prove Theorem 9.28. [Hint: Use the fact that any limiting point of a sequence of points from a compact set E belongs to E.]

6. Is it true that if all observations in the decay model are positive then the NLS exists?

7. Does the NLS estimate always exist for the quadratic regression depicted in Figure 9.15? [Hint: Use Theorem 9.28.]

8. (a) Reduce the NLS problem for the quadratic regression $Y_i = \beta + \beta^2 x_i + \varepsilon_i$ with an arbitrary n to a cubic polynomial solution and check your algorithm by running `nls`. (b) Use simulations to estimate the probability that there will be two local minima of the RSS and plot the probability as a function of σ similarly to Figure 9.16. (c) For those simulations when there are two local minima run `nls` to see what solution it converges to. [Hint: Use function `polyroot`.]

9.8 Optimal design of experiments with nonlinear regression

Sometimes values $\mathbf{x}_i$ in nonlinear regression (9.3) can be chosen by an experimentalist, as in many engineering or chemical applications. What values are optimal for the best estimation of unknown β? In statistics, this is addressed by the *optimal design of experiments*. The hanging wire example illustrates the concept.

9.8.1 Motivating examples

Example 9.31 *Hanging wire. The wire (or chain) is supported at the ends by poles of unit height at locations 1 and −1 (see Figure 9.17). The height of the* *wire $y = y(x)$ as a function of location, x, between the poles is modeled by an ordinary differential equation (ODE) $y'' = \beta\sqrt{1 + y'^2}$ with an unknown sagging parameter β that depends on wire's the physical properties. At what location x should the measurements of the height y be taken to get the maximum precision of the sagging parameter, β, estimated from the nonlinear regression?*

Solution. Today, the well-known solution of the ODE can be expressed in closed form as

$$y(x;\beta) = \frac{e^{-x\beta} + e^{x\beta}}{e^{-\beta} + e^{\beta}}.$$

Sometimes, this solution is called a *catenary* (Latin for "chain") solution. The function that governs the shape of a hanging flexible wire was on the mind of physicists and mathematicians for centuries. For example, Galileo claimed the wire's curve is parabolic. Many famous mathematicians contributed to the correct formula as we know it today, including Leibniz, Huygens, and Bernoulli.

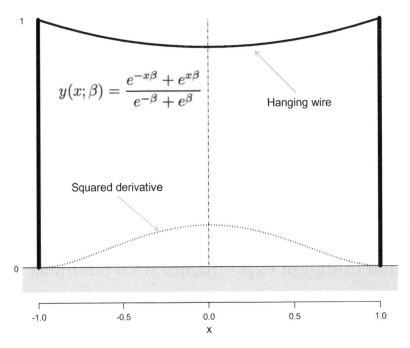

Figure 9.17: *The problem of a hanging wire. Where should measurements be made to estimate the sagging parameter, β, with maximum precision?*

Since just one measurement is sufficient to identify β, we rephrase the problem to finding x that yields the minimum variance of the nonlinear least squares (NLS) estimate, $\widehat{\beta}$. If h_* is the measured distance of the wire from the ground at point x_*, we write $h_* = y(x_*;\beta) + \varepsilon$, where ε is a random error with zero mean and variance σ^2 representing the measurement error. Since we deal with one measurement, the

NLS estimate of β is the solution to the equation $(e^{-x_*\beta} + e^{x_*\beta}) = (e^{-\beta} + e^{\beta})h_*$. From the theory of nonlinear regression, we know that the variance of the NLS estimate $\widehat{\beta}$ is approximated as $\mathrm{var}(\widehat{\beta}) = \sigma^2 \left\| \frac{dy}{d\beta} \right\|^{-2}$. Consequently, to make our estimate of β most efficient, we find x_* that maximizes the squared derivative of the catenary, $\|dy/d\beta\|^2$. Using elementary calculus or the symmetry argument, one can show that this maximum is attained at $x_* = 0$. Indeed, the squared derivative, $\|dy/d\beta\|^2$ takes maximum at $x_* = 0$ regardless of β. This point is called the optimal support point and gives the NLS estimate

$$\widehat{\beta} = \ln \frac{h_*}{1 - \sqrt{1 - h_*^2}}.$$

Thus, we conclude that the maximum precision is gained when the measurement is taken in the middle of the wire where sagging is maximum. To minimize the variance of $\widehat{\beta}$, several measurements of height may be taken at the same location, $x_* = 0$. Then we substitute h_* with the average height. □

In the previous example, the optimal measurement point was independent of the true parameter value. This is not typically it true, as in the examples below.

Example 9.32 *Hanging wire continued. As in the previous example, the wire is supported by two poles, but of different heights, H_1 and H_2. The generated ODE is the same, but the boundary conditions are different. Derive the solution $y(x; \beta, H_1, H_2)$ and formulate the problem of the optimal location of measurement x depending on which of the three parameters are known.*

Solution. The general solution to the ODE $y'' = \beta\sqrt{1 + y'^2}$ is

$$y = \frac{1}{2\beta} \left(e^{\beta x + C_1\beta} + e^{-\beta x - C_1\beta} \right) + C_2, \quad -1 \leq x \leq 1,$$

where constants C_1 and C_2 are derived from the boundary conditions $y(-1) = H_1$ and $y(1) = H_2$. After some algebra, one can show that the solution to the hanging wire problem can be written as

$$y(x; \beta, H_1, H_2) = \frac{1}{2\beta} \left(\rho e^{\beta x} + \frac{1}{\rho} e^{-\beta x} \right) + \frac{-e^{4\beta} + 2\rho H_1 \beta e^{3\beta} + 1 - 2H_2\beta\rho e^\beta}{2e^\beta \left(e^{2\beta} - 1 \right) \beta \rho},$$

where $\rho > 0$ is the solution to the quadratic equation $\left(e^{2\beta} - 1 \right) \rho^2 + 2\beta e^\beta \left(H_1 - H_2 \right) \rho + 1 - e^{2\beta} = 0$. It is easy to prove that this quadratic equation has a unique positive solution.

When the poles had the same heights, the optimal location of the measurement was in the middle regardless of β. Now we face a new problem: to find the optimal location, we must know β. If H_1 and H_2 were known, the criterion function is $\mathrm{var}(\widehat{\beta})$ and one measurement is sufficient to estimate the sagging parameter. However, if H_1 and H_2 are also unknown, at least three locations should be used. The criterion function may remain var$(\widehat{\beta})$ if only the sagging parameter is of interest, or may involve variances of H_1 and H_2 using the A- or D-optimality criterion, see the next section.

Example 9.33 *When to make temperature measurement in the cup of tea.* *Find the optimal time for temperature measurement in the cup of tea from Example 9.2.*

Solution. As follows from the previous example, the minimum variance of the NLS estimator occurs when the squared derivative of $T(t)$ with respect to γ is maximum – this is the general result when regression depends on a single parameter. Since $dT/d\gamma = -(100-T_{\text{room}})\gamma e^{-\gamma t}$, the squared derivative reaches maximum when $t^2 e^{-2\gamma t}$ reaches maximum. Multiplying by γ^2, we arrive at maximization of function $f(u) = u^2 e^{-2u}$ where $u = \gamma t$. Taking the derivative with respect to u yields the answer: $u = 1$. This means that the optimal time of measurement is when $\gamma t = 1$, i.e. when the temperature in the cup is $T_{\text{room}} + (100 - T_{\text{room}})e^{-1}$. For example, if the room temperature is $20°C$ the optimal time of measurement is when the tea has temperature $T_{\text{opt}} = 20 + 80e^{-1} = 50°C$.

9.8.2   Optimal designs with nonlinear regression

Now we turn our attention to a general case when the nonlinear regression model is defined by (9.3). Before proceeding to the optimal design, one must chose a metric to judge the effectiveness of the parameter estimation. Since the finite sample properties of the NLS are difficult to state, we appeal to asymptotic properties and formulate the optimality criteria in the language of the $m \times m$ Fisher information matrix, which, in the case of nonlinear regression, is proportional to $\mathbf{J'J}$, where $\mathbf{J} = \partial f(\mathbf{x};\boldsymbol{\beta})/\partial\boldsymbol{\beta}$. In vague terms, we want to chose the design points $\mathbf{x}$ such that matrix $\mathbf{J'J}$ is maximum, or equivalently when $(\mathbf{J'J})^{-1}$ is minimum. In what follows, the parameter vector $\boldsymbol{\beta}$ is held fixed and we concentrate our attention on the choice of $\mathbf{x}$. Two optimality criteria are popular:

1. D-optimality finds the design that maximizes the generalized variance (see Sections 3.5 and 4.1) as the determinant of matrix $\mathbf{J'J}$.

2. A-optimality finds the design that minimizes the trace of the inverse matrix which is proportional to the sum of the variances of the NLS estimator, $\text{tr}(\mathbf{J'J})^{-1}$.

It is easy to see that, for one parameter ($m = 1$), the two optimality criteria are the same.

It was proven that the optimal design for a nonlinear regression with m parameters is defined by m points $\{\mathbf{x}_{*k}, k = 1, 2, ..., m\}$ called *support points*. The support points may repeat, i.e. several measurements may be done at the same point, as in Example 9.31. To account for repeated measurements, it is convenient to introduce positive weight coefficients, $\{w_k, k = 1, 2, ..., m\}$ summing to 1, as the proportion of repeated support points in the design. Finally, the design matrix $\mathbf{J}'\mathbf{J}$ takes the form

$$\sum_{k=1}^{m} w_k \left(\left. \frac{\partial f(\mathbf{x}; \boldsymbol{\beta})}{\partial \boldsymbol{\beta}} \right|_{\mathbf{x}=\mathbf{x}_{*k}} \right) \left(\left. \frac{\partial f(\mathbf{x}; \boldsymbol{\beta})}{\partial \boldsymbol{\beta}} \right|_{\mathbf{x}=\mathbf{x}_{*k}} \right)'. \tag{9.52}$$

The optimal design points and the respective weights optimize either the D- or A-optimality criterion (it is assumed that the design matrix is nonsingular). The following theorem finds the optimal weights w_k when the optimal support points $\mathbf{x}_{*k}$ are held fixed.

Theorem 9.34 *The optimal weights for D-optimality are the same, $w_k = 1/m$. The optimal weights for A-optimality are proportional to the diagonal elements of the inverse matrix (9.52).*

Proof. To shorten the notation, express the weights as the diagonal $m \times m$ matrix $\mathbf{W}$ with w_k on the diagonal and derivatives $\partial \mathbf{f}/\partial \boldsymbol{\beta}$ at m support points as row vectors in the $m \times m$ matrix $\mathbf{J}$. Then design matrix (9.52) can be compactly written as $\mathbf{J}'\mathbf{W}\mathbf{J}$.

The D-optimality criterion maximizes $\det(\mathbf{J}'\mathbf{W}\mathbf{J})$. Equivalently, the optimal weights maximize the log determinant, $\ln\det(\mathbf{J}'\mathbf{W}\mathbf{J}) = \ln\det(\mathbf{W}\mathbf{J}\mathbf{J}') = \ln\det(\mathbf{W}) + \ln\det(\mathbf{J}\mathbf{J}')$. Since $\ln\det(\mathbf{W}) = \sum_{k=1}^{m} \ln w_k$ and $\ln\det(\mathbf{J}\mathbf{J}')$ does not depend on the weights, the optimization problem reduces to maximization of $\sum_{k=1}^{m} \ln w_k$ under constraint $\sum_{k=1}^{m} w_k = 1$. Using the Lagrange multiplier technique with function

$$\mathcal{L}(w_1, ..., w_m, \lambda) = \sum_{k=1}^{m} \ln w_k - \lambda \left(\sum_{k=1}^{m} w_k - 1 \right),$$

differentiation yields that the optimal weights in the D-optimality criterion are uniform, $w_{*k} = 1/m$.

The A-optimality criterion minimizes $\operatorname{tr}(\mathbf{J}'\mathbf{W}\mathbf{J})^{-1}$. We have

$$\operatorname{tr}(\mathbf{J}'\mathbf{W}\mathbf{J})^{-1} = \operatorname{tr}(\mathbf{J}^{-1}\mathbf{W}^{-1}\mathbf{J}'^{-1}) = \operatorname{tr}(\mathbf{W}^{-1}(\mathbf{J}'\mathbf{J})^{-1}) = \sum_{k=1}^{m} \frac{(\mathbf{J}'\mathbf{J})^{kk}}{w_k},$$

where $(\mathbf{J}'\mathbf{J})^{kk}$ denotes the (k, k)th diagonal element of matrix $(\mathbf{J}'\mathbf{J})^{-1}$. Again, using the Lagrange multiplier technique, it is easy to show that the optimal weights are $w_{*k} = \sqrt{(\mathbf{J}'\mathbf{J})^{kk}} / \sum_{l=1}^{m} \sqrt{(\mathbf{J}'\mathbf{J})^{ll}}$, which completes the proof. $\quad\square$

In the example below we apply the optimal design to a sinusoidal regression function. Although the parameters of the function can be estimated by linear least squares, the question of optimal location of measurements is not trivial.

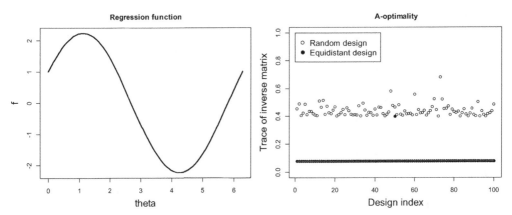

Figure 9.18: *The optimal design for a sinusoidal regression function. The number of measurements $n = 10$.*

Example 9.35 *Optimal design.* *An oscillation signal follows the sinusoidal function $f(a, b; \theta) = a \sin \theta + b \cos \theta$ of angle $\theta \in (0, 2\pi)$. (a) Find the optimal angles where the measurements should be made to optimally estimate the unknown amplitudes a and b. (b) Use simulations to illustrate the optimal design by comparing with random location and equidistant designs.*

Solution. See Figure 9.18 for illustration. (a) Since two parameters are involved, we seek two optimal angles, θ_1 and θ_2. Find the derivative

$$
\begin{bmatrix} \sin \theta_1 \\ \cos \theta_1 \end{bmatrix} \begin{bmatrix} \sin \theta_1 \\ \cos \theta_1 \end{bmatrix}' + \begin{bmatrix} \sin \theta_2 \\ \cos \theta_2 \end{bmatrix} \begin{bmatrix} \sin \theta_2 \\ \cos \theta_2 \end{bmatrix}'
$$

$$
= \begin{bmatrix} \sin^2 \theta_1 + \sin^2 \theta_2 & \sin \theta_1 \cos \theta_1 + \sin \theta_2 \cos \theta_2 \\ \sin \theta_1 \cos \theta_1 + \sin \theta_2 \cos \theta_2 & \cos^2 \theta_1 + \cos^2 \theta_2 \end{bmatrix}.
$$

with the determinant $D(\theta_1, \theta_2) = (\sin \theta_1 \cos \theta_2 - \sin \theta_2 \cos \theta_1)^2$. The D-optimality criterion maximizes function D. Taking the derivative of this function, we arrive at the solution. The optimal NLS estimates for a and b are obtained if the signal is measured at angles (θ_1, θ_2) with $\theta_1 = \theta < 3\pi/2$ and $\theta_2 = \theta + \pi/2$; angle $\theta \in (0, 3\pi/2)$ may be any.

Now we find the optimal design using the A-optimality criterion, the minimum of the trace of the inverse matrix. Some algebra gives that the trace of the inverse matrix is $2(\cos^2 \theta_2 - 2 \cos^2 \theta_2 \cos^2 \theta_1 + \cos^2 \theta_1 - 2 \sin \theta_1 \cos \theta_1 \sin \theta_2 \cos \theta_2)^{-1}$ with the minimum equal to 1. As in the above case the minimum is achieved at the same $\theta_1 = \theta$ and $\theta_2 = \theta_1 + \pi/2$ where $0 < \theta < 3\pi/2$. Thus, we conclude that the two criteria lead to the same optimal angles. (b) Let n measurements be made, say, $n = 10$. The optimal design suggests making five measurements at $\theta_2 = \theta$ and 5 measurements at $\theta_1 = \theta + \pi/2$. The random location design picks n random locations on $[0, 2\pi)$ and the equidistant design picks locations as $\theta_i = 2i\pi/n$, $i = 1, 2, ..., n$. The right plot in Figure 9.18 depicts the trace of the 2×2 matrix $(\mathbf{J'J})^{-1}$, where the first column of $\mathbf{J}$ is $\sin \theta_i$ and the second column is $\cos \theta_i$. The random and equidistant designs lead to a much larger sum of variances for a and b.   □

In the above examples, the optimal design was independent of the unknown parameter vector. If this is not the case, we face a problem: to find the optimal support points one needs to know vector $\boldsymbol{\beta}$, which is unknown. Two solutions may be suggested: (a) choose a reasonable parameter value from previous experience to find the support points $\mathbf{x}_{*k}$, or (b) at each iteration use the current value of $\boldsymbol{\beta}$ to adapt the design on the go ([32]).

Problems

1. (a) Prove that the NLS estimate in Example 9.31 is found from the equation $(e^{-x_*\beta} + e^{x_*\beta}) = (e^{-\beta} + e^{\beta})h_*$, where $0 < x_* < 1$ and $0 < h_* < 1$.

2. Suppose that in Example 9.31, the height of the poles is unknown and subject to estimation along with the sagging parameter. Where are the optimal location of the measurements? [Hint: Reduce the problem of estimation to one parameter.]

3. (a) Find the relationship between the sagging parameter and the length of the wire computed as $\int_{-1}^{1} \sqrt{1 + (dy/dx)^2}dx$. (b) Find the limiting wire position when $\beta \to 0$. What is the length of the wire? (c) What is the optimal position of the height measurement to estimate the length of the wire with maximum precision?

4. Suppose that in Example 9.32, the sagging parameter is known but H_1 and H_2 are not. What are optimal positions, x_1 and x_2? Write an R function with the β value as an argument. Use both A- and D-optimality criteria.

5. Using Theorem 9.34, find the optimal weights w_{*1} and w_{*2} when $m = 2$.

6. Find two optimal times of measurements using the A- and D-optimality criteria in the cup of tea Example 9.33 where estimates of T_{room} and γ are known. [Hint: Write an R program with `Troom` and `gamma` as arguments and use `contour` with the `par(mfrow=c(1,2))` option.]

7. Reconstruct Figure 9.18.

9.9   The Michaelis–Menten model

The Michaelis–Menten model is a popular statistical model in many application fields. It can be found in biology and chemistry applications to model the kinetics of interaction between two agents. It has a hyperbolic shape and relates y and x as an increasing function with an asymptote,

$$y = \frac{\alpha x}{\beta + x}. \tag{9.53}$$

Two types of numerical analyses are discussed here: the traditional nonlinear regression analysis relying on iterations and the exact numerical solution of reducing the normal equation to a polynomial.

9.9.1   The NLS solution

We use *Puromycin* data on the velocity (treated) of a chemical reaction $y = Y_i$ as a function of concentration x_i from an example provided by Bates and Watts [7], p. 269. The NLS is supposed to find the global minimum of the residual sum of squares (RSS)

$$S(\alpha, \beta) = \sum_{i=1}^{n} \left(Y_i - \frac{\alpha x_i}{\beta + x_i} \right)^2.$$

What follows is the analysis of the numerical properties of function S in terms of its minimum.

Starting values

Two starting values may be suggested to begin iterations. First, we can start `nls` with the initial guess at $\beta_0 = 0$ and optimal $\alpha_0 = \overline{Y}$ because it minimizes $\sum (Y_i - \alpha)^2$. Second, the starting values can be found from a linear regression by observing that $Y \simeq \alpha x/(\beta + x)$ can be replaced by $Y(\beta + x) \simeq \alpha x$ and $Yx \simeq \alpha x - \beta Y$. Thus, the initial guess for α and β can be obtained from regression of $Y_i x_i$ on x_i and Y_i (no intercept). The R code is found in function `michm`. This function fits the *Puromycin* data and computes all solutions of the normal equation as the roots of the polynomial (see below). `job=0` plots four simulated samples with parameters specified as arguments of the function; `job=1` estimates the Michaelis–Menten model by `nls` and produces a plot (not shown).

Statistical testing

The parameters alpha (`a`) and beta (`b`) are estimated with high precision but one should not be impressed with low p-values because they reflect how close these parameters are to zero. A more interesting hypotheses to test is if the relationship between Y and x is either (a) linear (when $\beta \to \infty$ and $\alpha = c\beta$ where c is a constant), or (b) constant (when $\beta \to 0$). Indeed, in case (a) the Michaelis–Menten model turns into the slope model

$$\lim_{\beta \to \infty} \frac{c\beta x_i}{\beta + x_i} = c \lim_{\beta \to \infty} \frac{x_i}{1 + x_i/\beta} = cx_i.$$

Which of the three statistical tests discussed in Section 9.4.3 is applicable for testing these hypotheses? The Wald and score tests involve computation of the Jacobian, which, in the limit ($\beta \to 0$ or $\beta \to \infty$), turns into a deficient matrix. In contrast, the LR test requires computation of the RSS for the limiting cases, which is a fairly simple task:

$$\widehat{S}_0 = \min_{\alpha} S(\alpha, 0) = \min_{\alpha} \sum_{i=1}^{n} (Y_i - \alpha)^2 = \sum_{i=1}^{n} (Y_i - \overline{Y})^2,$$

$$\widehat{S}_{\infty} = \min_{\alpha,\beta} \lim_{\beta \to \infty} S(\alpha, \beta) = \min_{c} \lim_{\beta \to \infty} \sum_{i=1}^{n} \left(Y_i - \frac{c\beta x_i}{\beta + x_i} \right)^2 = \min_{c} \sum_{i=1}^{n} (Y_i - cx_i)^2$$

$$= \sum_{i=1}^{n} Y_i^2 - \frac{\left(\sum_{i=1}^{n} Y_i x_i \right)^2}{\sum_{i=1}^{n} x_i^2}.$$

Then, according to the LR test, if the relationship between Y and x is a slope regression model, $Y_i = cx_i + \varepsilon_i$, or constant, $Y_i = c + \varepsilon_i$, we should have $n \ln(\widehat{S}_{\infty}/\widehat{S}) \simeq \chi^2(1)$ and $n \ln(\widehat{S}_0/\widehat{S}) \simeq \chi^2(1)$, respectively, where $\widehat{S}$ is the minimum RSS in the Michaelis–Menten model. As follows from the output of the call `michm(job=1)` both hypotheses are rejected (see `p-value slope` and `p-value const`).

9.9.2 The exact solution

The Michaelis–Menten model enables reducing the solution of the normal equation to finding the roots of a polynomial (computational algorithms for solution of a polynomial are well developed). The exact solution can be used to study computational problems of nonlinear least squares, as will be demonstrated below. Express α through β using the linear least squares,

$$\alpha = \frac{\sum Y_i x_i / (\beta + x_i)}{\sum x_i^2 / (\beta + x_i)^2}, \tag{9.54}$$

and consider S as a sole function of β. Then $dS/d\beta$ can be reduced to a $4(n-1)$ degree polynomial. Specifically, after some algebra, it is easy to show that the normal equations are equivalent to

$$\sum \frac{Y_i x_i}{(\beta + x_i)^2} \sum \frac{x_i^2}{(\beta + x_i)^2} - \sum \frac{Y_i x_i}{\beta + x_i} \sum \frac{x_i^2}{(\beta + x_i)^3} = 0,$$

along with (9.54). Note that instead of using `nls`, we could find a solution of the normal equation using the `uniroot` function , but we want to find all solutions. Multiply both sides of the above equation by $\prod (\beta + x_i)^4$. Then the normal equation becomes equivalent to the polynomial equation

$$P_{2(n-1)}(\beta) Q_{2(n-1)}(\beta) - R_{3(n-1)}(\beta) H_{n-1}(\beta) = 0 \tag{9.55}$$

of order $4(n-1)$, where the involved polynomial, are defined as

$$P_{2(n-1)}(\beta) = \sum_{i=1}^{n} Y_i x_i \prod_{j \neq i}^{n} (\beta + x_j)^2, \quad Q_{2(n-1)}(\beta) = \sum_{i=1}^{n} x_i^2 \prod_{j \neq i}^{n} (\beta + x_j)^2,$$

$$R_{3(n-1)}(\beta) = \sum_{i=1}^{n} Y_i x_i \prod_{j \neq i}^{n} (\beta + x_j), \quad H_{n-1}(\beta) = \sum_{i=1}^{n} x_i^2 \prod_{j \neq i}^{n} (\beta + x_j)^3.$$

The polynomial equation (9.55) has $4(n-1)$ roots; some of them may be complex and some of them may be real. There are stable numerical algorithms to find all roots of a polynomial – see function `polyroot` in R. We are looking for real positive roots.

Algebra of polynomials

To solve the polynomial equation (9.55) with `polyroot`, one has to find $4(n-1)$ coefficients which are derived from product and sum of elementary polynomials involved in the definition of P, Q, R and H. In fact, the degree of the polynomial (9.55) is $4(n-1)-1$ because the coefficient at $\beta^{4(n-1)}$ is zero. In the following, we accept the specification of the polynomial of the $(n-1)$ degree as an n-dimensional array of the coefficients **z**, with ascending power, $z_1 + z_2 x + ... + z_n x^{n-1}$. For example, to find the roots of the quadratic equation $x^2 - 3x - 1 = 0$ we call `polyroot(z=c(-1,-3,1))`. As a reminder, a polynomial of the $(n-1)$ degree has $n-1$ roots, some of them may be real and some of them may be complex. A polynomial of an odd degree has at least one real root. Three simple R utility functions are used to find the coefficients of the polynomial (9.55): (1) `polsum` returns the sum of two polynomials, (2) `polmultk` returns the product of a polynomial times an elementary polynomial of the type bx^k, and finally (3) `polmult` uses the previous two functions to compute the coefficients of the product of two general polynomials. This algebra of polynomials is used to find all real roots of equation (9.55).

Function `MMpolyn` computes the coefficients of the polynomial (9.55) using the utility functions described above and finds $4(n-1)-1$ roots using the `polyroot` function. All complex roots are brushed off by command `r=Re(r[abs(Im(r))` `<1e-07])` where `Re` and `Im` extract the real and imaginary parts of a complex number.

False solution

As we learned from Section 9.7, the NLS estimate may not exist or the normal equation may have several solutions leading to a possibly false NLS estimate. The polynomial equation (9.55) realized in the R function `MMpolyn(x,y)` finds all positive solutions of the normal equation for β with the respective slope coefficient using the least squares formula (9.54).

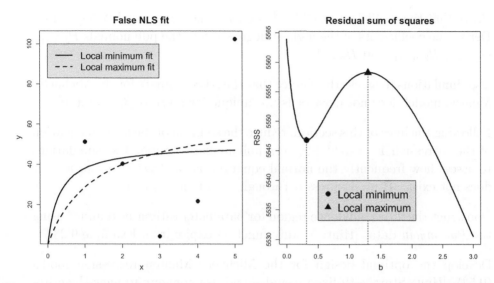

Figure 9.19: *For the data depicted in the left plot, the NLS does not exist and the* nls *produces a false estimate about* 0.3. *The RSS has one local minimum and one local maximum, but there is no global minimum.*

For the majority of data, the NLS estimate in the Michaelis–Menten regression model exists and is unique. However, this is not always the case. For example, function michm(job=2) offers data for x and Y ($n = 5$) for which the nls does not converge starting from a=100,b=3 with an error message. If the starting values are a=mean(y),b=0, as suggested above, iterations converge to a false local minimum (Solution 1) with RSS=5546.824 and b=0.3168224 (see below). The MMpolyn function returns two positive solutions of the normal equation: one being the point of local minimum and another b=1.2948885 being the point of local maximum with RSS = 5558.292 (Solution 2).

In Figure 9.19, the data with local minimum and maximum fits are shown at left and the RSS function of β with local minimum and maximum at right. Since the RSS at infinity, $\widehat{S}_\infty < \widehat{S}$, the NLS estimate is false: the nonlinear least squares solution for this data does not exist.

Problems

1. Start the michm(job=1) with the two starting values discussed above. Does nls converge or produce an error message?

2. (a) Modify function michm to repeat the analysis for the untreated velocity y=c(67,51,84, 86,98,115,131,124,144,158,160). Note that the concentration repeats once at 1.10. (b) Test the hypothesis that the asymptote parameter α is the same for treated and untreated data. [Hint: Combine the data using a dummy variable.]

3. Prove that the coefficient at $\beta^{4(n-1)}$ is zero in the polynomial (9.55). [Hint: Derive the coefficients of the highest degree of β in the polynomials $P_{2(n-1)}$, $Q_{2(n-1)}$, $R_{3(n-1)}$, and H_{n-1}.]

4. Use simulations to assess the chance that the NLS estimate for the Michaelis–Menten model does not exist or is not unique for given α, β, n, and σ^2.

5. Following the lines of this section, reduce the solution of the normal equation of the regression $Y_i = \alpha e^{i\beta} + \varepsilon_i$ to a polynomial solution. Use simulations to assess how frequently the normal equation has at least two solutions or does not exist. [Hint: Express α through β and let $x = e^{\beta}$.]

6. Construct the 95% confidence region for parameter estimates α and β using the *Puromycin* data. [Hint: Modify function `expgr` from Example 9.23.]

7. Develop the optimal design for the Michaelis–Menten regression model (9.53). [Hint: Start with fixed α and β and use `contour` to view the criterion function. Then use simulations to find an optimal adaptive design.]

Chapter 10

Appendix

10.1 Notation

Typically, we use boldface to indicate vectors and matrices. A small bold letter is typically used for vectors and a capital bold letter for matrices (normal font is used for scalars). For example, $\mathbf{a}$ is a vector, $\mathbf{A}$ is a matrix, and a is a scalar. The same convention works for Greek letters (following tradition, we use Greek letters for unknown parameters, subject to estimation); the hat over the symbol indicates that it is an estimator. For example, if μ is an unknown scalar parameter, $\widehat{\mu}$ indicates that it is an estimator of μ. Likewise, if $\boldsymbol{\theta}$ is an unknown parameter vector, $\widehat{\boldsymbol{\theta}}$ indicates its vector estimator. Finally, if $\boldsymbol{\Omega}$ is an unknown matrix (such as the covariance matrix), $\widehat{\boldsymbol{\Omega}}$ indicates that it is a matrix estimator.

Uppercase letters are used for random variables and the corresponding lowercase letters are used to indicate arguments of the density probability function (pdf), or compactly, density, and cumulative distribution function (cdf), or when using integration. For example, if X is a random variable, we use $f_X(x)$ or $f(x)$ to denote its density.

10.2 Basics of matrix algebra

The goal of this section is to review the basics of matrix algebra used in statistics. There are several excellent books on matrix algebra with statistical emphasis such as Searle [91], Graybill [42], Schott [89], and Haverville [54]. A concise problem–solution discussion of matrix algebra and calculus can be found in the book by Abadir and Magnus [1]. These books contain much more detail on matrix algebra.

Advanced Statistics with Applications in R, First Edition. Eugene Demidenko.
© 2020 John Wiley & Sons, Inc. Published 2020 by John Wiley & Sons, Inc.

10.2.1 Preliminaries and matrix inverse

Modern statistics processes large data. It is often convenient to deal with the object as whole rather than with its elements, especially when the number of elements is large. Therefore, matrix algebra and matrix calculus become indispensable tools for statistical computations. Vectors and matrices of the same size can be added or subtracted. For example, if $\mathbf{a}$ and $\mathbf{b}$, $\mathbf{A}$ and $\mathbf{B}$ are 4×1 vectors and 3×4 matrices, respectively, then the ith element of vector $\mathbf{a} + \mathbf{b}$ is $a_i + b_i$ and the (i, j)th element of matrix $\mathbf{A} - \mathbf{B}$ is $a_{ij} - b_{ij}$ (we say that matrices are of the same order if they have the same dimensions). We may use notation $\mathbf{A} = \{a_{ij}, i = 1, .., m, j = 1, .., n\}$ to indicate the elements of the $m \times n$ matrix. Sometimes the matrix dimensions are shown as a superindex, $\mathbf{A}^{m \times n}$. If the number of rows and columns are the same $(n = m)$ we say that the matrix is a square matrix. An $n \times 1$ vector $\mathbf{a}$ may be viewed as a matrix with single column. We assume that vectors are column vectors unless otherwise specified. Sometimes, we represent a matrix by a collection of vector columns, $\mathbf{A}^{m \times n} = \left[\mathbf{a}_1^{m \times 1}, \mathbf{a}_2^{m \times 1}, ..., \mathbf{a}_n^{m \times 1} \right]$.

The identity matrix is a special square matrix which has 1 on the main diagonal and 0 elsewhere. We use notation $\mathbf{I}$ to indicate the identity matrix; sometimes we use a subscript to indicate the size of the identity matrix, e.g. $\mathbf{I}_3$ is the 3×3 identity matrix. The (i, j)th element of the matrix product, $\mathbf{AB}$ is the sum of the product of the ith row of matrix $\mathbf{A}$ and the jth column of matrix $\mathbf{B}$. Succinctly, if $\mathbf{A}$ is $m \times n$ and $\mathbf{B}$ is $n \times k$, we have $(\mathbf{AB})_{ij} = \sum_{k=1}^{n} a_{ik} b_{kj}$ following the rule "row-by-column." For the matrix product to exist, the "neighboring" dimensions should be the same. Generally, square matrices of the same size do not commute, $\mathbf{AB} \neq \mathbf{BA}$. Hence, we need to distinguish the matrix multiplication at left from the matrix multiplication at right. Symmetric matrices commute if they share eigenvectors (see the next section). The associative rule holds, $\mathbf{A(B + C)} = \mathbf{AB} + \mathbf{AC}$. It is easy to see by the element-wise examination that $\mathbf{I}_m \mathbf{A} = \mathbf{A}$ and $\mathbf{AI}_n = \mathbf{A}$.

The matrix *inverse* is one of the most important concepts of matrix algebra. The inverse matrix is denoted $\mathbf{A}^{-1}$ and it may exist only for square matrices. By the definition of the inverse matrix $\mathbf{A}^{-1}\mathbf{A} = \mathbf{AA}^{-1} = \mathbf{I}$. Not all square matrices have an inverse matrix. A matrix without an inverse is called *singular*, or *deficient*. Reversely, a matrix with an inverse is call *nonsingular*. A square matrix $\mathbf{A}$ is singular if and only if the vector columns of the matrix are linearly dependent $\lambda_1 \mathbf{a}_1 + \lambda_2 \mathbf{a}_2 + ... + \lambda_n \mathbf{a}_n = \mathbf{0}$, where $\lambda_1, \lambda_2, ..., \lambda_n$ are scalars (not all zeros) and $\mathbf{0}$ is the $n \times 1$ null vector.

Transposition operation reflects a matrix by $90°$ across the main diagonal (rows become columns and columns become rows). In other words, the (i, j)th element of the transposed matrix $\mathbf{A}'$ is $\{A_{ji}\}$, where $'$ is the transposition sign (sometimes ^{T} is used). If $\mathbf{A}$ is $m \times n$ then $\mathbf{A}'$ is $n \times m$. A square matrix $\mathbf{A}$ is symmetric if $\mathbf{A}' = \mathbf{A}$. Four important properties hold: (a) $(\mathbf{AB})' = \mathbf{B}'\mathbf{A}'$, (b)

$(\alpha \mathbf{A} + \beta \mathbf{B})' = \alpha \mathbf{A}' + \beta \mathbf{B}'$, (c) $(\mathbf{A}^{-1})' = (\mathbf{A}')^{-1}$, and (d) $(\mathbf{AB})^{-1} = \mathbf{B}^{-1} \mathbf{A}^{-1}$.

The *rank* of a matrix is the maximum number of linearly independent vector columns, we write $\text{rank}(\mathbf{A}) = k$ to indicate that the rank of a $m \times n$ matrix $\mathbf{A}$ is k. We say that an $n \times m$ matrix is of full rank if $\text{rank}(\mathbf{A}) = \min(n, m)$. A matrix inverse exists if and only if it has full rank. If $\mathbf{A}$ is a nonsingular matrix then $\text{rank}(\mathbf{AB}) = \text{rank}(\mathbf{B})$. The following formulas are useful: $\text{rank}(\mathbf{A}) = \text{rank}(\mathbf{A}')$, $\text{rank}(\mathbf{A}) = \text{rank}(\mathbf{A}'\mathbf{A})$ and $\text{rank}(\mathbf{AB}) \leq \min(\text{rank}(\mathbf{A}), \text{rank}(\mathbf{B}))$. As follows from the first property, the number of linearly independent columns is the sane as the number of linearly independent rows.

If $\mathbf{a}$ is a vector column, then $\mathbf{a}'$ is a vector row. The product $\mathbf{a}'\mathbf{b}$ is called the scalar (or dot) product, where $\mathbf{a}$ and $\mathbf{b}$ are $n \times 1$. The reader may see other notations for the scalar product, such as $\mathbf{a} \cdot \mathbf{b}$ or $(\mathbf{a}, \mathbf{b})$. When $\mathbf{a}'\mathbf{b} = 0$, we say that $\mathbf{a}$ and $\mathbf{b}$ are orthogonal vectors in R^n, and we write $\mathbf{a} \perp \mathbf{b}$. If $\mathbf{a} = \mathbf{b}$ we have $\mathbf{a}'\mathbf{a} = \|\mathbf{a}\|^2 = \sum_{i=1}^{n} a_i^2$, where $\|\mathbf{a}\|$ is the norm of vector $\mathbf{a}$. The *Cauchy inequality* states that $(\sum a_i b_i)^2 \leq (\sum a_i^2)(\sum b_i^2)$ or in the vector form

$$(\mathbf{a}'\mathbf{b})^2 \leq \|\mathbf{a}\|^2 \|\mathbf{b}\|^2.$$

The equality takes place when vectors are collinear, or algebraically when $\mathbf{a} = \lambda \mathbf{b}$. The proof is easy: Consider a quadratic function of λ expressed through the squared norm of the vector, $\|\mathbf{a} - \lambda \mathbf{b}\|^2 \geq 0$. From elementary algebra, we know that a quadratic equation has one or no roots if and only if the discriminant $(\mathbf{a}'\mathbf{b})^2 - \|\mathbf{a}\|^2 \|\mathbf{b}\|^2 \leq 0$ which proves the Cauchy inequality.

The system of vectors $\{\mathbf{a}_1, \mathbf{a}_2, .., \mathbf{a}_n\}$ from R^m is called *orthogonal* if $\mathbf{a}_i \perp \mathbf{a}_j$ for $i \neq j$ ($n \leq m$). If the vector columns of matrix $\mathbf{A}^{m \times n}$ are orthogonal, then matrix $\mathbf{A}'\mathbf{A}$ is diagonal. An orthogonal system of vectors with unit length is called *orthonormal*; then matrix $\mathbf{A}$ is called orthonormal and $\mathbf{A}'\mathbf{A} = \mathbf{I}_n$.

The Cauchy inequality implies the *triangle inequality*,

$$\|\mathbf{x} - \mathbf{y}\| + \|\mathbf{y} - \mathbf{z}\| \geq \|\mathbf{x} - \mathbf{z}\|.$$

If $\mathbf{1} = (1, 1, ..., 1)'$, then $\mathbf{a}'\mathbf{1} = \mathbf{1}'\mathbf{a}$ is the sum of the elements of vector $\mathbf{a}$. As follows from the Cauchy inequality, the squared sum of the vector elements is less than or equal to the squared norm of the vector times its dimension, $(\mathbf{a}'\mathbf{1})^2 \leq n \|\mathbf{a}\|^2$. It is easy to see that $\mathbf{1}\mathbf{1}'$ is a matrix with all elements equal 1. The Cauchy inequality enables the definition of cosine of the angle, θ, between vectors:

$$\cos \theta = \frac{\mathbf{a}'\mathbf{b}}{\|\mathbf{a}\| \|\mathbf{b}\|}.$$

The angle is $90°$ if $\mathbf{a} \perp \mathbf{b}$, and $0°$ or $180°$ if the vectors are colinear. In fact, $\cos \theta$ is the correlation coefficient between data vectors $\mathbf{x}$ and $\mathbf{y}$ with the origin shifted to the center points $\bar{x}\mathbf{1}$ and $\bar{y}\mathbf{1}$.

The following formula holds

$$(\mathbf{I} + \mathbf{vv}')^{-1} = \mathbf{I} - \frac{1}{1 + \|\mathbf{v}\|^2} \mathbf{vv}', \tag{10.1}$$

where $\mathbf{I}$ is the identity matrix and $\mathbf{v}$ is a vector of the same dimension. This formula belongs to the family of general *dimension-reduction* formulas:

$$(\mathbf{D} + \mathbf{AB})^{-1} = \mathbf{D}^{-1} - \mathbf{D}^{-1}\mathbf{A}(\mathbf{I} + \mathbf{BD}^{-1}\mathbf{A})^{-1}\mathbf{BD}^{-1}, \qquad (10.2)$$

$$(\mathbf{D} + \mathbf{EFE'})^{-1} = \mathbf{D}^{-1} - \mathbf{D}^{-1}\mathbf{E}(\mathbf{E'D}^{-1}\mathbf{E} + \mathbf{F}^{-1})^{-1}\mathbf{E'D}^{-1}, \qquad (10.3)$$

$$(\mathbf{D} + \mathbf{B})^{-1} = \mathbf{D}^{-1} - \mathbf{D}^{-1}(\mathbf{D}^{-1} + \mathbf{B}^{-1})^{-1}\mathbf{D}^{-1}, \qquad (10.4)$$

$$(\mathbf{D} + \mathbf{B})^{-1}\mathbf{B} = \mathbf{I} - (\mathbf{D} + \mathbf{B})^{-1}\mathbf{D}, \qquad (10.5)$$

assuming that the inverse matrices exist. For example, (10.1) follows from (10.4) letting $\mathbf{D} = \mathbf{I}$, $\mathbf{B} = \mathbf{vv'}$. As a special case of matrix identity (10.2), we obtain

$$(\mathbf{A} + \mathbf{ab'})^{-1} = \mathbf{A}^{-1} - \frac{1}{1 + \mathbf{b'A}^{-1}\mathbf{a}}\mathbf{A}^{-1}\mathbf{ab'A}^{-1}. \qquad (10.6)$$

A square matrix $\mathbf{N}$ is called *idempotent* if $\mathbf{N}^2 = \mathbf{N}$. These matrices emerge in connection with the chi-square distribution. Obviously, the identity matrix is idempotent. It is easy to show that, if $\mathbf{N}$ is an idempotent matrix, then $\mathbf{I} - \mathbf{N}$ is an idempotent matrix. Let $\mathbf{X}$ be any $n \times m$ matrix such that $\mathbf{X'X}$ exists (this implies that rank$(\mathbf{X}) = m \leq n$). Then matrix

$$\mathbf{P} = \mathbf{X}(\mathbf{X'X})^{-1}\mathbf{X'} \qquad (10.7)$$

is called the *projection matrix*. This matrix emerges in the linear model; see Chapter 8. The projection matrix is an idempotent matrix,

$$\mathbf{P}^2 = \mathbf{X}(\mathbf{X'X})^{-1}\mathbf{X'X}(\mathbf{X'X})^{-1}\mathbf{X'} = \mathbf{X}(\mathbf{X'X})^{-1}\mathbf{X'} = \mathbf{P}.$$

The trace of a square matrix frequently emerges in statistics. For example, the trace of the covariance matrix is the total variance. By definition, the trace of a square matrix is the sum of the diagonal elements $\text{tr}(\mathbf{A}) = \sum A_{ii}$ and therefore can be viewed as a linear operator. The matrix norm of a rectangular matrix can be defined via the trace as $\|\mathbf{X}\| = \sqrt{\text{tr}(\mathbf{X'X})}$, although this is not the only definition of the matrix norm.

The following properties of the matrix trace hold: (1) $\text{tr}(\mathbf{A} + \mathbf{B}) = \text{tr}(\mathbf{A}) + \text{tr}(\mathbf{B})$, (2) $\text{tr}(a\mathbf{A}) = a \times \text{tr}(\mathbf{A})$, (3) $\text{tr}(\mathbf{A'}) = \text{tr}(\mathbf{A})$, (4) $\text{tr}(\mathbf{AB}) = \text{tr}(\mathbf{BA})$, (5) $\text{tr}(\mathbf{XX'}) = \text{tr}(\mathbf{X'X}) = \sum x_{ij}^2$. For example, the fourth property implies $\text{tr}(\mathbf{ab'}) = \mathbf{a'b}$, where $\mathbf{a}$ and $\mathbf{b}$ are vectors of the same length. In particular, $\text{tr}(\mathbf{aa'}) = \|\mathbf{a}\|^2$. The following inequality for matrices, as a version of the Cauchy inequality, holds: $\text{tr}^2(\mathbf{AB}) \leq \text{tr}(\mathbf{A}^2)\text{tr}(\mathbf{B}^2)$, where $\mathbf{A}$ and $\mathbf{B}$ are symmetric matrices of the same order.

R programming

R has a versatile built-in matrix algebra. Below are listed some major matrix/vector operations.

Math.	$\mathbf{a'b}$	$\mathbf{Ab}$	$\mathbf{AB}$	$\mathbf{A}^{-1}$	$\|\mathbf{a}\|^2$	$\mathbf{a'}$	$\mathbf{A'}$
R	sum(a*b)	A%*%b	A%*%B	solve(A)	sum(a^2)	t(a)	t(A)

Recall that `nrow(A)` and `ncol(A)` return the number of rows and columns. By default, vectors in R are vector columns. For instance, if `a=1:10` then it is assumed that `a` is a 10×1 vector. Matrix multiplication is distinguished from element-wise multiplication by surrounding * with % sign. For example, if A and B are matrices of the same size then `A*B` will be the matrix of the same size with the element-wise products, $a_{ij}b_{ij}$. Analogously, if `a` and `b` are vectors of the same size `a*b` means a vector with elements as the products of the corresponding elements of vectors `a` and `b`. This explains why the scalar product $\mathbf{a'b}$ is computed as `sum(a*b)`. Alternatively, we can compute the scalar product as `t(a)%*%b` which returns a 1×1 matrix. Consequently, if `M` is a 2×2 matrix then `(t(a)%*%b)*M` will produce an error. However, using `sum(a*b)*M` will be just fine. As a word of caution: there is function `rank` in R but it does not compute a rank of a matrix. One can use `det` to test if a rectangular matrix has full rank; see the next section.

10.2.2 Determinant

The determinant of a square matrix naturally arises in linear algebra in association with the volume of the parallelepiped. Moreover, the determinant of the covariance matrix is called the *generalized variance*. We use notation $|\mathbf{M}|$ to indicate the determinant of matrix $\mathbf{M}$. Some authors use notation $\det(\mathbf{M})$. A square matrix is singular if and only if its determinant is zero, $|\mathbf{A}| = 0$. It is a well known fact of linear algebra that if $\mathbf{A}^{m \times m}$ is the matrix of a linear transformation $R^m \to R^m$ and T is a body in R^m with the unit volume, then $vol(T_A) = |\mathbf{A}|$, where T_A is the transform of T upon $\mathbf{A}$ and $|\mathbf{A}|$ is the absolute value of the determinant (a less ambiguous notation is $|\det(\mathbf{A})|$).

If $\mathbf{A}$ and $\mathbf{B}$ are square matrices of the same order, then

$$|\mathbf{AB}| = |\mathbf{A}||\mathbf{B}|. \tag{10.8}$$

Other formulas to remember, $|\mathbf{A'}| = |\mathbf{A}|$ and $|\mathbf{A}^{-1}| = 1/|\mathbf{A}|$. The following general formula is useful

$$|\mathbf{D} - \mathbf{CA}^{-1}\mathbf{B}| = \frac{1}{|\mathbf{A}|}|\mathbf{D}||\mathbf{A} - \mathbf{BD}^{-1}\mathbf{C}|,$$

where $\mathbf{B}$ and $\mathbf{C}$ may be rectangular matrices. In a special case, $|\mathbf{D} - \mathbf{vw'}/a| = (a - \mathbf{w'D}^{-1}\mathbf{v})/a|\mathbf{D}|$. Applying this formula with $\mathbf{D} = \mathbf{I}$, $\mathbf{w} = -\mathbf{v}$ and $a = 1$ to the matrix in (10.1), we obtain $|\mathbf{I} + \mathbf{vv'}| = 1 + \|\mathbf{v}\|^2$.

10.2.3 Partition matrices

Sometimes we partition/split a matrix into several parts and consider submatrices. This consideration is relevant when two groups of predictors in a linear model are considered, or when the parameter vector is split into two parts, parameters of interest and nuisance parameters. For example, we may partition $\mathbf{X}^{n \times m} = \left[\mathbf{U}^{n \times m_1}; \mathbf{W}^{n \times m_2} \right]$, where $m_1 + m_2 = m$ meaning that matrix $\mathbf{U}$ consists of the first m_1 vector columns of matrix $\mathbf{X}$ and $\mathbf{W}$ consists of the last m_2 vector columns of matrix $\mathbf{X}$. Then $\mathbf{XX}' = \mathbf{UU}' + \mathbf{WW}'$ and

$$\mathbf{X}'\mathbf{X} = \begin{bmatrix} \mathbf{U}'\mathbf{U} & \mathbf{U}'\mathbf{W} \\ \mathbf{W}'\mathbf{U} & \mathbf{W}'\mathbf{W} \end{bmatrix}.$$

Let an $m \times m$ matrix be partitioned as follows (we do not assume the matrix is symmetric):

$$\mathbf{M} = \begin{bmatrix} \mathbf{M}_{11} & \mathbf{M}_{12} \\ \mathbf{M}_{21} & \mathbf{M}_{22} \end{bmatrix}. \tag{10.9}$$

Then the determinant of the matrix can be expressed via the determinants of the submatrices as follows:

$$|\mathbf{M}| = |\mathbf{M}_{22}| \left| \mathbf{M}_{11} - \mathbf{M}_{12}\mathbf{M}_{22}^{-1}\mathbf{M}_{21} \right| = |\mathbf{M}_{11}| \left| \mathbf{M}_{22} - \mathbf{M}_{21}\mathbf{M}_{11}^{-1}\mathbf{M}_{12} \right|. \tag{10.10}$$

Matrix $\mathbf{M}$ is symmetric if and only if $\mathbf{M}'_{11} = \mathbf{M}_{11}$, $\mathbf{M}'_{22} = \mathbf{M}_{22}$, and $\mathbf{M}_{21} = \mathbf{M}'_{12}$. Sometimes, we need to invert a partitioned symmetric matrix. The following formula for the block matrix inverse holds:

$$\begin{bmatrix} \mathbf{M}_{11} & \mathbf{M}_{12} \\ \mathbf{M}'_{12} & \mathbf{M}_{22} \end{bmatrix}^{-1} = \begin{bmatrix} \mathbf{M}_{11}^{-1} + \mathbf{F}\mathbf{E}^{-1}\mathbf{F}' & -\mathbf{F}\mathbf{E}^{-1} \\ -\mathbf{E}^{-1}\mathbf{F}' & \mathbf{E}^{-1} \end{bmatrix} = \begin{bmatrix} \mathbf{A}^{-1} & -\mathbf{A}^{-1}\mathbf{B}' \\ -\mathbf{B}\mathbf{A}^{-1} & \mathbf{M}_{22}^{-1} + \mathbf{B}\mathbf{A}^{-1}\mathbf{B}' \end{bmatrix} \tag{10.11}$$

where

$$\mathbf{E} = \mathbf{M}_{22} - \mathbf{M}'_{12}\mathbf{M}_{11}^{-1}\mathbf{M}_{12}, \quad \mathbf{F} = \mathbf{M}_{11}^{-1}\mathbf{M}_{12}, \quad \mathbf{A} = \mathbf{M}_{11} - \mathbf{M}_{12}\mathbf{M}_{22}^{-1}\mathbf{M}'_{12}, \quad \mathbf{B} = \mathbf{M}_{22}^{-1}\mathbf{M}'_{12}.$$

In a special case when $\mathbf{M}_{11}$ is a scalar we obtain

$$\begin{bmatrix} a & \mathbf{b}' \\ \mathbf{b} & \mathbf{A} \end{bmatrix}^{-1} = \begin{bmatrix} \frac{1}{a} + \frac{1}{a^2}\mathbf{b}'\mathbf{E}^{-1}\mathbf{b} & -\frac{1}{a}\mathbf{b}'\mathbf{E}^{-1} \\ -\frac{1}{a}\mathbf{E}^{-1}\mathbf{b} & \mathbf{E}^{-1} \end{bmatrix}, \tag{10.12}$$

where $\mathbf{E} = \mathbf{A} - \frac{1}{a}\mathbf{b}\mathbf{b}'$.

The block matrix inverse emerges in the multivariate normal distribution discussed in Section 4.1 when considering the conditional distribution. See also Example 10.7.

Problems

1. Deduce the triangle inequality from the Cauchy inequality. Write R code to verify the triangle inequality. [Hint: Use `runif` to generate vectors and use dimension as an argument of your R function.]

2. A three-horse problem: (a) The first horse pulls the weight North and the second horse pulls the weight East. The third horse is the weakest. Can she neutralize the first two horses? (b) The same conditions apply but the second horse pulls the weight South-East.

3. Derive (10.1) from (10.2) and (10.3).

4. Prove that $\mathbf{X}'\mathbf{X}$, $\mathbf{X}\mathbf{X}'$, and $0.5(\mathbf{A} + \mathbf{A}')$ are symmetric matrices. Write R code to verify the result. [Hint: Use `t` to transpose the matrix when checking the symmetry, use `runif` to generate random matrices, suggest a numeric criterion to check the symmetry.]

5. Prove that $\mathbf{A}^{-1}$ is a symmetric matrix if $\mathbf{A}$ is a symmetric matrix. Write R code to verify the result. [Hint: Suggest a numeric criterion to check the symmetry.]

6. Verify formulas (10.2) through (10.5) assuming that the matrices are diagonal.

7. Derive (10.6) from (10.2).

8. (a) The straight line in R^m is defined as the set of points $\{\mathbf{x} \in R^m : \mathbf{x} = \lambda\mathbf{a}, -\infty < \lambda < \infty\}$ where $\mathbf{a}$ is a nonzero direction vector. Project $\mathbf{y} \in R^m$ onto the line. Express the projection point using an idempotent matrix. (b) Generalize the previous problem to a linear subspace defined as $\{\mathbf{x} \in R^m : \mathbf{x} = \mathbf{A}\lambda\}$, where $\mathbf{A}$ is an $m \times k$ matrix of full rank $k < m$ and $\lambda \in R^k$ is arbitrary.

9. Can an idempotent matrix be nonsingular?

10. Prove that $\mathbf{A}'\mathbf{A} = \mathbf{0}$ implies $\mathbf{A} = \mathbf{0}$.

11. True or false: $(\mathbf{A} + \mathbf{B})^2 = \mathbf{A}^2 + 2\mathbf{A}\mathbf{B} + \mathbf{B}^2$, where $\mathbf{A}$ and $\mathbf{B}$ are symmetric matrices of the same order. Give a counterexample if it does not hold in the general case. Formulate the condition under which the equality holds. Write R code to verify the result. [Hint: Use `rnorm` to generate matrices and suggest a numeric criterion to check the equality.]

12. Prove that the sum of diagonal elements of matrix $\mathbf{A}\mathbf{a}\mathbf{a}'$ is equal to $\mathbf{a}'\mathbf{A}\mathbf{a}$, where $\mathbf{a}$ is a vector.

13. The R function `berndet` estimates the probability that a $n \times n$ matrix with iid Bernoulli distributed random variables is singular. (a) Find the probability that `det=0` when $n = 2$ in closed form and check via simulations. (b) Determine how the Bernoulli probability (`prob`) affects the probability that the determinant is 0.

14. The same as above but the elements of the matrix have a continuous distribution. Is it true that the random matrix is nonsingular with probability 1? Write R code to test this conjecture. Use `nExp`, `m` and `epsilon` as arguments (set the default values 10,000, 10, and 10^{-8}, respectively). Generate `nExp` matrices using `runif` and save their determinants in array `all.det`. Return `mean(abs(all.det)<epsilon)`. Repeat the same experiment with `rnorm`. Use a `for` loop and `apply` and compare their efficiency.

15. Derive the determinant of a matrix when $\mathbf{M}_{11}$ is scalar, similarly to the matrix in (10.12), using formula (10.10).

16. Is formula (10.11) beneficial for a large matrix inverse? In other words, what is faster to invert, say, a $1,000 \times 1,000$ matrix using `solve` or to partition the matrix in four 250×250 matrices and use formula (10.11)? Write R code to simulate 1000 random $1,000 \times 1,000$ matrices using the `for` loop and compare the inverse time by printing out `date()`.

17. $\mathbf{A}$ is a 10×6 matrix of full rank ($\text{rank}(A) = 6$). Find $\text{tr}(\mathbf{I} - \mathbf{A}(\mathbf{A}'\mathbf{A})^{-1}\mathbf{A}')$. Write R code to verify your analytic result by generating 60 matrix elements using `rnorm` function. [Hint: Use `sum(diag(M))` to compute $\text{tr}(\mathbf{M})$.]

18. Modify the R function `block.inv` to check formulas (10.10) and (10.12) by computation. [Hint: Create random matrices and compare direct computation with partition formulas.]

10.3   Eigenvalues and eigenvectors

A nonzero vector $\mathbf{p}$ is called an eigenvector (or a characteristic vector) of a symmetric $m \times m$ matrix $\mathbf{A}$ if $\mathbf{Ap} = \lambda\mathbf{p}$ for some scalar λ, which is called the eigenvalue (or the characteristic root). Although eigenvalues and eigenvectors can be defined for asymmetric matrices as well, we usually deal with symmetric matrices in statistics. Since eigenvectors multiplied by a scalar are eigenvectors, we usually apply the restriction $\|\mathbf{p}\| = 1$. However, even if this restriction applies, eigenvectors are not unique: if $\mathbf{p}$ is an eigenvector, $-\mathbf{p}$ is an eigenvector as well. Since equation $(\mathbf{A}-\lambda\mathbf{I})\mathbf{p} = \mathbf{0}$ has a unique solution $\mathbf{p} = \mathbf{0}$ if $|\mathbf{A}-\lambda\mathbf{I}| \neq 0$ for $\mathbf{p}$ to be a nonzero solution, we need to have $|\mathbf{A}-\lambda\mathbf{I}| = 0$. It immediately follows from the above equation that matrices $\mathbf{A}$ and $\mathbf{QAQ}^{-1}$ have the same set of eigenvalues for any nonsingular matrix $\mathbf{Q}$ due to property (10.8). $|\mathbf{A}-\lambda\mathbf{I}|$ as a function of λ is called the characteristic equation, a polynomial of the mth order.

By the fundamental theorem of algebra, there are m roots of the polynomial. That is, there are m eigenvalues for any $m \times m$ matrix. There may be complex roots, but, for a symmetric matrix $\mathbf{A}$, which is what we usually deal with in statistics, all roots/eigenvalues are real. Some eigenvalues may repeat just as the roots of the polynomial. For example, if the characteristic equation takes the form $(\lambda - 1)^2(\lambda - 2) = 0$ there are three roots, $\lambda_{1,2} = 1$, $\lambda_3 = 2$. If $\lambda_1 \neq \lambda_2$, the corresponding eigenvectors are orthogonal. Indeed, let $\mathbf{A}$ be symmetric, and $\mathbf{Ap}_1 = \lambda_1\mathbf{p}_1$, $\mathbf{Ap}_2 = \lambda_2\mathbf{p}_2$ where $\lambda_1 \neq \lambda_2$. Multiply the first equation by $\mathbf{p}_2'$ and the second equation by $\mathbf{p}_1'$ and subtract,

$$\mathbf{p}_2'\mathbf{Ap}_1 - \mathbf{p}_1'\mathbf{Ap}_2 = \lambda_1\mathbf{p}_2'\mathbf{p}_1 - \lambda_1\mathbf{p}_1'\mathbf{p}_2. \tag{10.13}$$

But $\mathbf{p}_2'\mathbf{p}_1 = \mathbf{p}_1'\mathbf{p}_2$ and $\mathbf{p}_2'\mathbf{Ap}_1 = \mathbf{p}_1'\mathbf{Ap}_2$. To prove the latter, use the fact that transposition of a scalar is the same scalar: then, using the formula $(\mathbf{AB})' = \mathbf{B}'\mathbf{A}'$, we have $\mathbf{p}_2'\mathbf{Ap}_1 = (\mathbf{p}_2'\mathbf{Ap}_1)' = \mathbf{p}_1'\mathbf{A}'(\mathbf{p}_2')' = \mathbf{p}_1'\mathbf{Ap}_2$. Then equation (10.13) implies $0 = (\lambda_1 - \lambda_2)(\mathbf{p}_1'\mathbf{p}_2)$. But, since $\lambda_1 - \lambda_2 \neq 0$, we finally obtain $\mathbf{p}_1'\mathbf{p}_2 = 0$. In geometric notation, $\mathbf{p}_1 \perp \mathbf{p}_2$. If two eigenvalues are the same (repeat) we can arbitrarily select two *orthonormal* (orthogonal unit length) eigenvectors. In general, for any $m \times m$ symmetric matrix, the set of eigenvectors constitutes m pairwise orthogonal vectors of unit length. Since eigenvectors have unit length, we have $\lambda = \mathbf{p}'\mathbf{Ap}$. A matrix is nonsingular if and only if none of its eigenvalues is zero.

Theorem 10.1 *If $\mathbf{A}$ is a $m \times m$ symmetric matrix with eigenvalues $\lambda_1, ..., \lambda_m$, then $tr(\mathbf{A}) = \sum \lambda_i$ and $|\mathbf{A}| = \prod \lambda_i$.*

In R, the eigenvectors and eigenvalues are computed by function `eigen`. For example, this R function is used in Section 4.1.1 to generate multivariate normally distributed random vectors.

10.3.1 Jordan spectral matrix decomposition

Let $\mathbf{A}$ be a $m \times m$ symmetric matrix with eigenvalues $\lambda_1, \lambda_2, ..., \lambda_m$ and corresponding eigenvectors $\mathbf{p}_1, \mathbf{p}_2, ..., \mathbf{p}_m$. The Jordan spectral matrix decomposition expresses the matrix as a linear combination of matrices of the type $\mathbf{pp}'$, or specifically, $\mathbf{A} = \sum_{i=1}^m \lambda_i\mathbf{p}_i\mathbf{p}_i'$. Combine eigenvectors into a $m \times m$ matrix $\mathbf{P} = [\mathbf{p}_1, \mathbf{p}_2, ..., \mathbf{p}_m]$ and let $\mathbf{\Lambda} = \text{diag}(\lambda_1, \lambda_2, ..., \lambda_m)$. Then the Jordan spectral matrix decomposition (spectral matrix decomposition) can be expressed in compact form as

$$\mathbf{A} = \mathbf{P}\mathbf{\Lambda}\mathbf{P}'. \tag{10.14}$$

Matrix $\mathbf{P}$ is orthonormal meaning that the vector columns are pairwise orthogonal and have unit length, or in matrix form $\mathbf{P}'\mathbf{P} = \mathbf{P}\mathbf{P}' = \mathbf{I}$, where $\mathbf{I}$ is the $m \times m$ identity matrix. Multiplying by $\mathbf{P}'$ and $\mathbf{P}$, we arrive at the matrix diagonalization

$\mathbf{P'AP} = \mathbf{\Lambda}$. In words, any symmetric matrix can be reduced to a diagonal matrix upon an orthonormal transformation. The Jordan spectral decomposition is frequently used in statistics to diagonalize quadratic forms, see the next section.

Definition 10.2 *Function of a matrix. Let* $\mathbf{A}$ *be a symmetric matrix with the spectral decomposition* $\mathbf{A} = \mathbf{P\Lambda P'}$ *and* f *be a scalar function,* $R^1 \to R^1$. *Define a matrix function as* $f(\mathbf{A}) = \mathbf{P}f(\mathbf{\Lambda})\mathbf{P'}$, *where* $f(\mathbf{\Lambda}) = \mathrm{diag}(f(\lambda_1), f(\lambda_2), ..., f(\lambda_m))$.

For example, define $\mathbf{A}^{1/2} = \mathbf{P\Lambda}^{1/2}\mathbf{P'}$, $\mathbf{A}^{-1} = \mathbf{P\Lambda}^{-1}\mathbf{P'}$ and $\mathbf{A}^{-1/2} = \mathbf{P\Lambda}^{-1/2}\mathbf{P'}$, assuming that all eigenvalues are positive. It is easy to prove that matrix functions with the same f share the eigenvectors with eigenvalues $f(\boldsymbol{\lambda})$, where $\boldsymbol{\lambda}$ is the vector of eigenvalues of matrix $\mathbf{A}$. Consequently, the eigenvalues of the inverse matrix, $\mathbf{A}^{-1}$, are the reciprocals of the eigenvalues of matrix $\mathbf{A}$.

The matrix square root is applied in statistics to generate m-variate normal random variables with specified covariance matrix $\mathbf{\Omega}$; see Section 4.1.1. Another, less computationally intensive algorithm to generate multivariate normal random variables is to use the Cholesky triangular decomposition, $\mathbf{\Omega} = \mathbf{T'T}$, where $\mathbf{T}$ is a nonsingular upper triangular matrix (elements below the diagonal are zero).

10.3.2   SVD: Singular value decomposition of a rectangular matrix

Rectangular matrices and square asymmetric matrices also may be expressed as the product of orthogonal matrices, but then we need two orthogonal matrices, $\mathbf{U}$ and $\mathbf{V}$. Specifically, for any $n \times m$ matrix $\mathbf{X}$, the following decomposition holds (without loss of generality we may assume that $n \geq m$):

$$\mathbf{X} = \mathbf{UDV'}, \tag{10.15}$$

where $\mathbf{U}$ and $\mathbf{V}$ are $n \times m$ and $m \times m$ orthogonal matrices, referred to as left and right eigenvectors, respectively, such that

$$\mathbf{U'U} = \mathbf{I}, \quad \mathbf{V'V} = \mathbf{I}, \tag{10.16}$$

and $\mathbf{D}$ is the $m \times m$ diagonal matrix. Decomposition (10.15) is called the singular value decomposition (SVD). In R, the SVD is computed using function svd; it returns a list with three components: \$u, \$v and \$d. A couple of comments: (1) as follows from (10.16), the vector columns of matrices $\mathbf{U}$ and $\mathbf{V}$ have unit length and are orthogonal to each other; (2) matrices $\mathbf{U}$ and $\mathbf{V}$ are not orthogonal. In general, $\mathbf{UV'} \neq \mathbf{0}$ and $\mathbf{UV} \neq \mathbf{0}$.

Connection to spectral decomposition of matrix $\mathbf{X'X}$

Let $\mathbf{X}$ be a $n \times m$ matrix with the SVD decomposition (10.15) and let $\mathbf{X'X} = \mathbf{P\Lambda P'}$ be the Jordan spectral matrix decomposition. It is easy to see that the two decompositions are connected through the following equalities:

$$\mathbf{\Lambda} = \mathbf{D}^2, \quad \mathbf{P} = \mathbf{V}. \tag{10.17}$$

Indeed, as follows from (10.15) and (10.16), we have

$$\mathbf{X}'\mathbf{X} = (\mathbf{UDV}')'\mathbf{UDV}' = \mathbf{VDU}'\mathbf{UDV}' = \mathbf{VD}^2\mathbf{V}'.$$

This means that, in decomposition (10.15), we may choose $\mathbf{P} = \mathbf{V}$ and $\mathbf{\Lambda} = \mathbf{D}^2$, so that $\mathbf{X}'\mathbf{X}$ can be written as $\mathbf{P\Lambda P}'$, as we sought to prove. In words, the eigenvalues of $\mathbf{X}'\mathbf{X}$ are squared values of matrix $\mathbf{D}$ and eigenvectors of $\mathbf{X}'\mathbf{X}$ are the right eigenvectors of the SVD of matrix $\mathbf{X}$.

Problems

1. (a) Prove that the eigenvalues of the projection matrix (10.7) are either 0 or 1. (b) Prove that the eigenvalues of any idempotent matrix are either 0 or 1. Verify this fact by writing R code using `rnorm` to generate random matrices and then using `eigen` to compute eigenvalues.

2. Prove that the trace of a symmetric matrix equals the sum of its eigenvalues, and use R to demonstrate. [Hint: Use spectral matrix decomposition.]

3. (a) Prove that the square root of matrix $\mathbf{I} - n^{-1}\mathbf{11}'$ is the same matrix. (b) Prove that the square root of any idempotent matrix is the matrix itself.

4. Prove that the rank of matrix $\mathbf{X}$ equals the number of nonzero eigenvalues of matrix $\mathbf{X}'\mathbf{X}$. Write R code to verify this result numerically using `eigen` and `rank`.

5. Find all eigenvectors and eigenvalues of matrix $\mathbf{aa}'$, where $\mathbf{a}$ is a nonzero $m \times 1$ vector.

6. (a) Is it true that eigenvalues of the sum of two symmetric matrices of the same dimension are the sums of the eigenvalues of the two matrices? Provide a counterexample if this is not true. (b) Formulate conditions under which it is true. Write R code to demonstrate.

7. Prove that eigenvalues of the $m \times m$ matrix $\mathbf{A} + \rho\mathbf{I}$ are $\{\lambda_i + \rho, i = 1, 2, ..., m\}$ where $\{\lambda_i\}$ are eigenvalues of matrix $\mathbf{A}$.

8. Prove that $\mathbf{P\Lambda}^{-1}\mathbf{P}'$ is an inverse matrix of $\mathbf{A} = \mathbf{P\Lambda P}'$ using spectral decomposition (10.14).

9. Prove that the eigenvalues of matrix $\mathbf{X}'\mathbf{X}$ are nonnegative.

10. (a) Define $\exp(\mathbf{A})$ and $\ln(\mathbf{A})$, where $\mathbf{A}$ is a symmetric matrix. (b) Prove that $\exp(\mathbf{A}) = \sum_{k=0}^{\infty} \mathbf{A}^k/k!$. (c) Derive matrix series expansion for $\ln(\mathbf{A})$, where $\mathbf{A} > \mathbf{0}$. Verify your results numerically in R.

11. Let $\mathbf{N}$ be a symmetric idempotent matrix. Show that (a) $(\mathbf{N} + \mathbf{I})^{-1} = \mathbf{I} - 0.5\mathbf{N}$, (b) $\exp(\mathbf{N}) = \mathbf{I} + (e - 1)\mathbf{N}$, and (c) $\ln(\mathbf{N} + \mathbf{I}) = \mathbf{N}\ln 2$. Check your results numerically in R. [Hint: Use projection to generate $\mathbf{N}$.]

12. Let the maximum eigenvalue of symmetric matrix $\mathbf{A}$ be less then 1. Prove that the following matrix series expansion holds: $(\mathbf{I} - \mathbf{A})^{-1} = \sum_{k=0}^{\infty} \mathbf{A}^k$. Verify this identity numerically in R by generating a random symmetric matrix with eigenvalues less than 1.

13. Write R function with arguments m and n as dimensions of a random matrix $\mathbf{X}$ to check that `eigen` and `svd` produce the same result (10.17).]Hint: Use `rnorm` to generate a random matrix.]

14. Prove that the set of nonzero eigenvalues of matrices $\mathbf{X}'\mathbf{X}$ and $\mathbf{XX}'$ are the same. [Hint: Use SVD.]

10.4 Quadratic forms and positive definite matrices

10.4.1 Quadratic forms

There are positive and negative numbers. For any two numbers, one may say which number is greater. How about matrices? Can we say which matrix is greater for any pair of matrices? The answer is not obvious because a matrix has several elements – some elements may be greater and some elements may be smaller. The concept of a positive (or nonnegative) definite matrix, as a generalization of a positive (nonnegative) number, is very important in statistics because it allows one to judge the performance of estimators by comparing their covariance matrices. Also the quadratic form emerges as the variance of a linear combination of random variables: if $\mathbf{U}$ is a random vector with covariance matrix $\mathbf{A}$, the variance of the linear combination, $\mathrm{var}(\mathbf{x}'\mathbf{U}) = \mathbf{x}'\mathbf{A}\mathbf{x}$, is a quadratic form of $\mathbf{x}$. The quadratic form is a natural choice as an estimator of the variance, see Section 6.6.

Definition 10.3 *The quadratic form, sometimes referred to as the ordinary quadratic form, is a function of the vector argument, $\mathbf{x}^{m \times 1}$, and is defined as*

$$Q(\mathbf{x}) \overset{\text{def}}{=} \mathbf{x}'\mathbf{A}\mathbf{x} = \sum_{i=1}^{m} \sum_{j=1}^{m} x_i x_j A_{ij},$$

where $\mathbf{A}^{m \times m}$ is a symmetric matrix. The general quadratic form is defined as $Q(\mathbf{x}) = a + \mathbf{b}'\mathbf{x} + \mathbf{x}'\mathbf{A}\mathbf{x}$, where $\mathbf{b}$ is a $m \times 1$ vector and a is a scalar.

For example, for $m = 2$, we have

$$\mathbf{x}'\mathbf{A}\mathbf{x} = x_1^2 A_{11} + 2x_1 x_2 A_{12} + x_2^2 A_{22}. \tag{10.18}$$

The general quadratic form can be expressed as the ordinary quadratic form plus a constant by *completing the square*,

$$a + \mathbf{b}'\mathbf{x} + \mathbf{x}'\mathbf{A}\mathbf{x} = c + (\mathbf{x} + \mathbf{d})'\mathbf{A}(\mathbf{x} + \mathbf{d}), \tag{10.19}$$

where $c = a - 0.25\mathbf{b}'\mathbf{A}^{-1}\mathbf{b}$ and $\mathbf{d} = 0.5\mathbf{A}^{-1}\mathbf{b}$, assuming that $\mathbf{A}$ is a nonsingular matrix. As an immediate corollary, we obtain that, if $\mathbf{A}$ is a positive definite matrix, the minimum of the general quadratic form is attained at $\mathbf{x} = -0.5\mathbf{A}^{-1}\mathbf{b}$.

Theorem 10.4 *Diagonalization of the quadratic form. Any quadratic form can be diagonalized as* $\mathbf{x}'\mathbf{A}\mathbf{a} = \sum_{i=1}^{m} \lambda_i z_i^2$, *where* $\{\lambda_i\}$ *are the eigenvalues and* $\{z_i\}$ *are the components of vector* $\mathbf{z} = \mathbf{P}'\mathbf{x}$, *where* $\mathbf{P}$ *is the matrix of eigenvectors of matrix* $\mathbf{A}$.

Proof. Indeed, as follows from the Jordan spectral decomposition (10.14), we have

$$\mathbf{x}'\mathbf{A}\mathbf{x} = \mathbf{x}'\mathbf{P}\mathbf{\Lambda}\mathbf{P}'\mathbf{x} = (\mathbf{P}'\mathbf{x})'\mathbf{\Lambda}(\mathbf{P}'\mathbf{x}) = \mathbf{z}'\mathbf{\Lambda}\mathbf{z} = \sum_{i=1}^{m} \lambda_i z_i^2.$$

10.4.2   Positive and nonnegative definite matrices

There is a close connection between the quadratic form and the eigenvalue. Namely, the minimum and maximum eigenvalue can be derived through minimization and maximization of the quadratic form under the constraint

$$\lambda_{\min} = \min_{\|\mathbf{x}\|=1} \mathbf{x}'\mathbf{A}\mathbf{x} = \min_{\mathbf{x}} \frac{\mathbf{x}'\mathbf{A}\mathbf{x}}{\|\mathbf{x}\|^2}, \quad \lambda_{\max} = \max_{\|\mathbf{x}\|=1} \mathbf{x}'\mathbf{A}\mathbf{x} = \max_{\mathbf{x}} \frac{\mathbf{x}'\mathbf{A}\mathbf{x}}{\|\mathbf{x}\|^2}, \quad (10.20)$$

where $\mathbf{A}$ is a symmetric matrix. To prove, consider function $f(\mathbf{x}) = \mathbf{x}'\mathbf{A}\mathbf{x}/\|\mathbf{x}\|^2$. The following formula for the differentiation of the quadratic form holds (see more detail in the next section):

$$\frac{\partial(\mathbf{x}'\mathbf{A}\mathbf{x})}{\partial \mathbf{x}} = 2\mathbf{A}\mathbf{x}.$$

The first-order condition for the minimum or maximum is $\partial f/\partial \mathbf{x} = \mathbf{0}$. Applying elementary differentiation rules, one obtains

$$\frac{\partial f}{\partial \mathbf{x}} = \frac{2\mathbf{A}\mathbf{x} \times \|\mathbf{x}\|^2 - 2\mathbf{x} \times \mathbf{x}'\mathbf{A}\mathbf{x}}{\|\mathbf{x}\|^4} = \mathbf{0},$$

which implies $\mathbf{A}\mathbf{x} = \lambda\mathbf{x}$, where $\lambda = \mathbf{x}'\mathbf{A}\mathbf{x}/\|\mathbf{x}\|^2$. Thus, the maximum or minimum of f is attained when $\mathbf{x}$ is an eigenvector. Consequently, the maximum and minimum eigenvectors yield the maximal and minimal values of $\mathbf{x}'\mathbf{A}\mathbf{x}/\|\mathbf{x}\|^2$, respectively. The formulas (10.20) imply that the quadratic form takes nonnegative values if the minimum eigenvalue is nonnegative, see below.

Now we are ready to generalize the concept of positiveness to matrices. This definition is a generalization of the following elementary observation: number A is positive if and only if $x^2 A = xAx$ is positive for every nonzero x.

Definition 10.5 *We say that a $m \times m$ symmetric matrix* $\mathbf{A}$ *is positive definite (denote as* $\mathbf{A} > \mathbf{0}$*) if the corresponding quadratic form is positive, i.e.*

$$\mathbf{x}'\mathbf{A}\mathbf{x} > 0 \quad \forall \mathbf{x} \neq \mathbf{0}. \tag{10.21}$$

We say that $\mathbf{A}$ *is nonnegative definite if* $\mathbf{x}'\mathbf{A}\mathbf{x} \geq 0$. *We write* $\mathbf{A} > \mathbf{B}$ *if matrix* $\mathbf{A} - \mathbf{B}$ *is positive definite. Likewise,* $\mathbf{A} \geq \mathbf{B}$ *means matrix* $\mathbf{A} - \mathbf{B}$ *is nonnegative definite.*

For any pair of numbers, one can say which is smaller, but some matrices are incomparable. That is why we say that Definition 10.5 establishes a partial order among symmetric matrices. As follows from representation (10.19), if $\mathbf{A}$ is a positive definite matrix, then

$$\min_{\mathbf{x}} \left(a + \mathbf{b}'\mathbf{x} + \mathbf{x}'\mathbf{A}\mathbf{x} \right) = a - \frac{1}{4}\mathbf{b}'\mathbf{A}^{-1}\mathbf{b}, \tag{10.22}$$

and $a + \mathbf{b}'\mathbf{x} + \mathbf{x}'\mathbf{A}\mathbf{x} \geq 0$ if and only if the right-hand side of (10.22) is nonnegative.

Representation (10.20) gives rise to the following criteria.

Theorem 10.6 *The following conditions for symmetric matrix* $\mathbf{A}$ *are equivalent. (1) Matrix is positive definite. (2) The quadratic form* $\mathbf{x}'\mathbf{A}\mathbf{x}$ *takes positive values for all* $\mathbf{x} \neq \mathbf{0}$*., (3) All eigenvalues are positive (*$\lambda_{\min} \geq 0$*).*

The following properties hold: (1) If $\mathbf{X}$ is a $n \times m$ matrix, then $\mathbf{X}'\mathbf{X}$ and $\mathbf{X}\mathbf{X}'$ are nonnegative definite matrices. Moreover, if $\mathbf{X}$ has full rank ($m < n$), then matrix $\mathbf{X}'\mathbf{X}$ is positive definite. (2) If $\mathbf{X}$ is a $n \times m$ matrix of full rank ($m \leq n$) and $\mathbf{A}$ is a positive definite $n \times n$ symmetric matrix, then $\mathbf{X}'\mathbf{A}\mathbf{X}$ is a positive definite matrix. (3) If $\mathbf{A}$ and $\mathbf{B}$ are nonnegative definite matrices, then $\mathbf{A} + \mathbf{B}$ is nonnegative definite matrix. Moreover, if $\mathbf{A}$ is positive definite and $\mathbf{B}$ is a nonnegative definite matrix, then $\mathbf{A} + \mathbf{B}$ is positive definite. (4) If $\mathbf{A}$ and $\mathbf{B}$ are symmetric matrices, then $\mathbf{A} \leq \mathbf{B}$ is equivalent to $\lambda_{\min}(\mathbf{B} - \mathbf{A}) \geq 0$.(5) If $\mathbf{A}$ is a positive (nonnegative) definite matrix then all its main submatrices are positive (nonnegative) definite. (6) If matrix $\mathbf{A}$ is symmetric and positive definite, then $\mathbf{A}^{-1}$ is symmetric and positive definite. (7) If $\mathbf{A}$ and $\mathbf{B}$ are positive definite matrices and $\mathbf{A} < \mathbf{B}$, then $\mathbf{A}^{-1} > \mathbf{B}^{-1}$. Similarly, if $\mathbf{A} \leq \mathbf{B}$ then $\mathbf{A}^{-1} \geq \mathbf{B}^{-1}$. (8) If $\mathbf{A}$ and $\mathbf{B}$ are $m \times m$ symmetric matrices and $\mathbf{A} \leq \mathbf{B}$, then $A_{ii} \leq B_{ii}$ for each $i = 1, 2, ..., m$. Consequently, $\mathrm{tr}(\mathbf{A}) \leq \mathrm{tr}(\mathbf{B})$. (9) Let $\mathbf{A}$ and $\mathbf{B}$ be symmetric $m \times m$ matrices and $\mathbf{X}$ be a $n \times m$ matrix. If $\mathbf{A} \leq \mathbf{B}$, then $\mathbf{X}'\mathbf{A}\mathbf{X} \leq \mathbf{X}'\mathbf{B}\mathbf{X}$. Moreover, if $n = m$ and $\mathbf{X}$ is a nonsingular matrix, then $\mathbf{X}'\mathbf{A}\mathbf{X} \leq \mathbf{X}'\mathbf{B}\mathbf{X}$ implies $\mathbf{A} \leq \mathbf{B}$. (10) An idempotent matrix is a nonnegative matrix because its eigenvalues are nonnegative (either 0 or 1). (11) If $\mathbf{A}$ is a positive definite matrix, then $\mathbf{A} - a\mathbf{1}\mathbf{1}'$ is positive definite if and only if $a < 1/(\mathbf{1}'\mathbf{A}^{-1}\mathbf{1})$. (12) If $\mathbf{B}$ is a $n \times m$ matrix of full rank m, then, for any $m \times p$ matrix $\mathbf{A}$, we have

$$\mathbf{A}'\mathbf{A} - \mathbf{A}'\mathbf{B}(\mathbf{B}'\mathbf{B})^{-1}\mathbf{B}'\mathbf{A} \geq 0.$$

This follows from the fact that matrix $\mathbf{I} - \mathbf{B}(\mathbf{B}'\mathbf{B})^{-1}\mathbf{B}'$ is idempotent and property (9).

The following example proves that the conditional covariance matrix in the multivariate normal distribution is positive definite; see Section 4.1.

Example 10.7 *Positive definite submatrix.* *Prove that if matrix $\mathbf{M}$ is positive definite and partitioned as in (10.9), then matrix $\mathbf{M}_{11} - \mathbf{M}_{12}\mathbf{M}_{22}^{-1}\mathbf{M}'_{12}$ is positive definite.*

Solution. From Property 6, we conclude that matrix $\mathbf{M}^{-1}$ is positive definite, so from Property 5, we infer that matrix $(\mathbf{M}_{11} - \mathbf{M}_{12}\mathbf{M}_{22}^{-1}\mathbf{M}'_{12})^{-1}$ is positive definite as well. Again using Property 6, we finally deduce that $\mathbf{M}_{11} - \mathbf{M}_{12}\mathbf{M}_{22}^{-1}\mathbf{M}'_{12}$ is positive definite. $\qquad\square$

For a positive definite $m \times m$ matrix $\mathbf{A}$, the *contour* of the quadratic form is defined as $E_{\mathbf{A}}(\rho) = \{\mathbf{x} \in R^m : \mathbf{x}'\mathbf{A}\mathbf{x} = \rho\}$, an ellipsoid for any $\rho > 0$. When $\rho = 1$ the ellipsoid is called *characteristic*. Similarly, define a *level set* as $V_{\mathbf{A}}(\rho) = \{\mathbf{x} \in R^m : \mathbf{x}'\mathbf{A}\mathbf{x} \le \rho\}$. If matrix $\mathbf{A}$ is positive definite the contours and level sets are bounded. In contrast, if $\mathbf{A}$ is not positive definite (some eigenvalues are negative), the level sets are unbounded: $\|\mathbf{x}_k\| \to \infty$ for some $\mathbf{x}_k \in V_{\mathbf{A}}(\rho)$. In particular, for $m = 2$, the contour is a hyperbola if $\mathbf{A}$ has one negative and one positive eigenvalue. The level set is related to the *confidence region* in statistics; see Section 7.8.5.

Problems

1. Find the condition under which the general quadratic form is positive for every $\mathbf{x}$. [Hint: Use representation (10.19).]

2. Prove that any quadratic form can be expressed as the sum of squares upon orthogonal transformation.

3. Prove that the following matrices are nonnegative definite: (a) $\mathbf{a}\mathbf{a}'$, (b) $\mathbf{X}'\mathbf{X}$, (c) $\mathbf{X}\mathbf{X}'$, (d) $\mathbf{I} - n^{-1}\mathbf{1}\mathbf{1}'$, (e) $\mathbf{X}(\mathbf{X}'\mathbf{X})^{-1}\mathbf{X}'$, and (f) $\mathbf{I} - \mathbf{X}(\mathbf{X}'\mathbf{X})^{-1}\mathbf{X}'$. Use R to verify by generating a random matrix $\mathbf{X}$.

4. Prove that matrix $\mathbf{X}'\mathbf{X}$ is positive definite if the $n \times m$ matrix $\mathbf{X}$ has rank m.

5. Let $\mathbf{A}$ be a symmetric matrix. (a) Prove that $\mathbf{A} \le \mathbf{A} + \mathbf{1}\mathbf{1}'$. (b) Prove that $\mathbf{A} \le \mathbf{A} + \mathbf{X}'\mathbf{X}$. (c) Prove that $\mathbf{A} \le \mathbf{A} + \mathbf{X}\mathbf{X}'$. Write R code to verify these inequalities through random matrices (use dimensions m and n as arguments of your function).

6. (a) Prove that $\mathbf{A} \le \mathbf{B}$ implies $\operatorname{tr}(\mathbf{A}) \le \operatorname{tr}(\mathbf{B})$. (b) Prove that $\mathbf{A} < \mathbf{B}$ implies $\operatorname{tr}(\mathbf{A}) < \operatorname{tr}(\mathbf{B})$.

7. (a) Does $\mathbf{A} < \mathbf{B}$ imply $|\mathbf{A}| < |\mathbf{B}|$? (b) Does $\mathbf{A} \leq \mathbf{B}$ imply $|\mathbf{A}| \leq |\mathbf{B}|$? Use R to verify.

8. Are conditions $\mathbf{A} \leq \mathbf{B}$ and $\lambda_{\max}(\mathbf{A}) \leq \lambda_{\min}(\mathbf{B})$ equivalent for nonnegative definite symmetric matrices?

9. Prove that (a) $\mathbf{A} \leq \mathbf{B}$ and $\mathbf{C} \leq \mathbf{D}$ implies $\mathbf{A} + \mathbf{C} \leq \mathbf{B} + \mathbf{D}$, (b) $0 < \mathbf{A} \leq \mathbf{B}$ implies $\mathbf{B}^{-1} \leq \mathbf{A}^{-1}$, and (c) $\mathbf{A} \leq \mathbf{B}$ implies $\mathbf{X}'\mathbf{AX} \leq \mathbf{X}'\mathbf{BX}$.

10. Write R code that generates many random symmetric matrices $\mathbf{A}$ and $\mathbf{B}$ using `runif` and estimate the probability that $\mathbf{A} < \mathbf{B}$ as a function of $n = 1, 2, 3, 4, 5$. [Hint: Use $\mathbf{MM}'$ to generate symmetric matrices.]

11. Find the minimum of quadratic function $f(\mathbf{x}) = 1 - 2\mathbf{a}'\mathbf{x} + \mathbf{x}'\mathbf{Ax}$, where $\mathbf{a} = (1, 0, ..., 0)$ and $\mathbf{A} = \mathbf{I} + \rho\mathbf{11}'$, $\rho > 0$. Use the `contour` function to verify your answer. [Hint: Use identity (10.1).]

12. Does $\mathbf{A} < \mathbf{B}$ imply that the ith eigenvalue of matrix $\mathbf{A}$ is less than the ith eigenvalue of matrix $\mathbf{B}$ assuming that the eigenvalues are arranged in descending order? Use R with random matrices to test this conjecture.

13. (a) Prove that if $\mathbf{0} < \mathbf{A} < \mathbf{B}$, then $V_{\mathbf{B}}(\rho) \subset V_{\mathbf{B}}(\rho)$ for every $\rho > 0$. (b) Prove that if $V_{\mathbf{B}}(1) \subset V_{\mathbf{B}}(1)$, then $V_{\mathbf{B}}(\rho) \subset V_{\mathbf{B}}(\rho)$ for every $\rho > 0$.

14. Modify function `sb60` to illustrate $\mathbf{A} \leq \mathbf{B}$ geometrically by plotting contours, where $A_{11} = 3, A_{12} = A_{21} = 1, A_{22} = 1$ and $B_{11} = 4, B_{12} = B_{21} = 2, B_{22} = 2$.

10.5 Vector and matrix calculus

Estimation problems in statistics typically are reduced to optimization problems with vectors or matrices. For example, to estimate a multivariate statistical model with unknown covariance matrix, we differentiate the log-likelihood function with respect to the covariance matrix. Moreover, to find the Fisher information matrix or Jacobian, we differentiate a vector with respect to a vector.

10.5.1 Differentiation of a scalar-valued function with respect to a vector

Let $f(\mathbf{x})$ be a real-valued function of a vector argument $\mathbf{x} = (x_1, x_2, ..., x_m)'$. The derivative (sometimes call the gradient) is an $m \times 1$ vector,

$$\partial f/\partial \mathbf{x} = (\partial f/\partial x_1, \partial f/\partial x_2, ..., \partial f/\partial x_m)'.$$

Sometimes, especially in physics and engineering applications, the gradient is denoted as ∇, then we write $\nabla f = \partial f/\partial \mathbf{x}$.

The ordinary calculus rules for the sum, product, and chain apply. (1) Differentiation of a linear combination: $\partial(\alpha f(\mathbf{x})/\partial\mathbf{x}+\beta g(\mathbf{x})) = \alpha\partial f/\partial\mathbf{x}+\beta\partial g/\partial\mathbf{x}$. (2) Differentiation of a product: $\partial(f(\mathbf{x})g(\mathbf{x}))/\partial\mathbf{x} = \partial f g(\mathbf{x})/\partial\mathbf{x}+f(\mathbf{x})\partial g/\partial\mathbf{x}$. (3) Chain rule: $\partial(H(f(\mathbf{x}))/\partial\mathbf{x} = (dH/df)(\partial f/\partial\mathbf{x})$, where H is a scalar-valued function of a scalar argument. (4) $\partial(\mathbf{a}f(\mathbf{x})/\partial\mathbf{x} = \mathbf{a}\,(\partial f/\partial\mathbf{x})'$, where $\mathbf{a}$ is a constant vector. The following examples illustrate differentiation.

$$\frac{\partial}{\partial\mathbf{x}}(\mathbf{a}'\mathbf{x})= \mathbf{a}, \;\; \frac{\partial}{\partial\mathbf{x}}\left\|\mathbf{x}\right\|^2 =2\mathbf{x}, \;\; \frac{\partial}{\partial\mathbf{x}}\mathbf{x}'\mathbf{A}\mathbf{x}=2\mathbf{A}\mathbf{x}, \;\; \frac{\partial}{\partial\mathbf{x}}e^{-\frac{1}{2}\mathbf{x}'\mathbf{A}\mathbf{x}} = -\frac{1}{e^{\frac{1}{2}\mathbf{x}'\mathbf{A}\mathbf{x}}}\mathbf{A}\mathbf{x}.$$

Using these formulas, one can prove that $\partial\left\|\mathbf{x}\right\|/\partial\mathbf{x} = \mathbf{x}/\left\|\mathbf{x}\right\|$.

Example 10.8 *Cauchy inequality.* *(a) Find the minimum of $f(\beta) = \left\|\mathbf{y}-\beta\mathbf{x}\right\|^2$ where $\mathbf{y}$ and $\mathbf{x}$ are $n \times 1$ vectors and $\mathbf{x} \neq \mathbf{0}$. (a) Derive the Cauchy inequality.*

Solution. (a) Find the derivative

$$\frac{df}{d\beta} = \frac{\partial}{\partial\beta}\left\|\mathbf{y}-\beta\mathbf{x}\right\|^2 = \left(\frac{\partial\left\|\mathbf{y}-\beta\mathbf{x}\right\|^2}{\partial(\mathbf{y}-\beta\mathbf{x})}\right)' \left(\frac{\partial(\mathbf{y}-\beta\mathbf{x})}{\partial\beta}\right) = -2(\mathbf{y}-\beta\mathbf{x})'\mathbf{x}.$$

Setting the derivative to 0 gives $\widehat{\beta} = \mathbf{y}'\mathbf{x}/\left\|\mathbf{x}\right\|^2$ and, after some vector algebra, one obtains $\min_\beta\left\|\mathbf{y}-\beta\mathbf{x}\right\|^2 = \left\|\mathbf{y} - (\mathbf{y}'\mathbf{x})/\left\|\mathbf{x}\right\|^2\mathbf{x}\right\|^2 = \left\|\mathbf{y}\right\|^2 - (\mathbf{y}'\mathbf{x})^2/\left\|\mathbf{x}\right\|^2$. (b) Since the left-hand side is nonnegative, the right-hand side is nonnegative as well. This proves the Cauchy inequality, $(\mathbf{y}'\mathbf{x})^2 \leq \left\|\mathbf{y}\right\|^2\left\|\mathbf{x}\right\|^2$.

10.5.2 Differentiation of a vector-valued function with respect to a vector

Now $\mathbf{f}(\mathbf{x})$ is a $p \times 1$ vector-valued function of a $m \times 1$ vector argument. This function can be viewed as a map of R^m to R^p, or symbolically $\mathbf{f}:R^m \rightarrow R^p$. The derivative of $\mathbf{f}$ with respect to $\mathbf{x}$ is a $p \times m$ matrix, sometimes called the Jacobian, $\mathbf{J}^{p\times m} = \partial\mathbf{f}/\partial\mathbf{x}$. For example, the Jacobian was used in the multivariate delta method (Section 3.10.3) and in the Gauss–Newton algorithm for nonlinear regression (Section 9.3). The (i,j)th element of matrix $\mathbf{J}$ is $\partial f_i/\partial x_j$ where $i = 1, 2, ..., p$ and $j = 1, 2, ..., m$.

The following rules apply ($\mathbf{x}$ is a $m \times 1$ vector). (1) Differentiation of a linear combination: $\partial(\alpha\mathbf{f}(\mathbf{x})+\beta\mathbf{g}(\mathbf{x}))/\partial\mathbf{x} = \alpha\partial\mathbf{f}/\partial\mathbf{x}+\beta\partial\mathbf{g}/\partial\mathbf{x}$, where α and β are scalars and $\mathbf{f}$ and $\mathbf{g}$ are $p\times1$. (2) Differentiation of a matrix factor: $\partial(\mathbf{A}(\mathbf{f}(\mathbf{x}))/\partial\mathbf{x} = \mathbf{A}(\partial\mathbf{f}/\,\partial\mathbf{x})$, where $\mathbf{A}$ is a constant $k\times p$ matrix. (3) Differentiation of the scalar product: $\partial(\mathbf{f}'(\mathbf{x})\mathbf{g}(\mathbf{x}))/\partial\mathbf{x} = (\partial\mathbf{f}/\partial\mathbf{x})'\,\mathbf{g}(\mathbf{x}) + (\partial\mathbf{g}/\partial\mathbf{x})'\,\mathbf{f}(\mathbf{x})$, where $\mathbf{g}$ is the $p \times 1$ function. (4) Differentiation of the quadratic form:

$$\frac{\partial(\mathbf{f}'(\mathbf{x})\mathbf{A}\mathbf{g}(\mathbf{x}))}{\partial\mathbf{x}}=\left(\frac{\partial\mathbf{f}}{\partial\mathbf{x}}\right)'\mathbf{A}\mathbf{g}(\mathbf{x})+\left(\frac{\partial\mathbf{g}}{\partial\mathbf{x}}\right)'\mathbf{A}\mathbf{f}(\mathbf{x}),$$

where $\mathbf{A}$ is a $p \times p$ symmetric matrix. In particular, using the above rules, we obtain $\partial\mathbf{x}'\mathbf{A}\mathbf{x}/\partial\mathbf{x} = 2\mathbf{A}\mathbf{x}$ and $\partial(\mathbf{y} - \mathbf{x})'\mathbf{A}(\mathbf{y} - \mathbf{x})/\partial\mathbf{x} = -2\mathbf{A}(\mathbf{y} - \mathbf{x})$.

10.5.3   Kronecker product

The Kronecker matrix product emerges in statistics when studying tabular data, such as the analysis of variance (ANOVA) or repeated measurements. Matrix calculus is needed to derive the ML estimator of a multivariate model and particularly when an unknown covariance matrix is involved. See Section 4.2.3 and Example 6.89 for application of the Kronecker product to the derivation of covariance matrices.

Let $\mathbf{A}$ and $\mathbf{B}$ be any $n \times m$ and $p \times k$ matrices. The Kronecker matrix product is defined as a $(np) \times (mk)$ block matrix,

$$\mathbf{A} \otimes \mathbf{B} = \begin{bmatrix} a_{11}\mathbf{B} & a_{12}\mathbf{B} & \cdots & a_{1m}\mathbf{B} \\ a_{21}\mathbf{B} & a_{22}\mathbf{B} & \cdots & a_{2m}\mathbf{B} \\ \vdots & \vdots & \ddots & \vdots \\ a_{n1}\mathbf{B} & a_{n2}\mathbf{B} & \cdots & a_{nm}\mathbf{B} \end{bmatrix}.$$

Each block is a matrix with the (i,j)th element of matrix $\mathbf{A}$ multiplied by matrix $\mathbf{B}$. Use `kronecker(A,B)` in R to compute the Kronecker product $\mathbf{A} \otimes \mathbf{B}$. If `a = c(1,3)` and `B=matrix(1:6,ncol=2)` commands `kronecker(a,B)` and `kronecker(B,a)` return 6×2 matrices. The following lists several properties of the Kronecker product: (1) $\mathbf{A} \otimes (\mathbf{B} + \mathbf{C}) = \mathbf{A} \otimes \mathbf{B} + \mathbf{A} \otimes \mathbf{C}$ and $(\mathbf{A} + \mathbf{C}) \otimes \mathbf{B} = \mathbf{A} \otimes \mathbf{B} + \mathbf{C} \otimes \mathbf{B}$. (2) $\mathbf{A} \otimes (\mathbf{B} \otimes \mathbf{C}) = (\mathbf{A} \otimes \mathbf{B}) \otimes \mathbf{C}$, and generally, $\mathbf{A} \otimes \mathbf{B} \neq \mathbf{B} \otimes \mathbf{A}$. (3) $(\mathbf{A} \otimes \mathbf{B})(\mathbf{C} \otimes \mathbf{D} = \mathbf{A}\mathbf{C} \otimes \mathbf{B}\mathbf{D}$. (4) $\mathbf{a} \otimes \mathbf{b}' = \mathbf{a}\mathbf{b}' = \mathbf{b}' \otimes \mathbf{a}$, where $\mathbf{a}$ and $\mathbf{b}$ are vectors. (5) $(\mathbf{A} \otimes \mathbf{B})' = \mathbf{A}' \otimes \mathbf{B}'$. (6) $(\mathbf{A} \otimes \mathbf{B})^{-1} = \mathbf{A}^{-1} \otimes \mathbf{B}^{-1}$. (7) $\text{tr}(\mathbf{A} \otimes \mathbf{B}) = (\text{tr}\mathbf{A})(\text{tr}\mathbf{B})$. (8) $|\mathbf{A} \otimes \mathbf{B}| = |\mathbf{A}|^p|\mathbf{B}|^n$, where square matrices $\mathbf{A}$ and $\mathbf{B}$ have dimensions $n \times n$ and $p \times p$, respectively.

10.5.4   vec operator

The vec operator transforms a $n \times m$ matrix $\mathbf{A}$ into a $(nm) \times 1$ vector, $\mathbf{a} = \text{vec}(\mathbf{A})$ by stacking vector columns. In R, if $\mathbf{A}$ is a matrix, this transformation is done by `as.vector(A)`. Reversely, if $\mathbf{a}$ is a $(nm) \times 1$ vector, then `matrix(a,ncol=m)` returns a $n \times m$ matrix.

The following formula is instrumental for many matrix problems, such as differentiation of a matrix with respect to a matrix:

$$\text{vec}(\mathbf{A}\mathbf{B}\mathbf{C}) = (\mathbf{C}' \otimes \mathbf{A})\text{vec}(\mathbf{B}). \tag{10.23}$$

Some properties of the vec operator: (1) $\text{vec}(\mathbf{a}\mathbf{b}') = \mathbf{a} \otimes \mathbf{b}$ and $\text{vec}(\mathbf{a}\mathbf{a}') = \mathbf{a} \otimes \mathbf{a}$. (2) $\text{vec}(\mathbf{A}\mathbf{B}) = (\mathbf{B}' \otimes \mathbf{I}_m)\text{vec}(\mathbf{A}) = (\mathbf{B}' \otimes \mathbf{A})\text{vec}(\mathbf{I}_n) = \text{vec}(\mathbf{I}_p \otimes \mathbf{A})\text{vec}(\mathbf{B})$, where matrices $\mathbf{A}$ and $\mathbf{B}$ have dimensions $m \times n$ and $n \times p$, respectively. (3) $\text{vec}(\mathbf{A}) = (\mathbf{I} \otimes \mathbf{A})\text{vec}(\mathbf{I}) = (\mathbf{A}' \otimes \mathbf{I}_m)\text{vec}(\mathbf{I}_m)$, where $\mathbf{A}$ is an $m \times n$ matrix. (4) $\text{tr}(\mathbf{A}'\mathbf{B}) = (\text{vec}(\mathbf{A}))'(\text{vec}(\mathbf{B}))$, where matrices $\mathbf{A}$ and $\mathbf{B}$ are $m \times n$. (5) $\text{tr}(\mathbf{A}'\mathbf{A}) =$

$\|\mathrm{vec}(\mathbf{A})\|^2$. (6) $\mathbf{x}'\mathbf{A}\mathbf{x} = (\mathbf{x} \otimes \mathbf{x})'\mathrm{vec}(\mathbf{A}) = \mathrm{vec}'(\mathbf{A})(\mathbf{x} \otimes \mathbf{x})$. (7) $\mathrm{tr}(\mathbf{ABCD}) = (\mathrm{vec}(\mathbf{D}'))'(\mathbf{C}' \otimes \mathbf{A})\mathrm{vec}(\mathbf{B}) = \mathrm{vec}'(\mathbf{D})(\mathbf{A} \otimes \mathbf{C}')\mathrm{vec}(\mathbf{B}')$.

Problems

1. Find the minimum and maximum of function $f(\mathbf{x}) = \mathbf{b}'\mathbf{x}/(1+\mathbf{x}'\mathbf{A}\mathbf{x})$, where $\mathbf{A}$ is a positive definite matrix, by setting the derivative to zero. [Hint: Reduce the optimization to a function of one variable.]

2. Apply the Kronecker product in combination with the vec operator to solve the Lyapunov equation $\mathbf{AX} + \mathbf{XB} = \mathbf{C}$ for the $n \times m$ matrix $\mathbf{X}$, where $\mathbf{A}$, $\mathbf{B}$, and $\mathbf{C}$ are given $n \times n$, $m \times m$, and $n \times m$ matrices, respectively.

10.6   Optimization

Only in rare statistical problems can an estimator be found in closed form, such as the solution to the linear least squares problem. Otherwise, one needs to apply iterations. For example, iterations are required to find the ML or the NLS estimate, as discussed in Sections 6.10.6 and 9.3.

Optimization is a frequent task in statistics – many methods of statistical inference are reduced to solving nonlinear equations or minimization (or maximization). Optimization problems can be roughly divided into two classes: convex and nonconvex optimization. Convex optimization is uncommon in practice but has beautiful mathematical properties and serves as a benchmark model for many optimization algorithms. However, the majority of optimization problems are nonconvex, including maximization of the log-likelihood function. Even the log-likelihood of a linear model is not concave (see below). Nonconvex functions may have several minima, complicating the optimization problem dramatically because the existing calculus-based criteria can only check if the minimum is local. Regarding the domain, an optimization problem may be classified into unconstrained or constrained; if a constraint is expressed as an equation, it can be handled with Lagrange multiplier technique discussed at the end of the section.

Optimization means either minimization or maximization. From a methodological perspective, minimization is equivalent to maximization because if f is the minimum, then $-f$ is the maximum and vice verse. Indeed, often we minimize the negative log-likelihood function because the formulas become simpler. Thus, to avoid obvious restatements, we will discuss minimization problems. We recommend classic books on optimization for further reading, such as Ortega and Rheinboldt [78], Polak [81], Dennis and Schnabel [33].

One must remember that existing minimization algorithms, gradient algorithms included, find a local minimum. Since, in practice, there is no guarantee that the criterion function to minimize, such as the residual sum of squares for a nonlinear regression, has one local minimum, the problem of the global minimum is difficult to address and still awaiting its solution. The challenge of finding a constructive criterion for the global minimum is explained by the fact that calculus studies local features of function, but global minimization depends on global properties. It is a good practice to start iterations from several different guesses to confirm that they converge to the same point. Of course, this does not prove whether the found minimum is the global one.

10.6.1 Convex and concave functions

Definition 10.9 *A set D in R^m is called convex if the segment connecting every pair of points from the set belongs to the set as well: for every $\mathbf{x}_1, \mathbf{x}_2 \in D$, we have $\lambda \mathbf{x}_1 + (1 - \lambda)\mathbf{x}_2 \in D$ for every $0 \leq \lambda \leq 1$. Function $f(\mathbf{x})$ defined on a convex set D is called convex if $f(\lambda \mathbf{x}_1 + (1 - \lambda)\mathbf{x}_2) \leq \lambda f(\mathbf{x}_1) + (1 - \lambda)f(\mathbf{x}_2)$ for all $0 \leq \lambda \leq 1$ and $\mathbf{x}_1, \mathbf{x}_2 \in D$. Function $f(\mathbf{x})$ on D is called concave if $f(\lambda \mathbf{x}_1 + (1 - \lambda)\mathbf{x}_2) \geq \lambda f(\mathbf{x}_1) + (1 - \lambda)f(\mathbf{x}_2)$. Function is called strictly concave (convex) if the inequalities are strict for $0 < \lambda < 1$ and $\mathbf{x}_1 \neq \mathbf{x}_2$.*

An important property of a convex function is formulated below.

Theorem 10.10 *A convex function defined on a convex set has at most one local minimum.*

Consequently, if a convex function has a local minimum this minimum is the global minimum. Note that a convex function may have no minimum or minimum, such as a linear function or e^x on $(-\infty, \infty)$.

Checking the convexity of a function following Definition 10.9 is difficult. Fortunately, calculus applies and then convexity reduces to checking if the Hessian matrix is positive definite . The vector and matrix calculus discussed in the previous section becomes very useful here.

Theorem 10.11 *Let $f(\mathbf{x})$ be a twice differentiable function defined on an open convex set $D \subset R^m$. Define the Hessian as the $m \times m$ matrix of second derivatives, $\mathbf{H}(\mathbf{x}) = \partial^2 f / \partial \mathbf{x}^2$. The function f is convex if the Hessian is a nonnegative definite matrix on D. The function is strictly convex if the Hessian is a positive definite matrix.*

Note that the set must be open to comply with the differentiation rule that the point belongs to the set along with a neighborhood. Theorem 10.10 does not assert that the minimum exists, a sufficient condition is provided below.

Theorem 10.12 *Let $f(\mathbf{x})$ be a continuous function and let there exist $\mathbf{x}_0$ such that the level set $D = \{\mathbf{x} : f(\mathbf{x}) \leq f(\mathbf{x}_0)\}$ is a bounded set (it is contained in a ball of finite radius). Then there is at least one local minimum and the global minimum is reached.*

Combined with Theorem 10.10, one can state that if there exists such $\mathbf{x}_0$ that a continuous function $f(\mathbf{x})$ is convex on the level set $\{\mathbf{x} : f(\mathbf{x}) \leq f(\mathbf{x}_0)\}$ then there exists a unique local minimum which is also the global one. Three types of functions with their contours are illustrated in Figure 10.1. Contours of a convex function look like concentric ellipses; contours of a unimodal function look like concentric ellipses with a dent, and multimodal (at least two local minima) function has unconnected contours.

Example 10.13 *Quadratic function.* *(a) Prove that the quadratic function* $f(\mathbf{x}) = \mathbf{x}'\mathbf{A}\mathbf{x} + \mathbf{b}'\mathbf{x} + c$, *where* $\mathbf{A}$ *is a positive definite matrix, is a strictly convex function on* R^m. *(b) Prove that the residual sum of squares,* $\|\mathbf{y} - \mathbf{X}\boldsymbol{\beta}\|^2$ *is a strictly convex function of* $\boldsymbol{\beta} \in R^m$ *if matrix* $\mathbf{X}$ *has full rank.*

Solution. (a) From the previous section, we know that

$$\frac{\partial^2 \mathbf{x}'\mathbf{A}\mathbf{x}}{\partial \mathbf{x}^2} = 2\mathbf{A}. \tag{10.24}$$

Thus, the Hessian matrix takes the form

$$\frac{\partial^2 f}{\partial \mathbf{x}^2} = \frac{\partial^2}{\partial \mathbf{x}^2}(\mathbf{x}'\mathbf{A}\mathbf{x}) + \frac{\partial^2}{\partial \mathbf{x}^2}(\mathbf{b}'\mathbf{x}) + \frac{\partial^2}{\partial \mathbf{x}^2}c = 2\mathbf{A}$$

and is positive definite by the given condition. Therefore, from Theorem 10.11, we conclude that $\mathbf{x}'\mathbf{A}\mathbf{x} + \mathbf{b}'\mathbf{x} + c$ is a strictly convex function.

(b) Express $\|\mathbf{y} - \mathbf{X}\boldsymbol{\beta}\|^2 = \boldsymbol{\beta}'(\mathbf{X}'\mathbf{X})\boldsymbol{\beta} - 2\mathbf{y}'\mathbf{X}\boldsymbol{\beta} + \|\mathbf{y}\|^2$, and using the fact that matrix $\mathbf{X}'\mathbf{X}$ is positive definite if $\mathbf{X}$ is a matrix of full rank, we reduce the problem to case (a).

10.6.2   Criteria for unconstrained minimization

Existence criteria

The global minimum of a continuous function f defined on a convex set D may not exist if D is not a compact set such as the entire space R^m. The following is a sufficient criterion based on the concept of existence level (see details in Demidenko [25]). Note that finding a point of local minimum does not mean the global minimum exits.

Theorem 10.14 *Criterion for global minimum.* *Let* ∂D *denote the boundary of* D. *The upper existence level for minimization of* $f(\mathbf{x})$ *on* D *is defined as*

$$\overline{f}_E = \inf_{\mathbf{x} \in \partial D} f(\mathbf{x}). \tag{10.25}$$

If there exists $\mathbf{x}_0 \in D$ *such that* $f(\mathbf{x}_0) < \overline{f}_E$, *then the level set* $\{\mathbf{x} \in D : f(\mathbf{x}) \leq f(\mathbf{x}_0)\}$ *is compact and the global minimum on* D *exists.*

In words, $\overline{f}_E$ is the infimum of f on the boundary of D. If D is unbounded, we consider sequences of points from D that converge to a point on the boundary.

Example 10.15 *Existence level for the linear model.* *Find the existence level of the residual sum of squares for the linear model and show that the global minimum exists.*

Solution. The residual sum of squares is $S(\boldsymbol{\beta}) = \|\mathbf{y} - \mathbf{X}\boldsymbol{\beta}\|^2$, where $\boldsymbol{\beta} \in D = R^m$, the entire space. We show that the existence level of the residual sum of squares, $\overline{S}_E = \infty$ if matrix $\mathbf{X}$ has full rank. It suffices to show that $S(\boldsymbol{\beta}_k) \to \infty$ when $\boldsymbol{\beta}_k \to \infty$. Use the linear model identity,

$$\|\mathbf{y} - \mathbf{X}\boldsymbol{\beta}_k\|^2 = \left\|\mathbf{y} - \mathbf{X}\widehat{\boldsymbol{\beta}}\right\|^2 + (\boldsymbol{\beta}_k - \widehat{\boldsymbol{\beta}})'(\mathbf{X}'\mathbf{X})(\boldsymbol{\beta}_k - \widehat{\boldsymbol{\beta}})$$

where $\widehat{\boldsymbol{\beta}}$ is the least squares solution. Denote as $\lambda_{\min} > 0$, the minimum eigenvalue of matrix $\mathbf{X}'\mathbf{X}$. Then

$$\|\mathbf{y} - \mathbf{X}\boldsymbol{\beta}_k\|^2 \geq \left\|\mathbf{y} - \mathbf{X}\widehat{\boldsymbol{\beta}}\right\|^2 + \lambda_{\min}\left\|\boldsymbol{\beta}_k - \widehat{\boldsymbol{\beta}}\right\|^2 \to \infty$$

as $\boldsymbol{\beta}_k \to \infty$. Since $\overline{S}_E = \infty$, the level set for any $\boldsymbol{\beta}_0$ is compact and the global minimum of the residual sum of squares exists.

Uniqueness criteria

Convexity (concavity) is a fragile property: a monotonic transformation of f, such as logarithm, or a one-to-one transformation of the minimization domain ($\mathbf{x}$) ruins convexity (concavity). The following example emphasizes that the concavity is a rarity in practical optimization.

Example 10.16 *Log-likelihood is not concave.* *Show that the log-likelihood function for the linear model is not concave.*

Solution. The log-likelihood function for the linear model $\mathbf{y} \sim \mathcal{N}(\mathbf{X}\boldsymbol{\beta}, \sigma^2\mathbf{I})$ takes the form $l(\boldsymbol{\beta}, \sigma^2) = -\frac{1}{2}n\ln\sigma^2 - \|\mathbf{y} - \mathbf{X}\boldsymbol{\beta}\|^2/\sigma^2$ defined on the half space $R^m \times (0, \infty)$ with the $(m+1) \times (m+1)$ Hessian,

$$\mathbf{H}(\boldsymbol{\beta}, \sigma^2) = \frac{1}{\sigma^6}\begin{bmatrix} -\sigma^4\mathbf{X}'\mathbf{X} & \sigma^2\mathbf{X}'(\mathbf{y} - \mathbf{X}\boldsymbol{\beta}) \\ \sigma^2\mathbf{X}'(\mathbf{y} - \mathbf{X}\boldsymbol{\beta}) & \frac{1}{2}\sigma^2 n - \|\mathbf{y} - \mathbf{X}\boldsymbol{\beta}\|^2 \end{bmatrix}. \tag{10.26}$$

Matrix $\mathbf{H}$ is not always negative definite because the $(2,2)$ block element may be positive for large σ^2. This means that the log-likelihood function is not concave everywhere on the half space. $\square$

The following is the main result of unconstrained nonconvex optimization. In words, (a) the gradient at the point of a local minimum must be zero, and (b) if the Hessian matrix is positive definite, then it is a local minimum (if the Hessian matrix is negative definite it is local maximum).

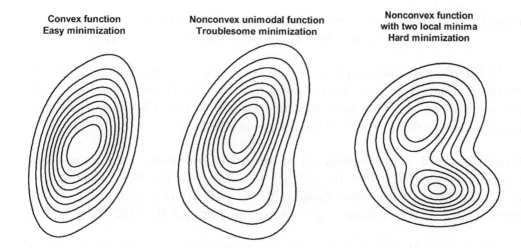

Figure 10.1: *Three types of functions to minimize. Contours of a convex function encompass a convex set, these functions are easy to minimize. Contours of a nonconvex function have a dent, gradient iterations may not converge. Contours of a nonconvex function with two local minima have two concentric contour sets. Iterations may converge to a false local minimum.*

Theorem 10.17 *Unconstrained nonconvex minimization.* *A necessary condition for the local optimum (minimum or maximum) is $\partial f / \partial \mathbf{x} = \mathbf{0}$, sometimes referred to as the first-order condition. A sufficient condition for the minimum. If in addition, the Hessian matrix, $\partial^2 f / \partial \mathbf{x}^2$, evaluated at the point $\mathbf{x}$ where $\partial f / \partial \mathbf{x} = \mathbf{0}$, is positive definite, then $\mathbf{x}$ is the point of local minimum.*

Sometimes, the point where the first derivative (gradient) turns into zero is called a *stationary point*. A stationary point may be of three types:

1. Point of local minimum, the Hessian matrix is positive definite, all eigenvalues are positive.

2. Point of local maximum, the Hessian matrix is negative definite, all eigenvalues are negative.

3. *Saddle point*, i.e. neither a point of minimum nor maximum, the Hessian matrix has positive and negative eigenvalues.

Convexity and level sets are closely related, see Figure 10.1. If a function is convex its level sets are convex and contours look like concentric ellipses. If a function is nonconvex but unimodal, the level sets may be not convex and contours look like ellipses with dents. If a function has multiple local minima, the level sets may be disconnected; see more discussion in Demidenko [24].

Definition 10.18 *A function to minimize is called unimodal if the Hessian is positive definite at the points where the gradient vanishes.*

Unlike convexity, unimodality remains upon monotonic function or argument transformation. Note that, for a convex function, the Hessian is positive definite everywhere, but, for a unimodal nonconvex function, it is sufficient to have Hessian positive definite only where the gradient is zero.

Example 10.19 *Transformation of unimodal function.* *Let $f(\mathbf{x})$ be a unimodal function. Show that (a) $g(f(\mathbf{x}))$ is a unimodal function, where g is a monotonic function such that $dg/df > 0$. (b) $f(\mathbf{T}(\mathbf{x}))$ is a unimodal function, where $\mathbf{T} : R^m \to R^m$ is a one-to-one transformation such that $|\partial \mathbf{T}/\partial \mathbf{x}| \neq 0$.*

Solution. (a) By the chain rule, we obtain $\partial g(f(\mathbf{x}))/\partial \mathbf{x} = (dg/df)(\partial f/\partial \mathbf{x}) = \mathbf{0}$, so that $\partial f/\partial \mathbf{x} = \mathbf{0}$. Further,

$$\frac{\partial^2 g(f(\mathbf{x}))}{\partial \mathbf{x}^2} = \frac{d^2 g}{df^2} \left(\frac{\partial f}{\partial \mathbf{x}}\right) \left(\frac{\partial f}{\partial \mathbf{x}}\right)' + \frac{dg}{df} \frac{\partial^2 f}{\partial \mathbf{x}^2}.$$

The first term at the right-hand side vanishes because $\partial f/\partial \mathbf{x} = \mathbf{0}$. But since $dg/df > 0$ and $\partial^2 f/\partial \mathbf{x}^2$ is a positive definite matrix the Hessian $\partial^2 g(f(\mathbf{x}))/\partial \mathbf{x}^2$ is a positive definite matrix as well.

(b) Again, using the chain rule the gradient is zero where $\partial f(\mathbf{T}(\mathbf{x}))/\partial \mathbf{x} = (\partial \mathbf{T}/\partial \mathbf{x})' (\partial f/\partial \mathbf{T}) = \mathbf{0}$. But $\partial \mathbf{T}/\partial \mathbf{x}$ is a nonsingular matrix, implying $\partial f/\partial \mathbf{T} = \mathbf{0}$. Now find the Hessian. By the product rule, we have

$$\frac{\partial^2 f(\mathbf{T}(\mathbf{x}))}{\partial \mathbf{x}^2} = \left(\frac{\partial}{\partial \mathbf{x}} \left(\frac{\partial \mathbf{T}}{\partial \mathbf{x}}\right)'\right) \frac{\partial f}{\partial \mathbf{T}} + \left(\frac{\partial \mathbf{T}}{\partial \mathbf{x}}\right)' \frac{\partial^2 f}{\partial \mathbf{T}^2} \left(\frac{\partial \mathbf{T}}{\partial \mathbf{x}}\right).$$

Again, the first term vanishes and therefore the Hessian is a positive definite matrix because $\partial^2 f/\partial \mathbf{T}^2$ is positive definite and $\partial \mathbf{T}/\partial \mathbf{x}$ is a nonsingular.   □

The definition of the unimodal function given above is supported by the following fact [25].

Theorem 10.20 *Fundamental theorem of nonconvex minimization.* *Let the function $f(\mathbf{x})$ approach the existence level $\overline{f}_E$ from below and let its Hessian be positive definite at the points where the gradient vanishes. Then, f has a unique local minimum which is the global minimum.*

Below we formulate a constructive global criterion.

Theorem 10.21 *Global criterion.* *Let $f(\mathbf{x})$ be a continuous function and let $\mathbf{x}_0$ be such that f is unimodal on a bounded nonempty level set $D = \{\mathbf{x} : f(\mathbf{x}) < f(\mathbf{x}_0)\}$. Let $\mathbf{x}_*$ be a local minimum on D. Then $\mathbf{x}_*$ is the global minimum of f.*

Proof. Since f is unimodal on D, we have $f(\mathbf{x}_*) < f(\mathbf{x}) < f(\mathbf{x}_0)$ for all $\mathbf{x} \in D$. If $\mathbf{x} \notin D$, then $f(\mathbf{x}) \geq f(\mathbf{x}_0) > f(\mathbf{x}_*)$. This means that $\mathbf{x}_*$ is the global minimum of f. □

The following, as a continuation of Example 10.16, illustrates an application of Theorem 10.17.

Example 10.22 *Linear model.* *Show that the log-likelihood function for the linear model is unimodal unless* $\mathbf{y}$ *is a linear combination of vector columns of* $\mathbf{X}$, *or in vector form* $\mathbf{y} = \mathbf{X}\boldsymbol{\beta}$ *for some* $\boldsymbol{\beta}$.

Solution. The gradient of l at the local maximum turns into zero, $\partial l/\partial \boldsymbol{\beta} = \mathbf{0}$ and $\partial l/\partial \sigma^2 = 0$ with the ML solution $\widehat{\boldsymbol{\beta}} = (\mathbf{X}'\mathbf{X})^{-1}\mathbf{X}'\mathbf{y}$ and $\widehat{\sigma}^2 = \left\|\mathbf{y} - \mathbf{X}\widehat{\boldsymbol{\beta}}\right\|^2/n$. Evaluate Hessian (10.26) at the ML solution

$$\mathbf{H}(\widehat{\boldsymbol{\beta}}, \widehat{\sigma}^2) = -\frac{1}{\widehat{\sigma}^6} \begin{bmatrix} \widehat{\sigma}^4 \mathbf{X}'\mathbf{X} & \mathbf{0}' \\ \mathbf{0} & \frac{1}{2}\widehat{\sigma}^2 n \end{bmatrix}.$$

Since $\mathbf{X}'\mathbf{X}$ is positive definite and $\left\|\mathbf{y} - \mathbf{X}\widehat{\boldsymbol{\beta}}\right\| > 0$ because $\mathbf{y}$ cannot be expressed as a linear combination of $\mathbf{X}$, matrix $\mathbf{H}(\widehat{\boldsymbol{\beta}}, \widehat{\sigma}^2)$ is negative definite, and therefore l is a unimodal function.

10.6.3   Gradient algorithms

Iterative algorithms based on the linear approximation are routinely used to solve equations and optimization problems. Newton's algorithm is the root of many gradient algorithms.

Newton's algorithm for one variable

The goal is to devise an iterative algorithm for solving the equation

$$f(x) = 0, \quad -\infty < x < \infty \tag{10.27}$$

by replacing f with its linear approximation at each iteration. Let x_k be the current approximation to the solution of equation (10.27) at iteration k. Using the Taylor series expansion of the first order around x_k, we approximate $f(x) \simeq f(x_k) + (x - x_k)f'(x_k)$. Now replace the original equation with the linear equation $f(x_k) + (x - x_k)f'(x_k) = 0$. Solving this equation for x generates an updated value

$$x_{k+1} = x_k - \frac{f(x_k)}{f'(x_k)}, \quad k = 0, 1, ..., \tag{10.28}$$

where x_0 is the starting point/initial guess. This algorithm is called the Newton's algorithm. It is easy to see that if Newton's algorithm converges, then the sequence generated by (10.28) converges to the solution of (10.27). Indeed, if $\lim_{k\to\infty} x_k = x_*$, then taking the limit at both sides of (10.28) when $k \to \infty$ yields $x_* = x_* - f(x_*)/f'(x_*)$ that implies $f(x_*) = 0$. Hence x_* is a solution of (10.27).

The reader may see a problem with iterations (10.28) when $f'(x_s) \simeq 0$. Indeed, Newton algorithm may fail to converge. Below is a sufficient criterion for success of Newton's iterations.

Theorem 10.23 *Newton's algorithm (10.28) converges to the solution x_* monotonically from below if (a) $f'(x) > 0$ and $f''(x) < 0$ or (b) $f'(x) < 0$ and $f''(x) > 0$ for all $x < x_*$ starting from $x_0 < x_*$.*

In words, the algorithm monotonically converges if the function is increasing and convex to the left of the solution. Clearly, the choice of the starting value may be crucial for convergence.

Example 10.24 *Newton's algorithm.* *Solve the equation $e^{-ax} = bx$ by Newton's algorithm, where a and b are fixed positive numbers.*

Solution. Rewrite the equation as $f(x) \overset{\text{def}}{=} e^{-ax} - bx = 0$. Then Newton's iterations (10.28) take the form

$$x_{k+1} = x_k + \frac{e^{-ax_k} - bx_k}{ae^{-ax_k} + b}, \quad k = 0, 1, 2, \ldots$$

Our function satisfies the conditions of Theorem 10.23 because $f'(x) = -ae^{-ax} - b < 0$ and $f''(x) = a^2 e^{-ax} > 0$ for all x. Starting point $x_0 = 0$ is to the left of the root because $f(0) = 1 > 0$. See the R function nLINEXP, which solves the equation $e^{-ax} = bx$ for user-defined a and b. A comment regarding the relative difference for convergence. Our choice means that if x_k is close to zero, then iterations stop if the absolute difference is small, $|x_{k+1} - x_k| < \varepsilon$. If $|x_k|$ is far from zero, then the criterion is close to the relative difference $|x_{k+1} - x_k| / |x_k| < \varepsilon$. In other words, this criterion combines absolute and relative difference between iterations in one formula. It took five iterations to converge to the solution $x = 1.1746$ with $f(x) = 1.2 \times 10^{-5}$.

Newton's algorithm for several variables

Generalization of Newton's algorithm to several variables is straightforward. Write the system to be solved as

$$\mathbf{g}(\mathbf{x}) = \mathbf{0}, \tag{10.29}$$

where $\mathbf{g}$ is a m-vector valued function of a $m \times 1$ vector, the system of m equations with m unknowns. As in the case of one equation, at each iteration, replace the original nonlinear system with a linear system using the approximation

$\mathbf{g}(\mathbf{x}) \simeq \mathbf{g}(\mathbf{x}_k) + \mathbf{H}_k(\mathbf{x} - \mathbf{x}_k)$, where $\mathbf{H}_k = \left.\frac{\partial \mathbf{g}}{\partial \mathbf{x}}\right|_{\mathbf{x}=\mathbf{x}_k}$ is the $m \times m$ matrix of derivatives. Under assumption that matrix $\mathbf{H}_k$ is nonsingular, Newton's iterations take the form

$$\mathbf{x}_{k+1} = \mathbf{x}_k - \mathbf{H}_k^{-1}\mathbf{g}(\mathbf{x}_k), \quad k = 0, 1, 2, \ldots \tag{10.30}$$

As in the case of one equation, if iterations converge to $\lim \mathbf{x}_{k\to\infty} = \mathbf{x}_*$, we have $\mathbf{g}(\mathbf{x}_*) = \mathbf{0}$ and $\mathbf{x}_*$ is a solution of (10.29).

Optimization algorithms

Consider an unconstrained minimization problem $\min_{\mathbf{x}} f(\mathbf{x})$. Instead of finding $\mathbf{x}_*$ where function f takes minimum, find a stationary point. That is, solve a system of equations, the first-order conditions, see Theorem 10.17,

$$\partial f/\partial \mathbf{x} = \mathbf{0}. \tag{10.31}$$

Now apply Newton's iterations (10.30) to (10.31) letting $g = \partial f/\partial \mathbf{x}$. In a more general form gradient algorithms for minimization take the form

$$\mathbf{x}_{k+1} = \mathbf{x}_k - \rho_k \mathbf{H}_k^{-1} \mathbf{g}_k, \quad k = 0, 1, 2, \ldots, \tag{10.32}$$

where $\mathbf{g}_k = \left.\frac{\partial f}{\partial \mathbf{x}}\right|_{\mathbf{x}=\mathbf{x}_k}$, $\rho_k > 0$ and $\mathbf{H}_k = \mathbf{H}(\mathbf{x}_k)$ is a positive definite matrix for each $\mathbf{x}$ (several choices of $\mathbf{H}$ are listed below). We make several comments.

1. $\rho_k \leq 1$ is called the step length (in the nls() function it is called step factor) and typically $\rho_k = 1$ as in the parent algorithm (10.30). Parameter $\rho_k \leq 1$ avoids overshooting to ensure that the value of the function decreases from iteration to iteration (see the next comment).

2. As follows directly from the parent algorithm (10.30), $\mathbf{H}_k$ may be

$$\mathbf{H}(\mathbf{x}) = \frac{\partial^2 f}{\partial \mathbf{x}^2}, \tag{10.33}$$

the Hessian matrix. Then the gradient algorithm is called the Newton–Raphson algorithm. The problem with this algorithm is that the Hessian may not be positive definite unless function f is convex. If $\mathbf{H}_k$ is positive definite, then there exists a positive step length ρ_k such that the value of the function at the next iteration is smaller than at the present iteration (of course, if we are not at the stationary point, $\mathbf{g}_k \neq \mathbf{0}$). Several choices of $\mathbf{H}$, other than the Hessian, are used in statistical algorithms: the Fisher and empirical Fisher scoring algorithms for the ML problems (Section 6.10.6) and the Gauss–Newton algorithm for nonlinear regression (Section 9.3). When $\mathbf{H}_k = \mathbf{I}$ the gradient algorithm (10.32) is called the *steepest descent* algorithm.

Now we formulate conditions for convergence of the gradient algorithm (10.32).

Theorem 10.25 *Let $f(\mathbf{x})$ be a continuously differentiable function to minimize and $\mathbf{x}_0$ exists such that the level set $S_0 = \{\mathbf{x} : f(\mathbf{x}) \leq f(\mathbf{x}_0)\}$ is bounded. Let the step length ρ_k be chosen such that the value of the function decreases from iteration to iteration, if $\mathbf{g}_k \neq \mathbf{0}$. Let $\{\rho_k\}$ be bounded from below, $0 < \rho < \rho_k$, and elements of the positive definite matrix $\mathbf{H}(\mathbf{x})$ be continuous function of $\mathbf{x}$. Then, at each limiting point of the sequence $\{\mathbf{x}_k\}$ generated by the gradient algorithm (10.32), the gradient of function $f(\mathbf{x})$ is zero (stationary point).*

Conditions of this theorem are important: (1) If the level set is not bounded, the iterations may diverge like in the case of $f(x) = e^{-x}$. (2) If ρ_k are not bounded from below, e.g. $\rho_k = O(1/k)$, iterations may converge too soon, where the gradient is not zero. (3) If matrix $\mathbf{H}$ is not positive definite at each iteration, there may be no ρ_k that decreases the value of the function. For example, no such ρ_k may exist for the Newton–Raphson algorithm. On the other hand, if matrix $\mathbf{H}$ is positive definite such as in Fisher scoring or Gauss–Newton algorithms, this never happens.

10.6.4   Constrained optimization: Lagrange multiplier technique

Some problems in statistics lead to an optimization under constraint, such as the minimum variance estimator under the unbiasedness constraint or the likelihood ratio test under the null hypothesis constraint. The Lagrange multiplier technique reduces constrained optimization to a solution of an unconstrained system of equations. See Figure 10.2 for an illustration.

Theorem 10.26 *Constrained minimization. A necessary condition for the minimum of function $f = f(\mathbf{x})$ under the constraint $g(\mathbf{x}) = 0$ is*

$$\frac{\partial \mathcal{L}}{\partial \mathbf{x}} = \frac{\partial f}{\partial \mathbf{x}} - \lambda \frac{\partial g}{\partial \mathbf{x}} = \mathbf{0}, \tag{10.34}$$

referred to as the first-order condition, where $\mathcal{L}(\mathbf{x}, \lambda) = f(\mathbf{x}) - \lambda g(\mathbf{x})$ is called the Lagrange function and λ is the Lagrange multiplier. A sufficient condition for the minimum is that if, in addition to the first-order condition, $\lambda > 0$ and the Hessian matrix of the Lagrange function at the minimum,

$$\frac{\partial^2 f}{\partial \mathbf{x}^2} - \lambda \frac{\partial^2 g}{\partial \mathbf{x}^2}, \tag{10.35}$$

is positive definite, $\mathbf{x}$ is the point of a local minimum of $f(\mathbf{x})$ under constraint $g(\mathbf{x}) = 0$.

Comments. (1) The optimal $\mathbf{x}$ and λ are found from solving the system of $m + 1$ equations, the first m equation, being (10.34) and the $(m + 1)$ equation being the constraint, $g(\mathbf{x}) = 0$. Thus, the optimization problem is reduced to solving a system of $m+1$ equations for $m+1$ unknowns. (2) Alternatively, the constraint may be written using equation $-g(\mathbf{x}) = 0$. The choice is dictated by condition (10.35): Matrix $\partial^2 g/\partial \mathbf{x}^2$ should be negative definite (function g is concave) to ensure that (10.35) is a positive definite matrix provided the first term is positive definite and $\lambda > 0$. For example, if $f(\mathbf{x}) = \mathbf{x}'\mathbf{A}\mathbf{x}$, where $\mathbf{A}$ is a positive definite matrix, to find the maximum eigenvalue the Lagrange function is written as $\mathbf{x}'\mathbf{A}\mathbf{x} - \lambda(1 - \|\mathbf{x}\|^2)$ to comply with (10.35), which yields the positive definite matrix $2\mathbf{A} + 2\lambda\mathbf{I}$. (3) Similarly to unconstrained minimization, the minimizer $\mathbf{x}_*$ based on the Lagrange multiplier technique yields a local minimum that may be different from the global minimizer. However, if function f is convex and function g is concave, $\mathbf{x}_*$ is the point of the global minimum.

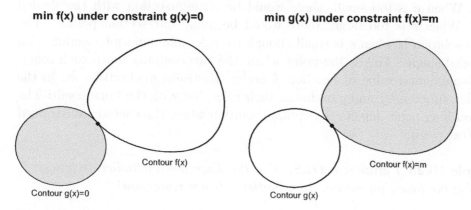

Figure 10.2: *Illustration of the optimization duality theorem. The two formulations lead to the same answer (the touching point): constraint and minimization functions interchange.*

Example 10.27 *Minimum of a quadratic function. Minimize the quadratic function $\mathbf{x}'\mathbf{A}\mathbf{x} + \mathbf{b}'\mathbf{x} + c$ under the constraint that the coordinates of vector $\mathbf{x}$ add to zero, where $\mathbf{A}$ is a $m \times m$ positive definite matrix, $\mathbf{b}$ is a fixed $m \times 1$ vector and c is a fixed scalar.*

Solution. In vector notation the condition that the coordinates of vector $\mathbf{x}$ add to zero is written as $\mathbf{x}'\mathbf{1} = 0$, where $\mathbf{1}$ is the $m \times 1$ vector of ones. Then the Lagrange function takes the form $\mathcal{L}(\mathbf{x},\lambda) = \mathbf{x}'\mathbf{A}\mathbf{x} + \mathbf{b}'\mathbf{x} + c - \lambda(\mathbf{x}'\mathbf{1})$. The necessary condition for the minimum is given by $\partial\mathcal{L}/\partial\mathbf{x} = 2\mathbf{A}\mathbf{x} + \mathbf{b} - \lambda\mathbf{1} = \mathbf{0}$. Express $\mathbf{x}$ through λ as $\mathbf{x} = 0.5\lambda\mathbf{A}^{-1}\mathbf{1} - 0.5\mathbf{A}^{-1}\mathbf{b}$. Substituting this expression into the constraint equation $\mathbf{x}'\mathbf{1} = 0$ find $\lambda = \mathbf{1}'\mathbf{A}^{-1}\mathbf{b}/\mathbf{1}'\mathbf{A}^{-1}\mathbf{1}$. Finally, the minimizer takes the form $\mathbf{x} = 0.5\mathbf{A}^{-1}(\lambda\mathbf{1} - \mathbf{b})$. The sufficient condition for the minimum is fulfilled here because $\partial^2\mathcal{L}/\partial^2\mathbf{x} = 2\mathbf{A}$ is a positive definite matrix.

Theorem 10.28 *Optimization duality theorem. Let $f(\mathbf{x})$ and $g(\mathbf{x})$ be two functions. The following six optimization problems are equivalent, meaning that*

they have the same necessary condition expressed as a Lagrange multiplier, λ: (1)
$\min_{g(\mathbf{x})=\mathbf{0}} f(\mathbf{x})$, *(2)* $\max_{g(\mathbf{x})=\mathbf{0}} f(\mathbf{x})$, *(3)* $\min_{f(\mathbf{x})=m} g(\mathbf{x})$, *(4)* $\max_{f(\mathbf{x})=M} g(\mathbf{x})$, *(5)*
$\min_{\mathbf{x}}(f(\mathbf{x})+\lambda g(\mathbf{x}))$, *(6)* $\min_{\mathbf{x}}(g(\mathbf{x})+\lambda f(\mathbf{x}))$.

It is obvious that the necessary condition for the optimum (10.34) for problems
1–4 is the same, but the Lagrange multiplier may be different. Note that, in
the first four problems, λ is derived from the constraint, but, in the last two
problems, λ is held fixed; therefore, these problems are reduced to unconstrained
optimization.

Figure 10.2 illustrates the optimization duality theorem between optimization
problems (1) and (3) using contours. The shaded shape represents the level set
of the constrained function. Consider the left plot: Contours $f(\mathbf{x}) = c$ increase
with c. When c is too small, there would be no intersection with the shaded
shape. When c is too large, there would be many intersection points. The
optimal solution is when c is small enough to make minimum intersection with
the shaded shape. This is the point where the two contours touch each other.
Let this minimum value of function f under constraint $g(\mathbf{x}) = 0$ be m. In the
right plot, functions f and g exchange their roles, but with the same result: The
same touching point depicts the optimal solution where the contour constrained
is $\{\mathbf{x} : f(\mathbf{x}) = m\}$.

Example 10.29 *Penalized OLS. Use the Lagrange multiplier technique to
show that the following optimization problems (ridge regression),*

$$\min_{\|\boldsymbol{\beta}\|^2=\rho} \|\mathbf{y} - \mathbf{X}\boldsymbol{\beta}\|^2, \qquad \min_{\|\mathbf{y}-\mathbf{X}\boldsymbol{\beta}\|^2=\eta} \|\boldsymbol{\beta}\|^2, \qquad \min_{\boldsymbol{\beta}} \left(\|\mathbf{y} - \mathbf{X}\boldsymbol{\beta}\|^2 + \lambda \|\boldsymbol{\beta}\|^2 \right),$$

*are equivalent. Namely, all solutions have the same form $\widehat{\boldsymbol{\beta}} = (\mathbf{X}'\mathbf{X}+\lambda\mathbf{I})^{-1}\mathbf{X}'\mathbf{y}$,
but λ is problem specific.*

Solution. The Lagrange function for the first problem is $\mathcal{L}(\boldsymbol{\beta},\lambda) = \|\mathbf{y} - \mathbf{X}\boldsymbol{\beta}\|^2 -$
$\lambda(\rho - \|\boldsymbol{\beta}\|^2)$. The first-order condition is $-2\mathbf{X}'(\mathbf{y} - \mathbf{X}\boldsymbol{\beta})+2\lambda\boldsymbol{\beta} = \mathbf{0}$ which yields
the solution $\widehat{\boldsymbol{\beta}}$, where λ is found from the solution of the equation

$$\mathbf{y}'\mathbf{X}(\mathbf{X}'\mathbf{X}+\lambda\mathbf{I})^{-2}\mathbf{X}'\mathbf{y} = \rho.$$

Under assumption that $0 < \rho < \mathbf{y}'\mathbf{X}(\mathbf{X}'\mathbf{X})^{-2}\mathbf{X}'\mathbf{y}$ the solution for λ exists, positive, and is unique because the left-hand side of the above equation as a function of
λ is strictly decreasing for $\lambda \geq 0$. The sufficient condition is met because $\lambda \geq 0$ and
the left-hand side of (10.35) is matrix $2(\mathbf{X}'\mathbf{X}+\lambda\mathbf{I})$, which is positive definite. The
Lagrange function for the second problem is $\mathcal{L}(\boldsymbol{\beta},\lambda) = \|\boldsymbol{\beta}\|^2 - \lambda(\eta - \|\mathbf{y} - \mathbf{X}\boldsymbol{\beta}\|^2)$,
which after differentiation yields $\widehat{\boldsymbol{\beta}} = (\mathbf{X}'\mathbf{X}+(1/\lambda)\mathbf{I})^{-1}\mathbf{X}'\mathbf{y}$. $\square$

The Lagrange multiplier technique can be applied to multiple constraints.
Below, we formulate the main result when constraints are linear.

Theorem 10.30 *Minimization under multiple linear constraints. A necessary condition for the minimizer,* $\mathbf{x}_*$*, of function* $f = f(\mathbf{x})$ *under linear constraints* $\mathbf{Lx} - \mathbf{p} = \mathbf{0}$*, where* $\mathbf{L}$ *is a* $k \times m$ *matrix of rank* $k < m$ *and* $\mathbf{p}$ *is a fixed* $k \times 1$ *vector takes the form* $\partial \mathcal{L}/\partial \mathbf{x} = \partial f/\partial \mathbf{x} - \mathbf{L}'\boldsymbol{\lambda} = \mathbf{0}$*, where the Lagrange function is* $\mathcal{L}(\mathbf{x}, \boldsymbol{\lambda}) = f(\mathbf{x}) - \boldsymbol{\lambda}'(\mathbf{Lx} - \mathbf{p})$*, and* $\boldsymbol{\lambda}$ *is the* $k \times 1$ *vector of Lagrange multipliers. A sufficient condition for the minimum is that the Hessian,* $\partial^2 f/\partial \mathbf{x}^2$*, evaluated at the solution is positive definite.*

Example 10.31 *Constrained minimization. Minimize quadratic function* $f(\mathbf{x}) = \mathbf{x}'\mathbf{Ax} + \mathbf{b}'\mathbf{x} + c$ *under multiple linear constraints* $\mathbf{Lx} = \mathbf{p}$*, where* $\mathbf{A}$ *is a* $m \times m$ *positive definite matrix,* $\mathbf{L}$ *is a fixed* $k \times m$ *matrix of rank* $k < m$ *and* $\mathbf{p}$ *is a fixed* $k \times 1$ *vector.*

Solution. The Lagrange function takes the form $\mathcal{L}(\mathbf{x}, \boldsymbol{\lambda}) = \mathbf{x}'\mathbf{Ax} + \mathbf{b}'\mathbf{x} + c - \boldsymbol{\lambda}'$ $(\mathbf{Lx} - \mathbf{p})$. Differentiate with respect to $\mathbf{x}$ to get the necessary condition for the minimum, $\partial \mathcal{L}/\partial \mathbf{x} = 2\mathbf{Ax} + \mathbf{b} - \mathbf{L}'\boldsymbol{\lambda} = \mathbf{0}$. Solve for $\mathbf{x}$ keeping other variables fixed, $\mathbf{x} = 0.5\mathbf{A}^{-1}(\mathbf{L}'\boldsymbol{\lambda} - \mathbf{b})$. To find the vector of Lagrange multipliers, use constraints $\mathbf{Lx} = \mathbf{p}$ which gives $0.5\mathbf{LA}^{-1}(\mathbf{L}'\boldsymbol{\lambda} - \mathbf{b}) = \mathbf{p}$. Find the optimal lambdas, $\boldsymbol{\lambda} = (\mathbf{LA}^{-1}\mathbf{L}')^{-1}(\mathbf{LA}^{-1}\mathbf{b} + 2\mathbf{p})$. Substituting $\boldsymbol{\lambda}$ back into the expression for $\mathbf{x}$ we arrive at the final solution $\mathbf{x} = 0.5\mathbf{A}^{-1}[\mathbf{L}'(\mathbf{LA}^{-1}\mathbf{L}')^{-1}(\mathbf{LA}^{-1}\mathbf{b} + 2\mathbf{p}) - \mathbf{b}]$. Since $\partial^2 f/\partial \mathbf{x}^2 = 2\mathbf{A}$ is a positive matrix, the sufficient condition for the minimum is fulfilled.

Problems

1. (a) Prove that linear function $f(\mathbf{x}) = \mathbf{a}'\mathbf{x}$ is convex and concave at the same time. (b) Prove that $f(\mathbf{x}) = g(\mathbf{a}'\mathbf{x})$ is a convex function where $\mathbf{a}$ is a fixed vector and $g(z)$ is a convex function on R^1. (c) Prove that a linear combination of convex functions with positive coefficients is a convex function.

2. (a) Prove that, if f is a convex function, then $-f$ is a concave function. (b) Is it true that if $f(\mathbf{x})$ is a convex function and g is an increasing function, then $g(f(\mathbf{x}))$ is a convex function? If true than prove; if false then provide a counterexample.

3. (a) Provide an example of a convex function defined on the entire real line that has no minimum. (b) Prove that function $S(\beta) = \sum_{i=1}^{n}(Y_i - e^{-\beta x_i})^2$ does not reach its minimum on the real line if $Y_i \leq 0$, $i = 1, 2, ..., n$.

4. Is function $f(x_1, x_2) = x_1^2 + x_2^2 - 3x_1x_2 - x_1$ a convex function? [Hint: Use Theorem 10.11.]

5. (a) Is it true that if function $f(\mathbf{x})$ is convex, then $f(\mathbf{x}) - \mathbf{a}'\mathbf{x}$ is convex as well? (b) Does addition of a linear function change the concavity of a function?

6. (a) Are functions $\log f(\mathbf{x})$ and $e^{-f(\mathbf{x})}$ convex in Example 10.13? (b) Are they unimodal?

7. (a) Is function $\log\|\mathbf{y} - \mathbf{X}\boldsymbol{\beta}\|^2$ convex? (b) Is function $\|\mathbf{y} - \mathbf{X}\boldsymbol{\beta}\|$ convex? (c) Are functions in (a) and (b) unimodal?

8. Prove that the existence level of $S(\boldsymbol{\beta})$ from Example 10.15 is not ∞ when $\mathrm{rank}(\mathbf{X}) = m - 1$, but the global minimum exists.

9. (a) Prove that the existence level of the negative log-likelihood function for linear model from Example 10.16 is ∞. (b) Prove that the global maximum of $l = l(\boldsymbol{\beta}, \sigma^2)$ exists.

10. Run function nLINEXP from Example 10.24 with different arguments a,b,x0, including negative values for a and b. Draw a conclusion on convergence.

11. Following Example 10.24, develop Newton's algorithm to solve the equation $x^a/(1 + x^a) = 1 - cx$ on $x > 0$, where $a > 0$ and $0 < c < 1$ are parameters. Write and test the R code.

12. Write R code to solve the system of equations $e^{-ax} + e^{-by} = 2$ and $e^{-ax-by} = 1$, where a and b are positive parameters, using Newton's algorithm. [Hint: Find the exact solution to check the code.]

13. Write R code to minimize function $(e^{-ax} + e^{-by} - 2)^2 + (e^{-ax-by} - 1)^2 + (x-y)^2$ using the gradient algorithm (10.32), where a and b are positive parameters. [Hint: Use the result of the previous problem.]

14. Write R code to minimize function $\|\mathbf{y} - \mathbf{X}\boldsymbol{\beta}\|^2$ using the steepest descent algorithm with a small $\rho_k = \rho_0 = \texttt{const}$ ($m = 2$). Test this algorithm in terms of convergence for different ρ_0 (plot contours of the least squares criterion function). Devise an appropriate stopping criterion. [Hint: Generate uniformly distributed $\mathbf{x}_1$ and $\mathbf{z}$ and let $\mathbf{x}_2 = \mathbf{z} + \kappa\mathbf{x}_1$, where $\kappa > 0$ controls the linear dependence.]

15. Derive the Cauchy inequality by maximizing $(\mathbf{a}'\mathbf{x})^2$ under constraint $\|\mathbf{x}\|^2 = c$ treating $\mathbf{a}$ as a fixed vector using the Lagrange multiplier technique.

16. Find the minimum of $\sum_{i=1}^n x_i^2$ under constraint $\sum_{i=1}^n x_i = 1$ and $x_1 = x_n$ using Theorem 10.30.

17. Show how the solution in Example 10.27 can be derived from the solution in Example 10.31.

Bibliography

[1] Abadir, K.M. (2005). The mean-median-mode inequality: counterexamples. *Econometric Theory* 21:477–482.

[2] Abadir, K.M. and Magnus, J.R. (2006). *Matrix Algebra.* Cambridge, England: Cambridge University Press.

[3] Abramowitz, M. and Stegun, I.A. (eds.) (1972). *Handbook of Mathematical Functions with Formulas, Graphs and Mathematical Tables.* New York: Wiley.

[4] Allmaraz, M., Bangerth, W., Linhart, J.M., Polanco, J., Wang, F., Wang, K., Webster, J. and Zedler, S. (2013). Estimating parameters in physical models through Bayesian inversion: A complete example. *SIAM Review* 55:149–167.

[5] Amrhein, V., Greenland, S. and McShane, B. (2019). Retire statistical significance. *Nature* 567:305–307.

[6] Bamber, D. (1975). The area above the ordinal dominance graph and the area below receiver operating characteristic graph. *Journal of Mathematical Psychology* 12:387–415.

[7] Bates, D.M., Watts, D.G. (1988). *Nonlinear Regression and Its Applications.* New York: Wiley.

[8] Benford, F. (1938). The law of anomalous numbers. *Proceedings of the American Philosophical Society* 78:551–572.

[9] Billingsley, P. (1995). *Probability and Measure,* 3ed. New York: Wiley.

[10] Bliss, C.I. (1939). The toxicity of poisons applied jointly. *Annals of Applied Biology* 26:585–615.

[11] Boos, D.D. and Stefanski, L.A. (2011). P-value precision and reproducibility. *American Statistician* 65:213–221.

Advanced Statistics with Applications in R, First Edition. Eugene Demidenko.
© 2020 John Wiley & Sons, Inc. Published 2020 by John Wiley & Sons, Inc.

[12] Borja, M.C. (2013). The strong birthday problem. *Significance* 12:18–20.

[13] Brams, S.J. and Kilgour, D.M. (1995). The box problem: To switch or not to switch. *Mathematics Magazine* 68:27–34.

[14] Brinton, E.A., Eisenberg, S. and Breslow, J.L. (1990). A low-fat diet decreases high density lipoprotein (HDL) cholesterol levels by decreasing HDL apolipoprotein transport rates. *Journal of Clinical Investigation* 85:144–151.

[15] Browne, R.H. (2010). The t-test p value and its relationship to the effect size and P(X>Y). *American Statistician* 64:30–33.

[16] Brazzale, A.R., Davison, A.C. and Reid, N. (2007). *Applied Asymptotics: Case Studies in Small-Sample Statistics*. Cambridge, England: Cambridge University Press.

[17] Casella, G., and Beger, R.L. (1990). *Statistical Inference*. Belmont, CA: Duxbury Press.

[18] Cobb, G.W. (1978). The problem of the Nile: Conditional solution to a changepoint problem. *Biometrika* 65:243–251.

[19] Cohen, J. (1994). The earth is round (*p* < .05). *American Psychologist* 49:997–1003.

[20] Cox, D.R. and Oakes D. (1984). *Analysis of Survival Data*. Boca Raton: Chapman & Hall.

[21] Cramér, H. (1946). *Mathematical Methods of Statistics*. Princeton, NJ: Princeton University Press.

[22] David, F.N. (1938). *Tables of the Distribution of the Correlation Coefficient*. London: Biometrika Office.

[23] Demidenko E. (2000). Is this the least squares estimate? *Biometrika* 87:437–452.

[24] Demidenko, E. (2006). Criteria for global minimum of sum of squares in nonlinear regression. *Computational Statistics & Data Analysis* 51:1739–1753.

[25] Demidenko, E. (2008). Criteria for unconstrained global optimization. *Journal of Optimization Theory and Applications* 136:375–395.

[26] Demidenko, E. Williams, B.B and Swartz, H.M. (2010). Radiation dose prediction using data on time to emesis in the case of nuclear terrorism. *Radiation Research* 171:310–319.

[27] Demidenko, E., Sargent, J. and Onega, T. (2012). Random effects coefficient of determination for mixed and meta-analysis models. *Communications in Statistics–Theory and Methods* 41:953–969.

[28] Demidenko, E. (2013). *Mixed Models. Theory and Applications with R.* Hoboken, NJ: Wiley.

[29] Demidenko, E. (2016). The p-value you can't buy. *American Statistician* 70:33–38.

[30] Demidenko, E., Glaholt, S.P., Kyker-Snowman, E., Shaw, J.R. and Chen, C.Y. (2017). Single toxin dose-response models revisited. *Toxicology and Applied Pharmacology* 314:12–23.

[31] Demidenko, E. (2017). Exact and approximate statistical inference for nonlinear regression and the estimating equation approach. *Journal of Scandinavian Statistics* 44:636–665.

[32] Demidenko, E. (2018). Optimal adaptive designs with inverse ordinary differential equations. *International Statistical Review* 86:169–188.

[33] Dennis, J.E. and Schnabel, R.B. (1983). *Numerical Methods for Unconstrained Optimization and Nonlinear Equations.* Englewood Cliffs, NJ: Prentice-Hall.

[34] Doob, J.L. (1953). *Stochastic Processes.* New York: Wiley.

[35] Gelman, A. and Stern, H. (2006). The difference between 'significant' and 'not significant' is not itself statistically significant. *American Statistician* 60:328–331.

[36] Gelman, A. and Carlin, J.B. (2013). *Bayesian Data Analysis.* Boca Raton: Chapman & Hall.

[37] Ghahramani, S. (2018). *Fundamentals of Probability with Stochastic Processes*, 4th ed. New York: Chapman and Hall/CRC.

[38] Giavarina, D. (2015). Understanding Bland Altman analysis. *Biochemia Medica* 25:141–151.

[39] Gill, J. (2008). *Bayesian Methods: A Social and Behaivioral Sciences Approach.* Boca Raton: CRC Press.

[40] Goldberg, J.L. (1991). *Matrix Theory with Applications.* New York: McGraw-Hill.

[41] Goodman, W. (2016). The promises and pitfalls of Benford's law. *Significance* 13:38–41.

[42] Graybill, F.A. (1982). *Matrices With Applications in Statistics.* 2d ed. Duxbury.

[43] Greene, W.H. (2012). *Econometric Analysis.* 7th ed. Boston: Prentice Hall.

[44] Guenther, W.C. (1972). On the use of the incomplete gamma table to obtain unbiased tests and unbiased confidence intervals for the variance of the normal distribution. *American Statistician* 26:31–34.

[45] Feller, W. (1967). *An Introduction to Probability Theory and Its Application.* New York: Wiley.

[46] Fieller, E.C. (1932) The Distribution of the index in a normal bivariate population. *Biometrika* 24:428–440.

[47] Finney, D.J. (1971). *Probit Analysis.* Cambridge, England: Cambridge University Press.

[48] Fisher, R.A. (1970). *Statistical Methods for Research Workers,* 14th ed. Edinburgh: Olyver and Boyed.

[49] Fisher, N.I. (1995). *Analysis of Circular Data.* Cambridge, England: Cambridge University Press.

[50] Fitzmaurice, G.M., Laird, N.M. and Ware, J.H. (2011). *Applied Longitudinal Analysis,* 2nd ed. New York: Wiley.

[51] Forsyth, A.R. (1959) *Theory of Differential Equations.* New York: Dover.

[52] Halmos, P.R. and Savage, L.J. (1949). Application of Radon-Nikodym theorem to the theory of sufficient statistics. *Annals of Mathematical Statistics* 20:225–241.

[53] Harlow, L.L., Mulaik, S.A. and Steiger, J.H. (eds.) (1997). *What If There Were No Significance Tests?* Mahwah, NJ: Lawrence Erlbaum Associates.

[54] Harville, D.A. (2008). *Matrix Algebra from a Statistician's Perspective.* New York: Springer.

[55] Hill, T.P. (1995). A statistical derivation of the significant-digit law. *Statistical Science* 19:354–363.

[56] Hollander, M, Wolfe, D.A. (1999). *Nonparametric Statistical Methods,* 2d ed. New York: Wiley.

[57] Hosford, S.R., Dillon, L.M., Bouley, S.J., Rosati, R. Yang, W., Chen, V.S., Demidenko, E. Morra, R.P. and Miller, T.E. (2017). Combined inhibition of both p110α and p110β isoforms of phosphatidylinositol 3-kinase is required for sustained therapeutic effect in PTEN-deficient, ER+ breast cancer. *Clinical Cancer Research* 23:e1-2918.

[58] Huber, P.J. and Ronchetti, E.M. (2008). *Robust Statistics*, 2d ed. New York: Wiley.

[59] Ivanov, A.V. (1997). *Asymptotic Theory of Nonlinear Regression*. Dordrecht: Kluwer.

[60] Johnson, N.L., Kotz, S. and Balakrishnan, N. (1995). *Continuous Univariate Distributions*, vol. 2, 2d ed. New York: Wiley.

[61] Kagan, A.M, Linnik, Y.V and Rao, C.R. (1973). *Characterization Problems in Mathematical Statistics*. New York: Wiley.

[62] Kalbfleisch, J.D. and Prentice, R.L. (2002). *The Statistical Analysis of Failure Time*. New York: Wiley.

[63] Karagas, M.R., Tosteson, T.D., Morris, J.S., Demidenko, E. Mott, L.A., Heaney, J., and Schned A. (2004). Incidence of transitional cell carcinoma of the bladder and arsenic exposure in New Hampshire. *Cancer Causes and Control* 15:465–472.

[64] Karian, Z.A, Dudewicz, E.L. (2011). *Handbook of Fitting Statistical Distributions with R*. Boca Raton: CRC Press.

[65] Kolmogorov, A.N. (1956). *Foundations of the Theory of Probability*, 2d ed. New York: Chelsea.

[66] Kolmogorov, A.N. and Fomin, S.V. (1961). *Elements of the Theory of Functions and Functional Analysis*. Rochester, NY: Graylock Press.

[67] Kotz, S., Balakrishnan, N. and Johnson, N.L. (2000). *Continuous Multivariate Distributions*, vol. 1, 2d ed. New York: Wiley.

[68] LeCun, Y., Léon B,. Yoshua, B. and Haffner, P. (1998). Gradient-based learning applied to document recognition. *Proceedings of the IEEE* 86:2278–2324.

[69] Lehmann, E.L. and Romano, J.P. (2005). *Testing Statistical Hypothesis*, 3d ed. New York: Springer.

[70] Loeve, M. (1977). *Probability Theory*. Berlin: Springer.

[71] Mardia, K.V. and Jupp, P.E. (2000). *Directional Statistics*, New York: Wiley.

[72] McGraw, K.O. and Wong, S.P. (1992). A common language effect size statistic. *Psycological Bulletin* 111:361–365.

[73] McShane, B.B. and Gal, D. (2017) Statistical significance and the dichotomization of evidence. *Journal of the American Statistical Association*
112:885–895.

[74] Miller, S. (2015). *Benfords's Law: Theory and Applications.* Princeton, NJ:
Princeton University Press.

[75] Mood, A.M., Graybill, F.A. and Boes, D.C. (1974). *Introduction to the
Theory of Statistics.* 3d. ed. New York: McGraw-Hill.

[76] Newcomb, S. (1881). Note on the frequency of use of different digits in
natural numbers. *American Journal of Mathematics* 4:39–40.

[77] O'Hara, J.A., Khan, N., Hou, H., Wilmot, C.M., Demidenko, E., Dunn,
J.F. and Swartz, H.M. (2004). Comparison of EPR oximetry and Eppen-
dorf polarographic electrode assessments of rat brain PtO2. *Physiological
Measurement* 25:1413–1423.

[78] Ortega, J.M. and Rheinboldt, W.C. (2000). *Iterative Solution of Nonlinear
Equations in Several Variables.* Philadelphia: SIAM.

[79] Parzen, E. (1962). On the estimation of a probability density function and
the mode. *Annals of Mathematical Statistics* 33:1065–1076.

[80] Pazman, A. (1993). *Nonlinear Statistical Models.* Dordrecht: Kluwer.

[81] Polak, E. (1971). *Computational Methods of Optimization.* New York: Wiley.

[82] Rao, C.R. (1973). *Linear Statistical Inference and Its Applications*, 2nd ed.
New York: Wiley.

[83] Ross, S.M. (1993). *Introduction to Probability Models.* 5th ed. San Diego:
Academic Press.

[84] Rosenwald A. et al. (2002). The use of molecular profiling to predict survival
after chemotherapy for diffuse large-B-cell lymphoma. *New England Journal
of Medicine* 346:1937–1947.

[85] Rubinstein, R.Y. and Kroese, D.P. (2011). *Simulation and the Monte Carlo
Method.* New York: Wiley.

[86] Sargent, J.D., Wills, T.A., Stoolmiller, M., Gibson, J. and Gibbons, F.X.
(2006) Alcohol use in motion pictures and its relation with early-onset teen
drinking. *Journal of Studies on Alcohol and Drugs* 67:54–65.

[87] Schervish, M.J. (1995). *Theory of Statistics.* New York: Springer.

[88] F.L. Schmidt and J. Hunter (1997). Eight common but false objections to the discontinuation of significance testing in the analysis of research data. In: *What If There Were No Significance Tests?* ed. by L.L Harlow, S.A. Mulaik, and J.H. Steiger. Lawrence Erlbaum Associates.

[89] Schott, J.R. (2005). *Matrix Analysis for Statistics*, 2d ed. Hoboken, NJ: Wiley.

[90] Schuemie, M.J., Ryan P.B., DuMouchel, W., Suchard, M.A. and Madigan, D. (2014). Interpreting observational studies: why empirical calibration is needed to correct p-values. *Statistics in Medicine* 33:209–218.

[91] Searle, S.R. (1982). *Matrix Algebra Useful for Statistics*. New York: Wiley.

[92] Shaffer, J.P. (1991). The Gauss–Markov theorem and random regressors. *American Statistician* 45:269–273.

[93] Serfling, R.J. (1980). *Approximation Theorems of Mathematical Statistics*. New York: Wiley.

[94] Shao, J. (2008). *Mathematical Statistics*, 2d ed. New York: Springer.

[95] Silverman, B. W. (1986). *Density Estimation for Statistics and Data Analysis*. London: Chapman and Hall.

[96] Smith, R.J. (2018). The continuing misuse of null hypothesis significance testing in biological anthropology. *American Journal of Physical Anthropology* 166:236–245.

[97] Stolley, P.D. (1991). When genius errs–Fisher, R.S. and the lung-cancer controversy. *American Journal of Epidemiology* 133:416–425.

[98] Tate, R.F. and Klett, G.W. (1959). Optimal confidence intervals for the variance of a normal distribution. *Journal of the American Statistical Association* 54:674–682.

[99] Torgo, L. (2011). *Data Mining with R. Learning with Case Studies*. Boca Raton: CRC Press.

[100] Trafimow, D., and Marks,M. (2015). "Editorial," *Basic and Social Psychology* 37:1–2.

[101] Wang, H-H and Kuo, C-H. (1998). On the frequency distribution of interoccurence times of earthquakes. *Journal of Seismology* 2:351–358.

[102] Warner, S.L. (1965). Randomized response: A survey technique for eliminating evasive answer bias. *Journal of the American Statistical Association* 60:63–69.

[103] Wasserstein, R.L. and Lazar, N.A. (2016). The ASA's statement on p-values: Context, process, and purpose. *American Statistician* 70:129–133.

[104] Wasserstein, R.L., Schirm, A.L. and Lazar, N.A. (2019) Moving to a world beyond "p < 0.05". *American Statistician* 73:1–19.

[105] Wilks, S.S. (1943). *Mathematical Statistics*. Princeton: Princeton University Press.

[106] Wolfe, D.A. and Hogg, R.V. (1971). On constructing statistics and reporting data. *American Statistician* 25:27–30.

[107] Wu, C-F. (1981). Asymptotic theory of nonlinear least squares estimation. *Annals of Statistics* 9:465–696.

[108] Zacks, S. (2014). *Examples and Problems in Mathematical Statistics*. Hoboken: Wiley.

[109] Zou, G.Y. (2007). Exact confidence interval for Cohen's effect size is readily available. *Statistics in Medicine* 26:3054–3056.

Index

A-optimality, 802

amplitude, 803

annihilator matrix, 630, 632

apply R, 21, 109, 238, 361, 373, 389, 503, 649

arsenic, 330

asymptotic normality, 479, 484, 486, 491, 492, 512, 757, 761

asymptotic unbiasedness, 95, 362, 374

attr R, 776, 789

AUC, 295, 299, 305, 544, 621

bandwidth, 313, 325, 331, 340, 342, 474, 488, 787

Bayes, 170, 180, 181

Bernoulli, 2, 5, 19, 26, 29, 45, 74, 96, 135, 177, 191, 353, 360, 416, 423, 437, 439, 455, 457, 464, 477

bias, 177, 359–361, 366, 367, 372, 374, 388, 392, 448, 469, 471, 488, 571, 783

biased estimator, 368, 447

binomial probability, 27, 34, 319, 353, 528, 533, 593

binomial proportion, 133, 420, 455, 573, 577, 599, 605

binormal ROC, 297, 305, 547, 624

bivariate cdf, 150

bivariate density, 335

bivariate distribution, 7, 169, 207, 210, 214, 240, 265, 338, 342, 463

bivariate pdf, 151, 194

BLUE, 379, 402, 405, 450, 639, 640

bootstrap, 307

cancer, 184, 306, 307, 317, 330, 343, 345, 536, 614, 702, 717, 731, 733, 784

cancer rate, 344

catenary, 800

Cauchy, 50, 52, 53, 130, 147, 220, 519, 762, 763, 813

Cauchy distribution, 50, 130, 146, 220, 500, 502, 503, 595, 647, 762, 777

Cauchy inequality, 52, 391, 437, 438, 448, 458, 513, 813

cdf, 43, 150

ceiling R, 30, 60

center of gravity, 7

central limit theorem (CLT), 72, 74, 94, 104, 205, 648, 665

central moment, 13, 50, 61, 69

Advanced Statistics with Applications in R, First Edition. Eugene Demidenko.
© 2020 John Wiley & Sons, Inc. Published 2020 by John Wiley & Sons, Inc.